CONCISE ENCYCLOPEDIA OF
TRAFFIC & TRANSPORTATION SYSTEMS

ADVANCES IN SYSTEMS, CONTROL AND INFORMATION ENGINEERING

This is a new series of Pergamon scientific reference works, each volume providing comprehensive, self-contained and up-to-date coverage of a selected area in the field of systems, control and information engineering. The series is being developed primarily from the highly acclaimed *Systems & Control Encyclopedia* published in 1987. Other titles in the series are listed below.

ATHERTON & BORNE (eds.)
Concise Encyclopedia of Modelling & Simulation

FINKELSTEIN & GRATTAN (eds.)
Concise Encyclopedia of Measurement & Instrumentation

MORRIS & TAMM (eds.)
Concise Encyclopedia of Software Engineering

PAYNE (ed.)
Concise Encyclopedia of Biological & Biomedical Measurement Systems

PELEGRIN & HOLLISTER (eds.)
Concise Encyclopedia of Aeronautics & Space Systems

SAGE (ed.)
Concise Encyclopedia of Information Processing in Systems & Organizations

YOUNG (ed.)
Concise Encyclopedia of Environmental Systems

NOTICE TO READERS

Dear Reader
If your library is not already a standing order/continuation order customer to the series **Advances in Systems, Control and Information Engineering**, may we recommend that you place a standing order/continuation order to receive immediately upon publication all new volumes. Should you find that these volumes no longer serve your needs, your order can be cancelled at any time without notice.

ROBERT MAXWELL
Publisher at Pergamon Press

CONCISE ENCYCLOPEDIA OF
TRAFFIC & TRANSPORTATION SYSTEMS

Editor
MARKOS PAPAGEORGIOU
Technische Universität München
Munich, Germany

Series Editor-in-Chief
MADAN G SINGH
UMIST, Manchester, UK

PERGAMON PRESS

Member of Maxwell Macmillan Pergamon Publishing Corporation
OXFORD • NEW YORK • BEIJING • FRANKFURT
SEOUL • SYDNEY • TOKYO

UK	Pergamon Press plc, Headington Hill Hall, Oxford OX3 0BW, England
USA	Pergamon Press, Inc, Maxwell House, Fairview Park, Elmsford, New York 10523, USA
PEOPLE'S REPUBLIC OF CHINA	Maxwell Pergamon China, Beijing Exhibition Centre, Xizhimenwai Dajie, Beijing 100044, People's Republic of China
GERMANY	Pergamon Press GmbH, Hammerweg 6, D-6242 Kronberg, Germany
KOREA	Pergamon Press Korea, KPO Box 315, Seoul 110-603, Korea
AUSTRALIA	Maxwell Macmillan Pergamon Publishing Australia Pty Ltd, Lakes Business Park, 2 Lord Street, Botany, NSW 2019, Australia
JAPAN	Pergamon Press, 8th Floor, Matsuoka Central Building, 1-7-1 Nishi-Shinjuku, Shinjuku-ku, Tokyo 160, Japan

Copyright © 1991 Pergamon Press plc

All rights reserved. No part of this publication may be reproduced, stored in any retrieval system or transmitted in any form or by any means: electronic, electrostatic, magnetic tape, mechanical, photocopying, recording or otherwise, without permission in writing from the publishers.

First edition 1991

Library of Congress Cataloging in Publication Data
Concise Encyclopedia of traffic & transportation systems / editor, Markos Papageorgiou. — 1st ed.
   p. cm. — (Advances in systems, control, and information engineering)
   Includes index.
   1. Transportation. 2. Transportation engineering. 3. Traffic engineering. I. Papageorgiou, M. (Markos),
1953– II. Title: Concise encyclopedia of traffic and transportation systems. III. Title: Traffic & transportation systems. IV. Series: Advances in systems, control & information engineering.
TA1145.C58   1991
388'.03—dc20   90–22479

British Library Cataloguing in Publication Data
Concise encyclopaedia of traffic & transportation systems
1. Transport
I. Papageorgiou, Markos
380.5

ISBN 0–08–036203–6

Printed and bound in Great Britain by BPCC Wheatons Ltd, Exeter

CONTENTS

Honorary Editorial Advisory Board	vi
Foreword	vii
Preface	ix
Guide to Use of the Encyclopedia	xi
Alphabetical List of Articles	xiii
An Introduction to Traffic and Transportation Systems	xv
Articles	1
List of Contributors	621
Subject Index	633

HONORARY EDITORIAL ADVISORY BOARD

Chairman
John F Coales CBE, FRS
Cambridge, UK

Editor-in-Chief
Madan G Singh
UMIST, Manchester, UK

D Aspinall
UMIST, Manchester, UK

K J Åström
*Lund Institute of Technology
Sweden*

A Bensoussan
INRIA, Le Chesnay, France

P Borne
*Institut Industriel du Nord
Villeneuve D'Ascq, France*

A W Goldsworthy OBE
*Jennings Industries Ltd
Victoria, Australia*

J Lesourne
*Conservatoire National
des Arts et Métiers,
Paris, France*

P A Payne
UMIST, Manchester, UK

A P Sage
*George Mason University
Fairfax, VA, USA*

Y Sawaragi
*Japan Institute for Systems
Research, Kyoto, Japan*

G Schmidt
*Technische Universität München
Germany*

B Tamm
*Tallinn Technical University
USSR*

M Thoma
*Universität Hannover
Germany*

R Vichnevetsky
*Rutgers University
New Brunswick, NJ, USA*

FOREWORD

With the publication of the eight-volume *Systems & Control Encyclopedia* in September 1987, Pergamon Press was very keen to ensure that the scholarship embodied in the Encyclopedia was both kept up to date and was disseminated to as wide an audience as possible. For these purposes, an Honorary Editorial Advisory Board was set up under the chairmanship of Professor John F Coales FRS, and I was invited to continue as Editor-in-Chief of both the Systems & Control Encyclopedia Series of Supplementary Volumes and the Advances in Systems, Control and Information Engineering Series of Concise Encyclopedias. This involved me personally editing a series of Supplementary Volumes updating the original Encyclopedia and also arranging for the editing of a series of Concise Encyclopedias being developed from it on specific subject areas. The Honorary Editorial Advisory Board helped to select a series of subject areas which were perceived to be appropriate for the publication of Concise Encyclopedias and chose the most distinguished experts in those areas to edit them. The Concise Encyclopedias contain some updated and revised articles from the Main Encyclopedia and many that are totally new, reflecting recent advances in the subject and gaps in the original Encyclopedia.

It gives me great pleasure to commend to the reader the second volume in the Advances in Systems, Control and Information Engineering series, the *Concise Encyclopedia of Traffic & Transportation Systems*, edited by Professor Markos Papageorgiou. This Concise Encyclopedia provides a comprehensive up-to-date view of this field which is of vital importance to all of us.

Madan G Singh
Series Editor-in-Chief

PREFACE

Transportation of people and goods has been necessary for society for thousands of years. In modern times, transportation by road, rail, air and sea has become a fundamental component of human activity. A great number of technical, economic, environmental and organizational problems related to traffic and transportation have been ingeniously resolved in the past and perhaps an even higher number of problems will have to be overcome in the future. Recently, it has been realized more and more that problems connected to transportation, traffic congestion and delay cannot be resolved by simply extending the available infrastructure. Efficient use of existing facilities is a feasible alternative which becomes possible by the application of concepts and methods provided by systems and information theory. Besides, recent developments in digital computer and communication technology provide the necessary practical tools for satisfactory and cheap solutions.

The Concise Encyclopedia of Traffic & Transportation Systems focuses mainly on the systems and control aspects of this vast field of scientific activity providing a concise, comprehensive and up-to-date overview. Moreover, a number of historical, social, economic and organizational issues are addressed. The volume comprises 118 articles written by 135 authors from 16 countries, providing an overview and a first reference to models, control methods and practical aspects of current and future road, rail, air and maritime traffic and transportation systems, with particular emphasis being placed on efficient utilization of available infrastructure. The subject of traffic and transportation systems is an interdisciplinary one, joining traffic and transportation engineering, control and systems science, management and operation of systems, computer science and information and communication technologies to mention but a few of the most important related fields. As a consequence, contributors to and potential readers of this Concise Encyclopedia originate from a broad spectrum of professional disciplines and include academics, operators, managers, engineers and graduate students.

In choosing the titles of the articles, an effort has been made to ensure that each title begins with the most significant keyword. The reader is guided to further reading by bibliographies and cross-references to other related articles within the volume. Also, an introductory article provides an overview of the articles in the main section of the Encyclopedia, and this may be consulted in order to find articles that fall into particular subject areas. It should be noted that some areas of interest (or aspects thereof) are covered by overlapping articles that reflect the point of view of the discipline and the practice of the country of the corresponding authors. This was felt to be a necessary approach in order to cover various needs and aspects of traffic and transportation systems. The subject areas covered in this Concise Encyclopedia offer an interesting linguistic reflection of the use of synonyms with American or British emphasis. An effort has been made to standardize terms, but where specific applications are mentioned or general use is biased against the "standard" term, the more commonly used term has been adopted.

This volume would not have been possible without the contributions of the numerous authors who agreed to write for it. Their names appear at the foot of the articles as well as in the alphabetical List of Contributors. Professors Günther Schmidt and Madan Singh suggested that I undertake this work and they made many useful suggestions which are gratefully acknowledged. At Pergamon Press, Colin Drayton and Peter Frank provided invaluable support for the successful completion of the work and Wayne Davies and Keith Mansfield contributed much to its editing.

Markos Papageorgiou
Editor

GUIDE TO USE OF THE ENCYCLOPEDIA

This Concise Encyclopedia is a comprehensive reference work covering all aspects of traffic and transportation systems. Information is presented in a series of alphabetically arranged articles which deal concisely with individual topics in a self-contained manner. This guide outlines the main features and organization of the Encyclopedia, and is intended to help the reader to locate the maximum amount of information on a given topic.

Accessibility of material is of vital importance in a reference work of this kind and article titles have therefore been selected, not only on the basis of article content, but also with the most probable needs of the reader in mind. An alphabetical list of all the articles contained in this Encyclopedia is to be found on pp. xiii and xiv.

Articles are linked by an extensive cross-referencing system. Cross-references to other articles in the Encyclopedia are of two types: in-text and end-of-text. Those in the body of the text are designed to refer the reader to articles that present in greater detail material on the specific topic under discussion. They generally take one of the following forms:

...as fully described in the article *Road Network Control*.

...or on simulation (see *Signal Control at Individual Junctions: Phase-Based Approach*).

The cross-references listed at the end of an article serve to identify broad background reading and to direct the reader to articles that cover different aspects of the same topic.

The nature of an encyclopedia demands a higher degree of uniformity in terminology and notation than many other scientific works. The widespread use of the International System of Units has determined that such units be used in this Encyclopedia. It has been recognized, however, that in some fields Imperial units are more generally used. Where this is the case, Imperial units are given with their SI equivalent quantity and unit following in parentheses. Where possible, the symbols defined in *Quantities, Units, and Symbols* published by the Royal Society of London have been used.

Most of the articles in the Encyclopedia include a bibliography giving sources of further information. Each bibliography consists of general items for further reading and/or references which cover specific aspects of the text. Where appropriate, authors are cited in the text using a name/date system as follows:

...as was recently reported (Smith 1988).

Jones (1984) describes...

The contributors' names and the organizations to which they are affiliated appear at the end of all the articles. All contributors can be found in the alphabetical List of Contributors, along with their full postal address and the titles of the articles of which they are authors or coauthors.

The Introduction to this Concise Encyclopedia provides an overview of traffic and transportation systems and directs the reader to many of the key articles in this work.

The most important information source for locating a particular topic in the Encyclopedia is the multilevel Subject Index, which has been made as complete and fully self-consistent as possible.

ALPHABETICAL LIST OF ARTICLES

Air Traffic Control: An Overview
Air Traffic Control Near Airports
Air Traffic Control: Trends
Air Traffic Simulators
Airborne Collision Avoidance
Automated Guideway Transit Systems and Personal Rapid Transit Systems
Automatic Landing Systems
Automatic Train Control: Protection Principles
Automatic Train Control: Safety and Reliability
Broadcasting Communication Systems
Car-Following Models
Cellular Communication Systems
Container Terminal Management
Corridor Control Systems
Data Links in Aeronautics
Data Processing in Air Traffic Control
Detectors for Road Traffic
Discrete-Time Point Processes: Applications to Road Traffic
DRIVE
Evaluation of Traffic Control Systems
Expert Systems Approach to Rail Traffic Control
Expert Systems Approach to Road Traffic Control
Flight Activity in General Aviation
Flight Activity: Prevision
Flow Variables
Flow Variables: Estimation
Freeway
Freeway Capacity: Reliability and Control
Freeway Control: An Overview
Freeway Network Control
Freeway Traffic Modelling
Fuel Conservation: Air Transport
Fuel Conservation: Road Transport
Fuel Conservation: Sea Transport
High-Speed Railroad: Modelling and Simulation
High-Speed Railroad Networks in Europe
High-Speed Railroad Networks in Japan
High-Speed Railroad: Systems Approach
IFAC Working Group on Transportation Systems
In-Vehicle Equipment for Future Traffic Control Systems
Incident Detection
Intelligent Traffic Control Systems
Kinematic Wave Theory
Magnetic Suspension Railroad Systems
Maintenance and Reliability of Traffic Control Systems
Marine Fleet Planning and Scheduling
Marine Propulsion Plants: Control
Merging Control
Mobile Communication
Navigation Control of Ships
Navigation Systems, Integrated
Network Modelling and Control: Store-and-Forward Approach

On-Ramp Control: Coordinated Time-of-Day Strategies
On-Ramp Control: Coordinated Traffic-Responsive Strategies
On-Ramp Control, Local
On-Ramp Control of Freeway Networks
On-Ramp Control: Realization Principles
Optimal Routing Applications
Origin–Destination Matrix: Dynamic Estimation
Origin–Destination Matrix: Static Estimation
Parking Control Systems
Prediction of Traffic Flow
Priority Intersection: Modelling
Queuing Theory Applications
Railroad Electronic Signalling
Railroad Systems: Active Control
Railroad Systems: Line Supervision and Control
Railroad Systems: Train Driving Control
Road Network Control
Road Network Signal Setting: Equilibrium Conditions
Road Networks: Dynamic Equilibrium Models
Road Traffic: An Introduction
Road Traffic Control, Demand-Responsive
Road Traffic Control: Progression Methods
Road Traffic Control Systems in Japan
Road Traffic Control: TRANSYT and SCOOT
Road Traffic Monitoring Equipment
Route Guidance, Collective
Route Guidance, Individual
Safety of Road Traffic
Safety of Road Traffic: Intervention and Evaluation
Ship Automation Control
Ship Dynamics: Modelling
Ship Positioning: Adaptive Control
Ship Rudder Roll Stabilization
Ship Stabilization: History
Ship Steering: Model-Reference Adaptive Control
Shortest-Path Algorithms
Signal Control and Traffic Assignment
Signal Control at Individual Junctions: Phase-Based Approach
Signal Control at Individual Junctions: Stage-Based Approach
Simulation of Urban Traffic: Software Environments
Simulation Programs, Macroscopic
Simulation Programs, Microscopic
Social Issues of Ship Automation
Social Issues of Transportation
Speed Limitation on Freeways: Traffic-Responsive Strategies
Traffic Assignment
Traffic Assignment, Dynamic
Traffic Assignment, Stochastic
Traffic Control Modes
Traffic Control Systems: Architecture
Traffic Control Systems: Trends

Traffic Management Systems
Traffic Office Management
Transportation Management: Systems Engineering Approach
Transportation Modelling
Transportation Planning: Activity-Based Approach
Transportation Planning: Microscopic Approach
Transportation Systems: Trends
Underground Railroad Modelling and Control
Underground Railroad Modelling and Control: Discrete-Event Approach
Underground Railroad: Organization of Operations
Vehicle Guidance by Computer Vision
Vehicle Monitoring and Control for Buses, Trolleys and Streetcars
Vessel Traffic Services: Management from Shore
Video Sensors
Visual and Instrument Flying Rules

AN INTRODUCTION TO TRAFFIC AND TRANSPORTATION SYSTEMS
by Markos Papageorgiou

The articles of the *Concise Encyclopedia of Traffic & Transportation Systems* may be classified into ten main groups which will be briefly outlined.

1. General Aspects

The first group comprises articles that are mainly nontechnical in character. In *Social Issues of Transportation*, traffic and transportation systems are reviewed with respect to their impact on society, whereas more particular socioeconomic and historical views are provided in *Social Issues of Ship Automation* and *Ship Stabilization: History*. Organizational issues are addressed in *Traffic Office Management* and *Transportation Management: Systems Engineering Approach*.

Some insight into current research organization and coordination is given in the articles *DRIVE* and *IFAC Working Group on Transportation Systems*.

Two further very important subjects belonging to this section are covered in *Evaluation of Traffic Control Systems*, *Safety of Road Traffic* and *Safety of Road Traffic: Intervention and Evaluation*. Evaluation methods are aimed at analyzing the impact of the introduction of advanced traffic control schemes on the basis of a multiplicity of criteria including economic and environmental assessment, safety considerations and so on. Each year, thousands of human beings pay with their lives because of society's need for mobility; increasing safety is therefore a fundamental concern in the framework of traffic and transportation research.

Some basic structural notions of traffic control systems are defined in *Traffic Control Modes*.

Finally, the present status of research and the future needs and trends both in traffic control systems and in general transportation systems are analyzed in *Traffic Control Systems: Trends*, *Air Traffic Control: Trends* and *Transportation Systems: Trends*.

2. Road Traffic Measurements and Communications

Road traffic control began with the police giving right of way to different traffic streams or pedestrians at urban intersections, mainly for safety reasons. Human traffic controllers were soon replaced by traffic lights and the evolution of the corresponding technology has followed the general development of electrical and electronic equipment: from analog to digital systems, from simple logic to intelligent devices, from local systems to global hierarchical structures, from simple sensors to computer vision.

Realization of a closed-loop system for traffic control implies availability of sensors, controllers, communication links between different control hierarchy layers and devices for traffic message display. The state of the art of the corresponding road traffic control technology is presented in *Detectors for Road Traffic* and *Road Traffic Monitoring Equipment*, whereas structural hardware aspects are included in *Traffic Control Systems: Architecture*. A most important practical view is given in *Maintenance and Reliability of Traffic Control Systems*.

Particular attention is paid to the new technologies of *Video Sensors* based on video cameras and image processing techniques, of *In-Vehicle Equipment for Future Traffic Control Systems* as they are needed for individual route guidance and of *Mobile Communication* and *Cellular Communication Systems*.

Finally, problems, methods and operational experience related to communication means between control systems and drivers are reviewed in *Broadcasting Communication Systems*.

3. Traffic Modelling and Simulation

Any scientific interest in process control begins with modelling (i.e., comprehension of the behavior of the processes). An initial surprise for new researchers entering the field of traffic flow modelling is the realization that—despite the involvement of drivers with different individual behavior—traffic flow can be viewed from a macroscopic point of view as a fluid with particular well-defined characteristics. However, unlike water flow, traffic flow consists of several substreams with prespecified origin–destinations and individual optimization strategies for route selection in a road network. Consequently, the mathematical description of traffic flow appears to be a more complex task than the macroscopic modelling of other fluids.

Consideration of traffic modelling usually starts in a microscopic framework. The article *Car-Following Models* describes the modelling of the motion (position and speed) of a vehicle in terms of the motion of the preceding vehicle. Car-following models applied to a number of successive vehicles are capable of describing the movement of a long string of vehicles. Questions of stability of the string and creation of congestion are of particular interest in this context. By introducing some additional modules describing lane changing and overtaking behavior, it is possible to simulate traffic flow on a multilane road stretch or even on a road network including traffic lights. A review is provided in *Simulation Programs, Microscopic*, whereas some particular road traffic phenomena are addressed in *Priority Intersection: Modelling*.

In contrast to microscopic models, the definition of aggregated traffic flow variables (see *Flow Variables*) such as traffic density (vehicles per kilometer) and traffic volume (vehicles per hour) leads to a macroscopic description of traffic flow. Under homogeneous traffic conditions in space and time, traffic density is related to traffic volume by a relationship known as the fundamental diagram. In contrast to the flow characteristics of other fluids, the fundamental diagram provides maximum flow at a critical density value. If density is further increased (e.g., due to entering traffic), traffic volume decreases and a more or less severe congestion results. This and further related issues are discussed in *Kinematic Wave Theory*.

Analogous to the modelling of fluids, macroscopic models of traffic flow make use of a conservation equation and a momentum equation. Macroscopic models are capable of describing the dynamic evolution of traffic flow on long roads and traffic networks. A review of theoretical macroscopic models and the most popular simulation packages is provided in *Freeway Traffic Modelling* and *Simulation Programs, Macroscopic*.

The particular use of macroscopic models for purposes of prediction is described in *Prediction of Traffic Flow*, while the article *Fuel Conservation: Road Transport* offers some modelling methods for fuel consumption.

An overall view of human–machine communication in the context of traffic simulation is provided by *Simulation of Urban Traffic: Software Environments*.

Finally, a more global view of transportation phenomena is provided in the articles *Transportation Modelling*, *Transportation Planning: Activity-Based Approach* and *Transportation Planning: Microscopic Approach*.

4. Road Traffic Control

The fundamental notions and problems related to road traffic control by traffic lights are given in *Road Traffic: An Introduction*.

Considering a single intersection with known demands and turning rates, the control problem consists of specifying the green and red phases of competing substreams so as to minimize a given criterion (e.g., total delay). Fixed-time control systems use historical data in order to specify optimal, time-of-day-dependent plans for traffic lights. Traffic-reponsive systems use current traffic data provided by detectors in order to specify appropriate lengths of green/red phases and traffic cycles at each intersection so as to minimize the extent of congestion. Fixed-time isolated junction control is treated in *Signal Control at Individual Junctions: Stage-Based Approach* and *Signal Control at Individual Junctions: Phase-Based Approach*, whereas real-time methods are included in *Road Traffic Control, Demand-Responsive*.

Interconnections between intersections arise due to the fact that traffic served at one intersection represents demand for the adjacent intersections. Coordinated control strategies are presented in the articles *Road Network Control*, *Road Traffic Control: TRANSYT and SCOOT* and *Road Traffic Control Systems in Japan*.

The article *Road Traffic Control: Progression Methods* considers methods that aim at maximizing the number of vehicles that travel on a green wave along a principal axis of a road network. Some innovative approaches to road traffic control are presented in *Expert Systems Approach to Road Traffic Control* and *Discrete-Time Point Processes: Applications to Road Traffic*, whereas particular efforts regarding public transport and parking are reviewed in *Vehicle Monitoring and Control for Buses, Trolleys and Streetcars* and *Parking Control Systems*, respectively.

Drivers modify their routes according to signal settings in order to minimize their individual travel times. This, however, changes the traffic data for which the signal settings were optimally calculated and eventually renders the signal settings nonoptimal. The interconnection between road traffic control and the driver's reaction is the principal subject of *Road Network Signal Setting: Equilibrium Conditions* and *Signal Control and Traffic Assignment*.

5. Freeway Traffic Control

Freeway traffic became problematic much later than road traffic. Nevertheless, modern freeways in most urban areas are seriously congested on a daily basis during rush hours. Even interurban freeway drivers experience considerable delays due to congestion, particularly on weekends, holidays and so on. Congestion on freeways appears a paradox to people who are not familiar with traffic flow characteristics. Drivers often spend several hours in long freeway queues although there is no apparent reason for a breakdown (no accident, no traffic lights). The answer to this paradox lies partially in keeping a safe distance between cars and partially in the dynamic driver behavior when acting as distance regulators.

For a definition of the term freeway see *Freeway* and for an overview of control problems and measures see *Freeway Control: An Overview*.

Freeway traffic control calls for high-level traffic flow information which cannot be provided directly by traffic detectors. Consequently, software devices are required that are capable of extracting the necessary information from measurement data in real time as reported in *Flow Variables: Estimation*.

An important prerequisite for traffic-responsive control and increased safety on freeways is the automatic detection of incidents. The problem to be resolved in this context is connected to the identification of an incident by the use of traffic data provided by adjacent detector stations (see *Incident Detection*).

Three classes of freeway traffic control measures aimed at increasing the efficiency and safety of the traffic system can be distinguished.

The first class is variable messages such as speed limitation, no overtaking and congestion warning, which aim at homogenizing traffic flow on a freeway axis. This usually leads to more stable behavior of the traffic flow and may increase traffic throughput and the critical density value, as reported in *Speed Limitation on Freeways: Traffic-Responsive Strategies* and *Freeway Capacity: Reliability and Control*.

The second class is ramp metering which, by the use of

traffic lights at on-ramps, is aimed at operating traffic flow near its maximum value, avoiding overload due to excessive demands. Although some delay may be caused at waiting ramp queues, the overall time may be decreased due to the optimal operation of the existing infrastructure. The practical problems related to this important freeway control measure are addressed in *On-Ramp Control: Realization Principles*. Various approaches to ramp metering are presented in the series of articles *On-Ramp Control, Local*, *On-Ramp Control: Coordinated Time-of-Day Strategies*, *On-Ramp Control: Coordinated Traffic-Responsive Strategies* and *On-Ramp Control of Freeway Networks*.

The third class of freeway control measures consists of variable route recommendation signs aimed at distributing traffic flow in a freeway network so as to minimize delays and to optimally utilize the existing infrastructure, as reported in *Freeway Network Control*.

6. General Traffic Networks

The next group of articles comprises items related to network flow that are not necessarily connected to a specific traffic type (road, freeway or mixed). An important problem of general traffic networks is the identification of origin–destination matrices from traffic flow measurements. Consider a traffic system receiving n traffic flows and distributing them to m exit links according to the vehicle destinations. The distributing system may be an intersection, a road network, a freeway network or any traffic network with unknown origin–destination rates. The corresponding problem is to identify nm matrix elements using traffic flow measurements. This task may be performed off line or in real time and the corresponding methods are reviewed in *Origin–Destination Matrix: Static Estimation* and *Origin–Destination Matrix: Dynamic Estimation*.

Another interesting problem of network traffic is traffic assignment (see *Traffic Assignment*). Consider an origin–destination pair (A, B) connected by more than one alternative route. Assume a traffic volume q arriving at node A and destined for node B. Let a route 1 be shorter than a route 2 and let travel times increase monotonically with increasing volume. The user optimum principle states that travel times on alternative routes used by drivers having free route choice for their travel from A to B are equal. In other words, if q is assigned to links 1, 2 according to the user optimum, then the corresponding travel times are equal. This is the "natural" traffic distribution (without control). On the contrary, a system optimum assignment implies optimization of an appropriate criterion (e.g., minimization of total travel time). It is interesting to note that the user optimum assignment is generally different from the system optimum assignment. Assignment problems (both static and dynamic) for general traffic networks are of particular importance for several areas of traffic engineering. Dynamic aspects (both one-day dynamics and day-to-day dynamics) and stochastic aspects of traffic assignment are addressed in *Traffic Assignment, Dynamic*, *Traffic Assignment, Stochastic* and *Road Networks: Dynamic Equilibrium Models*.

The application of methods from adjacent fields to important traffic network problems is reported in the articles *Queuing Theory Applications*, *Network Modelling and Control: Store-and-Forward Approach*, *Shortest-Path Algorithms* and *Optimal Routing Applications*, whereas further general network problems and methods are included in *Intelligent Traffic Control Systems* and *Traffic Management Systems*.

Particular problems of mixed networks including both freeways and urban streets are addressed in *Corridor Control Systems*.

A further interesting field related both to traffic assignment and to shortest-path algorithms arises from future vehicle navigation systems. The initial problem here is to find the shortest path through a network for a vehicle with a declared destination. This problem may be readily solved if network traffic conditions are accurately predicted for a future period corresponding to the trip duration. However, if more and more equipped vehicles obtain recommendations on route choice, the problem becomes more complex because decisions on route choice may change the predicted network conditions. Traffic assignment algorithms may appear more appropriate in this case. Problems and methods relating to this interesting new area are reviewed in *Route Guidance, Individual* and *Route Guidance, Collective*.

7. Automated Vehicle Traffic

The present capacity and safety of roads and freeways is connected to the corresponding efficiency of human distance regulators. For example, the given reaction time of drivers in case of emergency requires a corresponding speed-dependent safety distance from the preceding vehicle. However, if distance regulation is performed automatically, the distance between vehicles may be reduced for the same speed values and this may have a dramatic positive impact on road capacity. At the same time, traffic safety may be increased if the electronic equipment works more efficiently and more reliably than human drivers. Extensive research work is currently under way and many engineering (as well as other) problems will have to be resolved before this idea becomes reality, as reported in *Vehicle Guidance by Computer Vision*.

Particular problems of automatic merging control are reviewed along with suitable control strategies in *Merging Control*.

Public transport systems are the main subject of *Automated Guideway Transit Systems* and *Personal Rapid Transit Systems*.

8. Rail Transport

Reliability and safety are two items of central concern in rail transport as reported in the series of articles *Railroad Systems: Line Supervision and Control*, *Automatic Train Control: Safety and Reliability* and *Automatic Train Control:*

Protection Principles, whereas modelling and control aspects in the classical sense are included in *Railroad Systems: Train Driving Control* and *Underground Railroad Modelling and Control*.

However, problems underlying railroad transport cannot always be formulated so as to match the well-founded available mathematical tools. The application of innovative approaches, such as expert systems design and discrete-event methods, is envisaged in *Expert Systems Approach to Rail Traffic Control* and *Underground Railroad Modelling and Control: Discrete-Event Approach*.

Particular attention is paid to recent developments for high-speed trains both from the design–theoretical and the operational point of view, as in *High-Speed Railroad: Modelling and Simulation, High-Speed Railroad: Systems Approach, High-Speed Railroad Networks in Europe* and *High-Speed Railroad Networks in Japan*. There is also an article devoted to magnetic suspension trains (see *Magnetic Suspension Railroad Systems*). Further new trends are reported in *Railroad Systems: Active Control*, whereas some organizational and hardware aspects are covered in *Underground Railroad: Organization of Operations* and *Railroad Electronic Signalling*, respectively.

9. Air Traffic Control

Hand in hand with economic growth, the growing importance and availability of leisure time and technological progress in aviation, the demand for air transportation has grown considerably worldwide, and this growth has been generally stronger than that for the more "classical" transport modes of bus, car and rail. More precisely, global air traffic of scheduled services grew from 500 billion passenger kilometers in 1971 to almost 1600 billion passenger kilometers in 1987. This means that the average annual demand rose by about 70 billion passenger kilometers (see *Flight Activity: Prevision, Flight Activity in General Aviation*).

Air traffic control systems are aimed at providing a safe and efficient flow of aircraft at airports and in the airspace. Particularly with high-density traffic and under poor visibility conditions, a system of centralized ground control is necessary to maintain safety and efficiency. The major requirement is for a means of surveillance of the traffic, but the system also requires communications, navigation facilities, weather information, rules and procedures. The form of air traffic control systems has varied greatly with time, weather, technology and political conditions. The article *Air Traffic Control: An Overview* provides an overall view, whereas more particular items of air traffic control are addressed in *Air Traffic Control Near Airports, Air Traffic Simulators, Visual and Instrument Flying Rules* and *Airborne Collision Avoidance*.

A further topic is aircraft automatic landing in bad weather conditions. The first automatic landing of a civil aircraft with passengers on board took place in London on 10 June 1965 involving a Trident operated by the British airline BEA. Developments since this first automatic landing, current problems and future trends are discussed in *Automatic Landing Systems*.

Communication problems, methods and systems in the context of air traffic control are reviewed in *Navigation Systems, Integrated, Data Links in Aeronautics* and *Data Processing in Air Traffic Control*. Finally, fuel consumption aspects are discussed in *Fuel Conservation: Air Transport*.

10. Sea Transport

The last group of articles is devoted to problems and methods related to sea transport including both traffic aspects and modelling and control tasks for individual ships. Some traffic aspects of sea transport are presented in *Vessel Traffic Services: Management from Shore*, whereas traffic organization issues are covered by *Marine Fleet Planning and Scheduling* and *Container Terminal Management*. Fuel consumption is discussed in *Fuel Conservation: Sea Transport*.

The modelling and control of ships or parts of ships viewed as a dynamic process is considered in the articles *Ship Automation and Control, Navigation Control of Ships, Marine Propulsion Plants: Control, Ship Dynamics: Modelling* and *Ship Rudder Roll Stabilization*.

Particular emphasis to the application of modern adaptive methods to ship control problems is given in *Ship Steering: Model-Reference Adaptive Control* and *Ship Positioning: Adaptive Control*.

11. Other Classifications

The attempted classification of the articles of this Concise Encyclopedia into the ten groups given may appear reasonable from an application point of view, but it does not consider a number of methods and aspects that are common to more than one traffic system such as fuel consumption, navigation, obstacle avoidance, organization, communications, modern control methods, modelling approaches, simulation and many further topics of common interest to several traffic and transportation areas. The interested reader may consult the subject index in order to discover different application areas of one and the same method or aspect. The interrelationships between methods and applications are one of the major reasons why an alphabetical list of articles is used instead of any grouped ordering.

Air Traffic Control: An Overview

Air traffic control (ATC) is a system whose basic objective is to provide safe and efficient flow of aircraft at airports and in the airspace. In clear weather with limited numbers of aircraft, individual pilots can see and avoid one another for safety and space themselves in queues for efficient flow of traffic. However, with high traffic density and in poor visibility, a system of centralized ground control is necessary to maintain safety and efficiency. This is the ATC system.

The major requirement is a means for surveillance of the traffic, but the system also requires communications, navigation facilities, weather information, rules and procedures. The form of ATC systems has varied greatly with time, traffic density, weather, technology and political conditions.

1. History of ATC in the USA

Attempts to reach international agreement on general rules for air traffic started as early as 1910, but were unsuccessful until the International Commission for Air Navigation was developed at the Versailles peace conference in 1919. Although the USA failed to sign the treaty, it followed many of the concepts developed by the Commission. In 1929 the first air traffic controllers used checkered flags to clear aircraft for takeoff at busy airports. The first radio-equipped tower was opened at the Cleveland Municipal Airport in 1930. In 1935 the principal airlines opened three airway traffic control centers at Newark, Chicago and Cleveland. The government assumed responsibility for their operation in 1936 and rapidly expanded the number of centers.

These early centers used a teletype network for relaying information between airports and those ground stations with radio telephone facilities. By 1946, the number of centers had expanded to 24 and has stayed at about that level ever since. The jurisdiction of airport control towers was expanded to include control over aircraft making approaches under instrument conditions.

During the early 1950s, remote communications facilities were developed to provide direct pilot-to-controller communications. In the late 1950s primary radar was introduced, which allowed controllers to see aircraft positions in real time, albeit without altitude or identity information (which still had to be obtained by radio communication). Secondary radar was introduced in the 1960s. In the 1970s the secondary radar was upgraded to provide 4096 identity codes plus automatic altitude reporting. Computer data processing and computer-generated displays were also added to both the en route centers and the terminal approach facilities, which by now had been located remote from the airport control tower.

Upgrading of the data processing has continued with the addition of new features such as conflict alert, which provides the controller with a warning when it appears to the computer that two aircraft will come dangerously close together. Also, an improved form of secondary radar called mode S will soon be implemented. With this improvement will come two-way, digital, data-link communications.

The development of the ATC system in other countries has followed a similar pattern. Mode S, for example, has already received approval from the International Civil Aviation Organization (ICAO).

2. Modes of Operation

Four modes of ATC system operation are identified as a function of the surveillance capability available. These modes follow the historical development of ATC surveillance and are sometimes associated with the generations of ATC. The first mode was communications only. In this mode, the controller establishes the position and altitude of each aircraft from a pilot report. Traditionally, the information was recorded on strips of paper (one per aircraft) and stacked sequentially on a board which was updated by the controller. Aircraft were cleared in order down an airway or onto an approach path based on a report on his progress by the pilot ahead.

The second mode was primary surveillance radar (PSR). In this mode, the location of aircraft can be seen on the radar screen, but the controller has to identify the radar target and obtain altitude from pilot reports. Traditionally, the identity and altitude information is written on a small piece of plastic, which is pushed along by a controller to keep it close to its associated radar target. The piece of plastic is called a shrimp boat because of its characteristic shape. Because the controller can see the location of each aircraft, he can "vector" an individual aircraft along an arbitrary course by commanding heading and altitude assignments to the pilot over the communications radio channel. This technique is almost always the method used for merging and spacing of aircraft into and out of high-density terminal areas. Should the radar fail, the controller is obliged to revert to the communications only mode, relying on pilot reports to establish aircraft location.

The third mode was secondary surveillance radar (SSR). In this mode, identity and altitude information are automatically reported and available for computer processing. The display is computer generated, with a data block next to each aircraft symbol replacing the old shrimp boat. Data processing also provides velocity information, display options and many other features that make data readily available to the controller. The

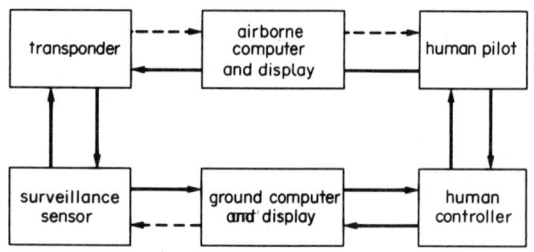

Figure 1
Information flow in air traffic control systems:
——, SSR; - - -, mode S

screen no longer has to be horizontal (to support the shrimp boats). The computer also provides controller prompting when aircraft approach control boundaries, violate their altitude clearance or are threatened by other traffic. Should the SSR mode fail, the controller reverts to the PSR mode.

The fourth mode is mode S data link. In this mode, communications can be sent and received via the surveillance system. It will be possible to have messages generated automatically and data requested and transferred automatically between air and ground without the intervention of pilot or controller. In each of the first three modes, the actual control of traffic was always carried out by the human controller and transmitted by voice commands. With mode S it will be possible to effect control by the computer, with clearances issued automatically via the data link.

The four modes are compared in Fig. 1, where the flow of information is shown schematically. The upper boxes represent airborne components, and the lower boxes represent ground components. For mode S with data-link operation, information can flow in all the channels. With SSR, the dashed line is lost as a communications channel and the airborne display for data-link information does not exist. Although limited data can come down through the surveillance channel, all the data going up have to go directly through the voice radio link between controller and pilot. With the primary radar mode, the down link through the surveillance channel is lost and there is no computer-generated display of down-link information. In the communications only mode, the only elements remaining are the pilot and controller, who pass information by the voice radio link.

3. Typical Flight Through the ATC System

The actual separation of traffic within the ATC system has always been carried out by human controllers. The airspace is divided into sectors and within each sector one controller has legal responsibility for the separation of all flights which he has accepted within his sector. As a flight proceeds from one sector of airspace to another, this responsibility is handed from one controller to another. The various controllers who handle a flight and their area of responsibility are as follows:

(a) ground controller, responsible for airport surface traffic;

(b) tower controller, responsible for takeoff and landing traffic within 8 km of airport;

(c) departure controller, responsible for departing aircraft in transition to en route flight;

(d) center controller, responsible for en route traffic within a sector of the center; and

(e) approach controller, responsible for arriving aircraft in transition from en route flight.

The ground and tower controllers are located in the airport tower, where they have visual contact with the majority of traffic they control. The approach and departure controllers are usually located together with their surveillance displays near the prime airport in their terminal area. The center controllers are housed in a remote location somewhere within the geographic area they serve.

The boundaries of the US centers are shown in Fig. 2. The airspace of each center is subdivided into about 20 sectors. The sector controllers of a single center sit side by side within the center facility. They can effect a hand off by simply talking to one another. The coordination of a hand off between sector controllers in adjacent centers requires intracenter communication over land lines or through the computer system.

To understand how the elements work together, consider a typical airline flight. A flight plan which includes the route, destination, cruise altitude and characteristics of the aircraft is filed in advance and stored on the ATC computer. Before takeoff, a clearance is sent to the departure airport tower for relay to the flight. Departure may be delayed until the en route centers and the destination airport are ready to accept the flight. Once the flight has been cleared, it is handed off from controller to controller as it progresses. At each hand off, the pilot is given the radio frequency of the next controller who has assumed separation responsibility.

The location and altitude of the aircraft are picked up at regular intervals by beacon and radar. On the basis of such data, the computer presents moving points on each controller's display screen that reflect the position and progress of each aircraft. Each point also carries information showing the identity, altitude and speed of each plane. In addition, the computer projects the path of each plane and gives visual as well as audible alarms if it expects two planes to come too close together. In general, alarms will be issued if lateral spacing becomes less than 8–16 km or vertical separation becomes less than 1000 ft (300 m), although these parameters vary with factors such as the altitude and speed of the aircraft. The controller advises the pilots of necessary changes in direction, altitude and speed to reach their desired destinations without conflict.

As each plane moves to the edge of a controller's sector, its supervision is transferred to the air traffic controller in the next sector, who may be at the same or another center. After such a hand off, the pilot is asked to contact the new controller by radio on a different, preassigned frequency. Thus, an aircraft will first be under the control of an airport control tower, then, in turn, under departure control, a series of sector controllers in several different centers, approach control and, finally, the control tower at the destination airport. Meanwhile, computers at the different centers pass along information about the plane to each controller who is to handle the flight.

If at any point a controller is unable to accept the aircraft into his sector, the aircraft has to be delayed until that controller is ready. An individual controller is capable of handling a maximum of about ten flights within his sector. As traffic density increases, it is necessary to reduce the size of the sector and increase the number of controllers. It would be desirable to automate the process, but that has yet to be accomplished.

4. Secondary Surveillance Radar

Secondary surveillance radar (SSR) as used in ATC is an outgrowth of the World War II military identification friend or foe (IFF) system. It is also referred to as the ATC radar beacon system (ATCRBS). It was originally introduced into ATC as a supplement to primary surveillance radar (PSR), the purpose of which was to provide data pertaining to the location and movement of aircraft.

The PSR tracks passive targets, whereas the SSR obtains tracking data by interrogating radar beacon transponders aboard the aircraft, which are equipped to provide identity and altitude information in their response (see *Data Processing in Air Traffic Control*). SSR employs two antennas for interrogating aircraft with pulse-coded signals within a range which may vary from one site to another and may be as large as 200 nautical miles (370 km). One of the antennas is highly directional and transmits a signal for the purpose of eliciting replies from transponders. The other is omnidirectional and transmits a signal, known as the sidelobe suppression (SLS) signal, to provide a reference amplitude which can be used to suppress replies that might be stimulated by the side lobes associated with the directional beam. Both antennas transmit their signals, multiplexed in time, at a frequency of 1030 MHz.

The directional antenna rotates at a rate that varies from station to station. A long-range en route SSR generally has a slow rotation rate (typically $\omega = 0.63\,\text{Hz}$) whereas a shorter-range terminal SSR has a faster rate (typically $\omega = 1.26\,\text{Hz}$). The beam width of the directional antenna is in the region of 3°. The sector over which a transponder will reply to the directional antenna signals is determined by the relative amplitudes of the omnidirectional and SLS signals. There should be a minimum of four to eight replies to interrogations per main beam passage for each interrogation mode, of which there are two types interlaced in an unspecified periodic pattern (i.e., the pattern may vary from one station to another). It is specified that aircraft above 5000 m must be able to transmit 1200 replies per second and aircraft below 5000 m must be able to transmit 1000 replies per second

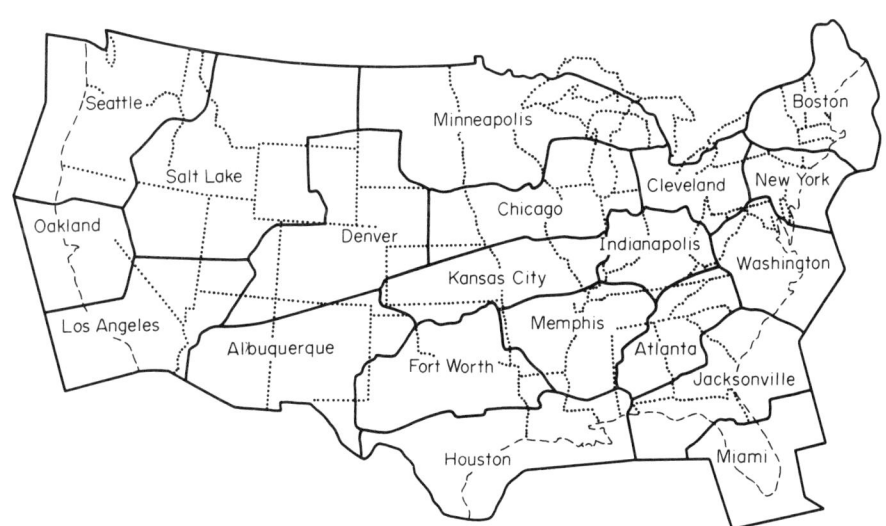

Figure 2
Boundaries of ATC centers in the continental USA

for a 15-pulse coded reply and, further, that the maximum interrogation frequency of any SSR shall be 450 interrogations per second.

Three pulses, designated in the sequential order of their transmission as P_1, P_2 and P_3, form the interrogation signal. The first, P_1, is transmitted by both antennas. The second, P_2, is the control pulse transmitted only by the omnidirectional antenna. The third, P_3, is transmitted only by the directional antenna.

The maximum power in the main beam of the directional antenna is recommended to be at least 24 dB above that in the strongest side lobe, and the power of the signal transmitted by the omnidirectional antenna is supposed to be at least 9 dB below the main beam maximum. Thus, an aircraft can determine whether it is receiving a main beam or a side-lobe signal by comparing the amplitudes of P_1 and P_2. The transmission of P_1 by both antennas is intended to aid in discriminating against interrogation signals which are multipath reflections from buildings or other structures rather than direct line-of-sight transmissions.

The time intervals between the P_1 and P_3 pulses are of discrete lengths to designate any of six corresponding modes. Two of the modes are relevant to ATC: mode A (8 μs) requests identification and mode C (21 μs) requests altitude.

Reply signals are transmitted at a frequency of 1090 MHz in a format consisting of two framing pulses, designated as F_1 and F_2 (see *Data Processing in Air Traffic Control*).

The transmission time of a reply is determined by the range propagation delay of a valid P_3 pulse; that is, one from a P_1, P_2, P_3 reply sequence for which the relative pulse amplitudes are in a relationship appropriate to a main beam transmission. Thus, when the SSR interrogator receives the transponder reply delayed again by the range propagation, it can determine the range to the transponder from a measurement of the total time between the transmission of the P_3 pulse and reception of the first framing pulse in the reply (which it recognizes by the 20.3 μs interval between the F_1 and F_2 pulses) after removing the fixed calibrated delays of the transponder. The SSR also estimates the azimuth of the transponder by beam splitting on all of the replies that have the same range delay.

Each SSR in a particular region transmits a different interrogation sequence pattern which can be used to identify it. The interrogation rate of any SSR is required to differ from that of any other by at least five interrogations per second. In addition, some transmit at a fixed pulse repetition interval (PRI), while others transmit in one of several possible staggered patterns.

5. Mode S

The development of selectively addressed SSR (mode S) was started in the early 1970s in the UK under the name ADSEL (MacKellar and Evans 1979) and in the USA under the name discrete address beacon

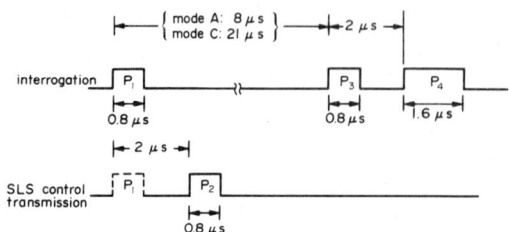

Figure 3
SSR interrogation waveform with mode S all-call pulse added

systems (DABS) (Drouilhet 1974). After a letter of agreement was signed between controlling agencies in the two countries in 1974, the two systems were developed cooperatively using a common signalling format. The advantages of selective addressing are improved accuracy and performance, plus the provision for data transmissions as part of the interrogation and reply formats.

To support these improvements, it is necessary to reduce the number of interrogations, which requires that reliable position information be extracted from each reply. This involves the use of a monopulse bearing measurement which is similar to the technique used in tracking radars, namely, using sum and difference antenna patterns and the ratio of signal amplitude in each pattern to locate the aircraft precisely in azimuth. To provide conformity with existing SSSR equipment, mode S operates on the frequencies of 1030 MHz and 1090 MHz with a compatible signal waveform. This is achieved simply by adding a fourth pulse, P_4, 2.0 μs after the standard P_1–P_3 combination presently used in SSR, as shown in Fig. 3.

This particular waveform with the P_4 pulse is defined as the all-call interrogation. It is used to interrogate all ATCRBS-equipped aircraft and for initial acquisition of the addresses of all mode S equipped aircraft for storage in a roll-call file. The SSR transponders recognize the first three conventional pulses, P_1, P_2 and P_3, but ignore the P_4 pulse, whereas Mode S transponders recognize the presence of P_4 as a request for discrete address. If there is no P_4 pulse present, as from an SSR interrogation, the mode S transponder replies in the conventional SSR mode. Once the ground sensor has acquired a mode S aircraft on roll call, the transponder is locked out from future all-calls, except under well-defined operational rules, and replies only when specifically addressed with its unique 24-bit discrete address code.

The discrete interrogation, as distinct from the all-call interrogation, is accomplished by again taking advantage of the SSR waveform. If an SSR transponder receives a P_2 pulse 2 μs after receiving the first pulse at the same relative amplitude or within 9 dB, it automatically suppresses and will not reply. Mode S uses this scheme to cause SSR transponders to suppress while interrogating on mode S. This is accomplished by radiating P_1 and P_2 in the main beam at the same amplitude. Mode S transponders recognize the P_1–P_2 combination as a discrete interrogation and wait 2 µs for the data block that follows. The data block contains either a 56-bit surveillance-only interrogation asking for the altitude of the aircraft, or a combination of surveillance and a 56-bit general data link message—a total of 112 bits.

There is also a third format that replaces both the 56-bit surveillance and the 56-bit data combination with an 80-bit data-only message. This latter format is used when long transmissions of data are required and is termed the extended length message (ELM). Up to 16 ELM segments can be transmitted sequentially to any suitably equipped aircraft while it is in the main beam, requiring only a single reply to acknowledge receipt. This allows large quantities of data to be transferred without excessive replies.

The modulation used for the uplink data block is differential phase shift keying (DPSK); a phase shift of 180° during a 0.25 µs bit space represents a binary 1; no phase shift in the bit space is 0.

The mode S reply format uses a pair of double pulses as a preamble to the data block containing either the 56-bit surveillance-only reply or the combination of 112-bit surveillance and data message or ELM reply. The modulation used on the downlink is pulse position modulation (PPM), where the position of the pulse determines whether it is a binary 1 or 0. The bit rate for all uplink formats is four megabits per second and one megabit per second for the downlink.

Mode S provides a position accuracy about six times as good as that of SSR, plus a reliable digital data link. Implementation is expected to have taken place by the early 1990s.

6. Backup Assurance

While the ground-based ATC system as described above is intended to prevent conflicts between aircraft, there are many instances when, for one reason or another, two aircraft fail to maintain separation. Research has been continuing since the mid-1960s to develop an airborne collision avoidance system (CAS) which would provide a backup to the primary ground-based ATC system when two aircraft, for whatever reason, find themselves on a near collision course with only just enough time remaining to make an escape maneuver (Bulford 1972, 1978) (see *Airborne Collision Avoidance*). The traffic alert and collision avoidance system (TCAS) is an outgrowth of the 20 years of CAS development and shows considerable promise for near-term implementation. TCAS is built around the mode S transponder.

The mode S transponder performs the same basic functions as the mode C transponder; that is, it automatically transmits aircraft identity and altitude data when triggered from the ground or the air. The principal advantage of the mode S transponder is its discrete, or selective, address capability. This selective address capability results from the fact that signalling formats which provide more than 16 million available codes are utilized, compared with 4096 for mode C equipment. This means that each mode S unit can be assigned a permanent call number.

In use, the mode S transponder transmits a periodic "squitter" signal which informs all TCAS-equipped aircraft in the area of its identity. This identification process establishes data-link communications for relaying TCAS advisories and collision avoidance information.

TCAS represents a development of the beacon collision avoidance system (BCAS) (Welch and Orlando 1980). The major technical improvement is the use of a directional antenna that permits reliable operation in high-density terminal areas. Like BCAS, TCAS determines the position of other aircraft in the vicinity by interrogating their transponders and analyzing the replies. It then presents appropriate traffic advisories and conflict resolution advisories to the pilot on his cockpit display.

In addition, TCAS will generate collision avoidance advisories for all conflicts involving aircraft with SSR altitude-reporting transponders. Initially, the equipment will provide only vertical escape maneuvers and the pilots will be advised either to descend or to climb. Horizontal maneuvers are a possibility for later versions. Moreover, when two TCAS aircraft are in conflict, the avoidance maneuvers will be coordinated via the mode S data link.

TCAS also communicates its position and intended collision avoidance maneuvers to mode S equipped airplanes using the crosslink feature. Proximate aircraft can be displayed on a VDU relative to the position of the receiving aircraft. Resolution advisories are typically generated with only 30 s remaining until closest approach.

Another approach to backup assurance is a cockpit display of traffic information (CDTI). CDTI differs from TCAS primarily in the longer range over which the CDTI display shows proximate aircraft. CDTI would also show a map background on the display so that the pilot could monitor the ATC operation as it applied to traffic in his vicinity. Although there has also been considerable research on CDTI (Connelly 1975), it is expected that TCAS will be implemented first.

Bibliography

Alexander B 1969 *Air Traffic Control Advisory Committee Report*. US Government Printing Office, Washington, DC

Bulford D E 1972 Collision avoidance: An annotated bibliography, FAA-NA-72-41. Federal Aviation Administration, Washington, DC

Bulford D E 1978 Collision avoidance: An annotated bibliography, FAA-NA-78-8. Federal Aviation Administration, Washington, DC

Connelly M E 1975 Applications of the airborne traffic situation display in ATC. *AGARD Conf. Proc.*, Vol. 188. Advisory Group for Aerospace Research and Development, Neuilly-sur-Seine, France

Drouilhet P R 1974 Provisional signal formats for the discrete address beacon system, MIT Lincoln Laboratory ATC-30, FAA-RD-74-62. Federal Aviation Administration, Washington, DC

Gilbert G A 1973 *Air Traffic Control: The Uncrowded Sky*. Smithsonian Institution Press, Washington, DC

Karp S, Haroules G G, Klein L (eds.) 1973 Special issue on aeronautical communications. *IEEE Trans. Commun.* **21**(5)

Kayton M, Fried W R 1969 *Avionics Navigation Systems*. Wiley, New York

MacKellar A C, Evans A J 1979 ADSEL—An improved form of secondary surveillance radar. *ICAO Bull.* **34**(3)

Orlando V A, Drouilhet P R 1980 DABS: Functional description, MIT Lincoln Laboratory ATC-42A, FAA-RD-80-41. Federal Aviation Administration, Washington, DC

Welch, J D, Orlando V A 1980 Active beacon collision avoidance system (BCAS), MIT Lincoln Laboratory ATC-102, FAA-RD-80-127. Federal Aviation Administration, Washington, DC

W. M. Hollister
[Massachusetts Institute of Technology,
Cambridge, Massachusetts, USA]

Air Traffic Control Near Airports

The increase in the volume of air traffic since the mid-1970s has led to a congestion of the air space, particularly in airport areas. This condition leads to interrupted climbs and descents, wasted time, and increases in both delay and fuel burnt resulting in increases in both operating costs and work load for pilots and control services. In turn, this means possible adverse consequences for safety. Such a situation will worsen in the future since an increase of 5% per year of total flights controlled seems to be realistic in spite of the so-called energy crisis (see *Flight Activity: Prevision*).

Short-term control in an airport area is exercised basically through area navigation. This allows good use of runway capacity while ensuring the desired separation between aircraft. However, this does not guarantee a satisfactory decrease in fuel consumption and delays because of the brevity of the control time span, the necessity to modify the ground track (i.e., the ground projection of the flight path), the low altitude and, consequently, the high-fuel-consumption flight evolutions and holding patterns. Hence, the need to optimize the use of available system capacity leads to the consideration of an extended zone of control.

Such an extended zone of control, combined with an accurate prediction of future flight paths, should allow aircraft to approach the runway on uninterrupted descents at idle thrust (leading to a reduction in fuel consumption) and along given ground tracks (giving a controller workload reduction). By considering the flux of aircraft entering the system, it should then be possible to ensure a "global" optimization of the traffic.

1. Analysis of the Problem

Before considering the control problem itself, a global analysis should be made of the objectives and constraints. The system constituted by air traffic management in a zone of convergence is particularly complex due to the variety of objectives that must be satisfied, the number of constraints inherent in the system, the variety of functions to be performed (e.g., management, sequencing, scheduling, conflict detection and resolution) and the modelling problems associated with variations in aircraft performances.

Broadly, the problem is to guide to the runway a flow of aircraft which enter the zone in a quasirandom fashion with respect to altitudes, speeds and arrival times.

In addition to constraints related to individual operating conditions of aircraft and to the constraints regarding procedures, an air traffic control (ATC) system must satisfy the safety standards which lay down a minimum separation between any two planes at any instant.

The overall objective is to minimize the global transit cost for all aircraft in the system. This global cost reflects the capability of the system to reduce both delays (i.e., indirect operating costs) and total consumption of fuel. The two objectives of minimization of transit time and of consumption are somewhat conflicting; minimum transit time for an individual aircraft requires the aircraft to proceed at the maximum allowable speed. Such a procedure may decrease some of the indirect operation costs, but at the expense of a significant increase in the fuel consumption. Alternatively, the minimum fuel trajectory requires the aircraft to descend at low speed. In practice, some compromise between these two policies is looked for by airlines, by weighting time spent and fuel burnt. The corresponding strategy may vary from one company to another and may be summarized by the definition of a "preferred profile" for each type of aircraft (see *Fuel Conservation: Air Transport*).

If the total cost of transit for all aircraft in the zone from a global point of view is considered, with the same criteria, the solution can be analyzed as a compromise between

(a) the minimization of the sum of the delays encountered by all the aircraft (a delay is the difference between the transit time actually experienced by an aircraft from its entry into the zone to landing and the nominal time it would have spent if alone in the zone and descending on its preferred profile); and

(b) the minimization of the global quantity of fuel burnt by all aircraft (it is important to note that the fuel consumption is dependent not only on time spent, thrust and speed, but also on altitude and, thus, for a given delay, fuel consumption depends on whether the delay occurs in level flight or in idle-thrust descent in the case of a level tracking on its altitude).

Any control procedure aimed at minimizing cost and/or fuel consumption will be efficient (i.e., will bring appreciable savings) only if the extent of the zone is sufficiently large. For this reason, the zone of control considered in this article, called the zone of convergence, is larger than the current airport area; for practical purposes, the radius of this zone varies from 100 nautical miles (185 km) to 200 nautical miles (370 km) around the airport of interest. In western Europe, a 100 nautical mile zone could be implemented on a national level. A 200 nautical mile zone implies international cooperation, in some cases, but its potential savings will be appreciably higher.

2. Dynamic Scheduling

The scheduling technique is essentially based on landing sequencing and on control of cruise and descent speed profiles. The landing sequence and the descent profiles have to be chosen in order to minimize the predefined criterion (i.e., the total transit cost for all aircraft from their entry into the zone to touchdown).

One of the difficulties associated with scheduling lies in its dynamic nature: each time an aircraft enters the zone a new optimal sequence has to be computed for the aircraft in the zone. This requires the design of an efficient optimization technique compatible with real-time operation.

There are two basic ideas underlying scheduling policy. First, it may be possible to improve the landing sequence since the minimum time separations between two landing aircraft depend on the types of plane involved. Second, it is possible to control the descent speed of aircraft, in order to improve the traffic flow: a fast descent may decrease the waiting time of following aircraft and a slow descent allows delays to be absorbed with reduced consumption, the waiting time being spent at idle thrust.

2.1 Sequencing Constraints

The safety standards at the runway or, more specifically, on the last common track of the instrument landing system (ILS) are defined in terms of longitudinal separation. This separation $s_l(i,j)$ between two aircraft, the jth aircraft following the ith aircraft, depends on the couple considered, since vortex phenomena imply a greater separation behind a jumbo jet than behind a medium-haul aircraft. To this longitudinal separation in nautical miles corresponds a time separation for landing $s(i,j)$. This depends on the longitudinal separation, the length of the last common track and the speeds of the two planes. The range of

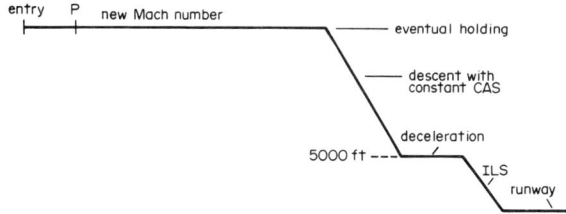

Figure 1
Classical aircraft descent procedure

speeds and the differences in length of the last common track of the routes lead to an important range of separations. The ratio between extreme values may reach a value of three. Because of this range of separations, optimization of the landing sequence may be efficient and improve the runway occupancy, thereby reducing potential delays.

2.2 Descent Procedure

The descent procedure considered here is a classical one and is compatible with present navigation and control equipment. An aircraft descends at idle thrust along a "Mach–CAS" profile; that is, the first part of the descent is at constant Mach number and the second part, below "transition altitude," is at constant conventional air speed (CAS). This descent is uninterrupted until approach altitude, which can vary between 3000 ft (900 m) and 5000 ft (1500 m) for different airports. The aircraft then decelerates at this altitude to its approach speed and intercepts the glide path. Further along this path, the aircraft decelerates to landing speed. This is shown in Fig. 1.

If the route is sufficiently long there is the possibility of en route speed control. In this case, after entry into the zone, the aircraft keeps its cruise Mach number for 4 min and adopts the new Mach given by the control (at point P in Fig. 1) for the last part of its cruise. The holding pattern, when it exists, takes place before descent. High-altitude holding returns better fuel consumption figures than low-level holding and also leads to a better occupancy of air space.

2.3 Descent Speed and Cost

For different speed descents with the same idle-thrust setting, the slope increases with speed. Hence, the beginning of descent must occur earlier for a low-speed descent than for one at high speed. This augments the phenomenon of fuel consumption reduction with decreasing speed. Considering two trajectories between the zone entry point and the runway, one flown at high speed and the other at low speed (see Fig. 2), the low-speed cruise gives lower fuel consumption and has a shorter flight path length, which can lead to substantial fuel savings. For example, if a typical medium-haul aircraft is considered, the fuel consumption over a distance of 200 nautical miles (370 km) between entry

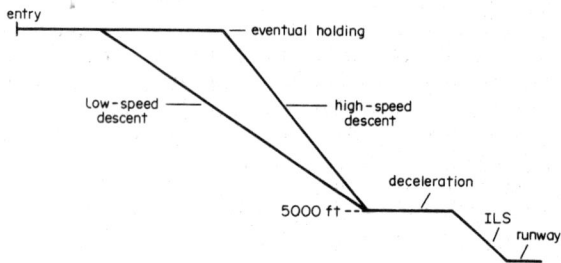

Figure 2
Alternative speeds of descent

and touchdown can range from 810 kg for optimum fuel profile to 1130 kg for the highest acceptable speed, an increase of over 30%.

Computing the cost associated with different speed profiles, fuel cost increases with speed while cost associated with transit time decreases. Hence, for the combined cost, a curve is obtained as a function of speed or as a function of transit time with the minimum corresponding to the preferential profile (see Fig. 3).

The choice of the aircraft speed has to be made from within the range of feasible speeds allowed by the performance manual of the aircraft. Maximum and minimum speeds correspond respectively to minimum and maximum transit times without holding; any increased transit time implies a holding time. It can also be noticed that for any time between the minimum and maximum transit times, the cost corresponding to a trajectory with no holding is a minimum, in the sense that any trajectory corresponding to the same transit time but flown at higher speed and including a holding pattern has a higher cost.

In the scheduling algorithm it is assumed that, for each aircraft, the choice of possible speeds has to be made from a finite set of speeds. This assumption is made with two objectives:

(a) to make the transmissions between control and pilot easier (the definition of possible speeds with 10 knot (19 km h^{-1}) intervals seems reasonable), and

(b) to preserve the combinatorial and discrete nature of the sequencing problem.

Practically, six profiles have been defined as possible choices for each type of aircraft: six descent speeds when only the descent speed is controlled and six Mach–CAS values for the case where en route speed control is carried out.

2.4 Sequencing

Each aircraft descends at its own chosen speed and, on reaching the holding point, holds if necessary. Each aircraft then lands in the order it arrived. This can be represented by first-come, first-served at runway sequencing (FCFSRW). The landing sequence is based on the landing times for aircraft descending with preferential speed profile (see Fig. 4a). If holding is necessary it is done at low altitude.

This policy has been used as a reference to compare the cost obtained by the following proposed scheduling procedures and to demonstrate the efficiency of these new scheduling methods.

The possibility of modifying the FCFSRW landing order by shifting aircraft forward or backward in the sequence can be considered. On the basis of cost minimization, however, a constraint must be put on this modification in order to avoid the phenomenon of indefinite delay for "cheap" aircraft (with low fuel consumption) and of priority for "expensive" (high fuel consumption) aircraft. This constraint may be expressed in terms of a maximum waiting time but, to be efficient, the maximum value has to be modulated as a function of traffic density. For this reason the concept of maximum position shifting (MPS) is preferred. A constraint is put on the number of positions an aircraft can be shifted backward or forward. Moreover, this form of constraint has the advantage of limiting the number of feasible solutions, and hence the number of

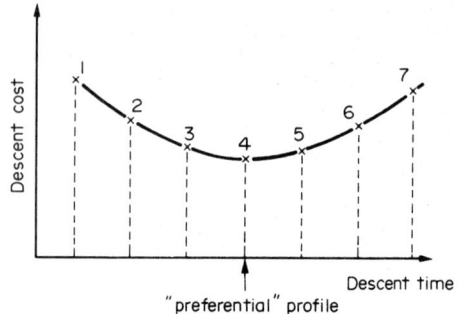

Figure 3
Cost of various descent speeds (1–7)

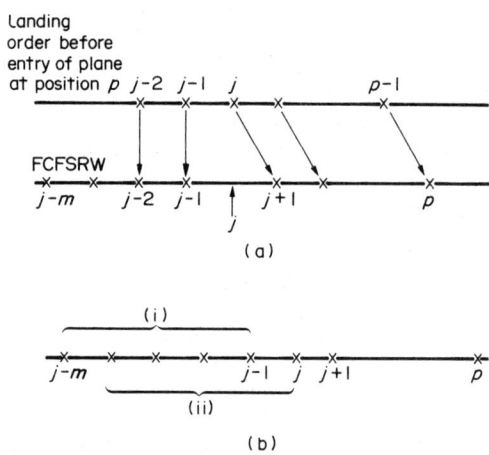

Figure 4
The problem of sequencing: (a) FCFSRW and (b) reordering planes to minimize the total transit cost

solutions to explore, and the solution can be found in real time.

The value of the MPS parameter is chosen to be an integer between one and four. An MPS equal to four means that an aircraft that occupies position n in an FCFSRW sequence may occupy positions $n-4$ to $n+4$ in the new schedule.

2.5 Optimization

Each time a new aircraft enters the zone, the problem of finding the "best" landing sequence and the "best" descent profiles to minimize a generalized cost criterion must be solved.

This optimization applies to all aircraft in the zone that have not initiated descent; that is, 1 min before descent (or holding when it exists), the highest point of the descent profile is given to the aircraft and cannot be subsequently changed. In the case where en route speed control is applied, the cruise and descent speed profile is fixed 1 min before cruise Mach-number change (i.e., 3 min after zone entry), but the landing time may be changed throughout the holding pattern until 1 min before holding (or descent).

When the pth aircraft enters the zone its landing number in the FCFSRW order can be determined on the basis of its potential landing time on the preferential profile descent with no delay. Thus, an optimal permutation of landing order under the constraint of maximum position shifting for all aircraft must be found. If the FCFSRW order of the entering aircraft is j and the maximum position shift is m, the possible positions for this plane are given by the increasing sequence of integers beginning with $j-m$ and ending with j, so that scheduling for all aircraft from $j-m-1$ to j must be considered. This optimization is achieved in several steps (see Fig. 4b):

(a) Determine the sequence of the $m+1$ aircraft preceding j, with j landing last, and determine the profiles for those aircraft (there being n possibilities for each), including aircraft j, in order to minimize their total transit cost (see (i) in Fig. 4b).

(b) On the basis of the new sequence, repeat the preceding scheme considering the $m+1$ aircraft preceding $j+1$, with aircraft $j+1$ landing last (see (ii) in Fig. 4b).

(c) Repeat until aircraft p is the last to land.

So, at each of the $p-j+1$ steps of this optimization, the algorithm has to select one among $(m+1)!$ permutations of aircraft and, for each of the $m+2$ aircraft, it has to select one among the n possible descent profiles that can be assigned to an aircraft.

The objective is to minimize the total transit cost (time plus fuel) under the constraint of the landing separations, which are given in the form of a time-separation matrix. The number of possibilities is $(m+1)! n^{m+2}$ at each of the $p-j+1$ steps. With six possible profiles ($n=6$) and a maximum value of position-shifting parameter m of four, the number of possibilities is approximately 5.6 million. Of course, if one of the aircraft concerned has attained the point after which it cannot be reordered, the optimization applies to the subsequent aircraft only and the number of cases is reduced.

An optimization program suitable for solving this problem within the computing-time constraints compatible with real-time use has been developed. The branch-and-bound algorithm is used. This essentially consists of separating the set of solutions into subsets for which the evaluation of a lower bound of the criterion is possible and exploring at each step the branch with minimum lower bound. The efficiency of such an algorithm depends mainly on the quality of both the separation principle and the evaluation of the lower bound for the criterion. In the present case, the algorithm performances are most satisfactory and lead to an average computing time of 0.5 s on a typical minicomputer to obtain the optimum solution each time a new aircraft enters the system.

The derivation of optimality and constraints equations, as well as details concerning the branch-and-bound resolution algorithm, are given in Sect. 5.

3. Operational Considerations and Constraints

3.1 Control–Pilot Communication

The scheduling program is run every time a new aircraft enters the zone. The transmission of the type of aircraft, its weight, flight level and speed allows the optimization program to select a landing sequence and a speed for all aircraft still cruising in the zone. The schedule is recomputed at each new entry, but the controller has to communicate the result to each aircraft only once; in fact, the controller transmits descent parameters to an aircraft only when these are firmly decided and the aircraft cannot be reordered any more (i.e. 1 min before descent or eventual holding). The parameters to be transmitted are holding duration, top of descent (i.e., the point where the descent has to be initiated) and speed of descent.

It has been deliberately supposed that transmissions take place without a data link; this is unnecessary since the small number of parameters to be transmitted will not significantly increase the workload for both pilot and control.

3.2 Aircraft Separation and Conflict Resolution

In the sequencing process, the separation constraint between aircraft is taken into account only for the approach and landing phases, and takes the form of the time-separation matrix. It ensures standard separation of all aircraft between the end of their descent and the runway.

When the optimization is performed, the speed profiles for the aircraft and their times of landing are known, and their trajectories can be predicted: it is then necessary to verify that they are nonconflicting. A conflict prediction algorithm is implemented and the solutions proposed for any resolution are based either on flight-level change or earlier descent. The associated speeds are, in all cases, computed so that the aircraft land at the scheduled time and the landing order remains unchanged.

3.3 Trajectory Prediction and Tracking Precision

(*a*) *Prediction and aircraft modelling.* The sequencing algorithm uses, as the data for each aircraft, the predicted time of landing and the cost (i.e., predicted consumption) associated with all possible predefined speed profiles. These data must be available for computation very shortly after the entry of an aircraft into the zone—as soon as its type, flight level, cruise Mach number and weight are known.

Generally, models of aircraft that include aerodynamics and propulsion characteristics are well known as the flight mechanics equations but, to meet the requirements of computation time, it is not feasible to integrate this model on-line. Therefore, the model used should be of some "aggregate" form. The kind of representation used can be precomputed times and consumptions for each type of aircraft, and can be for different entry levels, aircraft weights and winds. The method used here is the representation proposed by Benoit and Swierstra (1981) in the form of polynomial approximations of time, distance for descent and fuel consumption as functions of speed, given for different weights and flight levels. The meteorological conditions (pressure, temperature and, most importantly, winds) must also be introduced.

The accuracy of such polynomial approximations is sufficient for the scheduling purpose, but the main difficulty lies in wind prediction. Several experiments have been conducted to try to evaluate the error on final time between a precomputed trajectory and a real trajectory (Imbert *et al.* 1976, Pelegrin and Imbert 1977). All experiments have indicated that the error is due mainly to the poor accuracy of wind conditions prediction and, especially, of altitude wind profile prediction.

The results show that the errors in the final arrival times are too large to be consistent with an efficient scheduling strategy. If the separation times taken into account for the scheduling must include the uncertainty in arrival times, the runway capacity decreases considerably.

If the efficiency of the scheduling algorithm relies heavily on the capability to predict an aircraft trajectory, the problem can be reversed and it can be considered as relying on the capability of an aircraft to fly a predefined trajectory and reach the runway at a prescribed time, with or without control intervention. In order to ensure the final-time accuracy, the scheduling function must then be associated with some trajectory tracking algorithm.

(*b*) *Trajectory tracking.* It is clear that the accuracy of trajectory tracking relies on the technological aids available, either for trajectory data transmission or for actual tracking.

The feasibility of four-dimensional control of the actual position of the aircraft with respect to the desired position can be envisaged in the near future. This assumes the capability to accurately measure the position of an aircraft and the generation of the corresponding control commands. This function could be ensured by onboard computers of flight management system (FMS) type which have been developed by several avionics manufacturers.

A less sophisticated solution, compatible with current technology or in the future with small-aircraft equipment, relies on speed and heading corrections along the descent trajectory (Imbert *et al.* 1976, Pelegrin and Imbert 1977). Taking into account the control load and the pilot task, the number of corrections has been limited to three: two speed adjustments during level flight and descent, and a heading correction before ILS interception. The correction magnitude is computed on the ground on the basis of the recorded errors between theoretical and actual positions of aircraft, and the new value of speed or heading is transmitted to the pilot by the standard present transmission means.

The proposed solution has been tested in simulation and in the real environment on commercial flights. The error in final arrival time, which could be 1 min in the worst cases without any correction, can be reduced to the order of 10 s with correction.

4. *Example of Results*

Different samples of traffic on different real extended terminal areas have been checked in simulation in order to investigate the potential benefits of the proposed type of control.

A traffic sample is characterized by the distribution between different aircraft classes and by its density: the traffic density may be defined with reference to the runway capacity, which corresponds to the average number of aircraft that may land on the runway per hour. Since the time separations depend on the aircraft types, the capacity is a function of distribution and of the separation matrix. An average runway capacity is about 36 aircraft per hour. The traffic density is defined as a percentage of the runway capacity.

The samples for which the results are presented in this article include nine types of aircraft, ranging from a twin-engine light transport aircraft (Aerospatiale N 262) to wide-bodied long-range aircraft (McDonnell Douglas DC10 and Boeing 747).

Figure 5 shows the results obtained for different samples: they have been generated by considering the same sequence arriving at zone entry gates at a rate ranging from 60% to 125% of the runway capacity. The

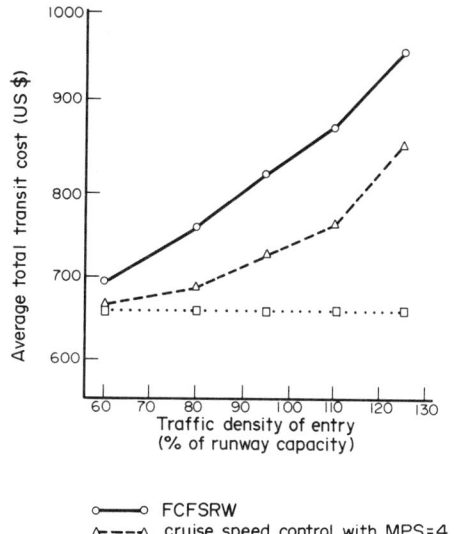

Figure 5
Comparison of costs of landing using different control procedures in the mid-1980s

zone considered was a fictitious zone with twelve entry gates located 200 nautical miles (370 km) from the runway and convergence points located at 100 nautical miles (185 km) and 60 nautical miles (110 km) from the runway. The average total transit cost was plotted from entry to landing for different policies: the FCFSRW procedure and optimization with cruise speed control and maximum position shifting parameter equal to four. The reference is the average minimum transit cost per aircraft, that is, the transit cost on preferential profile with no delay. This minimum may certainly not be achieved for saturated or oversaturated traffic; however, this value has been adopted in the absence of any measure of the reachable minimum.

The FCFSRW procedure can be considered as an approximation of the current practice. Nevertheless, it is clear that this procedure and the assumptions underlying it (i.e. uninterrupted descent and generalized use of preferential profiles in particular) constitute an optimistic view of the real situation as far as overall transit costs and fuel conservation procedures are concerned. This obviously implies that the system efficiency is somewhat underevaluated.

From these results and all the results obtained for different traffic samples and different zones (Imbert and Comes 1978, Imbert et al. 1979, Imbert 1980, 1982), some conclusions concerning the benefits expected from dynamic scheduling may be deduced. First, as already noted, the computing time is certainly compatible with real-time implementation, with an average of 0.5 s for each new computation. Second, an obvious conclusion is that the optimization is more effective in saturated conditions of traffic, since there is no major problem of waiting time for traffic density under 80%. Third, efficiency is increased only for a sufficiently large zone. The sequencing algorithm is applied when an aircraft enters the zone to all aircraft still in cruise. If the length of the routes is too small, aircraft are in descent when entering the zone or begin their descent very soon after this so that the number of aircraft in cruise is not sufficient for efficient sequencing. Fourth, the length of the different routes also have to be "homogeneous." A zone with very long and very short routes may lead to a kind of "locking" of the system. If an aircraft arrives in the zone very shortly before landing, it is not possible to reorder the other aircraft that are already descending and the new plane has to wait. In this case, the result of dynamic scheduling, with its hypothesis of uninterrupted descents, can be worse than the result of present policy.

5. Mathematical Treatment

When the pth aircraft enters the zone, if its FCFSRW number is j, the following problem must be solved $p-j+1$ times: find the sequence and the speed profiles of N aircraft, and the speed profile of the $(N+1)$th that minimize the total transit cost, under the runway separation constraint.

The criterion for these $N+1$ aircraft is

$$C = \sum_{i=1}^{N+1} \alpha_i t_i + \beta_i C_i \quad (1)$$

where i refers to the present order of the $N+1$ aircraft, t_i is the transit time in the zone from entry to runway including eventual holding time, α_i is the cost per unit time (fuel excluded) for the ith aircraft, C_i is the total fuel consumption of the ith aircraft in the zone and β_i is the unit fuel cost (see *Fuel Conservation: Air Transport*).

The criterion may be separated into two parts, one corresponding to transit itself, the other to the eventual holding pattern. Defining t^t_{ik} as the transit time of the ith aircraft on the kth speed profile ($1 \leq k \leq k_{max}$), c_{ik} as the corresponding consumption, t^h_i as the holding time and δ_i as the holding fuel consumption per unit time, then

$$C = \sum_i \alpha_i(t^t_{ik} + t^h_i) + \beta_i(c_{ik} + \delta_i t^h_i)$$

$$= \sum_i (\alpha_i t^t_{ik} + \beta_i c_{ik}) + \gamma_i t^h_i$$

$$= \sum_i C_{ik} + \gamma_i t^h_i \quad (2)$$

where C_{ik} is the total transit cost on kth profile and γ_i is the total holding cost per unit time.

The problem is to find the landing position j and the speed profile for each aircraft i. Define

$$y_{ijk}: \quad 1 \leq i \leq N, \quad 1 \leq j \leq N, \quad 1 \leq k \leq k_{max}$$

where

$y_{ijk} = 1$, iff aircraft i lands in the jth position and descends with kth profile, where the permutations (ijk) satisfying this condition are to be provided,

and

$$y_{ijk} = 0 \quad \text{in all other cases} \tag{3}$$

Note that

$$\sum_{jk} y_{ijk} = 1 \tag{4}$$

$$\sum_{ik} y_{ijk} = 1 \tag{5}$$

In addition, define

$$Z_k: 1 \leq k \leq k_{max}$$

where

$Z_k = 1$ iff aircraft $(N+1)$ descends with kth profile

and

$$Z_k = 0 \quad \text{otherwise} \tag{6}$$

Note that

$$\sum_k Z_k = 1 \tag{7}$$

In the following, the subscript i will correspond to a particular aircraft and the subscript j will correspond to a landing position. The separation constraints are easier to write if the holding time is defined corresponding to a landing position, rather than to a particular aircraft.

Let t_j^h be the holding time of aircraft landing in the jth position where

$$1 \leq j \leq N+1 \tag{8}$$

and let t_0 be the initial time (i.e., the instant when the previous aircraft has landed), S_{0i} the separation time between this aircraft and aircraft i, S_{il} the separation time between aircraft i and l, S_{iF} the separation time between aircraft i and $N+1$, and t_{ik} the possible time of arrival at the runway of aircraft i descending on profile k (with no holding). With these definitions, the landing time of an aircraft in the jth position is

$$\sum_{ik} y_{ijk} t_{ik} \tag{9}$$

and the separation time between aircraft landing in positions j and $j+1$ is

$$\sum_{ilkk'} y_{ijk} y_{l(j+1)k'} S_{il} \tag{10}$$

So, the $N+1$ separation constraints may be expressed as

$$t_0 + \sum_{ik} y_{i1k} S_{0i} \leq t_1^h + \sum_{ik} y_{i1k} t_{ik}$$

$$t_1^h + \sum_{ik} y_{i1k} t_{ik} + \sum_{ilkk'} y_{i1k} y_{l2k'} S_{il} \leq t_2^h + \sum_{lk'} y_{lk'} t_{lk'}$$

$$\vdots$$

$$t_{N-1}^h + \sum_{ik} y_{i(N-1)k} t_{ik} + \sum_{ilkk'} y_{i(N-1)k} y_{lNk'} S_{il} \leq t_N^h + \sum_{lk'} y_{lNk'} t_{lk'}$$

$$t_N^h + \sum_{ik} y_{iNk} t_{ik} + \sum_{ik} y_{iNk} S_{iF} \leq t_{N+1}^h + \sum_k Z_k t_{(N+1)k} \tag{11}$$

The problem is then to find the y_{ijk} and the Z_k minimizing C under the constraints in Eqn. (11).

The resolution is a branch-and-bound technique; the principle is to branch the set of solutions into subsets, with an evaluation at each separation of a lower bound of the criterion. In this case, the separation is based on the allocation of a landing position to an aircraft and on the choice of a particular profile.

ALGORITHM

Step 1: The solutions are separated into N subsets, subset i corresponding to aircraft i in the first position.

Step 2: Each of those subsets is separated on the basis of all possible profiles for this first aircraft.

Step 3: The subset of solutions, with aircraft i with the kth profile landing in the first position, is divided into subsets corresponding to plane l in the second position $(l \neq i)$.

This repeats until the profile of the $(N+1)$th aircraft.

The tree generated for the case where $N = 5$ and with six possible speeds per aircraft has about 5×10^6 nodes. Some of these can be eliminated by considering MPS constraint or criterion properties. At each node of the tree, a lower bound of the criterion must be evaluated for all solutions of the subset it represents. The criterion may be written as

$$C = \sum_i C_{ik} + \sum_j \sum_{ik} (\gamma_i y_{ijk}) t_j^h + t_{N+1}^h y_{N+1} \qquad (12)$$

In the first term, if the profile of aircraft i is not chosen at the considered node, the minimum cost for this aircraft is taken (preferential profile).

For the second term representing waiting-time cost, a lower bound of the waiting time has to be evaluated for nonassigned positions. To evaluate a lower bound of the waiting time of an aircraft in the jth position the first inequality is used where it appears:

$$t_{j-1}^h + \sum_{ik} y_{ijk} t_{ik} + \sum_{ilkk'} y_{i(j-1)k} y_{ljk'} S_{il} \leq t_j^h$$

$$+ \sum_{ik} y_{ijk} t_{ik} \qquad (13)$$

The quantities in the first term are replaced by their minimum value, the additional quantity of the second term is replaced by its maximum value.

The waiting times or their lower bound correspond to a position and not to an aircraft, but the waiting-time costs are known per aircraft; a lower bound of the total waiting cost is obtained by multiplying the waiting times (in order of decreasing values) by the costs (in order of increasing values).

Hence, at each node, a lower bound of the criterion may be evaluated for all the solutions it includes. This value has to be compared with all node evaluations already achieved and the algorithm starts from the node with the lowest value.

The convergence properties of the algorithm are related to the quality of evaluation and separation. The computing time and the core occupancy depend on the number of explored nodes; at each step it is necessary to store the extreme nodes with the corresponding value of the criterion.

The convergence of the algorithm with the predefined separation and evaluation techniques are most satisfactory and lead to the possibility of real-time implementation.

6. Scheduling for Several Airports

Practical results have pointed out the potential benefits of dynamic scheduling of traffic in a convergence zone. The principle is that with an enlargement of the zone it is possible to use the speed as control variable in order to improve the traffic flow and optimize the landing sequence.

An extension of this principle is coordination between several airports. Good as the sequencing algorithm may be, it has to deal with randomly entering aircraft. Taking into account the trajectory of aircraft further "upstream," it is possible to take their departure time as a control variable and to have them wait before taking off rather than hold before landing.

The problem for a set of airports is then to find a takeoff and landing sequence for each of them, in order to minimize a global cost criterion for the transit of all aircraft. The high coupling between airports and the existence of "external" traffic (i.e., aircraft with outside airports as origin or destination) lead to a highly complex optimization problem. If, theoretically, the same algorithms can be used for this problem, the computing-time constraints require the use of parallel computing and of suboptimal solutions. The results obtained so far show the feasibility and the benefits of such a solution.

See also: Airborne Collision Avoidance

Bibliography

Benoit A, Swierstra S 1981 Simulation of air traffic control operation in a zone of convergence—aircraft performance data. *Rapport Eurocontrol Doc.* 812031-2

Imbert N 1980 Application de l'algorithme d'ordonnancement du trafic aérien à la zone de Londres Heathrow. *Rapport CERT–DERA* 1/7240

Imbert N 1982 Algorithme de detection et de resolution de conflits entre trajectoires d'avion. *Rapport CERT–DERA* 2/7240-02

Imbert N, Comes M 1978 Gestion stratégique de trafic en zone de convergence ordonnancement dynamique et résultats. *Rapport CERT–DERA* 6-7/7172

Imbert N, Filleau J B, Fossard A 1976 Etude de la précision de navigation en zone terminale. *Rapport CERT–DERA*

Imbert N, Fossard A J, Comes M 1979 Gestion à moyen terme du trafic aérien en zone de convergence. *Proc. IFAC/IFORS Symp. Comparison of Automatic Control and Operational Research Techniques Applied to Large Systems Analysis and Control.* pp. 155–62

Pelegrin M, Imbert N 1977 Accurate timing in landings through air traffic control. *AGARD Conf. Proc.* 24. North Atlantic Treaty Organization, Neuilly-sur-Seine, France

N. Imbert
[CERT–DERA, Toulouse, France]

Air Traffic Control: Trends

Air traffic is divided into two main categories: controlled traffic and uncontrolled traffic. The controlled traffic includes en route and terminal area traffic. Air traffic routes are specified by way points, identified by radio navigation aids; however, planes may use other routes when under radar control. En route planes use on-board receivers or an autonomous inertial navigation system (INS), of which the many possibilities include long-range radioelectric navigation transmitters or, more recently, navigation satellites.

The controlled traffic is regulated by air traffic control (ATC) centers which normally know the position of each aircraft flying under instrument flying rules (IFR) (see *Visual and Instrument Flying Rules*). They give instructions to the planes (change of altitude, authorizations to descend, modification to their route) in order to ensure a smooth flow of traffic and to avoid risk of collision. Although planes are highly automated, the same is not yet true for ATCs and delays ocur in dense areas at certain periods of the year (e.g. the Mediterranean in summer time). Improvements are urgently needed and will require the use of automatic data links between ATCs and planes.

1. General Situation

Airspace is divided in controlled airspace and uncontrolled airspace; uncontrolled airspace is limited to below flight level (FL) 195 (FL195 means an altitude of 19 500 ft (5850 m) read on an altimeter set at 101.35×10^3 Pa).

Flights are divided into two categories according to whether they are using

(a) visual flying rules (VFR) or

(b) instrument flying rules (IFR).

This implies a distinction concerning the type of meteorological conditions. They are

(a) visual meteorological conditions (VMC) and

(b) instrument meteorological conditions (IMC).

Flights in uncontrolled airspace will not be considered here. VRF flights in controlled areas must be performed in VMC. Air vehicles operating under VFR may penetrate into airways (crossing them or using them) if they are flying in VMC. If using VFR, they must fly at a prescribed FL, normally "odd plus five levels," when flying routes with bearings between 0° and 179° and "even plus five levels," when flying routes between 180° and 359°. This means that they can use levels such as FL 55, 75, 95, ..., 135, ..., 195 for routes between 0° and 179° and FL 45, 65, ..., 145, ..., 185 for routes between 180° and 359°.

Note that IFR traffic as well as VFR traffic may change their flight levels. IFR vehicles may do this after authorization by ATC (see below), but ATC may ignore the presence of VFR planes in the airways since they do not have control over their flight paths. In fact, it is recommended that VFR planes entering airways are equipped with radio and inform ATC about their position and FL; in many countries there is no requirement for them to be equipped with a transponder before FL 120 although it is mandatory above this level. (A transponder is an on-board transmitter which is interrogated by secondary surveillance radars (SSRs).

The replay message includes the aircraft identification and, according to the type of transponder, it may also indicate the FL of the plane.)

In contradiction with what is generally thought, when flying in controlled airspace in VMC (good visibility), collision avoidance is the responsibility of the pilot, whether using VFR or IFR.

Considering IFR flights in controlled airspace, both VFR and IFR air vehicles may be present. For IFR, authorization to go on the flight is given by ATC. Additional information such as local meteorological conditions or terminal airfield situation are provided both for IFR and VFR, and a code number may be assigned to the plane if it is equipped with a transponder. Normally, planes are located by the ATC from radio reports by the crew when passing over beacons or way points and, in dense areas, by primary surveillance radar (PSR) and SSR.

The clearance is transmitted by the airport control tower which gives all the necessary data to start the flight: the first way point (see Sect. 1.1) to be reached; the altitude or FL; and other data for the flight after the first way point (plus additional information concerning the radio communication frequencies to be used). After departing the terminal area, the plane is under en route traffic control.

ATC assigns and controls the routes (and altitude or FL) in such a way that collisions are avoided and routes are as close to the routes requested in the flight plan as possible. The airspace is divided into regions (e.g., there are five regions over France); each region is divided into sectors (there are between eight and 12 sectors in a region), and each sector is monitored by one or two controllers.

1.1 Routes

It could be surprising that traffic is voluntarily concentrated on routes. This is the best way to monitor the motion of many planes in a three-dimensional space. Routes belong to controlled airspace; they are defined by the two points at the extremities of each leg (a leg is a straight line). These points, called way points (see Fig. 1), could be

(a) a radio beacon such as hf Omnirange for automatic direction finder (ADF) or vhf Omnirange (VOR);

(b) the intersection of two radials of VORs;

(c) the bearing and the distance to a VOR (polar local coordinates); or

(d) the longitude/latitude.

Routes are often designated by a number or a name; sometimes way points have a name coded with five letters; they are called reporting points when the crew must report passing over the way point (see Fig. 2). Additional information, such as minimum altitude or FL on the route, distance between the two way points and so on, are often indicated on maps.

ATC could have a precise position when the plane is flying in a space covered by radar but it can only rely on

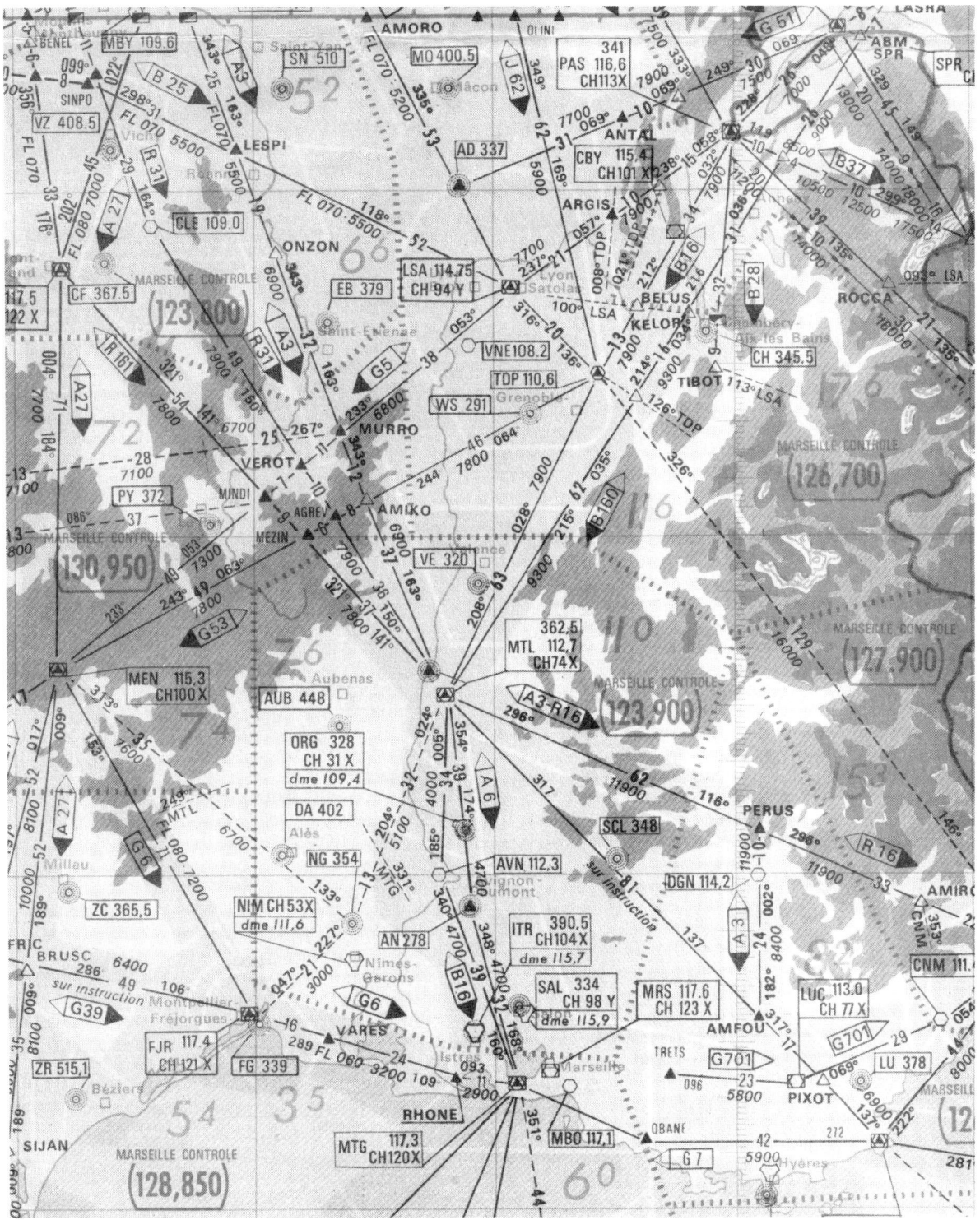

Figure 1
Map showing airways, beacons and way points in Southern France

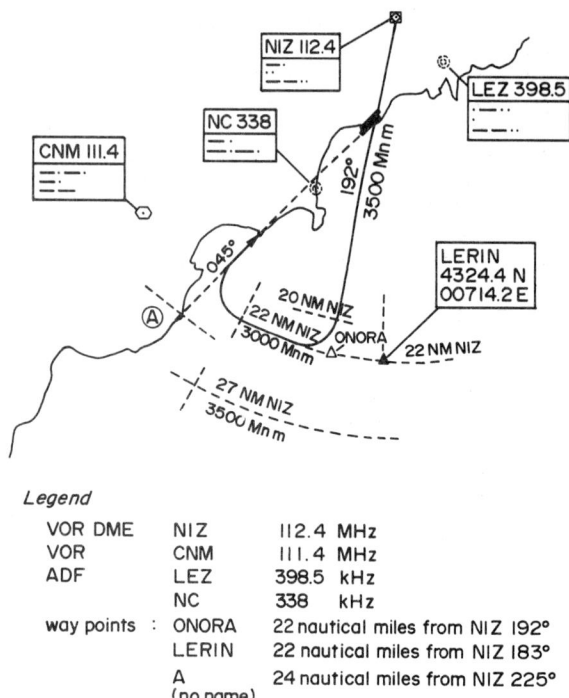

Figure 2
Map showing way point information near Nice-Côte d'Azur (LFMN)

the reports from the crew where there is no radar coverage. This is why the plane must have on-board navigation and the crew must report its position at either prescribed locations or times. The various long-range and short-range navigation aids are summarized in Table 1.

1.2 Terminal Area

Beacons are more dense in a terminal area and the power of the transmitters is small in order to avoid interference between beacons (these frequencies may be reused in other airports some 100 km away).

Navigation is controlled by approach controls. If, in a given period of time, more planes arrive than the maximum capacity of the runway(s), a holding pattern is provided (or sometimes two, with one in the proximity of the runway and one further away); this consists of two half-turns (which must be flown in 1 min) and two straight legs (which must be flown in 1 min or 1.5 min according to the speed of the plane). The total pattern is flown in 4–5 min (see Fig. 3).

All major airports are equipped with an instrument landing system (ILS), at least on one direction of a runway (note that a runway corresponds to two possibilities of landing with routes differing by 180°).

The ILS is, in fact, an ensemble of two transmitters which define a two-plane space (see Fig. 4): one, called the localizer, is a vertical plane containing the axis of the runway; and the other, called the glide slope, is an oblique plane containing horizontal lines which intercept the ground at about 300 m from the threshold of the runway. The intersection of the two planes defines an imaginary axis which represents the correct approach trajectory to fly. The slope of this axis lies between 2.5° and 3°. The localizer transmitter sends a Morse code (three letters) for identification.

According to the equipment in the plane, the qualification of the crew and the equipment on the ground, the landing is classified into one of three categories (CAT):

(a) CAT I, with vertical visibility 60 m and horizontal visibility 800 m;

(b) CAT II, with vertical visibility 30 m and horizontal visibility 400 m; or

(c) CAT III, with vertical visibility 0–5 m and horizontal visibility 75 m (CAT III runways are equipped with high-intensity lights on the axis of the runway).

Automatic landings are performed when conditions are CAT III. Vertical visibility may be 0 m (the A310 and A320 are so certified). However, airlines operators use CAT III A with height of decision (HD), which is the height at which the pilot is able to check the correct alignment of the plane on the axis of the runway. HD is 20 ft (6 m) under the wheels for the A320. If the alignment is correct, then the plane stays in the automatic landing mode up to the touchdown and during the braking phase, until it reaches 60–50 knot (110–90 km h^{-1}). If the pilot is unable to verify correct alignment, then he/she takes over.

Approach control monitors the plane after it has been handed off from en route control; tower stays with the plane until the runway is cleared. Then the crew contacts ground control for clearance to the ramp.

2. Navigation Systems

2.1 Technical Data on Mode S Radar

Mode S will soon be the type of SSR normally used for most of Europe and the USA. This is the combination of SSR and an automatic data link between the radar site and the plane.

At present, SSR is used with transponders in mode A and mode C. SSR is monopulse radar which excites the transponder in the plane. In mode A, the return message (which allows proper location of the plane) conveys the identity of the plane (a set of four octal digits); in mode C, the return message contains the FL. When mode C is used, mode A and mode C are, alternately, sent in order that the full identification of the plane is acquired (see Fig. 5). This system is called the air traffic control beacon system (ATCRBS). Mode S uses the same frequencies as ATCRBS for incoming and outgoing messages, and mode A, mode C and mode S are compatible.

The fundamental difference between mode S and ATCRBS is the manner of addressing aircraft or selecting which aircraft will respond to an interrogation. In ACTRBS, the selection is spatial (i.e., aircraft within the mainbeam of the interrogator respond). As the beam sweeps around, all azimuths are interrogated, and all aircraft within line-of-sight of the antenna respond. In mode S, each aircraft is assigned a unique address code. This is a 24-bit address (which allows about 20 million different addresses to be assigned: a permanent address can thus be allocated to a plane when it is delivered).

Selection of which aircraft is to respond to an interrogation is accomplished by including the address code of the aircraft in the interrogation. Each such interrogation is thus directed to a particular aircraft.

Narrow-beam antennas will continue to be used, but primarily for minimizing interference between sites and as an aid in the determination of aircraft azimuth.

An interrogator calls only those targets for which it has surveillance responsibility (this prevents saturation) and appropriate timing of interrogations avoids overlaps in answers (e.g., calls ordered according to the distance of the plane to the radar site).

There is provision in the messages for conveying additional data; this means that an automatic data link between ATC and plane potentially exists: 56 of the 112 bits of the interrogation or reply message are reserved for that purpose. Extended-length messages are also provided in up to 16 80-bit message segments.

The mode S–ATC interface is particularly simple in the case of an isolated mode S sensor interacting with a

Table 1
Navigation system comparisons

Navigation	Method used: coordinates provided	Coverage provided	Status of the system
Navstar global positioning system	Spherical ranging: three-dimensional position three-dimensional velocity Precise time	Global ($24\,h\,d^{-1}$)	Six satellites in 12 h orbits, available worldwide $1-4\,h\,d^{-1}$
Transit	Doppler shift: longitude latitude	Global except at the poles (periodic fixes only, typically $\frac{1}{2}$–2 h apart)	Five satellites in polar "birdcage" orbits; more than 10 000 sets in use, with 80% civilian users
Loran type C or D	Hyperbolic ranging: longitude latitude	Regional, coverage about 10% of earth	Eight loran C chains with 34 transmitters cover about 10% of the earth
Omega	Hyperbolic ranging: longitude latitude	Essentially global: 88% coverage by day, 98% by night	Eight transmitting stations in operation worldwide
Automatic direction finder	Goniometry on omnidirectional beacon	100 nautical miles around the beacon, not accurate in case of storm (lightning)	Worldwide
VOR/DME Tacan	Lighthouse signal and spherical ranging: heading slant range	Line of sight along present air routes	More than 1000 transmitters in operation; at least 250 000 users
ILS/microwave landing system (MLS)	Beam Steering: heading elevation range	Line of sight: 17–35 nautical miles, available only at properly equipped airports	Hundreds of systems operating worldwide; more than 120 000 domestic users
Inertial navigation	Integrating accelerometers: three-dimensional position three-dimensional velocity	Global with periodic updates	Thousands of self-contained units in use on civilian and military planes and ships

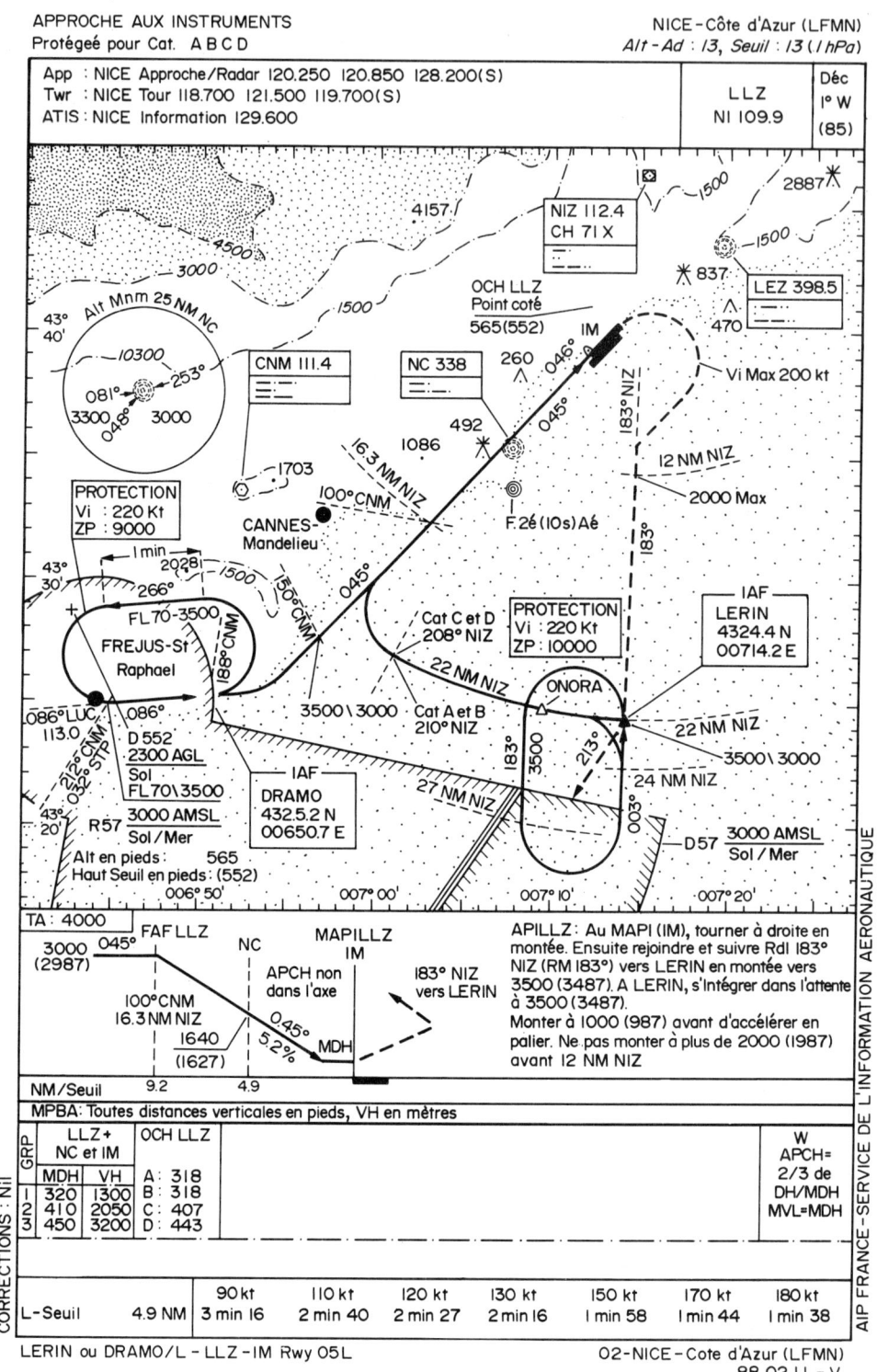

Figure 3
Approach to Nice-Côte d'Azur (as Fig. 2), runway 05R (actual magnetic bearing 046°, R means right) instrument landing system (ILS) with two possibilities: from the LERIN or DRAMO stackings (note that these two stackings are defined by way points which do not correspond to a radioelectric beacon)

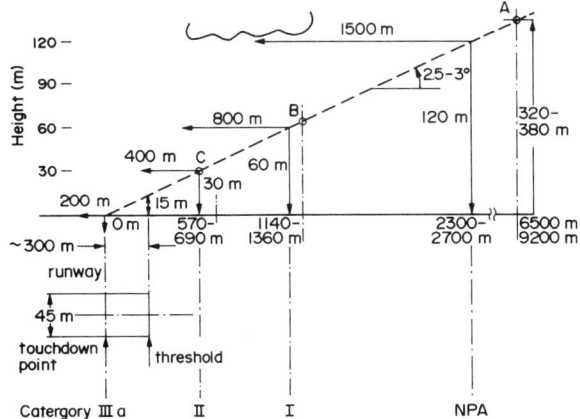

Figure 4
ILS and visual conditions for landing

single control facility (e.g., a sensor at an airport interconnected only with the local terminal tracking facilities (Tracon)). In this situation, the sensor provides surveillance data to the Tracon, and operates as a relay point for data-link messages between aircraft and ATC.

In general, however, each sensor is capable of providing a surveillance and communication service for more than one facility, and in turn each control facility may receive data from more than one sensor. This capability of greater connectivity permits control facilities to take advantage of multiple coverage to maintain surveillance and data-link service in the event of an equipment or link failure at a particular sensor. Surveillance boundaries between adjacent sensors are determined primarily by coverage geometry; these will not be the same as the control boundaries between adjacent ATC facilities, which are determined by air traffic flow patterns.

In general, a control facility will use the data from only one sensor to maintain its track on a particular aircraft. For an en route facility, data on the same aircraft may be available from another sensor as an instantaneous backup. Typically, the data from the sensor designated as primary would be used, as presumably this sensor would have best coverage in a particular region of airspace. The control facility may use any sensor which has an aircraft in its track file for the transmission of standard ground-to-air data-link messages to that aircraft.

2.2 Global Positioning System

Across oceans or over nondense areas, planes may use navigation systems based on satellites. The most popular one is the American global positioning system (GPS). Recently, the USSR has unveiled the Glonass system and has offered the possibility of using it to other countries. The main characteristics of Glonass are given in Table 2.

The principle of this system is to measure the distances from a point on the earth or in the sky to several satellites, the trajectories of which are precisely known. The "point" may be a plane, a ship or any ground vehicle down to a single man. A minimum of three satellites are required and, in fact, four are necessary to determine the altitude to a high degree of accuracy.

The distance measurement is derived from the time delay of an electromagnetic signal (travelling at the speed of light). Each satellite broadcasts information that gives the exact time at which the signal was transmitted and the exact location of the satellite. The ground receiver registers the arrival time and compares it with the time the signal was transmitted. A computer derives latitude, longitude and altitude from these data.

2.3 Navstar System

The Navstar system (GPS) will comprise 21 satellites (18 on line, three on standby to replace a failed satellite). The trajectories are circular at an altitude of 17 600 km. The orbits are coordinated so that, from any point on the earth, when one satellite sets below the horizon another rises to replace it. To do so, two orbital planes are used: 63° for the first seven satellites to be launched and 55° for the 11 others which are due to be

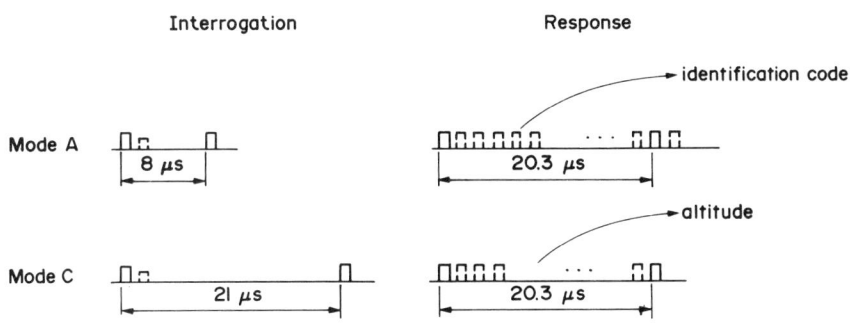

Figure 5
Message modes A and C

Table 2
Main characteristics of Glonass system

Number of satellites	24 (3 standby)
Number of orbits	3
Type of orbit	19.1 km circular
Period of rotation	11.25 h
Inclination	64.8°
Frequency band	$(1602.5625/1615.5) \pm 0.5$ MHz
Type of measurements	noninterrogative
Accuracy	
coordinates (x, y)	± 100 m
altitude (z)	± 150 m
velocity components	± 0.15 m s^{-1}

launched by the early 1990s (three spare satellites will also be launched). Each satellite is updated by signals transmitted by six tracking ground stations.

The computed accuracy, when the 21 satellites are operational, will be of the order of 30 m in x and y coordinates (latitude and longitude) and 40 m for z (altitude above the geoïd). (These accuracies are those which can be reached all over the world. US defense users may have a more accurate coding system which leads to smaller errors, at least for x and y coordinates.)

By January 1989, only seven satellites were in operation leading to large uncovered areas on a 24 h cycle (some highly populated areas such as Los Angeles are covered only 7 h per day). When four satellites are in sight, the computed accuracy is reached and no drift is noticed (the atomic clocks which are on board the satellites have an accuracy of 10^{-12}). Their expected life is five years; however, some Navstar satellites are still operational after nine years (the first satellite was launched on February 22, 1978).

Navstar is a quasiautonomous means of navigation; for long routes Navstar has a greater accuracy than inertial navigation (the accuracy is constant with regard to duration of flight or location); neither the GPS nor the Glonass system allows any other capability than navigation of the plane. Communications (voice or data) must be carried out by another satellite system.

2.4 Communication Navigation Surveillance–Air Traffic Management

In many parts of the world, in the European Community (EC) in particular, airline carriers are becoming increasingly aware of the inability of existing air traffic management (ATM) services to cope with the traffic estimated for the very near future. Saturation levels are already reached in many terminal zones and in certain sectors of flight information regions (FIRs).

In 1983, it was decided by the International Civil Aviation Organization (ICAO) that recommendations for future air navigation systems (FANS) be drawn up. These recommendations call for a fully integrated communication navigation surveillance (CNS) and ATM system. At present, the EC is studying the implementation of such a system; in North America, a plan for the development of an advanced automation system is underway.

Technologies currently available, such as satellite communications, satellite navigation, digital communications and, in the very near future, artificial intelligence (in particular, expert systems), are not yet integrated but public and commercial uses have been demonstrated. The future CNS–ATM system will include public communications both for passengers and airline operators. The difficulty comes more from political decision than from technical gaps.

The CNS–ATM fully integrated system can be divided into three interrelated domains.

(a) *Communication subsystems between air mobiles and ground stations.* At present, all communications are exchanged on vhf or hf channels from the air, but they can be relayed by satellite channels from ground stations. Ground stations providing various services, such as en route traffic, TMA traffic approach, company administration and public telephone services, must be fully integrated, at least for protocol message exchange policies.

The integrated CNS–ATM system will use communications satellites with direct links between air mobiles and satellites. However, mode S SSR will also allow communication between air mobiles equipped with mode S transponders and radar sites, and from these, with any other ground stations.

Message compatibility is mandatory. It would not seem wise to abandon mode S communication when communication satellites are fully operational because of the general trend to achieve a certain degree of redundancy in critical phases such as approaches in dense traffic areas. According to the Radio Conference for the Mobile Services, Geneva, 1987, the 1545–1559 MHz and 1646.5–1160.5 MHz frequency bands are allocated for public communication with air vehicles; and the 1593–1594 MHz (ground to air) and 1625.5–1626.5 MHz (air to ground) bands are allocated for terrestrial satellite communications. (Note that to take into account the Doppler shift due to satellite speed, each communication system must have a bandwidth of 6 kHz.)

(b) *Navigation subsystem.* This must be a four-dimensional navigation system (three position dimensions plus time). It may include a combination of on-board equipment and ground or satellite equipment. For example, the Navstar can be used by itself or it can be used to reset an inertial navigation system. It is important to evaluate the maximum navigation error both on-board and at the ground station, and in some areas or airways, when minimum accuracy equipment is required. This is the case on high-density traffic routes over the North Atlantic. Obviously, radar coverage on such routes is not achievable, even in the distant future. Several systems exist (see Table 1) but their accuracy varies according to the region in which the plane navigates and according to random parameters such as radio transmission perturbations. Solar activity, which is unpredictable except on a few minutes basis, may degrade the precision of navigation suddenly.

The navigation subsystem will normally receive more than one signal coming from navigation ground-based or satellite-based transmitters. It must derive from the data received the best estimate (Kalman filter) given by each system and then derive the 3 standard deviation (SD) error ellipse. The navigation subsystem must be able to transmit the position (and the 3 SD error zone) computed on board to ground stations, via satellite if necessary. The ground en route center must correlate the data received from the mobile with its own estimation and, in case of disagreement, must automatically carry out a dialog with the mobile.

(c) Surveillance subsystem. The EC is presently defining the requirements for the surveillance specification of the CNS–ATM fully integrated system. It includes the monitoring and controlling of planes throughout the EC airspace as a priority, but must be capable of extension to oceanic and low-density traffic in other countries.

The development of airborne collision avoidance systems must be included in the study both when used in an autonomous way by aircraft and when used in cooperation with ATC centers (see *Airborne Collision Avoidance*).

It is obvious that the subsystem surveillance is highly interrelated with the communication and navigation subsystems. The combination of these must lead to the derivation of positional information which can be assimilated to a pseudoradar traffic data presentation. A local representation of the air situation may be presented on board.

It is clear that the next technical step is not entirely a matter of ATC: it also concerns the aircraft. The aeronautical community, including the airline operators, is aware of the deficiency (the lack of an automatic data link between the aircraft and ATC) and have called for urgent internationally agreed decisions in order to control air traffic, which is increasing by 6–8% each year. Peak periods correspond to a demand that can be twice the average arrival traffic. However, safety must be maintained or increased: no compromise is acceptable in this matter.

Once an automatic data link is provided, the traffic can be controlled much more smoothly for many reasons.

(a) The "saturation" of the frequency of the approach control center will no longer appear because messages will be transmitted in a much shorter time in both directions.

(b) The stress on the crew will be kept reasonable because the radio will not be perpetually blaring. At present, in dense terminal areas, the radio is speaking continually: if between ten and 12 aircraft are under control by the approach control center, only 8–10% of the messages concern a given aircraft.

(c) It will be possible to reach the maximum landing rate of the airport due to the development of four-dimensional navigation.

3. Four-Dimensional Navigation

As stated previously, instructions given by the ATC to the planes consist of clearances to go to the next way point at a given FL or a given altitude, or to begin descent and reach a given way point at a prescribed level, or to proceed to climb, and so on.

No time is assigned to pass over a way point, but ATC often asks the crew for an estimation of this time. This is the way the en route navigation is performed. In the terminal area, the dialog is about the same; however, time becomes a much more important parameter. Two cases must be considered.

In the first case, the rate of incoming traffic is less or equal to the maximum landing rate for the period of time considered (the landing rate depends on the QFU (magnetic heading) of the runway in use (ILS or not) and on the instantaneous meteorological conditions such as horizontal and vertical visibilities, heavy rain, lateral wind and icing conditions). Aircraft are controlled directly to the runway through the requested paths or under radar guidance.

In order to sequence the aircraft smoothly, the controllers must have a correct knowledge of the speeds or, more precisely, of the ground speeds. To optimize the traffic, it is clear that the prescribed time of passing over beacons must be assigned to the aircraft, which must be equipped (perhaps with software stored in the flight management system (FMS) or flight management guidance computer (FMGC)) that ensures the beacon will be flown over at the correct time. This is four-dimensional navigation, which is shared by ATC and the crew.

In the second case, the rate of arrival of incoming traffic is temporarily higher than the maximum landing rate. Therefore, aircraft are "stored" in holding patterns (stacking) which must be flown in 4 min or 5 min according to the speed of the aircraft. The horizontal pattern is the same for all aircraft, but their altitudes differ by 1000 ft (300 m) (see Fig. 3). As time passes, aircraft exit the stack from the bottom. The remaining aircraft descend one level each time the lowest circuit is cleared. This is a common procedure.

However, 4 min or 5 min is a large delay with regard to the delay between two landings (between 80 s and 180 s according to meteorological conditions, runway equipment, etc). A velocity control and/or a precise location of the plane in the lowest circuit make the optimization of the escape rate from the lowest circuit to the final approach descent possible. In dense areas, the en route control and the terminal area control are tied so that the optimum landing rate can be achieved through permanent coordination between these two services.

In the early 1980s, it was shown that three corrections (two speed and one heading) in the terminal area might reduce the dispersion of the touchdown by a factor of three. In a basic document, the procedure reported simulation tests performed over more than 2000 flights. In the first batch of simulation tests, the procedure was the one described in the airport approach chart (i.e., without any speed correction or heading correction). Errors due to human behavior (setting of the route, altitude or speed) or to weather parameters (winds) were introduced in a random way according to a prescribed distribution. Dispersion of the touchdown times was computed as 27 s for 1 SD.

In the second batch of simulation tests, a prescribed time of touchdown was attached to the plane when passing the border of the TMA some 60–100 nautical miles ahead (trajectory length) and, on a permanent basis, the ground computer checked the error between the "real" plane and the "ideal" plane (i.e., the plane that will land at the prescribed time). The "top of descent" was determined by the ground computer (it may differ slightly from the one of the ideal plane in order to compensate for slight errors already accumulated). At about the middle of the descent, a speed correction was radioed to the pilot. A second speed correction was transmitted when the plane leveled off at 10 000 ft (3000 m) (this is a frequent procedure in the airport approach). A third correction concerns the heading to be taken for the ILS beam interception. Errors were introduced in a similar way (both human errors and winds). The dispersion of touchdown instants was computed as 9 s for 1 SD.

This type of control which requests the acceptance by the crew of speed controls has been tested in flight on commercial flights (Air-Inter) and a precision of less than 10 s has been reached after a flight (approach phase) of 15 min to 20 min. Langley (Knox and Imbert 1986) checked this procedure and came to the same conclusions. Later, Eurocontrol in Brussels developed the procedure and the algorithms to be used in the ground computer. A great number of simulations have been carried out and some tests in flight have also been performed (see *Fuel Conservation: Air Transport*).

Nowadays, the rapid extension of on-board FMS or FMGC gives a new push to this type of control, as does the large increase in computing facilities in the ground control sites. It is worth noting that the ground computers must have models of the planes; a sensitive study by CERT, Toulouse, France has shown that simple models (sixth order) are sufficient (i.e., the exact mass of the incoming plane is not necessary for assigning the time of the landing).

In fact, the airline speed profile and some airline rules are necessary for the computation of the prescribed time of arrival (touchdown). It is estimated that 400 models cover all the types of aircraft and company rules. From these models, the ground computer assigns the optimal time of touchdown for each incoming plane (the criterion to be used is a mixture of the "regularity of landing sequence" and "minimization of the approach time"; for the latter parameter it could be a global minimization for the approach zone). The crew may enter this time in the on-board computer and, if the program exists, the corrections no longer have to be radioed. The error between the ideal plane and the real plane can be sent to an index on the air speed indicator or to the flight director if heading corrections are accepted.

Finally, once again, the use of an automatic data link will ensure more safety plus frequent automatic control from the flight profile computed on the ATC computers (correlation between the ideal and the real aircraft).

See also: Data Links in Aeronautics

Bibliography

Carel O 1986 The selection of future curl air navigation systems. *AGARD Conf. Proc.* **140**

Knox C E, Imbert N 1986 Ground-based time guidance. National Aeronautics and Space Administration, Technical Paper No. 2616. NASA, Washington, DC

M. J. Pélegrin
[ONERA, Toulouse, France]

Air Traffic Simulators

Since the 1960s, the Ecole Nationale de l'Aviation Civile (ENAC, the French National Civil Aviation School) has been using a simulator to train air traffic controllers and to conduct experiments. Convinced of the efficacy of such a system, the French government has requested the Service Technique de la Navigation Aérienne (STNA) to develop a new simulation system.

1. Air Traffic Control

Air traffic control (ATC) is an internationally standardized service provided to general air traffic in the airspace allotted to it (see *Air Traffic Control: An Overview*). Its objectives are to prevent aerial collisions and aircraft colliding with obstacles while taxiing on the ground, to speed up and regulate the circulation of air traffic, and to alert the emergency services.

This service is provided by specialized entities such as the control tower, approach control centers and en route control centers (five in France, two in the UK, around 20 in the rest of Europe and 20 in the USA), and is carried out by air traffic controllers. The job of the air traffic controllers is to know the current air traffic situation, extrapolate it and regulate the traffic in order to avoid collisions and "traffic jams."

Figure 1
An SCU

Airspace is divided into basic volumes known as control sectors. This division entails an additional workload of information, negotiation and coordination between the controllers of the sectors through which the flight passes. Depending on the entity and the density of traffic, up to three controllers may be responsible for a single control sector. The workstation needed by the team to control its sector is called a space control unit (SCU, see Fig. 1).

The airport tower controller manages the ground and its immediate environs; in most cases the controller can see the aircraft. In contrast, the approach or en route controller does not actually see the aircraft, but builds up a picture of the traffic using data supplied by flight plans filed by the pilots or airlines, the radar picture, if any and radio messages exchanged with the pilots and telephone messages exchanged with other controllers. In some countries, controllers are assisted by computerized systems. These provide the appropriate flight plan and radar information at the relevant time and in a suitable form.

The initial flight plan processing system (IFPS) processes the original flight plan filed by the pilot. In particular, it contains the name and type of aircraft, its equipment, the time of departure, the desired route described as a set of points to be flown over, and the optimum cruising level for passenger comfort and fuel consumption. The system produces a machine flight plan principally giving the projected time and height of passage for each of the reference points, as well as listing the sectors through which the flight passes.

The radar data processing system (RDPS) follows aircraft using position measurements provided by radar. The positions calculated enable, in correlation with the flight plans, regular positioning and identification of aircraft on the SCU radar screens.

The flight data processing system (FDPS) updates times and heights of passage for the machine flight plans transmitted to it by the IFPS, using data supplied by the RDPS and the controllers. It prints out paper strips giving flight progress within the sector and displays the lists of aircraft relevant to the sector in the short term on interactive terminals, indicating the status of each aircraft: entering, leaving or coordinated (see *Data Processing in Air Traffic Control*).

There is usually one FDPS and one RDPS per control center, with several centers sharing an IFPS.

2. Types of Simulator

2.1 Classification Based on Application

(a) Training. This term is used in its widest sense, covering several activities: initial training to communicate practical know-how, and advanced training to maintain practical know-how and familiarize the trainee with a new working environment.

Controllers get their practical training either "on the job" or on a simulator. The advantages of using a simulator are obvious: the safety of the operational system is increased by reducing the amount of on-the-job training and training is more efficient as a full, progressive training plan can be rigorously adhered to.

Quantitatively, the biggest use of air traffic simulators is for initial training purposes. Their use in advanced training is still very restricted.

(b) Studies and experiments. These may be, for example, to examine the possibilities resulting from a reorganization of part of the airspace, to evaluate the efficacy of new work methods or tools, or to test workstation and dialog ergonomics.

2.2 Classification Based on Entity Simulated

There are several types of air traffic simulator, depending on the type of control center and the equipment to be simulated, and they are all fundamentally different.

(a) Airport control simulator. Until recently, the most frequently used technique for simulating ATC consisted of remote-controlled models moving about in reduced-scale scenery. Nowadays, there are airport control simulators that use a wide screen to show realistic synthesized images of the ground and moving objects as the controller sees them from the top of the control tower.

(b) Approach and en route simulator. The simulation of approach or en route control centers is less recent. It should be remembered that in these centers the controller does not actually see the aircraft. It is therefore necessary to reproduce the data that has to be dealt with using media and formats as close as possible to those found in operations centers.

There are two types of simulator for these centers, depending on whether the center is equipped with radar to follow traffic. There are radar approach and en route control simulators, and nonradar simulators, also referred to as procedure control simulators.

2.3 Classification Based on Type of Training

(a) Global simulation. This enables simultaneous practice of all the work procedures implemented by a

controller. It is, therefore, necessary to reproduce the whole of the working environment.

(*b*) *Analytical simulation.* This allows the trainee to concentrate on certain work processes. It is not necessarily important to reproduce the whole of the controller's working environment.

2.4 Classification Based on Simulation Technique

(*a*) *Arithmetic simulation.* For some studies, especially preliminary studies of the possibilities resulting from a reorganization of airspace, arithmetic simulators can be used. These model traffic flow and control work. There are no human operators intervening during the simulation; they run at accelerated speeds and only produce data files or paper printouts.

(*b*) *Real-time simulation.* For other activities, real-time simulators that reproduce the controller's environment are used. The environment is not fully modelled and requires the use of human operators, such as dummy pilots simulating aircraft and dummy controllers simulating neighboring sectors. There are three possible methods to produce a real-time simulator.

First, it is possible to use an "off-the-peg" simulator of the type marketed by a number of industrial companies throughout the world. They are intended for initial training purposes and generally have to be adapted to local needs.

Second, it is possible to include the simulation functions and data in the operational computer-aided control system itself. This is the option chosen in the USA.

Third, it is possible to develop a "made-to-measure" simulator, whose control functions are designed to be identical to those of the operational system. This is the option that has been preferred by the French government since 1972.

3. Example of a National Training Simulator: ELECTRA

ELECTRA is a real-time simulation system for en route and approach control centers, implementing global and analytical simulations for initial or advanced training and experimental purposes. The system is under development, being specifically designed to meet the requirements of the French civil aviation.

3.1 Functions

This simulator offers two main types of function: simulation, running functions enabling the implementation of simulation scenarios by faithfully reproducing the controller's working environment; and administrative functions such as entering and controlling all the data required to run simulations; planning the use of the simulator; and supervising simulator operation.

(*a*) *Simulation running functions.* A simulation scenario, or exercise, contains all the information needed by the system to reproduce the environment of a controller: aircraft flight plans, definition of weather, condition of ground equipment and so on. Different types of exercise can be run, each corresponding specifically to one of the missions listed in Sect. 2.

Global simulation exercises are training oriented. Experienced or trainee controllers practise at work positions identical to real control positions in terms of both equipment and operation; that is, they have the same layout, the same graphic display consoles (radar and television screens), the same data input methods (currently touch-sensitive screen and trackerball), the same printout method and the same methods of communication (radio for the pilots of aircraft being controlled and telephone for the controllers of neighboring sectors).

The rest of the controller's working environment is fully simulated using operators, working on specialized computer workstations, and computer modelling (see Sect. 3.3). "Dummy pilot" operators replace real pilots. They pilot simulated aircraft in accordance with the instructions of the trainee controller, and they send the trainee the same requests and make the same responses as real pilots. "Dummy controller" operators represent controllers of airspace adjoining the airspace controlled by the trainee. In particular, they negotiate the conditions for the transfer of aircraft from the trainee's airspace to a neighboring sector in the same way as occurs in ATC. These operators are aided by system models, which give indications as to the behavior to adopt; that is, a pilot model for the dummy pilot and a controller model for the dummy controller.

Other models enable the system to achieve total environment realism, in particular

(i) the aircraft model, reproducing aircraft behavior in response to pilot commands;

(ii) the weather model, modelling meteorological phenomena affecting ATC, such as altitude, winds, temperatures, visibility and storms; and

(iii) the radar model, reproducing the radar system used to supply information on aircraft trajectories to control positions.

These models not only reproduce behavior of the person, object or phenomenon modelled under normal conditions but also in situations where there is a malfunction or breakdown; for example, pilot and aircraft behavior during a breakdown in cabin pressurization, or a partial or total breakdown of aircraft equipment.

Trainees can work alone or under instructor supervision. The instructor uses a specialized computer workstation enabling him to

(i) control the running of an exercise, using such functions as start, freeze (i.e., pause) and unfreeze, taking a position at a particular point in time of the exercise, replay and redisplay;

(ii) modify the exercise by adding or removing a flight, altering a flight, or setting off an event relating to an aircraft (e.g., breakdown) or to the environment (e.g., storm); and

(iii) follow the trainee's work by consulting the measures taken during the exercise, thereby enabling the quality of the trainee's work to be evaluated (e.g., measuring the increases in length of flight trajectories resulting from the trainee's actions and listing any breaking of ATC regulations).

Analytical simulation exercises are for use in initial training. They enable acquisition of the basic data and procedures required for ATC: sector geography, study of aircraft performance and behavior, and methods of resolving conflicts between aircraft. These exercises run on individual workstations, do not require operators and use a limited environmental realism (the aircraft model is fully realistic, but the weather, pilot and neighboring controller models are simplified).

Experiments are very similar to global simulation exercises. However, more measures are taken during an experiment than during a global simulation exercise; it is not unusual for them to be specifically programmed. The users in this case are not trainees but qualified controllers and the instructors are replaced by experimenters.

(b) *Administrative functions.* These enable trainee, instructor and experimenter users to plan utilization of the simulator well in advance, and to have access to a complete and up-to-date range of sets of exercises that can be run under reliable conditions.

Exercise preparation is the construction of exercise scenarios for the benefit of users. This is carried out on graphics workstations by instructors, this time functioning as exercise designers, aided by a range of tools such as databanks of air traffic archives taken from real operations systems, weather situation databanks, and graphic display and design tools.

Experiment preparation is similar to exercise preparation, except that the experiment designer may need to specify and take delivery of specialized software and ensure that new equipment is available.

Management consists of keeping all the data used by the system up to date and planning system utilization. This function takes into account requests for training courses from training officers, resolves conflicts between different requests and enables the equipment and personnel required to run the courses to be forecast well in advance.

Supervision enables monitoring of system operation and allows all possible measures to be taken to ensure maximum availability of system services, especially in the case of malfunction or breakdown.

Management and supervision operations are carried out by the simulator administrators on graphic workstations.

3.2 System Organization

The simulation service is available at ENAC and in several control centers. For each site, there is a local simulation service, the use of which is independent of the other sites. They provide initial and advanced training through planned courses, each course comprising several sessions. Training can also be carried out on an informal basis using positions that have not been reserved for a course.

Practically all the management and supervision work is centralized at ENAC. However, each local simulation service has a certain amount of autonomy to manage its own specific data and supervise its own center.

3.3 Computer Techniques

(a) *Quality requirements.* The classification of quality into factors arrived at by J. A. MacCall will be considered. One of the most important factors that ELECTRA must satisfy is *correctness*; that is, it must be adequate for its purpose. This is a particularly stringent requirement, since the primary objective of ELECTRA is to be a training simulator that places the controller in a highly realistic environment. In order to achieve this objective, several measures have been implemented.

(i) A large number of environment simulation models have been developed (see Sect. 3.4).

(ii) Exercises have been prepared using real flight plan and trajectory archives taken from the national archive service.

(iii) The software for the operations systems aiding ATC will be included in the software for ELECTRA with practically no modification, along with their databases. In order to facilitate *maintainability*, a single version of each program will then be maintained, this version being used in both simulation and operation. The technical problems caused by the use of operations software are discussed later in this section.

Another essential factor for maintaining realism is *flexibility*: the system reproduces an operational organization of control work that is constantly evolving and changes in the real environment must be reproduced without delay. This is obtained by employing operations software as mentioned previously and using data to describe the organization of the control environment.

A further particularity of this system is that it is to be used at ENAC and in training centers, implying simultaneous use of several courses and exercises, as well as necessitating *expandability* of the system in terms of power, in order to cope rapidly with major unforecast increases in trainee controller intake. This requirement is taken into account by the hardware and software architectures described in this section.

It follows that ELECTRA is a complex system, not only in terms of technical design, but also in terms of its implementation: its *usability* must, therefore, be carefully maintained. The fact that ELECTRA employs operational software and data, new versions of which are constantly coming into existence, results in a large number of highly technical management tasks. It is not realistic to train more than a small number of specialists to carry out these tasks; the same applies to technical supervision work (monitoring equipment operation). These considerations have led to the centralizing of both administration work and hardware architecture at ENAC, thereby minimizing the amount of simulator equipment for monitoring in the control centers. The hardware architecture is described in this section.

As mentioned previously, ELECTRA is a major computer system supporting both real-time and management jobs. It therefore requires a considerable financial investment which must be lasting. Its estimated useful life is about 20 years and the current rate of progress in computer technology is such that it is difficult to envisage maintaining the operational level of such a system over such a period without changing the individual computers. The economic vulnerability of computer manufacturers is also a risk to be taken into account. As a result, *portability* is a prime quality factor for ELECTRA. This is achieved by applying an industrial policy to products and standards, the methods of selection being explained later in this section, and stating requirements to project software developers (priority to development on standard operating systems, or else isolate code specific to a type of hardware or basic software).

The remaining classical quality factors—reliability, testability, efficiency, integrity, reusability and interoperability—although still important, are no more critical for ELECTRA than for other systems.

(b) *Software architecture.* It has already been seen that a training simulator must be able to run several exercises simultaneously and that the need for realism for training purposes demands the use of software taken directly from computer-aided ATC systems (RDPS, FDPS, IFPS). The main technical difficulty of ELECTRA is in the meeting of both of these requirements. The resolution of this problem was used as the basis for defining the software architecture.

Operations control programs are generally designed to cope with a single type of traffic and a single environment: reality. Operations software cannot be reconfigured "from the inside" as this would result in significant modifications and force the production of "simulation" versions of the programs. In consequence, changes in operation functions could not be taken into account within a reasonable time period; both maintainability and adaptability would be poor. Nor is it possible to resolve the configuration problem by increasing the number of computers. An exercise uses at least one of each of the FDPS, RDPS and IFPS subsystems, and each subsystem normally uses one computer; however, the long-term objective is to be able to run up to 45 exercises simultaneously.

The chosen solution was to develop a virtual machine manager (VMM). This manager is capable of running a number of subsystems that normally use all the facilities of a computer within a single physical machine. These subsystems are then said to use the facilities of a "virtual machine" (i.e. completely transparent from the point of view of the applications software, which "thinks" it has all the computer for itself). Therefore, when configuring for multiple exercises, it is possible to install a number of operations subsystems in one machine, the number being limited only by the physical computing power available. Another basic software program is being developed to manage communications between virtual machines (whether or not they co-exist in the same physical machine), and between virtual machines and their peripherals. This is known as a communications supervisor.

A simplified diagram of the ELECTRA software architecture is given in Fig. 2.

(c) *Hardware architecture.* The ELECTRA hardware architecture was designed to

(i) maximize centralization of technical means, and

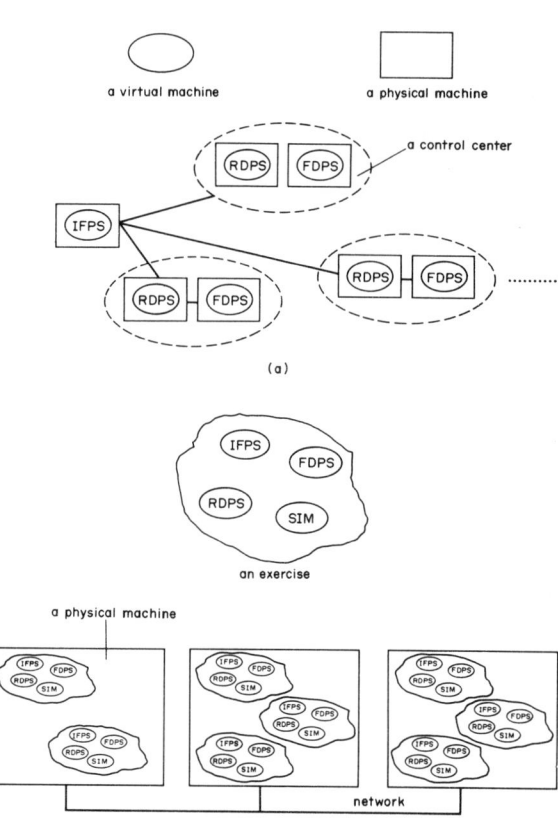

Figure 2
ELECTRA software architecture: (a) operational control system and (b) multiple exercise simulation system (simplified)

(ii) minimize the effects of a possible increase in required size and errors in estimating the calculating power needed.

Point (ii) led to the adoption of local networks as the means of communication between processing units on the same site, rather than one-to-one linkups. This means that it is always possible to add another processing unit to the network without altering the architecture logic.

The architecture comprises a central site, located at ENAC, and several local sites (the control centers and ENAC).

The central site is where the ATC system simulations are run and where the central management and supervision work is carried out. The data and exercise banks are located here. These functions run on 32-bit minicomputers identical to those used in operational control centers; their number depends on the simulation power required. These computers communicate with each other via a local network and are fully interchangeable (i.e., any management, supervision or simulation job for any site can be run on each and any of these machines). This flexibility is made possible by the communications supervisor previously mentioned. The network also includes the shared printers, disks and tape drives, and a master workstation which deals with system supervision and initialization. The central site is linked to outside systems such as the Centre d'Etudes de la Navigation Aérienne (CENA), for the transfer of experimental data, and air navigation centers, for the reception of flight plan archives and data representing the functioning of computerized control systems.

Hardware reliability is obtained using redundancy. A minicomputer is permanently kept as a backup or for tests. An electronic switching bank links each minicomputer to all the remote centres. Should a minicomputer break down, the supervising station will detect it and reconfigure the switching bank to replace the ailing computer by the backup machine. In addition, the disk drives have hardware protection and the supervising station has a warm backup (i.e., ready at all times to take over from the first, should it cease to function properly).

Each local site comprises

(i) a server linked to workstations by Ethernet, the workstations providing not only the man–machine graphic interfaces, but also the traffic and control environment simulation functions, and the exercise and local data preparation functions; and

(ii) a SCU group equipped with peripherals such as radar screens, digitatrons (touch input devices), strip printers and telephone/radio consoles.

Each local site is linked to the central site by

(i) transmission lines specifically designed to carry data for transferring environment and ATC simulation information requiring high-speed data transfer, an X25 protocol being used, and

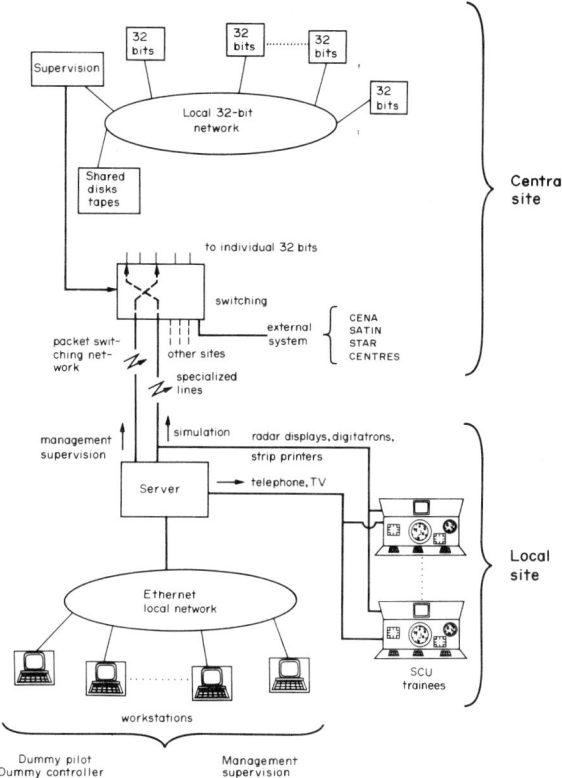

Figure 3
ELECTRA hardware architecture

(ii) a public or private packet switching network for less rigorous management and supervision messages.

A simplified diagram of the ELECTRA hardware architecture is given in Fig. 3.

(*d*) *Technical selections*. The method of specifying operational requirements (work concepts and organization) comes from the MERISE method, using its ability to describe conceptual characteristics of objects (conceptual data model equivalent to the entity/relation model) and work organization (organizational processing model).

The functional specification method of the software is SA-RT, which allows classical description by process and data flow, topped off by a process synchronization description mechanism useful for the real-time portion of the system.

The precise method of design is *object-oriented design*, the main advantages of which are greater ease in reusing software modules and natural data encapsulation.

The real-time language selected is ADA, because of the openings it offers into a future software components industry (packages) and because of its characteristics, which enable implementation of just a part of the requirements of object-oriented design (private types): ADA does not allow inheritance.

The *database* management system chosen is of the *relational* type. Despite the performance speed problems that the relational method causes, it has gradually become the standard for database management, mainly because of the existence of the standard data management language SQL, but also because it allows nonprocedural transactions based on the semantic data structure (close to the user) and not the implementation structure (too close to the design). This makes prototyping easy. Performance should soon be improved by increasingly powerful hardware.

The hardware and operating systems chosen are 32-bit minicomputers using their native DOS, for those machines transferred directly from the operational system (in which case they are the same as the machines used in real control work), and UNIX graphic workstations, for all purely simulation software. UNIX was chosen because it has become, for all practical purposes, the industry standard. ELECTRA, which is a system with average real-time requirements, will have to cope with the lack of task priority in UNIX.

3.4 Modelling Techniques

The models used by the system to reproduce the controller's environment must satisfy the following requirements:

(a) reproduce the behavior of the object, person or phenomenon modelled as the controller sees it during control work;

(b) give performance in line with the requirements of a real-time system; and

(c) be easily updated, so that changes in the behavior of the real object be reflected without delay by the model.

The objective of the aircraft model is to provide a model of the behavior of the aircraft in response to pilot commands. An aircraft can be modelled in three different ways (according to the classification made by the European Organisation for the Safety of Air Navigation).

The first type of model describes the aircraft as a solid and reproduces its movement in both translation and rotation. These models use flight mechanics equations. The polars and engine curves must take into account all actions taken by the pilot affecting the aircraft (e.g., rudder, wing flaps, tail flaps and engine). This type of model is used in flight simulators for pilot training.

Similar to the first, but simpler, the second type of model describes the aircraft as a point: only its movement in translation is reproduced. The corresponding advantage is that only a subset of the pilot's actions need be taken into account.

The third type of model concentrates on describing changes in certain aircraft parameters (e.g., vertical speed) as variable functions (e.g., of altitude and mass). It does not use flight mechanics but, rather, empirical functions. This type of model is used in ATC, both for computer-aided operations and simulation work. For example, the model used by ELECTRA supplies for each aircraft type

(a) speeds as a function of the flight phase (climbing, descending or cruising) and the airline that operates the plane;

(b) vertical speeds as a function of the speed, mass, flight phase, altitude and temperature of the aircraft;

(c) the angles of landing and takeoff as a function of mass;

(d) acceleration and deceleration values as functions of flight phase and altitude;

(e) the effect of the wingflaps and airbrakes on aircraft behavior; and

(f) performance limits (e.g., maximum speed and ceiling).

The pilot and controller models simulate the behavior of these operators (respectively pilots and controllers of neighboring sectors) when events occur relating to the trainee's practice sector. Each event is described in terms of its constituent elements: set off of the event, the series of operations carried out by the trainee and the simulated operators, and messages exchanged between these operators and the trainee. The set off of the event (unless this is done by the trainee) and the actions carried out by the simulated operators are reproduced using algorithms.

A simple weather model allows reproduction of weather phenomena affecting the controller. The precision looked for here is the macroscopic view of the weather conditions that the controller receives through communication with pilots and controllers, and observing the aircraft navigating. This model comprises several models using simple empirical equations enabling simulation of changes in important parameters such as temperature and wind varying with altitude, of certain specific phenomena (e.g., storms) and of weather conditions specific to an area. It allows rough testing, using simple rules, of the consistency of the weather parameters defined in a scenario. The time aspect is not usually taken into account, except for certain specific phenomena, such as storms.

4. Future Developments

4.1 Artificial Intelligence

Artificial intelligence (AI) techniques seem particularly promising for simulation work.

Knowledge-based systems work with concepts such as rules and facts, thereby enabling programming in a manner very similar to the way in which the experts whose knowledge is to be implemented express themselves. It is, therefore, in their flexibility, when the rules to be described are complex and constantly changing, that these techniques have an advantage over procedural techniques.

The inconvenience of using these techniques is that they generally give slower performance than classical techniques and are more expensive, due to the computing power and memory required. Furthermore, these techniques are still in their infancy, with the risk inherent in all new technology. It is envisageable that, in the early 1990s, AI techniques will be used in air traffic simulation systems for the following four applications.

(a) *Simulation of controller environment.* This is the simulation of the other controllers in liaison with the trainee controller during the simulation run. This environment, which reflects the work organization and functional logic of the operational control system, changes so much in time and space that AI is probably the only solution compatible with the maintainability requirements of ELECTRA.

(b) *Computer-aided teaching.* For a large number of specific and well-defined teaching points, it can be useful for a school and its trainees to run exercises in a machine-guided mode, without instructor supervision. The machine continually checks the quality of the trainee's work and provides guidance, using programmed teaching expertise, by suggesting activities (replay, revision of specific themes, new exercises). As in all situations where there exists informal, constantly evolving expertise, the use of AI techniques is appropriate for this.

(c) *Simulation of pilot behavior.* The joint use of AI and vocal techniques should, in the long run, allow complete automation of the dummy pilot function; that is, no operator required (see Sect. 4.2).

(d) *Automatic revision of exercises.* The exercises are training scenarios referring to data describing airspace structure, aircraft characteristics and so on. When this data is modified by air traffic organizations and aircraft manufacturers, it is usually without obtaining prior authorization from simulator operators. It is the simulator system manager's job to keep all the exercises playable without changing the teaching content. This work requires perfect knowledge of all the dependence links between objects and the action to be taken when some of the data is modified. This is highly specialized expertise, justifying an AI approach.

4.2 Voice Recognition

Getting together all the dummy pilot and dummy controller operators needed to run an exercise (usually between 1 and 4) is often difficult. In the early versions of ELECTRA, attempts are made to reduce the number of operators by increasing their efficiency using an appropriate man–machine interface and by the use of models included in the software. In the future, the use of voice input and synthesis techniques should enable certain dummy pilot operators to be replaced. The choice of dummy pilots for the initial application of vocal techniques is due to the fact that controller–pilot dialog in real situations is relatively standardized, using a set phraseology, which should make the work of voice input and synthesis easier.

An initial single-speaker system (i.e., requiring recognition of the voice) recognizing series of words has, therefore, been developed for pilot–controller (or dummy pilot–controller) communication by the CENA. It has a 100-word vocabulary, a 95% recognition for words, and a 70% recognition for phrases or control instructions.

A new prototype system of the same kind, PAROLE, is also being developed by CENA with a target performance of a 300-word vocabulary and a 95% recognition for phrases. The capabilities of PAROLE would allow it to replace the dummy pilot in initial training exercises. However, it is unrealistic in the medium term to envisage a voice input system for controller–controller (or dummy controller–trainee) dialog, as the vocabulary used is very large and communication is in natural language, not a standardized code.

See also: Air Traffic Control: Trends

C. Dujardin and R. Fondacci
[Service Technique de la Navigation Aérienne, Paris, France]

M. S. Redon
[Ecole Nationale de l'Aviation Civile, Toulouse, France]

Airborne Collision Avoidance

During the late 1950s, the worldwide growth in air traffic resulted in an increased number of airborne collisions. The Air Transport Association of America (ATA) initiated the consideration and development of an airborne collision warning and avoidance system in cooperation with manufacturers.

The idea presented by the ATA was the design and standardization of airborne equipment for detecting collision threats and supporting pilots in solving conflicts. Collision threats are very rare events occurring once per 5 000 000 to 10 000 000 flight hours. Thus, it was felt that a collision avoidance system (CAS) would rarely alert the pilot without significant increase of cockpit workload.

Collisions were found to occur in all airspaces, with or without air traffic control (ATC). Airlines flying worldwide therefore focused their interest on the design of an airborne system protecting their own aircraft everywhere and in all phases of flight. From this point of view, no changes in ATC functions were expected; the design of such a system at that time was

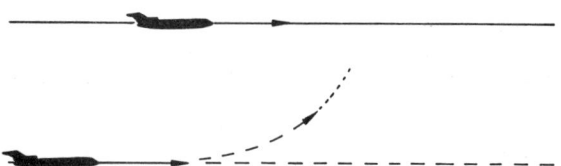

Figure 1
Maneuvering aircraft changes the calculated τ

believed to be only a matter of developing adequate measurement techniques, signal processing and data processing, and displays on board the aircraft.

The extensive work that has occurred since the 1950s, however, has shown this task to be a considerable challenge, with serious problems to be met.

1. Basic Requirements

1.1 General Functions and Technical Solutions

Collision avoidance requires the detection of a collision threat at a time τ before collision. The time τ must be long enough to allow the pilot to pick up the indication, react and solve the conflict by a maneuver. Therefore, τ must be between 20 s and 30 s. The detection of a potential threat requires the measurement of the range, direction and relative velocity of the intruding aircraft, and a prediction of the flight path of the intruder over the next τ. If the predicted flight paths of the two aircraft will be closer than a specified minimum horizontal miss distance (e.g., 200 m) and minimum altitude difference (e.g., 100 ft (30 m)), then a collision threat exists.

The following different technical systems would meet these requirements.

(a) A fully cooperative system, in which every aircraft would broadcast its position, heading, horizontal speed, altitude and vertical speed, and every aircraft would receive and process that data, is difficult to introduce, because the first equipped aircraft gets no protection at all.

(b) A completely autonomous system, such as a three-dimensional radar on board, would measure, track and process range, direction and elevation of the intruder without any cooperation. Such a system delivers benefits even for the first equipped aircraft, but would be extremely expensive.

(c) It is possible to have a combined system, mixing both of these extreme technical solutions in combination with other systems.

1.2 Maneuvering Aircraft

Aircraft maneuvers have a significant influence on the principle and design of CASs. For example, if two aircraft are flying parallel at a vertical distance of 1000 ft (300 m), apart then the calculated τ is infinite (see Fig. 1). However, if the lower aircraft initiates a climb rate of 3000 ft min^{-1} (30 m s^{-1}) or more, then τ will change very rapidly from infinity to 20 s or less within a few seconds of vertical acceleration.

Recognition of such a threat by the aircraft above requires accurate and frequently updated position data or measurement, and a quick-following tracking. It will be very difficult, or perhaps impossible, to detect such a vertical acceleration phase and to interpret it early enough as the establishment of a threat. This problem and its operational aspects illustrate that, in such a situation, pilots in both proximate aircraft should be prevented from sudden vertical acceleration in the direction of the other aircraft.

This consideration suggests that full protection of both aircraft requires CASs in both aircraft. A single system may deliver only limited protection. As such, autonomous systems lose much of their attraction.

1.3 Resolution Advisories

Position data of proximate or even threatening aircraft processed in an on-board CAS may be presented to the pilot on a graphic display showing the traffic situation. This enhances the "traffic awareness" of the pilot and significantly supports visual detection and conflict solution.

As known from collision warning systems on ships, such a graphic presentation of the traffic situation is difficult to interpret, particularly with respect to the selection of a proper escape maneuver (see Fig. 2). Under instrument meteorological conditions (IMC), the pilot needs some computer assistance for successful conflict resolution. Unfortunately, this need introduces a new problem into the area of collision avoidance, because any misinterpretation of an originally safe traffic situation by the CAS may create unnecessary or even dangerous resolution advisories (RAs).

If the introduction of a CAS reduces the number of critical air misses (one per 10^5 flight hours) or collisions (one per 2×10^7 flight hours) by at least an order of magnitude, then a CAS may only cause an extremely small number of critical air misses or collisions by giving false RAs. The demonstration of the quality of the system, therefore, requires extensive flight tests and investigations representing millions of flight hours. Trials and investigations of this extent cannot be afforded by airlines, associations or states, and may only be achieved on an international basis. This explains why past system proposals have not really succeeded.

An additional, but related, problem is that the maneuvering aircraft will increase the rate of false or even dangerous RAs. According to US statistics (Federal Aviation Administrtion 1987), most (43%) near midair collisions reported by pilots occur when one aircraft is flying level and the other is climbing or descending. The probability of maneuvers correlating with critical situations depends on the ATC procedures in the airspace; this suggests that tests and investigations of CASs are necessary in all types of airspaces.

2. History of CAS Development

Except for one, all systems proposals made in the past have been made on a cooperative basis. Acknowledging *a priori* the problems of measuring the relative bearing and, in particular, the elevation angle of a proximate aircraft on board a CAS-fitted aircraft, the proposed systems measured only the range of proximate aircraft and obtained altitude information by transmission of pressure altitude information incremented by 100 ft (30 m) steps.

2.1 CAS Systems Initiated by ATA

On this basis, ATA specified only vertical RAs to be delivered by the system (Air Transport Association of America 1967). Figure 3 illustrates schematically the specified relations between range R, reported altitude h, derived relative speed $\dot{R}$, (i.e., range rate or closing speed) and vertical speed $\dot{h}$, and the calculation of RAs by algorithms called the CAS logic.

The system design was based on experience taken from modified Tacan systems for station keeping of military aircraft, developed by Collins, and Bendix and Sierra Research, and from a traffic monitoring system developed by McDonnel-Douglas. Essential elements of this system, usually called EROS (elimination of range zero system) are the allocation of time slots to each aircraft in the system and synchronization by a particular ground station. In Fig. 4, the time-sharing scheme shows 2000 time slots, the first for the ground station and the other 1999 available for aircraft in the coverage of the ground station. Every aircraft is given a time slot and transmits a range pulse and an altitude pulse. The distance between range and altitude pulses represents the altitude–time code of the transmitting aircraft. Presuming synchronization of all aircraft, the receipt of the range pulse indicates the run time and, thus, the range to proximate aircraft (one-way ranging). $\dot{R}$ can be taken from the Doppler shift of the carrier frequency (about 1600 MHz) or derived from range. Within 3 s, every aircraft transmits this signal in its slot once and all other aircraft fitted with the system obtain all the information about other aircraft necessary for collision avoidance. Full information about further details and degraded versions in areas without ground-station coverage and synchronization may be found in Form (1983).

In Fig. 4, the very low update rate of the system was determined by the epoch of 3 s, limiting the accuracy of $\dot{R}$ and $\dot{h}$. McDonnel-Douglas, and Bendix and Sierra Research developed the system in cooperation, and a final flight test was made by the Martin Marietta Corporation for the ATA (Martin Marietta Corporation 1970). The required expensive infrastructure on the ground (ground stations) and on board (high-stability frequency sources) finally prevented the ATA and the Federal Aviation Administration (FAA) from introduction of this type of system.

2.2 Nonsynchronous Techniques

In 1970, the Radio Corporation of America (RCA) presented the Secant system (Parson 1972) which, in a

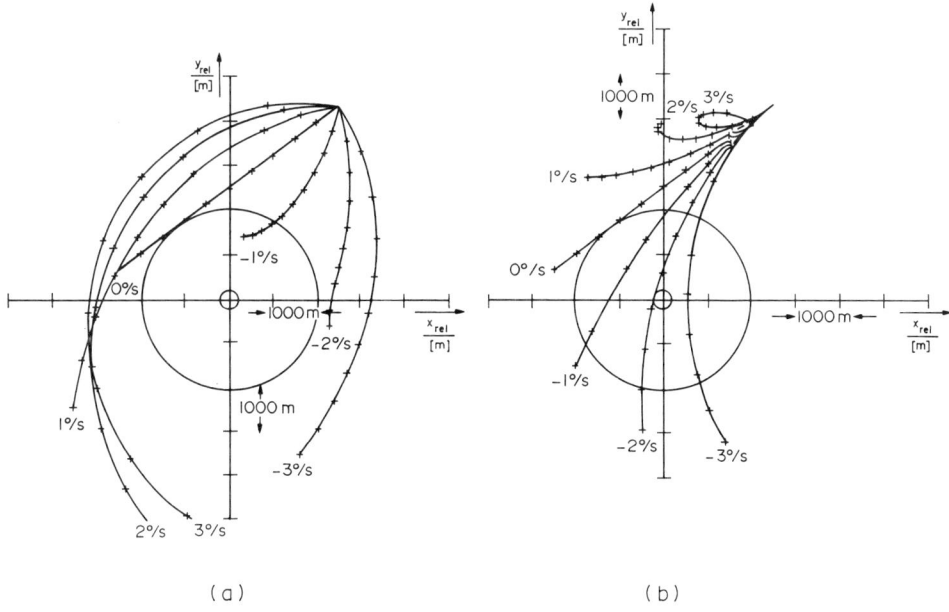

Figure 2
Relative position tracks of a threatening aircraft indicated on a cockpit display: (a) own aircraft turns only and (b) threatening aircraft turns (parameter is the rate of turn in degrees per second)

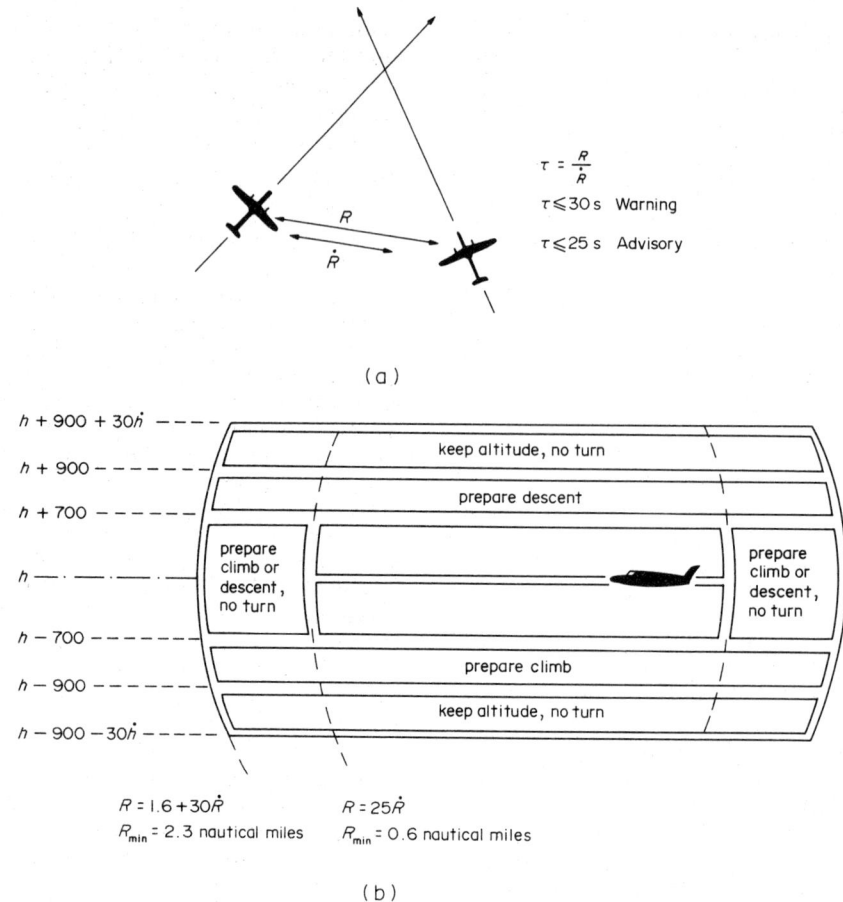

Figure 3
ATA requirements and calculation of τ: (a) in the horizontal plane and (b) in the vertical plane with the RAs delivered if an intruder flies in that sector

similar manner to distance measuring equipment (DME), uses a random interrogation–reply technique without synchronization and ground stations. A Secant-fitted aircraft transmits interrogation pulses (of 1 μs) in randomly selected time intervals (1 ms ± 0.25 ms) and uses statistical selection of one of two carrier frequencies; all proximate Secant-fitted aircraft within range respond with a reply pulse. A sequence of reply pulses from a proximate aircraft carries pressure altitude information on that aircraft using binary frequency modulation. Depending on the number of proximate aircraft, the number of replies to an interrogation may be large. Short pulses (1 μs) and a correlation technique provide proper selection of the aircraft and a range resolution of 150 m. Additional window tracking techniques improved range resolution to 5 m and the accuracy of $\dot{R}$ to 7.5 m s^{-1}.

In a second-phase development, the system was enhanced by a direction-finding antenna for relative bearing measurement. The system was tested by simulation (Los Angeles Basin traffic model) and flight tested by the US Navy in 1972. The results showed 90% detection probability, good range accuracy (average 22 m, with standard deviation 14 m) and precise $\dot{R}$ (average 1.75 m s^{-1}, with standard deviation 6.5 m s^{-1}).

Based on the same frequency band of about 1600 MHz, Honeywell Incorporated developed a

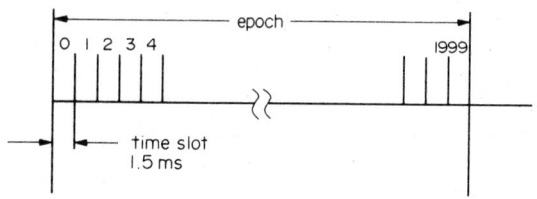

Figure 4
Epoch and time slot for ground station and aircraft in the EROS system

broadband interrogation–reply system called AVOIDS (Honeywell Incorporated 1971) which also followed ATA specifications. Originally designed for helicopter applications in the US Army, this system interrogates by two pairs of very short pulses. The length of a pulse is 0.1 µs; the distance of pulses is 0.5 µs in a pair. The interval between the two pairs of an interrogation indicates the pressure altitude of the interrogating aircraft. Another AVOIDS-fitted aircraft replies only if its own altitude does not differ by more than ±650 ft (200 m). This significantly reduces the rate of replies in the system. By a slow variation of the altitude coding of its own interrogations, the altitude of other proximate aircraft can be found out without extra altitude reports. Obviously, this system delivers good results, corresponding to ATA specifications.

2.3 Autonomous Systems

In 1971, Cyned Incorporated developed an airborne primary radar to be installed on the rudder within a small radome (Anon 1971). This system detects aircraft within a range of 0.75 nautical miles (1.4 km) and within an elevation of ±12.5°, and indicates relative aircraft position on a display in the cockpit without altitude information. Due to its limited capabilities, the system is called the proximity warning indicator (PWI).

Tests were also carried out using airborne weather radar equipment and transponders (developed by the Bendix Corporation and Motorola), but without satisfying ATA requirements (Anon 1975). A major reason for this is the very limited coverage given by weather radar.

2.4 Cooperative Systems Based on ATC Transponders

In 1968, G. B. Litchford presented proposals that incorporated the conventional ATC transponder into collision avoidance. The purpose of this was the detection of proximate aircraft not yet linked to a CAS system. Altitude reporting transponder equipage is mandatory for every aircraft flying under ATC in IMC. Aircraft flying without control may have transponders. The equipage of general aviation aircraft in the USA is about $\frac{2}{3}$, of which only 50% report altitude, which is a necessary piece of information for threat detection without unacceptable numbers of false alarms.

The interrogation of transponders of proximate aircraft by an airborne CAS and the receipt of replies has some fundamental problems.

(a) The additional interrogation of transponders by aircraft fitted with CASs and the replies elicited by these interrogations will increase the radio load in the surveillance radar frequency channels (1030 MHz, 1090 MHz).

(b) Additional blocking of transponders will be created (35–100 µs after each interrogation).

(c) In dense traffic, one interrogation will elicit many overlapping (synchronously garbled) replies, which might not be decoded.

These problems explain why the first proposal in this direction, called beacon CAS (BCAS), had two modes of operation:

(a) active interrogation in low-density areas, possibly without radar coverage; and

(b) a passive "listen-in" technique without interrogation, in which the BCAS only receives and evaluates replies elicited by radar interrogations (Litchford 1979).

During the 1970s, the FAA supported these sorts of proposals by a BCAS program in cooperation with MITRE and the Lincoln Laboratory (Massachusetts Institute of Technology). MITRE met the synchronous garble problem by establishing the "whisper–shout" technique, in which a sequence of rising interrogations and side-lobe suppression pulses are sent. The sequence interrogates the nearest aircraft first, blocks them during the next stronger interrogation and so on. The Lincoln Laboratory introduced a sectorized interrogation technique as an additional solution. Finally, the FAA developed the full BCAS program, in which a combination of the active and passive operation modes allows application in areas of low and high traffic density. This concept became very complex and expensive when the FAA added the ground-based collision avoidance system ATARS. This was an outgrowth of the discrete address beacon system with data link which began development by FAA, MITRE and the Lincoln Laboratory in 1972. The ATARS element in this concept should provide protection of aircraft in very-high-density areas (0.3 aircraft) per square nautical mile (0.1 aircraft km^{-2}). However, this full BCAS/ATARS progam was terminated because of the growing compatibility problems related to the complex combination of active/passive BCAS and ATARS, and was replaced in 1981 by the threat alert and collision avoidance system (TCAS) program.

3. Mode-S-Related CASs

3.1 General

With respect to the number of airborne collisions in the world, the International Civil Aviation Organization (ICAO) emphasized the need for better surveillance and ATC on the ground, and for an independent CAS in the air (International Civil Aviation Organization 1971). Following this recommendation, the development of an improved secondary surveillance radar (SSR) was pushed in the USA and in the UK. The main requirements were the elimination of synchronously garbled replies, reliable tracking and the introduction of a selective data link by a discretely addressing SSR. These system developments were initially called DABS in the USA and ADSEL in the UK, but later became known as mode S (selective) in both countries and by the ICAO, who standardized mode S as Annex 10 to their Standards and Recommendations (SARPs).

SSR mode S will be introduced into US airspace during the 1990s and other countries will follow, overcoming the deficiencies of SSR and of using the data link. The development of collision avoidance also requires an air-to-air data link for the coordination of two aircraft fitted with CASs in a conflict situation. Acknowledging this need, the ICAO considers only CAS systems proposals that incorporate mode S transponders providing the air-to-air crosslink capability.

In 1988, the USA established a rule requiring that TCAS II be fitted to all passenger aircraft with more than 30 seats. The present bill (April 1989) sets a deadline for installation of the end of 1992. The secondary radar improvement and collision avoidance system (SICAS) panel of the ICAO is considering three generic systems called airborne collision avoidance systems (ACASs) I, II and III.

(a) ACAS I provides traffic advisories (TAs) as an aid to initiate "see-and-avoid" action, but does not include the capability for generating resolution advisories (RAs). The ICAO standards for ACAS I only ensure compatible operation with other ACAS configurations and limit radio interference of the system. ACAS I is nearly identical to the TCAS I plans of the FAA.

(b) ACAS II provides vertical RAs in addition to TAs. The SICAS panel finished draft SARPs on ACAS II in 1989 and have proposed a preliminary introduction of ACAS II to get the experience of millions of flight hours necessary for final design and worldwide standardization. The draft ACAS II SARPs require some features not yet identical with flight-tested TCAS II equipment. ACAS II will require an on-board mode S transponder for coordination.

(c) ACAS III provides vertical and horizontal RAs, in addition to TAs. ACAS III has not yet been fully discussed by the ICAO, but the FAA is pushing the development and testing of the system in cooperation with the Bendix Corporation, because pilots prefer horizontal maneuvers for conflict resolution. The development of on-board direction-finding techniques for this purpose promises an accuracy of about 2°. ACAS III will require an on-board mode S transponder for coordination.

In principle, the ACAS should protect its own aircraft in every phase of flight and in every traffic density. Some functional limitations, however, must be acknowledged. For example, in all types of ACAS, interference with the SSR requires limitation of power and of the rate of ACAS interrogations. The rate of TA and RA needs some limitation of the warning time τ in the lower airspace and in even lower altitudes by a sensitivity control algorithm in the ACAS. Other operational limitations include the use of only vertical resolution maneuvers in the ACAS II system.

The Radio Technical Commission of Aeronautics of America (RTCA) has developed TCAS I guidelines (Radio Technical Commission of Aeronautics of America 1983a) minimum operational performance standards (MOPS) for TCAS II (Radio Technical Commission of Aeronautics of America 1983b) and a draft for MOPS for TCAS III. An improved version of MOPS for TCAS II aligned to the ICAO SARPs is expected in the early 1990s.

3.2 ACAS II Functions

Figure 5 shows the generic ACAS functions. The transmitter periodically radiates whisper–shout sequences to elicit altitude reporting replies from proximate aircraft equipped only with mode C transponders. As an option, mode A may also be interrogated. Receipt of the squitters (transponder address information randomly transmitted ($\sim 1\,\text{s}^{-1}$) by mode S transponder) of proximate aircraft equipped with mode S transponders will create discretely addressed interrogations to that aircraft. The surveillance establishes the tracks of all proximate aircraft with mode C or mode S transponders and submits these data to the logic. The ACAS logic is divided into two boxes. Both provide a range and an altitude test. If the projected flight path of an intruder passes both tests, then a TA or even an RA will be delivered to the pilot by a display or synthetic voice in the cockpit.

Some additional provisions are made for the solution of a multiple aircraft conflict. If an RA is indicated, then a coordination process will be established via the mode S crosslink to the threat aircraft if it is equipped with ACAS II or III and the RA will be submitted to the mode S transponder, which indicates that in the next reply to a mode S ground station an RA broadcast will be sent by the transmitter for receipt by ground stations that are not operating in mode S.

The mode S transponder receives coordination information from other ACAS aircraft and sensitivity level control (SLC) commands from mode S ground stations. Normally, the sensitivity of ACAS will be controlled by the altitude of its own aircraft or by manual settings by the pilot.

The limitation of interference requires information about the number of ACAS-equipped aircraft in the vicinity. Therefore, the ACAS transmitter radiates a broadcast every 10 s indicating the presence of ACAS equipment to other aircraft.

Figure 6 illustrates a retrofit version of a display, which shows the traffic situation around its own aircraft graphically if one proximate aircraft was identified to be a potential threat. The position of that potential threat is indicated by a yellow point; the positions of other proximate aircraft are white colored. If the aircraft becomes a real threat and creates an RA, then the color changes to red and the eyebrows advise the pilot to climb as for example, in Fig. 6, with a vertical rate of 15 000 ft min^{-1} (75 m s^{-1}).

Airborne Collision Avoidance

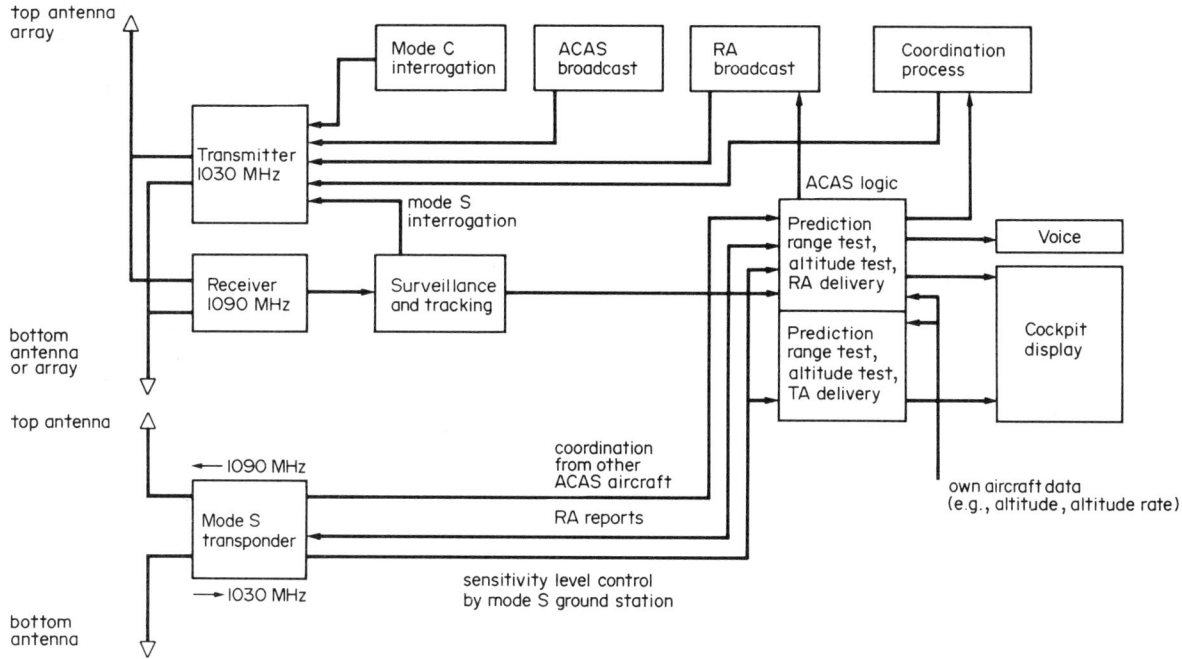

Figure 5
Generic ACAS functions in combination with a mode S transponder

3.3 TCAS II Flight Testing

Some results may be drawn from around 2000 flight hours flight tests and in-service evaluation (by the FAA, Piedmont Airlines, United Airlines, Air France and British Airways), and radar data analysis in Germany (by Technische Universität Braunschweig, representing about 2000 flight hours) and British airspace (by the Royal Signals and Radar Establishment, representing about 100 000 flight hours).

TA were delivered about once per 1–3 flight hours, depending on airspace and aircraft. Although this rate was high, pilots felt TA would enhance traffic awareness. Pilots could detect visually 67% of altitude-reporting potential threats, against only 25% without altitude reports. Only 33% of the TA against altitude-reporting intruders would be followed by an RA (once per 20–40 flight hours). About 10% of these advised the pilot to cross the altitude of the intruder (once per 400–500 flight hours). In these cases, the pilot was reluctant to follow, because the intruder might level off after the RA delivery. A very small percentage of these cases led to induced critical air miss situations. Further tests and evaluation during preliminary introduction will determine whether the required target level of safety (once per 5 000 000 flight hours) for ACAS-induced critical air misses can be met or whether further improvements of the system are required.

See also: Navigation Systems, Integrated; Visual and Instrument Flying Rules

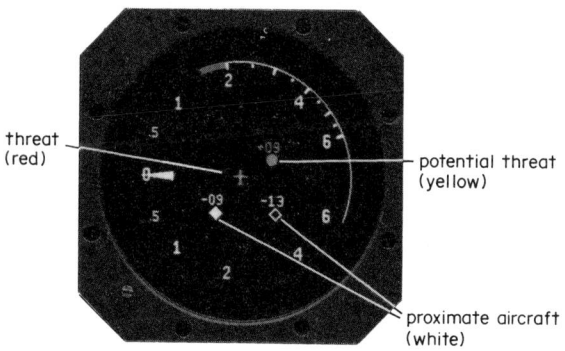

Figure 6
Retrofit version of a traffic and resolution advisory display integrated into vertical speed indicator (courtesy Bendix-King)

Bibliography

Air Transport Association of America 1967 ANTC Report No. 117
Anon 1971 FAA plans test of low-cost radar proximity warning unit. *Aviation Week and Space Technology* September
Anon 1975 The airborne radar navigator (Bendix multi-mode radar). *Business and Commercial Aviation* September

Federal Aviation Administration 1987 *Selected Statistics Concerning Pilot Reported Near Midair Collisions 1983–1986*. FAA, Washington, DC
Form P 1983 Kollisionsschutz in der Luftfahrt. Eine Studie für den Bundesminister für Verkehr. Technische Universität Braunschweig report
Honeywell Incorporated 1971 *Midair Collisions, the Problem—the Solution*, Report No. ASD 132-1. Honeywell Inc., Minneapolis, MN
International Civil Aviation Organization 1971 *ICAO Aeronautical Conf*. ICAO, Montreal
Litchford G B 1979 An SSR-based, air derived CAS can complement today's ATC system. Report on collision avoidance. *ICAO Bull. Spec.* March
Martin Marietta Corporation 1970 *Collision Avoidance System Test Flight and Evaluation Program*, Vols. 1 and 2. Martin Marietta Corp., Baltimore, MD
Parson I L 1972 SECANT, a solution to the problem of midair collisions. *International Federation of Airline Pilots Association Symp.*, Frankfurt
Radio Technical Commission of Aeronautics of America 1983a *TCAS I Functional Guidelines*, RTCA Document No. 184. RTCA, Washington, DC
Radio Technical Commission of Aeronautics of America 1983b *Minimum Operational Performance Standards (MOPS) for TCAS II*, RTCA Document No. 185. RTCA, Washington, DC
White F C 1968 Airborne collision avoidance systems development, introduction. *IEEE AES* **4**(2)

P. Form
[Technische Universität Braunschweig,
Braunschweig, FRG]

Automated Guideway Transit Systems and Personal Rapid Transit Systems

Since the end of the 1960s, a few countries, for example the USA, France, FRG and Japan, have been undertaking research to develop new modes of urban transportation which could contribute to resolving the problems raised by increasing use of the automobile within cities: great consumption of space and energy, traffic jams, accidents, pollution, and noise. One solution consists in the improvement of existing public transport and another in the adaptation of new automation technologies to new types of system. These new systems are the automated guideway transit (AGT) systems; the personal rapid transit (PRT) systems belong to this group, but with additional operational characteristics and increased use of computers.

1. General Characteristics of AGT Systems

AGT can be defined by the entire automation of the vehicles and the absence of a driver. Most of the other characteristics of AGT systems are consequences of this choice: for example, the traffic on an exclusive guideway and the relatively small capacities of the vehicles. To carry a given passenger flow, an AGT system can use smaller vehicles circulating with higher frequencies than in classic systems, first because there is no problem of multiplication of drivers which would increase the operating cost, and second because the response time of automatic devices is less than that of human beings, so intervals between vehicles can be reduced. Of course this comparison is relative and the capacity of an AGT passenger unit can vary between two passengers and 200 passengers (or even more) according to the kind of system, but for large passenger flows, AGT systems have smaller vehicles than classic metros, and for medium or low passenger flows, corresponding AGT systems have smaller vehicles than streetcars or buses.

The smaller size of vehicles and the higher frequency (or smaller headway between vehicles) have two consquences: first, greater attractiveness of the system because passengers wait only a short time and second, the substructure is cheaper than those for bigger vehicles. The substructure represents a significant part of the cost of the system because an exclusive guideway is necessary, and this means that the guideway must be protected from surrounding traffic. The systems are usually designed so that the guideway can be built at grade, elevated, in a cutting or underground; the choice between these four possibilities depends on factors which are independent of the technology: for example, the distribution of space inside the city.

As a consequence of the technological choices, AGT systems are characterized by their favorable impact with regard to environment problems.

(a) The consumption of space is small, relative to the number of passengers transported.

(b) The energy consumption is either comparable with or less than that of classic public transportation systems (bus, metro), which is about 100 W h (or between 0.015 kg and 0.030 kg of crude-oil equivalent) for one passenger transported 1 km, with an average vehicle occupancy of 20%. This means that, on average, the energy consumption is about one-fourth that of automobiles in the city.

(c) As a consequence of the use of electrical energy, AGT systems introduce no air pollution into the city where people are living, and very limited pollution outside the cities (if we consider that electricity production could also be, in some cases, a source of pollution) because of the low energy consumption. This is important since the automobile traffic produces large quantities of toxic fumes (CO, NO_x, lead, hydrocarbons, etc) which have as bad an impact on the health of the inhabitants as on the environment (acid rains, damage to forests, deterioration of buildings), and which will not be completely suppressed by the new regulation of exhaust fumes. The catalysts, for example, will reduce the emission of some pollutants, but will not reduce the emission of CO_2 at all.

(d) The noise level is reduced.

(e) The safety level is very high because the traffic is on an exclusive guideway and because of automation. This is very important if we consider that in France, for example, each year more than 150 000 people (among them pedestrians and cyclists) are injured by road traffic in cities.

Compared with other public transport systems, AGT systems are characterized in the following way.

(a) The average speed and the regularity of service due to the exclusive guideway are both satisfactory, which is not the case with buses or even streetcars, which can be stopped or slowed down by traffic jams.

(b) As already mentioned, the waiting time is reduced because of smaller intervals between vehicles, due to automation. This is also the case during peak hours.

(c) The use of modern technologies permits increased comfort in the vehicles as well as in the stations. The stations and vehicles of most AGT systems are even designed to be able to accept handicapped people in wheelchairs.

(d) The absence of drivers gives great flexibility for the insertion of vehicles. When necessary, in case of exceptional events, it is possible to operate the system 24 h d^{-1} (see *Merging Control*).

Details of many AGT systems currently in use or in construction are given in Table 1 and some details of characteristics of some systems in use are given in Table 2.

2. *Classification of AGT systems*

2.1 *American Classification*

In the mid-1970s, a classification born in the USA was in widespread use there, as well as being in use in other countries: the concepts used were shuttle and loop transit (SLT), group rapid transit (GRT) and personal rapid transit (PRT).

An SLT system is the simplest type of AGT system. The vehicles in this system may be of various sizes and travel on a fixed path which may have provision for several stations, but few or no switches. Vehicles may travel as single units or coupled together as trains to accommodate heavier passenger flow. In a shuttle system, vehicles move back and forth over a single guideway, while in loop transit they move over a closed path. There are numerous SLT systems currently in operation.

GRT differs from SLT systems in network and operational complexity, since it is designed more to serve groups of travellers with similar origins and destinations. For this reason, GRT has switching capabilities which allow for branch routes, and off-line stations so that vehicles on the main line are not delayed by those waiting at stations. There are two examples of GRT systems currently in public use (see Table 1).

PRT systems are characterized by small vehicles, usually carrying fewer than six people travelling together by choice. The headway, or time interval between the arrival of successive vehicles, is very short (usually less than 3 s), and the guideways are smaller and less obtrusive than for SLT or GRT. Plans for PRT systems call for a broad range of service policiers and require a high degree of technical sophistication. To date, there are no PRT systems in public use and, for cost reasons, all the development programs have been stopped, although a few prototypes have been successfully tested. It is this reason why the classification between SLT, GRT and PRT has lost a large part of its significance. It is, nevertheless, important to know these definitions, since the term of PRT is sometimes used instead of GRT, or even, in rare cases, instead of AGT.

In the USA, AGT systems, especially those of small or middle size, are very often called "people movers" (there are, for example, many airport people movers), or "downtown people movers" when they are built in the center of a city. However, as the term people movers can have a broader meaning, "automated people movers" is less ambiguous.

2.2 *Classification with Regard to Function*

There are three main types of application for AGT systems: hectometric, urban and suburban.

The concept of a hectometric system (which could also be called a system for short-to-middle distances) is often used in France and results from the following two factors.

(a) The need for AGT systems for sites that are rarely longer than 2–3 km, such as in big "closed complexes" (e.g., exhibition parks, airports), in connections between lines of public transport or between stations and parking, and in transfer lines towards lines with heavy traffic.

(b) The existence of technologies, such as mechanical active guideway, that are well adapted to this application but not appropriate to or less efficient for longer lines (see Sect. 9).

Specific urban systems adapted to lines many kilometers long, with interstations usually shorter than 1 km and a maximum speed that rarely exceeds 80 km h^{-1}, represent an important application case for AGT. Here, systems are sometimes called automatic mini metro, automatic light metro or automatic metro.

For specific applications, such as pole-to-pole connections, links between city and airport, and fast ringways, where the lines are not necessarily much longer than those of specific urban applications but where the interstations are a few kilometers long, there is a need for suburban systems with the following characteristics:

Automated Guideway Transit Systems and Personal Rapid Transit Systems

Table 1
46 of the AGT system lines around the world

System	Manufacturer	Country of manufacture	Application	Country of application	Year of opening	Line length (km)	Number of stations	Number of vehicles	Suspension	Motorization	Observations
VAL	Matra	France	Lille 1	France	1983	13.2	18	48×2	pneumatic tires	dc motor	
			Lille 1bis	France	1989	12	18	35×2			
			Jacksonville	USA	1989	1	3	2			
			Chicago	USA	1991	4.5	6	13			
			Taipeh	Taiwan	1991	11.5	13	102			
			Antony/Orly	France	1991	7.2	4	8×2			
			Toulouse 1	France	1993	10	15	29×2			
Miami DCM	Westinghouse	USA	Miami	USA	1986	3.8	9	12	pneumatic tires	dc motor (ac collection)	
Las Colinas APM	Westinghouse	USA	Las Colinas	USA	1989	1.9 (single)	4	4			with one agent
Various	Westinghouse	USA	10 Airports	USA and others	1971–1990						GRT
ALRT/ICTS	UTDC	Canada	Vancouver	Canada	1986	21.4	15	114	steel wheels	linear motor	
			Detroit	USA	1987	4.7 (single)	13	13			
			Toronto	Canada	1985	7	6	24			
Morgantown people mover	Boeing	USA	Morgantown	USA	1971	6.6	8	73	pneumatic tires	dc motor (ac collection)	GRT
Airtrans	Vought	USA	Dallas Airport	USA	1974	20 (single)	14	52	pneumatic tires	dc motor (ac collection)	
KNT	Kawasaki	Japan	Kobe	Japan	1981	9.3 (single)	9	72	pneumatic tires	dc motor (ac collection)	
NTS	Niigata (bought licence)	Japan	Osaka	Japan	1981	6.9	8	60	pneumatic tires	dc motor (ac collection)	
Vona	Nippon Sharyo	Japan	Omiya	Japan	1983	12.7	13		pneumatic tires	dc motor	
			Yokohama	Japan	1989	10.8	14				
			Sakura	Japan	1983	4.1	6				
M-Bahn	AEG	Germany	Berlin	Germany	1989	1.6	3	2×2	magnetic levitation	linear motor	
H-Bahn	Siemens/Duewag	Germany	Las Vegas	USA	1991	1.9	4	6	rubber tires	dc motor (ac collection)	suspended vehicles
			Dortmund	Germany	1984	1.1	2	2			
Poma 2000	SGTE	France	Laon	France	1989	1.5	3	3	pneumatic tires	cable	
SK	Soule	France	Villepinte (Paris Nord)	France	1986	0.3	2	16	rubber tires	cable	semicontinuous
			Vancouver	Canada	1986	0.15	2	12			
			Yokohama	Japan	1989	0.7	2	25			
			Noisy le Grand	France	1992	0.45	2	20			
			Paris (Gares Lyon–Austerlitz)	France	1994	0.6	2	25			
Wedway	TGI	USA	Orlando (Disneyworld)	USA	1975	1.4 (single)	1	30×5	rubber tires	linear motor	
Otis-Shuttle	Otis	USA	Houston Airport	USA	1981	2.2	5	6×3	air cushion	cable	
		USA	Tampa	USA	1985	0.7	2	2			
			Serfaus	Austria	1985	1.3	4	2			
			Sun City	South Africa	1986	1.7	3	3			
			Narita	Japan	1992	0.3	2	4			
Maglev	GEC	UK	Birmingham	UK	1984	0.6	2	2×2	magnetic levitation	linear motor	
Aeromovel	Sur Coestler	Brazil	Djakarta	Indonesia	1989	3.2	6		steel wheels		pneumatic

(a) a relatively high speed of about 150 km h^{-1} and sometimes higher,

(b) small curve radius to facilitate insertion into urban areas at the end of the line,

(c) a good acceleration and deceleration capability, and

(d) a design well adapted to aerial guideway, without excluding, of course, an insertion at grade or underground—this last solution has to be avoided as far as possible, but can sometimes be necessary on limited sections, for example, for the end of a line in a dense urban area, where the noise level must be very low.

The existence of such applications justifies the development of new magnetic levitation systems with middle-range speeds (see *Magnetic Suspension Railroad Systems*).

2.3 Technical Classification

There are two kinds of AGT systems: passive guideway systems and active guideway systems. Most are passive guideway systems; that means, the vehicle carries its own motorization device (as in classical systems), which is generally a dc electric motor but sometimes an ac motor or a linear motor. In this case, transfer of information between guideway and the vehicle is necessary, to transmit the speed diagram to the motor; information transfer can be effected similarly to the technique used for modern classical metros which are driven automatically but with the presence of a driver on board, who assures the performance of some residual functions. Therefore, in AGT systems, more information has to be transmitted so that these residual functions of the driver can be executed by automatic control equipment: the order of departure at each station; the state of on-board equipment, so as to prevent breakdown; and, possibly, the order for automatic pushing of the next vehicle, in case of total breakdown.

Some AGT systems are called active guideway systems because the motorization is not in the vehicle but on the guideway. In a mechanically active guideway, a cable transmits the propulsion power as well as the speed diagram from dc motors on the ground (at some stations) to the vehicle. In an electrically active guideway, inductors of linear motors placed on the guideway provide the propulsion power in relation to an armature mounted in the vehicle; the speed diagram is determined by the distribution of the inductors along the guideway.

Table 2
Examples of AGT system characteristics

	KNT	VAL	Airtrans	Morgantown	H-Bahn	Poma 2000	SK first generation	Miami DCM
Max. capacity of vehicles								
seated	24	22[a]	16	8	8–24	12	6–6	16
standing	51	82	24	13	13–67	21	6–9	96
total	75	104	40	21	21–91	33	12–15	112
Size of vehicle								
length (m)	9.1	12.7	6.4	4.72	3.45–11.64	4.9	2.75	11.9
width (m)	2.35	2.06	2.14	2.04	2.3	2.2	1.4	2.8
height (m)	3.15	3.25	3.05	2.68	2.3	2.9	2.51	3.35
Weight empty (kg)	9000	14 600	6350	3900	3200–8100	3500	950	12 500
Type	SLT	SLT	GRT	GRT	SLT	SLT	Semi continuous	SLT
Possibility of off-line stations	no	no	yes	yes	no	no	no	no
Min. headway (s)	90	60	18	15	60	variable	15	100
Passengers per lane per hour (crush load max)	18 000	25 000 (2 passenger units)	12 000	5000	1260–5460[b]	5000	3600	5200
Speed (km h^{-1})								
normal		60	27	36	60	35	20	43
max	70	80	27	48	60	40	25	43
Max. grade (%)	10	7	7.8		7.5 dc 15 linear	15	10	10
Switching				on-board	on-board			
Country	Japan	France	USA	USA	FRG	France	France	USA
First application	Kobe	Lille	Dallas Fort Worth	Morgantown	Dortmund	Laon	Villepinte	Miami

a Jump seat folded   b More if vehicles coupled

Another type of electrically active guideway consists of a continuous winding on the guideway. In active guideway systems there is no need for equipment to transmit the speed information to the vehicle, because motors and control equipment are on the guideway. This simplifies matters, but equipment is still necessary to ensure the collision avoidance function; in a mechanical active guideway, this equipment can be simplified because the cable contributes to keeping the right interval between vehicles.

3. AGT Systems: Main Components

3.1 Guideway

The guideway is made of either concrete or steel. Most of the guideways are designed to be built for aerial, grade and underground, but there can be specific configurations for aerial only guideways: for example, the guideway of the H-Bahn is made of a suspended box beam which protects the rolling bogie truck of the vehicles.

3.2 Support and Lateral Guidance System

Many AGT systems use rubber pneumatic tires for rolling as well as for lateral guidance, but it is also possible to use railway wheels and rails. German systems and some hectometric systems usually use steel wheels covered with special synthetic rubber. Air cushions and magnetic levitation are also utilized for AGT systems, in some specific cases.

3.3 Motor

The use of electrical energy has already been mentioned. An additional advantage is the reduction in vibration and the limitation of jerk during acceleration or deceleration. The use of the linear motor avoids grip problems on slopes or in icy conditions. Until recently, the linear motor was considered less efficient than rotating motors, but evaluations carried out at the laboratories of the Institut de Récherche des Transports (INRETS) proved that a new linear motor with a U-shaped armature invented by M. Guimbal of St Etienne has improved characteristics. This new motor, initially designed for high-speed systems, can also be adapted for AGT systems. In addition to the linear motor, which can be used either with an active guideway (inductors on the guideway, armature in the vehicle) or with a passive guideway (inductors in the vehicle, armature on the guideway), the mechanical active guideway (with cables) also provides a good grip of the wheels on steep slopes or on ice. Some systems propelled by rotating motors are provided with heating equipment for the running surfaces, to be used exceptionally in case of frost.

3.4 Braking System

In most cases, choppers provide regenerative electrical braking at high speeds and braking is provided by a mechanical brake at low speed. Emergency braking is mechanical and designed in a fail-safe manner.

3.5 Switching System

For some rare systems, the switching is on-board, or "active"; this means that two movable parts mounted on the vehicle carry guidewheels which can be moved either to the right-hand side or to the left-hand side against the guide rail at each switch. This kind of switching allows very short intervals between vehicles. Other types of switching equipment are more classical and "passive"; that is, the moving part is on the guideway.

3.6 Electrical Auxiliaries

The following functions are currently provided on the vehicles: control and monitoring of normal and emergency lighting, of ventilation and heating, of door status, of the compressor air tanks and brake circuit, and of the status of the mechanical coupling bar and electrical connector between the various vehicles of a passenger unit.

3.7 Command and Control System

The command and control systems are designed to operate in a fully automatic mode normally. They usually consist of the hardware and software necessary to provide the following functions:

(a) automatic vehicle protection (AVP) (see *Automatic Train Control: Protection Principles*),

(b) automatic vehicle operation (AVO),

(c) automatic vehicle supervision (AVS) (see *Railroad Systems: Line Supervision and Control*), and

(d) manual backup mode operations.

The AVP functions perform all automatic action necessary to provide safe operations regardless of malfunctions. The AVO functions perform all automatic vehicle operations that are not related to safety. The AVS functions provide a capability to change modes of operation. Manual backup mode operations are utilized in the event of a malfunction which cannot be safely restored in an automatic mode.

The AVP, AVO and AVS functions are performed by control system equipment distributed at three locations: at the control center, along the wayside and on board the vehicle. Control center equipment consists of control consoles, display panels, computers and data communications equipment.

Wayside control equipment consists of, for example, wayside control and communications units, dwell operation control unit, data transmission units, transmission lines and vehicle detectors. This equipment is located both in the stations and along the guideway. Wayside control and communication units (WCCU) perform all wayside safety functions such as collision avoidance and overspeed protection, and they provide all wayside uplink and downlink signalling functions.

Dwell operation control units (DOCU) manage station operation, including dwell control. Data transmission units (DTU) provide communication between the control center and all wayside equipment. Transmission lines are provided for two-way link between wayside and vehicle equipment. Visual signals are located at particular points along the guideway to indicate platform door status, departure authorization and switch status during manual backup operations.

Vehicle on-board control system equipment includes redundant AVP and AVO electronics and power supplies.

3.8 Safety Devices

Because of the absence of a driver, new equipment has to be developed, for example to avoid a passenger becoming wedged in the doors and dragged along by the departing vehicle. One solution is the construction of doors close to the platform, whose opening is synchronized with the opening of the vehicle doors.

A new tendency is the use of microprocessors, also for safety functions.

4. American Systems

The USA is the first country where AGT systems have been put into operation. About 15 AGT systems were constructed during the 1970s, most on relatively short lines, except for the Vought Airtrans system at Dallas Fort Worth Airport and the Boeing system at Morgantown, which operate on larger networks. This is due to the existence of specific sites which were more appropriate than urban areas for the experimentation of this new technology: airports, attraction parks, universities and so on. Among these "first generation" systems, which also include a few monorails, only the Westinghouse system is operating on several lines, essentially in airports (at the end of the 1980s, 12 lines were in operation around the world with two others under construction) and has been able to evolve in order to be installed also in urban areas, although only in two cases in Miami and Las Colinas (see Fig. 1).

At the end of the 1970s, the downtown people mover (DPM) project was launched in order to equip the city centers of the big American towns with AGT systems. Because of supplies restriction at the beginning of the 1980s, this project was reduced and, to date, only Miami, Detroit and Jacksonville are equipped. The so-called Miami Downtown Component of Metrorail, which makes a loop connected to the rail transit system, is the only DPM system completely manufactured in the USA. Among its technical characteristics, are the 600 V ac current collecting, by means of four rails (including one for earthing), two of them being also utilized to send frequencies for command and control purposes, and the automatic train protection system which utilizes two microprocessors in a checked-redundant way to achieve safety.

For the new implementation of this Westinghouse system in Las Colinas (a new town between Dallas and Fort Worth airport), a few improvements have been achieved:

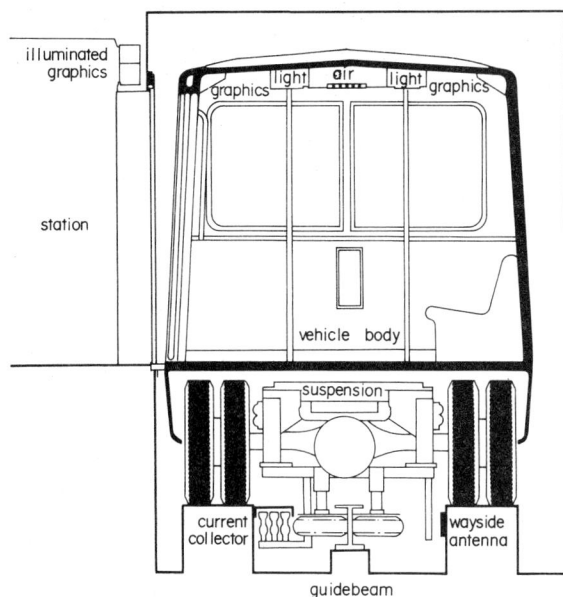

Figure 1
Cross section of a Westinghouse vehicle and guideway

(a) a lighter vehicle, manufactured from composite materials, and

(b) a new switching device, based on the horizontal pivoting of a guideway portion.

5. Japanese Systems

After the USA, Japan is the first country where AGT systems have been put into operation, this time for urban transportation, but in the specific environment of new urbanization in artificial islands.

The Kobe new transit and the new tram system in

Figure 2
New tram system in Osaka

Osaka (see Fig. 2) were both completed in 1981, and both of them connect a new port island with a rail transit system, by means of a line about 6 km long. Since 1981, nine other AGT systems have been constructed in Japan (the operation of the two last ones is foreseen in the early 1990s), if three automatic monorails are included.

Comparing the development of AGT systems in Japan with that of other countries, the choice of the monorail concept for some urban applications, and some similarities between the other systems, although they are manufactured by different constructors, can be noted. These similarities are:

(a) the type of suspension, with pneumatic tires or filled tires;

(b) the guidance by means of tires that are either central or on the sides;

(c) the frequent use of aerial concrete guideways;

(d) the size of the vehicles (a standardization is foreseen in order to go further in this direction);

(e) the coupling in trains comprising a few vehicles (usually more than two);

(f) the headway, usually more than 90 s, is more than most of the systems in other countries (which use smaller trains); and

(g) the power supply, which is very often ac or otherwise dc.

Sometimes in Japan, for psychological reasons, an agent remains in the vehicles during the first years of operation, although this is not necessary for the functioning of the system.

6. VAL System

A short time after the two Japanese systems, the VAL went into operation in Lille: in 1982 for the first stations and in 1983 for the entire 13 km first line. The VAL could qualify as the first fully automtic (light) metro in the world for the following reasons:

(a) the environment of the line was typically urban, with the crossing of a city center, of two suburbs and with the connection of a hospital center and a university; and

(b) the length of the line and its capacity were higher than those of all the other AGT systems that were functioning at this time.

VAL is now the most successful AGT system for various applications:

(a) the light automatic metro—two first lines are in operation at Lille with a third line foreseen, a first line is in construction in Toulouse, and lines are foreseen in other towns such as Rennes and Bordeaux;

Figure 3
VAL vehicle (Lille type)

(b) an automatic metro with a higher capacity—project of Taipeh (Taiwan);

(c) the DPM in Jacksonville (USA);

(d) an airport automated people mover in Chicago; and

(e) the AGT system in connection with a rail transit system—in construction between Antony and Orly airport.

Among the characteristics of this system, the following can be noted (see Figs. 3, 4):

(a) the width of the vehicle, which can vary with the type of application—2.06 m for French projects where the lines are partly in tunnels (narrow vehicles minimize the costs) and 2.56 m for American projects;

(b) the number of vehicles per train, which also varies—a single unit for American projects, married pairs in Lille with a possibility of doubling in the future, and longer trains for high-capacity metros;

(c) the pneumatic-tires suspension with an original axle bogie comprising two suspension wheels, four guiding wheels and two switching rollers;

(d) a dc electrical motor, with thyristor chopper or GTO chopper for the new generation already in operation on the second line in Lille;

(e) the transmission of information between vehicle and guideway (see Fig. 5) with a so-called "ground sheet" similar to that of the Paris metro, including lines which cross each other every 300 m (the speed reference is given by the distance between two crossings);

(f) the collision avoidance which is based on the fixed blocks principle and ultrasonic detectors (each block comprises a normal speed programme and a stopping programme);

Automated Guideway Transit Systems and Personal Rapid Transit Systems

(g) the automatic coupling and the automatic pushing of a failed vehicle (this occurs very rarely because most of the breakdowns are avoided by means of redundancies, telemonitoring and telecontrols, but the propulsion is nevertheless designed so that a vehicle can push the next one out of order, full of passengers, even up a 7% slope);

(h) the pneumatic secondary suspension by means of air bags, which enables a correct levelling of the vehicle floor in relation to the platform (together with station lifts, this is one of the arrangements that permits the access of handicapped people, even in wheelchairs); and

(i) the existence of platform doors.

7. Linear Motor Systems with Passive Guideway

A few AGT systems propelled by a linear induction motor have been developed. Passive guideway, contrary to active guideway, means that the vehicle

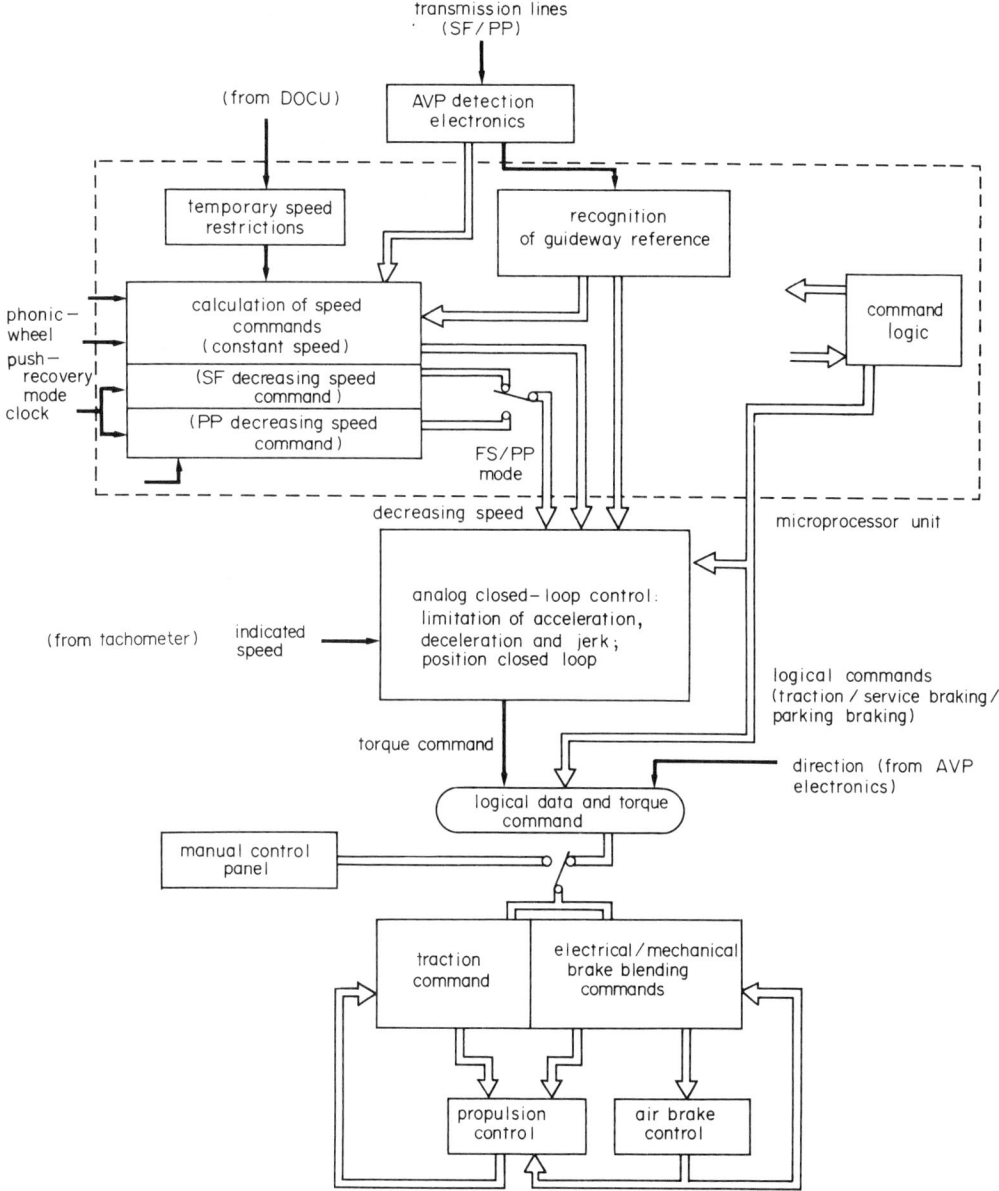

Figure 4
Block diagram of the on-board speed regulation equipment of the VAL system

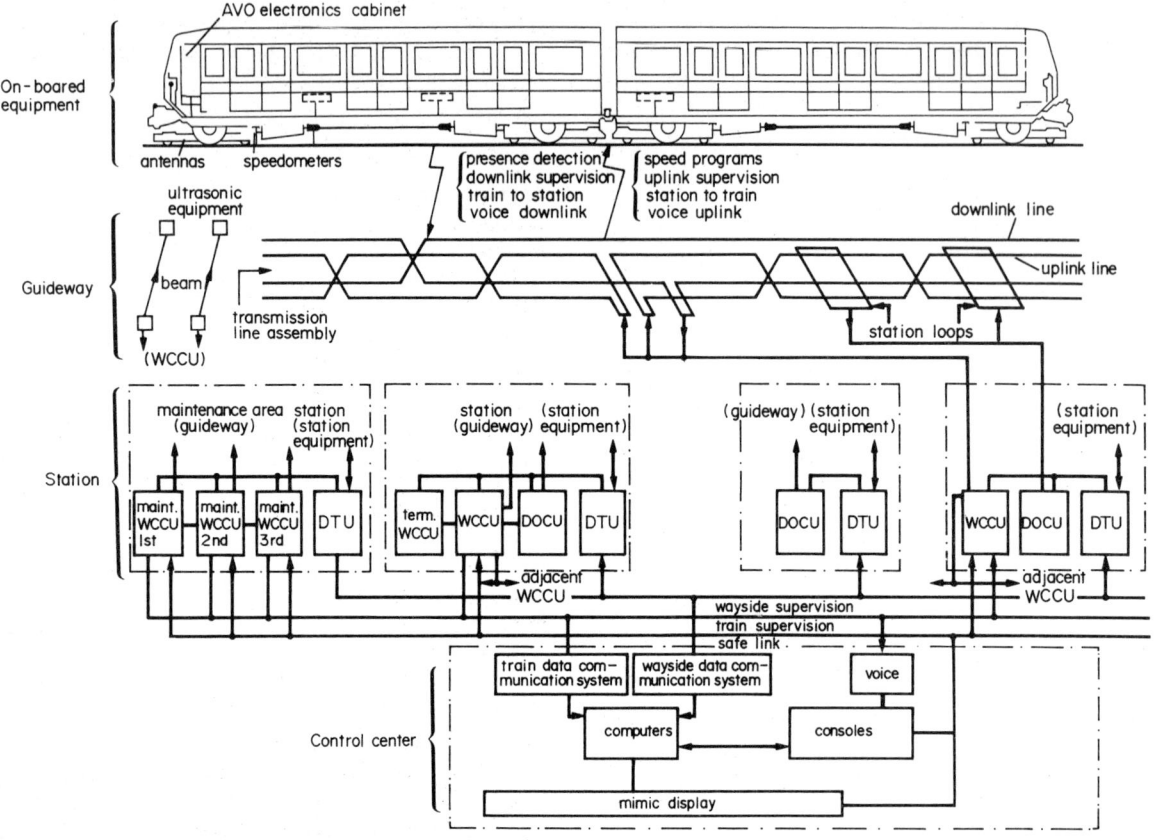

Figure 5
Location of on-board equipment and block diagram of control for the VAL system

carries the active part of the linear motor (the inductor) and the reaction rail is placed all along the track. In this case, an asynchronous linear motor is always used.

7.1 Advanced Light Rail Transit/Skytrain

The system developed by the Urban Transport Development Corporation (UTDC) as an intermediate capacity transit system (ICTS) is now called advanced light rail transit ALRT. It was built as a DPM in Detroit, where it is called Detroit central automated transit system (CATS) and in Vancouver where it is called Skytrain.

The main technical characteristics are the following ones:

(a) a flat asynchronous linear induction motor associated with wheel suspension;

(b) new types of bogies with mechanically steerable axles, with small nonindependant wheels (diameter 0.46 m). The advantages of such a bogie are the low level of the floor and the relatively smaller height of the vehicle, but improvements have still to be made to reduce noise and wear;

(c) command and control ensured by the German device SELTRAC utilizing microprocessors working with a redundancy and a comparison of results: two microprocessors on board and three microprocessors on the ground;

The Skytrain of Vancouver could be considered as the first AGT system with the wheel/rail technology (see Fig. 6).

7.2 Linear Motor Car

A prototype of a linear motor car or LM1 has been constructed in Japan, with the main purpose of reducing the section of the tunnels. The suspension consists of steel-wheeled bogies.

7.3 Magnetic Levitation Systems

Two systems have been studied for suburban applications, with cruise speeds of between 150 km h^{-1} and 200 km h^{-1}:

(a) HSST, designed in Japan with a flat linear motor, and a prototype of which is foreseen in Las Vegas; and

(b) STARLIM, studied in the context of a French–German cooperation, with an improved U-shaped linear motor.

A small low-speed system called Maglev is operating with a flat linear motor as the Birmingham International Airport transit link (see *Magnetic Suspension Railroad Systems*).

8. Electrical Active Guideway Systems

8.1 Flat Linear Motors

The Wedway system is in operation at Disneyworld in Orlando and at the Houston airport. This system supported by small wheels (diameter 0.25 m) with rubber tires is characterized by a low speed (20 km h^{-1}) but good reliability.

UTDC has recently constructed a prototype of a so-called low-capacity transit system (LCTS).

8.2 Long-Stator Linear Motors (Continuous Winding on the Track)

The German M-Bahn system utilizes magnetic levitation with permanent magnets which also serve as a reaction device for the linear propulsion, but numerous small wheels are still necessary for the following functions (see Figs. 7, 8):

(a) to support a small part of the vehicle weight and to regulate the air gap value of the magnets by means of a mechanical device;

(b) for lateral guidance; and

(c) for switching.

The guideway and its three-phase winding is subdivided into sections being fed up by a pulse-controlled inverter located on the ground.

The operating control system is of a hierarchical structure: organization (with central traffic control center), operation and process.

8.3 U-Shaped Linear Motors

Two systems equipped with the French U-shaped linear motor have been studied: the Telebus, the prototype of which has been functioning since 1976 on the St Etienne exhibition ground, and the Transville, the development of which began in 1989.

Transville utilizes an original low-pressure air cushion, with several circular air bags under each vehicle. The guidance of the U-shaped reaction element of the motor is made by vertical and horizontal small wheels circulating in a caisson at the lower part of the concrete guideway. There is a split in the upper part of the caisson for the anchor of the motor reaction element of the vehicle.

9. Mechanical Active Guideway

A few AGT systems with cable-hauled vehicles have been developed, essentially in France and the USA.

9.1 Shuttle Systems

A shuttle system is the simplest configuration. Two stations are served by one or two vehicles. In this last case the crossing takes place at the middle of the line. The guideway can be constructed with a single lane, except for the crossing area, in order to minimize the infrastructure costs. Cable propulsion is a good solution for shuttle systems when the speed does not exceed 40 km h^{-1}.

Figure 6
ARLT Vehicle

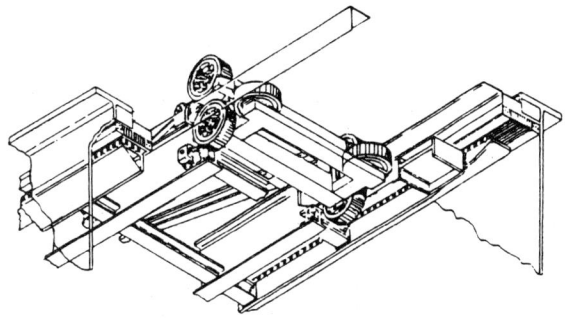

Figure 7
M-Bahn guide rollers on a levitation frame

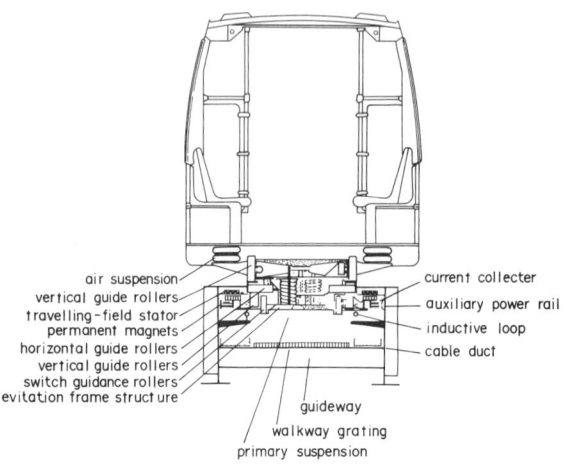

Figure 8
Cross section of M-Bahn vehicle and guideway

Figure 10
SK system at Villepinte

The Metro-Shuttle 6000 with rubber-tired vehicles is operating in Las Vegas with supported cabins and in Memphis with suspended cabins.

The OTIS-Shuttle is characterized by the suspension by means of air cushion, with circular air bags under the vehicles. Low-pressure air is supplied to the system by standard commercial centrifugal blowers. The thin film of air developed between the rubber air pad and the guideway running surface provides for a smooth riding, non-traction-dependent vertical support mechanism. Three systems are operating around the world, and a fourth is in construction in Japan.

9.2 Systems Serving more than Two Stations

The OTIS-Shuttle can also serve more than two stations, but in this case with only one vehicle and a longer waiting time (there is only one cable loop).

The Poma 2000 (see Fig. 9) is characterized by the existence of several cable loops; at each station the vehicle can go automatically from one loop to the following one by means of a special grip. The middle-sized vehicles are mounted on pneumatic tires. The first application in Laon (30 000 inhabitants) is characterized by:

(a) a relatively short line (1.5 km) but with many curves and an important slope (average slope 7%, maximum slope 13%); and

(b) the steel guideway, partly at grade, partly aerial.

The equipment managing the automatic pilot function is divided into two categories:

(a) the automatisms performed by a distributed data processing system that ensure the non-safety-related commands and centralized management of the alarms sent to the control stations; and

(b) the safety-related data processing system which controls the automatisms by triggering emergency braking or by inhibiting departure in the event that anomalies are detected. This is a SACEM-type system (i.e., it is based on the coded single-processor principle developed by the RATP (Paris Transport Authority)).

9.3 Semicontinuous Systems

These systems can also serve more than two stations. The vehicles do not come completely to rest in the stations but the passengers enter and exit while the vehicle continues at a very low speed past the platform. This favors a reduction in the service interval, and allows the choice of smaller vehicles, usually called cabins.

The Soule-Kermadec (SK) has been the most successful one with an application on short lines in Paris-Nord/Villepinte (see Figs. 10, 11), in Vancouver and in Yokohama. There are now other projects in Paris. If we except the two American GRT (Airtrans

Figure 9
Poma 2000 vehicle

Figure 11
SK vehicle in station

and Morgantown), the SK is the only example of AGT system consisting of cabins circulating with a short interval, not exceeding 20 s. Its main technical characteristics are the following:

(a) acceleration and deceleration of the cabins by means of belts supporting two wheels braked with a force proportional to the weight of the cabins;

(b) a mechanical device to weight the cabin;

(c) an original grip with progressive clutch, to ensure the cabin remains fixed on the cable without shaking;

(d) the complete passivity of the vehicles without any energy on board, the doors also being actuated by mechanical devices located on the ground—this favors the safety on board;

Figure 12
C-Bahn prototype

(e) the collision avoidance ensured both by the cable and a block system; and

(f) original safety devices in station, for example at the end of the embarkation platform.

9.4 Ropeways

In the context of this article, cable-hauled systems where vehicles are supported on wheels have been described, but it could be possible (in some cases) to utilize, for urban applications, systems more directly derived from ropeways, in which cables also ensure the suspension. There is one example in New Orleans with the Mississippi aerial river transit system derived from a Pomagalski single-cable ropeway.

10. Other Systems

10.1 Classical Steel Wheel Suspension Systems

The Docklands system in London is fully automated but an agent remains in the vehicles, and drives in the case of a breakdown. A possible evolution is the fully automatization of classical metros, without any agent on board. There is already the example of the D-line of the Lyon metro. The command and control system called Metro à Grand Gabarit de l'Agglomération Lyonnaise (MAGGALY) is based on moving variable blocks, and the code single-processor principle (SACEM type): the safety processing is carried out by a single processor (or group of processors), the data redundancy can itself detect the failures and errors generated by interference and disturbances.

10.2 Monorail Systems

In addition to the fully automated monorails already mentioned for American and Japanese systems, we can note the development of the German H-Bahn with vehicles suspended on bogies circulating in a box-beam. This system has found only one application on a short line at the Dortmund University but it is interesting to note that numerous other projects studied with suspended vehicles since 1970 have been completely abandoned.

10.3 Other Technologies

An example of an alternative technology is the Aeromovel with pneumatic active guideway, studied in Brazil and built in Djarkarta (Indonesia). The vehicles are propelled by low air pressure circulating in a caisson at the lower part of the concrete guideway.

11. PRT Systems

Although research carried out on this subject did not lead to commercial applications, it could give indications for the development of less ambitious AGT systems. Typically, PRT systems studied during the 1970s were:

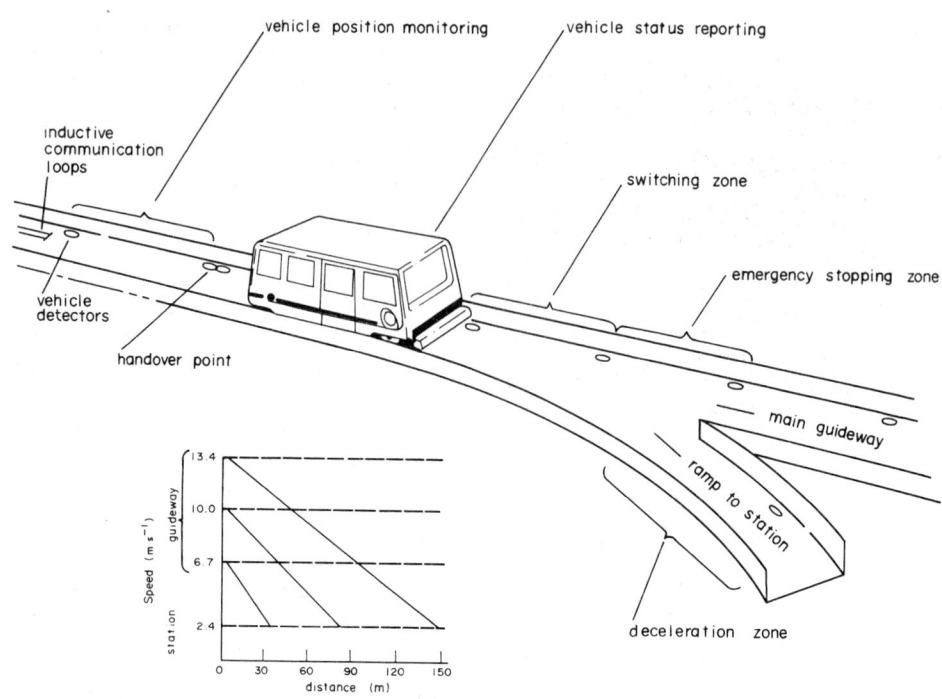

Figure 13
Guideway control and monitoring for the Morgantown people mover

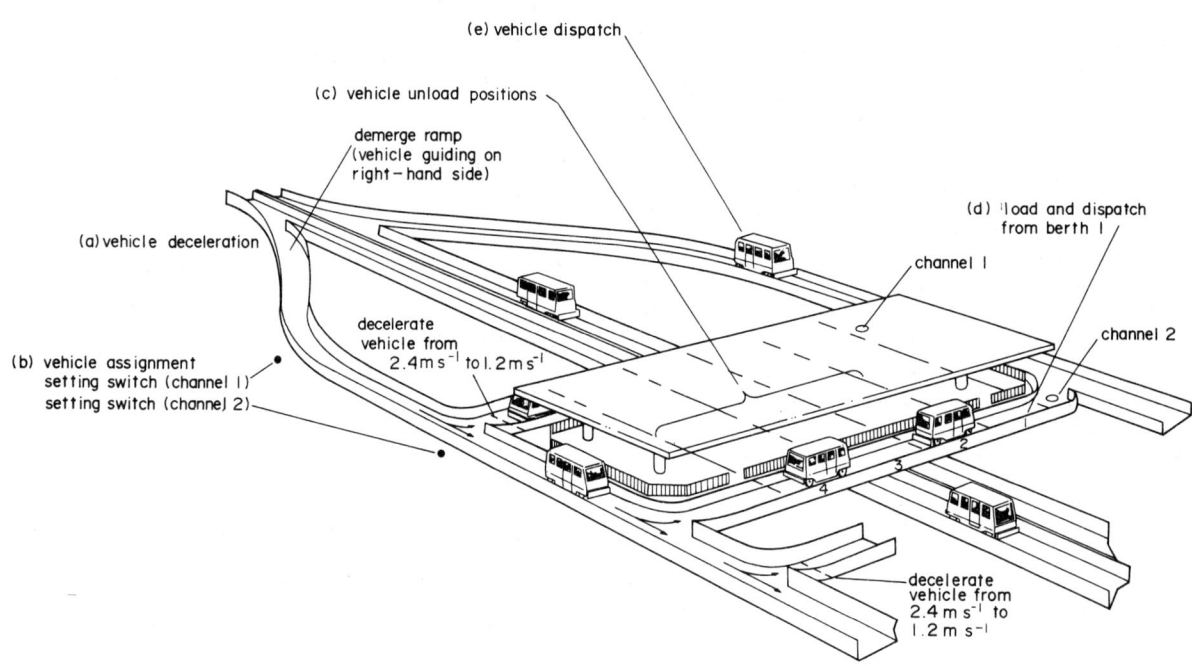

Figure 14
In-station vehicle management for the Morgantown people mover

Figure 15
Morgantown people mover vehicle

(a) CVS (computer controlled vehicle) in Japan,
(b) TTI-OTIS in the USA,
(c) Cabinentaxi in Germany (as one version of the C-Bahn family studied with suspended vehicles as well as with supported vehicles, as shown in Fig. 12), and
(d) the first version of Aramis in France.

Aramis has evolved to a non-PRT version, with larger cabins (ten places), and an application was foreseen on the Petite Ceinture in Paris, but this project was cancelled, essentially for cost reasons. Among the technical characteristics, the following can be retained:

(a) the immaterial coupling into sets of vehicles with a variable composition and "rendezvous," the distance of 0.3 m is measured by means of ultrasonic detectors;
(b) the command and control system of the SACEM type, with variable moving blocks; and
(c) the reluctance effect motors associated with GTO converters.

Figures 13 and 14 show some characteristics of the Morgantown people mover.

The Morgantown people mover is sometimes considered as a PRT because of the exisxtence of off-lines stations, although it is in fact a GRT (see Fig. 15 for the vehicle).

In Fig. 13, the time between vehicle positions is monitored; if out of tolerance the software will detect this, report it and effect appropriate controlling action. Vehicle status reporting covers the vehicle identity, current location, destination and switch state, speed and performance level, door status, braking condition and anomaly status. In the switching zone, the vehicle is commanded to switch from the main guideway if the destination is this particular station. In the emergency stopping zone, the vehicle is commanded to stop if the switching is not confirmed. In the deceleration zone, the vehicle is commanded to profile from guideway speed to station entry speed. At the handover point, the vehicle control and monitoring is transferred from one station to the next.

The in-station vehicle management proceeds in the following sequence (see Fig. 14):

(a) the vehicle is decelerated from guideway speed to 2.4 m s^{-1};
(b) it is assigned a channel for unloading by a setting switch (e.g., switch first left if assigned to channel 1 and second left if assigned to channel 2);
(c) it stops at the forwardmost available berth and is unloaded;
(d) it is loaded and dispatched; and
(e) it accelerates to main guideway speed along the station exit ramp.

12. Future Developments

Automated guideway transit systems using new automation technologies can provide good solutions to the problems of transportation within cities. In recent years, several systems have been put into operation in Europe, the USA and Japan. More extensive usage of these systems will depend not only on the technology involved, which is continually improving, but also on political choices involving the quality of urban life.

PRT systems which represent the highest level of automated guideway transit are currently judged too complex and expensive to be implemented in cities.

See also: High-Speed Railroad Networks in Europe; High-Speed Railroad Networks in Japan; Underground Railroad Modelling and Control: Discrete-Event Approach

Bibliography

Anagnostopoulos G 1981 *Interim Assessment of VAL AGT System.*, TSC, Cambridge, MA; Institut de Réchèrche des Transports, Arcueil, France
Automated People Movers II 1989 Proc. 2nd Int. Conf. Miami, FL
David Y, Heddebaut M, Stuparu A, Soulas C 1989 Compte-rendu du congrès ASCE "Automated People Movers." INRETS CRESTA/LTN, Arcueil, France
Gerland H, Soulas C 1981 *Europe, Japan: Development of AGT Systems*. TEC, Paris
International Transit Compendium 1983 Automated Guideway Transit, Vol. 4N1.ND Lea, Washington, DC
Ludwig N, Gerland H, Soulas C 1981 *USA, Canada: Urban Automatic Guided Systems*. TEC, Paris
Soulas C 1988 Transports guidés au sol—Solutions nouvelles—Application au transport de voyageurs. *Tech. Ingénieur* **A996**, 1–22
Soulas C 1989 Transports guidés urbains et innovation technologique: La recherche de solutions nouvelles. *Transp. Urbains* **66**, 5–16

C. Soulas
[INRETS, Arcueil, France]

Automatic Landing Systems

The first automatic landing of a civil aircraft with passengers took place in London on 10 June 1965 with a Trident operated by the British airline BEA; the era of automatic landing had begun. A few years later on 9 January 1969, during a scheduled flight between Lyon and Paris Orly, the French domestic carrier Air-Inter landed a Caravelle in category III visibility conditions; that is, 200 m runway visual range and a ceiling lower than 100 ft (30 m) (see Sect. 1.4). These two "firsts" should not be quoted without associating them with the names of the aircraft manufacturers who effected all-weather automatic landing systems: Hawker-Siddeley (now British Aerospace) on the British side for the Trident and Sud-Aviation (later Aérospatiale) on the French side for the Caravelle.

Since then, the large American constructors Boeing, Douglas and Lockheed have equipped their aircraft, and several hundreds of thousands of automatic landings have been performed, 1–2% of which have been in category III conditions, without any significant incident or accident being reported. Although this result is still modest from a statistical point of view, it is a positive indication that the high level of safety sought for can be achieved by the systems involved in automatic landing.

It must be remembered that regularity has always been one of the most important objectives pursued by air transport. Automatic landing has appeared as the surest way of achieving it, even in conditions of very bad visibility. This is, therefore, the primary motivation, but it will be seen later that there are other reasons for developing this system.

1. Regulatory Aspect

There are many regulations—depending on the country defining the requirements—to be met before operational approval or certification for automatic landing is granted, in particular in conditions of bad visibility: in the USA, AC20-57A–AC120-28B; in France, CTC25-3; in the UK, BCAR papers 575–742; and in Europe, JAR.

Particular mention should be made of the UK regulations, which have allowed the prolific concepts of average risk and particular risk to be introduced, and thus allow the problem of the safety of this system to be dealt with from a probabilistic point of view.

1.1 Average Risk

Accident statistics for the major commercial carriers show about three accidents per million flight hours, with one of these accidents occurring at landing. From this statistical observation, which only concerns manual landings, an average risk of 10^{-6} per landing can be deduced and the objective for automatic landing is to be ten times better than this, giving an average risk objective of 10^{-7} per landing. It is a risk on an average, which means that it is acceptable to take a greater risk from time to time, provided that the product of the value of this risk and the frequency of occurrence of the circumstances that led to its being taken complies with the objective of 10^{-7}. Mathematically, this is written as

$$\sum_1^n f_i p_i \leqslant 10^{-7}$$

where n is the number of landings, p_i the value of the risk for the ith landing and f_i the frequency of repetition of the circumstances encountered during the ith landing.

1.2 Particular Risk

This is where the concept of the maximum risk allowed on a given day intervenes. For example, why take a 10^{-5} risk when a safer action exists, which would be not to land but to go round for a further circuit for which the risk is known not to exceed 3×10^{-6}? This value, therefore, constitutes the upper limit of the risk that must not be exceeded on a given day during an automatic landing.

1.3 Design Heights

In addition to the risk objectives defined in Sects. 1.1 and 1.2, regulations have defined design heights for bad visibility landings.

Decision height (DH) is the height at which a missed approach procedure must be initiated unless the pilot has acquired adequate visual references to enable a safe landing to be guaranteed. The final safety precaution during a bad visibility landing can be seen here because, in the end, the pilot must be able to see in order to land. It should be noted that this precaution becomes less important and even disappears as the DH is decreased towards zero.

Alert height is the height after which it is considered that if a single failure occurs, it is safer to continue the landing than to go round for an additional circuit. This height explicitly introduces the concept of redundancy (see Sect. 2.2).

1.4 All-Weather Landing Categories

Landing conditions are categorized according to the runway visual range (RVR) and the ceiling that generally corresponds to it (95% of the cases) and which determines the DH:

Category I	RVR $\geqslant$ 800 m
	DH $\geqslant$ 200 ft
Category II	800 m > RVR $\geqslant$ 400 m
	200 ft > DH $\geqslant$ 100 ft
Category III A	400 m > RVR $\geqslant$ 200 m
	100 ft > DH $\geqslant$ 0 ft

B   200 m > RVR ⩾ 50 m
    100 ft > DH ⩾ 0 ft
C   RVR = 0 m
    DH = 0 ft

As far as category III is concerned, the classification described above corresponds to the International Civil Aeronautical Organisation (ICAO) regulations. More recently, the European regulations, JAR, have preferred to define category III as follows:

Category III divided into
  100 ft > DH ⩾ 50 ft and 50 ft > DH > 0 ft
Category III with DH = 0

2. Concepts

The concepts governing the design of an automatic landing system largely result from the confrontation of the undesirable events likely to occur with the previously defined safety objectives. This is not the only aspect: the operational aspect, in particular the role of the crew, intervenes just as strongly in the design concepts, but this subject will not be dealt with here.

In very bad visibility conditions, less than 200 m, the undesirable events can be classified into two categories:

(a) operation without a failure: the system operates correctly but its performance is downgraded; and

(b) operation with a failure: here a distinction can be made between loss of the system and undetected failure.

It should be noted that these are conventional types of events, which are considered when dealing with the safety of systems in general. For example, Fig. 1 shows a possible distribution of the 10^{-7} average risk among each of these events.

2.1 Performance

Figure 2 illustrates the different phases of an automatic landing. The aircraft is guided in elevation and azimuth by the glide and localizer radio beams with a flare maneuver which is initiated by the radio altimeter at 50 ft (15 m); the aircraft is then slaved to follow a vertical velocity program with the radio altimeter on the longitudinal axis.

As far as safety is concerned, the important parameters are touchdown point longitudinal distance, lateral position, vertical speed at touchdown, landing gear transverse speed component, and pitch and roll attitudes. Each of these parameters is controlled by the autopilot via servo loops, and the performance expressed as deviations from a target value will depend on the quality of the loop, on one hand, and on the level of disturbance, in particular the disturbance due to atmospheric turbulence, on the other. Taking as an example the parameter "lateral position of the aircraft at touchdown," which shows a distribution centered on the runway center line (see Fig. 3), the problem is to reduce scatter and, thus, standard deviation (SD) so that the sum of the crosshatched areas is less than the risk allocated to this parameter (2.5×10^{-8}).

This is a classical servosystem problem involving, in particular, the compromise that has to be found between accuracy and stability. It should be noted that increasing the performance of the aircraft tracking on the localizer is not the only way of reducing the risk; the level of acceptable disturbance can also be decreased which, in practice, leads to prescribing wind limits that must not be exceeded.

2.2 Redundancy

Redundancy is the technique used to reduce the probability of the event "loss of the system." However, as the resulting risk depends on altitude and visibility conditions, the concept of redundancy is required only for systems claiming very low values of DH.

Thus, the following conditions apply.

(a) Down to DH ⩾ 50 ft (15 m), a single autoland channel is sufficient. It is even thought that for commuter-type light aircraft, even automatic landing is not necessary and that automatic guidance is sufficient down to 50 ft, with the pilot taking over manually at this height.

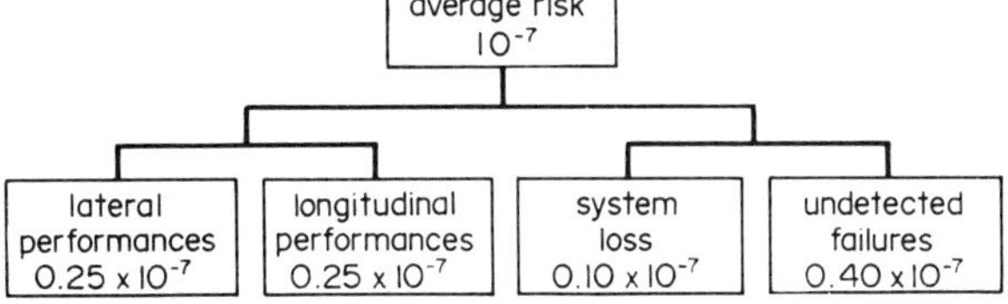

	average risk 10^{-7}		
lateral performances 0.25×10^{-7}	longitudinal performances 0.25×10^{-7}	system loss 0.10×10^{-7}	undetected failures 0.40×10^{-7}

Figure 1
Possible distribution of the 10^{-7} average risk for automatic landing in very bad visibility conditions among the undesirable events that may occur

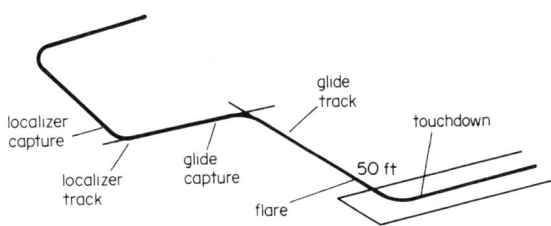

Figure 2
Phases of an automatic landing

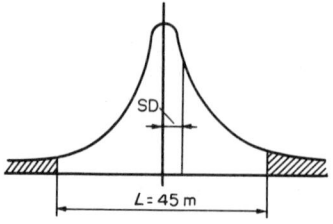

Figure 3
Distribution of the parameter "lateral position of the aircraft at touchdown"

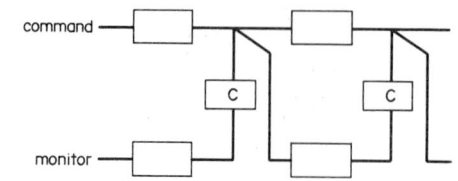

Figure 4
In-line monitoring (C is comparator)

(b) For DH < 50 ft, it becomes necessary to have a redundancy that enables the landing to be continued after failure of the first channel. This redundancy can be simple considering the risk time, which is about 30 s, and the mean time between failures (MTBF), about 250 h, of an automatic landing channel. Furthermore, this redundancy is not necessarily another autopilot—it could be a head-up display (HUD) type device—though there is a definite trend towards full automation for very low DHs.

2.3 Monitoring

The other event to be taken into consideration from the safety point of view is undetected failure of the autopilot channel in control. Different types of monitoring are possible.

In-line monitoring consists of monitoring the channel in control by a servomodel channel that has a monitoring function only and comparing them with one another (see Fig. 4). They are compared at several points in order to avoid accumulating the tolerances of each of the channels. This type of monitoring is used on aircraft such as the VC10 and Concorde.

Vote monitoring involves selecting, and thus voting for, the command channel from several channels. This can be mechanical, by summing the servomotor forces as used on the Trident and the Boeing 747, or electronic, by selecting the mean or midvalue, as on the A300, L1011 and DC10 (see Fig. 5).

The techniques mentioned previously imply channel redundancy, which requires extreme vigilance at the common points which exist at the design stage and at the production stage. External monitoring eliminates this difficulty. It consists in checking that the movements of the aircraft around the center of gravity and those of the center of gravity itself are acceptable. It is, therefore, overall monitoring, independent from the control loop as a matter of principle and very satisfactory from the safety point of view. A difficulty exists, however, in that it is sometimes hard to distinguish between a normal movement of the aircraft resulting from turbulence and an abnormal one caused by a failure. This technique was applied on the Caravelles.

3. Means of Demonstration

Before authorizing the use of an automatic landing system in commercial operation, proof of its safe operation must be provided. This demonstration is made in two stages: first by the aircraft manufacturer, with a certification of airworthiness, then by the airline, with operational certification.

3.1 Certification of Airworthiness

Certification of airworthiness is obtained on the basis of three complementary demonstrations.

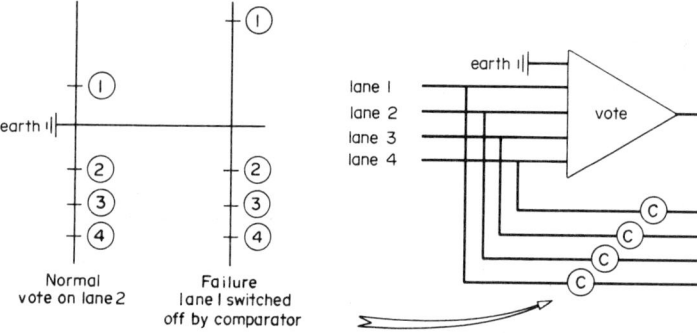

Figure 5
Electronic vote monitoring

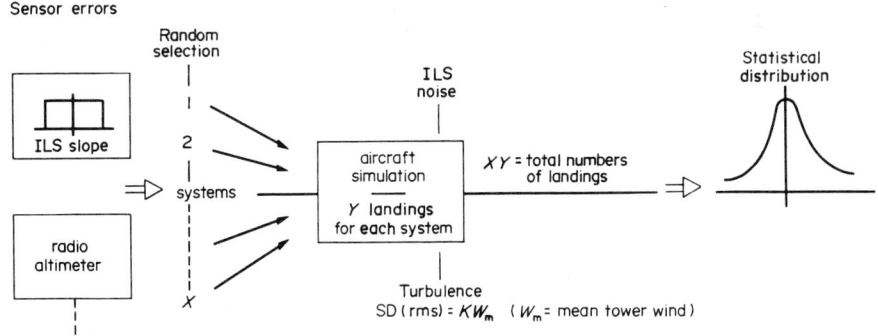

Figure 6
Monte Carlo simulation of the parameter "longitudinal scatter of the touchdown point" (ILS is instrument landing system)

(a) *Safety assessments.* These involve selecting from diagrams those failures that could have significant consequences. The most frequently used method consists in listing the elementary failures that could affect each of the components and grouping them together according to their consequences, thus forming global failures, the probability of which is then assessed. Another method consists in listing the undesirable events and searching for the elementary failures likely to produce them. This method is better known as the fault tree method.

(b) *Simulation.* Simulation of the aircraft and its systems is used for two purposes: performance and failure assessment.

For performance, tens of thousands of automatic landings can be performed in accelerated time with the computing capacities of modern computers. Provided that the effects of each of the causes of scatter are suitably proportioned, the statistical distributions on the main parameters can be obtained.

A method in general use is the Monte Carlo method. Figure 6 illustrates this process, considering the parameter "longitudinal scatter of the touchdown point." The advantage of this method is that it not only enables compliance with the average risk (10^{-9}) to be demonstrated with a certain confidence factor, but also allows the maximum value of each of the parameters known to the crew (e.g., wind speed at the control tower, glide slope, weight) to be determined. For wind speed, for example, all that has to be done is to increase the mean wind speed value to a maximum value which generates a risk of 3×10^{-6}, with the other parameters evolving according to their own probability density distribution.

For failure, the effects of the failures resulting from the safety assessments are studied. These tests imply the presence of pilots and, therefore, of a flight deck. This flight deck is mobile and is associated with a visual simulation of the outside world. It enables actual flight conditions to be reconstituted, making it possible to simulate failures and observe the subsequent behavior of the aircraft, taking the pilot's reactions into account, and therefore to pronounce judgement on the gravity and consequences of each failure. The criticality of the consequences is then judged in terms of the probabilities of occurrence of each failure according to Fig. 7.

In the field of automatic landing, as in many other aeronautical areas, accelerated and real-time simulation proves to be an extremely powerful tool which not only allows an almost exhaustive investigation into problems but also enables the number of full-scale demonstrations that will eventually have to be made on the actual aircraft to be considerably reduced. To achieve this, a high level of quality must be reached in performing the simulation; in fact, it is this level of quality that sets the limits of the method.

(c) *Flight tests.* These tests cover performance and failure.

For performance, in average conditions, about a hundred test landings are made, during which the weight, center of gravity and wind speed ranges are fully swept and several landing runways used. The essential purpose of this is to check that there is no large disagreement between simulation and flight as far as the mean values and SD values of the main parameters involved in touchdown are concerned.

In limit conditions, about fifty test landings are performed to assess the behaviour of the system qualitatively.

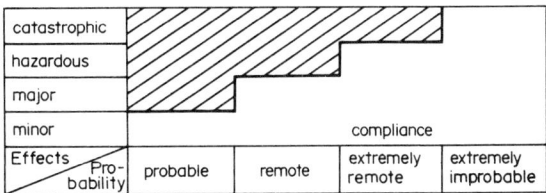

Figure 7
Probabilities of occurrence of failures affecting automatic landing

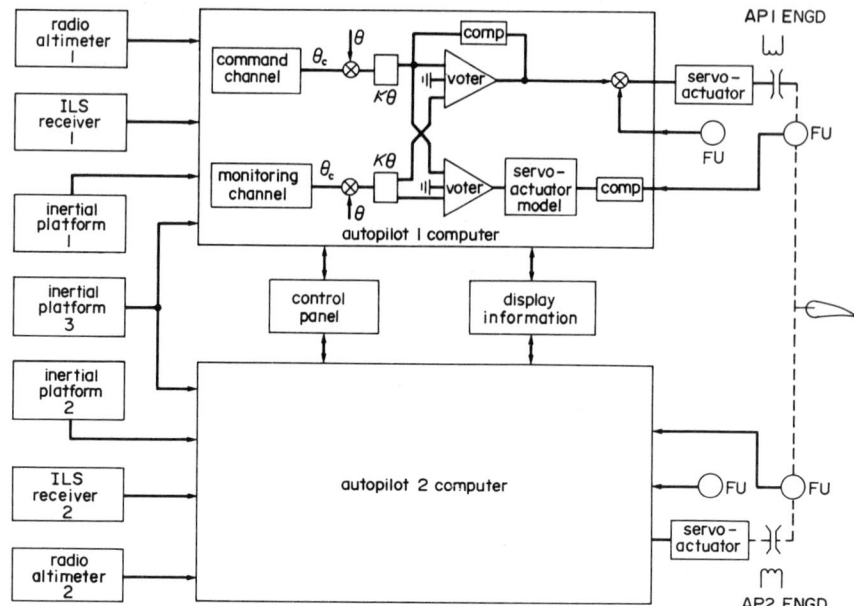

Figure 8
An automatic landing system (comp, comparator; θ, pitch attitude; θ_c, pitch attitude command; $K\theta$, pitch attitude gain; FU, follow up; AP1, first autopilot; AP2, second autopilot; ENGD, engaged)

About thirty go-arounds have to be added to these landing tests to demonstrate the capability of the aircraft to take off again whatever the altitude, even in the event of an engine failure.

For failure, the tests again essentially involve the most unfavorable cases resulting from the simulator testing. They are also used to check that the devices indicating normal and abnormal operation are well suited to the special conditions of blind approach.

3.2 Operational Certification

Certification of airworthiness is acquired by the aircraft manufacturer. It is part of so-called type certification which is generally sufficient to enable the operators—the airlines—to use the aircraft with the required level of safety. As far as all-weather landing is concerned, the certification authorities impose an additional condition in requiring the operator to demonstrate the correct operation of the system in airline service so as to grant final authorization for operation with bad visibility. This additional certification comprises

(a) a crew training program completed by a certain number of approaches in really bad visibility conditions with an instructor on board;

(b) an equipment maintenance program with which the airlines must comply; and

(c) at least a hundred automatic landings in clear weather to check that the system performance is still consistent with the performance achieved during airworthiness certification.

It is only when all these conditions have been met that operators are authorized to reduce their worst-weather criteria (meteorological minima) for landing. They must then carry out this reduction progressively and selectively according to the qualifications of specific crews.

4. Example of a System

Figure 8 shows an example of an automatic landing system. The system comprises the following.

(a) *Sensors*. These are instrument landing system (ILS) receivers that transmit information on aircraft deviations from the ILS radio beam, radio altimeters, which allow the system gain to be changed and are particularly used for the guidance law for the flare maneuver, and inertia systems, which provide attitude and acceleration information.

(b) *Autopilots*. Each of the two autopilots is self-monitored by duplication of the channels and three-party voting upstream of the power loop. In the normal configuration, autopilot 1 is in operation and autopilot 2 is in synchronized standby condition. In the event of the failure of autopilot 1, which is self-detected by its internal monitoring, autopilot 2 automatically takes over, thus making the system fail operational.

(c) *Control and display devices*. These inform the crew that the landing is being performed normally, but also provide warning information in the event of faulty operation.

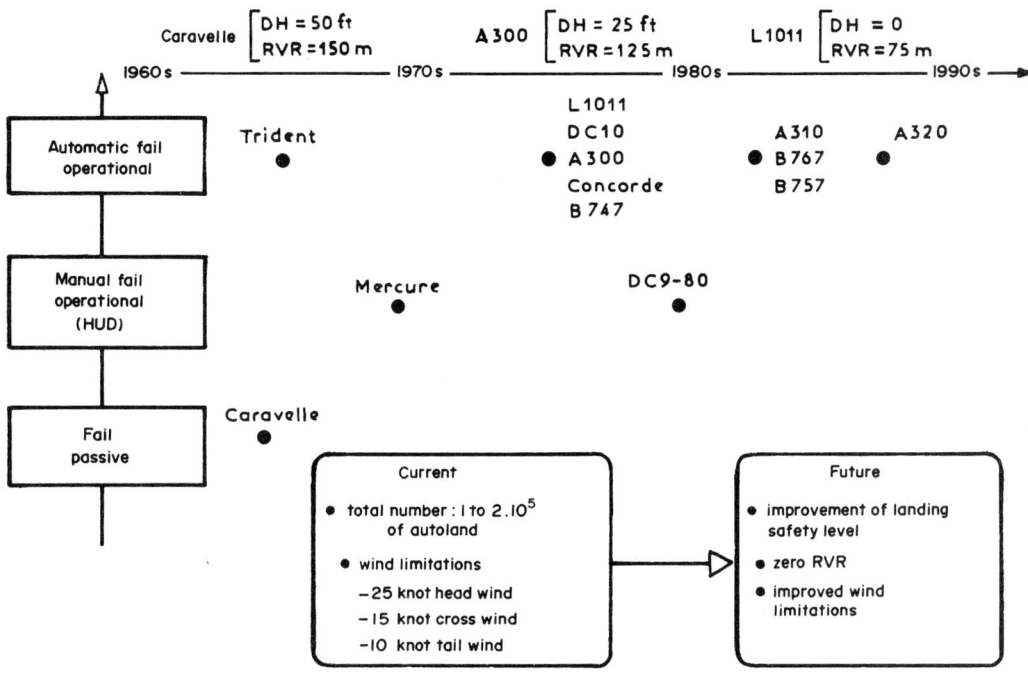

Figure 9
Meteorological minima for automatic landing system designs used in various aircraft

5. Future Developments

Since 1969, the first year in which a civil aircraft with passengers landed automatically in really bad visibility conditions, the use of automatic landing has developed considerably, with the objective of reducing meteorological minima in order to increase flight regularity. Figure 9 shows the meteorological minima that aircraft built since then have reached according to the type of system used. There is a constant trend towards lower minima.

It is important to indicate another use of automatic landing. Most systems are limited to values of mean wind speed announced by the control tower, which do not allow these systems to be used in very rough weather; thus, the pilots have to manage by themselves. An interesting course would be to promote automatic landing in all-weather conditions, including very high wind. This course has been opened up by Lockheed on the L1011, which is equipped with a system called "direct lift control" that acts directly on the lifting surfaces and enables the aircraft to contend with turbulence and wind-shear far more efficiently.

See also: Air Traffic Control Near Airports; Airborne Collision Avoidance; Visual and Instrument Flying Rules

Y. Negre
[Aérospatiale, Toulouse, France]

Automatic Train Control: Protection Principles

Guided transportation systems—from high-speed trains through conventional metros to downtown people movers (DPMs)—require safety and monitoring procedures to prevent accidents and to ensure a reliable service.

In 1989, more than 80 metros were in operation worldwide, all of which used automatic systems of varying sophistication, from basic protection up to comprehensive automation. All of these transport systems have common aims and the basic principles used are similar, even if the means of achieving them varies.

By convention, the complete automatic train control (ATC) system is considered as comprising several functional subsystems.

(a) *Automatic train protection (ATP)*. This is aimed at avoiding collisions and derailments (particularly those whose origin is not mechanical); it is, after the driver, the basis of train movement safety.

(b) *Automatic train operation (ATO)*. This is aimed at assisting, or even taking the place of, the driver and generally improves safety.

(c) *Automatic line supervision (ALS)*. This is aimed at managing the train movements on the line, but also participates in safety.

Automatic Train Control: Protection Principles

1. Basic Principles

Train protection is aimed at avoiding collisions and derailments. This takes on particular importance for DPMs, metros or rapid transit systems since the headway between vehicles (trains) is reduced.

The collision aspect mainly covers collisions with other trains. (Protection against collisions with unforeseen obstacles normally takes the form of protection for the line itself; for example, sensitive barriers, which transmit an automatic stop signal to the train if an abnormal situation is detected, or a contact detection bar in front of the train.) To protect trains from colliding with each other requires that their positions must always be known. This is generally achieved by means of track circuits. A track circuit consists of a portion of the track that is insulated at both ends (mechanically or electrically), into which an alternating or pulsing current is fed; the signal indicating that the circuit is engaged is transmitted when a train axle shunts this circuit. Another method of detection is based on input–output of the train or its axles viewed through the occultation of a beam or the action of a pedal.

The most frequently used method of protecting successive trains on the same track is the block procedure: the track is divided into sections, known as blocks, each with one or several track circuits whose minimum length is the stopping distance required by the train at its maximum allowed speed. A train cannot enter a block if the block is already engaged by another train.

During manual driving, permission to enter a block is given by a signal at the entry to the block. This then forms a spacing signalling system, whose layout depends on the rolling stock performance and the track profile. Again, varying levels of sophistication for the processing logics can be used, to suit the level of performance required.

The minimum possible headway between trains depends on the rapidity with which each block is "opened" after the passage of a train; that is, how closely the blocks follow it. At the limit, the distance between trains may be only slightly more than the stopping distance (continuous moving block). In general, the safety distance is calculated assuming that the leading train can stop instantaneously; this assumption is not the same as that used for road traffic.

Certain lines include branches; in addition, secondary tracks are needed for the storage and maintenance of trains. The junction point must, therefore, also be protected and this is achieved using a route signalling system. This systems monitors the position of the switch points and checks the compatibility between the routes that have been selected (interlocking logic). On all metros, these signalling systems use either electromechanical or electronic circuits (see *Railroad Electronic Signalling*).

Finally, the dynamic characteristics of the rolling stock and the track geometry (e.g., curves, cant) mean that certain speed limits must be respected; this is the function of the speed signalling system (speed indicators or signal settings).

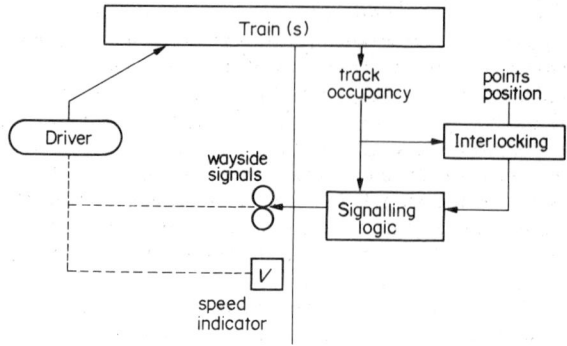

Figure 1
Basic train protection system

Figure 1 shows this basic protection system. All the data from the signal systems are displayed either

(a) along the track (generally by signals with at least two colors representing "free track" and "stop"); or

(b) on board the train (cab signal), the information being transmitted to the train from the track, the transmitter being either the track itself (the rails) or a cable laid alongside it.

The cab instruments can give similar information to the wayside signals or can be more sophisticated, indicating the target speed and the target distance (the distance at which the target speed must be respected). The target speed is zero when stopping at a signal is required. These more sophisticated systems are generally associated with automatic protection or automatic driving systems (see *Railroad Systems: Train Driving Control*; *Underground Railroad: Organization of Operations*).

2. First Level of Automated Control

A number of measures are normally taken to ensure safety in the event of a driver fault or error.

2.1 Driver's Vigilance

The driver's vigilance is constantly checked by making it necessary to regularly press (every few seconds to a minute, depending on conditions) a vigilance monitor or deadman's handle (e.g. the VACMA, an automatic watch with action check). If the driver does not fulfill this action, the train automatically stops. This is a simple, on-board system but, when a track-train transmission link is used, a check is also made to ensure that the driver has correctly acted on the caution signals transmitted to the train.

2.2 Automatic Train Stop

The train is automatically stopped if it runs through a stop signal. Former versions of this system were mechanical (trip stop): a pad on the ground was lifted if the signal was set to stop. As the train ran over this pad, it triggered an emergency braking system. More modern versions use an electromagnetic link that repeats the status of the signal that has been run through in the train. The signal transmitted can either be used by the on-board system to stop the train as it runs through the signal or it can be used for a local speed check. In the second case, a speed measurement time base is generated (e.g. by pedals) and a "go" signal is transmitted to the downstream track coil only if the speed is below the acceptable limit.

3. Continuous Overspeed Protection

Continuous overspeed protection control is used worldwide. It is normally considered as a safety feature although, when used in conjunction with manual driving, it is really only an aid to the driver, who retains control. Depending on the system employed, this feature uses different principles which allow different levels of performance (headway between trains, speeds) to be achieved; Fig. 2 shows the general system.

A visual or audible signal to the driver can be provided to indicate that the real speed is getting close to the limiting speed or that the speed limit has just been reduced. In this second case, an additional aid can be provided in the form of a target speed display.

3.1 Constant Speed Sections

This is the most common system. Speed limit signals are transmitted to the train for each section of track. In

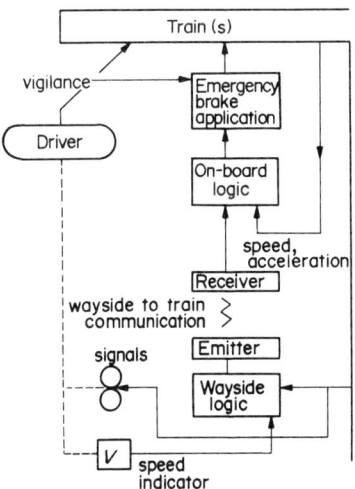

Figure 2
Continuous overspeed protection system

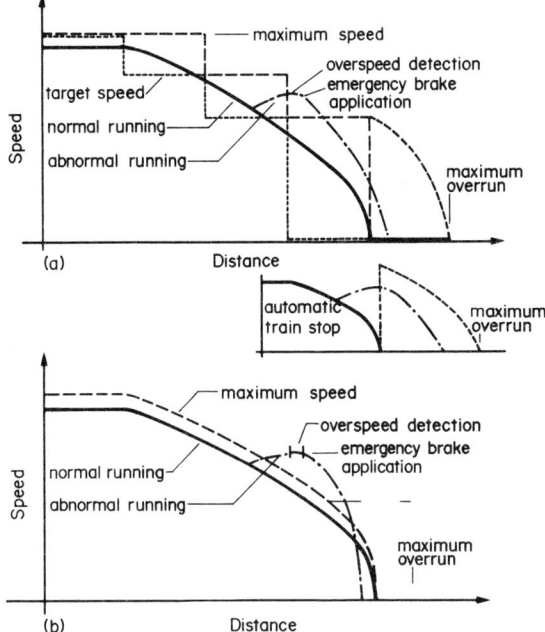

Figure 3
Automatic train protection: (a) discrete and (b) continuous overspeed protection

the train, the signal received is compared with the real speed measured from the rotation of the wheels (though care must be taken to eliminate errors due to slip and adjustments are provided to compensate for wheel wear). If the speed is too high, the emergency braking system is triggered and, generally, held until the train stops.

3.2 Constant Speed Sections and Deceleration

During braking phases, protection by constant speed sections is relatively unsatisfactory, since it means that for safety a section must be kept free upstream of the previous train (overlap). A deceleration monitoring system is therefore added.

3.3 Full Continuous Overspeed Protection

To achieve optimum performance, the normal braking curve must be more accurately "enveloped." This is carried out by constantly transmitting continually varying data determined from the distance covered.

A second solution consists of an on-board computer which calculates the overspeed envelope from data transmitted from the trackside. Continuous transmission is not necessary, but local transmitters can be mounted at given points alongside the line.

Figure 3 represents constant speed sections and continuous overspeed protection with, in each case, an illustration of abnormal operation followed by a trip of the system. An automatic emergency stop due to running through a stop signal is also shown. It can be seen that the higher the number of speed sections (i.e., the finer the cutting out of track circuits or information transmission sections), the closer the upstream train can be run to the leading train; that is, the service may be more frequent.

4. Example

The rapid transit lines (RER) of Paris are equipped with a classical block system and wayside signals with ATP of the type described in Sect. 2, allowing 150 s headways between trains 220 m long, with dwell times up to 50 s due to peak exchanges in the stations of inner Paris.

Traffic demand increased significantly during the 1980s, leading to the development of a new system to improve the line capacity. SACEM (driving, operation and maintenance aiding system) provides full continuous overspeed protection (cf. Sect. 3.3). In order to obtain shorter headways (120 s), the track circuits have been divided into subcircuits at every important station entrance, giving discrete moving blocks (steps of 40 m). Finally, a cab signal has been installed on the trains and wayside signals are cancelled ahead of trains whose on board systems are active. Wayside signalling acts like a downgraded system when SACEM is not available or when trains are not equipped.

SACEM was put into service in May 1989 on RER line A, increasing the capacity of the line by 25%. It is an example of advanced technology applied to railways; the functions—even the vital ones—are implemented on microprocessors.

See also: Automatic Train Control: Safety and Reliability; Railroad Systems: Train Driving Control

Bibliography

Blakey R 1987 Computer takes over underground. *Railw. Gaz. Int.* **143**, 675–8
David Y 1987 Systèmes de commande de fonctions de sécurité par microprocesseur en service dans les transports terrestres. *Rech., Trans., Sécurité* 15, 53–6
Perrin J-P (ed.) 1989 *IFAC Proc. Int. Symp. Control, Computers, Communications in Transportation '89.* Pergamon, Oxford
Revue Générale des Chemins de Fer **6**, 1990 Special issue on SACEM
Rowbotham A J 1989 Increased safety and efficiency through new signalling developments. In: *Developing World Transport.* Grosvenor, London, pp. 191–2
Savarzeix M, Auclair J P, Lazard G, Bridou J 1985 Poste à commande informatique. *Rev. Gen. Chem. Fer.* **104**, 489–504
Stalder O 1987 Bahn 200 prompts SBB resignalling. *Railw. Gaz. Int.* **143**, 669–71

J-P. Perrin
[RATP, Vincennes, France]

Automatic Train Control: Safety and Reliability

It is the function of any company that manages a metro system to provide a continuous transport service in complete safety. This definition includes two basic concepts:

(a) Human safety is a particularly sensitive aspect, since the effects of even a minor accident in a tunnel can very quickly become serious.

(b) The availability of the transport system, in socio-economic terms, is fundamental for a large city.

However, the two concepts are not independent: the necessity of offering a continuous service means that safety must be constantly maintained.

Safety can be defined as the probability of not having an accident. This may appear obvious; nevertheless, it expresses the fact that at the overall transport system level, operational safety is dependent on a large number of factors.

Consider, as an example, the probability of a collision between two trains. This results from the probabilities of a failure in the train control function and in the train protection function (signalling system, overspeed protection), combined with the probability of two trains closing on each other under potentially dangerous conditions. If, to increase the capacity of a line, it is desired to increase the number of trains and therefore reduce the headway, the automatic train protection provided must be strengthened; it may even be necessary to replace the human driver by an automatic driving system.

It becomes obvious, therefore, that so-called train safety systems are not the only systems to contribute to overall safety. For example, the use of an automatic line supervision system to manage the headway between trains also tends to increase overall safety.

1. Safety in Relation to Availability and Reliability

Once all of the functions that contribute to safety in varying degrees have been identified, it becomes necessary to compensate for nonavailability of any one of the functions in a way that ensures that the overall safety level is maintained.

Again the two train example will be considered. If one of the train-driving or protection automatic systems becomes unavailable, it will be possible to continue operation with the remaining functions, but the train movement conditions will become limiting. For example, the headway will have to be increased, the speed limit lowered or both. In this way, a continuous service is maintained with the same overall safety level, but the quality of service offered suffers.

The availability of a given function depends on the structure of the system that fulfills it, on the reliability of the components within the system and on the ease of maintenance. One solution to ensure that a function is always available is to build active or passive redundancy into the system that provides the function. However, this is not always possible for cost reasons. Another way of achieving high availability is to build a very reliable system whose cost, although high, is frequently less than that of duplicated less-reliable systems.

To ensure an even better result, the system must be very easily maintainable: it must be possible to identify quickly the defective subassembly and change it to allow the function to be operated correctly as soon as possible. The subassembly is then repaired off-line and has no effect on availability.

Overall safety mainly depends on the set of automatic train protection systems, frequently called safety systems, since these are specifically designed and manufactured to have an extremely low probability of dangerous failure. The two types of design that can be used in these systems are

(a) a fail-safe design, where any fault whatsoever always imposes stricter limits on the system than would have existed had the fault not occurred; and

(b) a design based on probabilistic analysis of failures such that the overall probability of a failure that could reduce safety remains below a fixed value deemed to be satisfactory.

In both cases an exhaustive analysis is essential. This means that the systems must be simple (or separable into simple subassemblies) and that the number of possible faults must be small.

These two constraints are themselves conducive to high reliability. Moreover, the exhaustive analysis of possible faults which must be carried out as part of the safety studies allows optimum technical options to be chosen to improve reliability. It is, therefore, reasonable to conclude that a safety system is reliable by its very nature (which does not necessarily mean that it has high availability).

It is essential that no maintenance operation can make a safety system unsafe. It is therefore essential that maintenance is easy and that the tests after corrective work are simple and exhaustive, and are capable of being carried out by highly qualified personnel. These criteria must be allowed for right from the design stage, so that operational safety is guaranteed not only when the system is put into service, but throughout its lifetime.

In most cases, safety systems are designed to allow the maintenance activities to be divided into three levels.

(a) The first level consists in changing complete subassemblies (with no adjustments being required). This allows the defective unit to be quickly returned to service.

(b) The second level is where the defective module (e.g., PCB, relay) within the subassembly is identified and changed (if possible without requiring adjustment). The subassembly is then tested to ensure that it operates correctly and is returned to the first maintenance level.

(c) The third level consists of a centralized laboratory, fully equipped with instruments and analysis and test equipment. This laboratory is manned by highly qualified personnel who repair, adjust and check the modules before returning them to the second level.

2. Impact of Technological Developments

Up to the 1960s, the protection of trains was limited to trackside spacing and speed signals, and to route protection. The systems consisted of fail-safe electromechanical assemblies. The rapid development of electronics from 1960 onwards allowed automatic systems offering higher performance to be developed (e.g., cab signals, continuous overspeed protection, see *Automatic Train Control: Protection Principles*). Circuits that use discrete components (e.g., semiconductors, resistors, capacitors, inductors) are still designed on fail-safe principles, although this is more difficult than with conventional electromechanical technology, since all possible faults for each component must be known and the effects of each failure studied individually.

With the development of large-scale integrated (LSI) circuits and, in particular, microprocessors and microcomputers throughout the 1970s, the development of systems with even higher performance became possible. However, fail-safe principles can no longer be applied with complex components, for which it is impossible to establish an exhaustive list of all possible faults and their effects. It therefore became necessary to use a new procedure, probabilistic analysis, to guarantee the operational safety of these systems. The obvious objective was to attain a level of safety at least equal to that already existing (to maintain overall safety). However, a major difficulty arose with these new technological facilities; that is, to demonstrate the safety of the design and the integration of the software.

However, experience has shown that these technologies have considerable advantages.

(a) They are more reliable, since the mean time between failures (MTBF) for a VLSI circuit is roughly the same as for a single transistor.

(b) They are more maintainable, since self-test software can easily be associated with the operational software, or a special program system can be installed to monitor, detect and diagnose faults automatically; this same system can identify the faulty subassemblies and transmit the information direct to the repair center.

(c) They are more available, because of the two advantages in (a) and (b), but also because, during transmission and processing, the digital data are less sensitive than analog data to the many sources of interference that exist along a metro line.

However, the disadvantage of these new technologies, both for conventional electronics and even more so for microelectronics, is that their lifetime is relatively short. This is not due to any inherent weakness leading to a rapid reduction in reliability, but to the fact that the very rapid development in this field means rapid obsolescence.

Concerning these developments, far from simply replacing each other, the various technologies and principles must be considered as complementary to one another. It is, therefore, the task of specialists to determine the best combination of methods to provide the most cost-effective solution; that is, the solution that while minimizing costs guarantees the safety and continuity of service characteristics of any mass-transit system such as a metro.

See also: Automatic Train Control: Protection Principles; Maintenance and Reliability of Traffic Control Systems; Safety of Road Traffic

Bibliography

Barwell F T 1973 *Automation and Control in Transport.* Pergamon, Oxford
Ferbeck D 1981 Le système VAL appliqué au métro de Lille. *Rev. Gen. Chem. Fer* **24**, 48–52
Frank K H 1980 Automatisierung der Hamburger U. Bahn. *Nahverkehrs-praxis* **11**, 12–16
Freehafer J 1975 Design philosophies in automatic control. *Railw. Gaz. Int.* **132**, 35–9
Maxwell M W 1979 Towards a fully automated London Underground. *Railw. Gaz. Int.* **135**, 12–16
Perrin J P 1972 New possibilities offered by automatic driving for the operations of metro lines. *Proc. 5th IFAC World Cong.* Pergamon, Oxford
Perrin J P 1980 20 ans pour une mutation, l'automatisation du métro de Paris. *Nouv. Autom.* **47**, 28–30
US Congress 1976 *Automatic Train Control in Rail Rapid Transit.* US Government Printing Office, Washington, DC
Weber O 1982 Is it time to change concepts of safety installations? *Rail Int.* **22**, 41–5
Yonezawa K, Takemura S 1976 Saporo subway and its computer total system. *Jpn. Rail. Eng.* **16**, 55–126

C. Hennebert
[RATP, Paris, France]

Broadcasting Communication Systems

At the beginning of the 1970s, various radio stations in Europe started to broadcast special information for motorists. The experience gained and the information concerning the international traffic in the holiday periods started to be exchanged within the European Broadcasting Union (EBU), a professional association of the broadcasting organizations having importance at national level. At that time, the EBU created within its Radio Programme Committee a working party that existed until the end of 1988, and which established the procedures for making the announcements and, also, for exchanging the relevant information. This also required coordination with the traffic authorities, the police and the automobile clubs. For this purpose, three Eurotravel conferences were organized by the EBU—in Geneva (1980), Grado (1984) and Copenhagen (1988)—involving these partners of the broadcasters in the process of optimizing the use of the broadcasting medium for traffic information and traffic guidance.

On the technical side, the Technical Committee of the EBU studied the possibilities for simplifying the tuning mechanism of new radio receivers to such programmes giving the traffic announcements. This led, in 1974, to the recommendation of the Autofahrer Rundfunk Information (ARI) system that used a 57 kHz subcarrier in a very-high frequency/frequency modulation (vhf/FM) multiplex signal for identification and, additionally, an announcement signal and an area code. That system was implemented in the FRG, Austria, Luxembourg and later also in Switzerland with the effect that by the early 1990s the large majority of car receivers in these countries are equipped for evaluating at least the identification during search tuning.

In other European countries, this solution was not considered to be satisfactory from a programming point of view, simply because it required the broadcaster to dedicate one program service for the insertion of traffic announcements. The general practice was, on the contrary, that certain strategic information concerning long-distance travelling was inserted in one programme, while local traffic information was given in those programmes that had a certain relevance in a part of the country. Ideally, it was desirable to have an identification system for vhf/FM broadcasting that permitted simplification of tuning to all programs, always finding the best frequency and also identifying the particular program on a display. The listener still should have the choice to select a timing mode that would accept only stations giving traffic announcements. This aim resulted in the development of the radio data system (RDS). Its specification was published by the EBU in 1984 and, since 1987, the RDS system has been implemented in the large majority of European countries. Its use outside Europe has also begun. In the context of traffic communication, RDS contains the same possibilities already offered by the ARI systems in 1974. As a consequence, ARI will be phased out and RDS, as a European solution, will be used instead. Because RDS uses digital data signals that are inaudibly inserted into the FM radio programme on the same subcarrier position (i.e., 57 kHz) used by ARI, a number of new possibilities for traffic communication will be offered or still need to be developed for the RDS system. In this concept, the European Commission is involved with the DRIVE project that in the time-frame 1989–1992 will lead to the specification of a new protocol concerning the RDS-traffic message channel (RDS-TMC) (see *DRIVE*).

1. RDS-TMC

Within the RDS, a TMC appears to offer a new possibility for broadcasting information to drivers on the actual traffic situation and, at the same time, for giving advice and recommendations concerning, for example, alternative routes and departure times. So, RDS-TMC has to be seen in the context of traffic control as well as in the context of route guidance and in-vehicle information. Furthermore, it forms one part of the telecommunications network within the European system architecture.

In the RDS-TMC, messages are transmitted in digital form and must be reconstructed in the vehicle using a special RDS-TMC receiver unit that contains a microcomputer with sufficient memory for data storage and processing, a database for the standardized international message set (some 2000 messages) and the respective national location sets (some 130 000 locations per country), a speech synthesizer, and the necessary interface for the RDS radio and the display, which has, preferably, to be integrated within the car dashboard and will, perhaps, at a later stage also provide navigational aid (see Fig. 1). Because this can be done in the language of the driver's choice, the system is not language dependent and can, therefore, be introduced throughout the whole of western Europe. Messages can be presented directly to the driver or integrated into an autonomous navigation aid such as the CARIN, EVA or AUTOGUIDE systems. The possible success of the RDS-TMC will depend on the quality of the broadcasting features as well as on the accuracy and timeliness of the traffic messages: in some countries, very sophisticated traffic monitoring systems that include facilities for automatic incident detection already exist or are under construction, such as the

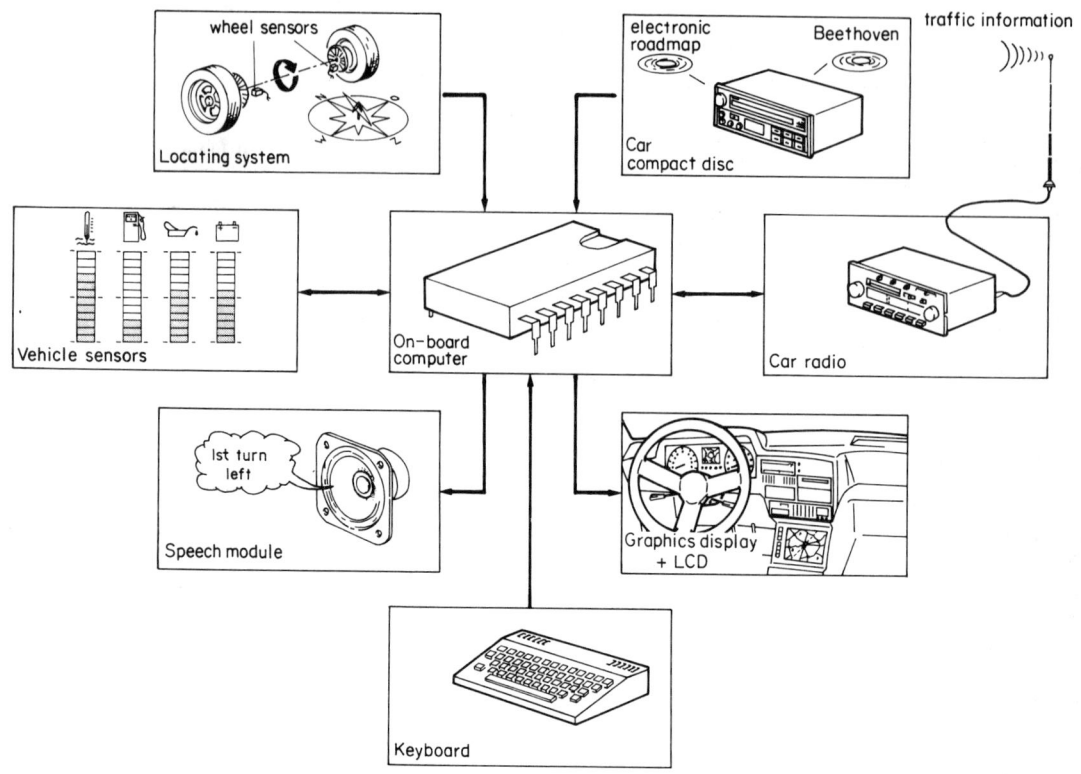

Figure 1
The development of new means for traffic communication involving broadcasting systems is to be seen in the context of a new powerful information processing environment: concentrated around the on-board computer, the car radio will only be one of the components to be considered; another important component will use an electronic roadmap (CD-ROM) for car navigation with updating for the traffic situation via RDS-TMC (courtesy of Philips N.V.)

ARIAM system in the FRG (see Figs. 2, 3 and 4), but very often the necessary data collection will still need to be done using traditional methods.

Driver information through the RDS-TMC is expected to increase transport efficiency by reducing, or even in some cases preventing, major traffic jams. Information on actual disturbances or bottlenecks should also help to prevent accidents.

2. Description, Status and Trends of RDS-TMC

At present, RDS is a *de facto* broadcast standard that was developed by the broadcasting organizations that cooperate on a European level within the EBU. This standard was published by the EBU in 1984 and it became a recommendation of the CCIR (an organization of the International Telecommunications Union (ITU) dealing with the worldwide radiocommunication standards) in 1986. The RDS specification of the EBU is now being considered by TC 107 of CEN/CENELEC for adoption as a European standard. It allows the transmission of a continuous stream of digital data inaudibly and in parallel with a normal vhf/FM stereo radio program. Like the ARI system, the data is superimposed on the stereo-multiplex signal with a sub-carrier frequency of 57 kHz.

The data is transmitted in "groups" which contain four "blocks" of 26 bits each. Each block comprises an information word of 16 bits, and a checkword of ten bits that allows for the recognition and the correction of transmission errors. Ten different groups have been defined by the EBU—group 8 being reserved for a TMC—but each of them contains the same information on program identification, program type and identification of traffic channel in the first two blocks. So, for further information, only the two information words of blocks 3 and 4 and, additionally, three bits which are undefined in block 2 are available. This means that one 8-type group has a maximum capacity for traffic information of 35 bits. With a total data rate of 1187.5 bits per second—equivalent to 11.4 groups per second—this leads theoretically to a total capacity for coded traffic messages of 400 bits per second if messages of group 8 only are sent. In practice, however, the RDS system fulfills in vhf/FM broadcasting, as a primary task, automated tuning functions which absorb by far the largest part of the data transmission capacity so that no more than one or two group 8 messages can be inserted for the TMC function during each second. Thus, the feasible message rate per minute will probably be limited to 15 to 30 messages, a figure also dependent on the repetition of the message required to ensure a high reception reliability in the mobile reception mode of the TMC receiver and on the number of type-8 groups that will be necessary for the transmission of one message.

Depending on the way it is finally agreed that traffic messages should be coded, 35 bits may be enough for the coding of complete traffic messages in most cases. However, a number of exceptions are expected where sequences of two or three groups will be needed for the transmission of one message.

For the definition of a standard traffic message code, several efforts are underway. A working group of the Conference of European Ministers of Transport (CEMT) has preliminarily agreed a format for a two-sequence traffic message, but some broadcasters and equipment manufacturers prefer one-sequence standard messages in order to increase capacity and are working on an agreement for this. At the same time, the European Commission is concerned to increase the coordination between the different parties involved, also expecting that the EBU, still responsible for the RDS specification will agree to help with the identification of a preliminary TMC protocol on which field tests can be implemented. At a later stage (after the tests are

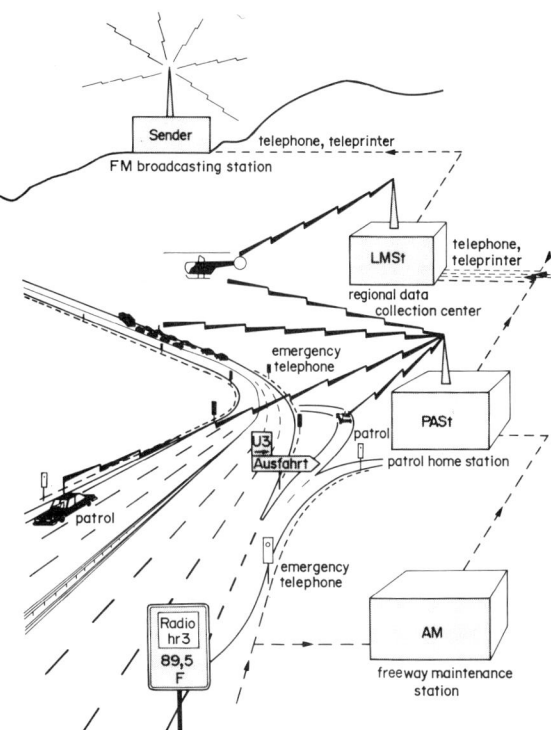

Figure 3
The collection of traffic information involves several information collection centers which are all interconnected. This results, of course, in a certain delay of the messages distributed. The delay will be longer if the informtion is collected manually and shorter if automated incident detection systems are used (courtesy of Heusch-Boesefeldt GmbH)

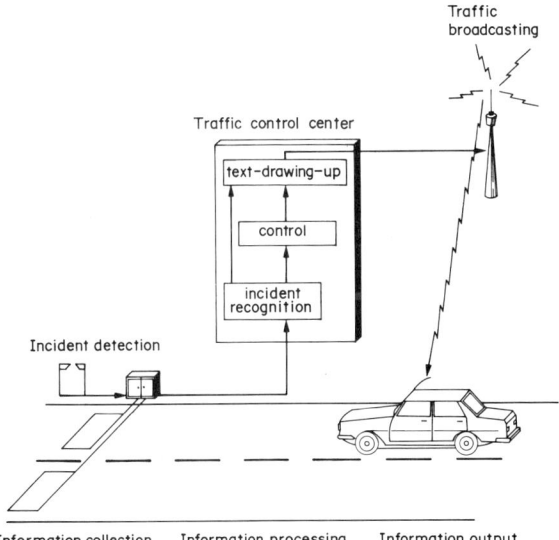

Figure 2
The broadcasting traffic service requires information collection, which can be automated, and information processing by the traffic authorities and the broadcasters (courtesy of Heusch-Boesefeldt GmbH)

completed in 1991), standardization of RDS-TMC within CEN/CENELEC TC 107 will then most probably become a necessity.

Most European broadcasters will introduce RDS in the years 1987–1991, mainly as a tuning aid for a new generation of vhf/FM radios. The first phase of RDS will generally not include the RDS-TMC feature. For the countries that have not yet introduced ARI, RDS immediately offers a possibility of identification of traffic broadcasts and also traffic announcements. In order to enable traffic messages to be inserted over TMC, it will be necessary to agree with broadcasters, preferably within the EBU, the definition of a communication protocol for the generation of RDS-TMC data groups that would enable traffic control centers to create a link to the RDS encoders used at the broadcast transmitters.

The main contribution of DRIVE in this particular area will be to encourage the adoption of RDS-TMC as a system for Europe and to assist the harmonization of the different efforts. Furthermore, DRIVE will investigate the possibility of including additional information

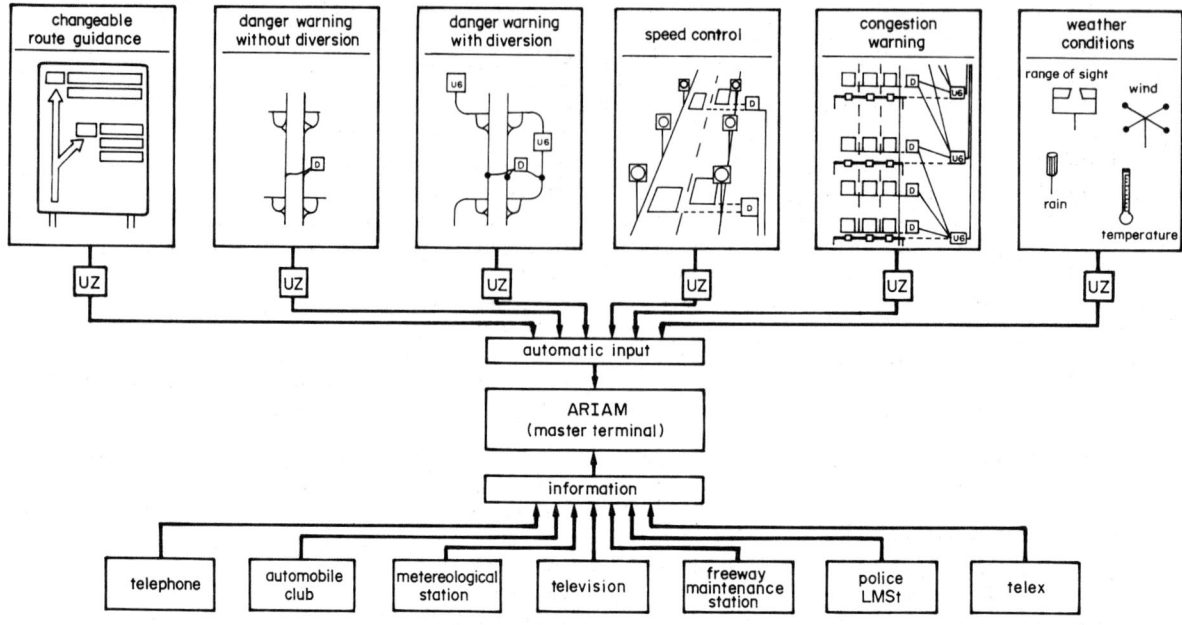

Figure 4
In the Frankfurt/Munich area in the FRG the ARIAM system is used for the collection and distribution of traffic information interfacing with traffic authorities, automobile clubs and broadcasters (courtesy of Heusch-Boesefeldt GmbH)

in the TMC, which might be useful or even necessary for traffic management purposes, but is not included in the current proposals.

If the RDS turns out to be an efficient system, it is not sufficient to have only quick and comprehensive transmission of messages. The collection of actual traffic data has at least the same importance, so the investigation of and, later on, the introduction of the RDS infrastructure should be accompanied by the investigation and installation of infrastructure for measuring traffic flow and, especially, for incident detection. Furthermore, there is also a strong need for algorithms and software for information analysis and forecasting, and for information management in general and especially for RDS-TMC. Also, the international exchange of messages in the context of heavy international traffic in holiday periods requires standardization of a message-exchange protocol.

See also: Mobile Communication

Bibliography

European Broadcasting Union 1984 Specification of the Radio Data System RDS for VHF/FM broadcasting. Technical Document 3244. EBU, Grand Saconnex, Switzerland

Eurotravel 1988—Traffic information before the year 2000 *EBU Rev.* **40** (2), 27–34

Giesa S, Everts K 1987 ARIAM—Car-driver-radio information on the basis of automatic incident detection. *Traffic Eng. Control* June, 344–8

C. Bielefeldt
[The MVA Consultancy, Woking, UK]

D. Kopitz
[European Broadcasting Union, Geneva, Switzerland]

Car-Following Models

In the field of traffic modelling, car-following models, also called microscopic models, are used to describe the behaviour of single vehicles in a network or, more precisely, the behavior of the driver–vehicle system in a stream of interacting vehicles. The main applications of such models are:

(a) to obtain a better understanding of the driver–vehicle system behavior which can lead to the development of new safety devices, and

(b) to provide the basic component of microscopic simulation models which are now widely used, for instance to test or improve new traffic control strategies (see *Simulation Programs, Microscopic*).

Car-following models consist of differential difference equations giving the acceleration of a vehicle with respect to the behaviour of the preceding ones; they are often guessed at, but control and estimation theory is also used to derive the structure of an optimal control model, based on the assumption that the driver performs car-following by minimizing an appropriate quadratic function. Car-following theory was first proposed by Reuschel (1950) and Pipes (1953) and was greatly extended by Herman *et al.* (1959). Among the various syntheses dealing with this subject are Gazis (1974), Gerlough and Huber (1975) and Bekey *et al.* (1977).

1. Main Car-Following Models

1.1 General Relation

The general form of car-following models can be represented by the expression:

$$\text{response}(t+T) = \text{sensitivity} \times \text{stimulus}(t) \quad (1)$$

where T is the reaction time of the driver–vehicle system. Although an exact description of this stimulus–response reaction would be very complicated, it has been proved that rather simple continuous differential difference equations could give a very good approximation of the phenomenon.

For the various models described, the response will always be the acceleration (or deceleration) of the following car, the stimulus being, in most cases, the difference in velocity between the lead car and the follower.

1.2 Simple Linear Car-Following Model

This model is mathematically expressed by:

$$\ddot{x}_{n+1}(t+T) = a[\dot{x}_n(t) - \dot{x}_{n+1}(t)] \quad (2)$$

where n is the number of the lead car, $n+1$ the number of the follower, T the time lag and a the sensitivity coefficient (per second). Chandler *et al.* (1958) obtained car-following data to validate this model at the General Motors Technical Center. They showed experimentally that T was approximately 1.5 s and a was approximately $0.37\,\text{s}^{-1}$. This very simple model is not very satisfactory, as will be seen in Sect. 3 in its implications for steady-state flow, but its stability can be studied easily using Fourier analysis.

The stability of a traffic model must be studied to verify that a change in velocity by the lead vehicle of a stream of cars will not be amplified by successive vehicles in the stream until a collision occurs. There are two types of stability. Local stability considers the response of a vehicle to the change in motion of the vehicle immediately in front of it, while asymptotic stability deals with the propagation of a fluctuation through a platoon of vehicles.

For the linear model, it can be demonstrated that local stability is guaranteed if $aT \leq \pi/2$, while a line of traffic is asymptotically stable only when $aT < 0.5$ (Herman *et al.* 1959).

1.3 Nonlinear Car-Following Models

In Eqn. (2), a is a constant; this means that the response to a given difference in velocity between the two vehicles is independent of their spacing, which is not realistic. Hence, Gazis *et al.* (1959) developed the following model:

$$\ddot{x}_{n+1}(t+T) = a_0 \frac{\dot{x}_n(t) - \dot{x}_{n+1}(t)}{x_n(t) - x_{n+1}(t)} \quad (3)$$

which shows the response to be inversely proportional to the spacing; a_0 is another sensitivity coefficient (in meters per second).

Pipes (1953) derived a car-following model based on the assumption that the acceleration for the following vehicle is proportional to the driver's perception of the rate of change of the visual angle θ (see Fig. 1), and

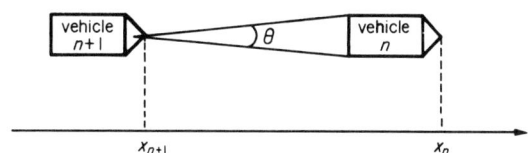

Figure 1
The visual angle model

arrived at the following nonlinear equation:

$$\ddot{x}_{n+1}(t+T) = a_0 \frac{\dot{x}_n(t) - \dot{x}_{n+1}(t)}{[x_n(t) - x_{n+1}(t)]^2} \quad (4)$$

A further analysis of the driver's behavior shows that, for a given difference in velocity and a given spacing, the response is more important the higher the velocity of the vehicle. Finally, these considerations lead to the following generalized form for nonlinear car-following models:

$$\ddot{x}_{n+1}(t+T) = a_0 \dot{x}_{n+1}^m(t+T) \frac{\dot{x}_n(t) - \dot{x}_{n+1}(t)}{[x_n(t) - x_{n+1}(t)]^l} \quad (5)$$

where l and m are constants.

1.4 Other Car-Following Models

(a) *Three-car model*. Several attempts have been made to build models including a second lead vehicle, such as the linear model:

$$\ddot{x}_{n+2}(t+T) = a_1[\dot{x}_{n+1}(t) - \dot{x}_{n+2}(t)] + a_2[\dot{x}_n(t) - \dot{x}_{n+2}(t)] \quad (6)$$

but vehicle experiments showed that the stimulus provided by the second lead car (vehicle number n in this case) was not very significant.

(b) *Helly model*. With the preceding car-following models, the spacing between two vehicles driving at the same speed remains constant; Helly (1961) supposed that the driver would seek to minimize both the velocity difference and the difference between his actual headway $(x_n - x_{n+1})$ and a desired headway. A simplified version of this model is:

$$\ddot{x}_{n+1}(t+T) = c_1[\dot{x}_n(t) - \dot{x}_{n+1}(t)] + c_2[x_n(t) - x_{n+1}(t) - D] \quad (7)$$

where c_1 is the velocity control parameter, c_2 the headway control parameter and D the desired headway.

This model is used in the SITRA-B microscopic simulation program (Gabard et al. 1982), where the desired headway is taken as equal to:

$$l_n + \tau_{n+1}\dot{x}_{n+1}(t)$$

with l_n being the length of vehicle n and τ_{n+1} the time headway for vehicle $n+1$.

(c) *Asymmetric models*. In the previous models, the assumption was made that the amplitude of the response was the same for acceleration and deceleration, but it is well known that the capabilities of cars and drivers' reactions are greater for deceleration than for acceleration. The following model has been tested:

$$\ddot{x}_{n+1}(t+T) = a_+[\dot{x}_n(t) - \dot{x}_{n+1}(t)] \quad (8a)$$

for positive relative velocity, and

$$\ddot{x}_{n+1}(t+T) = a_-[\dot{x}_n(t) - \dot{x}_{n+1}(t)] \quad (8b)$$

for negative relative velocity. This is an extension of the linear model; fitting of data to the model showed an average weighting factor of 1.1 for a_-/a_+.

In other models, deadbands have also been introduced in order to simulate driver perception thresholds.

(d) *Gipps model*. More recently, Gipps (1981) proposed a new car-following model based on the assumption that each driver sets limits to his desired braking and acceleration rates.

The model has two components, which cover acceleration and braking separately. For acceleration:

$$\dot{x}_{n+1}^a(t+T) = \dot{x}_{n+1}(t) + 2.5\, a_{n+1} T \left[1 - \frac{\dot{x}_{n+1}(t)}{V_{n+1}}\right]$$

$$\times \left[0.025 + \frac{\dot{x}_{n+1}(t)}{V_{n+1}}\right]^{1/2} \quad (9)$$

where $\dot{x}_{n+1}^a(t+T)$ is the maximum speed to which vehicle $n+1$ can accelerate during the time interval $(t, t+T)$, V_{n+1} is the desired speed for vehicle $n+1$ and a_{n+1} is the maximum acceleration for vehicle $n+1$. For braking:

$$\dot{x}_{n+1}^b(t+T) = \beta_{n+1} T$$
$$+ \left([\beta_{n+1}T]^2 - \beta_{n+1}\left\{2[x_n(t) - l_n - x_{n+1}(t)]\right.\right.$$
$$\left.\left. - \dot{x}_{n+1}(t)T - \frac{\dot{x}_n(t)^2}{\hat{\beta}}\right\}\right)^{1/2} \quad (10)$$

where $\dot{x}_{n+1}^b(t+T)$ is the maximum safe speed for vehicle $n+1$ with respect to vehicle n, β_{n+1} is the most severe braking the driver of vehicle $n+1$ wishes to undertake (<0), l_n is the effective length of vehicle n and $\hat{\beta}$ is the estimate of β_n used by the driver of vehicle $n+1$.

Finally, in any given circumstances, the speed adopted by vehicle $n+1$ is the minimum of $\dot{x}_{n+1}^a(t+T)$ and $\dot{x}_{n+1}^b(t+T)$.

This model is used in MULTSIM, a program for simulating vehicular traffic in multilane arterial roads (Gipps 1986), in which special attention was devoted to the modelling of the structure of lane-changing decisions.

2. Models Derived from Optimal Control Theory

In the approach of deriving a model from optimal control theory (Bekey et al. 1977), the follower-vehicle dynamics is represented by the expression:

$$x_{n+1}(t+\Delta t) = \mathbf{A}_{n+1} x_{n+1}(t) + \mathbf{B} u(t) \quad (11)$$

where $x_{n+1}(t)$ is the two-dimensional state vector of the follower (position $x_{n+1}(t)$ and velocity $\dot{x}_{n+1}(t)$), $\mathbf{A}_{n+1}$ is a 2×2 coefficient matrix, $\mathbf{B}$ is a 2×1 matrix and $u(t)$ is the control exerted by the driver (acceleration or deceleration). A similar representation is used for the lead vehicle dynamics (but without the control term $\mathbf{B}u(t)$), and the expression of the quadratic performance criterion to be minimized by the follower is:

$$j = \frac{1}{2} \sum_{i=0}^{\infty} \{[x_n(t_i) - x_{n+1}(t_i) - \alpha \dot{x}_{n+1}(t_i)]^2 q_1$$
$$+ [\dot{x}_n(t_i) - \dot{x}_{n+1}(t_i)]^2 q_2 + r u^2(t_i)\} \quad (12)$$

where q_1, q_2 and r are weighting factors and α is a constant; in this criterion, the first term represents the deviation of the vehicle spacing from a desired spacing (which is proportional to the follower velocity), the second term represents the relative velocity error, and the third term represents the cost of control.

With the supplementary hypothesis that the leader and follower car dynamics are identical, it can be shown that the solution of this optimization problem leads to the following general structure of control $u(t)$:

$$u(t) = c_1 [\dot{x}_n(t) - \dot{x}_{n+1}(t)]$$
$$+ c_2 [x_n(t) - x_{n+1}(t) - c_3 \dot{x}_{n+1}(t)] \quad (13)$$

It is particularly interesting to note that this structure is very similar to the one used by Helly (1961) (see Sect. 1.4).

This approach leads to deterministic models; a stochastic approach can also be considered, by introducing observation and state noises in the process; Kalman filtering is then used to derive the solution of this stochastic control problem.

3. From Microscopic to Macroscopic Models

Gazis et al. (1959) demonstrated that it was possible to derive equations of traffic stream flow (i.e., macroscopic models) directly from the microscopic models. This is done by integration of the preceding equations in a steady-state formulation, neglecting the time lag T; the expression obtained gives the velocity as a function of the density, which leads to a flow equation, using the well-known relation $q = uk$, where q is the flow (vehicles per second), u is the steady-state velocity of the traffic stream (meters per second) and k is the steady-state density (vehicles per meter) (see *Flow Variables*).

For example, integration of the linear model,

Eqn. (2), leads to the following traffic stream model:

$$q = a \left(1 - \frac{k}{k_j} \right) \quad (14)$$

The integration constant is determined by using boundary conditions; here, k_j represents the "jam" density (when velocity $u=0$). The sensitivity coefficient a can also be determined by solving the equation for a known condition. In this case, the final equation is:

$$q = q_m \left(1 - \frac{k}{k_j} \right) \quad (15)$$

where q_m is the maximum flow. Or, in terms of velocity:

$$u = q_m \left(\frac{1}{k} - \frac{1}{k_j} \right) \quad (16)$$

As stated previously, this macroscopic model cannot be considered as satisfactory; velocity does not approach infinity when k approaches zero and the $q=f(k)$ curve is not a straight line but a convex curve.

Better results are obtained with integration of the nonlinear car-following models; Table 1 shows the traffic stream model equations resulting from the integration of Eqn. (4) for various values of l and m, which have been verified by observations of vehicle flow. The case of noninteger values of m and l has also been examined.

4. Application of the Car-Following Theory

4.1 Microscopic Simulation Models

Car-following models are the basic component of microscopic simulation; among the various simulation

Table 1
Flow models derived from nonlinear car-following models (u_m = velocity at maximum flow, u_f = velocity at free flow, k_0 = density at maximum flow)

l	m	Flow equation
0	0	$q = q_m \left(1 - \dfrac{k}{k_j}\right)$
1	0	$q = u_m k \ln\left(\dfrac{k_j}{k}\right)$
1.5	0	$q = u_f k \left[1 - \left(\dfrac{k}{k_j}\right)^{1/2}\right]$
2	0	$q = u_f k \left(1 - \dfrac{k}{k_j}\right)$
2	1	$q = u_f k \exp\left(-\dfrac{k}{k_0}\right)$
3	1	$q = u_f k \exp\left[-\dfrac{1}{2}\left(\dfrac{k}{k_0}\right)^2\right]$

models, microscopic ones are used when it is necessary, for instance, to know precisely the position of the vehicles in a network, to differentiate between different types (cars, buses, lorries). The main application field of such models is to test and assess traffic control strategies: adaptive, real-time strategies for cars, priority strategies for buses, and so on.

They have also been used to predict noise level and pollutant emissions at signalized intersections, and to test the efficiency of safety devices intended to decrease rear-end collision risks in a stream of vehicles; in this latter study, a car-following model was used to extend experimental results obtained with two equipped cars to a simulated file of vehicles.

4.2 Automatic Car-Following

The main objective of the studies connected with the derivation of car-following models from automatic control theory was to achieve highway automation, in order to increase both highway capacity and highway safety (Fenton 1970). These studies will probably become very topical again, with the development of on-board electronics and the related research projects sponsored by the European Community (see *DRIVE*).

See also: Priority Intersection: Modelling; Simulation Programs, Microscopic

Bibliography

Bekey G A, Burnham G O, Seo J 1977 Control theoretic models of human drivers in car following. *Hum. Factors* **19**(4), 399–413
Chandler R E, Herman R C, Montroll E W 1958 Traffic dynamics: Studies in car-following. *Oper. Res.* **6**, 165–84
Fenton R E 1970 Automatic vehicle guidance and control—A state of the art survey. *IEEE Trans. Veh. Technol.* **19**(1), 153–61
Gabard J F, Henry J J, Tuffal J, David Y 1982 Traffic responsive or adaptive fixed time policies? A critical analysis with SITRA-B. *Proc. Int. Conf. Road Traffic Signalling*. Institution of Electrical Engineers, London, pp. 89–92
Gazis D C 1974 *Traffic Science*. Wiley, New York
Gazis D C, Herman R C, Potts R B 1959 Car-following theory of steady state flow. *Oper. Res.* **7**, 499–505
Gerlough D L, Huber M J 1975 Traffic Flow Theory: A Special Report. Special Report No. 165. US Transpsoration Research Board, Washington, DC
Gipps P G 1981 A behavioural car-following model for computer simulation. *Transp. Res. B* **15**, 105–11
Gipps P G 1986 A model for the structure of lane-changing decisions. *Transp. Res. B* **20**, 403–14
Helly W 1961 Simulation of bottlenecks in single-lane traffic flow. In: Herman R C (ed.) *Theory of Traffic Flow, Proc. Symp. Theory of Traffic Flow*. Elsevier, Amsterdam, pp. 207–38
Herman R C, Montroll E W, Potts R B, Rothery R W 1959 Traffic dynamics: Analysis of stability in car-following. *Oper. Res.* **7**, 86–106
Pipes L A 1953 An operational analysis of traffic dynamics. *J. Appl. Phys.* **24**, 274–81

Reuschel R 1950 Fahrzeugbewegungen in der Kolonne bei gleichformig beschleunigtem oder verzögertem Leitfahrzeug. *Z. Österr. Ing. Arch. Ver.* **95**, 52–62; 73–7

J. F. Gabard
[ONERA, Toulouse, France]

Cellular Communication Systems

Modern society is characterized by extensive mobility of its members, using various means of transportation. Two different reasons induce people on the move to communicate. The needs of navigation and motion safety dominate in rail, sea, air and space communications. In road communications, to be dealt with in this article, more general (but less vital) needs for information prevail, so mobile road services are relatively underdeveloped.

For decades, land-mobile radio (LMR) communication existed in the simple form of dispatch systems, coordinating the motion of vehicles and the work of their crews. Dispatching is necessary in various rescue and relief operations, law enforcement, utilities, transportation, repair services, and so on. Until the beginning of the 1980s the great majority of LMR mobile stations have belonged to small, autonomous dispatch networks. Usually, the dispatch center communicates by voice with a closed group (fleet) of vehicles, equipped by trained personnel. As a rule, direct interconnection of mobiles with the public telephone network is not required.

Since the 1980s conventional public telephony has been extended to mobile vehicles and persons in road, rail and sea transportation. In large-area multizone networks (see Fig. 1), sophisticated technology is indispensable, although less vital private or general information is being transferred. Users (subscribers) of public LMR systems are untrained and cannot cooperate to make the system more efficient. Also, an average public conversation is much longer than a typical dispatch message.

The radio spectrum is a scarce and nonrenewable resource; it is clearly impossible to assign a dedicated channel to every LMR user. For efficient spectrum utilization in large networks, two methods have been introduced: trunking (transferred from telephony) and the use of innovative cellular structures.

The radio spectrum assigned to an LMR system and its zones is subdivided into channels, less numerous than their potential users. Thus, the pool (multiplex) of channels in each zone has to be time-shared among the users; a user gets intermittent access to the system by seizing a free channel. This principle of trunking ensures easier access availability and equalizes the loading of channels from the trunk. This happens at the cost of providing each user with automatic access to any of the whole pool of channels. In the case of occupancy

Cellular Communication Systems

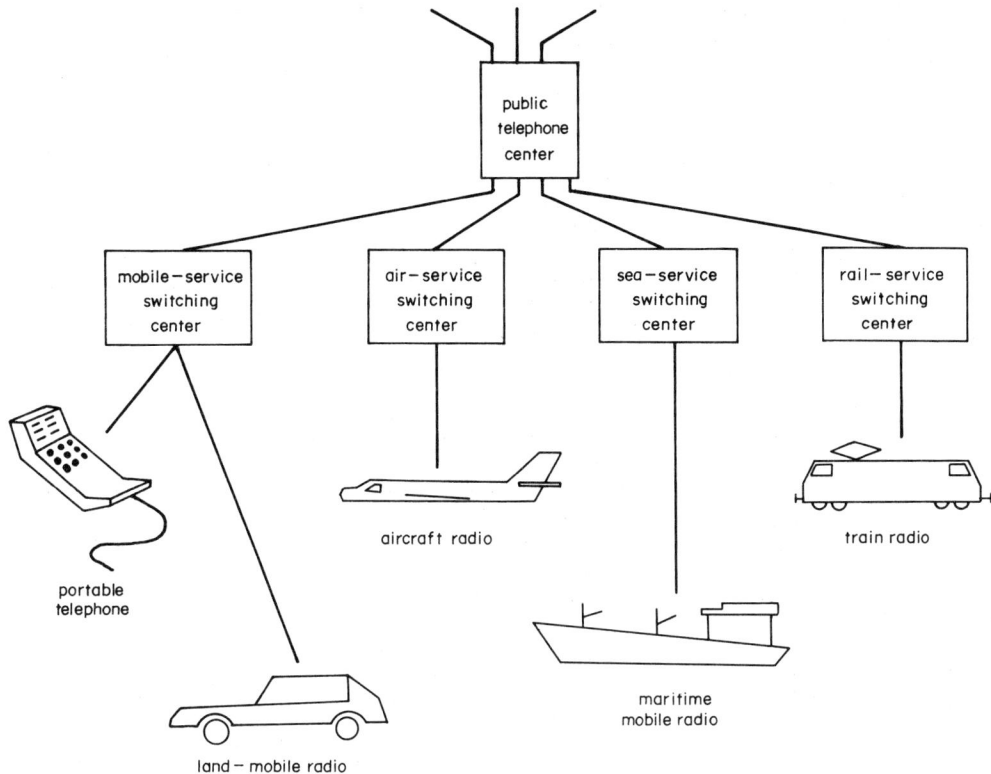

Figure 1
Large-area telephone network structure

of all channels, call attempts are blocked. Also, collisions of users trying to seize a channel are probable and have to be mitigated.

The cellular (small zone) structure of the network permits the reuse of radio channels in properly spaced cells. The growth of mobile radio (as well as broadcasting) networks relied for decades on increasing the power and the range of transmitters—only recently has this trend been reversed. Subdividing the system area into small cells was made feasible by the advent of the necessary integrated microelectronics and system techniques.

The cellular infrastucture takes care of mobiles roaming over the whole area. Two-way contact with them is automatically maintained, in spite of the poor quality of intracell radio links. An average connection with a mobile or between mobiles contains intercall cable links. Thus, the radio space is less occupied and the number of mobiles (the system capacity) can be increased substantially.

Mobility of LMR users presents many novel problems that do not occur in stationary telecommunication networks. A mobile user's access to the system is only temporary. Nevertheless, his location must be known to within one cell in order that he can be reached from the cellular system or from outside.

The integrity of a cellular LMR system is maintained at the central switch level, controlling cable links to base stations. Active connections from or to roaming mobiles should not be interrupted at the boundaries of cells; this requires automatic switching of radio mobile station to another channel in the neighboring cell.

1. Cellular Systems

1.1 Basic Concepts and Models

Cellular LMR systems cover large areas, subdivided into small cells, each equipped with a fixed base station (BS) in cell site (CCIR 1986, Lee 1988). Within each cell, a set of trunked radio channels transmits two-way information flow to and from mobile stations (MSs); any roaming MS can be active on the whole system area. For the whole network, several disjoint channel sets are allocated. BSs are interconnected with the switching centers (SCs), ensuring access to the outer telecommunication network.

Structures of cellular LMR networks are customarily modelled by uniform planar lattices. A spatially regularized array of points represents the sites of BSs; the numbering of points represents assignments of the

Cellular Communication Systems

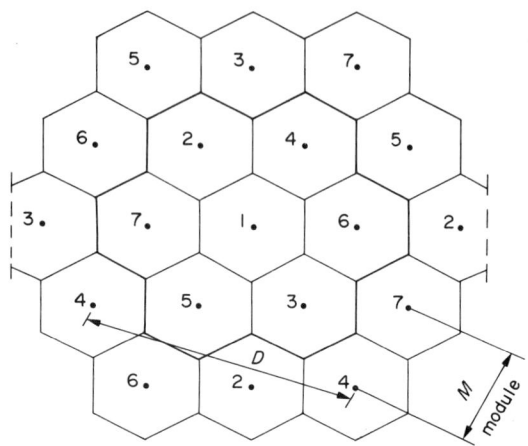

Figure 2
Sketch of a regular lattice: the cluster consists of seven hexagonal cells ($N=7$) numbers denote sets of frequency channels (reused)

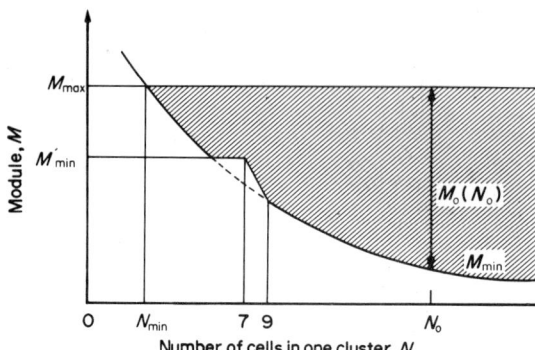

Figure 3
Admissible variation of the $M(N)$ values in the hatched area: regular lattice is represented by one point (M_o, N_o); the adjacent-channel constraint in small clusters is also shown

channel sets. All contiguous hexagonal cells with different assignments constitute a cluster, the second level in the network hierarchy of cell, cluster and area.

The geometry of a regular lattice (see Fig. 2) is described by two distances: the module M (least spacing between BSs) and the reuse (coordination) distance D (least spacing between cochannel BSs). The integer N, the third parameter of the regular lattice, determines the number of cells in one cluster, equal to the number of disjoint channel sets in the network.

1.2 Structural Relations

The study of regular lattices relies on conventional assumptions: hexagonal cells, identical stations, omnidirectional propagation, and stationary noise (cf. CCIR 1986 a, b, c). Three parameters of the model structure determine its regularity and efficiency. The array of lattice points constitutes equilateral triangles of side M; this pattern minimizes the number of BSs covering a given area.

With the fixed reuse distance D, cochannel interference in the network is minimized. Regular geometry of the lattice is feasible with some values of N only. Spectral regularity is achieved by the linear numbering (arithmetic progression modulo N) of cell/channel sets. Optimum lattices exhibit geometric and spectral regularity.

Complex topological analysis leads to the simple structural equation of any regular lattice (Wojnar 1988):

$$D = MN^{1/2} \qquad (1)$$

Only two parameters in Eqn. (1) are independent. Also, the values of N ensuring completely regular geometry are given by:

$$N_o = K^2 + l^2 + kl, \quad k \geq 1; \ l \geq 0 \qquad (2)$$

and k and l are integers. The following values are of interest for LMR (excluding future microcellular structures):

$$N_o = \mathbf{3}, 4, \mathbf{7}, 9, 12, \mathbf{13}, 16, \mathbf{19}, \mathbf{21}, 25, 27, \ldots \qquad (3)$$

The bold values constitute a small-density sequence of "rhombic" numbers, leading to optimum structures.

The dimensioning of cellular structures is affected by two radioengineering constraints: the intracell service range d_s and the intercell interference range d_i. The coverage criterion is:

$$M_{max} = d_s \sqrt{3} \qquad (4)$$

The cochannel compatibility, with disjoint d_s and d_i ranges, imposes:

$$D_{min} = d_s + d_i \qquad (5)$$

The theoretical lower bounds on lattice parameters result from Eqns. (1), (4) and (5):

$$M_{min} = N^{-1/2}(d_s + d_i) \qquad (6)$$
$$N_{min} = 3d_s^{-2}(d_s + d_i)_2 \qquad (7)$$

The potentialities of lattice structures are best visible in the $M(N)$ graph (See Fig. 3). All points (N_o, M_o) representing regular structures are located at vertical sections $N_o = $ constant, within the hatched area limited by the bounds given in Eqns. (4) and (6). The transition to real, deformed structures is simple: given N_o, the difference $M_{max} - M_{min}$ indicates the admissible deformability of the array of BS locations (Wojnar 1988).

Evidently, increasing the cluster size N_o improves the structural deformability of the real topography of cell sites at the cost of less efficient spectrum usage: trunked channel sets in the cells have to be diminished with growing N.

1.3 Performance Assessment

The primary objective of mobile radiocommunication is to ensure information flow to and from mobile subscribers. Thus, network performance should be expressed in terms of radiotraffic intensity (using the term traffic in the sense of information flow only) and of network capacity (i.e., the maximum number of subscribers adequately served by the system).

For assessing radiotraffic intensity A, measures from the theory of telephone traffic (Bear 1976) are adopted. The relevant intensity unit of 1 erlang corresponds to the permanent activity (during the observation time) of one transmission facility (e.g., the radio channel). The traffic (calls) is generated by a group of subscribers S much larger than the set of channels C. Occupancy of all channels blocks the access of subscribers to the network. Quality of service is measured by the proportion B of calls blocked and lost where $\doteq$ denotes proportionality and $|$ separates the required condition.

The three traffic-describing quantities A, B, C are interrelated through Erlang's lost-call formula $B(A,C)$ (Bear 1976). It is a nonlinear function of two variables. In the region of major interest ($C \geq 5$), an approximate proportionality:

$$A \doteq C\,|\,B = \text{constant} \quad (8)$$

is roughly valid in any LMR cell and also in the whole network (cellwise additivity of traffic).

In a cellular LMR network over a fixed area, the total number of cells n depends on the mean cell size through:

$$n \doteq M^{-2} \quad (9)$$

Additivity of traffic combined with Eqns. (8) and (9) leads to:

$$A \doteq M^{-2} N^{-1} \quad (10)$$

Thus, network performance improves with the lattice congestion (decreasing M). Less known advantages of diminishing the cluster size (decreasing N) are also revealed.

The network capacity S also depends on the same lattice parameters:

$$S \doteq M^{-2} N^{-1} \quad (11)$$

The efficiency of radio-spectrum utilization by a cellular network is customarily measured in erlangs $\text{km}^{-2}\,\text{MHz}^{-1}$. This measure behaves like traffic intensity in Eqn. (10), provided that the network area and the spectrum allocation are fixed.

In summary, substantial growth of a cellular network can be achieved by lattice congestion. For this purpose, cell-splitting techniques, both omnidirectional (BS site in the cell center) and sectorial (BS site in a cell corner), have been elaborated. Advantages of sectorial splitting with directional BS antennas are analyzed by Wojnar (1988). First, the coordination distance D in the sectorial structure, given by Eqn. (5) with omnidirectional cells can be diminished to:

$$D_{\min} = d_i - d_s \quad (12)$$

Consequently, the cluster size N can be decreased and the number of channels in each cell increased. Sectorial splitting produces essential gains in network performance.

2. Specific Applications

2.1 Radio Dispatching in Road Transportation

The professional dispatching service (sometimes known as the private dispatching service) has been the only kind of civilian LMR, paralleling combat radio nets in military applications. Fleets of vehicles with personnel are controlled by small and simple autonomous networks. Routine system parameters are utilized: very-high frequency (vhf) < 300 MHz, analog frequency modulation (FM) or phase modulation (PM), simplex operation and channel width 25 kHz.

With strongly diversified needs and system structures, it is justifiable to speak about a fuzzy class of systems. Nevertheless, the typical characteristics of dispatch radio traffic can be determined. An average conversation between the dispatcher and a mobile user is short (10–20 s) and contains more or less formalized commands and messages. It is then feasible to accommodate tens or even hundreds of MSs per channel or per dispatcher. In novel large networks, dispatch posts are multiplied and multichannel trunks with automatic selection of a free channel are provided for. However, dispatch systems often operate at high loads so that trunking becomes less effective. A dispatcher typically "stays on the air" most of the time; thus, the dispatch center can be the bottleneck of the system. Preemptions, signalling channels, various queuing techniques and access methods are in use to alleviate the congestion. The characteristics of trunking efficiency and the estimates of delays in overloaded networks can be found in CCIR (1986b).

Evidently, ever-growing road transportation calls for adequate communications supply. It is then necessary to radically increase spectrum utilization efficiency by very special measures. First, advantages of scale can be attained by aggregating separated networks into cellular regionwide (e.g., in the USA and Japan) or even countrywide (e.g., in France and Sweden) trunked systems, with at least several tens of channels (CCIR 1986a, b, c).

The countrywide system in France, Radiocom 2000, is unique in merging dispatch and public radiotelephone functions. The system uses blocks of channels in two vhf and uhf bands; the maximum of 256 channels can accommodate up to 20 000 dispatch-type MSs or 7000 radiotelephone subscribers or any combination of these two services. This versatility will be followed in the future Pan-European system (see Sect. 2.4) and in satellite systems (see Sect. 3.3).

2.2 Dispatching by Radio-Date Transmission

The Swedish nationwide dispatch system called Mobitex was introduced in 1987. The infrastructure comprises main and area exchanges as well as BSs in cellular configuration. When completed, the system will serve up to 40 000 MSs, using around 150 BSs with about 1000 channel equipments and 4000 dispatch posts throughout the country (CCIR 1986 a, b, c).

Mobitex handles primarily packetized text and data; mobile terminals have access to the user's subsystem and to national telex and data networks. Auxiliary voice links are interconnected to the public telephone network; call holding time is restricted. Automatic roaming between base stations is provided for.

Mobitex is a "store-and-forward" system without real-time contact between two terminals. Text or data packets to and from the terminals are transferred through network nodes (exchanges and BSs) with acknowledgement or error correction at each node. The maximum length of a packet is 512 characters; with radio transmission speed 1200 bits per second the packet occupies a few seconds of channel time. It is thus possible to accommodate up to 1000 mobiles per channel.

Mobitex radio channels of standard width are flexibly used. One dedicated (national) channel is used by all BSs with time division. Each BS has a few traffic channels for text/data or for speech. Stored-text emergency messages can be transmitted from mobiles.

The mobile station is composed of the digital equipment (e.g., visual display, keyboard, printer, various sensors, measuring and data-collection facility) and of the application software. Simple asynchronous character-oriented terminals are also admitted.

The fixed terminals communicate with the Mobitex area exchange via a fixed line and a standard modem. There are many possibilities of designing a user-specific application subsystem. Very important is the instant access to the user's host computer system, as well as to its databases. Among various user categories, emphasis is laid on transportation, utilities, emergency services, and so on. Mobitex reflects the needs of modern automatic and computerized management, especially in companies and institutions covering large areas by mobile personnel.

2.3 First-Generation Public Systems

Many problems of large-area roaming and of ample system capacity have been solved by cellular techniques. During the 1980s, first-generation public networks with analog speech transmission have been introduced in well-developed countries. Unfortunately, radio transmission standards differ and are noncompatible (CCIR 1986c).

The following operational (but not technological) features are typical in the existing public networks:

(a) automatic setting up of duplex radioconnections, to and from MSs independently of their localization (with the exception of large buildings, tunnels, mountain gorges and similar obstructions);

(b) continuous roaming of MSs with automatic registration of location;

(c) automatic transfer ("handover") of calls in progress through cell boundaries;

(d) telephonelike numbering and tariffication principles;

(e) continuous control of radio-link quality in order to approach the telephony standards;

(f) provision for ample system signalling, with digital sequences redundantly coded;

(g) efficient utilization of the radio spectrum thanks to cellular structure; and

(h) simple manipulations of subscribers' sets, designed for nonqualified users.

The principle of privately owned telephone subnetworks (private automatic branch exchanges) is not easily transferable to radiotelephony. This problem, solved in the dispatch system Mobitex, remains unsolved in first-generation public systems.

With the introduction of the first public LMR systems, the demand was only partly solved. In the largest urban agglomerations, the capacity of public networks was rapidly exhausted. Very soon modifications of the cellular structure (sectorial splitting) were indispensable (e.g., in the USA and Sweden). More radical has been the Japanese approach: in the new version of the public system, channels have been halved (12.5 kHz wide instead of 25 kHz wide). Loss of transmission quality has to be compensated by automatic selection of intracell channels.

Many other problems of public radiotelephony remain to be solved. Energy-saving portable stations have to be introduced for personal communication. On the other hand, LMR systems have to absorb various nonvoice services, now commonplace in stationary telecommunication with digital transmission. Next-generation public LMR systems will follow this trend.

2.4 Second-Generation Pan-European System

The first-generation cellular mobile systems are operating under various standards and lack international roaming. The European Community countries, preparing for their union in 1992, decided to develop and to introduce simultaneously a Pan-European cellular public system. The system, groupe speciale mobile

(GSM), is expected to remove the deficiencies of the present systems and to ensure novel services and facilities (Kittel 1988).

Enhanced potentialities of the GSM system originate mainly from the digital transmission of voice and nonvoice waveforms, combined with diversified digital processing of control and other useful signals. The functional architecture of the system (see Fig. 4) is rather conventional, with extra registers of subscribers, necessary for long distances when national networks are interconnected.

By 1989, the GSM system standard had been completed (c. 4000 pages) and verified by laboratory and field experiments aided by computer simulation. Emphasis has been laid on strict standardization of intracell radio channels; the adopted eightfold time-multiplexed channel of 200 kHz width corresponds to eight standard analog channels.

Source coding of speech begins with the conventional pulse coded modulation (PCM) digitizing: 8000 samples per second, eight bits each, permit transparent interconnection with stationary digital networks. Linear-prediction vocoders reduce the transmission rate to 13 kbytes per second, but robust channel coding increases it again to around 35 kbytes per second per speech signal. The resulting speech quality, under worst radio-propagation disturbances, amounts to 3.5 mean-opinion score (MOS) measured subjectively by a trained team on the 5-point scale—vs 2.0 MOS in first-generation analog systems.

Digital transmission resists cochannel interference; thus, the reuse distance D and the cluster size N can be diminished with a large increase in network capacity and efficiency (see Sect. 1.3). Presence of active speech during a conversation in the GSM system is detected and serves for introducing discontinuous radio transmission. The energy saving obtained decreases the size and weight of handheld MSs.

Exploiting the experience in military LMR, the GSM will ensure confidential transmission of information and protection of a subscriber's identity. Besides, to achieve protection against fraudulent use of the MS, a mandatory authentication procedure utilizing an intelligent, credit-card-like module is provided for.

The most important advantage of digital technology in the GSM system will be the direct interconnection to the IDN and ISDN stationary networks. (IDN is a stationary network with digital speech transmission, and ISDN is a network with integration of digital voice and nonvoice services.) Following the precedent in the Mobitex network, various text and data terminals will be included in some MSs for novel value-added services. Establishing user-specific subsystems will lead to the absorption of some larger dispatch systems.

Many innovations in the GSM system rely directly on digital techniques, partly exceeding the achievements in stationary public networks and in military radio systems. Novel GSM solutions in the inherently difficult wireless medium offer not only quality at the cost of complexity, but also two other unprecedented merits: immense size and versatility. Also, a step towards personal communication will be made.

3. Future Trends

3.1 Microcellular Networks

Mobile communication is growing very rapidly and concepts of new services are proliferating. In order to reduce the complexity of cellular LMR systems, simpler networks with microcellular structure have been submitted.

First, there is the telepoint, a wireless equivalent of public payphones with one-way access. The user of a telepoint set will be able to make calls from anywhere within the range of a BS (c. 50 m in cities and up to 200 m over open ground). Thus, a dense array of microcells has to cover the network area (e.g., 500 BSs are planned for Greater London). For this purpose, 40 uhf channels within a 4 MHz band have been allocated in the UK, with transmitter power limited to 10 mW. The radio interface has to comply with the cordless telephony standard CT2. (Wireless connection of the handset with the telephone set facilitates short-range mobility within housing, gardens and so on.)

In CT2, the digital channels are separated in frequency by frequency difference multiplexing (FDM). Speech to and from the user is accommodated on the same channel thanks to time-division duplexing. Each 2 ms period of digital speech is packetized and transmitted in just under 1 ms. The gaps between packets are used for transmission in the opposite direction and the spectral efficiency increases. The expensive duplexer filter in the conventional MS is replaced by a simple synchronous switch. The set of trunked 40 channels is dynamically managed: at the beginning, the MS scans all channels and chooses the most quiet one. In the case of interference, interchannel handover is automatic.

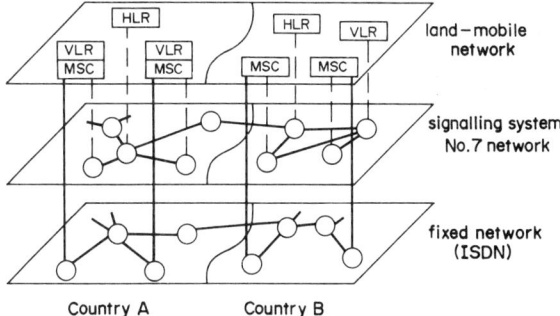

Figure 4
Architecture of a land mobile system showing the home location register (HLR), the visitor location register (VLR) and the mobile-service switching center (MSC)

In the telepoint system, the user cannot use his set on the move and cannot accept incoming calls. To overcome this drawback, a paging receiver could be incorporated into a telepoint MS (see *Broadcasting Communication Systems*).

Regular two-way communication in a microcellular network is also feasible and some concepts (beyond the scope of this article) have been published (Institute of Electrical and Electronics Engineers 1988). Local wireless communication in office buildings over factory terrains, construction sites, and so on is sought after, with a mixture of dispatch and general-type traffic. Since microwaves over 1 GHz are contemplated, large building attenuation has to be allowed for. Standardization of such systems is not very crucial and has not, to date, been accomplished.

3.2 Integration of Services

The main use of the existing LMR systems (except Mobitex) is speech communication. The ever increasing role of text and data has imposed the integration of services in stationary ISDN networks; it is a natural progression to extend this to the mobile communication field.

First-generation analog cellular systems are capable of providing simplified data links with modems in voiceband channels. In second-generation systems, the digital channels will absorb data transmission as an integral part of the system. It should be noted that large-area tactical military radio systems (e.g., Ptarmigan in the UK) have been service integrated since the late 1980s.

The ISDN stationary networks, inaugurated in 1988–1989 in several countries, are based on detailed standards, issued by the Comité Consulta International des Télécommunications (CCITT); the basic channel of 144 kbytes per second supports two speech streams and one data stream. Signalling system no. 7 facilitates future inclusion of mobile services.

Two categories of nonvoice services are envisaged for future LMR networks interworking with ISDN: the teleservices (teletext, fascimile, short text messages) and the bearer services (asynchronous and synchronous data transmission). In the GSM system, for instance, asynchronous and synchronous circuit-switched data up to 9.6 kbytes per second will be transmitted from and to ISDN through 12-kbytes-per-second radio channels. Also, packet-switched data with the same maximum speed will be transmitted from and to dedicated data networks, with rate adaptation if necessary.

The radiocommunication and the ISDN concepts have to merge in the GSM mobile station. The user's access to different services requires different terminals and a set of different MS configurations. The core of the MS handles speech communication and teleservices. A non-ISDN termination will be needed for nonintegrated data transmission over modems. An ISDN termination will support bearer services, with rate adaptation to cope with the higher speed in the stationary ISDN channels.

A large library of data communication protocols is necessary to ensure access to the various services and proper processing of data. The ever increasing role of software is characteristic for nonvoice communications.

3.3 Mobile Satellite Communications

The existing synchronous-orbit Inmarsat satellites satisfy the radiotermination needs in global sea and air navigation; they also establish maritime communication for oceanic routes. Currently, some cellular systems reach offshore and compete for short-range fixed and mobile communication.

A new generation of powerful Inmarsat satellites, with digital 16-kbytes-per-second transmission, could also complement and extend mobile communication to remote sparsely populated areas, not covered by cellular networks. There are, however, indications that several countries are considering introducing "anticellular" mobile satellite systems as a nationwide service in the 1990s (Institute of Electrical and Electronics Engineers 1988). For the time being, an L-band of 1.5–1.6 GHz has been allocated for this purpose.

The economy of land-mobile satellite service depends on solving some fundamental problems:

(a) satellite and fixed earth equipment has to be shared with other systems,

(b) compact and cost-efficient mobile terminals need to be developed, and

(c) compatibility with other LMR services should be maintained.

Research in this field is actively conducted in the USA, Canada and Japan with the objective of expanding mobile communication to rural, unpopulated and coastal areas.

In Japan, the experimental mobile satellite system is equipped with the multibeam test satellite V. Microwave propagation research is underway, as well as transmission and application experiments. In particular, digital voice communication, paging and low-speed nonvoice services are studied. Three types of nonvoice communications are employed: real time, store and forward (see Sect. 2.2) and message box (storing messages for mobile terminals at the satellite earth station).

The radio link (of *c*. 36 000 km length) is organized along the single-channel-per-carrier scheme, with in-channel control signalling. Robust channel coding counteracts severe signal fluctuations in mobile communication. Total digital transmission speed amounts to 330 kbytes per second.

Research in the USA and Canada has been summarized at the mobile satellite conference in Pasadena California, 1988. Emphasis was laid on versatile satellite system usage for dispatch as well as on a public radiotelephone service (paging excluded). Because of different traffic characteristics in these services, diversified operation procedures were proposed (queuing of dispatch messages belonging to the same system).

With respect to the spectrum scarcity, frequency reuse between the satellite antenna beams and short channel setup procedures will be introduced. The control center for the mobile satellite system will be equipped with a dedicated signalling channel, for call processing and monitoring, and for network administration (management and billing).

Satellite systems will also be of interest for developing countries. Absence of an established road and telecommunications infrastructure can be compensated for by temporary or stationary, fixed and mobile satellite links. They will also be cost effective for road and rail transportation.

Acknowledgement

It is with deep regret that we announce the death, on 15 September 1989, of the author Professor A. H. Wojnar. The contributions of Professor Dr H. S. Hahn and Dr R. Pelka in completing the manuscript and checking the proofs are acknowledged.

See also: Data Processing in Air Traffic Control; Mobile Communication

Bibliography

Bear D 1976 *Principles of Telecommunication-Traffic Engineering*. Peregrinus, Stevenage, UK
Comité Consultatif International des Radiocommunications 1986a General aspects of cellular systems. Report No. 740, Vol. 8
Comité Consultatif International des Radiocommunications 1986b Multi-channel mobile systems for dispatch traffic. Report No. 741, Vol. 8, pp. 110–24
Comité Consultatif International des Radiocommunications 1986c General principles of public land mobile telephone systems. Report No. 742, Vol. 8, pp. 127–48
Institute of Electrical and Electronics Engineers 1988 *Proc. Vehicular Technology Conference*. IEEE, New York
Kittel L (ed.) 1988 *Proc. Digital Cellular Radio Conference*. Hagen, FRG (papers 1a–7c)
Lee W C Y 1988 *Mobile Cellular Telecommunications System*. ITT Defense Communications Division, Nutley, NJ
Wojnar A H 1988 Analysis and synthesis of cellular structures for mobile radio systems. *Proc. EEE European Conference EUROCON 88*. Stockholm, pp. 354–7

A. H. Wojnar ✠
[Warsaw Academy of Technology,
Warsaw, Poland]

Container Terminal Management

In the field of integrated transport, particular attention is paid to the development and use of information and telecommunication technologies, both for management of intermodal terminals (networks and other facilities) and for organization of the services supplied. For the management systems of intermodal integrated transport, the key issue concerns the integration of information flows related to harbor activities; in fact, most transport modes and systems converge onto the harbor facilities. In harbor systems, the container terminals, representing the main connection between sea carriage and road and railroad transport, are playing an ever-increasing role. This has been confirmed in recent years by the remarkable developments of harbor facilities (in terms of size and complexity) and of the management of the volume of container transit and related management and monitoring operations. On these grounds, the use and the enhancement of information systems and automation in the whole of the terminal operating system, and for connection with other existing systems, has become increasingly important.

1. Basic Systems Category

When container terminal systems were first realized, information technology applications were limited to the traditional electronic data processing (EDP) management procedures. All problems relating to on-line control of the container handling cycle remained open because of technological reasons and costs. From the mid-1970s, however, when information and automation technologies were already offering good opportunities and their use was spreading (thanks to the drop in their costs), new architectures of container terminal management systems integrated with on-line management of the handling operation cycle were developed.

A rough classification of the existing systems, which are nearly always different from one another in terms of the integration level attained and the embodied technology level, may be:

(a) systems making use of information technology for the recording (tracking) of containers and related information, with the purpose of following the container throughout the terminal operating cycle ("passive" systems); and

(b) on-line control systems for the whole container operating cycle, from the expected arrival of the container to its leaving the terminal, in order to optimize the process effectiveness ("active" systems).

Past experiences have shown that the introduction of information and automation technologies in the field of container terminal management (just as in any production process) causes problems in organization and in correctly weighing the necessary investments. For these reasons, the analysis and definition of the requirements of a system for container terminal management must be properly integrated by effective planning, action scheduling and defining of a suitable organizational model.

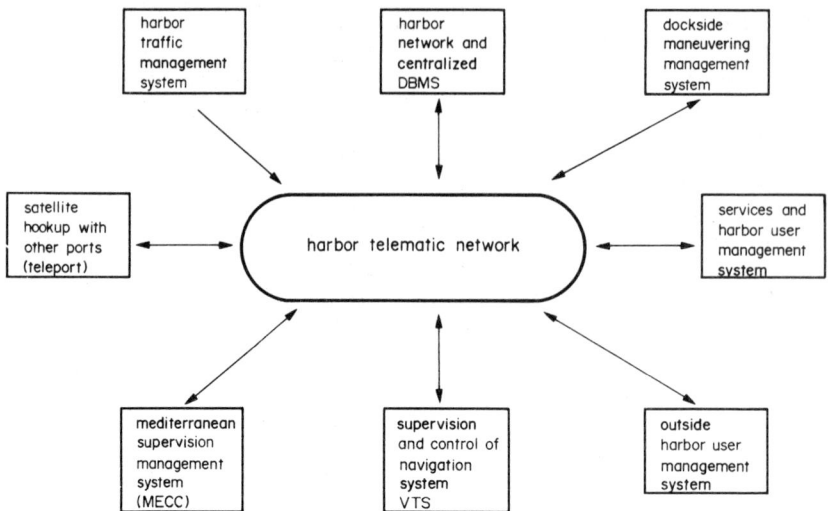

Figure 1
Harbor information system layout

2. General Objectives

The introduction of information systems for container terminal management aims to:

(a) improve the services offered to the terminal users through the rationalization of all processes of production, information, documentation and container-handling management;

(b) improve operational reliability by providing more flexibility with regard to changing demand;

(c) reduce operational costs and increase economic efficiency; and

(d) integrate terminal information with other systems in the field of goods transport, particularly sea traffic control systems (see *Vessel Traffic Services: Management from Shore*).

This last goal derives from a unitary view of a harbor system, intended as a multimodular system with reciprocal integration by a telecommunication network for data and information exchange so as to manage the transport cycle in its entirety.

Some actions have been taken to establish single modules of this network, such as the capability of information exchange among different harbors, the monitoring of sea traffic (e.g., EC projects such as EVHA, COST 301, EASI), and the integration of individual and specific information systems (e.g., projects GINA, ADEMAR, SOFI).

Figure 1 shows the logical scheme of a possible harbor telecommunications network, where each block represents a given information system (the container terminal is one module of the harbor system).

3. Current Systems

The information management systems that were installed during the 1980s in the major harbors of the world can be considered as belonging mainly to the passive category. Such realizations are based mainly on computerized management procedures whose major features may be summarized as

(a) the collection of a large mass of data from operational points,

(b) the management of large databases,

(c) the provision of graphic interfaces as decision-making tools for operators, and

(d) the automated management and editing of documents.

From an operational viewpoint, once the system has acquired the necessary information at the starting point, it will "follow" the path of each container and collect data directly at the operating nodes. Data collection is usually performed by means of lengthy data entry operations at the operating cycle node and, in some cases, in parallel with the carrying out of the container operations.

The main goals that may be attained through a container terminal management system of the passive type are:

(a) rationalization of operating procedures and related information flows;

(b) "soft" reorganization of the operating environments by introducing information technology;

(c) support to the operators, in terms of control and management of the system faults; and

(d) rationalization of administrative document editing.

3.1 Architecture and Functions

The general architecture of a passive-type container management system is organized on only one star-shaped information level: a central mainframe is able to manage the main database and the communications network through terminals distributed at various operating nodes of the system for input–output operations and control messages.

Figure 2 shows schematically the main functions of this kind of system. The characteristics that lead to functional deficiencies of this system are as follows:

(a) lack of feedback from terminal operating areas (the system is not reactive); and

(b) relative poorness of programming procedures, tending only to organize the terminal work and to control the final results collected by the operators.

3.2 Advantages and Disadvantages

The advantages of a passive-type system may be summarized as follows:

(a) use of less sophisticated but widely tested technologies;

(b) "soft" organizational impact through gradual conversion processes;

(c) elimination and/or rationalization of manual editing; and

(d) high flexibility in system management, especially in the case of unexpected events.

The disadvantages of the system are

(a) low reliability of data collection,

(b) low degree of decision-making autonomy,

(c) no real-time management of production resources (e.g., means, stocking yards), and

(d) critical connections with other systems inside and outside the same harbor.

4. Advanced Terminal Management Systems

The convergence of computing (on-line data collection tools) and telecommunication (integration of off-line systems) technologies has permitted the establishment of more sophisticated system architectures, showing higher levels of automation. This has allowed the achievement of on-line control of the whole operational cycle of container handling (an active system), as well as optimization of terminal management and production activities.

The technological developments have been quite rapid in the following areas directly concerning container terminal management systems:

(a) hardware/software through multilevel architectures, distributed databases, human–machine interfaces, etc.;

(b) communication systems through interconnected local networks, worldwide transmission, etc.; and

(c) very sophisticated programmable logic controller (PLC) automation systems, sensors, actuators, identification systems, etc.

The availability of this technology and, more importantly, the poor improvement in production effectiveness obtained by implementing passive systems has brought about many attempts to change to active-type systems. This change is still being made in many harbors and is being developed gradually, with particular care not to upset the current organization frames.

Starting from the passive system configuration and functions and the related actual organization, more and more advanced automation levels are being introduced in order to

(a) speed up container handling (mainly on board);

(b) reduce operating costs, through rationalization of resources and reorganization of all auxiliary activities to support the exploitation of resources; and

(c) increase real-time data exchange among the different users.

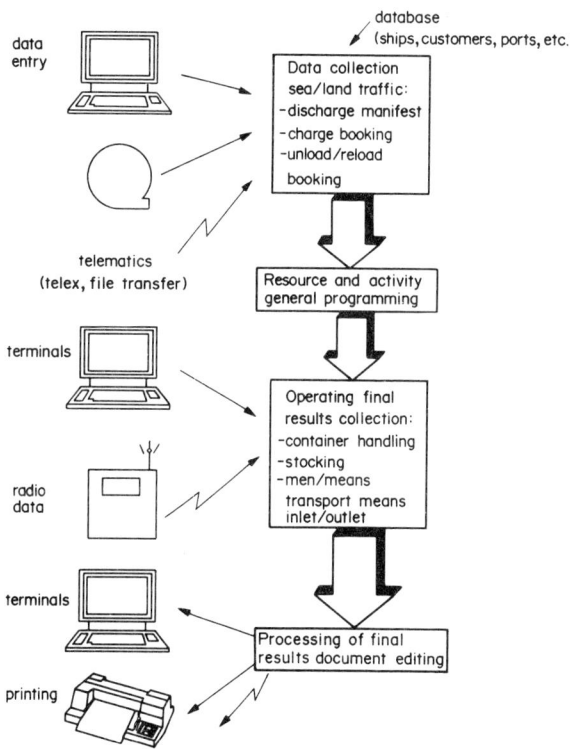

Figure 2
Passive system functions

These characteristics make terminal management systems true on-line control systems, enabling them to perform feedback actions on the whole handling cycle. The term active, besides describing this capability, stresses the dynamic and modular features of this kind of system.

4.1 Architecture and Functions

The architecture of an active system is generally organized on at least three levels, each one being intended to manage distinct functions.

(a) *Upper level.* This is the data processing center (the heart of the passive system), which performs the traditional EDP functions (e.g., administration, invoicing, data registration), as well as the operating ones, such as

(i) traffic data collection,

(ii) activity and resource planning,

(iii) resource management,

(iv) maintenance,

(v) arrival and departure management, and

(vi) database management.

From the functional point of view, the activity of ship planning might be associated with this level, in spite of the fact that it requires the use of dedicated workstations.

(b) *Intermediate level.* This is, in practice, a dedicated subsystem to manage single activities at terminal operating nodes. Based on working program outputs from the upper level, this subsystem performs the following main functions:

(i) working instructions processing and transmission to the various operating nodes;

(ii) interfacing with other possible automation systems at the operating nodes;

(iii) control and supervision of the whole operating management system (e.g., stocking yards, means);

(iv) fault diagnosis and advisory in real time; and

(v) processing of documents and final results transmission.

(c) *Local level.* Possible automation subsystems for implementation and control of some peculiar activities are located on this level, such as

(i) automotive drive of the lifting facilites,

(ii) automatic identification of yard positions,

(iii) container automatic positioning and lifting,

(iv) automatic identification of containers and transport means,

(v) survey of transport means inlet and run, and

(vi) automatic weighing.

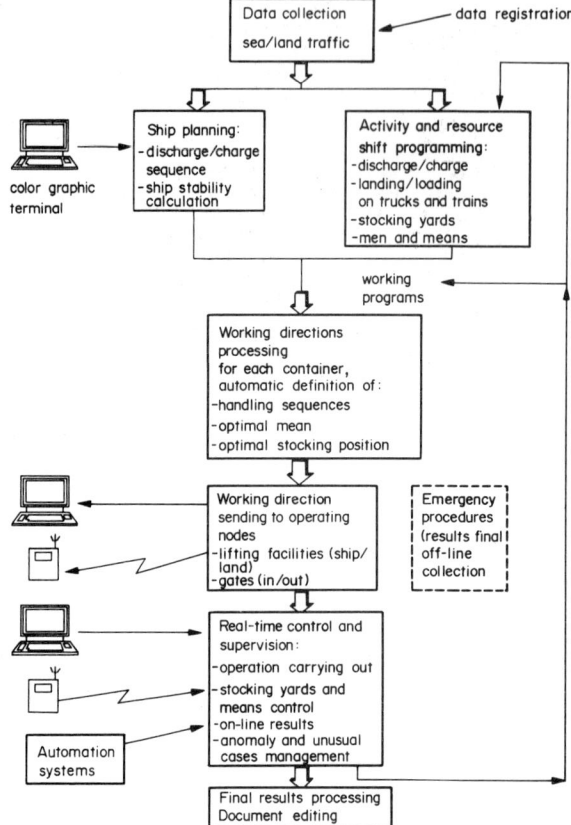

Figure 3
Active system functions

Figure 3 shows schematically the main functions of an active system. As can be seen, the operating management subsystem provides for real-time control of the activities programmed at the upper level. Through real-time feedback from the operating field, this subsystem enables the transfer to the upper level of information on the development of the phases of operation and facilitates immediate action to remedy possible faults.

4.2 Advantages and Disadvantages

Active systems give the following advantages:

(a) improvements in production effectiveness,

(b) optimization of all resources involved in the operating cycle (e.g., areas, means, time),

(c) increase of operations control capability,

(d) increase of reliability of information/data exchange, and

(e) capability of communicating with other systems.

The disadvantages of these systems are

(a) higher investments, because of a more sophisticated technology;
(b) lower global flexibility, owing to more detailed and sophisticated programming;
(c) stronger organizational impact, due also to the consequent need of personal skill;
(d) higher requirements in terms of whole system reliability; and
(e) higher needs of operation supervision and emergency procedures.

5. Future Developments

From the point of view of the services supplied, the main efforts are aimed at increasing the efficiency of

(a) ship operations (ship-to-shore and shore-to-ship container handling) as regards programming each activity in advance, advanced automation of embarking and landing phases, real-time control and planning of available resources (e.g., stocking yards, means);
(b) land operations (container loading and unloading on trucks and trains) as regards programming and control of terminal access gates, rationalization of stocking yards and internal handling, thus eliminating "useless handling" as much as possible; and
(c) terminal working organization by means of on-line procedures and tools for container yard inventory control, and real-time management of handling means.

When considering the terminal as a node of an intermodal and integrated transport system, the main effort is aimed at developing a telecommunications network between the various active parties, such as shipping agents, customs offices, trucking companies, and sea and harbor services. The principal problem is to do with cargo data transmission between terminals involved in set trade routes in order to obtain storage plans well in advance of the arrival of the ship, so that terminal planning may be made easier.

In reference to the application of information and telecommunications technologies to container handling operations, there is a need for further research in different fields of container management systems. The areas for potential change include

(a) scanning the container and making a record of its identification code (the best known is the barcode, the potential use may be by video camera or transponder);
(b) using simulation tools for evaluating container yard layout, operations models, crane analysis and allocation, and resource planning; and
(c) using advanced data transmission systems for displaying operational information in the control stations and the results of the operation itself.

See also: Marine Fleet Planning and Scheduling

Bibliography

Bianco L, La Bella A (eds.) 1988 *Freight Transport Planning and Logistics*. Springer, Berlin

Bonsall P, Bell M 1987 *Information Technology Applications in Transport*, Topics in Transportation Series. VNU, Utrech, Netherlands

Daskin M S 1985 Logistics: An overview of the state of the art and perspectives on future research. *Transp. Res. A.* **19** (5/6), 383–98

Ettorre J J 1987 New intermodalism: Joint ventures or megacarries. *Handling Shipp. Manage,* **28** (4), 22–6

Guiducci G, Ambrosino G 1990 Container terminal management systems: Problems and operational experiences. *Proc. ATS'90 Automation in Transportation Systems*, Trieste, pp. 666–78

Institute of Management Services 1985 New Computer applications in ports and the maritime industry. *Proc. Annual Conf. Shipping and Port Operations*. IMS, Rotterdam

G. Ambrosino and G. Guiducci
[Automa, Genova, Italy]

Corridor Control Systems

Growth in urban traffic demand, especially during the peak commuter periods, has caused overloading of high-speed, high-capacity freeways. This has caused flow breakdown, where the benefits of high-level designs are lost and speeds are reduced to a "crawl". In an attempt to alleviate this, freeway operating agencies have effectively implemented a reduction in demand on the freeways by metering access at on-ramps. This has resulted in improved freeway flow to the detriment of flow on parallel surface streets. There is clearly a need for integrated management of the freeways and surface roads in a corridor. Since these components interact, their controlling authorities should exchange information and, preferably, cooperate in their operations to provide a better overall system, instead of attempting to operate independently of one another. However, the overall effects of ramp controls are seldom even evaluated, so that system optimization is rarely even considered.

Transportation corridors are characterized by essentially one-dimensional networks. A corridor often develops between two simply linked traffic generators as additional links are forged between them. A corridor may exist to link a major generator and another major or satellite generator. Alternatively, a corridor may just evolve as a central generator grows in a single predominant direction due to geographic constraints such as water, mountains or environmentally sensitive lands. It may also develop as land-use activities move in to take advantage of a major interurban highway.

For the purposes of traffic control, a corridor is defined as a longitudinal network consisting of at least one freeway plus one or more viable and essentially parallel alternative routes, such as frontage or arterial roads. To qualify as a corridor within the context of

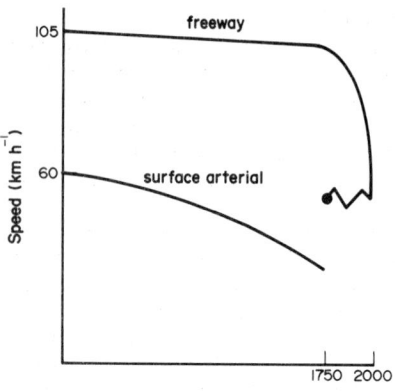

Figure 1
Typical speed–flow relationships for freeway and surface arterials

traffic control, a significant number of trips must have more than one major alternative route or else there will be only independent deterministic links. In addition, where corridor traffic controls are appropriate, there is usually quite frequent access between the parallel links.

1. Freeways

High-speed multilane freeways originally provided a newfound opportunity to serve high volumes of traffic at high-service levels. They later became victims of their own success, attracting even higher volumes of traffic, until demand exceeded freeway capacity, especially on urban freeways and during the peak commuter periods (see *Freeway*).

1.1 Freeway Characteristics

Their high-level design and controlled access allows freeways to carry more traffic at higher speeds than similar numbers of lanes of surface streets. In addition, freeway speeds are less sensitive to traffic demands than speeds on other highways for most of their range of flows. This is illustrated in Fig. 1, which compares the speed–flow relationship per lane of freeway to that per lane of a stretch of urban arterial roadway between traffic signals. Since surface arterials are generally controlled by traffic signals, each freeway lane can carry about the same volume of traffic as a signalized arterial with two lanes in one direction. Doubling the number of lanes approximately doubles the capacity of the arterial and, in addition to this, there are usually flanges at the signalized intersections which provide temporary additional lane(s) for turning maneuvers. The combined effects of these additional lanes and flanges tend to balance the movement time lost due to the signal, where the effective green light time is typically less than half of the cycle time. Therefore, the surface arterial curve of Fig. 1 can be considered to represent the speed–volume relationship for two lanes of signalized surface arterial in one direction having a reasonably good progression of offsets between adjacent signals.

1.2 The Case for Freeway Controls

The robust high-speed freeway speeds shown in Fig. 1 provided a false sense of security. In cases where urban freeways were not yet congested in the peak periods, they generally attracted sufficient land use to increase their volumes up to the range where speeds do drop precipitously and flow breakdowns occur. While freeways provide excellent quality of service in terms of speed and safety under free-flow conditions, the service level drops in a number of ways under queuing conditions. Speeds are much lower and the forced flow that occurs after breakdown in flow results in higher accident rates, often due to the sudden transition from upstream free flow to the lower speed and higher density of traffic immediately downstream. In addition to this, there are relatively few opportunities to escape to alternative routes under the limited access conditions. This is especially serious when incidents reduce the capacity and high volumes of traffic are trapped on the freeway.

After it was identified that freeways have the volatile characteristic shown in Fig. 1, it was generally resolved that it is better to operate in the free-flow regime with its relatively higher quality of service, than risk flow breakdown by allowing flows to reach the ultimate capacity of the freeway.

1.3 History of Freeway Controls

Controls on freeway volumes started in the early 1960s as isolated closures and metering of individual onramps. Pioneering freeway control systems were implemented in Chicago and Detroit (May 1963). While the freeways clearly operated better, there was a negative impact on the users who were metered off and, indeed, on the quality of service for all users of the surface routes onto which the metered users were dumped (May 1963). While these surface arterials do not have the precipitous drop in service that the freeways do, they are quite sensitive to increased flows from the outset, and especially to the relatively large volumes that freeway controls can force onto them.

1.4 Effects of Restricting Freeway Access

The controlling of freeway access signalled the beginning of involuntary route changes caused by the denial of access to large numbers of drivers at selected key locations. Unlike the equilibria that had evolved from totally free user path selections, planned major capacity reductions were now implemented with a view towards improving the global picture at the expense of relatively few users. It became axiomatic that it was desirable to restrict freeway flows to the free-flow range. However, this simple axiom does not, by itself, provide any alternatives, as it does not suggest which flows should be metered off the freeway and where they should be sent. It is not even clear that metering is always, if ever, a good thing.

Benefits and disadvantages of metering freeway ramps are discussed at some length in Yagar (1989). The benefits include

(a) minimizing an aggregate objective function such as total travel time to all users,

(b) efficient use of the total capacity in a corridor,

(c) discouraging paths with high societal costs,

(d) introducing order into the path selection jockeying of commuters,

(e) reducing the variance in trip times in a corridor, and

(f) public acceptance.

Disadvantages include

(a) encouraging longer trips,

(b) the transfer of land values,

(c) favoring through traffic over the metered local traffic,

(d) tampering with an evolved status quo, and

(e) the cost of operating the system.

1.5 The Case for Corridor-Wide Considerations

Successful planning should incorporate as global an evaluation as possible of the likely effects of freeway controls. Therefore, the evaluation should be expanded to include the effects of controls on other parts of the network. As a practical compromise between ignoring the concerns of the many drivers whose routes change in response to freeway controls and expanding the problem to an unmanageable size, control of urban freeways is expanding to the corridor level. This article deals with the effects of rerouting, which distinguishes corridor management from pure freeway control. While the latter is not able to evaluate the impacts caused by rerouting, it has received most of the attention to date, perhaps, because it is straightforward to model. Thus, it is still widely applied.

1.6 Freeway Models

In order to predict the performance of a system under various operating conditions and to possibly optimize its operation, a representative model of the system is required. FREQ (FREeway Queuing), the most celebrated of the freeway models, provides a macroscopic representation of the spillback effects of queues at various locations on a stretch of freeway as a function of the volumes that enter and leave at each on- and off-ramp. FREQ has had numerous applications and has become the representative state-of-the-art model. Gibson (1981) listed a number of other freeway models, as well as some FREQ derivatives along with their general characteristics. These models can test the local effects of various sensor locations, ramp designs and so on. Some models, particularly the derivatives of FREQ, evaluate the effects of high occupancy vehicle (HOV) lanes or ramp metering schemes, and even optimize according to various criteria, for given ramp origin–destination (OD) flow patterns. However, they are not able to evaluate the effects of altered flow patterns on the surface street subsystem.

Despite this, FREQ has often been confused as being a full corridor model, with the capability of representing the reallocation of trips that are metered off the freeway, perhaps because it accepts a set of ramp OD patterns. The additional information provided by these ramp OD patterns, compared to simple ramp volumes, does not enhance analysis by FREQ significantly and the term OD pattern has served to fuel the misapprehension that FREQ can evaluate the true corridor-wide impacts of freeway ramp controls.

The practical state of the art, as defined by the overwhelming majority of applications, is still freeway-oriented optimization, with special and ad hoc considerations for unacceptable impacts, such as queue spillbacks from metering stations to signalized intersections, rather than general corridor optimization. For example, the objective function on the Hanshin Expressway (Inoue et al. 1988) in Japan is to maximize the number of vehicles allowed onto the freeway. There are two interesting aspects to this objective function.

(a) It is more an economic- than a traffic-based optimization, as it maximizes the number of access fares collected.

(b) It is stated in terms of indirect surface street benefit, virtually to the exclusion of freeway concerns, as it maximizes the number of vehicles taken off the surface streets, essentially transferring the storage of queuing and congestion onto the freeway.

While the objective of the Hanshin control algorithm seems to be the antithesis of those that would optimize quality of service on the freeway, the common characteristic of both types of algorithm is that they are suboptimizations evaluating only on the freeway and ignoring the critical driver's path selection process.

2. Corridor Considerations

Corridor-wide implications of freeway controls are generally quite complex. Since each driver has an individual OD pattern, the impacts of freeway ramp controls on traffic demands at other corridor locations are not easy to predict. Since freeway controls affect large numbers of vehicles, whose effects interact and are superimposed in the corridor network, predicting the demand shifts can be quite difficult, requiring the assistance of models. This is perhaps the reason why analyses and optimizations have emphasized the freeway, usually to the exclusion of delays and other effects on surface streets. Clearly, the various assorted benefits and disadvantages of freeway ramp metering discussed in Sect. 1.4 are difficult to quantify. Nevertheless, there are some quantifiable effects such as queuing and travel time on surface arterials. These are of interest to traffic operations managers and can aid policy makers. However, they add an additional dimension to prediction models, as they introduce the elements of predicting trip paths under conditions of queuing and congestion to what were previously much simpler models. Since path changes occur in large numbers when freeway ramps are controlled, they must be considered when evaluating alternative freeway control schemes.

With daily recurring congestion due to excessive but predictable demands, the drivers' path selections can be considered to converge to a predictable equilibrium for any given control pattern, as the number of vehicles is large enough to mask random individual driver decisions. In addition, while the flow patterns vary with the time of day, they are generally treated as repeating from day to day. These can, therefore, be modelled for testing control strategies for recurrent incident-free conditions. Nonrecurrent major incidents are more complex, and usually involve detection and dynamic decision making.

2.1 Incident-Free Conditions

Development and testing of integrated strategies for the management of freeway–arterial corridors has historically been based on calculations, approximations and expert intuition. Since the end of the 1960s, efforts have been devoted to the development of models that would evaluate schemes such as sets of ramp metering rates provided by the traffic engineer and, in some cases, even recommend metering rates if given the freedom to do this by the engineer. This latter feature was accomplished by the CORQ (CORridor Queuing) model (Yagar 1975), using an inherent iterative bootstrapping technique to combine the features of traffic assignment and corridor queuing.

CORQ is a traffic-assignment-based model (see *Traffic Assignment*). The traffic assignment capability allows CORQ to vary traffic flows on competing routes as their relative speeds and travel times change. CORQ models the queuing that occurs on congested roads and takes this into account when routing trip demands through a corridor. This allows it to predict the buildup and dissipation of queues in a network as the demands and capacities vary with time, such as in a peak period. CORQ has been applied in testing strategies in Ottawa and Toronto.

Some of the key elements that CORQ has had to represent in bringing traffic assignment and queuing into an operational model are (Yagar 1976):

(a) the cost of queuing on appropriate links;
(b) queue spillback; and
(c) the respective capacities of merging links whose capacities depend on one another's flows and queues, which are yet to be determined by the traffic assignment procedure.

A study by Van Aerde *et al.* (1987) of potential models for evaluating the net effects of peak period corridor controls found that two models have illustrated their ability to predict the evolution of time-varying flows and queues in a network: CORQ and CONTRAM (Leonard and Gower 1982). Although CORQ was found to be robust to alternative strategies, predicting flows and queues that would result as drivers responded to these measures, it was not very user friendly. CONTRAM also has the capability to predict flows and queues for time-varying situations, but must assign and reassign the demands in small packets of typically five vehicles, and does not have the capability to treat freeways. A number of other models were found to consider freeways, but these are not sensitive to the dynamic queue buildup and dissipation which occurs in peak urban commuting periods.

CORQ has since been overhauled to greatly reduce its input data requirements and to improve its output leading to a simplification of user effort. CORQ had required the user to obtain complete OD tables. This, in turn, required sampling of drivers' travel patterns throughout the network, and a careful synthesis and calibration of a sequence of OD matrices. The new version, CORQ2, allows the user to merely feed in the flows and queues on the network links for a series of time slices, each typically 15 min in duration, and supplement this with selective trip routing information (Yagar 1988). The latter is obtained at critical times and locations, from strategically selected drivers, whose trip paths might change significantly in light of any traffic management schemes that are to be considered. The CORQ2 preprocessor creates the OD tables automatically from this basic information.

While CORQ required that demand–travel time relationships be represented in terms of piecewise-linear components, CORQ2 derives appropriate continuous relationships from simple basic information; that is, free speed, link capacity and type of link.

CORQ2 also has a look-ahead feature. CORQ had previously assumed that a driver knows the congestion levels in a network at all times, but does not know the future states of the network (i.e., cannot look ahead in time). However, a regular commuter's knowledge of the downstream characteristics of a corridor relate more closely to the time of day when that commuter would be at the downstream locations(s) than to the present time when the path selection must be made. CORQ2 incorporates this "preknowledge" into its representation of drivers' path selections.

2.2 Incident Conditions

When an incident occurs the time-of-day plan gives way to an incident response, and all of the capacity of the corridor should be employed in order to keep the overall queue buildup as small as possible. Since drivers' preknowledge of typical nonincident conditions is of little, if any, use here, communication is of paramount importance. The sequence of events is detection followed by response.

(a) Incident detection. Methods that have been used to report incidents in the past include reports by the motoring public (call boxes, CB radios), surveillance vehicles and cameras, and traffic detectors (see *Incident Detection*).

The modern method employs traffic presence detectors embedded in the roadway, one in each lane. These can provide vehicle counts and traffic occupancy or density a number of times per second, from which traffic conditions can be deduced and compared to normal conditions. A second set of detectors placed about 10 m downstream can provide vehicle speeds as well. By repeating this detection system at key points, usually before and after interchanges, and about 500 m apart on long stretches between interchanges, it is possible to obtain continuous information on the state of the freeway system.

When an incident occurs, the traffic conditions upstream and downstream of the incident are quite different. It typically takes 1 min for a detection algorithm to identify and confirm this. The detectors are often supported by a set of overhead cameras, about 1 km apart, which can scan and zoom in on the reported incident area to pinpoint the location and the type of problem. Video imaging techniques are being developed, whereby an overhead camera provides pixels of information that are processed to indicate the presence of vehicles. While this type of system could ultimately replace the present detector and surveillance camera system, it is still in its infancy (see *Video Sensors*).

(b) Response. This consists of removing the incident and implementing appropriate traffic management measures. The former involves dispatching the appropriate response team, which might consist of police, ambulance, firefighters, tow truck and maintenance crews, and so on. The latter involves selection of the appropriate traffic control plan, which will depend on the severity of the incident, the time of day and other traffic conditions. Once the optimum response has been selected it must be implemented. This usually involves rerouting of drivers either by giving them information relating to the incident and alternative routes or by controlling access at key points. Information is usually provided by radio or changeable message signs. Control is exercised by signals or barriers, such as the commands of a police officer.

A number of in-vehicle routing systems are being developed, notably in London, Berlin and Tokyo (see *Route Guidance, Collective*). These will have the capability of monitoring traffic in a whole urban area and providing information and advice to drivers. The state of the art in these systems is summarized by French (1988). Van Aerde and Yagar (1988) have described a technique for modelling in-vehicle route guidance systems.

See also: Freeway Network Control; On-Ramp Control: Coordinated Traffic-Responsive Strategies; Signal Control and Traffic Assignment

Bibliography

French R 1988 Improvement of urban traffic conditions by vehicular navigation and route guidance. In: Yagar S (ed.) *Management and Control of Urban Traffic Systems.* Engineering Foundation Press, New York, pp. 327–36

Gibson D 1981 Available computer models for traffic operations analysis. *Transp. Res. Board. Spec. Rep.* **194**, 12–22

Inoue N, Toshiharu H, Takishi M 1988 Traffic control system on the Hanshin Expressway—Further developments. In: Yagar S (ed.) *Management and Control of Urban Traffic Systems.* Engineering Foundation Press, New York, pp. 149–60

Leonard D R, Gower P 1982 User guide to Contram version 4. British Transport and Road Research Laboratory report No. SR735

May A 1963 Development and evaluation of Congress Street Expressway pilot detection system. *Highw. Res. Rec.* **21**

Van Aerde M, Yagar S 1988 Dynamic integrated freeway/traffic signal networks: A routing based modelling approach. *Transp. Res. A* **22**(6), 445–55

Van Aerde M, Yagar S, Ugge A, Case E R 1987 A review of candidate freeway-arterial corridor traffic models. *Transp. Res. Rec.* **1132**, 42–50

Yagar S 1975 CORQ—A model for predicting flows and queues in a road corridor. *Transp. Res. Rec.* **533**, 77–87

Yagar S 1976 Applications of traffic flow theory in modelling network operations. *Transp. Res. Rec.* **567**, 65–9

Yagar S 1988 Generating partial origin–destination tables for streamlined application of traffic models. *Transp. Res. Rec.* **1194**, 135–8

Yagar S 1989 Metering freeway access. *Transp. Q.* **43**(2), 215–24

S. Yagar
[University of Waterloo, Waterloo, Canada]

D

Data Links in Aeronautics

A plane may fly in either visual flying rules (VFR) or instrument flying rules (IFR) (see *Visual and Instrument Flying Rules*). In the first case, there is no minimum requirement for radio communication or radio navigation. However, most planes have one or two vhf radios and some radio navigational aids. In the second case, two vhf communicators, two vhf Omnirange (VOR) beacons and at least one automatic direction finder (ADF) are mandatory. A transponder, which is a receiver/transmitter on radar frequencies, is mandatory to take off from, to land at or to transit above certain airports. As will be discussed, a transponder carries data.

It is recognized that data transmission between airline operators, military bases and, in the very near future, air traffic control (ATC) centers is a necessity.

1. General Situation

All data concerning the navigation of the plane in IFR conditions, from the time the plane is ready to have the engines started to the time when the plane has reached its parking space at the destination, are transmitted by vhf. This type of data will not be considered further.

Radio navigation is provided by various means (see *Air Traffic Control: Trends*). Each of these (e.g., a VOR beacon) is identified by its frequency and an identification code which, in the VOR case, is two or three letters regularly transmitted (every minute) in Morse code.

Transponders are radar receivers on a prescribed frequency, that of the secondary surveillance radar (SSR). This radar covers the upper levels (roughly above 10 000–15 000 ft (3000–4500 m)) in developed countries. When a plane receives the radar signal, it answers by transmitting a signal that is rigorously delayed with regard to the incoming signal—in order to give a precise measurement of the distance to the ground site—and which contains data in mode A and mode C.

In mode A, a code (four digits, 0–7) identifies the plane: this code is given through the vhf communication channels prior to take off by the control tower, or by the en route control or approach centers when it is necessary to use a code. When the identity of the plane and the distance appear at the radar site, the bearing and the ground speed are derived from these data by an on-site computer.

In mode C, the coded flight level (FL) of the plane is retransmitted. FL is the pseudoaltitude, expressed in hundreds of feet, read on an altimeter set at 1.0135×10^5 Pa, irrespective of the present barometric reference pressure (e.g., FL 290 means that an altimeter set at 1.0135×10^5 Pa indicates 29 000 ft (8700 m)).

Due to the fact that the message transmitted by the on-board transponder comprises only 12 bits (4096 possibilities), it is impossible to transmit both the identification of the plane and its FL. Messages are interlaced, twice identification (mode A) and once FL (mode C) if the transponder has this mode.

In mode S a longer message is available. This mode will soon be in application. In dense traffic zones, such as in the vicinity (50–100 nautical miles (95–190 km)) of major airports, saturation has already appeared; the answering message of several aircraft may overlap and are consequently not usable. In mode S systems, the possibility of selective calls is provided. Each plane may be called when it is necessary.

2. Mode S System

The requirements are as follows.

(a) It must provide selective interrogation in addition to general calls as they exist in modes A and C. A plane can be addressed individually, thus the superposition of transponder messages in dense areas is avoided.

(b) It must give some degree of collision avoidance.

(c) It must allow data transmission from ATC to planes (and back), from planes to company technical centers (and back) and from meteorological offices to planes (and back, in order that planes following the same routes may benefit from the experience gained by planes that are ahead).

(d) It must be compatible with modes A and C.

The solutions are as follows.

(a) A permanent unique address is provided for each plane: a 24-bit message allows 16 million addresses.

(b) Data messages are 56 bits or 112 bits long, with a possibility of grouping messages up to 1280 bits (ground to air).

(c) Transmission rates of 4 Mbyte s^{-1} (ground to air) and 1 Mbyte s^{-1} (air to ground) are used.

(d) As for the threat avoidance collision system (TCAS), it will be possible to use the on-board transponder with mode C or S (if available). Each plane equipped with such a transponder periodically broadcasts interrogation codes. The range is about 20 nautical miles (40 km). It collects the response distance (accurate data), bearing (nonaccurate) and altitude (very accurate). An on-board computer detects and proposes evasive maneuvers to avoid collision. This will be a vertical maneuver which is radioed to the other plane and to ATC (e.g., for an altitude difference of 750 ft (225 m) a 20 s warning will be given). (Note that this system only works when potentially colliding planes are equipped with mode S transponders and, at least one, has the interrogator possibility plus the on-board computer.)

In 1989, the experience gained with this equipment led to one warning for each 3 h flight period, one conflict resolution for each 50 h flight period, one air miss for each 100 000 h flight period and one collision for each 10^6 h flight period.

3. Communication Navigation Surveillance System

This is a European project which started in 1989. It will provide the following services.

(a) *Communications*. Three services are forecast. These are ATC which may include coded data (messages) and voice, aeronautical operational communication (AOC) for the companies' use (e.g., automatic data transmission concerning engines or meteorological conditions) and aeronautical administrative communication (AAC) for the companies' use (e.g., flights, organization, passenger reservations).

(b) *Navigation*, a factor of merit giving the expected maximum error is required (i.e., required navigation performance capability (RNPC)). In dense zones, automatic reporting systems will also be provided.

(c) *Surveillance*. This must be provided for traffic en route and for collision avoidance.

The need for an automatic data transmission, if possible at any time during the flight (i.e., even over oceans or unpopulated areas), is mandatory: the link must transmit to the ground all the data, both air to ground and ground to air, needed to achieve a safe and optimal flight. A full compatibility between the various means of transmission (radar Mode S, relay satellites or vhf channels) must be achieved. To comply with the traffic increase, the implementation of the automatic data link becomes a necessity before 1995.

See also: Data Processing in Air Traffic Control

Bibliography

Aviation Safety 2nd Int. Symp. 1986 Cepadues Edition, Toulouse, France
Aviation Safety 3rd Int. Symp. 1988 Cepadues Edition, Toulouse, France
Cox M E 1986 Possible contributions from SSR mode S data link to the conduct of efficient aircraft operations. *AGARD Conf. Proc.* **410**, 25

Pélegrin M 1990 Data transmission between planes and control centers. *Prepr. IFAC 11th World Congr.* IFAC Secretariat, Laxenburg, Austria

M. J. Pélegrin
[ONERA, Toulouse, France]

Data Processing in Air Traffic Control

In order to ensure a safe, orderly and expeditious flow of air traffic, many different data are required. In the particular case of air traffic control (ATC), data refers to facts pertaining to the flight of an aircraft or to the air traffic structure in general (see Fig. 1). The data needed in ATC are classified according to the needs for information of the different air traffic services. With reference to this classification, data processing is defined as a systematic sequence of operations performed on data. The method of processing depends on the technical standard and equipment of the responsible air traffic services unit. In this article, the data processing described is based on the ATC equipment of Western Europe.

1. Classification of Data

1.1 Air Traffic Management Data

The purpose of the air traffic management service is to ensure an optimum flow of air traffic to or through areas with traffic demand at times exceeding the available capacity of the ATC system (see *Air Traffic Control: An Overview*). In order to ensure an optimum flow of air traffic, it is necessary that the air traffic management service achieves a balance between air traffic demand and ATC capacity. This requires, of course, that the air traffic management service must have at its disposal three categories of information:

(a) data representing the air navigation infrastructure (e.g., description of the air route network, airspace organization in terms of flight information regions, upper information regions, terminal maneuvering areas, ATC sectors, location of airports and navigational aids, ATC procedures including those affecting the profiles of aircraft, performance tables of various categories of aircraft);

(b) data on planned flight operations, the analysis of which gives a picture of the expected traffic demand; and

(c) ATC capacity values.

1.2 Preflight Planning Data

Preflight planning data are based on all information essential to an intended flight. The preflight information service provides preflight information verbally, in writing or by presentation of preflight information documents for self-briefing.

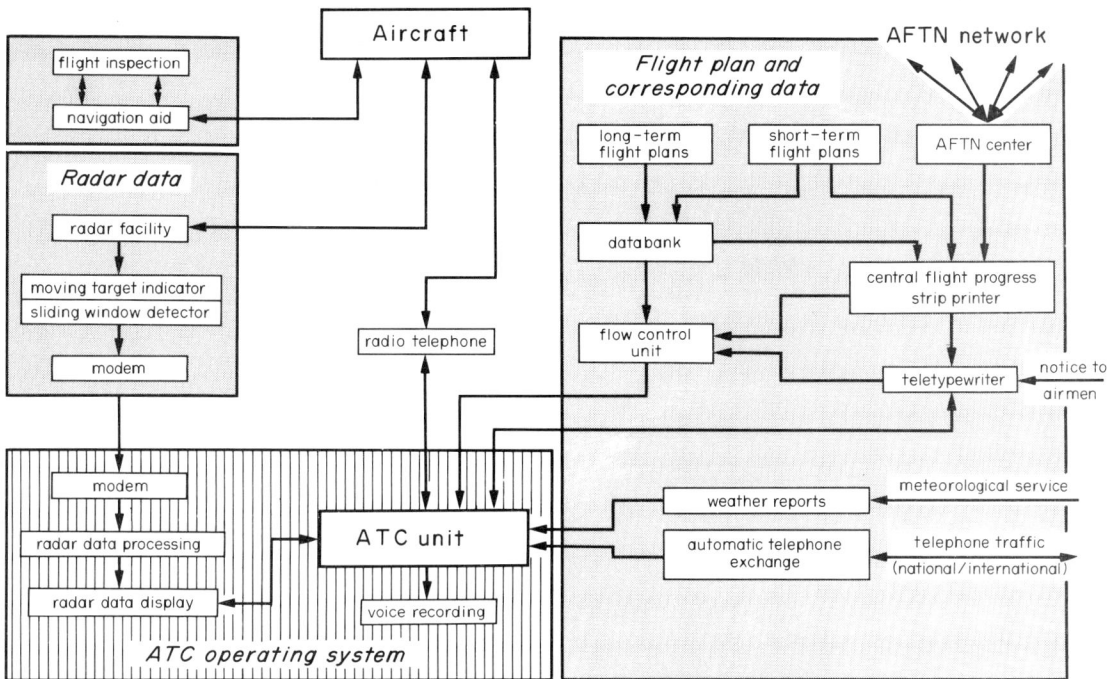

Figure 1
Data structure and processing in ATC

1.3 Flight Plan Data

The flight plan is a compilation of specified data about the planned flight of an aircraft to be submitted to ATC. Its purpose is to inform the relevant ATC units enabling them to supervise the flight within the scope of ATC as well as to provide flight information and alerting services.

1.4 Radar Data

Radar is a radio detection device that provides information on range, azimuth and/or elevation of objects. In the context of ATC, the primary radar system uses radio signals reflected at the aircraft. Secondary surveillance radar (SSR) uses ground interrogators and airborne transponders to determine the position of an aircraft in range and azimuth. When agreed modes and codes are used, height and identity are also transmitted. An electronic display of radar-derived information is given, continuously showing the positions and movements of the aircrafts.

1.5 Meteorological Data

Metorological data is discussed in Sect. 6.

2. Processing of Air Traffic Management Data

The basis of air traffic management in Western Europe is the operation of a databank by Eurocontrol. This databank must accept, update, process and analyze data of a planned flight operation on the air navigation infrastructure and the limits within which the ATC system is to accommodate the traffic. It has the function of providing the interested parties with counts and listings of traffic demand, together with the results of its comparison of demand with the available ATC capacity. This gives an indication of where and when traffic demand will reach or exceed specific thresholds. In particular, the databank is able to

(a) accept and update planned flight data provided by operators;

(b) examine the planned flight data received and eliminate any duplication that might exist;

(c) accept and update information provided by states on airspace organization and related assessments of ATC capacity;

(d) collate, analyze and compare available air traffic demand and ATC capacity data, so as to identify where and when demand is likely to reach or exceed capacity;

(e) transmit information on expected critical situations to the relevant air traffic management units in Europe at the earliest possible time;

(f) provide air traffic management units in Europe with information on expected traffic flows and other traffic information as requested; and

(g) provide, when requested by an air traffic management unit, information on possible alternative routings.

The data concerning the air traffic services environment is extracted from aeronautical information publications and stored in the databank. In particular, 45 000 airport pairs, 40 000 routes between airport pairs, 5000 way points, 1300 airports and 500 ATC sectors can be stored. Using this data a route catalog is prepared.

The route catalog contains standard routings; that is, descriptions of the routes between departure and destination aerodromes. If there is more than one standard route between two aerodromes, they will be stored together with an indication of which aircraft operators prefer which route. Similarly, one of the routes may be mandatory due to a traffic orientation scheme and this information is also stored with the route.

Aircraft operators are asked to provide to the Databank, at the earliest possible time, details of all instrument flying rules (IFR) flight operations planned before midnight of the day of actual operation and intended to operate within the databank trial area. Subsequent revisions should be made known as they occur.

This data will mainly be collected in the form of planned flight data (i.e., timetable information), but may also be in the form of flight plans or repetitive flight plans.

The data on planned flight operations are used to create flight data records and are allocated to routes from the route catalog. A flight data record is a unique combination of flight identification, aerodrome of departure and aerodrome of destination. Each flight data record may have several pages due, for example, to the use of different aircraft types on different days of the week or the use of different routes on weekdays and at weekends. The databank is capable of storing traffic demand consisting of approximately 60 000 flight data records.

When the data has been properly assembled into flight data records and flight data record pages, a profile is constructed. From this the estimated arrival times and flight levels at the various points on the route, and the entry times to the various ATC sectors the aircraft will enter are calculated. These data are then stored on the databank ready for interrogation.

The databank may be queried in order to count or list the described flight data associated with a geographical parameter. The types of geographical parameter that may be specified are

(a) points (navigation aid or way point);
(b) traffic flows, which are specified as two points, traffic crossing both points in the order in which they are specified being counted (listed);
(c) aerodromes; and
(d) count areas (e.g., ATC sectors, groups of sectors, terminal maneuvering areas).

Requests for counts result in an output in the form of the number of aircraft per hour for the time and geographical parameters specified. The number of aircraft per hour means the number of aircraft crossing a point, entering a traffic flow, arriving at or departing from an aerodrome, entering a predefined count area and so on. Listings give details of the individual flights counted.

3. Processing of Preflight Planning Data

Preflight information will be based on all information essential to the intended flight. Aeronautical information service units are connected to a "notice to airman" databank, from which summaries and bulletins may be retrieved for delivery to pilots and/or operators, or for preflight information via telephone. Summary means all current notice to airman data of a country or flight information region. Bulletin means a selection of notice to airman data for flights within an established area or along an established route including the aerodrome of departure, the aerodrome of destination and the alternate aerodromes. A bulletin contains the temporary information effective at the date of the bulletin and the advance information that will become effective on the date of the bulletin or within the following two days (it takes into account area/route and flight level as well as flight rules such as the permanent information that has not yet been incorporated into relevant aeronautical information publications and/or charts). Furthermore, detailed information is included about aerodromes, en route radio and navigation facilities, air traffic regulations and air traffic services procedures, regional navigation warnings and available supraregional informations.

4. Processing of Flight Plan Data

4.1 Flight Plan

The flight plan is an internationally recognized document which, for ease of transmission and understanding on a worldwide basis, is prepared in a standard format (see Fig. 2). The types of flights that are required to submit flight plans are also agreed upon internationally and are outlined by the International Civil Aviation organization (ICAO) (Rules of the Air, Annex 2). Accordingly, a flight plan shall be submitted in the event of

(a) any IFR flight;
(b) any visual flying rules (VFR) flight within controlled airspace;
(c) any VFR flight during nighttime within controlled airspace;
(d) any acrobatic flight within controlled airspace or in close vicinity to an aerodrome that is controlled by ATC services;

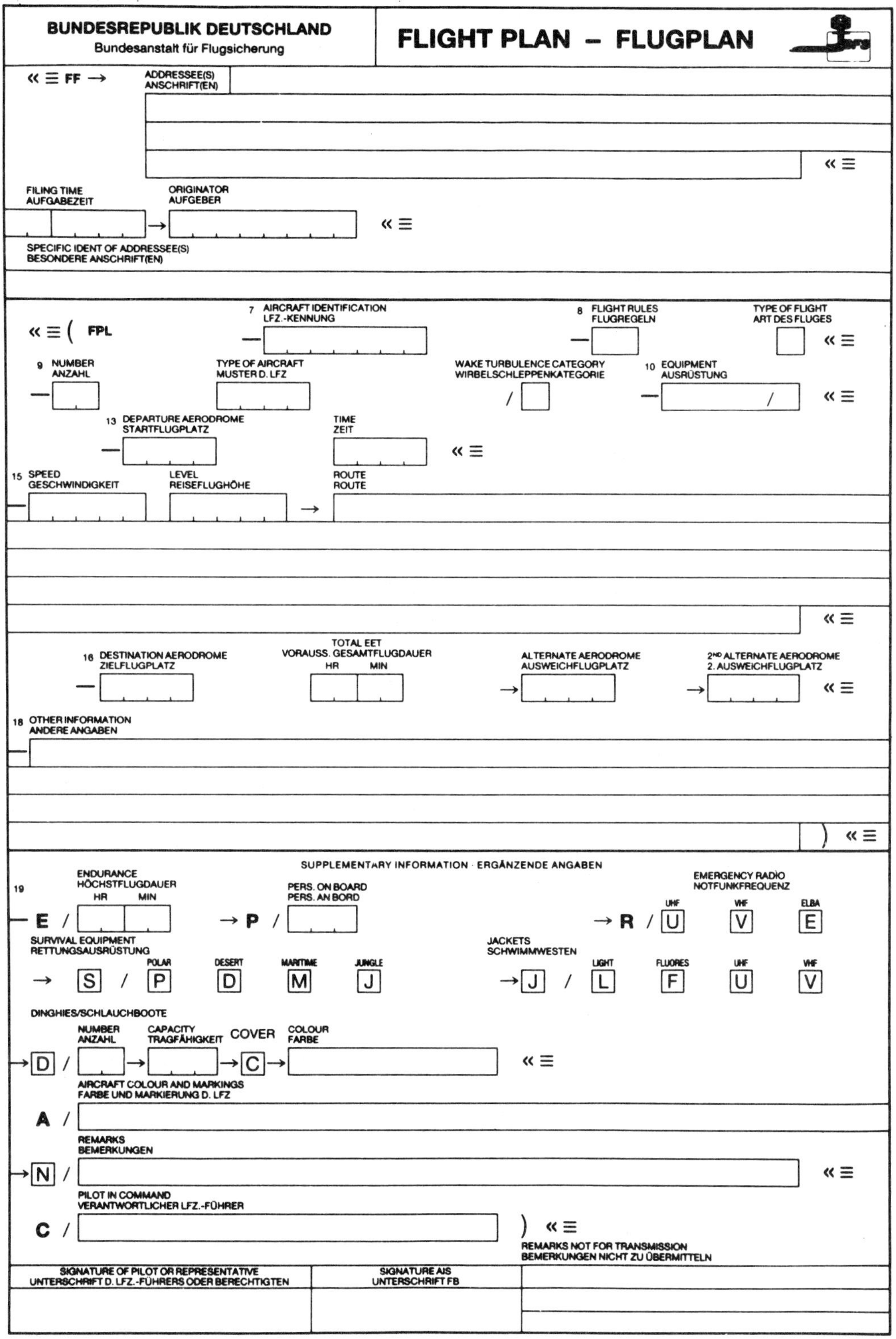

BUNDESREPUBLIK DEUTSCHLAND
Bundesanstalt für Flugsicherung

FLIGHT PLAN – FLUGPLAN

« ≡ FF → ADDRESSEE(S) ANSCHRIFT(EN) « ≡

FILING TIME AUFGABEZEIT → ORIGINATOR AUFGEBER « ≡

SPECIFIC IDENT OF ADDRESSEE(S) BESONDERE ANSCHRIFT(EN)

« ≡ (FPL

7 AIRCRAFT IDENTIFICATION LFZ.-KENNUNG

8 FLIGHT RULES FLUGREGELN

TYPE OF FLIGHT ART DES FLUGES « ≡

9 NUMBER ANZAHL

TYPE OF AIRCRAFT MUSTER D. LFZ

WAKE TURBULENCE CATEGORY WIRBELSCHLEPPENKATEGORIE /

10 EQUIPMENT AUSRÜSTUNG / « ≡

13 DEPARTURE AERODROME STARTFLUGPLATZ

TIME ZEIT « ≡

15 SPEED GESCHWINDIGKEIT

LEVEL REISEFLUGHÖHE →

ROUTE ROUTE « ≡

16 DESTINATION AERODROME ZIELFLUGPLATZ

TOTAL EET VORAUSS. GESAMTFLUGDAUER HR MIN

→ ALTERNATE AERODROME AUSWEICHFLUGPLATZ

→ 2ND ALTERNATE AERODROME 2. AUSWEICHFLUGPLATZ « ≡

18 OTHER INFORMATION ANDERE ANGABEN) « ≡

SUPPLEMENTARY INFORMATION · ERGÄNZENDE ANGABEN

19 ENDURANCE HÖCHSTFLUGDAUER HR MIN – E /

PERS. ON BOARD PERS. AN BORD → P /

EMERGENCY RADIO NOTFUNKFREQUENZ → R / UHF U VHF V ELBA E

SURVIVAL EQUIPMENT RETTUNGSAUSRÜSTUNG → S / POLAR P DESERT D MARITIME M JUNGLE J

JACKETS SCHWIMMWESTEN → J / LIGHT L FLUORES F UHF U VHF V

DINGHIES/SCHLAUCHBOOTE

NUMBER ANZAHL CAPACITY TRAGFÄHIGKEIT COVER COLOUR FARBE

→ D / → → C → « ≡

AIRCRAFT COLOUR AND MARKINGS FARBE UND MARKIERUNG D. LFZ

A /

REMARKS BEMERKUNGEN

→ N / « ≡

PILOT IN COMMAND VERANTWORTLICHER LFZ.-FÜHRER

C /) « ≡

REMARKS NOT FOR TRANSMISSION BEMERKUNGEN NICHT ZU ÜBERMITTELN

SIGNATURE OF PILOT OR REPRESENTATIVE UNTERSCHRIFT D. LFZ.-FÜHRERS ODER BERECHTIGTEN

SIGNATURE AIS UNTERSCHRIFT FB

Figure 2
Flight plan

(e) any gliding above clouds;

(f) any manned free balloon or airship flight;

(g) any ascents of unmanned free balloons with a total weight balloon cover and ballast of more than 0.5 kg, any ascents of bundled free balloons and any mass ascents of unmanned free balloons;

(h) any flight crossing international borders; and

(i) any flight within restricted areas.

The listed cases may vary a little in interpretation by the contracting member states of the ICAO, but the intent of the corresponding rules are applied worldwide.

As a result, if a flight is planned, the pilot-in-command or the responsible person fills in the flight plan prior to takeoff. The pilot in-command may file a flight plan during the flight if called for by special circumstances unknown before takeoff. This procedure is necessary because the application of ATC depends on knowledge of the present position of the aircraft and the intentions of the pilot in command. Therefore, the flight plan contains the data applicable to the planned flight in accordance with the flight plan form. The details that the pilot or the pilot's representative are required to insert on the flight plan are as follows (see Fig. 2):

(a) aircraft identification;

(b) flight rules and type of flight;

(c) number and type of aircraft, and wake turbulence category;

(d) equipment on the aircraft (e.g., radio equipment, radio navigation equipment for en route and approach navigation, SSR transponder equipment);

(e) departure aerodrome and estimated off-block time;

(f) speed and requested cruising level, routing;

(g) destination aerodrome and total estimated elapsed time, alternative aerodromes;

(h) other information (i.e., any information that is significant for the handling of the flight by ATC); and

(i) supplementary information (e.g., the endurance of the aircraft, the number of persons on board, the available emergency radio frequencies, the type of survival equipment carried, the type of life jackets carried, the number, capacity, type and color of the dinghies carried, the color of the aircraft including any significant markings, any possible additional information regarding the survival equipment, the name of the pilot in command).

The flight plan containing the listed information should be filled out up to 24 h before the estimated off-block time. However, it must be filed early enough so as to allow a period of not less than 30 min between the acceptance of the flight plan by the aeronautical information service (the responsible authority) and the estimated off-block time. For flights into areas with air traffic flow restrictions, the flight plan should be filed not later than 3 h before the estimated off-block time, unless a repetitive flight plan has been submitted. This does not affect the regulations prescribing filing of a flight plan for certain flights at an earlier time due to other reasons.

4.2 Flight Progress Strip

To prepare the flight plan data for the needs of the ATC services, all flight plans that the aeronautical information service accept are transmitted to a central flight progress strip print system by the Aeronautical Fixed Telecommunication Network or other national telecommunication networks (see Fig. 1). The Aeronautical Fixed Telecommunication Network is an integrated worldwide system of aeronautical fixed circuits. These are provided as a part of the aeronautical fixed service for the exchange of messages and/or digital data between aeronautical fixed stations having the same or compatible communications characteristics.

All transmitted flight plans will be concentrated into data blocks. The data blocks make a databank. The databank consists of a static part and a dynamic part. All repetitive flight plans and the nonactivated flight plans are stored in the static part. The activated flight plans are stored in the dynamic part. The relevant data of a defined flight (e.g., aerodrome of departure, estimated time of departure) will be extracted from the dynamic part and automatically transmitted to the respective ATC unit (see Fig. 1).

The flight process strip printer, part of the equipment of the ATC units, completes printing of the flight progress strip automatically (see Fig. 3). The flight progress strip is a strip of paper and is used by the air traffic controller for displaying aircraft data of an approaching flight, such as aircraft identification and type, SSR equipment, departure aerodrome, desired destination aerodrome and cleared flight level, time of departure, and times over way points, as well as flight times from the last and to the next way point. In the middle of the strip, marked by a frame, is the related way point. The areas around the frame give "from" and "to" information of the routing. The manner in which these flight progress strips are displayed to the controller depends to a large degree on the tasks that are performed by the respective ATC facility. Normally, for en route control, flight progress strip boards are used, upon which the flight progress strips can be displayed under geographical headings. These are usually related to the framed en route reporting points previously mentioned. This enables the controller to plan the traffic flow within their area of responsibility. An approach control unit which has a smaller area of

Data Processing in Air Traffic Control

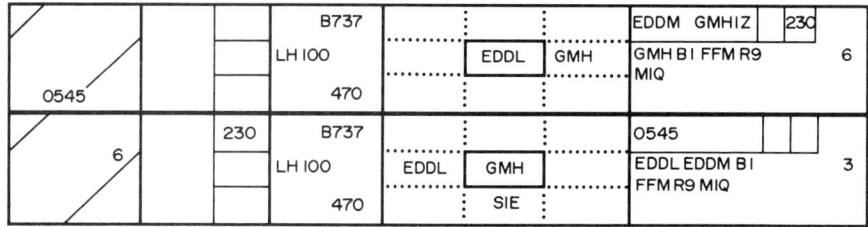

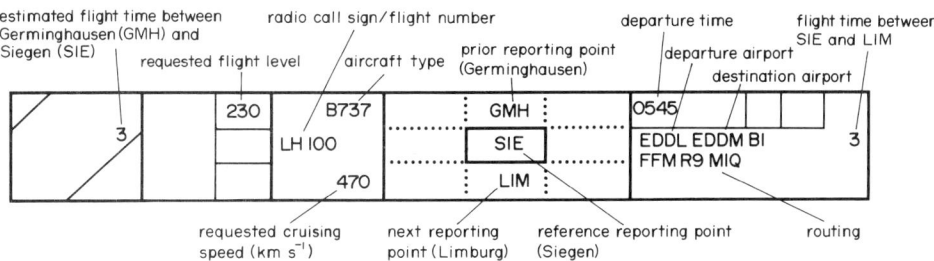

Figure 3
Flight progress strip

responsibility would only need a simple display with the flight progress strips sequenced under the headings of the holding facilities (stacks) serving the aerodrome. Even ATC systems that are radar based provide the controller with flight progress strips, not only to assist in planning tasks, but also to provide a basis for the procedural control of air traffic in the event of a radar failure. In this regard, it is considered important to recognize the fundamental role of a flight data display, particularly when it is used in conjunction with a radar display. The flight data display or the flight progress strip board is the planning tool. It displays advance information to the controller and is updated by the radar data.

5. Processing of Radar Data

5.1 Primary Radar

In addition to the flight progress strip, radar data are the most important information about the real position of an aircraft to be controlled by ATC. The term radar is the short form of radio detection and ranging. Its basic functions can be described as the detection of the presence of objects and the determination of their direction and range by means of radio echoes. It is based on the principle that electromagnetic energy, propagated by a transmitter at the speed of light, can be directed onto a reflecting object and, thus, the distance of the object can be calculated by measuring the time a pulse of the radio energy takes to travel to the object and back again. By using an antenna that emits and receives a narrow beam of energy, and by rotating this beam through 360°, the direction from which this energy is being reflected can also be determined. This technique is known as primary radar. For the described purposes, a radar set consists of

(a) a transmitter which produces the radio energy pulses;

(b) an antenna which radiates the energy and collects the echo;

(c) a receiver which decodes the reflected energy;

(d) an amplifier and processor which amplify the energy (signal), separate it from clutter and transform it into video form; and

(e) an indicator in the form of a cathode ray tube on which the returned and transformed signal can be displayed.

According to the different requirements of ATC, the following types of radar exist:

(a) en route radar, for controlling aircraft en route;

(b) airport surveillance radar, for controlling aircraft within the terminal maneuvering area;

(c) precision approach radar, for controlling precision approaches of aircraft; and

(d) airport surface detection equipment (or airport surface movement indicators), for controlling movements of aircraft on the ground (i.e., on the runway, taxi ways and the apron area).

These four groups exemplify the very different characteristics required of ground-based radar equipment. A small number of characteristics implied by the operational requirements indicate differences in the technology required. They are

(a) data renewal rate (e.g., antenna rotation rate),

(b) resolution capability,

(c) mean power requirements,

(d) range performance, and

(e) complexity of processing.

The most important disadvantages of the primary radar technique for the purposes of ATC are as follows.

(a) The height of the aircraft is unknown and must be requested by radio telephone.

(b) At first step, correlation between the flying target (the aircraft) and the displayed radar target is not possible. For exact identification, the aircraft is asked to perform a special procedure turn.

5.2 Secondary Surveillance Radar

To overcome the disadvantages of the primary radar technique, most of the en route radar facilities are combined with SSR. Indeed, from the point of view of controlling aircraft, the introduction of SSR has been the most significant advance since the application of primary radar. The SSR system includes ground-based and airborne elements. The ground station emits pulses of reply frequency energy via the directional beam of a rotating antenna using a frequency of 1030 MHz. When the antenna beam is pointing in the direction of an aircraft, airborne equipment, known as a transponder, detects the emitted interrogation signal and modulates its own transmitter in response. This reply, on a frequency of 1090 MHz, is detected by the ground station and is processed by a plot extractor. The latter equipment measures the range and bearing of the aircraft, decodes the replies of the aircraft to determine its identity and flight level, and passes the data to radar displays at an ATC center. The use of an airborne transponder permits the transponder reply frequency to be different from the ground transmitter frequency, thereby avoiding the problems of clutter returns experienced by primary radar. The presence of the transponder also enables the reply signal to be modulated, so that the additional data of identity and flight level can be communicated by the aircraft.

5.3 Interrogation Signal

The ground antenna used has two principal beams: the interrogate beam, which has a high gain and narrow main lobe with low side lobes, and the control beam, which is broader but has a lower peak gain (see Fig. 4). The essential feature is that the gain of the control beam is greater than that of the interrogate beam in all directions, except that of the narrow main lobe. The control beam is used to prevent the aircraft from replying to signals from the interrogate beam side lobes. The signal transmitted by the ground station is conventionally called an interrogation.

Two pulses (P_1 and P_3) are transmitted via the interrogate beam of the antenna and the spacing of these two pulses will determine the data content of the transponder reply (see Fig. 5). A further pulse (P_2) radiates from the control beam 2 µs after the P_1 pulse. By comparing the relative strengths of pulses P_1 and P_2, the airborne transponder can determine whether the signal originates from the main lobe or a side lobe. It can therefore determine whether a reply is required. By this process responses to side lobes are suppressed. This facility is known as interrogator side-lobe suppression. The pulses P_1 and P_3, both of which are transmitted via the interrogate channel, can adopt a number of different spacings. All identification pulses have a duration of 0.8 µs.

5.4 Reply Signal

The reply signal transmitted by the aircraft in response to an interrogation is inserted between two pulses called framing pulses designated F_1 and F_2 and spaced 20.3 µs apart. The data pulses are designated A, B, C and D with a suffix 1, 2, or 4 giving a total number of 12 (see Fig. 6), and are spaced in increments of 1.45 µs. The pulse in the middle, the X pulse, is reserved for future use. In addition, a final pulse, the special position interrogation (SPI), can be added to the reply pulse train 4.35 µs after F_2 in case of a mode 3/A reply. The SPI is usually transmitted with manual control by the pilot on voice communications request of ATC to aid in identifying a particular aircraft. The 12 pulses that are used give 4096 permutations and communicate the reply data. The principal interrogation mode is mode 3/A, the common civil/military mode. This mode is used for general identification with the identity number of the aircraft composed of the octal value of the reply pulses in the order ABCD. By international

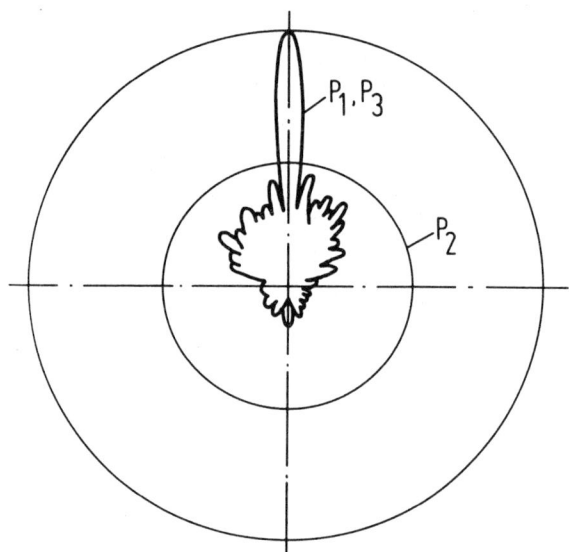

Figure 4
Antenna diagram with side lobes

Data Processing in Air Traffic Control

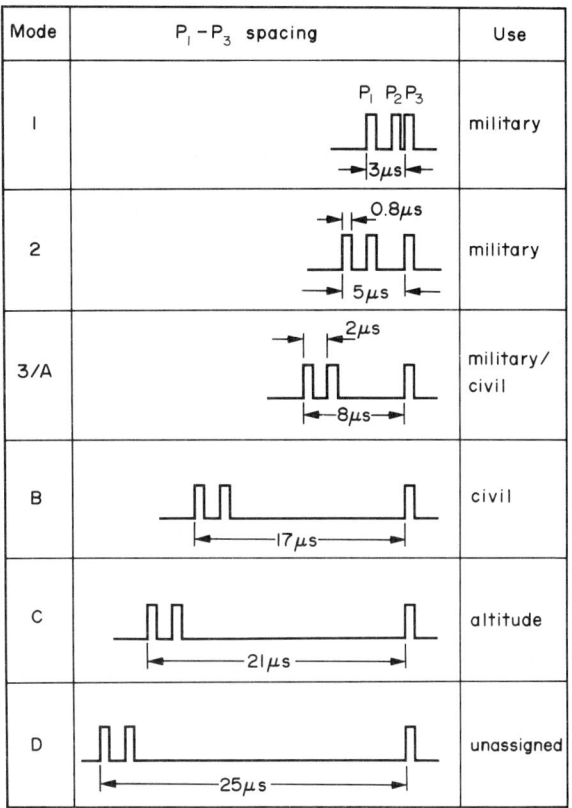

Figure 5
The use of SSR interrogation modes

agreement, specific sets of identity numbers can indicate the type of flight, or the destination or origin of the aircraft. Three particular codes are universally used to indicate emergency conditions: 7700 in emergency cases, 7600 for radio failure and 7500 in case of hijacking.

The next most important mode is mode C. This mode is used to communicate to the ground the height of the aircraft as indicated by its aneroid barometer. Only 11 pulses are used in a mode C reply (pulse D_1 is omitted),

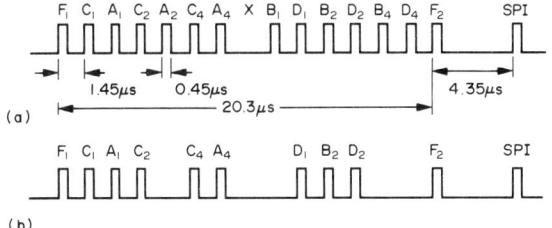

Figure 6
SSR reply: (a) format and (b) example of code 5273 with pulse values of $A_n = 1000n$, $B_n = 100n$, $C_n = 10n$ and $D_n = n$, where $n = 1, 2, 4$

but the 2048 permutations resulting are sufficient to indicate the height in 100 ft (30 m) increments from -1000 ft (-300 m) to 121 000 ft (36 300 m). The reply is initially encoded on eight internal reply pulse positions in a Gray code providing barometric altitudes in 150 m increments. The 150 m interval is then decomposed into 30 m intervals by a second Gray code, using three internal reply pulse positions. Since the height reported is obtained from the measurement of local air pressure, it is necessary to compensate for the normal changes in barometric pressure by adjusting the airborne instrument to the local ground barometric pressure as reported by the ATC facility on the ground. Above a transition altitude, normally around 5000–6000 ft (1500–1800 m), height measurement is based on a standard barometric pressure of 1.01325×10^5 Pa regardless of surface atmospheric pressure. This process removes the necessity for long distance aircraft to change their ground barometric reference pressure frequently as their journey progresses. Above the transition level, the mode C reply data do not relate to height above sea level but to flight level in steps of 100 ft (30 m). This provides a safe separation of all aircraft on the basis that their barometrically measured flight levels differ by at least 1000 ft (300 m). Above FL 290 (a flight level of approximately 29 000 ft (8700 m)) the separation widens to 2000 ft (600 m).

5.5 Moving Target Indicator

Before the signals determining the position of an aircraft (azimuth, range and height) can be transmitted, certain steps of signal processing must be carried out. The first step is to remove clutter. Therefore, coherent moving target indicators or pulse Doppler processing are used. Clutter, resulting from radar returns from the ground or sea, is essentially at zero Doppler velocity. Hence, if a notch filter is placed at zero Doppler frequency (with the width of this filter equal to the width of the clutter spectrum), the clutter is eliminated. In effect, moving target indicator processing puts a notch filter or rejection band at zero Doppler velocity and multiplies by the radar pulse repetition frequency. A two-pulse moving target indicator canceller simply involves subtracting the echo of the present transmission from the preceding return. If the scatterer is not moving (such as with a clutter patch), the phases and amplitudes of the two pulses are the same, and thereby cancel. Pulse Doppler processing involves the coherent processing of a large number of pulses, typically eight or more. These pulses are Fourier analyzed to obtain the return signal spectrum. Generally, the signal echoes having zero Doppler velocity are interpreted as clutter and ignored; the echoes having non-zero Doppler velocity are interpreted as indicating a moving target. Other techniques can be applied, depending on the state of the respective radar system.

5.6 Sliding Window Detector

The next step that should be performed is to decide whether a moving target is an aircraft or a false alarm.

This decision is made by a plot extractor. Normally, a two-threshold form of plot extractor is used. The decision to be made is whether the "first threshold of detection" has been crossed; that is, does the radar output amplitude exceed a level that constitutes an event other than noise. Many signal processors include this decision-making circuit but, if not, it has to be incorporated into the plot extractor.

The second decision is made by various methods. The most common is the sliding window technique. In this method, arrangements are made to store the decision on whether the first threshold has been crossed for every range increment of a pulse repetition period. It is usual to provide at least two, and there may be as many as four, range increments per resolution cell, but the decision is expressed in terms of a 1 if a resolution cell contains a first threshold crossing and a 0 if it does not. At the next interpulse period the same resolution increment is examined and its 1 and 0 state is stored with its predecessor. Storage equal to the number of pulses per beam width is usually provided. As soon as the count reaches a predetermined value a plot leading edge is declared. A target must be present. The counting goes on until another criterion is reached; that is, where the number of cells occupied falls into a defined level out of the predetermined value range. This is equivalent to a decision, stating that even if a leading edge has previously been declared, there are now insufficient detections to declare target presence; that is, a trailing edge has been found. Therefore, where the plot starts and where it finishes are defined. The antenna position and a digital expression of the azimuth at which these decisions have been made are also stored. Thus, within the plot extractor, there are the following data:

(a) plot start azimuth (leading edge),

(b) plot finish azimuth (trailing edge),

(c) plot range (resolution cell position), and

(d) plot presence (leading edge followed by trailing edge).

The thereby successfully detected targets should be continuously registered in the form of target lists. The target lists are then transmitted by coaxial cable to the respective ATC center. The capabilities of displaying the radar data on a radar display depend on the stage of development of the computer-based ATC system, on the different philosophies used to control the aircraft and on the different needs of information as a result of the special ATC service provided (see Fig. 7).

5.7 Radar Data Displaying

Most advanced computer-based ATC systems providing an en route ATC service in Europe have the capacity to carry out the following main functions.

(a) They can provide the controller with a pictorial representation of all aircraft current positions in the form of a composite radar picture. This is based on the explained plot messages which contain the measured aircraft position relative to the radar location, the SSR code (if available) and certain other technical information.

(b) They can provide radar tracking; that is, plot messages for a specific target received in subsequent radar scan periods from a specific radar are chained with each other and are subject to local radar tracking. This is a filtering process that produces a calculated position and speed of the target for the current scan, and an estimated position for the next scan (see Fig. 8).

(c) In order to warn controllers of potentialy lost separation, a short-term conflict function may be available that uses the track speed vector to extrapolate the target trajectories to determine those pairs of targets for which an infringement of horizontal and vertical separation must be expected within the next 2 min (see Fig. 9).

(d) Intermixed with radar plot messages, weather contour vectors are also received from certain radar stations or directly transmitted from weather stations.

To handle all the prepared data, advanced ATC centers may be equipped with working positions that support the following kinds of display and input devices:

(a) synthetic dynamic display, to show the traffic situation;

(b) touch input display, to queue text messages related to aircraft and labels of input menus;

(c) electronic data display, for presentation of messages upon controller request;

(d) display control panel, providing function keys used to control the filtering and data selection functions of the synthetic dynamic display; and

(e) rolling ball, used to make position designations on the synthetic dynamic display.

6. *Processing of Meteorological Data*

The activities of the meteorological service for aeronautical applications are regulated in compliance with the provisions of the International Civil Aviation Organization (ICAO) and the World Meteorological Organization. Their main functions are

(a) to provide aeronautical personnel with meteorological documentation and briefing, including all meteorological information and forecasts necessary for flight planning and operation of aircraft;

(b) to provide air traffic services units with meteorological reports, forecasts and warnings required for the control of air traffic and for transmission to aircraft in flight;

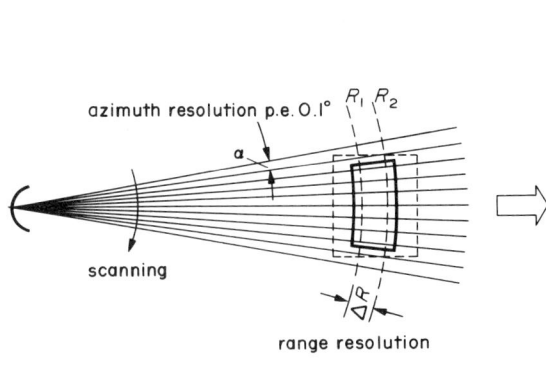

Figure 7
Operating method of the sliding window detector

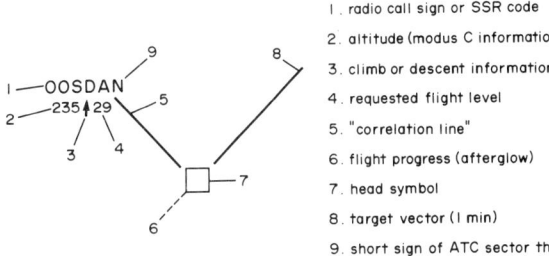

Figure 8
Short sign of ATC sector the aircraft is using

(c) to maintain a weather observation and reporting service, meeting the requirements of air navigation;

(d) to provide continuous meteorological observation and to disseminate warnings; and

(e) to carry out scientific work on meteorological problems of an aeronautical nature and to provide climatological data (see Fig. 1).

The meteorological forecast service for aeronautical applications is performed by aeronautic meteorological offices. The forecast service is provided, according to local agreement, verbally and/or in writing. The documentation is issued in tabular and/or pictorial form and includes

(a) aerodrome forecasts for aerodromes of departure and destination, and alternative aerodromes as available;

(b) prognostic charts of en route weather conditions (significant weather, upper winds and temperatures, tropopause heights, maximum wind); and

(c) other documentation on special request.

Information on meteorological conditions affecting flight operations may be provided (as far as operationally possible) individually or collectively, via telephone or teletype, in a form adjusted to the specific

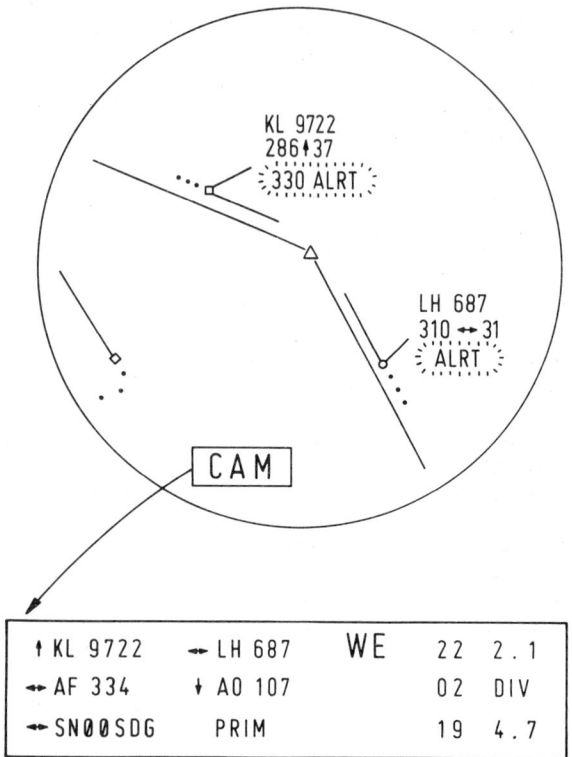

Figure 9
Presentation of a conflict warning on a radar traffic display (synthetic dynamic display); CAM is a conflict alert message

requirements. Meteorological forecasts via telephone are kept on tapes at most aeronautic meteorological offices. Requests for meteorological forecasts are obtained from the appopriate aeronautic meteorological office (individual briefing), via automatic telephone responders (automatic briefing) or via view data (automatic briefing).

7. Radio Telephony

Radio telephony may also be understood as a method of data processing in ATC. All of the aids at the controller's disposal ultimately call for the use of a speech circuit to issue an executive instruction to the pilot of the aircraft for the safeguarding of the flight. It is generally standard practice for a controller at an operating position to have a personal radio telephone circuit, discreet to the controller and to the pilots of aircraft within the airspace for which the controller is responsible. The frequencies used are normally in the vhf/uhf band for airport and en route control. For communications over long distances, high-frequency speech is used. Independent of the state of development of the ATC facility, all individual data (information) may be transmitted in this way, mentally processed by the pilot and controller.

See also: Data Links in Aeronautics

Bibliography

Baur E 1985 *Radartechnik*. Teubner, Stuttgart, Germany
Brookner E 1977 *Radar Technology*. Artech House, Dedham, MA
Bundesanstalt für Flugsicherung 1989 *Aeronautical Information Publication*. Bundesanstalt für Flugsicherung, Frankfurt, Germany
Cole H W 1985 *Understanding Radar*. Collins, London
Eurocontrol 1984 *Central Data Bank of Air Traffic Demand*. Eurocontrol, Brussels
Eurocontrol 1986 *The Maastricht Automatic Data Processing System*. Eurocontrol, Maastricht, The Netherlands
Field A 1985 *International Air Traffic Control*. Pergamon, Oxford
Mensen H 1989 *Moderne Flugsicherung*. Springer, Berlin
Stevens M C 1988 *Secondary Surveillance Radar*. Artech House, Dedham, MA

H. Mensen
[Technische Universität Berlin, Berlin, Germany]

Detectors for Road Traffic

This article is primarily concerned with four types of detectors for road traffic: ultrasonic Doppler-type detectors, loop-type detectors, vehicle profile classifier detectors and image sensors.

1. Ultrasonic Doppler-Type Detectors

As a method of detecting the exact velocity of moving vehicles, equipment exists that measures velocity using microwaves, but its use is restricted by law. However, ultrasonic Doppler-type detectors have the great advantage of so-called "easy handling" because they are not restricted by this legislation. Also, they can be processed in a suitable way to have an accuracy equivalent to that of the radar type detectors.

A method of measuring the velocity of moving cars by ultrasonic Doppler waves is already known; however, because it requires a complicated logical operation and also has the unsolved problem of fluctuations in the speed of sound due to the atmospheric temperature, as well as the effect of the wind, it is thought to be difficult to put it to practical use. However, with the establishment of a method of Doppler-signal control, a method of compensating for the atmospheric temperature, a method of operation of the speed, a method of installing an ultrasonic transceiver, and so on, the

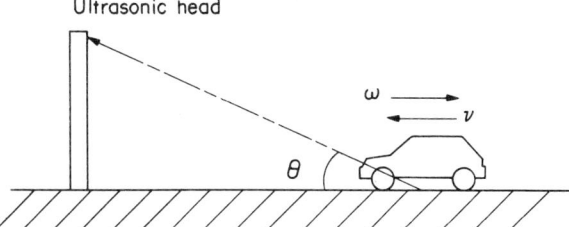

Figure 1
Outline of installation of an ultrasonic Doppler-type detector

ultrasonic Doppler-type detector has confirmed its practicality as a speed detector. Also, it is now applicable to other uses, such as measurement of occupancy rates and traffic volumes (see *Flow Variables*).

1.1 Principle of Action

(a) *Principle of the Doppler effect.* The Doppler effect (on a sound wave) is a phenomenon that affects the frequency of a sound wave reflected from a body moving at some speed relative to the emitting body.

When the sound wave of a particular frequency is reflected by the body moving relative to the emitter, it experiences a change of frequency in proportion to the relative velocity of the emitter and the reflecting body. From the difference between the frequency of the sound source and that of the reflected wave, the relative velocity of the two bodies can be measured. This difference in frequency is referred to as the Doppler frequency f_a.

As shown in Fig. 1, an ultrasonic beam with frequency ω is projected towards the progressive direction of a vehicle from a transmitting head which is a sound source (referred to as an ultrasonic head), and a reflected wave with frequency v is received at a receiving head.

(b) *Influence of atmospheric temperature.* The value of f_a, which is obtained while a car is moving in an ultrasonic beam at a certain velocity, is a function of its velocity, but it is also influenced by the velocity of sound. Thus, f_a cannot be obtained as an exact function of the velocity without corrections for the temperature on the velocity of sound. Therefore, this equipment uses a temperature sensor at the operation process stage to perform the correction for the velocity of sound and the operation of velocity.

(c) *Influence of wind.* A sound wave is carried away by the wind, affecting the velocity and the direction of propagation. However, considering the ordinary velocity of the wind, a target accuracy can be fully obtained without influence from the wind velocity.

1.2 Summary of Equipment

(a) *Structure.* The equipment consists of the ultrasonic head and the signal process section, as shown in Fig. 2.

The ultrasonic head includes two independent heads (i.e., the transmitting head and the receiving head) and is distinguished by a united structure including a temperature sensor. The temperature sensor has a dual structure to avoid the influence of rain, direct rays and so on.

(b) *Accuracy.* The accuracy of the velocity measurements, gives results within 10% of the real velocity of the vehicles. Considering the influence of a 10 m s^{-1} wind, this will be 0.17% in the case of a vehicle travelling at 30 km h^{-1} and 0.32% in the case of a vehicle travelling at 100 km h^{-1}, while, for a 25 m s^{-1} wind, this will be 0.7% in the case of a vehicle travelling at 30 km h^{-1}, and approximately 1.1% for a vehicle travelling at 100 km h^{-1}.

Similarly, the influence of the temperature, in theory, causes inaccuracy because the velocity of ultrasonic propagation is influenced by the atmospheric temperature, so a sensor is mounted for the continuous measurement of temperature and the velocity is compensated for with the measured value. As a result, the influence of the atmospheric temperature due to the tolerance of velocity measurement is held down to within 0.5%. As stated previously, there are doubts about its accuracy in the case where an ultrasonic Doppler-type detector is used for speed control; however, the detector is thought to be fully available for the collection of speed data and for speed warning (including speed indication).

2. Loop-Type Detectors

Loop-type detectors have been used in many detection situations in traffic control, as well as the ultrasonic types. Loop-type detectors utilize the characteristics of a loop coil buried in a road surface, namely that its electric fixed number fluctuates due to the approach of metallic objects (vehicles) and so it detects their existence and passing. That is, as shown in Fig. 3, while

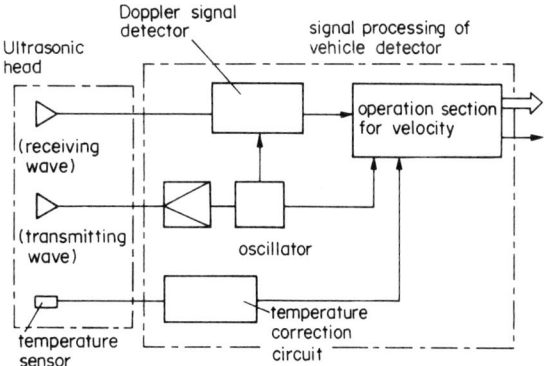

Figure 2
Block diagram of an ultrasonic Doppler-type detector

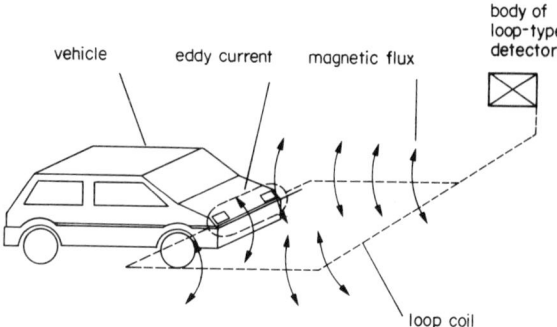

Figure 3
Principle of operation of a loop-type detector

an alternating current is being sent to loop coils buried in a road surface, an ac magnetic field of the same frequency is produced near the road surface. If a metallic object enters this ac magnetic field, then magnetic induction causes eddy currents in the metallic object. This results in a change in the impedance of the loop coil at the input side. First, this change of impedance is detected, then the object (car) is detected. In this case, it can be said that the loop coil and the vehicle act as a transformer; that is, if the loop coil is the primary winding, the metallic part of the car is the secondary coil, which is being shorted by the metallic resistance. The primary winding part is being combined with the secondary winding part by extremely low mutual inductance, which becomes zero while any vehicle is present on the loop coil. As the car approaches the loop coil, the mutual inductance becomes high, causing the impedance to change at the primary coil side of the transformer. This change of impedance is manifested as changes in resistance, inductance and phase. All of the above quantities are available for the detection of vehicles; however, for the most part, a detecting method based on the change in inductance is being put to practical use. The change in inductance in the loop coil is also caused by temperature, rain and the transformation of a road surface, so in the long run, the change caused by these factors is sometimes more than that caused by the penetration of vehicles into the locality. Therefore, this must be compensated for. Practical equipment measures the passage of vehicles, only taking into account the amount of inductance changed during a short time to produce its output. The change in inductance caused by the passage of vehicles varies with the type of vehicle. The system is more sensitive to vehicles with low bodies and less sensitive to higher bodied vehicles.

3. Vehicle Profile Classifier Detectors

A vehicle profile classifier detector classifies and totals the volume of traffic passed per vehicle profile based on the information from loop-type detectors and ultrasonic detectors (loop type, ultrasonic type and a combination of both types measure the height of floor surface, the height of vehicles and the length of vehicles, respectively). There are four classifications—large buses, large trucks, small trucks and cars—based on combinations of height, length and floor surface (see Fig. 4).

This equipment has an operating function for recording occupancy rate and velocity as secondary information, including recording the volume of traffic per vehicle profile, and also has a function for the transmission of this information to a distant place. Moreover, this equipment is available as terminal equipment for the collection of wider traffic information, such as a road plan, traffic safety, maintenance and management of a road, preservation of the environment, effective and practical use of a road, smooth running of vehicles on a road and so on.

3.1 Principle of Operation

This equipment can measure the height of three classifications of vehicle with its ultrasonic-type detector, the height of two classifications of vehicle floor surfaces with its loop-type detector and the length of two classifications of vehicle with the combination of output information from both units, and it classifies vehicles into four profiles according to the following flow and classification standard.

(*a*) *Measurement of vehicle height using an ultrasonic-type detector*. This measures the height of a vehicle using the ultrasonic transmission time from the emission of an ultrasonic audible sound at an ultrasonic transceiver to its return after having been reflected from the vehicle. The height classifications are: greater than 2.0 m, between 1.5 m and 2.0 m, and less than 1.5 m.

(*b*) *Measurement of the height of floor surface using a loop-type detector*. Inductance of the loop coil buried under the road surface is affected when a vehicle passes over it. The change in inductance varies with the floor area of a vehicle and with the height of the base of a vehicle (referred to as the height of the floor surface). That is, the change of inductance is a function of the distance between a loop coil and the base of a vehicle, so the height of the floor surface can be measured using the change of inductance. Thus, large trucks and buses can be distinguished from small trucks and cars.

(*c*) *Measurement of the length of vehicles using a combination of loop-type and ultrasonic-type detectors*. To obtain two classifications of vehicle length (4.75 m or longer and shorter than 4.75 m) from the condition of the overlap of the output signal from each detector, a loop coil and an ultrasonic transceiver are being utilized.

4. Image Sensors

As a method of collecting information about a stream of vehicles in a traffic control system, loop-type detectors or ultrasonic-type detectors have usually been

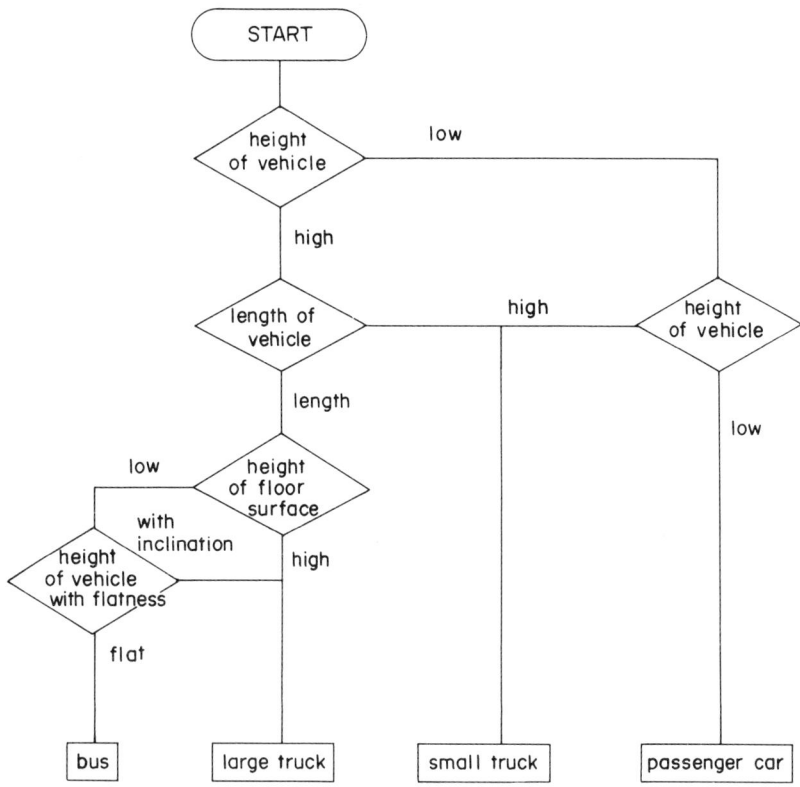

Figure 4
Flowchart of classification of vehicle profiles

employed. To measure parameters of a stream of vehicles, such as the number of vehicles, the velocity of vehicles, the percentage of occupation time, and so on, detectors first detect vehicles that have passed a point on the ground, then their output signal is processed by a central processing unit. As such, the detectors mentioned previously measure a stream of vehicles on a certain point, while image sensors can optically detect the condition of a stream of vehicles in the spatial range on a road to collect directly the overall information about a stream of vehicles. The equipment is classified into two sorts by its method of detecting the optical information: one employs a secondary sensor, the CCTV camera, and the other employs the equipment with a primary sensor, the minute photoelectron alley. The equipment with such an image sensor can detect vehicles in a field of view, in order to measure the information about a stream of vehicles which is necessary for traffic control, such as the number of vehicles passing, the velocity at a point and the average velocity. Also such information can be collected in real time, so incidents can be detected early to support disaster prevention.

CCTV cameras can be switched to give a controller supervisory control if it is necessary to obtain more detailed information from a picture (see *Video Sensors*).

4.1 Construction

A CCTV system consists of the CCTV camera for imaging a spatial range on a road and the picture processing section which processes the data for the image, and abstracts the distinctive signal of vehicles from the data to compute information about a stream of vehicles, such as the number of vehicles, the velocity of the vehicles, the percentage of occupancy of the road and so on (see Fig. 5). A minute photoelectron alley system consists of a primary sensor camera, which detects picture information on plural detecting points set ahead with a photoelectron alley, and the data processing section, which processes the analog information and abstracts the information about vehicles from that in order to compute the information about a stream of vehicles, such as the number of vehicles, the average velocity of vehicles and so on (see Fig. 6).

4.2 Principle of Operation

(*a*) *CCTV system.* As shown in a fundamental construction diagram, a video signal from a CCTV camera is input to the picture processing section, which performs

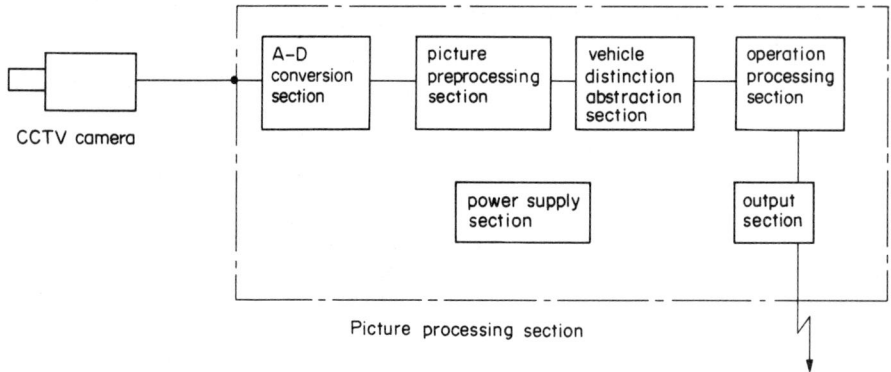

Figure 5
Fundamental construction diagram of a CCTV camera system

analog–digital (A–D) conversion and abstracts the distinctive signal of vehicles in the measuring region in the picture according to the vertical difference abstraction method. This system measures the changed degree of each picture element in the direction of scanning line to decide whether or not there is a vehicle for each scanning line. To be certain, this system compares an A–D converted picture element with each signal level of other picture elements away from the element, for k picture elements through the delay circuit, and outputs the changed degree of the vertical direction. Then, based on the changed degree, the distinctive signal abstraction section determines whether or not there is a vehicle.

After such recognition of vehicles, the operation processing section measures the volume of traffic, the velocity of vehicles, the percentage of space occupancy and the distance between vehicles based on the fundamental elements; that is, the measuring time and the positions of existing vehicles (corresponding to numbers of scanning lines), and outputs the results in the form of data suitable for the system.

(b) *Minute photoelectron alley system.* Minute photoelectronic sensors in a camera are able to receive the optical signal, through their lenses, at a detecting point set ahead on a road. The change in current of each photoelectronic sensor, caused by the passing of vehicles, is amplified to be output to the processing section. The output of the camera is made A–D conversion, then is compared with the last output level. Based on the changed degree and the direction of change, the image is removed and the vehicles abstracted. To compute the velocity of vehicles, two photoelectronic sensors are positioned at a detecting point with a certain space between them so that the velocity of a vehicle can be computed from the difference between the time difference of the two sensors detecting the vehicle. Based on such information, the volume of traffic, the velocity of vehicles at a point, the average velocity of vehicles in space, a traffic snarl and so on, can all be measured and then output in the form suitable for the system.

4.3 Features

Features of such equipment with an image sensor are the following.

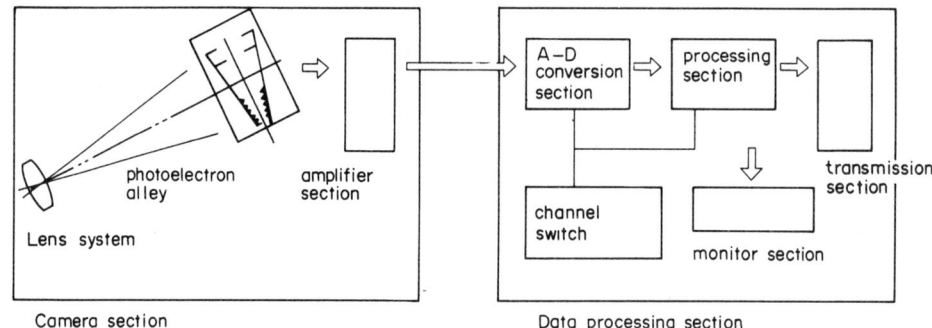

Figure 6
Fundamental construction diagram of a minute photoelectron alley system

(a) They can measure information about a stream of vehicles that cannot be obtained at usual detectors for vehicles.

(b) Through measurement in space, they can measure directly information about a stream of vehicles that cannot be obtained directly from usual detectors for vehicles.

(c) They can operate anywhere so as not to cause a lowering of accuracy, although usual detectors cannot operate without lowering the accuracy at some places where an abnormal stream of vehicles is found.

(d) In the case of a CCTV system, controllers can effectively use a CCTV camera for observing a stream of vehicles.

See also: In-Vehicle Equipment for Future Traffic Control Systems; Road Traffic Monitoring Equipment; Video Sensors

J. Kamata and T. Oda
[Matsushita Com. Ind. Co.,
Yokohama, Japan]

Discrete-Time Point Processes: Applications to Road Traffic

Traffic flows are intrinsically stochastic and dynamic. Deterministic models can be used as approximations to reality, when the detailed nature of the problem is not very important (e.g., in a macroscopic analysis). On the other hand, stochastic stationary models allow for the evaluation of equilibrium behavior only. Models of this kind have been proposed for the evaluation of traffic behavior at a roundabout (Pearce 1987), for the long term evaluation of on-line control strategies (Darroch et al. 1964, Lin 1982) and for the evaluation of queuing behavior at road junctions (Kimber and Daly 1986).

Neither deterministic nor stationary stochastic models are suited for the on-line control of traffic flows. In fact, in this case it is necessary to take into account both the fluctuations and the variations over time of traffic flows. The development of stochastic dynamical models can allow for more accurate estimation of traffic flow behavior, and hence hopefully for better control strategies.

1. Discrete-Time Point Processes

A few basic facts about dynamical point processes and their martingale representation can be briefly reviewed, substantially following the development of Segall and Kailath (1975). The use of point processes and martingale theory, as opposed to more traditional approaches such as Markov chain theory, is motivated (quoting Brémaud 1981)

"by the need for a dynamical model which takes the information dynamics into account in a direct and effective manner and by the necessity of controlling a system on line on the basis of the data collected at the time of the implementation of the control. The ideal tool in this respect is *martingale theory*, which indeed fully acknowledges the existence of information patterns that increase with time ... the martingale point of view is adapted to problems of a dynamical nature where the goal is not to assess different strategies, but to find the best one."

DEFINITION 1. *A point process* $\{T(n)\}$, $n = 0, 1, \ldots$ *is a real positive-valued stochastic process such that*

$$T(0) = 0$$
$$T(n) < T(n+1) \quad \forall n \text{ s.t. } T(n) < \infty$$

Point processes are the natural candidates for the description of occurrence times; that is, times at which a particular event takes place (e.g., a vehicle arrival, a departure, etc.). Associated with each point process there is an integer-valued stochastic process which acts as a counter of the number of the events which occurred up to a time instant t.

DEFINITION 2. *Given a point process* $\{T(n)\}$, $n = 0, 1, \ldots$ *a counting process (or unit jump process) is a stochastic process* $\{Y(t), t \in \mathbb{R}^+\}$ *s.t.*

$$Y(t) = \begin{cases} n & \text{if } t \in [T(n), T(n+1)); n \geq 0 \\ +\infty & \text{if } t \geq T(\infty) = \lim_{n \to \infty} T(n) \end{cases}$$

In many situations, different counting processes are superimposed (e.g., in the case of the arrival process in a multilane highway—the process relative to each lane can be thought of as a counting process).

DEFINITION 3. *A k-variate stochastic process* $\{Y(t)\}$ *is a multivariate (k-variate) counting process if its ith component* $N_i(t)$, $i = 1, \ldots, k$ *is s.t.*

$$N_i(t) = \mathbf{1}_{\{T(n) \leq t\}} \mathbf{1}_{\{Z(n) = i\}}$$

where $\mathbf{1}_{\{A\}}$ is the indicator function of the event A, $\{T(n)\}$ is a point process and $\{Z(n)\}$ is a sequence of $\{1, 2, \ldots, k\}$-valued random variables defined on the same probability space as $\{T(n)\}$.

Loosely speaking, $T(n)$ records the time at which the nth "change of state" of $\{Y(t)\}$ takes place, while $Z(n)$ indicates which of the k components of $\{Y(t)\}$ "jumped." It is supposed that at any time at most one of the components of $Y(t)$ can have a jump.

It is assumed that the system under consideration can be modelled as a stochastic (generally multivariate) process $\{X(t)\}$ called the internal state process. In traffic situations, the internal state may be, for example, a queue length. It is also assumed that, associated with

101

the internal state process, there is an observation process $\{Y(t)\}$. This is the situation when a vehicle-detecting device records vehicle passages on a road.

DEFINITION 4. *The internal history is the collection of σ-algebras β^t-generated by the observation and internal state processes:*

$$\beta^t = \sigma(X(s), Y(s), s \in [0, t])$$

DEFINITION 5. *The observed history is the collection of σ-algebras $\mathcal{F}^t$ generated by the observation process:*

$$\mathcal{F}^t = \sigma(Y(s), s \in [0, t])$$

For discrete-time point processes it is particularly convenient to introduce a $\{0, 1\}$-valued stochastic process $\{y(t)\}$ defined as

$$y(0) = 0$$
$$y(t) = Y(t) - Y(t-1) \quad \forall t \geq 1$$

which represents the "jumps" of the process $\{Y(t)\}$. It is easily seen that $y(t)$ admits the following representation:

$$y(t) = r(t; \beta^{t-1}) + w(t)$$

where $r(t;)$ is defined as the stochastic rate of $y(t)$ and takes the form:

$$r(t; \beta^{t-1}) = E(y(t) | \beta^{t-1})$$

The process $\{w(t)\}$ is a martingale difference with respect to β^{t-1}; that is,

$$E(w(t) | \beta^{t-1}) = 0$$

Notice that $w(t)$ is not a gaussian noise as in the more familiar case of Wiener processes.

In many cases, the internal state process is chosen "wide enough" to contain all the relevant information about the evolution of the system under inspection. In particular, it is often assumed that the internal history contains the observation history. Thus, in the following, the reference to β and $\mathcal{F}$ will be dropped and it will be assumed that

$$\beta^{t-1} = \sigma(x^t, y^{t-1})$$

with $x^t = \{x(0), x(1), \ldots, x(t)\}$ being the internal state sequence, and $y^{t-1} = \{y(0), y(1), \ldots, y(t-1)\}$ the observation sequence. It is easily seen that both the observation and the internal state processes admit a martingale representation

$$x(t+1) = s(t; y^{t-1}; x^t) + v(t) \quad (1a)$$

$$y(t) = r(t; y^{t-1}; x^t) + w(t) \quad (1b)$$

with

$$s(t; y^{t-1}; x^t) = E(x(t+1) | \beta^{t-1}) \quad (2a)$$
$$r(t; y^{t-1}; x^t) = E(y(t) | \beta^{t-1}) \quad (2b)$$

and v and w being two martingale difference processes. Explicit forms for the processes s and r will be given in the following sections for specific traffic situations.

2. Filtering for Dynamical Point Processes

The main objective of building an accurate model of the evolution of traffic characteristics is to obtain tools for their prediction and estimation. Estimation and prediction can be performed in a dynamic fashion based on the observations collected. The mechanism can be substantially derived from the application of Bayesian (least squares) estimation procedures.

Considering the following quantities

$$\hat{s}(t|t-1) = E(s(t; y^{t-1}; x^t) | \mathcal{F}^{t-1})$$
$$\tilde{s}(t|t-1) = s(t; y^{t-1}; x^t) - \hat{s}(t|t-1)$$
$$\hat{r}(t|t-1) = E(r(t; y^{t-1}; x^t) | \mathcal{F}^{t-1})$$
$$\tilde{r}(t|t-1) = r(t; y^{t-1}; x^t) - \hat{r}(t|t-1)$$
$$v(t) = y(t) - \hat{r}(t|t-1)$$

the first two represent, respectively, the Bayesian estimate of s based on the first $t-1$ observations and the error of the estimate, respectively; analogously for $\hat{r}$ and $\tilde{r}$; $v(t)$ is the so-called innovation process, the difference between the actual and the predicted observation.

Assume that $y(t)$ is a binary vector such that

$$y^T(t)y(t) \leq 1 \quad (3)$$

This causes no loss of generality when the state space of the observation process is discrete.

Based on these definitions, the optimal (in a least-squares sense) estimate, usually called the "filter," of the internal state vector at time $t+1$ on the basis of all the observations collected up to time t is given by the formula:

$$\hat{x}(t+1|t) = \hat{s}(t|t-1) + E(\tilde{s}(t|t-1)\tilde{r}^T(t|t-1)$$
$$+ v(t)w^T(t) | \mathcal{F}^{t-1})$$
$$\times (\text{diag}\{\hat{r}(t|t-1)\} - \hat{r}(t|t-1)$$
$$\times \hat{r}^T(t|t-1))^{-1}v(t) \quad (4)$$
$$\hat{x}(1|0) = E(x(1))$$

where $\text{diag}\{a\}$ represents a diagonal matrix with the ith diagonal element equal to the ith component of a. The derivation of this formula can be obtained as in Betrò et al. (1983).

The actual computation of the filter is obviously committed to the possibility of computing the expectation in Eqn. (4). It should be observed that in order to compute $E(v(t)w^T(t)|\mathcal{F}^{t-1})$ the correlation between $x(t+1)$ and $y(t)$ should be taken into account; thus, it cannot be expected that such an explicit computation could be carried out by manipulation of the above formula, but instead further modelling assumptions need to be formulated.

A particularly important special case for the computation of filters is that of linearity, in which

$$x(t+1) = S(t; y^{t-1})x(t) + v(t) \quad (5a)$$
$$y(t) = R(t; y^{t-1})x(t) + w(t). \quad (5b)$$

The filter, then, is easily computed as

$$\begin{aligned}\tilde{x}(t+1|t) =\ & S(t; y^{t-1})\hat{x}(t|t-1) \\ & + \{E(x(t+1)y^T(t)|\mathcal{F}^{t-1}) \\ & - S(t; y^{t-1})\hat{x}(t|t-1)\hat{x}^T(t|t-1)R^T(t; y^{t-1})\} \\ & \times \{\text{diag}\{R(t; y^{t-1})\hat{x}(t|t-1)\} - R(t; y^{t-1}) \\ & \times \hat{x}(t|t-1)\hat{x}^T(t|t-1)R^T(t; y^{t-1})\}^{-1} \\ & \times \{y(t) - R(t; y^{t-1})(\hat{x}(t|t-1))\} \quad (6)\end{aligned}$$

which again, as is obvious, contains an expectation that can be computed only after the introduction of more detailed modelling assumptions.

3. A Simple Model

This section is devoted to the presentation of a first class of models that fits into the general framework discussed in the previous sections. Full descriptions of these models were given by Baras et al. (1977, 1979a, b, 1985) and Baras and Dorsey (1981).

The simplest model introduced by Baras et al. (1979b), will be considered in some detail in order to enlighten the approach. A single, isolated intersection of two one-way single-lane streets is considered. The intersection is controlled by a two-phase traffic light and, on each lane, a detector is located at some distance upstream the stop line, enabling up to N vehicles to queue up before the queue extends beyond the detector.

Let the process $\{x(t)\}$ represent the queue evolution process, where the queue is defined as the set of vehicles between the stop line and the detector. It is convenient to define $\{x(t)\}$ as an N-vector, with binary components such that

$$x_k(t) = \begin{cases} 1 & \text{if the queue length is } k \text{ at time } t \\ 0 & \text{otherwise} \end{cases} \quad (7)$$

In such a way, the filter given by Eqn. (4) gives a complete description of the probability mass function of $x(t)$.

Let the scalar process $y(t)$ be defined as

$$y(t) = \begin{cases} 1 & \text{if an arrival is detected at time } t \\ 0 & \text{otherwise} \end{cases} \quad (8)$$

It is therefore assumed that one vehicle at most can arrive at time t. This assumption limits the power of the model as it uses a necessarily thin discretization of the time. Observe however that Eqn. (3) is trivially fulfilled.

A third discrete-time point process of interest in the evolution of the system is the departure process $y_d(t)$, which is not observed and which can be assumed as binary valued due to the thin time discretization:

$$y_d(t) = \begin{cases} 1 & \text{if there is a departure from the stop line at time } t \\ 0 & \text{otherwise} \end{cases}$$

In order to derive the explicit form of $s(t; y^{t-1}; x^t)$ in Eqn. (1a) and of $r(t; y^{t-1}; x^t)$ in Eqn. (1b), it is necessary to point out the probabilistic dependencies. Roughly speaking, the processes $y(t)$ and $y_d(t)$ are assumed to be dependent on the past history through $x(t)$ only. In particular, the number of vehicles that depart from the stop line depends on whether the queue is empty or not. Moreover, the queue at time $t+1$ is assumed to depend on the past history through the queue at time t, $x(t)$.

From these dependency assumptions and from Eqn. (2) and the definition of $x(t+1)$, it immediately follows that

$$s_k(t; y^{t-1}; x^t) = P\{x_k(t+1) = 1 | x(t)\}$$

and that

$$r(t; y^{t-1}; x^t) = E(y(t)|x(t)) = P\{y(t) = 1 | x(t)\}$$

If the matrix $S(t)$ is now introduced whose elements are given by

$$S_{hk}(t) = P\{x_k(t+1) = 1 | x_h(t) = 1\}$$

it immediately follows from Eqns. (1a) and (7) that

$$x(t+1) = S(t)x(t) + v(t) \quad (9)$$

The elements of $S(t)$ are immediately found to be

$$\begin{aligned}S_{k,k-1}(t) &= \mu_k(t)[1 - \lambda_k(t)] \\ S_{k,k+1}(t) &= \lambda_k(t)[1 - \mu_k(t)] \\ S_{k,k}(t) &= [1 - \mu_k(t)][1 - \lambda_k(t)] \\ S_{k,h}(t) &= 0 \quad |k-h| > 1\end{aligned}$$

where

$$\lambda_k(t) = P\{y(t) = 1 \mid x_k(t) = 1\}$$
$$\mu_k(t) = P\{y_d)t) = 1 \mid x_k(t) = 1\}$$

From Eqns. (1b) and (7), it follows that

$$y(t) = \lambda^T(t)x(t) + w(t)$$

that, together with Eqn. (9), completes the definition of the system in the linear form of Eqn. (5). In order to compute $\hat{x}(t+1 \mid t)$, it is now necessary to specify $E(x(t+1)y^T(t) \mid x(t))$ on the basis of the modelling assumptions. Recalling Eqns. (7) and (8) it follows that

$$\begin{aligned}
E(x_k(t+1)y(t) \mid x(t)) \\
&= P\{x_k(t+1) = 1, y(t) = 1 \mid \mathbf{x}(t)\} \\
&= \lambda_{k-1}(t)[1 - \mu_{k-1}(t)]P\{x_{k-1}(t) = 1 \mid x(t)\} \\
&+ \lambda_k(t)\mu_k(t) + P\{x_k(t) = 1 \mid x(t)\}
\end{aligned}$$

Recalling that $P\{x_k(t) = 1 \mid x(t)\} = \hat{x}_k(t \mid t-1)$, the above derivation makes it possible to recursively compute the optimal estimate of the queue length through Eqn. (6).

In Baras et al. (1979b), two other models are presented. These models take into account the fact that, for each vehicle velocity, detectors can give the occurrence time and that, on each lane, another detector might have been installed in order to detect departures too. However, in order to deal with the increased complexity of the situation, the models become more tangled.

4. A General Model

Consider now a more general model, presented in detail by Betrò et al. (1986, 1987a, b). As previously, it is assumed that a detector is placed on a link approaching an intersection, at a distance from the stop line such that it is unlikely for the queue to extend beyond the detector. The detector returns, at integer time instants $t = 1, 2, \ldots$, the count of vehicles that have arrived during the time interval $[t-1, t)$; for the sake of simplicity it is assumed that no detection error occurs, although this case can also be easily accomplished (Betrò et al. 1987a).

Let the process $\{a(t)\}$ represent the detector output at time $t+1$ and assume that it has state space $0, \ldots, K_a$ and depends on a parameter (which might actually be a set of parameters), being itself a random process, say $\{\lambda(t)\}$. For the sake of simplicity the state space of $\{\lambda(t)\}$ will be taken as finite, with K_λ states $\lambda_1, \ldots, \lambda_{K_\lambda}$. The interest is in inference about the queue evolution $\{q(t)\}$, where $q(t) \in \{0, \ldots, K_q\}$ is the number of vehicles between the detector and the stop line at instant t, on the basis of traffic counts provided by $\{a(t)\}$. If the process $\{d(t)\}$, $d(t) \in \{0, \ldots, K_d\}$, represents the departure process at the stop line, then we may write

$$q(t+1) = q(t) + a(t) - d(t) \tag{10}$$

The following reasonable assumptions will now be introduced.

ASSUMPTION 1. $d(t) \leq q(t)$; this rules out the possibility that more departures than vehicles in the queue can take place and, in particular, that arriving vehicles immediately cross the stop line.

ASSUMPTION 2. $\lambda(t)$ depends only on $q(t), \lambda(t-1)$, in the sense that its distribution for each instant t is determined once $q(t)$ and $\lambda(t-1)$ are given. Dependence on the queue allows the influence that queue extension has on the arrival process to be taken into account, while dependence on $\lambda(t-1)$ considers the interactions within the traffic flow.

ASSUMPTION 3. $a(t)$ depends, in the sense of Assumption 2, only on $\lambda(t)$; this gives the meaning of $\lambda(t)$ as the "parameter" of the distribution of $a(t)$.

ASSUMPTION 4. $d(t)$ depends only on $q(t)$; notice that dependence on time t is not ruled out, so that, for example, the condition $d(t) = 0$ may hold if, at t, the traffic light at the intersection is red.

In order to write down the system dynamics in the form of Eqn. (5), consider the $(K_q + 1) \times K_\lambda$ state vector $x(t)$ with binary components

$$x_k(t) = 1 \leftrightarrow q(t) = i(k), \quad \lambda(t) = \lambda_{l(k)} \tag{11}$$

where $0 \leq i(k) \leq K_q$ and $\leq l(k) \leq K_\lambda$ are uniquely determined by the equation $l(k) + (K_\lambda + 1)(q(k) - 1) = k$; and the observation vector $y(t)$ with binary components

$$y_j(t) = 1 \leftrightarrow a(t) = j, \quad j = 0, \ldots, K_a$$

Then, by Eqn. (10) and Assumption 2:

$$\begin{aligned}
E(x_k(t+1) \mid y^{t-1}, x^t) \\
&= P\{x_k(t+1) = 1 \mid y^{t-1}, x^t\} \\
&= P\{\lambda(t+1) = \lambda_{l(k)} \mid q(t+1) = i(k), \lambda(t)\} \\
&\quad \times \{q(t+1) = i(k) \mid q(t), \lambda(t)\}
\end{aligned}$$

from which it is easily seen, recalling Eqn. (11), that

$$E(x(t+1) \mid y^{t-1}, x^t) = \mathbf{S}(t)x(t)$$

where

$$S_{k_1 k_2} = P\{\lambda = \lambda_{l(k_1)} \mid q(t+1) = i(k_1), \lambda(t) = \lambda_{l(k_2)}\}$$
$$\times P\{q(t+1) = i(k_1) \mid q(t) = i(k_2), \lambda(t) = \lambda_{l(k_2)}\} \tag{12}$$

with

$$P\{q(t+1) = i(k) | q(t), \lambda(t)\}$$
$$= \sum_{a(t)} P\{q(t+1) = i(k) | q(t), a(t)\} P\{a(t) | \lambda(t)\}$$
$$= \sum_{a(t), d(t)} P\{q(t+1) = i(k) | q(t), a(t), d(t)\}$$
$$\times p\{a(t) | \lambda(t)\} p\{d)t) | q(t)\},$$
$$q(t) + a(t) - d(t) = i(k)$$

where $p(\ )$ denotes the probability mass function.

For what concerns the observation process $\{y(t)\}$, by Assumption 3:

$$E(y_j(t) | y^{t-1}, x^t) = P\{a(t) = j | \lambda(t)\}$$

so that, introducing the $K_a \times (K_q + 1) K_\lambda$ matrix

$$\mathbf{R}_{jk} = P\{a(t) = j | \lambda(t) = \lambda_{l(k)}\} \quad (13)$$

we have

$$E(y(t) | y^{t-1}, x^t) = \mathbf{R}(t) x(t).$$

Looking at Eqns. (1) and (2), it turns out that $x(t+1)$ and $y(t+1)$ can be written in the linear form of Eqn. (5) and hence the filter $\hat{x}(t+1|t)$ can be expressed in the form of Eqn. (6). It remains to evaluate the term $E(x(t+1) y^T(t) | \mathcal{F}^{t-1})$, for which we can proceed as follows:

$$E(x_k(t+1) y_j(t) | y^{t-1}) = P\{x_k(t+1) = 1, y_j(t) = 1 | y^{t-1}\}$$
$$= \sum_{q(t), \lambda(t)} P\{\lambda(t+1) = \lambda_{l(k)} | q(t+1) = i(k), \lambda(t)\}$$
$$\times P\{q(t+1) = i(k) | q(t), a(t) = j\} \quad (14)$$
$$\times P\{a(t) = j | \lambda(t)\} p(q(t), \lambda(t) | y^{t-1})$$

which shows that $E(x(t+1) y^T(t) | \mathcal{F}^{t-1})$ can be expressed in terms of $p(q(t), \lambda(t) | y^{t-1})$, that is, in terms of the filter at time $t-1$, $\hat{x}(t|t-1)$. According to Eqn. (6), this gives the filter $\hat{x}(t+1|t)$ a fully explicit recursive form.

5. An Example

In order to give some hints about the construction of a model for actual situations, the following further modelling assumptions can be introduced.

ASSUMPTION 5. *The parameter $\lambda(t)$ represents the expected number of vehicle arrivals during a unit time interval; it is assumed that its evolution is independent of $q(t)$ and that it can take only two values λ_L, λ_H with $\lambda_L < \lambda_H$. It is further assumed that two quantities α, $\beta \in [0, 1]$ are known s.t.*

$$\alpha = P\{\lambda(t) = \lambda_H | \lambda(t-1) = \lambda_H\}$$
$$= 1 - P\{\lambda(t) = \lambda_L | \lambda(t-1) = \lambda_H\}$$
$$\beta = P\{\lambda(t) = \lambda_H | \lambda(t-1) = \lambda_L\}$$
$$= 1 - P\{\lambda(t) = \lambda_L | \lambda(t-1) = \lambda_L\}$$

ASSUMPTION 6. *Arrivals during a time unit are assumed to be distributed as a binomial random variable with mean $\lambda(t)$:*

$$P\{a(t) = k | \lambda(t)\} = \binom{K_a}{k} \left[\frac{\lambda(t)}{K_a}\right]^k \left[1 - \frac{\lambda(t)}{K_a}\right]^{K_a - k}$$

with $k = 0, 1, \ldots, K_a$.

ASSUMPTION 7. *The departure process also follows a binomial distribution:*

$$P\{d(t) = k | q(t)\} = \binom{d_{\max}(q(t))}{k} \mu^k (1-\mu)^{d_{\max}(q(t))-k}$$

with $k = 0, 1, \ldots, d_{\max}(t)$, $d_{\max}(q) = \min\{K_d, q\}$, and $\mu \in (0, 1)$; it is also reasonable to assume that $K_d \leq K_a$.

In order to simplify the treatment of the particular situation that occurs when the queue reaches its maximum allowable extension, it is supposed here that $K_q = \infty$; all the formulae of the previous section carry over to this case without any theoretical difficulty.

It is easy to see that $|q(t+1) - q(t)| \leq K_a$; as a consequence, the matrix s in Eqn. (12) can be represented as a block K_a-diagonal matrix, where each block is a 2×2 matrix of the form

$$\begin{bmatrix} \phi(\lambda_L)(1-\beta) & \phi(\lambda_L)(1-\alpha) \\ \phi(\lambda_H)\beta & \phi(\lambda_H)\alpha \end{bmatrix}$$

The generic block in position $(i+h, i)$ (with $i \geq 0, 0 > h \leq K_a$) is characterized by

$$\phi(\lambda) = \sum_{k=h}^{\min\{K_a, h+K_d, h+i\}} \binom{K_a}{k} \left(\frac{\lambda}{K_a}\right)^k \left(1 - \frac{\lambda}{K_a}\right)^{K_a-k}$$
$$\times \left[\frac{d_{\max}(i)}{k-h}\right] \mu^{k-h} (1-\mu)^{d_{\max}(i)-k-h}$$

where $\lambda \in \{\lambda_L, \lambda_H\}$.

Similarly, the $(i-h, i)$ block is characterized by

$$\phi(\lambda) = \sum_{k=h}^{\min\{K_d, i\}} \binom{K_a}{k-h} \left(\frac{\lambda}{K_a}\right)^{k-h} \left(1 - \frac{\lambda}{K_a}\right)^{K_a-k-h}$$
$$\times \binom{d_{\max}(i)}{k} \mu^k (1-\mu)^{d_{\max}(i)-k}$$

and, for the (i, i) block,

$$\phi(\lambda) = \sum_{k=0}^{\min\{K_d, i\}} \binom{K_a}{k} \left(\frac{\lambda}{K_a}\right)^k \left(1-\frac{\lambda}{K_a}\right)^{K_a-k}$$

$$\times \binom{d_{\max}(i)}{k} \mu^k (1-\mu)^{d_{\max}(i)-k}$$

The **R** matrix in Eqn. (13) is given by

$$R_{jk} = \begin{cases} \binom{K_a}{j} \left(\frac{\lambda_L}{K_a}\right)^j \left(1-\frac{\lambda_L}{K_a}\right)^{K_a-j} & \text{if } k = 1, 3, \ldots \\ \binom{K_a}{j} \left(\frac{\lambda_H}{K_a}\right)^j \left(1-\frac{\lambda_H}{K_a}\right)^{K_a-j} & \text{if } k = 2, 4, \ldots \end{cases}$$

with $j = 0, 1, \ldots, K_a$.

Finally, the quantity $E(x_k(t+1) y_j(t) | y^{t-1})$ in Eqn. (14), is given by

$$(1-\beta) \sum_q \sum_{d=0}^{\min\{K_d, q\}} \mathbf{1}_{\{i(k) = q+j-d\}} \binom{K_a}{j} \left(\frac{\lambda_L}{K_a}\right)^j \left(1-\frac{\lambda_L}{K_a}\right)^{K_a-j}$$

$$\times \binom{d_{\max}(q)}{d} \mu^d (1-\mu)^{d_{\max}(q)-d}$$

$$\times P\{q(t) = q, \lambda(t) = \lambda_L | y^{t-1}\}$$

$$+ (1-\alpha) \sum_q \sum_{d=0}^{\min\{K_d, q\}} \mathbf{1}_{\{i(k) = q+j-d\}}$$

$$\times \binom{K_a}{j} \left(\frac{\lambda_H}{K_a}\right)^j \left(1-\frac{\lambda_H}{K_a}\right)^{K_a-j}$$

$$\times \binom{d_{\max}(q)}{d} \mu^d (1-\mu)^{d_{\max}(q)-d}$$

$$\times P\{q(t) = q, \lambda(t) = \lambda_H | y^{t-1}\}$$

if $\lambda_{l(k)} = \lambda_L$ and by

$$\beta \sum_q \sum_{d=0}^{\min\{K_d, q\}} \mathbf{1}_{\{i(k) = q+j-d\}} \binom{K_a}{j} \left(\frac{\lambda_L}{K_a}\right)^j \left(1-\frac{\lambda_L}{K_a}\right)^{K_a-j}$$

$$\times \binom{d_{\max}(q)}{d} \mu^d (1-\mu)^{d_{\max}(q)-d}$$

$$\times P\{q(t) = q, \lambda(t) = \lambda_L | y^{t-1}\}$$

$$+ \alpha \sum_q \sum_{d=0}^{\min\{K_d, q\}} \mathbf{1}_{\{i(k) = q+j-d\}} \binom{K_a}{j} \left(\frac{\lambda_H}{K_a}\right)^j \left(1-\frac{\lambda_H}{K_a}\right)^{K_a-j}$$

$$\times \binom{d_{\max}(q)}{d} \mu^d (1-\mu)^{d_{\max}(1)-d}$$

$$\times P\{q(t) = q, \lambda(t) = \lambda_H | y^{t-1}\}$$

if $\lambda_{l(k)} = \lambda_H$.

See also: Underground Railroad Modelling and Control: Discrete-Event Approach

Bibliography

Baras J S, Dorsey A J 1981 Stochastic control of two partially observed competing queues. *IEEE Trans. Autom. Control* **26**, 1106–17

Baras J S, Dorsey A J, Levine W S 1979a Estimation of traffic platoon structure from headway statistics. *IEEE Trans. Autom. Control* **24**, 553–9

Baras J S, Dorsey A J, Makowski A M 1985 Two competing queues with linear cost and geometric service requirements: The μc-rule is often optimal. *Adv. Appl. Probab.* **17**, 186–209

Baras J S, Levine W S, Dorsey A J, Lin T L 1977 Advanced filtering and prediction software for urban traffic control systems. Department of Transport Final Report Contract DOT-05-60134

Baras J S, Levine W S, Lin T L 1979b Discrete time point processes in urban traffic queue estimation. *IEEE Trans. Autom. Control* **24**, 12–27

Betrò B, Schoen F, Speranza M G 1983 Modelling and optimization of stochastic traffic flows. Istituto Applicazioni Matematica e Informatica Report No. 83.8 (Milan)

Betrò B, Schoen F, Speranza M G 1986 Markov models for traffic evolution at an intersection. *Methods Oper. Res.* **53**, 365–73

Betrò B, Schoen F, Speranza M G 1987a Dynamic estimation of queue behaviour in urban traffic. *EJOR* **31**, 368–75

Betrò B, Schoen F, Speranza M G 1987b A stochastic environment for the adaptive control of single intersections. In: Gartner N H, Wilson N H M (eds.) *Transportation and Traffic Theory*. Elsevier, New York, pp. 217–32

Brémaud P 1981 *Point Processes and Queues—Martingale Dynamics*. Springer, New York

Darroch J N, Newell G F, Morris R W J 1964 Queues for a vehicle-actuated traffic light. *Oper. Res.* **12**, 882–95

Kimber R M, Daly P N 1986 Time-dependent queueing at road junctions: Observation and prediction. *Transp. Res. B* **20**, 187–203

Lin F 1982 Predictive models of traffic-actuated cycle splits. *Transp. Res. B* **16**, 361–72

Pearce C E M 1987 A probabilistic model for the behavior of traffic at a roundabout. *Transp. Res. B* **21**, 207–16

Segall A, Kailath T 1975 The modeling of randomly modulated jump processes. *IEEE Trans. Inform. Theory* **21**, 135–43

B. Betrò
[Istituto Applicazioni Matematica e Informatica, Milan, Italy]

F. Schoen
[Dipartimento di Scienze dell'Informazione, Milan, Italy]

M. G. Speranza
[Dipartimento di Matematica e Informatica, Udine, Italy]

DRIVE

The transport sector is, and will continue to be, a very important part of the European economy. It represents more than 6% of the gross national product (GNP) and more than 10% of the average family budget is devoted to transport. Road transport represents between 80% and 85% of the total passenger distance travelled in Europe. Furthermore, car ownership has steadily increased during the 1980s. Although there are now some 120 million motor cars in Europe, the rate of car ownership still lags substantially behind that of the USA (330 as compared with 550 cars per 1000 inhabitants), indicating that further growth can be expected. The negative impacts of road transport on both human safety and the environment are significant.

(a) Every year in the European Community (EC) around 55 000 people are killed on the roads, 1 700 000 are injured and 150 000 are permanently handicapped. The financial cost of this is estimated to be more than ECU 50 billion (US$ 65 billion) per year; the social cost in terms of human misery and suffering cannot be measured.

(b) The total cost of operating vehicles in the EC, including time spent driving, which for an average European represents 2.5 years of life, is estimated to be ECU 500 billion (US$ 650 billion) per year. There is an enormous potential for savings simply through improved routing and congestion reduction.

(c) Vehicle emissions are a significant element of total environmental pollution. Emissions are estimated to cost Europe between ECU 5 billion (US$ 6.5 billion) and ECU 10 billion (US$ 13 billion) per year.

Current trends indicate an average growth rate in demand for road travel in excess of 3% per year which, if continued, implies a 50% increase in traffic between 1990 and 2000. Demand for international road travel is increasing at double this rate. The aim of the EC of achieving the Internal Market by 1992 is giving extra impetus to the forces creating demand.

Existing approaches such as traffic management schemes, civil engineering improvement, engine management technology and EC directives on vehicle standards are important but have their limitations. However, innovations and cost reductions in information technology, telecommunications and broadcasting offer new opportunities for effective solutions to many of these problems.

1. EC Initiative

In December 1984, a European Council resolution on road safety invited the European Commission to propose an appropriate research action. As a result, during 1985, background studies into various aspects of current and future developments in the domain of road transport informatics were conducted on behalf of the European Commission to investigate their potential for application to road and vehicle safety. The results of these studies, together with road transport research initiatives started within the framework of EUREKA, confirmed the view of the European Commission of the urgent need for a strategic program in this domain. Their proposal for a European Council regulation for an EC research and development program in the field of road transport informatics and telecommunications (European Council of Ministers 1987) was made in July 1987 and was named DRIVE (dedicated road infrastructure for vehicle safety in Europe). Following favorable opinions from the Economic and Social Committee, from the European Parliament and from the Council of Ministers, DRIVE was formally adopted in June 1988 as an EC research program of three years' duration and with a budget of ECU 60 million (US$ 80 million).

DRIVE has three primary objectives:

(a) to improve road safety;

(b) to maximize road transport efficiency and to engineer breakthroughs in this field; and

(c) to contribute to environmental improvements.

DRIVE envisages a common European road transport environment in which drivers are better informed and "intelligent" vehicles communicate and cooperate with the road infrastructure itself. The program follows a top-down systems approach to the research and overall design of traffic management and safety systems, which will represent a significant advance over those systems currently available.

If the innovations and cost reductions in information technology, telecommunications and broadcasting are brought together to provide integrated advanced communications, control and information systems, they will enable new, more flexible and responsive forms of traffic management and safety systems to be created to the benefit of all road users. DRIVE therefore seeks to create favorable conditions for the development of the integrated road transport environment (IRTE), through the precompetitive and collaborative research and development in the field of information technology and telecommunications applied to road transport. This is known as road transport informatics (RTI).

DRIVE will entail

(a) research, development and assessment of a range of RTI technologies;

(b) the evaluation of strategic choices of candidate systems; and

(c) a significant amount of standardization work.

Specifically DRIVE aims to achieve the following results:

(a) the identification of the best choice of systems on the basis of economic and technical criteria;

(b) the best strategy for their implementation;

(c) the specification of performance and compatibility standards that will enable industry to develop the necessary equipment and systems;

(d) the provision of directives and guidelines to which industrial products and intelligent European road transport infrastructures should conform; and

(e) the design and, if necessary, implementation of pilot schemes to assess the performance of equipment and systems.

The importance of harmonized European standards must be stressed, for without them difficulties are put in the way of international road travel and non tariff barriers hinder industry. Additionally, common standards will result in a unified European market for RTI products which will help to bring down the cost of equipment.

DRIVE brings together road users, research institutions, providers of broadcasting and telecommunications services, industry and road transportation authorities. It has developed and will maintain close links with other European actions in the domain, notably those carried out under EUREKA, such as Prometheus, Carminat and Europolis, and COST. In particular, DRIVE will involve EC action with regard to standardization and common functional specifications for advanced infrastructure systems. Such cooperation is essential for supporting the close-to-market activities of European industry and for ensuring that incompatibilities and unnecessary duplication of effort do not occur.

2. The Workplan

In order to meet the objectives of DRIVE, a detailed workplan was prepared by the European Commission, with the participation of all sector participants, and was adopted by the DRIVE management committee (which consists of representatives from all the member states). The DRIVE workplan defines exactly the scope and content of the program in a series of tasks. It has been used as the basis for the public call for proposals from which all current DRIVE projects originated. The workplan comprises six chapters (see Fig. 1).

(a) *Chapter 0: management tasks*. This relates to the overall management of the program by DRIVE central office, including information exchange and project control. It is also concerned with the external environment through consensus formation with the participants and with promoting acceptance of European RTI standards.

(b) *Chapter 1: systems approach*. This relates to the strategic direction and coherence of the entire research program, ensuring a top-down approach accompanied by iteration with the projects carried out under the other chapters to secure adjustments to preliminary

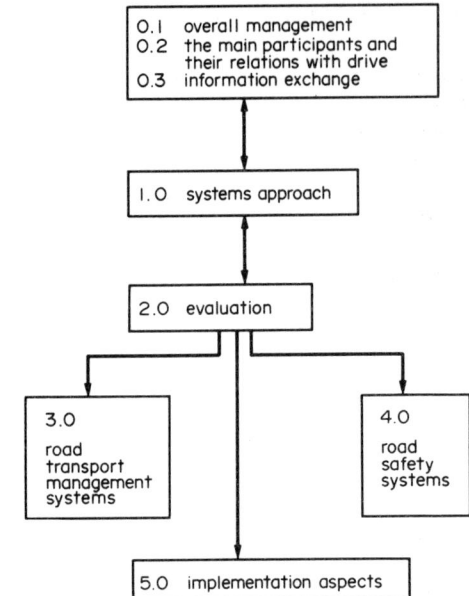

Figure 1
The six workplan chapters

definitions of system scope and feasibility (see *Transportation Management: Systems Engineering Approach*). Chapters 1, 2 and 5 are complementary "horizontal" activities.

(c) *Chapter 2: evaluation*. The ideas involved in this chapter are to develop tools and criteria against which to evaluate the results of the research carried out under the program; to support this with modelling of various "base" road traffic system characteristics against which scenarios of changed behavior arising from the use of RTI systems and services can be assessed; to perform evaluations of results; and to use these assessments to make recommendations for standards, for common systems, for implementations, for demonstrations and for further research (see *Evaluation of Traffic Control Systems*).

(d) *Chapter 3: road transport management systems*. The core research and development activities are in this chapter and chapter 4. Chapter 3 relates to the development of functional specifications and system design for RTI-based road transport management taking into account existing developments and the aim of realizing system integration (see *Traffic Control Systems: Architecture*).

(e) *Chapter 4: road safety systems*. This relates to the investigation of the new safety systems and services offered by the use of RTI; and to the definition of functional specifications and of system designs and standards for them, taking into account the need to provide improvements for all road users. It also relates to the definition of safety performance specifications for all new RTI systems and their testing (see *Safety of Road Traffic*).

(f) Chapter 5: implementation aspects. This relates to the standardization (economic and financial, legislative and regulatory) and social issues affecting Europe-wide implementation of RTI systems and services.

3. The Public Call

3.1 Conditions for Participation

The workplan formed the basis for the public call for proposals for specific projects to be carried out under the rules established by the framework program of the EC. There are three essential features of these rules.

First, projects are carried out by means of shared cost contracts, under which contracts will be expected to bear a substantial proportion of the costs, which should normally be at least 50% of the total expenditure. Alternatively, in respect of universities and research institutes, the EC may contribute up to 100% of the additional expenditure involved.

Second, proposals, which were submitted in reply to an open tender published in the official journal of the EC on 2 July 1988, generally involve the participation of at least two independent partners who are not established in the same member state. At least one of the partners is normally an industrial concern. The closing date for proposals was 17 October 1988. A supplementary call took place in April 1989.

Third, organizations from non-EC European countries (notably the European Free Trade Association (EFTA)) may participate in the program where agreements allow and where there would be mutual advantage; in such cases, the EC would expect the organizations concerned to make a full financial contribution.

3.2 Results

As a result of the call, more than 200 proposals were received totalling about US$ 730 million worth of effort. Following the evaluation of the proposals, 61 consortia started work in January 1989 while another 12 projects began in September 1989. More than 450 different participants are involved from all member states and from the EFTA countries.

Research institutes, industry and service providers participate in the program with an almost equally shared effort. Most projects are vertically integrated in terms of their participants (i.e., they include all participants typically involved with the introduction of a new technology from research establishments through manufacturing industry to service providers and leading edge users). The total effort during the 1989-1991 period amounts to 12 500 person months.

4. Projects

As a result of the public call, the retained proposals span a representative spectrum of advanced applications in such areas as

(a) two-way communication systems (groupe speciale mobile (GSM), infrared, microwave),

(b) integrated traffic control (strategies, applications, artificial intelligence),

(c) environmental control strategies,

(d) road safety (user-machine interfaces, anticollision systems, behavioral aspects),

(e) advanced public transport and freight systems,

(f) evaluation techniques and strategies, and

(g) systems integration and management techniques.

Within the program, projects have been arranged into four groups on the basis of their areas of research and on the degree of cooperation required between them. This cooperation and grouping of projects was envisaged in the set of tasks described in the DRIVE workplan. The four groups are shown in Fig. 2. The existing projects have succeeded in covering tasks from more than one chapter of the DRIVE workplan so that the program is vertically integrated as originally envisaged. Thus, group 1 covers tasks from chapters 1, 2 and 5 of the workplan, group 2 covers tasks from chapters 2, 4 and 5, group 3 covers tasks from chapters 1 and 3, and group 4 covers tasks from chapters 1, 3 and 5. In this way, a single project can address strategies, and simulation and technological questions.

4.1 Group 1: Evaluation and Modelling

To achieve the goal of optimizing RTI applications in Europe, the DRIVE program is financing the development of suitable tools to simulate and evaluate the effects of RTI implementation.

As a result of the first call, a large number of good proposals were received from European universities, research centers, and industrial and consultancy firms to undertake the workplan tasks on traffic, transportation and evaluation modelling. Mergers were negotiated between some of the proposals in order to have the strongest possible consortia and to have good technical integration. The projects in this area dovetail together and, where relevant, different simulations and

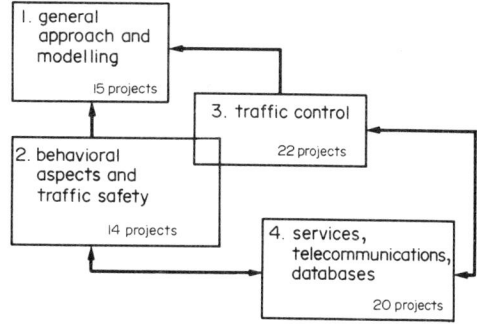

Figure 2
The four project groups

evaluations will be compatible across this part of the program; in particular, in transportation modelling and in evaluation guidelines.

In transportation modelling, an innovative approach is carried out in order to forecast and simulate the effects of RTI on transport demand, traffic performance and the environment. The seven models under development will be dynamic in character and will be based on existing research on perception, and driver and household behavior. These models will also explicitly include the effect of drivers having information about the road environment so that the consequences of partial and imperfect information can be modelled directly (see *Transportation Modelling*).

The guidelines for the evaluation of RTI systems and trials will be prepared by two consortia working respectively on field trials, and evaluation methods and criteria.

At this stage, work in this area of the program focuses on the consequences of introducing RTI on users and society at large.

4.2 Group 2: Behavioral Aspects and Traffic Safety

The DRIVE program offers a new approach to improve road safety through the application of RTI. This will be very welcome as the experience of the most motorized countries shows that the traditional ways of attacking the traffic injury problem have been pushed to their limits and further investment in them now yields very little benefit.

Projects in this area will research the following:

(a) hazard and accident data analysis, particularly in relation to RTI systems;

(b) safety for vulnerable road users;

(c) behavioral changes due to the introduction of RTI systems;

(d) requirements for user–machine interfaces and collision avoidance systems;

(e) data recording using a vehicle journey recorder, and systems for automatic policing; and

(f) impact and implementation studies.

DRIVE projects in this area include both theoretical and experimental work. The objective of the research is to define functional specifications and requirements for the relevant RTI systems, perhaps developed elsewhere in DRIVE. For systems that have been developed elsewhere, the effect on safety is one of the factors to be considered. Impact studies are necessary from the human, social and economic points of view, not only to see the positive effects of RTI systems, but also to discover to what extent they may have an adverse effect on safety (e.g., by reducing driver concentration and/or by driver ignorance of warning signals).

The safety problems of the most vulnerable road users—pedestrians and cyclists—need special attention in this new RTI approach.

It is a challenge for researchers and developers in this area to contribute to a reduction in the loss of human resources caused by inadequacies in the present traffic system.

4.3 Group 3: Traffic Control

The projects in the field of traffic control aim to improve the efficiency and, partly, the safety of the road network by using RTI facilities. One quarter of all the projects received in the call for proposals focused on this area.

The retained projects will contribute to the integrated road transport environment (IRTE) by developing either system components or by working on the integration of those components into a single system. For example, one project will define the functional requirements and specifications for the IRTE and will recommend international standards. Another deals with the integration of dynamic route guidance and traffic control systems (see *Signal Control and Traffic Assignment*). The other projects in this area deal with the more detailed components of the IRTE.

Traffic demand projects include

(a) the development of strategies for demand management,

(b) management of off-street parking (an important element of urban traffic control), and

(c) investigation of smart cards which could be used for automatic debiting.

Traffic control projects concern both urban and extraurban areas. In summary, there are projects concerning

(a) the development of separate traffic control subsystems, such as for tidal flow or for tunnel control (although each subsystem must be integrated into the overall IRTE);

(b) the improvement of traffic signal control by making use of better origin/destination information;

(c) using expert systems to advise on appropriate traffic control decisions (see *Expert Systems Approach to Road Traffic Control*).

(d) the use of artificial intelligence (AI) for data acquisition, traffic condition intepretation and prediction (later projects will study other possible applications for AI);

(e) the definition of system architectures and the building of prototypes for new road condition and weather monitoring systems (this project will also make recommendations for the use of this information in traffic management);

(f) freeway ramp control (see *On-Ramp Control of Freeway Networks*); and

(g) cooperative motorway driving.

Congestion control is really a special type of traffic control. Projects in this area concern

(a) alternative methods for general incident detection (see *Incident Detection*),

(b) the use of computer vision for general incident detection (see *Video Sensors*), and

(c) strategies that should be adopted to prevent congestion from occurring.

4.4 Group 4: Services, Telecommunications, Databases

Group 4 contains a diverse set of projects concerned with the telecommunications and information flow aspects of the IRTE and the special needs of fleet operators. This concern extends to the sources of information as well as to the use of the information; so projects include information sources, such as positioning systems and digital maps, and information users, such as driver information systems (see *Road Traffic Monitoring Equipment*). All the projects in group 4 contribute directly or indirectly to the realization of the main DRIVE objectives: improvement of road safety, increased efficiency and better air quality.

The projects in group 4 can be arranged into five clusters.

(a) Public transport cluster. Two projects will develop strategies for integrated public transport management and information systems, as well as for passenger information, vehicle scheduling and control systems for public transport. The goal of these projects is to arrive at a more efficient and more attractive public transport system, which will encourage more journeys to be made on public transport in preference to in private cars.

(b) Freight management cluster. Two projects will develop strategies for integrated freight and commercial fleet management and will set out the functional characteristics of an RTI-based pan-European system of road freight operation in order to optimize efficiency, and economic and other types of performance.

(c) Digital maps and databases cluster. Three projects will study a road database management structure and will develop digital maps. These databases and digital maps will be of major importance for the operational launching of route guidance systems. These systems will have a significant positive influence on traffic efficiency (shortest route), traffic safety (avoidance of secondary accidents in case of congestion) and reduction of pollution (less vehicle kilometres travelled and less congestion) (see *Route Guidance, Collective*).

(d) Information and broadcasting systems cluster. Three projects will deal with the full scope of operations in traffic information from gathering to dissemination, while another project will study the use of electronic cards for portable travel and transport information (see *Broadcasting Communication Systems*).

(e) Communications technologies, systems and architecture cluster. Three projects will examine in detail communications technologies that can be used in RTI systems. One will determine the suitability of different cable systems while another will develop recommendations for a European standard for microwave links in RTI applications. The third project will not only review, analyze, develop and demonstrate basic data acquisition and communication techniques for use in RTI systems, but will also evaluate and assess the suitability of these techniques.

For communications systems, three projects will be particularly concerned with the application of specific communications technologies to particular RTI systems. One concentrates on the use of cellular radio as the communications link for route navigation and driver information. Another will develop and test a microwave link for automatic two-way data communication between vehicles and the roadside to be used for traffic monitoring and pricing applications. It comprises a coherent set of tasks that runs right across the DRIVE program, from strategy through system specification, design and fabrication to eventual testing in the field. One of the major tasks involves the strategic context of overall demand management of traffic, within which the more detailed considerations of vehicle identification and automatic debiting will be implemented. The third project in this area will evaluate the economic and operational advantages of satellite-based vehicle location systems.

For communications architecture, a project will examine in detail the telecommunications and information processing infrastructure for RTI systems, and will consider how existing communications systems can be used as well as telecommunications network structures. It will also consider the integrity aspect of data, extraurban signalling and communications systems, and the overall framework of communications standards, developing protocols for vehicle–infrastructure communications.

All these communications technologies and systems will constitute important elements of traffic control and of guidance systems and will, as such, contribute to the improvement of traffic efficiency (e.g., microwave communications for cost assignment) and of traffic safety (e.g., microwave communications for anticollision systems), and to the reduction of air pollution.

5. System Integration and Consensus

Research results achieved in DRIVE projects must lead to effective systems integration. Given the complexity involved, this will be impossible unless an overall and consistent systems engineering approach is adopted from the outset. To support and help guide the operation of this approach among the projects, the DRIVE workplan foresaw the need for a set of horizontal management tasks.

In addition, as a complement to technical systems engineering, there are a range of other tasks that have collectively been called consensus formation. In essence, these tasks are concerned with ensuring continual external feedback to DRIVE projects throughout the life of the program with the ultimate goal of recommending a preferred implementation strategy for RTI systems.

The most important contribution DRIVE is set to make is to bring about consensus within Europe on the functional characteristics of the future RTI infrastructures. Irrespective of the quality of the research and development undertaken, DRIVE will not be successful without broad agreement between the sector participants that the results of the projects actually meet their requirements and will be adopted widely in the market place.

Mechanisms to seek and obtain consensus at the working level are built into the DRIVE program. These include deliverable handling procedures and regular concertation meetings where all DRIVE projects participate. However, at the strategy level, it is also necessary to involve major participants who are not directly involved in the DRIVE program but who are nevertheless affected by the evolution of advanced RTI systems. These participants must be brought together with senior strategists from the organizations that are involved in DRIVE. The task of ensuring this consensus has been assigned to a DRIVE project—the Systems Engineering and Consensus Formation Office (SECFO). The lead contractor is a major European vehicle producer. The other partners represent the information technology industry, public transport, freight operators, motorists' organizations and touring clubs, telecommunications equipment producers, traffic control engineers and road transport authorities.

6. Future Developments

The work initiated under DRIVE is only the first, albeit important, step towards the introduction of an IRTE in Europe. The program is expected to contribute to the achievement of an improved traffic management and road safety system. The main success of the program so far is that it has brought all sector participants together to work for the achievement of a better traffic environment in Europe. The next step must be to build on the current work, verify the results and move towards implementation of the appropriate systems.

Bibliography

Commission of the European Communities 1987 Report No. COM(87)351 final. Commission of the EC, Brussels

DRIVE 1988 The DRIVE workplan. Report No. DRI 100. DRIVE, Brussels

DRIVE 1989 The DRIVE programme in 1989. Report No. DRI 200. DRIVE, Brussels

F. Karamitsos
[DRIVE, Commission of the European Communities, Brussels, Belgium]

E

Evaluation of Traffic Control Systems

Evaluation is an integral part of the development and application of traffic control systems. There are four key roles expected of an evaluation framework:

(a) assistance in indicating how existing system designs should be modified as part of an iterative research and development process;

(b) identification of viable systems as candidates for implementation, thereby supporting the decision process;

(c) assistance in controlling the amount of resources that can be allocated to the development of each potential system; and

(d) analysis of the efficiency of a system after implementation.

For the technical assessment of systems there are three major requirements.

(a) The consequences of new technologies must be recognized as early as possible.

(b) A comprehensive assessment of all consequences is needed.

(c) The results of the assessment must be decision relevant (i.e., different possible options of the technology have to be presented to the decision maker).

These tasks demand two different approaches before and after implementation.

(a) *Ex ante* evaluation is an analysis of the efficiency of a system or the ranking of a set of systems with respect to their relative efficiency as a decision support before their implementation.

(b) *Ex post* evaluation is the analysis of the efficiency of a system after its installation.

This spectrum of requirements causes high expectations of the complexity and quality of the evaluation methods. However, the evaluation methodology for traffic control systems is not so far advanced that there could be a formalized procedure for the evaluation of the whole complexity associated with the impacts of control systems. The quality of an evaluation depends, therefore, on the expertise of the evaluation staff and their ability to grasp the evaluation problem in its decision environment and to include all relevant factors in the analysis.

1. Alternatives, Evaluators and Objectives

The value of a control system can only be viewed in the context of its decision environment, its decision makers and their objectives.

Candidates of possible control system alternatives for evaluation can be defined at different levels of integration; that is, the level of subsystems (e.g., navigation), single systems (e.g., route guidance), integrated systems (e.g., urban traffic control) and integrated control environments (e.g., passenger and freight transport control). Therefore, it becomes obvious that the consideration of isolated systems might neglect the synergetic and/or compensating effects that occur at an integrated level of control.

Different actors have to be considered in the evaluation as part of a decision process. These are people or groups of people who are affected by a candidate and/ or are actual decision makers. The evaluators are those actors whose preference structure with respect to the different objectives is actually considered in an evaluation production.

The objectives expected of a control system are defined in the context of the decision environment and the actors involved. With an objective (e.g., to improve traffic safety) both a criterion (e.g., traffic accidents) and an orientation of preference (e.g., minimize) with respect to that criterion have to be specified. There is also a need to identify a measure of effectiveness (e.g., accidents per vehicle kilometer) with each criterion to indicate the degree to which it has been achieved. This physical measure has then to be subjected to a valuation; for example, in monetary terms (e.g., cost per accident) or utility (e.g., in terms of percentage points) which is then dependent on the evaluator's preferences. The summing up of the valued measures then leads to the decision criteria (e.g., cost–benefit ratio) depending on the evaluation method applied. The valuation of the effects and their synthesis is a crucial step in an evaluation process, which has to be performed by the decision maker or by the decision maker in close cooperation with the evaluating staff.

Some effects engendered by control systems cannot be assessed in physical quantities so that intangible effects have to be treated as part of the evaluation task.

The first step of an evaluation process is to define the decision environment by

(a) formulating the problem,

(b) specifying the actors and the objectives, and

(c) identifying the candidates for evaluation.

2. Evaluation as Part of Systems Analysis

Studies by Quade and Boucher (1974) showed that high-level evaluation methods are part of a systems analysis approach which tries to describe a problem in

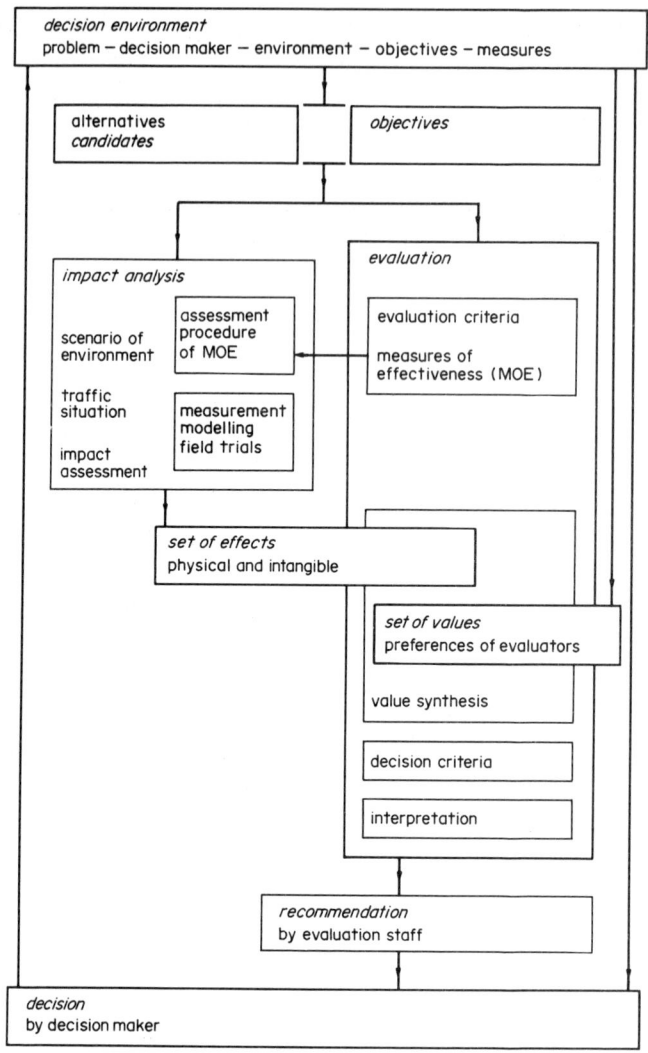

Figure 1
Evaluation as part of systems analysis

relation to the total structure of objectives, costs and benefits. Systems analysis is

"a systematic approach to help a decision maker choose a course of action by investigating his full problem, searching out objectives and alternatives and comparing them in the light of their consequences, using an appropriate framework—as far as possible analytical—to bring expert judgement and intuition to bear on the problem."

Systems analysis is "not a method or technique, nor . . . a fixed set of techniques," but the generic term for the systematic approach "to search out, examine and evaluate possible courses of action." This clarifies that evaluation has to be considered as an iterative process that includes the possibility of redefinition of the objectives and assumptions and the development of alternative candidates that will yield the objectives.

Figure 1 shows the position of the evaluation of traffic control systems in the decision process and the stages of the evaluation, as well as the interrelated activities and results.

Four consecutive stages are relevant for the evaluation of traffic and transport control systems:

(a) building of scenarios of the future decision environment for the evaluation candidates,

(b) measurement and/or modelling of traffic situations,

(c) assessment of physical and intangible impacts per evaluation criterion through measures of effectiveness, and

(d) evaluation of the impacts and synthesis of values for the impact components.

These stages allow a separate interpretation and decision on their respective results in order to achieve a high degree of acceptance by the actors involved in the evaluation process.

A relatively high degree of acceptance can be obtained for the building of scenarios for future transport environments and for the modelling of traffic performance. The assessment of impacts is achieved with varying accuracy depending on the measures of effectiveness, as intangibles such as driver comfort indicate. The lowest common denominator is reached in the fourth stage, where values are assigned to the measures of effectiveness, a process that always contains subjective judgements.

3. Classical Methods

The last step mentioned in Sect. 2 is, in particular, a discriminating factor in selecting the appropriate evaluation method. The framework of classical evaluation methods is summarized with respect to their possible objectives, cost and benefit structure in Fig. 2. It refers to the evaluation methods of

(a) cost–benefit analysis (CBA),

(b) cost-effectiveness analysis (CEA), and

(c) decision analysis (DA).

These allow a comprehensive evaluation of the impacts considered.

3.1 Cost–Benefit Analysis

CBA to evaluate public investments is based on the welfare theory. Investments are judged by the criterion of compensation; that is, whether those favored by a measure can bring the disadvantaged back to their previous benefit level by payments of compensation and still keep benefits for themselves. This leads to the decision criterion of CBA, where a measure is reasonable if the expected benefits (product of increase of goods b_j and price p_j) are greater than the expected costs (product of invested factors c_k and costs p_k), that is

$$E = \sum_j b_j p_j - \sum_k c_k p_k$$

The computational determination of the changes in welfare using this equation indicates the problems associated with CBA:

(a) all benefits are valued in monetary units;

(b) not all changes in benefits and costs can be measured in market prices; and

(c) the advantageousness of a candidate is judged by a single objective, which maximizes economic efficiency exclusively.

To be able to consider the different times of investments and of occurrence of costs and benefits, the monetary effects are discounted in the year in which the decision is to be made. Then, either the differential decision criterion

$$D = \sum_t \left(\sum_j b_{jt} p_{jt} - \sum_k c_{kt} p_{kt} \right) (1+i)^{-t}$$

or the ratio decision criterion

$$Q = \frac{\sum_t \sum_j b_{jt} p_{jt} (1+i)^{-t}}{\sum_t \sum_k c_{kt} p_{kt} (1+i)^{-t}}$$

is used, where i is the discount rate and t the number of years of the period of analysis. A candidate is considered to be economically reasonable if

$$D > 0 \text{ or } Q > 1$$

If there are several alternative candidates, they are ranked according to the decreasing cost–benefit ratio q.

Three points should be noted about CBA. First, its expressiveness increases with the share of monetary effects on the total impacts. Second, the choice of interest rate has a strong influence on the accounted costs and benefits and, therefore, decides the ranking of candidates. Third, to increase the transparency of

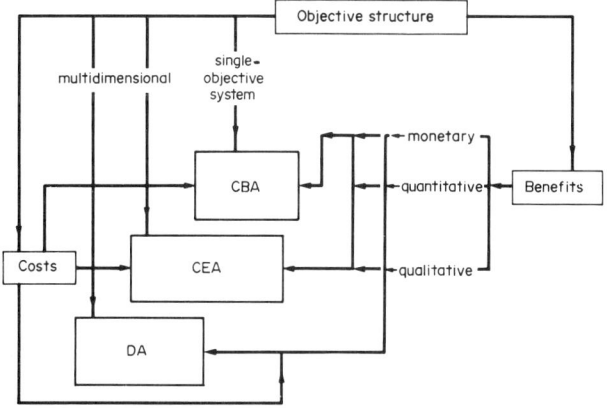

Figure 2
Framework of classical evaluation methods

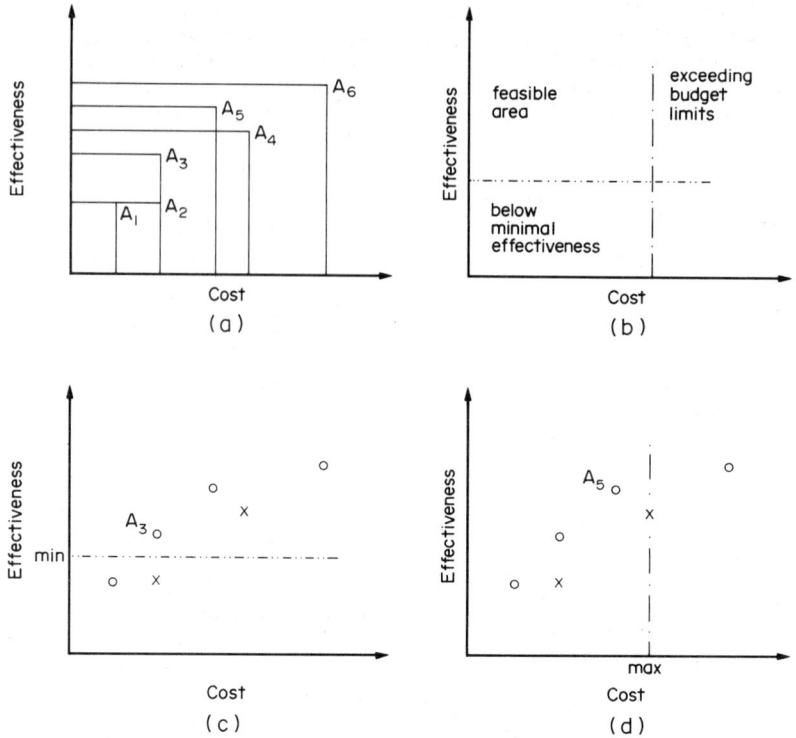

Figure 3
Decision criteria of CEA: (a) comparison of pairs, (b) feasible area, (c) fixed effectiveness and (d) fixed budget restriction (after Quade 1974)

the results, generally only capital and operating costs are treated in the cost section, while all other costs are treated as negative benefits.

3.2 Cost-Effectiveness Analysis

CEA was originally developed to analyze the effects of investments in the military sector (Quade and Boucher 1974). Because there are often no market prices for the social effects of public investments, the physical interpretation of benefits in the CEA is considered satisfactory. Capital and operating costs are accounted for in monetary terms.

In the high-level form of CEA, a problem is studied and evaluated as part of a systems approach by relating it to the total structure of objectives, benefits and costs. The low-level CEA assumes a fixed set of objectives.

The CEA approach allows for a multidimensional objective system, in which monetary, quantitative and qualitative effects can be considered, although a discounting of effects is only possible for the quantitative and not the qualitative effects. The effectiveness of a candidate can then be summed up into a one-dimensional expression by summarizing the target scores for the individual effects, which are weighted and transformed for this purpose according to the preference structure of the evaluators (generally using a scale from 0 to 100 points). Two decision criteria are used:

(a) preference for the candidate with the highest effectiveness, if the costs of the candidates are equal; and

(b) preference for the candidate with the least cost, if the effectiveness of the candidates is equal.

Taking these principles into consideration, the least advantageous candidate can be excluded, while the remaining candidates are selected by applying the following restrictions (see Fig. 3):

(a) definition of minimal effectiveness requiring at the same time the minimization of costs; and

(b) definition of maximal cost, requiring at the same time the minimization of the effectiveness.

The ratio of nonmonetary effectiveness (e.g., percentage points) to monetary costs is a one-dimensional index which permits a relative ranking of alternative candidates, but not an assessment of the absolute advantageousness of one candidate in the system.

Due to the subjective elements of the summing up of the effects into a one-dimensional effectiveness per candidate, it is often recommended to perform only an effect analysis; that is, to present the decision maker with only the matrix of quantitative and qualitative effects of the individual evaluation candidates.

Thus, CEA allows for a multidimensional objective system in which discounting of effects is only possible for the quantitative and not for the qualitative effects. The decision criterion is a one-dimensional index (e.g. 1 to 100 points per unit money), which can only be interpreted as a relative figure, so that more than one candidate has to be analysed with CEA, because only the relative advantage between candidates can be assessed.

3.3 Decision Analysis

In DA, the separate treatment of benefits and costs is abandoned. Monetary costs (capital and operating costs) are treated as effects in the same way as benefits. Costs are, therefore, elements of the multidimensional objective function.

Zangemeister (1976) formulated the following stages of a DA. The initial stage encompasses the formulation of the objective system in relation to the candidates to be evaluated. The objectives are structured in a treelike hierarchy with the higher branches expressing the objectives and their direction of preference while the lowest branches represent indicators (i.e., the operational measures of effectiveness which can be expressed in monetary, quantitative or qualitative terms).

In the next stage a target score is established for each measure of effectiveness using a specific assessment procedure (e.g., via traffic modelling). In order to depict the target scores as a single dimension, they are transformed into target values.

Finally, the target values of the individual measures of effectiveness are weighted according to the preferences of the respective evaluator (e.g., with weights 1 to 100 in the case of cardinal scales), allowing for their summation into the decision criterion, which is a one-dimensional index. The candidate that achieves the highest utility (e.g., percentage points) is then considered as the best alternative.

Thus, DA allows for a multidimensional system which includes capital and operating costs. Three points should be noted: first, the decision criterion is a dimensionless utility, which can only be interpreted as a relative figure, so that more than one candidate has to be included in the DA.

Second, the necessary transformations for the target scores of the measures of effectiveness and for their weighting in the value synthesis include a highly subjective element of judgement by the evaluators.

Third, because of the multidimensional objective system, monetary-valued criteria are not the only criteria considered in the evaluation, which is an important feature for the analysis of public investments.

3.4 General

Figure 4 gives a short form of the decision criteria employed in the three classical evaluation methods.

If the evaluation concentrates on certain impacts (e.g., under the assumption that other impacts are neglected or considered to be constant), sectoral studies are performed such as the following.

(a) *Cost analysis*. The cheapest alternative of equally valued alternatives is selected.

(b) *Benefit analysis*. For a given budget, the alternative that achieves the highest benefit with respect to a specific objective is selected.

(c) *Deficiency analysis*. A test is conducted to find which candidate falls short of generally recognized standards.

(d) *Compatibility test*. The measures are tested with respect to their compatibility with possible negative side effects (e.g., environmental impact studies).

Figure 2 clearly shows that, in the case of CBA, only a limited spectrum of impacts is covered. Experience shows, however, that with a monetary-based evaluation, a better agreement between different decision makers can be reached than with the other methods. It is, therefore, common to perform a CBA for the monetary impacts and then to incorporate it into a comprehensive multicriteria analysis such as CEA or DA. Generally, it is recommended to perform sensitivity analysis to test the stability of the evaluation results.

4. Measures of Effectiveness

The consequences of traffic and transport control systems will have an impact on a wide spectrum of the public. Keeney (1988) refers to such problems, which typically have environmental, social, health and safety implications, as problems of public interest. The affected parties should be able to participate in the decision-making processes.

The set of objectives used to evaluate the control system alternatives in public decision problems is the basis for the analysis of the said control system alternatives. The task is, therefore, to identify and structure the objectives. The preferences for the objectives will be different depending on the involved stakeholders. The final objective system has then to encompass all relevant aspects articulated by the different participants.

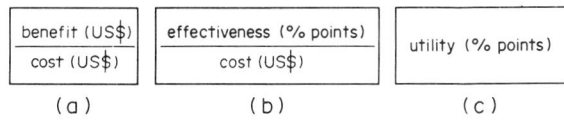

Figure 4
Short form of decision criteria for (a) CBA, (b) CEA and (c) DA

Participants or stakeholders affected by control systems can be categorized according to the following levels of involvement:

(a) individual level (directly affected), such as
   (i) private users (pedestrians, cyclists),
   (ii) car drivers (with distinct trip purposes),
   (iii) entrepreneurs (road haulage companies, tourists, taxis, services),
   (iv) emergency services, and
   (v) other transportation modes (transit systems, buses);
(b) public level (indirectly affected), such as
   (i) remaining participants in traffic (pedestrians, cyclists), and
   (ii) local residents;
(c) institutional level (public domain), such as
   (i) road transport authorities (planning and operation),
   (ii) government and local administrations, and
   (iii) international institutions;
(d) industrial level, such as
   (i) vehicle manufacturers and their suppliers, and
   (ii) traffic control equipment industry.

It is recommended to involve these stakeholder groups early on in the decision process so as to obtain their contribution for the selection and structuring of the objectives.

A crucial point for the assessment of the impacts of control systems is to find measures of effectiveness for the objectives that are sensitive to the changes in the effects induced by the control systems. This is closely linked with the design of the assessment procedures according to which these measures are determined (e.g., by modelling the effected taffic situations and their impacts on the energy consumption of vehicles).

The following list gives a typical set of objectives, criteria and measures of effectiveness:

(a) increase traffic safety, assessed by
   (i) accident rate by severity and road type,
   (ii) fatalities and injured persons per period of time, and
   (iii) damage to vehicles, infrastructure and buildings;
(b) improve transport efficiency, assessed by travel time savings by private trip purpose, in commercial transportation, in public transport and in service systems;
(c) reduce energy consumption, assessed by fuel consumption per kilometer, and by motor vehicle (car, truck) and road element;
(d) improve environment, assessed by
   (i) vehicle exhaust emission rate per component (CO, NO_x, CH, soot),
   (ii) vehicle noise emission levels per environment type,
   (iii) urban qualities, and
   (iv) visual barriers;
(e) improve driver comfort, assessed by
   (i) travel information density per road type, and
   (ii) stress frequencies per traffic situation and road type;
(f) reduce costs, assessed by
   (i) monetary costs of investment and operation for infrastructure,
   (ii) monetary costs for operation of traffic control systems, and
   (iii) monetary costs for on-board equipment.

This range of criteria stresses the importance of the fact that the objectives have to be independent of each other and that the expected spectrum of possible impacts has to be defined in the spatial and temporal dimensions at the beginning of the evaluation process.

Changes in the structure of costs and benefits have to be considered over time to be able to discount them for a defined time horizon. This means that the lifespan of control installation and equipment has to be known in order to be able to make capital and operating costs comparable (e.g., via annual costs). For the determination of the benefits, this means that an estimation of the future transport demand is necessary either by traffic estimation models or by scenario building.

5. Problems and Future Developments

To preserve the highest transparency in the presentation of the evaluation results, it is important to assess the set of physical and intangible effects separately from the set of values that are assigned to these effects. This two-stage approach leaves open the possibility of considering the different preference structures for the different effects of the individual participants involved in the evaluation process. In this way, the subjective element of the evaluation is separated from the objective impacts for which a higher degree of acceptance can be attained. This approach also considers the possibility of only presenting the decision maker with the set of calculated effects and allowing a decision to be made on the basis of personal preferences.

A major task of the evaluation is the detection and isolation of eventual small changes in the impacts caused by the control systems. This demands detailed assessment procedures of how to determine the target scores for the measures of effectiveness. These have to be formulated in close contact with the design of the experiment in the case of measurements or with the modelling of the traffic performance due to the control function.

The evaluation requires a design of the future decision environment either by scenario building or by forecast methods and, therefore, contains an element of relative uncertainty. The impact analysis can either be based on a before-and-after or a with-and-without control measure approach. This provides a reference level for the considered impacts which is a prerequisite for the application of CBA, while multicriteria analysis, such as CEA and DA, requires the analysis of at least two alternative candidate systems, because they only allow for relative judgements.

A typical problem associated with the evaluation of control systems is that hybrid situations have to be considered where, for example, only part of the vehicles or only part of the infrastructure is equipped with the control functions during the phase of system introduction. In addition, due to the nonmandatory character of some of the information and recommendations provided by the control systems, their usage and acceptance by the drivers are parameters of high relevance in the impact analysis.

A wide spectrum of intangible impacts will have to be considered, but this can hardly be handled by the analytical evaluation methods. To comply with these requirements, a team of experts can be formed from different disciplines, who can assist in the evaluation process with their experience and expertise. These experts can also serve as a panel for eventual ad hoc evaluations during the developmental phase of alternatives and for the interpretation of evaluation results for the decision makers.

The multidimensional implications of traffic control systems have to be reflected in the selected objective functions. The resulting integrated road transport environment for traffic control requires a hierarchical approach with respect to the different transportation planning and traffic operation levels, as well as to the different levels of the local, area, national and international involvement of the affected actors.

Numerous evaluation studies for traffic control systems have been performed. Due to the complexity of the possible impacts of traffic control systems, it has to be stressed that the application of the classical formalized evaluation methods can only serve as an information basis and as an assistance to the decision maker who must take responsibility for the actual decision.

See also: Safety of Road Traffic; Safety of Road Traffic: Intervention and Evaluation; Transportation Management: Systems Engineering Approach

Bibliography

Commission of the European Communities—DRIVE 1990 Evaluation process for road transport informatics. Project No. V1036-EVA. Technische Universität München, Munich, Germany

Everts K, Keller H, Zackor H 1987 Wirtschaftlichkeit von Zielführungssystemen unter Berücksichtigung des Entwicklungsstands. *Schriftenr. Forsch. Strassenbau Strassenverkehrstech.* **502**

Geiss H, Keller H, Pollmann P, Zackor H 1983 Wirtschaftlichkeitsuntersuchungen von Verkehrsleitsystemen. *Strassenverkehrstech.* **27**, 37–42, 92–7, 114–17

Hanusch H 1987 *Nutzen-Kosten-Analyse.* Vahlen, Munich, Germany

Kenney R L 1988 Structuring objectives for problems of public interest. *Oper. Res.* **36**, 396–405

Miser H J, Quade E S 1970 *Handbook of System Analysis.* Wiley, New York

Quade E S, Boucher W J 1974 *System Analysis and Policy Planning.* Wiley, New York

Zangemeister C 1976 *Nutzwertanalyse in der Systemtechnik.* Munich, Germany

H. Keller
[Technische Universität München,
Munich, Germany]

Expert Systems Approach to Rail Traffic Control

The application of computers to rail traffic control is promoted in order to make railroad systems more efficient and safe. It is very difficult, however, to implement the preparation of a train diagram and the restoration of rail traffic on a computer, because defining the necessary conditions numerically is very complicated. Therefore, these tasks still depend heavily on the flexible information-processing and decision-making abilities of human experts. The new approach using artificial intelligence techniques has been applied to real-world scheduling. It is based on a human expert's knowledge and emulates that expert's problem solving process on a computer. The ESTRAC project is the first application of knowledge engineering technology to the restoration of rail traffic and three expert systems have been developed since 1982.

1. Restoration of Rail Traffic

For commuter lines, rail traffic is centrally controlled by a computer system, and also by a small number of human experts called dispatchers. In normal situations, rail operation can be controlled automatically by a computer system according to a planned diagram. When rail traffic is disrupted, however, the dispatcher adjusts the train schedule (e.g. by moving an overtaking operation or suspending a rail service) to restore rail traffic as soon as possible. Such restoration work is called train schedule adjustment, for which the dispatcher must carry out the following four types of jobs: information gathering, problem setting, problem solving and instruction sending.

1.1 Information Gathering

When an accident or other disruption occurs, the dispatcher can find where and which trains are delayed by the alarm from the computer-aided rail traffic control system. The first action to be taken by the dispatcher is to find the cause of traffic disorder and to predict the time when trains can run normally. Therefore, it is

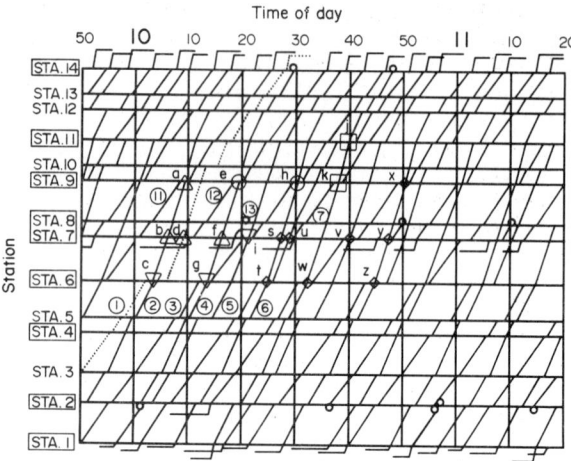

Figure 1
Train diagram: circled numbers are train numbers and points a–z are overtaking operations

necessary for the dispatcher to gather information concerning details and the scale of the disorder directly from the train crew or the station staff over the telephone. While traffic disorders vary from minor ones lasting a few minutes to major ones lasting several hours, they can be classified into the following types:

(a) delay in arrival/departure of one or more trains at a station;

(b) speed limitation in a particular section of a line during a certain period; and

(c) closure of a particular section of a line during a certain period.

1.2 Problem Setting

From the information gathered, the dispatcher must judge the type of disorder taking place. It is also necessary to predict the time when trains can run normally. The problem of train schedule adjustment can be settled on the basis of the judgment and prediction. However, even skilled dispatchers find it difficult to be exact in this prediction. Taking prediction errors into consideration, the dispatcher usually assumes 2–3 different cases, and problems are set depending on each case.

1.3 Problem Solving

After setting the problem, the dispatcher solves it by trial and error using a train diagram and prepares an adjustment plan. The train diagram is a kind of chart that draws the locus of each train on a two-dimensional plane of time and distance, and shows the arrival and departure times of the train at each station (see Fig. 1). This diagram is useful in that the relationship between each train schedule can be shown visually. On the train diagram, the dispatcher simulates manually the travel of trains that are related on the schedule, and then decides how the schedules of these trains should be adjusted. To solve the problem, the dispatcher repeats the simulation of train travel and the decision about schedule adjustment by trial and error until a suitable adjustment plan is arrived at. In this work, there are many different methods of schedule adjustment with a tremendous number of combinations. Using different methods in combination, a skilled dispatcher can solve the train schedule adjustment problem within a short time.

1.4 Instruction Sending

After solving the problem, the dispatcher must send the instruction for changes from the original train schedule to the relevant people as soon as possible. While the instruction is simply sent by telephone to the train crew and the station staff, the dispatcher must enter commands via the console on the computer-aided control system to change the train schedule stored in it. For the dispatcher to manually enter commands within a limited time can be difficult.

2. Train Schedule Adjustment Problem

In train schedule adjustment, the dispatcher's major concern is the arrival and departure of each train at each station. Train travel, therefore, is taken as a discrete-event system of the arrival at and departure from each station of each train. In this case, train operation is restricted by the operating conditions and the schedule. Restrictions due to the operating conditions are:

(a) the train travel conditions (differ according to each type of train), with
  (i) a minimum value of travelling time between two stations, and
  (ii) a minimum value of stopping time at each station; and

(b) the operating conditions at each station, with
  (i) a minimum value of the headway at arrival at or departure from each station,
  (ii) a minimum value of the time required for turning back at a station, and
  (iii) a maximum number of trains that can run at a time between two stations.

Restrictions due to the schedule are:

(a) the train schedule, with
  (i) the arrival and departure time at and from each station,
  (ii) the arrival and departure track used in each station, and
  (iii) the train-car operation; and

(b) the schedule at each station, with
  (i) the order of arriving trains at each station, and

(ii) the order of departing trains at each station.

The operating conditions are fixed because of physical restrictions on train cars and equipment. However, the schedule is a logical restriction determined by the person who prepared the train diagram and can be changed. Typical schedule adjustments applied by the dispatcher include a change of train order (i.e., shifting an overtaking operation), of the track to be used and of the train-car operation. Train-car operation can be changed at the station where the car turns back and starts again. Other options available to the dispatcher are suspending a train service, adding a train, switching two trains, and so on. Switching means replacing the schedule of two trains from an intermediate station where one of two trains starts. It is applicable when two trains are of the same type and have the same destination.

The purpose of train schedule adjustment is to restore rail traffic to normal operation as soon as possible and to minimize any inconvenience to passengers. This problem has conflicting objectives. While the railroad company would tend to place greater emphasis on minimizing the scope of traffic disorder and the number of adjustments from the original schedule, the passengers' primary concerns would be the extended travel time and the degree of crowding. The dispatcher generally evaluates his adjustment plan in the first approximation by calculating the total delay time (i.e., by adding up the delays of each train at each station). If additional time is available, he may work further to optimize other factors, such as the time when rail traffic is restored and also the number of schedule adjustments. In practice, however, evaluation of an adjustment plan itself is a difficult problem as the scale of traffic disorder, time of day, peculiar situations of the line and so on must be taken into consideration.

The mathematical programming of a computer, for example by the branch-and-bound method, is applicable for obtaining a train order at each station that minimizes the total delay time. In train schedule adjustment for real-world railroad networks, however, adjustment methods other than order change should also be taken into consideration, and other objectives are also involved. Therefore, train schedule adjustment is essentially a large-scale combinatorial problem and it cannot be solved by mathematical programming within a practical computation time because of combinatorial explosion.

3. Expert Systems Approach

The expert systems approach is based on the analysis of a dispatcher's problem-solving process by applying knowledge engineering technology. The ESTRAC project is the first application of knowledge engineering technology to the train schedule adjustment problem, and three expert systems have been developed since 1982. An outline of each system is given in Table 1.

The first version, ESTRAC-I (Araya et al. 1983), employed a production system that implemented heuristic rules to change train order at a station on a computer. ESTRAC-I could determine suitable train order at each station within a practical computation time. In addition, the train order that was determined by ESTRAC-I was evaluated by the branch-and-bound method to minimize the total delay time. Although ESTRAC-I supposed a small and simple railroad network, it introduced the new technology of knowledge engineering to the train schedule adjustment problem.

The second version, ESTRAC-II (Araya and Fukumori 1984), implemented on a computer the skilled dispatcher's know-how for schedule adjustment of real-world railroad networks. To extract the know-how, Araya and Fukumori developed a computer-aided system consisting of a graphics terminal to display a train diagram and a mouse to change the train schedule. They then took a train diagram used in a real-world railroad network and introduced typical traffic disorders. They had a skilled dispatcher prepare an adjustment plan and explain the logic used. They recorded the history of schedule change and his explanation, and studied these recordings in detail. ESTRAC-II employed a production system that effectively represented the dispatcher's know-how and applied it correctly to adjust the schedule. Although ESTRAC-II could prepare an adjustment plan that only applied the dispatcher's know-how for changing train order and train-car operation, it was made clear that the knowledge engineering technology would be applicable to the train schedule adjustment of large and complicated railroad networks.

Table 1
Outline of the ESTRAC system

	ESTRAC-I	ESTRAC-II	ESTRAC-III
Years	1982–1983 (I)	1984–1985 (II)	1986–1987 (III)
Railroad network	small, simple	large, complicated	large, complicated
Schedule adjustments	train order	train order, train-car operation	train order, train-car operation, switching, etc.
Remarks	introduction of artificial intelligence	knowledge aquisition	partial simulation

The latest version, ESTRAC-III (Komaya and Fukuda 1989), is based on the analysis of the skilled dispatcher's preparation process for the adjustment plan that was acquired through the development of ESTRAC-II. Features of ESTRAC-III are:

(a) partial simulation to collectively transact events related to each other is defined and is used as a basic unit of simulation;

(b) the basic commands to control execution of the partial simulation are defined; and

(c) knowledge of human experts is classified into the strategic knowledge of problem solving and the if ... then rules of schedule adjustment on local conditions.

In ESTRAC-III, the collective transaction of arrival and departure of one train at a station is defined as a partial simulation. If any overtaking is involved, the arrival and departure of the passed and the passing trains are defined to be transacted in one partial simulation. The partial simulations can be executed separately from each other. It is very easy, therefore, to understand the operation of trains that are related on the schedule.

The partial simulation is executed by receiving a basic command. At the station receiving the basic command, a judgment is made on the conditions and schedule of the station and trains subjected to the partial simulation as to whether or not the partial simulation is possible, and the result of this judgment is sent back. If the partial simulation is judged possible, then it is executed. In ESTRAC-III, three basic commands are defined according to the station where the partial simulation is executed, the station where the basic command is received, the next station, or the preceding station. Using three basic commands, the train operation is simulated by the sequence of a partial simulation in a specific order, although the conventional discrete-event simulation is performed by using a timer in the order which the event occurs in real-world systems. In railroad networks, many trains run at a time, and it is impossible for human experts to recognize whole events in order of time, as the discrete-event simulator does. The basic commands play a very important role in performing the partial simulation in the order that the skilled dispatcher prepares the adjustment plan.

Through the acquired preparation process for the adjustment plan of the skilled dispatcher, it was supposed that the skilled dispatcher recognizes only necessary events in a suitable order. For duplicating the skilled dispatcher's preparation process for the adjustment plan, the strategic knowledge describes to which station and in what order the three basic commands should be sent. The if ... then rules describe how the schedule should be adjusted for local conditions. By repeating the partial simulation and the schedule adjustment based on if ... then rules in a specific order

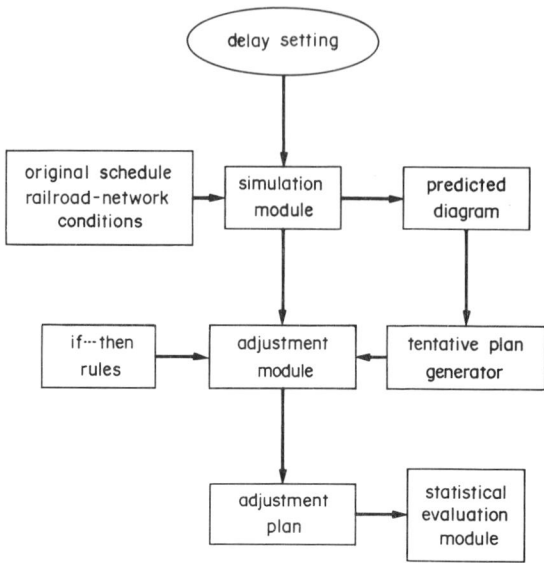

Figure 2
Process flow of the ESTRAC-III system

that is governed by the strategic knowledge, ESTRAC-III prepares the adjustment plan that is functionally equivalent to that of the skilled dispatcher working manually.

4. ESTRAC-III

The ESTRAC-III system is designed to cope with the delay in arrival/departure, and with the speed limitation in a certain section of a line. ESTRAC-III can prepare the adjustment plan that involves many different methods of schedule adjustment such as changing train order and train-car operation, suspending a train service, adding a train, and switching two trains. The process through which ESTRAC-III prepares the adjustment plan is shown in Fig. 2.

When a delay occurs, the dispatcher enters the delay information via the console. The system then simulates train operation that would result from running the trains according to the original schedule by the sequence of partial simulation. This simulation is performed to predict the scope of disorder and to identify overtaking operations that cannot take place on time. In this simulation, therefore, the partial simulation is performed in such an order that only the disturbed operation of trains is involved. Next, the system automatically generates several tentative plans for shifting overtaking operations by a domain-specific heuristic method. For each tentative plan so developed, the partial simulation is performed and if ... then rules are subsequently applied to adjust the original schedule for local conditions. By repeating steps of this kind in a specific order, the system prepares an adjustment plan.

Finally, a quantitative assessment of the adjustment plan is calculated. ESTRAC-III is thus able to display to the dispatcher an adjustment plan for each tentative plan, along with a quantitative assessment of its value.

The ESTRAC-III system uses two sets of if ... then rules for schedule adjustment. One set belongs to the train and governs the shifting of an overtaking operation, changing train-car operation, switching two trains, and suspending a train service. The other set belongs to the station and covers shifting an overtaking operation, changing the track to be used, and adding a train. For example, for each train, the rule for shifting an overtaking operation can be expressed in the following form (e.g., from station K (scheduled) to station $K+1$):

if the passing train arrives at station K delayed by a period of T seconds or more
then shift the overtaking operation to station $K+1$

The rule for shifting an overtaking operation (e.g., from station $K+1$ (scheduled) to station K) is:

if the train arriving first at station K is type S_1 (a local) and the train arriving second at station K is type S_2 (an express) and the difference in departure times is less than T seconds
then move the overtaking operation ahead to station K

5. An Operation Example

Consider the application of ESTRAC-III to a large and complicated railroad network operated in the real world; Fig. 1 shows a train diagram of that network from 9:50 to 11:20. This line operates six types of trains with an average total of 40 trains running in both directions outside the rush hours. Trains can overtake

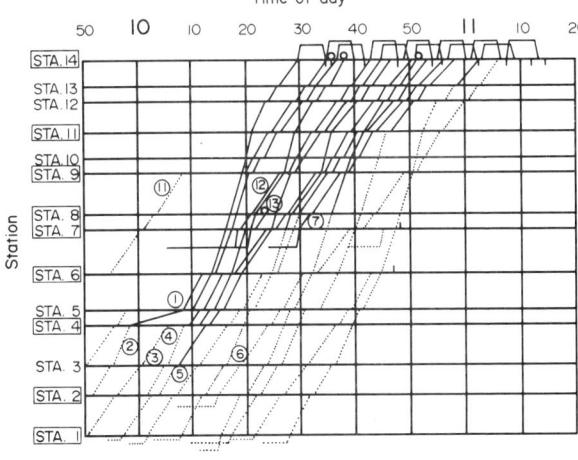

Figure 3
Predicted train diagram after a typical traffic disorder has been introduced

Table 2
Five tentative plans

Plan	c	g	i
1	station 6	station 4	station 4
2	station 6	station 4	station 6
3	station 6	station 4	station 7
4	station 6	station 6	station 6
5	station 6	station 6	station 7

at nine of the 14 stations (displayed in boxes). To improve clarity, the down-bound trains are not shown in Fig. 1.

On this train diagram, a typical traffic disorder has been introduced (i.e., train 1 has been delayed from departing for 10 min at station 5). Figure 3 shows the effect of the delay as performed by the simulation module. In this case, since most of the schedule adjustments are applied to up-bound trains, ESTRAC-III is designed to simulate the operation of up-bound and delayed trains only. In Fig. 3, the dotted line indicates trains running on time, while the solid line indicates delayed operation.

Through the predicted diagram shown in Fig. 3, overtaking operations s-z in Fig. 1 can take place on time; however, overtaking operations a-k are delayed. Trains 1-5, delayed at station 5, are interlinked by overtaking operations a-i. Of these nine overtaking operations a-i, three—c, g and i—are crucial in this situation. Once these are decided, the remainder of the overtaking operations and the other schedule adjustment can be determined automatically by successively applying the if ... then rules mentioned previously. ESTRAC-III then generates tentative plans from the possibilities, to which stations the overtaking operations c, g and i can be shifted, respectively. In this case, tentative plans are listed in Table 2.

For each tentative plan, ESTRAC-III can prepare a full adjustment plan. In this example, there are five adjustment plans. Table 3 shows the comparative figures for these five plans. Though ESTRAC-III can calculate other criteria, only the essential ones are

Table 3
Quantitative assessments of adjustment plans

Plan	Total delay delay time (h:min:s)	Number of delayed trains	Number of schedule adjustments
0	8:30:40	13	(0 0 0 0)
1	4:19:10	8	(5 1 1 0)
2	4:20:20	8	(5 1 1 0)
3	5:34:30	10	(5 1 0 0)
4	4:52:20	9	(5 1 1 0)
5	5:39:20	10	(5 1 0 0)

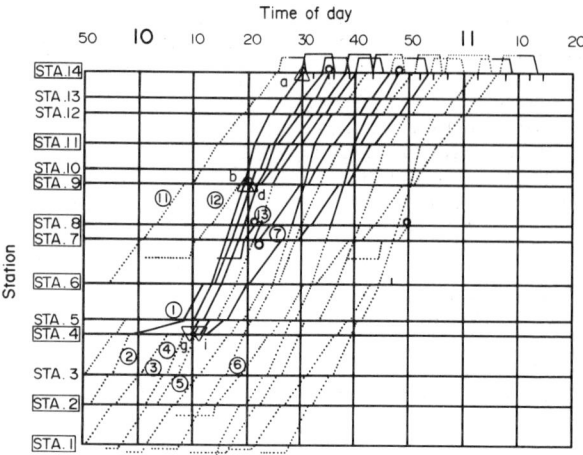

Figure 4
Predicted train diagram for plan 1

listed in Table 3. In Table 3, plan 0 denotes the case in which trains will be operated without any schedule adjustment (see Fig. 3) and, from right to left, the four figures in the number of schedule adjustments column indicate the number of changes in train order and train-car operation, switching two trains, and suspending a train service, respectively.

As shown in Table 3, ESTRAC-III can prepare an adjustment plan that involves many kinds of methods of schedule adjustment in combination, and a different adjustment plan for each tentative plan. In this case, plan 1 is selected as the optimum from the number of schedule adjustments and the total delay time, and the skilled dispatcher also prepares the same adjustment plan manually. Figure 4 shows the predicted diagram for plan 1 and Table 4 shows the complete list of the schedule adjustments.

Through the experience of applying ESTRAC-III to other disturbed situations, it has been confirmed that adjustment plans prepared by ESTRAC-III are functionally equivalent to those of a skilled dispatcher working manually. The software of ESTRAC-III is written in the computer language C, and runs on an engineering workstation. The system prepares a single adjustment plan in about 5 s. It is believed, therefore, that ESTRAC-III can support the dispatcher efficiently

Table 4
Results of schedule adjustments involving plan 1

Shift overtaking b from station 7 to 9
Shift overtaking d from station 7 to 9
Shift overtaking a from station 9 to 14
Change car operation of train 11 and train 2 at station 14
Shift overtaking g from station 6 to 4
Shift overtaking i from station 7 to 4
Switching train 3 and train 7 at station 7

and effectively in disturbed situations and that the expert systems approach is effective to support the decisions of human experts in this field.

See also: Expert Systems Approach to Road Traffic Control; Railroad Systems: Line Supervision and Control

Bibliography

Araya S, Abe K, Fukumori K 1983 An optimal rescheduling for online train traffic control in disturbed situations. *Proc. 22nd IEEE Conf. Decision and Control*. IEEE, New York, pp. 489–94

Araya S, Fukumori K 1984 ESTRAC-II: An expert system for train traffic control in disturbed situations. *Proc. 6th European Conf. Artificial Intelligence*. Elsevier, Amsterdam, pp. 23–32

Komaya K, Fukuda T 1989 ESTRAC-III: An expert system for train traffic control in disturbed situations. *CCCT '89*. Pergamon, Oxford, pp. 147–53

K. Komaya and T. Fukuda
[Mitsubishi Electric Corporation, Amagasaki, Japan]

Expert Systems Approach to Road Traffic Control

Knowledge-based systems (KBSs) and, in particular, expert systems (ESs) have been developed in the field of artificial intelligence (AI) research with the aim of emulating human problem-solving behavior in complex real-world tasks. The potentiality of such systems and their possibilities of application to transport problems has received considerable and increasing attention within the transport engineering community (Logie and Neffendorf 1984, Hendrickson et al., 1985, Bonsall and Kirby 1986, Yeh et al. 1986, Wentworth 1987). A variety of areas in this field can benefit from this approach. These include, among others, infrastructure design, transport planning, safety and maintenance, structures and equipment, vehicle scheduling, and traffic monitoring and control. For example, as regards air traffic control problems, several applications of AI techniques and ESs have been identified and widely discussed by Goslin (1987).

Because of increasing transport demand, increasing congestion and problems of safety and environmental degradation, road traffic control, especially in urban areas, is likely to be the transport sector where the benefits of ESs are potentially the most relevant. However, it is also the most challenging for assessing feasibility, effectiveness and usefulness of such systems, given the number of demanding requirements placed on this relatively young and still evolving technology. The exploration of this field is a new ongoing research direction. Nevertheless, some promising lines of developments are already appearing.

Table 1
Comparison of a conventional program and a KBS

Conventional programs	KBS
Defined, complete sequence of operations	Collection of rules of thumb
Order of execution precisely stated	Order of execution not specified prior to solving the problem
Amalgamation of information and control	Separation of knowledge and control
Correct data must be supplied in order to obtain correct results	Provision of data provide plausible solutions (i.e., valid in most cases, but not always)
All required data must be provided in input	Can operate with incomplete data
Provide only answers	May disclose their functioning and provide justifications for their reasoning
User unfriendly	User friendly
May be difficult to use and understand	Transparency of knowledge representation
	Transparency of dialog
May be difficult to modify	Incremental growth capability

1. ESs and Their Function

Knowledge-based ESs are problem-solving programs modelled on the reasoning of human experts. They are designed to incorporate the expert's judgement, rules of thumb, heuristics and reasoning strategies, and to provide knowledge advice about the application task. Such classes of systems are rather different from conventional computer programs both in the principles organizing the program architecture and in the way in which the information is used. ESs reason by manipulating symbolic, nonnumeric information (i.e., a logical representation of the problem domain) and are based on a clear and explicit separation between the knowledge they store (knowledge representation) and the mechanisms for manipulating such knowledge (inference and reasoning strategies). Furthermore, they are provided with some (limited) form of self-knowledge; that is, they are able to keep a representation of their internal structure and functioning. This allows them to control their reasoning and to reconstruct inference paths (e.g., for the purpose of explaining and justifying solutions to the user). There is a set of distinguishing features that characterizes these kinds of systems. Table 1 contains a summary of the main differences between conventional algorithmic programs and knowledge-based (expert) systems.

ESs typically include:

(a) a knowledge base, which contains a model of domain knowledge;

(b) an inference engine, which embodies reasoning methods to act on the knowledge stored in the knowledge base and to make deductions;

(c) an explanation module, which allows ESs to provide justifications for its reasoning steps;

(d) a knowledge acquisition module, to enable ES developers to build, modify and extend the knowledge base; and

(e) an input–output interface, which allows the user (or other application programs) to interact with the system.

The inference engine acts by performing the reasoning task using the knowledge stored in the knowledge base. Given a set of data (problem data and current state of solution) and some stated goals to be reached (problem solution) the inference engine looks for applicable rules or pieces of knowledge in the knowledge base, activates them and performs associated actions, the main kind of action being the deduction of new facts (data) which are then used for subsequent deductions. This process goes in cycles, until either the solution is met (success) or no more knowledge (rules) is applicable to current data (failure).

The various implementations of KBSs are all more or less sophisticated elaborations of this basic scheme. Classes of application cover different typical problem areas of engineering including design, planning, data interpretation, prediction, fault diagnosis, monitoring and control.

2. Application to Road Traffic Engineering and Control

Traffic control systems include data collection through roadside sensors, computer-based processing of collected traffic data and flow control and guidance through facilities such as traffic signals and variable message signs. Technical advances in computer science offer traffic engineers an entire set of programming options and control tools to implement the various strategies required to cope with the large range of situations and variations in traffic behavior. Software systems based on mathematical modelling and optimization are largely in use to provide support in a variety of tasks including, for example, traffic analysis, signal timing and evaluation of alternative traffic management strategies. Benefits from the application of ES techniques in order to achieve better efficiency and safety in traffic management operations are expected in two different, complimentary directions: first, to overcome some limits of current traffic control technology and to extend the range of situations the control system is able to provide adequate responses for; and second, to provide operators of traffic control centers with better support for managing all the system facilities and for coping better with the increasing flexibility of available technology.

2.1 Motivations

There are a number of problems in current traffic control systems for which ES technology is expected to provide a positive contribution. The most relevant issues include the following.

(*a*) *Managing congestion.* Conventional traffic control methods (i.e., vehicle-actuated, demand-responsive and fixed-time signal control systems) perform well in normal conditions; that is, in undersaturated situations or in saturated conditions, provided that saturation occurs for quite limited time intervals. The increasing levels of congestion and the ways patterns of congestion expand in a network require complex control strategies for monitoring, solving or even preventing the occurrence of congestion that are beyond the capabilities of current systems.

(*b*) *Responding to incidents and critical events.* Current control systems also have to face recurrent problems with critical events which break down smooth traffic flows and cause hazards, such as accidents, road obstructions, poor weather conditions and special events. Strategies are needed to recognize such events rapidly from traffic flow data and to recover the occurring reduced capacity promptly.

(*c*) *Dealing with conflicting constraints and goals.* Problems do not occur in isolation. While well-designed systems or experienced operators are able to respond effectively to single problems, they do not cope well with problems occurring simultaneously or with conflicting constraints and goals (e.g., limiting traffic to reduce pollution vs satisfying increasing demand in an area). Strategies and techniques are needed to enable control systems (and operators) to evaluate alternatives and to solve conflicting situations following adaptively the changing operating conditions and control objectives.

(*d*) *Dealing with incomplete and unreliable information.* Because of both economical and technical factors (i.e., cost of sensors, problems of installation and maintenance, etc.), generally, not all sections of the road network are provided with data collection equipment. Thus, only a partial view of the whole traffic in the network is available to the control systems for analysis and decision. Furthermore, failure of sensors or of data transmission equipment results in frequent losses of reliability of collected traffic data. Current systems do not perform well when faced with uncertainty or incompleteness of information. Techniques are needed for making inferences about traffic conditions on the basis of incomplete information when sensors are not present, functional or properly operating.

2.2 Basic Features

In all the problems in Sect. 2.1, the role of the human operator is still crucial and the operator's experience and knowledge considerably affect the quality of the control. A basic assumption behind the current interest in ES techniques is that the analysis and decision processes involved in such problems are hardly of algorithmic nature; therefore, heuristics and simulated human problem-solving approaches have more chances of success. Several features of typical ES techniques are considered to provide potentially adequate answers to attack the problems in Sect. 2.1. Among them the following are recognized as most relevant by many authors.

(*a*) *Heuristic programming.* Foremost among such features is, obviously, the original reason for research into expert systems; that is, the possibility of capturing human expertise and making it available and usable by a computer system. More heuristic knowledge into computer programs means, in principle, an extended capability of modellng problem-solving strategies when real-world problems do not conform to rigid, analytical assumptions, and less computational effort when searching for solutions to problems.

(*b*) *Transparency and cooperativeness.* A second reason is related to the improved interaction, between the computer and the user, allowed by ES technology. This relates, essentially, to two things: first, the transparency of the system—as relevant domain knowledge is mostly encoded declaratively in the knowledge base and the knowledge base is separated from the program that uses it during the problem-solving process (the inference engine), the information embedded in the application is more easily understood by the user and more readily manipulated and modified; and second, the interactiveness of the system—for example, the capability of explaining the reasoning sequence that led to results. Together, such capabilities can contribute to increase the usability and the acceptance of computer systems.

(*c*) *New software techniques.* A third aspect is related to the distinctive programming techniques used in ES applications. Data-driven programming, for example, is one such useful technique. In rule-based systems, rules can be executed whenever patterns in the current data state (e.g., in the data set currently acquired from the outside world) make it possible. This relieves programmers of the task of explicitly anticipating all the eventualities that programs might encounter and when, and in which order, such eventualities might occur. The ability of handling rules with incomplete, imprecise and inconsistent data is another useful characteristic of knowledge-based programming. A number of techniques exist for combining degrees of belief and disbelief in propositions formulated on database patterns and for modelling imprecise knowledge processing.

(d) New insight in modelling real-world systems. A last interest is rooted in some novel techniques offered by AI to model complex aspects of real-world phenomena. The ability of representing time and temporal reasoning processes (e.g., allowing a computer system to infer temporal relationships between events and to reason accordingly) is one such technique for which advantageous application of ESs to dynamic, real-world systems control is recognized by many authors (Rijckart *et al.* 1988).

2.3 Research issues

Current and expected applications of ESs address a variety of problem areas, including off-line tasks (e.g., rephasing signals at an intersection, and updating a library with a new signal plans) and on-line surveillance and control operations, isolated intersections, arterial roads and network areas, urban and extraurban environments. Conventional control technology (i.e., signalized networks, variable message signs, etc.) is mostly considered (although the application of ES or, more generally, of AI techniques to more advanced control systems including on-vehicle route guidance and driver information systems, is being studied in the framework of the European project PROMETHEUS, Pro-Art subprogram).

(a) Off-line applications. Although, strictly speaking, this is not part of the control problem, planning, designing and updating the operation of traffic signals is a problem area where the use of ESs can have a lasting influence on the performance of an overall control strategy. A number of manual procedures and of software tools are currently available to assist technicians in designing suitable phase distributions, optimizing signal plans, evaluating the effectiveness of a strategy, and so on. Even if a considerable part of the task is carried out by the computer models, the ultimate solution of such problems remains largely dependent on the experience the traffic engineer has with the problem. This is a factor that remarkably affects the final results.

Most work in this area addresses the above problems at the level of isolated intersection, while little attention has been paid to coordination and network control problems. Two examples of such applications are the systems presented by Kirby and Montgomery (1987) and Zozaya-Gorostiza and Hendrickson (1987). Following quite similar approaches, these systems explore the use of knowledge-based techniques for traffic signal design for isolated intersections. Such a problem has both an algorithmic part, concerned with calculating the optimum timing for a defined sequence of stages, and a nonalgorithmic part, which involves choosing the most appropriate sequence under a series of design constraints (e.g., geometric constraints, safety constraints, country- or site-specific regulations, and technical constraints related to the equipment used). Both systems, therefore, combine the application of ES techniques, to capture the expertise and engineering judgements for individuating an acceptable sequence of stages for the signals, and the use of conventional algorithms for calculating signal timings for the designed sequences (e.g., Webster's formula, in the program of Zozaya-Gorostiza and Hendrickson, for estimating user delay). Essentially, the advantages such systems demonstrate are that:

(i) the use of heuristics for the search of possible phase distributions helps in avoiding a complete enumeration of all possible sequences—this can cut down the search space of solutions enormously, thus making it possible to deal with complex junctions (i.e., large numbers of movements) that, otherwise, would lead to combinatorial explosion of the search process;

(ii) ES techniques help in making the design decision process more transparent and, therefore, the final result more effective—that is, they may help in analyzing possible solutions when the selection of the best alternative involves multiple competing figures of merit (e.g., minimum delay vs acceptable delay and short average queues); and

(iii) the programs are quite easily extensible and adaptable to different users and situations because most of the design knowledge is encoded in the knowledge base (e.g., they can be adapted with respect to the various national regulations of different countries).

(b) On-line applications. The application area where the use of ES techniques may become more advantageous is, of course, that of real-time control of traffic flows. The basic requirements for an ES at this level, include (Ambrosino *et al.* 1990):

(i) a real-time database, where traffic flow and data related to other subsystems (e.g., road and weather monitoring systems) are collected from roadside sensors and continuously updated at any cycle of operations;

(ii) a data completion and integration functional level, to assure consistency and completeness of the information stored in the real-time database—at this level, the control system should be able to detect erroneous and incomplete data (e.g., due to lacking or defective detectors) to infer missing information and to integrate information coming from different subsystems (e.g., traffic counts from inductive loops and additional traffic parameters extracted from closed-circuit television images);

(iii) a data analysis and interpretation functional level, to identify the current traffic state based on the information stored in the database, to promptly recognize possible tendencies toward critical, near future, developments (e.g., the onset of a local congestion), and to diagnose the causes of detected problems or trends toward problems (e.g., a particular approach to an intersection as the cause of an expanding congestion); and

(iv) a decision and control functional level, to decide what action should be implemented in order to respond to detected or foreseen problems (e.g., altering the phase of the traffic lights at a given location, overriding the timing of the current signal plan, etc.)—at this level, the control system should be able to reconcile possible competing constraints and goals (e.g., favoring one approach or an area instead of another) and to estimate the effects of possible alternative actions to solve the identified problem(s).

These requirements could, in principle, be considered both at the level of isolated intersection control measures and of major section routes, freeways or urban networks control, as well.

3. Examples of On-Line Applications

3.1 Isolated Intersections Control

The background objective in the area of isolated intersections control is to assess the use of ES techniques for implementing strategies for intelligent intersection control; that is, for adaptive control and interleaving of merging streams at the intersection, under the usual general requirements of flow optimization and safety. In general, such systems are seen, in a broader perspective, as building blocks of an overall decentralized network control system (see *Signal Control at Individual Junctions*: *Stage-Based Approach*).

One such system is presented by Chang and Tang (1989). The prototype expert system, called intelligent traffic signal control software (ITSS) is part of a larger system, called INTEL, which, besides intersection microcomputer and controller subsystems, includes an image-analysis component for traffic surveillance at the intersection. ITSS performs control operations by continuously running a phase and timing design process, based on the data transmitted by the data capture equipment. The design process is carried out in real time and consists of a heuristic part (phase sequence design) and a numerical part (timing optimization). The overall process can be regarded, essentially, as an on-line transposition of the knowledge-based signal setting design approach outlined in Sect. 2.3. The knowledge-base for phase design (consisting of nearly 400 rules, in the prototype version) contains a set of design constraints related to geometric, traffic and control strategy factors. Timing optimization is based on traffic forecasting and platoon dispersion models. Phase and timing are altered gradually. The phase design process proceeds in accordance with a sequence of steps in which the different sets of constraints are brought to bear: identification of the dominating phase; superimposition of geometric, vehicle operation and safety constraints; determination of minimum delays for movement arrangement; maximization of capacity by consolidating the number of phases; and determination of phase distribution and order of execution.

Radwan and Goul (1989) developed another system which is representative of this category of applications. The signal control for isolated intersection (SCII) project follows a quite different approach based on a strategy of continuous adjustment of the signal cycle, operating in a pretimed signalization mode (involving a four-phase sequence in a prototype version tested through simulation). Essentially, this involves a procedure to decide on-line whether the current cycle length should be maintained or modified. The heuristic criteria for such a decision are based on procedures for evaluating the level of service of an intersection, these are used in traffic engineering practice in the US *Highway Capacity Manual*. Contrary to the previous system, no phase sequence design operations are performed on-line. The level of service of the intersection for the current cycle is continuously evaluated using the actual volumes contained in the real-time database and utilizing user-supplied thresholds for deciding about modifications of the cycle length. A traffic forecasting model is also used to include future volume information in the decision process.

3.2 Arterial Road and Urban Networks Control

An ES for coordinated control of traffic on arterial roads or urban network areas is seen as an operator's associate acting, essentially, at two levels (Bell 1988).

(a) There is the operational and tactical level; that is short- and medium-term on-line control of traffic flows on major road sections and network areas. At this level, the ES should embed sufficient traffic engineering knowledge to shift the focus of attention and to locate areas where problems occur, to set up priorities to prevent propagation and to solve detected problems, to deliver advice accordingly (or even to implement decisions automatically) and to follow implemented actions in order to assess their effectiveness with respect to the proposed control policies.

(b) There is also the strategic level; that is, long-term, off-line assessment, upgrading and adaptation of control strategies with respect to changes and variations in traffic behavior. This requires a background activity of continuous updating and analysis of historical traffic data collected on-line and the integration and use of traffic assignment and planning models on the advice of the system and operator requests.

Most work underway in the area of urban network control concentrates on traffic congestion processing.

An experimental system for urban congestion surveillance based on expert system techniques, called SAGE, was set up in Paris (Foraste' and Scemama 1987). This system operates with the aim of analyzing traffic in congested situations and of suggesting appropriate remedy actions. The control strategies handled by the ES are essentially heuristic and consist of practices usually adopted in traffic control centers; that is, gating and metering demand to hold traffic upstream of specific oversaturated locations or favoring particular routes. SAGE includes some of the above major operational steps of an ES for on-line urban traffic network management:

(a) from traffic data collected by sensors—and roughly quantified, for the purposes of ES reasoning into free flow, precongested, congested and unknown—the system has (limited) capabilities of inferring missing data (congestion levels in some links where detectors are not present, in accordance with the above clsssification);

(b) congested routes are analyzed and road stretches and intersections which are likely origins of observed developed congestions are deduced;

(c) control actions which are susceptible to providing an appropriate response to detected problems (i.e., plan selection at given intersections) are proposed for implementation; and

(d) continuous monitoring of implemented control measures is assured by indicating, if necessary, those ones becoming inadequate.

SAGE follows a quite straightforward (and classical in AI terms) rule-based approach. It consists of three major components:

(a) the fact base—a working memory containing both permanent data (network structure, possible control actions) and temporary data (state of the links, causes of congestion, proposed control measures);

(b) the rule base—the encoding of the empiric traffic engineering knowledge to make a diagnosis about traffic in the network, to propose possible control actions and to assess their adequacy; and

(c) the inference engine—activating and controlling the reasoning by matching rules with the facts in the fact base, by selecting applicable rules and executing them.

The system performs, essentially, according to the data-driven programming technique; that is, it uses incoming traffic data (acquired every 3 min) to forward-chain its rules in order to deduce all the possible conclusions, given the current network situation, and to elaborate possible responses.

3.3 Freeway Control

While sharing the same basic requirements and objectives of applications to arterial road and urban network control, ongoing research in the area of freeway control addresses the problems related to safety and incident management more directly (see *Freeway Control: An Overview*). Monitoring traffic flows in order to predict or detect the occurrence of incidents, managing the traffic flow approaching the incident, and devising and implementing appropriate plans for response and recovery are typical tasks for which the application of ESs is deemed to enhance the performance of current technology. In particular, benefits are expected in the task of recognizing the type of incident, reasoning about its severity and helping the operators in elaborating the most appropriate strategic plan for response, on-site management and clearance. Chin-Ping Chan (1989), Faghri and Demetsky (1989) and Lakshminarayanan and Stephanedes (1989), among others, are authors who have tackled these problems.

One system is presented by Cuena (1989) for (almost) real-time traffic prediction, evaluation and policy recommendation in an urban access freeway. Implementation and testing of this system is planned in Madrid and Barcelona. The system is being developed to be installed on top of a conventional monitoring and control system (loop detectors, real-time database, variable message signs for lane control and traffic lights for access control). Essentially, it consists of three interacting knowledge bases: the prediction knowledge base, for the estimation of possible, short-term future situations; the interpretation knowledge base, for the identification of problems in present and predicted situations; and the control knowledge base for the recommendation of possible control actions at a local level (e.g., speed recommendation on a road section, traffic metering at an access point) and at a general level (e.g., introduction of a reversible lane in a zone). The distinguishing feature of this system is represented by its ability to make qualitative predictions of short-term traffic situations based on the use of a qualitative model of traffic flows. This model is derived from a macroscopic model of multilane traffic flows existing in the literature (based on the continuity equation and a dynamic relationship between mean speed and traffic density) adapted for the purposes of the ES, through an AI representation technique known as qualitative modelling.

4. Future Developments

Application of ES technology to road traffic monitoring and control is a young research area. Several researchers share the view that, in the whole field of transport theory and practice, ES technology is where other computing technologies such as simulation and traffic modelling were years ago. More coordinated and full-scale research efforts are deemed necessary, since most of the existing studies and experiences have been carried out at a quite preliminary and conceptual level. Also, extensive field testing of the prototypes realized is still needed. Already, some major developments are being undertaken within the DRIVE programme (projects V1015, V1026, V1055) supported by the European Community, from which a deeper insight into research and development problems is expected (see *DRIVE*).

Several issues appear to be relevant for the near future developments in the field.

(a) Theoretical and design issues. Knowledge, of course, is the essence of any ES. The extent to which traffic engineering knowledge can be formalized and structured into a knowledge base for efficient search and use by the ES is a key factor for the success of this class of applications. In particular, further research is needed to understand how to achieve an appropriate balance between general, site-independent traffic engineering knowledge and particular, site-specific knowledge and expertise. Several other basic research issues appear to have a prominent position in short-term research concerns. The development of self-learning capabilities into the system knowledge base and a fuller utilization of temporal reasoning techniques are two examples.

(b) System architecture. Owing to the novelty of the approach, most research work has concentrated on "pure" AI/ES programs. Little attention has been paid to the problems of achieving efficient integration of ES components into present-day traffic control systems technology and to related standardization aspects. Distributed systems technology is also expected to have a key role at an architectural level. Relevant topics for future research work include concurrent processing technology, distributed knowledge bases, cooperative ESs and problem-solving architectures. Some of these issues are actually arguments of basic research and development in AI.

(c) Software tools. Serious applications in the traffic control area place a number of demanding requirements on the software technology used for developing and implementing systems. These include the ability to cope with real-time constraints, an efficient integration of symbolic and numeric processing, and the robustness of the software tools. Some of these issues appear to be an order of magnitude beyond the capacity of current AI technology. More achievements are expected from ongoing research and development in the field.

See also: Expert Systems Approach to Rail Traffic Control

Bibliography

Ambrosino G, Bielli M, Boero M, Mastretta M 1990 Artificial intelligence techniques for urban traffic control. *Transp. Res. A* (in press)

Bell M C 1988 The fundamental issues of an expert system for urban traffic control. *Proc. Colloquium Applications of Expert Systems in Road Transportation*. Institution of Electrical Engineers, London, pp.7/1–6

Bonsall P W, Kirby H R 1986 The role of expert systems in transport. In: Bonsall P W, Bell M (eds.) *Information Technology Applications in Transport*. VNU Science Press, Utrecht, The Netherlands, pp.353–82

Chang A T S, Tang R Y 1989 A prototype expert system for traffic signal control in real-time. *Proc. 1st Int. Conf. Applications of Advanced Technologies in Transportation Engineering*. American Society of Civil Engineers, New York, pp.183–8

Chin-Ping Chan E 1989 Expert systems applications for freeway incident management. *Proc. 1st Int. Conf. Applications of Advanced Technologies in Transportation Engineering*. San Diego, CA, pp.147–52

Cuena J 1989 AURA: Second generation expert system for traffic control in urban motorways. *Proc. 9th Int. Workshop on Expert Systems and Their Applications*. EC2, Avignon, France, pp.145–56

Faghri A, Demetsky M J 1989 TRANZ: A prototype expert system for traffic control in highway work zones. *Proc. 1st Int. Conf. Applications of Advanced Technologies in Transportation Engineering*. San Diego, CA, pp. 165–70

Foraste' B, Scemama G 1987 Intelligent traffic surveillance system. *Compendium of Technical Papers, Institute of Transportation Engineers 57th Annual Meeting*. Institute of Transportation Engineers, New York, pp. 333–7

Goslin, G D 1987 Identification of artificial intelligence applications in air traffic control. *Transp. Res. A* **21**, 27–38

Hendrickson C T, Rehak D R, Fenves S J 1985 Expert systems in transportation systems engineering. Department of Civil Engineering. Research Report. Carnagie-Mellon University, Pittsburgh, PA

Kirby, H R, Montgomery F O 1987 Towards a rule-based approach for traffic signal design. *Civ. Eng. Syst.* **4**, 20–6

Lakshminarayanan N M, Stephanedes Y J 1989 Expert system for strategic response to freeway incidents. *Proc. 1st Int. Conf. Applications of Advanced Technologies in Transporation Engineering*. American Society of Civil Engineers, New York, pp. 153–8

Logie M, Neffendorf H 1984 Expert systems in transportation. *PTRC Summer Annual Meeting*. Planning and Transport Research and Computation, Brighton, UK

Radwan A E, Goul M 1989 Knowledge-based system development in real-time traffic control. *Proc. 1st Int. Conf. Applications of Advanced Technologies in Transportation Engineering*. San Diego, CA, pp.189–92

Rijckart M J, Debroey V, Bogaerts W, 1988 Expert Systems: The state of the art. In: Mitra G. (ed.) *Mathematical Models for Decision Support*. Springer, Berlin

Wentworth J 1987 Advisory (expert) systems—An assessment of opportunities in the federal highways administration. *Public Roads*. Federal Highways Administration, Washington, DC

Yeh C, Ritchie S G, Schneider J B 1986 Potential applications of knowledge-based expert systems in transportation planning and engineering. *Transp. Res. Rec.* **1076**, 58–65

Zozaya-Gorostiza C, Hendrickson C T 1987 Expert system for traffic signal setting assistance. *J. Transp. Eng.* **113**(2), 108–26

G. Ambrosino, M. Boero and M. Mastretta
[Automa, Genoa, Italy]

M. Bielli
[IASI-CNR, Rome, Italy]

Flight Activity in General Aviation

Flight activity in general aviation represents, particularly in the USA but also in some European countries, a large part of the entire flight activity of civil aviation. Therefore, it is of prime importance to describe developments in this sector by means of a detailed analysis of significant traffic-relevant data such as the number of aircraft, flight movements or hours flown. The task is then to identify parameters capable of adequately explaining this development. With the help of different prognosis methods, it is possible to make forecasts about the different segments of general aviation.

1. Definition of General Aviation

Civil aviation and general aviation cannot be seen as one and the same. Civil aviation is subdivided into numerous different transport demand segments with various travel purposes. The allocation of some segments of civil aviation to general aviation is not uniform in all countries; it is often based on convention. Generally, for the European market, all commercial and noncommercial aviation for the transportation of passengers and cargo that is not scheduled or charter air traffic is catagorized as general aviation.

General aviation can be subdivided into the following categories according to use:

(a) private business aviation,
(b) other commercial aviation (e.g., spray flights with helicopters or offshore transportation),
(c) training flights by commercial flight schools or noncommercial aviation clubs, and
(d) private aviation.

In contrast to European countries, there exists a distinct definition in the USA concerning the segments of civil aviation belonging to general aviation. This definition expresses that all civil aviation activity, except that of air carriers certificated in accordance with the Federal Aviation Administration (FAA) parts 121 and 127 (commercial air carriers), part 123 (air travel clubs) or part 135 (commuter air carriers and air taxi operators), belongs to general aviation. The European convention generally counts air taxi operators as general aviation.

2. Development of General Aviation

Generally, there are considered to be two main segments of general aviation regarding flight activity. The first is private aviation with aircraft, mostly single-engine or multiengine piston aircraft, often used for leisure and, particularly in the USA, for business travel purposes. The second important segment is that of training flights. All other segments have only niche functions. Therefore, most of the following explanations deal with only these two categories of general aviation.

A worldwide analysis and forecast of the flight activity of general aviation is not useful, because the regional distribution of the world general aviation fleet shows that only a small number of countries are centers of general aviation activity. Comparison of the existing general aviation aircraft fleets shows the enormous leading position of the USA. More than 80% of general aviation aircraft are licensed in the USA and Canada. The leading countries of Europe, such as Germany, the UK and France, together only have about 6% of the world fleet. For the following analysis and forecast of general aviation, it is therefore sufficient to consider the USA and, as an example for European general aviation, Germany.

As a consequence of the great demand for general aviation in the USA, most of the manufacturers of general aviation aircraft can be found there. Manufacturers in Europe did not play an important role during the period of great increase in the growth rates of general aviation (up to the early 1980s). More recently, European manufacturers have played a very innovative role in the technical development of new general aviation aircraft, but the leading position still belongs to the American manufacturers. Thus, it will suffice to analyze the production rates of the important American manufacturers, who are organized in the General Aviation Manufacturers Association (GAMA). During the period from 1970 up to 1987, they produced 168 945 units (79% single-engine piston aircraft, 15% multiengine piston aircraft and 6% turboprops and jets) for the worldwide general aviation market.

An overview of the yearly production rate of the GAMA manufacturers in this period is given in Fig. 1, which shows that the production rate and turnover steadily increased up to 1978, with more than 17 000 units produced in that year. However, the production rate decreased rapidly between 1980 and 1983, falling to 2691 aircraft per year, and this decline continued to 1987, with little more than 1000 units produced in that year. Because the production of the more expensive multiengine and turboprop aircraft did not decline at the same high rate as that of single-engine aircraft, the turnover stabilized after 1983 at between US$1.26 billion and US$1.68 billion.

The export of general aviation aircraft has been very important to the American manufacturers. The US

Flight Activity in General Aviation

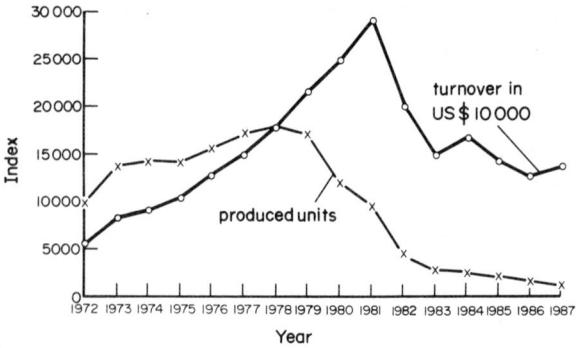

Figure 1
Production rate and value of GAMA general aviation aircraft shipments in the period 1970–1987

export rate between 1975 and 1983 was 28% by worth. In more recent years, this rate has decreased to less than 15%, showing that the US export market has declined faster than the US home market.

There are 220 000 aircraft in the active general aviation fleet in the USA. Most of these aircraft—about 195 000 units—are single-engine or multiengine piston aircraft. In the period since 1980, the number of aircraft in the active fleet has been nearly constant, with a growing trend of enlarging the turboprop aircraft fleet. Between 1980 and 1986, the numbers of single-engine piston aircraft declined by 4000 units, while the turboprop fleet increased from 6200 to 9800 units.

While the size of the aircraft fleet has remained constant in recent years, flight activity has rapidly declined in the USA, from 200 h flown per general aviation aircraft in 1980 to about 150 h in 1985. Whereas the number of turbine aircraft hours flown has increased, piston aircraft, in particular, have been used less and less in the USA.

In the German market up to 1982 there was also a steady increase of licensed general aviation aircraft, mostly single-engine models with a maximum takeoff weight of 2 t, there being 5703 units in that year.

After a decrease of about 250 units (4%), there has been a stagnation since 1986. Comparable with the development in the USA, the flight activity of these aircraft sank from 120 h flown per aircraft in 1980 to 100 h flown in 1985. The use of these aircraft was substantially less in Germany than in the USA, however, and the decline in flight activity per aircraft was not so high.

3. Factors Influencing General Aviation

The question arises of which factors have influenced the development of general aviation flight activity or aircraft fleets in the USA and in Europe. Numerous studies by offices of aviation policy, by universities and by the industry have shown that economic factors, as well as aircraft prices and operating costs, are responsible for explaining the recent decline.

The FAA has pointed out that the general aviation industry has, to date, failed to respond to the current economic recovery, one of the most robust of the postwar period. Historically, the economic cycle of the general aviation industry has closely paralleled that of the national economy. The analysis of European, especially German, general aviation shows the same development. Up to the early 1980s, the real family net income per month was the economic factor with the greatest influence on general aviation flight activity in noncommercial private aviation and in the training segment. The commercial and noncommercial business aviation segment was influenced in the same way by the national growth product.

The lack of development of general aviation in the USA and Europe since 1980 shows that, although the economic factors mentioned previously are of importance, other factors must have increased their degree of influence, perhaps even to a higher level than that of the economic factors.

The theories about the effects of new factors are diverse. Some note high aircraft prices and the availability of low-cost alternatives, such as ultralights, or the general change of leisure time interests, which may be shown in the fundamental change in the taste and preferences of the population. Others hypothesize that high operating costs or changes in tax laws are responsible for the depression of the general aviation industry.

From these theories, as well as from analysis of the European market, one thing has become certain: the costs involved (i.e., the operating and, particularly, the total cost per hour flown along with the purchase costs of the aircraft) have become the second most important influencing factor.

Figure 2 shows as an example the development of the total costs of a typical single-engine piston aircraft operated in the FRG. Comparisons with other European countries and the USA show that the trend is nearly the same. Between 1970 and 1979 the costs

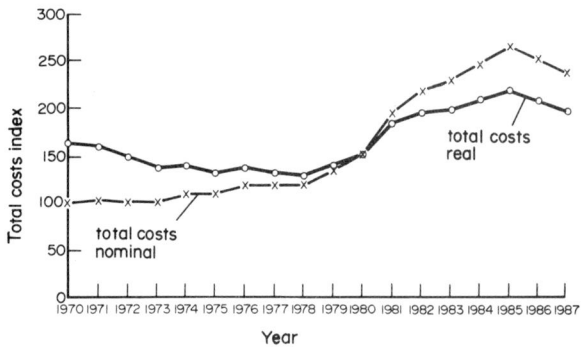

Figure 2
Development of the total costs index of a typical single-engine piston aircraft in Europe

(inflation rate reflected) were nearly constant, but they then began to climb sharply up to 1985. In the following period, a process of cost declining in small steps can be noticed, but costs remain on a high level. Reasons for this development are the enormous increases in aircraft prices, maintenance costs and fuel price.

Since 1987, the increase of costs has come to a standstill without reactivating general aviation activities; therefore, it must be assumed that there are other reasons beyond the influencing factors known. In the USA, it has been realized that changing tastes could upset the fundamental economic equations that have held for many years for the industry. If this is occurring, then falling prices and operating costs, accompanied by economic growth, may not be sufficient to revive the market. As a nation becomes wealthier, households can afford to pay the higher price for specialized items and a proliferation of varieties generally takes place. This intensifies the competition in specific areas of the market. During the recent strong economic recovery, recreational flying in conventional aircraft has been rapidly declining, while the demand for relatively expensive cars, homes and boats has been expanding. This lost market may be difficult to recover, even if the economic forces shift in favor of aviation.

4. Forecast Methods and Results

The reason for this lost market, which cannot be quantified, must also be taken into consideration for establishing prognoses for the general aviation market in the USA and in Europe. First, it is necessary to know what instruments are available for forecasting. The most usual methods for forecasting aviation items are

(a) trend extrapolation,

(b) forecast of saturation,

(c) the method of traffic models,

(d) regression analysis, and

(e) qualitative estimation procedures.

Trend extrapolation only takes into consideration the analyzed traffic data. Future trends are derived from developments in the past. The basis of this method is the often incorrect assumption that all components that have influenced development in the past will do so with the same form and force in the future.

The forecast of saturation method can only be used if the form of the trend and other logical relationships are giving indications that the traffic data, in this case primarily the aircraft part, will reach a saturation point in the future. This method very seldom gives a satisfactory forecast.

The method of using traffic models derives from the forecasting of urban traffic. Some studies have come to the conclusion that this method is, for several reasons, not applicable for the prognosis of flight activity in general aviation.

A method often used is that of meanwhile (multiple) regression analysis as an econometric model. With the help of computer software and long-duration historical databases, it is possible to explain one traffic factor, such as aircraft movements, by several logical and independent variables, the so-called explaining variables. The result is a function that is given in the linear form by

$$y = a + \sum_{i=1}^{n} b_i \, x_i$$

where y is the traffic factor to be forecast, a and b_i are regression coefficients delivered by the computer software, and x_i are the explaining variables. The disadvantage of this mathematical form of forecast is that only quantified variables can be used; however, all those factors with influence on the traffic factor that are not included in databanks cannot be considered.

The consideration of nonquantified variables is only possible by the method of estimation. With the help of an expert knowledge of the air traffic market that is to be forecast, it is possible to develop, discuss and verify the possible forecast. This method has a second great advantage: the results of a multiple regression analysis can be checked critically and, when the occasion arises, be modified.

The forecasts for general aviation by means of the discussed methods with consideration of the analyzed influential factors show a stagnation of the flight activity and the aircraft fleet for the USA as well as for Europe. The FAA forecasts that up to the year 2000, the hours flown will only increase by 0.4% on average, resulting in an estimated 36.6×10^6 h flown in the USA in 1998. During the 1960s and 1970s, the average annual growth rate of hours flown was about 6.0%. The number of single-engine piston aircraft hours flown is forecast to decline from 22.0×10^6 h in 1987 to 21.3×10^6 h in 1999. The active general aviation fleet is forecast to decline from 1987 to 1992 and then to grow slowly for the remainder of the forecast period. The population of active aircraft is forecast to increase only slightly over the 12-year period with a decline of 0.2% between 1987 and 1992 and with a growth of 0.2% between 1992 and 1999. The number of active single-engine piston aircraft is expected to decline at an annual rate of 0.4%.

For Germany and, in the long run, for most European countries, a stagnation or a small increase in the flight activity and in the existing fleet can be expected. Nevertheless, for this slight positive development, the expected economic boom is necessary to support general aviation up to 2000. The expected small increase in activity and fleets will immediately be disturbed if the national income does not increase as positively as forecast by economic institutes or if an overproportionally high increase of operating costs occurs to hinder general aviation. In these circumstances, it must be expected that the negative trend of the early 1980s will continue, a forecast result that is not only valid for European countries with general aviation activities, but also for the USA.

See also: Air Traffic Control; Flight Activity: Prevision; Social Issues of Transportation

Bibliography

Desel U 1988 Analyse der Flugaktivität im nichtgewerblichen Motorflugverkehr und Schulflugverkehr der Bundesrepublik Deutschland als Grundlage für eine Abschätzung seiner künftigen Entwicklung. *Veroeff. Verkehrswissenschaftliches Inst. RWTH Aachen* **42**
Federal Aviation Administration 1982 *General Aviation Activity and Avionics Survey*, Annual Summary Report. FAA. Washington, DC
Federal Aviation Administration 1983 *General Aviation Pilot and Aircraft Activity Survey*. FAA, Washington, DC
Federal Aviation Administration 1988 *FAA Aviation Forecast, Fiscal Years 1988–1999*. FAA, Washington, DC
General Aviation Manufacturers Association 1988 *General Aviation Statistical Databook*. GAMA, Washington, DC
Seeg H.-G. 1979 Bestimmung und Analyse der Einflussfaktoren für die Nachfrage nach Verkehrsleistung im Geschäftsreise und Werkverkehr in der Bundesrepublik Deutschland. Dissertation, TU Karlsrühe, Germany
Vahovich S G 1978 Income and cost impact of general aviation hours flown by individual owners. *Transp. Res.* **12**

U. Desel
[Niedernhausen, Germany]

Flight Activity: Prevision

The entire air transportation system consists of many components, the main parts being military and civil aviation, the latter being of interest in this article. Civil aviation consists of commercial air transportation, with scheduled and charter air services, and noncommercial or general aviation. Whereas general aviation can be regarded as a means of flying primarily for the private and leisure purposes of individuals (though business travel by company-owned aircraft also belongs to general aviation), it is commercial air transportation that serves the general public as a transport mode providing interregional, international and intercontinental services. This article concentrates on describing the structure and development of demand for scheduled and charter services of the commercial air transportation system.

1. History of Global Demand

Hand in hand with economic growth, the growing importance and availability of leisure time and technological progress in aviation, the demand for air transportation has grown considerably worldwide, and has generally outstripped the growth of the more "classical" modes of bus, car and railroad.

As shown in Fig. 1, global air traffic of scheduled services grew from 500 billion passenger kilometers in 1971 to almost 1600 billion passenger kilometers in 1987. This means that on average the annual demand rose by about 70 billion passenger kilometers. Since passengers have used airplanes for travelling increasing distances, the demand as expressed in passenger kilometers has grown faster than the demand expressed by passenger volume; the average growth rate for passengers carried was 5.5% as compared with 7.0% for passenger kilometers. The number of passengers carried on scheduled services exceeded 1 billion for the first time in 1987 and has doubled since 1974.

Although the main task of airlines is passenger transport, airfreight and airmail also play a growing role in this passenger-oriented transport system. More than one quarter of all tonne kilometers performed belonged to speed-sensitive freight and mail (see Fig. 1).

In some areas of worldwide air transportation scheduled services are supplemented by charter services, primarily serving holiday travellers. Total international nonscheduled passenger kilometers performed worldwide reached 157 billion in 1987, representing almost 20% of the total international travel on scheduled and nonscheduled services. Travel between the Mediterranean region and European countries constitutes the largest international charter market, the

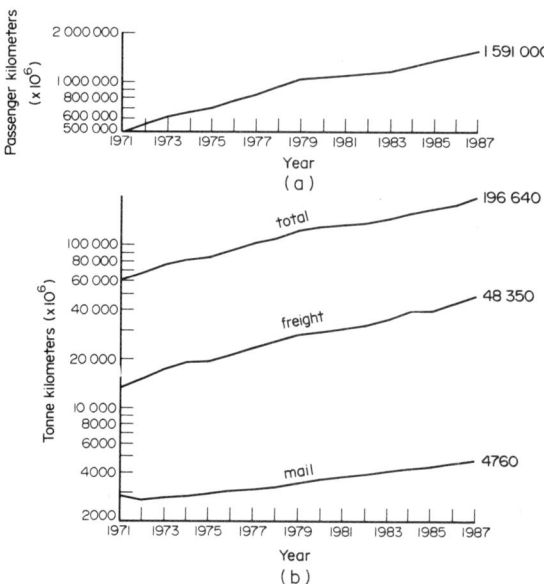

Figure 1
ICAO states total scheduled services and revenue traffic 1971–1987 in (a) passenger kilometers and (b) tonne kilometers (after ICAO (1988), 1987 data are provisional)

volume of nonscheduled traffic being comparable to that of scheduled traffic. Since charter traffic has not succeeded in gaining worldwide coverage, it has not grown as fast as scheduled traffic; during 1977–1987 the average growth rate of passenger kilometers performed on charter services was 3.9%.

2. Structural Characteristics

Global demand figures are primarily of statistical value for analyzing and forecasting air traffic; structural characteristics such as regional, relational or travel purpose data have more significance.

Regional or national markets vary widely in size, travel structure, network density and traffic development patterns. Figure 2 shows the development of total tonne kilometers performed by scheduled services of airlines registered in the International Civil Aviation Organization (ICAO) states of each region. Approximately 40% of the total tonne kilometers volume handled by airlines of the contracting states of ICAO (numbering almost 160) is accounted for by the carriers of North America, the most important air traffic region, with 31% being handled by the airlines of Europe and only 29% by the airlines of all the other countries. However, regional air traffic growth has been greatest in Asia and the Pacific, with its traffic share increasing from 13.5% in 1978 to almost 19% in 1987, while at the same time the traffic share of North America and Europe dropped from 76% to 71%.

Because of the generally greater distances between urban areas in North America, domestic air transportation is a more important part of the total national transportation system than it is in European countries. Scheduled domestic traffic within the USA (58%) and the USSR (21%) accounts for almost 80% of all domestic traffic in the world. On international services in 1987, US airlines carried 18% of the total international tonne kilometers traffic, while European airlines had a 37% share. The two international city pairs with the highest scheduled passenger traffic are London and New York and London and Paris, which had traffic flows of about 2.2 million passengers each in 1986.

In 1987, the total commercial air transport fleet of all scheduled and nonscheduled carriers of ICAO contracting states (except China and the USSR) exceeded 10 000 aircraft for the first time. There have been important changes in the composition of airline fleets, with the number of jet aircraft increasing from almost 6000 in 1978 to 7640 in 1987, rising from 68% to 76% of the fleet. While the number of turboprop aircraft also increased, the number of piston engine aircraft dropped to 610, constituting only about 6% of the total fleet. These fleets were operated by approximately 900 commercial air carriers worldwide, of which around 500 performed international services. Of the total number of international carriers, there were almost as many nonscheduled as scheduled airlines.

In order to carry the total demand of passengers and freight, airlines performed more than 14 million flights in 1987, 13 million of which were scheduled services. The total distance covered on these flights was more than 13 billion km, which corresponds to flying around the earth more than 300 000 times.

3. Objectives and Problems of Forecasting

In the aviation industry, the political and scientific communities, and among the general public, there is a growing consensus about the need for objective evaluation of future trends in air transportation. Transport technology research and development and transport policy and planning need estimates of the future demand for transport services and estimates of the consequences of anticipated political measures. As long as established political objectives are followed, especially that of freedom of choice of transport mode, and demand-oriented infrastructures are provided, then it is necessary to have a quantitative analysis of what is meant by demand for different economic scenarios as well as for different transport strategies.

The task of forecasting is not to make such statements about the long-term development of transportation which participants of the planning process (politicians, planners and members of the public who may or may not be affected) have to make judgements

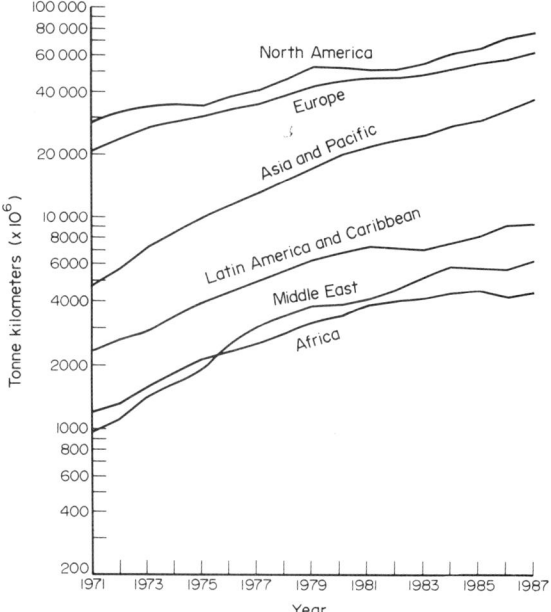

Figure 2
Long-term regional trends: total tonne kilometers performed by scheduled services of airlines registered in ICAO states of each region 1971–1987 (after ICAO (1988))

on, but rather to elaborate on relations (i.e., functions or chains of arguments) between the demand and the influencing factors and to comment on them and the significance of the results. Forecasts are to be if–then forecasts, with results conditioned by the strength or weakness of both the input and the underlying relationships. It is the responsibility of politicians to draw conclusions regarding the realization of projects and, if conditions are favorable, the type of investment involved.

There is no unique solution to the problem of how to bridge the gap between the risk and uncertainty of forecasting and the necessity of making forecasts available for planning purposes. One possibility for reducing the significance of the task is to treat forecasting as a continuous effort and, thus, take account of the newest data and methodology developments.

The adequacy of forecasts depends on the availability of methods and data. The type of forecasting method and the type of transport demand data needed depend on

(a) the timescale of the study (i.e., short term, medium term or long term);
(b) the geographical scale of the study (i.e., studies on a whole network or just on a single corridor);
(c) whether the change being considered is radical or has only a marginal impact on the transport services offered or on the social and economic climate;
(d) whether one (i.e., the air mode) or several modes of transport or category of travellers is involved;
(e) whether the demand is already evident to a greater or lesser extent at the time that the study is made;
(f) the degree of detail needed in the demand data, taking account of the standard of results anticipated or the methods used (e.g., aggregate or disaggregate); and
(g) the user of the forecast, be it a government agency, an airport or an airline.

The output of the forecast also depends on the user.

(a) Government agencies responsible for transport infrastructure planning may require global demand forecasts, as well as airport-related forecasts of both passengers and aircraft movements.
(b) An authority responsible for running and planning an airport needs short-, medium- and long-term forecasts of peak hour and annual numbers of passengers, freight and aircraft movements, and of characteristics derived from these, such as employees, visitors and parking spaces.
(c) An airline requires travel flow forecasts of passengers by type and of freight transport flows, as well as the shares between competing airlines and modes.

Because of the uncertainty about future traffic demand, a range of forecasts ("high" and "low" forecasts) are often generated rather than a single "most likely" forecast. It is important in the planning process to consider the consequences of the range of possible traffic developments. The greater the range the greater the risk associated with large capital investments and the greater the need for plans having maximum flexibility, with investment programs structured over a period of time.

4. Forecasting Methodology

Forecasting air transport demand typically means extrapolating the past development of demand directly or extrapolating relationships between demand and influencing factors that have been observed and calibrated in the past and which are believed to represent future demand in the same way. The differences in forecasting applications lie in the varying complexities of extrapolation and of formulating functional relationships.

There exists, for example, a rather complex relationship between air transport demand and supply, with strong feedback effects of supply on demand. For the purpose of simulating this relationship for forecasting, it must be broken up into several straightforward functions which are dealt with consecutively. In the first phase, it is assumed that there will be no capacity restraint in the transport system, so that the projected demand can be handled. In following phases, this must be reexamined by forecasting system capacities, by comparing traffic volumes with capacities and, in the case of shortcomings, by introducing transport strategies in order to reestimate the demand and repeat the sequences.

The development of air transport demand depends on many factors, the forecast of which is partly risky or requires a complex procedure. Because of this, it is appropriate and, with respect to interpreting the results, even seems necessary to state the input forecasts, that is the development of relevant air transport factors. Since these factors are intercorrelated in different ways, it must be ascertained that their forecast is internally consistent. Worthy of being stated are those factors that influence the transportation variable to a great extent and that are, at the same time, controversial in their estimated values.

For estimating future air transport demand, many hypotheses regarding the development of factors influencing the demand have to be stated. They belong to four categories:

(a) general political and economic conditions,
(b) sociodemographic and socioeconomic conditions,
(c) transport political conditions, and
(d) travel behavior.

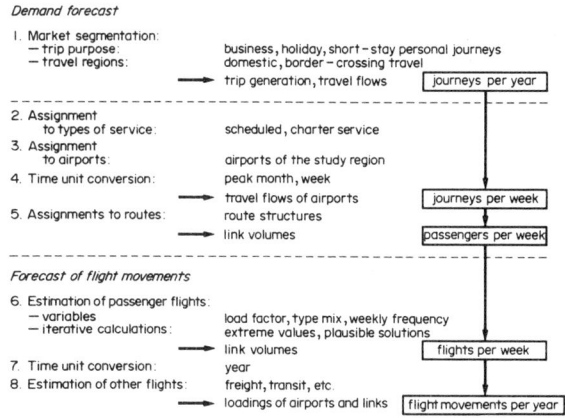

Figure 3
Working steps of an air transport forecast study

Relatively few premises are quantifiable and as such few direct variables enter the demand and supply functions; most others are qualitative in nature and, thus, limit the range of validity of the forecast results.

If the objective of the forecast is to estimate the number of aircraft movements at airports or in the airspace, it is essential first to estimate the future demand for air services and then, on this basis, the aircraft movements. Figure 3 shows in a simplified way the main steps of an air transport forecast for both travel flows and aircraft movements (Wilken *et al.* 1981). Of the eight phases shown, phase 1, the demand analysis and forecast, and phase 6, the derivation of flight movements, are the main phases of the study; however, for calculating airport and route loadings, the other steps are carried out (phases 2–5 and 7–8).

Reflecting the factors that cause people to travel (e.g., holidays, meeting friends and relatives, seeing places, and selling goods and services), the demand should be analyzed and forecast by trip purpose. Since official passenger statistics do not differentiate flows by purpose (but by type of service) special surveys, which are normally carried out by airline companies for market research purposes, have to be relied on. These trip-purpose-specific data must be regarded as an essential base for forecasting, because travel intensities and developments and spatial structures vary significantly with trip purpose.

In addition, market regions have to be identified that vary in modal substitution potential and destination attractiveness. On short-range traffic relations, the air passenger flow is only part of the total demand between two regions, which should be taken into account by forecasting both the total flow and the modal split.

Data are often not available for following a trip-purpose-specific and multimodal approach. Alternative and more global methods have, therefore, been developed, especially by the ICAO and airline associations such as the International Air Transport Association (IATA).

In general, forecasting methods may be subdivided into three broad categories: quantitative methods, qualitative (judgemental) methods and decision analysis techniques. Quantitative approaches can be further classified into time-series analysis techniques and functional methods such as regression and econometric analyses or simulation techniques. Qualitative methods include expert interview and Delphi techniques. Decision analysis primarily incorporates market research approaches.

Econometric theory has progressed a great deal, so that econometric methods have been widely used for forecasting the global air transport demand of a country, of a region or between regions. Most econometric models of air passenger demand include measures of market size or spending power ("income") and price as explanatory variables. A typical demand function is

$$PAX = a \cdot GDP^b \cdot YIELD^c$$

where PAX is the passenger volume, GDP the gross domestic product of a country, YIELD the average revenue per passenger kilometer, and a, b and c are parameters to be calibrated. In an ICAO global study (International Civil Aviation Organization 1985) of world scheduled air traffic, the income elasticity of the GDP variable $b = 1.8$ and the price elasticity of the YIELD variable $c = 0.8$.

Large variations in the values of b and of c have been found in case studies, indicating both weaknesses in model specification and differences in underlying demand situations regarding, for example, trip purpose and modal split characteristics of markets under study.

Trip-purpose-specific data are available for the German air transport market and have been used for forecasting. Nine out of ten origin–destination passengers on German domestic services are business travellers. Factors causing and influencing travel can therefore be found in the production and distribution processes of the German economy. An econometric relationship can be calibrated that determines trip rates in relation to the macroeconomic productivity as expressed by the GDP per employed person (see Fig. 4).

Similar functional relationships have been calibrated for the air travel demand of border-crossing business. For private travel, demand functions are more complex since nonquantifiable factors such as attitude or household constraints influence travel intensity. One approach is to extrapolate travel intensity and propensity rates for homogeneous travel behavior groups of the population.

If aircraft movements have to be forecast, trip-purpose-specific demand has to be assigned to scheduled and charter flight services, to origin–destination relations and to routes served in order to take account of corresponding passengers. The results of these assignment phases are passenger volumes on network

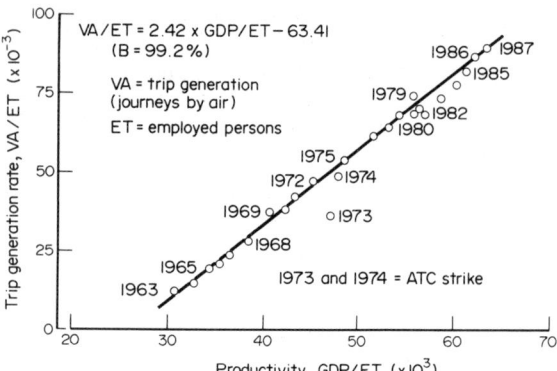

Figure 4
Relationship between domestic air transport demand in the FRG and macroeconomic productivity

links, which serve as a base for estimating flight movements on these links. Three key variables have to be specified for each link and service type:

(a) load factor (passengers per flight),

(b) frequency of services (number of flights per week or year), and

(c) distribution of aircraft types by seat capacity class offered on any one link type.

The frequency is both an input and, after having checked with limiting lower and upper values of the variables involved, an output variable. The variables are interrelated by the following sequence of functions:

$$SC_c = PAX_c/LF_p$$
$$\downarrow$$
$$SC_c = SCF_p \cdot WF_p \rightarrow WF_c = SC_c/SCF_c \quad (1)$$
$$\downarrow$$
$$SCF_c = SC_c/WF_p$$

with

$$LF_{min} < LF_p(LF_c) < LF_{max} \quad (2)$$
$$WF_{min} < WF_p(WF_c) < WF_{max} \quad (3)$$
$$SCF_{min} < SCF_p(SCF_c) < SCF_{max} \quad (4)$$

where PAX is the weekly passenger volume per network link as a result of the demand forecast (input), SC is the seat capacity to be offered on the link, LF is the average load factor per flight expressed as a percentage, SCF is the average seat capacity per flight (aircraft type), WF is the weekly frequency (which corresponds to the number of flights to be offered in future on the link), and c and p indicate calculated and postulated values, respectively.

For the derivation of plausible values of WF_c, it is necessary to go through a sequence of calculations of estimating SCF and WF values by employing previously derived limiting values and Eqns. (1–4). It should be noted that the LF and SCF values that enter the equations are average values, the distributions of which have to be estimated after having derived plausible WF results.

A final comparison of airport and airspace loadings with corresponding capacities will either verify the working hypothesis that the forecast demand will be met by sufficient capacity or, in the case of flight movements exceeding future system capacities, cause an iterative procedure of new estimates of flight movements and/or passenger travel flows.

5. Future Development Trends and Problems

Both the past development of air traffic (see Figs. 1, 2) and forecast model applications indicate that the demand for air traffic services will continue to grow. Even in the weak growth conditions of the general economy increasing demand has to be expected; according to income elasticities found in various studies, air travel will grow faster than the GDP. Another reason for assuming a continuing rise in air traffic is the low trip intensity rate in many countries: only a small, though growing, fraction of the total population travels by air. Since international specialization in industrial production and the exchange of goods and services between countries and continents will continue to increase, as will the marketing of resort areas worldwide, the demand for air traffic services in both business and recreational travel is bound to grow.

Caused by political pressures to liberalize international air transport regulations and by new and more economical aircraft of smaller size, airlines will open up new services and increase service frequencies so that the volume of flight movements may increase almost as quickly as passenger demand. This was not the case in the 1970s.

Although the future development of demand factors, such as national income, along with favorable political factors will cause further growth in air travel, there are in many countries, especially the USA and Europe, factors that have restraining effects. Lack of capacity reserve in airport infrastructure and air traffic control cause growing problems for accepting and handling additional air traffic. One of the most apparent indications that system capacity is inadequate is the growing number of flights delayed and the growing duration of these delays. If both these indicators grow faster than the air traffic itself, as has been the case, it becomes evident that traffic volumes will approach capacity levels.

At the same time, there is growing resistance to the realization of new airport infrastructures because the people affected are afraid of further environmental pollution and destruction of nature. Because of this fundamental conflict, governments will come increasingly under pressure to choose between irreconcilable objectives. Since the prospects for new runways and airports are rather small in the USA and in Europe, new ideas and policy measures regarding maximizing the utilization of existing infrastructures are needed. Methods of air traffic demand and capacity forecasting will have to adapt to these new political transport problems.

See also: Air Traffic Control: Trends; Flight Activity in General Aviation

Bibliography

Horn K W 1982 The Frequency of Air Travel. In: James G W (ed.) *Airline Economics*. Lexington Books, Lexington, MA, pp. 23–34
Horonjeff R 1980 *Planning and Design of Airports*. McGraw-Hill, New York
International Civil Aviation Organization 1985 Manual on Air Traffic Forecasting. Document No. 8991-AT/772. ICAO, Montreal
International Civil Aviation Organization 1988 Civil Aviation Statistics of the World, ICAO Statistical Yearbook. Document No. 9180/13. ICAO, Montreal
Ippolito R A 1981 Estimating airline demand with quality of service variables. *J. Transp. Econ. Policy* **15**(1), 7–15
Manheim M L 1979 *Fundamentals of Transportation Systems Analysis*, Vol 1. The MIT Press, Cambridge, MA
Taneja N K 1978 *Airline Traffic Forecasting*. Lexington Books, Lexington, MA
Thomson J M, Aurignac A, Wilken D 1976 *The Future of European Passenger Transport*. Organisation for Economic Co-operation and Development, Paris
Wilken D, Bachmann K, Urbatzka E, Focke H 1981 Der gewerbliche Luftverkehr der Bundesrepublik Deutschland 1975–1990–2000. DFVLR Report No. FB 81–36. DFVLR, Cologne, Germany

D. Wilken
[DLR, Cologne, Germany]

Flow Variables

The considerable increase in the amount of road traffic has led to the need for the development of theories, able to describe, explain and predict interactions between vehicles and flows. Deterministic or probabilistic in nature, these theories involve the use of a certain number of traffic variables and of characteristic relations and distributions. Some parameters relate to individual vehicles. Others refer to platoons of vehicles by means of distributions in time and space.

1. Microscopic Variables and Their Applications

1.1 Headway

The headway is the period of time between the instants at which the front ends of successive vehicles pass a given point on the road. Measurements allow the determination of empirical distributions of these headways under various traffic conditions. These distributions can be adjusted to theoretical distributions. Thus, in the case of a fluid flow of traffic, headway h follows an exponential distribution with a probability density function

$$f(t) = \text{prob}(h < t) = \frac{1}{h_m} \exp\left(-\frac{t}{h_m}\right) \qquad (1)$$

where h_m is the mean headway.

The log-normal probability density function, illustrated in Fig. 1 and given by

$$f(t) = \frac{1}{\sigma t (2\pi)^{1/2}} \exp\left[-\frac{1}{2\sigma^2}(\ln t - h_m)^2\right] \qquad (2)$$

where σ is the standard deviation of headways, gives a closer approximation for denser traffic.

A number of authors refer to a mixture of distributions for a given volume of traffic. Such an approach is justified on the assumption that a flow is made up of both "free" and "constrained" vehicles, each of these categories being represented by one of the components f and g of the mixture. In practice, the problem consists of "solving" the mixture from the knowledge of the empirical distribution of h. This means determining the components f and g of the free and the constrained distributions, such as

$$h(t) = bg(t) + (1 - b)f(t) \qquad (3)$$

where the constant b corresponds to the proportion of constrained vehicles in the traffic. Several statistical techniques can be used such as the method of moments, minimum χ^2 and dynamic clusters.

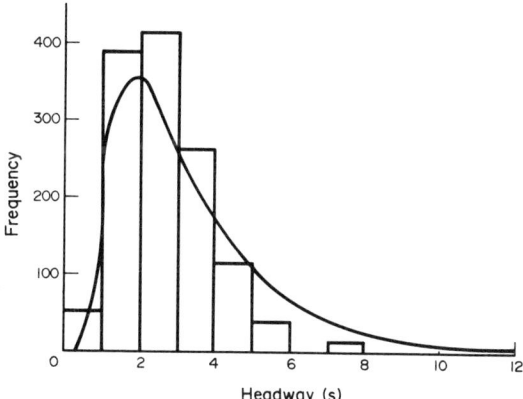

Figure 1
Empirical distribution of headway: shifted log-normal probability density function

1.2 Spacing

Spacing is the distance separating the front ends of successive vehicles. As in the case of headway distributions, theoretical distributions can be employed in representing empirical spacing distributions.

1.3 Applications

Reference to both headways and spacings can be particularly useful in conducting studies on road safety or in dealing with simulation problems. These variables provide a useful indication of the amount of traffic present on road infrastructures. Headway also allows the effect of traffic composition to be quantified by means of the passenger car unit coefficient, denoted by e and expressed in pcu. This coefficient corresponds to the number of passenger cars that each category of vehicles represents for given traffic conditions. It is estimated by the ratio

$$e \text{ (category/passenger car)} = \frac{\text{headway(category)}}{\text{headway(passenger cars)}} \quad (4)$$

1.4 Speed

Measurements of the instantaneous speed of a vehicle, recorded over the whole length of a trip, lead to a speed–time history. Such graphs yield useful parameters. They provide an indication of the quality of the traffic along a particular route by, for example, giving the proportion of stop time. They are also frequently used when studying fuel consumption and emissions of vehicles.

Conversely, by recording the instantaneous speed of all vehicles at a particular point along the road, the empirical distribution of these speeds can be plotted. Such a distribution is used in traffic simulation models, in particular, where it is often adjusted to fit an Erlang distribution for a given category of vehicles. The corresponding probability density function is given by

$$f(u) = \frac{c^k}{(k-1)!} u^{k-1} \exp(-cu) \quad (5)$$

where u is the instantaneous speed, c and k are constants, and k is a positive integer.

1.5 Acceleration and Acceleration Noise

Acceleration values are derived from speed profiles by differentiation and smoothing. In urban traffic, a characteristic feature of the distribution is the pronounced peak in the vicinity of the origin. This peak corresponds to the time spent by vehicles at nearly constant or zero speed. Although the shape of the distribution curves depends very much on the driving mode, these curves nevertheless tend to be symmetrical as a result of a similarity between acceleration and deceleration phases. Stop time being excluded, acceleration is fitted by a normal centered distribution. For a trip of duration T, the mean acceleration is given by

$$a_m = \frac{1}{T} \int_0^T a(t) \, dt = \frac{u(T) - u(0)}{T} \quad (6)$$

where $u(t)$ and $a(t)$ refer to the speed and acceleration of the vehicle at time t.

Acceleration also provides a measure of traffic quality: the acceleration noise σ_a. This variable depends on driver behavior, road characteristics and traffic conditions. It is defined by

$$\sigma_a^2 = \frac{1}{T} \int_0^T [a(t) - a_m]^2 \, dt \quad (7)$$

Acceleration noise increases with the level of traffic congestion.

1.6 Safe Spacing and Instantaneous Speed

Studies on road safety have tried to investigate relationships between instantaneous speed u and spacing s. The total time necessary for the driver to avoid a rear-end collision is assumed to be the sum of the reaction time t_r, before the beginning of the braking phase, and the braking time. The distance travelled during the reaction time is proportional to u, while during braking it is proportional to u^2. Thus, the minimum safe spacing has the following form:

$$s_{\min} = L + t_r u + \frac{u^2}{2a_{\max}} \quad (8)$$

where L is the vehicle length and $a_{\max}$ is the maximum deceleration. Experimental results confirm the validity of this relation.

2. Macroscopic Variables

Macroscopic variables define characteristics of platoons of vehicles on a given road section.

2.1 Flow

Flow (or volume) refers to the distribution of the vehicles in time. The average flow $q(t_1, t_2, x)$ between times t_1 and t_2 at a point x along the road is defined as

$$q(t_1, t_2, x) = \frac{n(t_1, t_2, x)}{t_2 - t_1} \quad (9)$$

where $n(t_1, t_2, x)$ refers to the number of vehicles passing the point x between the times t_1 and t_2. Average flow is determined on the basis of roadside counting.

According to some traffic theories, traffic flow is regarded as a continuous entity and the flow at time t and a point x on the road is defined by

$$q(x, t) = \lim_{\Delta t \to 0} q\left(t - \frac{\Delta t}{2}, t + \frac{\Delta t}{2}, x\right) \quad (10)$$

Such a definition does not apply to a discrete theory of traffic, for this limit would equal infinity or zero, according to whether a vehicle did or did not pass by at time t. Thus, this equation is a convenient simplification, and $q(x, t)$ and $q(t - \Delta t/2, t + \Delta t/2, x)$ are considered equivalent for small intervals Δt of about 20 s.

The link between macroscopic and microscopic approaches should be noted. For stationary or steady flows (i.e., without large variations of individual speeds) average flow is the inverse of the average headway.

2.2 Concentration

Concentration (or density) refers to the distribution of vehicles in space. Average concentration at time t on a section of road extending from point x_1 to x_2 is given by the ratio

$$\rho(x_1, x_2, t) = \frac{n(x_1, x_2, t)}{x_2 - x_1} \qquad (11)$$

where $n(x_1, x_2, t)$ refers to the number of vehicles present on the road section at time t. Concentration values can be directly determined, for example, by means of aerial photographs.

In continuous theories, concentration at time t at a point x on the road is defined by

$$\rho(x, t) = \lim_{\Delta x \to 0} \rho\left(x - \frac{\Delta x}{2}, x + \frac{\Delta x}{2}, t\right) \qquad (12)$$

Without repeating the argument given in Sect. 2.1 as to whether flow is regarded as a continuous or a discontinuous function, it should be noted that in the case of concentration, $\rho(x, t)$ and $\rho(x_1, x_2, t)$ are effectively equivalent for a Δx of 50–100 m. Thus, on a short specified length of road during a short specified period of time, a certain fractional quantity of vehicles (e.g., 0.001 vehicles) can pass by. In practice, in carrying out observations or in coming to any conclusions, considerations are limited to lengths of road and intervals of time that ensure these quantities are meaningful.

For steady flows of traffic, average concentration is the inverse of average spacing.

2.3 Occupancy

The occupancy variable is very commonly employed in road traffic operations and control. It is usually measured by inductive loops embedded in the carriageway which are sensitive to variations in magnetic field due to the passage of the metallic mass of vehicles. It is a dimensionless quantity, defined as the proportion of time during which vehicles are present above the loop. Occupancy τ, is directly related to ρ by the relation

$$\tau = (L + l)\rho \qquad (13)$$

where L is the average length of the vehicles and l is the length of the inductive loop. This is a very useful relation, since the direct measurement of concentration, unlike that of occupancy, is a complex and costly operation.

2.4 Mean Speeds

Time mean speed v_t at a given point on the road is defined as the arithmetic mean of the instantaneous speeds of the vehicles passing the given point during a given period of time.

Space mean speed v_s is a more useful variable in practice. It is defined as the arithmetic mean of the instantaneous speeds of the vehicles on a given length of road at a given instant.

These two concepts are different. For a steady flow of traffic (i.e., one in which flow, concentration and speed do not vary very much around their respective mean values q, ρ and v), Wardrop (1952) established the following relation:

$$v_t = v_s + \frac{\sigma^2}{v_s} \qquad (14)$$

where σ is the standard deviation of the speed distribution in space. In addition, mean headway h_m, mean spacing s_m and v_s are related by

$$h_m = \frac{s_m}{v_s} \qquad (15)$$

2.5 Basic Relation

For an isovelocic flow of traffic (i.e., flow in which all the vehicles run at the same speed) the mean or common speed v is referred to for a collection of vehicles. In this simple case,

$$q(x, t) = v\rho(x, t) \qquad (16)$$

Thus, average speed for any flow can be defined by

$$v(x, t) = \frac{q(x, t)}{\rho(x, t)} \qquad (17)$$

Given this relation, it can be shown that $v(x, t)$ corresponds to v_s.

It should also be noted that for a steady flow of traffic, $v(x, t)$ is the harmonic mean of the speeds v_i of the vehicles passing a point x on the road during a period of time:

$$\frac{1}{v(x, t)} = \frac{1}{n} \sum_{i=1}^{n} \frac{1}{v_i} \qquad (18)$$

Flow Variables

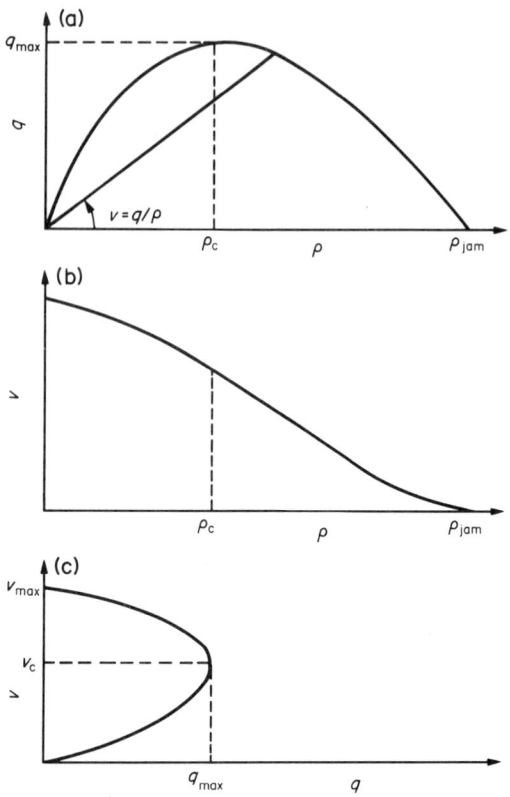

Figure 2
Fundamental diagram representations: (a) $q(\rho)$, (b) $v(\rho)$ and (c) $v(q)$ curves

Table 1
Main speed–density models

Model	Equation
Greenshields	$v = v_f(1 - \rho/\rho_{jam})$
Drew	$v = v_f[1 - (\rho/\rho_{jam})^{1/2}]$
Greenberg	$v = v_m \ln(\rho_{jam}/\rho)$
Underwood	$v = v_f \exp(-\rho/\rho_m)$
May	$v = v_f \exp[-\frac{1}{2}(-\rho/\rho_m)^2]$

v mean speed, v_m optimal speed, v_f free speed, ρ density, ρ_m optimal density, ρ_{jam} maximal density

2.6 Fundamental Diagram

A homogeneous and steady flow of traffic will be considered. At low concentrations, the average speed v of the flow is usually high. Vehicles can reach their desired speed. As concentration ρ increases, the degree of interaction between vehicles becomes greater and v decreases. This suggests that v is a monotonic decreasing function of ρ. This functional relation, namely

$$v = v(\rho) \quad (19)$$

corresponds to the hypothesis of the fundamental diagram.

Given the connection between the three variables q, ρ and v, such a diagram completely defines the traffic characteristics of a given section of road. The shape of this diagram is usually as shown in Fig. 2. The first part of the curve corresponds to free-flow traffic conditions. Flow increases with concentration up to a maximum value corresponding to a critical concentration ρ_c. This maximum flow corresponds to the capacity of the road section; that is, the maximum number of vehicles that can be driven through the section in a given period of time. If the concentration increases beyond ρ_c, the flow starts to decrease and vehicles interfere with one another more and more. Traffic is said to be saturated or congested. Flow tends to be unstable. There are two distinct concentration values for the same flow, according to whether traffic is free-flowing or congested. Thus, for example, zero flow can indicate that the road section is empty or, on the contrary, that concentration has reached its maximum value ρ_{jam}, as a result of a downstream blockage.

2.7 Calibration of the Fundamental Diagram

In practice, fundamental diagrams are derived from roadside measurements. Values of flow and occupancy are always fairly widely dispersed. The flow–occupancy (or flow–concentration) curve is fitted from these clusters by means of statistical regression techniques. Polynomial, logarithmic or exponential functions are frequently used, as shown in Table 1. Nonlinear functions often give a better calibration. Figure 3 illustrates an example of a generalized exponential model, fitted on the A13 motorway to the west of Paris.

The shape of the fundamental diagram depends on a number of factors, such as geometry of the road, nature of the traffic, traffic control practices and weather conditions.

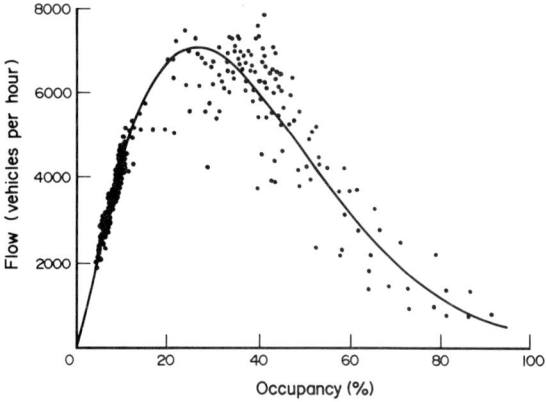

Figure 3
Generalized exponential model fitted to freeway measurements

Fundamental diagrams are well suited to the study of various traffic problems and they are particularly useful in connection with the design and subsequent operation of road developments.

See also: Detectors for Road Traffic; Road Traffic: An Introduction; Simulation Programs, Macroscopic; Simulation Programs, Microscopic

Bibliography

Drew D R 1968 *Traffic Flow Theory and Control*. McGraw-Hill, New York
Gazis D C (ed.) 1974 *Traffic Science*. Wiley, New York
Gerlough D L, Huber M J 1975 *Traffic Flow Theory. A Monograph*. US Transportation Research Board, National Research Council, Washington, DC
Haberman R 1977 *Mathematical Models: Mechanical Vibrations, Population Dynamics and Traffic Flow*. Prentice-Hall, Englewood Cliffs, NJ
Taylor M A P, Young W 1988 *Traffic Analysis. New Technology and New Solutions*. Hargreen, Australia
US Transportation Research Board 1985 *Highway Capacity Manual*, Special Report 209. US Transportation Research Board, Washington, DC
Wardrop J G 1952 Some theoretical aspects of road traffic research. *Proc. Inst. Civ. Eng.* **1**, 325–62
Whol M, Martin B V 1967 *Traffic Systems Analysis for Engineers and Planners*. McGraw-Hill, New York

S. Cohen
[INRETS, Arcueil, France]

Flow Variables: Estimation

Impressive innovations in the area of road traffic have been induced by developments in the field of informatics. Many advanced control concepts such as traffic-responsive area-wide control, dynamic route guidance, accurate trip planning, intelligent management of traffic facilities and numerous others that have so far appeared as sophisticated but somewhat artificial ideas are on the brink of realization. This is made evident by the many national and international programs in Europe, Japan, the USA and elsewhere. All of these advanced control or information systems have a common necessity; they need detailed and complete information about the current state of the traffic facility under consideration.

This kind of information can, however, only be directly obtained from measurements taken by expensive equipment such as television cameras or, more conventionally, by a dense sequence of loop detectors. Even in cases where investment in costly equipment seems reasonable, there is still considerable data processing effort required to transform the measurement data into traffic variables and parameters. This kind of expensive monitoring is restricted to a few critical sites such as tunnels, bridges or bottlenecks in freeways. For all other parts of a road network, the traffic surveillance system has to rely on incomplete samples of data which are provided by two types of sources: conventional loop detectors at a separation of several kilometers and, probably in the near future, vehicles serving as mobile sensors which from time to time transmit trip data (e.g., travel time) to the control center via beacons or other communication channels. Obviously, measurement data such as these from isolated detectors and vehicles do not by themselves provide a complete picture of the current traffic situation over the whole distance of a road section. This leads to questions of if and how this incomplete information from sample measurements can be completed by an appropriate data processing routine using the available record of the measurement time sequences and some knowledge of the rules and mechanisms of traffic flow phenomena to produce reliable estimates for the unknown state on the road section under consideration.

It is intuitively clear that a mathematical model that comprises knowledge about dynamic traffic phenomena, such as density wave propagation, shock waves and formation of congestion, may play a major part in the estimation of traffic variables along a road section. This is well known in control engineering where the problem is often to estimate the internal state of a system from few measurements. Therefore, a number of model-based estimation methods have been developed in the field of control theory (see Ljung and Söderström 1983 for a survey) which may be adapted to the problem of estimating the traffic state on a road section.

Since freeways are high-capacity roads carrying large traffic volumes, there is a particular interest in efficient surveillance schemes for these roads (see *Freeway*).

1. Model-Based Estimation

There are two important aspects that must be taken into account when choosing a model as a central part of an estimator. First, the model must be formulated such that it is able to reproduce the variables that are to be estimated. Second, however, if the model is too detailed, it will induce ambiguity when it is used to interpret the measurements to conclude the actual traffic state by causal reasoning. To be more specific, in a realistic scenario where measurements of traffic volumes and average speeds are taken from loop detectors at a separation of several kilometers, information about the maneuvers of individual vehicles within this road section cannot be expected, no matter how sophisticated the model used. The question of what kind of information can be extracted from given measurements can be answered from control theory by the concept of observability. A model that is inadequately detailed and complex will turn out to be not, or only weakly, observable from given measurements.

For the given problem of monitoring the state of a freeway section for information and control systems,

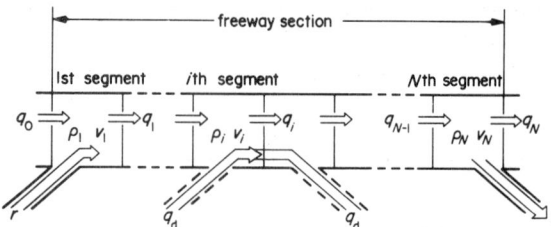

Figure 1
Discretized section of a freeway with macroscopic variables

aggregate information about the density profile along the section, the length of congestions and other macroscopic parameters is of interest, rather than detailed information about microscopic phenomena. Consequently, an appropriate model for the problem of state estimation will be a macroscopic dynamic model that describes traffic flow by aggregate variables such as section density, average speed and traffic volumes. A model that has proved to be most suitable for this purpose was first proposed by Payne (1971) and has been improved by several authors. As this model is described in detail in the article *Freeway Traffic Modelling*, it will be introduced only briefly.

Consider a section of freeway that, for convenience, is subdivided into several road segments with a length of about 500 m (see Fig. 1). With respect to this space-discrete configuration, $\rho_i(k)$ is the traffic density in segment i at time kT (vehicles per kilometer), $v_i(k)$ is the mean speed of the vehicles within segment i at time kT (km h^{-1}), $q_i(k)$ is the volume from segment i into segment $i+1$ during the time interval t given by $kT \leq t < (k+1)T$ (vehicles per hour), $r(k)$ and $s(k)$ are possible entering or leaving ramp volumes within the time interval t (vehicles per hour), $q_d(k)$ is the hypothetical flow (see Sect. 3) within the time interval t (vehicles per hour), and $w_0(k)$ and $w_N(k)$ are measurements of local average velocities at upstream and downstream ends of the section within the time interval t, respectively (km h^{-1}).

Using these variables, a simple balance for the vehicles within segment i at time $(k+1)T$ gives the following difference equation:

$$\rho_i(k+1) = \rho_i(k) + \frac{T}{\Delta_i}[q_{i-1}(k) - q_i(k) + r(k) - s(k)] \quad (1)$$

where Δ_i is the length of segment i. The boundaries of the section are chosen without loss of generality at possible on- and off-ramps. Then, $r(k)$ may have to be added for $i=1$, while $s(k)$ may have to be subtracted in the last segment where $i=N$.

The following difference equation for section average velocity which was formulated according to empirical observations has proved to be quite realistic:

$$v_i(k+1) = v_i(k) + \frac{T}{\tau}[V(\rho_i) - v_i](k)$$

$$+ \frac{T}{\Delta_i} v_i(k)[v_{i-1} - v_i](k)$$

$$+ v \frac{T}{\Delta_i \tau}\left[\frac{\rho_i - \rho_{i+1}}{\rho_i + \kappa}\right](k) \quad (2)$$

where τ is a time constant, κ is a density constant and v is a sensitivity factor. In Eqn. (2), the first bracketed term on the right-hand side reflects dynamic adaptation to the steady-state speed–density characteristic $V(\rho)$, the second term accounts for the convection of a speed gradient and the third term represents the driver's anticipation of a foreseen density gradient in the downstream direction.

Traffic volume will only be measured once in several kilometers (i.e., at the boundaries of the section) but can be expressed within the section according to the rules of hydromechanics as a product of density and velocity. Because of the allocation at the end of a segment, a weighted average seems to be appropriate:

$$q_i(k) = \alpha \rho_i(k) v_i(k)$$
$$+ (1-\alpha)\rho_{i+1}(k)v_{i+1}(k) \quad 0.5 \leq \alpha \leq 1.0;$$
$$i = 1, \ldots, N-1 \quad (3)$$

A dominating function within these dynamic model equations is $V(\rho)$ which in a general way may be given by

$$V(\rho) = v_f\left[\rho - \left(\frac{\rho}{\rho_{jam}}\right)^m\right]^l_{\rho_{jam}} \quad (4)$$

where v_f is the free velocity, ρ_{jam} denotes jam density and m and l are positive real numbers.

Equations (1–4) for $i=1, \ldots, N$ establish a nonlinear discrete-time model of order $2N$ for the dynamic phenomena of traffic flow within the section. The equations contain a number of parameters (τ, κ, v, weighting factor α and the four parameters of Eqn. (4)) which must be calibrated to bring the model close to reality (e.g., Cremer and Papageorgiou 1981).

Having defined a model for the process of traffic flow, the problem of estimation can be formulated. Suppose that the dynamic phenomena of traffic flow are adequately described by Eqns. (1–4) of the model for the whole range of possible density values (i.e., from zero density to full congestion at ρ_{jam}). It is assumed that time sequences of possible noise-corrupted measurements $\{q_0(k), q_N(k), w_0(k), w_N(k)\}$ for volumes and local average speeds are collected at both ends of the section. An estimation algorithm is then sought that, on the basis of the measured information and the model equations, produces reliable estimates $\{\hat{\rho}_i(k), \hat{v}_i(k)\}_{i=1,...,N}$ for the current density and mean speed values within the section.

The question of how the model can be used for the task of estimating the state within a freeway section must now be returned to. A quite simple and natural way would be to use the model as a simulation tool running in parallel with the real process. In this configuration, as shown in Fig. 2, the measurement data are inserted into the model equations appropriately as boundary values. This procedure has, however, a serious disadvantage: since the model uses a balancing of vehicles in Eqn. 1, the real densities would never be obtained unless the computation was started with the real density values which are normally unknown. Moreover, if the detector data are inaccurate and erroneous as they often are in practice, the model variables may diverge from the real values with increasing time, producing totally incorrect estimates. Therefore, a more efficient scheme has been developed in control engineering, as shown in Fig. 3. In this case, measurements of external variables are simply inserted into the model equations as inputs. Those measurements that are related to the variables within the section (here q_N and w_N) are compared with corresponding estimates ($\hat{q}_N$ and $\hat{w}_N$) giving an estimation error which is then used by a fairly sophisticated correction scheme to improve the estimated variables. If properly designed, this correction loop guarantees convergence of the estimates in spite of wrong initial values, measurement errors and model inaccuracies. This highly general configuration is applied by several well-known estimators, for example, by the Luenberger observer, various types of the recursive least-square estimator and the Kalman filter.

For the traffic flow model given by Eqns. (1–4), it has to be decided which measurements should be treated as simple inputs and which should be regarded as model reactions. This is not a trivial task since, depending on

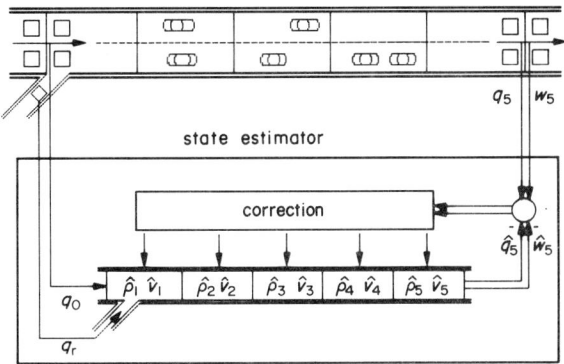

Figure 3
Configuration of an estimator with a correction feedback

traffic conditions, these variables may be more dependant on the situation inside the section or more on the conditions outside the section. Investigations have shown that for estimator design the best results are obtained by assigning the entering volume q_0 to be treated as an external input while all other measurements (w_0, q_N and w_N) should be treated as system reactions which are related to the state variables by the following equations:

$$w_0(k) = (1 + \alpha\varepsilon)v_1(k) - \alpha\varepsilon v_2(k) \qquad (5)$$

$$q_N(k) = [1 + (1-\alpha)\varepsilon]\rho_N(k)v_N(k)$$
$$\qquad - (1-\alpha)\varepsilon\rho_{N-1}(k)v_{N-1}(k) \qquad (6)$$

$$w_N(k) = [1 + (1-\alpha)\varepsilon]v_N(k) - (1-\alpha)\varepsilon v_{N-1}(k) \qquad (7)$$

These equations complete the description of the model.

2. Estimator Design and Experiments

The estimation scheme outlined in Sect. 1 was applied successfully by several researchers. First applications were reported in surveillance systems for short distances of critical road sections by Knapp (1972), Szeto and Gazis (1972), and Nahi and Trivedi (1973). In these and later investigations (e.g., Smulders 1987, Bhuori et al. 1989), only short sections with a length below 2 km were considered and the proposed estimation procedures were based on the Kalman filter approach. The applied models were less detailed than the one described in Sect. 1 and did not include the mechanisms of dynamic speed adaptation as given in Eqn. (2). Another approach that used the general feedback correction scheme of Fig. 3 was proposed by Maarseven (1983) by applying the martingale theory.

If the more elaborate model given in Sect. 1 is used, estimation of traffic variables can be performed successfully even for longer sections (i.e., when the distance between local detectors is increased). Cremer *et*

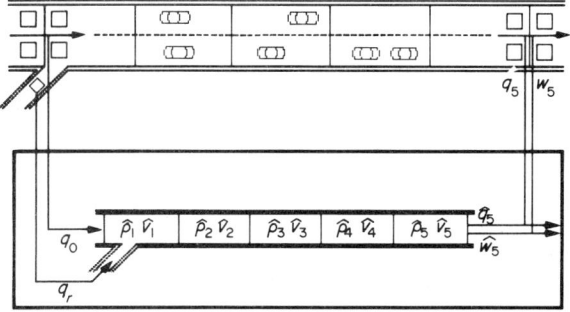

Figure 2
Open loop configuration of parallel model simulation

al. (1980) reported successful applications using measurement data from a freeway section where the upstream and downstream detector loops were 3 km apart. A typical result is shown in Fig. 4 where the time diagrams of actual and estimated mean speed in the middle of a 2.8 km section are shown. It can be seen from this figure that even in the critical case of a short congestion, the collapse and recovery of traffic flow is accurately estimated without significant delay.

Since the Kalman filter is a well-established tool in control engineering for the estimation of state variables, it is not surprising that it is used in most reported applications for estimating freeway traffic variables. However, there are some particular problems when applying it to this problem. The Kalman filter relies on the assumptions that the system under consideration is linear and that the noise terms which enter the state equations and measurement equations additively are random white and Gaussian noise processes. Under these assumptions and provided that the system state is completely observable from the measurements, it generates estimates that have minimum error variances.

In the case of freeway traffic flow, however, the system dynamics are nonlinear, as made evident by the model, and the noise terms that may be added to the state and measurement equations of the model on the right-hand side may not in reality be white Gaussian noise processes. Consequently, the conventional Kalman filter calculus cannot be applied directly. Instead, the extended Kalman filter must be used which accounts for nonlinearities by two modifications: first, it uses the nonlinear model within the filter equations and, second, for the computation of the correction loop, the model is linearized about the estimated state. However, for the extended Kalman filter optimality and, in some cases, even convergence of the estimates cannot be guaranteed.

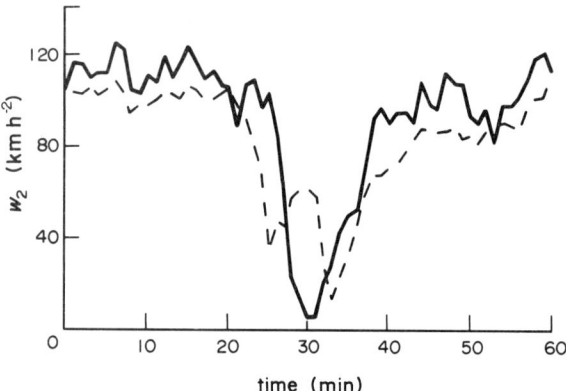

Figure 4
Real and estimated speed diagram inside of a 2.8 km freeway section: —— measurements, and --- estimates

Since the extended Kalman filter plays an important role in freeway surveillance systems, its equations will be briefly presented. First, for notational convenience, the $2N \times 1$ vector x of state variables ρ_i and v_i is introduced, where $i = 1, \ldots, N$. The right-hand sides of Eqns. (1) and (2) then comprise the function vector $f(x, u)$ where u denotes the external inputs, which are q_0, r and possibly s. Thus, Eqns. (1) with (3) and (2) with (4) may formally be represented by the vector difference equation

$$\mathbf{x}(k+1) = \mathbf{f}[\mathbf{x}(k), \mathbf{u}(k)] + \gamma(k) \quad (8)$$

Treating the measurement equations (Eqns. (5) (6) and (7)) similarly gives the abbreviated vector equation

$$\mathbf{y}(k) = \mathbf{g}[\mathbf{x}(k)] + \zeta(k) \quad (9)$$

with $\gamma(k)$ and $\zeta(k)$ being Gaussian white noise terms which account for model disturbances, model inaccuracies and measurement errors. The mean values and covariances of the noise terms are assumed to be given by

$$\begin{aligned} E\{\gamma(k)\} &= E\{\zeta(k)\} = 0 \\ E\{\gamma(j)^T \gamma(k)\} &= \mathbf{Q}(k)\delta_{j,k} \\ E\{\zeta(j)^T \zeta(k)\} &= \mathbf{R}(k)\gamma_{j,k} \\ E\{\gamma(j)^T \zeta(k)\} &= \mathbf{M}(k)\delta_{j,k} \end{aligned} \quad (10)$$

where $\delta_{j,k}$ denotes the Kronecker symbol. For the unknown initial state x_0, it is assumed that the estimation error has the covariance matrix

$$\mathbf{P}(0) = E\{(\mathbf{x}_0 - \hat{\mathbf{x}}_0)^T (\mathbf{x}_0 - \hat{\mathbf{x}}_0)\} \quad (11)$$

Furthermore, from the components of the vector functions $f()$ and $g()$ which are supposed to be continuously differentiable, the following Jacobian matrices are computed as representatives of a linearized system description:

$$\left. \begin{aligned} \mathbf{F}_k &= \frac{\partial \mathbf{f}}{\partial \mathbf{x}} \bigg|_{\substack{\mathbf{x} = \hat{\mathbf{x}}(k) \\ \mathbf{u} = \mathbf{u}(k)}} \\ \mathbf{H}_k &= \frac{\partial \mathbf{g}}{\partial \mathbf{x}} \bigg|_{\mathbf{x} = \hat{\mathbf{x}}(k)} \end{aligned} \right\} \quad (12)$$

The extended Kalman filter then determines suboptimal estimates by Eqns. (13–15). The filtermodel equation is

$$\hat{\mathbf{x}}(k+1) = \mathbf{f}[\hat{\mathbf{x}}(k), \mathbf{u}(k)] + \mathbf{D}(k)\{\mathbf{y}(k) - \mathbf{g}[\hat{\mathbf{x}}(k)]\} \quad (13)$$

(The second term on the right-hand side represents correction of the estimates according to the measurement output error $y(k) - \hat{y}(k)$.) The correction gain matrix $\mathbf{D}$ is computed for each time step

$$\mathbf{D}(k) = [\mathbf{F}_k \mathbf{P}(k) \mathbf{H}_k^T + \mathbf{M}(k)] [\mathbf{H}_k \mathbf{P}(k) \mathbf{H}_k^T + \mathbf{R}(k)]^{-1} \quad (14)$$

where the following 'covariance matrix' of the conditional estimation error is evaluated recursively:

$$P(k+1) = [F_k - D(k)H_k]P(k)F_k^T \\ + Q(k) - D(k)M(k)^T \qquad (15)$$

As can be seen from Eqns. (14) and (15), the correction gain matrix **D** is computed using the linearization matrices **F** and **H** evaluated for the values of the estimated state variables.

Since the estimated state may sometimes differ considerably from the real state, the correction mechanism may become less effective. Thus, optimality of the estimates is not guaranteed when applying the extended Kalman filter. In fact, it was shown by Cremer et al. (1980) for the extreme case where the whole freeway section is congested while the initial state of the filter model is assumed to be free flowing that the estimates of the extended Kalman filter did not even converge to the real variables. Instead, a less complicated correction scheme with a switching mechanism with two constant correction matrices D_1 and D_2 was shown to work better and to make the estimates converge to the real values.

3. Extensions

The model equations contain a number of parameters that have tacitly been assumed to be fixed. In real traffic, however, some of these parameters may vary with daytime, weather and other external conditions. For example, the parameters free velocity v_f and maximum density ρ_{max} define the shape of the speed–density curve and, thus, the capacity q_{max} of the freeway, which may be different for sunny or rainy weather, and morning or evening hours. Clearly, the road capacity is an important parameter for control purposes and it may be considered whether this estimation scheme could also be used to estimate these parameters. This problem has been investigated by several authors and it has been shown that slow variations of these parameters can be tracked by a suitably extended estimation algorithm (Szeto and Gazis 1972, Grewal and Payne 1976, Cremer and Schütt 1990).

The technique of this extension is quite straightforward. The state of the system is augmented by the model parameter p which is formally assumed to vary with time according to the simple relation

$$p(k+1) = p(k) + \gamma_p(k) \qquad (16)$$

where γ_p is a random variable with zero mean and known covariance. In this way, the system state variables are augmented by one for each parameter to be estimated and the dimensions of the involved matrices are increased accordingly.

It is obvious that the number of parameters to be additionally estimated on the basis of the same few measurements cannot become too high without affecting the quality of the estimates. According to the reasoning in Sect. 1, it must be expected that the observability properties of the model become worse and the estimation scheme may come under so much demand that it cannot unambiguously trace back any irregularity in the measurements to the correct reason within the augmented model. It has been shown by Cremer and Schütt (1990) that the most important parameters v_f and ρ_{max} can be estimated simultaneously with the traffic flow state inside a freeway section of about 3 km in length. This observation from simulation studies was substantiated by a systematic observability analysis.

Having the efficient tool of a model-based state estimator, the question might be asked as to whether it could be applied for early incident detection (see *Incident Detection*). Investigations have shown that this is not feasible without altering the model since the model is formulated for regular traffic phenomena and cannot treat irregular cases such as incidents. From this it is clear that incident detection capability can only be achieved by estimation techniques if a suitable extension is formulated for the model. From a macroscopic point of view, the most perceivable effect of an incident is a strong reduction of road capacity at the site of the incident. This could be modelled by a local change of v_f, ρ_{max} or of both. However, as these parameters enter the model equations in a nonlinear way and influence capacity less directly, another idea has proven to be more efficient and easier: the introduction of a hypothetical flow q_d which enters segment i and leaves segment $i+1$ as shown in Fig. 1. In the model equation, this is easily done by adding q_d to the density balance equation for segment i and subtracting q_d from the equation for ρ_{i+1}. The model has to be augmented by a simple equation similar to Eqn. (16):

$$q_d(k+1) = q_d(k) + \gamma_q(k) \qquad (17)$$

In normal conditions, this flow would be estimated as zero while after an accident the estimator would increase this flow drastically to account for the bottleneck caused by the incident as shown by the simulation results in Fig. 5 (Cremer 1981). Another promising approach using a multimodel estimation scheme has been reported by Green et al. (1977).

4. Future Developments

Complete and detailed information about the current state within a road network is a necessary prerequisite for many innovative driver information and traffic control systems. Model based estimation techniques which have been developed in the field of control engineering represent an efficient method of continuously estimating traffic variables which cannot be measured directly. The measurement data which are

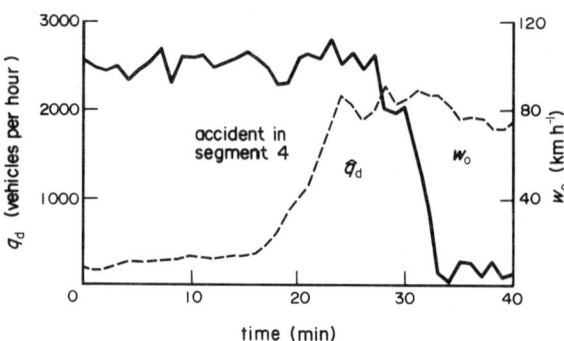

Figure 5
Incident detection via a fictitious flow q_d: —— local mean speed at the upstream detector and – – – estimate of fictitious flow q_d as incident indicator

required for reliable estimates of freeway traffic variables are conventional volume and speed measurements taken at a distance of three to four kilometers.

Though estimation methods for freeway traffic variables are well established on a simulation level, few field tests have been performed. In view of the current impact of control engineering and informatics on advanced road traffic systems, it is very likely that these methods will be tested extensively in the near future. Theoretical research is expected to focus on questions of how trip data delivered from individual vehicles serving as mobile sensors may be included to enhance the reliability of estimates and of how these methods can be transferred from the field of freeway traffic to urban streets and rural roads where traffic phenomena are different.

See also: Freeway Capacity: Reliability and Control; Prediction of Traffic Flow; Simulation Programs, Macroscopic

Bibliography

Bhouri N, Hadj-Salem H, Papageorgiou M, Blosseville J-M 1989 Estimation of traffic density on motorways using presence detector measurements. In: Husson R (ed.) *Prepr. IFAC/IMACS/IFORS Int. Symp. AIPAC.* IFAC, Laxenburg, Austria, pp. 325–30

Cremer M 1981 Incident detection on freeways by filtering techniques. *Proc. 8th Int. IFAC World Congr.* Pergamon, Oxford, pp. 96–101

Cremer M, Papageorgiou M 1981 Parameter identification for a traffic flow model. *Automatica* **17**, 837–43

Cremer M, Papageorgiou M, Schmidt G 1980 Einsatz regelungstechnischer Mittel zur Verbesserung des Verkehrsablaufs auf Schnellstraßen. *Forsch. Strassenbau Strassenberkehrstech.* **307**

Cremer M, Schütt H 1990 A comprehensive concept for simultaneous state observation. Parameter estimation and incident detection. *Proc. 11th Int. Symp. Transportation and Traffic Theory.* Elsevier, Amsterdam, pp. 95–111

Green C S, Hopt P K, Willsky A S, Gershwin S B 1977 Dynamic detection and identification of incidents on freeways: The multi model method. Massachusetts Institute of Technology Report No. ESL-R-766

Grewal M S, Payne H J 1976 Identification of parameters in a freeway traffic model. *IEEE Trans. Syst., Man Cybern.* **6**, 176–85

Knapp C H 1972 Traffic estimation and control at bottlenecks. *Proc. IEEE Systems, Man and Cybernetics Conf.* IEEE, New York

Ljung L, Söderström T 1983 *Theory and Practice of Recursive Identification.* MIT Press, Cambridge, MA

Maarseven M F 1983 A martingale approach to estimation and control of traffic flow on motorways. *Proc. IFAC/IFIP/IFORS Conf. Control in Transportation Systems.* Pergamon, Oxford, pp. 203–10

Nahi N F, Trivedi A N 1973 Recursive estimation of traffic variables: Section density and average speed. *Transp. Sci.* **7**, 269–86

Payne H J 1971 Models of freeway traffic control. *Simulation Council Proc. 1.* Simulation Council Inc., pp. 51–61

Smulders S A 1987 Modelling and filtering of freeway traffic flow. *Proc. 10th Int. Symp. Transportation and Traffic Theory.* Elsevier, Amsterdam, pp. 139–56

Szeto M, Gazis D C 1972 Application of Kalman filtering to the surveillance and control of traffic systems. *Transp. Sci.* **6**, 419–39

M. Cremer
[Technische Universität Hamburg-Harburg,
Hamburg, Germany]

Freeway

The term freeway is used to characterize a road traffic system with the following characteristics:

(a) a long road,

(b) unilateral flow,

(c) more than one lane (so that overtaking is possible),

(d) no amber lights along the mainstream, and

(e) interactions with adjacent roads limited to particular on-ramps and off-ramps.

Since freeways in American cities were the first traffic systems of this kind to be controlled, the term freeway is often used as a generic term for all traffic systems of the defined class. Highways (USA), motorways (UK), Autobahnen (Germany), autostrade (Italy) and autoroutes (France) are further examples of traffic systems belonging to this class.

For modelling and control purposes, freeway systems may be viewed as isolated axes with several on-ramps and off-ramps, as periurban ringways (e.g., the Boulevard Périphérique in Paris), as networks (see *Freeway Network Control*) or as mixed networks with surface streets (see *Corridor Control Systems*).

M. Papageorgiou
[Technische Universität München,
Munich, Germany]

Freeway Capacity: Reliability and Control

Control strategies that are used for preventing congestion on freeways can be divided into two classes. Strategies of the first class are directed towards the control of transport demand in order to avoid traffic volume on carriageways exceeding capacity, for example by metering traffic flow on entrance ramps or by modifying drivers' routes in order to divert traffic from the overloaded arcs of the network to those that have excess capacity. The second class includes strategies directed towards modifying drivers' behavior on the road when instability is approaching, with the aim of increasing the carriageway capacity. A crucial point in freeway traffic control is the detection of the approach of instability and the choice of the most suitable control strategy in each case. The theory of freeway reliability is a valid aid in solving both of these problems.

1. Reliability of Freeway Transport Systems

Consider a sequence of vehicles that pass, with constant flow rate Q, along a lane of a freeway in view of an observer located at the edge of the carriageway. If Q is high enough, vehicles proceed with small gaps between them, so that any variation in the speed of a vehicle causes a corresponding reaction in the following vehicle. In this case, the sequence of speeds recorded by the observer does not constitute a renewal process and the conditional means of speeds in the successive instants, given the preceding history of the process, form a random sequence that defines the level of the process over time.

Let $\ldots, t-1, t, t+1, \ldots$ be the instants in which vehicles pass before the observation point and let a_{t-1} be the shift of the speed v_{t-1} of the vehicle passing at the instant $t-1$ from the process level $\bar{v}_{t-1}$. This shift causes a reaction of the driver of the following vehicle and influences the level $\bar{v}_t$ at the instant t. It is assumed that the difference between $\bar{v}_t$ and $\bar{v}_{t-1}$ is proportional to a_{t-1}:

$$\bar{v}_t - \bar{v}_{t-1} = \lambda a_{t-1} \qquad (1)$$

where λ is a coefficient ranging between 0 and 1, and the shifts a_t are random variables normally and independently distributed with zero mean. With these hypotheses, it is easily verified that

$$v_t = v_{t-1} + a_t - (1-\lambda)a_{t-1} \qquad (2)$$

which shows that the sequence of the v_t is a realization of an integrated moving average process of the first order: its parameters are λ and the variance σ^2 of the a_t distribution. In this process, the relation between the levels $\bar{v}_t$ and $\bar{v}_{t-k}$ is given by

$$\bar{v}_t = \bar{v}_{t-k} + \lambda \sum_{j=1}^{k} a_{t-j} \qquad (3)$$

which indicates that the sequence of the $\bar{v}_t$ levels starting from the instant $t-k$ is the result of a random walk generated by the variable λa_t.

Any variation of the level $\bar{v}_t$ produces a corresponding variation of the density ρ of the traffic stream at the observation point, according to

$$\rho \bar{v}_t = q = \text{constant} \qquad (4)$$

Every traffic stream is characterized by a proper limit density ρ^*, at which stable flow is only possible if vehicles maintain a constant speed over time. In reality, vehicles undergo speed fluctuations due to many causes; therefore, when the speed level decreases to the value $\bar{v}^* = q/\rho^*$, flow becomes unstable. This instability is a random event, because whether or not it happens depends on the particular realization of the $\bar{v}_t$ process. The probability that this event does not occur during a time interval t is the reliability of the traffic stream relative to time t and can be calculated by studying the process in Eqn. (3).

2. Experimental Validation of the Reliability Theory

The results summarized in Sect. 1 were validated by analyzing the data recorded on two double-lane freeway carriageways, in Italy and in The Netherlands. It was seen from these data that instability occurs initially on the off-side lane and then reaches the near-side lane. This result enables the reliability of a carriageway to be identified with that of its off-side lane and allows the study of the speed process to be carried out on only this lane.

The long sequences of speeds recorded in different survey periods on the two freeways were broken down by a sequential procedure into successive realizations of homogeneous processes (Eqn. (2)), each of them characterized by different λ and σ^2 parameters and by different values of q and ρ. For each realization, the hypothesis of independent normal distribution with null mean of a_t was verified. It resulted in three relations.

First, the number N of speed values v_t belonging to a homogeneous process is an exponential random variable, whose parameter L is a linear function of the logarithm of q:

$$L = 0.0341 - 0.0039 \ln q \qquad (5)$$

Second, the λ parameter is a linear function of the logarithm of ρ:

$$\lambda = -0.126 + 0.274 \ln \rho \qquad (6)$$

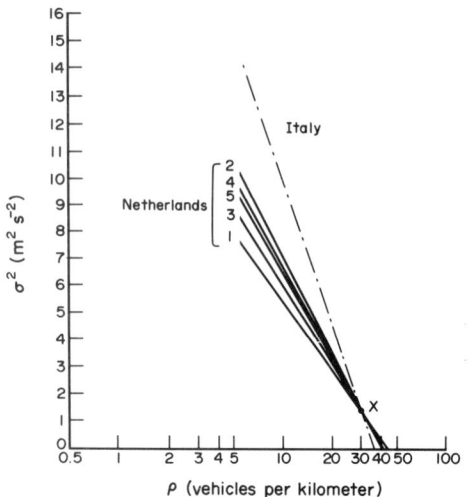

Figure 1
Relationships between σ^2 and ρ on the off-side lanes of two freeways

3. Calculation of Reliability

Consider a traffic stream, characterized by a flow rate q and by a given b, travelling along the off-side lane of a freeway. Its reliability ϕ relative to a time interval t can be obtained by executing a numerical simulation of the process given in Eqn. (3) in order to calculate the probability that the speed level $\bar{v}^* = q/\rho^*$ is not reached during the time t. In fact, repeating the simulation m times calculates ϕ as a ratio between the number of times in which $\bar{v}^*$ is not reached and m.

Values of ϕ relative to a time $t = 30$ min were calculated by this method for many values of q and b. Taking the regression of the ϕ values on the corresponding q and b values, the following equation was obtained:

$$\phi = 1 - 0.00407(q/1000)^{3.22}b \qquad (8)$$

4. The Use of Reliability in Traffic Control

The results reported in Sects. 1–3 can be used to detect the approach of instability on a freeway carriageway. Consider a cross section of a carriageway in which sequences of speeds and time gaps between vehicles travelling on its off-side lane are recorded. These sequences are broken down into successive subsequences of 50 vehicles and for each of these subsequences an estimate of λ and σ^2 is made in real time, along with that of q and ρ. The calculated values of λ and σ^2 vary from one subsequence to another, essentially because of errors of estimates and traffic density variations; however, b is generally the same for a certain number of successive subsequences, because drivers' behavior remains unchanged in a sufficiently limited time interval. Taking into account this fact, an estimation of b is made after the passage of the last vehicle of each subsequence, by calculating the linear regression of the σ^2 values relative to the last n subsequences on the corresponding $\ln \rho$, constrained by the condition that the regression line passes through the point having coordinates $\sigma^2 = 1.5 \text{ m}^2 \text{ s}^{-2}$ and $\ln \rho = \ln(30 \text{ vehicles per kilometer})$.

The choice of the value $n (>1)$ depends on the necessity of smoothing the differences between the successive values of b due to errors in σ^2 and ρ estimates. The greater the value of n, the more marked is the smoothing, but the greater is the risk of hiding (i.e., retarding the detection of) modifications in drivers' behavior. A value of $n = 10$ seems to be a good compromise between these two different needs.

The coefficients of Eqns. (5) and (6) were estimated using all the data recorded on the two freeways, because the equations obtained on both the freeways and in different survey periods were not statistically different. For this reason, the relationships in Eqns. (5) and (6) seem to be characteristics of the freeway transport system.

Third, the σ^2 parameter is a linear function of the logarithm of ρ:

$$\sigma^2 = a - b \ln \rho \qquad (7)$$

The estimates of a and b coefficients obtained on the two freeways, and even those estimates made during different survey periods on the Dutch motorway, are statistically different. Plots of Eqn. (7) using data obtained on the Italian freeway and on five different days on the Dutch freeway are shown in Fig. 1: all are straight lines that pass through the same point (x) having coordinates $\rho = 30$ vehicles per kilometer and $\sigma^2 = 1.5 \text{ m}^2 \text{ s}^{-2}$. The values of σ^2 are extremely low for values of ρ near the intersections of the regression lines with the axis of the abscissae. When ρ reaches the values corresponding to these intersections, circulation is possible only if speed shifts are practically nonexistent: they are then likely measures of limit density ρ^*.

It can be seen from Fig. 1 that ρ^* is different in different situations and decreases with the increase of the coefficient b in Eqn. (7); that is, with the increase of σ^2 values for the same values of ρ. σ^2 is also a parameter of the speed level process (Eqn. (3)); thus, Eqn. (7) is a fundamental characteristic of a traffic stream in order to determine its reliability.

In the same way, the flow rate is calculated after the passage of the last vehicle of each subsequence as an arithmetic mean of the values of q estimated on the last n subsequences. By substituting b and q values into Eqn. (8), the reliability of the traffic stream is calculated in real time after the passage of each subsequence.

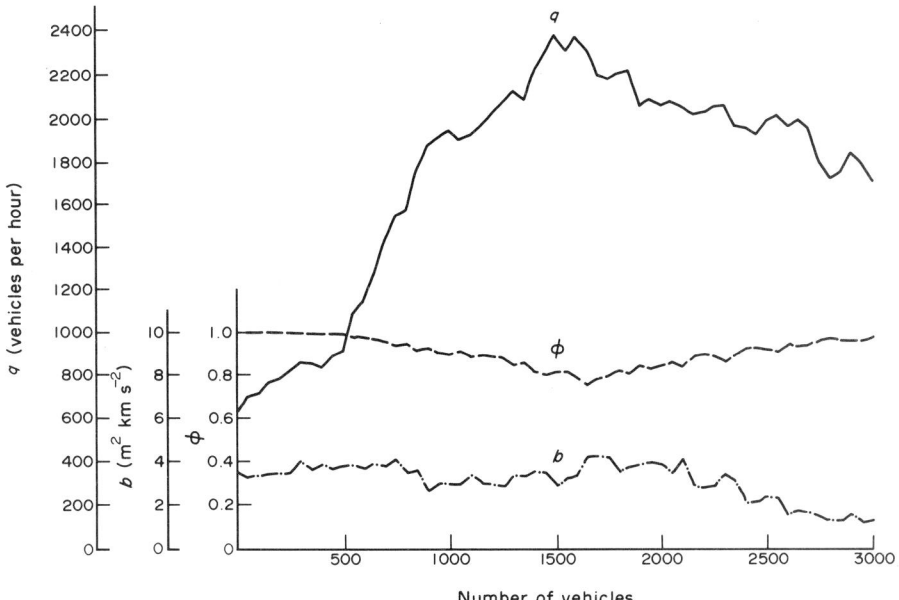

Figure 2
Sequences of q, b and ϕ values on the off-side lane of a Dutch freeway

5. Examples of Detection of Instability

Figure 2 shows, as an example of the application of the method described in Sect. 4, the succession of q, b and ϕ values calculated on a sequence of 3000 vehicles passing through a cross section of the off-side lane of a Dutch motorway. It can be seen that ϕ stays near 1 during a first period corresponding to the passage of 500 vehicles and then decreases progressively as a consequence of the increase of q, while b fluctuates

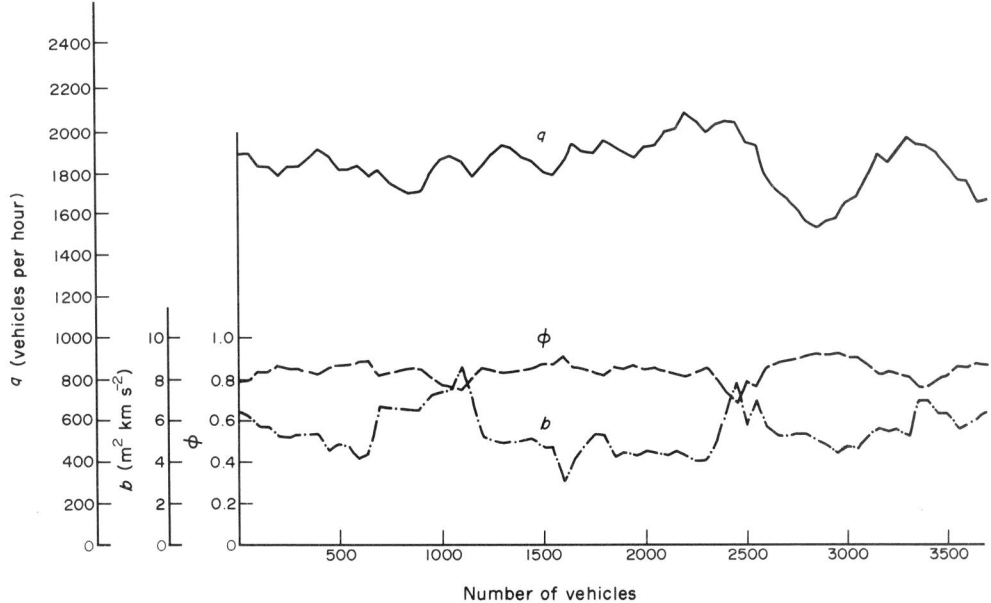

Figure 3
Sequences of q, b and ϕ values on the off-side lane of an Italian freeway

slightly around a mean value of $3.5 \text{ m}^2 \text{ km s}^{-2}$. After the passage of 1600 vehicles, the simultaneous increase of q up to 2400 vehicles per hour and of b beyond $4 \text{ m}^2 \text{ km s}^{-2}$ triggers off the instability, which continues for a long period. This is shown in Fig. 2 by the simultaneous decreases of q and b, a signal of instability. In fact, when in stable traffic conditions q decreases, speed fluctuations increase and ρ decreases, so that b remains constant on average.

In Fig. 2, the increase of ϕ during the period of instability is only the result of calculation and has no physical meaning, because ϕ is a characteristic of stable flow periods. Indeed, the progressive decrease of ϕ during the passage of the first 1600 vehicles implies a corresponding increase in the probability of instability, which actually takes place at $\phi = 0.75$. Therefore, when the values of ϕ calculated in real time by this method falls below a certain threshold, the approach of instability is detected and the intervention of a control strategy becomes necessary.

The choice of the threshold is not a theoretical problem, but a practical one. It depends both on the risk of instability that is desired and on the regularity of traffic flow: a value of 0.8 could be suggested.

Figure 3 reproduces analogous values of q, b and ϕ with reference to an Italian freeway. It can be seen that ϕ falls below 0.8 three times, after the passage of 1000, 2400 and 3300 vehicles, respectively. The corresponding temporary decreases of q and b indicate the triggering of instability, but this disappears quite rapidly because of the sufficiently low values of q and the consequent marked decrease of b. In this case, the drops in the value of ϕ are essentially due to high peak values of b and, because of the correspondent frequent triggerings of instability, values of q on the off-side lane generally remain below 2000 vehicles per hour.

By comparing Figs. 2 and 3, it can be deduced that the causes of instability are different in these two cases and, therefore, that two different control strategies are necessary when reliability approaches the threshold value. On the Dutch freeway, traffic demand has to be reduced so that the carriageway capacity is not exceeded. On the Italian freeway, drivers' behavior can be modified; for example, by imposing a speed limit or a particular utilization of the carriageway lanes, in order to cut down the high peak values of b and to increase the carriageway capacity (see *Speed Limitation on Freeways: Traffic-Responsive Strategies*).

See also: Prediction of Traffic Flow

Bibliography

Box J E P, Jenkins G M 1976 *Time Series Analysis: Forecasting and Control*. Holden Day, San Francisco, CA

Ferrari P 1988 The reliability of the motorway transport system. *Transp. Res. B* **22**, 291–310

Papageorgiou M 1983 *Application of Automatic Control Concepts to Traffic Flow Modeling and Control*. Springer, Berlin

Treiterer J, Myers J A 1974 The hysteresis phenomenon in traffic flow. *Proc. 6th Int. Symp. Transport and Traffic Theory*. Elsevier, New York, pp. 13–38

P. Ferrari
[Università di Pisa, Pisa, Italy]

Freeway Control: An Overview

Freeways have characteristics that require dedicated control systems and methods. (Most of what is stated in this article applies not only to freeways (for a definition of the term freeway see *Freeway*) but also to toll roads.) Speeds and densities are relatively high. As a consequence, disruptions in the traffic flow can rapidly propagate and increase in severity. Similarly, when imposing traffic measures, care should be taken that these do not lead to dangerous reactions by drivers.

Effective control requires a good understanding of traffic behavior as well as special hardware and software for monitoring and signalling. Development of both takes place interactively as specially developed hardware is needed to provide the basic traffic data for flow studies. Flow studies lead to a better insight into which data should be collected. Compared with traffic light control in cities, freeway control has proved difficult to realize. Only since the early 1970s have successful control systems very gradually become established. In this process of development, the human engineering aspect was long ignored to the detriment of the success of the systems. Behavioral science is as important a basis for traffic control as traffic flow theory.

1. Evaluation

1.1 System Requirements

There is no wholehearted public acceptance of freeway control systems. The major complaint is that the systems are ineffective and incapable of solving the problems. Only in a few cases are systems so successful that there is a public demand for them.

It is typical of the freeway control situation that very specific issues are dealt with; an overall systems approach incorporating various functions is rare. All control systems are used to improve on one or more of the following aspects:

(a) safety,

(b) throughput,

(c) energy consumption,

(d) pollution of the environment, and

(e) travel time predictability.

Priorities depend upon which of these aspects becomes of political interest. With time the importance placed on particular aspects shifts. This does not necessarily mean that the control systems and measures will also change. Most systems have a positive effect on more than one aspect and, in these cases, the control measures will remain practically the same. In the same way that traffic engineers have learned to include human engineering aspects in their system designs, they must now learn to meet the needs of political acceptability.

1.2 Problems

There is still a wide gap between academic research and operational applicability. Knowledge of traffic behavior is often confined to specific cases when, in practice, a generalization is needed. Research is difficult as a large proportion of the freeway network does not consist of long straight stretches of road without ramps. Monitoring is also a problem because algorithms take too long to discover any problems on the road.

Even when there is good insight into the traffic situation, there is still the problem of which measure to take. Solving a problem is sometimes more difficult than detecting it. Furthermore, most control measures will include advice or commands for drivers, which must be introduced gradually to allow learning to take place.

Systems engineering aspects are often underestimated; this can have indirect but very serious consequences. When systems are operational, the public expects them to work correctly. Supplying incorrect or no information will lead to drivers disregarding the system, making the system useless. However, some systems have been so successful that the drivers trust them blindly, with the result that when the system fails to warn the drivers they are in danger because of their inattentiveness.

Evaluation of effectiveness has only recently become common practice. Operational systems where research work is ongoing are few and these are commonly known.

2. Fog

Of all the adverse weather conditions, fog is one of the most interesting. Research has been carried out in California on traffic behavior in fog. These studies showed that when a safe speed was advised, a considerable percentage of drivers would not accept this speed, considering it too low. Thus, a low advised speed would cause two populations of drivers each with their own speed; this is a situation that should be avoided under fog conditions.

A differentiation can be made between two types of fog: valley fog and blanket fog.

Especially in mountainous areas, there are valleys where dense fog is common. These areas can be very dangerous in the dark when the change from good to bad visibility is abrupt and unexpected. This is a general problem for all roads where speeds are high. Fortunately, the location of these fog-prone sites is well known and allocation of fog detectors is simple.

Blanket fog, however, spreads out over low-lying areas and drifts around. It is possible to predict in which areas this fog will start to form, but a reliable detection system should have fog detectors on both sides of the freeway at intervals of 1–2 km. This type of fog can cover wide areas and the danger to traffic has not been well understood.

Accident statistics do not show that fog increases accident risk substantially. However, most severe accidents involve fog. The explanation for this is simple: fog accidents on a freeway are mainly caused by traffic running into slow-moving or standing vehicles because visibility is poor. Oncoming drivers crash into the jam caused by the original accident because visibility is still poor. This process repeats itself until adequate on-site slowdown warnings are given.

Most actions against fog accidents have been guided by the well-intentioned but incorrect idea that drivers should be told to drive more slowly. This idea is wrong for two reasons.

First, drivers will not slow down to a safe speed. In fog, drivers are misled by the environment as all contours are vague. The brain translates this vagueness as "far away." Thus, drivers feel that they are not driving as fast as they really are. Thus, any advice to drive slowly will, if followed, still lead to speeds that are too high. This also explains to some extent the Californian research results.

Second, only with severe speed differences (i.e., traffic jams) is there a real danger. Field tests have shown that when drivers in fog are shown an advised speed equal to the average speed, the variance of the speed distribution is reduced to one third. However, the impact of the signals when used as a warning against a traffic jam is, at the same time, reduced.

It can therefore be concluded that, although it is preferable to have slower driving and a very homogeneous speed distribution in fog, the priority consideration is to give an immediate warning against traffic jams whenever traffic is crawling or at a standstill. It is obvious from applications in The Netherlands that such an approach works perfectly. If freeway systems are only capable of giving simple information then they should never be used to warn traffic against blanket fog as people can see this fog clearly. The facility should be reserved for warning against slow traffic. Only adequate monitoring and signalling systems can prevent fog accidents; a general warning is not enough. The mechanisms needed will be described in Sect. 6.

Plans to reduce the severity of fog accidents by separating trucks from cars are difficult to carry out. In most cases, flow volumes are such that it is impossible to impose such rules.

3. Ice

During World War II, much research went into investigating the possibilities of reducing fog, by warming up the air and so on. Such schemes are not feasible for freeways. In the case of ice, however, these schemes are the only way to handle the problem. Warning against black ice is of little help as no average vehicle can manage on an icy surface. Therefore, research is geared towards an in-time prediction of ice-prone conditions such that maintenance crews can spray the road. Sand, salt and chemicals are used. The most effective spray is a mixture of water and salt as this solution has an immediate effect. As with fog, there are blackspots which can be easily identified with the help of infrared photography. At such sites, continuous scanning of the temperature, moisture and level of salination is easily possible. For the rest of the freeways a sparse network of detectors is sufficient. Accuracy can be improved if the temperatures of the ground and of the air are also monitored, as well as that of the cloudbase. However, such refinements still require tuning in by expert meteorologists.

4. Rain

During rain, accident risks increase threefold, though very little is known about traffic control to fight these risks. Rain reduces throughput by some 5%, a clear indication that drivers do tend to keep longer headways, although the reduction is not sufficient when compared with the loss of grip on the road. In addition, the reduction in throughput increases the risk of oversaturation with an increase of shock wave accidents. The only work known on freeway control to reduce the danger caused by rain is that in the field of stabilizing traffic flow mentioned later. Improvements seem to have come more from other directions such as antilock braking systems and open-asphalt road surfaces which do not form a film of water on the road.

5. Oversaturation

The more than linear increase of the accident rate with increase of flow has led to many studies on the effects of oversaturation and on measures to fight oversaturation. Oversaturation should be avoided by preventative rather than curative measures: ramp metering is one of the best known (see *On-Ramp Control: Realization Principles*).

Another method used to handle oversaturation is to dampen the shock waves. The first applications came from sites where advised speeds of $50\,\text{km}\,\text{h}^{-1}$ were shown to drivers in a queue of traffic. Although this caused some irritation because traffic was often unable to drive on, let alone drive at $50\,\text{km}\,\text{h}^{-1}$, the speed difference in the shock waves that always accompany oversaturation was reduced from 70–$80\,\text{km}\,\text{h}^{-1}$ to 50–$60\,\text{km}\,\text{h}^{-1}$. Throughput was 5% higher than when there was no advised speeds. Although speed–flow data show little or no change caused by such speed signs, the number of lane changes is more than halved.

It is desired to obtain a similar improvement in running traffic and this is possible, although it is very difficult to work out the algorithm that controls the advised speeds. This algorithm is very dependent on traffic behavior and, therefore, reacts differently in wet and dry weather conditions. This new method implies refined monitoring of the traffic. The advantages lie not only in the increased throughput and reduced accident risk, but also in the reduction of pollution and the improvement of travel time prediction.

5.1 Rerouting and Alternative Routes

Rerouting is a control measure on a different level used to avoid congestion, which is operational in various situations and in several forms. Typical examples can be found in France, Germany and the USA.

In France, especially during holiday periods, a practically permanent signing system can be used to let drivers follow alternative routes when the main routes are overcrowded. This scheme is not only for freeways. It is almost completely hand operated by the police who have an extensive nationwide monitoring network, as well as modern television and computer-network facilities to inform the public.

In Germany, a numbered alternative route starts at every off-ramp. Following these numbered signs brings drivers to a more downstream point of the freeway. This system is used by the police in emergencies. It is completely hand operated.

A much more advanced system is being implemented in Germany in which all of the freeway network is closely monitored with detector sites every 3–18 km depending on the traffic density. In cases of oversaturation, alternative routes can be advised from central control sites via radio and special changeable direction signs. These signs give information similar to that of the permanent direction signs and drivers are guided all the way along the alternative route. This scheme keeps traffic on the freeways. Interestingly, investigations have shown a demand from the public to receive clear information about alternatives of the same standard (i.e., there is a reluctance to leave the freeways and there is a reluctance to have to find the way unprepared).

In the USA, a totally different type of "rerouting" is common because freeways often have feeder roads running in parallel over a long distance. In these cases, it is, of course, attractive to reroute some traffic over these parallel facilities. Operational schemes vary from hand operated to more automated systems, although the more automatic systems only cover shorter stretches of freeway.

5.2 Tidal Flow

Alternate lane use is another option to cope with oversaturation. Although very common for a long time in urban areas, tidal-flow lanes for freeways have remained an option in the USA only. Normally, the extra lanes are reserved for a selected group such as high-occupancy vehicles (HOVs) or van pools. These so-called diamond lanes come in various forms and sizes, and much experience with geometry is now available. Control in the older systems is quite simple as signs are only operated at the entrance. In more modern systems, automatic monitoring and signalling to avoid secondary accidents are found.

6. Secondary Accidents

6.1 Importance

Accidents on freeways can be differentiated by their cause. The biggest single group is that of secondary accidents (i.e., accidents caused by a traffic jam). This traffic jam may be caused by an accident or by oversaturation. The relative proportion of secondary accidents increases rapidly with increasing flow. On most freeways, excluding remote ones, this proportion lies between 40% and 60%. From experience, it is known that once police are on site warning traffic to slow down the risk of secondary accidents practically disappears. The solution is therefore obvious: detect traffic jams and slow down traffic. One more aspect should be noted: the risk that a traffic jam causes a secondary accident is nearly twice as high in the first few minutes as the risk 10 min or more after the start of the jam. Therefore, detection and warning should be done quickly.

6.2 Detection

The development and introduction of good detection algorithms has suffered from two mistakes. The first was the rather academic attitude of most researchers who insisted that an algorithm should be developed to detect any stranded vehicle rather than a traffic jam. It is very difficult to detect a stranded vehicle when traffic volumes are so low that the traffic is hardly, if at all, affected by it. Knowing that secondary accidents take place mainly at higher flow volumes and, from accident statistics, that the risk a stranded vehicle causes to low-density traffic can be neglected, it is questionable why so much emphasis has been put on the perfect stranded-vehicle detector by researchers. The second mistake involved another attempt at perfection, that of the perfect queue detection algorithm. The forming of a traffic jam may take only approximately 10 s. However, it is such an outstanding phenomena that it is relatively simple to detect. Therefore, various algorithms can be used to detect traffic jams with up to 1 s difference in their operating times. Some, however, rely on information that is very stochastic in nature, such as occupancy (see *Flow Variables*). Occupancy is not only affected by speed, but also by vehicle length. However, vehicle length has little to do with traffic jams and, therefore, an algorithm based on the exponential smoothing of the speed of each vehicle should be preferred. This requires double loops, which makes the monitoring slightly more expensive. However, double loops do make it possible to check correct functioning as well as to fall back on a single-loop algorithm when one loop fails. A minor disadvantage is that it requires some tuning in as traffic behavior, especially in the cases of traffic jams forming, is somewhat site dependent.

Traffic queue detection should be reliable and fast. Reliability can be expressed in terms of, for example, percentage of queues missed and percentage of false alarms. As stated, it is not too difficult to detect queues, so the percentage of queues missed can be ignored. The percentage of false alarms is, however, high with some of the simpler single-loop algorithms. This is dangerous because it takes about 20 correct warnings of a traffic jam to regain the trust of someone who has been given a false warning.

All this holds for systems that make decisions based on information from a single site. The longer the distances between the sites, the greater the chance that a traffic jam will not be noticed. However, if detectors lie 150 m apart any traffic jam can be noticed as traffic behavior will differ sufficiently over at least one detector site.

Systems that work on the basis of a comparison of data from two successive sites have a much longer reaction time. However, these systems have their own assets; they can be used to obtain a good impression of the traffic flow stability. These systems are, therefore, the ones needed when stabilizing traffic (see Sect. 5).

Research into the detection of the single stranded vehicle will of course continue but there is little hope that it will lead to algorithms that can become operational in the near future.

6.3 Signalling

Signalling should bring traffic smoothly to a halt. This is best achieved by telling traffic exactly how to slow down; that is, by first slowing down to 70 km h^{-1} and then to 50 km h^{-1}, a pattern that is spontaneously followed when police alongside the road warn traffic against an accident.

7. Road Works

Road works are often the major cause of disruptions, delays and even accidents. The good use of signs is necessary to protect road workers. However, sign utilization is time consuming. In addition, it is desirable to make bottlenecks as short as possible but, with road works, constant adapting of the signs to the actual situation is needed and this is also time consuming. Sign usage is therefore mostly of the worst-case type, causing more disruption than necessary. Finally, roadworks can cause traffic jams, which are potential causes for secondary accidents the moment the jam becomes longer than the road works sign area. How often road works are needed is a policy issue; in the USA there is more of a tendency to leave damage without repair than there is in Europe. The tendency to keep the freeway service up to standard and the high traffic volumes handled have made The Netherlands a pioneer in freeway control systems for road maintenance. Based on cost–benefit figures, the Dutch have a scheme where, for example, roads that carry more than 60 000 vehicles average annual daily traffic (AADT) on 2×2 lanes get fully automatic control and signalling with signs over each lane, whereas with lower traffic volumes mostly hand-operated gantries with green arrow/red cross signs are used. These decisions are based on experience gained over many years with the fully automatic motorway control and signalling (MCS) system, which automatically warns against traffic jams and which, with the help of operators, can easily reserve lanes. This lane reservation is effected by showing a slanting arrow on one gantry (or bridge) over the lane closed further downstream accompanied by a 90 (km h^{-1}) over the free or open lanes, and showing a red cross plus 70s on the following gantry. If there is time (there is not in the case of emergency lane reservations), a trailer is placed just under this red cross to protect the maintenance crews. The effect of this arrangement is that, without any problems, some 1600–1800 vehicles per hour can merge from two lanes into one. The throughput is 5% higher than in those sites where semipermanent sign arrangements are used.

8. Enhanced Signs

The information given to traffic is, in all the examples that have been discussed, concise and to the point. There has been much discussion about the advantages and disadvantages of presenting drivers with additional information explaining why a certain measure has been taken or why a certain speed is being shown. In the case of rerouting, displaying "queue ahead" has been tried, but this is not very explanatory. As part of a European demonstration, explanatory signs have been used to give reasons why advisory speeds were shown. They gave noticeable improvements but were taken down after the demonstration. The problem of such enhancements is the high cost involved (full television coverage) in investigating what to show. However, such explanations are very much appreciated by the public and, where goodwill is needed, properly operated signs may, in an indirect way, pay off.

9. Future Developments

The facilities needed to improve the traffic situation in rain or oversaturation conditions at roadworks and elsewhere using traffic stabilization techniques, ramp metering and modern signalling are to a great extent the same. As a result, all of these systems will normally be incorporated into one overall system. Such systems pay off on busier freeways.

The modern rerouting systems as described in the example from Germany (see Sect. 5.1) require similar facilities, although the monitoring and signalling take place on a much less dense scale. Integration with an overall system is therefore possible to an extent where there is no duplication of equipment.

In freeway control, there is a big difference of attitude between the USA and Europe. In the USA, much more emphasis is placed on the rapid clearing of the road after an incident, whereas in Europe the focus is on protection against secondary accidents. These aims define the design of the system. Only fully automatic surveillance and signalling can help against secondary accidents, whereas for a rapid clearing of the road relatively simple but very sensitive monitoring linked with television surveillance must be used. These two systems are definitely not the same and, although there is some overlap, if the best features of both systems are required, the price comes close to that of two systems. This is exactly what happens in most tunnels where the features of one system are not good enough.

See also: Freeway Network Control; Incident Detection; Speed Limitation on Freeways: Traffic-Responsive Strategies

Bibliography

Institution of Electrical Engineers 1989 *Proc. 2nd Int. Conf. Road Traffic Monitoring.* IEE, London

Kroes J L de 1983 *An Evaluation of the External Effects of the Motorway Traffic Control System.* Rijkswaterstaat, The Hague, The Netherlands

Organisation for Economic Co-operation and Development 1981 *Traffic Control in Saturated Conditions.* OECD, Paris

Smulders S 1989 Control of freeway traffic flow. Thesis, University of Leiden

J. J. Klijnhout
[Rijkswaterstaat, Rotterdam, The Netherlands]

Freeway Network Control

In the densely populated regions of Western Europe and North America, there is often a close-meshed network of freeways or freeway-type roads to be found. A road network permits many origin–destination relationships over different, almost equal, routes. Due to the dynamic character of the road traffic and to the great variety of stochastic processes occurring in the traffic flow, the optimum routes can change very frequently within a close-meshed freeway network. Static signs indicating the best route to the nonlocal driver from the traffic policy point of view cannot adapt to extraordinary situations such as congestion, accidents or severe weather conditions. In these cases, dynamic traffic control measures are required in order to distribute the traffic streams within the road network. These measures are based on up-to-date information on traffic flow.

1. Route Guidance Measures

In principle, a distinction can be made between three different dynamic route guidance measures considered mainly for freeways.

1.1 Collective Control by Variable Message Signs

In collective control by means of variable message signs, partial streams (vehicle collectives) are influenced in their route choices by roadside installations (variable message signs) indicating to all drivers variable route recommendations for specific trip destinations at certain points of decision (freeway interchanges).

1.2 Collective Control via Traffic Radio Service

The most common method of controlling traffic stream distibution is by providing information to road users on the current traffic situation and on route recommendations via broadcast radio. Most Western European and North American countries possess the necessary technical and organizational facilities for information acquisition and distribution. The advantage of this method is that information on the traffic situation can be received not only during a trip, but also prior to the start of a trip. In case of disturbances on the originally desired route, the driver retains the decision whether to take an alternative route (perhaps recommended by the traffic radio service), postpone departure or select another transport mode (if the information was transmitted prior to the start of the trip). However, the inertia of the system (long time intervals between traffic radio service reports) and the vast amount of information given in these reports mean that this system is still unsatisfactory. A distinct improvement in the traffic radio service is anticipated with the introduction of the digital radio data system (RDS), which will be capable of storing and retrieving information on a selective basis (see *Broadcasting Communication Systems*).

1.3 Individual Control

Special on-board equipment (position-finding and navigation aids, on-board computers, stored road network data) will make individual route guidance easier. Unlike collective control methods, trip destinations entered by the driver prior to the start of the trip can be taken into consideration. A distinction must be made between autonomous (static) systems, which operate exclusively on the basis of data stored on board, and supported (dynamic) systems, which also include up-to-date information from roadside beacons in their route algorithms (see *Route Guidance, Individual*).

2. Objective Criteria for Network Control

The bases for developing control strategies for an alternative route guidance system in a freeway network are traffic policy objectives which are also valid for other control measures. Of particular importance are the minimization of travel time and the increase in safety and efficiency by avoiding trips in unstable traffic. In addition, greater driving convenience that can be achieved by alternative route guidance should not be underestimated. Most road users experience extreme discomfort and annoyance during trips in unstable traffic or even in delays due to congestion. Many drivers would accept detours, even if it meant slightly longer travel times, as long as they could maintain an acceptable speed on the alternative route.

3. The Control Model

Traffic-actuated control of a variable message sign system requires, first, qualified off-line-acquired basic data and, second, a careful analysis and forecast of current measurement values. The core of the control algorithm is generally the objective function which is capable of evaluating a certain number of alternative control states within the system. There are four control model elements.

3.1 Basic Data

(a) *Road network database*. The first step when developing a control model for a variable route guidance system is to define the control area and describe the model of the road network under study. Therefore, the network must be broken down, separately for each direction, into road sections of approximately constant lane and traffic conditions.

Section boundaries are mainly situated at access points and freeway interchanges, as well as at the transitions between different alignments (e.g., at lane narrowings or widenings and at major changes in the longitudinal gradient). The network model should be variable so that long-term and short-term changes in the network (e.g., new access points or roadworks) can be taken into account. Access points can be represented as point-shaped nodes (i.e., for road sections between entrance and exit ramps, no separate edge need be defined). The more extensive interchanges or three-way interchanges should, however, be broken down into subsections, since the connecting ramps within these interchanges are the road sections which are frequently prone to congestion.

The road section length and the main traffic parameters are required to describe each road section in the road network database. The traffic parameters can be combined to form a reference number which ultimately describes the capacity.

(b) *Anticipated traffic loads*. In order to control a variable route guidance system, it is important to know the anticipated traffic loads on individual road sections over longer periods of time. The required forecast period is therefore longer the more extensive is the control area (i.e., the longer are the travel times through the network). Since the extrapolation of current measurement values over a longer period of time entails a higher forecast risk, it is practical to base the forecast on anticipated values. These include standard time graphs of traffic loads for individual road sections in the network database. They can be derived from historic data material or from preinvestigations. The standard time graphs should be provided as daily time graphs for different days or types of days; for example, they could be differentiated by being assigned as workday traffic, Friday traffic, weekend traffic, holiday traffic on workdays or holiday traffic at weekends.

(c) Controllable traffic streams. In order to evaluate the alternative control states for variable message sign systems, it is not sufficient to know the traffic loads in the current system state. Consideration must be given to all the alternatives to the current control state of the variable message sign system defined as practical. This is based on assumptions of the percentages of traffic streams that can be diverted by the various control states. These can be derived, as a rule, from special studies of the origin–destination relationships in the road network. Furthermore, assumptions must also be made about the anticipated rate of compliance with the individual control states. Both parameters should be divided according to type of vehicle (e.g., private cars, trucks) and type of day, if possible.

(d) Fundamental diagrams. In order to assess the different control states of a variable message sign system, the anticipated alterations in traffic conditions caused by control plan changes must be determined. This is generally carried out using fundamental diagrams which show the stochastic interrelationships between the macroscopic traffic parameters traffic volume q, traffic density ρ and mean speed v. The traffic parameters are interlinked by the condition equation $q = \rho v$ (see *Flow Variables*).

As an example, Fig. 1 depicts the fundamental diagrams for two-, three- and four-lane carriageways and bottlenecks. These diagrams are based on the variable message sign system in the Rhine–Main area in Germany.

3.2 Traffic Condition Analysis

The current traffic conditions in a studied road network are continuously analyzed using on-line-acquired data. The aim is to be able to determine current travel times on the available alternative routes and to compare them with each other. A distinction is made between two basically different methodological approaches.

The first approach involves the determination of traffic densities on the individual road sections by comparing in-flowing and out-flowing traffic. This method measures the traffic volumes at the entrances and exits of each road section and continuously calculates the number of vehicles in the road section by addition or subtraction. It is very easy to detect whether the capacity of an individual section is exceeded, leading to an increase in travel times. However, consideration must be given to determining

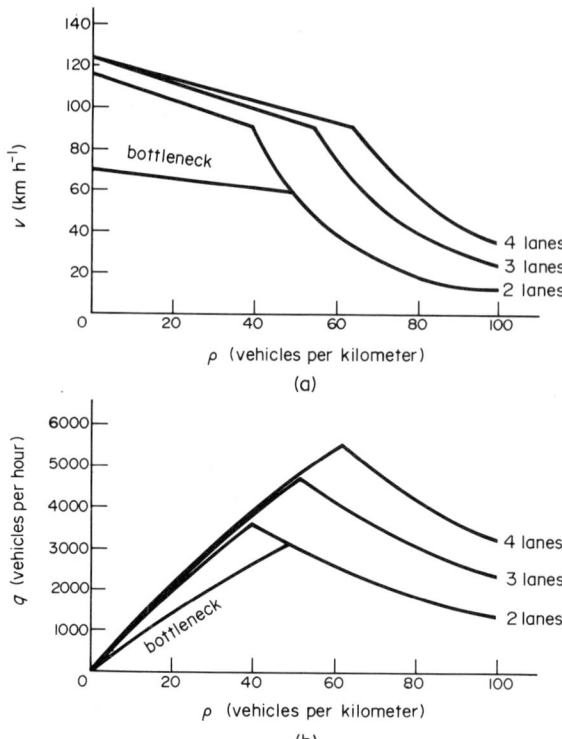

Figure 1
Examples of fundamental diagrams for (a) ρ–v and (b) ρ–q relations

section density through the ongoing addition and subtraction of in-flowing and out-flowing traffic, and this is prone to systematic measurement errors. Therefore, in the control model for the variable route guidance system at Dernbach/Koblenz in Germany which is based on this methodological approach, not only current measurement values but also forecast values are included to determine section density.

The second approach involves the determination of main traffic flow parameters at representative points within the road network. In this method, the local values of individual measurement points (or a combination of measurement points) are considered to be representative for a certain road section. The locally acquired mean speed is used as the main basis for calculating travel times. The difficulty with this method lies in the fact that the local speed is not necessarily representative of the speed profile of the assigned section, particularly at high traffic loads. Therefore, to compensate for this, the anticipated speeds derived from the given fundamental diagram at the measured traffic volume are used (taking into consideration the truck percentage).

3.3 Forecasting Traffic Conditions

Probably the most difficult part of a control model for a variable message sign system is forecasting the traffic flow. It is the basis for determining the objective parameters and, therefore, for evaluating alternative control states. Basically, two tasks result.

The first task involves forecasting the traffic buildup in the current control state. For a variable message sign system, which must not only react (i.e., as a possible measure after an incident has occurred) but also act (i.e., to achieve the optimum distribution of traffic on the available network as a preventive measure to avoid traffic disturbances), it is necessary to forecast the traffic buildup on individual sections of the road network under study. The required forecast horizon is dependent on the distance (or more precisely on the travel time) between the decision point (i.e., the place where it is possible to guide the road user to alternative routes) and the system exit or another decision point.

One possibility for forecasting traffic buildup is by extrapolating current measurement values. Since the forecast risk rises very quickly with increasing interval between the forecast and the measurement time, it is advisable to adapt the forecast values to the curve of anticipated values taken from the given standard time graph and to adjust them with the level of the standard time graph as the time horizon rises.

The second task involves forecasting the effects of alternative control states. In order to estimate the effects of control plan changes, it is initially necessary to calculate the diverting traffic volumes by assuming values for the divertable traffic percentages and for the anticipated rate of compliance. The forecast traffic volumes for the possible changes then result from the increase or decrease in the forecast values in the current control state by the divertable traffic volumes. The main forecast speeds for determining travel times can then be derived from the given fundamental diagrams.

3.4 Determination of the Optimum Control Plan

The optimum control plan is generally determined by an objective function that combines the various objective parameters to obtain one objective function value. Usually, the objective function is expressed as a pure cost function with objective parameters assessed in terms of money. The control plan incurring the lowest total cost is then selected.

The overall structure of a control model for a variable message sign system of dynamic route guidance in a freeway network is depicted in Fig. 2 in the form of a flow diagram.

4. System Structure

A system of dynamic route guidance in a freeway network consists basically of five subsystems (see Fig. 3).

4.1 Data Acquisition System

In order to control a variable route guidance system, current measurement values of the traffic flow on individual road sections of the studied road network are required. In general, measuring traffic volume and mean speed separately for private cars and trucks is sufficient. The data have to be ascertained in defined cycles, so that a reliable data basis is ensured and fast reaction of the system to changes in the traffic condition is possible. In practice, a measurement interval of 5 min has proved to be appropriate. Since for cost reasons the measurement point density of an extended road network is not generally sufficient to carry out automatic incident detection satisfactorily, the data acquisition system should also allow manual input of reports on roadworks, accidents, severe weather conditions and other particular events.

4.2 Data Processing System

Two levels have to be distinguished when processing data: local and central. At the local level (i.e., in the outstations at the individual measurement points), a plausibility check of the measurement values and a precondensation of the data are carried out. If

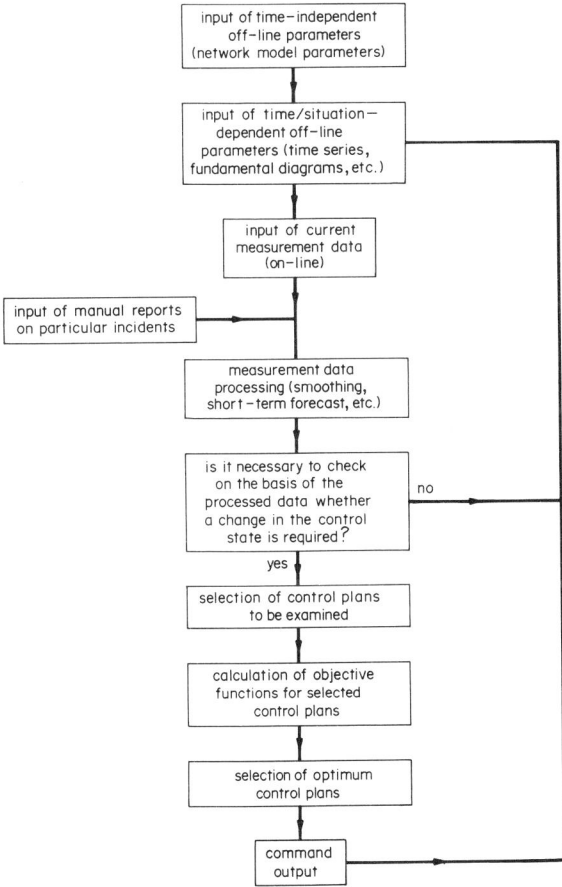

Figure 2
Flow diagram of a control model for a route guidance system

Freeway Network Control

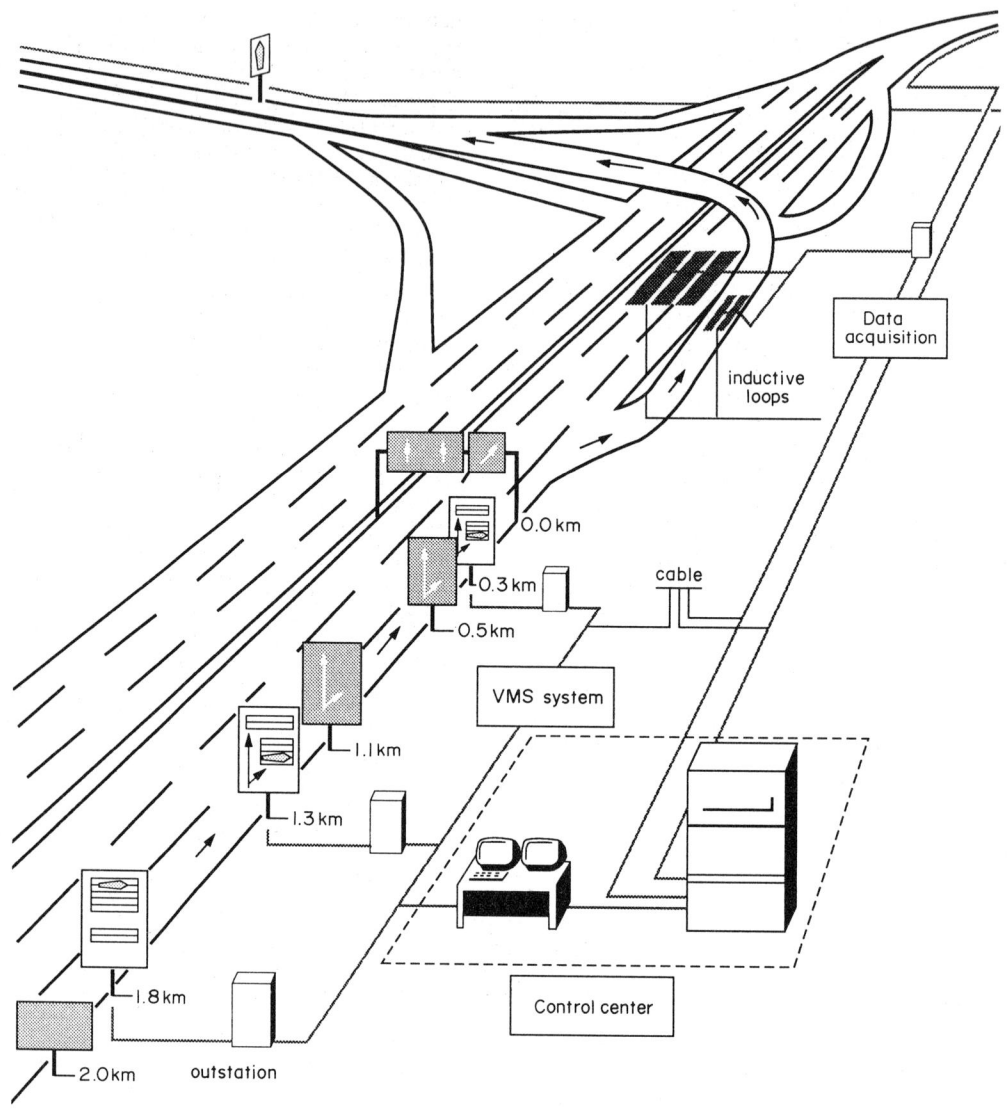

Figure 3
System structure for a route guidance system (courtesy Bavarian ministry)

necessary, first traffic condition analyses and prognoses can be performed at this point. The control algorithm for deciding which is the optimum control state is run at the central level, since from this point comprehensive information on the whole control area is necessary. In addition, reports on particular events that are entered manually are also collected and integrated into the decision process at this point.

4.3 Display System

As far as the collective systems of dynamic route guidance are concerned, communication with the driver takes place via variable message signs positioned at the decision points. Distinction can be made between two variants of display system.

(a) Substitutive system. In this system, the static destinations of the stationary signs are replaced by variable destinations, whereby the outer appearance of the guide signs does not change. It must be made certain that, in case of deviations, the new destinations at the decision points, which are written on the guide signs, can be refound on the alternative routes in order to ensure continuous guidance. The principle of substitutive signs applies particularly to nonlocal drivers, whose main orientation aids are the guide signs. The rate of compliance by local road users is, however, rather low

if changes are not underlined by supplementary indications (e.g., flashing beacons at guide signs).

(b) Additive system. Stationary signs remain unaltered and the system is completed by additional specially designed variable guide signs, which are neutral for normal traffic guidance, but if necessary indicate deviation recommendations for which reasons may be given by supplementary information (e.g., details on type and place of traffic disturbances). Due to the particular design of these variable message signs and due to the possibilities of informing the road users about the cause of a deviation, the local driver can be addressed better with the additive than with the substitutive display system.

4.4 Data Transmission System

Data transmission generally takes place via already existing data cables or new cables have to be laid. Since the already existing telecommunication cables often have only limited capacity reserves and new cabling is very cost intensive, wire-saving data transmission techniques (multiplex methods) are usually used in the available line network.

4.5 Energy Supply Systems

The possibility of supplying the measurement and display points with energy are mainly dependent on the local conditions and the available installations. In general, the display devices consume most energy in the roadside installations, whereas the electronic devices at the measurement points are extremely economical. The supply for the central installations does not generally constitute a problem, since these are usually put up in buildings with already existing energy supply installations.

5. Examples

By far the greatest proportion of variable message sign systems already installed or planned are situated in Germany. The main reason for this is that the freeway network in Germany has the required density and intermeshing for the practical use of such guidance systems. Alternative routes are often available, particularly for the incident-prone main routes for holiday traffic and for the highly loaded network parts in conurbation areas. Therfore, the responsible administrative bodies of Germany supported the development of dynamic variable route guidance systems in the early 1970s and have pressed for the realization of projects.

Table 1 gives an overview of the variable route guidance systems in Germany that are presently in operation or are being set up.

This overview shows that mainly systems based on the additive principle are installed. Valuable experience has been gained with these systems with regard to

Table 1
Variable route guidance systems in Germany

Area of variable message sign system operation	Year of inauguration	Size of area (km²)	Section length (km)	Detour factors	System principle	Number of variable message sign chains	Number of measurement points
Rhine–Main area near Frankfurt/Darmstadt/Wiesbaden	1974	1200	130	~1·6	substitutive	12 (at 6 intersections)	180
Rhineland between Koblenz and Cologne	1977 (manual operation) 1982 (automatic operation)	2500	150	~1·3	additive	4 (at 4 intersections)	30
Ruhr area between Wuppertal and Münster	1982	3000	250	~1·1	additive	7 (at 4 intersections)	66
Baden-Württemberg between Stuttgart and Walldorf	1990	3000	200	~1·0	additive	3 (at 2 intersections)	40
South Bavaria, North Munich	1992	100	35	~1·8	additive	6 (at 4 intersections)	30
North Bavaria between Nürnberg and Würzburg	1986 (manual operation) 1993 (automatic operation)	4000	300	~1·4	additive	5 (at 3 intersections)	110

the compliance of the deviation recommendations. Due to the overall positive effects obtained, it is planned to extend already existing systems and to carry out further projects.

See also: Corridor Control Systems; Freeway Control: An Overview; On-Ramp Control of Freeway Networks; Route Guidance, Collective

Bibliography

Bolte F, Reichelt P, Siegener W, Spies G, Zackor H, Ziegler M 1983 Substitutive oder additive Wechselwegweisung—Ein Systemvergleich. *Str. Autobahn* **3**

DRIVE 1989 *Requirements and System Specification for Dynamic Traffic Messages (VAMOS)*, DRIVE Project No. 1003. DRIVE, Brussels, Deliverables Nos. 3, 9, 10

W. Balz and H. Zackor
[Steierwald Schönharting und Partner,
Stuttgart, Germany]

Freeway Traffic Modelling

Macroscopic models of traffic flow describe traffic behavior on roads in terms of appropriate aggregated traffic variables. There is an interesting analogy between the mathematical description of traffic flow on roads and of water flow. Nevertheless, there are some particular phenomena in traffic flow that do not correspond to water flow, such as traffic flow instability, congestion and stop-and-go traffic. Macroscopic models of traffic flow are valuable tools for simulation (see *Simulation Programs, Macroscopic*), prediction (see *Prediction of Traffic Flow*), estimation (see *Flow Variables: Estimation*) and control strategy design (see *On-Ramp Control: Coordinated Traffic-Responsive Strategies*).

1. Preliminary Issues

The macroscopic description of traffic flow implies the definition of adequate flow variables expressing the average behavior of the vehicles at a specific location and time instant (see *Flow Variables*). By definition, the traffic density $\rho(x, t)$ is the number of vehicles per length unit (vehicles per kilometer), the space mean speed $v(x, t)$ (km h^{-1}) is the instantaneous average speed of vehicles in a length increment (Wardrop 1952) and the traffic volume $q(x, t)$ is the number of vehicles passing a specific location in a time unit (vehicles per hour).

The following equation is obtained as a direct consequence of the definition of these macroscopic or continuum variables:

$$q = \rho v \qquad (1)$$

This has a direct analogue in hydromechanics.

For safety reasons, distances between vehicles are adjusted by drivers according to the speeds of the vehicles. Hence, increasing density in a given freeway stretch, corresponding to shorter headways, leads to reduced mean speed. Several mathematical formulae have been proposed for the description of a speed-density relationship under homogeneous traffic conditions (e.g., May and Keller 1967). A fairly general formula satisfying appropriate boundary conditions is given by

$$V(\rho) = v_f \left[1 - \left(\frac{\rho}{\rho_{\text{jam}}} \right)^l \right]^m \qquad (2)$$

where $l(>0)$ and $m(>1)$ are real-valued parameters, v_f is the free speed and ρ_{jam} denotes the density for a traffic jam. Several $V(\rho)$ formulae proposed in the literature can be considered as special cases of Eqn. (2) by appropriate choice of m and l. For example,

$$V(\rho) = v_f \exp\left[\frac{-1}{a} \left(\frac{\rho}{\rho_c} \right)^a \right] \qquad (3)$$

results from Eqn. (2) with $l = a$, $\rho_{\text{jam}} = \rho_c (am)^{l/a}$ and $m \to \infty$.

Substituting $V(\rho)$ into Eqn. (1), a volume–density relationship $Q(\rho)$ is obtained which is broadly known as the fundamental diagram of traffic engineering. It can easily be shown that the $Q(\rho)$ relationship provides a maximum volume q_{max} at a critical density ρ_c. This is a particular characteristic that distinguishes traffic flow from any other flow process.

2. The Conservation Equation

Traffic on a long multilane freeway with several off-ramps and on-ramps will be considered. Mathematical models of the traffic flow are required to describe the dynamic evolution of traffic variables along the freeway.

Regarding traffic flow as a fluid with density $\rho(x, t)$ and volume $q(x, t)$, the fundamental equation for conservation of matter may be written as

$$\frac{\partial \rho}{\partial t} + \frac{\partial q}{\partial x} = r - s \qquad (4)$$

where $r - s$ is the exogenous on-ramp/off-ramp source term.

For some purposes, it is more convenient to consider difference equations. For this reason, the freeway is subdivided into N sections with lengths Δ_i ($i = 1, \ldots, N$) each having at most one on-ramp and one off-ramp. For a discrete time kT, where T is the sample time interval, the following space–time-discretized traffic variables will be introduced (see

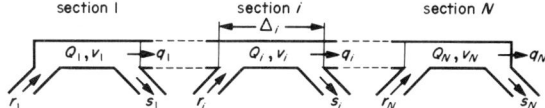

Figure 1
A freeway axis subdivided into sections

Fig. 1): $\rho_i(k)$ is the number of vehicles in the freeway section i at time kT divided by the length Δ_i of the section, $v_i(k)$ is the space mean speed of vehicles in the freeway section i at time kT, $q_i(k)$ is the number of vehicles leaving section i during the time period $[kT, (k+1)T]$, divided by T, and $r_i(k)$ and $s_i(k)$ are on-ramp and off-ramp volumes of section i (if any) respectively.

With these variables, a space–time-discretized form of Eqn. (4) can be given as follows:

$$\rho_i(k+1) = \rho(k) + \frac{T}{\Delta_i}[q_{i-1}(k) - q_i(k) + r_i(k) - s_i(k)] \quad (5)$$

Obviously, Eqns. (4) and (5) hold exactly. The conservation equation can be expanded into a complete model of traffic flow if traffic volumes are expressed in terms of traffic densities in these equations. This can be accomplished in different ways as described in Sects. 3–5.

3. An Algebraic Volume–Density Relationship

Assuming the fundamental diagram to be valid, even under nonhomogeneous traffic conditions, gives

$$q(x,t) = Q[\rho(x,t)] \quad (6)$$

In view of Eqn. (1), Eqn. (6) corresponds to the assumption

$$v(x,t) = V[\rho(x,t)] = \frac{Q[\rho(x,t)]}{\rho(x,t)} \quad (7)$$

Eqns. (4) and (6) constitute a complete traffic flow model for the nondiscretized case. Lighthill and Whitham (1955a, b) have shown that in order for this model to provide unique solutions, traffic variables should be allowed to have discontinuities (shock waves) in space and time (see *Kinematic Wave Theory*). More concretely, discontinuities may be developed by continuous wave forms due to the overtaking of slower waves by faster ones. At points of discontinuity, a shock wave moving condition

$$\frac{dx_s}{dt} = \frac{Q(\rho_2) - Q(\rho_1)}{\rho_2 - \rho_1} \quad (8)$$

replaces the use of Eqns. (4) and (6) which are valid elsewhere, x_s being the position of the shock wave and ρ_1 and ρ_2 ($\rho_2 > \rho_1$) being the density values upstream and downstream of the discontinuity, respectively.

Space and/or time discretization of this model is particularly difficult due to the existence of discontinuity waves (Michalopoulos *et al.* 1984).

4. Introduction of a Density Gradient Term

It may be assumed that mean speed adjusts to traffic density according to

$$v(x,t) = V[\rho(x + \Delta x, t)] \quad (9)$$

instead of as in Eqn. (7). For $\Delta x > 0$, Eqn. (9) suggests that drivers take into account density changes downstream when adjusting their speed. For small Δx, the following relation can be obtained by expanding Eqn. (9) in a Taylor series and neglecting higher-order terms:

$$v(x,t) = V[\rho(x,t)] + \left[\frac{\partial V}{\partial \rho}\frac{\partial \rho}{\partial x}\right]_{(x,t)} \Delta x \quad (10)$$

Payne (1971) proposed the space increment to be $\Delta x = 0.5/\rho$ based on heuristic microscopic considerations. Furthermore, it may be assumed that $\partial V/\partial \rho$ is approximately constant in which case the following equation can be obtained from Eqn. (10):

$$v(x,t) = V[\rho(x,t)] - \frac{\nu}{\rho(x,t)}\frac{\partial \rho(x,t)}{\partial x} \quad (11)$$

where $\nu = -0.5\, \partial V/\partial \rho\, (>0)$ is a constant parameter. Using Eqn. (1), the traffic volume is obtained as

$$q(x,t) = Q[\rho(x,t)] - \nu \frac{\partial \rho(x,t)}{\partial x} \quad (12)$$

which replaces Eqn. (6).

Eqns. (5) and (12) constitute a complete traffic flow model for the nondiscretized case. This model leads to unique solutions without the need for introducing discontinuities for traffic variables as was the case in Sect. 3. Hence, possible abrupt changes of traffic density always occur in finite lengths of space and time which corresponds more closely to real traffic phenomena.

In view of Eqn. (12), it becomes apparent that even under steady-state conditions, maximum traffic volume is not constant but depends on the density gradient. In particular, it is interesting to note that Eqn. (12) allows for q to become greater than q_{max} when ρ is sufficiently large and $\partial \rho/\partial x$ is negative.

In the discretized case, Eqn. (11) may be written as

$$v_i(k) = V[\rho_i(k)] - \frac{v}{\Delta_i} \frac{\rho_{i+1}(k)/\lambda_{i+1} - \rho_i(k)/\lambda_i}{\rho_i(k)/\lambda_i + \kappa} \quad (13)$$

where λ_i denotes the number of lanes in section i. The constant parameter κ has been added to limit the second term in case of very low density values. Traffic volume $q_i(k)$ between two sections may be obtained as a weighted sum of the traffic volumes corresponding to the densities of the two sections. Recalling Eqn. (1), this leads to

$$q_i(k) = \alpha \rho_i(k) v_i(k) + (1-\alpha) \rho_{i+1}(k) v_{i+1}(k) \quad (14)$$

Eqns. (5), (13) and (14) constitute a complete traffic flow model for the discretized case.

5. A Dynamic Mean Speed–Density Relationship

The two previous models contain stationary mean speed–density relationships (i.e., mean speed is supposed to adjust instantaneously to traffic density whenever a density change occurs). A more realistic description of traffic flow is expected if this hypothesis is removed by introducing a small time delay τ in Eqn. (9) as proposed by Payne (1971):

$$v(x, t+\tau) = V[\rho(x+\Delta x, t)] \quad (15)$$

Expanding the left-hand side of Eqn. (15) in a Taylor series with respect to τ and the right-hand side with respect to Δx, using the same assumptions as in Sect. 4 and rearranging terms gives

$$\tau \frac{dv}{dt} = V(\rho) - v - \frac{v}{\rho} \frac{\partial \rho}{\partial x} \quad (16)$$

where the arguments (x, t) are suppressed for convenience. dv/dt is the acceleration of an observer moving with the traffic stream:

$$\frac{dv}{dt} = \frac{\partial v}{\partial t} + v \frac{\partial v}{\partial x} \quad (17)$$

Combining Eqns. (16) and (17) gives

$$\frac{\partial v}{\partial t} = -v \frac{\partial v}{\partial x} + \left[V(\rho) - v - \frac{v}{\rho} \frac{\partial \rho}{\partial x} \right] \tau \quad (18)$$

where τ is an additional constant parameter. Equations (1), (4) and (18) constitute a complete traffic flow model for the nondiscretized case which describes dynamic phenomena (particularly in congested traffic) with higher accuracy than previous models.

Discretization of Eqn. (18) leads to the nonlinear difference equation

$$v_i(k+1) = v_i(k) + \frac{T}{\tau} \{ V[\rho_i(k)] - v_i(k) \}$$
$$+ \frac{T}{\Delta_i} v_i(k) [v_{i-1}(k) - v_i(k)]$$
$$- \frac{vT}{\tau \Delta_i} \frac{\rho_{i+1}(k)/\lambda_{i+1} - \rho_i(k)/\lambda_i}{\rho_i(k)/\lambda_i + \kappa} \quad (19)$$

where the constant parameter κ is again added in order to keep the last term limited when ρ_i becomes small.

In this way, a further dynamic traffic flow model is obtained which consists of Eqns. (5), (14) and (19). Further interesting aspects of macroscopic traffic flow modelling are given in Grewal 1974, Grewal and Payne 1976, Papageorgiou 1983, Papageorgiou et al. 1983, Babcock 1984, Kühne 1984, and Ross 1988.

6. Weaving Phenomena

Weaving phenomena occur when groups of vehicles are forced by the road layout to change lanes in one or both directions. Weaving leads to "energy losses" due to the corresponding acceleration or deceleration of groups of cars and should be appropriately included in macroscopic models.

6.1 Weaving due to Entering or Exiting Traffic

With reference to Eqn. (18), it is assumed that lateral traffic enters at point x with injection rate $r(x, t)$ (in vehicles per hour per kilometer) at a speed v_e. Since generally $v_e \leq v$, the rate of change $\partial v / \partial t$ is influenced by an additional term $(v_e - v) r/\rho$. Then, if it is assumed that the entering speed is given by $v_e = \delta' v$ ($0 \leq \delta' \leq 1$), the additional term becomes $-\delta v r/\rho$ ($0 \leq \delta \leq 1$). Eventually, Eqn. (18) may be changed to become

$$\frac{\partial v}{\partial t} = -v \frac{\partial v}{\partial x} + \left[V(\rho) - v - \frac{v}{\rho} \frac{\partial \rho}{\partial x} \right] / \tau - \frac{\delta v r}{\rho} \quad (20)$$

where δ is assumed to be a constant parameter depending upon the layout of the ramp (Papageorgiou et al. 1989). After space discretization, the constant parameter κ may be added to the denominator of the new term as was done for Eqn. (19). Hence, recalling $r = r_i/\Delta_i$, the discretized form of the additional term becomes $-\delta(T/\Delta_i) v_i r_i/(\rho_i + \kappa)$. A similar term has been proposed by Cremer and May (1986).

At this point, it is interesting to note that Eqns. (1), (4) and (18) have a perfect analogue in the description of unsteady water flow in open channels (e.g., Ligget 1975). The analogue of Eqn. (18), which is called the momentum equation, is derived from the conservation of momentum in water flow. The additional term to be included in Eqn. (18) is in case of lateral inflow and has the same structure as in the momentum equation of water flow.

At off-ramps, exiting cars generally decelerate leading to lower mean speeds in the corresponding freeway sections. Hence, the impact of off-ramp volumes may be described in the same way as for on-ramp volumes, with r being replaced by s.

6.2 Weaving due to Lane Drop

Lane changing due to a lane drop may be understood as a phenomenon similar to on-ramp traffic injection into the mainstream. Eventually, an additional term in Eqn. (19) may preserve the structure of the term introduced in Sect. 6.1. In this case, r_i should be replaced by the lane changing traffic volume which is roughly equal to $q_i(\lambda_i - \lambda_{i+1})/\lambda_i$ with $\lambda_{i+1} < \lambda_i$. However, if q_i is substituted using Eqn. (14) with $\alpha = 1$, the additional term $-\phi(T/\Delta_i)v_i(k)^2(\lambda_i - \lambda_{i+1})/\lambda_i$ is obtained, where ϕ is a new constant parameter. It is interesting to note that a similar term appears in the modelling of unsteady water flow in open channels. Of course, speed reduction due to lane drop is expected to be more pronounced for dense or congested traffic. For this reason, the additional term is modified to give

$$-\phi \frac{T}{\Delta_i} \frac{\lambda_i - \lambda_{i+1}}{\lambda_i} \frac{\rho_i(k)}{\rho_c} v_i(k)^2 \quad (21)$$

This term is added to the right-hand side of Eqn. (19) where $\lambda_{i+1} < \lambda_i$.

7. Limitations of the Models

Aggregate models describe traffic flow in one dimension (along the road axis). Although traffic flow along the road width is generally inhomogeneous (e.g., trucks and slow vehicles generally use the near-side lanes of a highway), traffic flow dynamics are described accurately enough by one-dimensional models under certain conditions. It is difficult to state these conditions precisely. However, it is possible to identify some traffic situations for which one-dimensional models will probably be inaccurate unless appropriate modifications are introduced.

One of these traffic situations is off-ramp congestion occurring when, for whatever reason, acutal off-ramp throughput is inferior to the amount of traffic wishing to exit. In this case, one or more near-side lanes of the mainstream may be blocked while traffic on off-side lanes may be fluid.

A second traffic situation that probably leads to model failure may occur when trucks are not allowed to use the outside lane. If the number and percentage of trucks exceeds certain limits, the lanes used by trucks may be blocked while traffic in the outside lane may be fluid. Finally, the same problem may arise in the case of special lanes dedicated to buses, taxis and so on.

Suitable extensions of the model equations may be necessary in these cases in order to facilitate adequate descriptions of traffic phenomena. Extensions towards two-dimensional traffic flow modelling are considered in Michalopoulos et al. (1984). The speed–density relationships of the various models presented in this article do not take into account the percentage of trucks in traffic. Different truck percentages do not require a structural change in the models, but results may differ quantitatively for different truck proportions. Cremer (1976) and Cohen (1981) have proposed suitable extensions of the fundamental diagram so as to take into account the actual percentage of trucks.

8. Validation

The mathematical models presented in Sects. 3, 4 and 5 constitute a hierarchy of models in the sense that the model in Sect. 3 may be viewed as a special case of the one in Sect. 4, for $\nu = 0$, and both models may be viewed as special cases of the one in Sect. 5, for $\tau = 0$. Hence, generally, each model is potentially more accurate than its predecessor. This has been confirmed

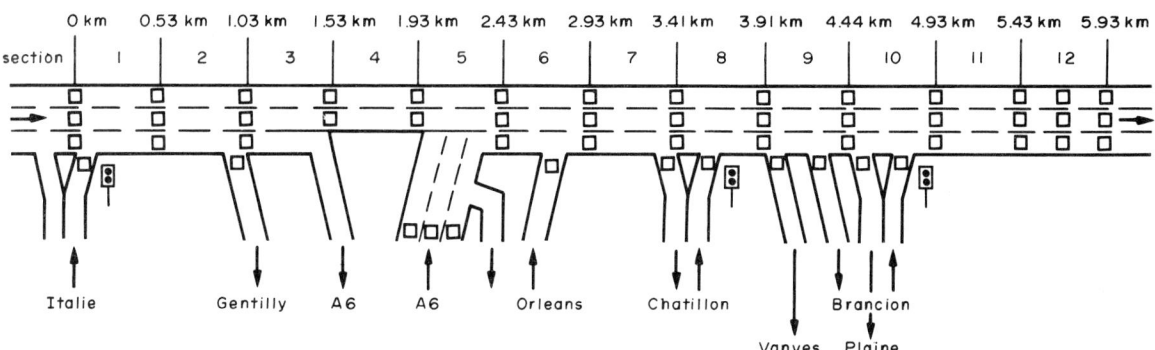

Figure 2
A stretch of the Boulevard Périphérique

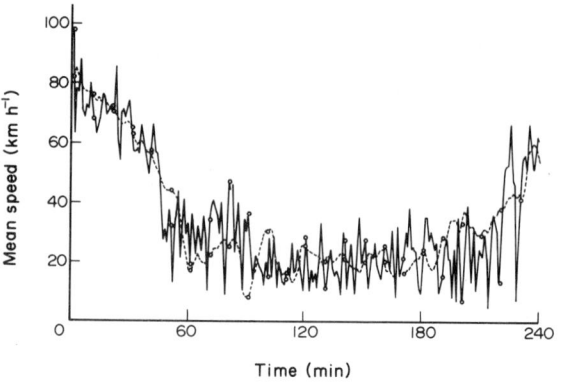

Figure 3
Mean speed trajectory measured (——) and calculated by the model (– – –) at an internal site

in some validation studies by comparison with real traffic data (Cremer and Papageorgiou 1981, Beskos et al. 1984, Papageorgiou et al. 1989).

Validation of the macroscopic model of traffic flow presented in Sect. 5 has been reported with excellent results for the Hollywood Freeway in Los Angeles (Payne et al. 1973), for a German Autobahn (Cremer and Papageorgiou 1981), for the Santa Monica Freeway in Los Angeles (Cremer and May 1986) and for the Boulevard Périphérique in Paris (Papageorgiou et al. 1990).

As an example, Fig. 2 presents a stretch of Boulevard Périphérique in Paris that has a length of 6 km and includes five on-ramps and six off-ramps. For application of the macroscopic model of Sect. 5, the stretch was subdivided into 12 sections of 500 m in length. The model parameters were calibrated so as to achieve accurate results for a representative set of different traffic conditions. The modelling equations were fed with measured boundary values and the traffic state inside the considered stretch was reproduced by the macroscopic model with high accuracy. Figures 3 and 4

present a time response of mean speed and traffic volume trajectories for an internal site of the considered stretch (section 3 in Fig. 2). It may be seen that calculations correspond fairly acurately to real measurements even under heavily congested conditions. Sensitivity investigations indicate that the macroscopic model is fairly insensitive with respect to moderate variations of all model parameters except for v_f and ρ_c; that is, the parameters of the fundamental diagram (Papageorgiou et al. 1990). The values of these two parameters depend upon the particular geometrical and traffic composition conditions. Therefore, they should be carefully calibrated for each particular application of the model.

See also: Car-Following Models; Kinematic Wave Theory; Prediction of Traffic Flow; Simulation Programs, Macroscopic

Bibliography

Babcock P S 1984 Improved dynamics and performance for the FRECON freeway simulation model Technical Document UCB-ITS-TD-84-1. Institute of Transportation Studies, University of California, CA

Beskos D E, Okutani I, Michalopoulos P G 1984 Testing of dynamic models for signal controlled intersections. *Transp. Res. B* **18**, 397–408

Cohen S 1981 Effet des poids lourds sur la capacité des autoroutes urbaines. Internal Report. INRETS, Arcueil, France

Cremer M 1976 A new scheme for traffic flow estimation and control with a two component model. *Proc. 3rd IFAC/IFIP/IFORS Int. Conf. Control in Transportation Systems*. Pergamon, Oxford, pp. 29–37

Cremer M, May A D 1986 An extended traffic flow model for inner urban freeways. *Proc. 5th IFAC/IFIP/IFORS Int. Conf. Control in Transportation Systems*. Pergamon, Oxford

Cremer M, Papageorgiou M 1981 Parameter identification for a traffic flow model. *Automatica* **17**, 837–43

Grewal M S 1974 Modeling and identification of freeway traffic systems. Ph.D. thesis, University of Southern California

Grewal M S, Payne H J 1976 Identification of parameters in a freeway traffic model. *IEEE Trans. Syst., Man Cybern.* **6**, 176–85

Kühne R D 1984 Macroscopic freeway model for dense traffic—stop–start waves and incident detection. *9th Int. Symp. Transportation and Traffic Theory*. VNU, Utrecht, The Netherlands, pp. 21–42

Ligget J A 1975 Basic equations of unsteady flow. In: Mahmood K, Yevjevich V (eds.) *Unsteady Flow in Open Channels*. Water Resources Publications, Fort Collins, CO, pp. 29–62

Lighthill M J, Whitham G B 1955 On kinematic waves I. Flood movement on long rivers. *Proc. R. Soc. London, Ser. A* **229**, 281–316

Lighthill M J, Whitham G B 1955 On kinematic waves II. A theory of traffic flow on long crowded roads. *Proc. R. Soc. London, Ser. A* **229**, 317–45

May A D Jr, Keller H E M 1967 Non-integer car-following models. *Highw. Res. Rec.* **199**, 19–32

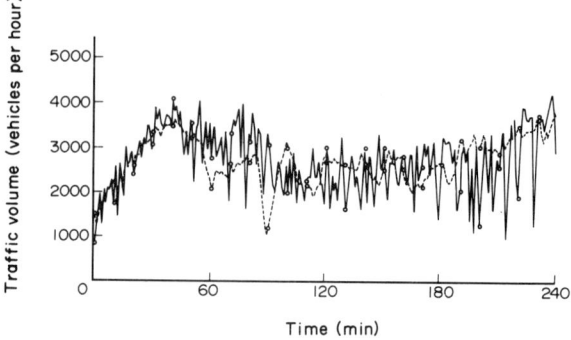

Figure 4
Traffic volume trajectory measured (——) and calculated by the model (– – –) at an internal site

Michalopoulos P G, Beskos D E, Lin J-K 1984 Analysis of interrupted traffic flow by finite difference methods. *Transp. Res. B* **18**, 409–21

Papageorgiou M 1983 *Applications of Automatic Control Concepts to Traffic Flow Modeling and Control.* Springer, Berlin

Papageorgiou M, Blosseville J M, Hadj-Salem H 1989 Macroscopic modelling of traffic flow on the Boulevard Périphérique in Paris. *Transp. Res. B* **23**, 29–47

Papageorgiou M, Blosseville J M, Hadj-Salem H 1990 Modelling and real-time control of traffic flow on the southern part of Boulevard Périphérique in Paris. I. Modelling. *Transp. Res. A* **24**, 345–59

Papageorgiou M, Posch B, Schmidt G 1983 Comparison of macroscopic models for control of freeway traffic. *Transp. Res. B* **17**, 107–16

Payne H J 1971 Models of freeway traffic and control. *Simulation Council Proc.* **1**, 51–61

Payne H J, Thompson W A, Isaksen L 1973 Design of a traffic-responsive control system for a Los Angeles freeway. *IEEE Trans. Syst. Man Cybern.* **3**, 213–24

Ross P 1988 Traffic dynamics. *Transp. Res. B* **22**, 421–35

Wardrop J G 1952 Some theoretical aspects of road traffic research. *Proc. Inst. Civ. Eng.* **1**, 325–62

M. Papageorgiou and G. Schmidt
[Technische Universität München,
Munich, Germany]

Fuel Conservation: Air Transport

In Western Europe, the average length of a civil aircraft flight is in the region of 300 nautical miles (550 km). Accordingly, cost-efficient control of the traffic calls for an integrated approach covering the entire flight and including the departure time among the control variables. In view of the present airspace structure and the level of coordination required, this can be seen only as a long-term goal. In contrast, the zone of convergence (ZOC) concept, which integrates the control of inbound traffic over an extended area including and surrounding a main terminal, could constitute a short-term air traffic control (ATC) contribution to economy in air transport.

1. Flight Economy and Transit Cost

1.1 Optimum Aim vs Present Situation

Civil aircraft operators aim to conduct their flights at minimum cost within the constraints resulting from the presence of the military (restricted airspace), from the existence of other traffic (air traffic control procedures and restrictions), from environmental constraints (proximity of residential areas, leading, in particular, to noise abatement) and possibly local or temporary policies. Without such restrictions, operation at minimum cost would mean direct uninterrupted climb, cruise at optimum altitude and speed and direct descent from cruise level to the runway without steps or delays, the whole flight being conducted from departure to arrival along the shortest air route. The present situation is far from this optimum and deviations are considerable (Benoît and Devry 1981, Renteux and Schroeter 1982).

In Western Europe, more specifically in the region covered by the Eurocontrol route charges system, the excess cost of flights (actual as against ideal) is estimated to be equivalent to at least 2×10^6 t of fuel per year: of this, half is for the excess route lengths and half is for the use of nonoptimum trajectories in the present air route network. Further, as the average length of a flight in this area is in the region of 300 nautical miles (550 km), even short delays have an appreciable influence on relative flight economy.

1.2 Equivalent Transit Cost

The direct cost C of a flight includes the fuel consumption and a term that increases with the flight duration t. If, for ease of discussion, an average value of the direct operating cost per unit time is properly defined and designated p, the direct cost of the flight can be expressed in linear form:

$$C = zf + tp \quad (1)$$

where f represents the cost of unit mass of fuel.

For comparison purposes, it is convenient to introduce an equivalent cost C_f expressed in mass of fuel in the form

$$C_f = z + t\alpha \quad (2)$$

where $\alpha = p/f$, the ratio of the unit flight time direct operating cost to the unit fuel cost, measures the importance of the time component in the direct cost. Obviously, for a given type of aircraft, this cost value will vary in accordance with the operator's policy. For airlines that are members of the International Air Transport Association (IATA), the cost is shared roughly equally, on average, between the fuel consumption and flight time components, although this ratio may vary appreciably from one type of aircraft to another (Cunningham 1982).

1.3 Selection of Flight Parameters

For a specific aircraft (identified by type, series, power plant and level of maintenance) connecting two cities along a given route, a set of parameters (in particular, power setting, speed components and cruise altitude) exists that minimizes the cost of the flight. The resulting relationship is illustrated in Figs. 1 and 2 for a short-haul aircraft representative of the Boeing B-737 class operating over 300 nautical miles (550 km), as measured along the actual route between the exit gate of the departure airport and the entry into the destination terminal.

These diagrams are self-explanatory. There is a preferential transit time which results from the optimum flight conditions, namely cruise altitude, climb, cruise and descent speed profiles. If the time of arrival and/or

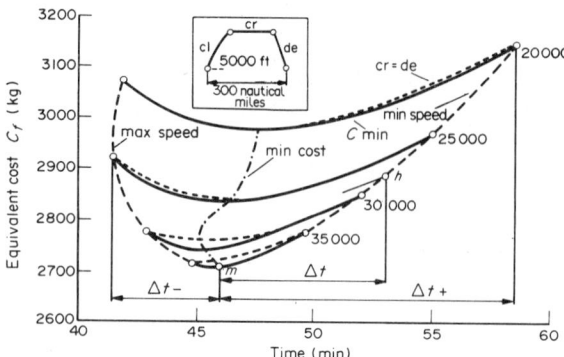

Figure 1
Cost–time relationship (climb–cruise–descent); figures at right-hand side denote altitude (in feet)

the cruise altitude are constrained, as a result of the traffic situation for instance, the flight should be conducted either along the envelope or at the intersection of the transit time value and the cleared cruise altitude, depending on the type and importance of the restrictions. With reference to the optimum optimorum (point m in Fig. 1), the range of the transit time would extend between approximately 5 min advance and 12 min delay, this maximum delay corresponding to an increase in cost of approximately 17%.

For shorter flights, the optimum cruise altitude normally takes an intermediate value and, as a result, plays a slightly different role in the cruise level allocation, as discussed by Benoît and Swierstra (1982a).

1.4 Integrated Zone of Control

It is obvious that, in Western Europe in particular, cost-efficient control of air traffic should cover the complete flight, integrating the departure time among the flight control variables (Benoît *et al.* 1977). As indicated, the potential benefits are considerable. Nevertheless, the historical development of ATC has been such that at present, even within one country, the airspace structure often prevents the control from covering the entire flight. This fact leads us to envisage an intermediate step, implying an appreciably less demanding level of control coordination and covering the whole of the traffic inbound for a given main airport.

2. Control of Inbound Flights

At present, when inbound aircraft are handed over to the approach control they are already relatively close to the runway (say 20–40 nautical miles (35–70 km)). The organization of the landing sequence is then made on the basis of the runway capacity available and with the aim of making full use of it in saturation periods. However, as this is done in a way practically independent of the upstream conditions, it leads necessarily to an appreciable excess cost, as estimated *a priori* and measured in actual environments.

2.1 Dynamics of Inbound Flights

In an extended area including and surrounding a main terminal, as depicted in Fig. 3, the inbound flights, except for a few connections from secondary airports, are essentially made up of the cruise or part thereof, the en route descent and the final descent and landing phases. For a given cruise altitude, the cost of a direct flight depends on the cruise and descent speeds selected (see Fig. 4, for a long-range, wide-bodied aircraft of the McDonnell Douglas DC-10 class).

Among all possible cruise–descent speed combinations only those corresponding (a) to the locus of minimum cost vs time transit and (b) to the smooth cruise–descent speed transition profile would be of practical interest, all the others leading to an unjustified excess cost.

It has been shown (Benoît and Swierstra 1980a, 1982b) that these two families of profiles are close to each other in terms of flight cost (or consumption), although their corresponding speed components may differ appreciably (see Fig. 5, for an aircraft of the McDonnell Douglas DC-10 class).

2.2. Control of Transit Time

Within the operationally acceptable speed range, the duration of the transit from entry to touchdown can be controlled to ensure (a) meeting a particular landing time window at (b) minimum cost. The range of control for direct flight increases with the cruise–descent flight extent. The term preferential refers to the optimum (i.e., lowest possible) cost of transit.

It is of interest to compare the transit costs when using cruise–descent speed control as against present practice (Benoît and Swierstra 1980b).

2.3 Comparison of Transit Time Control Procedures

When an aircraft arrives at an assembly point or possibly at a previous fix, it may be delayed to meet the

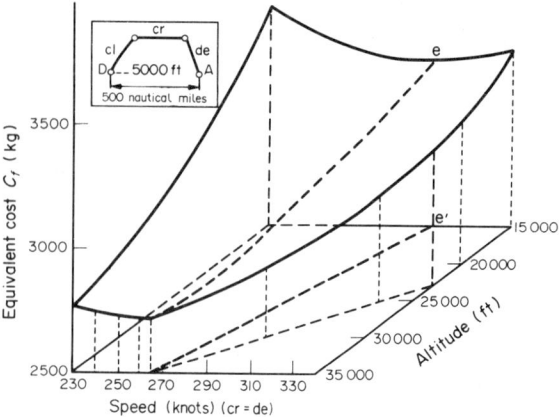

Figure 2
Cost–speed relationship (climb–cruise–descent)

Fuel Conservation: Air Transport

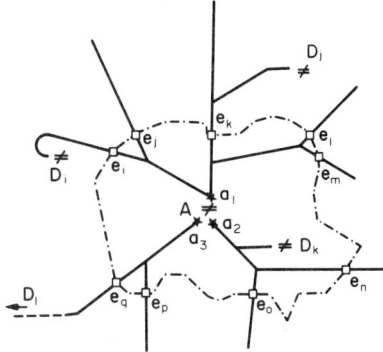

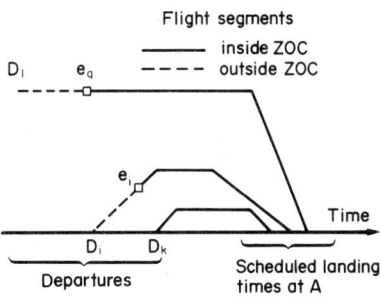

Figure 3
Schematic structure of a ZOC

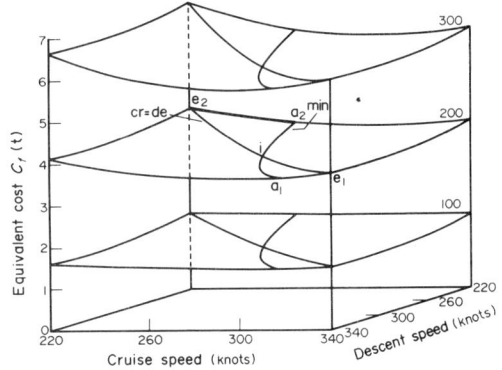

Figure 4
Cost as a function of speed for flight lengths of 100, 200 and 300 nautical miles, respectively, cruise + descent

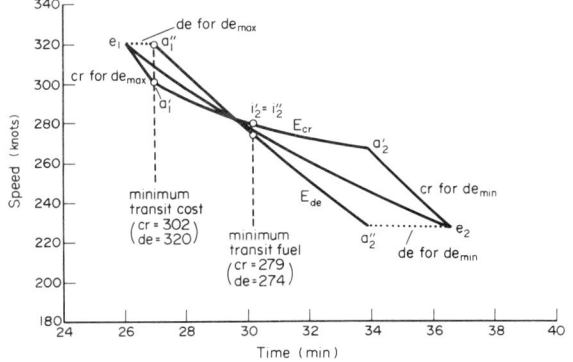

Figure 5
Minimum cost and smooth cruise–descent transition speed profiles (R zone 200 nautical miles; FL 300); E, envelope of fuel/cost–time profile

landing slots available at the runway. Currently, such delaying techniques may include:

(a) diversion at cruise level,
(b) holding at high altitude,
(c) path stretching at low altitude, or
(d) holding at low altitude, for example flight level (FL) 50. Here, FL is the pseudoaltitude, expressed in hundreds of feet, read on an altimeter set at 1.0135×10^5 Pa irrespective of the barometric reference pressure, so FL 50 means that an altimeter set at 1.0135×10^5 Pa indicated 5000 ft (1500 m).

The impact on flight cost of such procedures has been analyzed and compared with the effect of a technique based on the use of cruise–descent speed profile control in a variety of flight conditions. Obviously, the range of applicability and consequently the effects depend on the specific flight configuration (cruise altitude and cruise–descent speed control extent) and the performance characteristics of the aircraft concerned.

A sample of the results obtained is given in Fig. 6 for an aircraft of the McDonnell Douglas DC-10 class cruising at an altitude of 30 000 ft (9000 m) over a 150

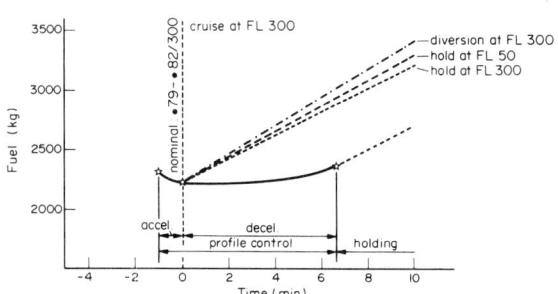

Figure 6
Comparison of transit time control techniques

nautical mile (280 km) cruise–descent segment. The diagram shows the fuel consumption associated with four control techniques (although often applied, path stretching at low altitude has not been explicitly considered, in view of its high cost) and the range in transit time which can be accommodated using cruise–descent speed control in cases of both advances and delays. The nominal, preferential profile used was recommended by a particular European airline.

The conclusions from the analysis made are as follows.

(a) The three conventional techniques considered are roughly equivalent in terms of fuel consumption, although holding at cruising level is slightly less penalizing than holding at low altitude which, in turn, requires rather less fuel than a diversion under cruise conditions. For this reason, the subsequent comparisons will be made with holding at high altitude.

(b) Conventional delaying techniques require appreciably more fuel than the application of transit speed profile control.

(c) When the imposed transit time can no longer be accommodated through the application of cruise–descent speed control, obviously one of the other techniques, preferably holding at high altitude, will have to be applied.

The comparison of cruise–descent speed control with holding at cruise altitude is shown in Fig. 7 for the same configuration as in Fig. 6.

When generalized to cover a variety of aircraft ranging from a short-haul lightweight (e.g., Fokker F-28) to

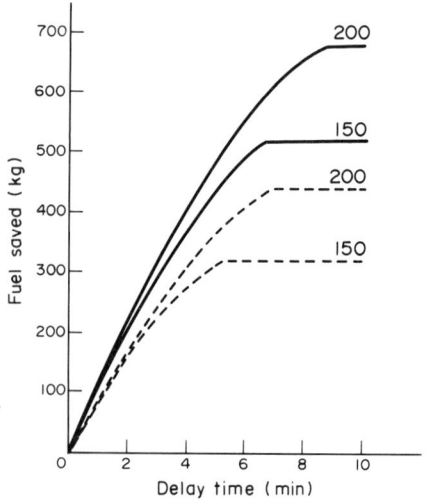

Figure 7
Fuel saving when using profile control instead of holding at cruise level: —— entry FL 300; --- entry Fl 350; figures on curves denote distance to touchdown (in nautical miles)

a long-range wide-bodied aircraft (e.g., Boeing B-747), the benefit of cruise–descent control over holding at high altitude is appreciable. For a cruise–descent extent of 150 nautical miles (280 km) (from entry to touchdown) and a 5 min delay imposed by ATC, the benefit is in the region of 20–25% of the total amount of fuel that would be required for the transit.

3. Dynamic Control of Aircraft for Minimum Cost Operation

3.1 Control in a ZOC

The ZOC concept aims at minimizing the overall transit cost for all aircraft operated in the area considered. The area covered is essentially anisotropic, as its extent varies along each route, reflecting the actual geography of the area (Fig. 3). At the boundaries, no particular requirement is imposed except for freedom from conflict as currently practised. In the zone, whenever possible, aircraft are scheduled in accordance with their preferential trajectories (optimum climb–cruise–descent speed profile and cruise altitude).

If the interests of individual aircraft conflict, ATC will aim for a global minimum cost solution. Procedurally, this is pursued through decoupling, as far as possible, of inbound and outbound traffic. For the aircraft inbound to the main terminal in the zone, ATC sets up a landing sequence based on both the "first come, first served" (FCFS) rule and the preferential transit times. Global cost minimization, considering all aircraft inbound to the terminal, could result from small modifications of the preferential landing order.

On the basis of the above considerations and using the cruise–descent transit speed as a means of controlling the transit time, the following procedure is proposed.

(a) For each aircraft entering the zone, a tentative landing time is computed, aiming at minimum transit cost for this particular aircraft. The associated speed profile, based on the smooth cruise–descent transition technique (Benoît and Swierstra 1982b) is transmitted to the aircraft immediately after entry.

(b) As long as the aircraft is not yet descending, the descent speed remains available as a control variable in the optimization process. To minimize the global cost in the zone, shifts in the landing sequence (limited to two positions) are considered. The resulting top-of-descent location and descent speed profile are sent to the aircraft, say 30–60 s before the transition from cruise to descent.

Initial investigations of the ZOC concept concentrated on cost minimization mainly through optimization of the landing sequence. A preliminary approach was based on a discrete number of possible transit speed profiles. The "branch-and-bound" optimization technique used was found to be compatible with real-time operation (Imbert 1980). Further, the initial computer program pointed to a number of difficulties which lead us to reconsider certain elements, in particular the ground–air control procedure and its implications for the optimization process at all levels of runway saturation.

With regard to present practice, it has become apparent that a simple organization of the landing sequence aiming to use the maximum available runway capacity duly combined with cruise–descent speed control as discussed in Sect. 2, would be responsible for the major part, say about 90%, of the potential fuel or cost savings.

Accordingly, the present developments make use of these findings, applying additionally an analytical technique based on aircraft trajectory and consumption performance data presented in an essentially parabolic polynomial form (PARZOC) (Benoît and Swierstra 1982e). The resulting programs are compatible with available microprocessor technology, and are suitable for on-line operation.

The concept has been exposed to a number of experiments aiming at assessing its compatibility with real-time and on-line operation in actual cockpit and/or ATC environments.

3.2 Concept Assessment

To test the compatibility of the proposed procedure with normal aircraft operation, two series of experiments were organized in cooperation with Lufthansa and Sabena, using the B-737, A-300 and DC-10 flight simulators (Benoît and Swierstra 1982c). The results indicated that the proposed procedure is adequate to control the time of arrival over an assembly point with a high degree of accuracy. When all the inaccuracies inherent in the ground and air coordinated control system, including human factors, are taken into account, the maximum discrepancy in arrival time after a 200 nautical miles (370 km) flight distance remains within 26 s. It is expected that over such a distance, one or possibly two updates could compensate for wind prediction errors.

In parallel, preliminary on-line experiments have been conducted during actual regular scheduled flights in cooperation with British Airways and the ATC authorities of Belgium and of the UK aircraft (Trident 3B and BAC 111) inbound to Brussels National, coming from Birmingham, London and Zurich, were given precise directive as to where to initiate the descent, the relevant decision being taken 80 nautical miles (150 km) from the runway. The meteorological information used was that available at the control center in Brussels. On entry to the ZOC, the mass of the aircraft was passed to the ground using the ATC voice channel. The integrated flight prediction system (IFLIPS) was used for the profile definition and prediction (Benoît and Swierstra 1982d).

For a series of some 20 flights, the maximum discrepancy observed at the 2000 ft (600 m) point was in the region of 20 s. These results were obtained without update or correction over an 80 nautical mile (150 km) flight distance.

Furthermore, the concept has been the subject of an ATC real-time simulation, using the facilities available at the Eurocontrol experimental centre. Controller reaction was extremely positive, the compatibility with operational life raised no essential difficulty and the tools offered appeared to reduce the human workload appreciably (Miller 1982).

Additional sources of information on the state of validation in 1986 are provided in the bibliography.

3.3 Compatibility with Advanced On-Board Equipment

The control technique proposed appears to be compatible with present on-board equipment. It does not require four-dimensional navigation or advanced on-board computing facilites. However, in the future, such facilities will become available in the cockpit and accordingly questions may arise regarding their impact on the ZOC concept.

The flight management system (FMS) is certainly suitable to advise the pilot on the most cost-effective transit procedure. However, these directives cannot take into account the traffic situation at the destination airport or any possible conflict resolution action envisaged by ATC. Therefore, when aircraft are inbound to a busy terminal, adequate coordination with ATC is essential. Accordingly, for such aircraft, it would be sufficient for ATC to transmit the ZOC directives in the form of required passing conditions (times and altitudes over way points), thus leaving it to the FMS to advise the pilot on how to implement the clearance in the most economical manner.

4. Fuel Burn in Extended Terminal Areas and Potential Savings

4.1 ZOCs of London and Brussels

To validate the fuel-saving potential of the ZOC concept, two data collections were organized during which radar tracks, flight plan information and meteorological data were recorded for a period of 2 h. One concentrated on flights inbound to London Heathrow, representative of one of Europe's highest-density airports (Fig. 8) and the other on Brussels National, which can be considered as a medium-density terminal. The density of the recorded traffic samples was such that the maximum number of aircraft heading simultaneously for the terminal was eight in the Brussels area and 18 in the London zone.

4.2 Assessment of Nugatory Fuel Consumption

Techniques have been developed for the accurate assessment of the amount of fuel used by an aircraft on an inbound flight (Benoît and Swierstra 1982d). The fuel consumption in the two zones for the traffic samples considered is presented in Fig. 9, which shows the amount of fuel consumed using the following four different transit procedures.

(a) Observed consumption: this constitutes an estimate of the evolution of the actual amount of fuel consumed by all aircraft in the traffic sample.

(b) Use of speed control: using the cruise–descent speed profile (Fig. 2) a minimum-cost profile is established for each aircraft on the basis of the actual entry conditions and the observed landing time (accordingly, the observed landing sequence established by the air traffic controller has been maintained). The results obtained represent an estimate of the fuel burn if this control method had been applied. It should be noted that some additional savings could have been expected if the landing sequence had been optimized at some earlier stage.

(c) Pilot-selected profile: in these investigations the "pilot-selected" profile constitutes an estimate of the profile recommended by the individual aircraft operator (i.e., the minimum-cost transit speed profile). The curve represents the fuel consumed if the aircraft had flown along this profile during its transit through the zone. The pilot-selected profile is extracted for each flight from the observed radar and meteorological data during the cruise phase and the upper part of the en route descent phase.

(d) Minimum-fuel profile: the minimum-fuel curve represents the total fuel consumption if all aircraft had flown on the minimum cruise–descent speed profiles without interference from ATC.

The results presented in Fig. 9 show that for the London zone the absolute minimum transit fuel is about 30% lower than the amount of fuel actually burnt. However, this comparison is purely theoretical and does not take account of practical aspects such as runway occupancy and possible conflicts along the route. Accordingly, it constitutes an upper bound to

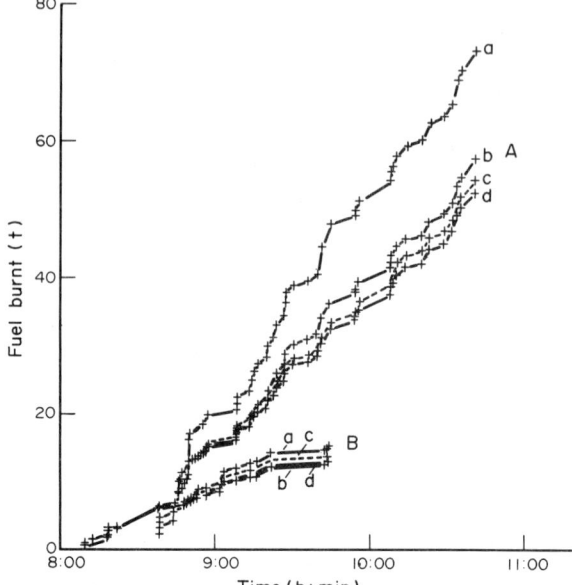

Figure 9
Fuel burnt in extended terminal areas: A, London Heathrow; and B, Brussels National. Transit speed profiles: a, actual observation; b, use of speed control; c, pilot-selected; and d, minimum fuel

the fuel saving that might be achieved. The same applies to the pilot-selected profile, on which the consumption would have been approximately 28% lower than the observed value. It indicates, however, that the pilots, if not hindered by ATC, do indeed tend to fly very economically; it was found that the average transit time using the pilot-selected speed profile was considerably lower than the corresponding time computed for the minimum-fuel profile, while the total consumption was only slightly higher.

The investigations summarized in this diagram serve principally to confirm that the application of cruise–descent speed control, as envisaged in the ZOC concept, may indeed make a major contribution to the economy of air transport. Using this technique, it is shown that if transit times and landing order are kept as observed, savings in fuel consumption of approximately 22% in the case of a high-density terminal and of approximately 12% in the case of a medium-density terminal can be obtained. These percentages refer to the total amount of fuel burnt by all inbound traffic in the zones considered. These estimates of potential savings in fuel consumption are in line with the theoretical results presented in Sect. 2.

5. Future Developments

Techniques compatible with real-time and on-line operations are currently available to provide ATC with accurate estimates of time, fuel consumption and cost

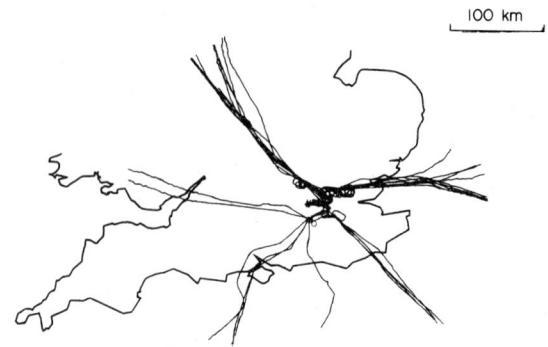

Figure 8
Map illustrating flights inbound to London Heathrow

of transit for any aircraft requested to follow a particular control pattern between two given positions. On the basis of such techniques, a global approach is proposed, integrating en route, approach and landing phases of inbound flights. At all times, expedition of traffic is ensured in the most economic manner, safety being considered as a prime constraint. To this effect, adequate use is made of the available on-board equipment, operational capabilities of the aircraft and the potential of tools available for conducting on-line the prediction, control and economy assessment of flight profiles.

In the present European context, the possible generalization of this approach from airport to airport may be considered only as the ultimate goal. In contrast, the concept of a ZOC could well be considered as an essential short-term contribution to air transport economy. The traffic is controlled over a large area including and surrounding at least a main terminal and extending over about 180–550 km. The relevant area is referred to as a ZOC. No asumption is required regarding the incoming traffic, except that it is, as usual, free of conflicts when arriving in the zone (i.e., when the transfer of control occurs).

The developments undertaken within the ZOC concept provide an adequate framework for realizing the following objectives:

(a) appreciable increase in safety;
(b) reduction of controller workload;
(c) maximum use of available ATC capacity, particularly at the runway;
(d) minimum flight operating cost for the overall traffic in the zone;
(e) as a consequence, consumption of fuel in line with operators' requirements; and
(f) no individual penalty resulting from the use of a particular type of aircraft.

See also: Air Traffic Control Near Airports; Fuel Conservation: Road Transport; Fuel Conservation: Sea Transport

Bibliography

Attwooll V, Benoît A 1985 Fuel economies effected by the use of FMS in an advanced TMA. *J. Navig.* **38**(1)
Attwooll V, Rickman D 1984 A study of conflicts within a zone of convergence. SPACDAR GEDARC-11. Eurocontrol, Brussels
Benoît a 1986 4-D control of current air carries in the present environment. Eurocontrol Report No. 862013. Eurocontrol, Brussels
Benoît A, Devry H 1981 Impact of excess route lengths on fuel consumption in a European air-route network. Eurocontrol Report No. 802021. Eurocontrol, Brussels
Benoît A, Swierstra S 1980a Optimum use of cruise/descent control for the scheduling of inbound traffic. *Proc. Int. Conf. Fuel Economy in the Airlines*. Royal Aeronautical Society, London
Benoît A, Swierstra S 1980b A minimum fuel transit procedure for the control of inbound traffic. Eurocontrol Report No. 802007. Eurocontrol, Brussels
Benoît A, Swierstra S 1982a The dynamic control of aircraft for minimum cost operation. *Proc. Int. Semin. ATC Contribution to Fuel Economy.* Eurocontrol Institute of Air Navigation, Luxembourg
Benoît A, Swierstra S 1982b Dynamic control fo inbound flights for minimum cost operation. *AGARD Conf. Proc.* **321**
Benoît A, Swierstra S 1982c A ground/air coordinated control procedure for minimum cost operation over an extended area. *Today and Tomorrow—Mini and Micro Computers in Airline Operations.* Royal Aeronautical Society, London
Benoît A, Swierstra S 1982d Available tools for the prediction, control and economy assessment of flight profiles. *Proc. Int. Seminar on ATC Contribution to Fuel Economy.* Eurocontrol, Luxembourg
Benoît A, Swierstra S 1982e Simulation of air traffic control operation in a zone of convergence (2): Aircraft (PARZOC) performance data. Eurocontrol Report No. 812031-2. Eurocontrol, Brussels
Benoît A, Swierstra S 1984 Air traffic control in a zone of convergence: Assessment with Belgian airspace. Eurocontrol Report No. 842009. Eurocontrol, Brussels
Benoît A, Swierstra S 1986 A simulation facility for assessing the next generation of 4-D air traffic control procedures. Eurocontrol Report No. 862017. Eurocontrol, Brussels
Benoît A, Swierstra S, Cox M, Storey J 1977 An evolutionary application of advanced flight path prediction capability to ATC. *Proc. Int. Conf. Electronic Systems and Navigation Aids.* SEE/GIEL, Paris
Benoît A, Swierstra S, De Wispelaere R 1986 Next generation of control techniques in advanced TMA: Automatic assistance for the controller/pilot dialogue. *AGARD Conf. Proc.* **401**
Cunningham P 1982 The impact of fuel prices on airlines' economy and operations. *Proc. Int. Semin. ATC Contribution to Fuel Economy.* Eurocontrol, Luxembourg
Imbert N 1980 Application de l'algorithme d'ordonnancement du traffic aérien à la zone de Londres Heathrow. Report No. CERT-DERA-1/7240, CERT, Toulouse
Miller W 1982 Adaptation of simulation tools to the study of fuel consumption problems at the Eurocontrol Experimental Centre. *Proc. Int. Seminar ATC Contribution to Fuel Economy.* Eurocontrol Institute of Air Navigation, Luxembourg
Renteux J L, Schroeter H 1982 Fuel saving in air transport, Eurocontrol Report No. 822007. Eurocontrol, Brussels

A. Benoît and S. Swierstra
[European Organisation for the Safety of Air Navigation, Brussels, Belgium]

Fuel Conservation: Road Transport

Road transport plays an important economic role in all developed countries and accounts for an appreciable proportion of the total consumption of hydrocarbons as a source of energy. The consumption due to road

Fuel Conservation: Road Transport

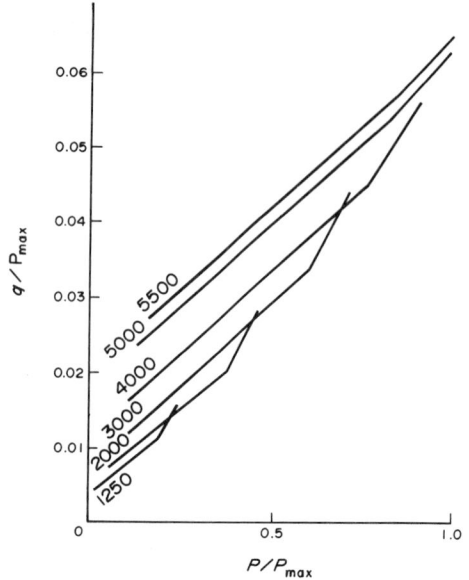

Figure 1
Curves of the specific rate of consumption of fuel as a function of engine power output for different engine speeds (in revolutions per minute)

transport has continued to increase over the years. For example, between 1973 and 1987 in France

(a) the consumption of petrol by road transport increased by 42%, and

(b) the total consumption of petrol as a fuel decreased by 25% and the proportion consumed by road transport increased from 25% to 47%.

Thus, the need to reduce oil imports has resulted in most countries undertaking actions to limit the consumption of fuel by road transport. Such actions are concerned with the vehicles, the traffic, the road systems and the drivers.

Different parameters affect the fuel consumption of road vehicles according to different equations.

1. Calculation of the Fuel Consumption of a Vehicle

1.1 Engine Fuel Consumption

The instantaneous fuel consumption of an engine is determined by its output power P and its speed N.

A typical example of the way in which this consumption varies is shown in Fig. 1. Measured values generally correspond to a linear variation:

$$q = aP + bN \qquad (1)$$

where q is the rate of fuel consumption by weight and a and b are constants characteristic of the engine.

Equation (1) applies to all engines fitted in road vehicles.

1.2 Vehicle Fuel Consumption

The power provided by the engine serves not only to propel the vehicle but also to operate certain items of auxiliary equipment.

(a) *Power absorbed by auxiliary equipment.* The total power absorbed by the auxiliary equipment (fans, pumps, compressors, etc.) can amount to 20% of the maximum power output of the engine.

Certain items of auxiliary equipment are mechanically driven by the engine and derive the necessary power directly from the engine shaft, while others (electric fans, lighting) are supplied with electricity from the alternator which has a maximum efficiency of 30–40%.

It has been calculated that for a car travelling in town at an average speed of 20 km h^{-1}, a consumption of 300 W of electrical power, which corresponds to an input of 1000 W mechanical power, results in a consumption of 2 l petrol per 100 km.

(b) *Propulsion of the vehicle.* Some of the power produced by the engine is absorbed by the transmission (gear box, drive axle) and some in overcoming the rotational inertias of the rotating parts on accelerating the vehicle, the balance being the propulsive power delivered to the wheels.

In order to simplify the calculations, it is assumed that the rotational inertias correspond to an external force and that the mechanical efficiency of the transmission depends only on the gear ratio being employed. This gives

$$P_m = \eta_i P_e \qquad (2)$$

$$P_m = F_{ext} v \qquad (3)$$

$$\left. \begin{array}{l} F_{ext} = RMg + \tfrac{1}{2}\rho A C_x v^2 + Mgs + I_i M\gamma \\ = F_1 + F_2 + F_3 + F_4 \end{array} \right\} \qquad (4)$$

where P_m is the driving power at the wheels (W), i is the index corresponding to the gearbox ratio, η_i is the efficiency of the drive train (dependent on i), P_e is the power provided by the engine (W), F_{ext} is the resultant of external forces (N), v is the vehicle speed (m s^{-1}), R is the rolling resistance coefficient of the tires (kg kg^{-1}), M is the mass of the vehicle (kg), g is the acceleration due to gravity (m s^{-2}), ρ is the mass per unit volume of air (kg m^{-3}), A is the frontal area of the vehicle (m^2), C_x is the aerodynamic drag coefficient, s is the road slope (m m^{-1}), I_i is the total inertia coefficient (dependent on i) and γ is the vehicle acceleration (m s^{-2}).

The total force F_{ext} is the sum of four external forces:

(i) F_1 is the tire rolling resistance,

(ii) F_2 is the aerodynamic resistance,

(iii) F_3 is the force due to gravity, and

(iv) F_4 is the inertial force, with

$$F_4 = \underbrace{M\gamma}_{\text{vehicle inertia}} + \underbrace{(I_i - 1)M\gamma}_{\text{rotational inertias}}$$

Numerical values of the different parameters given in Eqns. (2)–(4) together with the laws for how some of them vary, in particular R and C_x, are to be found in Roumegoux (1979) and Bonnetain and Roumegoux (1980).

(c) *Fuel consumption.* By using Eqns (1)–(4), the instantaneous consumption of a vehicle can be determined under given conditions.

Considering steady-speed consumption, when the vehicle is travelling at a constant speed ($\gamma = 0$), then:

$$q = \frac{a}{\eta_i} v(RMg + \tfrac{1}{2}\rho A C_x v^2 + Mgs) + bN \quad (5)$$

On taking account of the overall gear ratio given by $\delta_i = v/N$ and of the density of the fuel, the fuel consumed per 100 km can be calculated by dividing the two right-hand terms of Eqn. (5) by the speed v:

$$C = \frac{a'}{\eta_i}(RMg + \tfrac{1}{2}\rho A C_x v^2 + Mgs) + \frac{b'}{\delta_i} \quad (6)$$

Equation (6) shows how the fuel consumption per unit distance varies as a function of transmission efficiency and overall gear ratio, vehicle weight, streamlining of the vehicle as given by the product AC_x, air density, vehicle speed and slope of the road.

Considering fuel consumption for a given journey, there are, in general, two situations where the power delivered to the wheels of the vehicle is negative: namely, in certain descents and when the driver is braking. The engine is functioning as a brake in both of these situations; the set of curves shown in Fig. 1 no longer applies and the flow of fuel can be assumed to be constant. The same assumption can be made in the case of an idling engine. Thus, the total quantity of fuel consumed amounts to:

$$F = F_+ + F_- + F_0 \quad (7a)$$

$$F = F_+ + q_- t_- + q_0 t_0 \quad (7b)$$

where the subscripts $+$, $-$ and 0 refer, respectively, to the situations where the power output of the engine is positive or negative and/or where the engine is idling and where q_i and t_i refer, respectively, to the corresponding rates of consumption and their durations. The fuel consumption per unit distance is obtained on dividing Eqn. (7b) by the distance vt, where v is the average speed and t the total duration concerned:

$$C = \frac{F_+}{vt} + \left(\frac{q_- t_-}{t} + \frac{q_0 t_0}{t}\right) v^{-1} \quad (8)$$

The first term F_+/vt leads to an expression similar to that of Eqn. (6) to which a term corresponding to the positive variations of kinetic energy needs to be added. The second term introduces a variation of the fuel consumption as an inverse function of the average speed. Either of these terms can be preponderant or negligible depending on the traffic conditions. The final equation for the fuel consumption per unit distance is:

$$C = \lambda_1 v^2 + \lambda_2 + \lambda_3 s + \lambda_4 v^{-1} \quad (9)$$

where λ_1, λ_2, λ_3 and λ_4 are coefficients.

2. Fuel Consumption of the Traffic as a Whole

A number of different methods can be employed to determine the fuel consumption of the traffic as a whole, the choice of the appropriate method depending on the data available for defining the kinematics of the vehicles making up the traffic.

This fuel consumption is determined as the sum of the fuel consumption of each of the vehicles concerned (cars, buses, lorries, etc.), in turn determined in one of two different ways:

(a) according to the instantaneous kinematics of those vehicles (vehicle speed and acceleration, engine speed) using equations of the form of Eqn. (1); or

(b) according to the average values (average speed) using equations of the form of Eqn. (9).

The following examples are taken from Roumegoux (1979) and Bonnetain and Roumegoux (1980).

2.1 Effect of Traffic Lights at a Road Junction

The operation of traffic lights at a road junction can be optimized so as to minimize the total consumption of fuel by the traffic according to the average flow of vehicles on each of the roads making up the junction.

The results shown in Fig. 2 were obtained by referring to a microsimulation of the traffic and by determining the instantaneous fuel consumption of each vehicle.

The total fuel consumption of the traffic on a road depends on the flow of vehicles and the duration of the red phase (RP) of the traffic lights.

2.2 Urban Traffic Control

Attempts are made to improve the circulation of traffic in a town by reducing traffic jams and by increasing the steady flow of traffic, which results in an increase in the average speed of the vehicles. This usually leads to fuel savings. Thus, given that the average speeds are usually low, the term $\lambda_1 v^2$ of Eqn. (9), corresponding to the aerodynamic resistance, is negligible and the fuel

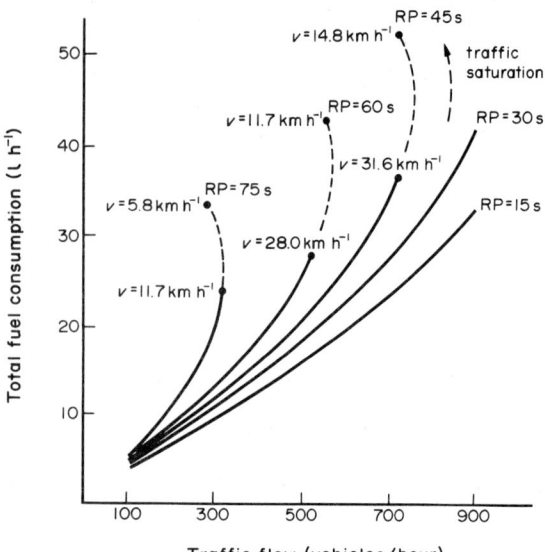

Figure 2
Total consumption of fuel per hour by cars travelling along a 600 m section of road containing traffic lights according to the flow of vehicles per hour and the duration of the RP (total duration of cycle: 90 s)

consumption per unit distance varies inversely to the average speed:

$$C = a + bv^{-1} \qquad (10)$$

where a and b are coefficients.

However, the overall effect of a traffic plan must be evaluated by taking account of the distances covered by the vehicles which are often greater because of the introduction of one-way streets. In the case of centralized control of traffic lights, it is also necessary to take account of the number of times vehicles are stopped. Thus, some investigators, often as a result of experimental studies, have established equations of the form:

$$C_{tot} = aD + bt + cS \qquad (11)$$

where C_{tot} is the total quantity of fuel consumed (l), D is the total distance covered (km), t is the total time spent on the road system (h), S is the total number of stops, and a, b and c are coefficients dependent on the type of vehicle. Such equations can be introduced into the computer programs used to control the traffic lights in a district.

Different studies aimed at evaluating the fuel savings due to the operation of urban traffic control systems show how such savings only amount to a few percent and there are often adverse effects made possible such as an increase in the amount of traffic as a result of the steadier flow of vehicles.

It should also be noted how it is difficult to evaluate these fuel savings since the equations employed, such as Eqn. (11), are not very exact. More exact equations cannot be employed since this would call for the collection of data that is difficult to obtain. For example, the fuel consumed due to vehicles being stopped, represented by coefficient c in Eqn. (11), depends on:

(a) the times for which the vehicles are at rest, which can be estimated fairly well according to the operation of the traffic lights provided that the traffic has not reached saturation point; and

(b) the kinetic energies involved in accelerating the vehicles again, which depend on the speeds reached following their being stopped.

This last point can be illustrated by referring to an example taken from a study of the effects of speed limits on the fuel consumption of vehicles circulating in towns (see Fig. 3). This study showed how the fuel consumption due to a stop increased as the square of the speed reached following the stop, which is in good agreement with the conversion to kinetic energy.

2.3 Effects of Road System Characteristics

On constructing a new road or on carrying out improvements to an existing road (e.g., by eliminating bends) outside urban areas, the effects of the geometrical characteristics of the road on the fuel consumption by the traffic are of concern. The additional consumption of energy for the construction of the road can be offset by the saving in the fuel consumed by the traffic, such that the cost of the former is soon defrayed.

In studying these effects, aggregate models of the fuel consumed by the traffic have been produced that take account of the average speed of the vehicles, the bends in the road and the slope of the road.

(a) Bends in the road. It has been shown theoretically how the fuel consumption per unit distance along a level road varies as follows:

$$C = av^2 + b \qquad (12)$$

with

$$a = a_0 + \lambda_1 \sum \alpha + \lambda_2 n$$

$$b = b_0(1 - nl) - \lambda_3 nr$$

where C is the average fuel consumption per vehicle (l per 100 km), v is the average vehicle speed (km h^{-1}), $\Sigma\alpha$ is the sum of the angles of the large-radius bends (rad km^{-1}), n is the number of small-radius bends (km^{-1}), l and r are the average length and radius of these small-radius bends (m), λ_1, λ_2 and λ_3 are constant coefficients for a given vehicle, and a_0 and b_0 are values of a and b on a straight road ($\Sigma\alpha = 0$, $n = 0$).

The bends are grouped into either of the following two categories:

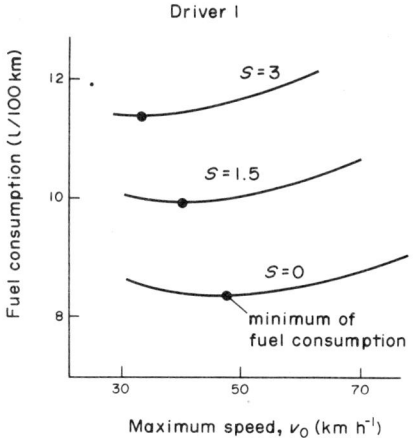
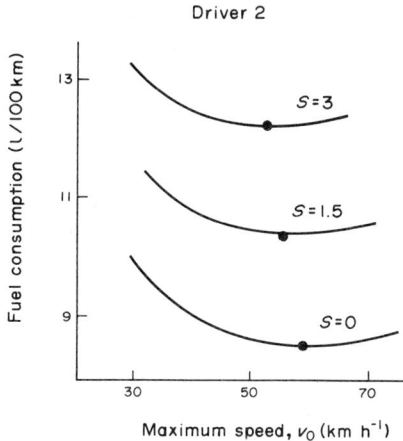

Figure 3
Fuel consumption as a function of the maximum speed v_0 and the number of stops per kilometre S, showing the effects of the type of driving

(i) small-radius bends, for which vehicles are obliged to slow down and then speed up again; and

(ii) large-radius bends, which vehicles can negotiate without slowing down but which give rise to an increase in the resistance to their forward motion proportional to the centrifugal force involved.

A distinction between the two categories of bend is made with reference to a limiting radius R_0, given by:

$$R_0 = v^2 \gamma_t^{-1} \qquad (13)$$

where γ_t is a centrifugal acceleration having a value of 2–3 m s^{-2}.

Measurements carried out on a number of vehicles with a number of different drivers on roads made up of various combinations of bends have confirmed the validity of Eqn. (12). The variations in fuel consumption are shown in Fig. 4.

It should also be noted how the type of driving can affect the fuel consumption giving rise to significant differences in fuel consumption on very winding roads (of up to 30%) with little change in average speed.

(*b*) *Slope of the road.* The slope of the road can have a significant effect on fuel consumption, particularly in the case of heavy vehicles. For example, in the case of a road having a 6% slope:

(i) a medium-sized car consumes 12 l of fuel per 100 km on mounting a slope and 1.5 l per 100 km on descending it, compared with a consumption of 6–7 l per 100 km when proceeding along a level road at a speed of 80 km h^{-1}; while

(ii) a 38 t lorry consumes 200 l of fuel per 100 km on mounting a slope at a speed of 25 km h^{-1} and 0 l per 100 km on descending it, compared with a consumption of about 38 l per 100 km when proceeding along a level road at a speed of 80 km h^{-1}.

In agreement with Eqn. (9), the fuel consumption varies in an approximately linear way with the slope of the road according to Fig. 5:

$$C = C_0 + ks, \quad s \geqslant s_1$$
$$C = C_1, \quad s < s_1$$

where k is proportional to the weight of the vehicle.

The limiting value of the slope s_1 (negative) corresponds to the case where the power delivered by the engine is cancelled out. This value depends on the characteristics of the vehicle and, in particular, on its weight (cancelling out of the first term of Eqn. (6)).

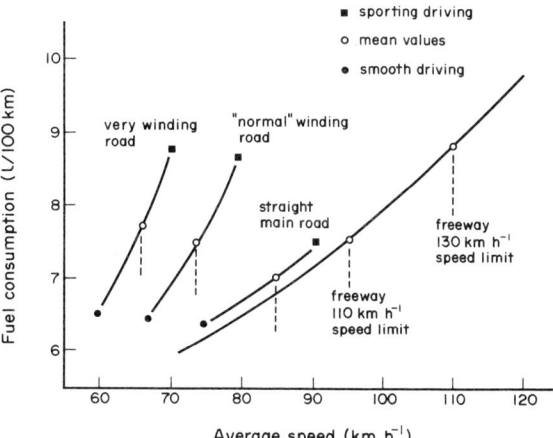

Figure 4
Fuel consumption per vehicle (car traffic) as a function of the average speed and the type of road (consumption corrected to correspond to travel along a level road)

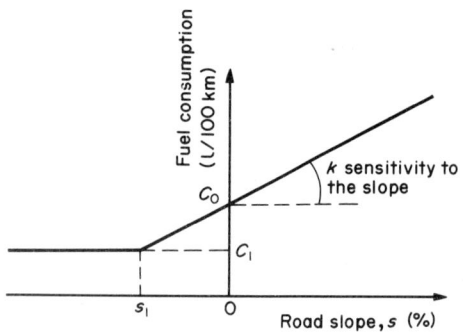

Figure 5
Simplified modelling of fuel consumption as a function of the slope of the road

The value of the fuel consumption C_1 depends on the vehicle speed as well as on the fuel supply system

(i) for a petrol-engined car supplied with fuel via a carburettor, $C_1 = q_-/v$ (see Eqn. (8)); and

(ii) for a lorry fitted with a direct-injection diesel engine, $C_1 = 0$.

The fuel consumption values show how it is necessary to take account of the average fuel consumption of the traffic for both directions of travel (mounting and descending the slope). The limiting value of the slope s_1 is then important:

(i) for a medium-sized car, $s_1 = -4\%$ to -5%; and

(ii) for a 38 t lorry, $s_1 = -1.5\%$.

Considering the coefficient k for the fuel consumption sensitivity to the slope:

(i) for a 1 t car, k is of the order of 0.9 l per 100 km for each percent of slope; and

(ii) for a 38 t lorry, k is of the order of 25 l per 100 km for each percent of slope.

3. Other Parameters Affecting Fuel Consumption

What has been said in Sects. 1 and 2 concerns well-maintained vehicles, engines operating under optimum thermal conditions, standard weather (no rain or wind) and average drivers, the last being rather difficult to define.

Most of these factors do not concern the traffic control systems as such directly, but their significance requires a brief mention, particularly when it is a matter of the experimental evaluation of fuel consumption values and variations.

3.1 Engine Adjustments

As a result of inquiries it is thought that the timing and carburettor settings of more than half of the petrol-engined vehicles on the road are out of adjustment, and experimental studies have shown how this maladjustment gives rise to an additional fuel consumption of 8% on average.

3.2 Type of Driving

"Sporty" drivers consume 25–30% more fuel than "calm" ones for an average speed which is practically the same in towns and only a little higher on winding roads.

3.3 Weather Conditions

The wind increases the consumption of fuel by the traffic, particularly where large vehicles are involved. The fuel consumption of a tractor plus semitrailer running on a freeway is increased by 10% as a result of a 10 km h^{-1} side wind.

3.4 Thermal Condition of Engines

The thermal condition of the engine on starting a journey is very significant. If the engine is started when cold, all the mechanical and viscous resistances are greater and the combustion efficiency of the engine is reduced.

The results of measurements carried out on a medium-sized car operating over a 4 km urban traffic cycle shows that (Fig. 6):

(a) for an ambient temperature of 0 °C, a car starting off with a cold engine (0 °C) consumes 40% more fuel than one starting off with a hot engine (~ 80 °C); and

(b) for an ambient temperature of 20 °C, this difference still amounts to 25%.

Many short journeys (less than 4 km) are undertaken in towns with cars starting off cold, and a traffic plan favoring the use of public transport will, accordingly, result in fuel savings.

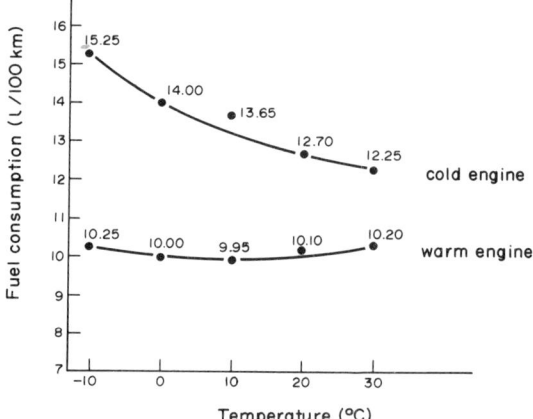

Figure 6
Effects of the ambient temperature on the fuel consumption of a medium-sized car starting off with either a hot or cold engine (European cycle, 4052 m)

4. Comments

The subject of fuel consumption by road transport involves a number of very different factors:

(a) road systems (ranging from small streets in towns to freeways),
(b) vehicles (two-wheeled vehicles, cars, buses, lorries, etc.),
(c) drivers (private or professional),
(d) traffic conditions (the extent to which there is a steady flow of traffic),
(e) amount of traffic,
(f) weather conditions, and
(g) organization of the transport between the different modes.

Fuel savings or, failing that, control of increases in fuel consumption are possible in most of the fields concerned. However, there are many constraints and some of these will modify the methods that can be employed in future. In particular, this is a matter of constraints related to the conservation of the environment.

A reduction in the emission of exhaust-gas pollutants by motor vehicles, for example by means of catalytic systems, will probably lead to an increase in fuel consumption for the same traffic conditions.

The limitation of the concentration of pollutants in the atmosphere, particularly in towns, will call for the provision of traffic control systems that will allow the flow of vehicles to be limited in some cases and better control over their speeds to be exercised in others.

See also: Fuel Conservation: Air Transport; Fuel Conservation: Sea Transport

Bibliography

Bonnetain Y, Roumegoux J P 1980 Le poids lourd, conception et fonctionnement. Information Note No. 18. Institut de Recherche des Transports, Bron, France.
Roumegoux J P 1979 Consommation d'énergie par la circulation routière. Information Note No. 14. Institut de Recherche des Transports, Bron, France

J. P. Roumégoux
[INRETS, Bron, France]

Fuel Conservation: Sea Transport

Increases in fuel prices occasioned by the oil crises of recent years have meant that fuel saving in sea transport systems has attracted serious attention. Various technical and operational aspects of the problem have been investigated. There are two basic approaches: one is inclined to making total sea transport systems more energy efficient, the other to improving fuel saving in individual transport units such as ships. Only the latter scheme is discussed in this article.

Each of the areas of sea transport, land transport and air transport has its proper domain of application, as clearly indicated in Fig. 1 (Akagi 1971). The diagram is known as the Kármán–Gabrielli diagram and shows the relation between the vehicle speed and HP/Wv. Where HP is the power consumption of the vehicle, W is the total weight of the vehicle including the payload, and v is the speed. The value of HP/Wv is directly related to the energy consumption per tonne kilometer, as shown in Sect. 2.

The most notable characteristic of water-surface transportation, as seen from Fig. 1, is its very low energy consumption per tonne kilometer. Inventions such as the hovercraft and the hydrofoil have certainly been successful in expanding the application limit of ships to higher speeds in some special cases, but the main trend in research and development of ships is towards strengthening the proper advantage of sea transport by further lowering the energy consumption.

1. Energy Saving as a Means of Cost Saving

Energy saving has been an important field of science and technology since the invention of the heat engine and the blast furnace in the eighteenth century. However, considering recent history, the most notable energy-saving measures were realized during the period 1975–1985 after the first oil crisis. Energy saving is certainly worthy of pursuit for its own sake because of its effect on the conservation of natural resources and

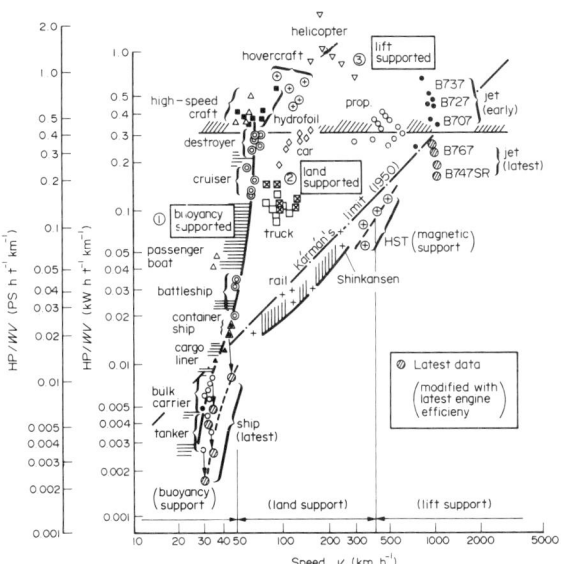

Figure 1
The Kármán–Gabrielli diagram (from Akagi 1971)

on protection of the environment. However, it should be noted that additional expenditure on energy saving is ordinarily expected to be redeemed in about two years in the industrial field, as exemplified in Fig. 2 (Ishigai and Akagi 1980).

The total transport cost of a ship includes not only the energy cost but also the capital cost, the crew cost, the research and development cost, and many others. The energy-saving cost must, therefore, not unreasonably burden other costs. The investment for marine power plant replacement (i.e., taking out the originally installed steam propulsion plant and installing a smaller-output diesel plant to bring about fuel saving, as explained in Sect. 5.1) as shown in Fig. 2 has a redemption time of seven years (Ishigai and Akagi 1980). Such a long redemption time indicates that fuel saving is very important for marine power plants and also that the situation of the market for marine transportation has changed drastically since the first oil crisis in 1973.

The rapid technological progress in marine engineering is tending towards the production of more energy-efficient ships and machinery at lower prices as time progresses. Estimates of price trends and the energy efficiency are, therefore, important factors in designing a ship. Several time constants, such as the engine price divided by the annual fuel cost, should always be taken into consideration. However, the methods of total cost optimization is beyond the scope of this article.

2. Energy Consumption Characteristics

A ship consumes energy not only for propulsion but also for cargo care, for maintenance work including painting, and for several other purposes. However, by far the greatest part of the energy is consumed as the fuel during navigation. The fuel consumption index e of a ship can be defined as follows (Akagi 1976):

$$e = \frac{\mathrm{HP}f(l/v)}{\mathrm{DW}l} = \frac{\mathrm{HP}}{\mathrm{DW}v}f \qquad (1)$$

where HP is the total output of the propulsive plant of the ship (PS, where 1 PS = 0.735 kW), DW is the deadweight capacity of the ship (t), l is the navigation distance (nautical miles), v is the speed of the ship (knots) and f is the specific all-purpose fuel consumption of the power plant of the ship $(\mathrm{g\,(PS)^{-1}\,h^{-1}})$.

Equation (1) indicates that the fuel consumption per tonne kilometer can be improved by reducing the factor HP/DWv and/or f of the power plant. The value of HP/DWv is closely related to the efficiency of the ship hull characteristics such as the size, and the speed and power of the ship. On the other hand, f is related mainly to the efficiency of the power plant. Generally suggested methods of improving these two factors are (Ishigai and Akagi 1980):

(a) improvement of ship hulls (long, slender hull form, weight reduction, etc.),

(b) improvement of propulsive efficiency (slow-turning propeller, etc.),

(c) improvement of power plants (improvement of specific fuel consumption, slow-revolution diesel, etc.),

(d) waste-heat recovery (turbogenerator system with exhaust economizer, etc.), and

(e) improvement and ship operation (optimization of ship speed, etc.).

These are shown schematically in Fig. 3. An example of a typical ship applying various means for fuel saving is also shown in Fig. 4.

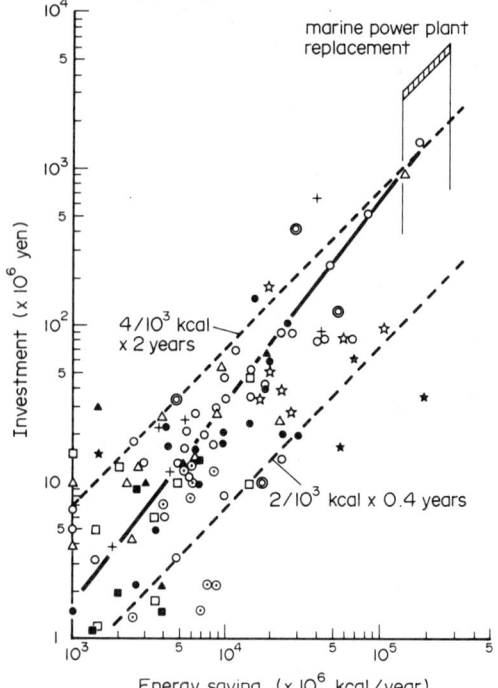

Figure 2
Comparison of energy saving investments between land and ships (power plant replacement)

- maritime fuel conservation
  - 1 improvement of performance of ship resistance and propulsive efficiency
    - reduction of ship resistance
      - improvement of ship hull form
      - improvement of ship hull paint
      - reduction of ship hull weight
      - reduction of air resistance
    - improvement of propulsive efficiency
      - large, slow-turning propeller
      - variable-pitch propeller
    - improvement of seakeeping qualities
      - improvement of ship hull form
      - improvement of steering system
  - 2 improvement of power plants
    - improvement of conventional power plants
      - improvement of slow-speed diesel engine
      - geared diesel system
      - waste heat recovery
    - new types of power plants
      - combined cycle plant
      - stirling engine
  - 3 improvement of ship operations
    - optimization of ship speed
    - optimization of voyage route
    - improvement of autopilot

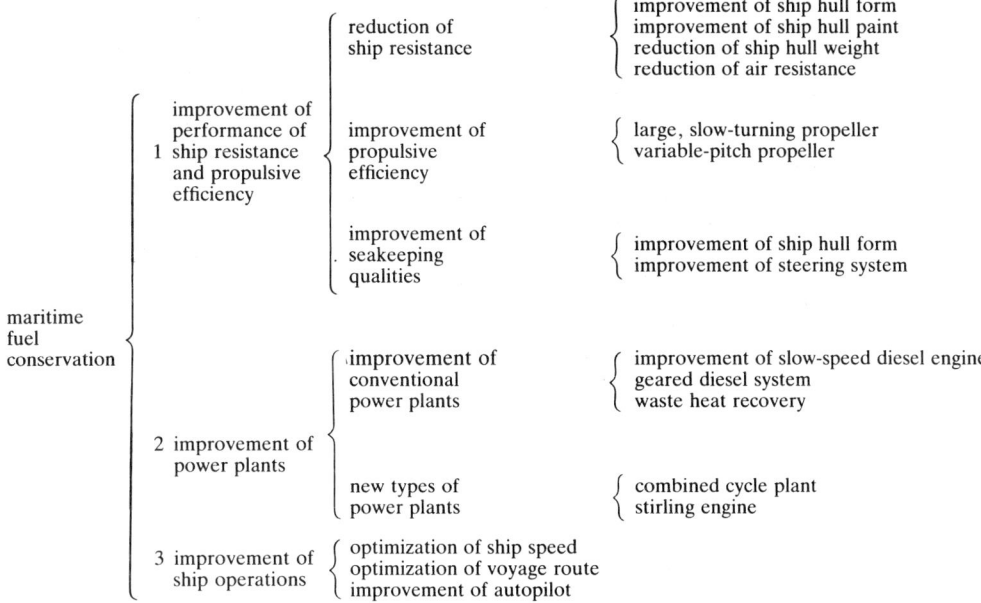

Figure 3
General scheme for maritime fuel conservation

3. *Fuel Savings Through Reduction of Ship Hull Resistance*

3.1 Effects of Ship Size

Ship size has increased with increasing worldwide production activity. During the period of high economic growth before the first oil crisis in 1973, ship size increased remarkably, especially the size of tankers which nearly doubled every five years and became as large as 500 000 t. The effect of ship size on fuel saving per unit transport of cargo (tonne nautical mile) is evident in Fig. 5, which shows the relation between HP/DWv and DW for actual oil tankers (Ishigai and Akagi 1980). The unit energy consumption of a 500 000 t ship decreases to one-quarter of that of a 50 000 t ship.

This example clearly explains the reason for the high rate of growth in the size of ships up until 1973. The ship size is among the most powerful factors affecting the fuel saving.

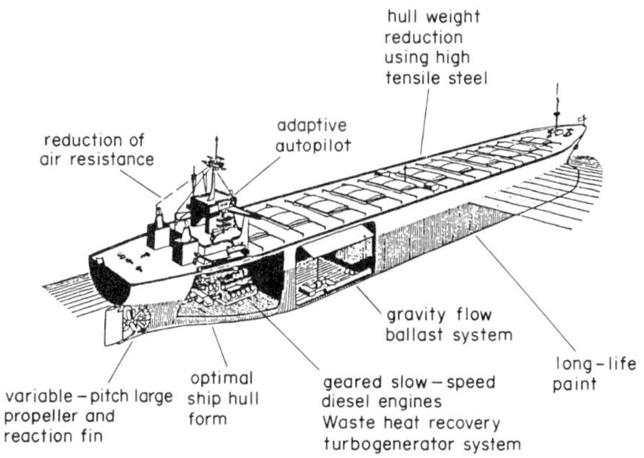

Figure 4
Typical ship with various means for fuel saving

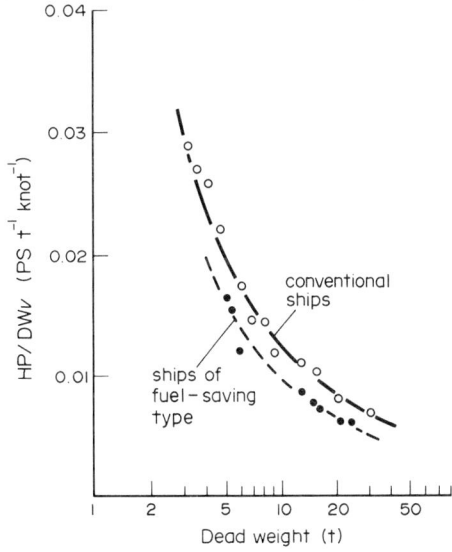

Figure 5
Evaluation of fuel conservation by enlarging ship size
(1 PS t^{-1} $knot^{-1}$ = 0.397 kW h t^{-1} km^{-1})

3.2 Effects Ship Speed and Hull Form

The optimal speed of a merchant ship is usually determined by economic criteria. As the fuel cost is drastically influenced by the ship speed, the effective approach to reduce fuel costs in a period of economic depression is to reduce the ship speed. On the other hand, to reduce fuel costs for a given speed, the development of ship hull forms with lower resistance has been progressing steadily.

The contribution of the ship hull form to the fuel saving can be seen in Fig. 6 for a container ship. In Fig. 6, the index HP/DWv from Eqn. (1) is modified to HP/Δv by taking the displacement of ships Δ instead of DW for representing the ship size. This index is more suitable for the case of a container ship (Nishikawa and Hashimoto 1983). It can easily be seen that Eqn. (1) is transformed into

$$e = \left(\frac{\text{HP}}{\Delta v}\right)\left(\frac{\Delta}{\text{DW}}\right) f \qquad (2)$$

In Fig. 6, the index HP/Δv is compared among four groups of ships (i.e., the older cargo ships, the new cargo ships, the early-generation container ships and the latest container ships). Despite the great increase in the container ship speed when compared with cargo ships, the unit fuel rate HP/Δv is not very much different. The distance between the straight lines represents the technological improvement in the hull form. Improvement in the latest design when compared with the older design is also evident.

The reason why container ships can attain such a high speed without an increase of HP/Δv can be explained by their inherent characteristics, which make them suitable for high speeds. The typical feature of a container ship is that it carries containers on its deck. This makes it possible to use a slender hull form which has a low resistance. The hull of the conventional cargo ship has a dual function; that is, it holds the cargo and, at the same time, supports the ship by its buoyancy. In contrast, the container ship is designed to separate these two functions into the "deck" and the "hull," respectively (see Fig. 7) (Ishigai and Akagi 1980). In general, the development of any system is inevitably accompanied by such a "functional separation." In a hydrofoil, which is faster than a displacement-type ship, these two functions are completely separated, as shown in Fig. 7 (Ishigai and Akagi 1980).

In Fig. 3, three other methods of reducing ship resistance are grouped together with hull-form improvement. They are the application of a high-performance antifouling hull paint to reduce the frictional hull resistance, the reduction of air resistance by using a small deck house, and the application of high-tensile steel to reduce the weight of the hull.

Mention should be made of the derating of the propulsion engine. Let R and DR be the subscripts for the rated output condition and the derated output condition, respectively. Then, from Eqn. (1):

$$\frac{e_{\text{DR}}}{e_{\text{R}}} = \left(\frac{\text{HP}_{\text{DR}}}{\text{HP}_{\text{R}}}\right)\left(\frac{v_{\text{R}}}{v_{\text{DR}}}\right)\left(\frac{f_{\text{DR}}}{f_{\text{R}}}\right)$$

and since HP is proportional to v^3:

$$\frac{e_{\text{DR}}}{e_{\text{R}}} = \left(\frac{v_{\text{DR}}}{v_{\text{R}}}\right)^2 \left(\frac{f_{\text{DR}}}{f_{\text{R}}}\right) \qquad (3)$$

For the steam power plant, $f_{\text{DR}}/f_{\text{R}}$ is greater than unity but, for the diesel plant, it is about equal to or (in the latest diesel plant) even slightly less than unity. Because the energy consumption ratio is proportional to the square of the speed ratio, as seen in Eqn. (3), derated navigation is a very powerful method of saving energy per tonne kilometer. This explains the tendency for favoring derated navigation when fuel prices are high and the shipping business is in depression, and also for favoring the diesel plant in the above-mentioned situation.

Unstable fuel prices make it necessary to design ships taking into consideration the variation in optimal ship speed due to the variation in fuel price. Selection of the type of propulsion engine and the type of propeller is one of the most important factors in such a consideration.

4. Improvement of Propulsive Efficiency

Equation (1) can be transformed as

$$e = \left(\frac{\text{HP}}{\text{DW}v}\right)f = \left(\frac{R}{\text{DW}v}\right)\left(\frac{1}{\eta}\right)f \qquad (4)$$

where R is the resistance of the ship hull and η is the propulsive efficiency. The fuel consumption index e is, therefore, improved by increasing η, which can be effected by reducing the number of revolutions of the propeller. This method was initiated with the turbine

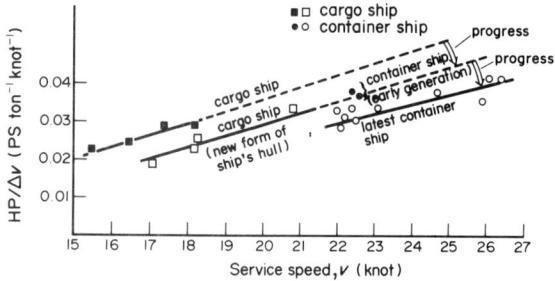

Figure 6
Improvement of ship hull form of container ships (1 knot = 1.852 km h^{-1}; 1 PS t^{-1} knot^{-1} = 0.397 kW h t^{-1} km^{-1})

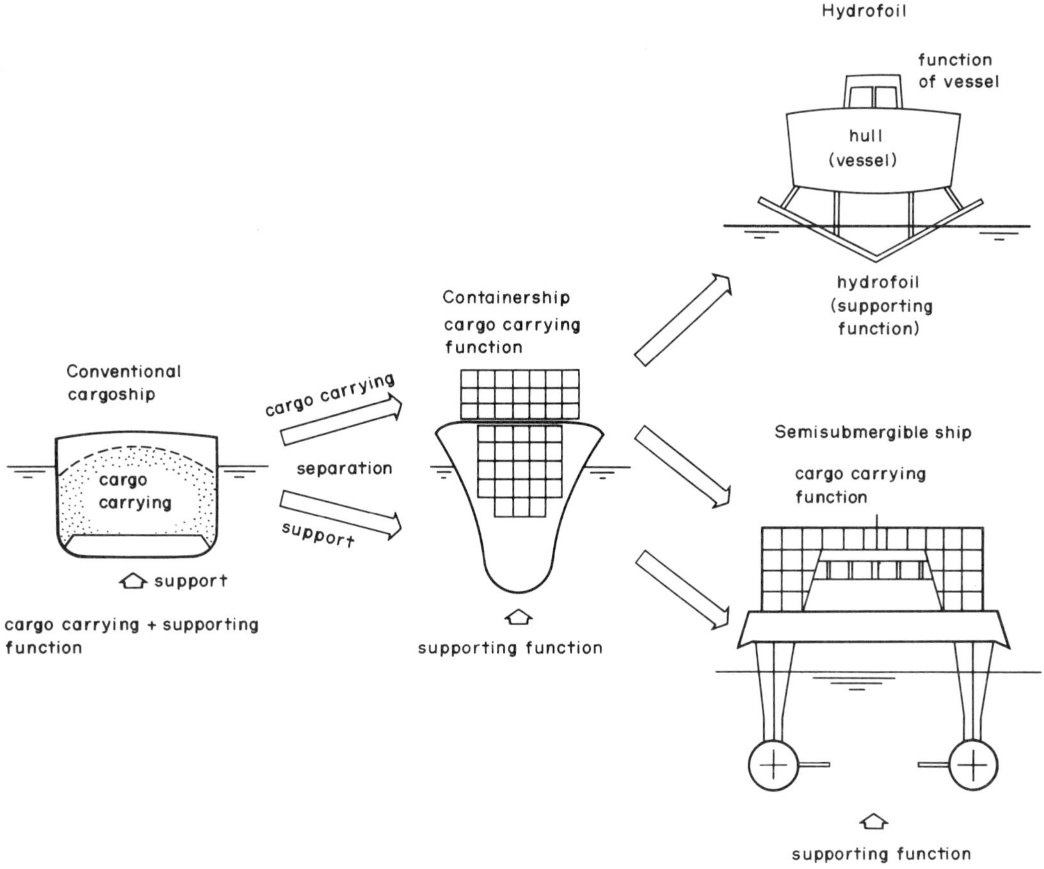

Fuel Conservation: Sea Transport

Figure 7
Functional separation in container ships and hydrofoils

engine, which was necessarily equipped with the reduction gear. The high performance of larger, slower propellers was, thus, demonstrated and brought about further research and development on propellers and reduction gears, which in turn brought about the development of the geared diesel and the long-stroke diesel.

Figure 8 shows the results of a case study (Kadoi 1980). In fact, when the number of revolutions is reduced from 120 revolutions per minute (rpm) to 60 rpm, the propulsive efficiency increases by 16%. This results in a power saving of 14–15%. The other measures to improve the propulsive efficiency are the applications of a variable pitch propeller, reaction fins (see Fig. 9), and so forth.

5. Improvement of Marine Power Plants

Another important method of fuel saving is the improvement of the specific fuel consumption f of the power plant, as indicated by Eqn. (1). Because of the decline in demand for marine transportation and the overcapacity of oil tankers after the oil crisis, it is undesirable to build very large ships as a means of fuel saving. Emphasis is therefore now being put on improving the fuel consumption of the power plant itself.

5.1 Energy Flow

Ships require some energy for lighting, heating, ventilation, cargo care, deck machinery and other purposes.

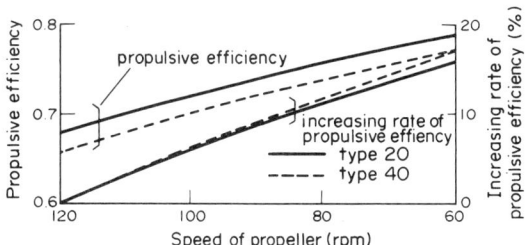

Figure 8
Relation between number of revolutions and propulsive efficiency of propeller

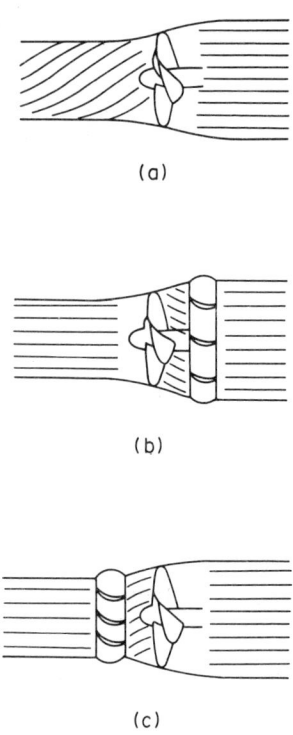

Figure 9
Reaction fin

However, the sum of all energy consumptions for purposes other than propulsion is not significant enough to make it possible to improve the overall power plant efficiency by cogeneration of heat and power. Even the waste heat from the main propulsion engine must, therefore, be converted into power. An example of a sea-going-condition diesel plant, typical in 1975, is shown in Fig. 10, in which 32% of the total thermal energy of fuel is used for the propulsion, only 2.8% for the heating service, and 1.7% for the electric power. The remainder is wasted as the thermal energy of the exhaust gas, the cooling water, the air cooler and so on (Akagi 1976). Consequently, the improvement of the efficiency of the main propulsive engine itself is the central technological issue.

5.2 Examples of Fuel-Saving Methods

(*a*) *Diesel engines.* Ever since its successful application to marine use at the beginning of the twentieth century, the diesel engine has been the most efficient heat engine. The introduction of supercharging to the slow-speed two-cycle diesel engine in 1955 further strengthened this position, and the latest success of static-pressure supercharge has brought its specific fuel consumption down to less than $130 \text{ g (PS)}^{-1} \text{h}^{-1}$ (Nagai and Suzuki 1982). The trend shown in Fig. 11 indicates that the rapid improvement in thermal efficiency is still taking place, but it is outside the scope of this article to go into details of the progress being made in this area.

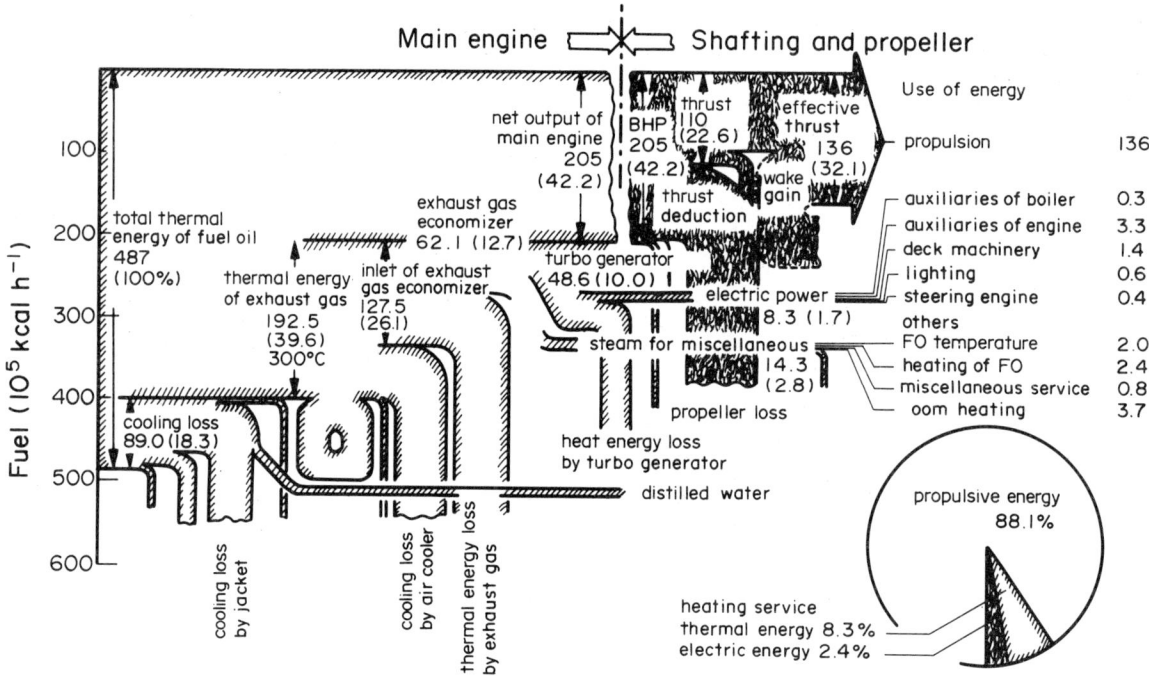

Figure 10
Energy flow of a marine power plant, typical in 1975

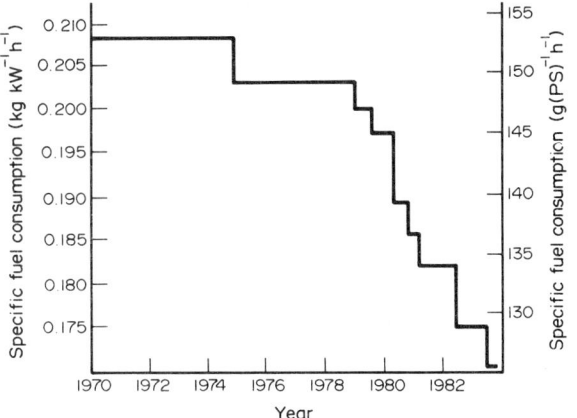

Figure 11
Trend in specific fuel consumption for large, slow-speed diesel engines

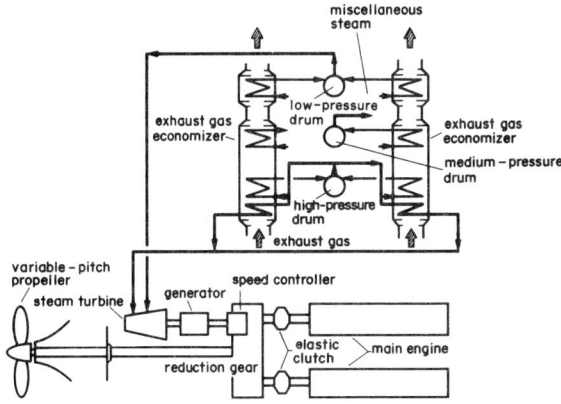

Figure 13
Combined system of turbogenerator, shaft generator and shaft driving

(*b*) *Recovery of waste heat.* The utilization of waste heat (from the exhaust gas, the turbocharged air and the cooling water of the main engine) is one of the most effective means of fuel saving. In fact, the electric power generation system by the exhaust gas of the main engine is usually adopted. Figure 12 shows a turbogenerating system with an exhaust gas economizer (Toda 1980). The exhaust gas of a main diesel engine is led to the exhaust gas economizer and steam is generated by its waste heat. High-pressure steam is led to the generator turbine which is connected to the electric generator and low-pressure steam is also used for the general services on the ship. The system shown in Fig. 13 is a more advanced turbogenerating system in which the turbine output is supplied to both the electric generator and the propulsion shaft through a reduction gear. In this system, the electric generator can also be driven by the propulsion shaft, when the heat gain from the exhaust gas is insufficient to meet the electricity demand.

(*c*) *Combined cycle plant.* Some typical combined gas–steam turbine plants are shown in Fig. 14. These are systems in which the gas turbine exhaust is sent to an exhaust gas boiler to generate steam for driving a steam turbine. These systems certainly have potential, but further study is required before their actual utilization for marine power plants will be possible.

6. Wind-Power Utilization

Regarding wind-power utilization, the good fuel-saving performance of Shin Aitoku Maru should be noted. This is a sail-rigged motor tanker built in 1980, with a dead weight of 1450 t, equipped with a 1.2 MW diesel engine assisted by two sails with a total area of 194.4 m². A 50% energy saving has been attained when compared with conventionally designed ships of the

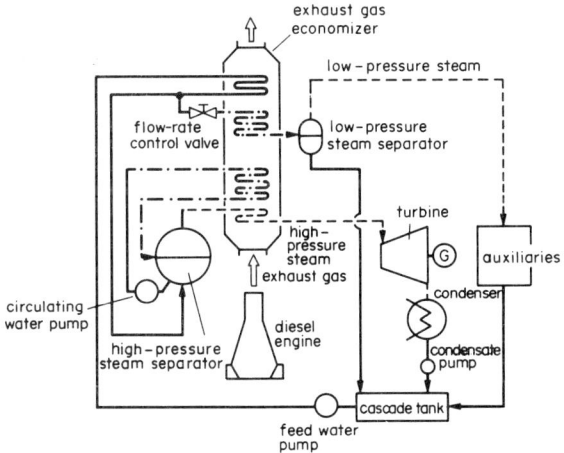

Figure 12
Schematic diagram of exhaust gas economizer/turbogenerating system

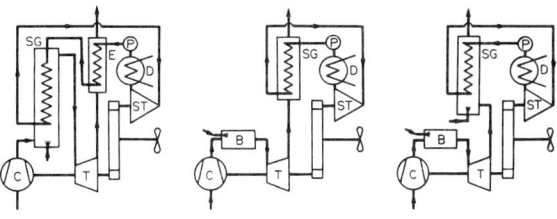

Figure 14
Schematic diagram of combined gas and steam turbine plants

same size. The stability, reliability and availability are all reported to be good.

See also: Fuel Conservation: Air Transport; Fuel Conservation: Road Transport

Bibliography

Akagi S 1971 *Transport Vehicle Engineering* (in Japanese). Corona Publishing
Akagi S 1976 Analysis and economization of energies in ships. *Bull. Mar. Eng. Soc. Jpn.* **4** (2), 77–88
Ishigai S, Akagi S 1980 Maritime fuel conservation. *Bull. Mar. Eng. Soc. Jpn.* **8** (4), 311–18
Kadoi H 1980 Reduction of power of ships resulting from improved propulsive efficiency (power saving by means of large, slow turning propeller). *J. Mar. Eng. Soc. Jpn.* **15**, 331–7
Nagai M, Suzuki S 1982 Examination of thermal efficiency improvements of marine diesel engines (in Japanese). *J. Mar. Eng. Soc. Jpn.* **17**, 1033–43
Nishikawa E, Hashimoto J 1983 Feasibility study of a catamaran ferry for domestic cargo shipping. *J. Mar. Eng. Soc. Jpn.* **18** 338–47
Toda K 1980 Economical waste heat recovery turbo generating plant for motor ship. *J. Mar. Eng. Soc. Jpn.* **15**, 365–72

S. Akagi
[Osaka University, Osaka, Japan]

High-Speed Railroad: Modelling and Simulation

Rail-guided ground transportation systems are currently of great interest as an alternative to air transportation, especially in areas with high population density and medium travel distances as in Europe and Japan. For example, Germany and Japan are developing new high-speed systems based on magnetic levitation (maglev) and are also investigating the potential of the conventional wheel–rail system (Jänsch 1988) (see *Magnetic Suspension Railroad Systems*).

At the desired velocity range of these systems (300–500 km h^{-1}), the system dynamics and its control play a dominant role (Verein Deutscher Ingenieure 1984, 1987). The complexity, costs, and risks involved in modern high-speed ground transportation systems (HSGT) require dynamic modelling and computer simulation at an early stage of the development. The advance into new velocity regimes, under stringent conditions of ride comfort, safety and economy of material, is a challenge for advanced computer-aided analysis and design (CAAD) tools.

A necessary fundamental requirement for the application of computer-aided methods is the availability of reliable mathematical models, also called computer or simulation models. In particular, simulation models are needed for:

(a) early design considerations such as approximate estimation of the dynamic behavior of alternative concepts based on simple models;
(b) design and optimization of system performance based on (improved, validated but still low-order) design models;
(c) evaluation of the performance of the final design (e.g., lateral stability, curving ability, ride comfort, and force levels); and
(d) precomputation of experiments and field tests in conjunction with acceptance testing, checking of performance specifications, and model verification activities.

Validated simulation models are not only required for effective and systematic CAAD but also for monitoring, diagnostics, and failure prediction of revenue systems.

1. Dynamic Modelling of Rail-Guided Vehicles

Figure 1 summarizes the main CAAD steps (software) as well as the supporting hardware and experimental activities. The fundamental first step is always the physical and mathematical modelling leading to the system equations.

Any vehicle, whether on rails or on roads, resembles in the first place a mechanical system consisting of bodies having mass and inertia properties. The individual bodies are linked together by interconnections with small or negligible mass giving rise to coupling forces or constraint relations. These types of systems are—in modern terminology—called multibody systems (MBSs), for which powerful methods have been developed during the 1970s and 1980s (e.g., Kortüm and Schiehlen 1985).

The most effective modelling strategy is to "disassemble" the vehicle into basic elements, to determine the parameters and equations for these elements and, finally, to synthesize (reassemble) the equations for the full vehicle. This strategy is usually called physical modelling as it is based on physical principles for the elements (see Fig. 2).

In contrast, deriving the equations from experiments (excitation and response measurements) of the complete (vehicle) system, is called identification. Naturally, since identification could be performed on the component level (e.g., for suspensions), both methods can be combined advantageously. Identification of equations and parameters for complete vehicles is still rudimentary, especially if the structure of the equations is of concern (Brockhaus and Doherr 1987).

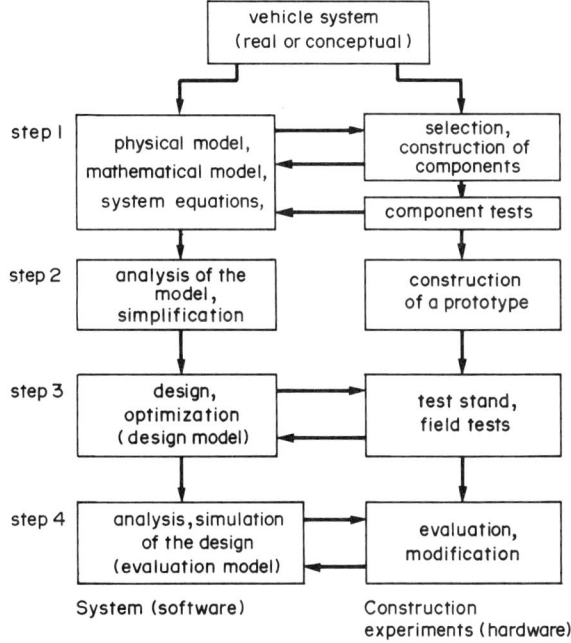

Figure 1
Main steps in CAAD

1.1 Vehicle Bodies

Continuing with physical modelling, the first task is to decompose a full vehicle (or even a train) into single bodies ("free-body diagram") and to define their interconnection elements. Once this major modelling task has been performed, the model can—in principle only—proceed safely. The well-established principles (axioms) of mechanics (e.g., the Newton–Euler equations of motion) can be applied to obtain the mathematical models, which turn out to be a set of nonlinear ordinary differential equations (plus, possibly, a set of algebraic relations).

Naturally, the purpose of investigation and the required accuracy of the model have a major influence on the resulting model and its degrees of freedom. Dukkipati and Amyot (1988) distinguish in this context as many as eight different model categories for railway investigations (Fig. 3).

The next problem which is also nontrivial is to establish the mass geometry parameters (i.e., each body mass and its inertia matrix). They may be computed from the drawing of the vehicle and its mass distribution (specific weight of the construction elements). However, in practice this may be tedious and ambiguous. Therefore, experimental facilities are used as well; for example, weighting the body to obtain its mass and performing rotational vibration experiments—using physical pendulums (i.e., gravity restoring, or torsional springs)—to obtain the inertia parameters.

A further complication arises when some of the vehicle bodies cannot be modelled with rigid degrees of freedom alone, but due to structural flexibility they have to be considered as elastic continua (Wallrapp 1989).

1.2 Suspensions

Suspensions connecting the vehicle bodies are also fundamental components of rail vehicles. Their main tasks are:

(a) to support the weight of the vehicle and to provide the necessary guidance along the track; and

(b) to isolate the vehicle body with its passengers or freight from disturbances such as track irregularities or external forces (e.g., wind gusts, load changes).

In order to fulfill their function, suspensions consist of restoring-force and energy-absorbing elements that provide stiffness and damping to the system. They are called flexible suspensions as their compliance creates coupling forces between contiguous vehicle bodies. Examples are springs, dampers or shock absorbers; more recently, active (feedback-controlled) actuators are considered in railways (yaw suspensions, vehicle-body tilting, or even between the two wheels of a wheel set) (see *Railroad Systems: Active Control*). Their mechanical, or more generally their mathematical, models are called coupling elements. What are required are the relations describing the force (or moment) characteristics with respect to displacement or velocity of these components.

Many of these suspensions are not mechanical devices; their effects may result from pneumatic, hydraulic, electric or magnetic principles. Bond graph methods may in many instances effectively assist the modelling work for such devices (Karnopp and Rosenberg 1983). Also, even if the structure of the force laws can be derived, their parameters quite frequently have to be determined with the aid of parameter-identification experiments. A typical force-displacement characteristic of a buffer model is shown in Fig. 4, where the hysteresis has a dominant influence on the dynamic behavior. Finally, the reader's attention is drawn to Morman and Giannopoulos (1982) who give a useful survey of the state of the art of modelling passive and active vehicle suspensions.

1.3 Wheel–Rail Interaction

The dynamic behavior of a railway vehicle is significantly affected by the interactive forces between wheel and rail. The forces acting in the contact region between wheel and rail turn out to be highly nonlinear phenomena. There are two main sources for these nonlinearities.

(a) If the wheel and rail are both considered as rigid bodies, the coordinates of a wheel set depends nonlinearly on the geometric profile of wheel and rail. Figure 5a depicts the fundamental kinematic problem of a wheel set on rails. The relations

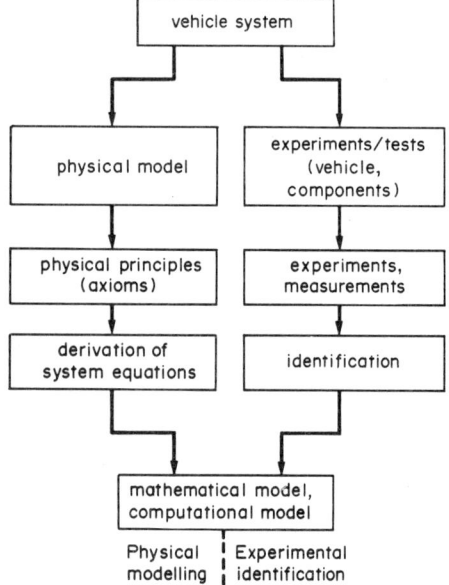

Figure 2
Physical and experimental modelling (identification)

High-Speed Railroad: Modelling and Simulation

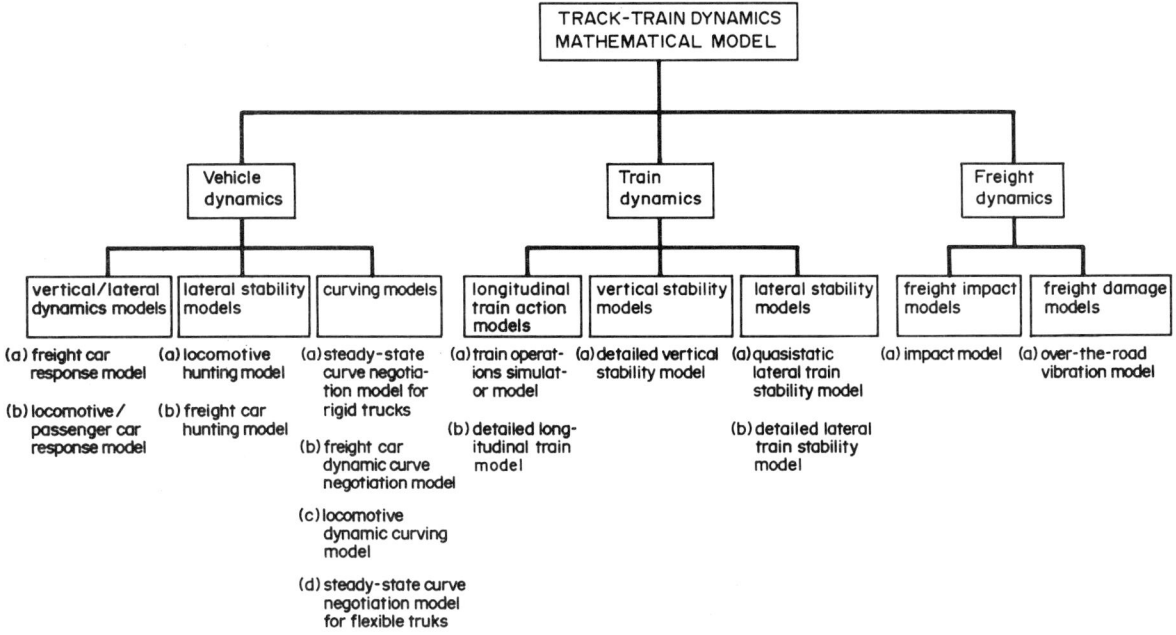

Figure 3
Model categories for railway dynamic investigaton (after Garg and Dukkipati 1984)

necessary to determine the location of the contact point and the relations among the wheel set coordinates can be computed from purely geometric considerations (see Fig. 5b).

(b) For the contact forces, the elasticity of the contacting material plays an important role, because of its influence on the size of the contact patch. According to Hertz's theory, the contact region is elliptical and the ratio of the semiaxes a/b can be calculated from a knowledge of the principal radii of curvature of the bodies of revolution. In order to calculate the actual size of the ellipse, the normal force is required as well. The resulting forces are most significantly affected by the so-called creepages, which are defined as the ratio of the relative velocities between wheel and rail and the vehicle speed v (see Fig. 6). Furthermore, the contact forces between wheel and rail depend on the adhesion conditions in the contact area, influenced by the surface roughness and also by the contamination due to humidity, dust, oil and so on. Both the contact geometry relations and the creep force–creepage relations need to be computed numerically with efficient software routines.

(i) Contact geometry programs fed with the mathematical shape functions of the wheel and rail profiles compute all required kinematical relations. As a result, the diagrams of Fig. 5b show some significant results for the contact angle δ_r, the roll angle ϕ and the contact point location at the wheel. Note the nonlinear history of the graphs when the contact changes from the tread to the flange of the wheel.

(ii) Programs based on Kalker's theory of rolling contact compute the creep forces between wheel and rail. Note that some of the geometrical relations computed in the step above are needed here.

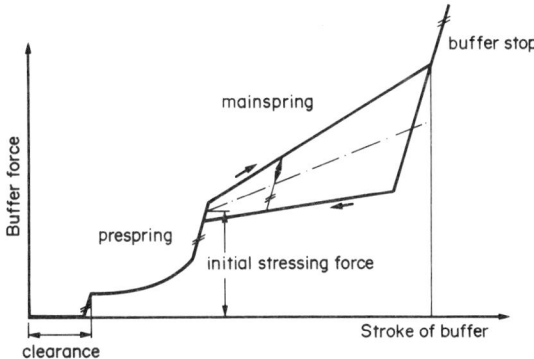

Figure 4
Typical force–displacement curve of a buffer model (after Horn *et al.* 1988)

189

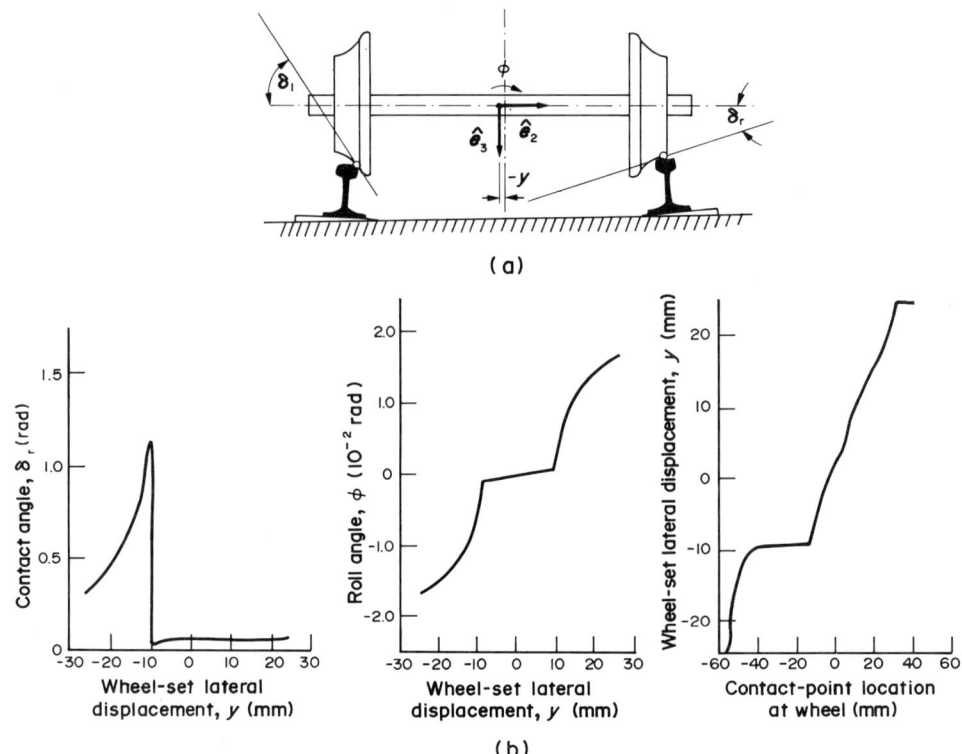

Figure 5
(a) The wheel–rail geometry problem and (b) some significant results

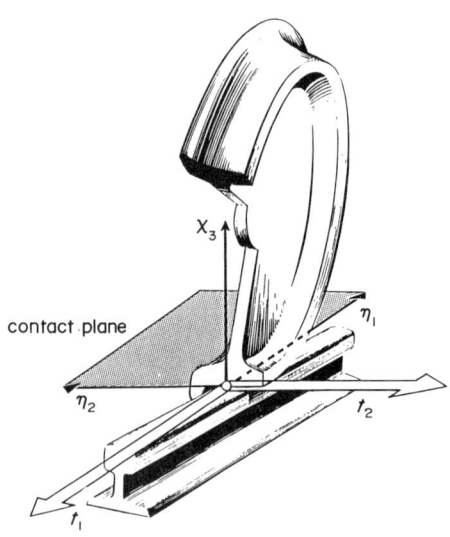

Figure 6
Definition of creep forces between wheel and rail

The results of this theory are usually represented in a nondimensional form:

$$t_{1,2} = t_{1,2}(\eta_1, \eta_2, \chi_3, a/b) \qquad (1)$$

where $t_{1,2}$ denotes the normalized longitudinal and lateral creep forces, η_1 and η_2 are the normalized longitudinal and lateral creepages, χ_3 denotes the normalized spin creepage and, finally, a/b represents the ratio of the semiaxes of the contact ellipse. This nondimensional form allows a helpful graphical representation with only four parameters instead of nine in the original representation. Figure 7 gives a demonstration of these relationships for some essential values of the four parameters of Eqn. (1).

Within a dynamic simulation of a wheel–rail vehicle with several wheel sets, efficient evaluation of these two steps is a necessity, because during time integration of the vehicle equations, both geometry and force computation have to be performed several times at every integration step. Therefore, attempts have been made to avoid the geometry programs to be evaluated at every time step and for every contact point. Also, the computer time for the creep-force computation could be reduced using combinations of linear creep-force relations and nonlinear force computation. In addition, approximations of creep-force laws based on curve fittings are useful (Duffek and Jaschinski 1982).

High-Speed Railroad: Modelling and Simulation

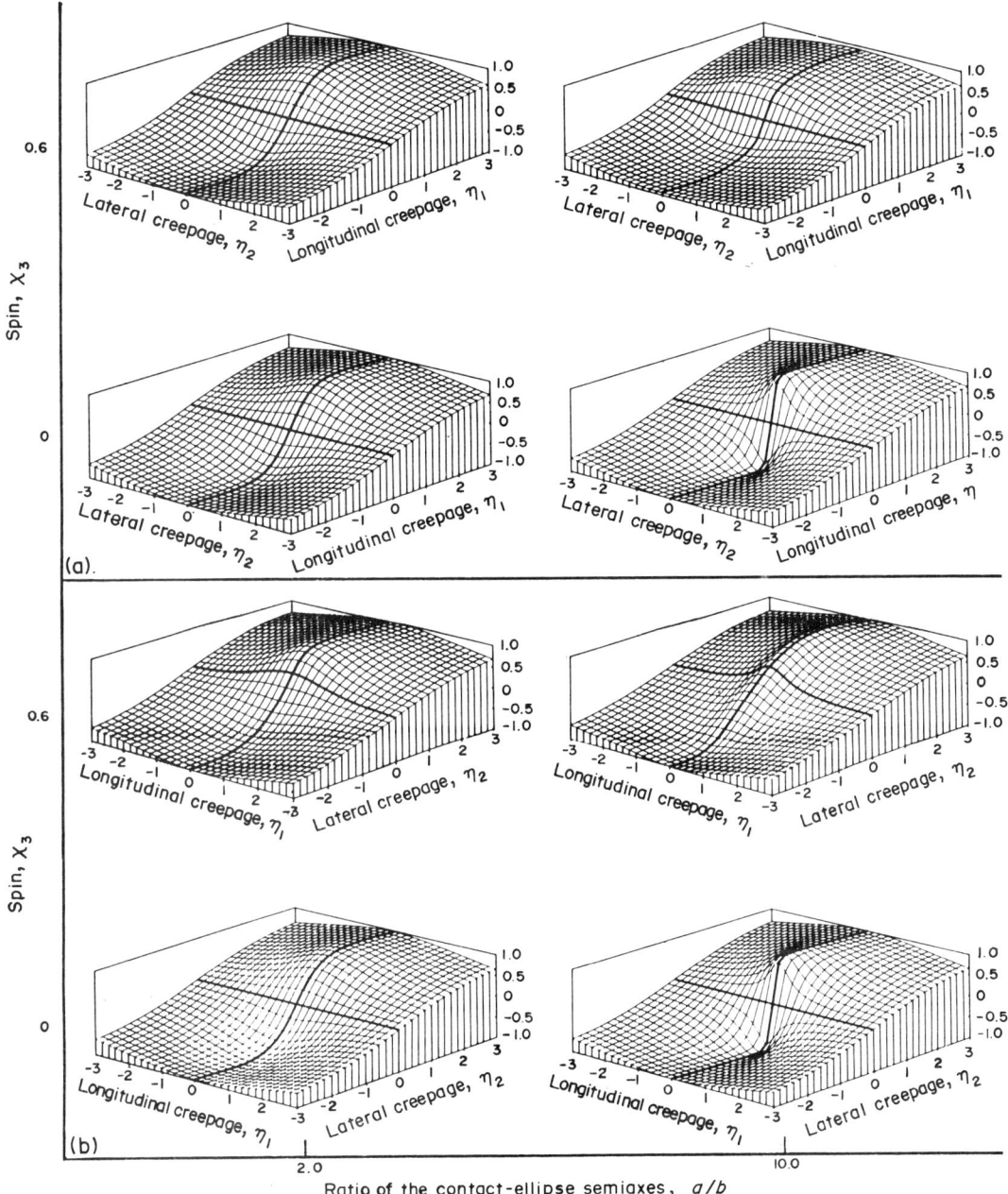

Figure 7
Selected results of creep force–creepage relations: (a) normalized longitudinal forces t_1 and (b) normalized lateral forces t_2

2. Railroad System Dynamics Simulation Software

With the modelling components as described in Sect. 1, individually specialized simulation models for railway vehicles or even for trains can, in prinicple, be built up. However, because the software development is expensive (in time as well as costs) and error prone, engineers are interested in using existing software packages. The most comprehensive survey of existing railway software is given by Dukkipati and Amyot (1988). Rather advanced is general purpose vehicle system dynamics software, which is based on multibody formalisms (Kortüm and Schiehlen 1985), since a wide class of vehicle or train models can be built up quite rapidly without being restricted to specialized models. In this context, MEDYNA is a multibody program that is especially adapted to the needs of railroad dynamics (Jaschinski *et al.* 1986); it will therefore be given as an example.

2.1 Modelling Capabilities of MEDYNA

MEDYNA handles small rigid-body motions relative to a large motion of one or more reference frames. Gravitational and gyroscopic effects of the individual bodies can be taken into account. The bodies can be kinematically connected in tree or closed loop configurations. Even small elastic deformations of the vehicle bodies are allowed.

A library of the most important coupling elements for modelling suspensions is available; it comprises linear, nonlinear, dynamic as well as active (i.e., feedback-controlled) force elements.

Another essential feature is the availability of specific railroad vehicle oriented models describing the primary lift and guidance suspension mechanics.

For wheel–rail vehicles, three model options are available:

(a) a general wheel–rail interconnection element with evaluation of the nonlinear contact geometry and contact forces;

(b) a general linear or quasilinear wheel–rail interconnection element with precomputation of the linearization parameters; and

(c) a nonlinear substructure describing a wheel set (with torsional flexibility of the axle) on a track element with precomputed tabulated geometry data and pointwise evaluation of the contact forces.

For magnetically levitated vehicles, linear and nonlinear analytical models for the magnetic levitation and guidance forces as functions of the air gap (equal to the relative motion of the vehicleborne magnets to the rails), magnetic flux, and voltage have been used.

For these models, MEDYNA establishes the dynamic equations, together with the holonomic constraints, automatically in the form

$$\mathbf{M}\ddot{p} + \mathbf{D}\dot{p} + \mathbf{K}p + \mathbf{C}_z f_{zd} = g(p, \dot{p}, r, u, t) \quad (2)$$
$$\mathbf{C}_z^T p = z$$

where $\mathbf{M}$ is the mass matrix, and $\mathbf{D}$ and $\mathbf{K}$ are the damping and stiffness matrices of linear interaction force laws and gyroscopic terms. Additionally, $\mathbf{K}$ also contains the effects of the preloads resulting from nominal constraints or applied forces. The term $\mathbf{C}_z$ is the constrained matrix, p is the total position vector of the unconstrained system, f_{zd} is the dynamic part of the constraint forces and z is the vector of kinematic excitations in the constraints. The vector g contains the generalized forces of nonlinear interaction laws, moving reference frames, effects of system inputs as external forces and kinematic excitations (vector u) and dynamic interaction laws (state vector r) (Wallrapp 1989).

The equations of motion are reduced by incorporating the constraint equations and transformed to the minimal (i.e., state-space) form. A special algorithm is used to determine the independent constraint relations and appropriate generalized coordinates y (also for closed loops). Finally, p is expressed in terms of y and z as

$$p = \mathbf{J}_y y + \mathbf{J}_z z \quad (3)$$

where $\mathbf{J}_y$ and $\mathbf{J}_z$ are constant Jacobian matrices of the free and locked modes of the MBS. The resulting state-space equations of the MBS are obtained in the following format:

$$\dot{x} = \mathbf{A}x + \mathbf{B}u + b(x, u, t)$$
$$x = [y^T \dot{y}^T, r^T]^T \quad (4)$$

where x is the state vector and $\mathbf{A}$ and $\mathbf{B}$ are the linear system and input matrices, respectively; in $b(x, u, t)$, all nonlinear and time-dependent terms resulting from nonlinear interaction forces, time-dependent inputs, moving reference frames and nonlinear substructures are summarized. Equation (4) describes the small body motion of the MBS relative to the moving reference frame under the assumption that the MBS starts from the static equilibrium.

2.2 Computational Methods

Special attention has been paid to the computational methods and numerical algorithms implemented in MEDYNA.

(a) Static analysis. For a given MBS configuration and given nominal applied external forces, the nominal constraint forces and the nominal interaction forces at all interconnections can be computed.

On the other hand, if the nominal interaction forces are specified but the equilibrium conditions are violated, the true equilibrium can be computed iteratively with the help of the Newton–Raphson method allowing large body motion. It should be noted that static analysis is required for correct linearization of the system, since the nominal loads (preloads) of the interconnections are required.

(b) Linear system analysis. Since the state equations are established, in the case of fully linear models, stability of the system can be concluded from the eigenvalues. In addition, all further linear analysis is based primarily on the eigenvalues and eigenvectors, as inputs to the MBS time-dependent external forces at certain body locations or time- or position-dependent

excitations (e.g., rail irregularities) at interconnections can be defined. Their components and possibly their time derivatives constitute the input vector $u(t)$.

Other components of the state vector, acceleration at specified vehicle locations or interaction forces can be chosen as output quantities. MEDYNA establishes the linear input–output equations in the form

$$\dot{x} = \mathbf{A}x + \mathbf{B}u$$
$$y_M = \mathbf{C}x + \mathbf{D}u \quad (5)$$

The input vector u for a vehicle with several axles can be obtained from the actual disturbance inputs with the aid of time delays; it can be given in the form of deterministic signals (e.g., sine waves) or as power-spectral densities (PSD).

Frequency domain analysis of Eqn. (5) starts from numerical computation of the transfer matrix

$$\mathbf{G}(j\omega) = \mathbf{C}(j\omega \mathbf{E} - \mathbf{A})^{-1}\mathbf{B} + \mathbf{D} \quad (6)$$

and yields frequency responses (Bode plots) or output spectral densities in the case of stochastic inputs. From the output densities, ride-comfort measures are computed by integration over frequency.

As an alternative, MEDYNA also allows the stochastic evaluation by covariance analysis. In this case the stochastic input process is represented by the output of a linear dynamic filter with white noise as input. Here, a series of Ljapunov matrix equations have to be solved numerically, yielding the desired covariance matrix, the elements of which can be used to obtain ride-comfort measures.

(c) *Nonlinear simulation via time integration.* The main tool for solving the usually nonlinear state equations is numerical integration. Several different integration routines have been tested. For most vehicle dynamics problems, Runge–Kutta methods of different order with error control and variable step size have proved to be efficient. Several Runge–Kutta methods are being used of various order with and without error control and with variable or constant step size including modifications by Merson, Shampine and Bettis; also included are multistep methods for stiff systems. These methods are very important in vehicle dynamics because of stiffness of the systems at low speeds and also when suspensions with Coulomb friction and hysteresis are modelled.

2.3 Program Handling and Preprocessing

MEDYNA is an integrated interactive program containing the model options and computational methods mentioned in Sects. 2.1 and 2.2.

(a) *Handling.* Referring to Fig. 8, the user is offered a menu of options, from which the desired input and computational routines are chosen. All options such as system definition, equation generation, analysis and evaluation are unified into one program. Therefore, the user can perform a complete system analysis and design task. Numerous built-in validitiy checks avoid, to a large extent, undesired program interruptions. Providing short or, if desired, comprehensive help texts, the program is self explanatory, which reduces the learning period substantially. A special data-organizational concept allows the session to be terminated at any input without loss of the previous data and computational results (restart capability). All model data and results are stored in one database. Through copying such a database and modifying it, new models are easily built up. As output media, the input terminal is available for hints and brief results and the printer and graphic terminal are available for detailed results. The program allows the definition of user-specific routines for nonlinear force laws of interconnections, system inputs, system outputs or measured wheel–rail profiles, which can be filled in by the individual user.

(b) *The ASKESE preprocessor.* For further processing of the data obtained from finite-element (FE) modelling for elastic bodies, a preprocessor for FE software

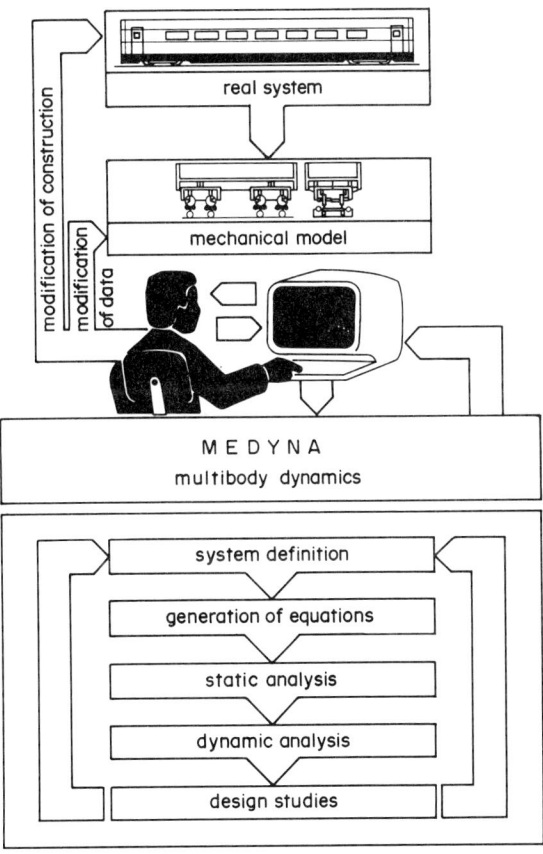

Figure 8
Schematic representation of the interactive multibody program MEDYNA

has been implemented. This processor transforms the system matrices (i.e., the mass, damping and stiffness matrices), generated from the FE program, into the modal matrices, as required for MEDYNA. In addition, the modal-shape functions are transferred as required to compute the modal loads. Through this interface, elastic degrees of freedom can be combined with rigid-body degrees of freedom for any part of the vehicle within MEDYNA.

3. Future Development

Mathematical modelling and simulation have become an essential part of CAAD of railroad systems. Modern software for complex railroad vehicles and trains has two key ingredients: a multibody formalism to build up easily vehicle models with various degrees of freedom and model assumptions; and the required wheel–rail interaction models, essentially a wheel–rail geometry and a creep force–creepage program. MEDYNA is a representative of such a tool and is currently the only tool to provide this combination as well as the required solution techniques.

Further development in this area needs to concentrate on the following issues:

(a) verifying and validating model assumptions;

(b) providing straightforward systematic techniques for obtaining the model parameters;

(c) speeding up the computational time by making use of advanced computer structures such as parallel processing; and

(d) reducing the effort for model definition and interpretation by making use of computer graphics and animation.

See also: Railroad Systems: Active Control

Bibliography

Brockhaus R, Doherr K F (eds.) 1987 Systemidentifikation in der Fahrzeugdynamik. Proc. Symp. Sicherheit im Luftverkehr. *DFVLR Mitt.* **22**
Duffek W, Jaschinski A 1982 Efficient implementation of wheel–rail contact mechanics in dynamic curving. *Proc. 7th IAVSD Symp.* Swets Zeitlinger, Lisse, The Netherlands, pp. 441–54
Dukkipati R V, Amyot J R 1988 *Computer-Aided Simulation in Railway Dynamics.* Dekker, New York
Garg V K, Dukkipati R V 1984 *Dynamics of Railway Vehicle Systems.* Academic Press, Toronto
Horn H, Jaschinski A, Sedlmair S 1988 Dynamic simulation of freight cars with nonlinear suspensions during curving. *Proc. 10th IAVSD Symp.* Swets Zeitlinger, Lisse, The Netherlands, pp. 161–8
Jänsch E 1988 Rad/Schiene Schnellbahn und Magnetschwebebahn im Vergleich. *ETR* **37** (5/6), 263–73
Jaschinski A, Kortüm W, Wallrapp O 1986 Simultation of ground vehicles with the multibody program MEDYNA, Symp. Simulation and Control of Ground Vehicles and Transport Systems. *AMD (Am. Soc. Mech. Eng.)* **80**, 315–41
Karnopp D C, Rosenberg R C 1983 *Introduction to Physical System Dynamics.* McGraw-Hill, New York
Kortüm W, Schiehlen W O 1985 General purpose vehicle system dynamics software based on multibody formalisms. *J. Veh. Syst. Dyn.* **14**, 229–63
Morman K N, Giannopoulos F 1982 Recent advances in the analytical and computational aspects of modelling active and passive suspensions. Computational methods in ground transportation. *AMD (Am. Soc. Mech. Eng.)* **50**, 75–115
Verein Deutscher Ingenieure (eds.) 1984 Dynamik schneller Bahnsysteme—Rad/Schiene und Magnetschwebetechnik. *VDI Ber.* **510**
Verein Deutscher Ingenieure (eds.) 1987 Dynamik fortschrittlicher Bahnsysteme. *VDI Ber.* **635**
Wallrapp O 1989 Entwicklung rechnergestützter Methoden der Mehrkörperdynamik in der Fahrzeugtechnik. Dissertation, Technical University of Berlin

W. Kortüm
[DLR, Wessling, Germany]

High-Speed Railroad Networks in Europe

Railroads, equipped with an obsolete infrastructure from the nineteenth century and burdened with numerous public duties, have been unable to compete with the dynamic development of road and air transport in the past few decades. The result has been a declining market share and large losses, compensated by governmental subsidies.

However, in terms of the environmental problems in Europe, railroads have considerable advantages: they cause only minimal noise and air pollution, they are safe and their energy consumption is comparatively low. Perhaps still more important, railroads are often underutilized—at a time when road and air traffic suffer from severe congestion.

In principle, railroads offer the basis for an advanced, high-speed, high-productivity transport system capable of moving even more goods and people along the most-travelled corridors.

1. The High-Speed Concept

1.1 Trip Time

Deficient speed is a key factor for the poor competitiveness of railroads. As a rule, private vehicles are superior on short distances and airplanes on long distances. In order to maintain superiority, at least on medium distances, maximum train speed must well exceed 200 km h^{-1}. Otherwise, the important door-to-door trip time (including access time), which is a very significant parameter for mode choice, is usually longer than for other modes. By exceeding this speed threshold, a

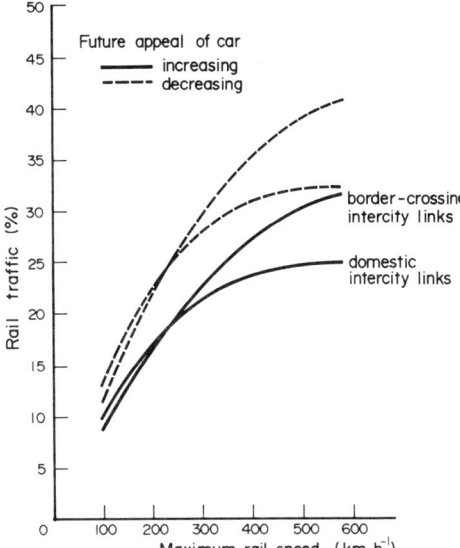

Figure 1
Predicted long-distance passenger transport market share in European corridors

change in travellers' mode-choice behavior in favor of rail travel can be expected.

Figure 1 displays the results of theoretical computations for the long-distance passenger market in Europe, based on mode-choice models. According to these calculations, a sensitive market reaction is expected for speeds up to about 400 km h^{-1} and even higher for European cross-border traffic (because of, on average, longer distances).

1.2 Experiences in France and Japan

Since 1981, when the high-speed track between Paris and Lyon began operation, the concept of high-speed trains has received much attention in Europe. The success of this service, its popularity with the public and its effect on the competitive situation has aroused considerable interest. The French national railroads (SNCF) had offered the *train à grande vitesse* (TGV), a completely new product, which was presented as an integrated service extending far beyond the high-speed line itself.

Two high-speed lines, the TGV southeast and the TGV Atlantic (under construction), are to form the main axes of a high-speed network that will serve three-quarters of the French population and will account for about 40% of total SNCF traffic.

The TGV southeast line has carried more than 100 million passengers. Actually, 20% of SNCF long-distance passengers are carried by TGV trains, at annual revenues of approximately US$600 million. The increase in passenger numbers between Paris and Lyon is more than 50% compared with 1980, the year before the high-speed link was opened.

The Japanese have the most experience with high-speed trains; the first Shinkansen was put into operation in 1964. Thus, there are especially comprehensive results of empirical investigations into the effects of high-speed services based on 25 years of operation.

From these results, it can be concluded that the Shinkansen has not only been a major success in the field of transport but has also stimulated the economic development of the regions served.

1.3 Future Potential

These examples demonstrate that higher train speeds can result in considerable effects from both commercial and economic points of view. Moreover, high-speed rail is not only a technical concept, but an evolutionary/revolutionary factor in the way people think about time, that could enable railroads to be seen in a new perspective. It could allow the railroad companies to make a strong "comeback" in public opinion and give new impetus to the notion of public service.

Consequently, in most of the industrialized countries, efforts have been intensified to develop high-speed trains, typically designed for speeds of about 250–300 km h^{-1}, and to provide the necessary infrastructure. Besides the French lines mentioned in Sect. 1.2, this has occurred mainly in Germany (Hannover–Würzburg, Mannheim–Stuttgart and some smaller sections), Italy (Rome–Florence), Austria and Switzerland.

Mainly in Germany and Japan, completely new systems are being developed that are based on magnetic levitation (maglev). These maglev systems are designed for cruise speeds of 400–500 km h^{-1}. As they are not compatible with wheel-on-rail systems, new tracks are needed (see *Magnetic Suspension Railroad Systems*).

2. Aspects of the European Transport Market

2.1 Traffic Demand

In planning high-speed services, not only domestic lines but also border-crossing links should be taken into consideration. While domestic demand shows a general tendency towards saturation, the cross-border transport market in Europe is increasing.

For example, in Germany, domestic passenger movements have increased by 50% as compared with 1960, while cross-border movements have trebled during the same period.

Forecasts for the Western European countries yield, on average, only a moderate 30% increase of traffic demand on domestic routes by the year 2010, due to the relatively moderate expectations on further development of gross national product (GNP) and the presumably minor changes of population. But for cross-border movements, the expected increase is twofold, above all as a result of the anticipated further progress towards European integration.

Unfortunately, it has to be stated that the national railroad companies neglected this growing market in the past, with the result that they have a market share of just 7%. This is less than cross-border bus traffic and also less than rail traffic on comparable domestic long-distance routes.

However, in recent years, some positive examples of international cooperation in the field of high-quality rail services have been proposed; namely, the channel tunnel, which is under construction, and the high-speed Paris–Brussels–Cologne rail project, which is being designed by the railways and governments concerned, with a final decision on construction expected soon. These links could form the first step on the way to a future common European high-speed rail network.

2.2 Major Axes of European Settlement

Considering the transport movements in Europe, it appears logical to consider those corridors with the highest total traffic movements as an appropriate basis for a network layout. Also, the particular advantages of high-speed trains are more apparent for the longer intercity movements.

As shown in Fig. 2, the spatial distribution of conurbations in Europe has several axes covering a wide area, such as the north–south corridors:

(a) Glasgow–Manchester–London–Paris–Lyons–Marseilles,

(b) Western Netherlands conurbation (Randstad Holland)–Brussels–Strasbourg–Zurich–Milan–Rome, and

(c) Rhine/Ruhr–Rhine/Main–Rhine/Neckar–Strasbourg–Zurich–Milan–Rome.

The following major axes exist in the west–east direction:

(d) Paris–Brussels–Cologne–Rhine/Ruhr–Hamburg, and

(e) Paris–Rhine/Neckar–Munich.

Characteristic data related to the axes and including all towns with a population of over 400 000 are given in Table 1.

A network that follows these axes would link the major conurbations in Central Europe and provide rapid connections with major peripheral areas. It connects 26 conurbations with a population of about 65 million. The network length is approximately 7000 km. The average distance between the conurbations is about 200 km; 90% of all such distances are greater than 120 km. These distances are best suited to high-speed trains.

Already this simple consideration demonstrates a high affinity between structural settlement features in Europe and the performance of high-speed trains.

3. Future Scenarios

In the following, results are given of model computations of demand and economy for European high-speed rail networks. The study was performed jointly by the German aerospace research establishment (DLR), the Institut National de Recherche sur les Transports et leur Sécurité (INRETS) and the Dutch center for transportation research (NEA). It was sponsored by the Commission of the European Community. The time horizon of the considered scenarios was the year 2010.

3.1 Network Configuration

The territories of 11 European countries are included in the considerations. In this area, different networks are treated which are all oriented at the European main axes but differ essentially in the degree of extension. The assumed extension reaches from a small high-speed network which mainly includes only sections that have already been constructed, are under construction or at least being planned, to very extended networks of more than 6000 km length.

Three of the investigated networks are selected and the results presented here. The networks—denoted A1, A2 and A3—are displayed in Figs. 3–5.

In scenario A1, it is assumed that as a minimum the existing high-speed links are complemented by new links under construction or planned, such as Hannover–Würzburg or Paris–Brussels–Cologne. In addition to these new links the plan shows several improved links as programmed by the national railroads. Hence, Fig. 3 virtually corresponds to actual high-speed railroad construction in Europe. In scenario A2, mainly the north–south connections are completed by new links. The network in scenario A3 has several additional links (e.g., two new east–west border-crossing links between Mannheim and Paris and between Lyons and Milan, and a fully improved London–Lille link). While the scenarios A1 and A2 are based on rather moderate and plausible assumptions, A3 appears to be realistic only if there is a strong growth in demand in the future.

In assessing the network layout, it should be taken into account that the study was to give a first orientation; an optimization of configurations was not included. In particular, it should be examined whether the Alps-crossing link and both the middle and south east–west sections would be better treated as predominantly improved instead of rebuilt. Also, more attention could perhaps be paid to an upgrading of the connections to Scandinavia and the Iberian peninsula.

However, disregarding such details, the applied hypotheses are well suited to giving useful information about the likely impact of high-speed rail networks on transport and economy, especially as a function of network extensiveness and also for particular routes.

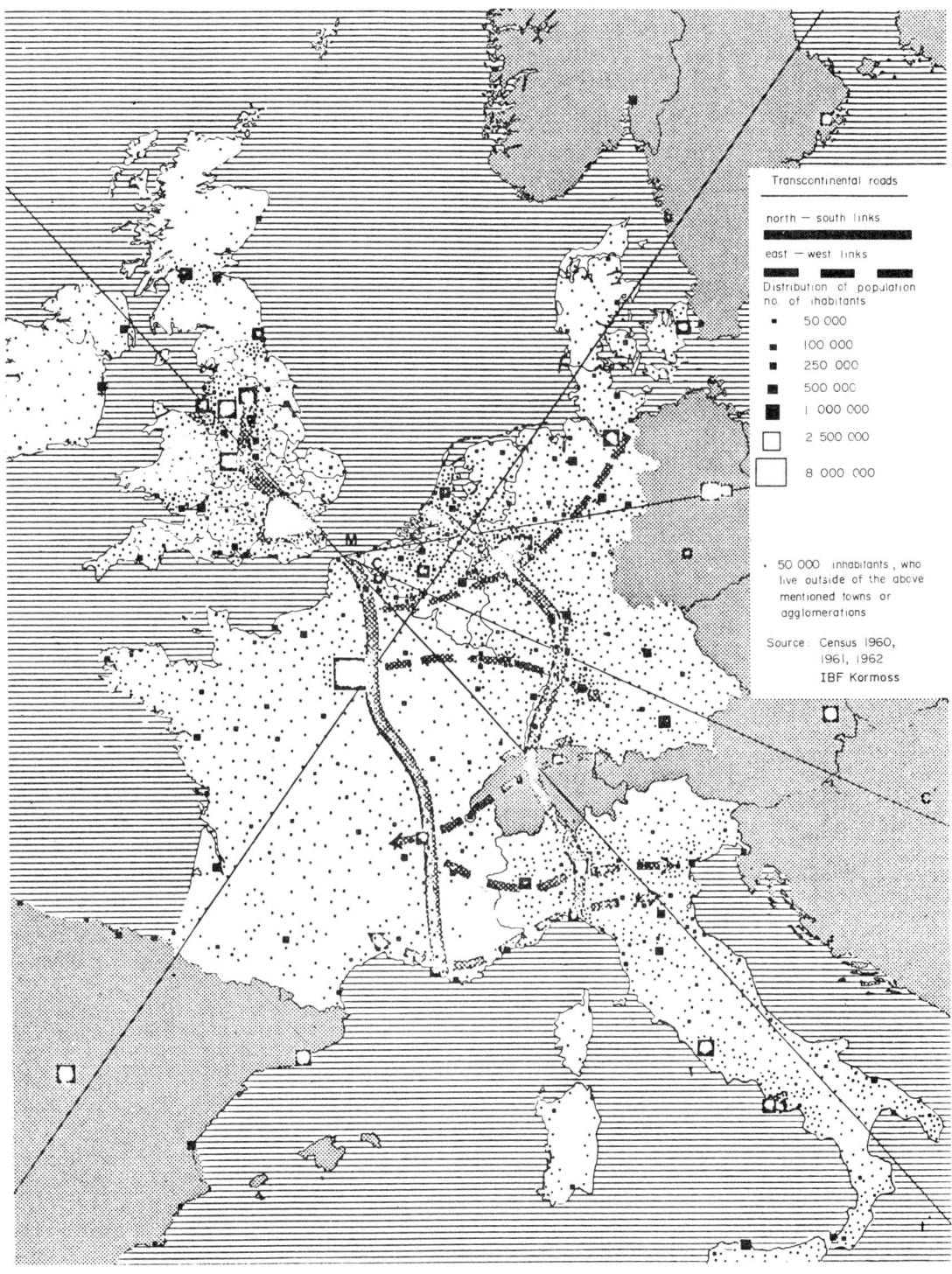

Figure 2
Distribution of conurbations in Europe showing several axes of settlement

Table 1
Corridor characteristics

Axis	Length (km)	Residents ($\times 10^6$)	Number of stations	Average intermediate distance (km)
(a)	2012	31.3	9	252
(b)	1722	14.1	12	157
(c)	1544	18.4	11	154
(d)	982	21.7	7	164
(e)	841	13.6	6	168

3.2 Estimated Demand

Demand was estimated by a passenger transport simulation model that comprises both transport and socioeconomic factors and their empirically derived interrelationships. The size of computations was extraordinarily large so that the results appear well founded and applicable—as far as future predictions can be made.

For model calibration, 3000 traffic relations were analyzed, in terms of supply and demand, separately

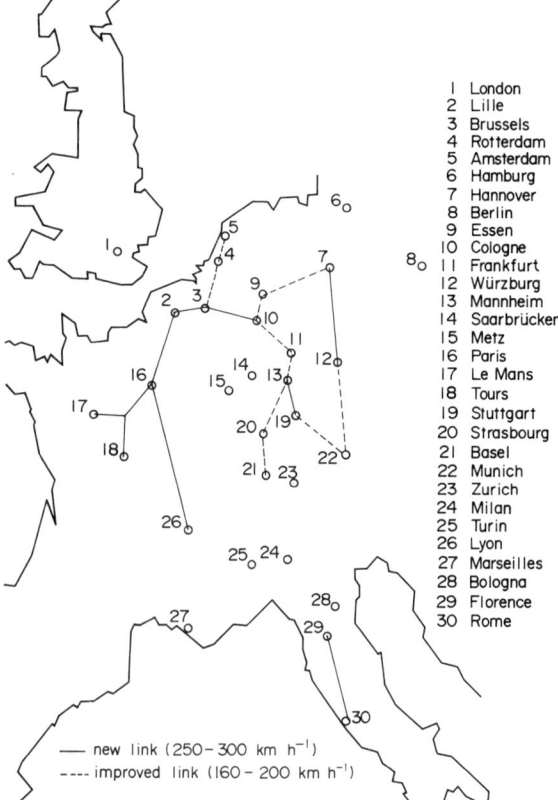

1 London
2 Lille
3 Brussels
4 Rotterdam
5 Amsterdam
6 Hamburg
7 Hannover
8 Berlin
9 Essen
10 Cologne
11 Frankfurt
12 Würzburg
13 Mannheim
14 Saarbrücken
15 Metz
16 Paris
17 Le Mans
18 Tours
19 Stuttgart
20 Strasbourg
21 Basel
22 Munich
23 Zurich
24 Milan
25 Turin
26 Lyon
27 Marseilles
28 Bologna
29 Florence
30 Rome

— new link (250–300 km h^{-1})
---- improved link (160–200 km h^{-1})

Figure 3
European high-speed railroad network: scenario A1

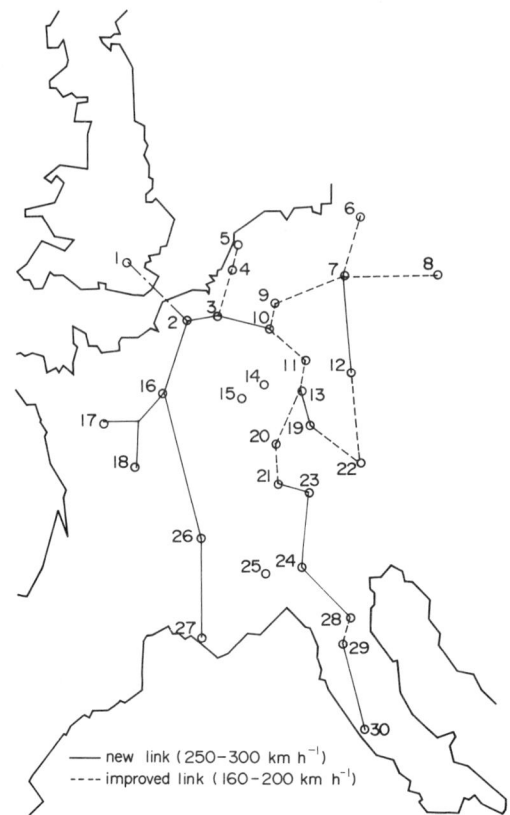

— new link (250–300 km h^{-1})
---- improved link (160–200 km h^{-1})

Figure 4
European high-speed railroad network: scenario A2 (legend as for Fig. 3)

for three trip purposes (business, vacation and other private reasons). As demand is strongly influenced by the socioeconomic environment, corresponding parameters were also applied (e.g., population, average household size, mean income, GNP, degree of urbanization and car availability). The data were used in a regionalized form (i.e., differentiated for 83 regions) and forecasts were made for the year 2010 on the basis of the most recent prognoses available.

Serving as a reference case, Fig. 6 shows the expected traffic flows in 2010 for a situation where no high-speed lines are put into operation. The calculated results for the traffic demand are displayed in Figs. 7–9. The thickness of lines corresponds to the number of traffic movements on the network sections.

The figures clearly show the considerable increase in traffic volumes on high-speed sections and also, to some extent, because of a spread effect, on other sections of the investigated network. As expected, total demand is highest in the scenario with the most extended high-speed network.

As the most important (aggregated) result of the computations, the considerable increase in rail traffic has to be emphasized, which amounts to:

(a) 40–90% on high-speed sections, and

(b) about 25% on other sections.

Furthermore, the relatively low traffic volumes yielded on the (all border-crossing) east–west routes in contrast to very high traffic volumes on the north–south axes have to be stressed as a structurally conspicuous result. It reflects the fact that, in the past, economic and sociocultural interconnections developed mainly within the national territories. However, the envisaged progress towards European integration will lead to a significant increase in the importance of east–west connections.

Nevertheless, the comparatively low demand on these sections generates major problems if cost-benefit considerations are based on the usually applied so-called territorial principle (i.e., if investment, expenses and revenues are allocated to the countries in accordance with their proportion of line). This problem has also influenced the decision on the Paris–Brussels–Cologne high-speed rail project. The anticipated high demand on the Paris–Brussels section makes good profitability likely there, whereas on the Brussels–

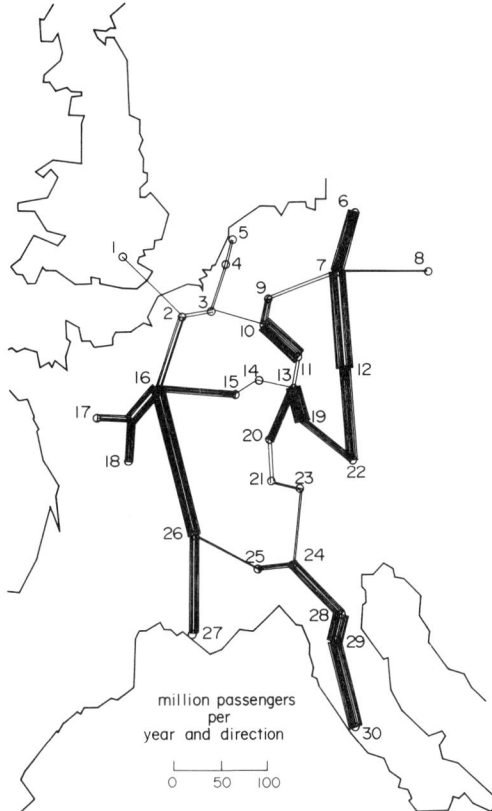

Figure 6
Predicted traffic flows in 2010 with no high-speed lines (legend as for Fig. 3)

Cologne section only a comparatively low demand can be expected. When basing the calculations on the territorial principle, the results are quite different for each country, varying from very satisfactory to unacceptable in terms of profitability.

The construction of a common European high-speed rail network inevitably requires European thinking by all parties concerned.

3.3 Economic Considerations

The investment required was estimated for each scenario by simplifying schematic assumptions about average cost per kilometer of track as a function of topography and population density. This method yields investments in the range US$15–50 million, according to the scenario considered.

This is of course a rough method, but calculating construction costs in advance and in great detail is extraordinarily complicated, and often nearly impossible, because the result is greatly dependent on the respective local situation.

If the estimate of investment costs is accepted as reliable, the resulting figures about the rentability are

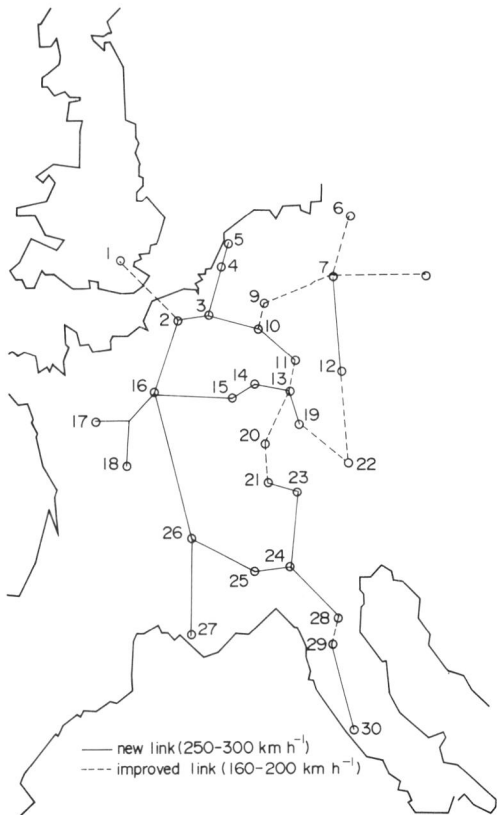

Figure 5
European high-speed railroad network: scenario A3 (legend as for Fig. 3)

encouraging. According to these calculations, it is expected that the annual net receipts (i.e., gross revenues minus operation costs including amortization) amount to 8.4–10.7% of investment cost.

As a meaningful result, the calculations show that the rentability tends to decrease with increasing network extension. Hence, the favoured strategy should be to invest in heavily frequented corridors but to limit investment in other corridors, with only minor upgrading of speeds to about 160 km h^{-1}.

The most immediate advantage of high-speed trains is the considerable trip-time reduction for travellers. The total reduction is estimated at 3×10^8 h annually or, more evidently when converted into monetary values, 7.6–10.4% of investment costs. The higher value again applies to the small network in A1.

This means the consumer surplus is of the same order of magnitude as the commercial indicator used above, or in other words, the social profitability is twice as high as the financial profitability.

The anticipated impacts on the overall economy also prove to be considerable. They are obviously highest in the most extended network in scenario A3. For this case, the most important effects ascertained are:

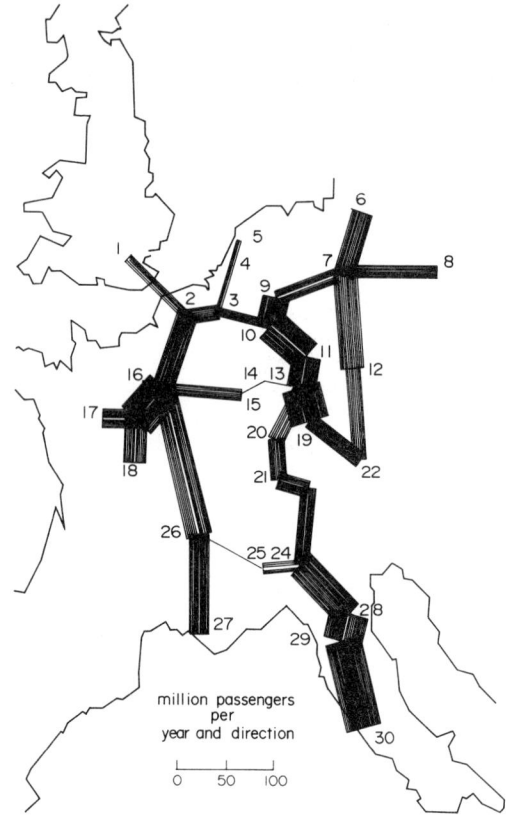

Figure 8
Predicted traffic flows in 2010 with scenario A2 (legend as for Fig. 3)

(a) an effect on employment, amounting to 2×10^6 man years in the construction phase;

(b) creation of 200 000 new working places in the operation phase, including both direct and indirect effects;

(c) an increase in the GNP of the countries concerned, of the order of US$5000 million;

(d) annual savings of between 2×10^9 l and 5×10^9 l oil compared with a situation where cars are the only means of transport; and

(e) a noticeable decline in accident numbers.

4. Future Development

Experience on the considerable effects of existing high-speed railroads and the described model computations about European high-speed rail networks clearly demonstrate that the high-speed concept enables railroads to be viewed in a new perspective.

Do railroads offer a bright future with clean, safe,

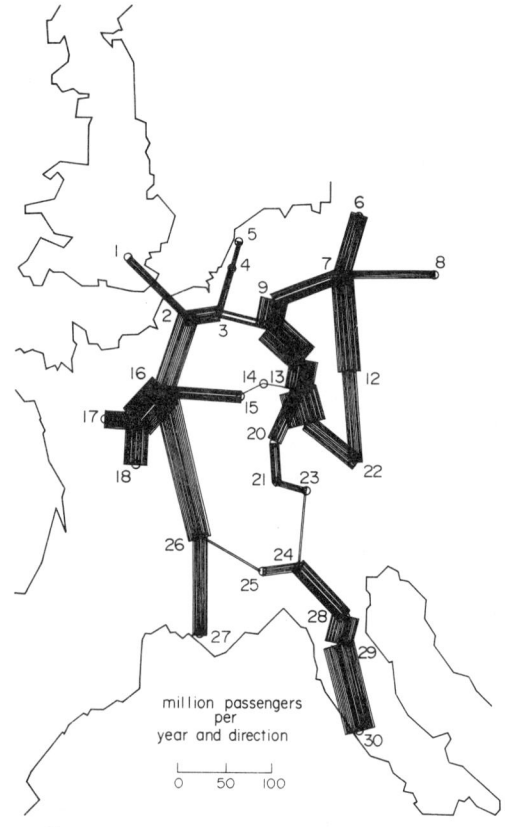

Figure 7
Predicted traffic flows in 2010 with scenario A1 (legend as for Fig. 3)

fuel-efficient technology and the ability to carry substantially more passengers and goods without making additional claims on precious land?

On the one hand, this can be affirmed without reluctance, all the more as there is much more technological potential for railroad innovation than is dealt with in this article, which focuses solely on high-speed railroads. However, obviously, in order to realize the potential, a much higher degree of market orientation and a radical change in the political understanding of the role of railroads is essential.

The experience of the Japanese national railroads (JNR) may be warning enough. For a long time the JNR was one of the most technologically advanced railroads attracting the highest passenger numbers. The profits of the Shinkansen high-speed lines, which on the main Tokyo–Osaka section made up a considerable 50% of revenue, were not used for the benefit of this market but to match losses on unattractive lines, thus avoiding their necessary closure. This caused irritation to many Shinkansen bullet-train passengers.

Furthermore, the JNR also agreed to construct high-speed lines which could not be profitable. For the last two lines, the Tohoku Shinkansen and the Joetsu Shinkansen, put into operation in 1982, their unprofitability was well evident in advance. Their implementation was politically motivated for reasons of regional planning.

The financial problems of JNR were not going to be solved in this way. By the time of its privatization in 1987, the conclusion from the tremendous loss makings had been drawn.

High-speed railroad networks have considerable potential for the benefit of the railroads, the natural environment and the economy, but their realization needs actions to be taken in accordance with the requirements of the market.

See also: High-Speed Railroad Networks in Japan; High-Speed Railroad: Systems Approach; Magnetic Suspension Railroad Systems

Bibliography

DLR, NEA, INRETS 1986 *Study on the Development of a High-Speed Rail Network in the European Community.* DLR, Cologne; NEA, Rijswijk; INRETS, Arcueil

Eberlein D 1981 Einsatzbereiche für Schnellbahnen unter besonderer Berücksichtigung der japanischen Erfahrungen. *Tagungsband Dtsch. Verkehrswiss. Ges. B* **60**, 271–94

Eberlein D 1985 Demand potentials for maglev passenger transport in Europe. *Proc. 1985 Conf. Maglev Transport.* Institute of Electrical Engineers of Japan, Tokyo, pp. 277–84

Eberlein D, Weber P J 1985 Compétition modale et vitesse ferroviaire optimale. *Les Aspects Socio-Economiques des Trains à Grande Vitesse.* La Documentation Française, Paris, pp. 82–92

Roundtable of European Industrialists 1984 *Missing Links.* REI, Brussels

Roundtable of European Industrialists 1987 *Keeping Europe Mobile.* REI, Brussels

Union Internationale des Chemins de Fer 1989 *Proposal for a European High-Speed Network.* UIC, Paris

Weber P J 1980 *Quantitative Abschätzung der Nachfrage nach Verkehrsleistungen Schneller Bahnsysteme um das Jahr 2000.* DLR, Cologne

D. Eberlein
[DLR, Cologne, Germany]

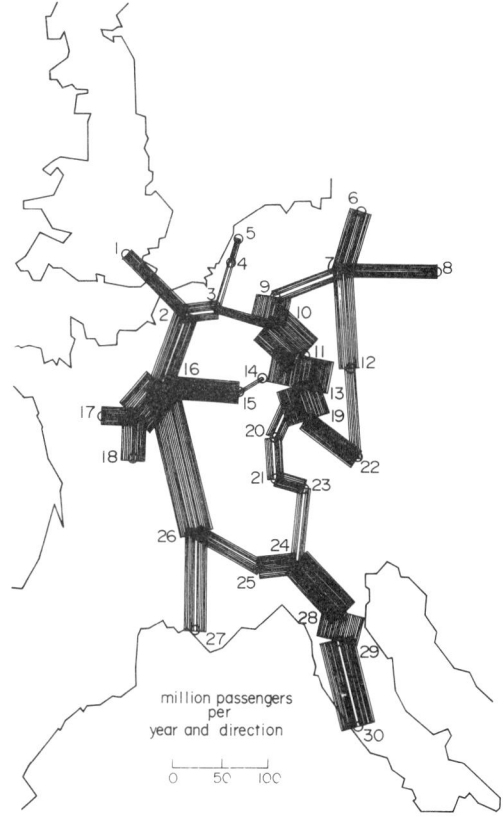

Figure 9
Predicted traffic flows in 2010 with scenario A3 (legend as for Fig. 3)

High-Speed Railroad Networks in Japan

Since the railroad was opened in Japan in 1872 the total service route length has grown to more than 30 000 km. In Japan, a large area of land is mountainous. The many bends and slopes make it difficult to increase the average speeds of trains, so the railroads have often lost out to airlines and expressways in high-speed transportation between cities and ports.

The population of Japan is concentrated in the cities along the Pacific corridor. With increasing economic growth the conventional Tokaido line was unable to cope with the increased demand for transportation. Consequently, Japanese national railroads (JNR) constructed the 516 km Tokaido Shinkansen line between Tokyo and Osaka in 1964. The Shinkansen railroad system was a new high-speed mass transportation system quite different from conventional railroad systems.

1. Outline of the Shinkansen Railroad System

The individual techniques used in the Shinkansen railroad are an extension of conventional railroad techniques, but the system itself is a new one and has enabled the safe, punctual and frequent operation of high-speed trains at lower costs than before.

The features of the Shinkansen railroad system are:

(a) the new line was constructed to avoid bends and slopes as much as possible so that the trains could run at high speeds constantly;

(b) the new train-safety system adopted is automatic train control (ATC) at a high level, making automatic operation possible to ensure safety;

(c) a fleet of high-speed trains is supervised through the centralized traffic control (CTC) system;

(d) two-level crossings to all roads are used for safety and access to the track is prohibited by law;

(e) all trains can run at the same speed with the same level of performance;

(f) both the production and the maintenance costs of the rolling stock are reduced by their standardization—use of the electric brake for stopping trains at high speeds is safe and makes routine maintenance easier;

(g) since all the rolling stock is driven by a motor and each axle load is equally heavy, the track suffers only slight damage and is inspected by inspection trains which are able to run at the same speed as the passenger trains;

(h) the train schedule, including six Hikari trains and four Kodama trains per hour, enables passengers to take a train at almost any time; and

(i) the reservation of seats is exclusively controlled by the central computer—this is convenient for passengers and ensures high reliability.

Between 1964 and 1988, the Tokaido and Sanyo Shinkansen lines carried 2500 million passengers, with a daily maximum of 282 trains and 1.030 million passengers, and maintained a high level of safety. The opening of the Tohoku Shinkansen and the Joetsu Shinkansen in 1982 brought the total Shinkansen route length up to 1810 km and the Shinkasen network is now the main artery of the transportation system in Japan.

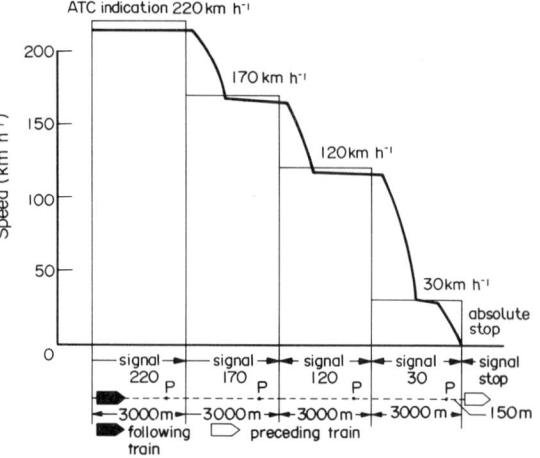

Figure 1
ATC system (when necessary, a signal for 70 km h^{-1} can be sent)

2. Advanced Control System for Reliable Operation

2.1 Control System

The control system for operating the Shinkasen railroad consists of the two main systems, ATC and the computer-aided traffic control system (COMTRAC). Furthermore, the CTC system controls a fleet of trains. The outline for the reliable operation of each system on the Shinkasen is as follows.

(*a*) *ATC*. This guarantees safe operation by automatically regulating speed and braking functions (see Fig. 1) Although the Shinkansen has no visual signalling equipment along the line, speed and brake commands are transmitted directly to the signal indication meter in the train cabin via electrical impulses sent over the rails.

The Shinkansen tracks have electrical circuit insulators installed at 3 km intervals. As a train passes over these insulators, the left and right rails are electrically linked by the wheels and axles of the train, thus indicating that it has entered that 3 km zone. Subsequent trains can then be instructed to maintain an appropriate speed. The ground-level ATC units automatically send out signals for speeds of 220, 170, 120, 70 or 30 km h^{-1}. If a train exceeds the indicated maximum speed, it is automatically slowed down. When approaching a zone in which a preceding train is running, or a station at which it is scheduled to stop, the train receives a stop signal. Speeds of 70 km h^{-1} and 120 km h^{-1} are used only when maintenance work is in progress and where there are tight bends.

All of these systems are fully automatic, and operations perfomed by ATC are chanelled through the CTC and then duly recorded.

(b) *COMTRAC*. This is the core of the Shinkansen control system (see Fig. 2). Utilizing a total of seven computers, it coordinates such operational information as departure and arrival times, track numbers and car configurations.

(c) *Centralized information control (CIC)*. This is a communications installation monitoring control system. When a mechanical breakdown occurs, such as in the train radiotelephone or some other part of the communications network, CIC swiftly and accurately locates the source of the trouble and assesses the extent of the problem and any possible repercussions for the entire Shinkansen network. Repair and maintenance operations can then begin without delay.

(d) *Centralized substation control (CSC)*. This monitors the supply of electric power to the Shinkansen. Electric power companies in eastern, central and western Japan supply electricity to the Shinkansen substations. Transmission lines send three-phase 77 kV and 275 kV electric power to transformers which then convert it into single-phase, 25 kV electricity. The power supply network is monitored via a system board, and a character display shows the state of equipment in unmanned substations and sectioning posts. Thus, rapid response to breakdowns and other problems is assured.

(e) *CTC*. This is the COMTRAC nerve-center that transmits instructions around the network. Relevant information and commands are sent from the CTC central controller by fiber-optic cable to controllers at each station. A vast amount of information is processed through this system, including routing, train numbers, stops, delays, train locations, wind velocity alarms, tunnel water levels and signalling.

(f) *Man–machine advanced processor (MAP)*. This consists of two computers that analyze data forwarded by the CTC. Information about the location of trains, the situation at various stations and other operational factors is then available for the dispatchers to retrieve on their terminals.

(g) *Program route control (PRC)*. This comprises three computers that monitor and control all aspects of the daily operation of the Shinkansen system. This covers departure and arrival times, selection of platforms and train carriages, and carriage configurations. Furthermore, PRC advises dispatchers if a train is delayed and can implement the automatic adjustments to the timetable generated by electronic data processing (EDP) if a major revision is necessary. One of the three computers can adequately handle the many tasks necessary, but a second one is always operated in parallel to constantly confirm the accuracy of data and decisions. The third computer is a backup designed to guarantee absolute safety in the event of an unpredictable failure in one of the other two.

(h) *EDP*. This is made up of two powerful computers that conduct basic planning for Shinkansen operations,

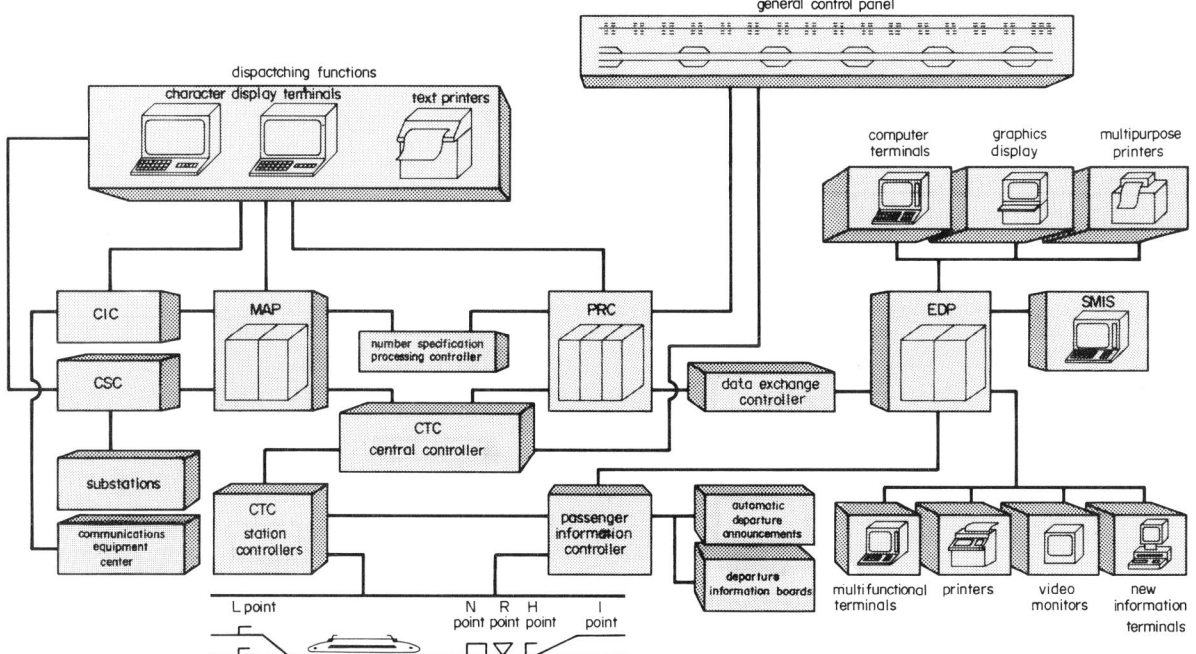

Figure 2
COMTRAC system

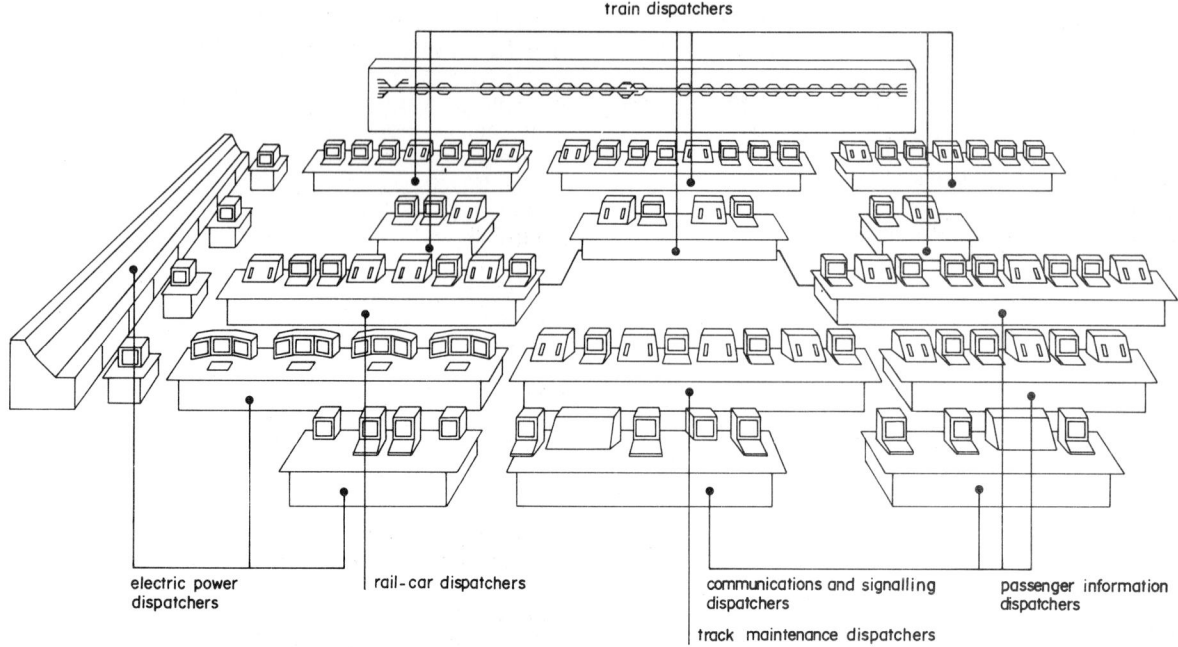

Figure 3
General control center

which entails coordinating daily schedules for both services and equipment, creating major timetable revisions and planning inspection activities. These plans are based on an analysis of the number of kilometers travelled and the number of hours of operation. To minimize problems stemming from train delays, EDP generates revised schedules and advises dispatchers of the likely impact on operations.

(i) *Shinkansen management information system (SMIS)*. Data from the systems described are channelled to SMIS, which coordinates information on Shinkansen operations. Data processed by SMIS is then forwarded to the Shinkansen management for analysis.

SMIS supplies information on vehicle use, performance and breakdowns. Track deterioration, line friction, and other problems are checked by a special inspection train and then the data are input and analyzed, providing information for use in visual inspections and maintenance of the entire network.

Passenger traffic trends, traffic volume by zone, passenger loads, sales results and other facts provide data essential for developing schedules for regular and holiday trains.

2.2 General Control Center
All facets of the operation of the system must be carefully monitored, including train movements and scheduling, maintenance of tracks, facilities and vehicles, the electric power supply and signalling systems (see *Railroad Electronic Signalling*).

In November 1987, a new general control center for the Shinkansen was opened. This facility is half the size of its predecessor but has 20 times the data processing capacity and four times the number of terminals, permitting faster and more reliable control (see Fig. 3). The many interrelated systems that ensure safe, punctual operation of the Shinkansen are coordinated by this center. The two main systems are ATC, installed along the tracks and in the trains, and COMTRAC, the centralized monitoring system.

(a) *Train dispatchers*. These dispatchers can monitor train movements via their video terminals and transmit instructions and assume control of trains during natural disasters, adverse weather conditions and other incidents. Using character and graphics displays, they make arrangements for necessary repair work and restoration of service in accordance with other dispatchers' information.

(b) *Rail-car dispatchers*. These dispatchers inform the engineers and conductors of changes in route or assignment and offer advice on steps to take if a rail car malfunctions. They also arrange for emergency rail car replacement and plan regular inspection and maintenance work and overhauls.

(c) *Electric power dispatchers*. While observing the CSC control board, these dispatchers ensure a stable supply of electric power from regional power companies to the entire Shinkansen network. They also give necessary instructions to maintain or restore power during interruptions to normal operation.

(d) *Communications and signalling dispatchers*. These dispatchers are responsible for monitoring the performance of such systems as ATC, CTC, the train radio-telephone, other devices in the communications equipment center and all signalling equipment. Should a breakdown occur, they initiate prompt action by informing technicians in the affected areas.

(e) *Passenger information dispatchers*. These dispatchers provide stations and conductors with a smooth, rapid flow of information relating to passenger services. In addition to providing information under normal operating circumstances, when delays occur they must promptly process and transmit data on train connections in order to minimize inconvenience to passengers.

(f) *Track maintenance dispatchers*. These dispatchers analyze any track abnormalities detected by the special inspection train, by monitoring devices installed in scheduled trains and by visual inspections conducted daily by maintenance workers. They then issue instructions for trains in the affected area to reduce speed or stop and for emergency repair work to be undertaken. These dispatchers are also responsible for coordinating regular maintenance activities, which are undertaken at night after scheduled runs are completed.

2.3 Safety and Maintenance Standards

(a) *Weather monitoring devices*. These have been placed at regular intervals along Shinkansen lines. On the Tokaido Shinkansen line, 35 anemometers have been located in order to record wind velocity, and 28 devices to measure and record the amount of rainfall. This equipment alerts dispatchers in the general control center when weather conditions pose a safety hazard. If, for example, the wind velocity or the precipitation exceeds safety standards, dispatchers transmit a signal to halt the trains.

Additionally, in areas of heavy snowfall, particularly in Shiga Prefecture, high-velocity water spraying devices have been installed that can damp snow to prevent it swirling up and accumulating in the slipstream of the Shinkansen.

The climate along the Tohoku Shinkansen is cold but snow seldom falls. Therefore, removal of snow piled on the tracks by snow plows on trains and viaducts capable of storing snow under the rail are the fundamental countermeasures. On the other hand, the Joetsu Shinkansen runs through the heavy-snow region of Japan, with snow drifts deeper than 4 m. Thus, the basic countermeasure is to sprinkle heated water from sprinklers along the tracks when snowfall is detected by automatic detectors.

(b) *Earthquake countermeasures*. Because the Japanese archipelago lies in an earthquake zone, earthquake countermeasures are necessary for safe operation of the Shinkansen. Seismographs, each with an effective range of approximately 20 km, are installed at the 25 transformer substations between Tokyo and Osaka, thereby covering the entire route. When an earthquake of intensity four or greater on the Japanese scale is detected (equivalent to 0.4 m s^{-2}), power is automatically shut down within a 20 km radius and trains are stopped. For rapid detection of earthquakes in the Pacific Ocean, a sensor system is linked to the electric power dispatcher's graphics display in the general control center.

For the Tohoku and Joetsu Shinkansen, seismometers are placed along the track as well as along the distant Pacific Ocean coastline where there is an earthquake zone. Using this equipment, it is possible to obtain information earler about an earthquake occurring in the Pacific Ocean, and so to reduce train speeds before the tremor reaches the inland Shinkansen lines. This is a double system that uses wires on the ground and a communications satellite in order to increase the reliability.

(c) *Safe construction*. The Shinkansen has been constructed with the utmost safety in mind. The Shinkansen track is completely separated from roads and expressways. The track itself is completely separated from conventional lines and is protected by a high fence, thus deterring unauthorized access.

To enable high-speed operation, the track is virtually straight and almost completely level. As a result, tunnels through mountainous areas tend to be long and are, therefore, equipped with advanced safety features. Lighting is installed at 15 m intervals, and emergency telephones can be found every 500 m. Fire-extinguishing equipment has also been installed at the entrances to and inside all tunnels.

(d) *Maintenance*. To ensure the highest safety standards, section teams responsible for track and electrical equipment maintenance are always on duty at various points along the Shinkansen line. These teams carefully monitor their respective areas and perform daily visual inspections. In addition, every ten days, a seven-car multipurpose inspection train passes along regular passenger routes at normal speeds. Utilizing on-board computers, it detects and measures any irregularities in both track and electrical facilities; this data is centrally processed and analyzed. According to the information resulting from the daily patrols and computer analysis, over 3000 maintenance workers perform both routine and special repairs after the regular train service ends at night.

In addition, recording devices are periodically mounted on scheduled trains to determine the actual level of riding comfort. Necessary adjustments are made if any abnormalities are detected.

Thorough inspection of rolling stock is conducted on a periodic basis to detect irregularities and spot potential problem areas. Based on rail-car usage, external examination of parts vital to operational safety is carried out and regular inspection of the bogies and other parts is performed at five rolling-stock depots. Major disassembling and overhauls are scheduled for at least every 36 months or 900 000 km (whichever is the sooner) and are performed at the workshops.

3. Recent Developments

The ATC signalling system on the Tohoku and Joetsu Shinkansens is very expensive. A version that makes use of transponders is under development that will eventually reduce costs.

The earthquake detection system mentioned in Sect. 2.3 is effective for the Tohoku and Joetsu Shinkansens. However, it does not provide enough warning for the Tokaido Shinkansen, so the urgent detection and alarm system (UrEDAS) is under development. This will be able to predict seismic intensity by detecting the p wave of an earthquake.

Since 1989, an extra Hikari train per hour has operated on the Tokaido Shinkansen. However, the demand is such that a further increase in the number of trains is needed. Use of the existing ATC system with fixed blocks makes it very difficult to add any further trains, so an ATC system with movable blocks is under development.

Leakage coaxial cables (LCX) were installed along the entire Tokaido Shinkansen line and each train has an antenna on the side of the vehicle body. The system came into service in 1989. It allows more stable transmission quality to be achieved and makes possible various services by connecting terminals inside the vehicle with a ground station. The system is used to meet the increasing demand for public telephones on trains, to relay radio broadcasts to passenger cars and to provide passengers with information relating to passenger services.

4. Planned Projects

4.1 Increasing Average Speeds

(a) *Technical problems for higher-speed commercial operation on the existing Shinkansen lines.* Sophisticated scheduling of trains is necessary to operate the higher-speed trains because the timetable on the Tokaido line is already overcrowded. From the environmental viewpoint, it is difficult to control the ground vibration induced by higher-speed trains because the structures are a little weaker on the Tokaido line than on the Sanyo, Tohoku and Joetsu lines. Even on the Sanyo, Tohoku and Joetsu lines, it is very expensive to control the low-frequency noise resulting from the air pressure wave produced when a higher-speed train passes through a tunnel.

With improving equipment, the present trains should be replaced by ones with higher speeds and lighter vehicles must be developed that can solve some of the problems of environmental conservation. These measures will allow the speeding up of the existing Shinkansen.

(b) *Latest tests.* The maximum speed attained on the Shinkansen is 319 km h^{-1} in 1979. However, for the next commercial speeding up on the Tohoku line (opened to service with a maximum speed of 240 km h^{-1}) tests were carried out attaining a speed of 275 km h^{-1} in 1988. The object of the test in 1988 was mainly to investigate the many technical measures to control the external noise at 275 km h^{-1}.

The new series 100 train for the Tokaido–Sanyo Shinkansen was tested at a maximum speed of 260 km h^{-1} in 1985 for the future speeding up on the Sanyo line, and the external noise of the series 100 was measured at the same time.

(c) *Commercial services.* As the result of these high speed tests, JNR increased the maximum speed on the Tohoku line from 210 km h^{-1} to 240 km h^{-1} in 1985 and on the Tokaido–Sanyo line from 210 km h^{-1} to 220 km h^{-1} in 1986. On both lines, JNR has accomplished the increase in speed with the existing rolling stock.

West Japan railroad company increased the maximum speed on the Sanyo line from 220 km h^{-1} to 230 km h^{-1} in 1989 using the new series 100N.

For the ATC system, the signal for 210 km h^{-1} was changed to 220 km h^{-1} or 230 km h^{-1} on board the train. The length of some ATC sections was adjusted.

(d) *Future developments.* After several high-speed tests, measures are possible for improving equipment and rolling stock that would allow the service to operate at a maximum speed of 300 km h^{-1}. Among the problems raised are the control of external noise and the reduction of operational costs for energy consumption and maintenance.

The improvement of equipment for speeding up the existing Shinkansen is technically very difficult and will need much time and expense. Therefore, the way to proceed should be to steadily introduce new, more lightweight rolling stock designed to have a low running resistance based on using new technology.

For the moment the target maximum speed is 270 km h^{-1} to be followed by 300 km h^{-1} in the next stage. The aim is to achieve a journey time of 2.5 h between Tokyo and Osaka, and also between Osaka and Hakata.

4.2 Environmental Conservation

It is difficult to solve the problems of external noise and vibration. The environmental standards for the control of external noise are that it be no more than 70 phon in residential areas and no more than 75 phon in commercial or industrial regions.

The external noise is made up from the noise from the current collection system, the rolling noise between wheel and rail, the aerodynamic noise from the surface of bodies and their projecting parts and the noise arising from the structures.

The external noise from the current collection system consists of the spark noise by contact loss, the sliding noise and the aerodynamic noise from pantographs. Among them the spark noise is the loudest. However, the spark noise problem can nearly be solved by a reduction in the number of pantographs and the connection of pantographs (in parallel) through an electric circuit. By connecting the pantographs in this way the collecting current flows from a pantograph that has lost contact to others, and the spark disappears.

These measures have already been carried out on the Tohoku line and are due to be introduced on the Tokaido–Sanyo line. They will become possible because of the improvement of switching the electric feeding system from the booster transformer (BT) system on the Tokaido line to the autotransformer (AT) system. The AT feeding system has already been adopted on the Sanyo line.

After reducing the spark noise, the aerodynamic noise of the pantograph comes to predominate at higher speeds. It is recognized that a device to deflect air currents from a pantograph would be effective in reducing the aerodynamic noise, so research into a noiseless shape for the pantograph is continuing.

The rolling noise between wheel and rail is decreased somewhat by setting side walls along the tracks, and smoothing the upper surface of the rail is effective for reducing the noise at source. This was carried out on the Shinkansen lines and the grinding technique is in the process of being improved.

It is difficult to reduce the aerodynamic noise from the surface of bodies and their projecting parts when running through the air at higher speeds. This noise increases more markedly than other noises over 250 km h^{-1}. Proposed measures are to reduce the sectional area of body, to smooth the body surface and to remove and/or shroud the projecting parts. The frontal shape of the series 100 train, which is 900 mm longer than the series 0 train, proved these measures to be effective. However, further efforts must be made to decrease the surface aerodynamic noise.

The ground vibrations induced by higher-speed trains are localized depending on the structural forms and the ground condition. Basically, reducing the weight of the rolling stock is effective in reducing the vibration.

The low-frequency noise resulting from the air pressure wave when a higher-speed train passes through a tunnel is generally reduced by changing the shapes of the tunnel ends. However, this is not always practical for the existing Shinkansen, so other lower-cost means are being sought.

4.3 Introduction of Shinkansen Trains on Narrow-Gauge Lines

To revitalize the old railroads, Shinkansen trains are being introduced on old lines. The gauge of the old lines is 1067 mm, but the gauge of Shinkansens is 1435 mm.

When passengers travel to locations along old lines, they must transfer from the Shinkansen line to the old line. This transfer is inconvenient, troublesome and time consuming; therefore, the development of a through train moving from the Shinkansen onto the old lines has long been desired.

It is necessary for the through operation to convert the existing 1067 mm gauge lines to 1435 mm, to combine the two systems by laying a second set of rails or to develop variable-gauge bogies like the Talgo train in Spain. However, it is difficult to develop variable-gauge bogies for high-speed multiple units. Instead, it should prove relatively easy to convert the existing lines to 1435 mm gauge or to lay a second set of rails. Either of these will be chosen depending on the natural conditions. For example, if there is a bypass line, the old line can be converted to 1435 mm gauge. Track maintenance will then be easier in the future.

Since the economic advantage of this idea consists in not converting the conventional cross sections, the cross section of tunnels, bridges or platform facilities can remain unchanged. Thus, it is technically necessary to run small-sized rolling stock on Shinkansen lines at higher speed. On the old lines, there are many small-radius bends so the wheelbase of the bogie must be shorter. The development of a bogie with a shorter wheelbase is being pursued so as to secure the running stability at higher speeds on the Shinkansen. Increasing the speed around bends is necessary to reduce the travel time on the old lines.

At the same time, lighter, smaller-bodied vehicles with smaller power equipment are to be introduced in a few years.

The new form of railroad will be attractive for small and medium cities (e.g., Yamagata) on the old railroad and near to the Shinkansen.

4.4 The Future

The Japanese high-speed network had been progressing steadily with the construction of the Tokaido, Sanyo, Tohoku and Joetsu Shinkansens. However, JNR was divided into JR groups in 1987, interrupting further progress. JR groups have been very cautious about the constructon of more Shinkansen lines.

The linear motor car Maglev, whose maximum speed is more than 500 km h^{-1} is under development and has entered the test phase for practical use. The idea of the Japan corridor has been proposed, whereby the distance between Tokyo and Osaka would be covered in about an hour by Maglev.

On the other hand, the new form of railroad which introduces the Shinkansen trains onto old lines is being realized. As a model case, the old line between Fukushima and Yamagata was selected.

Thus, it is very difficult to predict the future development of the Japanese high-speed network under the present variable conditions.

JR groups operating the Shinkansens have plans to increase speeds while meeting the environmental conditions. Research and development efforts of the

Railway Technical Research Institute are eagerly anticipated by JR groups.

The Central Japan railroad company plans to cover the distance between Tokyo and Osaka in 2.5 h at a maximum speed of 270 km h^{-1}, and the West Japan railroad company plans to link Osaka and Hakata in 2.5 h at a maximum speed of 300 km h^{-1}. The East Japan railroad company, operating at a maximum speed of 240 km h^{-1} on the Tohoku Shinkansen, is of the opinion that a commercial service operated at a maximum speed of 270 km h^{-1} should not present many technical difficulties.

See also: Automatic Train Control: Protection Principles; Automatic Train Control: Safety and Reliability; High-Speed Railroad: Modelling and Simulation; High-Speed Railroad Networks in Europe; High-Speed Railroad: Systems Approach

A. Mochizuki
[Japan Railroad Group, Tokyo, Japan]

High-Speed Railroad: Systems Approach

The history of railroads goes back over a hundred years and, in that time, technological progress and changes in organization methodology have, in general, been superimposed on preexisting structures. This approach has maintained a sense of continuity with the past and allowed the exploitation of consolidated experience of running railroads. However, the system in its entirety has grown up not with an integrated and unitary approach, but rather through a collection of subsystems linked by technological and procedural/organizational interfaces, often of a highly complex nature. This evolution was acceptable until the end of the 1960s. The external world with which the railroad interacted was relatively stable. Since then a fundamentally new situation has arisen; the extremely rapid evolution of communications technology is fast making the railways obsolete and it is no longer possible to follow a strategy of minor adjustments to existing equipment or low integration.

The high-speed (HS) project, created by the Italian railroad Ferrovie dello Stato (FS), is not only, therefore, simply the idea to develop a high-speed train but above all an HS system.

The different subsystems (train, contact line, track, fixed plants, etc.) have been laid out in an integrated fashion and each subsystem has been designed according to an integrated approach.

The term high speed has been introduced into the vocabulary of the railroad service to signify a threshold velocity of approximately 250 km h^{-1}. Above this speed, the rail transport system is differentiated in terms of its technical, commercial and performance aspects from the usual system using normal speeds.

This article sets out to describe, in a very general manner, some characteristic aspects of the HS system of the FS that distinguish the system from "normal" railway systems.

1. Basic Characteristics of the HS System

The HS system does not completely replace the present rail transport system but is intended to integrate it with a higher quality service with operating costs and times considerably reduced.

To meet its general objectives, the FS HS system has the task of:

(a) orienting all technical and organizational choices on the basis of cost–benefit evaluations;

(b) planning for rolling stock capable of travelling at a maximum speed of 300 km h^{-1} and with an operating speed of 275 km h^{-1}—with respect to performance, the rolling stock must in any case offer maximum comfort to passengers, be arranged to permit rapid and efficient servicing and be able to run on normal tracks;

(c) using plant and infrastructures that (for selected rolling stock) are not specific to the HS trains;

(d) using a signalling system that enables the high speeds to be reached on the relevant lines and, where necessary, integrates information, mechanisms and so on to solve problems regarding the high speeds;

(e) using a 3 kV feed system that has electric substation (ESS) catenary components that assure the power required and minimize the need for maintenance;

(f) organizing maintenance work on the basis of an exhaustive store of diagnostic information regarding the rolling stock, on the line and on fixed plants in order to optimize maintenance;

(g) integrating current running with a minimal logistic impact on the present arrangement of functions and stations, with adequate scheduling and service;

(h) fitting into the environment with respect to criteria of landscape protection and safety.

2. Division of the HS System into Subsystems

The HS system can be divided into a series of subsystems, not necessarily physical, that represent particular viewpoints from which the problems associated with the project are tackled and solved.

Figure 1 shows the general scheme representing the HS system divided into subsystems, and these, in turn, divided into components. Figure 2 shows the interconnections between various subsystems and the critical links.

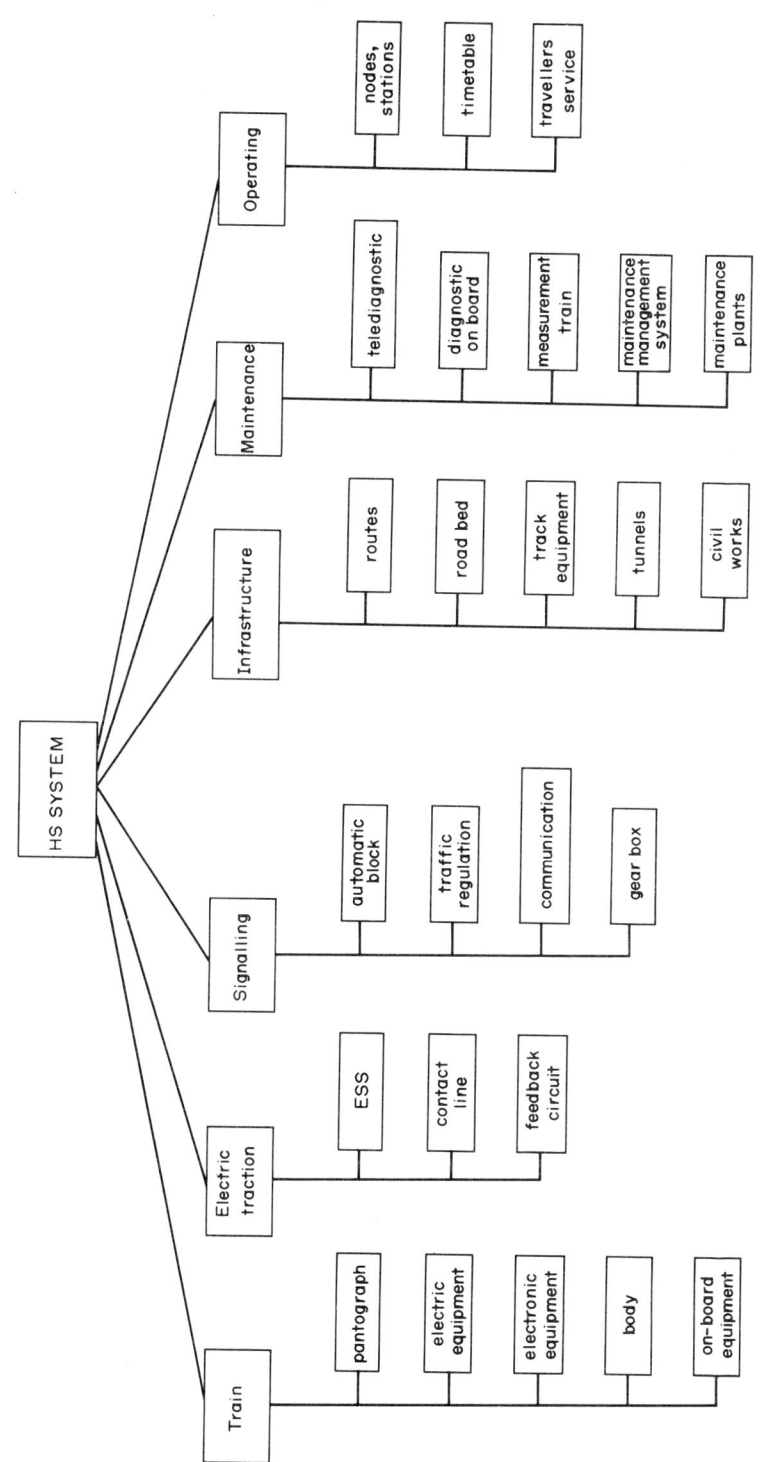

Figure 1
Division of the HS system into subsystems

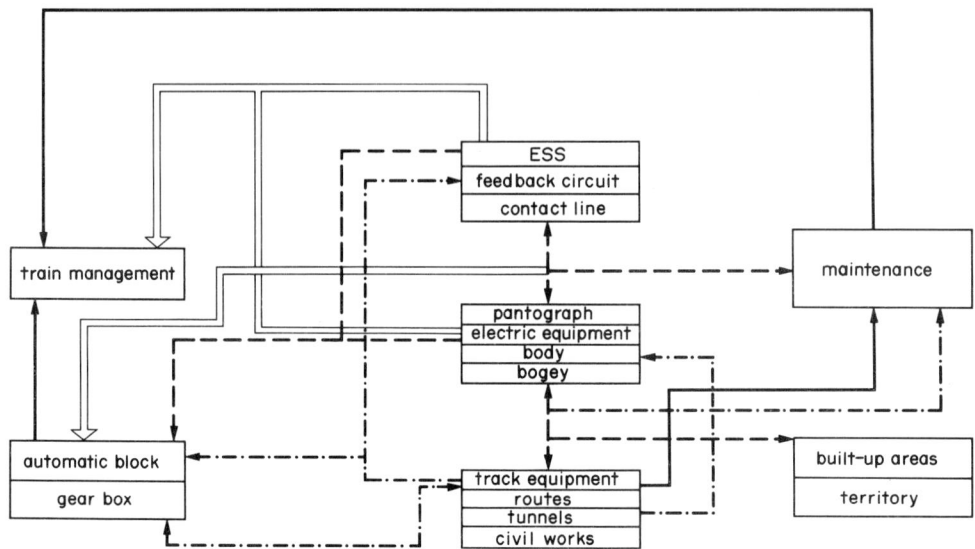

Figure 2
Interconnections between subsystems and the critical links

With reference to Figs. 1 and 2, the HS system and its subsystems are defined as follows.

(a) *HS system.* This is the set of everything that must be designed, realized, adapted, reused, prepared and tested in order to reach the general objectives.
(b) *Rolling stock.* This includes mechanical, electrical and electronic parts and the informatic components of the vehicles that make it possible to reach required speeds and standards of safety and comfort.
(c) *Infrastructures.* This includes everything that must be prepared to realize the work dedicated to the HS system;
(d) *Signalling.* This includes the equipment, philosophy and materials used for safe running of HS trains on the dedicated tracks, as well as the regulation of traffic and communications.
(e) *Electric traction.* This includes everything that is necessary to supply the power required for the HS engines (from the ESSs to the catenary).
(f) *Maintenance.* This includes methods, equipment plants and the organization of the maintenance services for the rolling stock, the line and the plants.
(g) *Running.* This includes everything involved with the running of trains and how this interferes with the traditional system at junctions and stations.

2.1 Rolling Stock
The rolling stock used for the HS line consists of ETR 500 series trains with a locomotive at each end and a variable number of towed vehicles.

The locomotive has four asynchronous traction engines fed with double power conversion. The first is at 3000 Vdc of the catenary, at 2800 Vdc with chopper. The second is at 1400 Vdc, at 1090 Vdc three-phase voltage amplitude, with variable current and frequency, with inverter.

The main characteristics of the train are its

(a) reversibility, in that there are locomotives at each end;
(b) versatility, in that each vehicle is independent and the number of vehicles that make up the train depends on the running requirements (different demand according to routes, schedule phases or periods of the week or year)—the service is reserved for passengers with first and second class seating and the maximum size of a train is 14 wagons;
(c) pressurization, in that every part of the train is pressurized as it needs to be protected from overpressures caused when passing other trains or on entry into tunnels;
(d) aerodynamics, in that particular care has been taken in the study of the aerodynamics of the train mainly to reduce resistance and the effects of high pressure for both normal running conditions and in the case of perturbations (crossings and tunnels); and
(e) high speed performance, which are a 300 km h^{-1} maximum test speed and a 275 km h^{-1} maximum running speed.

2.2 Infrastructures
There are two main aspects that distinguish the design of HS train infrastructures from that of normal-speed

traffic. These are the aerodynamic problems relating to the profile of the train and the tunnels and the reciprocal stresses between the train and the ballast, especially over bridges and viaducts.

From studies carried out by the Office de Recherches et d'Essais (ORE) de l'Union International des Chemins de Fer (UIC) offices on a European scale and by the FS it has been found that:

(a) the entry of a train into a tunnel generates pressure and decompression waves that propagate inside the tunnel and provoke variations in pressure that can be felt by people on land and by passengers; and

(b) the variation in pressure on a person 1.5 m from the track varies according to v^3 (where v is the velocity of the train) as the front of the train passes and according to v^4 as the rear of the train passes.

The parameters that most influence the generation of pressure waves are the relation between the section of the train and that of the tunnel and the velocity of the train. In contrast, the lengths of the tunnel and the train have virtually no effect.

As regards the design of bridges and viaducts for HS trains, account must be taken not only of the effect of the dynamic stresses of the load on the structure but also of the effects on the train induced by the dynamic deformations of the structure.

To this end, it is best to avoid the construction of similar works with a succession of elements with similar characteristics of rigidity.

Other considerations were made about the minimum curvature radius, the center distance of the tracks and the maximum gradient. The design specifications adopted are summarized in Table 1, in terms of maximum velocity.

2.3 Fixed Electric Traction Plants

The electrification of the HS lines has, as mentioned in Sect. 2.1, a nominal voltage of 3000 Vdc with a distance between ESSs of 12 km.

The ESSs are fed by an FS long-distance line, in turn connected to FS or ENEL (the national body for supplying electric power) long-distance lines, so as to create a grid system that limits the outside services to a maximum of two consecutive ESSs. This ensures continuous working, even if there is some reduction in power.

All the ESSs are remote controlled from a single control center.

The contact line consists of two contact wires and two messenger cables with automatic regulation, the lines and cables are sweepback, the length between poles is practically constant and the distribution of suspension wire is designed to reduce the difference in elasticity.

2.4 Regulation Plants

A station plant for passenger services has not been foreseen on the new HS lines. Therefore, the operating points are:

(a) communication points for access from one track to another,

(b) connections between the HS lines and existing lines, and

(c) shunting areas to give precedence and permit parking of traction engines and vehicles that are out of order.

The safety plants situated in working areas are not substantially different from those used in analogous situations for normal-speed rail traffic.

Separation between trains is by feed line circuit coded current automatic lock to permit a continuous signal to the locomotive. The minimum interval between trains has been fixed at a little more than 3 min, the separation distance (<14 km for trains travelling at $v = 250$ km h^{-1} and <17 km for trains travelling at $v = 300$ km h^{-1}) has been established on track circuits of about 1500 m. The signalling is two aspect and the sections are block—in order to maximize the theoretical potential of the line, according to safety standards, the sections are about 6 km long. Table 2 shows the connection between velocity and breaking distance.

The automatic lock is standardized to permit simple track circulation and parallel circulation.

The continuous locomotive signalling is that used on the traditional FS network based on the codes 75, 120,

Table 2
Relationship between velocity and breaking distance for HS trains

Velocity (km h^{-1})	Braking distance (km)
160	1.5
160–180	3
180–220	4.5
220–250	6
250–270	7.5
270–300	9

Table 1
Design specifications for HS lines

	Maximum velocity	
	300 km h^{-1}	250 km h^{-1}
Minimum curvature (m)	5.417	3.695
Maximum superelevation (cm)	10.5	11
Center distance between lines (m)	5.00	4.6
Maximum gradient	18‰ in the open	15‰ in tunnels

180 and 270 obtained from the 50 Hz and 178 Hz with which the track currents are fed. This makes it possible for the trains to circulate on normal lines. In addition to the basic system outlined above, five messages will be added for the present, giving a total of nine with a further possibility to arrive at a total of 20. The continuous locomotive signalling will be connected to an on-board speed control system.

All the new HS lines, together with part of those with interconnections, are managed by a multilevel system of circulation regulation supervision.

The managing nodes consist of the various central operational center substations. These regulate traffic circulation of functional line tracts with:

(a) remote command and remote control of peripheral stations;

(b) train describer;

(c) automatic programming of routes; and

(d) other functions.

The whole system is based on a network of computers and simulation models to solve conflicts.

2.5 Maintenance

The advanced characteristics of the materials and subsystems used for the HS train are such that maintenance is of particular importance. In fact, to assure HS train users of a service that meets their expectations, the service must assure a high standard of regularity, travelling comfort and, of course, safety. Therefore, materials and equipment must always function to a very high standard.

The maintenance of the line and fixed plants must meet more exacting standards of performance than traditional lines and therefore includes preventive maintenance with verification methods and techniques that do not adversely affect plants, on the ground and on board, and continuous verification of the state of deterioration of the track and of the coupling bogey/track, in order to reduce noise and improve travelling comfort.

The FS has undertaken two approaches:

(a) the realization of special trains specially equipped to verify the line infrastructures (line-testing trains); and

(b) the realization of a management program and control plan on the basis of the effective needs identified in the analysis of the data gathered during the line tests and during normal operation.

See also: High-Speed Railroad Networks in Europe; High-Speed Railroad Networks in Japan; Transportation Management: Systems Engineering Approach

Bibliography

Appun P, Gratzfeld P, Lossel W 1988 Drehstromantriebstechnik für Hochgeswindigkeitsfahrzeuge. *Ric. Nuove Tecnol. Trasp.*
Becker-Lindhorst K 1988 Bremseinrichtungen fur den Hochgeswindigkeitsverkehr. *Ric. Nuove Tecnol. Trasp.*
Buonanno M 1989 Ricerca per adeguare il progetto della catenaria della linea DD Roma–Firenze alle caratteristiche del nuovo schema ad alta velocita'. *Le Ferrovie nei Trasporti Degli Anni 2000.* Avenue Media, Bologna, pp. 172–5
Cacopardo D 1989 Per un nuovo ruolo strategico del transporto ferroviario: Il progetto alta velocita'. *Le Ferrovie nei Trasporti Degli Anni 2000.* Avenue Media, Bologna, pp. 44–52
Casini C 1988 I treni ETR 500 per il sistema AV delle FS. *Ing. Ferrov.* **43**, 7–27
Clementi G, Morin F, Tronconi P 1989 L'equipaggiamento elettrico di trazione del treno ad alta velocita' ETR 500 delle FS. *Le Ferrovie nei Trasporti Degli Anni 2000.* Avenue Media, Bologna, pp. 328–32
Ferrovie dello Stato 1987 *Capitolato ETR Y 500.* FS, Rome
Ferrovie dello Stato Sistema ferroviario italiano ad alta velocita'. *Ing. Ferrov.* **43**, 139–67
Focacci C 1989 L'esperienza italiana di binario senza massicciata. *Le Ferrovie nei Trasporti Degli Anni 2000.* Avenue Media, Bologna, pp. 132–8
Glockle H 1988 Aerodynamische Effekte beim Hochgeschwindigkeitsverkehr (Fahrzeug, Strecke, Tunnel). *Ric. Nuove Tecnol. Trasp.*
Morisi L 1988 L'azionamento elettrico delle motrici dell' ETR 500. *Ing. Ferrov.* **43**, 35–40
Orlandi D 1988 Tipi di armamento per l'alta velocita'. *Ing. Ferrov.* **43**, 440–7
Pafi E, Palma G 1989 La variazione della pressione nelle gallerie: Gli aspetti fisiologici. *Le Ferrovie nei Trasporti Degli Anni 2000.* Avenue Media, Bologna, pp. 82–4
Panagin R 1987 Comportamento in curva dei carrelli ferroviari ad assili ed a ruote indipendenti. Ing. Ferrov. **42**, 485–9
Ribeill G 1989 Le processus d'innovation dans les systemes ferroviaires, le cas de technologies de la grande vitesse. *Le Ferrovie nei Trasporti Degli Anni 2000.* Avenue Media, Bologna, pp. 53–60
Santoro F 1987 Velocita' ferroviaria e alta velocita'. *Ing. Ferrov.* **42**, 443–7

F. Belardinelli
[Ferrovie dello Stato, Rome, Italy]

G. Boccassi
[Automa, Genoa, Italy]

IFAC Working Group on Transportation Systems

Developments in transportation have, until recently, been uncoordinated. The main reasons for this have been the independent evolution of the organizations involved and the limitations on a holistic approach. Exchange of ideas between a scientific community with a long tradition in transport and the new community of aeronautics experts have happened only rarely. The information exchange between the theorists and those applying the theories has been poor, especially in such fast-developing areas as control theory, operations research and information theory.

In the early 1970s, international federations acting in such areas as computers, control, optimization, operations research and systems engineering recognized that it would be timely to establish groups for dealing with problems in transportation. They also decided that the transfer of knowledge should be improved and the gap between theory and practice should be reduced. The success of Japanese National Railways with the new Tokaido line may have influenced this development as this demonstrated that systems engineering applied to railways, and that the transfer of state-of-the-art technology on optimization methods, reliability approaches and so on to railroads would help to achieve not only the highest safety in the systems, but also the best economical success (Straszak 1981).

The needs of customers, the limitations of national infrastructures, particularly in urban areas, and the possibilities of new technologies, especially those of information and communication systems (telematics), all demand the coordination and integration of the different modes of transport. Furthermore, the growth of knowledge and tools in systems engineering stimulates the application of a systems approach. The current environmental problems and the possible impact of new solutions on the environment and on society force the consideration of a holistic viewpoint.

For these reasons, the Technical Committee on Systems Engineering (SECOM) of the International Federation of Automatic Control (IFAC) decided to establish the Working Group on Transportation Systems in 1973.

1. Organization of IFAC

On 12 September 1957, IFAC was founded as a multinational federation of national member organizations, each representing the engineering and scientific communities concerned with automatic control in its own country. The structure of IFAC is shown in Fig. 1.

The Working Group on Transportation Systems belongs to SECOM along with 7 other working groups.

SECOM is involved with applying systems engineering principles and methods to large-scale, complex systems in which technical, economic, social and scientific aspects and values are significant. In many cases, these systems include multidisciplinary features. Special consideration is given to the long-range planning and objectives of these systems. The systems analysis of interest to SECOM includes formulation, structuring, modelling, simulation, information handling, decision making, control, management, and methods for evaluation and testing of such complex systems.

The Aerospace Committee has a working group on the specification for automatic data links between air traffic control and planes and the Applications Committee has a working group on the control of vehicles and signals. Both of these groups are engaged to some extent in the field of transportation systems.

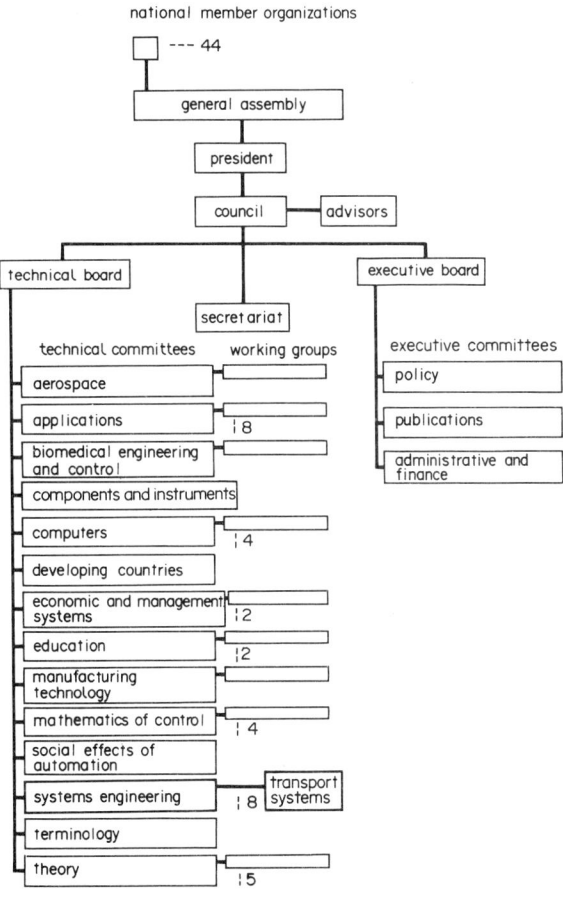

Figure 1
The structure of IFAC

IFAC has consultative status with the United Nations Educational, Scientific and Cultural Organization (UNESCO) and the United Nations Industrial Development Organization (UNIDO). The federation is a member of the Five International Associations Coordinating Committee (FIACC), which also comprises

(a) the International Federation for Information Processing (IFIP), 16 place Longemalle, CH-1204 Geneva, Switzerland;

(b) the International Federation of Operational Societies (IFORS), Technical University of Denmark, Building 321, DK-2800, Lyngby, Denmark;

(c) the International Association for Mathematics and Computers in Simulation (IMACS), Avenue de la Renaissance 30, B-1040 Brussels, Belgium; and

(d) the International Measurement Confederation (IMEKO), PO Box 457, H-1371 Budapest 5, Hungary.

Information about IFAC is available from the IFAC Secretariat, Schlossplatz 12, A-2361 Laxenburg, Austria.

2. Scope and Activities

The scope of the IFAC Working Group on Transportation Systems, which is connected with the scope of SECOM, includes the exchange of ideas and the dissemination of research results on large-scale, complex transportation systems. The technical scope includes system design, modelling, simulation, testing, evaluation and long-range planning, as well as systems analysis (including formulation, structuring, information handling and organization) of such systems. The approaches range from the application of automatic control theory to the consideration of social values and economic factors.

The objectives of the IFAC Working Group are to foster an interdisciplinary exchange of ideas, to encourage coordinated solutions for transportation problems, to stimulate discussions between managers and scientists, to inform management (including managers in developing countries) of what the methods can do and to motivate scientists to become more involved in these fields.

The intention of the IFAC Working Group is that the different areas of transportation, the different countries and the different fields of research, planning and operation should be represented by its members. At present, 11 countries are represented by 16 members. The chair has been held by

(a) Dr R. Genser, Austrian Federal Railways, from 1973 to 1979;

(b) Prof. R. E. Fenton, The Ohio State University from 1979 to 1987; and

(c) Mr J.-P. Perrin, RATP (Autonomous Administration of Paris Transport) from 1987 onwards.

Great importance is given to liaison with other groups. This is realized, for example, by having members in the IFIP Working Group on Transportation Systems (belonging to the IFIP Committee on Computer Applications in Technology), in the Airway Group of IFORS, on the Technical Committee on Safety, Security and Reliability of the European Workshop on Industrial Computer Systems (EWICS) and in the national or transnational associations in transportation.

The first event organized by the IFAC Working Group was the IFAC/IIASA Workshop on Optimization Applied to Transportation Systems, in Vienna, on 17–19 February 1976 (Strobel et al. 1977). The International Institute for Applied Systems Analysis (IIASA) was founded by the academies of science and the equivalent scientific organizations of 12 nations for multidisciplinary international research in Laxenburg, Austria, in 1972.

Beginning with the 5th IFAC Congress in Boston, Massachusetts, in 1975, the IFAC Working Group has had the responsibility for organizing the sessions on transportation systems at the IFAC Triennial World Congress. The proceedings of these IFAC congresses, as well as IFAC symposia, are published by Pergamon, Oxford.

The responsibility of the IFAC Working Group for the series of IFAC/IFIP/IFORS Symposia on Traffic Control and Transportation Systems started with the third symposia in Columbus, Ohio, in 1976. The main themes of these symposia have been

(a) control theory and optimization for traffic flow and vehicles (1970, Versailles, France),

(b) progress to aspects of previous symposium and the reliability approach (1974, Monte Carlo),

(c) the approach of using electronics and process computers in safety-related systems (1976, Columbus, Ohio),

(d) handling fuzzy information and automation in safety-related systems (1983, Baden-Baden, Germany),

(e) the systems engineering approach for national transportation systems and logistics (1986, Vienna), and

(f) the integration of computers, communications and control (1989, Paris).

The correlation between attendance and the quality of operational systems is evident. Furthermore, the attendance of people responsible for such systems has encouraged the flow of information beyond their own limited fields of expertise. For example, some railway administrations, not taking part at such meetings, have a time lag of 15 years or more in the application of advanced control or optimization methods.

Information transfer and stimulation are now taking place between different fields of application, such as aeronautics and railways, as well as between the areas of theory and practice. The amplification of ideas will occur if people in different fields continue to come together. For this to happen, an organization must have efficiency, flexibility and adaptability; sensibility towards new developments is a precondition.

Bibliography

Straszak A (ed.) 1981 *The Shinkansen Program*. International Institute for Applied Systems Analysis, Laxenburg, Austria

Strobel H, Genser R, Etschmaier M M (eds.) 1977 *Optimization Applied to Transportation Systems*. International Institute for Applied Systems Analysis, Laxenburg, Austria

R. Genser
[Austrian Federal Railways, Vienna, Austria]

In-Vehicle Equipment for Future Traffic Control Systems

In-vehicle equipment linked by mobile data communication to traffic control centers is prominent among advanced vehicle–highway technologies under development in Europe, Japan and the USA for applying computer and communications technologies to improve traffic efficiency, road capacity, safety and environmental conditions. In-vehicle equipment, which typically takes the form of navigation or route guidance systems (often called driver information systems) automatically keeps the driver informed of vehicle location, deduces the best routes to desired destinations taking into account current traffic and road conditions, and speaks or displays turn-by-turn instructions as the vehicle travels over the route. Such systems may also be able to direct the driver to the nearest service station that is open, the closest hospital, the retailer of a particular product and so on. Upon approaching a destination, the system may direct the driver to the nearest available parking space. Driver information systems thus extend individualized traffic management into the vehicle itself, and vehicles so equipped will spend less time on the road and/or travel shorter distances in performing their missions. This will help alleviate congestion by making more road capacity available for other vehicles, as well as reducing the risk of equipped vehicles being involved in accidents and reducing pollution.

1. Requirements and Functions

Driver information systems require in-vehicle equipment, a supporting infrastructure including traffic control centers with comprehensive real-time data on traffic conditions over wide areas and a communications link (preferably two-way) connecting the two. The in-vehicle equipment for typical driver information systems of the future is expected to include the elements shown in Fig. 1. The following features are necessary:

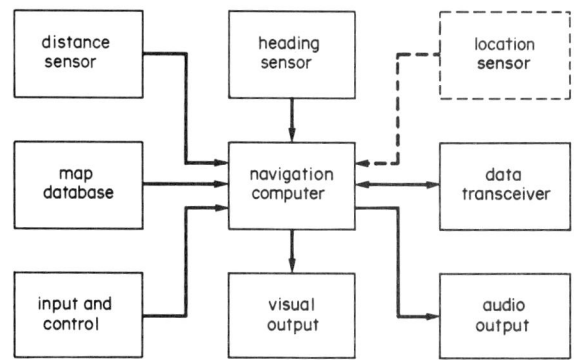

Figure 1
Typical components and subsystems of an in-vehicle driver information system (the location sensor, if included, may be a radionavigation receiver or a proximity beacon receiver)

(a) the means for automatically determining vehicle location with sufficient accuracy to identify the road travelled and each intersection approached;

(b) digital maps to identify and give the location, class, traffic regulations and address ranges for each road in areas where the vehicle operates, and which include the locations and descriptions of service stations, garages, parking, public buildings, hotels, restaurants, tourist attractions and other types of commonly used "directory" information;

(c) route guidance software and driver interface means to determine the optimum route and to present real-time route instructions to a specified destination; and

(d) a mobile communications transceiver for receiving traffic information and other variable data (e.g., available parking) for the area being travelled, and for transmitting travel times actually experienced over different road segments to the traffic control centers for augmenting data on local traffic conditions.

Vehicle location is determined by various combinations of dead-reckoning, map-matching, satellite-positioning and proximity-beacon technologies.

Dead reckoning is the process of calculating the location of a vehicle by integrating measured increments of distance and direction of travel away from a known initial location. Since distance and heading measurements always have some error, dead-reckoned vehicle location accuracy gradually decreases until the vehicle location is updated by map matching or other means.

Map matching is an artificial intelligence process that recognizes the location of a vehicle by matching its apparent path with the patterns of digital road maps stored in a computer file (e.g., CD-ROM) which is also used for map displays, route guidance, directory information and so on (French 1989). For example, each time a vehicle makes a turn whose location, direction and so on closely match those of a mapped turn, the vehicle is presumed to be at the mapped location. Dead reckoning with map matching is used in virtually all state-of-the-art vehicular navigation systems.

Satellite-positioning systems provide accurate location information when the satellite receiver of a vehicle has simultaneous reception of signals from several different satellites. However, high-frequency radio signals from satellites not directly overhead are often blocked by buildings in urban areas where driver information systems are most needed. Since dead reckoning is required to fill these gaps, the main use of satellite positioning in driver information systems will be to eliminate occasional situations where dead reckoning with map matching may fail due to extensive travel off of mapped roads, after ferry crossings and so on. The Navstar global positioning system (GPS), scheduled to be completed in the early 1990s by the US Department of Defense (US Department of Defense 1988), is the principal satellite system considered for use with driver information systems.

Proximity beacons are short-range-signal emitters that are strategically located at key intersections and the reception of their location-coded signals by passing vehicles confirms the location of the vehicles. Dead reckoning with map matching is used for navigating between beacons. The drawback of the proximity beacon of requiring extensive roadside equipment is partially offset by its usefulness as a communications link for traffic and other localized data.

Other alternatives for data communications between vehicular equipment and traffic data centers include broadcast subcarrier, cellular radio, land mobile radio (LMR) and mobile satellite. The broadcast subcarrier approach, which superimposes inaudible data on the sideband of regular commercial frequency-modulated (FM) radio stations for decoding by an inexpensive attachment to ordinary radio receivers, is limited to one-way communication. It is called the radio data system (RDS) in Europe where it is being tested for communicating traffic data. Various adaptations of cellular radio are also being tested in Japan and studied in Europe. The appeal of mobile satellites is limited by their large signal "footprints" which are ill-suited for communicating traffic data that is useful only within local areas. These and other mobile communications technologies useful for traffic and transportation applications are described in the article *Mobile Communication*.

2. Overall System Architecture and Functions

Future traffic control systems that use mobile data communication links to integrate in-vehicle equipment with traffic control centers will bring unprecedented capabilities for both collective and individualized traffic management. Figure 2 shows the linkage of in-vehicle systems with a traffic control center and the overall linkage of both with the traffic system which comprises the road system, all vehicles within the system and their respective drivers (Case and French 1987).

Link A represents the monitoring of traffic information (flow, density, queues) by vehicle detectors located throughout the road network. Link B represents traffic control information presented to the driver by conventional means (e.g., signal lights, changeable message signs, highway advisory radio). Link C represents monitoring vehicle location and input destination information by the in-vehicle system, and link D represents the flow of route guidance or other information provided to the driver by the in-vehicle system. Link E represents monitoring by the traffic control center of the location, destination, link travel times and so on of individual equipped vehicles, and link F represents the flow of traffic data or route guidance and other information from the traffic control center to the driver information systems of individual vehicles.

The links that are present and active determine the mode of operation for the navigation system. Without links E and F, the system can operate only in a static mode providing navigation information or route guidance developed on board based on historical traffic data. With the addition of link F, dynamic route guidance taking into account current traffic conditions is possible. With link E present, the link travel times and individual destinations of all equipped vehicles may be considered to enable interactive route guidance for the highest overall system efficiency.

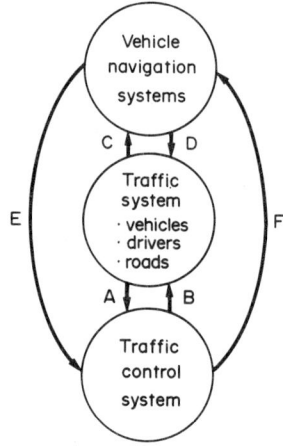

Figure 2
Information flows in an integrated traffic system

3. Potential Benefits

The benefits of driver information systems comprising in-vehicle navigation and route guidance equipment that receives real-time traffic datas from the infrastructure are in the form of reduced congestion, improved safety and fewer environmentally damaging consequences. The potential magnitude of expected benefits can be estimated by considering the traffic situation in the USA and by the results of various studies in Europe and Japan as well as in the USA.

The need for driver information systems is illustrated by a study that estimates that 2×10^9 vehicle hours of delay occurred on urban expressways in the USA in 1987 and that, without further countermeasures, the delay will increase to 11×10^9 vehicle hours in the year 2005 (Lindley 1989). The delay cost US$16 000 million in 1987 and will cost US$88 000 million in 2005, in terms of the 1987 value of the US dollar. Another estimate places the 1987 delay cost as high as US$42 000 million including insurance and other costs (Harris and Bridges 1989).

According to another US study, 6.9% of the distance travelled by noncommercial vehicles and 12.2% of the time spent in such travel is wasted due to poor navigation and route following skills, thus contributing to further congestion. The annual cost to individuals and to society of this excess travel was estimated at US$45 700 million (King 1986). Results of similar studies in Europe and Japan generally agree with those in the USA.

The number of highway-related fatalities worldwide exceeds 500 000 people annually. In the USA alone, there were 47 000 fatalities in 1988 and the total cost of all accidents was estimated at US$130 000 million (Betsold 1989). Without countermeasures, the number of accidents and their costs to society will increase because of the growth in traffic and the rapidly increasing fraction of older drivers.

Errors associated with manual navigation, such as improper trip planning and poor direction finding, may result in slow driving, erratic maneuvers, delay, and lost or confused drivers. Unforeseen events en route can negate a trip plan and require the driver to make midcourse corrections, sometimes without adequate information. Under such conditions, drivers typically slow down, make last-minute lane changes, travel in reverse on exit ramps, make illegal U turns and so on. Secondary consequences include incidents, congestion and increased exposure to accident risk (Lunenfeld 1989). Safety should thus be enhanced by driver information systems, because the driver can proceed with confidence to the destination without hesitating at decision points, or having to make lane changes or other abrupt last minute maneuvers to follow the proper route.

Although a number of potential safety impacts of in-vehicle navigation has been recognized and may be characterized in qualitative terms, relatively little quantitative information is available. In particular, there seems to be few, if any, research results or statistical studies that provide a sound measure of the expected safety benefits. As for safety disadvantages, there could be offsetting increases in accident risk if system displays and controls are not designed with safety in mind. There is much cautionary literature about the possible adverse effects of in-vehicle navigation systems. These views are based largely on generic human factor considerations as only sparse information from actual research with navigation systems is available.

Although direct statistics are not available to show what fraction of accidents may be due to poor navigation, an evaluation of the potential efficiency of various types of driving aids concluded that 3% of a particular group of accidents would have been avoided by use of a navigation aid (Fontaine *et al.*, 1989). This estimate was based on analysis of the police reports on 350 accidents in France and implies that navigation aids could save US$3900 million annually if US accident causes parallel those of the French study.

Navigation-based driver information systems also have potential for improving road safety through reduced exposure to accidents as a result of travelling shorter distances and/or for shorter periods of time. Although the relationship between accident frequency and total vehicle mileage is complex and not well understood, King (1986) estimated that US$4400 million in accident losses could be saved each year through improved route following in the USA. Combining this potential saving with that suggested previously for accident avoidance yields a total of US$8300 million that could be saved annually by the use of navigation aids (see *Safety of Road Traffic*).

King (1986) also concluded that US$10 000 million in vehicle operating costs was wasted annually in the USA and that US$31 300 million was wasted in the form of lost time. Combining these potential savings with the US$8300 million in potential safety benefits and the smaller of the estimates of 1987 congestion waste (US$16 000 million) yields a total potential saving of US$65 600 million annually. This potential saving would be accompanied by as much as a 10% reduction in the pollution caused by automobile traffic.

4. Worldwide Status

The development of driver information systems for dynamic route guidance has already progressed to the stage of large-scale pilot demonstrations in Europe. One demonstration on a more modest scale is underway in the USA and a second is being developed. Several demonstrations and trials involving communication of traffic data to in-vehicle navigation systems have also been carried out in Japan since the mid-1980s.

Traffic data centers, communication links and other infrastructure support required by driver information systems necessitate public sector involvement. The potential of infrastructure-supported vehicle route guidance for improving the management of traffic and for safe and efficient public road utilization attracted government sponsorship for early research in the USA during the 1960s (French 1986). European and Japanese government agencies then carried route guidance research forward after the USA aborted its pioneering effort in 1970 and pulled further ahead as the US government minimized its involvement in such civil matters during the 1980s.

Prometheus (program for a European traffic with highest efficiency and unprecedented safety) and DRIVE (dedicated road infrastructure for vehicle safety in Europe) are the major cooperative programs for developing route guidance and other forms of advanced vehicle–highway systems in Europe (Gillan 1989). Prometheus is an eight-year, US$750 million program originated by the automotive industry in 1986 under Eureka, the 19-nation organization for coordinating European industrial research. DRIVE is a cooperative government-industry progam for developing road transport informatics (RTI) by the 12-nation Commission of the European Communities which is due to complete its three-year, US$140 million initial phase by 1992. While DRIVE and Prometheus are still in the midst of comprehensive top-down analyses and precompetitive research projects spanning the entire range of intelligent vehicle–highway systems, loosely associated pilot projects are underway in West Berlin (LISB) and London (Autoguide) for large-scale field tests of in-vehicle route guidance supported by two-way infrared proximity beacons (Jeffery 1989).

Japan is well positioned for early exploitation of driver information systems because government agencies already have numerous traffic control centers with real-time traffic data for coordinating signal timing over wide areas, operating variable message signs and so on. With the addition of data communications links, these centers could support in-vehicle route guidance. Field tests and demonstrations of a two-way microwave proximity-beacon communication link by the RACS project of the Ministry of Construction (Shibata 1989) and of a cellular like 800 MHz radio approach by the AMTICS project of the National Police Agency (Tsuzawa and Okamoto 1989) are nearing completion as these two agencies vie to provide the infrastructure support. These two projects were expected to be consolidated by the end of 1990.

Japanese industry has already introduced the world's first autonomous navigation and driver information systems to become available as original automotive equipment (Shoji et al. 1988, Itoh et al. 1990). Future versions of these production systems, as well as of advanced prototype systems developed by virtually all automobile and major electronics manufacturers in Japan, need little more than communication links and software enhancements to become comprehensive infrastructure-supported driver information systems.

The USA is poised for a major push for advanced vehicle–highway systems including driver information systems largely as a result of the mid-1980s initiatives of the California Department of Transportation (Caltrans) and the subsequent efforts of Mobility 2000, an informal coalition of federal, state and local government officials, private sector representatives and university researchers. Mobility 2000 has strived since 1987 to coordinate scattered research efforts and to define and recommend a unified intelligent vehicle–highway systems (IVHS) program (Harris and Bridges 1989). Broad support has developed among transportation organizations and the US Congress is expected to consider legislation which could establish a 10-year, US$1000 million IVHS program, including advanced driver information systems (ADIS) linked to traffic control centers as a major program objective.

Pathfinder, a low-budget cooperative project of Caltrans, the Federal Highway Administration (FHWA) and General Motors to test the usefulness of a navigation system with in-vehicle traffic data, is underway in Los Angeles (Betsold 1989). Travtek, a cooperative effort proposed by the FHWA, the Florida Department of Transportation, the City of Orlando, General Motors and the American Automobile Association, will provide a more advanced form of dynamic route guidance with information on parking, tourist attractions and other services to the drivers of 100 equipped vehicles in the Orlando, Florida, urban area (Betsold 1989).

5. Future Developments

Although in-vehicle navigation and driver information systems are no panacea for traffic congestion and safety problems, they will assure more efficient utilization of available road capacity and, if designed with safety in mind, will reduce future accident rates. In so doing, the computer and communications technologies of the information age will permeate road transportation just as they have already permeated the laboratory, the factory and the office, and will bring benefits that should be no less dramatic.

Suitable technologies for the systems are available. However, considerable research is required to determine the optimum system architecture and the most cost-effective and desirable functions, and to obtain a sufficient understanding of the overall safety impacts to ensure that the systems are designed to maximize safety benefits. Other challenges include the establishment of standards and of new types of institutional arrangements necessary for systems that must allow an extensive exchange of information between the equipped vehicles and the traffic management infrastructure.

See also: Route Guidance, Individual; Vehicle Guidance by Computer Vision

Bibliography

Betsold R J 1989 Intelligent vehicle/highway systems for the United States—An emerging national program. *Proc. JSK Int. Symp. Technological Innovations for Tomorrow's Automobile Traffic and Driving Information Systems.* JSK Foundation, Tokyo, pp. 53–9

Case E R, French R L 1987 Vehicular navigation and route guidance systems: Technology and applications. *Proc. 57th Annual Meeting Institute of Transportation Engineers.* Institute of Transportation Engineers, Washington, DC, pp. 203–8

Fontaine H, Malaterre G, Van Elslande P 1989 Evaluation of the potential efficiency of driving aids. *Proc. VNIS '89 —Vehicle Navigation and Information Systems Conf.* Institute of Electrical and Electronics Engineers, Piscataway, NJ, pp. 454–9

French R L 1986 In-vehicle route guidance in the United States 1910–1985. *Proc. 2nd IEEE Int. Conf. Road Traffic Control.* Institution of Electrical Engineers, London, pp. 6–9

French R L 1989 Map matching origins, approaches and applications. *Proc. 2nd Int. Symp. Land Vehicle Navigation.* TÜV Rheinland, Cologne, Germany, pp. 91–116

Gillan W J 1989 PROMETHEUS and DRIVE: Their implications for traffic managers. *Proc. VNIS '89— Vehicle Navigation and Information Systems Conf.* Institute of Electrical and Electronics Engineers, Piscataway, NJ, pp. 237–43

Harris W J, Bridges G S (eds.) 1989 *Proc. Workshop Intelligent Vehicle/Highway Systems by Mobility 2000.* Texas Transportation Institute, College Station, TX

Itoh T, Tsunoda S, Hirano K, Tanaka J 1990 Navigation system with map-matching method. Society of Automotive Engineers Technical Paper Series No. 900471. SAE, Warrendale, PA

Jeffery D J 1989 In-vehicle route guidance: the future for automobile transport and traffic operation in Europe. *Proc. JSK Int. Symp. Technological Innovations for Tomorrow's Automobile Traffic and Driving Information Systems.* JSK Foundation, Tokyo, pp. 61–79

King G E 1986 Economic assessment of potential solutions for improving motorist route following. US Federal Highway Administration Report No. FHWA/RD-86/029

Lindley J A 1989 Urban freeway congestion problems and solutions: an update. *ITE J.* **59**, 21–3

Lunenfeld H 1989 Human factors considerations of motorist navigation and information systems. *Proc. VNIS '89 —Vehicle Navigation and Information Systems Conf.* Institute of Electrical and Electronics Engineers, Piscataway, NJ, pp. 35–42

Shibata M 1989 Road management in Japan and development of the road/automobile communication system (RACS). *Proc. JSK Int. Symp. Technological Innovations for Tomorrow's Automobile Traffic and Driving Information Systems.* JSK Foundation, Tokyo, pp. 29–37

Shoji Y, Horibe T, Kondo N 1988 Toyota Electro Multivision. Society of Automotive Engineers Technical Paper Series No. 880220. SAE, Warrendale, PA

Tsuzawa M, Okamoto H 1989 Advanced mobile traffic information and communication system—AMTICS. *Proc. VNIS '89—Vehicle Navigation and Information Systems Conf.* Institute of Electrical and Electronics Engineers, Piscataway, NJ, pp. 475–83

US Department of Defense 1988 *Federal Radionavigation Plan.* Report No. DOD-4650.4. US Department of Defense, Washington, DC

R. L. French
[R L French & Associates, Fort Worth, Texas, USA]

Incident Detection

In general, automatic incident detection constitutes a separate and independent branch of research and development due to the comparatively complex and difficult nature of the topic. However, when applied in practice, it must be considered as an integral element of a comprehensive traffic control system. Any dynamic control system (i.e., any control system reacting to the traffic flow) requires an accurate assessment of the prevailing traffic condition. Relevance to the current situation and the quality of the assessment are of decisive importance in terms of the effectiveness of the overall system. One important element is therefore the fast and acurate location and identification of any kind of traffic flow disturbance; that is, incident detection.

The importance and potential for improvement in safety, economic efficiency and environmental compatibility of incident detection may be underlined by some examples.

First, depending on the traffic conditions and the environmental situation, between 20% and 50% of all accidents on freeways are caused by preceding (primary) incidents.

Second, far more than 50% of these secondary accidents occur within 10 min of the first incident. This is corroborated by accident series on freeways which are often caused by minor primary incidents and which, in many cases, could be avoided if unprepared approaching drivers could be warned in time.

Third, US studies in the Los Angeles conurbation show that more vehicle hours of delay result from extraordinary and accidentally occurring traffic disturbances than from regularly occurring network overload during daily peak hours.

The consequence of such findings is that, particularly in highly motorized countries, control and warning systems that are able to react in time to disturbances and other influences are increasingly being planned or put into operation. Accompanying investigations confirm, in general, the success of such systems expressed, for example, by a significant reduction in accident rates.

1. Basic Principles
1.1 Traffic Control
Basically, all traffic control systems work according to the same principle. Their major functions are shown in Fig. 1.

Incident Detection

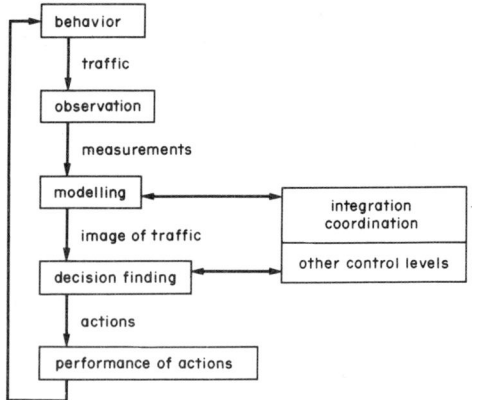

Figure 1
Basic functions of a traffic control system

(*a*) *Observation.* Different techniques are available for obtaining information about the current traffic situation in a given time–space window. They range from standard local measurements at individual sites by means of inductive loops to image processing or even to measurements by means of specially equipped vehicles; for example, the vehicles of modern electronic route guidance systems (see *Route Guidance, Individual*).

(*b*) *Traffic modelling.* The information gained from the observation process needs to be combined and extrapolated to create a complete and detailed image of the current traffic situation. This is carried out by means of fairly sophisticated models of the real traffic world, where the degree of detail is influenced by the information gained from the observation, on the one hand, and the required precision from the decision-finding stage, on the other.

(*c*) *Decision finding.* The current traffic situation has to be characterized and assessed in order to decide on appropriate actions. The decision depends on the individual local situation as well as on the overall strategy and the state of the network.

(*d*) *Performance of actions.* If an action has been chosen, it must be initialized and performed. The performance must be checked by the system. This applies to the switching of variable message signs as well as to the appropriate adaptation of a traffic signal control.

(*e*) *Coordination/integration.* Coordination and integration are mainly needed

(i) for the overall network control management level (correct adaptation/creation of the current image of the traffic situation, coordinated control measures),
(ii) for systems running at a parallel level (e.g., speed control on a freeway section), and
(iii) for local subsystems (e.g., traffic signals at the next intersection).

It is evident that the overall performance of the control system depends on each individual function and on well-coordinated cooperation between them. Vital factors that influence the performance with regard to incident detection are

(a) the degree of information produced by the sensor systems,
(b) the precision of measurements,
(c) the fault tolerance of system components,
(d) the type of traffic model (degree of detail and validation range),
(e) the reaction time of the model to the traffic behavior,
(f) the clear-cut separation of different traffic situations and incidents,
(g) the velocity of the decision-finding process,
(h) the adequateness of the chosen actions, and
(i) the velocity of initialization and performance of measures.

The importance of point (i) must not be underestimated since time savings up to several minutes, made possible by a powerful incident detection, are only significant with an efficient, delay-free conversion within the overall system.

As the points show, the essential factors for the capacity of the system are the traffic model and its capability to clearly delimit the different traffic conditions, as well as the strategy linked to it.

1.2 Incidents and Their Detection

Incident detection in a general sense comprises all activities related to spatial and temporal location of unforeseeable anomalies in traffic flow. Usually a distinction is made between automatic incident detection (AID) and nonautomatic incident detection (NAID). Whereas NAID implies human observation and report (e.g., incident location by patrols, emergency systems, citizens band (CB) radio or television monitoring), AID is detection by independent analysis of data directly measured on the road section. This can be carried out in on-line or off-line mode.

Less clear is the definition of the term "incident." In the literature, numerous partially different definitions can be found. This is one reason why the results of investigations or details about the efficiency of individual systems or strategies are difficult to compare. The definition of what is finally regarded as an incident and, therefore, as something to be detected is influenced by pragmatic aspects as well as by the application area itself (i.e., by the kind of traffic flow and the type of road section). Thus, varying "definitions" are used for freeway traffic on multilane carriageways and for urban traffic in signal-controlled networks.

Nevertheless, some universally valid statements can be made. A distinction must be made between the producing event and its consequences (i.e., the cause and the effect). The cause is the actual incident; that is, any change in the road space that occurs either in the environment of the lanes (e.g., counterflow lane, fog or parking lanes) or on the lanes themselves (e.g., accident or slow platoons). Normally, the causing events cannot be directly detected; they have to be deduced from their effects.

In any case, the effect of an incident is a change in the quantitative relationship of the macroscopic traffic variables (volume, density and speed) in the immediate neighborhood of the cause. This means that a different fundamental diagram is valid in this area for as long as the disturbance lasts.

An incident is, therefore, generally characterized by the transition of a traffic stream from one traffic condition to another. Such a transition is temporally and spatially limited by a shock wave. Depending on the relationship of inflow and capacity in the area, two basic types of incidents can be distinguished:

(a) incidents with spatially limited, invariable extension of their effects (inflow less than capacity delimited by stationary shock waves), and

(b) incidents with a spreading area of influenced traffic (inflow greater than capacity delimited by shock waves that spread out from the incident).

An example is shown in Fig. 2. In a simplified form it depicts the trajectories of individual vehicles as well as the spreading of an incident of type (b) at the distance–time level. It becomes clear that even after the actual incident (cause) has disappeared at time t_2, the tailback increases further (up to time t_3) and normal traffic flow is only restored at the incident site s_1 much later on (time t_4).

A quantitative description of incidents and their consequences is desirable in many cases (e.g., for assessing the urgency of resultant measures for more precise warnings within the framework of scientific objectives). The following parameters are used (terminology according to Euco-Cost 30 (1979) and Busch (1986)):

(a) core time is the duration of the incident (cause);

(b) total time is the period of time between incident occurrence and the end of all effects at the incident site;

(c) core area is the temporal–spatial area created by the incident and the tailback;

(d) total area is the temporal–spatial area showing disturbed traffic conditions upstream from the incident site;

(e) intensity is the relationship between total area and core area;

(f) extension is the length of incident site and tailback; and

(g) shock wave speed is the speed with which the end of the tailback moves away against the direction of travel.

The basic disadvantage of all such parameters is that their determination is usually very difficult and requires certain knowledge or assumptions about the incident itself.

1.3 Assessment Criteria

In order to assess the capacity of an AID method and to compare alternative methods or systems, various parameters are determined:

(a) detection rate (DR) is the proportion of incidents detected during a period of observation;

(b) false alarm rate (FAR) is the proportion of false alarms out of all alarms (on-line definition) or the proportion of false alarms out of all alarm tests in the case of incident-free conditions (off-line definition);

(c) mean time to detect (MTD) is the mean period of time passed between incident occurrence and detection;

(d) mean delay time (MDT) is the mean period of time passed between incident disappearance and the end of the alarm;

(e) detection index (DI) is the mean weighted detection quality. It considers the significance of a detection, the detection time and a weighting of type and severity of an incident (Busch 1986);

(f) efforts for realization is the number and type of measurement sites, data processing operations and so on. In general, the costs are used for the quantification of this parameter.

It must be noted that these parameters always have to refer to a period of observation and have to consider a complete monitoring system. This means, for example,

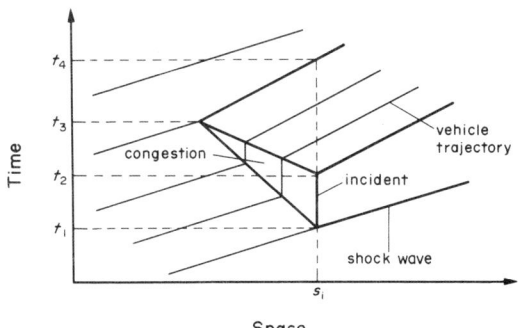

Figure 2
Space–time diagram of an incident situation

Table 1
Classification of AID methods (brackets indicate minor importance)

Algorithm type	Local		Section	
	Microscopic	Macroscopic	Microscopic	Macroscopic
Direct comparison	(×)	×		×
Temporal forecast	(×)	×	(×)	×
Spatial forecast			×	×
Model-based estimation				×
Pattern recognition			×	×
Cluster analysis		×		×
Probabilistic approach	(×)	×	(×)	×

that it is necessary to take account of the number of control sections (i.e., the measurement cross sections). It is equally necessary to characterize in detail the sample used (the number, type and severity of the incidents), since otherwise assessing or comparing the determined parameter values is impossible. The last problem is partly avoided when using the DI.

There is one particularity that should be noted when calibrating AID methods. There is a natural connection between the DR and the FAR of a method: the higher the aspired detection quality, the higher the FAR, since the method has to be tuned more sensitively. However, the error rate of the whole traffic control system is composed of the added rates of its components, so that for the AID module itself, only minor rates are acceptable (in practice, a FAR of 5% is often too high).

2. *State of the Art*

Studies dealing specifically with AID started around the mid-1960s in the USA. They were undertaken in connection with the planning of control systems for freeways in conurbations. Only later did AID become an independent research field.

These studies have mainly concentrated on freeway traffic (see *Freeway*). Little experience has been gained with other road types. This is due to the expected high safety gains on freeways by preventing secondary accidents, compared with application in innerurban areas where only improved traffic management (prevention/reduction of congestion and delay) can be achieved. Correspondingly, a number of different but generally valid methods are available for freeways, whereas for other road types (rural roads, urban roads and tunnels) only specific solutions for the local traffic situations exist. These are usually based on spot measurements (traffic volume, occupancy and, in exceptional cases, speed) and classify the current traffic situation by comparing the measured values with predefined thresholds (primarily, congestion is checked by means of high occupancy). The decision logic can then activate a different signal plan (tunnel, urban traffic control (UTC)) which is more applicable to the traffic situation.

Independently from the application field, the measurement equipment almost exclusively comprises inductive loops. Their features may, however, differ greatly depending on type and on the manner in which they were installed, ranging from simple loops installed over several lanes to frequently scanned, lane-wise installed pairs of loops, which allow a detailed differentiation of the vehicles.

Besides inductive loops, radar sensors and automatic image processing systems have been used (see *Video Sensors*). Radar sensors do not provide fundamentally different traffic information from inductive loops and the experience gained with image processing has been limited to a few pilot systems. Thus, the AID methods treated in the rest of this article are generally based on spot measurements of traffic flow by means of inductive loops. In keeping with available experience and realized systems, they refer to non-signal-controlled freeway traffic on multilane carriageways.

2.1 *AID Methods*

The methods suggested so far can be divided into two groups:

(a) local algorithms (i.e., observation at one single site), and

(b) section algorithms (i.e., observation at two adjacent sites with data comparison between them).

Both groups can be subdivided according to the type of input data they require into microscopic and macroscopic methods. While the microscopic methods need detailed single-vehicle data, the macroscopic ones work with aggregated information (e.g., average speed over 30 s intervals). The largest amount of data is, therefore, needed for the microscopic section methods.

The basic principles of the algorithms are listed in Table 1.

(a) *Direct comparison*. Measured values or values derived from them are directly compared to previously fixed threshold values. In order to eliminate the influence of stochastic variations of the test parameters during short measurement intervals, the calculation of smoothed mean values is used and this has turned out to be a simple but very effective method. The most meaningful test parameters are occupancy and mean speed, as well as appropriate combinations of both.

(*b*) *Temporal forecast.* Knowing how a test parameter changes over time until a specific point allows a forecast for a future time to be made and compared to the real value. The basic idea is the assumption that an incident produces a deviation of the parameter from its previous course beyond the limits set by statistical noise. Although advanced time-series analysis methods have been employed on a trial basis, for practical applications with short forecast time intervals, a simple determination of the smoothed mean value with linear trend corrections is sufficiently precise. The test parameters are the same as in the direct comparison.

(*c*) *Spatial forecast.* Naturally, this approach can only be used for sections. On the basis of the measurement values at the upstream cross section, a forecast for the traffic behavior expected at the downstream cross section is made and compared to the real measurement values. It is assumed that, in the case of undisturbed traffic, the arrival of single vehicles or, at least, of vehicle groups can be forecast for the following cross section, delayed by the travel time. If a greater deviation between forecast and measurement values occurs, an incident between the cross sections may be inferred. In this case, the necessary measurement values are speed and the number of vehicles.

(*d*) *Model-based estimation.* A simplified mathematical model is set up for the traffic system situated between the measurement cross sections. It is permanently calibrated with the help of current measurement values and error corrections so that it optimally reproduces the present traffic situation. With the underlying traffic model, a fairly precise statement can be made about the momentary situation within the real system, including a possible assessment of the location and severity of incidents. These methods use the Kalman–Bucy filter and take traffic volume and speed as measurement values.

(*e*) *Pattern recognition.* The underlying assumption is that, in the case of undisturbed conditions, traffic patterns occurring upstream may be rediscovered downstream delayed by a certain period of time. If this is not the case or if the estimated offset changes significantly, an incident has occurred. Empirical methods and correlation techniques are used to recognize the patterns and thus the offset. Good results are obtained with these approaches, particularly at low traffic volumes. The measurement values are mainly occupancy or density. For more complicated methods, detailed information on individual vehicles is also acquired with corresponding detectors.

(*f*) *Cluster analysis.* By means of cluster analysis (a multidimensional statistical method), sets of traffic parameters for disturbed and undisturbed situations are analyzed in an off-line period. These data are transformed in a suitable manner so that distinct groups (clusters) of different traffic conditions can be identified. The conditions that may be drawn up for the delimitation of these groups together with the transformation equations form the on-line applied AID method. The measurement values and test parameters used are similar to those used in the direct methods.

(*g*) *Probabilistic approach.* For any AID algorithm, it is possible to increase the meaningfulness of a current incident report by linking it with statistical certainty. To do this, the probability of an incident on the monitored section as well as the detection rates of the algorithm for disturbed and undisturbed traffic situations are determined in presurveys and updated from time to time during operation. The current probability of an incident (i.e., the statistical significance of an alarm) can then be determined by combining these single probabilities for the totality of the last messages of the algorithm.

2.2 Comparisons of AID Algorithms

A number of investigations have tried to compare different algorithms at a fairly general level. Four of these seem to hold more generally than the rest:

(a) Cook and Cleveland (1974) compared 19 algorithms (direct comparison and temporal forecast);

(b) Payne *et al.* (1976) compared 24 algorithms (direct comparison and temporal forecast);

(c) Levin and Krause (1979) compared 5 algorithms (direct comparison); and

(d) Leutzbach (1983), Busch (1986) and Fellendorf (1988) compared 12 algorithms (which cover the main part of Table 1).

In terms of results, the four conclusions of investigation (d) may be considered.

(a) None of the tested methods proved to be superior in all situations. This means that no generally applicable method is available at the moment.

(b) Most of the algorithms produced a high detection rate and short detection times at high traffic volumes and detector spacings up to 1 km. The reason for this is that these algorithms react to shock waves which move very fast at high volumes and do not diffuse within short distances.

(c) At low volumes most of the algorithms failed to produce satisfactory results. The reason for this is that the shock waves move slowly and diffuse quickly. Comparatively, the best results have been achieved by the pattern recognition technique.

(d) Increasing the intersensor distance enforced this high-volume tendency; that is, local shock wave detection was only applicable at high volumes (but long detection times). Section methods based on a suitable traffic model proved to be superior.

2.3 Systems in Operation

A number of traffic control systems using AID methods as an integral part are in operation. They all use different system configurations and varying AID approaches. A review of 21 European freeway surveillance systems nevertheless revealed some common features and a certain generality.

(a) Nearly all systems use inductive loop detectors as the main measuring equipment. Video observation is sometimes used as an additional source of information for the operator.

(b) The distance between detector stations varies (usually even within one system) from 100 m to 2000 m; the average distance is about 400 m to 600 m.

(c) The type of AID algorithm varies, but most methods use a comparison of measured (and sometimes smoothed) data with predefined thresholds. Mostly local algorithms are used for decisions on traffic conditions; later on the control measures are coordinated along the section.

(d) The variable mostly used is occupancy. One reason for this is that only one loop is needed. If available, mean speed is taken as the second parameter. No microscopic data or other more detailed information (e.g., statistical distributions) is used.

(e) The time interval used for the decision finding of the algorithm varies from 1 s to 2 min. Most systems use intervals of either 10–20 s or of 1 min.

(f) Most of the systems operate in an open-loop mode; that is, any control measure has to be activated by the operator (see *Traffic Control Modes*).

Point (f) in particular shows that, despite the increasing willingness of system operators to use AID, the confidence put in the method is relatively low and human control is still considered necessary.

3. *Future Developments*

The present situation can be characterized as follows:

(a) Different, varyingly powerful theoretical approaches exist whose hardware and software requirements are widely varying.

(b) Although AID methods are being more frequently integrated into traffic control systems, very simple methods are usually used which do not correspond to the state of the art.

(c) Studies and applications have been restricted almost without exception to extraurban freeways since this is where the greatest safety gains can be expected.

Thus, some global guidelines for future research and development can be derived. It seems necessary

(a) to improve the known algorithms with regard to their reliability and robustness in real traffic;

(b) to develop techniques for low-volume situations at long intersensor distances;

(c) to investigate the possibilities of surveilling freeway junctions, which often tend to be the origin of incidents on upstream sections (first approaches to this problem were undertaken by Leutzbach and Busch (1986));

(d) to improve and to accelerate the flow of information from the detection system back to the road user and to the relevant emergency systems; and

(e) to develop methods using new techniques of traffic observation arising from recent industry and scientific research.

In particular, point (e) offers a wide application field, which possibly provides a new quality of incident detection. For example, it seems possible to reduce the FAR and, at the same time, to increase the detection quality (up to the identification of the incident type) by using self-teaching methods from the field of artificial intelligence. Methods of automatic image processing which recognize congestions with high certainty are already known from pilot applications and will become more suitable for practical application with the increasing capacity of microprocessors. In the field of automotive research, in-vehicle computer systems will be developed that are supplied by different types of sensors and that can transmit complex measurement data e.g., on the course of the trip, to the roadside infrastructure and also to other vehicles. Such investigations are currently being performed on a large scale in two European research and development programs: DRIVE (dedicated road infrastructure for vehicle safety in Europe) and Prometheus (**program** for a **E**uropean **t**raffic with **h**ighest **e**fficiency and **u**nprecedented **s**afety) (see *DRIVE*).

Bibliography

Ahmed S A, Cook A R 1981 Point process models for freeway incident detection. *Proc. 8th Int. Symp. Transportation and Traffic Theory*. University of Toronto Press, Toronto, Canada, pp. 20–30

Bell M G H, Thancanamootoo B 1988 Automatic incident detection within urban traffic control systems. *Proc. Roads and Traffic 2000 Conf.*, Berlin, pp. 35–9

Böttger R 1979 Ein Verfahren zur messtechnischen Feststellung von Verkehrsstörungen auf Fernstrassen und Autobahnen. *Strassenverkehrstech.* **6**, 173–9

Busch F 1986 Automatische Störungserkennung auf Schnellverkehrsstrassen—Ein Verfahrensvergleich. Thesis, Technical University of Karlsruhe

Busch F, Fellendorf M 1990 Automatic incident detection on motorways. *Traffic Eng. Control* 221–7
Collins J F, Hopkins C M, Martin J A 1979 Automatic incident detection—TRRL algorithms HIOCC and PATREG. *Transp. Road Res. Lab. (U.K.), TRRL Suppl. Rep.* **526**
Cook A R, Cleveland D E 1974 The detection of freeway capacity-reducing incidents by traffic streams measurements. *Transp. Res. Rec.* **495**, 1–11
Cremer M 1981 Incident detection on freeways by filtering techniques. *Proc. 8th IFAC Congr.* Pergamon, Oxford
Euco-Cost 30 1979 Theme 6. Incident detection. European Community Report No. 30/94/79-XII/880/79
Fellendorf M 1988 Comparison of procedures for automatic incident detection. *Proc. Roads and Traffic 2000 Conf.*, Berlin, pp. 233–8
Hilgers C J 1980 *A Method for Calculating a Group of Algorithms for Automatic Incident Detection*. Rijkswaterstaat, The Hague, The Netherlands
Hobbs A A, Clifford R J 1989 AUTOWARN—A motorway incident detection and signalling system. *Institution of Electrical Engineers, Conf. Road Traffic Monitoring.* IEEE Report No. 299, pp. 167–71
Kühne R 1987 Freeway speed distribution and acceleration noise calculations from stochastic continuum theory and comparison with measurements. *Proc. 10th Int. Symp. Transport and Traffic Theory.* Elsevier, Amsterdam, pp. 119–37
Leutzbach W, Busch F 1983 Störfallentdeckung. Research project. Federal Ministry of Transport, Karlsruhe, Germany
Leutzbach W, Busch F 1986 Störungserkennung im Bereich von BAB-Knotenpunkten. Research project. Federal Ministry of Transport, Karlsruhe, Germany
Levin M, Krause G M 1979 Incident detection algorithms. *Transp. Res. Rec.* **722**, 49–58
Levin M, Krause G M 1982 Incident detection—A Bayesian approach. *Transp. Res. Rec.* **682**, 52–8
Payne H J, Goodwin D N, Teener M D 1976 Evaluation of existing incident detection algorithms. Federal Highway Administration Report No. FHWA-RD-75-39
Pfannerstill E 1989 Automatic monitoring of traffic conditions by reidentification of vehicles. *Institution of Electrical Engineers, Conf. Road Traffic Monitoring.* IEEE Report No. 299, pp. 172–5

F. Busch
[Siemans AG, Munich, Germany]

Intelligent Traffic Control Systems

Modern traffic control systems cover large urban areas with extremely complex and variable traffic patterns and traffic situations. Therefore, they must possess highly sophisticated capabilities to manage effectively the flow of traffic in a modern city with many intersections fed by a network of highways. Since the traffic control problem is different for city streets and for highways, the two problems will be treated separately. However, the same theoretical approach will be used for both cases (Saridis and Lee 1979).

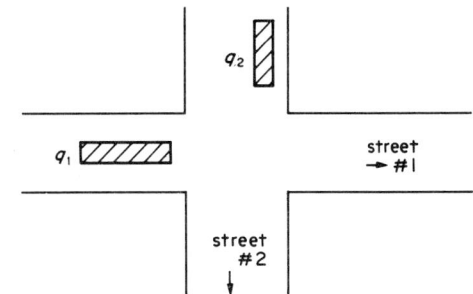

Figure 1
Simple intersection model

1. Urban Traffic Control

This section describes a hierarchically intelligent control procedure for the urban traffic management problem. On-line learning capabilities introduced for this task considerably reduce the need for extensive data communications and complex numerical computations. A hierarchically intelligent system is composed of three levels which are, in descending order, the linguistic organization level, the coordination level and the on-line control level. These levels assume responsibility for the management of traffic in city, neighborhoods and single intersections, respectively, in order of decreasing intelligence and increasing complexity.

The lowest level in the hierarchically intelligent system is a single intersection. Each individual intersection possesses an on-line controller which manages its signal cycles. A group of such intersections forms a coordination which is supervised by a coordinator. The traffic situations generally encountered at a single intersection of an urban area are divided into five categories:

(a) sparse traffic/off-line control,
(b) light traffic,
(c) heavy traffic,
(d) oversaturated traffic, and
(e) immobile traffic/incident.

For each mode, a discrete-time mathematical model is separately derived, thus markedly reducing the complexity of the system and the cost of management of the individual intersections. The general mathematical model for the state of the queues at the two-competing, one-lane, one-way street intersection of Fig. 1 is given by

$$q_s(k+1) = q_s(k) + r_s(k) - \frac{S_s(l)}{c} u_s(k), \quad s = 1, 2$$

$$\sum_{s=1}^{2} u_s(k) \leq c, \quad \forall k$$

where k refers to the discrete-time instance, that is, the cycle: $q_1(k)$ and $q_2(k)$ are the queue lengths at the intersection at the beginning of the kth cycle or, equivalently, at the end of the $(k-1)$th cycle, expressed in terms of the number of vehicles: $r_1(k)$ and $r_2(k)$ represent the numbers of vehicles arriving at the intersection during the kth cycle; $u_1(k)$ and $u_2(k)$ are the lengths of the "green" phases for the two competing streets during the kth cycle; l is the designator of the traffic mode taking values 1–5 for traffic situations (a–e), respectively; $S_s(l)$ is the maximum throughput of street number s while in mode l, expressed in a number of vehicles per cycle; and c is the full length of a cycle in seconds (see *Traffic Control Modes*).

The lengths of the green phases of the traffic-light cycles at each intersection serve as local control inputs. Typically, the control will be operating off-line when the flow of traffic is low (e.g., sparse traffic) or when the communication lines of the network are severed. A present phase pattern locally stored will be assigned automatically to each intersection. For the other four traffic situations, on-line control inputs will be adopted through optimization of appropriate performance functions.

For each of the four on-line-controlled traffic situations, various performance criteria are examined to meet the following performance objectives:

(a) maximum average flow,
(b) minimum queue length at the intersection, and
(c) minimum drivers' discomfort.

To accommodate the needs of each traffic situation, an appropriate performance objective is assigned to it. The on-line control inputs are computed on a microcomputer so that it can learn to optimize the performance functions associated with the proper objectives for each of the existing traffic situations.

Additional parameters entering the control function, such as the maximum throughput depending largely on the offset of the green phases between two intersections and the length of the cycles, are defined by the higher levels of the hierarchy for better coordination of the traffic flow.

The next level in hierarchical intelligent systems is the coordination level, which also possesses learning capabilities. Each coordinator is in control of a neighborhood of an urban area, composed of a number of interacting intersections. These neighborhoods are assigned according to zoning rules and uniformity of their designated population. Hence, control over any neighborhood is maintained independently of others. Coordination patterns for each neighborhood are assigned and monitored by the organization level.

The offset of the green phases between two intersections immediately affects the maximum throughput of traffic and may serve as the control variable of the coordinator. If the arrival rates at the neighboring intersections are modelled properly, the offsets may be set at every predetermined number of cycles to optimize the traffic flow through the neighborhood for each traffic situation.

A possible performance criterion assigned to each coordinator is the maximum throughput through the neighborhood. Feedback information from the local intersections, aggregated over the predetermined number of intersection cycles, is used to train the controller to minimize the performance criterion for the neighborhood for each control situation. Past experience is stored in the coordination computer for future use and training. Various weighting coefficients in the performance criterion may be adjusted by the coordinator itself to provide uniformity of traffic flow throughout the neighborhood. In a large urban environment, multiple levels of coordinators may be used.

The organization level, representing the highest level of the hierarchy, is centrally located and is operated by a mainframe computer. It provides an interactive link with human operators, accumulates and processes information on traffic situations across the network for different weather conditions and time of the day and week and assigns the values of vital parameters to the coordinators through learning, such as cycle lengths and performance criteria, for various traffic situations. It also serves as a link with the freeway system computers for collection of traffic information and coordination purposes. Since most of its functions can be performed almost off-line, it may be time-shared with other activities.

The functional structure and information flow of a hierarchically intelligent control system are shown in Fig. 2 and its computer implementation in Fig. 3. Simulation results have established the feasibility of such a hierarchically intelligent control of urban traffic systems (Saridis *et al.* 1981, Kashani and Saridis 1983).

2. Freeway Traffic Control

The objective of freeway traffic control is regulation of the number of vehicles allowed to enter the freeway through the local access ramps. Control of the flow in the main stream of the freeway is also another point of interest, although very few controllable parameters exist in this case.

The underlying approach proposed in this article is based on the principle of "increasing intelligence with decreasing precision" in the control hierarchy, which permits the use of advanced formal computer languages for high-level decision making on traffic patterns and performance. The main idea is to organize the complex problem of controlling the freeway flow into a hierarchical computer system consisting of interrelated computer-operated subsystems with lower degrees of complexity. The proposed system is endowed with learning and decision-making capabilities to adapt the controls onto the appropriate pattern in time and space. The on-line learning capability is one of the major differences between the proposed hierarchical systems (Saridis *et al.* 1981).

The higher-order continuous-time mathematical model of freeway traffic flow is given in terms of the dynamic speed equation

$$\frac{\partial u}{\partial t} = -u\frac{\partial u}{\partial x} - \frac{l}{T}\left[u - u_e(\rho) + \frac{\gamma}{\rho}\frac{\partial \rho}{\partial x}\right]$$

where x is the spatial distance along the freeway, increasing in the direction of flow, t is the time, $\rho(x, t)$ is the density of vehicles (vehicle per lane per kilometer), $l(x, t)$ is the number of lanes, $u(x, t)$ is the space mean speed (km h^{-1}), $q(x, t)$ is the flow rate (vehicles per hour) and $f(x, t)$ is the net flow entering the freeway (vehicles per hour), corresponding to on-ramp and off-ramp traffic (see *Flow Variables*). This formula suggests that the time rate of change of speed is given by three terms:

(a) $-u(\partial u/\partial x)$, which is called connection, the effect of vehicles from upstream arriving at this point;

(b) $[u - u_e(\rho)]$, which is called relaxation to equilibrium, the effect of drivers adjusting their speeds to conform to the spacing $(1/\rho)$; and

(c) $(\gamma/\rho)(\partial \rho/\partial x)$, which is called anticipation, the effect of drivers reacting to conditions downstream, for example slowing down in anticipation of higher densities downstream $(\rho/x > 0)$.

Parameters γ (s) and T (km^2 h^{-1}) are referred to as the relaxation and anticipation parameters, respectively. It should be noted that the choice of equilibrium density relation (u_e) and the parameters T and γ are critical to the functioning of the model. This model was derived by postulating terms that affect driver acceleration in different ways; the last two terms especially represent driver behavior (see *Freeway Traffic Modelling*).

The discrete-time mathematical model for a segment k of a freeway is given by the following three equations:

$$\rho_k(j+1) = \rho_k(j) - a_k[q_k(j) - q_{k-1}(j)]$$
$$+ \frac{a_k}{i_k}[r_k(j) - g_k(j)]$$

$$u_k(j+1) = u_k(j) - \frac{2\Delta t}{x_{k+1} - x_{k-1}}[u_k(j)][u_k(j) - u_{k-1}(j)]$$

$$- \frac{\Delta t}{T}\left\{u_k(j) - u_e(\rho_k(j))\right.$$

$$\left. - \frac{2\gamma}{\rho_k(j)}\left[\frac{\rho_{k+1}(j) - \rho_k(j)}{x_{k+2} - x_k}\right]\right\}$$

$$q_k(j+1) = \tfrac{1}{4}[\rho_{k-1}(j) + \rho_k(j)][u_{k-1}(j) + u_k(j)]$$

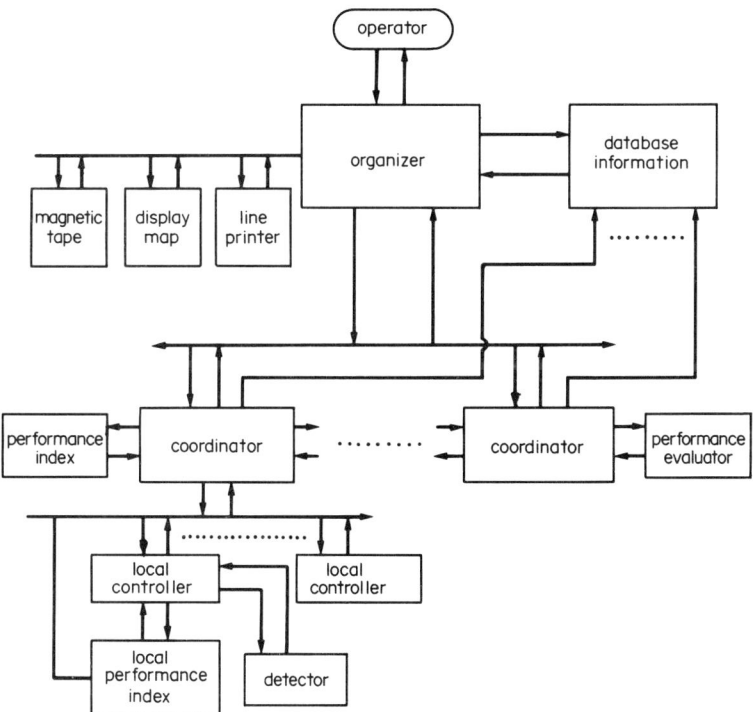

Figure 2
Functional structure and information-flow diagram of the proposed traffic control system

Intelligent Traffic Control Systems

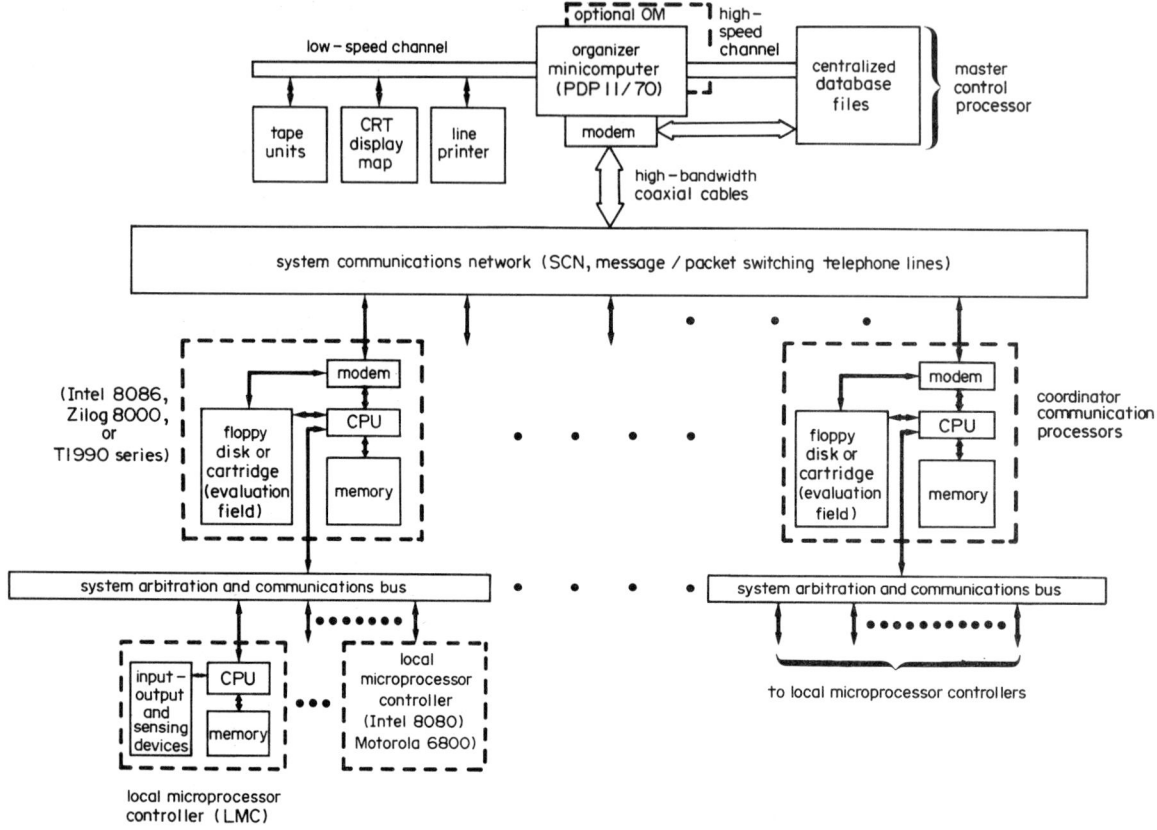

Figure 3
System architecture of the proposed computer communications network for traffic control

where $a_k = \Delta t/\Delta x_k$, $\Delta x_k = x_{k+1} - x_k$, and Δt is the full length of a cycle in seconds. Discrete-time models are easier to use on a digital computer and are therefore preferred to continuous-time models.

The hierarchically intelligent control of highway flow can be logically decomposed into three hierarchically arranged decision-making levels. This reduces the complexity of the problem and makes a solution for each main level easier. The decomposition is performed according to ascending order of complexity and required intelligence. The three main levels are

(a) the on-line control level,

(b) the coordination level, and

(c) the linguistic organization level.

The on-line control is the lowest and least complex level in the hierarchy. Practically speaking, its main job is to provide the timing signal for each individual entry ramp. It also acquires traffic information at each ramp through the traffic sensors and extracts features from these measurements. These features are sent to the coordination level for traffic coordination with the neighboring ramps. The features are also used to characterize the traffic conditions, such as rush hours (see *On-Ramp Control, Local*).

As noted, this level deals directly with the control problem of a traffic system component. The control law associated with each recognized traffic situation is used in this level to determine the individual timing sequence of the ramp. The control law mentioned can be thought of as optimizing an objective function to achieve a certain criterion, such as minimum waiting time or maximum traffic flow. Some of the parameters of this function are set by the coordination level which supervises a group of ramps.

The operation of the on-line controller can be summarized as follows:

(a) recognizing the traffic pattern through the sensors,

(b) relaying the information about the traffic situation to the coordinator,

(c) choosing the appropriate control function to be optimized with regard to the present situation,

(d) receiving some parameters of the chosen function from the coordination level, and

(e) optimizing the function and obtaining the timing sequence.

The main function of the coordination level is to coordinate the operation of a group of on-line controllers to improve the performance of the system. The traffic network is divided into several groups, each of which is coordinated by a coordinator. These groups can be urban road intersections, highway segments or an interacting combination of both.

The coordinator is in contact with both on-line controllers and the organization level. It receives the traffic task assignment from the organization level and the feature vectors containing information about the traffic situation from the on-line controllers. Then it chooses a global objective function which consists of the weighted sum of all the local objective functions under its command, or in mathematical terms

$$J_j(\alpha_i^j) = \sum_{i=1}^{n} (\alpha_i^j) J_i(u)$$

(for m coordinators each commanding n on-line controllers) where $0 \leq \alpha_i^j \leq 1$, $i = 1, 2, \ldots, n$, $j = 1, 2, \ldots, m$, $J_j(\alpha_i^j)$ is the global function chosen by the jth coordinator, $J_i(j)$ is the local objective function of the ith on-line controller and $(\alpha_i^j)^T = (\alpha_1^j, \alpha_2^j, \ldots, \alpha_n^j)$ is the coordination vector for the jth coordinator.

The coordination vector (α_i^j) is adjusted by optimizing the global objective function. By doing so, the coordinator can penalize a specific ramp (because the traffic on that section of highway is moving slowly), or it can reward another ramp. This can be done by assigning a large or small value to the ith position (corresponding to the ith on-line controller) of the coordination vector.

It should be noted that the optimization of $J_j(\alpha_i^j)$ and determination of (α^j) can be achieved through a learning scheme. The feature vector which characterizes the traffic pattern is matched with vectors stored in a database library. If such a situation has previously existed, then the appropriate coordination vector can be easily picked from past experience. If the traffic situation is a new pattern, then the appropriate coordination vector is obtained and then stored in the library for future reference.

The linguistic organization level is the intelligent level of the system. Its main purpose is to establish a communication medium between the traffic operator and the control system. The organization level receives input commands (e.g., typing a linguistic command on a keyboard terminal) and relays the necessary information to the coordination level for execution.

In general, the operator's commands can be expressed in strings of spoken words or typed characters. The command string is parsed by a bottom-up procedure. If it is recognized as syntactically valid, the parser output is mapped into a sequence of goals that determines the performance of the lower levels.

The main advantage of the proposed system over the existing methods is the capability to adapt the controls to the different traffic patterns with little computational effort. This flexibility is due to the learning characteristic of the system, which enables it to recognize similar recurring control situations and use the best previously obtained control law for them. It should be noted that traffic patterns are influenced by many factors such as the weather, the time of day and incidents. Therefore, such a system which can adapt the controls to the rapid changes of the flow is very desirable. Note also that the system can perform its operations in an environment for which no *a priori* information is available (Saridis *et al.* 1981).

See also: Expert Systems Approach to Road Traffic Control; On-Ramp Control of Freeway Networks; Road Network Control

Bibliography

Kashani H, Saridis G N 1983 Intelligent control for urban traffic systems. *Automatica* **19**, 191–7

Saridis G N, Kashani H R, Khotanzad A 1981 A Hierarchically Intelligent Control System for Urban and Freeway Traffic. Technical Report, TR-EE-81-19. Purdue University, IN

Saridis G N, Lee C S G 1979 On hierarchically intelligent controls and management of traffic systems. *Proc. Engineering Foundation Conf. Research Direction in Computer Control of Urban Traffic Systems*. American Society of Control Engineers, New York, pp. 209–18

G. N. Saridis
[Rensselaer Polytechnic Institute, Troy, New York, USA]

Kinematic Wave Theory

Early theoretical road traffic studies tried to establish analogies with physical laws of incompressible fluids. It was found that such analogies had certain limitations. For example, the behavior of a flow of traffic can only be compared with that of a continuous fluid for high concentrations. Despite these limitations, the consideration of a hydrodynamic theory of traffic, suggested by fluid flow analogies, has proved significant and has enabled several traffic control practices to be improved.

1. Basic Principles of Hydrodynamic Theory

Hydrodynamic theory is based on three fundamental principles:

(a) the continuous representation of variables,

(b) the law of conservation of mass, and

(c) the statement of fundamental diagram.

1.1 Continuous Representation

This principle is introduced by using an adequate definition of traffic variables (see *Flow Variables*). It consists of considering flow $q(x, t)$ and concentration $k(x, t)$ as continuous variables. The same applies in the case of mean speed $u(x, t)$, given by the relation

$$u(x,t) = \frac{q(x,t)}{k(x,t)} \quad (1)$$

This relation underlines the analogy with the flow of a fluid in a pipe: speed of flow also corresponds to the flow–concentration ratio.

1.2 Law of Conservation of Mass

Unlike the first principle, which gives rise to certain reservations, the second is closely related to real traffic conditions. The law of conservation of mass can be defined as follows. The variation in the number of vehicles on a road section (x_1, x_2) is equal to the difference between the number of vehicles entering the section at the point x_1 between times t_1 and t_2 and the number of vehicles exiting at point x_2 during the same period of time. In mathematical terms, this principle is expressed as

$$\int_{x_1}^{x_2}[k(x,t_2)-k(x,t_1)]dx = \int_{t_1}^{t_2}[q(x_1,t)-q(x_2,t)]dt \quad (2)$$

Provided that the flow and concentration functions can be differentiated, the law of conservation of mass implies the following "continuity equation":

$$\frac{\partial k(x,t)}{\partial t} + \frac{\partial q(x,t)}{\partial x} = 0 \quad (3)$$

1.3 Fundamental Diagram

The assumption of the fundamental diagram is expressed as follows. At any point along the road and at any time, speed is a decreasing function of concentration

$$u = u(k) \quad (4)$$

Given Eqn. (1), the law of traffic flow on a given section of road during a given period of time can be expressed in terms of an equation relating two out of three variables, namely flow, concentration and speed.

Experimental measurements confirm the assumption of the fundamental diagram. The flow–concentration relation can also be derived on the basis of microscopic traffic theories, in particular that concerned with the concept of car following (see *Car-Following Models*).

2. Kinematic Waves

Lighthill and Whitham (1955) have elaborated a theory concerning the appearance of kinematic waves in traffic flow. This theory is based on the three principles mentioned above. Thus, assuming that the fundamental diagram is applicable, the continuity equation can be written as

$$\frac{\partial k(x,t)}{\partial t} + q'(k)\frac{\partial k(x,t)}{\partial x} = 0 \quad (5)$$

Wave theory attempts to solve this partial differential equation. With certain limiting conditions taken into account, Eqn. (5) can be used to predict the concentration at any point on the road, at any time. If $q'(k)$ is a class C function, the continuity equation is a quasi-linear, partial differential one. Local solution yields a so-called regular solution of the form

$$k(x,t) = F\{x - tq'[k(x,t)]\} \quad (6)$$

where F is an arbitrary class C function. Resolution of the equation yields two integrals corresponding to

$$x - tq'[k(x,t)] = \text{constant} \quad \text{and} \quad k = \text{constant} \quad (7)$$

Kinematic Wave Theory

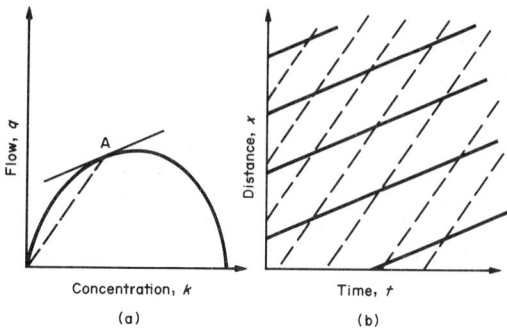

Figure 1
Speed and trajectories of waves and vehicles in (a) a flow–concentration diagram and (b) a space–time diagram

Curves $[x = tq'(k) + \text{constant}]$ are straight lines and can still be regarded as characteristics where concentration, flow and speed remain constant. These isoflow lines can therefore be regarded as wave fronts associated with the propagation of a flow q at a speed $q'(k)$. Propagation speed of kinematic waves corresponds to the slope of the tangent on the fundamental diagram, as illustrated in Fig. 1. In other words, an observer travelling at a speed corresponding at every point to the slope of the $q(k)$ curve at this point would not experience any variation in flow at his level.

It can be shown that the propagation speed of kinematic waves is lower than that of the vehicle travelling at the mean speed u of the flow. Waves are propagated either:

(a) forward, when concentration k is less than the critical concentration k_c; or

(b) backward, when k is greater than k_c.

This property leads to a distinction between two types of flow, namely free flow when k is less than k_c and congested flow when k is greater than k_c. These results should be compared with what happens in practice. Once traffic is congested, disturbances propagate from downstream.

3. Shock Waves

Suppose that there is a change in the state of flow. This could arise, for example, when a group of high-speed vehicles joins vehicles travelling at a lower speed. Traffic changes from a state indicated on curve $q(k)$ by a point A toward a state represented by point B, as illustrated by Fig. 2. Kinematic waves, travelling at the higher speed, join those being propagated at lower speed, so as to give rise to a shock wave. The propagation speed w of this shock wave is given by the slope of the chord AB joining the operating points A and B in the fundamental diagram:

$$w = \frac{q_B - q_A}{k_B - k_A} \tag{8}$$

The propagation of the shock wave under the conditions defined above can be shown on a space–time diagram. In region A, downstream of the shock wave, low-speed waves, drawn parallel to the tangent to point A, correspond to high flows. Conversely, in region B upstream of the shock wave, kinematic waves travelling at a higher speed are associated with lower rates of flow. The shock wave, separating the two zones, travels at an intermediate speed. In the space–time diagram, its slope is parallel to the chord AB. The dotted line shows the trajectory of a vehicle through two segments, respectively parallel to the chords OB and OA.

In theory, the shock wave corresponds to a discontinuity of the concentration function $k(x, t)$, passing suddenly from a value k_A to a value k_B. In practice, this change becomes apparent in the form of sudden disturbances in the flow, perceived by drivers. This process leads to reductions in flow and speed, at any point of the road section.

4. Specific Cases and Applications

4.1 Variations in Peak Flow

The wave theory provides information on the formation and the dissipation of peaks. On a uniform section of road, traffic first increases up to a very sharp peak and then decreases. Vehicles become involved in a shock wave on approaching the resulting traffic jam. The region where vehicles encounter the shock wave moves in the direction of the traffic. The speed of kinematic waves decreases as concentration increases. On the space–time diagram illustrated in Fig. 3, lines corresponding to the progress of the waves first diverge and then, after a decrease in flow and concentration, converge and finally come together. This gives rise to a shock wave. The amplitude of the disturbance constituting the shock wave first increases rapidly and then slowly decreases, while moving forward at a steady speed, as illustrated by Fig. 4a–h.

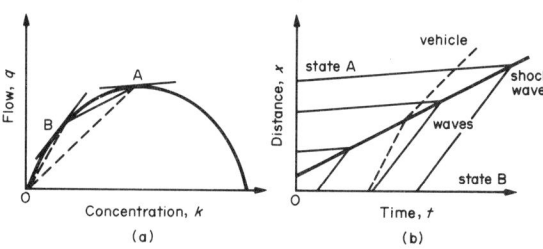

Figure 2
Relationship between (a) the flow–concentration diagram and (b) the space–time diagram

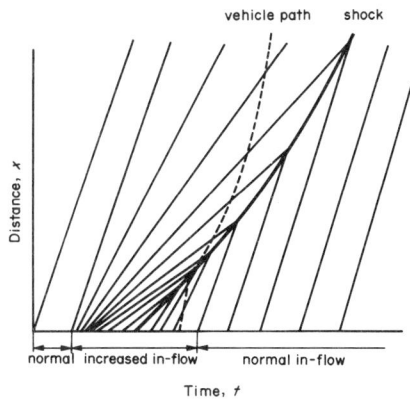

Figure 3
Behavior of a traffic peak

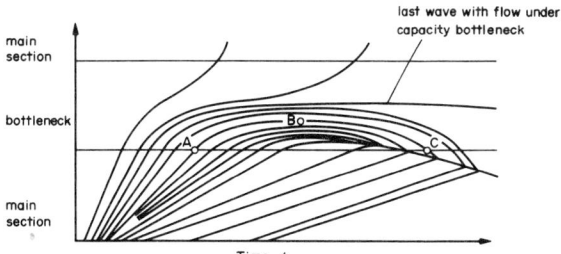

Figure 5
Shock wave in front of a traffic peak as it enters a bottleneck

4.2 Effects of Bottlenecks

A uniform subsection of a road constitutes a bottleneck when its capacity is lower than that of the main section. Two eventualities are considered, depending on the nature of traffic demand upstream of the bottleneck.

(a) If the volume is less than the capacity of the bottleneck, the shock wave is stationary and its speed is zero. Traffic flow is simply penalized by a reduction in its average speed.

(b) As demand increases, kinematic flow waves travel through the bottleneck, until the capacity of the bottleneck is reached. On the space–time diagram of Fig. 5, lines corresponding to the progress of these waves bend and change direction as they go through the bottleneck. They straighten again as waves progress along the main road section. When the capacity of the bottleneck is exceeded, the distortion of these lines, in a space-time diagram, is such that they go through previous positions and meet other incident waves. This gives rise to a shock wave as shown in Fig. 6. This shock wave is propagated back into the flow of vehicles upstream from the bottleneck until demand falls below the capacity of the bottleneck.

A better understanding of the effects of bottlenecks, on the basis of this wave theory, has been successfully used to improve traffic conditions in several road tunnels, especially in New York city.

4.3 Signallized Intersections

Hydrodynamic theory has been employed to study stop–start effects at signallized intersections. When traffic is interrupted during the red period, kinematic waves associated with maximum and minimum values of concentration are propagated in upstream and downstream directions. These waves can be represented by collections of parallel straight lines separated by the horizontal axis. When traffic is starting up at green, kinematic waves form a divergent collection of straight lines associated with the propagation of all possible values of concentration, from zero to the maximum possible. When these waves meet the first type of wave, shock waves occur as traffic starts up or stops.

4.4 Applications

There have been other recent applications of the theory. Reference can be made, in particular, to the

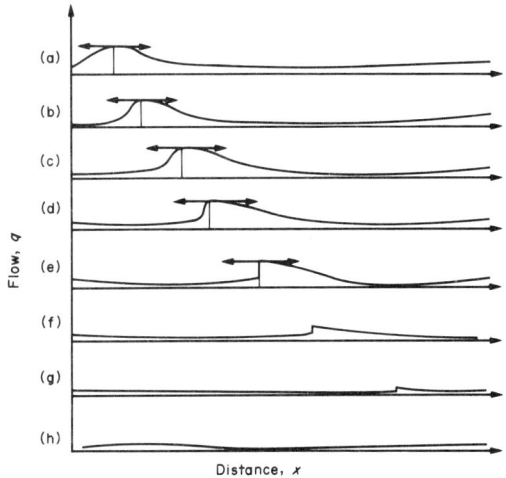

Figure 4
Rapid growth and slow decay of a shock wave

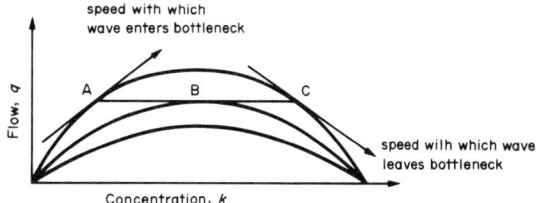

Figure 6
Reflection of a wave from a bottleneck in a flow–concentration diagram

233

development of algorithms for automatic incident detection on freeways. Determination of travel times on multilane roads is also based on this theory. Several macroscopic simulation models, for use in testing control strategies on urban freeways, have also been produced by making use of hydrodynamic concepts.

See also: Freeway Traffic Modelling; Prediction of Traffic Flow

Bibliography

Edie L C, Foote R S 1958 Traffic flows in tunnels. *Proc. Highway Res. Board* **37**, 334–44

Edie L C, Herman R, Rothery R (eds.) 1967 *Vehicular Traffic Science*. Elsevier, New York

Gerlough D L, Huber M J 1975 *Traffic flow theory. A Monograph*. Transportation Research Board, National Research Council, Washington, DC

Greenberg H 1959 An analysis of traffic flow. *Oper. Res.* **7**, 79–85

Haberman R 1977 *Mathematical Models: Mechanical Vibrations, Population Dynamics and Traffic Flow*. Prentice-Hall, Englewood Cliffs, NJ

Lighthill M J, Whitham G B 1955 On kinematic waves 11. A theory of traffic flow on long crowded roads. *Proc. R. Soc. London, Ser. A* **229**, 317–45

Richards P I 1956 Shock waves on a highway. *Oper. Res.* **4**, 42–51

S. Cohen
[INRETS, Arcueil, France]

Magnetic Suspension Railroad Systems

High-speed and urban magnetic levitation (maglev) vehicles were developed as alternatives to conventional railroad systems. For distances up to 500 km, high-speed maglev trains have been proven to be superior to airplanes with regard to energy consumption and travelling time. Mechanical and electrical modularization and decentralized hierarchical control architectures of the levitation and guidance systems (LGS) are outstanding characteristics of robust operational vehicles. Further key technologies are linear induction motors (LIM) for propulsion and linear induction generators (LIG) for on-board power generation.

Between 1960 and 1970, the necessity of tracked high-speed ground transportation systems for passenger travel as well as transportation of goods was recognized worldwide (Hochleistungs-Schnellbahn Studiengesellschaft 1971). The challenge was to find a ground-based system that could better compete with air transportation and which, in order to be accepted by the public, had to be efficient and economic on the one hand (see Fig. 1) and fast and attractive on the other (Rogg 1988). Different levitation concepts were investigated, but only two technologies have been pursued: electrodynamic suspension (EDS) and electromagnetic suspension (EMS).

1. Reasons for Using Maglev

The dynamic load on guideway and undercarriage for a given guideway increases quadratically with the velocity of the vehicle. For wheel-on-rail systems, the necessary effort with respect to accuracy and bedding of rails and guideway increases rapidly with velocity, resulting in costly consequences for guideway construction.

Equally, the operating expenses for maintenance and repair due to wear and sagging increase.

In addition, the noise level produced by a high-speed wheel-on-rail system is hardly acceptable in terms of environmental considerations (Gottzein 1984).

For magnetic guidance and levitation systems, contrary to wheel-on-rail vehicles, loads are evenly distributed and, therefore, extension gaps within the guideway are admissible (see Fig. 2).

For contactless guidance and levitation systems, contactless and friction-independent propulsion and brake systems will be used, as well as contactless power transfer.

Thus, new economically and ecologically advantageous solutions to guideway construction may be found. An elevated concrete—or steel—beam guideway represents a low-cost solution and the stiff guidance and levitation system allows steep banks, allowing small curve radii without any reduction of ride comfort. The possibility of combining vertical and horizontal radii and to realize steeper gradients results in a much higher flexibility in tracing. With better adaptation of the guideway to the terrain, it is possible to reduce the number of costly tunnels and bridges required for a given route.

2. Maglev Techniques

In the EMS technique, attractive forces are generated by controlled electromagnets on board the vehicle against ferromagnetic rails on the guideway.

EDS uses superconductive helium-cooled coil magnets which generate extremely high magnetic field

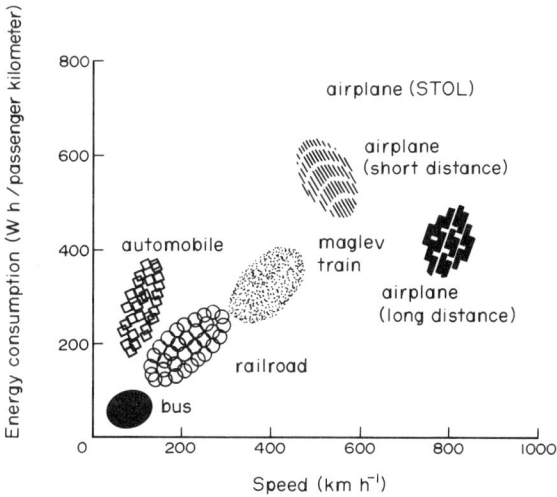

Figure 1
Comparison of primary energy consumption in different transportation systems

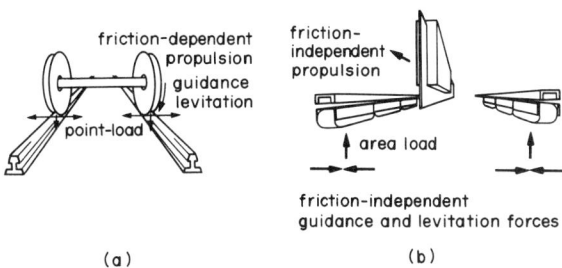

Figure 2
Characteristics of (a) wheel-on-rail and (b) maglev systems

Magnetic Suspension Railroad Systems

Table 1
Maglev vehicles developed for high-speed transportation between 1971 and 1989

Vehicle	Company/ institution	Country of origin	Year of origin	Speed (km h^{-1})	LGS principle	Central or decentral LGS	Number of persons carried	Short- or long-stator LIM	Remarks
Prinzipfahrzeug	MBB	Germany	1971	90	EMS[a]	central[a]	14	short[a]	
TR 02	KM	Germany	1971	164	EMS	central	8	short	
ROMAG TM	Rohr Ind.	USA	1971/1973		EMS	central	12/6	short	integrated levitation guidance and propulsion
ML 100	JNR	Japan	1972	60[b]	EDS[a]	central	4	short	
TR 04	KM	Germany	1973	253.2	EMS	central	20	short	
Test vehicle	University of Sussex	UK	1974		EMS	central	4	short	
EET 01	Seimens	Germany	1974	230	EDS	central	2	short	
HMB 2	TH	Germany	1974	36	EMS	central	4	long	iron-cored long-stator LIM
KOMET	MBB	Germany	1975	401.3[a]	EMS	central			propelled by hot-water rockets
HSST 01	JAL	Japan	1975	307.8	EMS	central		short	
KOMET M	MBB	Germany	1976	341	EMS	decentral[a]			propelled by hot-water rockets
ML 500	JNR	Japan	1977	512[a]	EDS	central		long	
HSST 02	JAL	Japan	1978	100[b]	EMS	central	7 (+2)	short	
TR 05	Transrapid EMS	Germany	1979	75	EMS	decentral	68	long	first operational maglev transport system, in service at the IVA, Hamburg[a]
ML 500 R	JNR	Japan	1979	201	EDS	central		long	
MLU 001-01/02/03	JNR	Japan	1980/1981/1982	400	EDS		32	long	400 km h^{-1} in 1 section operation
TR 06	Transrapid EMS	Germany	1983	412.6	EMS	decentral	192 (+4)	long	full-scale high-speed vehicle[a]
HSST 03	JAL	Japan	1984	30[b]	EMS		45	short	
TR 07	TH	Germany	1988		EMS	decentral		long	under test

Source: Rossberg (1983)
a Proof of concept b Nominal speed

forces. The coils induce currents in support rails, causing repulsive support or guidance forces, lifting the vehicle when the take-off speed is exceeded.

3. EMS Vehicle Development and History

The idea of magnetic levitation dates back to the beginning of the twentieth century. Proof of feasibility was given by Kemper (1935). Roughly from 1965 onwards, development of maglev for vehicles began in many industrialized nations (Gottzein and Rogg 1984). Test vehicles were developed in Germany, USA, USSR, Japan, the UK and Romania, but only Germany, Japan and the UK continued development of operational transportation systems, as shown in Tables 1 and 2.

In Germany, electromagnetic suspension systems based on controlled electromagnets have been under development since 1970. The feasibility of magnetically levitated vehicles was proved with the Prinzipfahrzeug in 1971. This made use of a centralized control concept (degree of freedom control) which proved an adequate solution for heave, sway, pitch, yaw and roll in the short, compact test vehicle. The linear induction motor chosen was a short-stator asynchronous motor, the stator on board the vehicle and an aluminium part of the guideway representing the squirrel cage. Problems of power transmission at high speed and the loss of payload, due to heavy stators and power conditioning units on board led to long-stator solutions with linear induction motors for propulsion and linear induction generators for power transmission.

In this version, the stator winding is mounted on the guideway. The levitation magnets on board the vehicle represent the "rotor." Proof of the feasibility of this concept was established by the vehicle HMB 2. To prove the high-speed capability of EMS a linear high-speed test stand was built to operate the test vehicle KOMET. On the 1200 m track in Manching near Munich, the KOMET, propelled by hot-water rockets, reached 401.3 km h^{-1}. The limiting factor was the length of the guideway, not the magnetic levitation.

For reasons of mechanical tolerances, mass and power budget, guideway disturbances and redundancy and safety aspects, a decentralized hierarchical control concept has been developed for the KOMET M (modular), and subsequently, modular levitation and guidance systems have been chosen for all full-scale vehicles (see Fig. 3). This allows for decentralization into autonomous control functions and electrical and mechanical separation of hardware modules. By elastic suspension of individual magnets, the masses to be accelerated are made smaller, the influence of the mechanical tolerances of vehicle and guideway on the magnet gap is nearly eliminated, excitation of structural modes by the single-magnet controllers is nearly impossible and malfunction of a single magnet has no influence on vehicle operation.

4. Basic Subsystems of EMS Vehicles

The basic subsystems of EMS vehicles are described here based on the full-scale vehicle Transrapid 06 (TR 06), which corresponds to an operational vehicle for high-speed transportation between cities and airports.

The TR 06 consists of two sections (see Fig. 4). Each cabin is carried by four magnet frames. The secondary suspension between cabin and magnet frames consists of pairs of pneumatic springs. The magnet frames themselves are interconnected by joints.

Each of the four magnet frames of one section carries four levitation and three or four guidance magnets on each side.

The autonomous subsystems consisting of a single magnet with individual suspension and single-magnet controller is called a "magnetic wheel" (Gottzein 1984). The vehicle is propelled by a synchronous long-stator motor, which also performs the braking.

For safety, the TR 06 is equipped with an autonomous mechanical emergency braking system and elastically suspended emergency skids for levitation and guidance.

The guideway is a continuously elevated single—or double—beam design, partly steel, partly concrete with a clearance height of at least 4.5 m.

The magnetic wheels are the core of the modular levitation and guidance system of the TR 06. Each section is equipped with 60 autonomous magnetic wheels. The components of a magnetic wheel are:

(a) an electromagnet with individual magnet suspension consisting of springs and dampers,

(b) the magnet current drivers,

(c) a gap sensor and accelerometer,

(d) a single-magnet controller, and

(e) command and telemetry interfaces and monitoring equipment.

For the TR 06, an overall availability of 99.5% during an 18 h service period is achieved by functional redundancy of magnetic wheels which allows for switching off magnets without interruption to service.

Table 2
Maglev vehicles developed for local traffic between 1971 and 1989

Vehicle	Company/ institution	Country of origin	Year of origin	Speed (km h^{-1})	LGS principle	Central or decentral LGS	Number of persons carried	Short- or long-stator LIM	Remarks
Transurban 03	KM	Germany	1974	73	EMS	central	12	short	
EML 50	JNR	Japan	1975	40	EMS	central		short	
Birmingham Maglev	People Mover Group	UK	1984	54	EMS	central	8	short	only maglev transport system in public service (Birmingham airport)[a]
M40-1/M70-2	Magnetbahn GmbH	Germany	1976/1980	47			40–53 70–110	long	combined permanent magnet/ wheel- levitation and guidance system

Source: Rossberg (1983)
a Proof of concept

Magnetic Suspension Railroad Systems

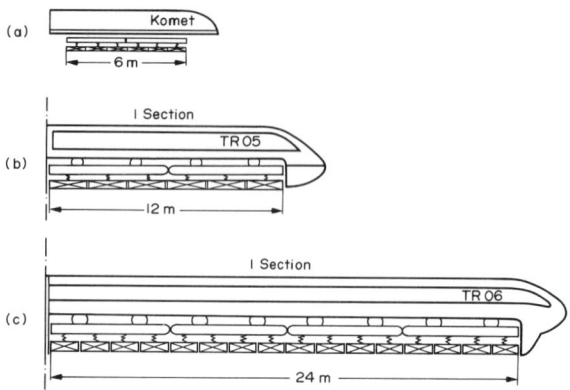

Figure 3
Maglev vehicles with modular levitation and guidance system, drawn to scale: (a) Komet M, 1976; (b) TR 05, 1979; and (c) TR 06, 1983/1984

5. State of the Art and Prospects

Development of maglev transportation systems is still being continued by Japan, the UK and Germany.

In Japan, competitive development of EMS and EDS is performed by Japanese airlines (JAL) and Japanese national railroads (JNR), respectively. The unmanned test vehicle ML 500 holds the world record for EDS vehicles with a speed of 512 km h^{-1}. The concurrent EMS development is a three-section vehicle MLU 001, designed for travel at 220 km h^{-1} and 32 persons (Masada 1988).

Birmingham Maglev is the only maglev transportation system in public service. It connects Birmingham International airport and the railroad station.

Design and development of the TR 06 in Germany was based on ten years of experience in maglev and, in particular, on test experience with eight vehicles, each designed to be a milestone towards an operational high-speed train for public service.

The main data of the TR 06 are summarized in Table 3. Tests of the TR 06 are performed on the Transrapid Versuchanlage Emsland (TVE) (see Fig. 5). The TVE guideway has all the features of an operational maglev guideway, such as high-speed straights, reversing loops and switches.

Based on the TR 06 experience, the status of high-speed maglev train development in Germany has been summarized by Rogg (1988) as follows:

(a) functioning of contact-free levitation on all guideway elements over the full speed range of up to 500 km h^{-1} with full-scale operational vehicles;

(b) high travelling comfort;

(c) excellent acceleration and braking ability as a result of acceleration sections of the long-stator motor and low vehicle weight (e.g., 0.5 t per seat);

(d) reasonable investment cost as a result of low guideway loads and flexible routing parameters (curve radii of 4000 m at 400 km h^{-1}; gradient up to 10%);

Table 3
Specifications of the TR 06

Length (m)	54.2
Width (m)	3.7
Height (m)	4.2
Empty weight (t)	102.4
Payload (t)	20.0
Total weight (t)	122.4
Number of seats	192

Figure 4
TR 06 on the TVE

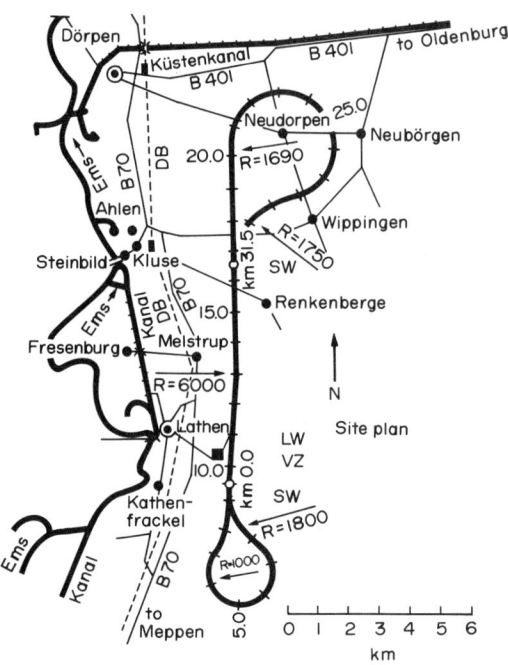

Figure 5
Site plan of the TVE (SW is a high-speed switch, LW a low-speed switch and VZ the test center)

(e) favorable operating costs as a result of levitation/guidance/propulsion and braking systems free from wear;

(f) low-energy consumption (e.g., 60 W h per seat kilometer at 400 km h^{-1} from a substation);

(g) high safety standards due to derail-safe vehicles;

(h) relatively low noise level, especially when entering densely populated urban areas at reduced speeds (e.g., maximum noise level at 200 km h^{-1} measured at 25 m distance is 77 dB (A)) and no mechanical vibrations as a result of contact-free ride technique;

(i) minimal land requirements (2000 m^2 km^{-1}); and

(j) totally negligible stray electromagnetic fields.

From 1983 onward, the TR 06 has been operated on the TVE. Up to mid-1985, it carried more than 20 000 people over a distance of more than 40 000 km. It holds the present world record for manned maglev trains of 412.6 km h^{-1}. From present results there is no reason to believe that the limitations of the magnetic LGS have been reached (Menden *et al.* 1989).

See also: High-Speed Railroad: Modelling and Simulation

Bibliography

Gottzein E 1984 Das "Magnetische Rad" als autonome Functionseinheit modularer Trag- und Führsysteme für Magnetbahnen. *Fortschr. Ber., VDI Z. Reihe 8* **68**

Gottzein E, Rogg D 1984 Status of high speed MAGLEV train development in the FRG. Report No. C 407/84. Institute of Mechanical Engineers, London

Hochleistungs-Schnellbahn Studiengesellschaft 1971 Bundesministerium für Verkehr, Studie über ein Hochleistungsschnellverkehrssystem. HSB Studiengesellschaft, Munich, Germany

Kemper H 1935 Trassierung mit räderlosen Fahrzeugen. German Patent No. 644,303

Masada E 1988 EMS-technique and state of applications in Japan. *10th Int. Conf. Magnetically Levitated Systems.* TÜV Rheinland, Cologne, Germany

Menden W, Mayer W J, Rogg D 1989 State of development and future prospects of the MAGLEV-systems Transrapid, M-Bahn and Starlim. *11th Int. Conf. Magnetically Levitated Systems.* IEE, Tokyo

Rogg D 1988 The development of high-speed magnetic levitation systems in the Federal Republic of Germany, objectives and present state. *10th Int. Conf. Magnetically Levitated Systems.* TÜV Rheinland, Cologne, Germany

Rogg D, Schulz H 1978 Systementscheidung bei der Magnetschwebe technik. *ETR Eisenbahntech. Rundsch.* **11**, 721–8

Rossberg R R 1983 *Radlos in die Zukunft.* Orell-Füssli, Zurich

E. Gottzein
[MBB Deutsche Aerospace, Munich, Germany]
W. Crämer
[Fachhochschule, Rosenheim, Germany]

Maintenance and Reliability of Traffic Control Systems

In the traffic control sector, notwithstanding the safety problems, the reliability of the systems is seldom tackled in a rigorous way. Taking as an example the area of traffic light control, few products are characterized by precise data on reliability; that is, by the average rates of various kinds of failures.

To highlight this failure, it is sufficient to make a comparison between the designing and maintaining of automation and control systems in manufacturing plants and in traffic control systems. No private enterprise would accept systems satisfying the maintenance and reliability requirements of most control systems installed on road intersections.

This failure is obviously less pronounced in the industrialized countries, but the gap between manufacturing automation and automation in traffic management is large everywhere. Consequently, higher costs burden both communities and transport authorities. In fact, the low reliability of the components used, their missing harmonization, the lack of systems for diagnosing and automatically signalling the failures and the weakness of the design with respect to maintenance problems are all leading to higher failure rates and more difficult maintenance with longer repair times. The presence of faults increases the danger and causes delays to the detriment of the community. Excessive failure rates and poor maintainability of the system bring about an increase of maintenance costs which are borne by the companies or authorities involved.

There are, however, signs pointing towards a change of direction. The first large traffic control system which, according to the literature, has been designed to conform to modern reliability criteria is in Tokyo (Inose 1976). Other systems have followed, among them an experimental project in Turin (Donati *et al.* 1984) and, especially worthy of mention, work in London, where a fault control center (FCC) has been implemented and is now in operation (Oastler 1985, Oastler and Palmer 1989).

The urban traffic light control system is considered as a representative example and reference case for this article. Design concepts and the technology of the subsystems evolved rapidly during the 1980s; nevertheless, the recommended guidelines and the conclusions reached should be applied both to existing and operating systems and to projects still in the development stage.

1. Types of Failures

In a traffic control system all kinds of failures can occur; an attempt to classify such failures on the basis of what can break down would have no significance here. Indeed, this would be a classification peculiar to a specific commercial product. Changing the product and consequently the hardware structure of the system means that the possible failures will also change.

This article examines the failures from the point of view of their effects rather than with reference to what breaks down; that is, the problem is tackled as the user sees it, following an approach that aims to formulate the general specifications for a supply order.

In order to establish a hierarchy for appreciating the level of operation of the system and also for determining the maintenance requirements, the following classification into four types of failures is given.

1.1. Dangerous Failures

The class of failures that cause dangers to the users includes

(a) opposing greens (a conflict situation existing when green lights are given simultaneously to two directions of traffic flow so that collisions can occur),

(b) missing red (the red light is missing for one traffic stream so that the user moving in that direction might think that the traffic light is blacked out),

(c) missing amber (this situation is particularly dangerous whenever the "removal phase" does not envisage an "all-red" interval of time),

(d) cycle stop on specific colors, and

(e) when one green time lasts less than the given minimum or extends for more than the (given) maximum period.

The presence of failed lamps (either red or amber) should not be considered as dangerous failure if the user has the possibility of seeing at least another traffic light, clearly visible, belonging to the same group of signals.

1.2 Apparent Failures

The failures that cause evident malfunctioning are ones that, being associated with evidence of faults, have negative influence on the traffic and increase the waiting times without originating an immediate danger for the users. They are, therefore, failures of all kinds in one of the following situations:

(a) signals blacked out on all approaches, and

(b) flashing amber.

These faulty states can be caused by a wide variety of failures which can hit the system at different points.

For example, a protection device is normally used against potentially dangerous faulty states which, when recognizing the faulty state, sets the system in flashing amber or in all blacked out. As a consequence, the faulty state considered here is frequently due to a failure that potentially belongs to the category described in Sect. 1.1 (e.g., the burning out of the lamps).

It can also happen that the protection device is released due to disturbances. In this case, the device for putting the system into operation need only be reset.

Further examples include other failures of the electronics of the controller, insulation failures of the electric supply of controller and lamps, and short circuits in the lamp holders of the traffic lights.

1.3 Degradation

Some failures do not originate conditions dangerous to the users or evidence of faults, but they can still cause a degradation of the control system performance and an increase of waiting times at the intersections for all vehicles (or for some categories of them). These failures are typical of modern computerized control systems which, equipped with self-diagnosis, recognize the faulty state and actuate a consistent strategy of degraded control.

For example, this category includes faults with sensors detecting both private and public vehicles that the control system is able to diagnose and, therefore, to actuate a new control strategy that excludes the use of the sensors that are faulty. Should, however, the self-diagnosis not be provided or be inefficient, then the failure of one sensor for a traffic light that is traffic actuated or has the task of assigning priority to public vehicles is a dangerous failure (see Sect. 1.1).

Another example is any fault concerning the central control system or the communications system between center and intersections. Obviously, these kinds of failures are recognized by the peripheral system, which automatically sets itself into a local degraded operating condition.

1.4 Failure with no Effect

For some components, active or standby redundancies have been provided so that the user neither perceives the fault nor bears any consequences.

This class of faults regards only the maintenance service which must actuate special diagnosis methods in order to recognize the failure and act as soon as possible so as to restore the redundancy conditions.

2. Indices of Operation

Considering the smallest functional unit of the control system, which typically corresponds to an intersection, the possible states of a traffic light control system are subdivided into four classes:

(a) in operation (state 0),

(b) dangerous failure (state 1),

(c) apparent failure (state 2), and

(d) degradation failure (state 3).

States 1, 2 and 3 correspond to the existence of failures connected with the categories given in Sects. 1.1, 1.2 and 1.3, respectively. The state of the whole control system is then described by the combination of the states of its elementary units (i.e., of the individual intersections).

The probability that each unit of the system is in any one of the four states describes the operational level of the system itself.

Given an observation period, each probability corresponds to the expected value of the ratio of the time during which the system remains in the state to the total time considered.

Denoting the probabilities of states 0, 1, 2 and 3 by P_0, P_1, P_2 and P_3, respectively, then

$$P_0 + P_1 + P_2 + P_3 = 1 \qquad (1)$$

P_0 is the index of the operational level; P_1, P_2 and P_3 are indices of unavailability.

These probabilities represent indices summarizing the characteristics of the plant reliability, of the maintainability and of the efficiency of the maintenance service.

The reliability is expressed by the failure rate or, more usually, by its inverse value—the mean time between failures (MTBF). With reference to the classification of failures introduced and to the functional unit, three different values $MTBF_1$, $MTBF_2$ and $MTBF_3$ are employed, corresponding to the average time interval between occurrences of two different failures that bring the system into the faulty states 1, 2 and 3, respectively.

The maintainability, like the reliability, is a characteristic dependent on the plant design. It is evaluated through the mean time to repair (MTTR); that is, the total corrective maintenance repair time required to complete a number of corrective maintenance operations, divided by the number of operations. To be precise, a specific MTTR corresponds to each specific failure. It is convenient to introduce three average values $MTTR_1$, $MTTR_2$ and $MTTR_3$, defined as the average time required to repair the "average" faults that bring the system into the faulty states 1, 2 and 3, respectively.

Both the fault-locating time and the true repair time contribute to the MTTR. To reduce the MTTR values, it is important to adopt automatic techniques of failure diagnosis, as well as to use modularization of the subsystems: the fault repair phase "in field" will then be reduced to the replacement of the module located by the self-diagnosis device.

The efficiency of the maintenance service can be expressed through two indices: the mean time to signal (MTTS) and the mean time to intervene (MTTI). The MTTS can be significantly reduced by using automatic fault diagnosis and signalling systems connected with the FCC. However, in the extreme case of a fault signal due to user complaints, the value of the MTTS can become very large. The MTTI depends essentially on the size of the maintenance service (as the number of maintenance crews is relative to the plant size), on the number of hours per day of operation of this service, on whether or not the service is operating during holidays and, not of least importance, on the communications system between crews and the FCC.

The sum of MTTR, MTTS and MTTI gives the mean down time (MDT); that is, the average time the plant remains in a faulty state after the failure has occurred. Obviously, with regard to the three types of faults possible, three values can be defined: MDT_1, MDT_2 and MDT_3.

With small acceptable approximations, the following relations hold:

$$P_1 = \frac{MDT_1}{MTBF_1} \qquad (2)$$

$$P_2 = \frac{MDT_2}{MTBF_2} \qquad (3)$$

$$P_3 = \frac{MDT_3}{MTBF_3} \qquad (4)$$

The indices P_0, P_1, P_2 and P_3 are the most appropriate ones for expressing the requirements of the community. It is the duty of the designer and of the operator to adapt the various items that affect the system operation in order to minimize the costs under the same final performance.

3. Indicative Values of Indices

3.1 Dangerous Failures

The index P_1 expresses the probability that the system is in a condition that is dangerous for road users. It should be made as small as the community requires and is willing to pay for.

Rodgers (1971) identified for each operation a risk level of 10^{-4} as the boundary between acceptable and unacceptable, and asserted that, for risk levels below 10^{-5}, the US public tends to ignore the risk and does not approve of additional expenditure to reduce the risk further.

The operation considered is the crossing of an intersection when traffic signals and rules are observed. There is the risk that the signal given by the traffic light control system is wrong and there is the question of what would originate the danger of a serious accident.

Considering the minor increase in costs that would nowadays make it possible to reach a high level of safety, the limits given by Rodgers and proposed again by Hulscher (1977) should be considered as a level of risk no longer acceptable.

The average user in a medium to large town for the home–work–home journey crosses about 30 intersections equipped with traffic lights every day. A risk of 10^{-4} means that roughly one event per year will be experienced. Even considering that crossing an intersection when its control system is in a dangerous faulty state does not necessarily mean involvement in an accident (though it does mean taking a serious risk), it must be concluded that one event per year per user is too high.

With a suggested figure of 10^{-5} or 10^{-6} in mind, estimation of a more reasonable limit will be considered. Protection against dangerous failures is obtained through a hardware card which detects the faulty state and sets the system into one of the not dangerous faulty states: flashing amber or all blacked out. This card must be designed with an adequate MTBF, typically higher than for the other parts of the equipment.

If as usual, no safeguarding of the protection is operative, it must be assumed that a protection failure is not signalled and is therefore not repaired until the dangerous failure against which the protection is designed to act occurs. It is only the appearance of this failure that leads to the discovery that the protection is not working. The MTBF_1 of the appearance of a dangerous failure is therefore given by adding the MTBF of the failure occurrence to the MTBF of the protection card.

Assuming that a dangerous failure should be quickly signalled, at least at the first accident and that the intervention of a police officer should be sufficient for switching the controller into the not dangerous, fault state 2, the MDT_1 in urban areas can be estimated to be about 1 h. Thus

$$\text{MTBF}_1 = \frac{\text{MTD}_1}{P_1} \qquad (5)$$

To be conservative, MTBF_1 can be replaced with the MTBF of the protection card. Assuming $P_1 = 10^{-4}$, it follows that $\text{MTBF} = 10\,000$ h. Given the simple logic structure of the card, this can be obtained at very low cost, without any kind of component sifting or card debugging. An $\text{MTBF} = 100\,000$ h, which reduces P_1 to 10^{-5}, can be obtained without too much difficulty, still at low costs, by conveniently sifting the components and by breaking in the card in order to eliminate possible failures caused when assembling it.

Without going deeper into the problem, it seems reasonable to conclude that modern traffic control systems should have values of P_1 not greater than 10^{-5}, with 10^{-6} as a goal for the not too distant future.

However, it must be emphasized that traffic light control systems (some of them built in the 1980s) are still operating in several countries without any special protection against state 1 failures. In these conditions, it is likely that they run at P_1 level between 10^{-3} and 10^{-4}. The most recent realizations do not, however, have P_1 values larger than 10^{-5}.

3.2 Apparent Failures

The range of acceptable values for P_2 depends essentially on the location of the traffic control system and on the importance of the traffic conditions that the traffic control has to manage.

It should be noted that, in low-traffic conditions, the blackout of the light controller does not cause traffic delays but raises the probability of accidents at the intersection, whereas a traffic increase causes a delay that grows quickly with traffic intensity, up to complete blockage of the intersection for medium–high traffic (i.e., for a normal situation in a large town).

This failure, therefore, causes an increase of costs for the public administration, as they are compelled to keep a group of signalmen in the service of the local police, ready to intervene as soon as a failure becomes known, the number of signalmen required being proportional to the average number of failures.

Finally, if the failure rate goes above levels considered reasonable, the public image of the administration will suffer.

To find a compromise value of the index P_2, the damage due to the faulty state must be compared with the higher costs to be faced in reducing P_2.

An accurate cost–benefit comparison valid for all countries cannot be performed, since the economic evaluation of the damage due to a failure gives results that are different from country to country, both because of the different labor costs and because of the different ways of perceiving some types of costs.

Nevertheless, at least for what concerns the most industrialized countries, the trade-off point is apparent. Given the present cost of electronics (low compared with the cost of labor), it is convenient to reduce P_2 towards the smallest possible value within the range of application of normal commercial products. That is, without requiring special products which in small series would be too expensive, convenience points towards higher-quality commercial products. At the same time, it is suitable to exert a continuous pressure on the control system manufacturers to obtain increasingly better products comparable to the present development of electronics in other sectors.

Traffic light controllers with a MTBF of 15 000 h are on the market. It is conceivable that values of about 25 000 h can be reached without an increase in costs.

There are highly reliable traffic lights protected against water and dust (IP 55). The reliability of the electric subsystem depends on the care taken during the realization rather than on the cost of the materials. The weak point is in the lamps which, besides having a nonnegligible infancy failure rate, have an average life of the order of 6000 h.

The problem of lamp failures is often tackled through

a planned replacement at a specific time during their life. This approach, without being a perfect solution, offers considerable advantages but is rather expensive.

The way to a radical solution seems to be active redundancy of the lights seen by each driver, together with local automatic monitoring of burnt-out lamps, repeated at the FCC.

The maintenance services are not efficient everywhere: for large towns an MDT of 24 h can be accepted. In London, a central FCC operates through an active maintenance service continuously. The most urgent faults are repaired within 5 h (Blase 1979, Oastler 1985, Oastler and Palmer 1989).

It is thought, however, that a threshold of convenience for the costs exists: the MDT should not be reduced below certain limits, especially in medium and small towns.

With reference to modern systems installed in large towns, typical values of P_2 are in the range 10^{-2}–10^{-3}. In fact, in a modern design (all causes considered and damage through accidents included), it is possible to remain at 1–2 failures per year (Donati *et al.* 1986). An MDT of 24 h gives P_2 within the acceptable range. Values close to 10^{-3} are considered to be very good.

3.3 Degradation

Very few existing systems use sophisticated control strategies that, as a consequence of failures, can operate in a degraded way. In most cases, degraded strategies are not provided; therefore, the failure of even a nonessential component puts the system in failure state 1 or 2.

Where degraded operating conditions are admitted, then failure state 3 exists. Since this state does not appear externally, it is perceived with difficulty by the user. Even when noticing a slowing down of the traffic, the user thinks of a traffic increase rather than a degradation of the strategies. The first visual effects of a degradation of the control strategies are an increase in car density and extension of queues, the same phenomena that appear whenever a traffic increase occurs.

This is one of the reasons why, in modern control systems of hierarchical type, the failure rate of the failure state 3 category is always fairly high. The other reason is the need for keeping a check on expenditures. The failure rate of the computers mostly used for these purposes (minicomputers or microcomputers) is fairly high. Obviously, the rate changes, depending on the system configuration, but MTBF values around 1000 h are normal. To these must be added stops for ordinary maintenance, stops due to bugs in the software, stops caused by disturbances on the telecommunications lines and stops because of faulty operation by the FCC. As this is the beginning of the use of sophisticated systems (such as those operating in several European towns—each one representing an innovative solution—which actuate coordinated control of public–private traffic, with priority given to public vehicles), the conditions of operation are rather difficult since public administrations are not willing to spend too much for innovative systems.

In addition, in keeping down the costs, the reliability of the central system is given up, which would at least require a "hot" (active) redundancy of all elements subject to frequent failures.

However, for economic reasons, the MDT is never too small because even if the failure is immediately made known, outside firms are entrusted with the repairing contracts and this does not always ensure a prompt intervention. Consequently, the index P_3 is in the range 5–10%.

As an example, the SIS system operating in Turin will be considered (Gentile and Mauro 1988). This has the purpose of ensuring the regularity of the surface public transport system interacting with the traffic light control Progetto Torino. On average, it does not succeed in keeping more than 90–95% of the assigned vehicles under control, because of a wide number of failures of on-board equipment and because of disturbances in telecommunications.

It is to be expected that the value of P_3 will decrease considerably in the near future, as soon as the next generation of control systems are in operation.

See also: Road Traffic Monitoring Equipment; Safety of Road Traffic

Bibliography

Blase J H 1979 Computer aids to large-scale traffic signal maintenance. *Traffic Eng. Control* **20**, 341–7

Donati F, Margaria A, Piglione M C 1986 Il sistema semaforico sperimentale "Progetto Torino". *Rend. Convegno Nazionale ANIPLA.*

Donati F, Mauro V, Roncolini G, Vallauri M 1984 A hierarchical-decentralized traffic light control system. The first realization "Progetto Torino". *Prepr. IFAC 9th World Congr.*, Vol. 2. IFAC, Laxenburg, Austria, pp. 1–6

Gentile P, Mauro V 1988 Experience on S.I.S., Torino public transport operation aid system. *Conf. Automatic Vehicle Location.*

Hulscher F R 1977 Reliability aspects of road traffic control signals. *Traffic Eng. Control* **16**, 98–102

Inose H 1976 Road-traffic control with particular reference to the Tokyo traffic control and surveillance system. *Proc. IEEE.* **64**, 1028–39

Oastler K H S 1985 Maintenance of traffic signals in London. *Traffic Eng. Control* **26**, 104–8

Oastler K H S, Palmer R F 1989 Revised arrangements for the maintenance of traffic signals in London. *Traffic Eng. Control* **30**, 114–20

Rodgers W P 1971 *Introduction to Safety System Engineering.* Wiley, New York

F. Donati and M. Vallauri
[Politecnico di Torino,
Turin, Italy]

Marine Fleet Planning and Scheduling

The increase in efficiency of transport system functioning is closely connected to the problem of improving and automating control processes in transportation systems. Under these conditions, the task of efficient planning when the influence of the environment is essential plays an important role in these problems. The solution to these problems is very important for transportation systems, which operate under complex nonstationary conditions.

The computer-aided system is designed for efficient planning and scheduling of marine fleets allowing for their interaction with railroad networks. The functional basis of the system is the task of short-range planning and scheduling of marine fleets, formulated as the problem of obtaining optimal coordinated solutions for various periods of planning. The complex of models which describe transportation processes and controlling effects with necessary detail has been developed using methods of nonlinear optimization, dynamic programming and dialog procedures.

The system referred to in this article was designed for the Vanino–Kholmsk ferry in the USSR. However, the methods, models and program support may be used to solve other practical problems that can be described by these models (e.g., the short-range planning of other kinds of marine fleets).

1. Statement of Requirements

The objects to be controlled in the computer-aided system for planning and scheduling of marine fleet (CASPLAS) are ships as a specific part of the technological system of combined freight transportation between ports.

In solving the task, it is necessary to have information about the ship dislocation on line, as well as the state of the ports (the quantity of freight), forecasts of freight flows to the ports and the hydrometeorological conditions. Irregularity of freight flow to ports and inaccuracy of transportation are also taken into account.

The functional basis of the system is the task of planning and scheduling marine fleet (TPSMF), which consists of determining optimal plans and schedules for every ship in a current plan period (month or quarter) and values of plan indexes describing the work of the ferry being preassigned (preset) for this schedule.

The effectiveness of a solution can be assessed by various criteria. The profit of the fleet for the current plan period (CPP) has been taken as a main criterion for the effectiveness of short-range planning, while the main resources may be considered as a constant quality.

The essential limitations of the TPSMF are the limited capacity of by-port railroad terminals, tracks and port tracks, the limited carrying capacities of marine fleets, and the necessity for realization of CPP indexes. Also, the technological limitations are important, such as the limited number of moorages in ports, the necessity for changing the working team schedule, the limited capacities of by-port railroad tracks and port tracks, the limited capacities of ships, the specialization of ships to transport specific kinds of freight, the individual characters of ships and the account of current ships and state of the ports, the effect of hydrometeorological conditions, the repair schedule of ships and so on.

In general, the TPSMF is determined as a problem of profit maximization from transportation of freight taking into account CPP plan realization, forecasting the arrival of cars to the port, hydrometeorological conditions, the state of the ships and ports and so on.

2. Computer-Aided Planning and Control

Great experience has been accumulated in the use of computers to control marine transport (Ardonin 1982, Cashman, 1983, *Container News* 1983, Schonknecht 1983). The development of information systems takes place stage by stage, as a rule, by increasing service and functional opportunities, permitting the performance of information-and-advisory and information-and-controlling systems on this basis. These systems are usually developed for only one kind of transport. The other kinds of interactive transports are users of these systems (*Container News* 1983).

The control processes for combined transportation of freight in transportation and technological systems by the scheme "from door to door" require the participants of these systems to exchange information and work out coordinated plans using computers, communication means and data transmission (Etschmaier 1986, Polyantsev and Kiselev 1987).

Consider the use of computers for controlling ferry transportation, for instance between the USSR and Bulgaria (Ilyichevsk–Varna), and between the USSR and Germany (Kleipeda–Mukran). To control ships at the Ilyichevsk–Varna ferry, the united dispatching center has been developed and equipped with the necessary communication means for exchanging information between control points in Bulgaria, on the Odessa railway lines, on-board the ferries and also in the coastal ferry complexes. The computer enables the optimum plans for the realization of annual volumes of freight transported to be calculated as well as solving the problems of ship scheduling and optimizing the timetable (Pritulsky and Sukolenov 1986a, b, Polyantsev and Kiselev 1987).

Accumulated experience has allowed the beginning of attempts to solve more complicated problems such as the development of the computer-aided system for international railroad–ferry transportation between the USSR and Germany (Polyantsev and Kiselev 1987). The computer-aided system "Ferry" includes both railroad and marine transportation and is characterized by the various interactions between interactive systems.

Ferry also provides automatic support of the system database, forecast and control transportation processes, the estimation and analysis of ferry operation (including mutual calculations for transportation and service), the official documentation of transportation processes and so on. The continuous character of the transportation process and a high degree of variability of external factors, have resulted in the need to produce a complex system of constant efficient planning of marine transportation. This system is based on the systematic inclusion of variations and additions into the plans, with allowances for the present situation, the regular increase of plans for a certain period of time and the constant relationship of plans of various structural units and plans with the different planning periods (Leviy 1971, Aven et al. 1983). In this connection, it is of particular importance to work out methods and models that allow the automation of the processes of continuous planning and control.

Some scientists suggest an approach that facilitates the process of composing the coordinated plans for different periods of planning (Moiseenko 1978, Aven et al. 1983, 1985). This approach has been considered as an example for the united annual plan, with the division into quarters and the quarterly plan divided into months. The tasks of annual and quarterly planning are usually solved independently; this results in great expenditure of labour to coordinate these plans. Taking into account the fact that the initial information for the first quarter is identical for annual and quarterly plans, some authors suggest a model that allows both tasks to be solved simultaneously. In the case when quarterly plan indexes must be corrected in the intermediate periods between corrections to the annual plan, the model of quarterly planning is used with flexible limitations to the annual plan. The suggested model represents the task of large-scale linear programming. In order to solve it, decompositional methods based on limitations are used. This approach is also used for solving the TPSMF. However, the models, being the basis of the TPSMF, differ from the models suggested by Moiseenko (1978) and Aven et al. (1983, 1985). For the TPSMF, the models are related to discrete representation of the transportation process, the necessity of forecast dynamics, difficulties in the formalization of a 24 h period of planning, and criteria of optimization that have situational character and depend on the current situation in the transportation process.

3. Formalization of the Problem

The dynamic nature of transportation processes, and the randomness of external (load and truck arrival and hydrometeorological conditions) and internal (ship and port conditions) factors necessitate the TPSMF being considered as a problem of multistage stochastic programming. However, making a numerical and analytical decision is difficult because of the way the problem is posed. If the role of accidental factors is unimportant, the procedure of continuous planning gives an approximate optimum decision (Pervozvansky and Gitsgory 1979).

The procedure of continuous planning takes place as follows. The period of quarter planning is divided into steps (Pervozvansky and Gitsgory 1979, Aven et al. 1985) which must presume complete information about the state of the system by the beginning of planning, when the realization of the efficient decision is demanded. Then, the optimization of the multistep model is made with allowance for current data and factor forecasting for the future. Because of the inaccuracies involved in forecasting the state, which is reached by the beginning of the next step, it will contain discrepancies compared with the calculation and the procedure of efficient-planning decision making is repeated using current data.

Therefore, the TPSMF is regarded as a task to be determined by dynamic programming, which utilizes forecasts of accidental factors. These forecasts should be slightly different from the realization of accidental factors. The accuracy of forecasts for the ferry is achieved by gathering information from the railroad network about the location of trains and cars travelling to the port, coordinating forecasts of car arrivals at the seaport between the ship company and the railroad and the collection and processing of statistical data about the effects of hydrometeorological conditions on fleet work.

Regarding the essential functional task, solved by CASPLAS, write $t = 1, T$ for numbers of the stages of planning which are in the current plan period. The duration of the t-stages, measured in some unit of time, can be different.

For the TPSMF, a number of stages $\tau_1 = \{1, \ldots, t_0\}$ with a 24 h period can be distinguished from a number of stages $\tau_2 = \{t_1, \ldots, T\}$ whose period is, as a rule, longer. The TPSMF can be formulated as:

$$\sum_{t=1}^{T} \phi_t(\mathbf{y}_t) \to \max_{\{y_t\}} \quad (1)$$

$$\sum_{t=1}^{T} f_t^j(\mathbf{y}_t) \leq B^j, \quad j = \overline{1, J}, \quad (2)$$

$$\mathbf{x}_t = A_t \mathbf{x}_{t-1} + \psi_t(\mathbf{y}_t), \quad \mathbf{x}_0 = \mathbf{x}(0); t = \overline{1, T} \quad (3)$$

$$\mathbf{x}_t \in X_t \subset \mathbb{R}^{n_t^1}, \quad t = \overline{1, T} \quad (4)$$

$$\mathbf{y}_t \in Y_t \subset \mathbb{R}^{n_t^2}, \quad t = \overline{1, T} \quad (5)$$

$$n_{t_2}^i \leq n_{t_1}^i \text{ for } t_1 \leq t_2, \quad i = 1, 2 \quad (6)$$

where t is a number of the stages of planning.

Equation (1) reflects the necessity of obtaining the maximum economic effect from the system. Equation (2) reflects the limitation of resources for the whole process. Equation (3) is an equation of dynamic process state. Equation (4) reflects the state limitations of the process, including the conditions of fulfillment of the plan index at the stages t. Equation (5) reflects the limitations on the choice of stage control.

The peculiarity of this mode is the presence of the operator of state clustering A_t, hence, the process at different stages t is characterized by the state vector x_t, changing in different $\mathbb{R}^{n_t^i}$. This mode allows submodels of the different details describing the process to be united. For the ferry mode, Eqns. (1)–(6) are a combination of two submodels: the problem of time planning ($t \in \tau_2$) and the problem of scheduling ($t \in \tau_1$). These problems involve different details of the process description. Combining them into one model (Eqns. (1)–(6)) ensures that none of the system qualities is lost and that all of the important processes are described in the necessary detail.

The problem has a high dimension and may be solved by decompositional approaches. The dynamic planning method is quite effective.

Consider a set of functions

$$F_t(s_{t-1}), \quad t = \overline{1, T}; \, s_t \in S_t$$

where s_t is the state vector of the system at the end of stage t, whose components are the variables x_t and variables which correspond to the general process limits given by Eqn. (2). $F_t(s_{t-1})$ is the maximal profit from the beginning of stage t to the end of stage T.

Following the optimization principle, recurrent equations can be written:

$$F_t(s_{t-1}) = \max_{y_t \in Q(s_{t-1})} [\phi_t(y_t) + F_{t+1}(G(s_{t-1}, y_t))],$$

$$t = \overline{1, T\text{-}1} \quad (7)$$

$$F_t(s_{t-1}) = \max_{y_t \in Q(s_{t-1})} \phi_t(y_t), \quad t = T \quad (8)$$

where $Q(s_{t-1})$ is the limit for the equation of y_t at stage t under the condition that at the beginning of stage t the system was at state s_{t-1}.

From the solution of these equations with original state S_0, values $\{y_t\}$ are determined, running into a maximum of efficiency being the solution of Eqn. (1). In the stages $t \in \tau_1$, calculations are made based on scheduling problems and in the stage $t \in \tau_2$ they are made based on time-planning problems. Various methods are used to avoid large-sized difficulties. An approximately optimal decision $\{y_t^0\}$ is determined and then additional limitations are introduced to $\{y_t\}$ and $\{s_t\}$; the area of $\{y_t^0\}$ is reduced (Bellman and Dreyfus 1962, Moiseev 1974).

To determine $\{y_t^0\}$, the method of modified Lagrange's function turns out to be effective (Gill et al. 1981), using the model of the calendar-planning problem. The model given by Eqns. (1)–(6) is used both in automatic and dialog modes. While solving the task, the user has various opportunites: using solutions obtained during previous calculations; giving and estimating any solution of the problem; to correct problem limitations; and giving various model interpretations by setting a calculation on the databases, corresponding to various modifications of the TPSMF problem.

4. Time Planning

Let the forecasts of freight arrival α_t, the hydrometeorological forecasts β_t, the state of system (x_{t_0}), and the achieved values of the CPP indexes $\{P_j^{t_0}\}$ with $j = \overline{1, J}$ be known at the end of stage t_0. Control variables are the mean intensity of ship operation, the quality of runs between the ports and the loading of the ships with freight at stage t.

The problem of time planning can be formulated as:

$$\text{PR} = \text{I} - \text{E} = \sum_{t \in \tau_2} [(\text{I}_t(y_t) - \text{E}_t(y_t, \beta_t)] \to \max_{\{y_t\}} \quad (9)$$

under the condition of fulfillment of the CPP $\{P_j\}$, with $j \in J$:

$$\sum_{t \in \tau_2} f_t^j(y_t) \geq P_j - P_j^{t_0}, \quad j \in J \quad (10)$$

and fulfillment of balance and technological limitations:

$$x_t^i = x_{t-1}^i + \alpha_t^i - \Psi_t^i(y_t), \quad x_0 = x_{t_0}A; t \in \tau_2; i \in I \quad (11)$$

$$0 \leq x_t^i \leq x^{-i}, \quad i \in I \quad (12)$$

$$0 \leq y_t^k \leq y_t^{-k}(\beta_t), \quad k \in K \quad (13)$$

where PR represents profit, I represents incomes, E represents expenditure and f, Ψ and y are known functions.

The application of the model in the process of solving Eqns. (7) and (8), leads to a modification of the efficiency function and problem indexes according to the demands of Eqns. (7) and (8).

5. Scheduling Marine Fleet

The problem of scheduling marine fleet consists of determining the start and end of service for every ship in port at stage $t \in \tau_1$, according to the recurrent Eqns. (7)–(8). This equation is:

$$F_t(s_{t-1}) = \max_{r_t \in Q_t(s_{t-1})} \sum_{t \in \tau_1} [\phi_t(r_t) + F_{t+1}(G(s_{t-1}, r_t))] \quad (14)$$

where r_t is the schedule at stage $t \in \tau_1$, $\phi_t(r_t)$ is the profit as a function of ship schedule and $Q_t(s_{t-1})$ represents the limitations in schedule.

An examination of Eqn. (14) as a discrete-programming problem allows the known methods of discrete programming to be applied in order to solve it

(Mikhalevich and Kuska 1983, Aven et al. 1985, Aven and Alexeychuk 1986).

These methods are used for designing and examining heuristic procedures to search for optimal and semi-optimal schedules. Scheduling process can be regarded as simulations of different kinds of ship operation taking account of the limitation of $Q_t(S_{t-1})$. Decision making in conflict situations is undertaken by means of different heuristic procedures, handling priorities of ships, their economic indexes, the states of the ports, and the hydrometeorological conditions.

The oriented graph of dynamic-system transition from one state to another is the basis for solving this problem. Ferry operation is divided into 12 stages including detention of ships, loading, unloading, wharf waiting, waiting for departure permission, and the motion of ships.

At every simulation step the following may be determined:

(a) the working index for period τ_1,
(b) the number of the ship state,
(c) the end of service time for this state,
(d) the recommended and maximum speeds,
(e) the quantity of freight for every ship in the current period, and
(f) the maximum capacity of the ships.

The programme support includes the dialog with the model. The simulation ends with the scheduling of ships for the period $t \in \tau_1$, the determination of the system clustered state (s_{t-1}) at the end of stage t_0, and the calculation of optimizational function variables.

6. Essential Functions of CASPLAS

CASPLAS performs the following main functions (Artynov and Vasilchenko 1988).

(a) Managing a database in an interactive mode—interactive and batch updating is permitted.
(b) Efficient planning of marine fleet—the system permits the calculation of the values of plan indexes for a given number of steps comprising a CPP. CASPLAS supports both an on-line mode of decision of the TPSMF and an interactive mode, which allows the dispatcher's experience to be utilized.
(c) Scheduling a fleet timetable—this is the decision step at which additional technological constraints (e.g., ship location) are taken into account. A main mode used in the scheduling of a fleet timetable is an interactive mode. The system obtains optimal plans and timetables (if they exist), evaluates them numerically and gives a message if a TPSMF cannot be solved as a whole for a selected timetable.
(d) Supporting an interactive operation mode—for controlling the solution process in an interactive mode, the system uses a dialog monitor which supports a hierarchical dialog scenario. For processing large bodies of information, new tools for displaying and manipulating information in tabular form, as well as tools for printing desired record forms, have been developed.
(e) The dialog monitor is used for controlling the process of calculation in a dialog mode—the controller can solve the necessary task and, with the help of commands, look through and correct the information tables with the subsequent usage of correct information in the course of subroutine work. Every subroutine realizes, as a rule, one or several simple functions of CASPLAS.
(f) The computer outputs the results to the printer.

The system was realized in PL/1 language.

6.1 Structure of CASPLAS

The structure of CASPLAS is shown in Fig. 1. The essential components of the system and the control communication between components are reflected in the figure.

Block 1.1 is the reading of the instruction. This programme reads the command from the display.

Block 1.2 produces the syntactical analysis of the command. A command is correct, if it is in the vocabulary of commands and macrocommands. The analysis of macrocommands and instruction operands is realized.

Block 1.3 is the monitor. A subroutine governs the system operation.

The second-level programmes process first-level commands, but they may include a regime of work for making the second level of dialog. In this case, the consumer will be in the environment of the processor of the first-level commands; hence, only the subroutine environment will be within reach. The essential components of the second are the following.

Block 2.1 is the TPSMF. The programs of this block allow the TPSMF to be solved in both the automatic and the dialog regimes. Also, in the environment of the TPSMF, there is the possibility of working with the database for the printing of results, and so on.

Block 2.2 contains the programs for working with the database. They allow the user to correct, in dialog mode, the information in the system database.

Block 2.3 contains the programs that simplify the operation in dialog mode. These subroutines, as a rule, are used in the environment of other second-level operations for the realization of the dialog. Also, the block 2.3 programs allow some system parameters to be corrected.

Block 2.4 contains the printing programs. These programs allow the information in the buffer to be printed as tables with the help of printing parameters.

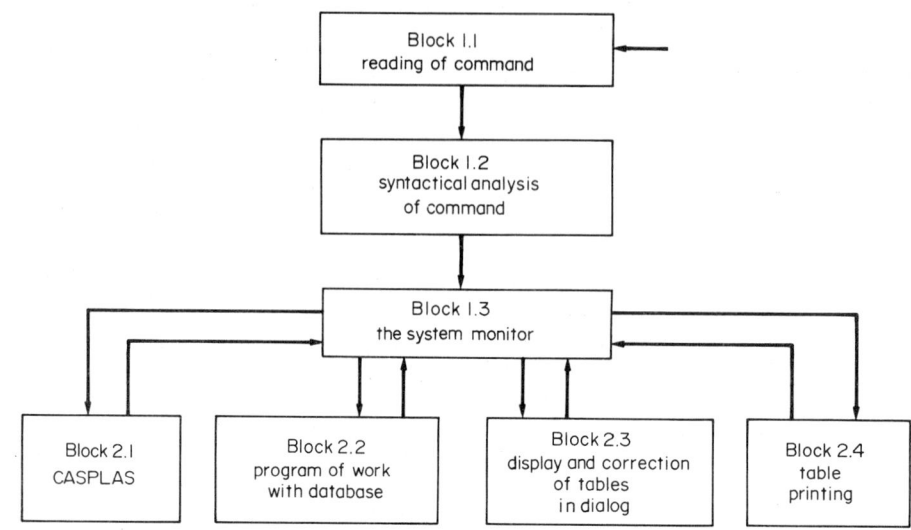

Figure 1
Control structure of CASPLAS

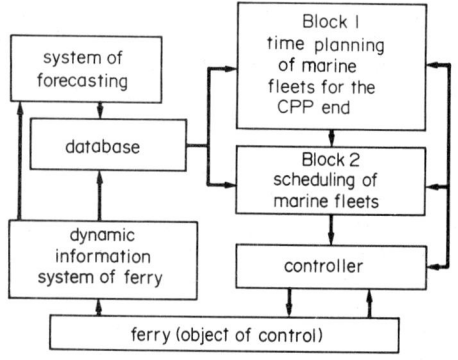

Figure 2
Functional structure of CASPLAS

6.2 Functional Structure

The functional structure of the system is given in Fig. 2. The essential components of CASPLAS, which solves the TPSMF and information relations between the components, are shown in the figure. The information environment is formed by the forecasting system and the dynamic-information tracking system for the transportation system on-board the ferry. The forecasting system forecasts the arrival time of the ships into the ports, as well as forecasting the hydrometeorological conditions. These forecasts cover different time periods, depending on the stage of solution of the TPSMF. The information system of ferry (see Fig. 2) reflects the current state of ports, ships and the fulfillment of plans and so on. Details of this information also depend on the solution stage.

Using the methods of decomposition for determining the optimal plans and schedule (Bellman's optimal principle), two blocks in the structure of continuous planning of the system must be distinguished: time planning marine fleets and scheduling marine fleets. Such decomposition of the original problem not only makes its solution possible, but allows each block to be interpreted more clearly.

Block 1, time planning marine fleets solves the optimization problem of plan determination of the indexes in stages of planning interval and evaluation of the CPP for the time remaining using clustered data.

The problem is solved on the basis of forecasting information concerning the arrival of freight and cars to the ports, the current state of the ports resources (i.e., ships) taking into account repair plans, the detention of ships and hydrometeorological conditions for the period τ, the clustered forecasts of hydrometeorological conditions and the arrival of freight in the period remaining.

The optimal plans for the transportation of freight, with details of the planning stages and suitable criteria are passed to block 2. Block 2 solves the problem of scheduling the movements of a ship for a period, taking account of the original ship location, the state of the ports and other technological limitations to their work. The capacity of the transportation levels of the plans, profits, and exploitation expenses are calculated for the chosen schedule.

This scheduling process is realized every day. It allows the deviation of the real working indexes from those planned to be taken into account and corrected, as well as coordinating the resolution of problems in different planning intervals (days, months or quarters) and coordinating the decisions of interacting transport systems.

Table 1
System parameters

Number of planning stages	2
Number of ships	9
Number of plan indexes	2
Number of ports	2

Table 2
State of the ports

Index name	Port 1	Port 2
Number of cars	120	150
Quantity of freight (t)	480	620

Table 3
Forecast of freight and cars

Stage number	Port 1		Port 2	
	Freight (t)	Cars	Freight (t)	Cars
01	6800	170	5200	130
02	8750	250	11 550	330

Table 4
Marine fleet plans

Index name	Index value
Car plan	1000
Freight plan	30 000

6.3 Information Support

The database is an essential information resource of CASPLAS, and includes the following:

(a) information, including the dialog-monitoring system indexes, the program-realization system indexes concerning the TPSMF decision method, and the information about ships, ports and stations; and

(b) operative information that must reflect the state of the transportation process up to the moment when the operative plan and schedule are calculated—fleet plans up to the end of the CPP, the work-time resources of the ships for the end of the CPP, the current state of the ports, forecasts of the arrival of freight and cars to the port, hydrometeorological conditions, the current location of ships and information about CPP dates and times.

In the computer-aided variant of the database, the database is worked on from the information system of the ferry (see Fig. 2). In the semiautomatic version, operative information in the database may be corrected in dialog mode.

6.4 Example of Operation

A brief description of the system in operation will now be given. The user is connected to the system by means of commands (or macrocommands). The information table is corrected and displayed. In these tables, there is information about the state of the system and the current calculation parameters. The results of such calculations on hypothetical data for the problem of time planning are given in Tables 1–6.

In Tables 1–4, there are parts of the system database for solving the problem of time planning—the amount of cars and freight in the ports, the forecast of freight and cars to the ports, car and freight marine-fleet plans to the end of the CPP, and so on.

An analysis of the solution to the problem is made using the information from Tables 5 and 6. Table 5 permits the evaluation of technical solution parameters—ship's load, turnrounds and economic parameters (income, expenditure and profit). Table 6 permits the evaluation of the dynamics of car arrivals at and departures from the ports. The satisfactory solutions are used by the dispatcher for the management of marine fleets.

7. Future Developments

The CASPLAS system is being used to solve the problems involved with the efficient planning and scheduling of fleet operations, accounting for the dynamics of freight flow to ports. These problems are formulated as the tasks of optimum coordinated decision making for various periods of planning.

Table 5
Analysis of solutions for ships (first stage of planning)

	Load (cars)					
	To port 1	To port 2	Turnrounds	Income	Expenditure	Profit
Ship 1	25	25	0,7	15.0	6.0	9.0
Ship 2	25	25	0,8	18.0	6.5	11.5

Table 6
Analysis of solutions for ports

Stage number	Initial state	Cars arriving	Cars departing	Resulting state	Limitations
Port 1					
stage 01	120	170	210	80	300
stage 02	80	250	210	120	300
Port 2					
stage 01	150	130	210	70	400
stage 02	70	330	210	190	400

Models and methods that allow the problem to be solved in the dialog regime have been worked out. A peculiarity of these models and methods is the coordinating of solutions obtained using different details and durations. The exploitation of the system has confirmed both its expendiency and efficiency. At present, CASPLAS is designed for computers such as the IBM360 and the IBM370. The development of the system is proceeding in the direction of personal computers, and automated communication with ships, ports and users of transportation systems. In order to expand the opportunities of the system and increase its efficiency, it is intended that the mechanism of information exchange between interacting transport systems and coordinated decision making on the efficient control of transportation processes be worked out.

Bibliography

Ardonin P A 1982 Microcomputer based on integrated management information system for transportation industry. *Mini and Microcomputers and Applications, Proc. ISMM Int. Symp.*, Paris

Artynov A P, Vasilchenko A I, Kurbatova Ye O 1988 Computer-aided system of the efficient planning and scheduling of ferry operation. *Prep.* Institute of Automation and Control Processes, USSR Academy of Sciences, Vladivostok, USSR (in Russian)

Aven O I, Alexeychuk A Ye 1986 Interactive computerized system for dynamic scheduling in transportation applications. *Prep. 5th IFAC Int. Conf. Container Transport Systems.* International Federation of Automatic Control, Dusseldorf, Germany, pp. 97–100

Aven O I, Alexeychuk A Ye, Lovetsky S Ye 1983 Computer-aided system of continuous planning of marine fleet. *Prep.* Institute of Control Problems, Moscow (in Russian)

Aven O I, Lovetsky S Ye, Moiseenko G Ye 1985 *The Optimization of Transport Flows*. Nauka, Moscow (in Russian)

Bellman R, Dreyfus S 1962 *Applied Dynamic Programming*. Princeton University Press, Princeton, NJ

Cashman J P 1983 Worldwide shipping information system. *Int. Association of Ports and Harbors, 13th Int. Conf.* IAPH, Tokyo pp. 28–38

Container News **18**(2), 1983 Computer systems changing intermodal shipping industry. 10–12; 14–17

Etschmaier M M 1986 Operational planning and control on transportation systems. *Prep. 5th IFAC Int. Conf. Container Transport Systems*. International Federation of Automatic Control, Dusseldorf, Germany, pp. 181–8

Gill P Ye, Murray W, Wright M 1981 *Practical Optimization*. Academic Press, New York

Leviy V D 1971 *Optimization of Fleet Planning*. Advertising bureau, MMF, Moscow (in Russian)

Mikhalevich V S, Kuksa A Ya 1983 *The Methods of Successive Optimization in Discrete Tasks of Optimum Resources Distribution*. Nauka, Moscow (in Russian)

Moiseenko G Ye 1978 Continuous planning and plan coordination with the various methods of planning. *Planning in Transport Systems—Models, Methods and Information Support*, Vol. 17. Institute of Control Problems, Moscow, pp. 49–58 (in Russian)

Moiseev N N 1974 *Elements in the Theory of Optimum Systems*. Nauka, Moscow (in Russian)

Pervozvansky A A, Gitsgory V G 1979 *Decomposition, Clustering and Approximate Optimization*. Nauka, Moscow (in Russian)

Polyantsev Yu D, Kiselev A V 1987 The improvement of control for transport—Technological systems by means of computers. *Express Information, Mar. Transp. Organ. Control* **3**, 1–13 (in Russian)

Pritulsky V S, Sukolenov A Ye 1986a Ferry operation control at the international ferry Ilyichevsk–Varna. *Express Information, Mar. Transp. Organ. Control* **5**, 1–12 (in Russian)

Pritulsky V S, Sukolenov A Ye 1986b The main principles and methods of the annual, quarterly and monthly planning of the international ferry Ilyichevsk–Varna. *Express Information, Mar. Transp. Organ. Control* **5**, 12–18 (in Russian)

Schonknecht K 1983 Automatische Informations Systeme in Seehafen. Ein Uberblick zum Entwicklungsstand in kapitalisticschen Seehafen. *DDR Verk.* **16**(7), 205–8

A. P. Artynov and A. I. Vasilchenko
[USSR Academy of Sciences, Vladivostok, USSR]

Marine Propulsion Plants: Control

Marine propulsion machinery has increased in complexity as steam driven plant has gradually given way to diesel or gas turbine powered systems. This change towards greater complexity has also been accompanied by an increase in both the difficulty and the number of control problems that need to be resolved. The demands that are placed on the control equipment have also increased. The use of digital computer control systems has further contributed to system complexity by virtue of the opportunities for advanced control and monitoring systems that need to be implemented if the computer is to be used in a cost effective and efficient way.

1. Gas Turbine Powered Marine Propulsion Plants

Gas turbine powered marine propulsion plants are typically found in modern warships, container vessels and some other types of large ships. The gas turbine is normally a version of an established and proven aero engine (e.g., a Rolls-Royce Olympus gas turbine engine) which has been marinized, that is, made suitable for operation in the salt-laden highly corrosive environment in which ships have to operate. Marinization involves modifying the engine to run on diesel oil instead of the kerosene burnt in aero engines, minimization of the effects of corrosion by the application of corrosive-resistant coatings, modifications to the properties and types of metals used in the construction of the engine components, and the installation of filters in the air intake to the engine and cleaning systems in the fuel supply.

Gas turbine designs can vary, but basically a typical engine consists of a gas generator that produces a gas flow which drives a power turbine. The gas generator comprises a compressor, a combustion chamber and a compressor turbine which is coupled to the compressor. Air is drawn into the gas generator, passing through the compressor, the combustion chamber and the compressor turbine, and then the high-temperature output flow from the gas generator passes into the power turbine, from where it is discharged into the atmosphere via the exhaust system.

The output from the power turbine is high speed, thus the gas turbine is connected to the propeller and its shafting via a reduction gearbox. The actual connection of the power turbine to the gearbox is made through a torque tube and a self-synchronizing clutch, the purpose of which is to enable the engine to be connected to and disconnected from the rest of the propulsion machinery.

The alternative to a gas turbine engine is a marine diesel. The main advantages of the gas turbine over the diesel are its power output–weight and power output–bulk ratios. The main disadvantage of the gas turbine is its high fuel consumption relative to the diesel engine. A gas turbine such as the Rolls-Royce Olympus has an output power of the order of 20 MW, thus providing the ship with a high-speed capability. However, the fuel consumption of such a high-power engine at part load is very poor. Typically, a modern naval vessel of 4000 t needs about 36 MW to achieve a full speed of about 15 m s^{-1}. However, for about 80% of the life of a ship, a cruising speed of about 9 m s^{-1} will not be exceeded, which coincides with a power requirement of about 6 MW. From typical fuel consumption vs part-load performance characteristics of a high-power gas turbine, it can be demonstrated that a doubling of fuel consumption can occur when operating at such part loads.

For reasons of fuel economy it is normal to use high-power gas turbines only for high speed or "sprint" maneuvering, and to use a second, more economical, low-power gas turbine running at full power for "cruise" maneuvering. A Rolls-Royce Tyne gas turbine which has a power output of about 4 MW can be used in this role. This results in an installation known as a combined gas or gas (COGOG) propulsion-plant arrangement, where either the sprint engine is in use or the cruise engine is operating. The two types of engines cannot however be used simultaneously to drive the propeller.

In some cases, a marine diesel is used for the cruise engine instead of the low-power gas turbine. This results in a combined diesel or gas turbine (CODOG) propulsion-plant arrangement. Sometimes, two marine diesels are installed instead of one, and while the diesels cannot be used to drive the system at the same time as the high-power gas turbine, it is normally possible to run either one diesel alone or both diesels simultaneously. It is also not uncommon in naval vessels to find twin shaft propulsion systems, where each shaft set consists of a sprint gas turbine, a cruise gas turbine engine (or diesel engines), a reduction gearbox, clutches and a propeller and associated shafting. It is also possible to have twin screw propulsion driven by a common engine system.

Gas turbines (and marine diesels) are unidirectional prime movers, thus it is necessary to provide some means of generating reverse thrust (e.g., to stop the ship quickly or for harbor maneuvering). Reverse thrust can be achieved either by the use of a reversing gearbox or by a controllable pitch propeller which allows the pitch angle to be reversed. A reversing gearbox is more complex and bulkier than a unidirectional gearbox. On the other hand, a controllable pitch propeller is less efficient than a fixed pitch propeller, but controllable pitch propellers improve the maneuverability of the ship and, for this reason, controllable pitch propellers tend to be fitted to naval ships and other types of vessels such as car ferries where maneuverability is important.

The remainder of this article focuses on gas turbine powered controllable pitch propeller propulsion systems and the associated control problems and available solutions.

2. Modelling and Simulation

Digital computer simulation now plays an important role in the design of ship propulsion systems. It is used in initial feasibility design studies, right through subsequent stages of design, to sea trials and beyond. Typically, computer simulation is used (a) to evaluate ship and propulsion-plant performance, (b) as an aid to propulsion-plant selection and development, (c) for control-system functional design and development, (d) for failure analysis, and (e) as an aid to control-system testing and commissioning.

The primary advantage of computer simulation is that it gives designers a flexible tool to predict steady-state and dynamic performance. It assists designers to establish and optimize gas turbine fuel-valve opening and closing times, propeller pitch stroking times, ship maneuvering performance and main machinery torque, thrust and rotational speeds under a wide range of maneuvering conditions.

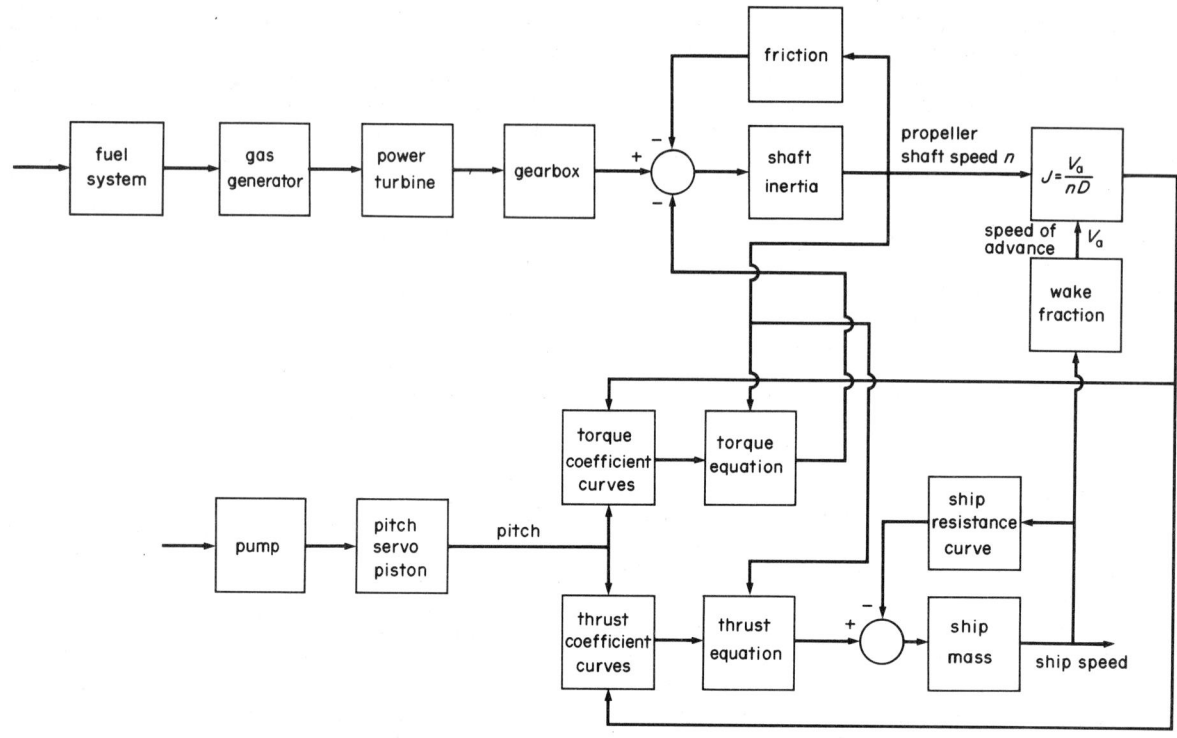

Figure 1
Interconnection of propulsion-system variables

Performance evaluation and control-systems development can account for a considerable portion of the overall ship design and development cost. The use of computer simulations can save both development time and cost. It allows control problems to be examined early on in the design of the ship, thus allowing systems and control design and development to proceed in parallel. The use of real-time distributed digital simulation facilities interfaced to control equipment, computer hardware and control-systems software, also allows a considerable amount of control-system testing and fault finding to be done prior to installation in the ship. This can considerably reduce commissioning time and cost. Such a simulation facility can also be used for training purposes, for developing operating procedures, and for developing fault finding and diagnostic routines.

The dynamics of a gas turbine powered controllable pitch propeller ship propulsion system are highly nonlinear, and performance characteristics are a function of operating conditions such as temperature, power, speed and so on. A realistic simulation model must include the nonlinear characteristics of the gas turbine, the propeller and the hull, the actuator nonlinearities and the main time constants of the system (which include the dynamics of the combined inertia of the gas turbine, propeller and propeller shaft, the nonlinear dynamics of the mass of the ship, and the nonlinear turbine torque dynamics). A block diagram showing the interconnection of system variables is shown in Fig. 1.

Linearized models can be extracted form the nonlinear simulation at various operating points if required, and these linear models can be used along with frequency domain control-system design techniques (such as Nyquist's or Bode's methods) to study the transient behavior and to produce linear control-system compensator designs to modify these characteristics. The nonlinear simulation can then be used for final tuning of these linear designs.

Sometimes it is easier to model the gas turbine using transfer functions, the parameters of which are scheduled to operating conditions. Linear models can also be used to model the ship propulsion system when the objective is to undertake perturbation studies such as those that might occur while trying to maintain constant speed against the effects of varying propeller loads arising from sea conditions.

3. Control Problems and Objectives

The basic control objectives for a gas turbine powered marine propulsion plant are (a) to develop the required thrust in the most efficient manner, (b) to achieve the desired degree of ship maneuverability, (c) to ensure that desired response characteristics to power-demand changes are achieved and (d) to minimize the effect of disturbances. All these objectives must be achieved without exceeding safe operating conditions. For example, the main propulsion control system must protect mechanical components such as the propeller shaft from overtorquing and overspeeding, while the engine control system must protect the gas turbine from overspeeding, surge, flame-out, and so on.

The gas generator part of the engine has requirements for fuel input limitation so that it cannot be run outside its permitted operating range. These limitations must override any fuel flow selected by the main propulsion control system external to the engine. In particular, the gas generator control must:

(a) provide scheduling of fuel flow during starting and run up to idling;
(b) prevent fuel flow exceeding maximum limiting values determined by the requirement to prevent engine surge during acceleration and the need to keep rotor speed, combustion temperature, exhaust gas temperature and power output within safe limits; and
(c) prevent fuel flow from being decreased below limiting values necessary to maintain combustion during deceleration.

A propeller operating in a seaway can encounter wide variations in load owing to the motion of the ship, for example, during pitching and heaving when the depth to which the propeller is submerged is changing. In general, the disturbance effect of the sea on propeller load is nonperiodic owing to an irregular sea. The effect of seaway, however, can produce gas turbine cycling; load torque oscillations of the order of 30% can produce variations of the order of 4% in power turbine inlet temperature and 6% in fuel flow, for very long period waves, but these variations tend towards zero as the period of the waves decreases. The variations in power turbine torque can attain peak values of the order of 6% and peak variation in engine speed can be of the order of 250 revolutions per minute (rpm). Some means of inhibiting the propulsion control system is therefore required in order to prevent hunting of the engine throttle resulting from sea conditions.

Gas turbine speed characteristics are dependent on the ambient temperature. As the temperature is decreased, the curves shift upwards resulting in increased torque for given gas compressor and power turbine speeds. The gas turbine torque-speed coefficient is also temperature dependent. Power changes resulting from air intake temperature changes therefore need to be controlled. For example, air intake temperatures may vary from $-30\,°C$ to $+40\,°C$, and these need to be controlled within $\pm 3\%$ of operating level power.

Control is made more difficult because the system is very nonlinear. The dynamics of the gas turbine are a function of its speed, and the parameters of the propulsion system are subject to large variations over the load range of the system. The gas turbine constitutes a multivariable system, as changes to the input (gas generator fuel flow) result in changes occurring in gas generator shaft speed, compressor delivery pressure, power turbine shaft speed, combustion temperature and exhaust gas temperature.

Propeller shaft speed and torque are both dependent on two input variables, gas generator fuel flow and propeller pitch angle, such that any change in fuel flow or pitch angle will result in changes to propeller shaft speed and torque, both of which must be maintained within safe operating limits. Thus, the propulsion plant is also a mulitivariable system in which a change in one input effects more than one output, and this interaction makes the control of the system more problematic. The propulsion system also displays nonminimum phase characteristics at small pitch angle (low fuelling rate operating conditions) and this generates further control problems.

The position of the controllable pitch propeller blades must be controlled by a hydraulic closed loop position control system. However, the rate at which the pitch angle is changed can have a significant effect on propeller shaft speed and torque and the thrust developed by the propeller. Unfortunately, there is no value of pitch angle rate of change that is suitable for all operating conditions and some means of controlling pitch rate is, therefore, required.

In addition to achieving the objectives already defined and dealing with the various problems and requirements identified, the control systems must also provide monitoring and display facilities for important plant parameters and alarm facilities to warn of dangerous plant conditions. The control system must be able to control the starting and stopping of each engine and the smooth transfer of load from one engine to another. In the case of naval vessels, control must also be available at two remote operating positions, the ship control centre and the bridge, and local manual control is required in the engine room. Each shaft set must be independently controllable, and the control system must normally "fail set," so that a control failure results in a machinery unit remaining in the condition existing immediatley prior to the failure.

4. Propulsion Control

A typical analog propulsion control system is shown in Fig. 2. With this system, control over the propulsion machinery is achieved through a mixture of open loop and closed loop control strategies (see *Traffic Control Modes*) and protection against overspeeding and overtorquing is achieved through the judicious choice of settings for a number of rate-limiting devices. A control system such as this was used in ships from the late 1960s onwards and would have been implemented in a centralized form using hard wired electronic modules.

The propeller pitch is controlled via closed loop control of the pitch angle. A propeller pitch angle demand signal is produced from a pitch angle schedule which is related to the power pitch control lever (PPCL). This demand signal is compared with a measurement of the actual propeller pitch angle, and the error determines the required position of the actuator, which then positions, via a hydraulic servo, the swash plate on a hydraulic pump to produce the fluid flow that adjusts the propeller pitch angle. When the error is zero, the actuator and swash plate are both set to zero, resulting in no fluid flow. The hydrauic pump, which is driven from the propeller shaft, is backed up by an electrically driven pump, which trips in when the fluid pressure falls below a set value (e.g., at low shaft speed).

Controllable pitch propellers have a high blade area–swept disk ratio, and this results in high reverse thrusts when the pitch angle approaches zero at high speeds. To reduce the reverse thrust during a high-power step, the pitch rate must be set to a level that would be unacceptable for low-power maneuvering. Thus, a pitch rate control loop is introduced into the system to protect the machinery from excessive torques and thrusts during maneuvers. The output of the achieved pitch angle measurement system is differentiated to give an achieved pitch rate signal. Owing to the necessity to vary the allowed pitch rate, the achieved pitch rate is compared with a pitch rate schedule which is related to propeller shaft speed, and by this means the system produces a maximum permitted pitch rate as a function of propeller shaft speed. Rates in excess of the schedule result in a reduction in pitch demand. Thus, at high shaft speeds the pitch rate is decreased (i.e., made slower) and at low shaft speeds the pitch rate is increased (i.e., made faster) towards the maximum allowed. This pitch rate control mechanism constitutes a self-adaptive feature of the control system, although it is never referred to as such.

The gas turbine fuel-demand signal is primarily determined from an open loop fuel schedule which is related to the PPCL. Another component of fuel demand is determined from a limited authority closed loop control on the propeller shaft speed. The controller in this speed control loop contains proportional plus integral plus derivative terms. The integral term has its authority limited to ±10% of the throttle position and operates on a long time constant providing an automatic trim facility to allow for changes in engine efficiency and hull resistance. The proportional and derivative terms have an authority limited to ±40% of throttle position, and a ±10% deadband which inhibits any output if the demanded and achieved shaft speeds are within ±10% of each other. This deadband is introduced to prevent shaft speed variations due solely to sea conditions being seen by the control system, where they cause the engine throttle to hunt. A minimum shaft speed control loop is introduced which maintains the shaft speed above a fixed datum required to maintain the operation of auxiliary systems driven directly from the propeller shaft.

A number of restrictions need to be incuded in the system to limit torque and thrust. The throttle demand signal obtained from the fuel schedule is restricted by the value of achieved pitch such that full power cannot be demanded unless full pitch has been achieved. This is brought about by comparing achieved pitch with the PPCL signal in a "lowest-wins" circuit before the fuel schedule. When full pitch is achieved, its signal goes high allowing the PPCL signal to pass directly to the

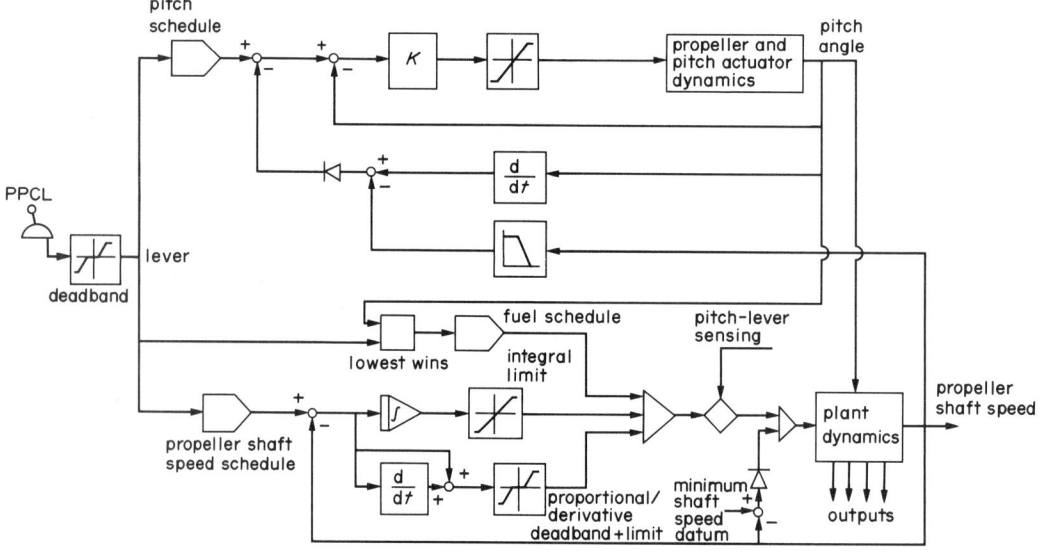

Figure 2
Analog propulsion control system where K is the controller gain in the pitch angle control loop

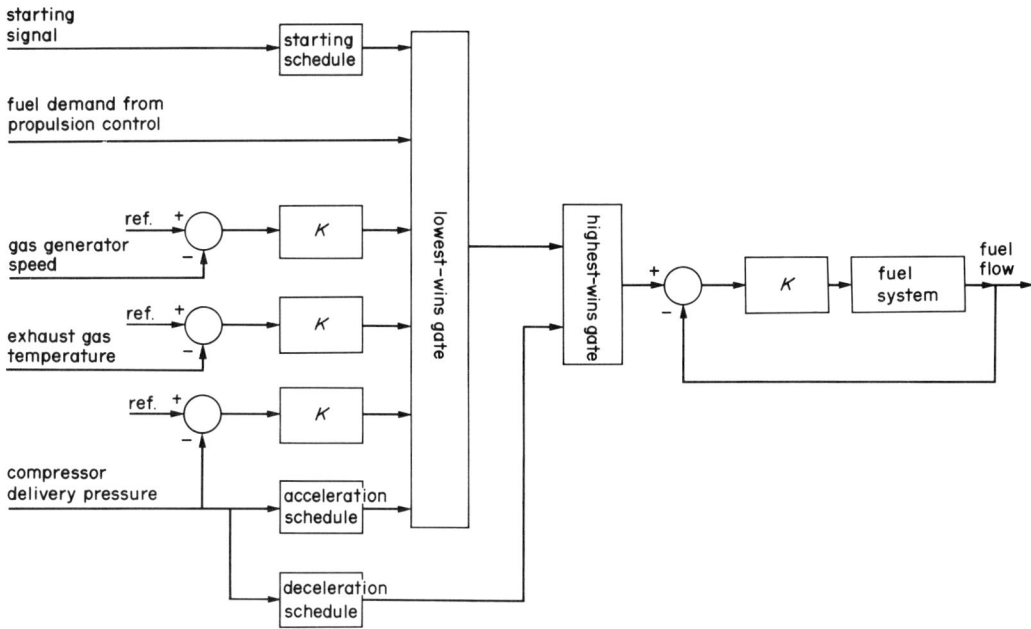

Figure 3
Engine control scheme

fuel schedule. No restrictions are placed on the shaft speed control loop and up to 40% of throttle can therefore be applied if the error is large enough. This allows rapid acceleration of the ship from the stationary position.

During maneuvers involving changes from ahead to astern or vice versa, the power needs to be reduced to a minimum while the pitch is changing sense in order to avoid overspeeding the shaft. Thus, the main fuel-demand signal is reduced to zero whenever the pitch angle and PPCL signals are in opposite sense, but the shaft speed is maintained at a minimum by virtue of the minimum shaft speed control loop.

5. Engine Control

The primary objective of the engine control system is to protect the gas turbine and to keep it within a safe operating range. A number of output variables can be used for this purpose. These include displacement of the actuator on the fuel valve, fuel flow, gas generator speed, power turbine speed, exhaust temperature and compressor delivery pressure. The best variable is compressor delivery pressure, but in practice a number of outputs are used.

The main fuel-demand signal comes from the propulsion control system but compressor delivery pressure needs to be used to control acceleration, because, if a stall occurs, compressor delivery pressure falls immediately. The gas temperature also rises quickly, but the speed remains constant and any increase in fuel demand resulting from a speed control loop attempting to achieve the desired speed would cause the temperature to increase further, rapidly causing an engine burnout. Limiting the increase in fuel flow as a function of compressor delivery pressure would, at the onset of stall, cause compressor delivery pressure and hence fuel flow to reduce, thus averting serious engine damage.

If flameout occurs at anytime, the speed and compressor delivery pressure will fall. The speed control loop will then demand extra fuel to achieve the desired level and the air–fuel ratio may increase to a level that is sufficient to cause fire or explosion in the hot exhaust trunking. A fuel flow scheduled to compressor delivery pressure can be used to reduce this risk.

A typical control system is shown in Fig. 3. The primary fuel demand comes from the propulsion control system, and a minor loop controls fuel flow in response to this demand. The fuel-demand signal can be reduced at a lowest-wins circuit by other control parameters, which include gas generator shaft speed, exhaust gas temperature and compressor delivery pressure, as well as the acceleration schedule and starting schedule. This allows fuel demand to be reduced as limiting conditions are approached. A further input can increase fuel flow at a highest-wins gate to ensure that flameout does not occur under deceleration conditions.

6. Digital Control

The propulsion and engine control systems that have been discussed in Sects. 4 and 5 are based on hard wired electronic modules, most of which are located at

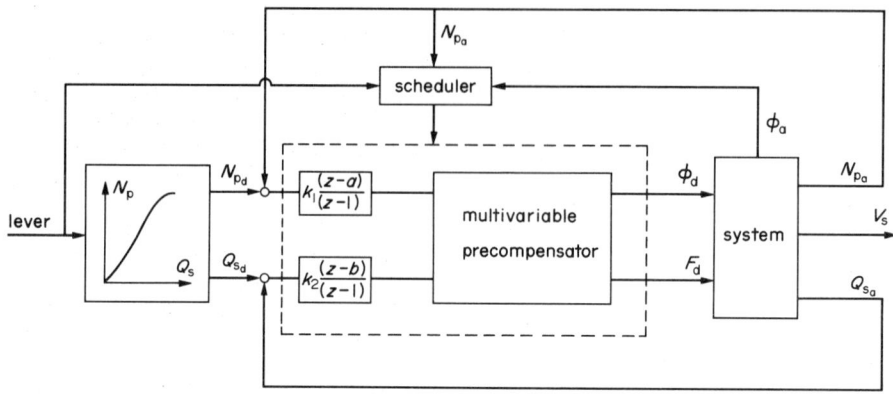

Figure 4
Gain scheduled multivariable control system: N_p is the propeller shaft speed; Q_s is the propeller shaft torque; V_s is the ship speed; F_d is the fuel demand; ϕ is the pitch angle; k_1, k_2, a and b are controller parameters and z is the z–transform operator

some central position. Digital computers have begun to replace these electronic control modules and this has resulted in more distributed forms of control, with each item of plant being controlled by a local microprocessor-based system, the activities of which are coordinated by a main central supervisory computer at the system level.

The main advantage of digital control over analog control lies in the possibility of obtaining better performance with enhanced fuel economy. This comes about because it is possible to operate the propulsion system and the gas turbine closer to their operating limits, thus bringing about performance improvements. Furthermore, because the dynamics of the propulsion system and the gas turbine vary with operating conditions, a fixed control strategy of the type described earlier must be a compromise. This is one of the main limitations of existing analog-type control systems; digital technology allows the parameters of the control systems to be adapted with relative ease to operating conditions. It also allows the implementation of the more complex control algorithms that result from using multivariable frequency domain design methods (e.g., Rosenbrock's inverse Nyquist array).

Thus, digtital technology should lead to improvements in both performance and fuel consumption when compared with analog systems by virtue of the improved stability and transient performance that can be achieved and maintained at all points in the power range. Additional improvements in performance will also be brought about by the implementation of condition monitoring systems.

A multivariable gain scheduled adaptive propulsion control system is shown in Fig. 4. With this system, closed loop control of propeller shaft speed and torque is achieved by manipulating the two input variables, gas turbine fuel demand and propeller pitch angle demand. A multivariable precompensator is implemented in order to reduce the interaction in the system, so that a change in one input only has a significant effect on one output variable rather than on both outputs.

Owing to variations in the plant parameters with power level, the coefficients in these compensators are scheduled to operating load conditions via achieved propeller shaft speed, achieved propeller pitch angle and the PPCL signal. Set points for the two control loops are determined from a schedule which is related to the PPCL.

Some control logic is required in order to achieve satisfactory performance. Pitch sensing is required to ensure that the fuel demand signal is set to zero when the pitch and lever are in opposite sense, and also to ensure that the pitch-demand signal is always in the correct sense during ahead to astern maneuvers. The rate limit on the signal applied to the gas turbine is less than that on a conventional analog control system and a fixed limit on the rate of change of pitch angle is included only during the initial phases of the ahead to astern maneuver. There is no need for a variable pitch rate.

The performance of this type of digital control system is much better than that of the conventional analog system. This improvement primarily arises from the fact that the restrictive rate limits used in the analog system have either been relaxed or removed, and because the transient performance of the control loops can be maintained by adapting the controller parameters to operating-load level. This means that closed loop control of propeller shaft speed and torque can always give the desired transient response characteristics and, by appropriate shaping of the responses, propeller shaft overtorquing and overspeeding can be avoided.

Further improvements in the performance of the control loops and simplifications to the control system can be achieved by impelementing high-gain feedback loops in conjunction with reference models which define the desired transient performance for each loop.

See also: Ship Steering: Model-Reference Adaptive Control

Bibliography

Beale G B, Gowans B J 1970 Transmission design for warships of the Royal Navy. *Trans. Inst. Mar. Eng.* **82**, 241–63
Birbanescu-Biran A 1976 Modelling of propulsion plants (linear analysis). In: Pitkin M, Roche J, Williams, T J (eds.) *Ship Operation Automation*. North-Holland, Amsterdam, pp. 183–92
Gillman D K 1979 The application of transfer function techniques to modelling ship control and navigation systems. *J. Navig.* **32**, 200–9
Kidd P T 1986 Design of a controller for a mulitivariable system with varying parameters using the extended Nyquist array. *Int. J. Control* **43**, 901–20
Kidd P T 1987 Ship propulsion control. In: Singh M G (ed.) *Systems and Control Encyclopedia*. Pergamon, Oxford, pp. 4267–71
Kidd P T, Munro N, Thiruarooran C, Winterbone D E 1985 Linear modelling of ship propulsion plants. *Proc. Inst. Mech. Eng., Part A* **199**, 53–8
Kidd P T, Munro N, Winterbone D E 1985 Development of an alternative control scheme for a naval propulsion plant. *Trans. Inst. Meas. Control* **7**, 46–56
Kidd P T, Munro N, Winterbone D E 1985 Design of compensation schemes for a nonminimum-phase multivariable plant. *IEE Proc. D* **132**, 75–85
Lutje-Schipholt R M 1975 Marinization of aero gas turbines. *Trans. Inst. Mar. Eng.* **87**, 1–16
O'Hare T L R, Holburn J G 1973 Operating experience with gas turbine container ships. *Trans. Inst. Mar. Eng.* **85**, 1–23
Patel R V, Munro N 1982 *Multivariable Systems Theory and Design*. Pergamon, Oxford
Rubis C J 1974 Gas turbine performance under varying torque propeller loads. *Int. Shipbuild. Prog.* **21**, 20–7
Standen G, Bowes J, Warsop J C 1975 Machinery installation in the type 42 destroyer. *Trans. Inst. Mar. Eng.* **87**, 229–45
Zeien C 1970 Application of the gas turbine to main propulsion of merchant ships. *Trans. Inst. Mar. Eng.* **82**, 193–209

P. T. Kidd
[Cheshire Henbury Research & Consultancy, Macclesfield, UK]

Merging Control

Intersection merging control is a dominating aspect of network control in fully automated rail or road transportation systems. At the merge points of the network, two segregated lanes intersect to allow vehicles from two different input lanes to share a common output direction. The basic objective of a merging control strategy is to guide the vehicles so that rear-end collisions of vehicles arriving at the merge point are prevented.

1. General Aspects

There are two fundamental aspects of the merging control problem, as illustrated in Fig. 1. First, since every vehicle changes from input lane A or input lane B to output lane C at the merge point, the sequence of vehicles in lane C has to be controlled appropriately. Second, with no proper motion control available, a vehicle on passing the merge point may collide with the vehicle immediately ahead in the common lane C. Therefore, the tasks connected with automatic merging can be attached on two interacting control layers.

First, there is the sequence assignment layer. Every time t_{ev} a vehicle enters the control region the merging sequence σ has to be updated. The sequence σ represents a prediction of the ordered set of identification codes of vehicles after having passed the merge point. Two consecutive vehicles within the ordered set σ are called "follower v" and "leader μc."

Second, there is the motion control layer, which ensures that:

(a) the actual time to go τ_v to the merge point of every single follower v will be greater than the actual time to go $\tau_{\mu c}$ of the leader μc according to the merging sequence σ as fixed by the sequence assignment layer; and

(b) when passing the merge point, the actual velocity of every vehicle v is identical to the prescribed merging velocity V_p. If $h(V_p)$ denotes the fixed time-headway reference in lane C, the major goals of the merging control strategy can be summarized as follows:

$$\tau_v(t) \stackrel{!}{\geq} \tau_{\mu c}(t) + h(V_p), \quad \forall \text{ pairs } \{v, \mu c\} \text{ within } \sigma \quad (1)$$

$$v_v[s(t) = 0] \stackrel{!}{=} V_p, \quad \forall v \quad (2)$$

where the exclamation marks above the equals sign and the greater than sign indicate that it must be made equal to or greater than, respectively, by a proper control strategy.

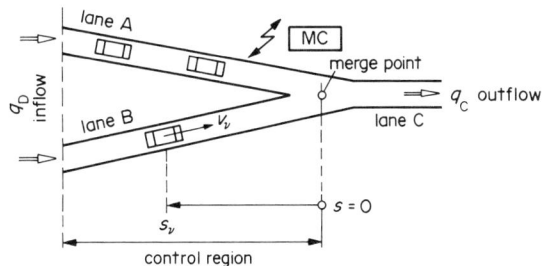

Figure 1
The merging problem

2. Cell-Following Methods

In order to simplify the design approach of both layers of a merging-control scheme, the following, partly restrictive assumptions can be made.

(a) In steady-state operation, vehicles occupy positions on the guideway within hypothetical equally spaced reference "cells" that move at constant velocity V_p (Godfrey 1968, Wilkie 1970); the size of the cells Δ must be chosen to satisfy desirable headway requirements.

(b) The moving reference cells (see Fig. 2), generated by an appropriate signal travelling along the guideway, are governed by overall network synchronization of position and movement.

(c) The moving boundaries of corresponding cells on the guideways of the two input lanes of the intersection are synchronized in relation to the merge point.

(d) Overall network scheduling ensures that traffic patterns entering the control region of a local merging controller (MC) contain a sufficient number of empty cells for resolving conflicts locally. One possible strategy is based on the preprogramming of the route of each vehicle and its arrival times at the merge points on its route (Stefanek and Wilkie 1973) and another strategy is the controlled bypassing of "exceeding" vehicles on alternative routes at the demerge points of the network (McGinley 1975, Caudill and Youngblood 1976).

(e) Merging control is a pure rearrangement: in order to accomplish merge, vehicles may have to be shifted from cell to cell.

(f) A vehicle has attained its steady-state position in a given cell before receiving a command to move to a new cell.

2.1 Sequence Assignment

As a result of these assumptions for a finite group of vehicles, the pattern of occupied and empty cells can be represented by a vector. Let a' and b' denote the binary vectors of the "initial states" within the assignment blocks (n' cells long) at the input lanes a certain distance ($n - n'$ cells) upstream of the merge point. If $a' \cap b' \neq 0$, identifying a merging conflict, the sequence assignment layer has to compute two command vectors m'_a and m'_b that cause the vectors a' and b' to be mapped into new assignment vectors $a = f(a', m'_a)$ and $b = f(b', m'_b)$ according to a "terminal state"

$$c = a \cup b \qquad (3)$$

with

$$a \cap b = 0 \qquad (4)$$

The command vectors m'_a and m'_b specify how many cells each vehicle in the assignment blocks must move in order to eliminate the conflict. These vectors can be computed off line by using a heuristic procedure (Whitney 1972) or by setting up a constraint integer programming problem (Sarachik and Chu 1975). The algorithms can be implemented via a combinatorial network or stored-state tables and transition diagrams.

2.2 Motion Control

Motion control is based on velocity and position error measurement with regard to the boundary of the assigned reference cells. It can be stated as a linear optimal "tracking problem" using a quadratic performance index. Ignoring the behavior of neighboring vehicles (Wilkie 1970) the feedback controller structure is simple, time invariant and independent of initial conditions. A modified approach including the error signals of the two nearest neighboring vehicles was presented by Sarachik and Chu 1975. Cell shift maneuvers can be executed by simply giving a corresponding step change in position error. In the case of merging low-speed vehicles ($v \to 0$) into the main-line stream ($v = V_p$) at an entrance ramp of a station, the problem must be formulated as a fixed-time, fixed-endpoint optimal control strategy. Merging vehicles always start from the same initial conditions and have a fixed time $\tau(0) = s(0)/V_p$ to reach an empty cell.

For cell-following methods, separate emergency control functions must be employed, overriding the normal headway and velocity regulation discussed above. Emergency (stopping) procedures have to be based on velocity difference and headway deviation measured with regard to the preceding vehicle. System breakdown caused by erratic behavior of vehicles leads to difficult synchronization problems.

3. Car-Following Methods

In order to avoid most of the restrictions featuring the operating conditions of cell-following methods, the following specifics of the process of automatically merging vehicles must be considered.

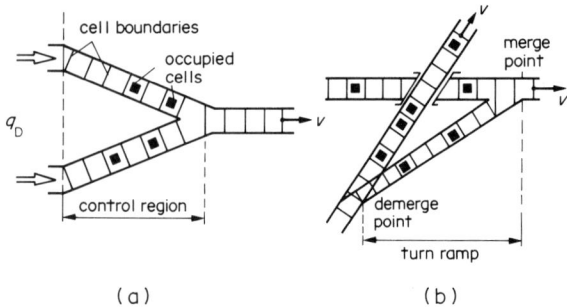

Figure 2
Moving cell scheme for two types of junction

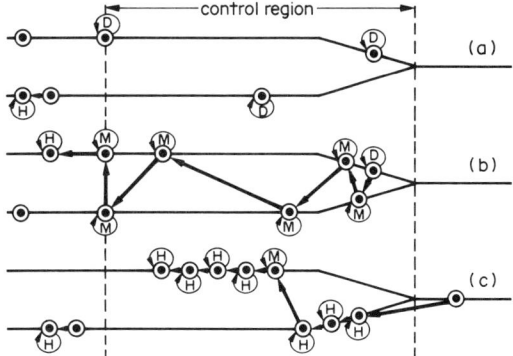

Figure 3
Typical coupling patterns: D means motion decoupled, M means coupled by merging conflict and H means coupled by headway conflict

First, a local merging controller has no control over the arbitrary arrival times t_{ev} and initial states of motion, $0 \leq v_v(t_{ev}) \leq 30 \text{ m s}^{-1}$ of the vehicles entering the control region. Each vehicle arrival has to be considered as an event. The actual number N of vehicles within the control region and consequently the order of the process vary arbitrarily with time.

Second, for a given sequence σ, the modes of motion control must be classified depending on the type of predicted conflicts (a)–(c) detailed below. In order to assure fail-safe behavior in any cases of detected conflicts, the states of motion of the vehicles involved must be coupled by an appropriate motion control strategy. Thus, the actual structure of the controlled process also varies with time. A graphical representation of the three possible basic coupling patterns within the controlled process of automatically merging vehicles is shown in digraph notation in Fig. 3. The dynamic model of the velocity controlled single vehicle is represented by the symbol ⊙; the functional relation of the different modes of motion control are marked by labelled lines.

The following modes of conflict may occur.

(a) If the estimated time to go τ_v and $\tau_{\mu c}$ of a consecutive pair $\{v, \mu c\}$ of vehicles in σ do not fulfill Eqn. (1), a merging conflict will be detected. It has to be solved by the local merging controller. In order to estimate the time to go, certain assumptions relating to the future motion of the vehicle have to be made. All reported merging control strategies (Athans 1969, Brown 1975) are based on the assumption that vehicles are travelling with constant velocity V_p to the merge point (i.e., $\tau_v(t) = s_v(t)/V_p$).

Due to the possible spectrum of entering velocities permitted, the actual vehicle trajectories do not remain sufficiently close to a reference trajectory $v(s) = V_p = $ constant.

(b) If the actual distance between two vehicles, travelling to the merge point in the same lane, becomes less than the reference value determined by the headway vs speed philosophy chosen (Posch 1982), a headway conflict will occur. In this case, motion control will be transferred from the local MC to the on-board headway controller of the follower vehicle v. In order to avoid persistent switching between the control laws, both merging and headway conflicts will only be solved by reducing the velocity of the follower v.

(c) If no merging or headway conflict exists, the non-conflict mode on the subordinate level of the local MC is in operation.

The frequency of occurrence of these types of conflict depends on the series of differing time gaps between the arrival of the vehicles at the control region (i.e., the intensity of the vehicle inflow).

3.1 Sequence Control
Because of safety requirements, the assignment of a merging sequence cannot be based on one given pattern, for example "first in first out" (Brown 1975), but has to be fixed with respect to the actual states of motion of each vehicle within the control region. Let α and β denote the numbers of vehicles already present in lanes A and B, respectively, at time t_{ev}. There exist $M = (\alpha + \beta)!/\alpha!\beta!$ possible merging sequences σ. In order to find a simple way to insert a newly entering vehicle at time $t = t_{ev}$ into an already specified sequence σ in the most reasonable way, consider the following. The lower level of the local merging controller has to initiate a control maneuver if the entering velocity of vehicles differs from the prescribed merging velocity. Assuming for the moment the absence of any disturbances, an appropriate non-clonflict-control law meeting the final velocity value V_p will generate "reference" trajectories $R(v, s)$ in the v–s plane with regard to a field of isochromes $T(v, s)$ corresponding to the values τ of the times to go; that is,

$$R(v, s) \Rightarrow T(v, s) \qquad (5)$$

Assuming that for $\tau = $ constant the functional relationship $v(\tau = \text{const}; s)$ is unique with respect to s (i.e., the field of isochrones has the basic structure shown in Fig. 4), a "natural" merging sequence σ_N can be deduced from the set of theoretically possible merging sequences.

The natural merging sequence σ_N results simply from the corresponding times to go of the isochrones related to the actual states of motion of the individual vehicles at a certain point in time. The sequence σ_N is defined by the ascending values of the times to go. A merging conflict is detected when the states of motion of two vehicles in different lanes belong to neighboring isochrones that do not satisfy Eqn. (1). In this situation, the motion control for the follower switches to the

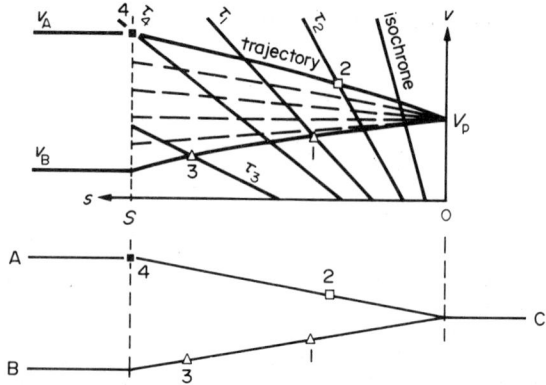

Figure 4
Field of isochrones and natural merging sequence for $t = t_{c4}$; entering sequence $\sigma_e = \{1, 2, 3, 4\}$ and merging sequence $\sigma_N = \{2, 1, 4, 3\}$

conflict mode of the control law to extend the time to go. In order to achieve minimal time delays during merging, the natural merging sequence has to be kept constant (Posch 1982). Minimum average time delay also gives the lowest absolute values of the mean vehicle acceleration. This, in turn, provides comfortable ride conditions.

The individual steps of the sequence assignment procedure for merging control have to be performed cyclically in real time. They can be summarized as follows:

(a) checking of number N, identification codes $\{v\}$ and states of motion of all vehicles in the control region at the actual instant of time t_c;

(b) prediction of the times to go τ of all N vehicles on the basis of the isochrones corresponding to the actual states of motion and appropriate reference trajectories;

(c) ordering of the identification codes according to ascending values of the time to go, leading to σ_N;

(d) detection of possible merging conflicts by comparison of successive values in the ordered set of times to go; and

(e) assignment of the appropriate motion control law depending on the result of conflict detection.

The structure of this merging-control algorithm will guarantee fail-safe behavior. Due to the cyclic check of the times to go related to the actual vehicle states, the merging sequence may change automatically if the speed of one vehicle in the control region decreases unexpectedly as, for example, caused by engine breakdown.

3.2 Motion Control in the Nonconflict Case

A heuristic control strategy can be developed (Posch 1982) by requiring that vehicle motion within the control region should be based on a constant value of acceleration. This requirement leads to simple formulae for the estimation of τ and comfortable ride conditions. The corresponding reference trajectories R of vehicle motion are defined by the following state differential equations:

$$R: \quad \dot{x} = \begin{cases} -\dot{s} = v & (6a) \\ \dot{v} = u_0 & (6b) \end{cases}$$

with

$$u_0 = (V_p^2 - v^2)/2s \quad (7)$$

The trajectories are parabolas going through the desired final values $s = 0$ and $v = V_p$ of Eqn. (2). If Eqn. (6) is considered as an idealized vehicle-motion model, then u_0 represents a final value control law for the nonconflict case. The corresponding field of isochrones T consists of straight lines according to

$$T: \quad v(t) = (2/\tau)s(t) - V_p, \quad \text{for } \tau = \text{const} \quad (8)$$

At a certain point in time, the time to go can be computed (i.e., predicted) from the formula

$$\hat{\tau}(t) = \frac{2s(t)}{v(t) + V_p} \quad (9)$$

providing the base for an assignment of the instantaneous merging sequence σ_N and the detection of merging conflicts.

3.3 Motion Control in the Conflict Case

In case of a merging conflict, the estimated times to go of a consecutive pair $\{v, \mu c\}$ in σ_N will contradict Eqn. (1). The existing conflict can be solved by synchronizing the motion of the follower v such that when leader μc passes the merge point with $v_{\mu c} = V_p$ the state of the follower v reaches the following final state:

$$v_v \stackrel{!}{=} V_p \quad (10a)$$

$$s_v \stackrel{!}{=} d(V_p) \quad (10b)$$

with

$$d(V_p) = V_p h(V_p) \quad (11)$$

where $d(V_p)$ is the safety distance. The specified final state can be reached by guiding vehicle v along an appropriately modified reference trajectory. Such a trajectory is again a parabola in the v–s plane connecting the final state, Eqns. (10) and (11), and a point $P_1[S_p, v_1]$ as depicted in Fig. 5. The point P_1 is defined by the intersection of the line S_p = constant and the line

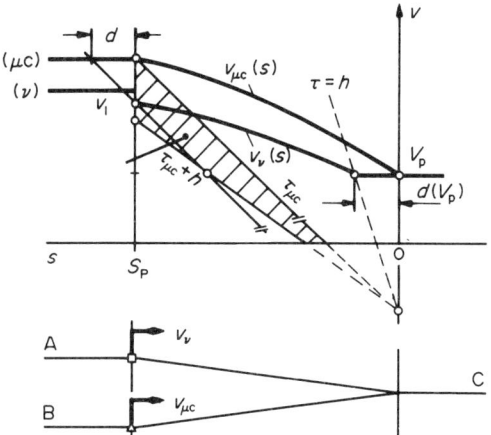

Figure 5
Modified reference trajectory in case of merging conflict

$\tau_{\mu c}(s_{\mu c}, v_{\mu c})$ shifted in parallel to the left by the safety distance $d(V_p)$. Figure 5 shows that this strategy may lead to a small discontinuous change of speed $v_1 - v_v(t_{ev})$ at the start of the control maneuver of follower v. With Eqns. (6)–(8), the modified reference trajectory can be defined by the following set of coupled differential equations:

$$\dot{x}_{\mu c} = \begin{cases} -s_{\mu c} = v_{\mu c} & (12a) \\ \dot{v}_{\mu c} = (V_p^2 - v_{\mu c}^2)/2s_{\mu c} & (12b) \end{cases}$$

$$\dot{x}_v = v_v = u_k \quad (12c)$$

with

$$u_k = \frac{V_p + v_{\mu c}}{s_{\mu c}}(s_v - d) - V_p \quad (13)$$

In the same way as in Sect. 3.2, u_k can be considered as a command value of the idealized vehicle motion described by Eqn. (12), with Eqn. (13) representing the final value control law for the conflict case.

4. Evaluation by Computer Simulation

Some results of a computer simulation study evaluating the merging-control scheme of Sect. 3 are now given. For the vehicles, a realistic fourth-order model of a typical motor vehicle was used. This nonlinear model with variable structure is based on measured steady-state engine and drag characteristics. The adaptation of the ideal command values u_0 and u_k to the velocity reference values w_0 and w_k of the speed-controlled vehicle was performed by a model-following approach (Posch 1982). With T_m indicating the dominant time constant of the controlled vehicle, Eqns. (14) and (15) are obtained for the nonconflict case and the conflict case, respectively:

$$w_0 := u_0 T_m + v \quad (14)$$

$$w_k := \dot{u}_k T_m + u_k \quad (15)$$

4.1 Evaluation of Motion Control

Figure 6 shows the time responses of two vehicles μc and v entering the control region simultaneously with maximum velocity $v(0) = 30 \text{ m s}^{-1}$. The length of the control region is 500 m, the prescribed merging velocity $V_p = 15 \text{ m s}^{-1}$ and the corresponding safety distance $d = 34$ m. According to the control law, w_0 and w_k, the speed of both vehicles is decreased smoothly to the final value V_p and the distance between them is increased steadily to the value $d(V_p)$. The control errors with regard to the prescribed final values are less than 1%.

As indicated by the time responses of the vehicle acceleration, the actual control laws smooth out the

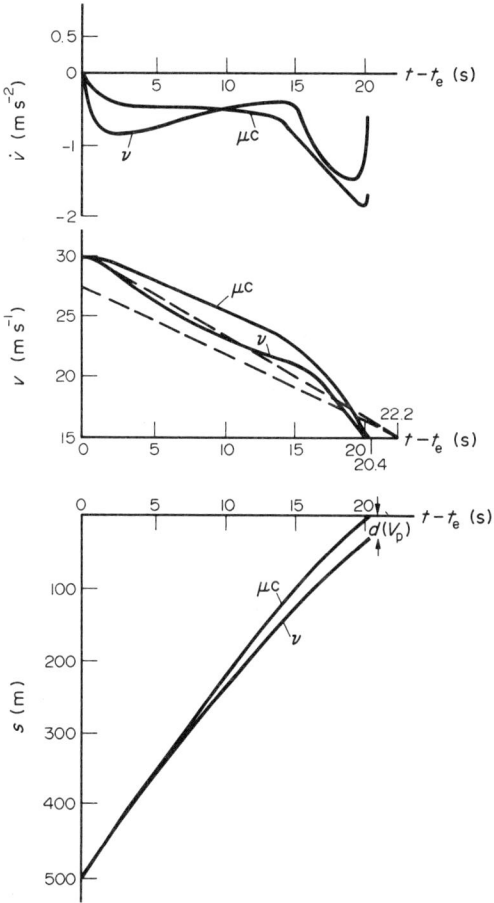

Figure 6
Time responses of two merging vehicles

Merging Control

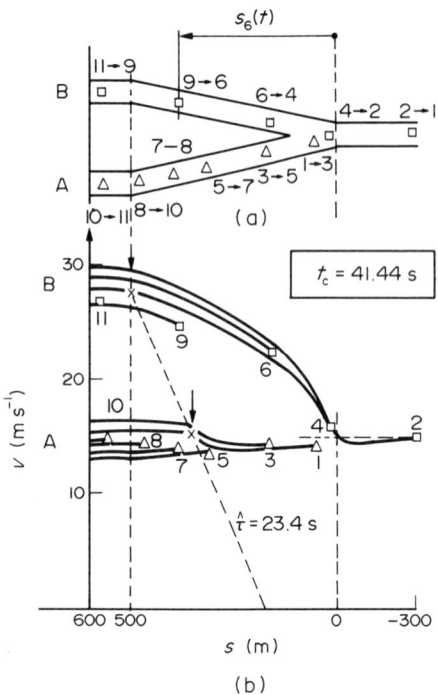

Figure 7
Snap shots of a computer-generated demonstration film: (a) bird's-eye view and (b) time displays of the individual trajectories

discontinuities existing in the idealized acceleration profiles as mentioned in Sect. 3.3 and in connection with Fig. 5. Due to the nonlinear behavior of the realistic vehicle model used in this study, the actual vehicle accelerations $\dot{v}$ differ to some degree from the constant value, as assumed for the basic design steps.

4.2 System Performance for Various Traffic Conditions

Depending on traffic volume, an MC must simultaneously serve up to $N=30$ vehicles (i.e., up to more than 100 vehicles during a 2–3 min time interval). Only a computer-generated demonstration film can give a comprehensive picture of the operation and the total performance of the described merging-control system. Figure 7 shows, for a selected example, one shot of such a film. The identification codes v in Fig. 7 correspond to the sequence in which the vehicles enter the control region. By introducing aggregate performance measures, the dynamic behavior of the merging process can be roughly evaluated. It turns out that useful aggregate measures are the time dependency of the total prolongation time of the individual vehicles in the control region and the average trajectory of all vehicles during a constant time interval.

(*a*) *Normal-flowing traffic.* This is characterized by steady-state inflow on both lanes A and B of

$$q_D = q_A + q_B < 1/h(V_p) \qquad (16)$$

and *no* initial vehicle congestion in the control region. In the example shown in Fig. 8, an average inflow of nearly 60% of $1/h(V_p)$ enters the control region. The average values of entering velocities at 28 m s^{-1} for lane A and 14 m s^{-1} for lane B. In spite of the large differences in entering velocities, the mean $\bar{v}$ and standard deviation $\bar{v} \pm D$ of the average trajectory in both lanes indicate comfortable ride conditions (see Fig. 8a). The total travelling time of the vehicles T_v is approximately equal to the travelling time predicted $\tau_{0,v}$ when entering the control region. Thus, the values of prolongation time $T_v - \tau_{0,v}$ are nearly equal to zero, as shown in Fig. 8b.

(*b*) *Temporarily oversaturated traffic.* As a rather extreme example, the simultaneous inflow of "dense packed" strings of vehicles on both lanes A and B is considered here. According to Fig. 9, all vehicles are assumed to enter the control region with merging velocity V_p and equidistant time headways. For an initial time span of 1 min, the inflow exceeds the maximum possible outflow by 80%; that is, $q_D = 1.8/h(V_p)$.

During this time interval, the average speed $\bar{v}$ in the control region drops to 2 m s^{-1} and the total travelling time is increased by up to 50 s. The vehicles are congested as depicted in Fig. 9a. Subsequently, inflow abruptly reduces to a value of $q_D = 0.8/h(V_p)$. As a consequence, total prolongation time gradually

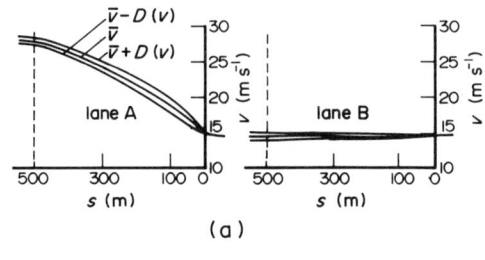

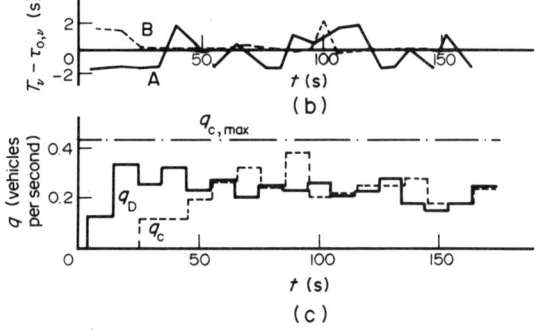

Figure 8
Merging in the case of normal-flowing traffic: (a) averaged trajectories, (b) total prolongation time and (c) flow

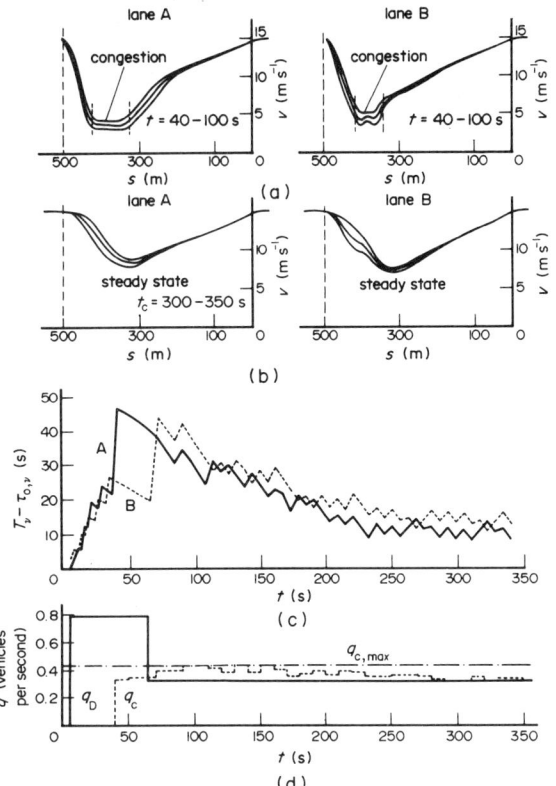

Figure 9
Merging in case of temporarily oversaturated traffic: (a) averaged trajectories for congestion, (b) averaged trajectories for steady state, (c) total prolongation time and (d) flow

decreases, but does not come back to a zero average of $T_v - \tau_{0,v}$. This phenomenon is due to the fact that the controlled merging process reaches a new steady state (outflow $q_C = q_D$), before all delayed, slow-moving vehicles have left the control region. The minimum average velocity within the control region cannot recover from the 8 m s^{-1} value, as shown in Fig. 9b. This, in some way, academic experiment demonstrates the necessity of a superimposed flow controller (Papageorgiou 1983). A recovery from the nonzero steady state of total prolongation time can only be achieved by a reduction of inflow $q_D(t)$ to zero value for a sufficiently long time interval. Further details related to the limitations, robustness of the designed merging controller and the required controller–vehicle communications are discussed in more detail by Posch (1982).

See also: Car-Following Models

Bibliography

Athans M 1969 A unified approach to the vehicle-merging problem. *Transp. Res.* **3**, 123–33
Brown S F 1975 Traffic merging under vehicle follower control. *Proc. Personal Rapid Transit III, Int. Conf. Denver, Colorado*. Audiovisual Library Service, University of Minnesota, Minneapolis, MN, pp. 257–67
Caudill R J, Youngblood J N 1976 Intersection merge control in automated transportation systems. *Transp. Res.* **10**, 17–24
Godfrey M B 1968 Merging delays in automated transportation systems. Ph.D. thesis, Massachusetts Institute of Technology, Cambridge, MA
McGinley F J 1975 An intersection control strategy for a short-headway PRT network. *Transp. Plann. Technol.* **3**, 45–53
Papageorgiou M 1983 *Applications of Automatic Control Concepts to Traffic Flow Modelling and Control*. Springer, Berlin
Posch B 1982 Ein Regelkonzept für die automatische Zusammenführung zweier Fahrzeugschlangen. Dr.-Ing. dissertation, Technische Universität München, Munich, Germany
Posch B, Schmidt G 1983 A comprehensive control concept for merging automated vehicles under a broad class of traffic conditions. *Prep. 4th IFAC Conf. Control in Transportation Systems*. VDI/VDE-Ges. Mess-u. Regelungstechnik, Dusseldorf, Germany, pp. 187–94
Sarachik P E, Chu K C 1975 Real-time merging of high-speed vehicular strings. *Transp. Sci.* **9**, 122–39
Stefanek R G, Wilkie D F 1973 Control aspects of a dual-mode transportation system. *IEEE Trans. Veh. Technol.* **22**(2), 7–13
Whitney D E 1972 A finite state approach to vehicle merging. *ASME J. Dyn. Syst. Meas. Control* **6**, 147–51
Wilkie D F 1970 A moving cell control scheme for automated transportation systems. *Transp. Sci.* **4**, 347–64

B. Posch
[Munich, Germany]

Mobile Communication

The operation of many forms of advanced traffic and transportation systems depends on mobile communication for exchanging information between in-vehicle equipment and traffic control centers, dispatch offices and other infrastructure installations. One-way (to vehicles) voice communication by means of highway advisory radio (HAR) using dedicated amplitude modulation (AM) frequencies in the USA and automatic radio information (ARI) using sidebands of frequency modulation (FM) stations in Europe has supplied traffic information to motorists since about 1970 (see *Broadcasting Communication Systems*). Two-way voice communication via land-mobile radio (LMR) has been used for dispatching commercial vehicles since about 1930, and radiotelephone services have seen some use for managing commercial vehicle operations during the 1980s. This article, however, will deal with mobile data communication which is rapidly supplanting voice for many applications. Mobile data communication is required by automatic vehicle location tracking systems and by future traffic control systems utilizing in-vehicle equipment (see *In-Vehicle Equipment for Future Traffic Control Systems*).

1. Background

For many years, the state of the art in mobile data communications was to connect telephone-type modems to LMR. This simple combination worked well provided that the data transmission needs were modest and the system was designed to cope with the harsh mobile-radio environment. Momentary fading, noise, distortion or interference, which is only a minor annoyance during voice communications, will either alter or destroy a data stream. The net result was that transmissions were limited to 300 bits per second and retransmission was used to control errors.

From these beginnings, mobile data communications has evolved to include packet-radio systems, an effective technology for transmission of data by radio networks interfaced to data devices and computers. Packet-radio systems are a derivative of methods pioneered on long-distance wired networks used by common carriers for both domestic and international data communications, and are also used with microwave and infrared data communications.

Extensive packet-radio research and development during the 1980s has resulted in a new resource to provide high acuracy, high throughput and a wide range of networking possibilities for data transmission by radio. Among the range of possibilities are radio transmission links at rates of 2400–9600 bits per second, bit error rates of one uncorrected or one undetected wrong bit in 10^{11} bits transmitted, with throughput efficiencies of greater than 75% using standard narrow land-mobile channels (Beeferman 1986). Data rates of 19 600 bits per second are expected soon. Much higher data rates (100 000–500 000 bits per second) are already achieved with microwave and infrared carriers.

Table 1
Characteristics of alternative mobile communication approaches

Approach	Characteristics
Broadcast SCA	one-way only voice and data low data rates extended area coverage includes ARI and RDS
Proximity beacon	one-way or two-way data only high data rates spot-area coverage
Inductive loop	one-way or two-way data only low data rates spot-area coverage
Land-mobile radio (dedicated)	two-way voice and data local-area coverage
Specialized mobile radio (SMR)	two-way voice and data extended-area coverage
Cellular radio	two-way voice and data local/extended-area coverage
Mobile satellite communicatons	one-way or two-way voice or data wide-area coverage
Meteor burst communications	two-way data only wide-area coverage involves time delays

2. Alternative Approaches

The principal communication systems identified for electronic transmission of information to and from vehicles are listed and broadly characterized in Table 1. These approaches, along with typical applications in traffic and transportation systems, are discussed individually below.

2.1 Broadcast SCA

Regulatory bodies allow commercial FM and television broadcast stations to transmit ancillary information on 53–99 kHz subchannels with a bandwidth of ±5 kHz. Called subcarrier authorization (SCA) in the USA, these subchannels carry analog or digital signals which may be detected and decoded by special attachments or design features of ordinary receivers. An early example is the automatic road information (ARI) system for local-area traffic advisories by voice transmission. The European radio data system (RDS), standardized by the European Broadcasting Union (EBU) in 1984, operates on the same principle for transmitting station and programming data for automatic tuning along with other information in digital form. Protocols for using RDS to convey traffic information to in-vehicle equipment, such as navigation systems or speech synthesis units which express the information in the language of the driver's choice, are being developed and tested in Europe (Davies *et al.* 1989).

Use of the SCA approach for area broadcast of digitized information to vehicles requires relatively little additional equipment other than a minor enhancement of the FM radio receiver that is already commonplace in vehicles. However, broadcast SCA has the disadvantage of being limited to one-way communication into vehicles and, thus, could not be used for location monitoring or status reporting. In addition, new uses have to compete with other subchannel uses such as background music services, stock quotation services, paging services, and so on. Moreover, the data rates are rather low. The European RDS, for example, provides a raw data rate of 1187.5 bits per second and a substantial fraction of this is reserved for control, error checking, and so on.

2.2 Proximity Beacon

Infrared and microwave proximity beacons, which have the advantages of high data rates in combination with spot coverage (i.e., typically limited to tens of meters) that permits messages to be tailored to highly localized needs, are strong candidates for a mobile data communications role in route guidance systems being developed and tested in Europe and Japan. In these densely populated geographic areas, the heavy infrastructure costs of installing beacons at close intervals throughout the road system and integrating them into overall communications networks do not seem as formidable as in the USA where, except for use as electronic signposts in obsolescent location monitoring systems for transit buses, proximity beacons have given way to other technologies.

Current examples of proximity beacons for mobile data communication are the infrared beacon used with the ALI-SCOUT route guidance system in Europe (von Tomkewitsch 1986) and the microwave beacon used with the road/automobile communication system (RACS) in Japan (Fukui et al. 1989). Both are used primarily for communicating routing and related traffic data to on-board driver information systems. The ALI-SCOUT beacon originally operated at 64 000 bits per second and a newer version is expected to operate at 500 000 bits per second. Thus, up to 32 kbytes of information may be exchanged while a vehicle is passing a beacon, even at highway speeds. RACS, which uses a 2.5 GHz microwave carrier, can handle similar data rates. In contrast, 247.2 kHz radio frequency radio beacons tested in an earlier phase of RACS experimentation, provided data rates of only 9600 bits per second.

It is anticipated that both ALI-SCOUT and RACS will be used to support fleet management applications by providing two-way data communication services as well as vehicle location reporting services. In addition to route guidance and traffic information transmitted to all equipped vehicles, the RACS microwave communications protocol provides for exchange of up to 25 kbytes of individually addressed information each time a vehicle passes a beacon. The packet data communication network connecting the beacons directs messages addressed to a particular vehicle to the beacons it is most likely to pass, based on the vehicle identification transmitted to previously passed beacons. In addition to data communications, the RACS beacons will also be able to transmit images (facsimile or still pictures) as well as voice messages by means of coding and signal compression.

2.3 Inductive Loop

Inductive loops, which are essentially proximity beacons in the form of radio antennas buried beneath the roadway, have the low data rates characteristic of radio-frequency proximity beacons (see *Detectors for Road Traffic*). They have the further disadvantage of being very expensive and awkward to install because traffic must be temporarily diverted from lanes being equipped. Maintenance is also expensive because of the wear and tear from traffic. Nonetheless, inductive loops were considered for communications in the early route guidance approaches (Rosen et al. 1970), and the California Department of Transportation (Caltrans) is currently experimenting with inductive loops for conveying information to on-board systems. Although similar loops are used fairly extensively for traffic monitoring in the USA, other countries (e.g., Japan) are turning more to overhead ultrasonic or video detectors for traffic monitoring purposes (see *Video Sensors*).

2.4 Dedicated LMR

Although individually owned and operated LMR systems for two-way voice communication have seen widespread use for fleet management within city areas the spread had been severely limited by the amount of the radio spectrum allocated for this purpose. New technologies such as trunking and digital addressing are now providing more traffic capability within the available frequencies. However, the greatest gains in dedicated land-mobile utilization are resulting from increased use of data communication as an alternative to voice communication.

Until recently, US Federal Communications Commission (FCC) rules limited nonvoice transmission by LMR to a maximum of 2 s without a break in the signal, and also required that such signals be secondary to voice transmission. Although the 2 s restriction has been eliminated and digital data made coequal with voice, shared voice and data communications is still a useful concept in the overall design and operation of LMR systems.

2.5 Specialized Mobile Radio

Specialized mobile radio (SMR) is a new class of LMR service that has been evolving since first authorized by the FCC in 1974. SMR has quickly emerged to become the preeminent provider of private land-mobile communications services, particularly in large metropolitan areas. According to the FCC, there were approximately 3000 trunked SMR operators in the USA serving some 670 000 mobile units by 1988.

Basically, SMR is a business and regulatory approach which permits different using entities to share common transmission facilities and frequencies. With one type of trunked SMR service, a user vehicle unit monitors a number of frequency channels for a digital code identifying a transmission addressed to the individual unit, and monitors channel use to select an available frequency for transmitting its own messages addressed to the user's dispatch station. With a second type of trunked system, the vehicle unit monitors a "control channel" which manages channel assignments.

SMR has many of the characteristics of services provided by common carriers in that base and repeater stations are owned and operated by an entrepreneur to serve many customers whose only investment is in the two-way units for their vehicles. In fact, the dispatcher typically uses a similar two-way unit rather than landlines to the SMR base station. In addition to eliminating the cost of a dedicated base station by paying either a flat fee or according to the amount of air time used, the SMR approach simplifies the usual problem of individual licensing for dedicated two-way radio systems, thus assuring easy access to radio service by virtually all business users or local agencies who would otherwise qualify for a dedicated license if frequencies were available. In recent years, FCC regulations have been further relaxed to permit SMR use by individuals, federal govenment agencies, and representatives of foreign governments.

Motorola Inc. has announced a scheme for tying individual SMR operators together with a hub-controlled packet network. This will eventually permit continuous communication coverage as well as location reporting over the 30% of the USA served by SMR. The CIDER project of DRIVE is exploring the possibility of an essentially similar pan-European network of trunked private mobile radio (PMR) for exchanging short messages between vehicles and information centers. Limited networks are alreaady in use in the Scandinavian countries (Mobitex) and in the UK (band III).

2.6 Cellular Radio

Cellular radio telephones have provided a major step forward for both business and personal voice communications in the vehicular environment. Cellular data communication is expected to broaden cellular applications during the next decade. However, the present scheme of using cellular technology exclusively for full-duplex individually addressed communications will limit the potential use of cellular radio in providing traffic and routing data for vehicular navigation and information systems. Effective adaptation of cellular radio to the data communications requirements of navigation and route guidance would require the establishment of a dedicated channel for repeatedly broadcasting the same map updates, traffic data, and so on, by all cell transmitters in the local area served by a cellular system. Equipped vehicles could have special receive-only units to detect the data for transfer to the on-board equipment.

In the meantime, commercial fleet management applications of conventional cellular telephones are growing, and have particular attraction for small vehicle fleets with only infrequent requirements for voice communications within local cellular service areas. Cellular data links are also being considered by a number of vehicle location monitoring systems integrators for localized applications. The first such system based on cellular radio was reported by Carter and Warburton (1985). Although the cellular approach has received relatively little attention for managing cross-country trucking operations in the USA, it is considered to have great potential for such applications in Europe where continuous geographic coverage at modest cost is expected once the groupe speciale mobile (GSM) digital cellular standard is implemented in the early 1990s. The SOCRATES project of DRIVE is currently investigating the possible use of GSM for traffic messaging services as well.

2.7 Satellite Communications

Unlike urban vehicle location monitoring, which can use short-range LMR or cellular telephone communication links, cross-country truck location monitoring requires long-distance communication links such as those characteristic of mobile satellite services. Thus, satellite communications have long been viewed as having great potential for cross-country trucking applications because of the essentially unlimited range. However, the wide area covered by satellites is not an advantage for traffic data communications because traffic data is not useful outside the subject local area.

Limited satellite communication services are already available for mobile applications and others are in the offing. Inmarsat, an international organization that has operated a global network of L-band (~ 1500 MHz) satellites since 1982, primarily for ship and aircraft communications, is introducing standard C service for land-mobile communications including two-way store-and-forward data/messaging (Bell 1989). Geostar (Locstar in Europe) is offering limited interim messaging services prior to completing its integrated satellite positioning and communication system. Qualcomm also offers satellite communications services for fleet opertations. Virtually all such services now available require that communications be routed through hub stations operated by the service provider.

The most comprehensive satellite communication service for land-mobile applications in North America is expected to be the Mobile Satellite Service (MSS) authorized by the FCC in 1986 with a spectral allocation of 27 MHz in the L-band. Most of the original contenders for the license to operate this service have joined together in the American Mobile Satellite Consortium (AMSC) which, along with Telesat Mobile, Inc. (TMI) of Canada, are jointly developing MSAT (Noreen 1989). MSAT will be implemented in 1993, approximately, with separate geostationary satellites for the USA and Canada such that the satellite from one country provides backup capacity for the other.

The basic services provided by MSAT will include mobile telephones (interconnected with the public switched telephone network), mobile radio dispatch, mobile radio services, aeronautical services and transportable telephone services. Ancillary services will include point-to-multipoint data distribution, paging, remote data collection, position location and new mobile services using alphanumeric terminals. Whereas most MSAT services will be accessed through hub stations, communications between fleet controllers and mobile units can be relayed directly through the satellite, thus circumventing terrestrial links entirely.

2.8 Meteor Burst Communications

Meteor burst communications, which have been used in various forms for specialized applications since the late 1970s, have now been developed to the point where they offer an alternative to satellites for the long-distance communication required for cross-country trucking operations. Meteor-scatter radio service, which allows a single vhf low-band base station to cover a radius as great as 1900 km, was originally offered to US trucking companies by the first FCC licensed operator, Transtrack Inc., and a second operator, the Pegasus Message Corporation, has an experimental license (Mickelson 1989).

Short-lived (a few milliseconds to several seconds) ionized trails created by dust-sized meteors slicing through the atmosphere at heights of 80–120 km reflect radio signal bursts. A base station constantly transmits a probe signal on one frequency and, on receiving the reflected signal, individual mobile units respond on another frequency when a response is necessary. The mobile units are typically equipped with a Loran-C receiver, keyboard, and message display. A network operations center coordinates system operation and is in telephone communication with trucking company offices to manage message traffic. With the approach used by Transtrack, as many as 32 characters can be sent per message packet, and sometimes more than one packet may be sent via a single meteor trail. The random process is said to provide approximately ten connections per hour with trucks within the reflection range of a base station.

3. Trends

As listed in Table 1 and discussed above, there is great diversity among available approaches potentially useful for mobile data communications. There are also substantial differences between the data communication requirements of traffic applications and commercial vehicle applications. Traffic applications generally require transmission of large amounts of highly localized data over short distances, whereas commercial vehicle applications require transmission of smaller amounts of data over great distances as well as relatively short distances. Consequently, no single approach is well suited for all applications.

State-of-the-art navigation and route guidance systems that require traffic data no longer require proximity beacons for updating vehicle location (see *In-Vehicle Equipment for Future Traffic Control Systems*). However, proximity beacons continue to be of interest because of their ability to transmit large amounts of traffic data highly specific to beacon location. This advantage at least partially offsets the disadvantage of the high infrastructure cost of the proximity beacon. Broadcast SCA, on the other hand, is very inexpensive to implement but has the disadvantage of very low data rates and being limited to one-way communication. Some form of cellular or trunked LMR network may eventually offer the best single approach for traffic applications. In the meantine, numerous approaches remain under investigation, and many are the subject of extensive field trials which will be useful in narrowing the selection.

Short-range communications trends for urban commercial vehicles include the expanding use of cellular telephone and trunked SMR. Although relatively high operating costs, overcrowding, and limited coverage in some areas are deterrents to greatly expanded cellular telephone use by commercial vehicles, some of these problems will be alleviated in the future by digital cellular telephones. The additional European problem caused by the lack of standardization among different countries will soon be eliminated with the implementation of the GSM digital cellular standard.

Satellite communications appear to be the present method of choice for cross-country vehicles, and may be expected to expand rapidly as more comprehensive satellite communication services become available. The role of meteor-burst communications for cross-country vehicles will depend very strongly on the selection of specialized applications and the development of transportation management techniques that are not unduly hampered by the time delays inherent to the approach. However, the rapid expansion of the capacity and service domain of LMR by trunking and networking appears to offer great potential as a future alternative to satellite and meteor-burst communications for cross-country trucking.

See also: Cellular Communication Systems; Data Processing in Air Traffic Control

Bibliography

Beeferman S 1986 Packet radio can be a cost-effective alternative to dedicated land lines for data communication networks. *Commun. News* **23**, 36–8

Bell J C 1989 Vehicular C^3N—a global satellite solution. *Proc. VNIS '89—Vehicle Navigation and Information Systems Conf.* Institute of Electrical and Electronics Engineers, Piscataway, NJ, pp. 403–8

Carter D A, Warburton R D H 1985 Using cellular telephones for automatic vehicle tracking. *Proc. RIN Land Navigation and Location for Mobile Applications Conf.*, Paper 28. Royal Institute of Navigation, London

Davies P, Hill C, Klein G 1989 Standards for the radio data system–traffic message channel. Society of Automotive Engineers Technical Paper Series, No. 891684. SAE, Warrendale, PA

Fukui R, Noji Y, Hashizume M 1989 Individual Communication Function of RACS: Road automobile communication system. *Proc. VNIS '89—Vehicle Navigation and Information Systems Conf.* Toronto, Canada, pp. 206–13

Mickelson K D 1989 Tracking 64 000 vehicles with meteor-scatter radio. *Mobile Radio Technol.* **7**, 24–38

Noreen G K 1989 MSAT: Mobile communications throughout North America. *Proc. 39th IEEE Vehicular*

Technology Conf., II. Institute of Electrical and Electronics Engineers, Piscataway, NJ, pp. 557–62

Rosen D A, Mammano F J, Favout R 1970 An electronic route guidance system for highway vehicles. *IEEE Trans. Veh. Technol.* **19**, 143–52

von Tomkewitsch R 1986 ALI-SCOUT—A universal guidance and information system for road traffic. *Proc. 2nd IEE Int. Conf. Road Traffic Control.* Institution of Electrical Engineers, London, pp. 22–5

R. L. French
[R L French & Associates,
Fort Worth, Texas, USA]

Navigation Control of Ships

Navigation control, also referred to as maneuvering of ships, can be defined as control of systems on ships that generate forces or moments of hydrodynamic, of aerodynamic or of mechanical nature, with the objective of reaching a desired position, speed, heading or course. There is a great variety of these systems. They vary not only in dimensions but also in principle of working, efficiency and so on. Typical systems are devices such as rudders, propellers, transverse thrusters, azimuthing propellers and water jets. The applicability of these devices is strongly dependent on the mission profile of the vessel. Modern control theory and simulation techniques are applied to optimize the performance of the navigation control systems.

1. Process of Maneuvering

Vessels can be divided into three main categories:

(a) the classical group of ships meant for the transportation of cargo or passengers from one place to another, over long distances with many different routes (e.g., tankers, freighters, tugs and cruise vessels),

(b) vessels following fixed predetermined routes (e.g., ferries, dredges, navy vessels and cable layers), and

(c) working platforms from which offshore activities take place (e.g., drilling platforms and diving support vessels).

These three categories require different control strategies.

During much the greater part of their lifetime, transporters have a required position which is not expressed in meters but in kilometers. Only during the entering of a harbor is a higher accuracy a necessity. In this case, use is made of tugs or, when a harbor is entered frequently, the vessel is installed with maneuver-enhancing devices such as bow thrusters and special rudders. The vessel is then maneuvered as category (b) vessels.

For category (b), it is desired to follow the selected route with a specified accuracy to maintain a predetermined speed. Within these constraints, the fuel consumption should be minimized. For ferries, for example, it is important to arrive at the expected time, but at the lowest cost.

For working platforms, the main task of which is to make it possible to perform work at a location, the required maneuvering accuracy is usually expressed in meters.

For transporters and ferries the reference system is the earth's coordinate system with the equator and the Greenwich meridian as references. Although sextants are still used, electronic position measurement is in common use. Examples are satellite navigation, Decca navigator marine receivers and Loran C. For platforms, the position reference system is a nearly fixed point or a system of points. Typical measurement systems are hydroacoustic measurement systems, taut wire systems or the Artemis system. New developments are the use of the global positioning system (GPS) and television-tracking systems. Of course, vessels exist that have a purpose where more types of position control are applied. A typical example is the supply vessel. It transports over longer distances and it maneuvers on-the-spot to unload near an off-shore platform.

It is evident that, although the three categories of vessel lead to different configurations of maneuvering devices and control systems, the process to be controlled is identical.

1.1 Process Description

The process in its simplest form can be described as the motion of a single-mass system under the influence of external forces. Normally, the study of navigation or maneuvering is limited to the motion in a horizontal plane—the flat water surface—though motions outside this plane certainly occur, influencing the working of propellers, for example, and generating forces and moments of secondary order which excite the mass to accelerate in the horizontal plane. When these out-of-plane motions—roll, pitch and heave—are neglected, the mathematical representation of the sway, surge and yaw motions can be described by three Euler equations:

$$\left.\begin{array}{c} m(\ddot{x} - \dot{y}\phi) = \sum X_0 \\ m(\ddot{y} + \dot{x}\phi) = \sum Y_0 \\ J\ddot{\phi} = \sum \phi_0 \end{array}\right\} \quad (1)$$

The coordinate system is shown in Fig 1. The mass m and the polar moment of inertia J normally include the coefficients of hydrodynamic forces and moment. These included components and those generated by the acceleration of the vessel and are often referred to as the added mass and added moment of inertia (Lamb 1945).

The right-hand side of each of Eqns. (1) represents three types of forces and moments:

(a) hydrodynamic pressures and shear stresses on the hull due to the motion of the vessel;

(b) forces and moments due to wind, waves and current or, in confined waters, due to the passage of nearby ships and changes of water depth, which can be called environmental influences of external forces; and

(c) forces and moments exerted by the maneuvering devices.

1.2 Hydrodynamic Pressures and Shear Stresses on the Hull

Forces and moments due to the motion of the ship result in a set of simultaneous equations. For example, the pure yawing motion generates forces such that swaying occurs and vice versa. This results in

$$
\left.\begin{array}{l}
m\ddot{x} + a\dot{x} = \sum X_1 \\[1em]
m\ddot{y} + b\dot{y} + c\phi + d\ddot{\phi} = \sum Y_1 \\[1em]
J\ddot{\phi} + e\dot{\phi} + f\dot{y} + g\ddot{y} = \sum \phi_1
\end{array}\right\} \quad (2)
$$

where only the first-order terms of the Taylor expansion have been taken into account. The coefficients b to g are known as the stability derivatives. The present state of the art does not allow the constants to be calculated. Quantitative analysis therefore has to be carried out by means of model tests in which a ship model is oscillated to determine its stability derivatives. With the stability derivatives known, it is possible to determine if the ship itself is stable, which means that it can keep a definite course. In the case of instability, a constant corrective action by means of the rudder is necessary. Response times to disturbances also follow from the same analysis.

1.3 Environmental Influences

The forces due to wind, waves and current are not only dependent on the vessel geometry but are also functions of heading and speeds. They can be expressed as $F_E = f(\phi, \dot{\phi}, \dot{x}, \dot{y})$. Consequently, a further coupling between the equations describing the motion occurs:

$$
\left.\begin{array}{l}
m\ddot{x} + a\dot{x} + X_E(\phi, \dot{\phi}, \dot{x}) = \sum X_2 \\[1em]
m\ddot{y} + b\dot{y} + c\phi + d\ddot{\phi} + Y_E(\phi, \dot{\phi}, \dot{y}) = \sum Y_2 \\[1em]
J\ddot{\phi} + e\dot{\phi} + f\dot{y} + g\ddot{y} + M_E(\phi, \dot{\phi}, \dot{y}) = \sum Q_2
\end{array}\right\} \quad (3)
$$

The forces and moments due to current are mainly of a constant nature over a relatively long period; those due to wind are not constant and can even have a gusting nature. The forces and moment due to waves vary greatly with time and can be represented by a series of sine functions. The frequency of these sine functions corresponds to the frequency of the wave components and has an order of magnitude of 0.1 Hz. The high frequency signals are neglected by the control system. The low frequency signals are estimated by specially designed adaptive filter methods (e.g., Kalman and extrapolation).

1.4 Maneuvering Devices

Except for sails, maneuvering devices are all based on hydrodynamic principles. The well-known rudder, of which many types and configurations exist, is almost always passive, which means that no mechanical energy is delivered to the rudder to generate forces. This means that forces and moments depend on the velocities in the water flowing to the rudder. Therefore, the relation between the control parameter—the rudder angle—and the rudder force is not a unique one but is dependent on ship velocities and propeller-induced velocities. The rudder forces can be expressed as $F_R = f(\delta, \dot{x}, \dot{y}, T)$, where δ is the rudder angle and T is the thrust of the propeller in front of the rudder.

All other maneuvering devices are based on the working of a propeller or impeller. In Fig. 2, possible maneuvering devices based on propeller action are shown. The forces exerted by the maneuvering devices depend on the amount of power absorbed by the propeller. For controllable-pitch propellers and azimuthing propellers, the rotational speed and the angular position also determine the forces. Since the working principle is the change of momentum of the water, the forces also depend on the velocity of the vessel. Summarizing, the forces can be represented by $F_M =$

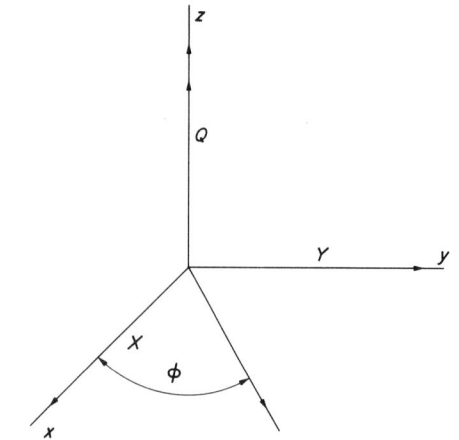

Figure 1
Coordinate system for sway, surge and yaw motions

$f(a, n, \varepsilon, \dot{x}, \dot{y})$, where a is the pitch of the propeller, n the rotational speed of the propeller and ε the angular position of an azimuthing propeller.

Thus, the total set of equations becomes

$$\left.\begin{array}{l} m\ddot{x} + a\dot{x} + X_E(\phi, \dot{\phi}, \dot{x}) + X_R(\delta, \dot{x}, \dot{y}, T) \\ \qquad + X_M(a, n, \varepsilon, \dot{x}, \dot{y}) = 0 \\ m\ddot{y} + b\dot{y} + c\phi + d\ddot{\phi} + Y_E(\phi, \dot{\phi}, \dot{y}) \\ \qquad + Y_R(\delta, \dot{x}, \dot{y}, T) + Y_M(a, n, \varepsilon, \dot{x}, \dot{y}) = 0 \\ J\ddot{\phi} + e\dot{\phi} + f\dot{y} + g\ddot{y} + M_E(\phi, \dot{\phi}, \dot{y}) + hX_R \\ \qquad + iY_R + hX_M + kY_M + Q_M(a, n, \varepsilon, \dot{x}, \dot{y}) = 0 \end{array}\right\} \quad (4)$$

where h to k are coefficients representing geometrical parameters describing the position of the maneuvering devices with respect to the center of gravity of the vessel. The settling times of pitch and rotational speed also need to be taken into account.

As can be seen, the coefficients of the equation terms containing the derivatives of the coordinates x and y can be influenced by the coefficients describing the performance of the maneuvering devices. A well-known example is the selection of coefficients for an autopilot, where the course stability can be strongly influenced. The autopilot controls the heading of the ship via the rudder angle. The desired heading is compared with the actual heading. The latter is obtained from the gyrocompass. Generally, a proportional-plus-integral-plus-derivative (PID) controller is applied. The operator has to adjust several control parameters depending on the environmental conditions. Since the introduction of microcomputers, self-tuning systems have also been applied. These controllers generally require a complicated controller parameter adaption mechanism and an identification system for ship parameters. A simpler approach for an autopilot is gain scheduling. For example, the control parameters are a function of the ship speed.

1.5 Prime Movers

All propellers or impellers need mechanical energy. For ships used as prime movers mainly diesel engines are in application. For fast navy vessels gas turbines are used. Steam turbines can be found in very large crude oil carriers. Most marine diesel engines are turbocharged and run at medium speed (~500 revolutions per minute (rpm)). The propellers are driven by a gear box or sometimes via an electrical shaft connection (generator-electric motor) or a fluid coupling.

The dynamic response of the pitch a or the rotational speed n of the propeller is dependent on the behavior of the prime mover. The rotational speed can be represented by

$$\dot{n} = \frac{1}{\Sigma J}[Q_c(f, N) - Q_p(n, a, \dot{x}) - W \cdot n] \quad (5)$$

where ΣJ represents the total moment of inertia, $Q_c(f, N)$ is the engine torque as a function of the fuel rack position f and the turboblower speed N, $Q_p(n, a, \dot{x})$ is the propeller torque as a function of shaft speed, of pitch and of ship speed, and W is frictional losses. The propeller pitch response is mainly determined by the capability of the engine. The diesel engine needs to be protected against overloading. In combination with a controllable pitch propeller, an optimum fuel consumption path as a function of the rotational speed can be defined. Load control systems are applied, which control the pitch such that this optimum fuel consumption path is followed (see Fig. 3). When the propeller is driven by two engines via a common gear box or a common electric motor, the engine loads need to be shared. This is achieved by a common governor controlling both engines or by a master–slave configuration.

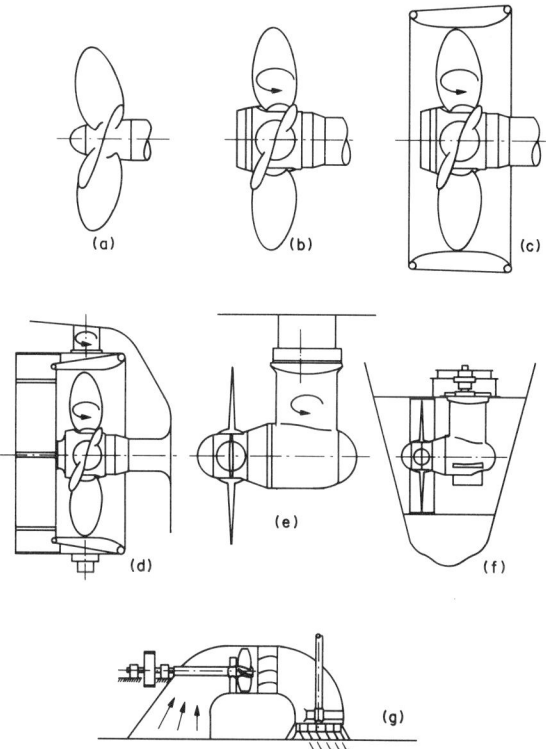

Figure 2
Maneuvering devices based on propeller action:
(a) fixed-pitch propeller, (b) controllable-pitch propeller, (c) propeller in duct, (d) propeller in rotatable nozzle, (e) azimuthing thruster with or without nozzle, (f) transverse thruster in tunnel, and (g) water jet with rotatable guide vanes

Navigation Control of Ships

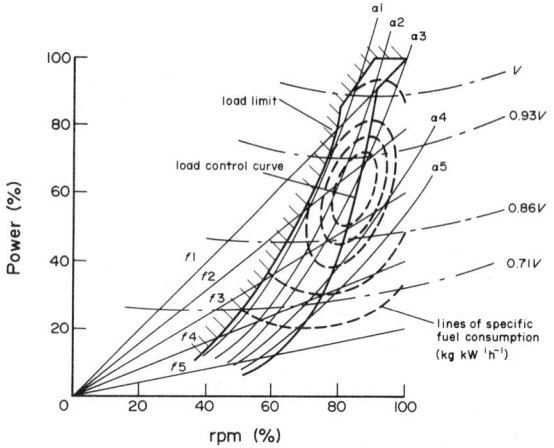

Figure 3
Optimum fuel consumption path for a diesel engine with a controllable pitch propeller (v is ship velocity; αk is pitch, with $k = 1$ being high pitch; fn is fuel rack content, with $n = 1$ being a full rack)

2. Control Systems

The control systems that are in operation range from the most advanced system on board dynamically

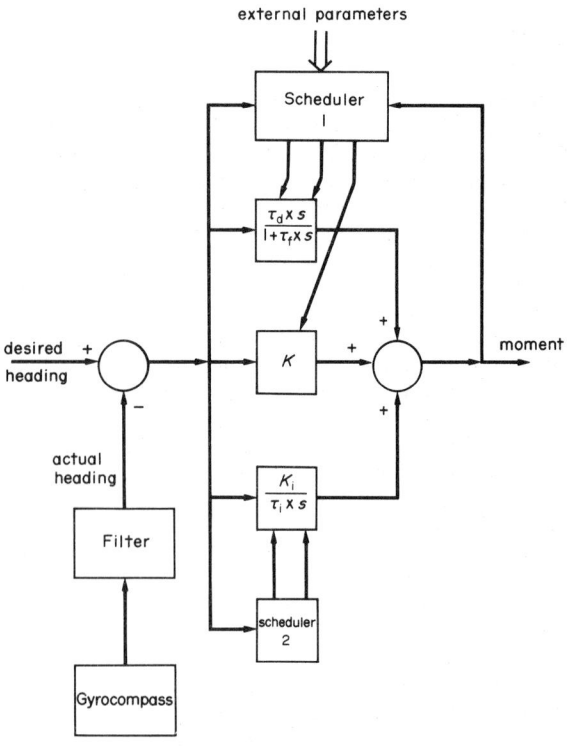

Figure 4
Autopilot with gain scheduling

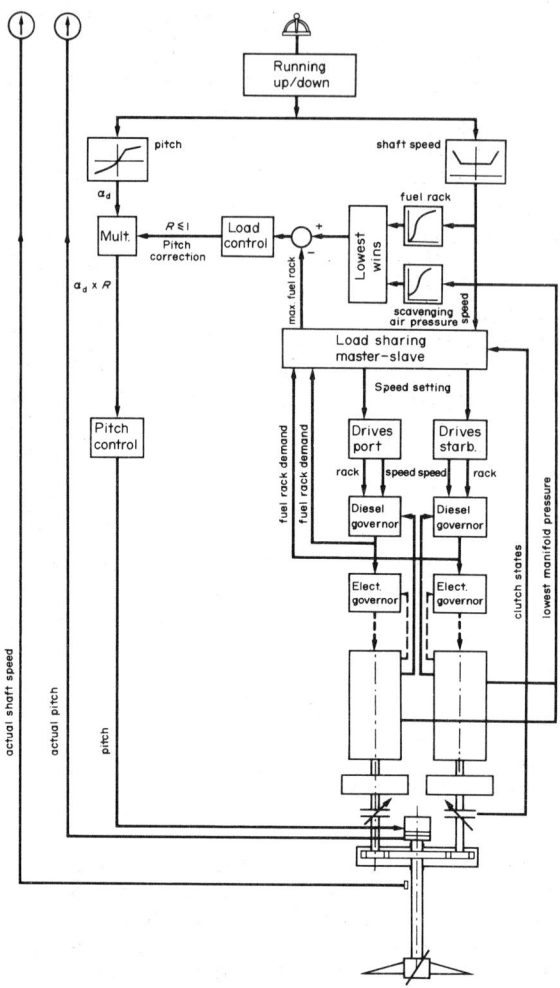

Figure 5
Propulsion control system for a twin diesel engine drive with controllable pitch propeller

positioned platforms, equipped with a number of maneuvering devices, to the system on board a single-propeller, single-rudder ship.

In the simplest case, the rotational speed of the diesel engine is controlled by a hydraulic–mechanical governor, based on Watt's fly balls, and a reversing gear mechanism. The commands for the governor and the reversing gear are transmitted via a lever on the bridge or in the engine control room. The lever range is from full astern via dead slow to full ahead. The rudder angle is controlled by the helmsman or by the autopilot (see Fig. 4).

The use of a controllable pitch propeller instead of the reversing gear gives another degree of freedom; that is, the pitch of the propeller. This feature allows for continuous and fast thrust control during maneuvering and for fuel optimization during a voyage.

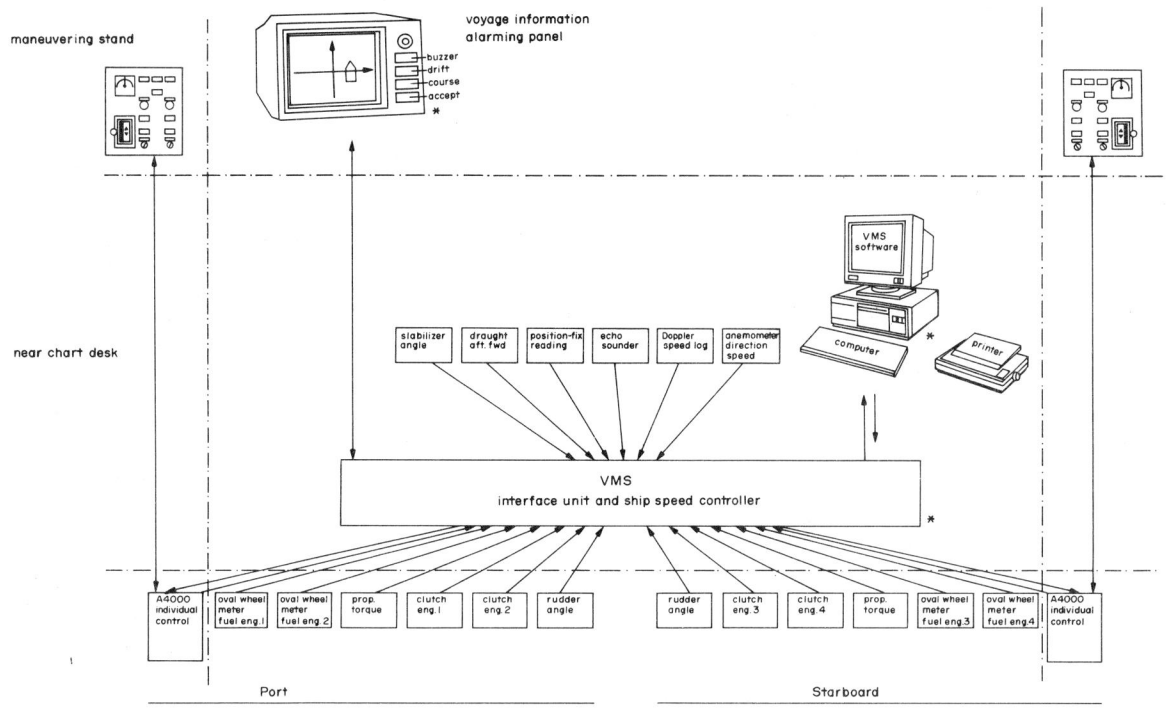

Figure 6
Voyage management system

For the transportation category of vessels, the classical combined pitch–rpm control in combination with a load control is mostly used (see Fig. 5). The commands are given by a single lever. The shaft speed is independently set by this lever. At low-power operation, the pitch is set simultaneously. At high-power operation, the load control suppresses the pitch in order to allow for optimum fuel consumption or to avoid overloading of the engine. In case shaft-driven generators are used, constant shaft speed is chosen, while the thrust is controlled by the pitch. Sometimes the electrical machinery can cope with large frequency deviations. In this case, during a voyage, the shaft speed is also controlled (Deichmann and Frey 1988).

For the fixed-route category of vessels, a voyage management system can control the forward ship speed. Especially for vessels with precise arrival times (e.g., ferries), it is beneficial to preselect the optimum route and to optimize the forward speed profile for the selected route. The voyage management system as depicted by Fig. 6 executes a continuous ship speed optimization.

Together with the increasing complexity and power density of prime movers goes the introduction of protection control, which is a substantial and integrated part of the maneuvering control. The additional degree of freedom of the controllable pitch propeller (the propeller pitch) can be used for automatic load running up and running down during acceleration and deceleration, for automatic load slow downs in case of a major malfunctioning of the machinery and so on. However, all of these protection systems require overriding systems, such as crash maneuvers, at the bridge, in order to obtain a safe navigation.

For the working platform category of vessels, more than two maneuvering devices (i.e., rudder and propeller) are installed. Generally, a single-lever maneuvering system will be applied in addition to the individual controls of the devices. Figure 7 shows a control system developed by LIPS ("LIPS stick").

By the single lever (joy stick), the longitudinal and transversal forces acting on the vessel are controlled, while the heading is automatically maintained by an autoheading controller. When the joy stick input signals are received from positional controllers and when the necessary position reference systems are installed, the system is called dynamic positioning (DP).

DP control systems are applied on vessels employed in difficult operational modes such as diving support, drilling operations, cable laying and single-point mooring. It is obvious that a high degree of safety is required for these operations. Hence, duplex computer systems and duplication of individual controls may be required,

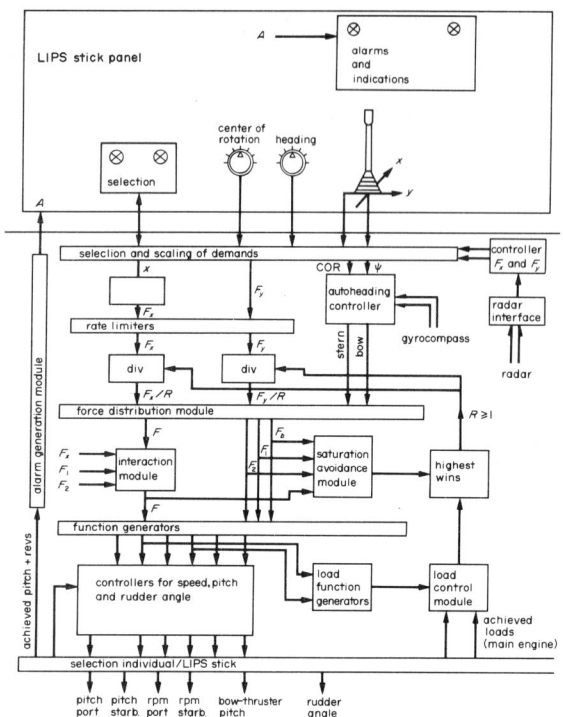

Figure 7
LIPS stick integrated maneuvering control system

depending on the notation of the classification society. The supply vessel serves as a typical example of where a simplex DP can be installed. During loading and unloading near an off-shore rig, long waiting times occur. The vessel has to be in the vicinity of the rig but, due to the pipelines and cables at the ocean floor, no anchor can be dropped. The positioning of a vessel (even with a joy stick) in adverse weather conditions is very tiring. The DP system suitable for such a task can be interfaced with the radar system on board. When the vessel is manually positioned at the selected point a "hold" mode is chosen. The radar will take the rig as the position reference. The control system uses the radar pulses and the heading flash to determine the distance to the rig and takes care to ensure smooth control of the propulsion devices and the rudders.

See also: Ship Automation and Control; Ship Positioning: Adaptive Control

Bibliography

Deichmann F, Frey D 1988 Auslegung der elektrischen Energieerzeugung und deren Auswirkung auf den Schiffsbetrieb bei Schiffen kleinerer Leistung. *Jahrb. Schiffbautech. Ges.* **82**

Lamb H 1945 *Hydrodynamics*. Dover, New York

Langelaar G, Wesselink A F 1983 Sense and nonsense in manoeuvring automation with single lever control systems. *5th LIPS Propeller Symp.* LIPS, Drunen, The Netherlands

Lough A 1985 Dynamic positioning. *Schip en Werf* **56**(7)

Van den Houten E 1989 The development of a voyage management system for fuel-efficient crossings. *7th LIPS Propeller Symp.* LIPS, Drunen, The Netherlands

A. Wesselink
[LIPS, Drunen, The Netherlands]

Navigation Systems, Integrated

For many years, different systems for navigation have been in operation in numerous vehicles. Modern transport aircraft, for example, are equipped with vhf omnidirectional range (VOR) receivers, distance measuring equipment (DME) interrogators and inertial navigation systems (INSs). In the case of vessels, radio navigation systems such as Decca, Omega and Loran C, as well as dead-reckoning systems which are based on velocity and heading, are used.

In general, only one of the different navigation systems on board a vehicle has been used as the active navigation system for a certain route. In aviation, for example, VOR navigation is applied in most cases for flying air routes over continents. In order to increase navigational accuracy and to improve the availability of navigation results, an integration of different navigation systems on board a vehicle can be carried out.

1. Principles

The position of a vehicle can be determined using the measurements from one or two navigation systems (e.g., VOR and DME). If there are additional systems, redundant measurements may be obtained. With a suitable adjustment procedure the most probable position can be determined. For example, the method of least-squares adjustment, developed by Gauss, can be used. However, this method is not normally used for the integration of different navigation systems; it is only used for special applications.

Normally, the term integrated navigation system means the combination of an autonomous on-board navigation system and a radio navigation system. Autonomous on-board systems are dead-reckoning systems such as Doppler navigation systems, INSs and simple systems which only use the heading and the velocity of the vehicle. The advantage of the integration of a dead-reckoning system and a radio navigation system is that systems with totally different error behavior are combined. The short-time accuracy of dead-reckoning systems is of a very high quality, whereas the errors of all radio navigation systems contain a systematic part which is superimposed by a relative high-frequency noise. Figure 1 clearly shows the different

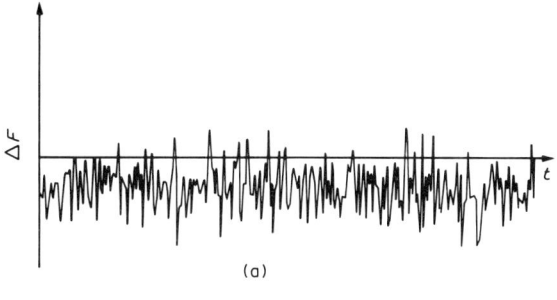
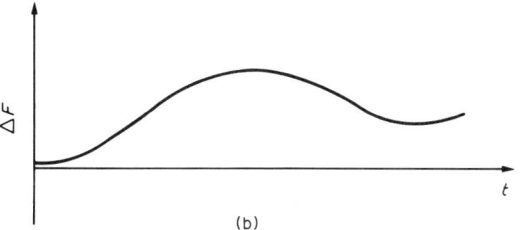

Figure 1
Error behavior of (a) VOR radio navigation, and (b) dead-reckoning INS

error behavior of a VOR radio navigation system and a dead-reckoning INS.

Using modern INSs, navigation accuracies can be achieved that are below 1 nautical mile per hour mission time. Similar accuracies are possible with Doppler navigation systems. Nevertheless, after a certain time, the navigation errors may reach a value that is too large for the mission in question. A solution for this problem can be obtained by aiding the dead-reckoning systems; that is, by using an integrated navigation system. This means that, in order to correct the dead-reckoning system, a comparison with other navigation systems is carried out. Therefore, the navigation error can be limited. Conventional methods are available for this process of comparing and carrying out corrections. The technique of Kalman filtering has been used increasingly to obtain the best correction for dead-reckoning systems. This technique is a further development of the method of least-squares adjustment proposed by Gauss.

Not only is the navigation accuracy of integrated navigation systems high, but the availability is also increased. When radio navigation systems are used, the navigation is sometimes interrupted because of propagation problems. In such cases, the navigation can be continued by navigating with the calibrated dead-reckoning system.

The development of integrated navigation systems has been pushed due to the availability of small and powerful computers. Because prices for computer hardware have decreased continuously, it can be assumed that, in the near future, many vehicles will be equipped with integrated navigation systems.

2. Kalman Filter

There are two formulations of the Kalman filter: time-continuous equations and time-discrete equations. For navigation applications, the time-discrete formulation of the Kalman filter is generally of importance.

Concerning the discrete formulation of the Kalman filter, certain equations have to be solved periodically (e.g., every minute). First, the prediction of the state vector from the previous iteration time to the present time instant has to be carried out. The following equation must be used:

$$x'_k = \phi_{k-1} x_{k-1} \qquad (1)$$

where x'_k and x_{k-1} are the state vectors for the time instants k and $k-1$, respectively. In connection with navigation systems, the state vector mainly contains elements representing the errors of a dead-reckoning system. With respect to an INS there are, for example, position errors, velocity errors and misalignment angles. The matrix ϕ represents the dynamic error model and can be derived from the system equations

$$\dot{x} = Ax + w \qquad (2)$$

in which A is the system matrix and the vector w contains noise inputs. This differential equation describes, for example, which position errors are built up in an INS as a result of misalignment errors of the gyros and noise contained in the sensor signals (i.e., in the accelerometers and gyros).

The next step of the discrete Kalman filter is the inclusion of measurements in order to correct the state vector. They are taken into account in the following manner:

$$x_k = x'_k + K_k(z_k - H_k x') \qquad (3)$$

where x_k and x'_k are state vectors with and without the inclusion of the measurements obtained for the time instant k, respectively. K_k is the gain matrix. The vector z_k contains the measurements; for example, the difference between the navigation results of an INS and a radio navigation system. H_k is the measurement matrix. This matrix relates the measurements to the corresponding elements of the state vector.

Besides the computation of the state vector, the covariance matrix P has to be calculated. This matrix gives an indication of the estimation accuracy that is obtained for the state vector. The diagonal elements of such a matrix are the squares of the standard deviations. Thus, these elements must always be positive. The off-diagonal elements are the cross covariances which describe the correlation between the individual errors. Concerning this matrix, a prediction in time is also necessary:

$$P'_k = \phi_{k-1} P_{k-1} \phi^T_{k-1} + Q_{k-1} \qquad (4)$$

where $\mathbf{P}_{k-1}$ is the covariance matrix for the time instant $k-1$ and $\mathbf{P}'_k$ for the time instant k without the inclusion of the measurements that are available at this time. The superscript T denotes the matrix transpose. The matrix $\mathbf{Q}$ takes into account noise inputs and can be derived from Eqn. (2).

In order to include the measurements as in Eqn. (3), the calculation of the gain matrix is necessary:

$$\mathbf{K}_k = \mathbf{P}'_k \mathbf{H}_k (\mathbf{H}_k \mathbf{P}'_k \mathbf{H}_k^T + \mathbf{R}_k)^{-1} \quad (5)$$

Matrix $\mathbf{R}$ is also a covariance matrix that gives information about the measurement noise. This information is important in order to use each measurement with a certain weight.

The following equation is necessary in order to include the last measurement with respect to the covariance matrix $\mathbf{P}$:

$$\mathbf{P}_k = (\mathbf{I} - \mathbf{K}_k \mathbf{H}_k) \mathbf{P}'_k \quad (6)$$

where $\mathbf{I}$ is the unit matrix and $\mathbf{P}'_k$ is the covariance matrix representing the estimation accuracy immediately before inclusion of the measurements. It can be seen from Eqn. (6) that each measurement increases the estimation accuracy.

At the beginning of the Kalman filter computations, some initial values have to be set. This concerns the state vector and the covariance matrix $\mathbf{P}$. Sometimes no measurements are available. In such cases, only the prediction of the state vector x and of the covariance matrix have to be carried out (Eqns. (2) and (4)).

3. A Simple System

A simple integrated system for aircraft navigation will be considered. This system consists of a simple dead-reckoning system that is aided by DME distance measurements.

The dead-reckoning system uses heading, true air speed (TAS) and the wind. Due to errors, the flight path determined by the dead-reckoning system will deviate from the true flight path. Therefore, this system is aided by DME distance measurements to different ground stations. The DME data are measured by one DME interrogator which is constantly switched over to frequencies of five different ground stations (multiple DME).

The Kalman filter used for the integration enables the estimation of DME biases that refer to the individual ground stations and to the DME interrogator. Because of the combination of TAS, heading and ground data (DME), it is also possible to estimate the wind components.

The position errors of the dead-reckoning system are caused by

(a) uncertain determination of TAS,

(b) uncertain heading and side slip angle, and

(c) unknown wind conditions.

Taking into account these error sources and also the biases of the DME ground stations, the following system equation is obtained:

$$\begin{bmatrix} \dot{f}_x \\ \dot{f}_y \\ \dot{k} \\ \Delta\dot{\theta} \\ \dot{W}_x \\ \dot{W}_y \\ \Delta\dot{D} \end{bmatrix} = \begin{bmatrix} 0 & 0 & v_x & v_y & 1 & 0 & \mathbf{0}^T \\ 0 & 0 & v_y & -v_x & 0 & 1 & \mathbf{0}^T \\ 0 & 0 & 0 & 0 & 0 & 0 & \mathbf{0}^T \\ 0 & 0 & 0 & 0 & 0 & 0 & \mathbf{0}^T \\ 0 & 0 & 0 & 0 & 0 & 0 & \mathbf{0}^T \\ 0 & 0 & 0 & 0 & 0 & 0 & \mathbf{0}^T \\ \mathbf{0} & \mathbf{0} & \mathbf{0} & \mathbf{0} & \mathbf{0} & \mathbf{0} & \mathbf{0} \end{bmatrix} \begin{bmatrix} f_x \\ f_y \\ k \\ \Delta\theta \\ W_x \\ W_y \\ \Delta D \end{bmatrix}$$

$$+ \begin{bmatrix} v_{rx} \\ v_{ry} \\ 0 \\ \theta_r \\ W_{rx} \\ W_{ry} \\ 0 \end{bmatrix} \quad (7)$$

The state elements f_x and f_y are the position errors in eastern and northern directions. For simplicity, the model is shown in an x–y coordinate system. k represents the scale factor error of TAS, $\Delta\theta$ the heading error, and W_x and W_y the wind velocity components in eastern and northern directions. ΔD is a subvector comprising the systematic errors (biases) of five DME ground stations.

The system matrix contains the velocity components v_x and v_y calculated from TAS and heading.

Random errors, such as turbulences, influence v_x and v_y, and are represented by the noise inputs v_{rx} and v_{ry}. The heading error has been modelled as a random-walk process. The power density of θ_r can be chosen in such a way that a certain heading drift may occur (e.g., 3 degrees h^{-1}). From Eqn. (7), it can be seen that the wind velocities are also modelled as random walks. A suitable power density for W_{rx} and W_{ry} may cause the wind components to change by 4 m s^{-1} in about 10 min.

The transition matrix ϕ and the covariance matrix $\mathbf{Q}$ for a certain time interval can be calculated from Eqn. (7). Different methods for doing this are available.

The covariance matrix of the measurement noise is one-dimensional and, thus, equals the standard deviation ($\mathbf{R} = \sigma_D$). There is a large number of DME data available which allows the determination of σ_D. A typical value for σ_D is 100 m. Tests of different organizations show that the standard deviation of the systematic error (bias error) of different DME ground stations is approximately 150 m.

As mentioned previously, initial values have to be set for the state vector elements as well as for their covariances. The values for the covariances may be chosen according to Table 1.

In order to consider the measurements, the matrix **H** has to be determined. In the case of a one-dimensional range measurement, the matrix reads

$$\mathbf{H} = \underbrace{\frac{\partial D}{\partial x}, \frac{\partial D}{\partial y}}_{\text{position}}, 0, 0, 0, 0, 1, 0, 0, 0, 0 \quad (8)$$
$$\quad\quad\quad 1.\,2.\,3.\,4.\,5.$$

A measurement is the difference between the measured DME distance and the distance calculated from the position generated by the dead-reckoning system and from the position of the respective DME ground station.

The first two elements of the measurement matrix are the derivatives of the DME distance D between the aircraft and the ground station that has just been used in eastern (x) and northern (y) directions. On the right-hand side of Eqn. (8), the 1 takes into account the systematic error of a ground station. This coefficient is an approximation of $1/\cos\gamma$. Therefore, Eqn. (8) is only valid for small elevation angles γ from the DME ground station to the aircraft. The position of the coefficient 1 in the measurement matrix changes according to the ground station (i.e., first position for first ground station, etc.).

Whenever an aircraft leaves the covering range of a ground station, a new one is selected. The elements of the state vector and the covariance matrix corresponding to the new station have to be reinitialized. The cycle time of the Kalman filter is chosen to be 2 s. Therefore, the DME station in use also has to be changed every 2 s.

In the case of this simple dead-reckoning system, the mechanization of the Kalman filter as a closed-loop filter has some advantages. In such a filter, the results of the estimation process are immediately used by the dead-reckoning system. One advantage is that no great position error will occur due, for example, to a strong wind. This aspect is also important with respect to the linearization of the measurement equation, especially concerning the first two elements.

In this case of time periods without DME observations, the navigation procedure continues by using the well-calibrated dead-reckoning system including the last estimate of the wind.

When the position of the aircraft is near a ground

Table 1
Initial values for the covariances

State vector element	Standard deviation
Position error f_x, f_y (m)	5000
Scale factor error k	0.03
Heading error $\Delta\theta$ (°)	3
Wind velocity W_x, W_y (m s^{-1})	40
DME bias error ΔD (m)	150

Table 2
Gyro classes

	Class		
	A	B	C
Bias drift (degrees h^{-1})	0.01	0.1	1
Random-walk drift (degrees h^{-1})	0.01	0.1	1
Acceleration-dependent drift (degrees h^{-1} g^{-1})	2	10	50

station, position errors have great influence on the first two elements of the measurement matrix. If the elevation angle between the ground station and the aircraft is greater than 10°, the measurements to that station are omitted.

This integrated navigation system has some important advantages.

(a) The position accuracy is better than 100 m (1σ). This is a great improvement in comparison with VOR/DME navigation which is used for short-range navigation.

(b) The navigation is not interrupted, if no DME signals are available.

(c) The wind is continuously estimated and, thus, available on board the aircraft.

4. Example Systems

Most modern aircraft are equipped with INSs. These systems can be aided by DME measurements in the same manner as described for the simple dead-reckoning system in Sect. 3. Special DME interrogators are used for this purpose. They are able to measure the distance of up to five different ground stations simultaneously. The position accuracy of such an integrated navigation system is approximately 30–50 m (1σ).

The availability of satellite navigation systems such as the Navstar global positioning system (GPS) opens the way to use a radio navigation system worldwide. For many reasons, it is sensible to combine such a system with an INS. However, the position accuracy of such an integrated system can only be slightly improved compared to pure satellite navigation. However, it must be considered that the velocity errors as well as the angle misalignment errors are also estimated. In connection with this, the question of which requirements concerning the accuracy of the INS are necessary is interesting. In order to answer this question, the accuracies of three classes (A–C) of INSs that are aided by GPS measurements are presented in Table 2. The three classes are characterized by the error behavior of the gyros. Rough estimates of the velocity errors as well as of the misalignment angles that are obtained with the three different systems are summarized in Table 3. In

Table 3
Errors of GPS-aided INS (1σ)

INS class	Velocity (m s^{-1})	Altitude (arc min)	Heading (arc min)
A	0.1	0.3	0.6
B	0.2	1.0	2.5
C	0.4	5.0	20.0

many cases, the less expensive INSs (systems B and C) are sufficient. The values given in Table 3 are results from computer simulations.

See also: Data Processing in Air Traffic Control

Bibliography

Brokof U, Hurrass K 1982 Simple integrated navigation systems. *Proc. Int. Congr. Institutes of Navigation.* Institut Francais de Navigation, Paris
Gelb A (ed.) 1974 *Applied Optimal Estimation.* MIT Press, Cambridge, MA
Kayton M, Fried K 1969 *Avionics Navigation Systems.* Wiley, New York
Latham R W 1974 Aircraft positioning with multiple DME. *J. Inst. Navig.* **21**(2)
Latham R W, Townes R S 1975 DME errors. *J. Inst. Navig.* **22**(4)
Rawlings R C 1981 The flight assessment and applications of DME/DME. *J. Navig.* **34**(1)

K. Hurrass
[Institut für Flugführung, Braunschweig, Germany]

Network Modelling and Control: Store-and-Forward Approach

Store-and-forward operation is very common in communication and transportation networks. It occurs when direct movement of information or goods is either unwanted or impossible, due to temporary unavailability of a link in a network. The latter case is particularly common in transportation networks during periods of congestion. Examples are traffic networks during periods of "rush hour," and air traffic networks during periods of high arrival and departure rates and/or reduced runway capacity due to inclement weather. The store-and-forward operation of transportation networks sets in by itself during periods of congestion, but in an unchecked way which generally produces undesirable consequences, such as excessive delays, "gridlock," waste of fuel, and safety exposures. Intervention by a controlling authority aims at mitigating if not altogether eliminating any combinations of such penalties of congestion, reflected into suitable objective functions. For example, unchecked adherence to airline departure schedules generated, for many years, excessive "stacking" of aircraft over major airports during periods of congestion. The desire for an improvement became a necessity in 1980 because of the airport controllers' strike in the USA. The resulting practice of holding the aircraft on the ground until a landing slot is available at their destination has contributed to fuel savings as well as operational safety. Still more could be achieved through proper scheduling and rescheduling of aircraft movements across the airspace and airport network in order to decrease the overall delay and inconvenience to air travellers. Similarly, the emerging applications of intelligent vehicle highway systems (IVHS) will require improved management of the store-and-forward operation of congested traffic systems.

1. Basic Formulation

Since the mid-1960s, the modelling and optimal control of congested transportation systems has been the subject of many studies, with particular attention paid to arterial networks of automobile traffic. A general characterization of such networks has been given by Gazis (1974a) and is shown in Fig. 1. The networks consist of arcs which handle time-dependent origin–destination requirements between certain nodes. Traffic streams associated with different origin–destination pairs may be viewed as different commodities, since elements of these streams are not interchangeable. There is a fixed travel time associated with each arc, as well as a capacity constraint. In addition, there is a storage capacity associated with each arc, which may be assumed as located near the end of the arc. Multiple such storage "bins" may be assumed in cases where

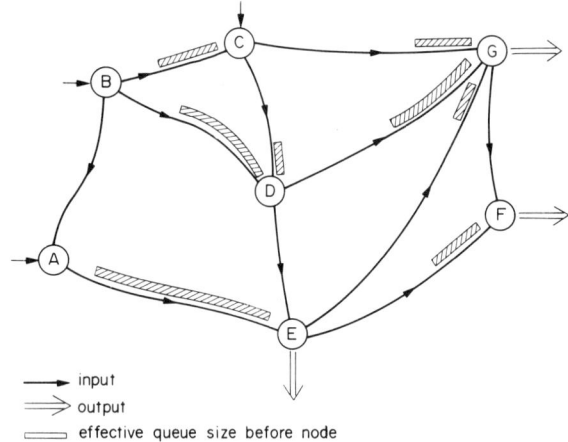

Figure 1
Schematic diagram of a transportation store-and-forward network (after Gazis 1974a)

traffic is separated into different directional movements. Traffic is assumed to travel at constant speed from node to node and then stored at the end of the arc until it is allowed to proceed to its destination. Service of the queues in the bins is first in first out (FIFO). There is usually sufficient storage capacity between bottlenecks to warrant an assumption of unlimited storage capacity. However, constraints on this storage capacity have been explicitly treated by Michalopoulos and Stephanopoulos (1977).

The optimal control of the network of Fig. 1 requires, in general, two operations:

(a) the (time-dependent) optimum allocation of a route to each unit of traffic from its origin to its destination; and

(b) the optimum "switching" at the nodes, determining the rate of discharge of the various queues (switching consists of allocating a certain percentage of time for discharging each of the queues at the node, in a way that satisfies the flow capacity restraints of the arcs downstream of the node).

It should be pointed out that the nodes of such networks are not to be identified with intersections of arterial networks which are usually the constraining elements of the network. The nodes of the network in Fig. 1 have infinite capacity and the traffic flow constraints are provided by the arc capacity limits. However, as pointed out by Gazis (1974a), the arterial intersection is subsumed in the above network definition. This can be seen in the case of an oversaturated intersection which is the simplest case of a store-and-forward operation of a congested system (Gazis and Potts 1965). The oversaturated intersection is one in which the cumulative demand during a rush period exceeds the possible cumulative service. Each leg of the intersection is associated with a saturation flow, which is the maximum flow obtainable per unit of green time. It is assumed that the flow across the intersection is equal to this saturation flow, after an appropriate start-up portion of the green phase, as long as there is a residual queue waiting to be served. The oversaturated intersection can be viewed as a network of the type shown in Fig. 1 if dummy links are added before the intersection to account for the saturation flow limits in the two competing directions (see Fig. 2). The dummy links have no storge capability, and capacity limits s_1 and s_2, where s_2 is less than s_1. Switching can be viewed as taking place either at node A or at node B, allowing alternating discharge of the two queues during percentages of times which roughly add up to 100% (minus a small percentage allowed for clearance of traffic before switching service from one queue to the other). The treatment of the optimal control of the oversaturated intersection is presented here as a representative methodology for optimal control of congested store-and-forward traffic networks.

2. The Oversaturated Intersection

Let x_1 and x_2 denote the queues along directions 1 and 2 of the intersection shown in Fig. 2a, $q_1(t)$ and $q_2(t)$ the time-dependent inputs, g_1 and g_2 the green times allocated to the two directions during a cycle c and s_1 and the s_2 the saturation flows. Then, x_1 and x_2 satisfy the differential equations

$$\left. \begin{array}{l} \dfrac{dx_1}{dt} = q_1(t) - \dfrac{s_1 g_1}{c} \\ \dfrac{dx_2}{dt} = q_2(t) - \dfrac{s_2 g_2}{c} \end{array} \right\} \quad (1)$$

It is further assumed that x_i is greater than zero as a result of oversaturation. The objective of optimal control is to minimize the delay (D) to the users of the intersection, defined by

$$D = \int_0^T (x_1 + x_2) dt \quad (2)$$

where T is the duration of the rush period, which is also minimized if both queues are exhausted at the same time. This defines the end conditions

$$x_1(T) = x_2(T) = 0 \quad (3)$$

The green times g_1 and g_2 are the control variables, which must satisfy the condition

$$g_1 + g_2 = c - L \quad (4)$$

where L is the total lost time during every cycle in order to clear the intersection after each green phase. In addition, g_i must be within certain bounds dictated by operational considerations, namely,

$$g_{min} \leq g_i \leq g_{max} \quad (5)$$

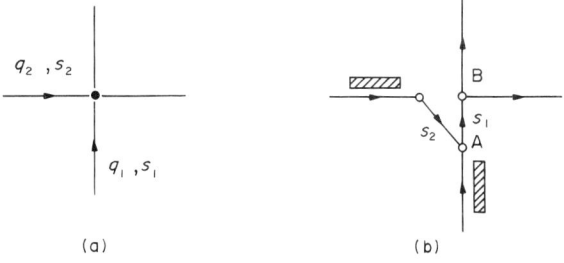

Figure 2
(a) Oversaturated intersection with time-dependent demands q_1 and q_2 exceeding capacity and corresponding saturation flows s_1 and s_2 and
(b) representation of the single oversaturated intersection as a store-and-forward network (after Gazis 1974a)

The above formulation casts the optimization problem in the framework of control theory. It can be solved by applying Pontryagin's maximum principle as was shown by Gazis (1964). The solution generally corresponds to values of the control variables that lie at the apexes of the admissible control domain. In the case of the oversaturated intersection, the optimal solution is a "bang-bang" solution. It consists of giving the maximum allowable green phase to the stream of traffic associated with the largest saturation flow and then switching, at an appropriate time, to maximum green phase for the stream associated with the smallest saturation flow. This control maximizes the aggregate cumulative service to the users of the intersection. This observation can be used as a guide for optimal control of other oversaturated systems, so long as the primary objective in their operation is to minimize the delay to the users.

Knowledge of the bang-bang nature of the solution permits the following approach to finding the optimal switching strategy using linear programming, also allowing a generalization for more complex systems.

Following Gazis (1974a), the obervation is used that the asymptotic behavior of the cumulative demand functions for the two competing streams is generally sufficient for optimization, and the functions are assumed to be given by

$$Q_i = \int_0^t q_i(z)dz = A_i + B_i t \quad (i = 1, 2) \quad (6)$$

as shown in Fig. 3. The aim is to minimize the delay D given by

$$D = \frac{1}{2}[(A_1 + A_2)\tau + (a_1 + a_2)T] \quad (7)$$

where a_1, a_2 and τ are three independent variables and T is a parameter of a parametric linear programming (PLP) problem. The green phases can be expressed as functions of these four quantities. Then, the constraints on the green phases, Eqns. (4) and (5), yield a feasibility domain for the PLP problem. The minimum T for which such a feasibility domain exists is the earliest end of the rush period. Thus, the optimal solution is obtained corresponding to the values

$$T = \frac{A_1 s_2 + A_2 s_1}{s_1 s_2 (1 - L/c) - (B_1 s_2 + B_2 s_1)} \quad (8)$$

$$\tau = \frac{(A_1 z - A_2 w) + u_{\min}(\lambda A_1 + a_2)}{(\lambda w + z)(u_{\max} - u_{\min})} \quad (9)$$

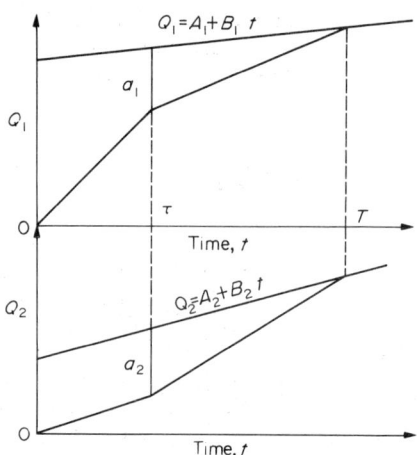

Figure 3
Linear programming solution of the problem of the oversaturated intersection (after Gazis 1974a)

where

$$\left.\begin{aligned} w &= b_1 \\ z &= b_2 - s_2\left(1 - \frac{L}{c}\right) \\ \lambda &= \frac{s_2}{s_1} \\ (u_{\min}; u_{\max}) &= \frac{s_1}{c}(g_{\min}; g_{\max}) \end{aligned}\right\} \quad (10)$$

The preceding discussion brings out the following key points regarding the optimal operation of congested store-and-forward networks.

(a) The optimal solution corresponds to a cumulative service which is a concave function of time. Several service patterns are maintained for various lengths of time, in a way that exhausts all queues at the same time. The sequence of these patterns is in a decreasing order of the corresponding throughput of the network, producing a concave cumulative service function.

(b) Traffic fluctuations on a time scale in comparison with the duration of the rush period, affect the optimal solution only to the extent that they influence the asymptotic behavior of the cumulative demand curves. It should be pointed out, however, that if such short-term fluctuations produce a dip in a cumulative demand which temporarily empties a particular queue before the end of the overall rush period, adjustments to the above solution are required for optimality. Whether or not such adjustments are appropriate will depend on the size of the dip in demand.

3. Systems of Intersections

The preceding observations can be used to devise a heuristic approach to the optimal control of an oversaturated store-and-forward network. The approach consists of constructing a cumulative service curve which is a concave function of time, exhausts all the queues simultaneously at the end of the rush period and consists of a sequence of sets of the control variable corresponding to the apexes of the admissible control domain. This approach can be illustrated by considering the system of three oversaturated intersections discussed by Gazis (1974a) and shown in Fig. 4.

Assume that traffic moves along the directions 1 to 3 without turning movements and that the cumulative demands are given by the equations

$$\left.\begin{array}{l} Q_1 = 1800 + 500t \\ Q_2 = 1400 + 300t \\ Q_3 = 1300 + 300t \end{array}\right\} \quad (11)$$

where the time t is given in hours. Also, assume that the saturation flows are $s_1 = 3600$, $s_2 = 2400$, and $s_3 = 2000$ vehicles per hour. The maximum green phase is assumed to be 60 s and the minimum 20 s, with 10 s allowed per cycle for clearing the intersection after switching the lights.

The admissible control domain is found to be the domain bounded by the hexahedral surface shown in Fig. 5. The coordinate axes γ_i of this figure are the fractions of the light cycle allocated to the green phases of the three directions. It is tacitly assumed that a common cycle c is used at all three intersections, hence

$$\gamma_i = \frac{g_i}{(c - L)} \quad (12)$$

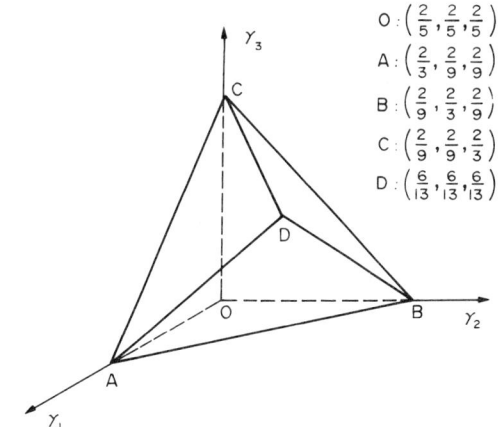

Figure 5
Admissible control domain for the system of intersections of Fig. 4: the variables γ_i are the fractions of each cycle allocated to the three traffic streams (after Gazis 1974a)

where g_i is the usable green phase for the ith stream and L is the total lost time for clearing the intersection after switching the light. For example, the apex A of Fig. 5 corresponds to $g_1 = 60$ s and $g_2 = g_3 = 20$ s. The usable green phase for the streams 2 and 3, corresponding to this apex, is limited to that allocated to these streams where they intersect stream 1. The intersection of streams 2 and 3 is not a bottleneck, and the specific allocation of green phases at that intersection has no influence on the performance of the intersection.

It may be easily ascertained that during the rush period at least one of the green phases must be equal to the maximum, in order to decrease the lost time for clearance. As a result, only the apexes A, B, C, and D need be considered in seeking a solution. Table 1 shows the service rates for the three streams corresponding to these four apexes, listed in order of decreasing total throughput of the system of intersections. The light settings corresponding to the four apexes are assumed

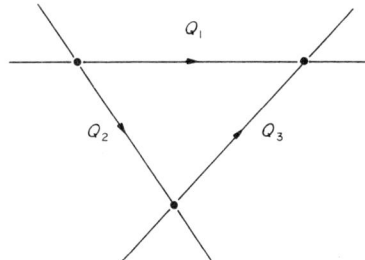

Figure 4
System of three interacting oversaturated intersections (after Gazis 1974a)

Table 1
Service rates corresponding to the apexes of the control domain for the system of intersections shown in Fig. 4

Apex		Stream 1	Stream 2	Stream 3	Total
D	green phase	60	60	60	$c = 130$
	service rate	1660	1110	920	3690
A	green phase	60	20	20	$c = 90$
	service rate	2400	530	440	2970
B	green phase	20	60	20	$c = 90$
	service rate	800	1600	440	2840
C	green phase	20	20	60	$c = 90$
	service rate	800	530	1330	2660

to be maintained for times T_A, T_B, T_C and T_D, and it is required that all queues be simultaneously exhausted at time $T_A + T_B + T_C + T_D$. Furthermore, to ensure a concave service curve, the sequence D, A, B, C was chosen as the optimal one. The optimal solution can now be found by solving the following linear programming (LP) problem.
Minimize

$$T = T_A + T_B + T_C + T_D \qquad (13)$$

subject to

$$A_{ij} x_j = b_i + c_i T \qquad (14)$$

where

$$A = \begin{bmatrix} 2400 & 800 & 800 & 1660 \\ 530 & 1600 & 530 & 1110 \\ 440 & 440 & 1330 & 920 \end{bmatrix}$$

$$b = \{1800, 1400, 1300\}$$

$$c = \{500, 300, 300\}$$

$$(15)$$

and the unknowns x_j satisfy the nonnegativity constraint

$$x_j \geq 0 \qquad (16)$$

The solution to this LP problem is

$$T_A = 0$$
$$T_B = 0.124 \text{ h}$$
$$T_C = 0.394 \text{ h}$$
$$T_D = 1.43 \text{ h}$$

It should be noted that an LP formalism can be used without knowledge of the exact form of the admissible control domain and, hence, the possible number of distinct control periods and corresponding service rates. The time axis may simply be divided into (equal or unequal) segments and the queue sizes on various links can be used as the unknowns of the LP problem, the objective being set at minimizing the overall delay to the users of the system. The solution to this LP problem is a suboptimal one, inscribed within the optimal solution and approaching this solution as the time segments are decreased. This approach will now be used to suggest an algorithm for the solution to the optimization problem of complex transportation networks.

4. Generalization to Complex Transportation Networks

Consider a network such as that shown in Fig. 1. The optimal control of such a network during periods of congestion entails both route control and queue management through switching at the nodes. The general solution to this optimization problem is rather complex, particularly for multicommodity networks. Here, the simplifying assumption is made that the route for each unit of traffic is predetermined and not under our control. In practice, several sets of such route choices can be assumed and the one corresponding to the least delay can be applied. It is further assumed that the various commodities are distributed roughly uniformly within each queue and so the requirement of FIFO service need not be considered explicitly. Finally, it is assumed that the predetermined paths of traffic can be roughly translated into percentages of turning movements which are expressed as functions of time. Then, the optimal control problem is formulated as follows.

Let $i, j, k, \ldots$ denote the nodes of the network and let

(ij) = arc between nodes i and j

$x_{ij}(t)$ = queue size along (ij) just in front of node j

$p_{ijk}(t)$ = percentage of discharge of queue x_{ij} proceeding towards k

$u_{ij}(t)$ = discharge rate of queue x_{ij}

r_{ij} = travel time along (ij)

Since the path of each traffic unit is predetermined, so is its minimum transit time, which is simply the sum of the travel times along the arcs of its path. It is then convenient to ignore this minimum transit time in the optimization problem and simply minimize the time spent while waiting in the queues before the nodes. Thus, the objective is to minimize, over the duration T of the rush period, the function

$$J = \int_0^T \sum_{i,j} x_{ij}(t) \, dt \qquad (17)$$

where the x_{ij} satisfy the state equations

$$\dot{x}_{ij}(t) = \sum_l p_{lij}(t - \tau_{ij}) u_{li}(t - \tau_{ij}) - u_{ij}(t) \qquad (18)$$

where l denotes the set of nodes discharging through i towards j. The control variables u_{ij} and state variables x_{ij} must, in general, satisfy the constraints

$$\sum_l p_{lij} u_{li} \leq c_{ij}, \qquad 0 \leq x_{ij} \leq X_{ij} \qquad (19)$$

where c_{ij} is the traffic flow capacity limit of arc (ij) and X_{ij} the storge capability along arc (ij) in front of node j. For entrance nodes, the sum over l of feed-forward traffic in Eqn. (19) is replaced by given input rates $q_i(t)$.

A discrete LP formulation of the above optimal control problem may be obtained by dividing a suitably long time period T, assumed to equal or exceed the

duration of the rush period, into n intervals δ, and using a difference approximation of the integral of Eqn. (17). The resulting formulation is as follows.
Minimize

$$J = \sum_{i,j,n} \frac{1}{2}(x_{ij,n} + x_{ij,n+1})\delta \qquad (20)$$

subject to

$$\frac{1}{\delta}(x_{ij,n+1} - x_{ij,n}) = \sum_{l} p_{lij,(n-s_{ij})} u_{li,(n-s_{ij})} - u_{ij,n} \qquad (21)$$

$$\sum_{l} p_{li,n} u_{li,n} \leq c_{ij}, \quad 0 \leq x_{ij,n} \leq X_{ij}, \quad u_{ij,n} \geq 0 \qquad (22)$$

where $s_{ij} = \tau_{ij}/\delta$ are assumed to be integers for all arcs (ij). In the above formulation, both $x_{ij,n}$ and $u_{ij,n}$ are considered as independent variables of the LP minimization of the objective function J, and the quantities $p_{lij,n}$ are constants, to be chosen by an appropriate scheme descending towards decreasing values of the objective function. The positivity constraints on $u_{ij,n}$ have been added in order to cast the problem in a standard LP form, although they may be redundant in practical cases.

A network that has only one origin and destination is a single commodity network, permitting the explicit treatment of turning movements as well as the rates of queue discharge within an LP framework. If control variables are defined as the quantities $u_{ijk}(t)$, equal to the rate of discharge from queue x_{ij} towards k, the problem is formulated as follows.
Minimize

$$J = \int_0^T \sum_{i,j} \left[x_{ij}(t) + \sum_k u_{ijl}(t) \right] dt \qquad (23)$$

subject to

$$\dot{x}_{ij}(t) = \sum_l u_{lij}(t - \tau_{lj}) - \sum_k u_{ijk}(t) \qquad (24)$$

$$\sum_l u_{lij} \leq c_{ij}, \quad 0 \leq x_{ij} \leq X_{ij} \qquad (25)$$

Discretization of Eqns. (23)–(25) leads again to an LP problem.

5. The IVHS as a Store-and-Forward System

An IVHS is intended to provide motorists with detailed instructions that will guide them from their origin to their destination with a minimum of inconvenience. To accomplish this, an IVHS must have detailed information about the intended destinations of the motorists, the current status of the system and reasonably accurate projections on expected demand within a control horizon. None of these three ingredients is easily available in present systems. In particular, the most ambitious projections for IVHS development do not anticipate an appreciable number of vehicles properly equipped for perfect communication with the system. As a result, it is likely that information about intended traffic movements will continue to rely on statistical information and sampling of actual inputs obtained from those vehicles that are properly instrumented. Traffic streams will continue to be controlled as continua, and individual instructions to motorists will be based on such control. The preceding formulation of store-and-forward networks can be used to achieve this type of control. It can be applied even in the cases when vehicles are passive targets detectable only through system sensors, but it will benefit greatly from additional information obtained from transponder-equipped vehicles (see *Route Guidance, Individual*).

In future IVHS implementations, the majority of vehicles can be expected to communicate with the system. In this case, a modification of the store-and-forward formalism will be appropriate, which, in effect, adds an additional discretization. Instead of using a continuous variable for a queue, the queue will be composed of individual units. In addition to increasing the complexity of the problem and the concomitant information processing requirements, such an operation will impose considerable requirements of data communication. However, the continued progress in computer and communications technology makes the prospect of a cost-effective implementation of such systems eminently feasible in the not too distant future.

See also: Optimal Routing Applications; Queuing Theory Applications

Bibliography

D'Ans G C, Gazis D C 1976 Optimal control of oversaturated store-and-forward transportation networks. *Transp. Sci.* **10**, 1–19

Gazis D C 1964 Optimum control of a system of oversaturated intersections. *Oper. Res.* **12**, 815–31

Gazis D C 1974a Modeling and optimal control of congested transportation systems. *Networks* **4**, 113–24

Gazis D C 1974b Traffic control—Theory and application. In: Gazis D C (ed.) *Traffic Science*. Wiley, New York, pp. 175–239

Gazis D C, Potts R B 1965 The oversaturated intersection. *Proc. 2nd Int. Symp. Theory of Road Traffic Flow*. Organization for Economic Cooperation and Development, Paris, pp. 221–37

Green D H 1968 Control of oversaturated intersections. *Oper. Res. Q.* **18**, 161–73

Kaltenbach M, Koivo H N 1974 Modelling and control of urban traffic flow. *Proc. Joint Automatic Control Conf.* University of Texas, Houston, TX, pp. 147–54

Michalopoulos P G, Stephanopoulos G 1977 Oversaturated signal systems with queue constraints: I, Single intersection; II, Systems of intersections. *Transp. Res.* **11**, 413–28

Pontryagin L S, Boltyanski V G, Gamkrelidze R V, Mishchenko E F 1962 *The Mathematical Theory of Optimal Processes* (translated by Trigoroff K N). Wiley, New York

D. C. Gazis
[IBM T. J. Watson Research Center, Yorktown Heights, New York, USA]

On-Ramp Control: Coordinated Time-of-Day Strategies

Ramp control can be defined as a method of improving overall freeway operations by limiting, regulating and timing the entrance of vehicles from one or more ramps onto the main line. In locations where freeway entrance ramps have adequate storage capacity or where the surrounding street network can accommodate additional traffic, ramp-control systems can provide substantial operational improvements under certain combinations of traffic demand and freeway capacity (Blumentritt *et al.* 1981). In order to carry out such effective ramp control, it is necessary to have an appropriate control strategy, coordinating control operations at related ramps under fluctuating traffic demand from time to time. This is the time-of-day strategy for coordinated ramp control.

1. Warrants for Ramp-Metering Implementation

Although on-ramp control is applicable to most freeways, it should not automatically be assumed that on-ramp control will be desirable and feasible for all freeways. Installation of freeway ramp-control signals may be warranted when the following occur.

(a) The expected reduction in delay to freeway traffic exceeds the expected delay to ramp users and the added travel time for diverted traffic and traffic on the alternative routes.

(b) There is adequate storage space for the vehicles that will be delayed.

(c) There are suitable alternative surface routes available having capacity for traffic diverted from the freeway ramps, and either:
  (i) there is recurring congestion on the freeway due to traffic demand in excess of the capacity, or
  (ii) there is recurring congestion or a severe accident hazard at the freeway on-ramp because of inadequate ramp-merging area.

2. Control Modes

On-ramp control modes can be organized into three basic categories: pretimed-control mode, local-actuated-control mode, and system-control mode (see *Traffic Control Modes*).

2.1 Pretimed-Control Mode

Any form of entrance-ramp metering that is not directly influenced by mainline traffic is called pretimed control. This does not imply the absence of vehicle detectors. In many cases, both demand and passage detectors are used to actuate and terminate each metering cycle. The individual metering rates used with pretimed control are wholly a function of past traffic observations, which may include origin–destination studies. When the set of rates has been established through a metering plan, metering operation depends solely on the time of day, the day of the week or on special events. Pretimed control can apply to any number of on-ramps. However, no interconnection with other ramps is used.

2.2 Local-Actuated-Control Mode

Local actuated control is directly influenced by the main-line traffic conditions during the metering period. The decision-making mechanism is based primarily on real-time, locally measured traffic conditions based on main-line detectors in the immediate vicinity of the ramp. No interconnection with other ramps is used and no attempt at system optimization is possible.

2.3 System-Control Mode

System control is the form of on-ramp metering in which real-time information on total freeway traffic conditions is used to control the entrance-ramp system. A significant feature of this class of metering is the interconnection that permits conditions at one location to affect the metering rate imposed at one or more other locations. Freeway traffic conditions, as reflected by detectors throughout the system, are analyzed at a central computer and metering rates for all ramps are established according to real-time metering (see *On-Ramp Control: Realization Principles*).

The principal components of a system-mode configuration are the controller, signal head(s), a cabinet, detections wiring, a data communications subsystem, a communications medium and a control computer. A diagram of these components is shown in Fig. 1.

3. Control Strategies

The control strategies are the algorithms followed in developing the rate at which the ramp signals are cycled (Blumentritt *et al.* 1981).

3.1 Pretimed-Control Strategies

A pretimed strategy is based on matching uniform demands with control measures that reduce freeway congestion. A preliminary control plan is developed which limits access by the desired amount, based on current traffic data. When the system becomes operational, traffic data that reflect the operational characteristics of the freeway under control are collected.

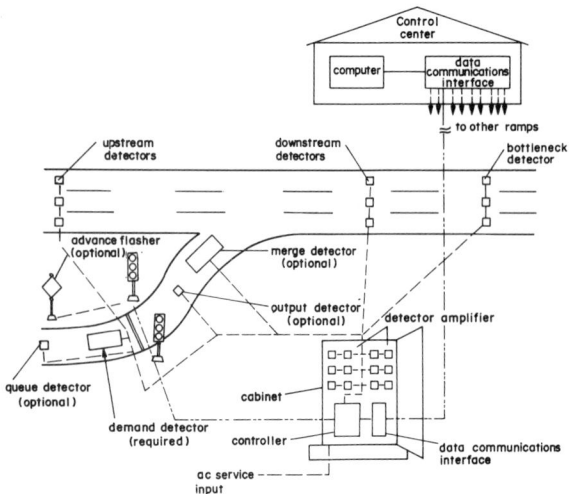

Figure 1
Principal hardware components of system-mode configuration

These data are then used as feedback to revise the metering rates for continuing control.

One procedure for calculating metering rates for the pretimed control is the integrated demand–capacity calculation. The calculation is practiced from upstream to downstream throughout the freeway section under control. More specifically, if traffic is congested within the freeway control section, then the on-ramp volume upstream of the bottleneck section is set so that the section volume is equal to or less than the bottleneck section capacity.

To accomplish this, engineers usually use a variable origin–destination matrix to determine the exit demands as a function of the upstream input volumes.

The linear-programming technique has been used in the formulation of the pretimed ramp-metering strategy ever since the early stages of freeway on-ramp control development (Wattleworth and Berry 1965). A detailed discussion concerning the linear-programming technique is given in Sect. 4.

3.2 Local-Actuated-Control Strategies

Local actuated control, being influenced by mainline traffic conditions, can respond to traffic conditions in the immediate vicinity of the ramp. Freeway speed, volume, density or occupancy can be used as a measure of the quality of flow. A typical local-actuated strategy limits the on-ramp volume to a desired value by correlation with the occupancy level of the adjacent mainline traffic. This occupancy level is determined by measuring the percentage of time that vehicles are over a point of detection. Table 1 is an example of occupancy levels with corresponding metering rates (see *On-Ramp Control, Local*).

3.3 System-Control Strategies

The system mode of control is the only control mode that can truly have a control strategy. Pretimed and locally actuated modes can be more accurately described according to a control plan rather than a control strategy.

Assume a control system with across-all-lanes detector counting stations placed at regular intervals and speed detectors at critical bottleneck locations. Thus, any given entrance ramp is to have a counting station both upstream and downstream of the ramp, together with a speed detector at a critical downstream bottleneck. The strategy would compare each of these three parameter measurements against a table of threshold values, and preassigned metering rates would be obtained for each of these measurements. A simple strategy is then to pick the smallest of the determined metering rates.

For example, assume a three-lane freeway section with a 1 min upstream flow of 95 vehicles, a 1 min downstream flow of 85 vehicles and a critical downstream bottleneck speed of 40 km h^{-1}. The associated parameters are given in Table 2, and the metering rates for upstream, downstream and bottleneck locations, respectively, are three, six and nine. Selecting the minimum metering rate of three, the next minute metering rate would be three vehicles per minute. Many other combinations of analytic procedures could be applied to achieve a strategy for assigning metering rates (see *On-Ramp Control: Coordinated Traffic-Responsive Strategies*).

3.4 Common Elements of Strategies

The main difference between the three mode strategies is time. The pretimed strategy is computed on the basis of a forecast of the average traffic conditions occurring during a specific time period. A dial is associated with the time period. Freeway operation resulting from the effect of metering during this time period is the consequence of this strategy.

The locally actuated mode operates in real time in response to freeway traffic conditions in the vicinity of each ramp. The freeway is to be divided into subsections delineated by the location of the main-line detector. The monitoring of local conditions is essentially

Table 1
Local actuated metering rates as a function of mainline occupancy

Occupancy (%)	Metering rate (vehicles min^{-1})
≤10	12
11–16	10
17–22	8
23–28	6
29–34	4
<34	3

Table 2
Control strategy parameters

Upstream		Downstream		Bottleneck	
1 min volume	metering rate	1 min volume	metering rate	speed (km h^{-1})	metering rate
91–100	3	91–100	3	65–80	15
81–90	6	81–90	6	49–64	12
71–80	9	71–80	9	33–48	9
61–70	12	61–70	12	17–32	6
<61	15	<61	15	0–16	3

continuous because the responses to occupancy change-over levels are immediate. The operation within each subsection is highly dependent on the operation of the immediate upstream and downstream subsections. The effectiveness of total system operation under local actuated control is a function of the propagation speed of congestive and clearing conditions of the downstream subsystem and also of the output of the upstream subsystem.

The system mode is an on-line real-time process that regularly calculates a new metering strategy according to existing traffic conditions. If, for example, the update interval is 1 min, then a calculation is made each minute to determine the metering rate for the next minute. As a result, feedback from the metering strategy is a continuous operation. Given this capability, a control algorithm is structured to provide a metering strategy that is compatible with the virtually continuous method of updating metering rates system-wide in real time.

4. Theoretical Aspects of the Control Strategies

Many theoretical investigations and empirical analyses have been reported with respect to freeway on-ramp control. Recognizing the need to develop a control strategy for a group of ramps as an interrelated system, recent research has been directed to the application of mathematical programming techniques and to the problem of ramp control. What follows are mathematical expressions of control strategies proposed by Kahng et al. (1984).

4.1 Extended Pretimed Traffic-Responsive Control

The extended pretimed traffic-responsive (EPT) control strategy has evolved from linear programming (LP) base pretimed control strategies (Wattleworth and Berry 1965, Wattleworth 1967). The major difference is that the EPT control determines the metering rate based on on-line traffic information and historical traffic data rather than based solely on historical data, so EPT control can respond to changing traffic situations. In EPT control the only real-time traffic data are traffic flow data from the first main-line section of the freeway.

Other traffic flow data needed for this control—namely, on-ramp and off-ramp flow data—are historical. Based on these real-time and historical data, EPT control determines metering rates to maximize the sum of input flows while preventing main-line congestion. In order to do this, EPT control uses the following steps.

For each control period, the central computer receives on-line measured traffic flow data of the first main-line section and processes this information to predict the first main-line traffic flow in the next control period using an appropriate method such as a historical-factor approach, a moving-average approach or an autoregressive approach.

By combining the predicted main-line traffic demand of the first section with the historical data for on-ramps and off-ramps, LP is used to generate optimal metering rates that maximize the sum of input flows within constraints of section capacities. Equations (1)–(4) give the formulations used in EPT control.

$$\text{maximize} \sum_{i=1}^{n} X_i \qquad (1)$$

$$\text{subject to} \sum_{i=1}^{n} A_{ij} \le C_j \quad j=1,2,\ldots,m \qquad (2)$$

$$0 \le X_i \le D_i \qquad (3)$$

$$X_{i,\min} \le X_i \le X_{i,\max} \qquad (4)$$

where X_i is the input rate from on-ramp i (i.e., the finalized metering rate for on-ramp i), n is the number of on-ramps, m is the number of freeway subsections, A_{ij} is the fraction of vehicles from on-ramp i going through subsection j, C_j is the capacity of freeway subsection j, $X_{i,\max}$ is the maximum metering rate for on-ramp i, and $X_{i,\min}$ is the minimum metering rate for on-ramp i.

Equation (1) simply states that the objective of the control is to maximize the total input rate from all ramps. Equation (2) defines the capacity constraints. Equation (3) states that the metering rate for any on-ramp cannot be negative and cannot exceed the demand. Equation (4) is for maximum and minimum metering-rate constraints. This formulation can be solved efficiently by the simplex method.

4.2 Extended Local Traffic-Responsive Control

Extended local traffic-responsive (ELT) control has evolved from local actuated control. The major difference between them is that ELT control has an extended view of the freeway, so that each on-ramp controller is aware of, and reacts to, the changing traffic situation at its neighboring on-ramps.

Based on the real-time traffic data collected from detectors on both the freeway main line and the on-ramps and off-ramps, ELT control determines the metering rate for each on-ramp to handle freeway bottlenecks at capacity while preventing main-line congestion and giving consideration to on-ramp queues. To accomplish this goal, the ELT control uses the following steps.

First, for each control period t, extended demand capacity analysis estimates the effective downstream main-line capacity on on-ramp i, $Cd_i(t)$, which is defined as the minimum capacity available at time t between on-ramp i and the next downstream on-ramp $(i+1)$.

Second, to balance out the queue length at each on-ramp, on-ramp i is allowed to ask for help from the upstream on-ramp $(i-1)$ to reduce its ramp metering rate. However, in the field, the exact queue length at the on-ramp is not usually available once the queue grows beyond the queue detector. Thus, the amount of help asked by on-ramp i from upstream on-ramp $(i-1)$ must be estimated from an equation such as:

$$\Delta X_i(t) = a[X_{i,\max} - X_i(t)] + b \qquad (5)$$

where $\Delta X_i(t)$ is the reduction in the metering rate of upstream on-ramp i at control period t, $X_i(t)$ is the metering rate at on-ramp i during control period t, $X_{i,\max}$ is the maximum metering rate at on-ramp i, and a and b are design parameters.

Third, the preliminary metering rate of on-ramp i for the next control period $(t+1)$ is set as the difference between the reduced effective downstream capacity and the on-line measured main-line traffic flow upstream of on-ramp i:

$$X_i(t+1) = [Cd_i(t) - \Delta X_{i+1}(t)] - Qf_i(t) \qquad (6)$$

where $Qf_i(t)$ is the on-line measured main-line flow immediately upstream of on-ramp i during control period t.

Fourth, a main-line occupancy check is made as a feedback mechanism of the ELT control. This operates at two levels. The first level compares each on-ramp upstream main-line occupancy level against the preset critical occupancy value as shown in Table 1. If this comparison indicates that the detector occupancy upstream of on-ramp i is greater than the critical occupancy associated with the location, then $X_i(t+1)$, calculated by Eqn. (6), is reduced to the minimum metering rate for on-ramp i. Otherwise, the $X_i(t+1)$ remains unchanged. The second-level occupancy checking allows on-ramp i to request help from upstream on-ramp $(i-1)$, in terms of a reduction in the metering rate of on-ramp $(i-1)$, if the main-line occupancy problems at on-ramp i persist for more than a user-specified number of control periods. In that case, the $X_{i-1}(t+1)$ is reduced to the minimum metering rate for on-ramp $(i-1)$. The revised preliminary metering rates are checked against minimum and maximum metering-rate constraints to be finalized.

4.3 Comparison of the Two Strategies

The two segment-wide traffic-responsive freeway control strategies mentioned above—EPT control and ELT control—were evaluated on a macroscopic dynamic freeway corridor simulation model (Kahng et al. 1984) using traffic data obtained over three days from the Santa Monica Freeway in Los Angeles. The data in Table 3 compare traffic-performance measures that result from the two control strategies. Both control strategies produced almost identical total travel distances in terms of vehicle kilometres for all three days.

However, total travel times, which indicate the effectiveness of the control, were different for the two controllers for all three days. The ELT control resulted in shorter total travel times (~0–7% less) compared with those from the EPT control.

5. Extension of Freeway Ramp-Metering Control to Corridor Control

As indicated in the comparison of traffic-performance measures in Sect. 4.3, freeway ramp-metering control cannot be evaluated without the traffic-performance

Table 3
Comparison of traffic performance measures

Day	Measure	ELT control	EPT control	Difference[a] (%)
1	TTD[b]	279 892	279 775	0.0
	FTT[c]	3537	3554	−0.5
	OWT[d]	777	849	−8.5
	TTT[e]	4314	4403	−2.0
2	TTD	281 326	281 318	0.0
	FTT	3624	3771	−3.9
	OWT	1031	904	14.0
	TTT	4655	4675	−0.4
3	TTD	262 593	262 587	0.0
	FTT	3232	3219	0.4
	OWT	92	372	−75.0
	TTT	3323	3591	−7.5

a Difference = 100 (ELT − EPT)/EPT b TTD = total travelled distance (vehicle kilometers) c FTT = freeway travel time (vehicle hours) d OWT = on-ramp wait time (vehicle hours) e TTT = total travel time (vehicle hours) statistics from related arterial surface streets. Therefore, the ramp-control strategies should be extended towards the direction of corridor control.

The objective of corridor control is to obtain an optimum balance between traffic demand and capacity within a corridor—a freeway and the system of roadways influenced by the freeway which accommodates travel demands along a predominate directional line in a portion of an urban area.

This is accomplished through the coordination of various control and driver information systems to facilitate the total movement of traffic on a freeway and the adjacent urban arterial streets. The basic concept of this type of control consists of a process whereby all routes in the corridor are constantly monitored and traffic is diverted from those facilities that are overloaded to alternative facilities that have excess capacity. Several corridor control projects have been proposed and are currently under development. However, these engineering projects are still in the research stage; consequently, corridor control strategies have not advanced to the point that they can fully take into account all the variables of freeway and urban street control. Freeway-corridor and arterial-network models have been developed for this purpose and have been used on a limited basis to assess various operational strategies as shown in Sect. 4 (Capelle 1979).

Here, one of the studies more directly related to the information systems on a freeway corridor is shown as an example of coordinated ramp-meeting strategies (Pretty 1972). The study deals with a freeway corridor with an alternative route with n choice points at which a vehicle may either enter the freeway or continue on the alternative route. A set of changeable information signs located at choice points can operate to indicate the quicker route to the next downstream ramp or choice point from each choice point according to the following inequality:

$$\frac{F_i + R_i + Q_i}{X_i} > \frac{S_i + R_{i+1} + Q_{i+1}}{X_i} \qquad (7)$$

where F_i is the freeway travel time (in minutes) from the ith entry ramp to the next entry ramp with metering in operation, S_i is the surface-street travel time (in minutes) from the ith choice point to the next choice point, R_i is the travel time (in minutes) from the ith choice point to the ith entry ramp, excluding a queuing delay on the ramp, X_i is the metering rate (in vehicles per minute) on the ith ramp necessary to prevent the downstream demand from exceeding capacity, and Q_i is the queue length on the ith ramp.

A set of signs could indicate the quicker route to the next downstream ramp or choice point from each choice point. If sufficient numbers follow the signs that suggest the use of the alternative route, then surface street and freeway travel times will reach a state where the travel times will be about the same for each route. The condition to attain this state of equilibrium is

$$p_i \geq 1 - \frac{60X_i}{D_i} \qquad (8)$$

where p_i is the proportion obeying the sign and D_i is the demand for the ith ramp (in vehicles per hour). From these equations, sign-switching decisions based on the queue length at each ramp can be derived easily. It is, of course, necessary to have detecting systems for ramp queues and travel times along the alternative route.

See also: On-Ramp Control: Coordinated Traffic-Responsive Strategies; On-Ramp Control, Local; On-Ramp Control of Freeway Networks

Bibliography

Blumentritt C W, Pinnel C, McCasland W R, Ross D W, Glazer J 1981 Guidelines for selection of ramp control systems. *Highw. Res. Board, Natl. Coop. Highw. Res. Program Rep.* **232**

Capelle D G 1979 Freeway corridor systems. *Proc. Int. Symp. Traffic Control Systems.* Institute of Transportation Studies, University of California, pp. 33–54

Kahng S J, Jenc C Y, Campbell J F, May A D 1984 Developing segmentwide traffic responsive freeway entry control. *Transp. Res. Rep.* **957**, 5–13

Payne H J, Meisel W S, Teener M D 1973 Ramp control to relieve freeway congestion caused by traffic disturbance. *Highw. Res. Rec.* **469**, 52–64

Pretty R L 1972 Control of a freeway system by means of ramp metering and information signs. *Highw. Res. Rec.* **388**, 62–73

Wang J, May A D 1973 Computer model for optimal freeway on-ramp control. *Highw. Res. Rec.* **469**, 16–25

Wattleworth J A 1967 Peak-period analysis and control of a freeway system. *Highw. Res. Rec.* **157**, 1–10

Wattleworth J A, Berry D S 1965 Peak-period control of a freeway system—Some theoretical investigations. *Highw. Res. Rec.* **89**, 1–25

Y. Makigami
[Ritsumeikan University, Kyoto, Japan]

On-Ramp Control: Coordinated Traffic-Responsive Strategies

Mathematical optimization theory provides some useful tools for efficient control strategy design. Application of optimization methods requires the traffic control problem to be formulated in the format of an optimal control problem consisting of three main components:

(a) a mathematical process model,

(b) some physical constraints, and

(c) a performance index to be minimized subject to the model equations and the constraints.

Since physical constraints and the performance index are expressed in terms of the variables of the model, it is the chosen mathematical model that mainly influences the computation and instrumentation effort required for the realization of the control strategy. Several control-oriented, deterministic, macroscopic models of freeway traffic flow have been proposed by various researchers since about 1970 (see *Freeway Traffic Modelling*). A good understanding of the features of each of these models and the relationships between them is of great importance for the development of efficient control strategies.

Total time spent by all drivers on a freeway system, including possible waiting times at the on-ramps, is a suitable performance index for freeway traffic control operation. This is because minimization of total time spent over a sufficiently long time horizon implies possible traffic congestion to be avoided and/or eliminated. In general terms, the freeway traffic control problem can be translated into the following general optimal control problem.

PROBLEM 1. *Select the on-ramp volumes from an admissible control region so as to minimize the total time spent on the freeway system over a specified time horizon subject to some mathematical model constraints and further physical/technical restrictions.*

The precise mathematical formulation of this problem depends on the traffic model used.

1. Problem Constraints

Ramp metering is implemented by installing common traffic lights at the on-ramps of the freeway. The underlying idea is to keep traffic density at values near the critical density (see *Freeway Traffic Modelling*) guaranteeing traffic operation at maximum traffic volume. In the case of nonrecurrent congestion, upstream on-ramp volumes should be reduced in order to enable a quick release.

Adjusting on-ramp volumes must be performed subject to several constraints. On-ramp volumes cannot be higher than the current demand at a given entrance ramp i:

$$r_i(k) \leq d_i(k) + \frac{l_i(k)}{T} \quad (1)$$

where r_i is the on-ramp volume, d_i is the arriving volume or demand volume and l_i is the nonnegative queue length described by the difference equation:

$$l_i(k+1) = l_i(k) + T[d_i(k) - r_i(k)] \quad (2)$$

where k denotes the discrete-time index and T denotes the sample time interval (see *Flow Variables*). Clearly, an entrance queue will be formed any time the volume of traffic permitted to enter the freeway is less than the volume desiring to use the ramp.

Due to the geometric characteristics of the entrance ramp, there is a maximum on-ramp volume value $r_{i,\max}$ which cannot be exceeded and hence

$$r_i(k) \leq r_{i,\max} \quad (3)$$

must always hold. On the other hand, too low rates of on-ramp volumes lead waiting drivers to judge the metering signal to be malfunctioning. Hence, a reasonable lower limit should be posed:

$$r_i(k) \geq r_{i,\min} > 0 \quad (4)$$

Finally, metering on-ramp volumes can lead to long entrance queues during the rush hours. In order to avoid interference with traffic in surface streets, a maximum queue length $l_{i,\max}$ should not be exceeded:

$$l_i(k) \leq l_{i,\max} \quad (5)$$

Substituting Eqn. (2) into Eqn. (5) gives:

$$r_i(k) \geq d_i(k) - \frac{1}{T}[l_{i,\max} - l_i(k)] \quad (6)$$

Summarizing Eqns. (1), (3), (4) and (6), the admissible control region for on-ramp volumes is given by:

$$\max\left\{r_{\min}, d(k) - \frac{1}{T}[l_{\max} - l(k)]\right\}$$
$$\leq r(k) \quad (7)$$
$$\leq \min\left\{r_{\max}, d(k) + \frac{1}{T}l(k)\right\}$$

2. Control Objectives

Several control objectives for freeway traffic have been proposed by various researchers. Minimization of the total time spent t_s by all drivers on the freeway system seems to be the most suitable one. The time t_s on the freeway includes total travel time and total waiting time at the on-ramps. Minimization of t_s on the freeway implies minimization of delays caused by congestions and is, thus, a suitable requirement in order to prevent or eliminate both recurrent and nonrecurrent congestions. The time t_s is given by the sum

$$t_s = t_t + t_w \quad (8)$$

where t_t is the total travel time and t_w is the total waiting time. The total travel time for a given time horizon Kt is given by the sum

$$t_t = \sum_{k=0}^{K} T\boldsymbol{\rho}(k)^T \boldsymbol{\Delta} \quad (9)$$

where $\boldsymbol{\rho}$ denotes the vector of traffic densities ρ_i in corresponding freeway sections and the Δ_i are the lengths of the sections. The total waiting time for the same time horizon is given by the sum

$$t_w = \sum_{k=0}^{K} T\boldsymbol{l}(k)^T \boldsymbol{e} \quad (10)$$

with $\boldsymbol{e}$ the unit vector of appropriate dimension. Minimizing t_s is equivalent to maximizing a weighted sum of total freeway output, as shown in Papageorgiou (1983):

$$(t_s \to \min) \leftrightarrow \left(S = \sum_{k=0}^{K} (K-k) \boldsymbol{s}(k)^T \boldsymbol{e} \to \max \right) \quad (11)$$

where $\boldsymbol{s}$ denotes the vector of exit volumes. From this, it becomes apparent that minimization of t_s corresponds to maximization of the weighted number of cars served during system operation. Hence, the objectives of Eqn. (11) benefit short trip drivers.

An alternative criterion sometimes utilized is the maximization of total travel distance:

$$d = \sum_{k=0}^{K} T\boldsymbol{q}(k)^T \boldsymbol{\Delta} = \sum_{k=0}^{K} T(\boldsymbol{\Delta \rho}(k))^T \boldsymbol{v}(k) \to \max \quad (12)$$

where the q_i and v_i are the traffic volumes and mean speeds of the sections i, respectively and $\boldsymbol{\Delta \rho}(k)^T = (\Delta_1 \rho_1, \ldots, \Delta_n \rho_n)$. Maximizing d is equivalent to maximum utilization of street capacity, even for the price of less cars being served. Hence, the objective of Eqn. (12) benefits long-trip drivers. Under certain conditions (e.g., equal section lengths and equal exit rates), Eqns. (11) and (12) are equivalent (i.e., they result in the same optimal control actions).

Minimization or maximization of overall freeway traffic quantities, such as t_s or d, may result in unequal distribution of waiting times in the individual on-ramps. There are two ways leading to an approximately uniform distribution of on-ramp queues. The first is by the requirement:

$$l_1(k) = l_2(k) = \cdots = l_n(k), \quad \forall k \quad (13)$$

The constraints of Eqn. (13) lead to identical queue lengths but because of their strictness they may lead to an unnecessary increase of the overall objective. It therefore seems more reasonable to augment the main control objective by adding (in the case of minimization) or subtracting (in the case of maximization) the quantity:

$$L = \sum_{k=0}^{K} \|\boldsymbol{l}(k)\|_Q^2 \quad (14)$$

where $\|\boldsymbol{l}\|_Q^2 = \boldsymbol{l}^T \boldsymbol{Q} \boldsymbol{l}$, and $\boldsymbol{Q}$ is an appropriately chosen weighting matrix. Minimization of the subcriterion given by Eqn. (14) is known to lead to an approximately uniform distribution in space and time.

3. Dynamic Nonlinear Open-Loop Optimal Control

Consider three classes of models for modelling freeway traffic: model B (see *Freeway Traffic Modelling*, Sect. 3), model E (see *Freeway Traffic Modelling*, Sect. 4) and the more complex model D (see *Freeway Traffic Modelling*, Sect. 5). Open-loop optimal control can be derived on the basis of any of these models by solving the following optimization problem.

PROBLEM 2. *Given an initial condition for the state variables and given predicted trajectories for demands $d(k)$, and exit rates $\gamma(k)$ $(k = 0, \ldots, K-1)$, find the on-ramp volume trajectories $r(k)$ $(k = 0, \ldots, K-1)$ which minimize a chosen performance index J subject to the equations of model B, E or D and subject to the constraints of Eqns. (2) and (7).*

Optimization Problem 2 can be solved by use of Pontryagin's maximum principle. However, numerical calculation of the optimal on-ramp volume trajectories is a nontrivial task.

3.1 Optimal Control Based on Model B

Problem 2 based on the equations of model B has been essentially considered by Kaya (1972), Tabak (1972) and Papageorgiou (1984) for minimization of the total time spent (i.e., t_s in Eqn. (8)). Unfortunately, none of these authors proposed a numerical algorithm for solution of the optimization problem. Also, in order to keep computational effort within certain limits, application of model B is accomplished for sections several kilometers long. However, in the case of congestion (i.e., the presence of overcritical density in a short freeway segment), the overcritical density would be dispersed all over the section, which may result in an inadequate description of traffic behavior. For this reason, a modification of the model equations is proposed in Papageorgiou (1984) for the case of congestion. A different approach using results of the kinematic wave theory was proposed by Greenlee and Payne (1977) for responding to incidents. They propose a subdivision of the freeway into three parts with moving boundaries: part 1, the congestion area; part 2, the part upstream; and part 3, the part downstream of the congestion area. The locations of the congestion boundaries X_1 and X_2 are considered as the state variables of the problem. A time-optimal control problem is then formulated with the terminal condition $X_1 = X_2$ (i.e., elimination of the congestion). Unfortunately, neither a solution algorithm for this time-optimal control problem, nor simulation tests of the proposed method are included by Greenlee and Payne.

3.2 Optimal Control Based on Model E

Bhouri et al. (1990) have proposed an optimal control scheme for ramp metering based on model E and reported its application to the Boulevard Périphérique in Paris. The approach aims at minimizing the total time spent t_s and utilizes a feasible search algorithm for numerical calculation of the optimal on-ramp trajectories.

3.3 Optimal Control Based on Model D

Optimal control Problem 2 was considered by Blinkin (1976) on the basis of model D in order to minimize a quadratic performance index attempting to balance on-ramp queues and freeway congestion. Cremer (1976) considered a similar optimization problem and used a nonlinear programming algorithm for its solution. An example with 16 freeway sections (8 km) and four on-ramps for a time horizon $K = 60$ (10 min) is provided in order to demonstrate application of the method. The computation time needed for the evaluation of the optimal control sequence was reported to be 7.5 min on a Telefunken TR 440 digital computer.

Finally, Papageorgiou and Mayr (1982) considered equations of model D in order to minimize deviations of traffic state from a desired situation. An example with 30 freeway sections (30 km) and six on-ramps for a time horizon $K = 120$ (30 min) is provided in order to demonstrate application of the method. The computation time needed for specification of the optimal solution was reported to be 20 s in a CYBER 175 digital computer. Simulation tests show that the method is capable of eliminating congestion (see Sect. 6).

4. Dynamic Linear-Quadratic Feedback Control

The well-known linear-quadratic (LQ) methodology may be applied to the freeway traffic control problem if the following simplifications are introduced into Problem 2:

(a) the nonlinear model equations are linearized around desired trajectories for the state and input variables;

(b) the inequality constraints given by Eqn. (7) are dropped from the problem formulation; and

(c) a quadratic index, penalizing deviations from some desired trajectories, is considered as the performance criterion.

Under these assumptions and for given desired trajectories, an optimal gain matrix trajectory $\mathbf{L}(k)$ can be calculated by solution of the well-known Riccati equation. The optimal on-ramp volumes are given by the feedback law:

$$r(k) = r_d(k) - \mathbf{L}(k)[x(k) - x_d(k)] \quad (15)$$

where x is the state vector provided by measurements, and $r_d(k)$ and $x_d(k)$ are prescribed desired trajectories. If r_d and x_d are steady-state constant values, the time horizon K in the performance index can be set to infinity, and we obtain a constant gain matrix $\mathbf{L}$.

The LQ problem for freeway traffic control has been considered by Yuan and Kreer (1968) and Kaya (1972) on the basis of model B, and by Knapp (1972), Isaksen and Payne (1973), Athans et al. (1975), Goldstein and Kumar (1982) and Papageorgiou (1984) on the basis of model D. Papageorgiou et al. (1990) have proposed an LQ-feedback law with integral parts.

The gain matrix $\mathbf{L}$ of the feedback law was found to have an overlapping structure which permits its calculation with decentralization techniques aiming at reducing the corresponding calculation effort. Moreover, the structure of $\mathbf{L}$ permits implementation to be based on an overlapping feedback information structure.

A further way of looking for (possibly nonlinear) feedback laws is to prescribe a feedback structure that appears reasonable, and to look for regulator parameters that are optimal in some defined sense. This suboptimal procedure was studied by Payne et al. (1973), Cremer (1978) and Looze et al. (1978).

5. Hierarchical (Multilayer) Control Structures

Multilayer control structures for freeway traffic have been proposed by several researchers. The overall control problem is vertically decomposed into a number of subproblems that interact with each other in a way so as to combine efficiency of open-loop control systems with computational feasibility and robustness of closed-loop control systems. The multilayer structure consists of an optimization layer, an adaptation layer and a direct control layer, hierarchically ordered as visualized in Fig. 1. The task of each control layer will be outlined separately.

The optimization layer plays a central role in the control hierarchy. An optimization problem (e.g., Problem 2) is solved here on-line under consideration of the overall freeway traffic conditions. As an optimization result, desired values for the traffic state x_d and the on-ramp volumes r_d are specified. The loss of performance caused by model inaccuracies and simplifications is partially eliminated by appropriate design of the two additional control layers. The optimization problem may be a steady-state problem or a dynamic one. It is solved on the basis of updated predictions of demands and origin–destination rates and further

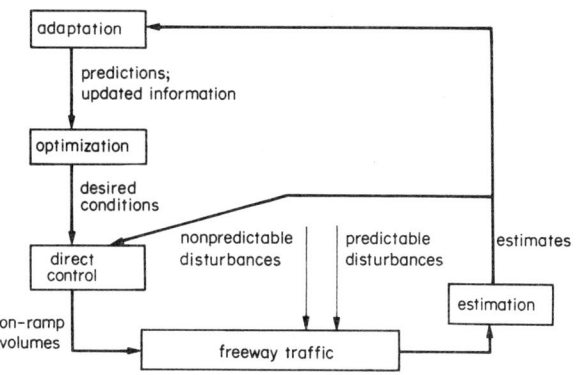

Figure 1
Hierarchical multilayer control structure

updated process information (initial conditions, occurrence of an incident, etc.).

The adaptation layer accomplishes information updating on the basis of both current and historical, stored measurements. It is mainly designed in a heuristic way. Updating is actuated periodically and/or whenever predicted traffic conditions are judged to deviate significantly from actual traffic conditions.

The direct control layer keeps traffic near the desired values given by the optimization layer (i.e., it suppresses the effects of model inaccuracies and slight unexpected disturbances). The direct control layer may consist of a tandem of decentralized or overlapping feedback laws (e.g., Eqn. (15)) which lead to better reliability properties and lower implementation costs.

The estimation task, also appearing in Fig. 1, is described elsewhere (see *Flow Variables: Estimation*).

Hierarchical control structures of this kind have been considered by Kaya (1972), Isaksen and Payne (1973), Athans *et al.* (1975), May (1976) and Papageorgiou (1984).

6. A Simulation Example

To illustrate the efficiency of different control schemes, consider a simulation example taken from Papageorgiou (1986). The example is of a hypothetical,

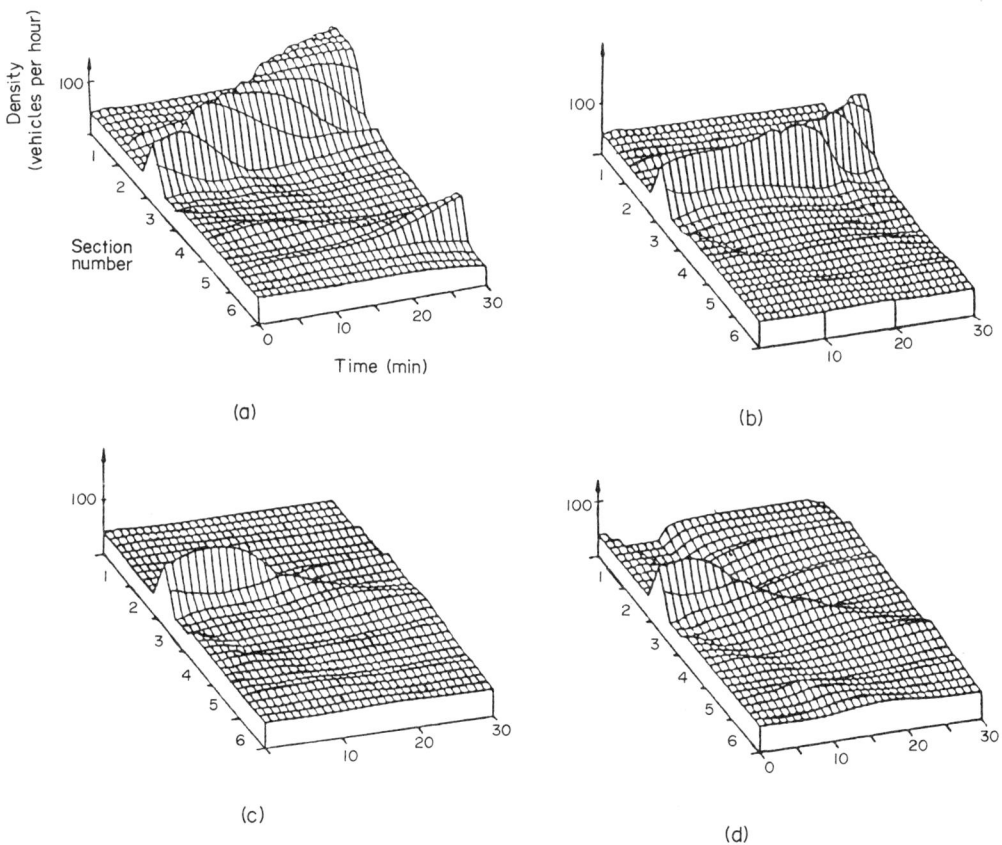

Figure 2
Traffic density evolution for a simulated congested freeway: (a) no control action, (b) LQ-feedback control, (c) nonlinear optimal control and (d) multilayer control

two-lane, 30 km freeway with six on-ramps and six off-ramps. The freeway is subdivided into six 5 km sections, each including one on-ramp and one off-ramp. Each freeway section is again subdivided into five 1 km parts. Application of model D to this freeway results in a state vector of the order 66.

Severe congestion is assumed to be present in section 3 of the freeway at time zero. The simulation is performed over a time horizon of 30 min for the following cases:

(a) no control action

(b) LQ-feedback control,

(c) nonlinear optimal control, and

(d) multilayer control.

Figure 2 depicts density evolution for the four cases. In the absence of any control action (case (a)), traffic flow in section 3 becomes unstable, and moreover recurrent congestion occurs in section 5 due to high demand. LQ control (case (b)) ameliorates the traffic situation, although it does not succeed in alleviating congestion completely. Nonlinear optimal control (case (c)) and multilayer control (case (d)) lead traffic flow back to normal conditions.

See also: On-Ramp Control: Coordinated Time-of-Day Strategies; On-Ramp Control of Freeway Networks

Bibliography

Athans M, Houpt P K, Looze D, Orlhac D, Gershwin S B, Speyer J L 1975 Stochastic control of freeway corridor systems. *Proc. 1975 IEEE Conf. Decision and Control.* Institute of Electrical and Electronic Engineers, New York, pp. 676–85

Bhouri N, Papageorgiou M, Blosseville J M 1990 Optimal control of traffic flow on periurban ringways with application to Boulevard Périphérique in Paris. *Proc. 11th IFAC World Congress.* International Federation of Automatic Control, Dusseldorf, Germany (in press)

Blinkin M Ya 1976 Problem of optimal control of traffic flow on highways. *Autom. Remote Control* **37**, 662–7

Cremer M 1976 A new scheme for traffic flow estimation and control with a two component model. *Proc. 3rd IFAC/IFIP/IFORS Symp. Control in Transportation Systems.* Pergamon, Oxford, pp. 29–37

Cremer M 1978 A state feedback approach to freeway traffic control. *Prep. 7th IFAC World Congress*, International Federation of Automatic Control, Dusseldorf, Germany, pp. 1575–82

Goldstein N B, Kumar K S P 1982 A decentralized control strategy for freeway regulation. *Transp. Res. B* **16**, 279–90

Greenlee T L, Payne H J 1977 Freeway ramp metering strategies for responding to incidents. *Proc. 1977 IEEE Conf. Decision and Control.* Institute of Electrical and Electronic Engineers, New York, pp. 982–7

Isaksen L, Payne H J 1973 Suboptimal control of linear systems by augmentation with application to freeway traffic regulation. *IEEE Trans. Autom. Control* **18**, 210–19

Kaya A 1972 Computer and optimization techniques for efficient utilization or urban freeway systems. *Proc. 5th IFAC World Congress.* International Federation of Automatic Control, Dusseldorf, Germany, paper 12.1

Knapp C H 1972 Traffic estimation and control at bottlenecks. *Proc. Int. Conf. Cybernetics and Society.* Institute of Electrical and Electronic Engineers, New York, pp. 469–72

Looze D P, Houpt P K, Sandell N R, Athans M 1978 On decentralized estimation and control with application to freeway ramp metering. *IEEE Trans. Autom. Control* **23**, 268–75

May A D 1976 A proposed dynamic freeway control system hierarchy. *Proc. IFAC/IFIP/IFORS 3rd Int. Symp. Transportation Systems.* Pergamon, Oxford, pp. 1–12

Papageorgiou M 1983 *Applications of Automatic Control Concepts to Traffic Flow Modeling and Control.* Springer, Berlin

Papageorgiou M 1984 Multilayer control system design applied to freeway traffic. *IEEE Trans. Autom. Control* **29**, 482–90

Papageorgiou M 1986 Freeway on-ramp control strategies: Overview, discussion and possible application to Boulevard Périphérique de Paris. Internal Report, INRETS, Arcueil, France

Papageorgiou M, Blosseville J M, Hadj-Salem H 1990 Modelling and real-time control of traffic flow on the southern part of Boulevard Périphérique in Paris, Pt 2—Coordinated on-ramp metering. *Transp. Res. B* (in press)

Papageorgiou M, Mayr R 1982 Optimal decomposition methods applied to motorway traffic control. *Int. J. Control* **35**, 269–80

Payne H J, Meisel W S, Teener M D 1973 Ramp control to relieve freeway congestion caused by traffic disturbances. *Highw. Res. Rec.* **469**, 52–64

Tabak D 1972 A linear programming model of highway traffic control. *Proc. 6th Annual Princeton Conf. Information Science and Systems.* Princeton, NJ, pp. 568–70

Yuan L S, Kreer J B 1968 An optimal control algorithm for ramp metering of urban freeways. *Proc. 6th IEEE Annual Allerton Conf. Circuit and System Theory.* University of Illinois, Allerton, IL

M. Papageorgiou
[Technische Universität München,
Munich, Germany]

On-Ramp Control, Local

Congestion on freeways has become a common phenomenon leading to delays, reduced traffic security, increased fuel consumption and severe air pollution. Freeway traffic flow can be influenced by on-ramp metering, variable speed limitation signs, variable route recommendations, information provided to the drivers and further variable traffic signs. A substantial amelioration can be achieved by applying the tools provided by automatic control theory and computer

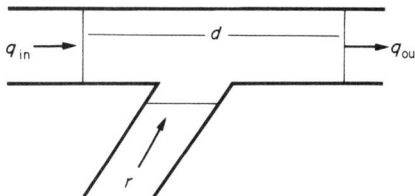

Figure 1
The traffic flow process

technology aiming at transforming traffic flow on freeways into a controllable, optimally operating system. Traffic-responsive on-ramp metering systems are currently operating on several freeways in the USA (Masher *et al.* 1975), the UK, France and the Netherlands. This article presents a comparative assessment of several local control strategies that were performed at an entrance ramp of the Boulevard Périphérique in Paris.

1. Local Ramp-Metering Strategies

The followng quantities (see Fig. 1) must be defined:

(a) O_{out} and O_{in} are the measured occupancy rates downstream and upstream, respectively;

(b) q_{out} and q_{in} are the measured traffic volumes downstream and upstream, respectively;

(c) r is the on-ramp traffic volume;

(d) r_{min} is the admissible on-ramp traffic volume; and

(e) Cap is the downstream bottleneck capacity.

1.1 Static Control

Static control consists in restricting geometrically the end of the ramp so as to force the cars to enter the mainstream in one lane. No traffic lights are utilized in this context, but the restriction is effected by the use of beacons so as to avoid violation by the drivers.

1.2 Fixed Time Control

Fixed time control is effected by ordinary traffic lights with a cycle of 40 s divided into a green phase of 25 s and a red phase of 15 s. The utilized cycle and phases were found to be adequate during real-world experiments performed on the same site at an earlier stage (Blosseville 1985).

1.3 Demand-Capacity Strategy

The demand-capacity strategy, which is extensively used in the USA (Masher *et al.* 1975, Koble *et al.* 1980) is based on measuring the volume (demand) q_{in} upstream of the merge area and comparing this demand with the capacity Cap of the bottleneck downstream of the merge area, as determined by historical data.

However, since the measurement of volume alone is insufficient to determine whether the freeway is congested or free flowing, occupancy O_{out} from the downstream detector stations is used. If the occupancy is above a preset threshold determined from historical data, congested flow is assumed to exist and the minimum metering rate r_{min} is used. If occupany is below the threshold value, the upstream volume is compared with capacity to determine the ramp-metering rate. More precisely, at each cycle k ($k = 1, 2, 3, \ldots$)

$$r(k) = \begin{cases} \text{Cap} - q_{in}(k), & \text{if } O_{out} \leq O_{cr} \\ r_{min}, & \text{otherwise} \end{cases}$$

1.4 Percent-Occupancy Strategy

The percent-occupancy strategy (Masher *et al.* 1975, Koble *et al.* 1980) which is in use in the USA, is essentially based on the same philosophy as the demand-capacity strategy; that is,

(a) measure upstream demand and add on-ramp volume so as to maintain downstream capacity; and

(b) in the case of congestion, switch to a minimum on-ramp volume value.

The difference between percent-occupancy and demand-capacity strategies is the following.

(a) Upstream demand is estimated using occupancy measurements. The main reason for this is that only one detector can be used (e.g., in the middle lane) in order to estimate the upstream demand according to a calibrated curve. However, in the experiment detailed in Sect. 2, all three detectors (for three mainstream lanes) have been used.

(b) Congestion is detected by the upstream detector (i.e., the method needs only one mainstream detector station).

The final form of the percent-occupancy strategy is depicted in Fig. 2. The critical value of the upstream

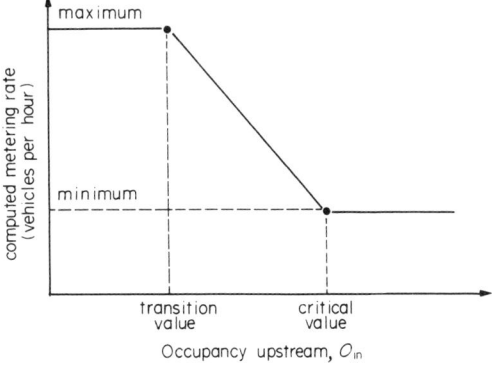

Figure 2
Percent-occupancy strategy

occupancy is specified by use of historical data and the transition value is found by trial and error in accordance with the historical on-ramp demand (Hadj Salem et al. 1988).

1.5 Demand-Capacity INRETS

This strategy is an ameliorated version of the demand-capacity strategy described in Sect. 1.3. Demand-capacity INRETS utilizes measurements from three mainstream detector stations in order to better estimate the degree of congestion and react accordingly.

Under free-flowing conditions and under conditions of severe congestion, this strategy reacts in the same way as demand capacity. There are, however, a series of (typically two) intermediate traffic situations where the ramp-metering rate is calculated by use of the following equation:

$$r(k) = \beta q_{out}(k) - q_{in}(k)$$

which means that:

(a) the capacity utilized in the demand-capacity strategy is now replaced by the actually measured downstream volume; and

(b) a parameter β is utilized that is valued at 1 for rather slight congestion and at 0.9 for stronger congestion.

1.6 ALINEA Strategy

All traffic-responsive strategies presented so far are based on a feedforward philosophy which is known to be particularly sensitive with respect to disturbances. On the contrary, ALINEA is based on a feedback philosophy. In other words, since the main aim of ramp metering is to maintain capacity flow *downstream* of the merge area, the control strategy should be based on *downstream* measurements. In fact, ALINEA was developed by application of well-known methods of the classical feedback theory—see Papageorgiou et al. (1989) for a detailed derivation—and obtains the form

$$r(k) = r(k-1) + K_R[\hat{O} - O_{out}(k)]$$

where $\hat{O}$ is a preset desired value (typically $\hat{O} = O_{cr}$) and K_R is a regulation parameter. This feedback law is simpler than other local metering strategies. It suggests a fairly plausible control behavior: if the measured occupancy $O_{out}(k)$ at cycle k is found to be lower (higher) than the desired occupancy $\hat{O}$, the term $\hat{O} - O_{out}(k)$ of the right hand side of the equation becomes positive (negative) and the ordered on-ramp volume $r(k)$ is increased (decreased) as compared with its last value $r(k-1)$. The feedback law acts in the same way for both congested and light traffic, so no switchings are necessary.

Note that the family of percent-occupancy strategies reacts to excessive occupancies only after congestion is created and has reached an upstream measurement location, whereas ALINEA reacts smoothly even to slight differences $\hat{O} - O_{out}(k)$ and may thus prevent congestion in an elegant way.

On the other hand, the family of demand-capacity strategies react to excessive downstream occupancy only after a threshold value is exceeded. Typically, and in contrast to ALINEA, the reaction of all previous control strategies to excessive occupancy is rather crude: on-ramp volumes are set equal to their minima values.

A value of $K_R = 70$ vehicles per hour was found to give good results in real-world experiments. Note that K_R is the only parameter to be adjusted in the implementation phase since no thresholds or other constants are included in the feedback law. Moreover, from theoretical considerations it is known that:

(a) results are insensitive for a wide range of K_R values;

(b) increasing (decreasing) K_R leads to stronger (smoother) reactions of the regulator and regulation times get shorter (longer); and

(c) for extremely high values of K_R, the regulator may have oscillatory, unstable behavior.

ALINEA requires only one detector station which measures occupancy O_{out} downstream of the merge area. The measurement location should be such that congestion originating from excessive on-ramp volumes should be visible in the measurements.

If the upstream traffic volume q_{in} is constant, then the feedback law is easily shown to lead to $O_{out} = \hat{O}$ in the steady state. In other words, whatever the value of a constant (and not measured) upstream traffic volume, the feedback law leads occupancy to its desired value.

Similarly, if upstream traffic volume q_{in} is perturbed around a constant or slowly varying average, the feedback law keeps downstream occupancy O_{out} close to $\hat{O}$ on average. On the other hand, rapid time variations of q_{in} around the average value are only slightly reduced by the control system.

1.7 Override Tactics

The same override tactics may be applied to all the traffic-responsive strategies mentioned. More precisely:

(a) if the queue of vehicles on the ramp becomes excessive, interference with surface street traffic may occur (excessive queue lengths may be detected by suitably placed detectors); and

(b) the on-ramp volumes calculated by the traffic-responsive strategies are limited from above and from below.

The on-ramp volume values calculated by the strategies may be transformed into a corresponding green/red phasing by utilization of a saturation flow value. Alternatively, on-ramp volumes may be implemented in a car-by-car mode.

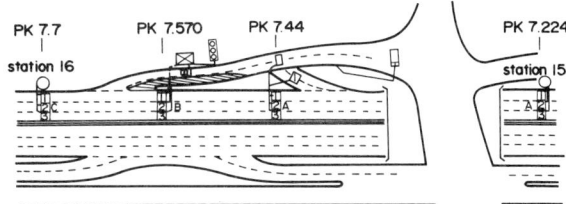

Figure 3
Functional diagram of the internal side of the southern part of the BP at Porte de Brancion

2. An experimental site

A functional diagram of the internal side of the southern part of the Boulevard Périphérique (BP) in Paris, including one on-ramp and one off-ramp is shown in Fig. 3. The total length of the stretch is equal to 1 km. The main features of the problems experienced at the Porte de Brancion site were regular and prolonged periods of congestion caused by high levels of demand. Congestion arose as a result of breakdown in the flow in the downstream section due to the high level of on-ramp volume.

Ramp metering is effected by ordinary traffic signals. All control functions are performed by a IBM-compatible microcomputer equipped with acquisition boards (Master-PC2).

3. Assessment of different strategies

Six strategies have been installed and tested on one on-ramp (Brancion) of the southern part of the BP. A systematic comparison of all candidate strategies was effected in a real-world study. From a one-year experiment, data collected during 13 d per strategy were selected for the comparative study. The comparison reported here is based on classical calculations, per strategy, of the following indexes:

(a) total travel time,

(b) number of vehicles served,

(c) congestion duration, and

(d) estimated number of diverted vehicles.

In order to perform a statistically relevant comparison, similarities between averages and deviations of demands were tested. Only demand measured during free-flow conditions was used to perform the tests. As a matter of fact, when the congestion reaches the upstream limit of the system, the demand is modified according to the control action. Both χ^2 and F tests proved that a statistical similarity at 5% level can be accepted (Hadj Salem *et al*. 1989).

Before reporting the detailed evaluation of the candidate strategies, the effect of the beaconing (geometrical restriction on the on-ramp) must be mentioned.

3.1 Static-Control Effect

Substantial benefits were obtained by physically restricting the ramp flow: beacons restricting the entering flow to one lane oblige the ramp users to enter the main flow in the right-hand lane only. Consequently, perturbations of the center and left-hand lanes are reduced along with a reduction of maximum ramp volume from 2000 vehicles per hour to 1600 vehicles per hour. This action, resulting in a smoothing effect on the mainstream flow, has allowed a reduction of the congestion

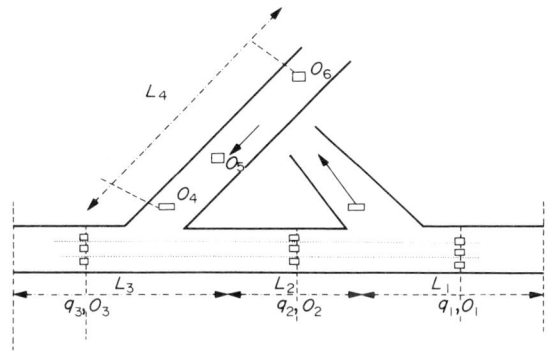

Figure 4
Diagram of the site used for the experiment

Table 1
TTT, TS and MTT per strategy for the peak period 7 am to 10 am

Control strategy	Time spent		Vehicles served		MTT	
	(vehicle hours)	(% benefit)	(vehicle kilometers)	(% benefit)	(s)	(% benefit)
Static regulation	421		16 463		92	
ALINEA	354	15.9	16 980	3.1	75	18.5
INRETS demand capacity	376	10.7	16 801	2.1	80	13.0
Simple demand capacity	407	3.3	15 143	−8.0	97	−5.4
Percent occupancy	438	−0.4	15 673	−4.8	100	−8.7
Fixed time	451	−7.1	16 116	−2.1	101	−9.8

Table 2
Congestion duration per strategy

Candidate strategy	Time spent in congestion per 6 h	
	(min)	(%)
ALINEA	53	14.7
INRETS demand capacity	81	22.5
Percent occupancy	103	28.6
Static regulation	108	30.0
Simple demand capacity	108	30.0
Fixed time	146	40.6

duration during the total peak period from 65% before beaconing installation to 24% afterwards. Because this action partially meets the general ramp-control objectives—regulating the ramp volume in order to reduce the insertion conflicts that generate main-flow perturbations—it has been considered as a control strategy. On the other hand, as this installation was maintained during the entire period of the experiment, this strategy has been taken as a reference in the remainder of the study.

3.2 Total Travel Time, Total Service and Mean Travel Time

The total tavel time (TTT) spent in the system was calculated as the sum of the travel time (TT) on the freeway (1 km around the ramp) and the waiting time (TW) on the ramp.

$$TTT = TT + TW$$

$$TT = \sum_{k=1}^{K} 3\alpha \, \Delta T [L_1 O_1(k) + L_2 O_2(k) + L_3 O_3(k)]$$

$$TW = \sum_{k=1}^{K} 2\alpha \, \Delta T L_4 [O_4(k) + O_5(k) + O_6(k)]/3$$

where $\alpha = 1.6$ vehicles per kilometer depends on the mean effective vehicle length (Bhouri 1987), $\Delta T = 40$ s is the cycle of the traffic lights, $L_1 = 0.375$ km, $L_2 = 0.125$ km, $L_3 = 0.5$ km and $L_4 = 0.4$ km and $O_1(k)$, $O_2(k)$, $O_3(k)$, $O_4(k)$, $O_5(k)$ and $O_6(k)$ are the occupancy rates ($\times 100$) measured on the BP and on the ramp, as shown in Fig. 4.

Total service (TS) is another index of effectiveness. Expressed in units of vehicle kilometers, it represents the number of vehicles served and is calculated as follows:

$$TS = \sum_{k=1}^{K} \Delta T [L_1 q_1(k) + L_2 q_2(k) + L_3 q_3(k)]$$

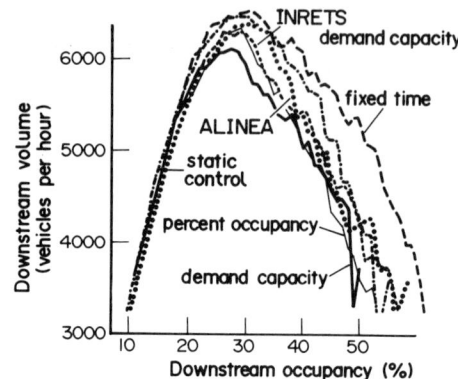

Figure 5
Downstream volume vs downstream occupancy for different control strategies

where $q_1(k)$, $q_2(k)$ and $q_3(k)$ are the mainstream volumes during the time horizon k.

Mean travel time (MTT) is given by TS/TT and values of TTT, TS and MTT per strategy are given in Table 1.

Due to the particular geometry of the experimentation site, it has been difficult to improve the rather optimal situation created with the static control. Table 1 shows that only two strategies (INRETS demand capacity and ALINEA) succeeded in improving the reference situation. ALINEA has given the best results: during the peak period (7 am to 10 am), the number of vehicles served is increased by 3% and the TTT is decreased by about 16%. Consequently, the MTT is reduced by 18.5%.

3.3 Congestion Duration

Congestion is determined by measuring the occupancy from the BP detector station installed downstream of the merge area. Congested flow is assumed to exist if occupancy is above the critical occupancy (31%). The congestion duration measured during the morning peak period ranges from 53 min to 146 min, expressing the great variability in the control efficiency (see Table 2). Fixed-time control proved to be the worst strategy (146 min). The feedback law ALINEA significantly improved the average situation obtained in the static control case, reducing the congestion duration by half.

Fundamental diagrams (see Fig. 5) present various

Table 3
Time spent during excessive queue override

Control strategy	Time (min)
ALINEA	17
INRETS demand capacity	37
Simple demand capacity	46
Percent occupancy	57

Table 4
Mean upstream demands (mainstream plus ramp) per strategy

Control strategy	Mean demand (vehicles per hour)	Difference (demand)	Mean on-ramp volume (vehicles per hour)	Difference (on-ramp)
Static regulation	4598		1108	
ALINEA	4849	251	1008	−100
INRETS demand capacity	4830	232	1008	−100
Fixed time	4521	−77	968	−140
Percent occupancy	4562	−36	918	−190
Simple demand capacity	4343	−255	947	−161

aspects regarding the strategy showing up some control details, especially in the congested situations: the control strategies do not act very differently during free-flow situations. On the contrary, the curves indicate that the strategies dealt differently with the congestion: fixed-time control allows 500 vehicles per hour more than the other strategies at a comparable occupancy rate. This means that during congestion and for this strategy, ramp users tend to accept shorter time headways during insertion. Furthermore, a 30% red-signal violation corresponding to this strategy, reinforces that interpretation. However, traffic conditions resulting from fixed-time control were much less stable than for other strategies and this results in long congestion duration as stated earlier. On the other hand, ALINEA leads to a more stable situation: the control is more consistent with the real conditions and red-signal violation is less frequent ($<10\%$).

An index corresponding to the time spent on excessive queue override (see Table 3), gives additional comprehension of the control strategies.

The control override is most frequently active during the occupancy strategy (17% of the total control period). As a matter of fact, entering volumes are kept high while the congestion has not reached the freeway upstream of the merge area. A major effect of this is the existence of violent congestion provoking frequent queues spilling back onto the surface road. Also, according to this index, ALINEA tends to react smoothly to high density and, hence, the excessive queue override arises for only 4.7% of the control period.

3.4 Diversion Aspects

So far, the efficiency of the control strategies was only approached in terms of measuring their effects within the controlled area. An attempt can now be made to show the control consequences outside this area (e.g., on diversion).

Direct measurement of the number of diverted vehicles would have demanded a complex and wide survey scheme including measurement capabilities not envisaged in the scope of this article.

In order to estimate the diversion effects, the following assumption was made: on the free-flow demand similarity basis, the differences that can be noticed upon upstream volumes (mainstream or ramp volumes) can be interpreted as a consequence of users diverting to avoid the congested area. More specifically, if one control strategy produces longer congestion on the mainstream than another, drivers will divert at off-ramps further upstream and will use alternative streets to reach their destination. This will result in lower accumulated upstream demand. On the other hand, on-ramp queues due to ramp metering may motivate some drivers to avoid the ramp (divert) and this again results in a smaller total volume entering by the ramp. Table 4 shows the differences between upstream demand volumes with respect to the tested strategies.

Only two strategies have proved to be superior to the reference case: the INRETS demand-capacity strategy and ALINEA. As a matter of fact, the two strategies each allow 100 ramp vehicles per hour fewer on average to enter the freeway than the static strategy. However, the noted upstream demand during the ALINEA control is 251 vehicles per hour higher than during the reference period.

Bibliography

Bhouri N 1987 Estimation de la densité dans une section d'autoroute. DEA Report, INRETS, Arcueil, France

Blosseville J M 1985 Stratégie adaptative de contrôle d'un accès à une autoroute. IRT Report, INRETS, Arcueil, France

Hadj Salem H, Blosseville J M, Davéc M M, Papageorgiou M 1988 ALINEA: Un outil de regulation d'accés isolé sur autoroute—Etude comparative sur site réel. INRETS Report No. 80. INRETS, Arcueil, France

Koble H M, Adams T A, Samant V S 1980 Control strategies in response to freeway incidents. Report No. FHWA/RD-80/005. Federal Highway Administration, Washington, DC

Masher D P, Ross D W, Wong P J, Tuan P L, Zeidler, Peracek S 1975 Guidelines for design and operating of ramp control systems. Standford Research Institute Report NCHRP 3-22, SRI Project 3340. SRI, Menid Park, CA

Papageorgiou M, Hadj Salem H, Blosseville J M 1989 ALINEA: A local feedback control law for on-ramp metering. *Prep. 70th Annual Meeting Transportation Research Board*, Paper No. 910659. TRB, Washington, DC

H. Hadj Salem and J. M. Blosseville
[INRETS, Arcueil, France]

M. Papageorgiou
[Technische Universität München, Munich, Germany]

On-Ramp Control of Freeway Networks

On-ramp control refers to measures taken to avoid or dissolve congestion by restricting the number of toll booths in operation and closing down on-ramps to adjust traffic inflow onto freeways. Since urban freeways form a complex network with many on-ramps, they require systematic on-ramp controls to decrease congestion efficiently. Linear programming (LP) is used for one of the inflow control methods.

The main objectives of on-ramp control are to reduce traffic congestion, traffic accidents and environmental pollution.

1. Principles of Freeway Traffic Control

There are two types of freeways: urban freeways and intercity freeways. Table 1 shows the differences between their characteristics based on pricing systems, networks and user conditions. These characteristics influence control principles (Sasaki 1969).

On urban freeways, the principles are considered as follows.

(a) Since users pay tolls in advance before entering the freeway, they may ask for a toll discount or reduced ticket if they are discharged from the freeway at an intermediate off-ramp to their destinations.

(b) Since the network usually extends both radially and circularly, it is not always possible to guarantee that the users are nearer to their trip destinations if they are discharged before their originally scheduled off-ramps.

(c) Since the off-ramps link up with urban streets which have inadequate capacity, it is difficult to discharge cars smoothly.

(d) Since users are sensitive to delay, administrators must maintain a smooth traffic flow whenever possible.

(e) Since there are many bottlenecks on urban freeways and the inflow demand is likely to exceed their capacities, natural congestion is very likely to occur.

(f) Since most users have homogeneous trip purposes and trip lengths, it is easy to estimate an origin–destination (OD) pattern between ramps. Then, by using the OD pattern and inflow volume from on-ramps, it is possible to predict the time when, and the link where, the natural congestion will occur.

(g) Since there are generally many on-ramps close to bottlenecks where natural congestion is predicted, it can be expected that natural congestion will be prevented if inflow volumes are controlled.

Table 1
Comparison of the characteristics of urban freeways and intercity freeways

	Urban freeway	intercity freeway
Pricing system	flat rate payment in advance	rate proportional to distance used deferred payment
Network		
geometry	radial and circular	extended linearly
ramp spacing	dense short interval	not so dense long interval
bottlenecks	many	few
off-ramps	linked with urban streets with limited capacity	linked with alternative highways with sufficient capacity
User conditions		
demand	large	not so large
trip purpose and length	homogeneous	heterogeneous
sensitivity to delay	sensitive	not so sensitive

From (a)–(c), it is apparent that discharge control should be kept to a minimum. Also, from (d)–(g), inflow control entails few problems. On urban freeways, inflow control on on-ramps should be introduced to prevent natural congestion before it occurs.

On intercity freeways, the principles are considered as follows.

(a) Since there are no serious bottlenecks on intercity freeways and inflow demand is not so large, natural congestion seldom occurs.

(b) Since the distance between ramps is long and a long time is usually taken to clear the obstacles after an accident has occurred, the accident congestion will develop in length.

(c) Since the users are not so sensitive to delays, it is not necessry to aim for perfect congestion prevention.

(d) Since the distances between ramps are long, the discharge control from the nearest upstream off-ramp can be expected to reduce the congestion more quickly than the inflow control on on-ramps far from the accident point.

(e) Since users pay tolls after leaving the freeway, there are few problems involving discharging tolls before the user's originally scheduled off-ramps.

(f) Since the network develops linearly with alternative highways, it is guaranteed that users approach their trip destinations when they are discharged before the originally scheduled off-ramps.

(g) Since the off-ramps link up with highways with sufficient capacity, it is fairly easy to discharge cars smoothly.

From (a)–(c), it is apparent that preventative controls against natural congestion need not be introduced, but induced controls should be introduced after accident congestion has already occurred. From (d)–(g), it is apparent that discharge control entails few problems. In intercity freeways, discharge control should be introduced to reduce accident congestion after it has already occurred.

2. Methods of Inflow Control

Inflow control is very effective in preventing cars from entering bottlenecks on freeways. Two methods of on-ramp control are described in the following. One is LP control and the other is sequential ramp closure (Sasaki and Myojin 1968, Myojin 1983).

2.1 LP Control

The control problem is stated as follows: optimize a certain linear function of inflow volume through each entrance ramp so that traffic flow, expressed as a linear function of inflow volume, is not above capacity at any point of the freeway.

Mathematically, this is described as the optimization of an objective function

$$F = a_1 u_1 + \ldots + a_i u_i + \ldots + a_k u_k \quad (1)$$

subject to

$$\left.\begin{array}{l} x_1 = Q_{11} u_1 + \ldots + Q_{i1} u_i + \ldots + Q_{k1} u_k \leq c_1 \\ \vdots \\ x_h = Q_{1h} u_1 + \ldots + Q_{ih} u_i + \ldots + Q_{kh} u_k \leq c_h \\ \vdots \\ x_m = Q_{1m} u_1 + \ldots + Q_{im} u_i + \ldots + Q_{km} u_k \leq c_m \end{array}\right\} \quad (2)$$

and

$$0 \leq u_1 \leq u_1^d, \ldots, 0 \leq u_i \leq u_i^d, \ldots, 0 \leq u_k \leq u_k^d \quad (3)$$

where u_i is the allowable inflow volume through on-ramp i ($i = 1, 2, \ldots, k$), u_i^d is the inflow demand volume through on-ramp i, x_h is the traffic volume estimated at section h on the freeway ($h = 1, 2, \ldots, m$), c_h is the traffic capacity of section h, a_i is a constant, Q_{ih} is a constant known as the influence factor, k is the number of on-ramps of the freeway and m is the number of sections of the freeway.

The freeway is divided into sections in each of which both the traffic capacity and the estimated traffic volume are uniform. In practice, on-ramps, off-ramps, merging points and diversion points divide the freeway into sections.

Q_{ih} is called the influence factor of on-ramp i on section h, since it represents the ratio of traffic volume appearing at section h to a single inflow through entrance ramp i.

Assuming that c_h and u_i^d are given in some way, two problems that remain are setting up an acceptable objective function and estimating influence factors.

(a) *Objective functions.* Two objective functions have been proposed.

If every $a_i = 1$, an objective function

$$F = u_1 + \ldots + u_i + \ldots + u_k \quad (4)$$

is obtained. Maximization of F means minimization of the inflow volume that is rejected entry by the control and maximum income for freeway systems with flat toll rates such as the Hanshin expressway in Japan. A certain portion of the excess traffic volume, $u_i^d - u_i$, waits to enter at on-ramp i until the next control time period and so the maximization of F is likely to lead to the minimization of waiting time at on-ramps.

The other objective function is set up by equating a_i with l_i, which is the mean use length of freeway by inflow from entrance on-ramp i to any destination off-ramp.

$$F = l_1 u_1 + \ldots + l_i u_i + \ldots + l_k u_k \tag{5}$$

Maximization of the objective function is equivalent to maximum service (in vehicle kilometers) supplied for freeway users.

By the second objective function, the shorter the mean use length of an on-ramp, the more likely the on-ramp is to be controlled. Further investigation, however, is needed in order to make a decision as to which function is the better one to use. The role the urban freeway has to perform in the area covered by the network, reasons why longer trips are preferable to shorter ones, the performance of traffic flows on freeways and the characteristics of the optimum solution are all matters to be investigated.

(b) Influence factors. For on-ramp i on section h of the freeway, this is given by Q_{ih}, the (i, h) component of the $k \times m$ matrix

$$\mathbf{Q} = (\mathbf{p}_1 \mathbf{R}_1, \ldots, \mathbf{p}_i \mathbf{R}_i, \ldots, \mathbf{p}_k \mathbf{R}_k)^{\mathrm{T}} \tag{6}$$

where $\mathbf{p}_i = (p_{i1}, \ldots, p_{ij}, \ldots, p_{in})$ and $i = 1, 2, \ldots, k$; $\mathbf{R}_i$ is $n \times m$ matrix whose components are 0 or 1; and n is the number of exit ramps of the network.

The row vector $\mathbf{p}_i$ has components p_{ij} equal to the ratio of a single vehicle entering through on-ramp i to exit off-ramp j.

$$p_{i1} + \ldots + p_{ij} + \ldots + p_{in} = 1 \quad (i = 1, \ldots, k)$$

It is supposed that $\mathbf{p}_i$ is known to the control.

$\mathbf{R}_i$ is called the route matrix, which expresses the routes from entrance ramp i to every exit ramp by entries of 0 or 1. It is written as

$$\mathbf{R}_i = \begin{pmatrix} r_{11}^i, \ldots, r_{1h}^i, \ldots, r_{1m}^i \\ \vdots \\ r_{j1}^i, \ldots, r_{jh}^i, \ldots, r_{jm}^i \\ \vdots \\ r_{n1}^i, \ldots, r_{nh}^i, \ldots, r_{nm}^i \end{pmatrix}$$

where

$$r_{jh}^i = \begin{cases} 1, & \text{if and only if the route from on-ramp } i \text{ to off-ramp } j \text{ passes through section } h \\ 0, & \text{otherwise} \end{cases}$$

The route matrix is easy to calculate. At most, one route is selected between each pair of entrance and exit ramps. The shortest one is selected if two or more routes exist between a pair of ramps and $r_{jh}^i = 0$ if there is no route between ramps i and j (i.e., j is unreachable from i). The reason why the shortest one is selected is that drivers are most likely to choose it under normal conditions.

It is clear that the influence factor Q_{ih} is given by the hth entry of row vector $\mathbf{p}_i \mathbf{R}_i$. $\mathbf{Q}$ is called the influence factor matrix.

(c) Vector form. Using matrix $\mathbf{Q}$ and inflow vector $\mathbf{u}$, traffic volume on every freeway section is shown by a row vector

$$\mathbf{x} = (x_1, \ldots, x_h, \ldots, x_m) = \mathbf{uQ} \tag{7}$$

where

$$\mathbf{u} = (u_1, \ldots, u_i, \ldots, u_k)$$

Eqn. (7) is another expression of the left-hand side of Eqn. (2).

Then, in matrix notation, LP control is stated as: optimize $F = \mathbf{ua}$, subject to $\mathbf{uQ} < \mathbf{c}$ and $0 < \mathbf{u} < \mathbf{u}^{\mathrm{d}}$, where $\mathbf{a} = (a_1, \ldots, a_i, \ldots, a_k)^{\mathrm{T}}$, $\mathbf{c} = (c_1, \ldots, c_h, \ldots, c_m)$ and $\mathbf{u}^{\mathrm{d}} = (u_1^{\mathrm{d}}, \ldots, u_i^{\mathrm{d}}, \ldots, u_k^{\mathrm{d}})$.

The inequality $\mathbf{uQ} < \mathbf{C}$ means that traffic flow is below capacity on every section of the freeway network. This means that traffic stagnation never occurs on the freeway.

(d) Queue-length constraint. LP control often leads to long periods of closure of some specified entrance on-ramps. The introduction of queue-length constraint was proposed to prevent the queue length on these entrance ramps from growing to disturb traffic flow on streets or on another freeway (Myojin *et al.* 1976).

2.2 Sequential Ramp Closure

Considering the lags of vehicles travelling between an entrance ramp and a certain section of freeway, traffic volume is expressed as

$$x_h(t) = u_1(t - t_{1h})Q_{1h} + \ldots + u_i(t - t_{ih})Q_{ih} + \ldots + u_k(t - t_{kh})Q_{kh} \quad (h = 1, 2, \ldots, m) \tag{8}$$

where $x_h(t)$ is the traffic volume at section h between t and $t + s$, t_{ih} is the travel time from entrance ramp i to section h, and $u_i(t)$ is the inflow through ramp i between t and $t + s$.

In sequential ramp closure, the ramp of minimum lag is, if necessary, closed first, and the one of second-minimum lag second, and so on, as is necessary. The cycle time of control is set to s, for example, $s = 5$ min.

Sequential ramp closure is better applied to those inflow demands that cause unsteady sharp changes, such as those in the "rush hour." However, it has no optimum criterion.

See also: On-Ramp Control: Coordinated Time-of-Day Strategies

Bibliography

Myojin S 1983 Traffic Control on urban expressway, LP control and sequential ramp control. *Proc. Seminar Highway and Urban Traffic Control Technology.* Taiwan Area National Freeway Bureau and China Road Federation, Taipei

Myojin S, Sakamoto H, Iwamoto S 1976 Some characteristics of ramp control on urban expressway network (in Japanese). *Proc. Jpn. Soc. Civ. Eng.* **247**(3)

Sasaki T 1969 Principles of traffic control on expressways (in Japanese). *Expressways and Automobiles* **12**(6), 27–32

Sasaki T, Myojin S 1968 Theory of inflow control on an urban expressway system. *Proc. Jpn. Soc. Civ. Eng.* **160**(12)

N. Inoue
[Fukuyama University, Fukuyama, Japan]

Y. Iida and T. Hasegawa
[Kyoto University, Kyoto, Japan]

On-Ramp Control: Realization Principles

Ramp metering is the practice of artificially limiting the number of vehicles entering a freeway in order to achieve some goal such as higher speeds on the freeway or redistributing traffic to lesser-used arterial streets. The overall objectives of freeway ramp metering and the details of the algorithms used to determine appropriate metering rates for individual on-ramps are discussed elsewhere (see *On-Ramp Control: Coordinated Traffic-Responsive Strategies*; *On-Ramp Control, Local*). This article describes how ramp metering is implemented at individual ramps after desired metering rates have been determined elsewhere.

Freeway ramp metering was first implemented on the Congress Street (now Eisenhower) Expressway in Chicago in September 1961. Since then a considerable body of experience has accumulated in North America, but no standard practice has evolved. The lack of a standard may be due to the fact that a wide variety of practices are effective.

1. General Principles

Ramp metering is usually effected with an ordinary two- or three-indication traffic signal which controls traffic flow on the ramp. Figure 1 shows a typical ramp-metering installation. In the US, the *Manual on Uniform Traffic Control Devices* requires a minimum of two signal faces for ramp-metering signals. Vehicles are required to stop at the signal on red and wait until the signal turns green. The green-plus-amber period is short, just long enough to allow passage of one or two vehicles, and the red period is varied to produce the desired metering rate. Amber indications are often omitted to imply that flow is limited to precisely one or two vehicles in each green period.

Best operation is achieved when one vehicle at a time is metered. A green period plus amber period of 3 s is appropriate. The maximum capacity that can be obtained with single-vehicle metering is about 900 vehicles per hour due to limitations on how quickly the metering signal can cycle.

Capacities up to about 1100 vehicles per hour can be achieved if two vehicles are allowed through in each green period. The two vehicles can be side by side (if the ramp is two lanes wide) or following one another. Two-abreast metering is generally considered slightly safer. If the two vehicles follow one another, the green period plus amber period should be 4.5–5.0 s. Three vehicles per green period produces no further increase in capacity and is generally believed to increase the frequency of rear-end accidents.

Some jurisdictions use a variant of two-abreast metering wherein vehicle release is alternated between the two lanes. The objective of this tactic is to remove the temptation for side-by-side vehicles to race for the lead as they merge into a single lane. However, jurisdictions that use simultaneous release generally observe cooperative behavior between the two vehicles; safety differences have not been demonstrated.

The minimum achievable ramp-metering rate is about 180 vehicles per hour with single-vehicle metering and twice that with double-vehicle metering. This limit occurs because the violation rate begins to rise steeply as red periods become longer than 15–20 s. If metering rates less than these are desired, full ramp closure should be used.

It is generally accepted that drivers are confused by ramp meters that sometimes pass single vehicles and at other times pass two or more vehicles, even if such operations occur at separate, well-defined times of day. Such operation is discouraged. However, no problems have been noticed where adjoining ramps use different metering styles so long as each ramp maintains a self-consistent mode of operation.

Figure 1
Typical single-lane ramp-metering installation

2. Geometry and Signing

There must be sufficient distance between the ramp-metering signal and the freeway to allow vehicles to accelerate to freeway speeds. On level terrain this requires at least 45 m for automobiles. Ramps that were originally designed under the assumption that vehicles would not have to stop may not have an adequate length available to allow ramp metering to be installed.

In addition to the acceleration distance from signal to freeway, there must be adequate storage space for vehicles in front of the signal. The amount of space required depends on the amount by which demand exceeds the metering rate, how long the excess demand persists and on driver tolerance for waiting. Since the willingness of drivers to wait depends on how attractive their alternative routes are, the queuing space required is highly dependent on local conditions.

Custom is to supply clear, definitive signing at the ramp-metering signal. "Stop here on red" and "one vehicle per green" are virtually standard.

Advance warning signs at the entrance to the ramp are required. Changeable message signs are most desirable; their message can be as brief as "meter on." Flashing amber indications accompanied by the fixed message "ramp metered when flashing" are less expensive and equally effective. Fixed signs with time-of-day indications such as "ramp metered from 6:30 to 9:00 am" are also acceptable.

3. Detectors

Although the overall metering rate is determined by systemwide considerations, there is considerable latitude in how the metering rate is achieved on each individual ramp. Ramp meters can operate successfully in a completely pretimed mode, but driver satisfaction is considerably enhanced when the ramp meter seems "responsive" (see *Traffic Control Modes*). Up to four vehicle detectors can be used on the ramp to produce locally responsive operation. Figure 2 shows how the four detectors are deployed on a typical freeway ramp. The checkin/demand detector is placed immediately in front of the metering signal in a position to detect any vehicles waiting for a green signal. In locally responsive operation, the metering signal dwells on red but a green signal is provided when the vehicle is detected or when the appropriate red period has elapsed, whichever comes last. Thus, a vehicle approaching a ramp-metering signal that has not been used recently receives an immediate green signal. This provides a high level of service under light traffic demand. The trend is to use two 2 m × 2 m loop detectors spaced 2 m apart for the checkin/demand detector. The loops are connected in parallel to a single detector to provide improved reliability and a longer detection area.

The checkout/passage detector is placed about 2.4 m beyond the ramp-metering signal. When a checkout/passage detector is used, the signal stays green until the checkout/passage detector indicates the vehicle has actually cleared the signal. This provides a longer green for those vehicles which are slow to react to the signal and prevents them from being cut off by the onset of a red signal. It also provides a shorter green period when the vehicle is quick to leave the signal, thus reducing any temptation for a second vehicle to squeeze through.

The short distance between the stop line and the checkout/passage detector can cause problems. Vehicles that are waiting for green at the metering signal may extend far enough beyond the stop line to cause actuation of the checkout/passage detector before the vehicles have been released. Moving the checkout/passage detector futher downstream is not an acceptable remedy since it introduces too much delay in the activation of the detector. However, logic that recognizes the premature activation of the detector and returns the metering signal to red when the activation of the checkout/passage detector *ceases* has been shown to be effective. Checkout/passage detectors are standard in new installations.

A queue detector is often placed upstream of the metering signal in a position to detect a queue before it spills back to interfere with arterial street traffic. When the queue detector is activated, the ramp-metering rate is increased, thus reducing the queue length. A queue detector can be used without any other detectors.

The least commonly used detector is the "merge detector". A merge detector is placed in the merge area in a location that will detect vehicles waiting to merge with the freeway main-line traffic. (Observation will generally indicate the exact location for this detector. A rule of thumb is to make the detector 24 m long, centered at the ramp "nose.") When the merge detector is activated, green is inhibited at the metering signal until the merge area is cleared. Merge detectors are rarely used without checkin/demand and checkout/passage detectors.

Detectors are often placed on the freeway main line.

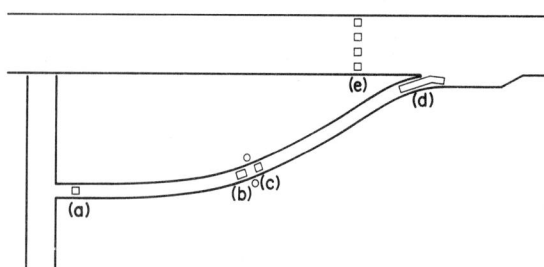

Figure 2
Deployment of detectors for responsive ramp metering: (a) queue detector, (b) checkin/demand detector, (c) checkout/passage detector, (d) merge detector and (e) main-line detectors (used to determine overall ramp-metering rates)

Such detectors are used to measure the main-line traffic performance from which an algorithm deduces the appropriate systemwide metering rates; except for determining metering *rates*, main-line detector information is not generally used in individual ramp operation. Algorithms to coordinate the release of vehicles from the ramp-metering signal with gaps in the main-line traffic were popular in the past but have largely been abandoned. Under heavy-flow conditions, there were rarely enough suitable main-line gaps detected to satisfy the required ramp-metering rates and vehicles had to be released even though no main-line gaps had been detected. Even under moderate-flow conditions, the benefits were modest; the main-line gaps often quickly disappeared, and the variability in acceleration of vehicles on the ramp meant the merging vehicle often failed to find the main-line gap, even if it still existed.

4. Start Up and Fail-Safe Operation

When ramp metering is not in operation, the ramp-metering signal is generally dark (although some successful systems allow it to display continuous green). When ramp metering is first turned on, the signal indicates green for about 1 min before the first red is displayed. (Signals that display green when not operating use amber or flashing amber before the first red.) In some systems, the queue detector, if present, is used to detect a 5–6 s gap which triggers the first red signal.

The most serious common malfunction is failure of the checkin/demand detector. The simplest safeguard is to limit the maximum red period to 15–20 s, even when there are no actuations of the checkin/demand detector. This is simple to implement but the 15–20 s default red is likely to be inappropriate. A better option is to time the detector actuations and, when none has occurred for about 2 min, fall back to time-of-day operation.

5. Priority Vehicles

In order to enhance the level of service provided by buses, they are usually exempted from queueing for the ramp-metering signal. There are two principal methods for exempting buses from ramp metering. Although each method is held to be the only effective method by some advocates, it seems that most reasonable methods work well.

One popular bus priority implementation is to provide a separate ramp, marked "buses only," which joins the regular ramp downstream of the ramp-metering signal. Thus, buses using the priority ramp completely bypass the waiting queue and metering signal. When this tactic is used, the priority buses do not stop; they travel unimpeded the length of their own ramp and merge with the metered traffic as it accelerates toward the freeway. There need be no detector

Figure 3
Ramp metering with car pool priority in adjoining lane

on the bypass ramp at all. If a detector is provided on the bypass ramp, it can be used to preempt the next cycle at the metering signal. This provides a gap for the priority buses to merge into and automatically includes them in the count of vehicles metered on the ramp.

The most common method of bus priority is to use two-lane ramps with one lane for ordinary traffic and the other reserved for priority buses. Some jurisdictions require buses to stop at the metering signal and wait for the next cycle, thus, in effect, metering both the ordinary traffic and the priority buses at equal rates. The priority buses have a shorter wait because there are fewer of them. Release can be simultaneous from both lanes or alternated. Other jurisdictions that use separate lanes for different classes of traffic on the same ramp allow the priority buses to pass the metering signal without stopping. When this is done, the metering signal is made inconspicuous from the priority lane and signing is provided to indicate that priority buses need not stop. The speed differential between the two lanes can be large, but no safety problems have been noted.

In many systems, car pools are given the same priority treatment as buses. Bus and car pool traffic is mixed on the priority lanes and the signing indicates "buses and car pools" instead of "buses only." Figure 3 shows a two-lane ramp-metering installation with priority treatment for car pools.

It is often difficult to implement priority treatments due to the difficulty in finding geometric arrangements that provide both a queuing area for regular vehicles upstream of the signal and a bypass or second lane for buses.

6. Exceptional Practice

Ramp metering is rare or unknown outside of the US and Canada. There is one ramp-metering installation in the UK; it is on the M6 motorway near Birmingham. The on-ramp consists of two lanes in the queuing area before the signal but narrows to one lane at the merge with the freeway main line. The control algorithm is highly responsive to conditions on the freeway. The metering signal is designed to operate more like a conventional traffic signal than a ramp meter in the American sense. The green interval conforms to the UK standards for minimum green period and is followed by an amber interval of ordinary duration. Platoons are usually eight or nine vehicles long. There is a queue detector at the upstream end of the ramp but no other detectors. The installation is reported to perform quite satisfactorily.

Some arterial street signals in Southern California are affected by the local ramp-metering rate. As the ramp metering rate is decreased, a historically based amount of traffic is assumed to be diverted from the freeway on-ramp to the arterial street. The arterial signal timing changes to reflect this increased total traffic. This technique is too new for its effectiveness to have been evaluated.

See also: Freeway Control: An Overview; On-Ramp Control: Coordinated Time-of-Day Strategies

Bibliography

Capelle D G, Basu S 1982 Freeway surveillance and control. In: Homburger W S (ed.) *Transportation and Traffic Engineering Handbook*. Prentice-Hall, Englewood Cliffs, NJ, pp. 786–93

Federal Highway Administration 1976 *Traffic Control Systems Handbook*. US Government Printing Office, Washington, DC

P. Ross
[United States Federal Highway Administration, McLean, Virginia, USA]

Optimal Routing Applications

The main goal in routing control is the assignment of an "optimal" route to all vehicles in a network, with respect to some cost criterion which is minimized. The intent is to get all the vehicles to their destinations as soon as possible or, equivalently, to clear the network of traffic, and the criterion selected is to reflect this intention. The criterion may, furthermore, reflect the cost or desirability of specific routes or the avoidance of congestion or delays at queues, and so on. The implementation considered is important both in the selection of the criterion and of the constraints imposed on the solution of the optimization problem.

In actual applications, the vehicle is usually expected to indicate its intended destination and the local computer at the intersection is to respond by specifying the route the vehicle is to follow. The statement may appear on a roadside display as the vehicle approaches or on a small indicator on the vehicle itself. The local or central computer is to base its decision on a number of parameters and variables, which it may originally know (e.g., network topology), measure locally (e.g., queue length at intersection) or receive as information transferred to it (e.g., downstream congestion). It will then decide, according to a given criterion which reflects total delays/costs in clearing congestion and minimizing travel, on the route to assign.

The concept of optimal routing in this context (i.e., providing a driver automatically with routing instructions at decision points in a road network) was initially researched in the USA in the early 1960s (Rosen *et al.* 1970). However, the whole area was dropped in the USA after the mid-1960s until recently, but was picked up and advanced by studies and implementation tests in both Japan and the FRG in the early 1980s (Braegas 1980). Of these two, the more ambitious was the Japanese approach which was evaluated in Tokyo, with route guidance and information units installed at 83 intersections (Hirose and Suzuki 1980, Tsuji and Kawashima 1981). The late 1980s produced a large amount of activity in Europe, due to the support of the Economic Community's Eureka program Prometheus (**pro**gram for **E**uropean **t**raffic with **h**ighest **e**fficiency and **u**nprecedented **s**afety). This has led to very large and ambitious route guidance experiments such as LISB in Germany and Autoguide in the UK. In Japan, experimentation has also developed through the AMTICS program.

1. Route Guidance

The basis of optimal routing for transportation applications is route guidance. The basis of all methods used in route guidance is a routing table. When an approaching vehicle indicates its intended destination to the roadside computer, a route (which is "best" by some measure) is assigned to the vehicle by checking through a routing table. (The term "computer" is used in a loose sense. It may be a microprocessor-based microcomputer or a special purpose digital circuit built with large-scale integration (LSI) technology.) The routing table for an intersection may either be preset (i.e., static, as in studies carried out in the USA including the work at Sperry), changeable from a central location under emergency conditions (as in the ALI system in Germany) or traffic dependent (i.e., dynamic, as in the CACS system in Japan). The dynamic aspect of CACS is indeed important since, in a congested urban network, it is this very aspect that provides the most benefit. Recent studies concur with this and provide some provisions in the roadside microcomputer for possible information exchange with a central computer facility. This large central computer would, presumably, either simulate traffic or compute an optimization algorithm with traffic-measurement-dependent parameters and, finally, update roadside routing tables. The CACS system essentially does this, providing an update every 15 min.

The advantages of dynamic route guidance are better handling of changes in traffic and emergency situations. These may be slightly offset by the more complicated hardware–software interface required (with the central computer) and the somewhat larger investment in the central facility. Modern approaches in decentralized/hierarchical control provide answers to these problems. Recent work has shown that, if the roadside computer is provided with some more computing power, dynamic route guidance may be carried out without resorting to a large central computing facility. A microcomputer at an intersection may be perfectly capable of updating local routing tables using approaches derived from recent advances in control.

Physical implementation of route guidance involves huge practical problems. However, intimately tied to these practical issues are more mathematical/theoretical questions.

2. The Model

Obviously, many models of different complexity may be used to represent flows in networks. These models can be stochastic or deterministic, and lumped or distributed. A simple deterministic model will be used to illustrate some of the issues related to decentralization and optimality tradeoffs. The model to be considered (Sarachik and Özgüner 1982) is that of a store-and-forward network as suggested by Gazis (1974) (see *Network Modelling and Control: Store-and-Forward Approach*). In this model, all delays due to congestion are associated with the queues that can build at the nodes of the network. Traffic is assumed to move between nodes at constant velocity. Thus, the time to pass from node to node is the sum of a constant delay for travel time and a portion associated with the waiting time in the queue. This eliminates the nonlinear dynamics that result when a density-dependent velocity profile is used. A similar model arises in data communications network except that the travel time between nodes is taken to be zero.

Consider a connected traffic network consisting of a single destination node O and N other nodes. All the nodes are connected by L directed branches or links and at each nondestination node n a queue of traffic $q_n(t)$ can form. Let $r_n(t)$ denote the rate at which traffic arrives at node n from outside the network and let $u_l(t)$ denote the rate of traffic flow along link l. (It is assumed that u_l is the value of the link traffic, originating at the node considered.) Each branch has a flow capacity C_l and the time delay in traversing link l is t_l. There are occasions when it will be clearer to use double subscripts to represent branch quantities as, for example, when link l is directed from node n to node k the notation u_{nk} or t_{nk} may be used. With the assumptions used here, any congestion delay encountered by traffic in traversing the network is associated with the time spent waiting in the queues. The dynamics of this

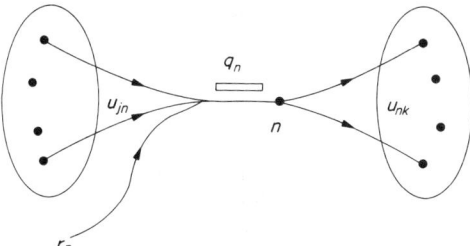

Figure 1
A single node of a network

system can be written as

$$\dot{q}_n(t) = r_n(t) + \sum_{j=1}^{N} u_{jn}(t - t_{jn}) - \sum_{k=1}^{N} u_{nk}(t) \quad (1)$$

or

$$\dot{q}_n(t) = r_n(t) + \sum_{l \in L_n^i} u_l(t) - \sum_{l \in L_n^o} u_l(t) \quad \text{for } n = 1, 2, \ldots, N \quad (2)$$

where L_n^i and L_n^o denote the index set of branches entering and leaving node n, respectively (see Fig. 1).

Since the queues can never become negative, each link has a limited capacity and traffic is not permitted to flow backwards on the directed links, the following constraints are imposed:

$$q_n(t) \geq 0 \quad \text{and} \quad 0 \leq u_l(t) \leq C_l \quad (3)$$

3. Decentralization

Decentralization implies that various controllers in the system are only allowed to measure certain outputs of the system and control certain inputs. This differs from the centralized case where a single controller may measure all variables in the system and control all inputs. The decentralized information structure often appears in large-scale systems, as traffic systems, where it may be impractical, unreliable and costly to centralize all inputs and measurements. Thus, the information structure associated with a network is the specific assignment of measurements to controls.

The question of deciding on an information structure must now be faced. Assumptions can be made about three types of structure:

(a) *Full centralization.* A central facility has access to all on-line measurements of variables and can inform the local controllers of the appropriate control action.

(b) *Full decentralization.* Local controllers can measure local variables (e.g., queue lengths, flows) and can, with a knowledge of the static characteristics of the network (capacities and costs of links), decide on a strategy. Obviously, any major changes downstream such as major deviations from statistical averages and accidents may make this unsatisfactory. However, consideration of a fully decentralized setup is important in that the same control strategy may be used by updating the changing parameters (e.g., link capacities or costs) at appropriate intervals.

(c) *Partial decentralizaetion.* This may occur in a number of ways. A regional coordinator may pass necessary aggregate information to the local decision makers or neighboring intersections may transmit aggregate information to each other. Furthermore, this may be done at coarser time intervals than used in local decision making.

The choice of which information structure to have and which specific piece of information to transmit is intimately related to, first, the optimality criterion assumed and, second, the cost of establishing the required information channels. In many optimal routing applications (specifically in communication networks), the "transported entity" and the "transmitted information" (regarding routing) utilize the same media. This, of course, is not the case in transportation networks.

4. Optimality

It must always be remembered that the specification of a cost criterion for calculation of an optimal control is essentially an intermediate step, a means of obtaining "good behavior" in the system. Therefore, different criteria may be considered and different "optimal" solutions may be found, the optimal control with respect to one criterion being suboptimal with respect to another. The choice of the final control to implement is an engineering decision made after considering the overall behavior of the system. This choice must be made after also considering the possible cost of adding new information channels and whether the computational burden imposed, be it in terms of time or computer facilities, is really worth the incremental enhancement in system behavior.

One performance measure by which the routing strategy is evaluated is the aggregate delay up to a specified time T. This is expressed as

$$J\{u(t)\} = \int_{t_0}^{T} \sum_{n=1}^{N} q_n(t) + \sum_{l=1}^{L} t_l u_l(t) \, dt \quad (4a)$$

where u denotes the vector with components u_l

($l=1, \ldots, L$). Controllers then compute u to obtain

$$\min_{u} [J\{u\} - J^*] \quad (4b)$$

where J^* is the steady-state cost.

Another possible performance measure is

$$J\{u(t)\} = \int_{t_0}^{T} \sum_{n=1}^{N} q_n^2(t) + \sum_{l=1}^{L} t_l (u_l(t) - u_l^*)^2 \, dt \quad (5)$$

where $\{u_l^*\}$ are the steady-state flows. The values of t_l (the travel times along links) are also the weights associated with travelling along that link.

The approach taken by Sarachik and Özgüner (1982) uses Eqn. (4) as an intersection level criterion and finds a decentralized control based on Eqn. (4b). A partially decentralized information structure is then assumed with estimates of t_l being supplied to each node from its neighbors. The criterion given by Eqn. (5) can also be considered with this type of embedding approach. The result is, of course, not optimal with respect to either of the criteria given in Eqns. (4) and (5) considered for the composite system in Eqn. (1). However, it is a feasible solution for the problem as simplified and restated.

Let $\gamma_l = C_l - u_l^*$ be the excess capacity of a link. The optimal solution is given by Sarachik and Özgüner (1982) as

$$u_i(t) = \begin{cases} C_i & \text{when } q_i > \sum_{l=1}^{i} \gamma_l (t_i - t_l) \\ 0 & \text{when } q_i \leq \sum_{l=1}^{i} \gamma_l (t_i - t_l) \end{cases} \quad (6)$$

Note that this implies the existence of a routing table at each intersection. The local measurement of the queue length is compared against values on this table to provide a decision regarding the use of each exit link.

One key feature of this approach is the estimation of the values of t_l and their transmittal. To be effective, these estimates have to be easy to calculate and lead to suboptimal solutions that are satisfactory.

5. Computational Complexity

The approach outlined in Sect. 4 is not necessarily optimal if a centralized implementation is possible. However, another problem not discussed so far is the computational complexity of obtaining an optimal solution. The decentralized approach presented evaded both this issue and the time delays in the system by considering small simple subsystem models and local optimization problems. A hierarchical optimization approach will be outlined to solve the general problem with queue length constraints and delays in the system.

The problem is defined in discrete time, so that the

queue length at node n at sampling time k is denoted as $q_n(k)$. Let $r_n(k)$ denote the rate at which traffic arrives at node n during the sampling interval. Let $r_n(k)$ denote the rate at which traffic arrives at node n during the sampling interval. Let $u_j(k)$ denote the rate of traffic flow along link j at sampling time k. The dynamics of the network can be represented by

$$q_n(k+1) = q_n(k) + \sum_{j \in L_n^i} u_j(k-t_j) - \sum_{j \in L_n^o} u_j(k) + r_n(k) \quad (7)$$

where, as before, L_n^i and L_n^o denote the index set of branches entering and leaving node n, respectively.

The cost function is described by

$$J = \sum_{k=0}^{K} \left[\sum_{i=1}^{N} Q_i q_i^2(k) + \sum_{j \in L} R_j (u_j^* - u_j)^2 \right] \quad (8)$$

where Q_i and R_j are weighting functions, u_j^* is the steady state for $u_j(k)$ along link j and L denotes the sets of all links. (This quadratic criterion, which is the discrete version of Eqn. (5), is picked simply to utilize an available control formulation. Its minimization implies an attempt to drive the queue lengths to zero and the link traffic to their steady-state levels.)

The overall system can be represented by a linear vector matrix difference equation with pure delays in the controls:

$$q(k+1) = \mathbf{A}_0 q(k) + \mathbf{B}_0 u(k) + \mathbf{B}_1 u(k-1) + \cdots$$
$$+ \mathbf{B}_m u(k-m) + v(k) \quad k = 0, 1, \ldots, K-1 \quad (9a)$$
$$q(k) = q_k \quad k = -m, -m+1, -m+2, \ldots, -1, 0 \quad (9b)$$
$$u(k) = u_k \quad k = -m, -m+1, -m+2, \ldots, -1 \quad (9c)$$

where q is the vector of queues at all the nondestination nodes, $\mathbf{A}_0$ is an identity matrix, $\mathbf{B}_j$ $(j=0,1,\ldots,m)$ are the control weighting matrices, $u(k)$ is the control vector, $u(k-j)(j=0,1,\ldots,m)$ are the delayed flow controls with pure delays on the links and v is the vector of external traffic added to the network.

The quadratic form has been chosen for analytical tractability. An approach due to Tamura (1972) will be applied to solve this optimization problem. In the context of a nonlinear programming problem, the cost function J given by Eqn. (8) can be considered as the primal problem which is minimized subject to

$$\mathbf{A}_0 q(k) + \mathbf{B}_0 u(k) + \mathbf{B}_1 u(k-1) + \cdots + \mathbf{B}_m u(k-m) + v(k)$$
$$= q(k+1) \quad k = 0, 1, \ldots, K-1 \quad (10a)$$
$$0 \leq q(k) \leq q_{\max} \quad k = 0, 1, \ldots, K \quad (10b)$$
$$0 \leq u(k) \leq C \quad k = 0, 1, \ldots, K-1 \quad (10c)$$

The constraints can be divided into two groups: Eqn. (10a) and Eqn. (10b, c). Suppose the optimal solution to the problem exists. The dual problem can be written;

in terms of the Lagrangian function $L(q, u, p)$ as

$$F(p) = \min_{q, u} [L(q, u, p)$$
subject to Eqns. (10b) and (10c)] (11)

where p, the dual variable (or Langrangian variable), is of the same dimension as the state vector $q(k)$. The Lagrangian is defined as

$$L(q, u, p) = J + \sum_{k=0}^{K-1} p^T(k) * [-q(k+1) + \mathbf{A}_0 q(k) + \mathbf{B}_0 u(k)$$
$$+ \mathbf{B}_1 u(k-1) + \cdots + \mathbf{B}_m u(K-m) + v(k)] \quad (12)$$

The algorithm is formulated within a two-level structure where on the first level, given initial values of $p(k)$, the Lagrangian L is decomposed using its additive separability and independent minimizations are performed for each queue and control variable to give, for $k = 0, 1, \ldots, K-1$,

$$q^i(k+1) = \text{sat}_1 [Q^{-1}(k^i(k) - \mathbf{A}_0 p^i(k+1)] \quad (13)$$

$$u^i(k) = \text{sat}_2 \left[-R^{-1} \sum_{j=0}^{m} \mathbf{B}_j p^i(k+j) \right] \quad (14)$$

where, for $a = 1, 2, \ldots, n$, the ath element of $\text{sat}_1[z]$ is given by

$$\text{sat}_1[z] = \begin{cases} q_{b, \max} & \text{if } w_b > q_{a, \max} \\ z_a & \text{if } 0 \leq z_a \leq q_{a, \max} \\ 0 & \text{if } z_a < 0 \end{cases} \quad (15a)$$

and, for $b = 1, 2, \ldots, r$, the bth element of $\text{sat}_2[w]$ is

$$\text{sat}_2[w] = \begin{cases} C_b & \text{if } u_b > C_b \\ w_b & \text{if } 0 \leq w_b \leq C_b \\ 0 & \text{if } w_b < 0 \end{cases} \quad (15b)$$

The value of $p(k)$ is generally not optimal and will be improved in the second level iteratively using

$$\nabla F(p)|_{p=p^*} = -q^*(k+1) + \mathbf{A}_0 q^*(k)$$
$$+ \mathbf{B}_0 u^*(k) + \mathbf{B}_1 u^*(k-1) + \cdots$$
$$+ \mathbf{B}_m u^*(k-m) + v(k) = e(k) \quad (16)$$

Further details of the algorithm with applications to specific networks are given by Ng (1984).

It can be observed that the optimal open-loop control for given initial conditions can be obtained. However, this is at the expense of extensive computation depending on how long the upper level iterations take to converge. Thus, this approach may not be useful for real-time implementation. Therefore, even if an optimal control is calculable, a suboptimal but computationally simpler solution may be preferable.

6. Further Research

A number of researchers have addressed theoretical/mathematical issues relevant to optimal routing. Some of these are in a communications network context. Basic work was carried out by Segall (1977) on the dynamic problem, proposing a centralized optimal control. The steady-state (static) problem was considered by Gallager (1977) and numerous other researchers. The single-destination, decentralized dynamic control problem was considered by Chu (1976), and Sarachik and Özgüner (1982). The multi-destination case has been receiving greater attention more recently (Özgüner and Sarachik 1979, Sarachik 1979, Casalino et al. 1983).

An attempt has been made here to discuss the interplay of the issues of decentralization, optimality and computational complexity in the context of optimal routing in traffic networks. The following points should be stressed.

It is possible to implement optimal routing using a partially decentralized approach. This approach is appealing since it does not require extensive local computational effort or extensive coordination. However, this approach may not be globally optimal if a centralized control implementation is possible.

A computationally attractive hierarchical optimization approach may also be used to solve for a globally optimal control. In spite of the numerical advantage it provides, this may still be too slow for real-time implementation. However, it does provide a benchmark against which suboptimal solutions may be compared.

See also: Queuing Theory Applications; Route Guidance, Collective; Route Guidance, Individual

Bibliography

Braegas P 1980 Function, equipment, and field testing of a route guidance and information system for drivers (ALI). *IEEE Trans Veh. Technol.*, **29**, 216–25

Casalino G, Davoli F, Minciardi R, Zoppoli R 1983 On the structure of decentralized dynamic routing strategies. *Proc. 22nd IEEE Conf. Dec. and Contr.* San Antonio, TX, pp. 472–6

Chu K C 1976 Decentralized real-time control of congested traffic networks. *IBM Research Report* No. RC-6101

Gallager R G 1977 A minimum delay routing algorithm using distributed computation. *IEEE Trans. Commun.* **25**(1), 73–84

Gazis D C 1974 Modelling and optimal control of congested transportation system. *Networks* **4**, 113–24

Hirose H, Suzuki N 1980 Testing and results of route guidance system in urban area. *J. Oper. Res. Soc. Jpn* **25**, 221–7

Ng H 1984 A study of optimal dynamic routing in traffic networks. M.Sc. thesis, The Ohio State University

Özgüner Ü, Sarachik P E 1979 Decentralized routing in congested multidestination traffic networks. *IEEE Int. Conf. of CCC*

Rosen D A, Mammano F J, Favout R 1970 An electronic route guidance system for highway vehicles. *IEEE Trans. Veh. Technol.* **19**, 143–51

Sarachik P E 1979 Clearing of congested multi-destination networks. *Proc. Engineering Foundation Conf. Research Direction in Computer Control of Urban Traffic Systems.*

Sarachik P E, Özgüner Ü 1982 On decentralized dynamic routing for congested traffic networks. *IEEE Trans. Autom. Control* **27**, 1233–8

Segall A 1977 The modelling of adaptive routing in data communication networks. *IEEE Trans. Commun.* **25**, 85–95

Tamura H 1972 Decentralized optimization for distributed-lag models of discrete systems. *Automatica* **11**, 593–602

Tsuji H, Kawashima H 1981 Testing and evaluating a pilot system for route guidance in Tokyo. *Proc. 8th IFAC World Congress*, Vol. 17. Pergamon, Oxford, pp. 132–8

Ü. Özgüner
[Ohio State University, Columbus, Ohio, USA]

Origin–Destination Matrix: Dynamic Estimation

Prediction of traffic evolution within a road network requires two sorts of information: the current traffic demand represented, for example, by volumes and density profiles within the network and the destinations of the various traffic flows, which are usually given in the form of the origin–destination (OD) matrix which contains flow values from each origin to every destination. While the actual density distribution provides the initial state for a prediction, the OD matrix represents basic information about the transition to future states. Thus, knowledge of the OD pattern of a traffic facility is an important prerequisite both for long-term planning and for short-term traffic management and control.

While traffic volumes can easily be measured, providing a basis for monitoring the current traffic state (see *Flow Variables: Estimation*), OD flows can only be observed directly with a great effort; for example, by taking individual interviews, by licence plate surveys or by costly image processing techniques (see *Video Sensors*). Although there will be some additional information available in the future from advanced route guidance and trip planning systems, these sample data will probably be insufficient to provide reliable OD estimates for a long time. For this reason, there has been considerable interest in developing methods by which missing OD information can be obtained indirectly using conventional volume measurements. First approaches involved static methods (see *Origin–Destination Matrix: Static Estimation*). These use accumulated traffic counts to set up a highly under-determined set of equations for the unknown OD flows. To complete this set, various additional assumptions are made and vague *a priori* knowledge is used to obtain estimates for the OD flows. For a further survey of these methods see Willis and May (1981) and Bell (1984).

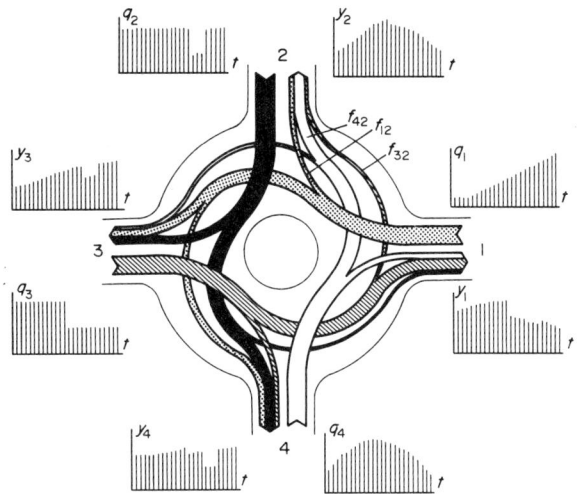

Figure 1
Roundabout with time diagrams of entry and exit flows

There are two major disadvantages with these methods. First, the additional assumptions are often not fulfilled, which leads to inaccurate if not unreliable estimates. Second, since these methods are designed to determine a constant OD matrix, they provide no efficient tool to track time-varying OD patterns as is necessary for traffic-responsive control systems. Attempts to apply these methods repeatedly at short intervals helped to track slow variations in simple cases, but did not overcome the general difficulties (Hendrickson and McNeil 1984, Nguyen et al. 1988).

In a pioneering work, Cremer and Keller (1981) showed that static methods waste measurement information by accumulating traffic counts. When counts are taken as short-interval time series, additional information can be extracted from the variation of the measurements over time which, in many cases, is sufficient to provide a unique, bias-free solution for the estimated OD flows. Further investigations have shown that a whole class of dynamic methods can be defined that follow this idea by applying well-known procedures from the fields of systems dynamics and estimation theory.

1. Basic Idea

A roundabout will be considered with four entries and four exits, as in Fig. 1. It is supposed that the proportions by which each entry flow splits according to the destinations of the three other exits are strictly constant over time (i.e., they are constant for each measurement interval). It is assumed that the sequences of short-time flow measurements depend on time according to the depicted time diagrams. Then, by the constant split ratios as indicated in Fig. 1 by the f_{ij} values, the exit flow sequences would sum up by superposition to the sequences given by the exit flow diagrams. Alternatively, given only the time diagrams of the entry and exit flows, the unknown splitting ratios can be inferred by simple comparison and analysis of the time diagrams. For example, the diagram of exit flow y_1 indicates by inspection that a major part of entry flow q_3, a medium part of q_4 and a small part of q_2 contribute to this exit flow.

However, entry flows show more of a random pattern than these clear-cut geometric forms. Moreover, the assumption of strictly constant splitting does not hold. This means that the unknown splitting ratios cannot be deduced simply by inspection of the time sequences of measured flows. Nevertheless, practically hidden under random noise, the output sequences bear information about the average splitting pattern within their time record and this can be disclosed by skilful signal processing procedures.

To formulate a mathematical model for the dynamic process of exit volume generation, which in this case is rather elementary, the following variables are introduced (see *Flow Variables*): $q_i(k)$ is the volume that enters entrance i during the time interval $(k-1)T \leq t < kT$ $(i = 1, \ldots, m)$, $y_j(k)$ is the volume that leaves exit j during the time interval $(k-1)T + \tau \leq t < kT + \tau$ $(j = 1, \ldots, n)$, $f_{ij}(k)$ is that part of volume $q_i(k)$ that leaves the intersection through exit j during the time interval $(k-1)T + \tau \leq t < kT + \tau$, and $b_{ij}(k)$ is the proportion of q_i forming $f_{ij}(k)$ (i.e., $b_{ij}(k) = f_{ij}(k)/q_i(k)$). T is a sampling interval defining short measurement periods. For an intersection, this interval should be chosen to be only a few minutes so that the time variations of the volumes are expressed representatively. The parameter τ denotes the average travel time a vehicle takes to pass from an entry to an exit. In the case of a network or a freeway, it may become necessary to choose a larger value of T and to introduce individual travel times τ_{ij} between entry i and exit j. (The presented procedures have to be modified appropriately to be applied in this case, as shown by Klaas (1986).)

From a balance of entering and leaving vehicles

$$y_j(k) = \sum_{i=1}^m f_{ij}(k) = \sum_{i=1}^m q_{ij}(k) b_{ij}(k) = \boldsymbol{q}^T(k) \cdot \boldsymbol{b}_j(k) \quad (1)$$

where $\boldsymbol{q}^T(k)$s is the $1 \times m$ vector of entry volumes $q_i(k)$ and $\boldsymbol{b}_j(k)$ is the $m \times 1$ column vector of split parameters $b_{ij}(k)$ $(i = 1, \ldots, m)$. Forming the $1 \times n$ vector $\boldsymbol{y}^T(k)$ of exit volumes and introducing the matrix $\mathbf{B}$ with elements $b_{ij}(k)$ (i.e., with columns $\boldsymbol{b}_j(k)$), the model equation (Eqn. (1)) can be written for all $j = 1, \ldots, n$ in condensed vector notation:

$$\boldsymbol{y}^T(k) = \boldsymbol{q}^T(k) \cdot \mathbf{B}(k) \quad (2)$$

From the definition of the split parameters b_{ij}, the following constraints are obtained:

$$0 \leq b_{ij}(k) \leq 1.0 \quad \text{for all } i, j \tag{3a}$$

$$\sum_{j=1}^{n} b_{ij}(k) = 1.0 \quad \text{for } i = 1, \ldots, m \tag{3b}$$

In addition, it is reasonable to assume that no car leaves the intersection through its entrance port, which gives a further condition

$$b_{ii}(k) = 0 \quad \text{for } i = 1, \ldots, m \tag{3c}$$

Equations (1) (or in condensed form Eqn. (2)) and (3) define the mathematical model of intersection flows. Equation (1) suggests that the model can be decomposed into n submodels for the individual exit flows. However, it should be noted that these submodels are coupled by constraints (Eqn. (3b)) which are imposed on the rows of the split coefficient matrix $\mathbf{B}$ while Eqn. (1) involves the columns of $\mathbf{B}$.

As mentioned, extended traffic facilities such as street corridors, freeways, road networks or public transport lines require more complicated models including individual travel times τ_{ij} and additional phenomena such as dispersion and nonlinear saturation effects. Basically, the following methods may correspondingly be applied to more complex models; however, as more complex models introduce additional sources of random uncertainties and weaken the correlation between entry and exit flow sequences, the quality of the estimates will deteriorate.

For most traffic facilities, the split pattern, as represented by the split coefficients $b_{ij}(k)$, varies less than the OD flows $f_{ij}(k)$ which also contain the fluctuations of the entry volumes $q_i(k)$. Therefore, dynamic estimation methods are formulated to identify the coefficients b_{ij}; the OD flows are then easily obtained by multiplication with the measured entry flows q_i.

2. Estimation Methods

Existing dynamic OD estimation methods follow a common principle which is shown in Fig. 2.

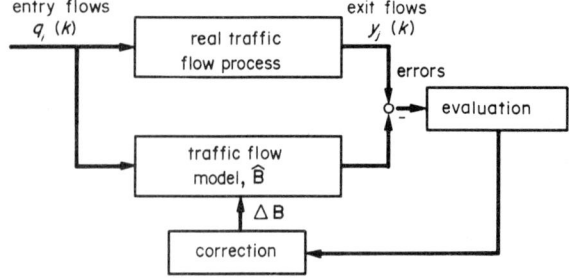

Figure 2
General principle of model-based estimation scheme

Measurements of entry and exit flows are collected from the real traffic system whose OD matrix is to be estimated. The entry flows are inserted into the model equations which, on the basis of tentative estimates for the OD parameters, determine estimates for the sequence of exit flows. These modelled exit flows are then compared with the real measurements and, from an evaluation of the exit flow estimation errors, corrections of the estimated OD parameters are deduced according to some correction scheme.

There are two classes of computational procedures. The methods of the first class consider the flow sequences of a whole observation period and try to adjust the OD parameters in order to minimize a function of the total error sequence (e.g., the mean quadratic error). These methods are quite useful for estimating constant mean values of OD parameters. The methods of the second class update the estimates recursively in each elementary sample interval whenever new measurements become available. Since these methods give new measurements more weight than old measurements, they are suitable to track time-varying OD patterns. As was shown by Nihan and Davis (1987), the two classes are interrelated and some methods may be formulated in both ways.

A least-squares optimization problem, which evolves in a natural way from the previous discussion, will be considered.

Given the sequences $q_i(k)$ and $y_i(k)$ of flow samples at the entries and exits of an intersection over an observation period of K sampling intervals, the problem is to find a constant matrix $\hat{\mathbf{B}}$ of estimates for the split parameters b_{ij} that minimizes the mean square of the prediction errors

$$J = \frac{1}{K} \sum_{k=1}^{K} \|\mathbf{y}(k) - \hat{\mathbf{y}}(k)\|_2 \to \min_{\hat{\mathbf{B}}} \tag{4}$$

where the estimates $\hat{b}_{ij}$ fulfill the constraints given in Eqn. (3).

For the present, the constraints given in Eqn. (3) will be ignored. Substituting $\hat{\mathbf{y}}$ in Eqn. (4) from Eqn. (2) (with $\mathbf{B}(k) = \hat{\mathbf{B}}$) and setting the first derivative of J with respect to $\hat{\mathbf{B}}$ equal to zero gives the ordinary least-squares solution

$$\hat{\mathbf{B}} = \left[\sum_{k=1}^{K} \mathbf{q}(k) \mathbf{q}^{\mathrm{T}}(k) \right]^{-1} \left[\sum_{k=1}^{K} \mathbf{q}(k) \mathbf{y}^{\mathrm{T}}(k) \right] \tag{5}$$

(For computational details see Cremer and Keller (1987), and Nihan and Davis (1987).) The matrix inverse on the right-hand side exists for $K \geq m$ provided no time sequence $q_i(k)$ is totally correlated with the other sequences $q_j(k)$. It is interesting to note that this solution was also obtained by Cremer (1983) using correlation techniques. In fact, the matrices within the brackets on the right-hand side of Eqn. (5) are the finite-interval cross-correlation matrices correlating $q(k)$ with $q(k)$ and $q(k)$ with $y(k)$, respectively.

If this solution fulfills the conditions given in Eqn. (3) (i.e., if the estimates $\hat{b}_{ij}$ lie in the admissible region of the parameter space), the optimization problem stated is solved. Unfortunately, this is not true in many cases. A possible procedure in this case is to use some kind of projection of this solution onto the admissible parameter subspace or, more simply, to apply truncation and normalization to fulfill the constraints afterwards. (Several possibilities are discussed in detail by Nihan and Davis (1987).) In all these cases, the modified estimates are not optimal in the sense stated previously. Another method that solves the constraint minimization problem by parameter optimization techniques was proposed by Cremer and Keller (1984); however, this method requires considerable effort and may create problems in real-time applications.

Suppose an estimated OD matrix has been computed on the basis of the last $k-1$ measurements. If these estimates are to be improved by the information contained within the new set of measurements arriving a sample time later at kT, the new terms $q(k)q^T(k)$ and $q(k)y^T(k)$ must be added to the right-hand side of Eqn. (5) and the matrix inversion must be carried out again. It has been shown by Nihan and Davis (1987) that, by applying the matrix inversion lemma, this matrix inversion can be avoided. According to this result, the computation of $\hat{\mathbf{B}}$ by Eqn. (5) can be reformulated into a recursive scheme where estimates are determined for each sample time interval yielding a time series of estimated matrices $\hat{\mathbf{B}}(k)$. This computational scheme consists of the following equations which are formulated for the jth column vector of $\hat{\mathbf{B}}$ for convenience:

$$\hat{b}_j(k) = \hat{b}_j(k-1) + d(k)[y_j(k) - q^T(k)\hat{b}_j(k-1)] \quad (6a)$$
$$d(k) = \mathbf{P}(k-1)q(k)[1 + q^T(t)\mathbf{P}(k-1)q(k)]^{-1} \quad (6b)$$
$$\mathbf{P}(k) = \mathbf{P}(k-1) - d(k)q^T(k)\mathbf{P}(k-1) \quad (6c)$$

where $\mathbf{P}(k)$ has the interpretation of the inverse correlation matrix given by the first bracket term on the right-hand side of Eqn. (5). To start this recursive scheme, initial values $\hat{b}_j(0)$ and $\mathbf{P}(0)$ are needed. The selection of these values is discussed by Nihan and Davis (1987) or, in a more general context, by Ljung and Söderström (1983).

The recursive computation scheme given in Eqn. (6) shows great similarity to the Kalman Filter equations which have been proposed as yet another method for OD parameter estimation by Cremer and Keller (1987) and by Nihan and Davis (1987). For this, the model equations (Eqn. (1)) are formulated in a slightly different way with state equations

$$b_j(k+1) = b_j(k) + w(k) \quad (7)$$

and measurement output equations

$$y_j(k) = q^T(k)b_j(k) + v_j(k) \quad (8)$$

Eqn. (1) takes the role of the measurement output equation with measurements $q_i(k)$ as time-variant system parameters. The variables $w(k)$ and $v_j(k)$ represent random white-noise terms with zero mean and covariances:

$$E\{w(k)w^T(l)\} = \begin{cases} \mathbf{W} & \text{for } k=l \\ 0 & \text{for } k \neq l \end{cases}$$
$$E\{v_j(k)v_j(l)\} = \begin{cases} r & \text{for } k=l \\ 0 & \text{for } k \neq l \end{cases} \quad (9)$$

Applying the Kalman Filter formalism to this problem yields the following set of equations:

$$\hat{b}_j(k+1) = \hat{b}_j(k) + d(k)[y_j(k) - q^T(k)\hat{b}_j(k-1)] \quad (10a)$$
$$d(k) = [\mathbf{P}(k-1) + \mathbf{W}]q(k)$$
$$\times [q^T(k)(\mathbf{P}(k-1) + \mathbf{W})q(k) + r]^{-1} \quad (10b)$$
$$\mathbf{P}(k) = [\mathbf{I} - d(k)q^T(k)][\mathbf{P}(k-1) + \mathbf{W}] \quad (10c)$$

This set produces conditional minimum variance estimates. Setting $\mathbf{W} = 0$ and $r = 1$ gives the recursive unconstrained least-squares estimator of Eqn. 6.

The first approach to dynamic OD estimation methods, which was proposed by Cremer and Keller (1981), belongs to this class of recursive methods. In this procedure, the new estimate is again computed by an equation of identical form to Eqns. (6a) and (10a) where the error gain vector $d(k)$ is computed by the very simple expression

$$d(k) = \gamma \Delta q(k) \quad (11)$$

where Δq is the vector of entry flow deviations from their respective mean values (i.e., $\Delta q_i(k) = q_i(k) - \bar{q}_i$) and γ is a constant which has to be appropriately chosen to guarantee stability and convergence.

It should be mentioned that modifications of the objective function (Eqn. (4)) (e.g., introduction of time-dependent weighting) may lead to other methods that are closely related to the methods presented in this article (e.g., the stochastic Gauss–Newton estimator presented by Nihan and Davis (1987)).

The recursive estimators of Eqns. (6), (10) and (11) do not implicitly fulfill the constraints given in Eqn. (3). As with the unconstrained least-squares estimator of Eqn. (5), several additional methods such as *a posteriori* truncation, normalization or projection onto the admissible space have to be carried out after each recursion to make the estimates consistent with the constraints.

3. Results

To demonstrate the efficiency of dynamic estimation methods, two representative results will be given. Further applications are reported by Cremer and Keller

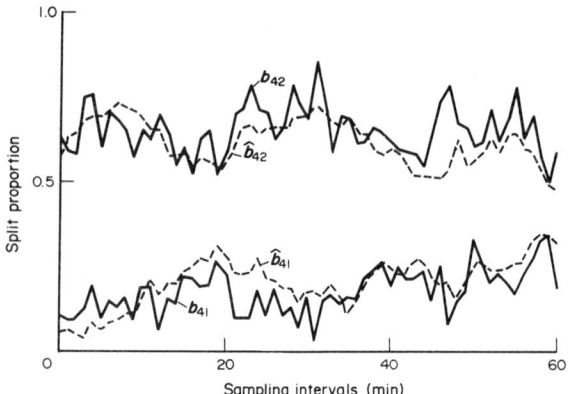

Figure 3
Estimation of time variable split parameter at an intersection: —— real measurements, --- estimates

(1981, 1984), Cremer *et al.* (1985), Keller and Ploss (1987), Luk and Besley (1987), and Nihan and Davis (1987).

For the first example, an experiment where measurements were taken from a four-entry four-exit intersection in Hamburg will be considered. The sample time T was chosen to be 90 s (the cycle time of the intersection) and the split parameters were estimated by the least-squares recursive formula with truncation and normalization. The time diagram for real and estimated parameters is shown in Fig. 3 for a 60 min observation period. It can be seen that the OD pattern actually changes with time and that the estimates are able to track these changes fairly well. (It should be noted that no use was made of knowledge of the different phases of the signalization and of the corresponding flows, which could make the problem solution much easier.)

For the second example, an experiment where synthetic measurement data were generated by simulation of a five-arm intersection with 20 unknown OD flows and nine linearly independent flow measurements from entries and exits will be considered. Estimates obtained by the nonrecursive constraint optimization method (Cremer and Keller 1984) were compared with results from the entropy maximization method of Van Zuylen and Willumsen (1980), which is considered one of the most efficient static methods. To obtain an unbiased comparison, no *a priori* information was used as is often done for static methods in the form of weights for the objective function or additional restrictions. It can be seen from Fig. 4, which shows the estimated flows vs the real flows, that dynamic methods obviously give better estimates.

These two examples and reported applications in the literature impressively emphasize that dynamic OD estimation procedures offer a great potential to estimate OD flows precisely and to track them if they are time variant.

4. Extensions

As stated previously, dynamic OD estimation procedures make use of the causal relationship between time records of entry and exit flows of a traffic system. The more these records are correlated, the better the estimation results. This recommends the application of these methods to small-scale systems such as intersections where there is a strong relation between entering and leaving flows.

If this causal relationship gets weaker in cases of extended networks due to various effects, it has to be expected that the potential of these methods will be diminished. The question that arises therefore, is whether benefit can also be obtained when applying these methods to more extended traffic and transportation facilities.

In the case of pure time delays between the ith entry and the jth exit, the presented methods may be applied quite successfully when the model is modified to take into account individual travel times τ_{ij} (Klaas 1986). Applications to freeways with many entries and many exits which can be regarded as a typical system with individual travel times were encouraging. The results were less satisfactory when dispersion and interruptions by signalized intersections deformed traffic flow patterns (Klaas 1986). If long-time averages are to be estimated, these problems can be avoided by increasing the sample time to the order of minutes and hours so that dispersion and other effects become negligible.

In urban networks, the strong impact of subsequent signalization on the flows as functions of time seemed to be a serious obstacle towards a successful application of dynamic methods, particularly when short-term variations were to be tracked. It was shown by Keller and

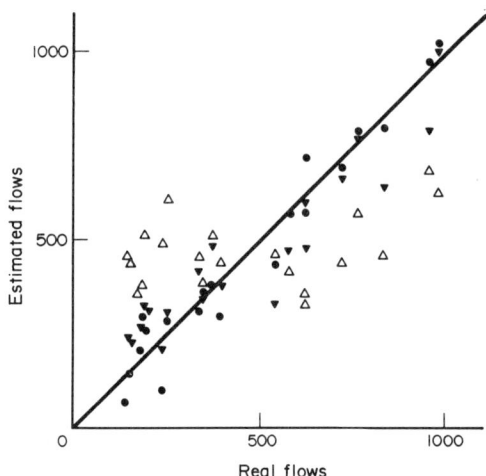

Figure 4
Comparison of estimates from several static and dynamic estimation procedures: △, Van Zuylen and Willumsen (1980); ▼, Cremer and Keller (1981); ●, Cremer and Keller (1984)

Ploss (1987) that, in such cases, a combination of dynamic and static methods may be efficient. According to their investigations, the performance of static methods when applied to a road network with many intersections is improved impressively when reliable estimates for OD flows at selected intersections are obtained from dynamic methods and used as additional information to support the static procedures.

5. Future Developments

According to international research activities aimed at equipping future vehicles with communications devices for route guidance, driver information purposes and traffic control functions, it can be expected that OD flow estimation in the future will be assisted by sample information from individual cars. How this additional information can be used properly in combination with available dynamic and static methods will be a promising subject of future research.

The innovative methods presented here formulate the generation and propagation of flows from each origin to each destination in a traffic system as a dynamic process and apply various estimation procedures from the field of system theory to this problem. The methods developed in recent years may be assigned to two categories of computational schemes. The first one contains procedures which make direct use of all measurements of a given observation period to determine estimates while the second category comprises recursive procedures where each new set of measurements from a single sample interval are processed recursively to improve the current estimates. Methods of the second class are especially suited when time variable OD patterns have to be tracked in real time for control purposes.

The most important approaches are presented and their properties are discussed and some results from applications with real and synthetic data are given to enable a comparison to be drawn between static and dynamic methods, which clearly underlines the superiority of the new methods.

Dynamic methods require the recording of flow rates as functions of time and since this is not performed by all counting devices, these methods, though very promising, have not been extensively tested. In view of current attempts to bring a new quality to road traffic by advanced information and control techniques, it is likely that widespread use of these methods will occur in the future.

See also: Origin–Destination Matrix: Static Estimation

Bibliography

Bell M H 1984 Log-linear models for the estimation of O–D matrices from traffic counts: An approximation. *Proc. 9th Int. Symp. Transportation and Traffic Theory.* VNU, Utrecht, The Netherlands, pp. 451–70

Cremer M 1983 Determining the time dependent trip distribution in a complex intersection for traffic responsive control. *Proc. 4th IFAC/IFIP/IFORS Symp. Control in Transportation Systems.* Pergamon, Oxford

Cremer M, Keller H 1981 Dynamic identification of flows from traffic counts at complex intersections. *Proc. 8th Int. Symp. Transportation and Traffic Theory.* University of Toronto Press, Toronto, pp. 199–209

Cremer M, Keller H 1984 A systems dynamics approach to the estimation of entry and exit O–D flows. *Proc. 9th Int. Symp. Transportation and Traffic Theory.* VNU, Utrecht, The Netherlands, pp. 431–50

Cremer M, Keller H 1987 A new class of dynamic methods for the identification of origin–destination flows. *Transp. Res. B* **21**, 117–32

Cremer M, Keller H, Klaas V, Ploss G 1985 Identifizierung der Herkunft-Ziel-Matrix von komplexen Verkehrsanlagen aus den Zeitverläufen von Querschittszählungen. Report No. Cr 69/1-1. Deutsche Forschungsgemeinschaft, Bonn

Hendrickson C, McNeil S 1984 Matrix entry estimation errors. *Proc. 9th Int. Symp. Transportation and Traffic Theory.* VNU, Utrecht, The Netherlands, pp. 413–30

Keller H, Ploss G 1987 Identification of O–D network flows from counts for urban traffic control. *Proc. 10th Int. Symp. Transportation and Traffic Theory.* Elsevier, Amsterdam, pp. 267–84

Klaas V 1986 Ein dynamisches Verfahren zur Bestimmung der Quelle-Ziel-Teilflüsse in Verkehrsanlagen mit Laufzeiten und Dispersion. Ph.D. thesis, Technische Universität Hamburg-Harburg

Ljung L, Söderström T 1983 *Theory and Practice of Recursive Identification.* MIT Press, Cambridge, MA

Luk J Y K, Besley M 1987 The box hill study—Queue length estimation in Scats. Internal Report. Australian Road Research Board, Vermont South, Victoria, Australia

Nguyen S, Morello E, Pallotino S 1988 Discrete time dynamic estimation model for passenger origin–destination matrices on transit networks. *Transp. Res. B* **22**, 251–60

Nihan N L, Davis G A 1987 Recursive estimation of origin–destination matrices from input/output counts. *Transp. Res. B* **21**, 149–63

Van Zuylen H J, Willumsen K G 1980 The most likely trip matrix estimated from traffic counts. *Transp. Res. B* **14**, 281–93

Willis A E, May A D 1981 Deriving origin-destination information from routinely collected traffic counts. Report No. UCB-ITS-RR-81-8. Institute of Transport Studies, Berkeley, CA

M. Cremer
[Technische Universität Hamburg-Harburg,
Hamburg, Germany]

Origin–Destination Matrix: Static Estimation

Origin–destination (OD) matrices have a number of uses, from the provision of on-line information to improve traffic control to the off-line design of traffic management schemes and the long-term planning and design of transportation systems. Therefore, the methods used to estimate trip matrices vary depending on their use: on-line traffic control systems require dynamic methods (see *Origin–Destination Matrix: Dynamic Estimation*) whereas design tasks call either for methods to estimate a representative trip matrix for a base year or for a future OD matrix for a target planning horizon. Traditional techniques for collecting OD information from, for example, home or roadside interviews tend to be costly, labor intensive and disruptive to trip makers. The need to develop low-cost methods to estimate the present and future OD matrices is apparent. This article considers a range of matrix estimation methods suitable for cases where only static data is available; that is, historic traffic counts or similar data collected over a period of time.

Traffic counts can be seen as resulting from a combination of a trip matrix and a route choice pattern. As such, they provide direct information about the sum of all OD flows that use the counted links. Traffic counts are very attractive as a data source because they are nondisruptive to travellers, generally available, relatively inexpensive to collect and automatic collection is well advanced. The idea of estimating trip matrices or demand models from traffic counts deserves serious consideration and the 1980s have seen the development of a number of practical approaches to tackle this problem.

A study area divided into N zones interconnected by a road network consisting of a series of links and nodes will be considered; as usual, these links may represent physical lengths of road, turning movements or a combination of both. The trip matrix for this study area consists of N^2 cells, or $N^2 - N$ cells if intrazonal trips may, as is often the case, be disregarded. A key element in the estimation of trip matrices from traffic counts is the identification of the paths followed by the trips from each origin to each destination. The variable p_{ij}^a is used to define the proportion of trips from zone i to zone j travelling through link a. Thus, the flow on a particular link a is the summation of the contributions of all trips between zones to that link:

$$V_a = \sum_{ij} T_{ij} p_{ij}^a \quad 0 \leq p_{ij}^a \leq 1 \quad (1)$$

The values of the indicator p_{ij}^a can be obtained using various trip assignment techniques ranging from simple all-or-nothing assignment to more advanced equilibrium assignment. Given all the p_{ij}^a and all the observed traffic counts V_a, there will be N^2 unknown values of T_{ij} to be estimated from a set of L simultaneous linear equations of the form of Eqn. (1) where L is the total number of traffic counts. In practice, the number of observed traffic counts is much less than the number of unknown values of T_{ij} and the problem is underspecified; that is, there will be more than one trip matrix that, when loaded onto the network, will reproduce the traffic counts. The solution space for the trip matrix must be reduced in order to select the most suitable one.

There are two basic approaches to provide this reduced solution space: structured and nonstructured methods. In the structured case, the modeller restricts the feasible space for the estimated matrix by imposing a particular structure which is usually provided by an existing travel demand model such as a gravity or direct-demand model. The nonstructured approach relies on general principles such as maximum likelihood or entropy maximization to provide the minimum of additional information required to estimate a matrix.

1. Route Choice and Matrix Estimation

Robillard (1975) classified assignment methods (see *Traffic Assignment*) for trip matrix estimation from counts under two main groups: proportional and nonproportional. Proportional methods assume that the proportion of drivers choosing each particular route between two zones is independent of the demand levels. The most common example is all-or-nothing assignment and in this case p_{ij}^a is defined as 1 if trips from origins i to destinations j use link a and 0 otherwise. Pure stochastic assignment methods, such as those of Burrell and of Dial, also fall into this group, but p_{ij}^a can take intermediate values between 0 and 1.

Nonproportional assignment techniques acknowledge the link between demand, flow levels and congestion and, therefore, recognize that the proportion of trips using each link does depend on link flows. Equilibrium and stochastic user equilibrium assignment methods are members of this group.

The advantage of proportional assignment methods is that they permit the separation of the route choice and matrix estimation problems. In contrast, nonproportional route choice requires the joint, or iterative, estimation of route choice and trip matrices so that both are consistent. It will be assumed that proportional assignment methods are a reasonable approximation to route choice.

2. Transport Model Estimation from Traffic Counts

The calibration of a gravity model from traffic counts was one of the first methods put forward for estimating trip matrices as by, for example, Holm et al. (1976). The basic idea is to postulate a particular form of gravity model and examine its assignment onto the network. In the case of interurban travel, the simplest gravity model would be

$$T_{ij} = \frac{bP_i P_j}{d_{ij}^2} \quad (2)$$

where P_i is the population of urban area i, d_{ij} is the distance between both areas and b is a constant for calibration, in this case the only one. The following equation results if a matrix of this kind is assigned onto the network:

$$V_a = \frac{\sum_{ij} p_{ij}^a b P_i P_j}{d_{ij}^2} = \frac{b \sum_{ij} p_{ij}^a P_i P_j}{d_{ij}^2} \quad (3)$$

Note that on the right-hand side of this equation the only unknown is b; the other values are provided by external data and a suitable route choice model. This model can be extended to include other trip generation/attraction factors (denoted as O_i and D_j) such as employment, industrial production, shopping floor area and a more general deterrence function f_{ij}. An example would be $f_{ij} = \exp(-\beta c_{ij})$ where c_{ij} is the (generalized) cost of travelling between i and j. If the gravity part of this model is designated by

$$G_{ij} = O_i D_j f_{ij} \quad (4)$$

and k journey purposes are allowed (or commodities if dealing with freight movements), then

$$V_a = \sum_k \sum_{ij} p_{ij}^a b_k O_i^k D_j^k f_{ij}^k = \sum_k b_k \sum_{ij} p_{ij}^a G_{ij}^k \quad (5)$$

where b_k are parameters for calibration as are β^k for the deterrence functions; the rest of the data is, once more, assumed available. The values of b_k and β^k may be estimated using a suitable method such as a least-squares technique; in this case, $V_a' = V_a + \varepsilon_a$ is postulated, where ε_a is an error term. A change of variable

$$X_k = \sum_{ij} p_{ij}^a G_{ij}^k \quad (6)$$

permits the writing of

$$V_a' = b_0 + \sum_k b_k X_k + \varepsilon_a \quad (7)$$

where b_0 is the intercept and may be deemed to depict the part of the flow not represented by the gravity model such as local or intrazonal traffic. The values of the parameters b_k can be calculated using a least-squares method, so as to minimize

$$S_0 = \frac{\sum_a (V_a' - V_a)^2}{V_a^*} \quad (8)$$

where $V_a^* = 1$ for nonweighted and $V_a^* = V_a'$ for weighted least-squares methods. The model is, in general, nonlinear but, if a value of β^k is transferred from another data set or at least different values are tried to obtain a best fit, it can be solved by a linear least-squares method.

The model in Eqns. (5) and (7) has at least one obvious deficiency; it is not constrained. A doubly constrained version of the gravity model can be introduced:

$$V_a = \sum_k b_k \sum_{ij} p_{ij}^a O_i^k D_j^k A_i^k B_j^k f_{ij}^k \quad (9)$$

where b_k is a scaling parameter that enables the use of different units for T_{ij}, O_i^k and D_j^k, and A_i^k and B_i^k are the balancing factors expressed as

$$A_i^k = \frac{1}{\sum_j B_j^k D_j^k f_{ij}^k} \quad (10)$$

$$B_j^k = \frac{1}{\sum_i A_i^k O_i^k f_{ij}^k} \quad (11)$$

Estimating this more conventional model from traffic counts represents a greater effort as the parameters for calibration are now A_i^k, B_j^k, β^k and b_k.

Tamin and Willumsen (1989) have generalized this approach combining features of the gravity and intervening opportunities models in a single model. The choice between gravity or opportunity approaches is decided empirically by allowing the estimation of parameters that control the global functional form of the trip distribution mechanism. Of course, other models (e.g., direct demand) can also be used in this type of estimation method. One interesting advantage of this approach is that, once a demand model is calibrated, it may also be used for forecasting purposes, provided future values for parameters such as O_i and D_j are available or estimable.

3. Nonstructured Approaches

Entropy-maximizing and information-minimizing techniques have been used as model-building tools in urban, regional and transport planning for many years, particularly after the work of Wilson (1970). It has been shown that, when used to derive models to estimate trip matrices from available data, the entropy-maximizing formalism generates a least-biased trip matrix consistent with the information available, represented as constraints to a maximization (of an entropy function) problem. In the case of the gravity model, the constraints represent trip-end and total cost information.

This idea was used by Willumsen (1981) to derive a model to estimate trip matrices from traffic counts. The problem can be written as

$$\text{maximize } S_1(T_{ij}) = -\sum_{ij}(T_{ij} \log T_{ij} - T_{ij}) \quad (12)$$

subject to

$$V_a - \sum_{ij} T_{ij} p_{ij}^a = 0 \quad (13a)$$

for each counted link a and to

$$T_{ij} \geq 0 \quad (13b)$$

The constraints given in Eqn. (13) replace the trip-end and cost constraints of the gravity model derivation. The use of Lagrangian methods permits the determination of the formal solution to this problem as

$$T_{ij} = \sum_a \exp -\mathcal{H}_a p_{ij}^a = \prod_a X_a^{p_{ij}^a} \quad (14)$$

where $\mathcal{H}_a$ are the Lagrange multipliers corresponding to the constraints (traffic counts) and

$$X_a = \exp -\mathcal{H}_a \quad (15)$$

An old matrix, a small-sample observed matrix or simply a matrix estimated (or cordoned off) from another study can be accommodated to some advantage. Let t be this prior matrix, sometimes called a target trip matrix; the objective function becomes

$$\text{maximize } S_2(T_{ij}/t_{ij}) = -\sum_{ij}(T_{ij} \log T_{ij}/t_{ij} - T_{ij} + t_{ij}) \quad (16)$$

subject to the constraints given by Eqn. (13). This objective function is convex and the term t_{ij}, being a constant, does not affect the optimum.

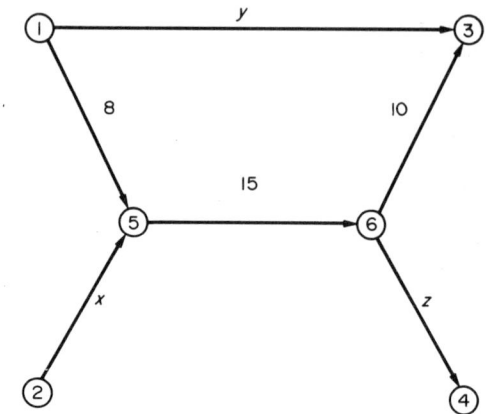

Figure 1
Simple network with observable flows

Using the same methodology and change of variables, the formal solution can be seen to be

$$T_{ij} = t_{ij} \sum_a \exp -\mathcal{H}_a p_{ij}^a = t_{ij} \prod_a X_a^{p_{ij}^a} \quad (17)$$

It is easy to see that, in the absence of prior information, the assumption that all OD pairs are equally likely to have trips (all t_{ij} equal 1) causes the model in Eqn. (17) to revert to Eqn. (14).

In order to illustrate the principles behind this approach consider the simple network depicted in Fig. 1. This network has two origins (1 and 2) and two destinations (3 and 4). The figure also depicts three observed flows {8, 15 and 10} and three, as yet, unobserved flows {x, y and z}. If it is assumed that only integer trips are possible then there are only six trip matrices that can reproduce the observed flows, as shown in Table 1. The entropy-maximizing formalism seeks to identify the most probable trip matrix consistent with the information available, in this case 3 traffic counts for 4 unknowns.

The values of the objective function $S(T_{ij})$ are also shown in Table 1. According to this, the most probable

Table 1
The six possible trip matrices for the flows in Fig. 1

	Destination nodes										
	3	4	3	4	3	4	3	4	3	4	3	4
Origin node 1	8	0	7	1	6	2	5	3	4	4	3	5
Origin node 2	2	5	3	4	4	3	5	2	6	1	7	0
$S(T_{ij})$	−11.07		−7.46		−5.98		−5.78		−6.84		−9.96	
$S'(T_{ij}/t_{ij})$	−5.79		−3.69		−3.70		−5.07		−7.22		−12.20	

trip matrix would be {5, 3, 5, 2} as it has the maximum entropy value. If a prior matrix is available, then the second objective function $S'(T_{ij})$ should be used. The value shown in Table 1 assumes a prior matrix {3, 2, 1, 3} is available; the most probable trip matrix is then {7, 1, 3, 4}. This illustrates how the estimated matrix depends on its prior matrix, if available. The method makes (minimal) use of the "structure" contained in the prior matrix to resolve the underspecification of the problem; it preserves the information contained in the prior matrix unless modified by the observations.

Obviously, it is not possible to calculate directly the entropy values of all possible matrices in larger, real problems. It should be noted, for example, that reducing the number of counts or increasing their absolute value augments the number of feasible trip matrices.

There are several possible algorithms for solving these models. The most widely used is the multiproportional approach as discussed by Lamond and Stewart (1981). This is, in essence, an extension of the biproportional method often used to gross up trip matrices when only data on future trip ends are available. In the present case, there is one balancing factor X_a for each traffic count and their calculation involves iterative estimation until the observed link flows are replicated within an acceptable tolerance. The power p_{ij}^a in the model (Eqn. (17)) ensures that each correction factor affects only OD pairs using the counted link.

The detailed analysis of this **m**aximum-entropy **m**atrix **e**stimation (ME2) model and that of a related approach, based on information-minimizing principles, are given by Van Zuylen and Willumsen (1980). Both models are practically equivalent and share most of their properties. The ME2 model will always reproduce the observations V'_a within a given tolerance provided the constraints define a feasible space; there must be at least one solution in nonnegative T_{ij}. An additional condition for the prior matrix **t** will be discussed.

It can be shown that minimizing the negative of the objective function (Eqn. (16)) is approximately equivalent to minimizing

$$S_3(T_{ij}/t_{ij}) = 0.5(T_{ij} - t_{ij})^2 / T_{ij} \quad (18)$$

This is an error like measure of the difference between the values of t_{ij} and T_{ij}. In effect, the negative of $S_2(T_{ij}/t_{ij})$ is also a natural measure of the difference between these cell values; it is zero when $t_{ij} = T_{ij}$ and increasingly positive as the difference increases. In this sense, the estimated matrix is that closest to the prior matrix that, when loaded onto the network, can reproduce the traffic counts.

The model can accommodate other sources of data provided they can be expressed as linear constraints. An example of this type may be information about expenditure or trip-length distribution for the study area. Information on the ticket sales of public transport systems for different zones or value ranges can also be incorporated as constraints. Moreover, the mathematical program can be written with a combination of equality and inequality constraints. Information on parking or, simply, on link capacities can be introduced in this form.

The solution to this extended optimization program is still a multiplicative model. Lamond and Stewart (1981) have shown how the multiproportional algorithm can be extended to handle inequality constraints; therefore, a similar solution method may be used for this extended model.

The ME2 model is multiplicative by nature. This means that if a cell in the prior matrix is zero, it will remain zero in the solution. This may be a source of problems if the cell in the prior matrix was zero by chance (i.e., because of the sampling rate adopted in the study) instead of representing an OD pair with no trips. One pragmatic solution to this problem, for very sparse prior matrices, is to "seed" the empty cells with a small value (e.g., 0.5 trips). The constraints, through the multiproportional or similar solution algorithm, will then ensure that some of these trips "grow" to one or more full trips while others regain a zero value.

4. Traffic Counts and Matrix Estimation

The nonstructured ME2 model requires a feasible solution space (Willumsen 1984). The conditions that must be met by the set of counts to be suitable for trip matrix estimation will be considered.

4.1 Independence

Not all traffic counts contain the same amount of information. For example, in Fig. 1, the flow in link 5–6 is made up of the sum of traffic links 1–5 and 2–5. Counting traffic on link 2–5 is then redundant and only two counts there can be said to be independent. Whenever a flow continuity equation of type "flows into a node equals flows out of a node" can be written, its counts will be linearly dependent. In this case, it will always be possible to describe one link flow as a linear combination of the rest.

4.2 Inconsistency

Counting errors and the fact that traffic counts are often obtained on different occasions (hours, days, weeks) are likely to lead to inconsistencies in the flows; that is, the expected flow continuity relationships will not be met. If the flow x in Fig. 1 turned out to be 6 instead of 7, the corresponding equation would be inconsistent and no trip matrix could possibly reproduce these flows. One way of reducing this problem is to allow an error term in the equations (as in Eqn. (7)) or to remove the inconsistencies beforehand. The latter approach was put forward by Van Zuylen and Willumsen (1980) who assumed that link flows are Poisson distributed and that the observations available are samples on these distributions. Maximum likelihood is then used to generate a model for producing improved and consistent estimates of the flows. It should be noted that model calibration from traffic counts, as discussed in Sect. 2.3, makes an explicit allowance for errors in the observed link flows, thus eliminating the need for removing inconsistencies in the counts.

In addition to these flow continuity conditions, there is a second source of inconsistencies: a mismatch between the assumed traffic assignment model and observed flows. In these conditions, there may be no trip matrix capable of reproducing the observed link flows using the assignment model. There are, therefore, conditions for flow level and path flow level consistency. The network in Fig. 1 will be considered with the three traffic counts plus the count $z = 5$, thus meeting flow continuity conditions. It is assumed that the route choice model assigns all traffic from 1 to 3 directly; that is, avoiding nodes 5 and 6. The flow on link 1–5 must, therefore, be destined to node 4. However, this is impossible as the volume on link 6–4 cannot be greater than 5. There is no feasible trip matrix that can satisfy these flows and the assumed route choice model.

The set of linear equations corresponding to this example is given by

(link 1–5)	$T_4 = 8$	(19)
(2–5)	$T_{23} + T_{24} = 7$	(20)
(5–6)	$T_{23} + T_{24} + T_{14} = 15$	(21)
(6–3)	$T_{23} = 10$	(22)
(6–4)	$T_{13} + T_{24} = 5$	(23)

Note that there are only three independent equations in this set. The solution to this system is $T_{14} = 8$, $T_{24} = -3$, $T_{23} = 10$ and, of course, $T_{13} = y$ (not in the system). This solution is useless as negative trips have no meaning. Clearly Eqns. (19) and (20) are each inconsistent with Eqn. (23) and the nonnegativity condition for trips. In simple problems such as this, inconsistencies can be ascertained by inspection but, in more complex networks, they can only be identified by means of row and column operations on the linear equations. For large systems, these operations are likely to be expensive in terms of computer requirements.

In this simplistic example, it is not difficult to see that the problem originates in the assumed single route between 1 and 3. If two paths were allowed, one direct and one via node 5, the inconsistency could be removed. Futhermore, the value of the resulting variable p_{13}^{56} cannot be arbitrarily chosen; in effect a feasible solution requires

$$\frac{3}{y+3} \leq p_{13}^{56} \leq \frac{8}{y+8} \qquad (24)$$

The fact that the path flow continuity conditions are not met reflects errors in assignment, whereas the link flow discontinuities reflect errors in traffic counts alone. To deal with the lack of consistency at the path flow level, an appropriate route choice model must be adopted. In general terms, consistency at the link flow level is a necessary but not sufficient condition for a feasible solution space. Consistency at path flow level is, however, a sufficient condition for link flow consistency and for a feasible solution space.

5. Improved Methods

ME2, probably because of its simplicity, relative efficiency and ease of programming, has been widely implemented and used, particularly in the UK. The model has, however, some known limitations.

One of these limitations is that the model tries to preserve the prior matrix and not just the relative proportions of the total trips in each cell. If a long period has elapsed since the prior matrix was produced, it is preferable to have a matrix with the closest proportions to the prior matrix and not the closest absolute values. This can be approximated by means of a general growth factor, for example,

$$\tau = \frac{\sum_a V_a}{\sum_a \sum_{ij} t_{ij} p_{ij}^a} \qquad (25)$$

which is then applied to the prior matrix before using the ME2 model. In this way, the structure of the prior matrix is preserved as much as possible. The estimation of τ in Eqn. (25) is only an approximation. Bell (1983) formulated a model that tries to preserve the structure of the prior matrix by adding a new constraint and thus modifying the mathematical program to the minimization of S_2 subject to Eqn. (13), to

$$\tau = \frac{\sum_{ij} T_{ij}}{\sum_{ij} t_{ij}} \qquad (26)$$

and to $T_{ij} \geq 0$. Bell used a Newton–Raphson method to solve this model with an iterative estimation for τ. This has advantages in terms of computer time and is also useful in tracing the effect of errors in the traffic counts through to the estimated trip matrix. However, this method requires considerably more memory and is, therefore, restricted to small- and medium-size networks.

A second limitation of ME2 is the fact that it considers the traffic counts as error-free observations on nonstochastic variables. In effect, the model gives complete credence to the traffic counts and uses the prior matrix only to compensate for the fact that they do not contain sufficient information for estimation purposes. Willumsen (1984) has suggested an approach to compensate for this difficulty by constructing a composite objective function:

$$\text{minimize } S_4 = \sum_{ij}(T_{ij}\log T_{ij}/t_{ij} - T_{ij} + t_{ij})$$
$$+ \sum_a \Phi_a(V_a \log V_a/V'_a - V_a + V'_a) \quad (27)$$

where V_a is now the "true" value of the traffic count at a, V'_a is the value of one observation of the flow made at a and Φ_a is a weighting factor that depends on the confidence attached to the observation V_a. The use of the Langrangian method leads to the solution

$$T_{ij} = t_{ij} \prod_a X_a^{p_{ij}^a} \quad (28)$$

and

$$V_a = V'_a X_a^{1/\Phi_a} \quad (29)$$

It should be noted that if Φ_a is very large (i.e., if a high weight has been assigned to the counts as they are believed to be very accurate), V_a tends to V'_a; in the limit with $\Phi_a = \infty$ the original model is used as $V_a = V'_a$. However, the smaller the value of Φ_a the greater the credence given to the prior matrix **t**.

Brenninger-Göthe *et al.* (1989) have shown that a very natural value for the weights Φ_a is the variance (or standard deviation) associated with the observations. They have further extended the model to consider weights attached to both the prior matrix μ_{ij} and the traffic counts Φ_a; thus, the new objective function becomes

$$\text{minimize } S_5 = \sum_{ij} \mu_{ij}(T_{ij}\log T_{ij}/t_{ij} - T_{ij} + t_{ij})$$
$$+ \sum_a \Phi_a(V_a \log V_a/V'_a - V_a + V'_a) \quad (30)$$

Cascetta (1984) and McNeil and Hendrickson (1985) have put forward versions involving generalized least-squares approaches as a variant to the original objective function S_1. One problem is that, under certain circumstances, these models may produce negative entries in the estimated trip matrix, particularly where the prior matrix originally had small values.

Maher (1983) proposed the use of a Bayesian approach to the trip matrix estimation problem which results in functional forms equivalent to the generalized least-squares method. A prior estimate of the trip matrix is updated in light of a set of traffic counts; both are assumed to be multivariate normally distributed variables with known covariances.

Spiess (1987) has put forward a maximum-likelihood model to solve the problem. He considered a specific formulation where, for each OD pair, t_{ij} is obtained by observing an independent Poisson process with mean $\Omega_{ij}T_{ij}$. This corresponds to the problem of taking a sample of an existing trip matrix with a sampling rate of $\Omega_{ij} < 1$. This generates a new objective function:

$$\text{maximize } S_6 = \sum_{ij}(t_{ij}\log(\Omega_{ij}T_{ij}) - \Omega_{ij}T_{ij} - \log t_{ij}!) \quad (31)$$

subject to the usual constraints. Separating the logarithm into the sum and discarding constant terms gives

$$\text{minimize } S'_6 = \sum_{ij}(\Omega_{ij}T_{ij} - t_{ij}\log T_{ij}) \quad (32)$$

This objective function is convex in T_{ij}; provided the set of constraints is consistent and the flows feasible then the existence of an optimal solution is assured. The solution may be obtained by any standard method for convex programming problems but Spiess (1987) has developed an algorithm that exploits some of the specific properties of this problem.

6. Treatment of Nonproportional Assignment

The models discussed so far are based on the assumption that it is possible to obtain the route choice proportions p_{ij}^a independently from the OD estimation process. Wherever congestion plays an important role in route choice, this assumption becomes questionable as the route choice proportions and the trip matrix become interdependent.

One way of tackling this problem is to adopt an iterative approach: assume a set of route choice proportions $\{p_{ij}^a\}$, estimate a matrix **T**, load it onto the network and obtain a new set of route choice proportions; repeat the process until route choice proportions and estimated matrices are mutually consistent.

A simple version of this was implemented by Holm *et al.* (1976) who extended the gravity model approach to include some features of capacity restraint assignment. They used the method of successive averages to obtain the proportion of trips using each link.

An improved approach has been adopted in the SATURN model (Hall *et al.* 1980) where the route choice proportions are estimated using the value Φ in the Frank–Wolfe algorithm (the optimum linear combination of accumulated and auxiliary flows) at each iteration. It is recognized that, in general, the path flows under equilibrium conditions are not unique and, therefore, the method is also a heuristic.

An alternative approach requires restating the original problem in terms of a three-dimensional matrix (origin, destination, route):

$$\text{maximize } S_7(T_{ij}/t_{ij}) = -\sum_{ijr} T_{ijr}(\log T_{ijr}/t_{ijr} - 1) \quad (33)$$

subject to

$$\sum_{ijr} T_{ijr}\delta^a_{ijr} - V_a = 0 \qquad T_{ijr} \geq 0 \quad (34)$$

where the index r indicates the route or path chosen, δ^a_{ijr} is 1 if route r between i and j uses link a, and 0 otherwise.

It is always possible, of course, to reconstruct the OD matrix $\{T_{ij}\}$ by aggregating the path flow matrices $\{T_{ijr}\}$. Again, the solution to this program is

$$T_{ijr} = t_{ijr} \prod_a X_a^{\delta^a_{ijr}} \quad (35)$$

and

$$T_{ij} = \sum_r T_{ijr} \quad (36)$$

The prior path flows may be calculated from the prior trip matrix as $t_{ijr} = t_{ij}/R_{ij}$, where R_{ij} is the number of paths between i and j. In this case, the path flows can take any value as they are not assumed unique. The Frank–Wolfe algorithm for equilibrium assignment is used to identify attractive paths (those selected in each all-or-nothing step), but not to define the strict proportions of the trip matrix using them. This is still only an improved heuristic scheme. A more general approach has been put forward by Fisk (1988) in which a maximum-entropy matrix estimation and user equilibrium assignment are combined as a single mathematical program. The computer requirements of such methods seem very high, thus limiting their practical value.

See also: Origin–Destination Matrix: Dynamic Estimation; Transportation Modelling

Bibliography

Bell M G H 1983 The estimation of an origin–destination matrix from traffic counts. *Transp. Sci.* **17**, 198–217

Brenninger-Göthe M, Jornsten K, Lundgren J 1989 Estimation of origin–destination matrices from traffic counts using multiobjective programming formulations. *Transp. Res. B* **23**, 257–80

Cascetta E 1984 Estimation of trip matrices from traffic counts and survey data: A generalised least-squares estimator. *Transp. Res. B* **18**, 289–99

Fisk C S 1988 On combining maximum entropy trip matrix estimation with user optimal assignment. *Transp. Res. B* **22**, 69–73

Hall M D, Van Vliet D, Willumsen L G 1980 SATURN—A simulation assignment model for the evaluation of traffic management schemes. *Traffic Eng. Control* **21**, 168–176

Holm J, Jensen T, Nielsen S, Christensen A, Johnsen B, Ronby G 1976 Calibrating traffic models on traffic census results only. *Traffic Eng. Control* **17**, 137–40

Lamond B, Stewart N F 1981 Bregman's balancing method. *Transp. Res. B* **15**, 239–48

McNeil S, Hendrickson C 1985 A regression formulation of the matrix estimation problem. *Transp. Sci.* **19**, 278–92

Maher M J 1983 Inferences on trip matrices from observations on link volumes: A Bayesian statistical approach. *Transp. Res. B* **17**, 435–47

Robillard P 1975 Estimating the O–D matrix from observed link volumes. *Transp. Res.* **9**, 123–8

Spiess H 1987 A maximum likelihood model for estimating origin–destination matrices. *Transp. Res. B* **21**, 395–412

Tamin O Z, Willumsen L G 1989 Transport demand model estimation from traffic counts. *Transportation* **16**, 3–26

Van Zuylen H, Willumsen L G 1980 The most likely trip matrix estimated from traffic counts. *Transp. Res. B* **14**, 281–293

Willumsen L G 1981 Simplified transport demand models based on traffic counts. *Transportation* **10**, 257–78

Willumsen L G 1984 Estimating time-dependent trip matrices from traffic counts. *Proc. 9th Int. Symp. Transportation and Traffic Theory.* VNU, Utrecht, The Netherlands

Wilson A G 1970 *Entropy in Urban and Regional Modelling.* Pion, London

L. G. Willumsen
[London, UK]

Parking Control Systems

Parking in city centers is becoming more and more difficult. Strategies to solve this problem consist of a whole bundle of appropriate measures. One of these is the automatic parking control or parking guidance system. The main task is to guide the driver who is looking for a parking space to one of the connected car parks which still has space available. The system covers the main road network of a city and indicates the best routes by variable guide signs corresponding to the current occupancy of car parks.

1. Background

The traffic burden and, simultaneously, its negative concomitants are still increasing. This affects, above all, city centers.

While roads leading to and roads inside urban areas are mostly developed in such a way that they can absorb the flowing traffic—although on the verge of capacity—serious difficulties arise when dealing with stationary traffic (parking) and especially with the search for parking space.

The significance of the correlation between flowing and stationary traffic becomes apparent when considering the nearly trivial fact that every trip with a car ends in a more or less long parking procedure. On average, a car is used only 2 h in every 24 h; for the rest of the day (i.e., for more than 90% of the time) it is parked (see Fig. 1).

This ratio shows that, while the traffic volume is permanently increasing, the already overloaded cities will have to deal with an immense parking problem if the development is left to the free interplay of forces and if appropriate measures are not taken.

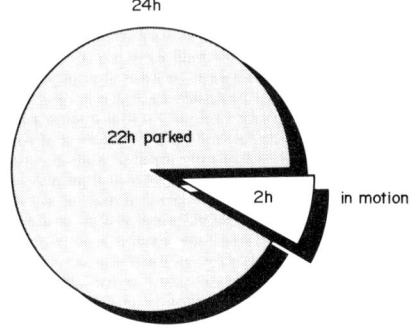

Figure 1
Ratio of time for cars parked and cars in motion

An effective strategy for the solution of parking problems must be assembled out of a whole bundle of measures.

In recent years, the aspects and weighting in the evaluation of the benefits and efficiency of traffic strategies have been changed. Consider as an example the setting up of new parking facilities. In the past, the accessibility of the decisive areas (destination) as well as the length of necessary walking distances from and to the parking place were used as the most important criteria. Nowadays, damage to the environment (by noise, exhaust or obstruction) and energy consumption play an important role. So, because of these reasons, the best-located place for the setting up of a new parking area cannot always be chosen. It follows that those parking areas that are near the destination are mostly overfilled, while at the same time there is still parking space available on newer parking areas which, however, are often less well located. This is the reason why

(a) there are queues in front of overcrowded car parks, and/or

(b) there is an unnecessary amount of traffic searching for other vacant parking places.

Negative concomitants connected with that are, for example:

(a) loss of time,

(b) increased traffic strain on the roads,

(c) disturbance to the flowing traffic,

(d) unnecessary damage to the environment by noise and exhaust, and

(e) unnecessary energy consumption.

Those defects can be remedied by early information and guiding recommendations for those searching for parking spaces.

Often enough, the implementation of traffic regulations—especially signposting—is insufficient because indication of parking space is often only announced right in front of the parking area. Such signposting is of little use for strangers because it only helps them in the last phase of their search. However, the problem is the same for those who know their way around. Although they do not need a car park guidance system, they also cannot find out before reaching the car park itself whether or not spaces are available.

So, the existing information in the form of static signposting is mostly insufficient. Real information should rather consist of the provision of indications as to where car parks with spaces available can be found and—with the help of a continuous guide sign system—in the provision of indications as to which route should be taken to get there so the drivers do not need to approach already overloaded car parks. However, this can only be achieved by a dynamic guidance system (variable message sign system) which can permanently adapt to changing conditions. Because of this consideration, the so-called automatic parking control system has been developed.

2. Operation

An automatic parking control system is a traffic control system which, corresponding to the current occupancy of all the connected car parks, can indicate optimum routes (as a rule the shortest ones) to car parks with spaces available. So, the main components of such an automatic parking control system are signposts which are developed as changeable message signs.

Figure 2 shows how an automatic parking control system works. Five multistorey car parks are shown in the figure, two of them still having spaces available. A driver arriving at intersection A finds a message sign with arrows pointing in two directions. If he goes straight on, he will get the information at intersection C that he should keep on going straight to the car park at the post office because all the car parks to the right and to the left are occupied. If, after a certain time, this car park at the post office is also overfilled, all the drivers arriving will get the information at intersection A to go to the left to the car park at the station where spaces are still available; the direction straight on will no longer be indicated. In principle, this handling is very plausible and encouraging for parking matters; consequently, some cities have already planned and installed this system.

An automatic parking control system consists of a number of single components and it is a requirement of the efficient use of such a system that these components work very precisely and absolutely faultlessly. As shown in Fig. 3, its main components are:

(a) the parking (ground-level places or multistorey car parks) with their entrance and exit counting mechanisms as well as systems capable of recognizing and collecting data on the current occupancy;

(b) the access roads, including traffic regulations and other controlling devices;

(c) the static and variable car park guidance along the roads as well as in front of the car parks; and

(d) the control center (possibly with subcenters) including the necessary hardware and software.

However, before an automatic parking control system can be installed or even planned, extensive preliminary investigations and surveys have to be carried out. It is the planner's task to analyze the existing conditions, to evaluate them and, at the same time, to become aware of the limits of such a system.

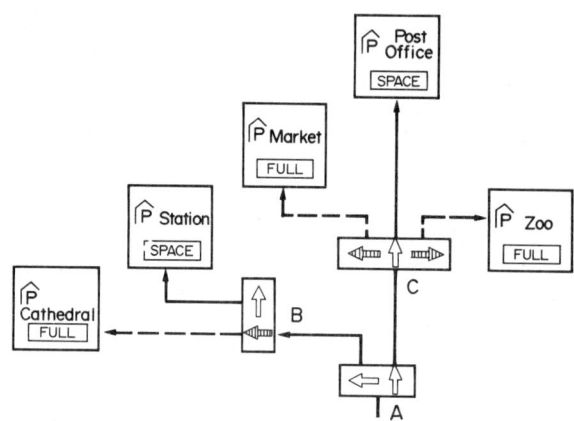

Figure 2
Principle of how a parking control system works

3. Requirements

3.1 Parking Supply and Demand

An essential, perhaps the most important, requirement for the effective use of an automatic parking control system is that a fairly good ratio between supply of and demand for parking spaces exists. The number of drivers searching for parking spaces should not exceed the number of given parking spaces. This should also apply to peak periods of stationary traffic. On the other hand, the installation of an automatic parking control

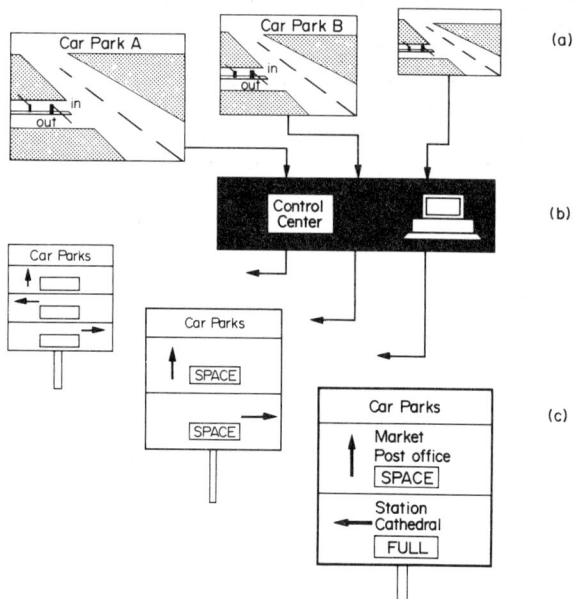

Figure 3
General configuration of an automatic parking guidance system: (a) data collection, (b) data evaluation and processing and (c) information display (variable guide signs)

system will not be useful if the supply of parking spaces in each car park is always higher than the demand. In areas where drivers can always be sure to find a parking space without an extensive search, the cheaper static car park guidance systems are definitely sufficient.

3.2 Allocation of Car Parks

Drivers searching for parking spaces always try to find places as near to their destinations as possible. Therefore, it is a requirement for an automatic parking control system that there be a supply of parking spaces near these destinations; the initial idea should be to guide the drivers searching for parking spaces and, above all, local drivers to the appropriate parking space as long as there are spaces available. Only when the maximum capacity of these well-located spaces is reached should a redirection to other parking areas ensue.

A parking area outside the destination centers can also be seen as well-located if it is near an efficient public transport stop to make a "park-and-ride" system possible.

Relating to the position of parking areas, situating entrances and exits away from main roads seems favorable. So, possible obstructions of the flowing traffic can largely be avoided, provided that there is enough space for possible queuing.

3.3 Structural requirements

In principle, every car park can be connected to an automatic parking control system regardless of whether it is a multistorey car park or a parking lot at ground level.

However, only those parking areas should be connected to an automatic parking control system which allow for reliable car detection. Therefore, it must be ensured that the structural situation only allows vehicles to come and go via entrances and exits or, in other words, uncontrolled use of the car park from all sides must be impossible. Entrances and exits have to be clearly separated so that vehicles entering and leaving can be registered continuously and reliably.

Furthermore, only those car parks should be connected to the automatic parking control system which provide a certain minimum capacity which can be fixed at a number of about 50 parking spaces.

3.4 Operational Requirements

To make as many drivers (who are searching for parking spaces) as possible willing to use such an automatic parking control system, absolutely reliable functioning of the system is required. Every consideration is intended to preserve and to increase the reliability and plausibility of such a system.

If a driver cannot find a parking space in a car park where one has been indicated vacant or if he can get a free space although a totally occupied car park has been indicated, the willingness to use this system will decrease very quickly.

The causes for misinformation are mostly to be found at the individual car parks. Therefore, before the installation of the automatic parking control system, it must be carefully examined to ensure that the vehicles entering and leaving have been registered and that the indication of the current occupancy is made to work faultlessly.

Another condition is that all those car parks which are to be connected to the automatic parking control system should have nearly the same tariffs if the parking areas are equally well located. In the case of car parks with a lower degree of acceptance, it is worth considering charging lower fees for the less attractive car parks.

3.5 Temporal Limits of Application

Principally, there are no restrictions to the temporal range of application. However, if the automatic parking control system is switched off in periods of lower traffic volume and, especially, on Sundays and on holidays, the automatic registration in the car parks must keep on working.

3.6 Spatial Limits of Application

Naturally, the application of automatic parking control systems is limited to areas with a short supply of parking spaces. In most cases, these are central shopping areas. On the other hand, other areas that also cause considerable traffic volume by their attractiveness can be taken into consideration. Among this group are

(a) airports,

(b) harbors near ferries,

(c) amusement parks,

(d) skiing areas,

(e) exhibition areas,

(f) historic places, and

(g) sports fields.

Here, at least two independent car parks are required.

4. Planning

4.1 General Remarks

In areas with intensive parking demand and where the requirements of Sect. 3 are fulfilled, a parking guidance system can help solve the parking problem. The planning can be divided into the following steps:

(a) development of routing plans (build-up trees) for each car park;

(b) coordination of all individual routing plans;

(c) determination of the position of variable message signs, types, and content of the messages;

(d) design of data collection, transmission and processing;

(e) structuring of control procedures into algorithms;

(f) development of software; and

(g) implementation and tuning.

4.2 Routing Plans

It is the object of graphical routing plans to describe the course of routes to the destinations in the form of a tree. The destinations (here, parking areas) correspond to the ends of the trunk. The routes (arterials) from the starting points (where the first information on how to get to the destination is given) up to the destination are identical to the routing-plan branches.

Single branches in the graphical routing plan are led together into the main branches which, in turn, merge into the trunk.

After the routing plans for the different destinations have been determined they must be superposed. While this procedure of superposing the graphical routing plans is only to be seen as an "addition" of the different routings, basic considerations have to be made before the course of the single-destination routing plans can be developed. Those considerations can lead to different routing-plan systems in single cities depending on the road network and traffic policy.

4.3 Message Signs

The signposts, their positions and their content result from the procedure of superposing the different routing plans. When developing their layout, it depends essentially on the plausibility of corresponding information about the best route to the destination. Because of the many ways information can be given, a large number of very different shapes and techniques exists for signposts and, especially, for variable signs.

Planning the guidance of vehicles in search of parking space is carried out from outside (city gates) to inside (center) with the help of the following types of signposts.

(*a*) *Static signposts as the first guiding element.* The first indication of parking areas should be given at important intersections before center or city areas are reached so that the drivers are able to take a first routing decision depending on destination area (e.g., shopping area, historical area or harbor).

(*b*) *Variable signposts.* After the first routing decision, the driver will find variable car park signposts (see Fig. 4) along the route which indicate the direction to take to parking areas with spaces available. These variable signposts are installed at every junction. The closer the driver gets to the possible parking space the lower will be the number of alternative routes.

(*c*) *Static signposts within the system.* When the driver is between the last variable message sign (last junction) and the car park, other static signposts might be necessary to keep him inside the system. This is also the case when he needs to turn away from the main direction somewhere between two variable signposts.

(*d*) *Variable information at the car park entrance.* Immediately at or in front of the respective car park, displays are necessary that indicate "spaces available" or "full."

Because of the fact that an automatic parking control system can only give advice or recommendations which cannot be compulsory for the driver, the standardization of the design of the different signpost types is very important. They should be easy to identify as a part of the system.

4.4 Control Procedures and Techniques

If drivers follow a variable sign which might even be reinforced by the indication "spaces available," only to reach a fully occupied car park, they will consider very carefully whether to trust those car park indications again.

If, the next time, a route is followed that, according to the car park guiding signs, leads only to occupied parking areas (i.e., according to the system this route should not be followed) and, nevertheless, the drivers find out that they can enter this car park because there are still spaces available, then they will probably be glad about having got a parking place, but will never believe car park guiding signs again. It is rare for technical defects to be responsible for these errors; most of the time they occur because signposts have changed their indications too late or too early. This means that the threshold value (the occupancy) of the parking area has been calculated wrongly.

A variable sign located at some distance from the parking area should not just switch to "full" when the car park is totally occupied, but a certain time before. Because a considerable number of cars might still be on the route between the last signpost and the car park it must, in the interest of the reliability of the system, be guaranteed that the driver who has just seen the variable message will, in fact, find a parking place.

The threshold value that triggers the switch from the indication of "spaces available" to "full" depends on various factors (e.g., the distance between the last variable signpost and the car park, and the expected percentage of drivers searching for parking space out of the total traffic queue). In each individual case, this can only be determined after a certain time of surveillance. After a period of experimental operation the thresholds should be adjusted.

On the other hand, when some cars leave a full car park again, the signpost may not simultaneously be switched back to "spaces available"; a "space available" indication must not be carried out until a sufficient number of spaces is available again.

The registration of vehicles entering and leaving a car park is carried out by induction loops or by digit emitters at the gates or at the ticket machines. If entrances and exits cannot be separated by the architectural design, dual indication loops are required to determine whether the registered vehicle is entering or leaving. For the detection of inflow and possible queues

Parking Control Systems

Figure 4
Examples of variable guide signs

in front of the entrances, indication loops have to be installed in the access roads.

The system can be controlled by different techniques. The older systems were switched by electro-mechanical devices; later, mainframe computers were used. Control is also possible by PC.

The control is based on an automatic, continuous computation of differences between entering and leaving vehicles. This is carried out for each individual car park. The switching of the parking guidance signs occurs in comparison with the threshold values.

5. Effects

In some cities, surveys have been carried out to quantify the effects of automatic parking control systems in relation to the degree of occupancy of certain car parks and the traffic volume in search of parking space.

In Aachen, before-and-after surveys were executed. Each survey included a number plate registration in combination with a time record and an inquiry of the car park users. The main results are listed below:

(a) a comparison of the before-and-after survey revealed that the number of car park users had generally increased by 10.5% in the investigated parking areas;

(b) in one unfavorably situated multistorey car park, the degree of occupancy climbed from 62.3% to 82.3%;

(c) 52.8% of the persons asked explained that they orientated themselves by the automatic parking control system, actually 73.9% of the strangers compared with 46.1% of those who were familiar with the locality; and

(d) in spite of the 10.5% increase of the number of parking events in the investigated parking areas, the proportionate volume of searching traffic referring to the totality of all persons asked had been reduced from 24.6% to 21.2%, which means a reduction of 13.8%.

The evaluation of these numbers should take into account the fact that, at that time, many of the drivers were still in a learning process (i.e., they only began to follow the automatic parking control system after they had first tried in vain to get a parking space by their own choice of route).

6. Outlook

First concepts that were concerned with an occupancy-oriented control of the traffic searching for parking space were carried out in 1963. The first automatic parking control system that was realized in 1971 as a pilot project system was the system in Aachen.

Meanwhile, a considerable number of cities in different countries have installed similar systems. New technology is used for control and variable message signing. For example, the Cologne park guidance system is one of the most modern and sophisticated. There are 27 car parks with about 11 500 places in total. The driver is guided by 231 variable signs. The system follows a new approach by indicating the number of available lots in each direction, thus integrating the driver more into the process (see Fig. 4).

The first results are rather good. The system is controlled by a mainframe and the data monitored in a control center. The technology for signalling is LCD. Park-and-ride information will be connected to the system.

Further activities concerning automatic parking control are carried out in the European research and development project DRIVE (see *DRIVE*). Here, the aspects of integration with other traffic control measures are also taken into consideration. Among others, a connection with in-vehicle information systems is considered.

See also: Route Guidance, Collective

Bibliography

Bernard N W, Hillen S M 1981 Compact urban traffic control—Evaluation of the Hull and Torbay systems. *Traf. Eng. Control* **5**, 281–8

Boesefeldt J, Kunze W 1982 Erfahrungen mit der Planung und dem Einsatz von Parkleitsystemen. *Strassenverkehrstechnik* **26**(4), 99–100

Derse K E 1971 Das Parkleitsystem der Stadt Aachen. *OECD Symp. Verbesserung der Verhältnisse in Städten durch Beschränkung des Individualverkehrs*, Cologne, Germany

Emde W, Heusch H 1974 Anwendungsbereich und Untersuchung von Auswirkungen eines Parkleitsystems. Forschungsauftrag des Bundesministers für Verkehr, Bonn

Hilton I C 1989 Parking access time: The pertinent information for users of public parking facilities. *Traf. Eng. Control*

Lennertz H, Philipps P 1976 Untersuchung zur Entwicklung eines Steuerungsmodells für Parkleitsysteme. Forschungsauftrag des Bundesministers für Verkehr, Bonn

Organisation for Economic Co-operation and Development 1980 *Road Research: Evaluation of Urban Parking Systems*. OECD, Paris.

Pieper F 1966 Automatische Wegweisung zu freien Parkplätzen. *Str.-Tiefbau* **3**, 236–8

Schneider H-W 1984 Sistemas de información al conductor desde el exterior, VII. *Congreso Nacional de Sequridad en el Trafico*, Barcelona, Spain

Schneider H-W 1988 *Parkleitsysteme, Verkehrsleittechnik für den Strassenverkehr*, Teil J. Springer, Berlin

Schneider H-W 1988 Erfahrungen mit der Planung und dem Einsatz von Parkleitsystemen. *Strassenverkehrstechnik und Verkehrssicherheit*, Strassen und Verkehr 2000, Band 4/2, Berlin

H.-W. Schneider
[Heusch/Boesefeldt, Aachen, Germany]

Prediction of Traffic Flow

In traffic control systems used for the regulation of urban networks, there is always a certain delay between the computation of traffic signal settings and their actual implementation on the network. Therefore, such systems can be appreciably improved by using predictive models.

Three specific types of prediction can be distinguished, depending on time scale.

(a) *Long-term prediction* (a few months or a year). This is mentioned only in passing; it is not actually used in control systems but can be used to forecast the required capacity of facilities such as freeways or tunnels, or to compute fixed-time coordination plans which are to be implemented over a long period.

(b) *Short-term prediction* (a few minutes). This is used in adaptive regulation systems including a package of coordination plans, from which a decision algorithm chooses, every 5 min or so, the one to be implemented; hence, the traffic variations have to be estimated a few minutes ahead.

(c) *Very-short-term prediction* (a few seconds to a minute). This is used by real-time regulation algorithms which are highly responsive to traffic variations.

In all cases, the parameters to be predicted are mostly traffic volumes, either through a road section or in a network. Various inputs and model structures are used, depending on which type of prediction is considered.

1. Short-Term Traffic Flow Prediction

Several kinds of predictive model based on techniques of time series analysis are used for short-term prediction. Some models rely on the periodicity of traffic volumes from one day or one week to another, and use historical data as inputs. Others rely on the stability of traffic volumes over a short period and use current-day measurements as inputs, while yet others use a combination of both kinds of input. All models are endogenous; that is, all inputs are previous values of the parameter to be predicted.

1.1 Linear Models Using Smoothed Information

These models are classified by reference to the predictive models developed in the USA for the second and third generations of urban traffic control systems (UTCS-2 and UTCS-3).

The second-generation models such as UTCS-2 (Ganslaw and Schaake 1973) or ASCOT (Ross *et al.* 1977) use both historical information and current-day measurements as inputs. The time unit of prediction is 5–15 min. For example, considering $f(t-1)$ the measured volume at time interval $t-1$, $m(t)$ the smoothed historical volume for time interval $t-1$ and $r(t) = ar(t-1) + (1-a)[f(t-1) - m(t-1)]$ the exponentially smoothed difference between measured and historical volumes, the UTCS-2 model will give the predicted volume at time interval t by

$$V(t) = m(t) + r(t) + d(t)$$

where $d(t) = b[f(t-1) - m(t-1) - r(t-1)]$ is the difference between the real and smoothed values of $f - m$, with a coefficient b computed by identification. The smoothed historical volume is obtained by fitting the Fourier series

$$m(t) = A_0 + \sum_{i=1}^{6} [A_i \cos(2\pi it/k) + B_i \sin(2\pi it/k)]$$

to a representative volume data set.

The main drawback of this kind of model is its poor responsiveness to abrupt changes in the traffic, related to the importance of historical information in the model. To avoid this drawback, the third-generation models such as UTCS-3 or similar algorithms (McShane *et al.* 1976) rely only on current-day measurements. The time unit of prediction is from one to a few traffic signal cycles, and the prediction is a two-step one. The UTCS-3 model is as follows. Considering $f(t)$ the measured volume at time interval t, $x(t) = ax(t-1) + (1-a)f(t)$ the exponentially smoothed volume and $d(t) = f(t) - x(t)$ the difference between the real and smoothed values, the predicted volume $V(t)$ is given by

$$V(t) = x(t-2) + bd(t-2)$$

where

$$b = \frac{(N-1)\sum_{t=1}^{N-2} d(t)d(t+2)}{(N-3)\sum_{N-1}^{N} d(t)^2}$$

The constant b is computed from a representative data set, and N is the number of data points used for this determination. This kind of model responds well to traffic changes, but always presents a certain inherent time lag.

Comparisons between the two generations of models (Kreer 1975) show that, generally speaking, the second-generation models are better than the third-generation ones, because of the use of historical data. On the other hand, third generation models do not need historical data and are more responsive to nonperiodic phenomena such as accidents and the weather. Figure 1 compares the prediction of a second-generation model with actual data.

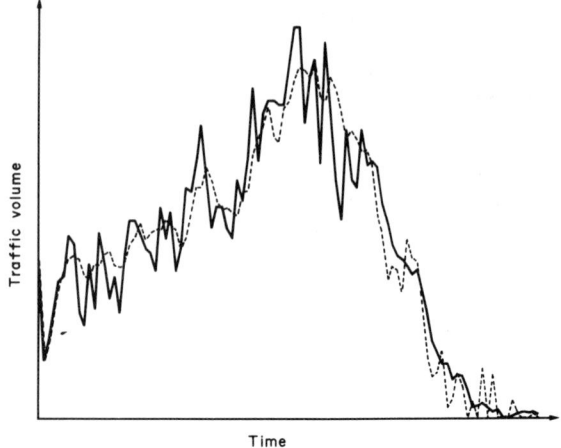

Figure 1
Prediction of a second-generation predictive model on a weekday with a time step of 6 min: ------ predicted traffic volume; ——— actual volume

1.2 Other Kinds of Model

Various other techniques have been applied to short-term prediction; the following are examples.

(a) General spectral analysis. Second-generation models use spectral analysis to compute smoothed historical volumes by fitting Fourier series. The technique is also used in a more general form (Nicholson and Swann 1974). The principle is to consider volume measurements on M days divided into N time intervals. The flow volume on day m at time interval k is defined as

$$V_m(k) = \sum_{i=1}^{k} C_{mi}\phi_i(k) + e_m(k), \quad m = 1, \ldots, M$$

where e_m is an independent expansion error and the $\phi_i(k)$ are a set of orthogonal vector functions representing the characteristic modes of functions V_m. Prediction is realized by computing, at time interval N' of day $M+1$, the coefficients $C_{M+1,i}$ from data $V_{M+1}(1), \ldots, V_{M+1}(N')$.

(b) Autoregressive integrated moving average (ARIMA). Box–Jenkins methods for time series analysis, also used for long-term prediction, have been applied to short-term prediction of volumes on freeways (Eldor 1977). The main interest of these methods is to provide the user with a whole class of models and to give him procedures for choosing the most appropriate one; classical schemes, such as exponential smoothings, can be considered as particular cases of the general method. The general ARIMA (p, d, q) is described by

$$\phi(B)\Delta^d f(t) = \theta_0 + \theta(B)a(t)$$

where $f(t)$ is the series value at time t and

$$\Delta f(t) = f(t) - f(t-1), \quad \Delta^d f(t) = \Delta[\Delta^{d-1} f(t)]$$
$$Bf(t) = f(t-1), \quad B^m f(t) = f(t-m)$$
$$\phi(B) = 1 - \phi_1 B - \phi_2 B^2 - \cdots - \phi_p B^p$$
$$\theta(B) = 1 - \theta_1 B - \theta_2 B^2 - \cdots - \theta_q B^q$$

θ_0 is the overall moving average constant and $a(t)$ is random noise. The predicted time series may be either $f(t)$ itself or the series

$$D(t) = f(t) - m(t)$$

where $m(t)$ is the smoothed historical volume at time interval t.

(c) Filtering methods. Sophisticated filtering methods, using the Kalman filter, have been proposed (Baras et al. 1979), resulting in accurate queue-length predictive models using only current-day measurements.

2. Very-Short-Term Prediction

The problem of very-short-term prediction, although it has some characteristics in common with short-term prediction, is very different. The most interesting available data are no longer previous measurements at the point studied but measurements at points situated upstream; hence, the models are exogenous and the main problem is to estimate travel time from upstream to downstream points, starting only from flow volume information.

Since the relations between flow volume and travel time cannot be precisely formalized in an urban network, no explicit model can be realized. The models used are therefore mostly linear or piecewise linear models (Lesort et al. 1982), fitted to a representative data set and eventually corrected on-line. A linear model can be defined as follows:

$$V_0(t+1) = \sum_{i=1}^{n} a_i V_1(t - \delta_i)$$

where $V_0(t)$ is the volume measured at time interval t at the downstream point, $V_1(t)$ is the volume measured at time interval t at the upstream point, $\delta_1, \ldots, \delta_n$ are possible values of travel time between two points and $a_1, \ldots, a_n$ are coefficients by linear regression for a representative data set. For a piecewise linear model, each piece of the model is representative of a certain range of travel times. The different pieces can all be defined as above, with different coefficients a_i for each piece of the model.

See also: Kinematic Wave Theory; Simulation Programs, Macroscopic; Transportation Modelling

Bibliography

Baras J S, Levine W S, Dorsey A J, Lin T L 1979 *Advanced Filtering and Prediction Software for Urban Traffic Control Systems*. Transportation Studies Center, University of Maryland, Baltimore, MD

Eldor M 1977 Demand predictors for computerized freeway control systems. *Proc. 7th Int. Symp. Transportation and Traffic Theory*. Institute of Systems Science Research, Kyoto, Japan

Ganslaw M J, Schaake J C 1973 A volume and speed predictor for the UTCS/BPS second generation software. Final report prepared for TRW systems. Houston, TX

Kreer J B 1975 A comparison of predictor algorithms for computerized traffic control system. *Traffic Eng.* **45**, 51–6

Lesort J B, Tardieu B, Lebacque P 1982 Short term traffic flow prediction in urban networks. *Proc. Int. Conf. Road Traffic Signalling*. Institution of Electrical Engineers, Stevenage, UK

Lieberman E B, McShane W R, Goldblatt R 1974 Variable cycle signal timing program. *Prediction Algorithms, Software and Hardware Requirements*, Vol 4, KLD, Huntingdon, WV

McShane W R, Lieberman E B, Goldblatt R 1976 Developing a predictor for highly responsive system-based traffic signal control. *Transp. Res. Rec.* **596**

Nicholson H, Swann C D 1974 The prediction of traffic flow volumes based on spectral analysis. *Transp. Res.* **8**, 533–8

Ross D W, Humphrey T L, Mahoney E E, Sandys R C, Williams G L, Wohnoutka R, Wong P J, Zeidler H M 1977 Improved control logic for use with computer-controlled traffic. Report NCHRP3. 18(1). Stanford Research Institute, Menlo Park, CA

J. B. Lesort
[INRETS, Arcueil, France]

Priority Intersection: Modelling

Most intersections in rural and even urban areas are not controlled with traffic signals. Safety is the main problem that engineers have to consider, but level of service is also worthy of consideration in order to aid them in the choice of the type of junction, the design of the layout or the economic appraisal. This level of service can be assessed with variables such as capacity, delay and queue lengths. Statistics, probabilities or simulation can help in the modelling of these variables for priority intersections (i.e., where users of a minor road have to give priority to users of a major road).

1. Capacity

The capacity of a priority intersection is the maximum possible flow outside of the minor road. So, it is not a single value like it is for a road section (see *Flow Variables*), but instead is expressed as a function of the major flow. The parameters of this function depend on geometric, environmental and behavioral characteristics.

1.1 Statistical Approach

The statistical approach consists of observing the variables under consideration over a number of selected intersections and then analyzing the data collected. Measurements have to be performed during periods of queuing to ensure that the observed entry flow is the maximum possible. The more significant work using this technique has been realized in the UK by the Transport and Road Research Laboratory (TRRL). After an extensive series of measurements, on a track and in the field, TRRL derived relationships expressing the capacity as a function of traffic and geometry for minor approaches to junctions and for roundabout entries. Their general shape is linear, slope and intercept being parameters affected by the geometry. The results have been summarized by Kimber (1988). Even if this empirical approach lacks an explanatory background model, the results obtained with such a number of observations as in the TRRL study are highly reliable and the expressions derived can be used as predictive relationships.

1.2 Basic Gap-Acceptance Model

In contrast, the probabilistic approach involves a vehicle–vehicle model of the system under study. In its basic form, this model implies that a minor flow gives priority to a major one that it has to merge into or to cross. The idea of the so-called discrete gap-acceptance model (see Fig. 1) is as follows: no vehicle of the minor flow can leave the give-way (or stop) line if the size of the available gap in the major flow is less than a given value t_c, which is called the critical gap. One vehicle can leave if the size is between t_c and $t_c + t_f$, where t_f is the time necessary for the second vehicle in the queue to reach the give-way line after the first one has left the queue (called the follow-up time or the move-up time). Two vehicles can insert into a gap in the range

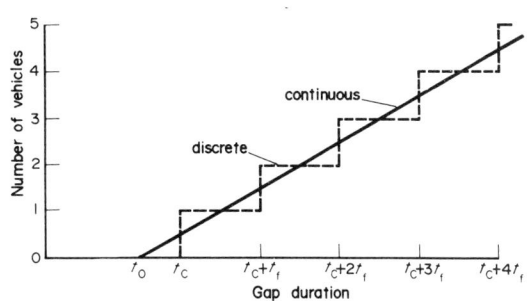

Figure 1
Number of vehicles able to insert into a gap

$[t_c+t_f, t_c+2t_f[$ and so on, so n vehicles can insert into a gap in the range $[t_c+(n-1)t_f, t_c+nt_f[$. This step function $n(t)$ (number of vehicles able to leave the queue during a gap of duration t) can be approximated with a switch line: no vehicle until a threshold t_0 and a noninteger number of vehicles varying linearly with t beyond t_0, with a slope equal to the saturation flow $1/t_f$: this is the continuous (or fluid) gap-acceptance model described, for instance, by Plank (1982). The correspondence between the two models occurs with $t_0 = t_c - t_f/2$.

It is necessary to distinguish the first gap encountered, called the lag, which is not bounded by two arrivals of vehicles in the major stream, but is opened by the arrival at the give-way line.

1.3 Parameters Involved

Critical gap and follow-up time are obviously not constant over the population (heterogeneity) and may also vary for the same driver, who can accept a gap shorter than another that he has previously rejected (inconsistency). The basic discrete model supposes homogeneity and consistency, but others avoid this simplification (see Sect. 1.5).

Many methods have been proposed to obtain values for the parameters of the gap-acceptance model from measuring the distributions of offered, rejected and accepted gaps. Miller (1971) and Hewitt (1985) have reviewed a dozen methods for processing such data. A bias appears however, due to the over-representation of cautious drivers, and so the observed percentage of accepted gaps of a given size is not the percentage of drivers prepared to accept such a gap. Ashworth (1968) showed that it can be corrected by subtracting a quantity $Q\sigma^2$ from the computed value, where σ is the standard deviation of the distribution (this is rigorous only for a normal distribution and random arrivals, but remains a good approximation for other cases). Possibly the most used method is named after Raff: it only uses the lags (this solution avoids the bias) and consists of determining graphically the intersection of the curves of accepted lags greater than t and rejected lags smaller than t. A better one is the probit method, well adapted to such data resulting from all-or-nothing answers. Maximum likelihood (Miller 1971) also gives accurate results. However, like the probit method, it implies a hypothesis on the distribution of t_c, generally normal or log normal.

The measurement of the follow-up time has led to few studies. Nonstandard regression on $n(t)$ can be considered, but the easy way is to measure the average headway when a queue is discharging. The follow-up time is often considered in the guidelines to be 60% of the critical gap.

The main factors affecting the value of t_c and t_f are, among others, the layout of the intersection (angles, visibilities, widths, etc.), the signing (stop or give way), the environment (rural or urban, weather and light conditions, etc.), the speed on the major road, the population (of drivers and of vehicles), the type of the maneuver and of the gap, and the impatience of the drivers.

1.4 Capacity Formulae

Given the distribution of available gaps, that is, the headway distribution $f(t)$ of the major flow Q (the models do not generally take account of the time necessary for a priority vehicle to cover its own length), it then becomes possible to derive a capacity formula. The capacity C can be expressed, over a large period, as the ratio of the mean number of vehicles inserting into a gap over the mean length of a gap:

$$C = Q \int_0^\infty f(t)\, n(t)\, dt \qquad (1)$$

(variables have to be expressed in homogeneous dimensions). With the two possible functions $n(t)$ mentioned above, it follows immediately that

$$C = Q \sum_{i=1}^\infty i \int_{t_c+(i-1)t_f}^{t_c+it_f} f(t)\, dt \qquad (2)$$

in the discrete case and

$$C = Q \int_{t_0}^\infty \frac{t-t_0}{t_f} f(t)\, dt \qquad (3)$$

in the continuous case.

Using the exponential distribution for $f(t)$ (i.e., random arrivals) leads to:

$$C = \frac{Q \exp(-Qt_c)}{1 - \exp(-Qt_f)} \qquad (4)$$

in the discrete case and

$$C = \frac{\exp(-Qt_0)}{t_f} \qquad (5)$$

in the continuous case.

The first of these two formulae is the basis of the German guidelines and of the late US Highway Capacity Manual, among others. Another usual distribution (which generalizes the previous one) is the regular-random distribution, which combines a proportion B of bunched vehicles with a uniform headway R, and the remaining $(1-B)$ vehicles driving free with a headway distributed according to a shifted exponential (R being the value of the shift). A particular case when $B = QR$ leads to the well-known Tanner's formula, derived through more tedious calculations and a rather different presentation, as the limit yielding infinite value for the delay (Tanner 1962):

$$C = \frac{Q(1-QR)\exp[-Q(t_c - R)]}{1 - \exp(-Qt_f)} \quad (6)$$

Other formulae, using either model and different headway distributions, can be found. Figure 2 shows that if the choice of a headway distribution is of a great importance, the continuous approximation of the discrete model introduces only slight differences in the numerical values.

1.5 Enhanced Models

Several authors have dealt with the introduction of heterogeneity or of inconsistency. Two papers have clarified the problem for the calculation of capacity. In the first, Plank and Catchpole (1984) consider the inconsistency of homogeneous drivers, provided that t_c is greater than t_f and range (t_c) is less than t_f. The general result is not always tractable but it simplifies for usual headway distributions. For example, using the regular-random distribution:

$$C = \frac{Q(1-B)\exp(\gamma R)d_c^*(\gamma)}{1 - d_f^*(\gamma)} \quad (7a)$$

with

$$\gamma = \frac{Q(1-B)}{(1-QR)} \quad (7b)$$

The terms d_c^* and d_f^* are the Laplace transforms of the distributions d_c and d_f of t_c and t_f. In the second paper, Catchpole and Plank (1986) allow for heterogeneity, but with a fixed value of t_f and some conditions on the ranges of the parameters. Again, the general expression is not always tractable. For the regular-random distribution it reduces to:

$$C = \frac{Q(1-B)\exp(\gamma R)}{[1 - \exp(-\gamma t_f)]\left[\sum_i a_i/d_{ci}^*(\gamma)\right]} \quad (8)$$

where the subscript i denotes the subpopulations of homogeneous drivers, in this case of a discrete set of classes of proportions a_i.

It can be shown, by this way or others, that the capacity C of a shared lane (i.e., occupied by vehicles that perform different maneuvers or that belong to different categories) is the harmonic mean of the capacities C_i calculated for the different sub-categories, weighted according to their proportions a_i (with exceptions such as a special case of regular headways):

$$C = \frac{1}{\sum_i a_i/C_i} \quad (9)$$

Many other improvements to the basic model have been proposed, such as distinguishing critical gap and critical lag, considering several major flows, introducing impatience and taking the flaring of a roundabout entry into account.

1.6 The Impedance Effect

Real intersections cannot, of course, be described by this simple interaction of two flows and it is necessary to account for different levels of priority. For instance, at a T junction (in a drive-on-the-right system) where the major arms are numbered 1 and 2 and the minor arm 3, left-turn vehicles from the major road (Q_{23}) have to give priority to the opposite straight-on vehicles (Q_{12}), but have priority over the left-turn vehicles from the minor road (Q_{31}): so Q_{12} is of level 1 (as are Q_{13} and Q_{21}), Q_{23} is of level 2 (as is Q_{32}) and Q_{31} is of level 3. The interaction between a stream of level 2 (or more for complex intersections) and one of higher level is called the impedance effect: vehicles queuing in Q_{23} block Q_{31}. The result obtained when applying a capacity formula has to be multiplied by a factor p_0 to obtain the capacity C_{31}, where p_0 is the probability for having no queuing vehicle in Q_{23}. The problem is knowing whether Q_{23} has to be considered in the major streams or not. The analytical solution is not known, and only approximate procedures are given (Brilon 1988 pp. 111–53). The exact result lies between the two possibilities.

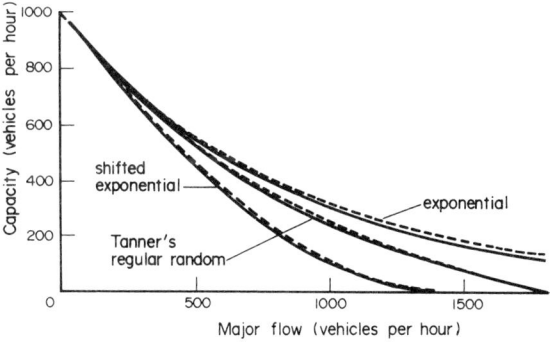

Figure 2
Capacity curves for the discrete (full line) and continuous (dashed line) models, with three different major headway distributions for $t_c = 6$ s, $t_f = 3.6$ s and $R = 2$ s

2. Delay

As with capacity, delay can be studied using the empirical way and an example is reported by Brilon (1988). However, most works on the topic involve queuing theory (see *Queuing Theory Applications*).

2.1 Steady State

A first step consists of studying the time spent in waiting for a gap of a sufficient duration. For an isolated vehicle arrived at a random instant, Adams established in 1936 the mean waiting time W_1 at the give-way line and its variance V_1, with Poisson arrivals in the main stream:

$$W_1 = \frac{\exp(Qt_c) - Qt_c - 1}{Q} \tag{10}$$

$$V_1 = \frac{\exp(2Qt_c) - 2Qt_c\exp(Qt_c) - 1}{Q^2} \tag{11}$$

However, the Pollaczek–Khintchine formula for obtaining the total delay W cannot simply be applied, for the problem is more complex, due to the fact that the service time is not the same for vehicles that have waited in the queue and for those that have not. For these systems, a formula exists, due to Yeo. The difference comes from the lag distribution encountered and from the consideration of the follow-up time (Daganzo 1977, Kimber *et al.* 1986).

Probably the most widely used formula is Tanner's formula (Tanner 1962). Under the assumptions seen in Sect. 1.3 and an exponential distribution for the minor stream Q', the total delay W per vehicle is (with the previous notation):

$$W = \frac{0.5E(y^2)/Y + Q'Y[1 - (1 + Qt_f)\exp(-Qt_f)]/Q}{1 - Q'Y[1 - \exp(-Qt_f)]} \tag{12a}$$

with

$$Y = E(y) + \frac{1}{Q} \tag{12b}$$

$$E(y) = \frac{A}{Q(1 - QR)} - \frac{1}{Q} \tag{12c}$$

$$E(y^2) = \frac{A - Qt_c(1 - QR) - 1 + QR - Q^2R^2 + 0.5Q^2R^2/(1 - QR)}{[Q^2(1 - QR)^2]/2A} \tag{12d}$$

$$A = \exp[Q(t_c - R)] \tag{12e}$$

In these expressions, y is the duration of a block in the main stream (i.e., major vehicles driving with headways such that a minor one cannot start).

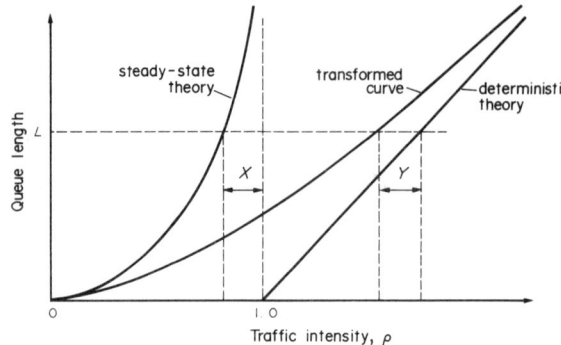

Figure 3
Illustration of the coordinate transformation for $L_0 = 0$ and a given value of t (after Kimber and Hollis 1979)

Various other solutions have been proposed, some of them being incorporated into national guidelines. One of the simplest, which is a good approximation for the Poisson case, is due to Siegloch (Brilon 1988):

$$W = \frac{\exp(-Q't_f)/(1 - \rho) - 1}{Q'} \tag{13}$$

where ρ is the traffic intensity $\rho = Q'/C$.

2.2 Nonstationary Conditions

All previous results are only valid under stationary conditions of traffic, and below capacity (and even for $\rho < 0.8$ for practical applications): beyond, delay and queue length become infinite. However, in real conditions, traffic may exceed capacity during limited periods of time. A deterministic approach to this problem consists of integrating the overcapacity during these periods to describe the growth and decay of the queue, but this theory does not consider queue due to random phenomena. TRRL has developed a method (Kimber and Hollis 1979) that combines probabilistic and deterministic approaches when traffic is given under the form of a flow profile (i.e., a series of short segments of time during which traffic is constant and may exceed capacity) and which avoids the long computations involved by the theory of transient states. It consists in determining the queue length L at the end of each time step, from the length L_0 at the beginning and the traffic intensity. In this method, a coordinate transformation of the curves corresponding to the steady state and the deterministic models is operated. Figure 3 illustrates the case for $L_0 = 0$ and a given t: the resulting transformed curve is such that the distances X and Y are equal. Delays can be derived, which can be expressed per unit time for the whole traffic or per arriving vehicle within a given period. Kimber and Hollis (1979) give the solutions for the M/M/1 and M/G/1 cases. The results are embodied in the computer programs PICADY and ARCADY, which assess delays and queue lengths from a capacity curve and a flow profile for priority junctions and roundabouts, respectively.

2.3 Geometric Delay

The geometric delay is caused by the fact that isolated vehicles have to slow down to perform their maneuver. For intersections operating under light or medium traffic, it can be much higher than the queuing delay. It has mainly been studied using empirical methods or a rough modelling of vehicles kinematics. Kimber *et al.* (1986) have, however, investigated the interrelation between geometric and queuing delays, within the gap-acceptance framework.

3. Simulation Programs

Since making probabilistic models more complex quickly leads to intractable equations, and simplifying them leads to inadmissible approximations, a useful alternative is simulation. Traffic at priority intersections is well suited to microscopic simulation, where individual behavior is modelled (see *Simulation Programs, Microscopic*). Simulation, which was initially a method for researchers to test a model or hypothesis, is now also a practical tool for engineers designing an intersection. Latest programs operate on microcomputers, are modular, enable the representation of any kind of layout, accept various headway distributions and time-varying traffic flows, and offer graphic displays or animation. Among the more recent, INSECT in Australia (which can also model roundabouts and signalized intersections), KNOSIMO in Germany and OCTAVE in France offer all or most of these characteristics.

4. Further Details

Most of the topics dealt with in this article are still being intensively researched. As the literature on the subject is very scattered, the interested reader will find useful surveys and detailed references on the most recent research in the proceedings of an international workshop on the subject (Brilon 1988).

See also: Road Traffic: An Introduction

Bibliography

Ashworth R 1968 A note on the selection of gap acceptance criteria for traffic simulation studies. *Transp. Res.* **2**, 171–5

Brilon W (ed.) 1988 *Intersections Without Traffic Signals, Proceedings of an International Workshop 16–18 March 1988, Bochum, West Germany*. Springer, Berlin

Catchpole E A, Plank A W 1986 The capacity of a priority intersection. *Transp. Res. B* **20**, 441–56

Daganzo C F 1977 Traffic delay at unsignalized intersections: Clarification of some issues. *Transp. Sci.* **11**, 180–9

Hewitt R H 1985 A comparison between some methods of measuring critical gap. *Traffic Eng. Control* **26**, 13–22

Kimber R M 1988 The design of unsignalized intersections in the UK. In: Brilon 1988, pp. 20–34

Kimber R M, Hollis E M 1979 Traffic queues and delays at road junctions. Transport and Road Research Laboratory Report No. LR909. TRRL, Crowthorne, UK

Kimber R M, Summersgill I, Burrow I J 1986 Delay processes at unsignalised junctions: The interrelation between geometric and queueing delay. *Transp. Res. B* **20**, 457–76

Miller A J 1971 Nine estimators of gap-acceptance parameters. In: Newell G F (ed.) *Proc. 5th Int. Symp. Theory of Traffic Flow and Transportation*. Elsevier, New York, pp. 215–35

Plank A W 1982 The capacity of a priority intersection—Two approaches. *Traffic Eng. Control* **23**, 88–92

Plank A W, Catchpole E A 1984 A general capacity formula for an uncontrolled intersection. *Traffic Eng. Control* **25**, 327–9

Tanner J C 1962 A theoretical analysis of delays at an uncontrolled intersection. *Biometrika* **49**, 163–70

G. Louah
[Centre d'Etudes Techniques de l'Equipement de l'Ouest, Nantes, France]

Queuing Theory Applications

The importance of queuing theory to the analysis of transport systems operation becomes obvious if it is considered not in terms of queuing, but in terms of congestion. For traffic congestion is simply queuing; that is, travellers or vehicles waiting to use some "server" such as a signalized intersection, a freeway bottleneck or a bus. In some cases, such as at stop signs or traffic signals, the queues are obvious and easily observed. The queues on urban freeways, however, are commonly so large that the public—and all too often even traffic analysts—do not recognize them as queues. In other cases, the queues are hidden from public view or located in places where they cannot be readily associated with a specific server. For example, inbound aircraft are sometimes directed to fly at reduced speed or to delay their takeoff from some other airport, tactics that reduce the costs associated with queuing, but displace rather than eliminate the queues (see *Air Traffic Control: An Overview*). In still other cases (e.g., scheduling trains on a busy railroad network), the object of control is to eliminate queuing, which really means reducing the probability of a queue occurring to some acceptable level (see *Underground Railroad Modelling and Control*).

1. Modelling

There is, of course, an immense body of literature on queuing theory but, surprisingly, little of it is useful for most of the problems encountered in transportation systems—particularly urban passenger systems, which are the primary topic of this article. The reason for this is that almost all of the literature is devoted to determining steady-state solutions for systems where the arrivals can be regarded as a stationary stochastic process; that is, systems in which the expected arrival rate $\lambda(t)$ does not vary with time t. A fundamental feature of urban passenger transport, however, is that the amount of travel varies over the day, with very high flows during morning and evening peak periods on weekdays (and in some locations on Sunday evenings), moderate-to-high flows during midday and much lower flows at other times. An obvious corollary is that on all but the most congested systems, the bulk of the congestion occurs during the peaks. This, however, leads to a major difficulty for analysts: peak period traffic is, almost by definition, not in a steady state, so the steady-state models that comprise the bulk of the queuing literature are, at best, approximations and are often totally inappropriate.

The fact that steady-state models cannot be used to analyze systems for which the expected arrival rate $\lambda(t)$ exceeds the average service rate or capacity, μ is obvious, so their use is ruled out for all urban freeway congestion and also for the analysis of many important signalized intersections. Whether they provide reasonable approximations in other situations is less obvious, though a good general rule is that it is unlikely in cases where the time-varying traffic intensity $\rho(t) = \lambda(t)/\mu$ is either near unity or changing rapidly.

One way to explore this problem further is to use the relationships that if the traffic intensity ρ is constant, the steady-state expected queue length is approximately $1/(1-\rho)$ and the time required for a queue length distribution to relax to a reasonable approximation of the steady-state equilibrium distribution from an initial queue length of zero is about $1/(1-\rho)^2$ service times. These relationships, obtained by approximations to approximations (Newell 1982 pp. 150–1, 239, 255) are not intended to be used for values of ρ much less than the values shown in Table 1 or to be accurate for all systems within even a factor of two, but the purpose at hand does not require great accuracy.

Table 1 lists the approximate equilibrium queue lengths and relaxation times for several values of ρ for mean service times $S = 1/\mu$ of 2 s and 2 min. The numbers were chosen for roundness rather than realism, but 2 s is a realistic service time for a highway lane or a pedestrian going through a doorway and 2 min is of the correct order of magnitude for a runway. Rough though the expected queue length and relaxation time estimates may be, their message is very clear. At $\rho = 0.5$, the queue length and the relaxation time are so small that the physical system can keep up with changes in the equilibrium queue length distribution, so the use of steady-state results seems quite safe. Beyond values of around $\rho = 0.9$, however, the situation becomes hopeless; there is no conceivable passenger transport system for which demand changes so slowly that the real queue length distribution would, in any way, resemble that given by the steady-state equations as ρ

Table 1
Approximate queue lengths, relaxation times and limits on the change in traffic intensity in 15 min

ρ	Expected queue length	Relaxation time $S=2$ s	Relaxation time $S=120$ s	$\Delta\rho$ in 15 min $S=2$ s	$\Delta\rho$ in 15 min $S=120$ s
0.5	2	8 s	8 min	56	0.9
0.75	4	32 s	32 min	7	0.1
0.9	10	3 min	3 h	0.5	0.01
0.95	20	13 min	13 h	0.06	10^{-3}
0.99	100	5 h	14 d	0.5×10^{-3}	10^{-5}
0.999	1000	23 d	4 years	0.5×10^{-6}	10^{-8}

changes from 0.95 to 0.99, a change that involves only a 4% increase in arrival flow, but a quintupling of the expected queue length. At these high traffic intensities, real queue lengths will obviously be significantly smaller than the steady-state equations predict.

For intermediate values of ρ, the situation is less obvious, but can be taken a little further by using the criterion that the physical system can keep up with changes in the equilibrium queue length distribution if $|d\rho(t)/dt| \ll \mu[1-\rho(t)]^3$ (Newell 1982 pp. 151–2). The last two columns in Table 1 are these limiting values of $d\rho(t)/dt$ multiplied by 15 min; if the change in ρ over a 15 min period is small compared to the values listed, it should be reasonably safe to use steady-state equations. The values show that this is very questionable when $\rho = 0.75$ for systems such as runways, where the average service time is measured in minutes, but remains reasonably safe up to $\rho \approx 0.9$ for systems with service times of only a few seconds.

For situations in which steady-state models are inappropriate, two alternative types of model are available, one deterministic and much used, the other stochastic and almost never used. The stochastic models are approximations of one sort or another, but seem to offer great promise for many of the problems encountered by transport analysts (for more details see Newell (1982)).

The deterministic models are usually presented in graphical form as shown in Fig. 1. Curve $A(t)$ represents the cumulative number of travellers that have arrived by time t and $D(t)$ represents the cumulative count of departures from the queue into the service. The queue length at any time t is simply the difference $Q(t) = A(t) - D(t)$. This conservation equation—which only states that anybody who has arrived must either have left or still be there—is all there is to the model, but the problem-solving power of such a simple idea is surprising. One feature which is particularly useful is that the area between the $A(t)$ and $D(t)$ curves is the total waiting time for all the customers who are delayed, measured in vehicle hours or some other convenient unit. Another is that if the queue discipline is first in, first out (FIFO), the horizontal distance between the curves is the waiting time of a particular customer.

For most transport applications, the service rate or capacity is fixed at some value μ or can be changed only by adding an additional server, so $D(t)$ will be a straight line or, at least, piecewise linear. This makes graphical solutions very easy: a straight edge is placed at slope μ and slid across the paper until it touches the $A(t)$ curve at some point t_0. Note that the implication is that to the left of the point of contact $A(t) = D(t)$ and there is no queue; the deterministic models do not supply any information about stochastic variations and actually imply that queues exist only when there is oversaturation (i.e., only in situations where $\lambda(t) > \mu$ throughout some finite time interval). Thus, they treat as negligible everything that steady-state queuing models are concerned with. This is the price that must be paid to have models that are no more complicated than the steady-state models, yet are capable of dealing with the peaking encountered in urban conditions.

Traveller counts are actually discrete, so the smooth curves $A(t)$ and $D(t)$ are approximations to step functions. In the case of $A(t)$, the smooth curve can also be thought of as the average of the data of many days or as the expected value of the step function that will develop on some future day. Thus, its slope $dA(t)/dt$ can be interepreted as the expected arrival rate $\lambda(t)$. It is also common to treat $D(t)$ as the expected departure curve, but this interpretation must be approached with caution. The problem that arises is that t_0, the time when the queue forms, is also a random variable and the $D(t)$ curves for individual days are straight lines beginning at the random points with coordinates $\{t_0, A(t_0)\}$. When $A(t)$ has the shape shown in Fig. 1, with a slope very close to μ over a fairly long time interval, the straight line drawn to represent $D(t)$ would seem to be a reasonable approximation of the expected departure curve, but if $A(t)$ curves sharply upward, the expected value of the departure curve will also be curved in the vicinity of t_0.

The $A(t)$ curve is more abstract than might be expected. For the model to work properly without modification, $A(t)$ must represent the number of travellers that would have reached the server by time t if there were no queue or, equivalently, the number that would have joined the queue if the queue occupied no space. For servers where the space occupied by the queue is negligible, this distinction seems pedantic, but for urban freeways—where queues are typically several kilometers long—it can be important.

2. Applications

2.1 Traffic Signals

Estimating the effectiveness of alternative intersection designs and traffic signal settings is the one area of urban transport analysis in which the use of queuing theory has become universal and almost unquestioned.

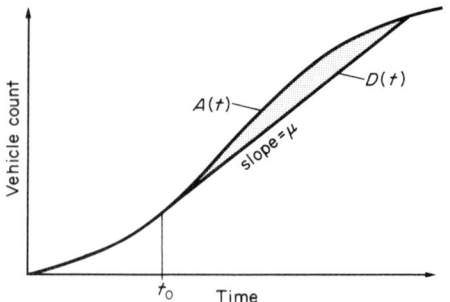

Figure 1
Arrival and departure curves, $A(t)$ and $D(t)$

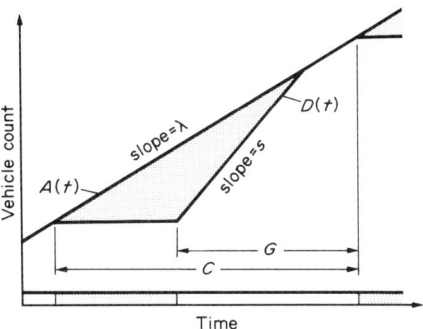

Figure 2
Arrival and departure curves for one signal cycle

The need is obvious: traffic signals by their very nature cause queuing delays and it is only sensible to try to keep these delays small.

A wide variety of equations is available for estimating the delay at pretimed traffic signals, but those in common use fit into two basic categories (Hurdle 1984a). In each of these, the first term is a deterministic estimate of the average delay per vehicle:

$$\frac{C(1-G/C)^2}{2(1-\lambda/s)} \quad (1)$$

where C is the cycle length, G is the effective length of the green interval, λ is the arrival rate and s is the saturation flow (defined as the rate at which vehicles are served while the light is green) (see *Road Traffic: An Introduction*). The expression in Eqn. (1) is simply the area of the triangle between the arrival and departure curves in Fig. 2 divided by the total number of arrivals during the cycle λC; it is sometimes called the uniform delay since this is what the delay would be if the headways between successive arrivals were all the same. Note that the uniform delay approaches half the length of the red interval as the signal approaches saturation (i.e., as $\rho = Gs/\lambda C$ approaches 1) and becomes meaningless when the signal is oversaturated (i.e., if $\rho > 1$), since the triangle ceases to be a triangle.

For the first set of equations, the remaining terms of the equations, derived from steady-state analytic models or simulations, represent the additional delay due to randomness in the arrival and service processes. Since most of this additional delay occurs when more vehicles arrive during some cycle than can be served and so spill over into the next cycle, it is frequently called overflow delay. For a steady-state traffic signal queue, the overflow delay approaches infinity as ρ approaches unity. (In the traffic signal literature, the traffic intensity ρ is generally called the degree of saturation and is represented by x or, in much of the US literature, by v/c.) For practical purposes, the estimate of the overflow delay can be regarded as negligible when $\rho < 0.6$ or even 0.7 and becomes totally unrealistic somewhere in the vicinity of $\rho = 0.9$ for the reasons discussed in Sect. 1.

Various versions of a second set of popular equations that originated with the TRANSYT computer program appear in such collections of procedures as the *Australian Road Research Report 123* (Akcelik 1981), the *Highway Capacity Manual* and the *Canadian Urban Capacity Guide*. These equations attempt to avoid the difficulty when ρ is near unity by effecting a smooth transition between the realistic overflow delays predicted by the steady-state equations for smaller values of ρ and those predicted for $\rho > 1$ by deterministic approximations. (The deterministic approximations predict that there is no overflow delay at all when $\rho \leq 1$, so the discrepancy between them and the prediction of the steady-state equations of infinite delay at $\rho = 1$ is dramatic.) These equations produce more realistic estimates than the steady-state models when ρ is near unity, but it should be noted that they predict the same delays for all values of $d\rho(t)/dt$, whereas the real delays will be larger when the arrival flow increases slowly than when the buildup is rapid. Their use is sometimes also suggested for oversaturated conditions, but this involves assuming an arrival pattern $\lambda(t)$ that will not often be encountered in real situations.

Both of the classes of models described assume that the cycle length is fixed, yet they are commonly used for traffic actuated signals—sometimes, as in the *Highway Capacity Manual*, with the delay modified by crude adjustment factors based on empirical data. Clearly, however, the operation of traffic-actuated signals is sufficiently different from that of pretimed signals to justify the development of models that explicitly recognize their characteristics. Both classes are also used, again sometimes with adjustment factors, to estimate delay at signals that are a part of a coordinated system. In this case, it is considerably more obvious that the answers produced are likely to be very wrong, but there is an alternative available. Computer models such as TRANSYT that analyze an entire network instead of a single signal have the capability of modelling the movement of platoons of vehicles through the system, so they can explicitly account for the fact that the arrival flow is not uniform, but cyclical. These models are deterministic simulations; they include stochastic effects, but estimate them in a manner similar to the models described previously, rather than by Monte Carlo simulation methods.

2.2 Freeways

All freeways in or near cities have locations where the peak arrival rates routinely exceed the roadway capacity; large queues form upstream from these bottlenecks unless entry is limited by some sort of control system. To see that this is unavoidable, it is only necessary to estimate the number of lanes that would be needed to accommodate the number of people who would like to drive home immediately after work from a central business district. In really large cities, it is not even worth considering whether it makes economic sense to build enough lanes to accommodate a demand that lasts for only a short time; it is not physically possible. Further away from major cities, of course, such queuing will not happen five days a week, nor in so many places, but that is not to say it will never occur. The classic example is a substantial part of Europe on the first and last days of school holidays.

Because of the extremely large delays involved, the analysis of freeway operation is another area in which the use of queuing theory is standard practice. For most purposes, the standard tools used for estimating freeway delays are deterministic models of the sort described in Sect. 1, though sometimes in the form of computer models rather than diagrams. These models are at their best in these conditions, since random variations in queue lengths can be expected to be small compared to the mean in situations where oversaturation for several hours and queues of several hundred vehicles are the norm. On the negative side, the routine use of deterministic models seems to make people forget that the processes involved are really stochastic.

Freeways are also subject to a great deal of nonrecurrent congestion caused by roadway maintenance, accidents or mechanical breakdown. A great deal can be gained for comparatively small expense by considering congestion effects when scheduling maintenance activities and the use of queuing theory for this purpose is standard practice for some agencies. A major objective is to schedule the work in such a way that queues caused by maintenance activities will vanish before recurrent congestion normally begins. Congestion caused by incidents is a more difficut problem since it cannot be known where or when it will occur, nor for how long the capacity of the road will be reduced. What can be done if the freeway management system has good data-gathering capabilities is to predict the future growth of the queue once it is discovered. Deterministic queuing models are very well suited for this purpose if the current state of the system is known with reasonable accuracy; the difficult part of the problem is to devise strategies for reducing the delay by diverting traffic to other routes.

Freeway management systems can also be designed to reduce recurrent congestion on the freeway by limiting entry. With the possible exception of control in the form of tolls, this will cause queues to form on the entrance ramps and possibly elsewhere in the system. Clearly the magnitude of the resulting change in total delay in the system must be estimated, both in order to design the details of the control scheme and to determine whether it is worthwhile. For this purpose the deterministic models are quite attractive. However, if a major goal is to allow high-speed bus transit without constructing busways or seriously decreasing the amount of capacity available for cars, the analysis becomes rather different. In this case, there would be obvious advantages if queuing could be completely eliminated from the freeway. This, however, is not a realistic objective; it is only possible to try to keep the probability of a queue forming within some acceptable level, a task for which a stochastic model (though not necessarily a queuing model) is needed.

There are some significant steps that are not, but could and should be, carried out in terms of freeway queuing analysis. Given that a system is going to have large queues, it seems obvious that they will be more disruptive in some locations than others. For example, a queue on a freeway is likely to be less disruptive than the same number of cars on town center streets. However, it might be even better to store these cars in an area built specifically for the purpose. Such queue management is sometimes part of control schemes for existing freeways, but is rarely, if ever, considered as part of area-wide planning efforts. It is a tool of those responsible for coping with current problems, but apparently not of those who try to prepare for the future. Unfortunately, this means that the amount of queue management possible is severely limited by the requirement that the overall system be changed only in very minor ways.

It is also comparatively uncommon for microeconomic or planning models (e.g., congestion pricing models, route assignment models, cost comparisons of alternate proposals) to recognize the type of queuing that occurs when $\rho(t) > 1$ for an extended period. In fact, such models oridinarily assume a steady state, though often without specifically saying so. That the conclusions reached can be very different when time-dependent queuing effects are considered is illustrated by Hurdle (1974, 1984b), D'Ans and Gazis (1976), Smith (1984), Daganzo (1985), Hurdle and Solomon (1986), Zawack and Thompson (1987) and Van Aerde and Yagar (1988).

2.3 Queuing to Meet a Schedule

A standard assumption of almost all queuing theory and transport analysis is that travellers appear according to some preordained schedule or distribution; if the cost of travel is too high they may choose to take a different route or to abandon the trip altogether, but they do not consider coming at a different time. Obviously, this is not what travellers do on their way to work; they must adjust their departure times in order to avoid being late. Thus, there is reason to think that morning peaking patterns might be systematically different from those in the evening and that morning travellers' responses to control schemes involving ramp metering or congestion tolls might be markedly different (Smith 1984, Daganzo 1985).

One conclusion reached by the majority of researchers is that arrival flows are likely to increase much more rapidly in the morning since travellers must compete for the scarce positions in the queue that will get them to work on time. A second is that, because there are not enough such positions, peak period travellers will not be able to reach their destinations when they want, but will have to settle for being either early or late. A third conclusion that might be made is that these models stretch conventional ideas of transport demand beyond its limits; there is a clear need for demand models that allow for cross elasticities between different times of day (see *Traffic Assignment, Dynamic*).

2.4 Queuing on Networks

Another area where a body of theory is beginning to appear is that of modelling flows on networks in which large, time-dependent queues form upstream from bottlenecks and extend into other links, changing the way in which these other links operate (Hurdle 1974, D'Ans and Gazis 1976, Zawack and Thompson 1987, and Van Aerde and Yagar 1988). The networks have grown from two nodes connected by two links (Hurdle 1974) to a standard network with multiple sources and sinks (Van Aerde and Yagar 1988). The four approaches given are radically different and, as might be expected in a field this new, each has obvious shortcomings, but they do illustrate the issues involved and suggest some possible approaches to the problem.

See also: Network Modelling and Control: Store-and-Forward Approach; Optimal Routing Applications.

Bibliography

Akcelik R 1981 *Traffic Signals: Capacity and Timing Analysis.* Australian Road Research Report No. 123. Australian Road Research Board, Victoria, Australia

Daganzo C F 1985 The uniqueness of a time-dependent equilibrium distribution of arrivals at a single bottleneck. *Transp. Sci.* **19**, 29–37

D'Ans G C, Gazis D C 1976 Optimal control of oversaturated store-and-forward transportation networks. *Transp. Sci.* **10**, 1–19

Hurdle V F 1974 The effect of queuing on traffic assignment in a simple road network. *Proc. 6th Int. Symp. Transportation and Traffic Theory.* Reed, Sydney, Australia, pp. 519–40

Hurdle V F 1984a Signalized intersection delay models—A primer for the uninitiated. *Transp. Res. Rec.* **971**, 96–105

Hurdle V F 1984b Equilibrium flows on urban freeways. *Transp. Sci.* **15**, 255–93

Hurdle V F, Solomon D Y 1986 Service functions for urban freeways—An empirical study. *Transp. Sci.* **20**, 153–63

Newell G F 1982 *Applications of Queuing Theory*, 2nd edn. Chapman and Hall, London

Smith M J 1984 The existence of a time-dependent equilibrium distribution of arrivals at a single bottleneck. *Transp. Sci.* **18**, 385–94

Van Aerde M, Yagar S 1988 Dynamic integrated freeway/traffic signal networks: A routing-based modelling approach. *Transp. Res.* A**22**, 445–53

Zawack D J, Thompson G L 1987 A dynamic space–time network flow model for city traffic congestion. *Transp. Sci.* **21**, 153–62

V. F. Hurdle
[University of Toronto, Toronto, Canada]

Railroad Electronic Signalling

Railroad signalling must be highly reliable and fail-safe. Electromagnetic relays with asymmetrical failure modes are used as the principal logic elements. Fail-safe technologies for electronic components such as a microcomputer whose failure mode cannot be uniquely predetermined have been established based on redundant structures or fail-safe logic circuits. Many kinds of new signalling systems utilizing the ability of high-speed, small-scale, low-power consumption and intelligence have been realized (Kalra 1979, Withehouse 1985).

1. Hierarchical Structure

A railroad signalling system has a hierarchical structure (see Fig. 1). The lowest level is a sensor and an actuator such as track circuits, conductor loops, signals and switches. The second level is a safety controller such as solid-state interlocking, electronic blocking, automatic train protection (ATP), automatic train operation (ATO) and level-crossing controller. Their objective is to ensure the train runs by controlling train routes, speeds, intervals and so on. Information on safety is transmitted between the train and the wayside. The third level is a transmission between stations and dispatching center, as with centralized traffic control (CTC). The highest level is a supervisory/management system such as automatic train supervision (ATS). Introduction of electronics has been advancing in every level (see *Railroad Systems: Line Supervision and Control*).

2. Architecture of the Fail-Safe Microcomputer

Several kinds of fail-safe architectures based on an off-the-shelf microcomputer have been adopted in railroad signalling. Their safety is ensured by error detection and diagnosis. Processing results of redundant modules are compared on a hardware or software basis in addition to self-checking by software in each computer. If any part of them fails, the control outputs are switched to safe-side state; for example, the power supply is cut off in the output circuit to give a "stop" signal. Redundant modules should be independent to avoid a double error in the same mode and the same instance. Several architectures are used in practice (see Fig. 2).

(a) *N*-version software is implemented into one computer and processing results of the versions produced by independent programming teams are compared (Fig. 2a).

(b) Application software and diagnosis one are implemented into one computer, which diagnoses both hardware and processing results.

(c) Two computers implemented with the same software are operated with synchronized clocks and data on their inner buses are compared by a fail-safe comparator for every machine cycle (Fig. 2b).

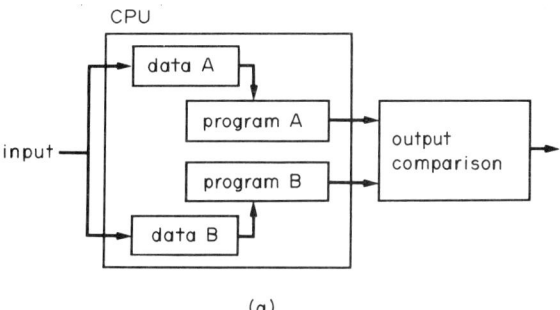

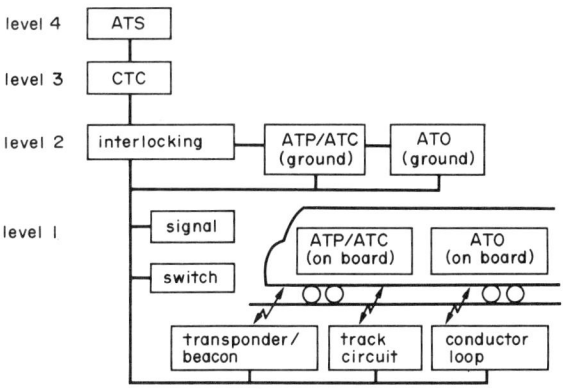

Figure 1
Hierarchical structure

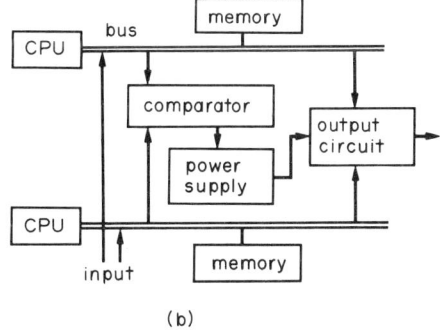

Figure 2
Architecture of a fail-safe computer: (a) with 2-version software and (b) two computers with bus-level comparison

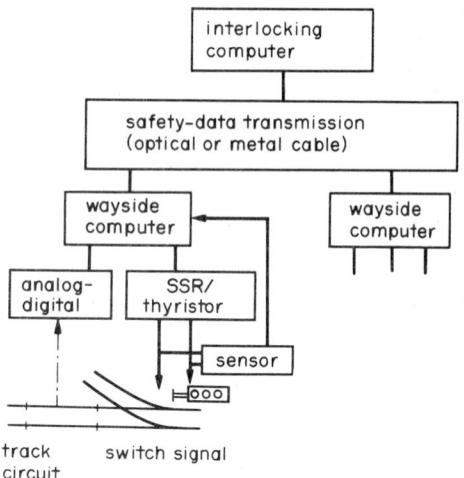

Figure 3
Structure of solid-state interlocking

(d) Two computers implemented with the same software are operated with independent clocks and processing results are compared on a task basis.

(e) Two computers operated with a time difference of T output their processing results to a comparator. The content of the comparator changing dynamically with agreement and disagreement is diagnosed on a software basis.

Software diversity is adopted by (a) and (b), while (c)–(e) do hardware diversity.

3. Solid-State Interlocking

Solid-state interlocking (see Fig. 3) controls signals and switches, and protects the trains against collision and derailment at a station (Cribbens 1987, Nakamura and Akita 1989). The kernel is a fail-safe interlocking microcomputer with triple modular redundancy or hot-standby structure. Wayside terminals which are also composed of a fail-safe microcomputer without back-up unit are installed around the station. They detect the train position by means of track circuits or wheel detectors, and actuate signal lamps and point motors through solid-state relays (SSR) or thyristors. Signal lamps and switches are checked for their failures by sensing the currents and voltages in the cables leading to such signalling devices. Permanent or transient errors of track circuits due to a rusty rail or an oil-contaminated rail are detected by continuously checking the voltage-level and the phase of the current received from the rail.

Safety data with a check code such as cyclic redundancy check (CRC) ensuring hamming distance of not less than 4 are transmitted periodically and error detection and diagnosis are done between the fail-safe computers. Construction cost of cables are considerably reduced.

4. Electronic Blocking System

An electronic blocking system is applied to low-traffic single track lines. Only one train, given an electronic token, is permitted to depart from any one of the adjacent stations. Several systems are in service (Birkin 1984, Sasaki 1986). One is equipped with a blocking computer with fail-safe characteristics at the control center. When the train driver makes a request for the token, by radio, by manipulating a console in the cab, the request is displayed on the console at the center. If the dispatcher gives the acknowledgement after voice communication with the driver, the token is sent from the computer by radio and the permission for departure is displayed in the cab (Fig. 4).

Another system is equipped with a blocking computer at every station. When the train driver requests the token for departure, by either radio or infrared transmission, the computer communicates with one at the adjacent station over line transmission, and produces "clear aspect" at the wayside signal. The dispatcher is unnecessary.

5. Level-Crossing Controller

The solid-state level-crossing controller detects the approaching train by means of track circuits or wheel detectors, and actuates crossing gates, speakers, flash lamps and train direction indicators (Kumagai 1987) (see Fig. 5). As a train detector, a crossed short-length loop coil installed between the rails is also used, which identifies the train by detecting the inductance change of the coil (Czehowsky 1988). The kernel of the controller is a fail-safe microcomputer. Sound alarms given by an oscillator and voice ones generated by a speech synthesis device are alternately issued from speakers. Flash lamps using light-emitting diodes (LEDs) and crossing gates are controlled through solid-state relays (SSRs). This controller has the merits of low cost, high reliability and being easy to maintain.

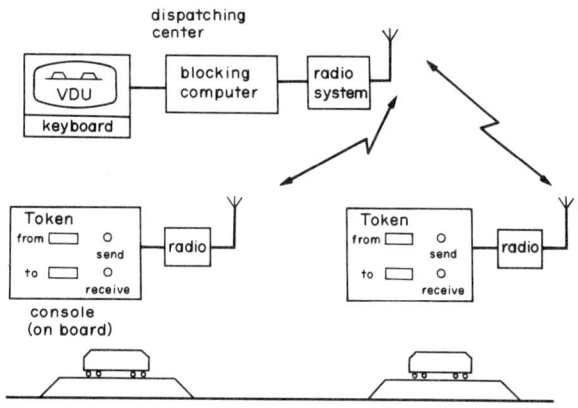

Figure 4
Electronic blocking system

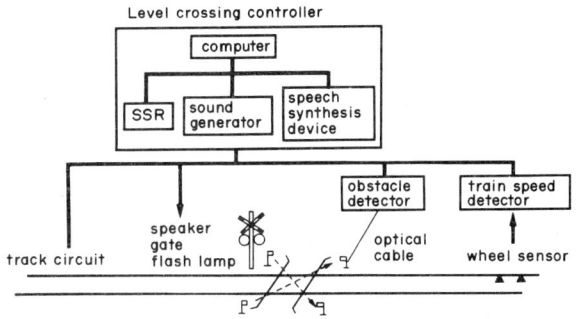

Figure 5
Solid-state level-crossing controller

Obstacle detection is done by an optical detector, a looped coil buried in the crossing road and so on. The optical detector, which is furnished with a microcomputer, identifies the obstacle when the light from the emitter to the optical sensor is interrupted. The looped coil recognizes the road vehicle by detecting the change in the inductance of the loop. Image processing utilizing infrared rays is under investigation.

To keep closed-gate duration constant regardless of train speed, a train speed detector can be added, which is composed of a fail-safe microcomputer and a pair of wheel sensors attached to the rail with intervals of tens of centimeters. The microcomputer calculates the train speed based on the running time of train between two sensors, each time the train passes over the sensors. It predicts the running time from the location of sensors to the level crossing, and determines the starting time for activating the level-crossing controller according to the train speed.

6. Transponders or Beacons

A transponder or beacon is used for safety-data transmission between wayside and train (Miyachi 1988) (Fig. 6). Two types of coils (an active type with a power supply and a passive type without one) are installed on the sleepers. The active type is connected with control equipment composed of a fail-safe microcomputer with over line transmission and variable data such as the presence of a stop signal ahead and distance from the preceding train can be transmitted to the train. The passive type is supplied with its power from on-board equipment over power transmission, and sends fixed data such as its location and the curve or slope ahead to the train. On-board equipment, the kernel of which is also a fail-safe microcomputer, communicates with the ground equipment via an antenna coil installed at the head of train. Train identification number, train speed and so forth can be sent to the ground.

Digital transmission is applied. A carrier frequency of several or tens of megahertz which is modulated by amplitude modulation (AM) or frequency-shift keying (FSK) is adopted for data transmission, while a signal of hundreds of kilohertz is used for the power transmission of several watts.

Transponders and beacons have been applied to automatic train control (ATC). Many train control systems are expected to be developed that will utilize the advantage of low-cost data transmission of transponders or beacons.

7. Advanced Train Control

A new-generation signalling system is being investigated in the USA, Canada, France, Japan and other countries (Bernard 1986, Detmold 1986). It is equipped with fail-safe computers on the ground and on the train, which communicate by means of radio transmission or telecommunications satellite. On-board equipment detects the position and speed of the train by itself. Several kinds of sensors are proposed. A Doppler radar or tachogenerator will be used for train speed detection, while train position will be detected based either on signals from global positioning satellites (GPS) launched in the USA or on accumulated pulses from the tachogenerator or Doppler radar (see Fig. 7). The output from tachogenerator or Doppler radar is calibrated by means of passive transponders installed at fixed positions along the railroad.

The equipment on the ground collects the positions and speeds of all trains and supplies each train with data on the distance from the preceding train. The on-board equipment determines the train speed based on the data from the ground, while train performance and topographical data are preset.

Active signalling devices such as track circuits along the right of way are unnecessary in this system, so construction and maintenance costs are reduced considerably. Further, the energy required for train operation is also reduced.

See also: Automatic Train Control: Protection Principles; Automatic Train Control: Safety and Reliability; Railroad Systems: Line Supervision and Control

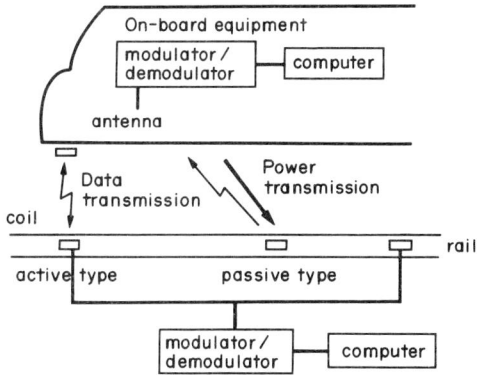

Figure 6
Structure of transponder

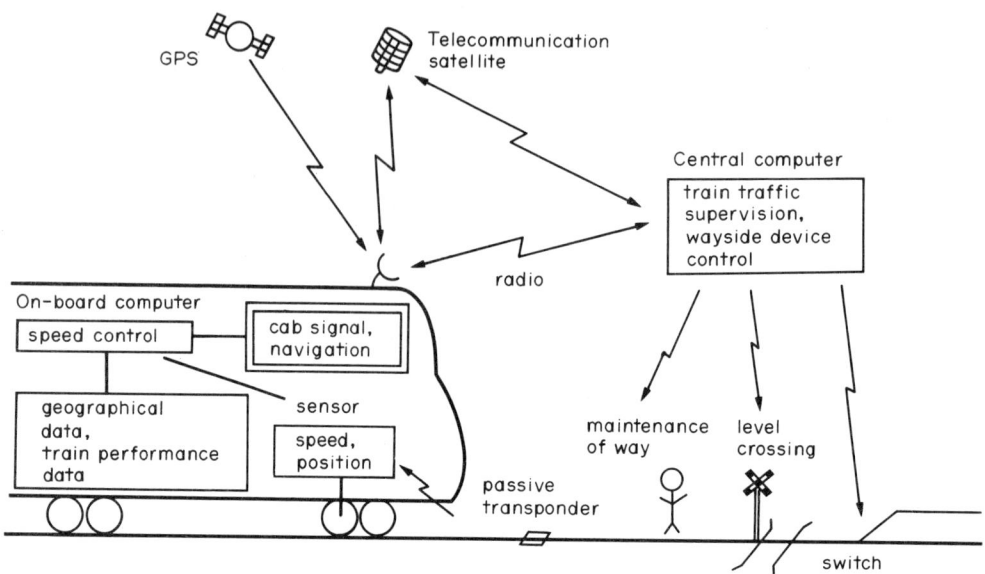

Figure 7
Advanced train control

Bibliography

Bernard R 1986 Ein universales System der Zuglaufsteuerng ASTREE. *Signal & Draht* **80**, 3–7
Birkin M S 1984 Low cost signalling by radio. *IRSE Proc.* 48–66
Cribbens A H 1987 Solid-state interlocking (SSI): An integrated electronic signalling system for mainline railways. *IEE Proc.* **134**, 148–58
Czehowsky J 1988 Fahrzeugsensoren als Schaltmittel im Gleis. *Signal & Draht* **80**, 31–9
Detmold P 1986 ATCS will close the control loop. *Railw. Gaz. Int.* January, 29–32
Kalra P S 1979 A green light for advanced train controls. *IEEE Spectrum* February, 44–9
Kumagai T 1987 Microelectronic alarm and protection control for level crossing equipment. *Jpn. Railw. Eng.* **103**, 18–20
Miyachi M 1988 The development of the speed verification type ATS with the transponder. *RTRI Q.* **29**(3), 103–6
Nakamura H, Akita K 1989 Architecture and safety of an electronic signalling system based on solid-state interlocking: SMILE. *Proc. IFAC Conf. CCCT'89.* International Federation of Automatic Control, Laxenburg, Austria
Sasaki T 1986 Development of the electronic blocking system. *Jpn. Railw. Eng.* **99**, 16–19
Welty G 1987 ATCS: On time, on target. *Railw. Age* June, 39–40
Withehouse W H 1985 New technology for signalling and telecommunications systems. *Int. Railw. Congr. Assoc./Int. Union Railw. Congr., Rail Int.* March, 123–31

K. Akita and H. Nakamura
[Railway Technical Research Institute,
Tokyo, Japan]

Railroad Systems: Active Control

Suspension systems for railway vehicles must be designed to accommodate the conflicting requirements of supporting the weight of the vehicle and guiding it along the track, and isolating the car body with its payload (passengers or freight) from disturbances, either track irregularities or externally imposed forces. These conflicting needs inevitably result in an engineering compromise, a design decision that must be taken from an overall systems viewpoint. Actively controlled suspension schemes have been under consideration for a number of years (Goodall and Kortüm 1983) and will often ease this engineering compromise, though they will usually be more expensive than a conventional (passive) suspension.

The concept of an actively controlled suspension involves the provision of the following subsystems:

(a) a set of actuators to replace or complement the passive suspension elements such as springs and dampers,
(b) a number of transducers or sensors by which the performance of the suspension may be monitored, and
(c) a controller to implement the chosen feedback law which will affect the actuators in accordance with information derived from the transducers.

Whereas a passive suspension can only store or dissipate energy by generating forces in response to local relative motion, an active system may inject energy, if needed, and may develop forces that are a combination of many variables, some of which may be remote.

1. Possible Uses for Active Suspension Technology

There are two important aspects when considering the potential application of active suspensions: the ways in which they may be applied to a vehicle (i.e., which dynamic modes are controlled) and the control techniques that can be used.

1.1 Vehicle Considerations

Railway vehicle suspensions are complex dynamic systems with many degrees of freedom. In theory, it would be possible to control all of these although, in practice, active control would be limited to a subset. When a set of actuators is fitted, it will influence a corresponding set of the degrees of freedom (e.g., fitting lateral actuators to the secondary suspension will mainly influence the lateral and yaw modes of the body). There will usually be a different objective associated with each of these.

In wheel set design, there is a well-known conflict between stability and curving. Independent wheels present no problems in terms of stability but do not generate any curving forces. Solid-axle wheel sets give curving action, but may be unstable above a certain critical speed. Consequently, control action that is able to free the wheels at critical frequencies, but that maintains the coupling in the steady state so as to ensure good curving, becomes a useful possibility.

Bogey design has to accommodate a number of requirements. The primary vertical suspension (from wheel sets to the bogey frame) is designed to be able to accommodate track twist without significantly unloading any of the wheels (too small a load leads to a derailment risk); at the same time, the primary deflection as the vehicle load changes from tare to fully laden is restricted by the mechanical design of the bogey. Generally, however, this is not a severe constraint for the bogey designer and so active control of the vertical primary is not an important application. The primary lateral/yaw suspension, however, is more significant. Since solid-axle wheel sets are nearly always used to give good curving, the primary suspension design and, in particular, the yaw restraint between wheel set and bogey is crucial in ensuring stability of the wheel set, although too stiff a restraint will degrade the curving performance. Hence, active control may be applied either to provide stabilizing forces to a softly connected wheel set or to relax an otherwise stiff suspension at low frequencies in order to improve the curving performance.

The secondary suspension between the bogey and body is the part that predominantly affects the ride quality. It is possible to improve the ride by designing softer springs, but this increases the suspension deflection and, in the extreme case, the vehicle body may move outside the loading gauge. In addition, parasitic effects, particularly friction, become more dominant and the expected improvements may not be achieved. The provision of secondary active control in the lateral and/or vertical directions can result in significant improvements in ride quality, especially since it is possible to implement absolute velocity or skyhook damping; this overcomes the design conflict between reducing the high-frequency transmissibility and maximizing the damping in the secondary modes with a passive damper. Secondary roll control (also known as body tilting) can also be used to improve the passenger environment by minimizing the steady-state accelerations in curves, usually with the objective of higher operating speeds. A number of other possibilities also exist. An example is active control that only acts at low frequencies to centralize the suspension within its total travel, either to compensate for load changes or to counteract centrifugal forces in curves, thereby "holding off" the suspension from the hard limits at the ends of travel. It should be mentioned that improved suspension performance can also be used to obtain an adequate quality of ride on a lower quality of track, which means that the extra capital investment in the rolling stock can be used to reduce the recurring maintenance costs associated with the track.

There are other smaller but quite important suspension functions that, for completeness, will be mentioned. On an electric train, the pantograph maintains contact with the overhead wire by exerting an upward force, which can cause problems at high speeds by setting up travelling waves. An actively controlled pantograph can ensure contact with a significantly lower force, hence reducing the problem. Another possibility relates to active control of the antivibration mounts on engines and auxiliary motor generators installed within the car body.

1.2 Control

Passive suspensions are restricted to generating forces in response to the local relative motion of a compliant element (spring) or a damping element. They also have fixed properties, even though their characteristics may be intentionally or unintentionally nonlinear. However, active suspensions can modulate the applied forces in response to many variables (local or remote) and can adapt to various levels of external forces and track irregularities, thus providing suspension characteristics that may appear simultaneously "soft" to track inputs and "hard" to external forces. Because of the added cost, energy expenditure, power requirements and complexity of a fully active suspension unit, various types of realizations have been considered.

The least complex systems are the so-called "adaptive" suspensions which intermittently adjust the damping characteristic—or, in a few cases, the spring characteristic—based upon external information such as speed, level of irregularities, and longitudinal or lateral acceleration level. Once a "best" level is selected, the suspension is totally passive and no change is made unless the monitored operating conditions alter to another discrete step. (It is suggested that, in order to avoid confusion with real adaptive

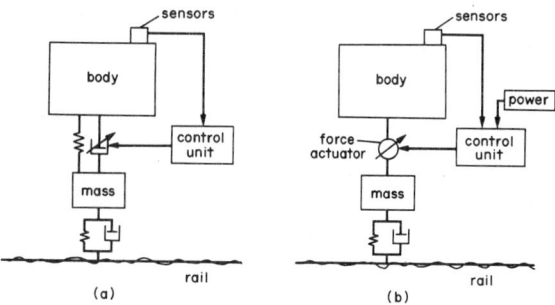

Figure 1
Schematics of (a) semiactive and (b) active suspensions

active suspensions, these types of suspensions are called "adjustable passive.")

Suspensions are termed semiactive if their parameters are varied in a continuous, real-time manner as a function of vehicle body motions. The damper (or spring) characteristic may be varied continuously between soft and firm settings, or to discrete values in between. The key feature of a semiactive suspension is the real-time continuous modulation of the characteristics (see Fig. 1a) as compared with the intermittent modulation of adjustable passive suspensions.

Fully active suspensions (see Fig. 1b) are characterized by (usually electronically) controlled actuators; their applied forces are functions of the actual state of the vehicle (body deflection, velocity or acceleration from any sensor location) and may require substantial external power. Sometimes, the control laws are altered as a function of vehicle system parameters or operating conditions. Simple versions may use gain switching or scheduling based upon vehicle speed, rail characteristics or curving scenarios. More complex schemes are envisaged that alter the control laws because of on-line identification of system parameters; these are called adaptive control systems or, especially if the feedback law is altered due to monitored vehicle system properties, variable structure control systems.

Finally, the most complex systems, in which external disturbances are measured before they excite the vehicle, are called preview control schemes.

1.3 Control Design Methods

There are many design methods for electronically controlled suspensions, the methodologies being divisible into five categories, emphasizing the state-space techniques.

(a) Classical feedback techniques. These usually start from a linearized time-invariant single-input single-output (SISO) description of the vehicle system, often in the form of transfer functions using the Laplace operator. Techniques such as Nyquist, Bode and root-locus are used to assess stability and performance. The feedback control laws that are synthesized in this way are usually simple transfer functions using proportional, integral and derivative (PID) terms. For vehicles, this results in displacement, velocity and, sometimes, body acceleration feedback from various degrees of freedom.

(b) Mode-decoupling techniques. An evident disadvantage of classical techniques is that only SISO systems are considered, whereas for vehicles it is obvious that SISO is frequently a crude approximation. A first step towards multi-input multi-output (MIMO) systems can be seen in decoupling control techniques. (These are sometimes incorrectly referred to as "modal control.") The linearized vehicle model is decoupled by setting up equations of motion as if the system has only one mode (e.g., heave, pitch, roll or yaw). Each mode (i.e., degree of freedom) is then designed individually using classical techniques. Actuator configurations are sought such that one or more control just a single mode. The resulting scheme is called decentralized control if it only needs local information and acts locally although, in more sophisticated realizations, the control law includes decoupling terms proportional to other modes in order to compensate for coupling terms in the equations of motion. Since vehicle disturbances usually excite more than one mode and since they are usually coupled (even in the linearized equations), the coordination of individual decentralized controllers has to be tested by simulation with the real coupled vehicle model.

(c) Linear state-space techniques. The modern control of MIMO systems rests heavily on state-space formulations of the linearized vehicle dynamics in the standard form

$$\left. \begin{array}{l} \dot{x} = Fx + Gu + Dw \\ y = Hx + v \end{array} \right\} \quad (1)$$

where x is the state vector (for vehicles usually consisting of z, the degrees of freedom, and $\dot{z}$, their time derivatives), u the control input and w the disturbance input vector, whereas y is the output vector (transducer signals). The matrices F, G, D and H are called the system, control input, disturbance input and output matrices, respectively. It is possible to use either full state feedback $u = -Cx$ or output feedback laws $u = -C_y y$, where C and C_y are gain matrices.

It is usual, as a first step, to use state-feedback laws (assuming that all states can be measured) and the main methods for design are

(i) pole placement,

(ii) modal control, and

(iii) Riccati design.

A major drawback with state-feedback techniques is that the full state vector may not be available, in which case state-estimate feedback is used. When these estimates involve statistical techniques, the state estimator is called a Kalman–Bucy filter; when the estimator is designed to have certain eigenvalues, it is known as a Luenberger observer.

It is also possible to produce a decoupled version of Eqn. (1) with the eigenvalues of the uncontrolled system on the diagonal of the system matrix. From this formulation, techniques are available for directly controlling each of the modes of the system. This can be contrasted with the mode-decoupling technique.

(*d*) *Nonlinear control techniques.* Whereas there are a number of analytical techniques for nonlinear systems, design techniques for nonlinear control laws are rather restricted. The following list identifies the most important:

(i) bilinear control,

(ii) bang-bang or contactor control, and

(iii) variable structure (sliding mode) control.

(*e*) *Neoclassical control techniques.* There have been many attempts to extend the strategies for control law design. One area, sometimes called expert-tuned design, involves developing the controller structure via basic considerations and then determining the free parameters in the design with the aid of formal optimization procedures. Another area under consideration is the extension of classical frequency domain techniques to MIMO systems. One important strategy is concerned with developing robust control laws (i.e., minimization of the sensitivity of the system with respect to uncertainties in the system parameters).

1.4 Actuators

There is little doubt that the most compact and effective form of actuation for achieving active control is servohydraulic. However, a complete system is complicated by the time all the components are incorporated and, of course, the overall efficiency is not good. Generally, the railway industry is unaccustomed to the level and quality of maintenance associated with servohydraulics, which must count against it. Electromechanical actuators are also a possibility, because high-efficiency ball nut/screw mechanisms are available for converting rotary to linear motion and, of course, modern electronic high-efficiency power amplifiers can be used to control either dc or ac motors. The technology is certainly one with which railway maintenance engineers will be more comfortable. Pneumatic actuators can also be used, but the compressibility of the air imposes a severe upper limit on the effectiveness of the control action; in addition, the efficiency is poor. Note, however, that many passenger vehicles already use a pneumatic actuator in the form of an airspring and, in many cases, this will include low-grade active control for self-levelling under changing loads, although the actual consumption of air and, hence, the overhead in terms of power is small. The only other possibility is to use electromagnetic actuators, which have high controllability and are rugged and simple.

The main restrictions are the limited possible displacement (unless they are used in the form of linear motors) and, possibly, the difficulty of installation since they are not particularly compact.

It is not necessary for the actuators to generate all the suspension forces; indeed, for reliability reasons, a passive backup is very desirable and so, generally, the actuators are fitted in parallel with existing passive springing. The actuator will be controlled to modulate the passive characteristics of the suspension on the basis of signals from the transducers.

2. Practical Developments

2.1 Ride Improvement

Japanese National Railways (in conjunction with industry) have been investigating control of secondary suspensions for a number of years, initially in the vertical direction using a 1/5th scale model, but more recently using lateral actuators fitted to a full size vehicle (Okamoto *et al.* 1987). In both cases, pneumatic actuators were fitted in parallel with normal passive springing and were controlled with servovalves (a number of studies around the world have looked at on–off control of pneumatic actuators, but have generally found the speed of response too restrictive). The controller for the lateral system had classically designed acceleration feedback loops driving the servovalves at each end of the vehicle. Field trials indicated an improvement of typically 40% in the peak–peak lateral acceleration at speeds of $120\,\mathrm{km\,h^{-1}}$ with a maximum force of 12 kN, although problems were encountered when negotiating curves due to excessive actuator force requirements.

A program of experimental work by British Rail has assessed a variety of actuator types in a number of secondary suspension applications (Pollard and Simons 1984). Two servohydraulic lateral actuators working in parallel with a passive airspring suspension achieved consistent reductions of around 40% in rms lateral acceleration. Force levels of 2.3 kN rms and a power consumption of 2 kW were recorded. It was noticeable that the energy in the measured acceleration spectrum at the rigid-body frequencies was very significantly reduced, although a slight worsening was experienced at higher frequencies as the bandwidth of the controller was reached. This highlights the important general conclusion that the active control bandwidth must be sufficiently high so that there is little energy in the input spectrum at frequencies for which this limit is reached. Another lateral active suspension used electromechanical actuators designed to fit into the space vacated by the lateral dampers on a standard railroad vehicle. The actuator used a dc motor driving a lead screw and was controlled by a high-efficiency power amplifier; if the control failed the actuator acted as an electric damper leaving the vehicle in a safe condition.

The British Rail work also considered two vertical suspensions with unusual actuator types. In the first case, "air pumps" were connected to the normal airsprings to give a servopneumatic system in which the airsprings were the actuators. The air pumps consisted of other small airsprings that were compressed by a dc motor–lead screw combination and ride improvements of 30–40% were achieved with a power consumption of 1.8 kW. The other type of vertical system used electromagnetic actuators comprising double-acting electromagnets producing a maximum force of 20 kN. These had the advantage that high-frequency control action was readily achieved as was borne out by experimental results. The vehicle fitted with these actuators was also used to experiment with different control approaches. Most of the other controllers used a classically designed scheme with feedback of absolute position and suspension displacement (often in addition to an inner feedback term), and these were incorporated into an integrated scheme which separately controlled the modes of vibration of the body. An optimally designed control system including a modal observer was also built and tested (Williams 1985). Theoretically, this offered great promise which was not fully realized, partly because the flexible body modes were excited by the controller (sometimes called the spillover effect). One alternative would have been to include these in the system model, but this would have significantly increased the controller and observer complexity, and so the problem was treated with narrow-band filters to reject the resonant frequencies.

2.2 Bogey Stability Control

High-speed operation of rail vehicles is limited by problems of ride quality, wheel and track wear, curve negotiation and hunting (lateral instability). At high speeds, the requirements to achieve satisfactory straight and curved track performance conflict to a degree and require a compromise in the choice of the suspension parameters; in general, curving performance is traded off with respect to straight track performance. A number of studies have shown that active control can significantly ease this tradeoff by employing restoring forces and torques which are functions of variables that cannot be achieved through passive components.

Sinha et al. (1978) demonstrated potential improvements through an active controller whose structure was partitioned into low-frequency actions which influence curving performance and higher-frequency actions which influence straight track performance. The relative displacements between bogey and body are used to modify curving performance, combined with bogey lateral velocity for straight track operation—these affect control actions generated in lateral force and steering torque actuators (located between bogey and body). Case studies have indicated reductions in rms acceleration and tracking errors of a factor of about five from an uncontrolled baseline vehicle design, while at the same time meeting specified curving performance. These improvements were predicted to need only modest levels of control power (~2 kW per bogey). While the controller parameters were selected based upon the performance at a given speed (by minimizing a quadratic performance criterion), the controller with these parameters was found to be effective in improving vehicle dynamic performance over a wide range of design speeds. Another study (Celniker and Hedrick 1982) used lateral actuators in pairs, and emphasized the significant power consumption and bandwidth requirements (64 kW and 7 Hz) needed to provide stability control as compared with ride control only (6.4 kW and 3 Hz). An experimental laboratory rig has been used to examine the hardware requirements for this type of system (Örley 1981). The rig was connected to a hybrid computer where the vehicle dynamics could be simulated in real time and the hydraulic actuators realistically tested. Götz (1983) studied the safety and reliability implications of such a system by means of qualitative analysis using redundancy techniques.

Unconventional forms such as magnetic guidance of bogeys have also been examined. For example, Caudill and Sweet (1982) proposed the use of linear motors attached to the bogey and/or body for propulsion, using them differentially to apply control torques; separate lateral acting motors are also employed and significant improvements in stability, curving and ride are predicted.

A significant practical development is being carried out in Germany (Leo 1984) in which a controlled electromagnetic coupling is used between the otherwise independent wheels of a wheel set. The general concept for operating these creep-controlled wheel sets is based upon shifting part of the creep from the wheel contact area to the coupling. This way, the longitudinal creep (causing wear) is diminished and the total creep reduced. The wheel set gear moment diminishes as the longitudinal creep is reduced, thus leading to effective stabilization of the bogey. If, however, the total creep is fully exploited, larger free lateral forces can be absorbed by increasing the lateral creep vectors, thus permitting the critical speed to be increased. Different control laws have been used to control the wheel set coupling; some of these are based upon extremely accurate measurement of the creep between the two wheels. The overall effect of the control scheme seems to be to energize the coupling such that the wheels are joined together in the low-frequency region, hence providing good curving performance, but they are decoupled at typical kinematic frequencies such that instability does not arise. Encouraging results have been obtained from tests on the German roller rig with two wheel sets in a test bogey for speeds up to 503 km h^{-1}, the maximum speed achievable with the rig. This development is combined with the use of fiber-composite bogey frames in order to reduce weight and achieve the required elasticity properties (Geuenich et al. 1985).

2.3 Bogey Steering Control

Rail vehicles are often required to negotiate tight curves, especially in urban transit systems. During curve negotiation, the performance of conventional vehicles generally suffers from increased wear at the wheels and rails, increased energy consumption, enhanced danger of derailment and objectionable noise generation. These problems are associated with large wheel–rail contact forces which develop due to angular misalignments of the wheel sets of conventional bogeys on curves. To improve curving performance, bogeys with softened suspension stiffnesses can be used, but this reduces the lateral stability of the bogey and thereby increases the tendency to hunt on straight track. Vehicles with bogeys that incorporate innovative suspension systems have been designed to steer their wheel sets into radial alignment without degrading dynamic stability. In these force-steered bogeys, linkages directly interconnect the wheel sets to the car body. In addition to providing the requisite stiffness for stability, the body–wheel set linkages measure the track curvature (by sensing the relative yaw angle between body and bogey frame) and steer the wheel sets into radial alignment. For negotiation of moderate-to-sharp curves, studies have shown that significant improvements in curving are possible, not only compared with normal bogeys, but also with self-steered bogeys which still rely on friction forces for steering (Jaschinski and Duffek 1983).

It is only a short step to achieve this type of steering control through active electronically controlled components. So far, only passive realizations consisting of purely mechanical parts have been reported in the literature. Frederich (1985), however, discourages such thoughts because of his expectations of poor cost effectiveness, and suggests novel concepts that discard the classical wheel set and use separate wheels including those with camber; these can be interpreted as mechanical realizations of feedback control.

The authors, however, believe that electronically controlled schemes should also be considered in order to achieve maximal performance with low effort, a relationship that depends upon the state of technology (e.g., electronic components vs mechanical realizations).

2.4 Body Tilting Systems

Of all the possibilities for active control, body tilting systems are in the most advanced state of development (Binnewies 1987). Strictly, they are for ride improvement. The most obvious method for achieving body tilt is via active control of the roll mode of the vertical secondary suspension and some studies have investigated the idea of using differential inflation and deflation of the airsprings from one side of the vehicle to the other. There are, however, three difficulties: first, the amount of tilt achievable is quite small (about 3°), second, the tilt center is low meaning that the top of the vehicle may move a long way as the body tilts and, third, the air consumption and speed of response of the pneumatic control may be a problem. Consequently, all recent implementations have had a separate mechanism for tilting. Three different mechanical configurations have been taken to a significant stage of development and these are shown in Fig. 2.

The tilting TALGO train of Spanish Railways uses pendular action, in which the vehicle bodies are suspended from airsprings on top of pedestals between the ends of the bodies (Lumpie 1977); this gives a high effective tilt center. Tilting is caused by a natural pendular effect, so the system is not strictly active.

The Italian-made Pendolino train (Losa 1987) has a nontilting bolster supported by coil springs from the bogey frame; the body is suspended from the bolster by means of inclined swing links which provide an effective tilt center somewhere around seat height. Servohydraulic actuators are fitted vertically between the bolster and the vehicle body in order to provide the tilt action.

British Rail's advanced passenger train (APT) (Halfpenny and Marshall 1977) used a tilting bolster suspended by inclined swing links from the bogey frame, with an airspring secondary suspension from the bolster

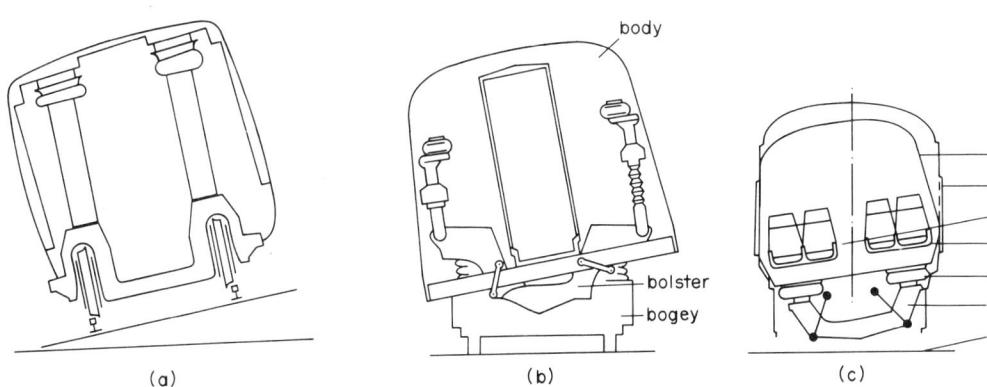

Figure 2
Tilt system configurations: (a) pendular, (b) nontilting bolster and (c) tilting bolster

to the vehicle body. Servohydraulic actuators operating laterally between the bogey frame and the bolster provided the tilt action. Electromechanical actuators, which offer potential for higher reliability and lower system cost, have also been successfully tested. A very similar mechanical arrangement is being used by Swedish Railways on their X15 test train (Lundgren 1988).

From an operational point of view, the TALGO train has been in service since the early 1980s. A single ETR401 Pendolino train has been used in revenue service by Italian Railways at speeds of up to 250 km h^{-1} and this forms the basis for their ETR450 production trains which have seen in excess of one million kilometers of service with good results. British Rail have operated a prototype tilting train in revenue service, but this is now discontinued. Swedish Railways have plans to introduce their concept into revenue service shortly.

In terms of control strategies, the TALGO train is fundamentally passive, although the normal airspring levelling action is disabled in curves to allow the body to swing freely (up to about 3.5°). All the other systems use measurements of lateral acceleration to compensate fully or partly for the centrifugal action, with tilt angles up to 9° or 10° possible, though not always used. One problem is that this signal measures not only the tilt angle, but also the lateral dynamic vibration as the vehicle moves along the track; this can cause unwanted high-frequency control activity which may degrade the ride. Filtering the signal is a possibility, but this will usually cause an unacceptably slow response when entering curves. Mounting the accelerometer on the vehicle body, which at least isolates it from the high lateral bogey accelerations, is also a possibility. However, this means that the secondary suspension dynamics are within the control loop and stability may be compromised, particularly for configurations in which the tilt mechanism is below the secondary suspension. To overcome this difficulty, the Pendolino uses a gyroscope from which the real tilt angle can be deduced, although an accelerometer is still needed for low-frequency measurements (considerable attention is paid in the control system design to reliability and redundancy). The APT control takes its acceleration measurement from a preceding bogey (i.e., preview control); this signal is filtered to remove the lateral dynamic movements, but the lag of the filter is arranged to compensate for the precedence time of the bogey at normal operating speeds. The solution of Swedish Railways appears to use a similar approach.

2.5 Others

In a practical sense, while there are clearly opportunities for other types of active control in railway vehicles, very little has been done recently in any area. The only work worth mentioning was carried out in the mid-to-late 1970s and investigated servocontrolled pantographs. British and German studies have been reported. The British work used a simple on-off hydraulic system with a constant force suspension for the contact head and achieved a much better force performance, although the actual contact performance was not significantly changed. The German system used hydraulic actuators on the upper and lower parts of the pantograph frame, which also provided attitude control of the head. Simulation studies, which were confirmed by only limited practical testing, predicted velocities of 360 km h^{-1} without loss of contact.

3. Future Developments

With the exception of tilting trains, the everyday use of active control for wheel–rail systems in revenue service is currently nonexistent. This situation is quite different in the automotive industry, where electronically controlled components are used for vehicle control (e.g., active suspensions and active four-wheel steering). This could lead to the conclusion that railway engineers and authorities are still more conservative, although the market forces are clearly very different between the two industries. Nevertheless, it is expected that actively controlled systems will, at some stage, be applied in rail systems for the following reasons.

(a) The potential of active control offers significant improvement in system performance, especially where the development potential of passive suspensions has been driven to its limit.

(b) Computer-aided control system design techniques with reliable and easy-to-use software are becoming widely available and will ease the design task.

(c) The experience of the aerospace and automotive industries with active electronic systems is spreading and should penetrate the rail industry.

(d) Hardware such as actuators, transducers, and microprocessors for the implementation of complex control laws is becoming both cheaper and more reliable.

(e) The competition between different modes of advanced transportation (Maglev vs rail, rail vs road, ground transportation vs air) will force industries to do their best with respect to comfort, reliability and travel speed.

The natural restraints on introducing active controls will be initially overcome in situations where it can be demonstrated that the same performance cannot be achieved by passive means (tilting trains are a good example); however, it must also be ensured that the active system is sufficiently reliable and/or there is a satisfactory passive backup in case of failure. It may be desirable to reduce the power requirements for active components by using semiactive strategies; naturally, any purely mechanical realization of an active control law is desirable if it is cost and weight efficient.

The continuing need to develop cheap and reliable hardware (in particular the actuators) must be emphasized. The implications of rapid advances in power semiconductor devices and the highly efficient control of power they offer should be studied. It is also possible that some of the criticism of active systems may be countered by using simple but rugged nonlinear controllers (e.g. bang-bang types), but only if they can offer the necessary performance.

It seems likely that the inevitable market demands which call for ever-increasing performance may force industry to adopt active solutions when passive designs can no longer meet these demands. Eventually, economic considerations must be paramount; any extra capital invested in the provision of active controls must yield benefits in the long run.

See also: High-Speed Railroad: Modelling and Simulation; High-Speed Railroad Networks in Europe; High-Speed Railroad Networks in Japan

Bibliography

Binnewies H 1987 Body-tilt passenger coaches. *Rail Int.* **18**(3), 21–5

Caudill R J, Sweet L M 1982 Magnetic guidance of conventional railroad vehicles. *J. Dyn. Syst., Meas. Control* **104**, 238–46

Celniker G, Hedrick J K 1982 Rail vehicle active suspensions for lateral ride quality and stability improvement. *J. Dyn. Syst., Meas. Control* **104**, 100–6

Frederich F 1985 Unbekannte und ungenutzte Möglichkeiten der Rad/Schiene Spurführung. *ZEV/Glas. Ann.* **109**, 41–7

Geuenich W, Gunther C, Leo R 1985 Fibre composite bogie has creep-controlled wheelsets. *Rail. Gaz. Int.* **141**, 279–82

Goodall R M, Kortüm W 1983 Active controls in ground transportation—A review of the state-of-the-art and future potential. *Veh. Syst. Dyn.* **12**, 225–57

Götz G 1983 Safety engineering for an active hunting controller for bogies in wheel/rail systems. *Proc. 4th IFAC/IFIP/IFORS Int. Conf. Control in Transportation Systems.* Pergamon, Oxford, pp. 325–31

Halfpenny D, Marshall J J 1977 Tilt technology. Paper presented to the Railway Division of the Institute of Mechanical Engineers. IME, London

Jaschinski A, Duffek W 1983 Evaluation of bogie models with respect to dynamic curving performance of rail vehicles. *Veh. Syst. Dyn.* **12**, 70–5

Leo R 1984 Erprobung schlupfgeregelter Radsatze auf dem Rollpfrufstand. *Proc. Dynamik Schneller Bahnsysteme: Rad/Schiene und Magnetschwebetechnik.* Verein Deutscher Ingenieure Berichte 510. VDI, Berlin

Losa P A 1987 The Pendolino body tilt control. *Energia Elettrica* **64**, 33–44

Lumpie D J C 1977 Experimental Talgo features pendular body-tilting. *Railw. Gaz. Int.* **133**(2), 58–9

Lundgren J 1988 High speed train class X2 for Swedish State Railways. Asea Brown Boveri Traction Document No. T88–117

Okamoto I, Koyangi S, Higaki H, Terada K, Sebata M, Takai H 1987 An active suspension for railroad passenger cars. *Proc. 1987 IEEE/ASME Joint Railroad Conf.* Institution of Electrical and Electronics Engineers, New York, pp. 141–6

Orley H 1981 Drehgestell hydraulisch gedämpft—Stellzylinder stabilisiert Laufwerke von Hochgeschwindigkeitszügen. *Z. Fluid* December, 24–6

Pollard M G, Simons N J A 1984 Passenger comfort—The role of active suspensions. *Proc., Inst. Mech. Eng., London* **198**(35), 1–15

Sinha P K, Wormley D N, Hedrick J K 1978 Rail passenger vehicle lateral dynamic performance improvement through active control. *J. Dyn. Syst. Meas. Control* **100**, 271–83

Williams R A 1985 Active suspensions—Classical or optimal? *Veh. Sys. Dyn.* **14**, 127–32

R. M. Goodall
[Loughborough University of Technology, Loughborough, UK]

W. Kortüm
[DLR, Wessling, Germany]

Railroad Systems: Line Supervision and Control

Automatic line supervision (ALS) is an automatic supervision and control system that acts on trains and ground installations, ensuring coherent operation and providing optimum line flow and general safety. Its basic function is train supervision. Information on train progress is sent to a central control room by a teletransmission system; these basic data are processed by a central computer which controls any corrective action.

ALS was initially developed in metro networks due to the density of traffic. It has extended to suburban and high-speed lines for similar reasons. Most of the following explanations are focused on rapid transit lines.

Train progress information basically comprises the number of each train and its geographical position. The choice of the number is determined by the train function (stations served, schedule, train composition); for each departure from a terminus, a number is allocated to the rolling stock available to carry out the journey. This number is given by the ALS in line with the theoretical daily schedule. In general, the geographical position of the train is given by the automatic train protection (ATP) in the form of a signalling block occupied by a train.

Given these basic data, the central computer "follows" the movement of each train from the terminus and, step by step, allocates the number of the train to each signalling section occupied. This "train describer" is a fundamental intermediate function of the ALS. It is used simultaneously by the operators in the central control room and by various automated ALS functions (train movement regulation, routing control, destination display control). For operator use, the train describer is usually displayed on large visual control panels which provide a geographical representation of the line with train number progression. Sometimes the visual control panel is simplified and only shows track circuit occupancy, the train number being given by a console; less often, tracking is displayed entirely by console.

1. Purpose of ALS

ALS is basically used to regularize traffic and maintain optimum line flow despite disruptions. The trains do not face the same type of risks as road traffic, as they travel within their own "private right-of-way." However, the nature of their mass transit role exposes them to other types of disruption; for example, extended dwell time following a passenger problem or a door failure (in a metro each door opens or closes every minute, presenting availability problems), and extended stopping time between stations after emergency brake operations by a passenger.

In a mass transit system, these disruptions would have a destabilizing effect on the line. In fact, the delayed train would be separated from the preceding train by an abnormally long period and the number of passengers awaiting this train downline would, therefore, be greater than usual, resulting in an increased dwell time, entailing an increase in delay. The phenomenon would, therefore, be cumulative.

This brief analysis shows that the phenomenon has to be eradicated as quickly as possible. The method to be implemented is, therefore, real-time data acquisition for all train movements to detect any delay as soon as it occurs, with immediate action on trains to regularize train headways.

Other functions concerning train movements and forming part of the overall control system are

(a) routing control,
(b) power distribution regulation,
(c) technical failure management,
(d) operation quality monitoring, and
(e) passenger information.

The ALS system is sometimes called automatic train supervision (ATS) by analogy with ATP and automatic train operation (ATO), but automatic line supervision is preferable, since this emphasizes its comprehensive ("line") nature.

2. Regulation of Train Movements

The regulation of train movements consists of regulating a line to ensure optimum line flow. It involves acting on trains both on the line and in the termini. The theoretical reference is the train schedule as far as it can be adhered to in the absence of disruptions. The

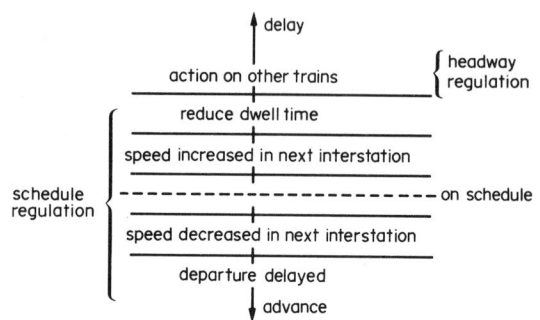

Figure 1
ALS regulation principle

importance to regulation efficiency of drawing up a good schedule is obvious. The establishment of schedules relies fundamentally on a good knowledge of traffic needs and driver working conditions.

Two levels can be defined: schedule regulation and headway regulation. Schedule regulation ensures that each train respects the theoretical run scheduled in the timetable; headway regulation regularizes the headway between trains. Schedule regulation is efficient for minor disruptions; however, in the event of larger or major disruptions, headway regulation becomes useful, since schedule regulation is no longer sufficient for stabilizing train runs (see Fig. 1).

The system then acts in the stations and termini on trains either side of the affected train (see Fig. 2).

(a) Downline holding consists of keeping trains in the stations downline from the station where the disruption is detected, to spread the ensuing delay over several trains and thereby bring about a better distribution of passenger load; this process, however, is not efficient when traffic is saturated (i.e., when trains are crowded with passengers).

(b) Upline holding consists of keeping trains in the stations upline of the disruption, so as to avoid a buildup of trains behind the affected train.

(c) Upstream and downstream regulation takes place at the terminus in the form of compensation: the

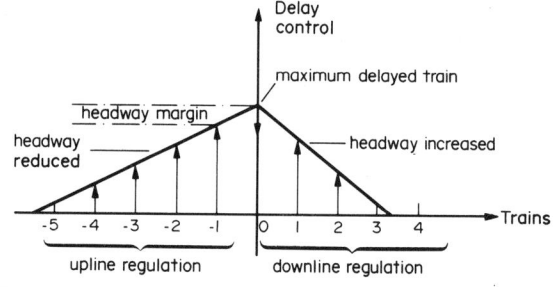

Figure 2
ALS headway regulation

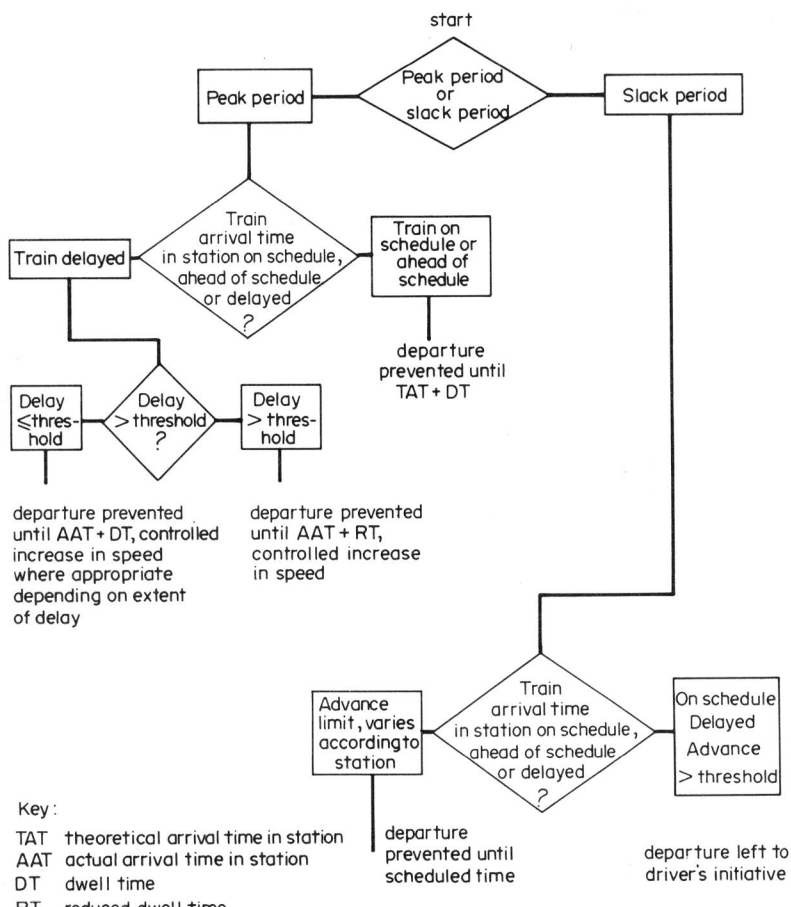

Figure 3
Example of dwell time regulation algorithm

schedule of trains leaving on one track is modified in accordance with the delay incurred by a train on the other track.

After regulation has been implemented, resulting in the delay of several trains, overtaking aims at returning these trains to the theoretical schedule. This is carried out either at the termini or on the line, usually outside peak periods.

Regulation acts upon train dwell time and on the travel time between two stations by means of ATO. However, dwell time cannot fall below a minimum that is governed by the nature of passenger interchange at the station involved (the passenger alighting time at least, supposing that the passengers waiting to board can wait for the following train). However, it may be extended in the event of a train being kept in the station; in such a case, the train remains at the platform with doors open to enable new passengers to board (see Fig. 3).

Travel time may vary between a minimum value, corresponding to maximum ATO–ATP performance, and a certain number of other higher values determined according to the needs of ALS. Operation at the minimum travel time results in high power consumption. It is, therefore, to be avoided as far as possible.

Headway regulation algorithms are quite simple in the case of a single disruption (see Fig. 2): downstream holding depends on the delay incurred by the affected train and decreases as one moves away from this train; downline regulation stops at the last station before the terminus. Upline holding corresponds to headway and decreases as one moves away from the affected train; upline regulation may be limited to the first train to leave. In the event of several disruptions, holding must interlink.

3. Routing Control Function

In certain cases, a train may follow several routes (on the line, in branch mode operation and in the terminus); moreover, emergency routes are always provided in case of lengthy train immobilization. Once the route of a train has been established, the corresponding points have to be controlled. The control acts upon the ATP, which guarantees the safety aspects of operations carried out over the section concerned.

The control of standard routes is usually automated on modern systems and, therefore, forms part of ALS. However, the control of emergency routes is always triggered by central control room personnel. This human intervention is indispensable, since the decision carries heavy responsibility and depends upon a multitude of parameters (e.g., type of incident, expected duration, staff at the scene, density of traffic) that only a human can comprehend; however, the automated systems can help to make the decision.

In general, routes are controlled by the central computer, according to the train describer (data concerning the number of the train and its geographical position). The teletransmission system sends this order to the local ATP equipment corresponding to the controlled route. A check of the correct implementation of the order is then retransmitted by the teletransmission system to the ALS (see Fig. 4). On a certain number of the most up-to-date North American systems (e.g., Atlanta, Washington), route control is carried out by the requesting train, in accordance with the number of this train and geographically determined points (see Fig. 5).

The structure in Fig. 4 is centralized while that in Fig. 5 is decentralized; however, in every case, there is always a central computer and a teletransmission system between local ATP equipment and the central computer. The decentralized structure offers certain protection against temporary central computer breakdowns, but at the price of an additional train–track link device.

4. Power Distribution Regulation Function

Power regulation consists of monitoring the electricity power supply to the line and, in the event of incidents or works, ensuring system reconfiguration to provide optimum power supply. Once reconfiguration decisions become complex, human intervention is required. The efficiency of power regulation relies on an overall and

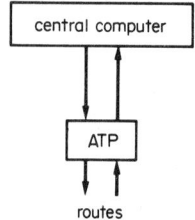

Figure 4
ALS interface with ATP: centralized method

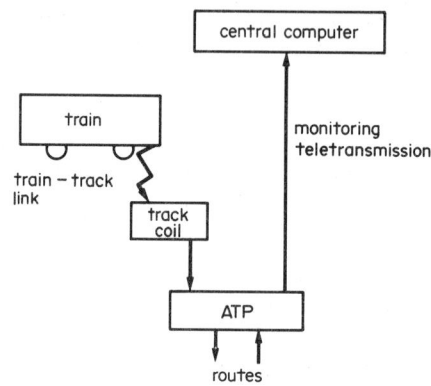

Figure 5
ALS interface with ATP: decentralized method

instantaneous picture of the situation; intervention, therefore, takes place from a central control room containing a visual display system (visual control panel or graphic consoles simultaneously providing raw information on the state of operation of the network and synthesized information for decision-making purposes). The central control room concerned with trains and power are sometimes merged.

In some cases, a discrimination must be made between direct power supply to the line and a power supply at a higher level; a transformation stage from a general (even national) supply to the private network of the transportation company may be required.

5. Technical Failure Management Function

The operation of a line not only depends on the ATO–ATP and power supply, but also on numerous other technical devices which have to be monitored and, where necessary, controlled from a central office.

Checks investigate the working order of the various items of equipment for trains and fixed installations (along the track, in tunnels or in stations), or of specific alarms, such as fire alarms.

Some of the dysfunctions can be easily detected and automatically registered for repair (autoreconfiguration, remote or local intervention); others need human diagnosis on site.

An interesting application is the control of interventions in case of fire and the resultant dispersion of smoke in the metro of Montreal.

In all cases, it is convenient to have a centralized recording system with a database, making it possible to distribute maintenance work and to establish statistics on mean times between failure and life cycles of the subsystems or components.

Of course, this centralization is necessary in the case of fully or partly unmanned systems: reconfiguration control of equipment in a train operated without a driver (e.g., the Lille metro) and control of escalators, fans and so on in stations operated without the permanent presence of a member of staff (e.g., the Atlanta and Lyon metros).

6. Passenger Information Monitoring Function

Operation monitoring consists of recording various parameters concerning train movements as they actually occur, processing these data to evaluate the quality of service, modifying the theoretical schedule if it proves to be inappropriate and storing the information for later investigations.

7. Passenger Information Function

Permanent information (e.g., structure of the network or of the line, stations served, schedules, fare system) is given on notice boards and does not form part of ALS. However, any temporary information concerning train movement does form part of the ALS.

In certain types of metro operation, trains entering a given station are not necessarily following the same route (as is the case with lines operated in branch mode, intermediate termini or trains running through certain stations without stopping); the passenger, therefore, has to be informed of the route to be taken by the train stopping at the platform. This information may be either oral or visual, and may be given on the platform, on the front of the train or inside the train. Control of this information is particularly crucial when the theoretical schedule can no longer be followed, owing to disruptions.

The most commonly used systems are

(a) a visual display of stations served, located on the platform and controlled by the ALS train describer system, and

(b) a visual display on the front of the train, directly dependent on the number of the train allocated by the ALS on departure from the terminus.

Finally, information concerning exceptional modifications to traffic as a result of incidents (e.g., interruption of operations in certain zones, lengthy train delays) is given both in trains and at stations.

This information (as with decisions involving the control of emergency routes described in Sect. 3) is provided by central control room staff, with the ALS playing only a supporting role. The staff may be helped by subsystems, such as automatic selection of stations at which the information has to be given, and composition of written or spoken (possibly prerecorded) messages.

8. Human Involvement

The very concept of ALS suggests completely automated control of the entire metro line; however, for the processing of situations involving exceptional disruption, it has already been pointed out that intervention by central office personnel is indispensable.

The general configuration of ALS is given in Fig. 6, that is, integral ALS with fully automated functions and human intervention limited to the presence of central control room personnel to deal with exceptional cases. It is operated on small transport systems (Morgantown, Dallas and Atlanta airports) or new simple operation metros (all trains following the same route): since 1981, Kobe, Lille, Miami, Vancouver, Detroit and Jacksonville have been opened to traffic; Osaka, London (Docklands) and Toronto are nearly fully automated but an attendant remains on board.

The other metros employ systematic human intervention for certain functions described in the ALS. The main reason for this is that automation is being introduced gradually, realistically taking account of technical, economic and human possibilities in accordance with already existing operations.

The great majority of metros have a central control room with an automatic train describer. Systematic human intervention is usually concentrated on the train movement regulation function.

The train movement regulation function may also be performed by drivers, who attempt to modulate the running of their trains so as to respect the schedule and who, in the event of disruptions, obey instructions from the central control room member of staff who has an overall and instant picture of the situation by virtue of the information provided by the train describer. The link between drivers and the central control room member of staff is via radiotelephone. Figures 7 and 8 give schematic representations of train movement regulation with and without systematic driver intervention, respectively.

9. Trends in Generalized Systems

The major progress made in the area of telecommunications during the late 1980s has brought new levels of capacity, facility and message integrity, leading to new

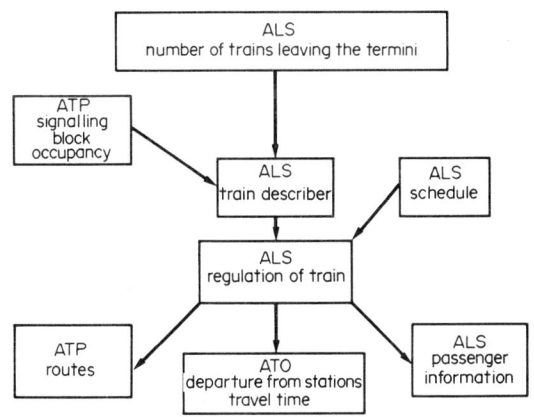

Figure 6
ALS: general configuration

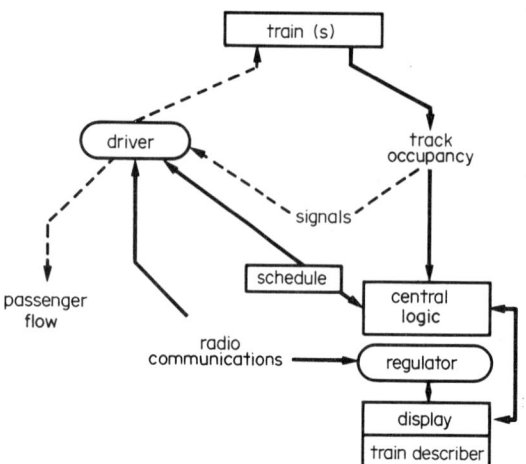

Figure 7
ALS with systematic intervention of driver in regulation function

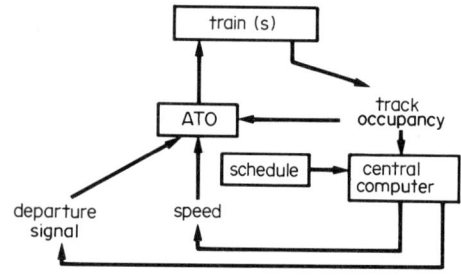

Figure 8
ALS with fully automatic regulation

concepts in ALS (even comprising a part of ATP) that are intended to cover large networks of railroads.

Cellular radio systems, digital transmissions, transponders and on-board computers enable vehicle identification and location that is sufficiently safe to allow the use of a centralized management control system for the fleet. Examples are the automatic train control system (ATCS), developed jointly by the USA and Canada, and the ASTREE project of the French railroads (SNCF).

See also: Automatic Train Control: Protection Principles; Railroad Systems: Train Driving Control; Underground Railroad: Organization of Operations

Bibliography

Blakey R 1987 Computer takes over underground. *Railw. Gaz. Int.* October, 675–8
Hooper A 1986 Centralized control of the Jubilee and Metropolitan lines. *IEE Colloq.* February
Perrin J-P (ed.) 1990 *Proc. IFAC Int. Symp. Control, Computers, Communications in Transportation.* Pergamon, Oxford
Perrin J-P, Beuchard P 1984 Automation in rail based urban transit systems. *Fr. Railw. Rev.* **2**(2), 101–11
Perrin J-P, Pascal J-P 1987 Les transports automatique de demain. *La Recherche, Suppl.* **190**, 18–29
Rowbotham A J 1989 Increased safety and efficiency through new signalling developments. *Developing World Transport.* Grosvenor, London, pp. 191–2
Stalder O 1987 Bahn 200 prompts SBB resignalling. *Railw. Gaz. Int.* October, 669–71

J.-P. Perrin
[RATP, Vincennes, France]

Railroad Systems: Train Driving Control

Automatic driving systems may have three objectives:

(a) aiding the driver;

(b) monitoring the driving (overspeed protection, automatic train protection (ATP)); and

(c) actually driving the train, partly or completely taking over from the human driver.

The purpose of these systems is to carry out repetitive tasks and to replace the driver completely when the schedule is tight (short headways).

In addition, automatic driving can also be classified by

(a) the safety level, depending on whether the human driver does or does not retain full responsibility for driving; and

(b) the amount of data transmitted from the trackside to the train (or from the train to the trackside).

Classification (b) partly depends on the technology employed. Microprocessor developments have led to the use of more "intelligent" systems capable of memorizing more data.

1. Basic Automatic Systems

Automatic systems may or may not be linked to an overspeed protection.

(a) Assigned speed or deceleration (Fig. 1). This is used for railroads and, more rarely, for metros with long

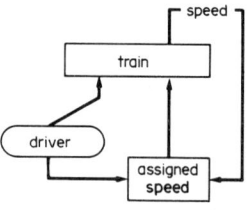

Figure 1
Block diagram of the assigned-speed system

interstation sections. The driver sets an assigned speed and a simple logic keeps the train at this speed, within given tolerances. This category also includes deceleration servosystems which allow this parameter to be controlled as a function of the line profile and the train load.

(b) *Programmed stop (Fig. 2).* This is an on-board system that is triggered by a local signal from the trackside. One or several braking curves are held in memory on the train and the train deceleration is then controlled to these curves.

(c) *Programmed running.* The normal running schedule is held in memory on the train and is automatically followed. If traffic is disturbed, the driver must resume control.

(d) *Optimized running schedule.* Signals received from the trackside are used to select, as a function of train control criteria, the most suitable program from several held in memory on board the train (see *Railroad Systems: Line Supervision and Control*).

(e) *Automatic driving (Fig. 3).* This term is used to describe systems that can completely take over from the driver; obviously, these can be used only in conjunction with overspeed protection. The complete system must be able to "drive" the train from one normal stopping point to the next, without the necessity of human intervention and respecting all safety conditions (train protection). By extension, it can include automatic door closing which, if all other conditions are satisfied, triggers the automatic departure of the train; the doors then automatically open (or are, at least, automatically unlocked) at the next stop.

The system permitting automatic train operation (ATO) is, in principle, a simple superimposition of overspeed protection and a programmed running schedule, where the most restricted parameter is used to drive the train. This method has the advantage of offering the possibility of using either automatic driving, required when traffic is dense, or manual

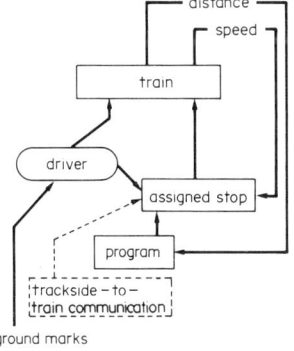

Figure 2
Block diagram of the programmed-stop system

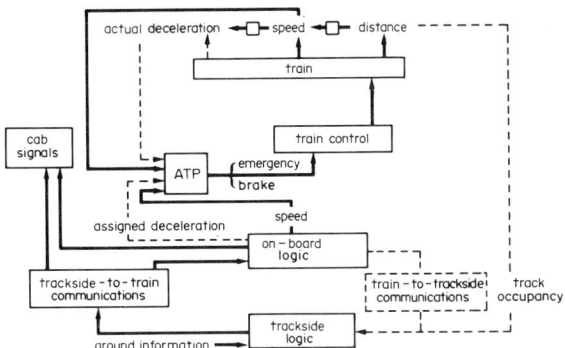

Figure 3
Block diagram of an automatic driving system

driving with overspeed protection. This second configuration allows drivers to keep in practice and also serves as a downgraded mode of operation if a breakdown occurs in the automatic driving system.

The speed signals given in interstation sections can either be the same as those used for overspeed protection (tight schedule), or can be reduced to save energy or for traffic regulation reasons. The deceleration and station stop phases are controlled either by trackside programs or by train-borne programs operated from ground information devices (most of the systems use track coils).

2. Full Automation

Some metro lines or similar facilities have been designed for completely automatic train control. The main difference between this system and automatic driving is the emphasis placed on maximum availability and the automatic processing of downgraded situations; for example, "docking" onto a train that has broken down in order to push it.

Some precautions also have to be taken regarding the safety of passengers during the movement of the train along the platforms.

Although these are the only differences, they do act as big constraints.

3. Driving Principles

In general, the trains are controlled by the gap (error) between the speed measured on board, by means of a tachogenerator or an electromagnetic device drive by the wheels, and the selected speed, which is either decoded from a message transmitted from the trackside or calculated on the train on the basis of data received from the trackside. Laws that are generally simple are used to transform this difference into a form suitable for traction motor or braking control.

The natural way of controlling motors is by the torque that they provide (i.e., the current they draw). This control system is linked to smoothing devices which limit acceleration and its derivative (jerk) to comfortable levels. On modern rolling stock, the brakes are normally electrically controlled (with an electropneumatic system for emergencies and very low speeds). They are frequently controlled by a servo-system during deceleration, even in manual driving so that the driver does not have to adapt action to the train weight (i.e., acceleration control instead of force control).

Figure 4 shows an example of an automatic driving process used in several metros. The running schedule is "written" along the trackside in the form of a twin transmitting cable (feed and return). Nodes in the cable are spaced such that the reference trip corresponds to a time t_0. Two sensors, one placed in front of the other, receive the electromagnetic signal. If the travel time over a sector is greater than t_0, the traction force is increased by a value proportional to the difference. If the travel time over a sector is less than t_0, a braking load is applied in the same way. The system illustrated uses fixed increments (stepped traction control T and braking control F). However, the same procedure can be used to provide continuously variable controls.

If the datum program is $x = V^2/2\gamma$, the control curves take the form $x_i = (V^2/2\gamma)(t_i^2/t_0^2)$. These are identified as T2, T1, F1, F2 and F3 on Fig. 4. An overspeed limit envelope curve is provided for ATP.

4. Classification by Type of Train–Track Transmission

There are several types of train–track transmission.

(a) Continuous transmission of elementary data may be carried out using the rails or a cable laid in the track. The most frequent example is transmission by modulation of the track circuits, which simply provides ATP.

(b) Continuous transmission of complex data may be carried out via the rails or a cable laid in the track. Generally, binary messages are transmitted and then decoded on board; a simpler method is to transmit superimposable frequencies which are then used on board to generate a certain number of functions. In certain cases, the cable arrangement allows running schedule data to be transmitted or geographical beacons to be identified.

(c) Local, point-by-point transmission may be used. In this case, the train-borne system needs to be more complex, since it must generate the ATP and ATO programs from data received, either instantaneously, if it is possible to transmit a high number of parameters, or by initializing calculations using programs held in memory on board.

(d) It is worth mentioning that research is being carried out on radio communication and transmission (uhf and microwaves).

5. Technological Developments

The first automatic driving systems were introduced in the 1960s (e.g., in Philadelphia, Leningrad, London and Paris). The technology was directly derived from that used for signalling systems (relays). The 1970s saw the development of several systems using electronic technology (e.g., Paris, San Francisco, Munich). Nevertheless, the safety aspects used conventional technology (in particular, tuning forks were used to guarantee transmission frequencies).

Microprocessors began to be introduced towards the end of the 1970s, initially for driving functions (the driving law) and then for safety functions using duplicated systems. The latest systems rely heavily on digital processing, which is more reliable and more stable than the analog technique. The introduction of thyristors into traction and braking controls has allowed much more flexible driving control to be achieved. Table 1 lists the locations of systems in existence in 1989.

See also: Automated Guideway Transit Systems and Personal Rapid Transit Systems; Automatic Train

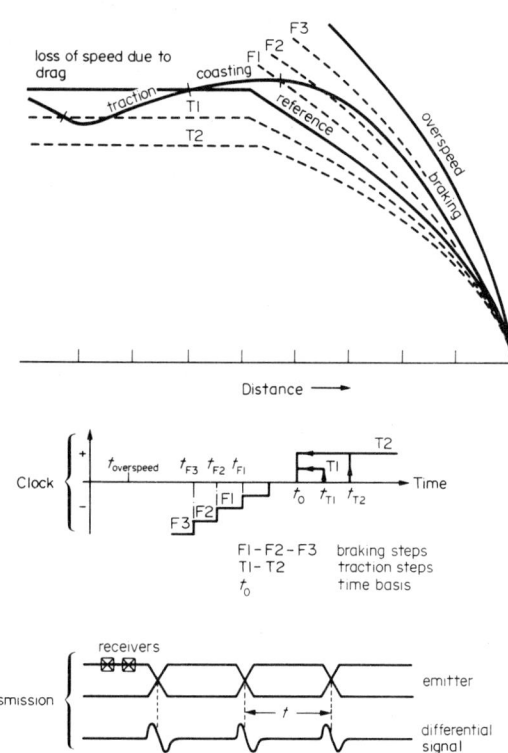

Figure 4
Example of an automatic driving process

Table 1
ATO subways in the world, 1989[a]

America			Asia			Europe		
	Venezuela:	Caracas (1)	Hong Kong		(3)	Germany:	Berlin	(1)
							Munich	(3)
	Brazil:	São Paulo (2)	Japan:	Sapporo	(1)			
		Rio de Janeiro (2)		Kobe	(*)	Austria:	Vienna	(3)
				Fukuoa	(1)			
	Chile:	Santiago (2)		Osaka	(+)	Spain:	Madrid	(1)
	Mexico:	Mexico City (8)	Singapore		(2)	Finland:	Helsinki	(1)
	Canada:	Montreal (4)				Netherlands:	Amsterdam	(1)
		Toronto (+)						
		Vancouver (*)				Italy:	Milan	(2)
	USA:	Philadelphia (1)				France:	Paris	(12)
		San Francisco					Lyons	(2)
		Washington (3)					Marseilles	(2)
		Atlanta (2)					Lille	(2*)
		Baltimore (1)						
		Miami (*)				UK:	London	(1+)
		Detroit (*)					(Docklands)	
		Jacksonville (*)					Glasgow	(1)
						USSR:	Moscow	(2)
							Leningrad	(2)
						Hungary:	Budapest	(1)

Source: *Metropolitan Railways in the World*, published by International Metropolitan Railways Committee (UTIP)
a Figures in brackets denote number of lines equipped; (*) fully automated line; (+) fully automated line with member of staff on board

Control: Protection Principles; Railroad Systems: Line Supervision and Control; Underground Railroad: Organization of Operations

Bibliography

Bartwell F T 1973 *Automation and Control in Transport*. Pergamon, Oxford
David Y 1987 Systèmes de commande de fonctions de sécurité par microprocesseurs en service dans les transports terrestres. *Recherche, Transports, Sécurité*, Vol. 15. INRETS, Arcueil, France
Perrin J-P (ed.) 1990 *Proc. IFAC Int. Symp. Control, Computers, Communciations in Transportation*. Pergamon, Oxford
Perrin J-P, Beuchard P 1984 Automation in rail based urban transit systems. *Fr. Railw. Rev.* **2**(2), 101–11
Perrin J-P, Pascal J-P 1987 Les transports automatique de demain. *La Recherche, Suppl.* **190**, 18–29

J.-P. Perrin
[RATP, Vincennes, France]

Road Network Control

The traffic increase in large urban areas calls for better management methods. New requirements are imposed on traffic-light control schemes. Congestion avoidance through better flow control is probably the most challenging goal as far as private traffic is concerned. On the other hand, improving the surface public transport by effective priority at intersections is commonly requested.

These requirements demand fast and efficient traffic-responsive schemes. Robustness and efficiency can only be reached by a closed-loop approach. To this end, the complex problem of urban traffic control must be decomposed in a suitable way into a series of reliable, smaller control problems. Then, feedback strategies are applied to each of them. An outline of the possible solutions is given in this article, together with the features and the results obtained in one implementation.

1. Statement of the Problem

The problem of network traffic control has generally been solved by setting suitable signal plans for each intersection. Plans can be fixed, time varying, or traffic actuated. They always rely on stationary or slow-varying traffic conditions, where suitable static coordination between intersections can be applied. Most existing traffic control schemes are based on these concepts (Strobel 1982).

Assume now that the following specifications are made for the control:

(a) absolute priority to selected public transport vehicles at every intersection (except when conflicts among them occur); and

(b) significant improvement of the mobility of private vehicles in all traffic conditions, including fast-varying traffic, partially oversaturated networks and consequences of capability variations (e.g., because of accidents).

Both requirements impose the need for a feedback solution. The feedback approach, in turn, requires that the network stability is analyzed together with the feasibility and robustness of the control system.

To this end, some methods common to the theory and practice of large-scale systems may be useful. Thus, a global approach to the problem is:

(a) finding a suitable decomposition of the control problem, in a hierarchical decentralized way;

(b) defining suitable functionals for all the resulting problems, together with the rules for interaction; and

(c) defining algorithms and methods for solving all the resulting problems.

Since the most attractive decomposition rule is the topological one (i.e., the network is composed of a series of intersections), the problem is quickly reduced to that of finding a robust feedback control for the intersections, with consistent rules for the interaction between intersections. To guarantee stability at the network level, interactions also occur with an upper level.

This global approach is assumed in this article. One feasible and comprehensive solution is suggested and outlined. This solution has been already implemented with very good results in an area of significant size in Turin. The applied strategies will be referred to as urban traffic optimization by integrated automation (UTOPIA) while the first implementation has the name of "Progetto Torino."

Given the specification for priority to public transport vehicles, a method for predicting the arrival times at intersections is needed. The related problems will not be dealt with in this article. Here, it is only assumed that such predictions are given. In Turin, an automated vehicle monitoring system is in operation: it supplies timely updates of the predicted arrivals, which are given with an advance time of more than 120 s and are improved as the vehicle approaches the intersection (Gentile and Mauro 1988). Other traffic-responsive network control methods have been developed (see *Road Traffic Control: TRANSYT and SCOOT*) (Henry et al. 1990).

1.1 Basic System Concept

The basic choice in the UTOPIA design has been the application of large-scale systems control theory to the

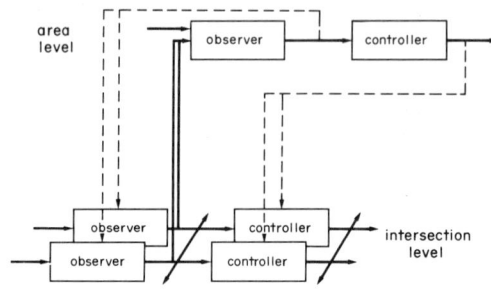

Figure 1
Decomposition of the wide-area traffic control problem into a set of smaller and interrelated subproblems, based on a topological rule. The subproblems are connected to the area-level controller and to each other by the intersections of the controlled network

whole traffic control problem. Thus, the overall problem has been decomposed into a series of interrelated, smaller problems that can be classified as belonging to two main classes: the intersection level and the area level (see Fig. 1).

(*a*) *The intersection level.* At this level, a series of "local" controls operate, one for each intersection. Every local control is further subdivided into two parts: the "observer" and the "controller."

The observer has the goal of updating the best estimates of the intersection state by making use of all the available data (traffic counts). The observer is based on a microscopic model of the intersection: state elements are groups of vehicles to be served for every link, and time discretization is in 3 s intervals. Thus, the state vector for the entire intersection is the composition of state vectors representing single links. These single-link state vectors are formed by the vehicles already in the link ordered by predicted arrival times at the stop lines and grouped in 3 s intervals. Thus, a link whose maximum length (in time) is 30 s, will be represented by a state vector of ten elements.

The controller outputs are the intersection signal settings to be applied to traffic lights. Every control optimizes a suitable functional adapted to the traffic situation in its area. Optimization is performed by making predictions at 120 s intervals. The optimal signal settings are applied to traffic lights for 6 s. Thus, a closed-loop control is obtained. It can be viewed as an open-loop feedback optimal (OLFO) or as an application of a rolling-horizon concept. If the optimization at the intersection level alone can neither guarantee optimality nor benefit at the town level, two important provisions are made.

(i) The "strong-interaction" concept is adopted: within the functional to be optimized at the intersection level, suitable terms are inserted to take into account the state of neighboring intersections.

Thus, a closed-loop capability of building up signal coordination is maintained.

(ii) The intersection controllers are constrained by limits given by the area level. Moreover, the weights in their functional are determined by the area level itself. In this way, the area level may establish preferred routes in town as well as imposing traffic-dependent criteria on the intersection level.

(b) *The area level.* The goal of the area level is to obtain a smooth traffic flow in every demand condition and all over the town. This goal is fulfilled by choosing suitable functionals for the intersection controllers (and by imposing constraints on them). At the same time, the area level may also support the local observers with some data.

The observer goal is that of predicting, in real time, the most important routes that will be taken by private traffic. The demand at the origin (flow levels) for these routes is also predicted by means of an observer based on a time-discrete model. Time discretization is done at 3 min intervals. The network of major roads is represented by a collection of oriented and connected "storage units" (see Fig. 2). Every storage unit may contain several real intersections. A series of fixed routes is superimposed on the network. States of the model are the number of vehicles in each storage unit belonging to every route.

The controller goal is that of optimizing the network functional, by acting on fictitious controls. Controls are average speeds and saturation flows within every storage unit. Suitable constraints, reflecting the actual ones, are imposed on both controls. Again, the control

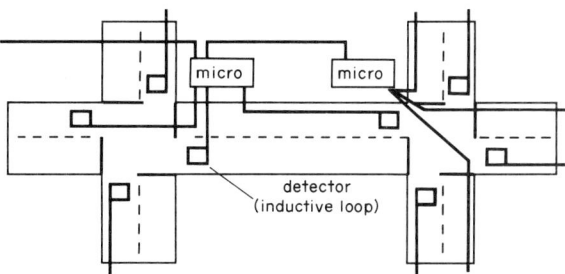

Figure 3
Layout of private traffic detectors at the intersections

is obtained by OLFO (or rolling-horizon) techniques; horizon time duration is 30 min. Control is recomputed when the observer detects a change in traffic conditions. Finally, the outcomes of the control are used to obtain the information and commands to be transferred to the intersection level.

1.2 Intersection Control Level

The intersection is fitted with loop detectors capable of counting vehicles: loops are installed at the exit from the intersection (see Fig. 3). Exit loops of one intersection are input loops to the next one. The optimal capability of predicting the arrivals to every stop line is thus obtained.

(a) *The local observer.* At the intersection level, two groups of variables have to be estimated in real time:

(i) a group of slow-varying parameters such as travel times (for every link), turning percentages and saturation flows; and

(ii) the intersection state (see Sect. 1.1).

A robust and correct estimation of the two groups is needed for efficient and robust closed-loop control.

State estimation is done by modified Kalman filtering of the traffic counts: state propagation is effected at every step (3 s) based also on the knowledge of the status of the traffic lights; state correction is then made after traffic lights have changed, by comparing flows propagated on exit detectors with the actual flows. Parameter estimation is also made via on-line filtering based on traffic counts and traffic-light status. An update is made every cycle and it is again based on the differences between predicted and actual counts.

The observer defined here provides efficient estimates of all the intersection variables and is well suited to closed-loop control. With the aim of increasing the robustness and stability of the observer, some information is used that is obtained from the area level. In particular, the area level transfers reference values for turning percentages, travel times and saturation flows.

(b) *The local controller.* By the definition of the state itself, predicted arrivals at the virtual stop line are determined at every instant for some tens of seconds.

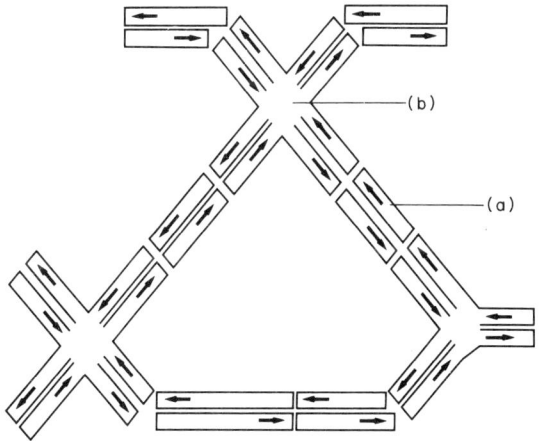

Figure 2
Model of the network at the area level: the main routes are represented by oriented and connected storage units which may contain more real junctions and represent (a) a piece of route or (b) the intersection of routes

This time horizon is not long enough to guarantee smooth control, due to the hard constraints of stage durations and the expressed goal of priority to public transport vehicles: thus a longer time horizon (120 s) is defined. Then, because there is no correct prediction of private car arrivals for the extra time, the signal setting is done by following different methods within the two logical parts of the horizon.

First, a suboptimal heuristic selection of signal settings for the extra-horizon (e.g., from 30 s to 120 s) is chosen by the evaluation of a cost function, taking into account the following: the probability of not providing public vehicles with priority (this cost is normally so high that safe green times are established); deviations from the signal setting as decided in the previous iteration (this ensures smooth behavior of the network); and deviations from the reference signal setting, as given by the area level (this ensures good coordination and helps in finding a stable and robust solution). The suboptimal signal setting obtained will never be actuated, but it will be used as follows. For the closed-loop time interval (e.g., the first 30 s) an optimal signal setting is computed by optimizing an additive functional having as cost elements the time lost by vehicles, vehicles stops and maximum queue lengths. For the intersection links, the same costs are seen at the adjacent intersections by the vehicles that left the actual one: time lost by the single public vehicle at the intersection, and cost of the needed modification of the open-loop policy (extra horizon).

In this functional, only the cost concerning the maximum queue lengths are not linear (quadratic); the search for the optimal solution is made by a branch-and-bound algorithm. The weights of the elements are chosen by the area level on the basis of network conditions; moreover the cost elements are associated with every link. By choosing the weights, the area level may fulfill its goal. Notice that the on-line signal coordination is automatically obtained by inserting the costs relating to the adjacent intersections into the functional (or, in other words, by imposing the strong-interaction concept on the network).

At the end, a signal setting is obtained that is optimal in the medium term (up to 20–30 s) and whose consequences have been evaluated for a longer time (up to 120 s). This signal setting will then be in operation for the next 6 s; after that time it will be computed again.

(c) *The local oversaturation problem.* Saturation occurs when the demand is higher than capacity for a significant time interval. If the saturation is predictable it is dealt with by the area level and its effects are minimized. However, some local oversaturations, which affect one or more intersections, may still arise. They can be the results of local, unpredictable changes in demand. In other cases, changes in the network parameters, due to accidents, traffic deviations or road-works could cause such effects.

The closed-loop control may well cope with such

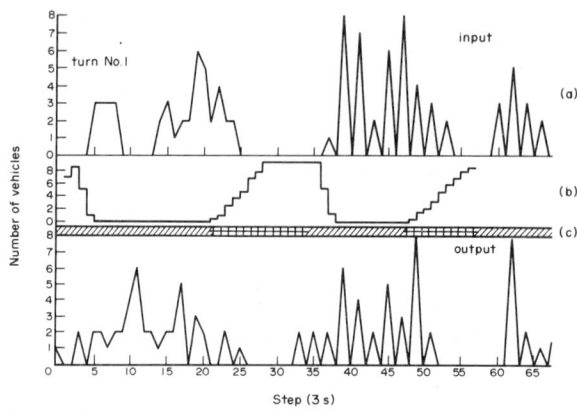

Figure 4
Information gathered directly by the local control (intersection no. 29) of Progetto Torino, under normal operating conditions: (a) private traffic arrivals foreseen at the stop line; (b) observed queues; and (c) actuated semaphoric stages as seen by the input link (dashed zones are green and crossed zones are red)

situations. However, in UTOPIA, some improvements to the control scheme had to be studied. The following provisions for dealing with the local oversaturation problem had to be introduced:

(i) giving the observer the added goal of detecting critical (oversaturation) conditions in specific links; and

(ii) weighting the number of vehicles released toward the adjacent intersection in the controller functional (an additional cost for receiver saturation)—as a consequence, in some cases it may be convenient to hold vehicles for a short time.

These two features will then be used to perform two actions. The first consists of minimizing oversaturation by increasing throughput. The second operates only if the first has not succeeded in decreasing the demand.

Both features require cooperation between adjacent intersections. In the first, the intersection that receives vehicles (the receiver) is involved. In the second, it is the one that sends vehicles to the saturated link (the sender).

Increasing throughput is obtained by relaxing constraints on green times in both the intersection and the receiver. Further on, the weights of the relevant queues are increased.

Decreasing demand is obtained at the cost of receiver saturation in the relevant links of the sender.

Both actions are driven by the saturation index, as estimated by the observer. The same index also re-establishes normal operating conditions.

(d) *UTOPIA.* The UTOPIA intersection control is very flexible and powerful. With the provisions for oversaturation it may deal with any real operating conditions. Concepts such as cycle, split and offset have no meaning. In Fig. 4, an actual, short-time, operation of one intersection is outlined to show that private traffic is accommodated in a correct way (the data for Fig. 4 have been taken from the Progetto Torino implementation, in normal operating conditions).

Some architectural remarks are worthy of mention. In any operating conditions, the intersection controller needs an efficient communication link with adjacent intersections. This is, indeed, always necessary for exchanging the forecasted signal setting before the optimization takes place. Moreover, in saturated conditions, requests for cooperative actions have to be transmitted. In the Progetto Torino implementation, point-to-point connections have been implemented.

For the best state and parameters estimation by the observer, a large number of loop detectors is needed. Moreover, the best loop position is as represented in Fig. 3. Every observer needs both the input loops and the output loops. This apparently doubles the wiring. The physical connection may be to the nearest controller. Data are then transmitted on the communication line.

1.3 Area Control Level

The area level goal is twofold: adding robustness in all conditions, by defining a medium-term reference scenario for the intersection level, while providing reference data for the local observers. The two goals are fulfilled, again, by the observers and controller.

(a) *Area level observer.* The observer has the goal of estimating and predicting the states of the already defined macroscopic model (see Sect. 1.1). To this end, it is assumed that the traffic is made up of a series of components.

First, there is a limited number of commuter-type routes having origin–destination attributes. Day-by-day similarity of the number of originated vehicles, departure times and assignment routes are assumed. The observer is then composed of two parts. One operates on a day-to-day basis and makes the optimal updates of the above-described data for all the traffic elements. The second runs in real time and detects, on the basis of events, the actual activation of the expected origins. Prediction can then be made by applying the filtered parameters to the measured origin flows.

Second, there is a small number of area-wide traffic phenomena that account for distributed traffic and are modelled by smooth (polynomial-type) time variations. In this case, the observer estimates the parameters of the time-variation law.

(b) *Area-level controller.* The on-line results given by the observer (time-varying traffic demand at the origins, assignment routes and predicted amplitude of the unmodelled flows in all main links) allow for the formulation of the following optimal control problem:

$$\min_{C} T_t = \min_{C} \int_{t}^{t+T} N \, dt$$

where T_t is the total travel time, C is the domain of feasible (area) controls, t is the current time and N is the number of vehicles in the network. Many simplifications can be made to keep the solution at an affordable level of complexity. As seen in Sect. 1.2, the outcome will not go directly to the traffic lights. The model is thus simplified. A series of storage units is considered, with a simple propagation law. The control variables are represented by a series of pairs α and β for every input–output relation of any storage unit.

The average speed and the capacity that must be given to adjust the traffic are represented by α and β, respectively. The set of α and β belonging to the same storage unit is constrained by realizability constraints. The problem is a (complex) deterministic optimal control problem for dynamic (large) systems. It is solved by a "feasible" method in discrete time, the time step being 3 min. The total horizon over which the functional is evaluated is 30–60 min. The area controller computes all the necessary parameters for the local controllers starting from the values of α and β obtained for every storage.

Other than the two functions of observer and controller, the area level performs all the operating functions for the complete system. Data collection from the local levels, data storage and retrieval in the database, as well as system diagnostics and maintenance, are all functions of the area level.

The experience of the Progetto Torino implementation has shown that:

(a) the system, due to the combined capabilities of area and intersection, is globally effective; and

(b) all the apparently complex algorithms have been put into levels of operation—larger areas could be dealt with.

1.4 Communication Network

Communication links must be established between adjacent local controllers and with the area controller. In the Progetto Torino, the resulting network is as shown in Fig. 5: a network of point-to-point physical connections at the local level with few connections to the area level. Telephonic-type lines have been used; throughputs are 1200 baud and 2400 baud (intersection and area). The software protocol provides for routing of all the messsages through the network up to the destination. The network protocol, which is resident in all nodes, is fault tolerant. Messages are rerouted in the case of failure of one of the lines.

2. Implementation and Results

2.1 System Implementation

The first realization has gone through various steps. In 1976, a series of studies aimed at designing the system were launched. The basic principles of the control

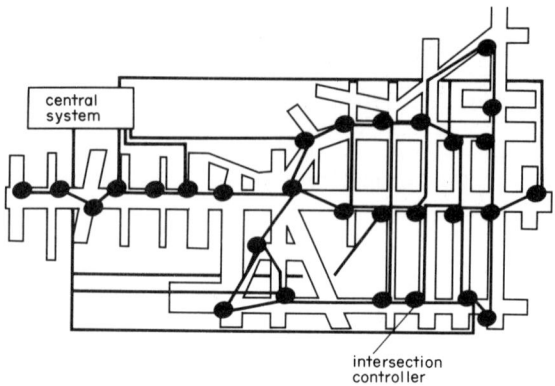

Figure 5
Progetto Torino communication network: only eight local controllers are directly connected to the area controller; the network is redundant and every node identifies on-line the best route for sending information towards the destination

strategies together with specifications for algorithm and architectures were obtained. The evaluation of benefits on private and public traffic was done via extensive use of microsimulation. Then the final area was chosen close to the center of Turin. It has a maximum length of 5 km, is 7 km^2 in area and comprises 39 intersections with traffic lights plus four controlled pedestrian crossings. Street-car line 10 crosses the entire area while the other two lines (1 and 12) partly share the same track. The area contains many critical intersections. Implementation began in 1981 and finished in early 1984. Architecture was composed of 33 intersection controllers, one area controller (three redundant computers), 400 loop detectors for private traffic and 110 detectors for public transport. Experimentation started in early 1984.

2.2 Experimental Results

In two weeks of 1985 and in a similar period in 1986, the experimental system was tested. To this end, travel times were measured by "floating" cars. In the meantime, the public transport travel times and priorities were measured. On some days, the system was actuating fixed-time policies. On other days, the UTOPIA strategies were in operation. The average benefits were as follows.

(a) The private traffic in 1985 improved by 9.5% (from 26.4 km h^{-1} to 28.9 km h^{-1}) and in 1986 by 15.9% (from 24.6 km h^{-1} to 28.5 km h^{-1}).

(b) In the same period, public transport obtained absolute priority on almost all the intersections. Only two intersections, having a very critical design, had a significant number of stops. The commercial speed improved, both in 1985 and 1986, by 20% (from 15.2 km h^{-1} to 18.3 km h^{-1}).

The benefits, in percentages, are significantly near to the ones forecast by microsimulation. The absolute values are different, due to the strong modifications in the traffic conditions. Moreover, detailed analysis of the algorithms and methods has also shown that the specifications were met.

See also: Road Traffic Control, Demand-Responsive; Road Traffic Control: TRANSYT and SCOOT

Bibliography

Donati F, Mauro V, Roncolini G, Vallauri M 1984 A hierarchical decentralized traffic light control system. The first realization. *IFAC 9th World Congr.*, Vol. II 11.G/A-1. International Federation of Automatic Control, Budapest

Gentile P, Mauro V 1988 Experience with SIS, Torino's public transport operation AID system. *Int. Conf. Automatic Vehicle Location in Urban Transit Systems*, Ottawa

Henry J J, Farges J L, Tuffal J 1990. PRODYN. *Proc. 6th IFAC–IFIP–IFORS Symp. Control, Computers, Communications in Transportation*. Pergamon, Oxford

Singh M G, Drew S A W, Coales J F 1985 Comparisons of practical hierarchical control methods for interconnected dynamical systems. *Automatica* **21**, 331–50

Strobel H *Computer Controlled Urban Transportation*. Wiley, Chichester, UK

V. Mauro
[MIZAR Automazione, Turin, Italy]

Road Network Signal Setting: Equilibrium Conditions

Since about 1950, mathematical models for transportation and traffic engineering have received great attention. They have been applied to, among others, the traffic signal setting problem, which plays an important role for the traffic management of urban networks.

Signal setting parameters are described by several decisional variables: green timing and scheduling for each junction, cycle time for each junction and offset for each pair of adjacent junctions. The flow pattern is described by the flow on each link (see *Road Traffic: An Introduction*).

Models for the optimization of a subset of these decisional variables, while assuming the others as fixed, have been proposed. They deal with three classes of problems: single-junction signal setting, interacting junction network coordination and traffic flow assignment. The first two problems assume the flow pattern is fixed (see *Road Network Control*; *Road Traffic Control: TRANSYT and SCOOT*; *Signal Control at Individual Junctions: Phase-Based Approach*; *Signal Control at Individual Junctions: Stage-Based Approach*). The third problem assumes the traffic signal setting parameters are fixed (see *Traffic Assignment*).

In real problems, however, delays at junctions and, therefore, user costs depend on signal setting. Thus, a change in the control strategy can have a redistributional effect on flows. Hence, control parameters should be chosen taking into account that a user, travelling from an origin to a destination, reacts to the adopted control strategy by choosing a new path such that individual cost is minimized (Allsop 1974, Smith 1985, Cantarella and Sforza 1987). For this reason, a combined evaluation of signal setting parameters and the flow pattern is needed to efficiently design a control strategy.

These considerations lead to the definition of models for traffic signal setting where user behavior is explicitly taken into account. These models can be referred to as equilibrium network traffic signal setting models and can be viewed as network design models where regulation parameters assume the role of design variables. A simplified formulation assumes that green times are the only signal setting variables. The computation of all of the previously mentioned signal setting variables can only be achieved through heuristic procedures that iteratively perform signal setting and traffic assignment.

1. Signal Setting and Flow Pattern Interaction

The equilibrium network traffic signal setting requires the following input parameters:

(a) network topology and travel demand;

(b) capacity, free-flow travel cost and cost–flow function for each link;

(c) crossing incompatibilities for each junction;

(d) saturation flow, lost time, minimum green time, maximum red time and maximum queue length for each approach; and

(e) distance and free-flow speed between each pair of adjacent junctions.

The decisional variables for the flow pattern and the signal setting are as described previously.

If the flow pattern is assigned, link flows as well as path choice are assumed independent of the adopted regulation. Under this hypothesis, traffic signal setting models can be used to compute regulation parameters. They can be formally expressed as

$$\min_{p} Z(p, q) \quad p \in P$$

where q is the assigned flow pattern, p is the vector of control parameters, P is the set of feasible vectors p and $Z(p, q)$ is the system performance index. This is a general model suitable for single-junction signal setting and network coordination. A widely used performance

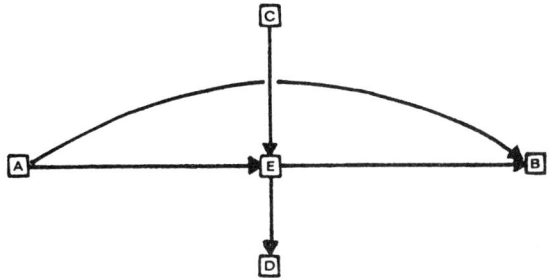

Figure 1
Example network

index is the total user delay (or travel time), expressed by

$$Z(p, q) = D(p, q) = \sum_i d_i(p, q) q_i$$

where $d_i(p, q)$ is the unitary delay at approach i. In the case of a single junction, another performance index, such as capacity factor, can be used.

Considering the simple network shown in Fig. 1, there are five nodes and five links (Dickson 1981). Two origin–destination pairs (A–B) and (C–D) are considered. Junction E is the only signalized junction. The parameters of such a simple problem (i.e., green times at junction E and link flows) can be graphically represented in a two-dimensional plane (p, q). Indeed, flows on paths A–B and A–E–B (i.e., q_{AB} and $q_{AE} = q_{EB}$) are complementary with respect to the demand value between A and B. Greens at approaches of junction E (g_{1E} on link A–E and g_{2E} on link C–E) are complementary with respect to the cycle time, which is assumed to be fixed. Moreover, flow on path C–E–D (i.e., $q_{CE} = q_{ED}$) is constant and equal to the demand value between C and D. Thus, the system has two degrees of freedom; that is, one pair of independent variables is needed to describe completely the problem (e.g., $q = q_{AE}$ and $p = g_{1E}$).

Figure 2 shows the contours of the system objective function $Z(p, q)$ in a plane (p, q), expressed by the total travel time or another system performance index. For value q_0, the function $Z(p, q_0)$ takes the minimum for the value p_0. These pairs (p_0, q_0) define a curve $p = \Gamma(q)$. Thus, the solutions of a traffic signal setting model correspond to the points of this curve, which expresses the optimal value of the regulation parameter for each flow value.

Alternatively, if traffic signal setting is fixed, the link flow pattern is found through a traffic assignment model according to some user-behavior criterion. This relationship can be formally expressed as $q = \phi(p)$. For example, in the case of deterministic symmetric equilibrium assignment with fixed demand, assuming a user behavior according to Wardrop's first principle (see *Traffic Assignment*), this relationship corresponds to the solution of the convex program:

$$\min_{q} I(p, q) \quad q \in Q$$

where Q is the set of feasible flow patterns and $I(p, q)$ is the integral travel cost; that is,

$$I(p, q) = \sum_{i} \int_{0}^{q_i} c_i(p, t) \, dt$$

where $c_i(p, t)$ is the cost–flow function for link i.

Figure 3 shows the contours of the integral travel cost $I(p, q)$. For value p_0, the function $I(p_0, q)$ takes the minimum for the value q_0. These (p_0, q_0) pairs define a curve $q = \phi(p)$. Thus, the solutions of a traffic assignment model correspond to the points of this curve, which expresses the equilibrium flow pattern for each regulation parameter set.

When both signal setting parameters and link flows are not fixed, their combined evaluation is necessary, requiring modelling the interaction between flow pattern and signal setting. This interaction can be seen as a game between two players: users and the traffic agency (Fisk 1984). The goal of the users is the minimization of their individual costs, according to Wardrop's first principle. The traffic agency chooses a control strategy to optimize a system performance index. This situation can be modelled by

$$\min_{p, q} Z(q, p) \quad p \in P; \sim q \in Q$$

From this viewpoint, the curve $p = \Gamma(q)$ represents a relation between equilibrium flow pattern q and signal setting p, and expresses the decision of the traffic agency. The curve $q = \phi(p)$ represents a relation between the same variables, but expresses the result of the users' choices. It is a constraint for the values of

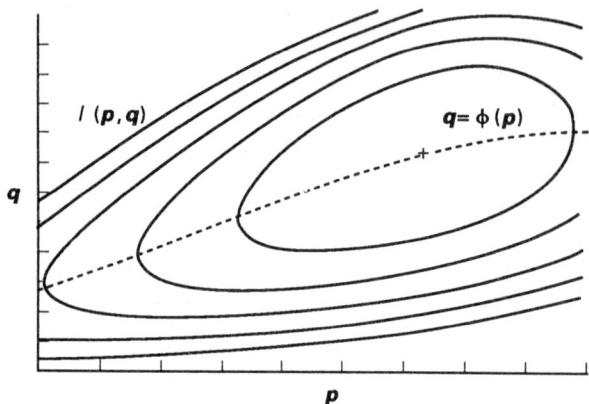

Figure 3
Traffic assignment: contours of function $I(p, q)$ and curve $q = \phi(p)$

variables p and q, and so a feasible pair (p, q) should be on this curve, which expresses the users' optimal points.

Figure 4 shows the contours of the function $Z(p, q)$ and the curve $q = \phi(p)$. The optimal solution is the point (p^*, q^*) of the curve $q = \phi(p)$ with the best value of $Z(p, q)$. A lower bound for the optimal objective function value $Z(p^*, q^*)$ is given by the value $Z(p_1, q_1)$. Generally, the solution (p_1, q_1) is not a feasible solution of the original model built for eqilibrium network traffic signal setting, since the set of user-behavior constraints has been removed. It is located on the curve $q = \theta(p)$, which expresses the system optimal points, at the intersection with the curve $p = \Gamma(q)$. This solution corresponds to an ideal situation where the users' route choices are also defined by the traffic agency to achieve the minimum system cost; that is, the users would

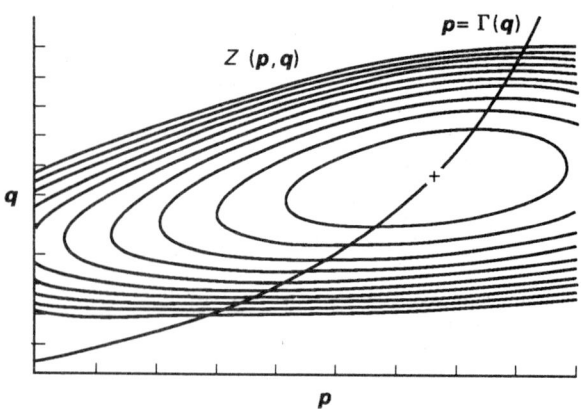

Figure 2
Signal setting: contours of function $Z(p, q)$ and curve $p = \Gamma(q)$

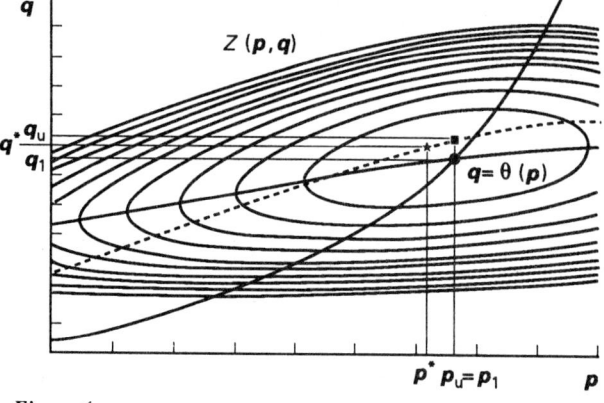

Figure 4
Equilibrium network traffic signal setting: contours of $Z(p, q)$ and curves $q = \phi(p)$, $q = \theta(p)$ and $p = \Gamma(q)$

adopt paths indicated by the traffic agency aimed at system optimization instead of choosing paths minimizing their individual cost, generating the so-called normative flow pattern.

Starting from the solution (p_1, q_1), a feasible solution (p_u, q_u) of the problem and, thus, an upper bound $Z(p_u, q_u)$ for $Z(p^*, q^*)$ can be easily found by setting $p_u = p_1$ and $q_u = \phi(p_1)$.

There are two main approaches to the solution of the equilibrium network traffic signal setting problem. The first is based on global optimization models. The second refers to iterative procedures which successively perform the stages of traffic signal setting and flow equilibrium assignment.

2. Global Optimization Models

These models assume $Z(p, q)$ as an objective function with a set of constraints expressing user behavior. The models thus obtained are continuous network design models (Magnanti and Wong 1984) in which regulation parameters play the role of design variables. These models assume that cycle time, green scheduling and link offsets are known parameters. Therefore, green times are the only design variables.

Thus, these models aim at the optimization of an objective function of flows and green times, where the flow pattern must satisfy a user-behavior criterion such as the widely used Wardrop's first principle. This requirement can be met in two different ways, leading to two-level or one-level optimization models.

In two-level models, the optimization is hierarchically carried out at two levels, assuming green times as the explicit decisional variables at the outer level and link flows as the implicit decisional variables at the inner level. With these definitions, assuming a symmetric equilibrium assignment, a two-level optimization model can be expressed by

$$Z(p^*, q^*) = \min_p Z(p, q^*) \quad p \in P$$

where q^* is such that

$$I(p, q^*) = \min_q I(p, q) \quad q \in Q$$

Hence, for each objective function evaluation at the outer level for a fixed p, the associated flow pattern is computed by performing an equilibrium assignment at the inner level. In this model, the user-behavior criterion is expressed by the optimization of the integral cost function. Hence, the equilibrium network traffic signal setting is performed, minimizing a total cost function with a flow pattern satisfying a user-behavior criterion. This approach appeared in Tan *et al.* (1979) who defined it as a hybrid optimization model, solved via an augmented Lagrangian method. Leblanc and Boyce (1986) proposed a new solution method based on heuristic generalization of techniques for bilevel linear programming to nonlinear problems.

In one-level models, both regulation parameters and link flows are explicit decisional variables. The user behavior is expressed by a set of constraints for which some formulations were proposed (Tan *et al.* 1979, Marcotte 1983). The model proposed by Marcotte adopted a variational inequality formulation for the user behavior constraints. Denoting the set of all flow extremal solutions (all-or-nothing flow patterns) by Q' and the link cost vector by $c(p, q)$, the model can be written as

$$\min Z(p, q) \text{ s.t. } p \in P; q \in Q$$
$$c(p, q) \cdot (q' - q) \geq 0 \quad \forall q' \in Q'$$

Marcotte suggested a solution approach that avoids the explicit enumeration of all extremal solutions in set Q'.

3. Iterative Procedures

An iterative procedure finds a feasible solution of the problem by sequentially solving an equilibrium assignment and a traffic signal setting until two successive flow patterns or signal settings are equal within a specified tolerance:

Compute an initial assignment

Repeat
  Compute traffic signal setting
  Compute equilibrium assignment
Until a convergence criterion is met

If this convergence criterion is met in a finite number of iterations, the final solution, which is a fixed point solution, can be said to be mutually consistent. This means that signal setting generates a set of link costs which determine a flow pattern such that signal setting is optimal.

In the case of the simple network of Fig. 1, the steps of an iterative procedure can be represented in a plane (p, q), between the curves $p = \Gamma(q)$ and $q = \phi(p)$ (see Fig. 5). Starting with an initial flow pattern q_0, the signal setting generates the point (p_1, q_0), with $p_1 = \Gamma(q_0)$. After the delay functions associated with these regulation parameters have been computed, the procedure finds a new flow pattern q_1, corresponding to the point (p_1, q_1), with $q_1 = \phi(p_1)$. Iterating this way, the procedure converges to the mutually consistent point (p_c, q_c), different from the optimal solution (p^*, q^*).

Allsop and Charlesworth (1977) introduced the computation of mutually consistent flow patterns and signal settings for a network of interacting junctions. In their procedure, traffic signal setting is performed using the TRANSYT procedure (see *Road Traffic Control:*

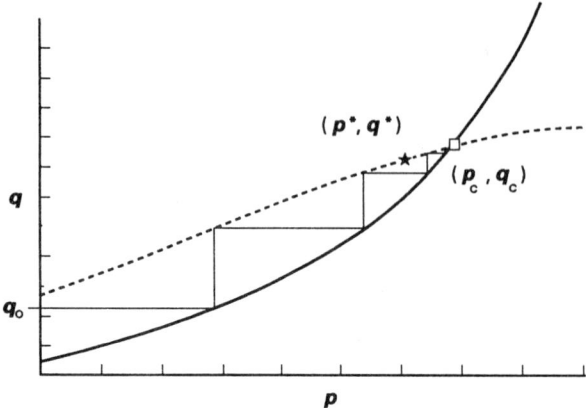

Figure 5
Steps of an iterative procedure for the example network in Fig. 1

TRANSYT and SCOOT). The cost–flow functions are estimated by evaluating delay costs for different values of flow, with the TRANSYT traffic simulation module, and then fitting these points with a polynomial function. The traffic assignment is performed through the deterministic equilibrium assignment (see *Traffic Assignment*).

Gartner *et al.* (1980) proposed two iterative procedures. The first assumes noninteracting junctions. In this way, offsets are not decisional variables and signal setting is performed by Webster's method. In the second, the signal setting stage is performed by using the MITROP method (see *Road Network Control*), which allows the evaluation of all the control variables. For this reason, a great computational effort, due to the solution of a mixed integer linear program, is required. In both cases, traffic assignment refers to the deterministic equilibrium model with fixed demand (see *Traffic Assignment*).

Catarella *et al.* (1986, 1991) proposed an iterative procedure, named ENETS, in which traffic signal setting is performed in two successive steps: green timing and scheduling at each junction, and signal coordination on the network. Green timing and scheduling at a single junction are based on a mixed binary linear program. This model is solved by a problem-oriented algorithm and also allows the green scheduling to be considered as a set of decisional variables (see *Signal Control at Individual Junctions: Phase-Based Approach*). Signal coordination for the whole network is performed by solving a discrete programming model with a total delay minimization objective function. This model is solved through a branch-and-bound algorithm which is efficiently designed for the model formulation (see *Road Network Control*). The traffic signal setting so performed allows the estimation of delay functions for movement links at the junctions. A traffic model, obtained with some simplifications from the TRANSYT traffic model, is used to evaluate delay costs for different values of flow. The resulting points are fitted by a generalized Webster two-term formula, which behaves better than a polynomial formula and is also appealing from a theoretical point of view. The flow assignment stage refers to the equilibrium model with fixed demand, as in other procedures.

It is worthwhile noting that iterative procedures do not necessarily converge to the optimal solution. Moreover, it has been noted that the final solution of an iterative procedure strongly depends on the initial assignment (Allsop and Charlesworth 1977, Tan *et al.* 1979, Marcotte 1983). Dickson (1981) showed that the system total cost, assumed as a performance index, can increase during the procedure, thus leading to a non-optimal solution.

Smith *et al.* (1987) analyzed the existence, uniqueness and stability of the solutions obtained by commonly used control strategies, such as total delay minimization and Webster's equisaturation method, assuming noninteracting adjacent junctions. They showed that these strategies can lead to multiple and/or nonstable solutions. Moreover, they analyzed the strategy proposed by Smith (1981a, b) which had been built up to avoid these conditions. It aims at the minimization of the objective function $S(p, q) = \Sigma_i d_i(p, q) s_i$ where unitary delay is weighted by saturation flow s_i instead of flow q_i.

Iterative procedures can be used also to compute the lower bound described in Sect. 1 for the value of the optimal solution. It can be found by solving a model in which constraints expressing the user behavior have been dropped. The relaxed model presents the same objective function both in signal setting and traffic assignment; thus, it can easily be solved by using an iterative procedure in which the assignment stage is performed to minimize the total instead of the integral cost. This corresponds to performing an iterative procedure between the curve $p = \Gamma(q)$ and $q = \theta(p)$ instead of $q = \phi(p)$ (see Fig. 6).

The upper bound described in Sect. 1 can be obtained by finding an equilibrium assignment q_u in which the cost–flow functions are derived from the signal setting p_1, relative to the previously defined lower bound. Sometimes this upper bound can be better than the mutually consistent solution of an iterative procedure and, thus, it is a good evaluation of the optimal solution. Alternatively, it can be used as a starting solution to avoid uncertainty about the starting solution of an iterative procedure. All of these defined solutions are depicted in Fig. 6.

4. Limitations and Future Developments

As stated in Sect. 2, global optimization models assume that green times are the only decisional variables. It is worth underlining that this approach to the problem

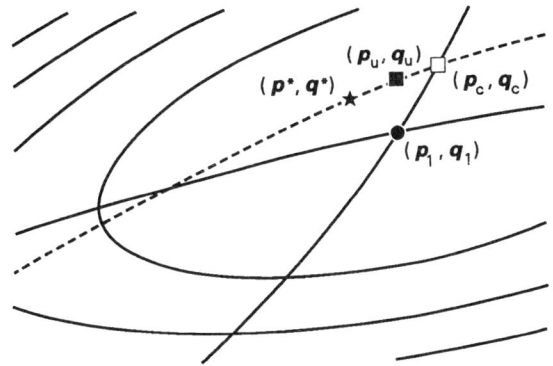

Figure 6
Lower and upper bounds of the optimal solution (enlargement of Fig. 4)

assumes that there is no interaction between adjacent junctions, thereby omitting the influence of signal coordination. This is a strongly limitative assumption for an urban traffic network where signal coordination plays a very important role in link cost definition and can greatly improve system performance.

Iterative procedures, however, allow a greater number of regulation parameters to be decisional variables, including green scheduling and offsets between adjacent junctions. They also allow the solution of medium-scale problems with an acceptable computational effort. For these reasons, real problems are usually solved via iterative procedures, in spite of their known limitations in finding the optimal solution.

Furthermore, it is worth noting the modular structure of an iterative procedure. This feature allows the use of methods already developed for solving the different problems comprising the general one, updating a single component if a new method is proposed for its solution or even changing a single model if a new approach is proposed for it.

From this viewpoint, research efforts can be devoted to finding more sophisticated delay-offset functions in which delay on a link generally depends on control parameters and the flow pattern of the whole network. In this case, the need for an asymmetric equilibrium model in the assignment stage arises (see *Traffic Assignment*).

Previous considerations indicate that equilibrium network traffic signal setting is an open research field both from the theoretical and the practical viewpoint. Further theoretical research work is needed to remove some of the simplifying assumptions made in global optimization methods, thus extending their field of application. On the practical side, further experimentation is needed to test the performance of iterative procedures so as to investigate their mathematical behavior in depth.

See also: Signal Control and Traffic Assignment; Traffic Assignment; Traffic Management Systems

Bibliography

Allsop R E 1974 Some possibilities for using traffic control to influence trip destination and route choice. *Proc. 6th Int. Symp. Transportation and Traffic Theory.* Elsevier, Amsterdam, pp. 345–74

Allsop R E, Charlesworth J A 1977 Traffic in a signal controlled road network: An example of different signal timings inducing different routing. *Traffic Eng. Control* **18**, 262–4

Cantarella G E, Cosentino M, Improta G, Sforza A 1986 Equilibrium network control system design. *Proc. 5th IFAC/IFIP/IFORS Int. Conf. Control in Transportation Systems.* Pergamon, Oxford

Cantarella G E, Improta A, Sforza A 1991 A procedure for equilibrium network traffic signal setting. *Transp. Res.* **A 25**

Cantarella G E, Sforza A 1987 Methods for equilibrium network traffic signal setting. In: Odoni A R, Bianco L, Szegö G (eds.) *Flow Control on Congested Networks.* Springer, Berlin

Dickson T J 1981 A note on traffic assignment and signal timings in a signal controlled road network. *Transp. Res. B* **15**, 267–71

Fisk C S 1984 Game theory and transportation systems modelling. *Transp. Res. B* **18**, 301–13

Gartner N H, Gershwin S B, Little J D C, Ross P 1980 Pilot study of computer-based urban traffic management. *Transp. Res. B* **14**, 203–17

Leblanc L J, Boyce D E 1986 A bilevel programming algorithm for exact solution of the network design problem with user optimal flows. *Transp. Res. B* **20**, 259–65

Magnanti T, Wong R 1984 Network design and transportation planning: Models and algorithms. *Transp. Sci.* **18**, 1–55

Marcotte P 1983 Network optimization with continuous control parameters. *Transp. Sci.* **17**, 181–97

Smith M J 1981a The existence of an equilibrium solution of the traffic assignment problem when there are junction interactions. *Transp. Res. B* **15**, 443–51

Smith M J 1981b Properties of a traffic control policy which ensure the existence of a traffic equilibrium consistent with the policy. *Transp. Res. B* **15**, 453–62

Smith M J 1985 Traffic signals in assignment. *Transp. Res. B* **19**, 155–60

Smith M J, Van Vuren T, Heydecker B J, Van Vliet D 1987 The interaction between signal control policies and route choice. In: Gartner N H, Wilson N H M (eds.) *Transportation and Traffic Theory.* Elsevier, Amsterdam

Tan H, Gershwin S B, Athans M 1979 Hybrid optimization in urban traffic networks. Massachusetts Institute of Technology report No. DOT-TSC-RSP-79-7

G. E. Cantarella
[Università di Reggio Calabria,
Reggio Calabria, Italy]

G. Improta
[Università di Napoli, Naples, Italy]

A. Sforza
[Università di Salerno, Fisciano, Italy]

Road Networks: Dynamic Equilibrium Models

Dynamic peak period models are concerned with how commuters adjust their travel decisions concerning route and departure time from day to day. This topic is very important given the urgent need to have operational models to describe how drivers will react to various policies (e.g., capacity expansion, changes in road signalling, information provision). However, a relatively small number of dynamic models can be found in the literature.

The problem can be described in a more general and formal way. There is a repetitive event which occurs at some exogenously given frequency. During an event, each individual of a given group has to make one or a series of decisions. The decisions of these individuals are interrelated, because the outcome associated with one individual's decision depends on the decisions made by the other individuals. The individuals can partially observe each other, partially forecast each other's decisions and be partially informed about the values of some exogenously given variables that affect the outcomes of the decisions they make. After each event, individuals are informed about the outcome of their own choices and will then use this information, as well as the information gathered about the system, to make a decision for the next event. There are two temporal processes. The first process occurs during each given event and the second from one event to the next. Each event is not instantaneous but spread over time, so that the individuals have the opportunity, at least partially, to observe the decisions of other individuals and the current state of the system before they make their own decisions. Moreover, an interval of time elapses between one event and the next so that each individual has the ability to process, at least partially, the various pieces of information gathered personally before the next event; learning can also take place so that the individuals may improve their heuristics event after event.

Examples of such situations are numerous. As a first example, consider recurrent professional meetings (e.g., the Transportation Research Board meeting each year in Washington). In this case, the decisions the individuals make are related to the papers they will present, to the topics they will discuss, to the people they will meet and so on.

Commuting is a second example. In this case, the drivers have to make daily decisions related to mode, departure time and route. The benefits of selecting a given route, for example, are obviously a function of the number of drivers selecting this route. This number is, in general, only partially known by the drivers by the time they make their decisions—indeed one route can become fairly congested after the individual has selected it. The outcome of each driver's decisions (e.g., the fact that the driver has experienced much or little congestion) will influence the decision on the next day. Various sources of uncertainty could make drivers' decisions more complicated.

Cocktail parties provide a third possible example. In this case, one of the decisions individuals have to make can be the quantity of alcohol consumed. Clearly, the individual's decisions are not independent of each other: each individual regulates personal consumption as a function of the instantaneous or cumulative individual and aggregate consumption. The benefits/costs are partially experienced instantaneously (the pleasure of drinking), partially experienced the next day (the headache) and possibly experienced in the long run (addiction). In addition, the benefits/costs associated with one event will influence the behavior on the next event.

1. The Dynamic Setting

The commuting problem in which drivers have to select their departure times and routes each morning will be considered. It will be assumed that each commuter selects the alternative that minimizes personal total cost. This cost is an increasing function of the travel time and the schedule delay, defined as the difference (in absolute value) between the actual and desired arrival times.

The reasons for the day-to-day adjustment process are explained by uncertainties that occur at both the system and the individual levels. If the traffic conditions were constant from day to day, there would be no need to develop a model that describes the day-to-day dynamics. However, this is not the case. Minor changes occur each day in the traffic networks. The commuters receive information (through personal experience or other sources) and adjust as best they can. These variations constantly modify the traffic conditions and introduce a constant source of uncertainty which implies that commuters' decisions are *a priori* very complicated.

1.1 Uncertainty at the System Level

Transportation systems are not in a constant state over time, but are subject to fluctuations from various sources. These fluctuations may occur because the road capacity changes from day to day; they are, in general, correlated over time. Poor visibility, precipitation and other adverse weather conditions reduce the safe driving speed and, hence, the effective flow capacity. Large-scale changes in work schedules, gasoline price changes and other shocks may create other sources of uncertainty.

1.2 Uncertainty at the Individual Level

The first and most obvious source of uncertainty at the individual level corresponds to the fact that the total demand for the transportation system may vary from day to day (elastic demand). The volume of traffic can also fluctuate severely if transit strikes or capacity shocks cause drivers to divert into the part of the network that is being studied. Another main point is that commuters' decisions to use one route or another depend on their perceptions of what the other commuters will do. This is a very difficult problem given that our expectations of what the other drivers will do depend on what we believe the other drivers' expectations of our decisions are and so on.

1.3 The Quality of Information

In terms of the quality of information available to the drivers, two extreme situations can be identified. In the first case (zero information), the drivers have no information about the traffic conditions prior to their departure time and route choice decisions; consequently, if they do not learn, they will not alter their decisions from day to day and the corresponding model is static. The deterministic user equilibrium (DUE) concept used in deterministic environments should be replaced in this case by the stochastic user equilibrium (SUE) concept. At the SUE, each commuter going from a given origin to a given destination cannot reduce personal expected total cost by changing departure time and/or route given all other commuters' decisions of departure time and route (see *Traffic Assignment, Stochastic*). The second extreme case (full information) corresponds to the situation where the individuals are perfectly informed about traffic conditions each day. In this case, the user faces a deterministic—but day specific—situation. The intermediate cases correspond to situations where individuals are partially informed about traffic conditions.

1.4 Motivation to Develop a Dynamic Model

From a strictly theoretical point of view, there is nothing to suggest the development of a dynamic model. The uncertainty about traffic conditions is a necessary but not a sufficient condition to justify the use of a dynamic framework. Indeed, for any of the situations discussed previously, the problem could be described within a static framework where each commuter is perfectly informed about the "rules of the game" (i.e., the exact performance law and the behavior of the other commuters) and is formally participating in an n-person game under uncertainty. This approach is very attractive from a theoretical point of view because it proposes a well-defined procedure which can be seen as a useful limiting case (Arnott *et al.* 1991).

There are three reasons why such a static procedure is not satisfactory from a behavioral point of view.

First, in the situations where individuals are not fully informed about the rules of the game, commuters will typically learn about the behavior of the other commuters and the conditions of the system from day to day, so that the corresponding model should obviously be dynamic. However, these rules are also changing over time so that perfect knowledge probably never occurs.

Second, even if the individual had this perfect knowledge of the rules of the game, the solution to the commuting problem remains very complicated from a mathematical point of view and would require the use of a large computer to be solved numerically. This procedure is not sound from a behavioral point of view because the cognitive costs to find the best alternative are excessively high and because the amount of information to process is unreasonably large (Bettman 1979). Extensive work in psychology has shown that in these situations individuals resort to heuristic methods (Kahneman *et al.* 1982). Indeed, given that the potential cost savings in finding a better alternative are of the same order of magnitude as the cognitive cost necessary to find it, the commuters will not, in practice, resort to a complicated optimization procedure but will follow simple rules. In our context, the method of trial and error from day to day is a reasonable heuristic. Therefore, a more satisfactory approach from the behavioral point of view, would involve a dynamic rather than a static model. It is more likely, with imperfect information, that each driver will try by trial and error to identify the best strategy to pursue. Each day some individuals may decide to stick to their previous choices. Other individuals may decide to review and modify their choices if they believe that by doing so they will reduce their cost. If the environment is constantly changing over time, the drivers will constantly modify their decisions as well by adapting to this evolving environment. If, however, the environment remains constant, it is expected that, after a sufficient period of time, the system will evolve towards a stationary distribution (characterized by the fact that the number of commuters selecting the various choices remains constant over time).

Finally, a dynamic model can be useful in situations in which a large change (predicted or not) in the transportation network has occurred, such as a capacity expansion or the construction of a new road. In this case, a dynamic model is useful to describe how the commuters readjust their travel decisions and collectively react over time.

2. Departure Time Model with Homogeneous Drivers

In the departure time model with homogeneous drivers, it is assumed that departure time is the only choice variable (i.e., there is only one route between the origin and the destination). The time-of-use choice involves an essential trade-off between travel time and schedule delay. Either commuters can use the facility at a popular time and experience large travel times and small schedule delays, or they can use the facility at a less attractive time, but experience less crowded conditions. This trade-off was first analyzed by Vickrey (1969) in the context of the morning rush hour.

In the simplest version of this model, N identical

commuters travel on a single road connecting a given origin (home) to a given destination (workplace) each morning. Along this road there is a bottleneck of finite capacity s. Commuters have to decide their departure times in order to minimize costs. This cost is assumed to be a function of travel time and schedule delay. DUE is obtained when no driver can modify departure time to decrease the cost incurred.

2.1 Model Specification

The performance of the transportation system is described by a simple deterministic queuing model. Travel time along the road is equal to 0 (without loss of generality) when there is no congestion. Congestion occurs when the arrival rate at the entrance of the bottleneck is larger than its capacity s. Assuming congestion starts at t_0 and still continues at time t, the queue length at t is

$$D(t) = \int_{t_0}^{t} r(u)\,du - s(t - t_0) \qquad (1)$$

where $r(u)$ denotes the departure rate of vehicles at time u from the origin (or the arrival rate at the entrance of the bottleneck). When there is no congestion, $D(t) = 0$. The travel time

$$T(t) = D(t)/s \qquad (2)$$

There are N commuters who travel each morning. The commuters' behavior is specified by assuming that the cost incurred by a driver departing at t is a linear function of travel time and schedule delays:

$$C(t) = \alpha T(t) + \beta \max\{[t_a - t - T(t)], 0\} + \gamma \max\{[t + T(t) - t_a], 0\} \qquad (3)$$

where t_a represents the official work start time.

2.2 Results

In equilibrium, all commuters incur the same level of cost C. The equilibrium distribution of departure rate $r_e(t)$ can be computed analytically; it is given by

$$r_e = s[1 + \beta/(\alpha - \beta)] \quad \text{for early arrivals} \qquad (4a)$$
$$r_e = s[1 - \gamma/(\alpha + \gamma)] \quad \text{for late arrivals} \qquad (4b)$$

The solution is represented in Fig. 1. Note that the equilibrium solution does not specify when a particular individual leaves home, but it specifies the distribution of departures. The aggregate total cost $TC_e = NC$, is given by

$$TC_e = N^2 \beta \gamma / [(\beta + \gamma)s] \qquad (5a)$$

In the social (or system) optimum, there is no

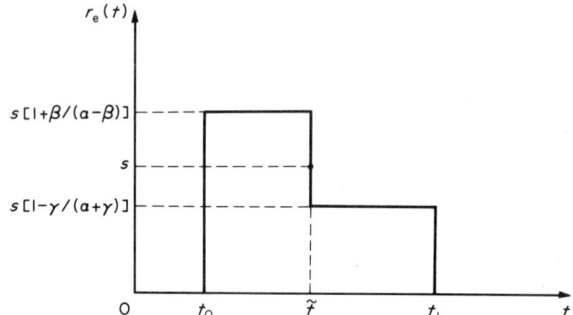

Figure 1
Equilibrium distribution of departure time $r_e(t)$: t_0 and t_1 are the beginning and end of congestion, and $\tilde{t}$ is the departure time for arrival on time

congestion and the total cost is reduced to

$$TC_0 = N^2 \beta \gamma / [2(\beta + \gamma)s] \qquad (5b)$$

The parameters of this model have been estimated by Small (1982).

3. Departure Time Model with Heterogeneous Drivers

The situation where N commuters are identical was discussed in Sect. 2. This is inconsistent with the fact that the value of travel time rises with income, that individuals have different workplaces and household constraints in choosing when to travel, and have idiosyncratic preferences for tavel time and schedule delay. A given individual k making a specific decision will perceive a cost associated with departure time t equal to

$$C^*(k, t) = C(t) + e(k, t) \qquad (6)$$

where $e(k, t)$ is a variable specific to the individual k and the alternative t. The commuter will select the departure time $t'(k)$ that minimizes the perceived cost:

$$t'(k) = \operatorname{argmin}\{C^*(k, t)\} \quad t \in [0, 1] \qquad (7)$$

where, without loss of generality, the set of possible values of t has been normalized to $[0, 1]$. The values of $e(k, t)$ are not observable so that the best that can be achieved is to model these variables as the realizations of independently distributed (i.d.) random variables, denoted by $\varepsilon(t)$. It is assumed that these random variables are absolutely continuous. Now, the model derived predicts the probability $P(t')$ that an individual $k (1 \leq k \leq N)$, selected randomly from a population of N commuters, chooses the departure time t'. These probabilities are given by

$$P(t') = \operatorname{prob}\{C^*(k, t') = \min C^*(k, t)\} \quad t \in [0, 1] \qquad (8)$$

In order to derive an explicit form for $P(t')$, it is necessary to make some supplementary assumptions on the random variables $\varepsilon(t)$. It is assumed that $\varepsilon(t)$ are independently identically distributed (i.i.d.) (across choices and individuals) according to the double exponential distribution

$$F(x) = \text{prob}\{\varepsilon(t) < x\} = \exp[-\exp(-x/\sigma)], \quad (9)$$

where σ represents the standard deviation (up to a multiplicative constant) of $\varepsilon(t)$. In this case, however, the set of choices is discrete (e.g., choices may correspond to a 5 min interval). The time intervals are labelled by $i = 1, \ldots, n$. For this specification of $\varepsilon(t)$, the choice probabilities are given by the multinomial logit model (e.g., McFadden 1974):

$$P(i) = \frac{\exp[-C(i)/\sigma]}{\sum_{j=1}^{n} \exp[-C(j)/\sigma]} \quad (10)$$

The system in Eqn. (10) is a nonlinear system of n equations with n unknowns. de Palma et al. (1983) have shown that, for the continuous approximation ($n \to \infty$), this system has a unique solution. The equilibrium departure time distribution is single peaked and there is at most one congestion period. Moreover, this solution $P(i)N$ converges, as $\sigma \to 0$, to the solution of the homogeneous case (see Eqn. (4)).

Other probabilistic models can be derived by making alternative assumptions on the random variables $\varepsilon(t)$. The best-known generalization of the logit model in the context of departure time models has been developed by Small (1987). The discrete choice model for ordered alternatives, introduced and estimated by Small, is an extension of the multinomial logit model which allows the description of a situation where the alternatives are ordered in a natural way (the departure times). This model no longer assumes independence but takes into account the fact that close alternatives are likely to be more correlated than alternatives that are distant.

4. Day-To-Day Adjustment Process for Departure Time Models

The model presented in Sect. 3 is static. It describes the behavior of a population of drivers and does not account for the day-to-day dynamic processes. However, for various distinct reasons discussed in Sect. 1, it is expected that the departure time distributions may vary from day to day. For an early work on the dynamics of peak period congestion, see Gaver (1967).

4.1 Specification of Framework to Develop a Dynamic Model

The dynamic models are *a priori* complex because the state of the system on day w depends on the state of the system during the M previous days. The $n \times 1$ column vector that represents the state of the system on day w is denoted by $X(w) = [X_1(w), \ldots, X_n(w)]$, $X_i(w)$ represents the number of commuters who have selected alternative i on day $w (i = 1, \ldots, n)$ and n is the number of alternatives available to the drivers (e.g., the total number of routes or the total number of feasible departure time intervals). The state of the system corresponds to the endogenous information acquired by the various commuters during a given day. It is also assumed that there is exogenous information $I(w)$ which is not the outcome of commuters' decisions (e.g., weather conditions, road damage, road signals).

These two sets of information are *a priori* extremely large. First, it is assumed that the commuters only have access to exogenous information concerning the current day. Second, it is assumed that all commuters remember perfectly the state of the system during M previous days and cannot remember it beyond M days. Therefore, the information available on day w will be limited to the number of individuals $x(w - j)$ selecting alternatives $1, \ldots, n$ on days $w - j (j = 1, \ldots, M)$ and to the exogenous information set $I(w)$ available on day w.

4.2 Dynamic Models of Departure Time: Basic Approach

To begin, the case where commuters base their decisions on the information gathered the previous day ($M = 1$) will be considered; that is, on day w the drivers base their decisions on the state of the system $X(w - 1)$ and on the exogenous information $I(w)$. Let $R_{ij}[X(w - 1), I(w)]$ denote the probability that a commuter having selected alternative i on day $w - 1$ decides to select alternative j on day $w (i, j = 1, \ldots, n, \; i \neq j)$. Moreover, it is defined that

$$R_{ii}[X(w-1), I(w)] = 1 - \sum_{j=1}^{n} R_{ij}[X(w-1), I(w)]$$
$$i = 1, \ldots, n; j \neq i \quad (11)$$

The evolution of the number of individuals selecting choice i on day w is given by

$$X_i(w) = \sum_{j=1}^{n} X_j(w-1) R_{ji}[X(w-1), I(w)] \quad i = 1, \ldots, n \quad (12)$$

This system of nonlinear difference equations is, in general, very difficult to solve. It is assumed that $I(w)$ is constant over time (and equal to I). The stationary state X^{st} of the system in Eqn. (12) is given by

$$X_i^{\text{st}} = \sum_{j=1}^{n} X_j^{\text{st}} R_{ji}[X^{\text{st}}, I] \quad i = 1, \ldots, n \quad (13)$$

If the transition probabilities are continuous functions of their argument X, each equation of the system in Eqn. (13) will correspond to a single-value continuous function from the simplex $\Sigma X_i = N$ to itself. Using Brouwer's theorem, it is assured that there exists a solution X^{st} to Eqn. (13). The formal proof of uniqueness of equilibrium is more difficult and has not yet been derived.

This model will now be specialized. Some individuals may find themselves satisfied and select the same choice on two consecutive days. Others may, for various reasons, decide to reconsider their choices. Reviewing may happen, for example, after a particularly bad experience or bad mood. The reviewing stage will be modelled as a stochastic process with Ω_i as the time-independent probability that a commuter having selecting alternative i reviews that choice on a given day. (Ben-Akiva and de Palma (1986) gives a computation of the reviewing rate based on an optimization procedure at the individual level.) An individual who decides to review choice i on day w then has a conditional probability denoted by $R_{j|i}[(w-1), I(w)]$ to select the alternative $j (j=1, \ldots, n)$ on day w; that is, it is assumed that commuters modify choice, from day to day, through two successive steps. First, during $[w-1, w]$, the present alternative i with a probability Ω_i is reviewed. Second, the alternative j with probability $R_{j|i}[(w-1), I(w)] (j=1, \ldots, n)$ is selected. Thus

$$R_{ij}[X(w-1), I(w)]$$
$$= \Omega_i R_{j|i}[(w-1), I(w)] \quad i, j, \ldots, n; j \neq i \quad (14)$$

The model can be simplified further by assuming that the alternative chosen by the commuter on day $w-1$ does not influence the alternative selected on day w, so that

$$R_{j|i}[X(w-1), I(w)]$$
$$= R_j[X(w-1), I(w)] \quad i, j = 1, \ldots, n; j \neq i \quad (15)$$

This hypothesis implies that there are no costs in shifting departure time and/or no intertemporal correlations in individual preferences from day to day. Note, however, that the commuters still have a strong memory of their past choices if the probability of reviewing is small enough. From now on, it is assumed that the environment is stable (i.e., $I(w)$ remains constant from day to day and equal to I). In such a case, the model would describe the day-to-day adjustment in a system after a change had occurred (e.g., road construction, capacity expansion or change in the official start hour). The specific functional forms of the conditional probabilities $R_{j|i}[(w-1), I](j=1, \ldots, n; j \neq i)$ remain to be specified. One simple way to proceed is to assume that, once the decision to review the choice has been made, the commuter is facing a static model similar to that described in Sect. 3; that is, a suitable parameterization of the conditional probabilities (see Eqn. (10)) is

$$R_i[X(w-1), I] = \frac{\exp[-C(i, w-1)/\sigma]}{\sum_{j=1}^{n} \exp[-C(j, w-1)/\sigma]} \quad (16)$$

The variable $C(i, w-1)$ represents the estimate that the individual makes about the cost on day w, based on personal knowledge of the cost on day $w-1$. Here, it is assumed that the static expectation which sets the average expected cost on day w is equal to the actual average cost on day $w-1$. The difference equation that describes the individual choices reduces to

$$X_i(w) = X_i(w-1)[1 - \Omega_i]$$
$$+ \frac{\exp[-C(i, w-1)/\sigma]}{\sum_{k=1}^{n} \exp[-C(k, w-1)/\sigma]} \sum_{j=1}^{n} X_j(w-1)\Omega_j$$
$$i = 1, \ldots, n \quad (17)$$

Note that when the reviewing rates are independent of the current alternative ($\Omega_i = \Omega, i = 1, \ldots, n$), then the stationary solution of Eqn. (17) can be written

$$X_i^{st}/N = \frac{\exp[-C^{st}(i)/\sigma]}{\sum_{k=1}^{n} \exp[-C^{st}(j)/\sigma]} \quad i = 1, \ldots, n \quad (18)$$

where $C^{st}(i)$ represents the cost associated with alternative i when the system is at the stationary state. The system in Eqn. (18) is exactly the same as the system in Eqn. (10), but now $P(i)$ is replaced by X_i^{st}/N. Therefore, the difference equation (Eqn. (17)) has a unique stationary solution in the case of a single bottleneck. It can be shown, moreover, that the system converges, for Ω small enough, to the stationary solution.

When the probability of reviewing depends on the alternative chosen and/or when the conditional probabilities depend on the initial choice selected, it is no longer true that the stationary solution coincides with the static one provided by Eqn. (18). Thus, in this case, the stationary state keeps some memory of the adjustment process that has led the system to it. de Palma and Lefèvre (1983) give different illustrations of this.

The day-to-day adjustment is illustrated by Fig. 2. The values of the parameters chosen are $\alpha = 6.4, \beta = 3.9, \gamma = 15.2, \Omega = 0.1, N = 10\,000$ vehicles and $\sigma = 1$ (Small 1982). In this simulation, it is assumed that commuters incur no penalty cost if their arrival times at the destination lie within the time interval [7:50 am, 8:50 am]. The initial state of this simulation is the stationary state of another simulation where $s = 4000$ vehicles per hour. On day 0, the capacity of the route is expanded to 6000 vehicles per hour. The individuals do not know this, but discover it by experiencing unexpected levels of congestion. Each day a fraction of the drivers modify their departure times in order to reduce their costs; it appears that after about 10 d of iterations the distribution of departure rates reaches its new stationary state.

Although this is certainly not the optimal procedure to reach the new stationary configuration, it is, however, interesting to note that such a simple procedure, where commuters do not anticipate the future (given that they assume that the traffic conditions on one day will be the same as on the previous day) and do not select the least-cost alternative (once they have decided to review their choices), converges reasonably quickly to the stationary solution. Note, also, that the difference equation (Eqn. (15)) will not converge when the reviewing rate Ω is too large or when the value of σ is too small.

4.3 Alternative Studies

Dynamic models of departure time have not been estimated empirically. Indeed, the data necessary to estimate day-to-day adjustment processes is very difficult and very costly to obtain so that Mahmassani and Chang (1986) have preferred to conduct microsimulations where they can compute the day-to-day adjustment patterns of hypothetical commuters. The users involved in the simulation make daily decisions concerning departure time and route choice according to some exogenously given rules. A special purpose traffic simulation program (e.g., Chang et al. 1985) models the resulting congestion levels and provides the actual travel time to the users. With this information the users select next iteration (day) travel time and route choice alternatives. Such procedures allow comparison of various day-to-day adjustment behaviors.

The simulation results were derived for a commuting corridor with a common destination which corresponds to a central business district (CBD). Two types of behavior were observed: convergence towards an equilibrium and lack of convergence (with, in particular, the occurrence of oscillatory behavior).

The mechanism advocated by Mahmassani and Chang is based on the concept of bounded rationality (Simon 1955, Day 1975). These authors assume that the drivers have a specific indifference interval that corresponds to the range of acceptable arrival times at their destination. If the actual arrival time is positioned in this interval, the user is assumed to have reached a desired satisfaction level and will not modify the subsequent choice (as long as the arrival time of the user still belongs to the indifference interval). Otherwise, a departure time equal to the average desired arrival time minus the expected travel time will be selected. Each day (each iteration) the commuters are assumed to readjust their perceived travel times according to some behavioral process that takes into account the experienced travel time and schedule delay (if any). In these adjustment processes the commuters can base their decisions on the travel time experienced on the last day they travelled (the case treated previously) or they can use the information accumulated over the previous days (see Sect. 5). The equilibrium provided by the model specified and that given by Vickrey (1969) are comparable to some extent. Here, however, the official work start time of Vickrey's model is replaced by an interval of time (the indifference interval) and β and γ are infinitely large given that no equilibrium will occur until the individuals are arriving within a given time interval. Note that this time interval is exogenously given in the work mentioned previously; it remains to be explored under what conditions the individual may interactively readjust the magnitude of this interval.

Another approach based on laboratory experiments was performed by Tong et al. (1987). The demand side consists of individuals participating in the experiments and the supply side is described using macroparticle traffic simulation models. The behavior of the participants was observed during several consecutive iterations and the parameters of various alternative behavioral models were estimated. Such experiments are important because they provide a better understanding of commuters' adjustment behaviors. In particular, Tong et al. were able to show that the bounded rationality concept is plausible in commuting behavior. Moreover, they could justify the assumption made by Ben-Akiva et al. (1984) that predicted travel time on day w is mainly based on the travel time experienced on day $w-1$. The precise way the predicted travel time is

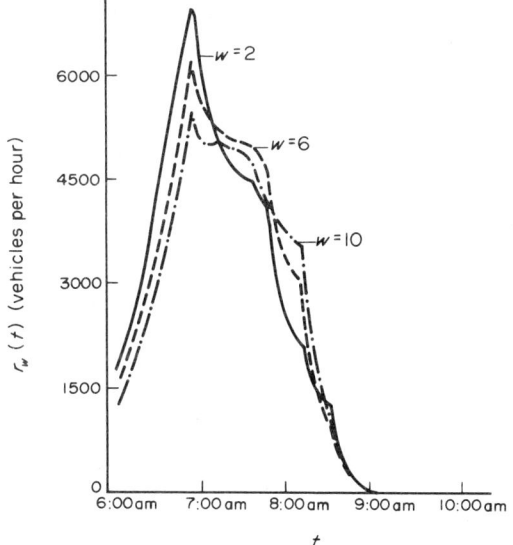

Figure 2
Transient distributions of departure time $r_w(t)$ for a 50% increase in capacity on day $w = 0$

computed is, however, different in the case of Tong *et al*. These authors argue that individuals base their predictions for the day w not only on the experienced travel time on day $w-1$ but also on the magnitude of the unexpected schedule delay experienced on day $w-1$. This corresponds to some type of self-correcting mechanism.

There is no doubt that simulation approaches and laboratory experiments correspond to a useful and necessary step. However, these dynamic models should eventually be calibrated on real data. This is an urgent issue given the growing interest in understanding the way commuters react over time. In particular, this question is very important in developing a sound model for forecasting the impact of information on road users.

5. Dynamic Models of Route Choice

The description of route choice models is mainly based on Horowitz (1984). For simplicity the models described only address the choice between two routes. The departure time choice treated in Sect. 4 is not discussed.

Horowitz considers three types of information gathering processes. In his model, the commuters base their decisions on the information they have gathered during several previous days. These processes express how individuals acquire information through learning and through personal experience in various contexts.

The first model he considers is

$$C_{iw} = \sum_{k=1}^{w} a_k C_{ik} + \varepsilon_{ik} \qquad (19)$$

where C_{ik} denotes the actual (measured) travel cost on route i on day k and C_{iw} represents the predicted travel cost on day w. The weights a_k are positive (or null) parameters which express how the past experiences influence the predictions and ε_{ik} is a random variable. Two limiting cases are of special interest. If the weights a_k are increasing with time, it implies that the individual places more value on recent rather than past information: this expresses the forgetting phenomenon. If, however, the weights a_k are decreasing with w, the users have developed some experience over time which is reinforced as time elapses.

An alternative formulation proposed by Horowitz assumes that the predicted travel cost C_{iw} is a weighted average of the predicted travel costs. Therefore

$$C_{iw} = \sum_{k=1}^{w} w_k (C_{ik} + \varepsilon_{ik}) \qquad (20)$$

The third model considered explicitly describes the drivers' information acquisition and assumes that the drivers can acquire information about road conditions only through personal experience. In this case, information will be necessarily individual specific, given that, even in identical conditions, individuals make different choices. Obviously, other situations can be considered where commuters can benefit at the same time from personal experience and from some exogenous source of information (through word of mouth and broadcasting).

Moreover, the model is completely defined if the traffic laws, as well as the individual choice rule (i.e., the function that predicts the number of users selecting each route as a function of the predicted costs), are defined.

The corresponding dynamic models correspond to nonlinear discrete-time equations. Conventional stability analysis and qualitative analysis of difference equations can be used to study their behavior. The details of these techniques are beyond the scope of this article but can be found, for example, in work by Gandolfo (1980) or Takayama (1985). Basically, these methods specify the conditions under which the system converges towards a stationary state, diverges or oscillates.

It is difficult to draw general conclusions based on this exploratory study. Convergence towards a stationary state when the probabilities of reviewing (see Eqn. (14)) are sufficiently small can be conjectured. This is because the major reason for the oscillatory behavior is that individuals make erroneous expectations. Therefore, if only a few individuals alternate route from day to day, the travel cost will tend to vary slowly as a function of time and, as a result, the predictions made by the individuals will tend to be more accurate. Note that the stationary state does not necessarily coincide with the equilibrium state given by the Wardrop principle (defined by equal travel cost on the two routes). This is because the behavioral model of route choice does not necessarily predict that the smallest cost route is selected by all individuals. Indeed, if this were the case, the errors in prediction would be more likely to modify stability. In general, when the system converges toward an equilibrium, its value depends on the prediction rule used by the individuals.

A variety of very simple deterministic adjustment rules are likely, however, to lead the system to equilibrium. For example, it may be assumed that the travellers who change their route necessarily select the cheapest route. For another adjustment process, the travellers who select another route only succeed in avoiding the worst route (the most costly route). Chu (1989) has shown for the two-route case that these processes lead to the Wardrop equilibrium. More generally, Smith (1984) has shown that, if the drivers only change routes in order to reduce their costs, the system will converge towards a Wardrop equilibrium. The proof presented by Smith is very general because it applies to a network; it is based on Liapunov's method, much used in economics and in stability theory.

6. Future Developments

The development of behavioral dynamic models in transportation is in its early stages. Although this article has focused on deterministic departure time or route choice models, combined models treating the departure time and route choices simultaneously have been studied by, among others, Ben-Akiva et al. (1986). Moreover, the stochastic approach cannot be ignored. It was initially developed by Alpha and Minh (1979); more recently, stochastic models have been studied from a theoretical and computational point of view by Cascetta (1989), and Cascetta and Cantarella (1989). Stochastic departure and route choice models are difficult to analyze; however, they may provide an adequate framework for studying the impact of information in an uncertain environment and, therefore, they may play an important role in the future.

Most of the literature uses the Markovian assumption, which assumes that the behavior of the users on day w can be predicted by the simple knowledge of their behavior on day $w-1$. When this hypothesis cannot be verified, a whole range of new models arises. It is difficult to select the best possible model and more research efforts need to be devoted to this problem.

The dynamic approach is justified by recognizing that individuals are not perfectly informed and do not have a perfect ability to choose. Indeed, in the present research, the behaviors of the individuals are exogenously given; the choice of the heuristic rules (or the quality of the heuristic rules) should be determined endogenously as in de Palma and Papageorgiou (1989). In their approach, the cognitive effort to improve heuristics and the time (and money) to acquire more information are modelled explicitly so that the quality of the heuristic rules can be described endogenously. It is believed that this approach can be applied to the modelling of the day-to-day adjustment process in transportation in order to provide sound theoretical justifications for using specific models.

Acknowledgement

Support from the National Science Foundation is gratefully acknowledged. Thanks go to Moshe Ben-Akiva and Qin Liu.

See also: Traffic Assignment, Dynamic; Traffic Assignment, Stochastic; Traffic Management Systems

Bibliography

Alpha A S, Minh D 1979 A stochastic model for the temporal distribution of traffic demand—The peak hour problem. *Transp. Sci.* **13**, 315–24

Arnott R, de Palma A, Lindsey R 1991 Does providing information to drivers reduce traffic congestion? *Transp. Res. B* (in press)

Ben-Akiva M, Cyna M, De Palma A 1984 Dynamic model of peak period congestion. *Transp. Res. B* **18**, 339–55

Ben-Akiva M, de Palma A 1986 Analysis of a dynamic residential location choice model with transaction costs. *J. Reg. Sci.* **26**, 321–41

Ben-Akiva M, de Palma A, Kanaroglou P 1986 Dynamic model of peak period traffic congestion with elastic arrival rates. *Transp. Sci.* **20**, 164–81

Bettman J R 1979 *An Information Processing Theory of Consumer Choice*. Addison-Wesley, Reading, MA

Cascetta E 1989 A stochastic process approach to the analysis of temporal dynamics in transportation networks. *Transp. Res. B* **23**, 1–17

Cascetta E, Cantarella G E 1989 Stodyn 2: A day-to-day and within-day dynamic stochastic assignment model. Internal report, Università di Napoli

Chang G-L, Mahmassani H, Herman R 1985 A macroparticle traffic simulation model to investigate peak period commuter decision dynamics. *Transp. Res. Rec.* **1005**, 107–21

Chu C 1989 A stability analysis of dynamic network equilibrium. Master thesis, Northwestern University

Day R 1975 Adaptive process and economic theory. In: Day R, Groves T (eds.) *Adaptive Economic Models*. Academic Press, New York

de Palma A, Ben-Akiva M, Lefèvre C, Litinas N 1983 Stochastic equilibrium model of peak period traffic congestion. *Transp. Sci.* **17**, 430–53

de Palma A, Lefèvre C 1983 Individual decision-making in dynamic collective systems. *J. Math. Sociol.* **9**, 103–24

de Palma A, Papageorgiou Y Y 1989 A model of rational choice behavior under imperfect ability to choose. Internal report, Northwestern University

Gandolfo G 1980 *Economic Dynamics: Methods and Models*, 2nd edn. Università di Roma, Rome

Gaver L R 1967 Headstart strategies for combating congestion. *Transp. Sci.* **2**, 172–81

Horowitz J L 1984 The stability of stochastic equilibrium in a two-link transportation network. *Transp. Res. B* **18**, 13–28

Kahneman D, Slovic P, Tversky A 1982 *Judgement Under Uncertainty: Heuristics and Biases*. Cambridge University Press, Cambridge

McFadden D 1974 Conditional logit analysis of qualitative choice behavior. In: Zarembka P (ed.) *Frontiers in Econometrics*. Academic Press, New York, pp. 105–42

Mahmassani H, Chang G-L 1986 Experiments with departure time choice dynamics of urban commuters. *Transp. Res.* **20**, 297–320

Mahmassani H, Chang G-L 1987 On boundedly rational user equilibrium in transportation systems. *Transp. Sci.* **21**, 89–99

Simon H 1955 A behavior model of rational choice. *Q. J. Econ.* **69**, 88–118

Small K A 1982 The scheduling of consumer activities: Work trips. *Am. Econ. Rev.* **72**, 467–79

Small K A 1987 A discrete choice model for ordered alternatives. *Econometrica* **55**, 409–24

Smith M J 1984 The stability of a dynamic model of traffic assignment—An application of a method of Lyapunov. *Transp. Sci.* **18**, 245–52

Takayama A 1985 *Mathematical Economics*, 2nd edn. Cambridge University Press, Cambridge

Tong C, Mahmassani H, Chang G 1987 Travel time prediction and information availability in commuter behavior dynamics. *66th Annual Meeting Transportation*

Research Board. Transportation Research Board, Washington, DC

Vickery W S 1969 Congestion theory and transportation investment. *Am. Econ. Rev.* **59**, 251–61

A. de Palma
[Northwestern University, Evanston,
Illinois, USA]

Road Traffic: An Introduction

Everyday experience of pedestrians, riders, drivers and passengers is that traffic conditions vary widely over the road system and that they are often far from satisfactory. Many people have ideas about how to make the traffic conditions better, but except in the simplest cases people often disagree about what they mean by better and, even if they agree about this, they often disagree about how to achieve it, or how much it is worth spending to do so. These disagreements stem partly from the differing, often conflicting, interests of different road users and of other affected groups and partly from the complexity of the road and traffic system itself. Better understanding of this system and how it works can help in improving traffic conditions on existing roads and in the design of new roads, and can also inform and clarify the discussion and resolution of conflicts of interest. The basic ideas of road traffic are outlined here in terms of the flow of vehicles along roads remote from junctions, the movement of vehicles and pedestrians in queues at road junctions of various types and at pedestrian crossings, and the movement of traffic in networks of roads.

In doing so, it is important to distinguish between two kinds of variation in traffic: systematic variation, as occurs for example on average over the day, week and year; and random variation, as occurs from vehicle to vehicle or minute to minute and between corresponding days of similar weeks. There is an associated distinction between deterministic analysis, in which systematic relationships are established and random variation is allowed for only in terms of its average effects, and stochastic analysis, in which random variation is studied explicitly. Another useful distinction is between macroscopic analysis, in which each stream of traffic is considered as a single entity without considering the vehicles or pedestrians individually, and microscopic analysis, in which the movements of individual vehicles and pedestrians are examined. Most macroscopic analyses are deterministic and most microscopic analyses are stochastic.

1. Vehicular Traffic Flow Remote from Road Junctions

On dual carriageways and on some multilane single carriageways, traffic travelling in the two directions can be considered as two separate streams between which there is little or no interaction in normal circumstances. On single carriageways where one or more lanes are available to traffic travelling in either direction, the streams of traffic travelling in the two directions interact through the use of the shared lanes. Such interaction is especially important on two-lane two-way roads, where overtaking is possible only by use of the lane used mainly by traffic travelling in the other direction.

Consider a stream of traffic travelling in one direction without interaction with opposing traffic. In macroscopic analysis the stream in the neighborhood of any point on the road is described by two parameters: the density or concentration k vehicles per unit length of road in the neighborhood of the point, and the flow q of vehicles per unit time passing the point. The corresponding average speed of traffic in the stream can be described either by the space-mean speed v_s, which is the average speed of vehicles present at a typical instant in the length of road over which k is measured, or by the time-mean speed v_t, which is the average speed of vehicles passing the point during a period over which q is measured. It can be shown (Wardrop 1952) that these two speeds are related by the equation

$$v_t = v_s + \sigma_s^2 / v_s$$

where σ_s^2 is the variance of the distribution of speeds of vehicles in the length of road over which v_s is defined. It can also be shown that

$$q = k v_s$$

so that speed, flow and density are not mutually independent (see *Flow Variables*).

There is a behavioral relationship between speed and density, in that drivers tend to drive more slowly when there is a lot of traffic about than when they have the road more or less to themselves. This leads to the observation that the space-mean speed is a decreasing function of density, which is shown as a linear function in Fig. 1, but need not be linear. The exact relationship can be affected by factors such as road layout and visibility. The intercept v_0 on the speed axis is called the mean free speed and is the average of the speeds at which the drivers in the stream would drive if they had the road to themselves. The intercept on the density axis is called the "jam" density and is the average density of the traffic when it is brought to a halt by congestion.

Equation (1) can be used to derive from the speed–density relationship the corresponding relationships between speed and flow and between flow and density. In Fig. 1, the linear relationship between v_s and k leads to the parabolic relationships between v_s and q and between q and k.

It can be seen that these relationships imply a certain maximum flow q_{max} in the stream and that any lesser flow can be attained either at higher speed and lower density or at lower speed and higher density than q_{max}.

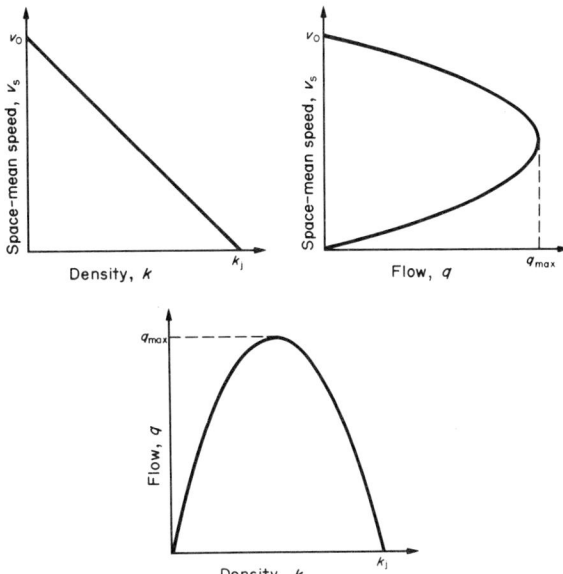

Figure 1
Relationship between density k, flow q and space-mean speed v_s of a stream of vehicular traffic when the speed–density relationship is linear, showing jam density k_j, mean free speed v_0 and maximum flow q_{max}

The former possibility is described as free flow and the latter as congested flow.

The maximum flow defined by Fig. 1 can be regarded as the theoretical traffic capacity of the length of road to which the curves apply. For practical purposes, however, the capacity of a length of road is usually defined in relation to a particular quality of traffic flow, or level of service, as being the highest flow consistent with the level of service remaining above a certain level specified, for example, in terms of a mean speed in the free-flow range. Flow has so far been defined and discussed in terms of vehicles, but it is a common observation that a given number of larger vehicles take up more of the capacity of a road than the same number of smaller vehicles. For this reason, each vehicle of any given type (e.g., motorcycle, coach, heavy lorry) on a given type of road is regarded as being equivalent to that number of typical passenger cars that would take up the same portion of the capacity of the road as does one typical vehicle of the type being considered. Traffic flow is then measured in passenger car units (pcu) per unit time, and the definition of the units implies that the capacity of the road in pcu is independent of the composition of the traffic. A vehicle of a given type may be equivalent to different numbers of pcu on different types of road and in different traffic conditions.

Even under free flow, drivers experience delay at all but the lowest flows, and the delay to a driver travelling along a length of road can be defined as the excess of the actual travel time over the travel time at the driver's free speed. Transition from free to congested flow can be observed upstream of any constriction that is sufficient to slow traffic down appreciably without bringing it to a halt, and can be analyzed by means of wave theory (see *Kinematic Wave Theory*).

In microscopic analysis of traffic flow, an important role is played by the headway between successive vehicles, which can be measured either as the time between the fronts of the two vehicles passing a given point or the distance between the fronts of the vehicles at a given instant. The headway is a random variable in that it is, in general, different for each pair of vehicles. Where the flow q is much less than the road can accommodate, so that overtaking is virtually unhindered, the headways h are usually approximately exponentially distributed, with probability density function $q \exp(-qh)$. Where traffic is heavier and overtaking is therefore inhibited, the vehicles in each lane tend to form bunches, each consisting of a leading vehicle which is free to travel at the driver's free speed, and following vehicles, whose drivers would like to travel faster and are probably looking for opportunities to overtake the leading vehicle. In such cases, the distribution of headways must contain a smaller proportion of very short headways than the exponential distribution and a larger proportion of headways representing close following in bunches. Various distributions more complicated than the exponential have been used to describe such situations. The proportion of vehicles that are following can be used as an indicator of level of service.

The movement of a following car relative to the car in front has been analyzed by means of car-following theory (see *Car-Following Models*). The basis for this theory is the supposition that if $x_n(t)$, $\dot{x}_n(t)$ and $\ddot{x}_n(t)$ are the position, speed and acceleration, respectively, of the nth car in a bunch at time t, then

$$\ddot{x}_{n+1}(t) = f[x_n(t-\tau) - x_{n+1}(t-\tau), \dot{x}_n(t-\tau) - \dot{x}_{n+1}(t-\tau)]$$

so that the driver of each car is regarded as accelerating or braking at time t in a manner that is determined by his speed and position relative to the car in front at a time τ previously. Various forms of the function f have been used (Gazis 1974). Such theory can be extended to lane changing and overtaking in terms of the positions and speeds of the next vehicles in front and behind in adjacent lanes.

Macroscopic relationships between speed, flow and density can also be applied to the total flow in both directions on two-way roads where there is interaction between flow in the two directions. The flow of traffic on two-lane two-way roads can also be analyzed by regarding the traffic travelling in each direction as a set of bunches separated by gaps, each bunch consisting of a moving queue of following vehicles waiting to overtake the leading vehicle in the opportunities provided by gaps in the opposing traffic stream.

In traffic engineering and the planning of new or improved roads, relationships between speed, flow and density, or average travel times based on car-following models, are used mainly to estimate how traffic conditions on a length of road can be expected to change if either the pattern of traffic or the layout of the road are altered, or what traffic conditions can be expected on a new length of road at any given traffic flow. Such estimates can be used to help to optimize the allocation of resources for road construction or improvement in a network and the timing of improvements to any given length of road in the light of a forecast rate of increase of traffic. In addition, they can be used to help in the detailed design of new roads and improvement schemes.

Microscopic analysis of traffic flow remote from road junctions is relevant not only to traffic conditions there but also to the way in which vehicles join queues that form at junctions or other bottlenecks at the ends of relatively uncongested stretches of road.

2. Traffic at Junctions and Pedestrian Crossings

Road junctions of all kinds except the most free-flowing grade-separated ones, and pedestrian crossings, have in common that some or all of the approaching vehicles and pedestrians cannot necessarily pass through the junction or over the crossing immediately they approach it. This means that queues of vehicles and pedestrians form. Queues of vehicles also form on roads remote from junctions at points where the flow of traffic is constricted severely enough to bring to a halt the traffic upstream of the constriction.

Traffic queues, like other types of queue, can be described and analyzed in terms of the arrival process (how traffic joins the queue) the queue discipline (how traffic behaves in the queue) and the departure process (how traffic leaves the front of the queue).

The arrival process can usually be described completely in terms of the distribution of the headways between the approaching vehicles or pedestrians and is thus similar for a wide range of traffic queues. The reciprocal of the mean headway, which equals the average approaching flow, is called the arrival rate.

The queue discipline describes how the order in which pedestrians or vehicles reach the front of the queue is related to the order in which they join the queue. Queuing pedestrians are usually assumed to wait side by side at the front of the queue ready to start crossing the road simultaneously when an opportunity occurs. When there are too many pedestrians for this to be possible, there is little error in assuming that they cross the road in the order in which they join the queue, even if this is not quite the case. Except for some two-wheelers, vehicles that queue in a single lane must reach the front of the queue in the order in which they join it. When vehicles queue in more than one lane, the question arises of whether the queuing in each lane should be analyzed separately, or whether the vehicles in two or more adjacent lanes should be considered as forming a single queue for purposes of analysis. They should be treated as a single queue if and only if enough of the approaching drivers have a free choice between lanes to keep the time taken to pass through the queue roughly the same in all lanes. If this condition is satisfied it follows that vehicles will reach the front of the queue in roughly the same order as they join it. If not, the traffic in different lanes should be treated as forming separate queues. Traffic in a single lane analyzed separately, or in two or more adjacent lanes treated as a single queue, is said to form a stream. Almost all traffic queues can therefore be analyzed on the basis that vehicles or pedestrians reach the front of the queue in the order in which they join it.

The departure process, unlike the arrival process and queue discipline, varies greatly between different kinds of traffic queue and will be discussed in relation to some of the most important kinds in turn. Several important features common to all kinds of traffic queue are discussed first.

2.1 General Behavior of Traffic Queues

For any traffic queue, a capacity can be defined as the average rate at which traffic can depart from the front of the queue so long as it never empties. The behavior of the queue then depends crucially on the ratio of the arrival rate to capacity. This ratio is called the traffic intensity. The notation will be used that q is the arrival rate, Q is the capacity (in the same units as q) and the traffic intensity $\rho = q/Q$.

As in the case of trafffic flow remote from junctions, different types of vehicle may take up different portions of the capacity of a vehicular queue, and the number of pcu to which a vehicle is equivalent in a queue is proportional to the average headway between successive vehicles of its own type departing without the queue becoming empty. Again, capacity measured in pcu per unit time is independent of traffic composition. Because, however, the behavior of a queue depends on the distribution of the headways between individual arrivals, and in a vehicular queue the individuals are vehicles rather than pcu, it is preferable to carry out all calculations with q and Q measured in vehicles per unit time, adjusting the capacity to these units according to the prevailing traffic composition.

If ρ remains steady for a time at a value less than unity, then the amount of traffic in the queue can be expected to approach and then vary randomly about an equilibrium value. In many cases where the traffic arrives randomly, this equilibrium value can be estimated theoretically in the form $\rho + C\rho^2/(1-\rho)$, where C is a parameter whose value, usually between 0.5 and 1, depends on the kind of queue. If ρ is only just less than unity, it follows that the equilibrium value is large, and the queue may take a long time to approach equilibrium. If ρ exceeds unity for a time then, apart from random fluctuations, the amount of traffic in the queue can be expected to increase steadily at rate $q - Q$ (see *Queuing Theory Applications*).

Time-dependent queuing theory (Kimber and Hollis 1979) has enabled the growth or decline of the queue towards equilibrium for $\rho < 1$ and the steady growth for $\rho > 1$ to be described by a single approximate analysis. According to this analysis, if L_t is the average amount of traffic to be expected in the queue at time t in a period of up to about 15 min beginning at $t = 0$ and during which q and Q remain constant, then L_t can be estimated as a function of q by one branch of a hyperbola whose equation in L_t and q has coefficients determined by C, L_0 and Q. The form of the resulting dependence of average amount of traffic in the queue on time is illustrated in Fig. 2 for various values of q relative to Q. Whether the queue tends to increase or to decrease depends on whether q is greater or less than the value q_0 for which the equilibrium amount of traffic in the queue is L_0. In each case, the mean delay to a vehicle or pedestrian joining the queue at time t is L_t/Q.

Because this method of estimation is approximate, its use in practice requires a little care, and advice about this is given by its originators. It is, however, especially useful because some of the queuing situations that are most important in practice are those in which a queue that has $\rho < 1$ before and after a peak period has $\rho > 1$ during the peak.

Over appreciable periods during which $\rho < 1$, the average rate at which traffic leaves the queue is approximately equal to the arrival rate unless the amount of traffic in the queue is initially very different from its equilibrium value. It is important to remember, however, that in the latter case and if $\rho > 1$, these two rates will differ and the arrival rate can be measured only upstream of the end of the queue.

Some important kinds of situations where traffic queues form are now discussed.

2.2 Priority Junctions

At a priority junction, vehicles in certain streams have to give way and wait for gaps between vehicles in certain other streams that have priority. Drivers of vehicles in a stream that must give way have to assess which gaps are acceptable to them. Theoretical analysis of this process of gap acceptance has led to various expressions for the capacity of a stream giving way and delay in that stream in terms of the characteristics of the traffic that has priority. The assumptions underlying such analyses are, however, rather restrictive, and the parameters in the resulting formulae are not clearly related to the layout of the junction. Empirical expressions for the capacities of the streams giving way have, therefore, been developed that are linear in the flows of traffic having priority and have parameters that depend explicitly on the layout of the junction and the intervisibility of the streams of traffic (see *Priority Intersection: Modelling*). This explicit dependence is helpful in junction design. Estimates of capacity obtained by either type of method can be used to estimate queue lengths and delay by the method outlined in Sect. 2.1.

2.3 Roundabouts

A roundabout is a junction at which each entering vehicle joins a circulating one-way stream of traffic in which it moves round the junction until it reaches the exit its driver wishes to take. Roundabouts operate most satisfactorily under the offside priority rule, whereby traffic entering the roundabout must give way to circulating traffic. Under this rule, provided that the exits from the roundabout and the circulating carriageway have sufficient capacity, queuing is confined to the entries. In these circumstances it has been found empirically (Kimber 1980) that the capacity of an entry can be estimated as a linear function of the circulating flow in which the parameters depend explicitly on the layout of the junction. This explicit dependence again makes the expression helpful in junction design, and the values of the capacity for the various entries can be used to estimate queue lengths and delay by the method outlined in Sect. 2.1.

Roundabouts operating to other priority rules have more complicated interactions between traffic streams and their operation often has to be assisted by signals unless the traffic is light.

2.4 Signal-Controlled Junctions

Control of traffic by signals enables conflicting traffic movements at a junction to take place in turn so that conflict is reduced or eliminated. There are two main types of operation of traffic signals. In fixed-time oper-

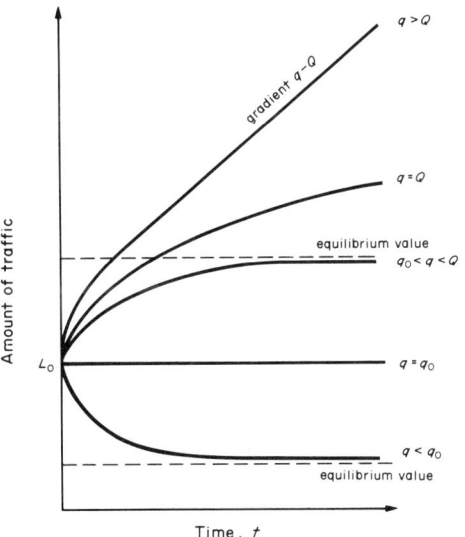

Figure 2
Form of dependence on t of average amount L_t of traffic at time t in a queue for which the arrival rate q and capacity Q remain steady and q_0 is the value of q for which the equilibrium amount of traffic in the queue is the initial amount L_0

ation different sets of movements are allowed to proceed in a prescribed cyclic order for predetermined lengths of time regardless of the traffic. One repetition of this process is called the signal cycle and its duration the cycle time. Alternatively, in vehicle-actuated operation, the changing of the signals is influenced by approaching vehicles through detectors and pedestrians through push buttons. Such operation may be cyclic, in that there is a normal order of changing of the signals which is varied only in the times between changes and by omitting green periods for sets of streams in which no traffic is waiting when their turn comes. Instead, it may be acyclic, in that each change of the signals is selected from all safe changes according to vehicle and pedestrian demand.

Cyclic vehicle-actuated operation approximates to fixed-time operation when there is enough traffic in enough of the streams approaching the junction, and the same is probably true of acyclic operation. Signal timings are constrained by safety in terms of fixed intergreen times between the green periods for conflicting streams, minimum green periods and a maximum cycle time. Within these constraints, the timings may be chosen to influence capacity and delay for traffic in different streams. In calculating timings it is usually assumed that vehicular traffic in a given stream passes the signal controlling it during a part of each cycle (the effective green time) at a constant rate (the saturation flow) so long as there is a queue in the stream. The effective green time is closely related to the actual green time of the relevant signal and the saturation flow to the layout of the part of the junction used by the stream (Webster and Cobbe 1966).

If a signal-controlled stream has saturation flow s and its effective green time forms a proportion Λ of all time, then its capacity is Λs. Signal timings can often be calculated by identifying a set of streams whose green periods and the intergreen times between them just cover the whole cycle time, and whose arrival rates are larger in relation to their saturation flow than those of the other streams. The effective green times for streams in the set are then chosen to make their traffic intensities equal. (The traffic intensity of a signal-controlled stream is often called its degree of saturation.) To maximize the capacity of these streams, the cycle time is given its maximum permitted value. If the common value of the traffic intensities in these streams is then substantially less than unity, the average delay per unit time at the junction can be roughly minimized by choosing a certain shorter cycle time. Several approximate formulae have been used for this purpose and for estimating delay in signal-controlled streams when vehicles arrive roughly at random. More comprehensive methods for calculating signal timings that are optimal in certain respects and some of their uses in junction design are dealt with in the articles *Signal Control at Individual Junctions: Phase-Based Approach* and *Signal Control at Individual Junctions: Stage-Based Approach*.

2.5 Systems of Signal-Controlled Junctions

When signal-controlled junctions are close together, traffic moves between them in well-defined bunches, which are called platoons, rather than at random, and advantage can be taken of this to coordinate the operation of signals at different junctions so that, as far as possible, green periods coincide with the arrival of platoons. This is relatively simple on a one-way arterial road with little turning traffic, substantially more difficult on a two-way arterial road and very complicated to achieve in appreciable measure in a network of streets. Signals whose operation is coordinated are called linked signals. Linking of signals at a substantial number of junctions in an area so that their coordinated operation is controlled by one or more computers is called area traffic control or urban traffic control.

Given the flow of traffic on each street, the turning movements at each junction, and the layout and signal cycle at each junction, various methods are available for calculating good signal timings for coordinated operation. Alternatively, with the help of an extensive system of traffic detectors, vehicle-actuated or traffic-responsive coordinated operation can be achived (see *Road Network Control*; *Traffic Control Systems: Architecture*).

2.6 Delay to Pedestrians

The average delay to a pedestrian waiting to cross a stream of vehicles in which the headways are exponentially distributed with mean $1/q$ and requiring a gap of duration at least c is $[\exp(qc) - 1 - qc]/q$. This increases more than proportionately to q, the arrival rate of the vehicles, which shows that the delay to pedestrians crossing a multilane stream of vehicles arriving randomly can be reduced by providing refuges so that they can cross one or a few lanes at a time.

Delay to a signal-controlled stream of pedestrians who obey the signals can be estimated by the same formulae as for signal-controlled streams of vehicles by assuming that the saturation flow of the pedestrians is high (e.g., several pedestrians per second for each meter of kerb along which they wait). The estimate of delay will be insensitive to this value unless the arrival rate of pedestrians is high. This will overestimate delay at sites where pedestrians cross through gaps between moving vehicles, but the overestimate is unlikely to be serious if the traffic intensity of the stream of vehicles is high.

Delay to pedestrians at other types of crossing depends on the regulations governing the crossing and the degree of compliance with them by drivers and pedestrians.

2.7 Other Measures of Performance

Delay or travel time is not the only indicator of performance of parts of the road system where queuing takes place. There are several reasons why the number of occasions on which vehicles stop and start during their journeys is important quite apart from the time lost in doing so. Starting from rest involves disproportionate fuel consumption, noise and exhaust emissions. The need to stop carries with it the risk of not doing so safely and is an irritation to the driver. For these reasons, it is useful to consider the number of vehicle stops in congested parts of the road system as a measure of performance. In queues where the vehicles arrive at random, the number of stops tends to increase with delay, so that a design that reduces delay is also likely to reduce the number of stops. However, this is by no means the case where traffic moves in platoons. Similar levels of delay may then be achieved with widely differing numbers of stops. Both these two quantities must then be considered in any optimization.

Fuel consumption is itself an important measure of performance and can be estimated quite well by function of travel time, stopped time and number of stops.

Safety is comparable in importance with reduction of travel times, and in some respects in human terms much more important, and there is an extensive literature concerning the ways in which particular features of road design and traffic engineering and control can affect safety. Empirical relationships established between accident risk and certain parameters of the design and operation of roads and their junctions lend themselves to explicit inclusion in optimization procedures. Other safety features are expressed as constraints on any optimization.

3. Particular Vehicle Types

Vehicular traffic consists typically of a mixture of different kinds of vehicles driven by different kinds of people for different reasons. Many of these differences can simply be regarded as contributions to the systematic and random variation in traffic, but some types of vehicles are sufficiently distinctive in their operational requirements to be considered specifically in traffic engineering and control.

3.1 Public Transport Vehicles

Because the ratio of the average number of passengers per public transport vehicle in a traffic stream to the corresponding average for cars is usually much higher than the pcu equivalent of a public transport vehicle, net benefit to passengers can in principle be achieved, even within a stream of given traffic intensity, by giving public transport vehicles priority over cars. For this and other reasons of transport policy, techniques have been developed for giving priority to buses in congested road networks (Webster and Bly 1976) and many of these apply with suitable adaptation to light rail vehicles.

Two methods require no special compliance by other road users and may not even be apparent to them: these are the calculation of signal timings to favor public transport vehicles, which is possible both at individual signal-controlled junctions and in coordinated operation, and selective detection of public transport vehicles by vehicle-actuated signals so that green can be extended or red cut short for the relevant stream. The latter can be integrated with public transport vehicle location and scheduling systems. Other methods require special road signs and markings and the compliance of other road users. The main ones are bus lanes, either in the same direction as other traffic or, to enable buses to avoid detours in one-way systems, in the opposite direction; segregation of light rail tracks; exemption from prohibited turns, usually opposed turns which other traffic has to make detours to achieve; and entry by public transport vehicle to areas to which car travellers have to walk from nearby streets or car parks. These other methods can be taken into account in various ways in the calculation of traffic capacity, delay and signal timings at affected junctions.

3.2 Pedal Cycles

After declining sharply as car ownership increased, pedal cycling has a new and continuing role in many countries and sympathy for the difficulties and danger encountered by cyclists in heavy traffic is widespread. The ideal arrangement for cyclists is a system of cycle routes separate from both motor traffic and pedestrians. Even where this can be achieved on a large scale, however, cyclists will still need to use parts of the general road system. A range of techniques exists for helping them to cross busy routes and negotiate difficult junctions and useful guidelines have been produced (Institution of Highways and Transportation 1983). These also cover the practicalities of allocating space to cyclists on routes shared with motor traffic or with pedestrians.

3.3 Emergency Vehicles

Traffic congestion is a source of risk to life, limb and property, not only through traffic accidents but also through the delay incurred by ambulances and fire tenders on rescue missions. Traffic control systems can be used to reduce such delays by providing successions of green signals on demand along previously identified routes and by giving emergency vehicles priority by selective detection at individual signal-controlled junctions in a similar way to public transport vehicles.

4. Traffic Management and Other Applications

Traffic engineering and control can be used to improve the traffic conditions resulting from the movement of a given flow of traffic along a stretch of road, or a given set of flows through a junction. The widespread application of such techniques throughout an area will, however, not merely tend to change conditions for the existing pattern of traffic, but will also lead to changes in the traffic pattern as drivers adjust their choices of route, or even of journeys, in the light of the altered traffic conditions. This leads on to the use of traffic engineering and control not just to provide for a pattern of traffic but deliberately to seek to influence it to produce improvements in accessibility and environment. This process is known as traffic management. When it is extended to include changes in public transport operation, possibly by rail as well as road, and parking policy it is sometimes known as transport (or transportation) system management.

In order to estimate the effects of sets of measures of traffic engineering and control that form traffic management schemes, it is necessary not only to analyze the flow of traffic along roads and the queues at junctions, but also to estimate how the routes taken by drivers in the road system concerned will be affected. For this purpose, use is made of a procedure known as traffic assignment, in which the road system is represented by a network of nodes and links, and the traffic conditions are represented by a set of relationships that express the travel time along each link as a function of the flow of trafffic on that and other relevant links. These relationships can be derived from the analyses described in Sects. 1 and 2. Given the relationships and a matrix specifying how many vehicles per unit time wish to travel from each point of entry to the network to each point of exit from it at any given time of day or week, the estimation of the routes that drivers will take and the consequent flow of traffic on each road and through each junction can itself be expressed as an optimization problem. Techniques for solving this problem in ways which allow for the variation of traffic conditions over time, especially in peak periods, have been incorporated into various models (Allsop 1983) designed to help traffic engineers to estimate the effects of area-wide traffic management schemes. Traffic assignment is also used at a wider and less detailed level to estimate the effects of major road construction and improvement on the routes used for longer-distance movement.

The basic ideas of road traffic that have been outlined here are used in various ways to develop theoretical and simulation models of traffic situations ranging from a simple bottleneck to a whole network of roads. These models in turn are used in the design and appraisal of schemes for building new roads, improving existing roads, or improving traffic conditions on the roads as they are (Institution of Highways and Transportation and Department of Transport 1987). In some cases, the models are used to forecast what conditions will be like if each of a set of possible options for change is implemented, and thus help to inform the choice between options. In other cases, the models can be used as the basis of an explicit optimization procedure, in order to calculate what is, in some specific sense, the best design for a scheme of a given type, or the best mode of operation for a given system. Advances in technology, especially information technology, can only increase the scope and effectiveness of application of these ideas.

Bibliography

Allsop R E 1983 Network models in traffic management and control. *Transp. Rev.* **3**, 157–82

Gazis D C 1974 *Traffic Science*. Wiley, New York

Institution of Highways and Transportation 1983 *Providing for the Cyclist.* IHT, London

Institution of Highways and Transportation, Department of Transport 1987 *Roads and Traffic in Urban Areas.* Her Majesty's Stationery Office, London

Kimber R M 1980 The traffic capacity of roundabouts. Departments of the Environment and Transport TRRL Report LR 942. Transport and Road Research Laboratory, Crowthorne, UK

Kimber R M, Hollis E M 1979 Traffic queues and delays at road junctions. Departments of the Environment and Transport TRRL Report LR 909. Transport and Road Research Laboratory, Crowthorne, UK

Transportation Research Board 1985 Highway capacity manual. Special Report 209. TRB, Washington, DC

Wardrop J G 1952 Some theoretical aspects of road traffic research. *Proc. Inst. Civ. Eng.* **1**, 325–78

Webster F V, Bly P H 1976 Bus priority systems. NATO CCMS Report 40. Transport and Road Research Laboratory, Crowthorne, UK

Webster F V, Cobbe B M 1966 Traffic signals. Road Research Technical Paper 56. Her Majesty's Stationery Office, London

R. E. Allsop
[University College London, London, UK]

Road Traffic Control, Demand-Responsive

Urban vehicular traffic, as an expression of human behavior, is variable in time and in space. The control of such traffic requires a high degree of adaptiveness to enable a suitable response to this variability. Ever since the inception of modern traffic signal controls, traffic engineers and signal system designers have attempted to make them as responsive as possible to prevailing traffic conditions. The premise always was that increased responsiveness must lead to improved traffic performance. This premise was applied to single intersection signals as well as to arterial and network signal systems. However, the extent to which traffic responsiveness is achieved depends on a variety of factors including strategy sophistication, hardware capabilities, surveillance and communication equipment, as well as operator capabilities (see *Traffic Control Modes*).

With the advent of computerized systems in the mid-1960s, many cities embarked on deploying centrally controlled and monitored traffic signal systems. Such systems offered significant advantages when compared with the previously used electromechanical devices. However, they also imposed rigidities which restricted the opportunities for traffic responsiveness. This was quite evident in the urban traffic control system (UTCS) experiments in Washington, DC (MacGowan and Fullerton 1979–1980), as well as in similar experiments that were conducted in Canada (Toronto Metropolitan Corporation 1974–1976) and in the UK (Holroyd and Robertson 1973). Various aspects of these rigidities persist.

The advent of advanced microprocessor technologies drastically changed the traffic signal control field and opened up new horizons and opportunities for demand-responsive control. It became feasible to develop much more sophisticated systems than before, systems that offer a great deal of responsiveness, work largely automatically and eliminate the need for operator intervention. However, the new technologies also require the development of new concepts and methods.

1. Prescription for Demand-Responsive Control

After a review of past experiments, a prescription for the development of an effective demand-responsive traffic control strategy was formulated as follows (Gartner 1982a, 1985).

(a) The strategy must provide better performance than off-line methods. Although this may seem self-evident, it was not always explicitly recognized in the development of previous responsive strategies; in some cases it was superseded by less relevant criteria such as main-street progression or variable cycle time.

(b) Development of new control concepts is needed and not merely an extrapolation of existing concepts. Effective responsiveness is not achieved by implementing off-line methods at an increased frequency. New methods that are better suited to the variability in traffic must be developed. For instance, the conventional notions of offset, split and cycle time, which are inherent in existing signal optimization methods, may be unsuited for demand-responsive control.

(c) The strategy must be truly demand-responsive (i.e., adapt to actual traffic conditions and not to historical or predicted values that may be far from the actual values).

(d) The strategy must not be arbitrarily restricted to control periods of a specified length but should be capable of frequent updating of plans, as necessary. Furthermore, it should not be encumbered by a rigid network model structure. Rather, it should be based on decentralized decision making.

While some of these principles were already incorporated in the on-line SCOOT method (Hunt *et al.* 1981), their full realization is only exhibited by more modern methods such as optimization policies for adaptive control (OPAC) (Gartner 1982b, 1983) or PRODYN (Henry and Farges 1989), both of which are based on dynamic programming. In this article, the methodological development and practical application of OPAC are described. In the first stage, a dynamic programming (DP) procedure was developed that served as a standard of performance for demand-responsive control, since DP is capable of generating optimal control strategies. Next, the procedure was simplified to replicate the performance of DP, while relinquishing its extensive computational requirements to make it suitable for on-line calculation. In the third stage, the procedure was further refined by introducing a rolling-horizon approach. In this way a practical strategy for demand-responsive control was obtained. Before describing the details of the method, intersection control modelling approaches should be reviewed.

2. Intersection Control Models

There is a tremendous variety of intersection control models, but only a limited number have been implemented in practice. The most common approach is to determine settings for a fixed-cycle light that minimize the average delay per car assuming constant arrival rates (Webster 1958, Miller 1963). Gazis and Potts (1965) obtained conditions for the optimal control of an intersection that becomes oversaturated for some finite length of time; the model was extended by Gazis (1964) to two intersections. Dunne and Potts (1964) developed time-varying control algorithms for an undersaturated intersection with constant arrivals which guarantee that, for any initial state, the system eventually reaches a limit cycle for which the equilibrium average delay per car is a minimum. In all these models, the control policy is not responsive to the dynamics of the traffic flow process since there is no real-time traffic flow information involved. The traffic flow process is represented by a statistical distribution (Poisson, Binomial, etc.) with the initial conditions being the initial queue lengths; alternatively, in the case of oversaturation it can be represented by a smooth function of predicted demand vs time. Obviously, none of these models can take advantage of the information contained in the actual arrival times of individual vehicles.

A dynamically self-optimizing strategy was proposed by Miller (1965). In this strategy, a decision whether to extend a phase is repeatedly made at very short fixed intervals h by the examination of a delay-based control function. This function estimates the difference in vehicle seconds of delay between the gain to the extra vehicles that will be allowed to cross the intersection during an extension of h s and the loss to the queuing vehicles on the cross street resulting from the extension (h is 1–2 s long). A similar approach was also proposed by van Zijverden and Kwakernaak (1969). Bang and Nilsson (1976) implemented and field tested Miller's strategy and showed that important gains can be obtained compared with fixed-time and vehicle-actuated control at isolated intersections. However, since this method has a very short projection horizon, with a correspondingly short optimization interval, it does not lend itself to implementation in a network of intersections and, furthermore, does not assure overall optimality of the control strategy. On the other hand, due to its capabilities for multistage decision making, DP is a more attractive technique for optimal dynamic

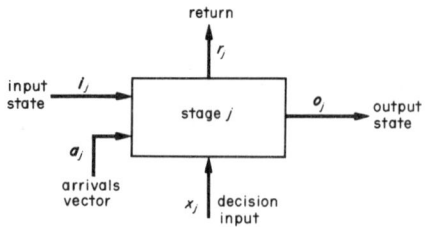

Figure 1
Typical stage of the DP process

signal control. Two DP models were proposed for optimal signal control. Grafton and Newell (1967) formulated a continuous-time model, whereas Robertson and Bretherton (1974) described a discrete-time model. The first model minimizes an infinite-horizon total discounted delay function. The second model minimizes the total delay aggregated over all intervals of a finite horizon. The DP optimization described in Sect. 3 employs the discrete-time version.

3. Dynamic Programming Optimization

DP is a mathematical technique used for the optimization of multistage decision processes. In this technique, the decisions (or control values) that affect the process are optimized in *stages* rather than simultaneously. This is done by dividing the original decision/control problem into small subproblems (stages) which can then be handled much more efficiently from a computational standpoint. DP is a systematic procedure for determining the combination of decisions that maximizes overall effectiveness, or minimizes overall disutility. It is based on the principle of optimality which was enunciated by Bellman (1957) as follows: "An optimal policy has the property that whatever the initial state and initial decision are, the remaining decisions must constitute an optimal policy with regard to the state resulting from the first decision."

Consider a single intersection with signal phases that consist of effective green times and effective red times only. All traffic arrivals on the approaches to the intersection are assumed to be known for a finite horizon length. The optimization process is decomposed into N stages, where each stage represents a discrete time interval (e.g., 5 s long). A typical stage j is illustrated in Fig. 1.

At stage j there is an input state vector i_j, an arrivals vector A_j, output state vector o_j, input decision variable x_j, economic return (cost) output r_j, and a set of transformations:

$$o_j = T_j(i_j, a_j, x_j)$$
$$r_j = R_j(i_j, a_j, x_j)$$

The state of the intersection is characterized by the state of the signal (green or red) and by the queue length on each of the approaches. Assuming a two-phase signal, the input decision variable indicates whether the signal is to be switched at this stage ($x = 1$) or remain in its present state ($x = 0$). The return cost output is the index of performance (total delay time) of the intersection which is to be minimized. The functional relationship between the input and output variables is based on the queuing-discharge process at the intersection (i.e., the inflows and outflows relative to the signal settings). The switching policy consists of the sequence of phase switch ons and phase switch offs throughout the horizon.

The recursive optimization functional is given by the following equation:

$$f_j^*(i_j) = \min_{x_j} \{R_j(i_j, a_j, x_j) + f_{j+1}^*(i_j, a_j, x_j)\}$$

The return at stage j is the queuing delay incurred at this stage and is measured in vehicle-interval units. The result of the optimization is a switching policy that minimizes the total delay over the horizon period for a given initial input state i_1. This policy identifies the sequence of switching decisions $(x_j^*, j = 1, \ldots, N)$ at all stages of the optimization process. An example of the demand-responsive control strategy calculated by this approach is shown in Fig. 2 for a 5 min horizon length. The signal is two phase and only two approaches are considered, A and B. The figure shows the arrivals on the approaches, the optimal switching policies and the resulting queue-length evolution. The signal timings appear as hatched (red) or blank (green) areas, including an all-red overlapping red interval at each switching point. The total performance index (PI) is 196 vehicle intervals. This is the best possible policy for the given arrival pattern.

4. Rolling-Horizon Optimization

The DP method for calculating demand-responsive control policies requires advanced knowledge of arrival data for the entire horizon period. This is usually beyond what can be obtained from available surveillance systems. Moreover, DP optimization requires an extensive computational effort and, since it is carried out backwards in time, precludes the opportunity for modification of forthcoming control decisions in the light of updated traffic data. Thus, the DP approach, while assuring global optimality of the control strategies, cannot be used in real time. Consequently, a simplified optimization procedure was developed that is amenable to on-line implementation, yet produces results of comparable quality. The procedure uses an optimal sequential constrained (OSCO) search and has the following basic features.

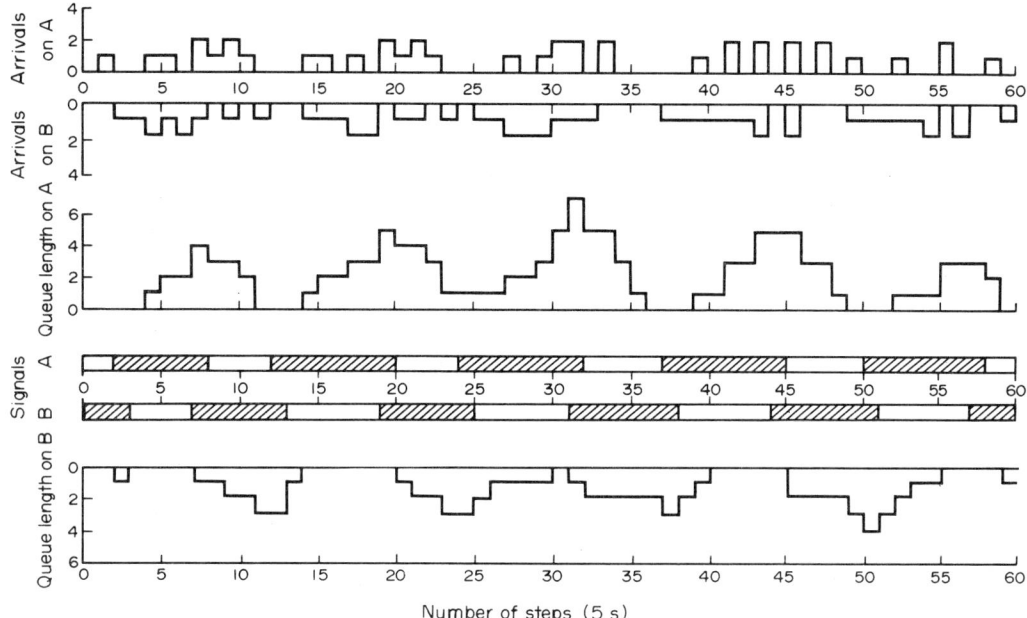

Figure 2
Example of a demand-responsive control strategy

(a) The optimization process is divided into sequential stages of T seconds. A stage length is in the range of 50–100 s.

(b) During each stage there is at least one signal change (switchover) and at most three phase switchovers.

(c) An objective function (total delay) is evaluated for all feasible switching sequences and the sequence generating the least delay is selected. For any given switching sequence at stage n a performance function is defined on each approach that calculates the total delay incurred during this stage on each approach (in vehicle intervals), as follows:

$$\phi_n(t_1, t_2, t_3) = \sum_{j=1}^{k} (Q_0 + A_j - D_j)$$

where Q_0 is the initial queue, A_j is the number of arrivals during interval j, D_j is the number of departures during interval j, k is the number of intervals in the stage and t_1, t_2 and t_3 are the possible switching times during this stage. The function ϕ measures the area enclosed between the cumulative arrivals and cumulative departures curves.

The optimal switching policies are calculated independently for each sequential stage in a forward manner (i.e., one stage after another). Since the calculation proceeds in parallel with the real-time information flow, this approach is amenable for use in an on-line system. The information and decision flow at a typical stage n is illustrated in Fig. 3.

Computational results show that the OSCO approach provides results that are very close (within 10%) to the original DP approach. Thus, the stage optimization can serve as a building block for demand-responsive decentralized control. However, the technique still requires future arrival information for the entire stage, which may be difficult to obtain in practice. To mitigate these requirements in such a way that only available flow data are used, a rolling-horizon optimization is introduced (Wagner 1977). From upstream detectors an advance flow information for the "head" of the stage is obtained. For the "tail" data from a model are used. An optimal policy is calculated for

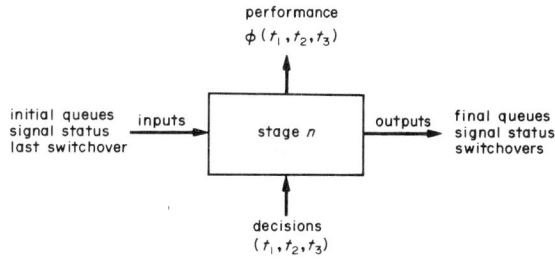

Figure 3
Information and decision flow at a typical stage of rolling-horizon optimization

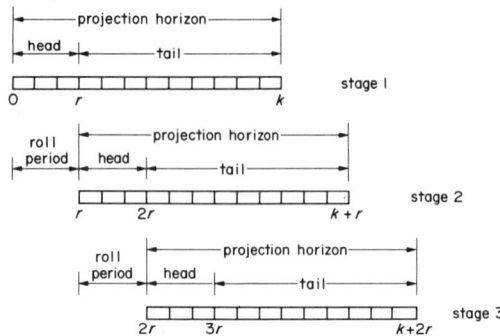

Figure 4
Rolling-horizon optimization process

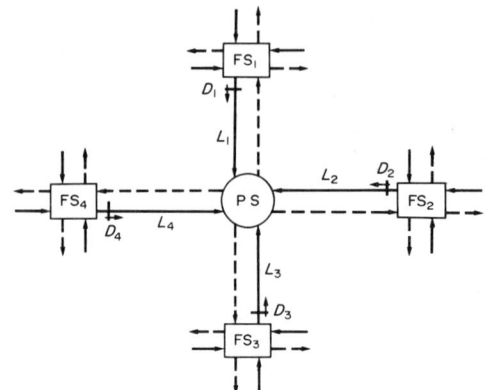

Figure 6
Envisaged information flow for the OPAC method applied to a network of signals

the entire stage, but implemented only for the head section. The horizon is then shifted forward, new flow data obtained for the entire stage (head and tail) and the process repeated. The process is illustrated in Fig. 4.

5. Implementation and Field Testing

The OPAC method was extensively tested using the NETSIM simulation model (Chen *et al.* 1987) (see *Simulation Programs, Microscopic*) and was also field tested in several locations by the Federal Highway Administration (Farradyne Systems Inc. 1989). Results have confirmed the operational capabilities of the method and have shown that significant improvements in performance can be obtained by its implementation. Average delays can be reduced by 30–50% when compared with fixed-time settings and by 5–15% when compared with actuated operation. Interestingly, the largest benefits occur during high volume/capacity conditions (e.g., see Fig. 5) when the intersection performance is most sensitive to the control input.

While the OPAC method was only applied for single-intersection control, it is also amenable for application in a network of signals. The analysis capabilities of OPAC would be used to structure the flows in the network so that coordination can be preserved on the one hand, while taking advantage of the ever present variations in flows on the other. This would require both local analysis capabilities at each intersection, as well as communication with adjacent controllers. The result would be a demand-responsive decentralized flexibly coordinated system. A diagram of the envisioned information flow in such a system is illustrated in Fig. 6.

6. Future Developments

On-line traffic control strategies should be capable of providing results that are better than those produced by off-line methods. Studies undertaken with OPAC and other methods indicate that substantial benefits can be achieved with truly responsive strategies. It was shown that OPAC offers a feasible and promising approach to real-time control. The strategy is designed to make use of readily available data, produces effective control policies and has manageable computational requirements. Of greater significance is the OPAC flow model. It considers the entire projection horizon and, therefore, can be used in a demand-responsive decentralized flexibly coordinated system. There are continuing research and development efforts in this area in the USA as well as in other countries.

See also: Road Network Control; Road Traffic Control: Progression Methods; Road Traffic Control: TRANSYT and SCOOT

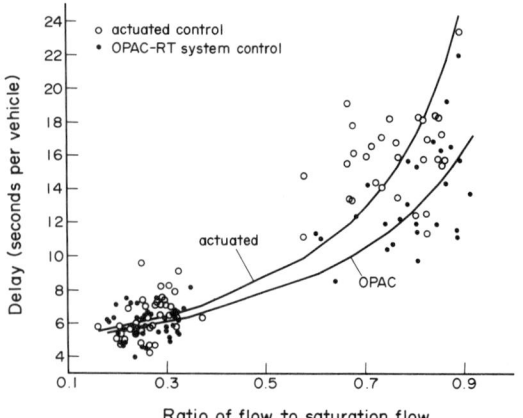

Figure 5
Results from field tests of the OPAC method

Bibliography

Bang K L, Nilsson L E 1976 Optimal control of isolated traffic signals. *ARRB Proc.* **8**, 16–24

Bellman R D 1957 *Dynamic Programming*. Princeton University Press, Princeton, NJ

Chen H et al. 1987 Simulation study of OPAC: A demand-responsive strategy for traffic signal control. In: Gartner N, Wilson N (eds.) *Transportation and Traffic Theory*. Elsevier, New York, pp. 233–49

Dunne M C, Potts R B 1964 Algorithm for traffic control. *Oper. Res.* **12**, 870–81

Farradyne Systems Inc. 1989 Evaluation of the optimized policies for adaptive control (OPAC) strategy. Report No FHWA-RD-89-135. Federal Highway Administration, Washington, DC

Gartner N H 1982a Prescription for demand-responsive urban traffic control. *Transp. Res. Rec.* **881**, 73–6

Gartner N H 1982b Development and testing of a demand-responsive strategy for traffic signal control. *Proc. 1982 American Control Conf.* pp. 578–83

Gartner N H 1983 OPAC: A demand-responsive strategy for traffic signal control. *Transp. Res. Rec.* **906**, 75–81

Gartner N H 1985 Demand-responsive traffic signal control research. *Transp. Res. A* **19**, 369–73

Gazis D C 1964 Optimum control of a system of oversaturated intersections. *Oper. Res.* **12**, 815–31

Gazis D C, Potts R B 1965 The oversaturated intersection. In: Almond J (ed.) *Proc. 2nd Int. Symp. Theory of Road Traffic Flow*. Organisation for Economic Co-operation and Development, Paris, pp. 221–37

Grafton R B, Newell G F 1967 Optimal policies for the control of an undersaturated intersection. In: Edie L C et al. (eds.) *Proc. 3rd Int. Symp. Theory of Traffic Flow*. Elsevier, New York, pp. 239–57

Henry J J, Farges J L 1989 PRODYN. *Proc., 6th IFAC–IFIP–FORS Symp. Transportation*. Pergamon, Oxford, pp. 505–7

Holroyd J, Robertson D I 1973 Strategies for area traffic control systems present and future. TRRL Report LR 569. Transport and Road Research Laboratory, Crowthorne, UK

Hunt R B, Robertson D I, Bretherton R D, Winton R I 1981 SCOOT: A traffic responsive method of coordinating signals. TRRL Report LR 1014. Transport and Road Research Laboratory, Crowthorne, UK

MacGowan J, Fullerton I J 1979–1980 Development and testing of advanced control strategies in the urban traffic control system. *Public Roads* **43** (2–4)

Miller A J 1963 Settings for fixed-cycle traffic signals. *Oper. Res. Q.* **14**, 373–86

Miller A J 1965 A computer control system for traffic networks. In: Almond J (ed.) *Proc. 2nd Int. Symp. Theory of Road Traffic Flow*. Organisation for Economics Co-operation and Development, Paris, pp. 200–20

Robertson D I, Bretherton R D 1974 Optimum control of an intersection for any known sequence of vehicle arrivals. *Proc. 2nd IFAC–IFIP–IFORS Symp. Traffic Control and Transportation Systems*. North-Holland, Amsterdam, pp. 3–17

Toronto Metropolitan Corporation 1974–1976 *Improved Operation of Urban Transportation Systems*, Vol. 1 (March 1974), Vol. 2 (November 1975), Vol. 3 (November 1976). TMC, Toronto, Canada

van Zijverden J D, Kwakernaak H 1969 A new approach to traffic-actuated computer control of intersections. In: Leutzbach W, Baron P (eds.) *Proc. 4th Int. Symp. Theory of Traffic Flow*. *Strassenbau Strassenverkehrstech.* **86**, 113–17

Wagner H M 1977 *Principles of Operations Research*, 2nd edn. Prentice-Hall, Englewood Cliffs, NJ

Webster F V 1958 Traffic signal settings. Road Research Technical Paper 39. Her Majesty's Stationery Office, London

N. H. Gartner
[University of Lowell, Lowell,
Massachusetts, USA]

Road Traffic Control: Progression Methods

Arterial progression schemes are widely used for signal coordination in the USA, as well as in other countries (Forschungsgesellschaft für Strassen-und Verkehrswesen 1981, Homburger 1982). Such schemes provide significant benefits in increased travel speed and smoothness of traffic flow. The conceptual basis for the progression design is that traffic lights tend to group vehicles into a "platoon" with more uniform headways than would otherwise occur. The platooning effect is prevalent on major streets, which have signalized intersections at frequent intervals. It seems desirable to encourage platooning so that continuous movement (or progression) of vehicle platoons through successive traffic lights is maintained. Signal timings are designed to maximize the width of continuous, uniform green bands in both directions along the artery at the projected speed of travel. Such designs work best when the main-street flow is predominantly through traffic and when the number of vehicles turning onto the main street is small.

Advances in optimization techniques and in computational capabilities have steadily increased the sophistication of arterial progression methods, replacing the manual design of time–space diagrams with computer-based methods. Two of the most widely used methods in the USA are PASSER-II and MAXBAND (Federal Highway Administration 1985, McShane and Roess 1990). Whereas PASSER-II uses a heuristic search procedure, MAXBAND employs a powerful mathematical programming technique, thus achieving greater versatility. It calculates the cycle time, offsets, progression speeds and order of left-turn phases (at each intersection) to maximize the weighted combination of the bandwidths in the two directions along the artery.

An alternative design for arterial progression is to use a delay-based signal network optimization method, such as the TRANSYT method (see *Road Traffic Control: TRANSYT and SCOOT*). However, many traffic engineers prefer to impose upon the TRANSYT solution some arterial progression scheme for the main streets in the network. The objective is to obtain a smoother flow of traffic on the principal arteries than is

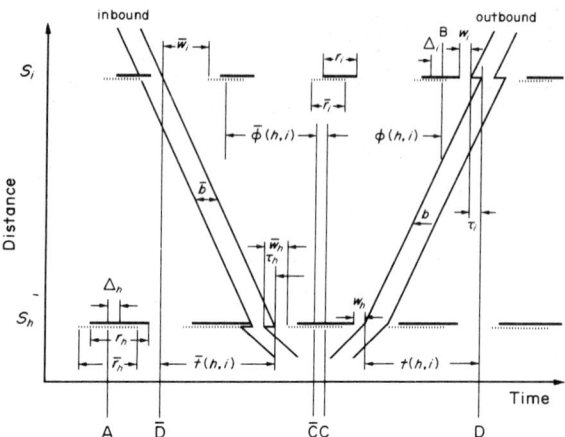

Figure 1
Time–space diagram for the MAXBAND model

achieved by TRANSYT settings alone. This has led to the development of "hybrid" versions in which the TRANSYT program is constrained by a bandwidth solution for the arterial streets (Cohen and Liu 1986, Liu 1988). Substantial benefits can also be gained by concurrent use of delay-based methods (such as TRANSYT) and bandwidth-based methods (such as MAXBAND) (Cohen 1983, Skabardonis and May 1985). These benefits accrue primarily due to the phase-sequencing capabilities of the progresssion methods.

1. Bandwidth Maximization

The basic, symmetric, uniform-width bandwidth maximization problem will first be formulated and then expanded by adding more decision variables. The generalized time–space diagram is shown in Fig. 1. An arterial with n signals is considered. The signal at node (intersection) i, where $i = 1, \ldots, n$, is denoted by S_i. All time variables are in units of the cycle time: b $(\bar{b})$ is the outbound (inbound) bandwidth; r_i $(\bar{r}_i)$ is the outbound (inbound) red time at S_i; w_i $(\bar{w}_i)$ is the interference variable which is the time from the right (left) side of red at S_i to the left (right) edge of the outbound (inbound) green band; $t(h, i)$ $[\bar{t}(h, i)]$ is the travel time from S_i to S_h outbound $[S_h$ to S_i inbound]; $\phi(h, i)$ $[\bar{\phi}(h, i)]$ is the internode offset which is the time from the center of an outbound [inbound] red at S_h to the center of a particular outbound [inbound] red at S_i; Δ_i is the intranode offset which is the time from the center of $\bar{r}_i$ to the nearest center of r_i (positive if the center of r_i is to right of the center of $\bar{r}_i$); and $\tau_i(\bar{\tau}_i)$ is the queue clearance time, an advance of the outbound (inbound) bandwidth at S_i to clear up turning-in traffic before arrival of the main-street platoon.

First, the directional interference constraints are introduced. These constraints make sure that the progression bands use only the available green time and do not infringe upon any of the red times. From Fig. 1 it can be seen that at each signal

$$\left. \begin{array}{l} w_i + b \leq 1 - r_i \\ \bar{w}_i + \bar{b} \leq 1 - \bar{r}_i \end{array} \right\} \quad (1)$$

Next, the loop-integer constraint is calculated. This constraint is due to the fact that the signals of the arterial are synchronized (i.e., they operate on a common cycle time). Starting at the center of the outbound red at S_h and proceeding along a loop consisting of the centers of: outbound red at S_i—inbound red at S_i—inbound red at S_h—outbound red at S_h, a point must be reached that is removed an integral number of cycle times from the point of departure. Summing algebraically the appropriate internode and intranode offsets along the loop

$$\phi(h, i) + \bar{\phi}(h, i) + \Delta_h - \Delta_i = m(h, i) \quad (2)$$

where $m(h, i)$ is the corresponding loop integer variable. This equation can be expressed in terms of the time variables defined earlier. In an arterial with n signals, there are $n - 1$ such constraints. For a symmetric progression it is necessary that $b = \bar{b}$. The following bandwidth maximization problem (BMP) is obtained.

BMP 1

Given: cycle time, splits
travel times (or speeds)
queue clearances

Find: bandwidth $b = \bar{b}$
offsets, interferences

To: maximize $b = \bar{b}$

Subject to: interference constraints
loop-integer constraints

Using the outputs of BMP 1 the final time–space diagram can be generated. An example of a symmetric progression is illustrated in Fig. 2. The basic problem can be solved by mixed-integer linear programming (Little 1966) or by heuristic combinatorial search optimization techniques (Morgan and Little 1964).

2. The MAXBAND Method

MAXBAND extends the basic maximization problem described in Sect. 1 to make it a more versatile design tool. The first extension concerns the directional weighting of the two bands. The traffic engineer may wish to favor one direction of traffic over the other (e.g., the inbound direction during the morning peak period and the outbound direction during the afternoon peak period). A "balanced" progression may be desirable during off-peak periods. Let k be the target ratio of inbound-to-outbound bandwidth (taken as the ratio

Road Traffic Control: Progression Methods

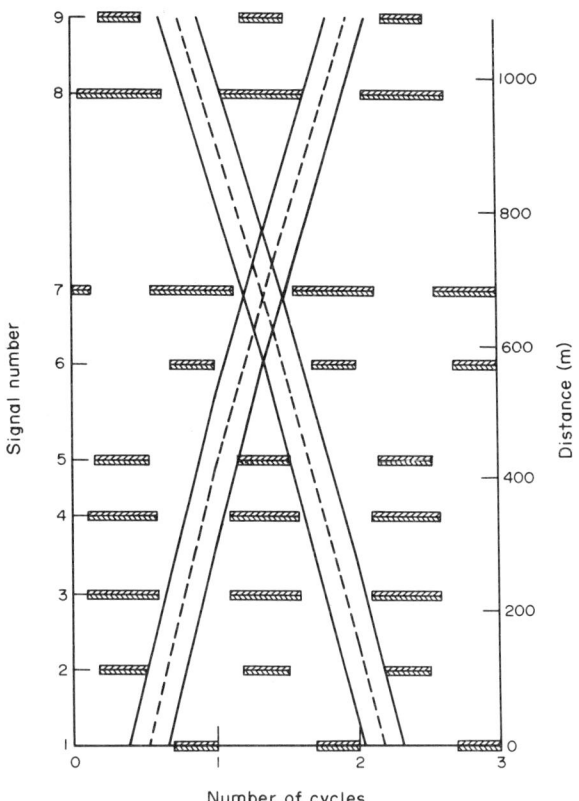

Figure 2
MAXBAND time–space diagram: symmetric progression ($k = 1.0$) and cycle time $C = 70$ s

of total inbound volumes to total outbound volumes along the arterial). The objective function and the ratio constraint can be set up as

$$\text{maximize } (b + k\bar{b})$$

subject to

$$\bar{b} \geq kb, \quad \text{if } k < 1 \text{ (outbound direction favored)}$$
$$\bar{b} \leq kb, \quad \text{if } k > 1 \text{ (inbound direction favored)}$$
$$\bar{b} = b, \quad \text{if } k = 1 \text{ (balanced progression)}$$

The first two inequalities can be replaced by a single inequality, called the bandwidth ratio constraint:

$$(1-k)\bar{b} \geq (1-k)kb \qquad (3)$$

The second extension is to let both the common signal cycle time C (s) and the link-specific progression speed $v_i(\bar{v}_i)$ (m s^{-1}) be optimizable variables (link i represents the road section between S_i and S_{i+1}). This introduces considerable flexibility in the calculation of the best arterial progression. Each of these variables is constrained by upper and lower limits: C_1 and C_2 are the lower and upper limits, respectively, on cycle length; e_i and f_i ($\bar{e}_i$ and $\bar{f}_i$) are the lower and upper limits, respectively, on an outbound (inbound) speed; and g_i and h_i ($\bar{g}_i$ and $\bar{h}_i$) are the lower and upper limits, respectively, on change in outbound (inbound) speed. For the outbound direction

$$C_1 \leq C \leq C_2 \qquad (4)$$
$$e_i \leq v_i \leq f_i \qquad (5)$$
$$g_i \leq v_{i+1} - v_i \leq h_i \qquad (6)$$

Corresponding expressions are obtained for the inbound direction. Another important decision capability that is offered by mixed-integer linear programming is to determine the order of the left-turn phase (if one is present) with respect to the through green at any signal. This is done by means of 0–1 decision variables. The variables for the phase-sequence decision are included in the loop-integer constraints.

Incorporating these extensions into BMP 1 yields a more versatile bandwidth maximization problem, BMP 2.

BMP 2

Given: splits, queue clearances
             target ratio of bandwidths
             limits on: cycle time, link speeds, changes in speeds

Find: cycle time, offsets, interferences
             bandwidths, b, $\bar{b}$
             link progression speeds
             left-turn phase patterns

To: maximize $b + k\bar{b}$

Subject to: cycle-time constraint
             bandwidth-ratio constraint
             interference constraints
             loop-integer constraints
             speed and speed-change constraints

To calculate the offsets, green splits need to be given or, alternatively, the user can provide traffic volume and capacity information for each intersection and the program will calculate the splits. BMP 2 is the current version of the MAXBAND program, which uses a mixed-integer linear programming code for the optimization (Little *et al.* 1981). MAXBAND was also extended for application to multiarterial closed networks (Cohen and Mekemson 1985). An example of a MAXBAND generated time–space diagram for $k \neq 1$ is shown in Fig. 3.

3. The MULTIBAND Method

A basic limitation of traditional progression schemes is that their design criterion (maximum bandwidth) does not depend on the actual traffic volumes on the arterial

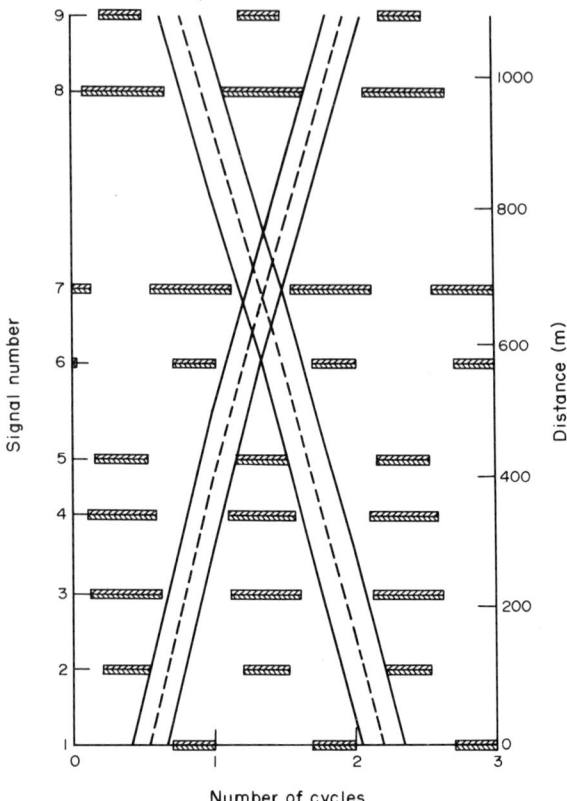

Figure 3
MAXBAND time–space diagram: band weight equals the total volume ratio ($k = 1:1.2$) and $C = 60$ s

to different traffic flow patterns. Referring to the geometry in Fig. 4, the following variables can be defined. Let $b_i (\bar{b}_i)$ be the outbound (inbound) bandwidth between signals S_i and S_{i+1}; there is now a specific band for each directional road section or link. Also, let $w_i (\bar{w}_i)$ be the time from the right (left) side of red at S_i to the center line of the outbound (inbound) green band; the time reference point at each signal is moved from the edges to the center line of the band.

The following constraints apply in the outbound directions at signal S_i:

$$\left. \begin{array}{c} w_i + \dfrac{b_i}{2} \leq 1 - r_i \\[6pt] w_i - \dfrac{b_i}{2} \geq 0 \end{array} \right\} \quad (7)$$

The pair of constraints can be combined as follows:

$$\dfrac{b_i}{2} \leq w_i \leq 1 - r_i - \dfrac{b_i}{2} \quad (8)$$

The same constraint exists at signal S_{i+1}, since band b_i must be constrained at both ends. Corresponding relationships exist in the inbound direction. These are the directional interference constraints. The loop-integer constraints given by Eqn. (2) remain unchanged and so do the travel time and speed-change constraints.

The bandwidth ratio constraint given by Eqn. (3) is also changed to reflect the multiband situation. For each pair of parallel links

$$(1 - k_i) \bar{b}_i \geq (1 - k_i) k_i b_i \quad (9)$$

where k_i is the target ratio of inbound to outbound bandwidth on section i (taken as the ratio of the corresponding volumes in each direction). There are now $n - 1$ such constraints.

In MULTIBAND the bands are link specific and can be weighted disaggregately to achieve desirable traffic

links and, therefore, is insensitive to variations in such volumes. The sum of bandwidths that are obtained for an arterial can be allocated in any desired ratio among the two directions of travel. A common practice is to apportion it according to the overall directional volume ratio k. However, because of turn-in and turn-out traffic, volumes are, generally, not constant in each direction. Consequently, the idea of a uniform platoon moving through all the signals does not always hold. The ratio of volumes on opposing road sections between each pair of adjacent signals may also be varying and a single adjustment parameter for the entire arterial (k) cannot adequately reflect this diversity.

MULTIBAND is a new progression method that is designed to overcome these limitations by incorporating in the bandwidth optimization procedure a systematic traffic-dependent criterion (Gartner *et al.* 1990, 1991). MULTIBAND assigns a different bandwidth to each directional road section of the arterial. The bandwidth can be individually weighted with respect to its contribution to the overall objective function. Thus, a method is obtained that is sensitive to varying traffic conditions and the progression scheme can be tailored

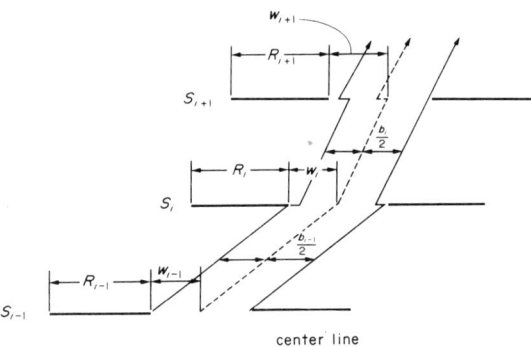

Figure 4
Geometric relations for the MULTIBAND model

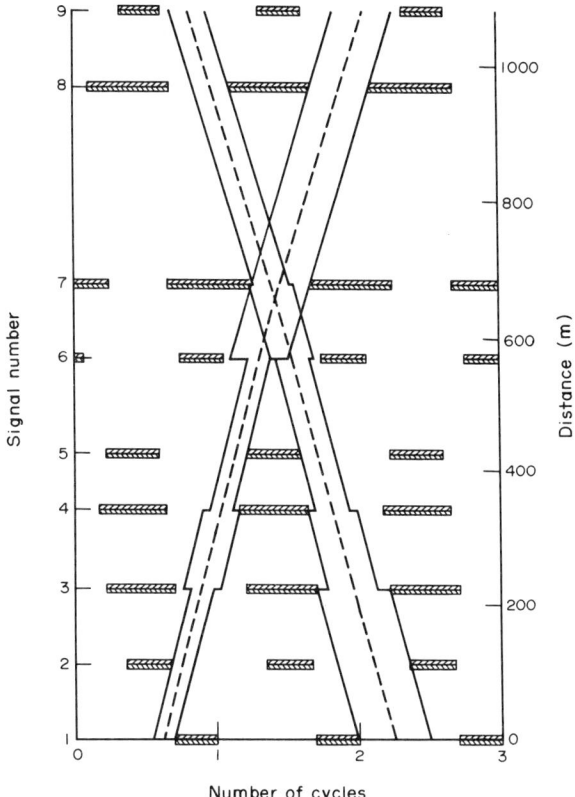

Figure 5
MULTIBAND time–space diagram: band weights equal to (total volume/ratio)^4 and $C = 68$ s

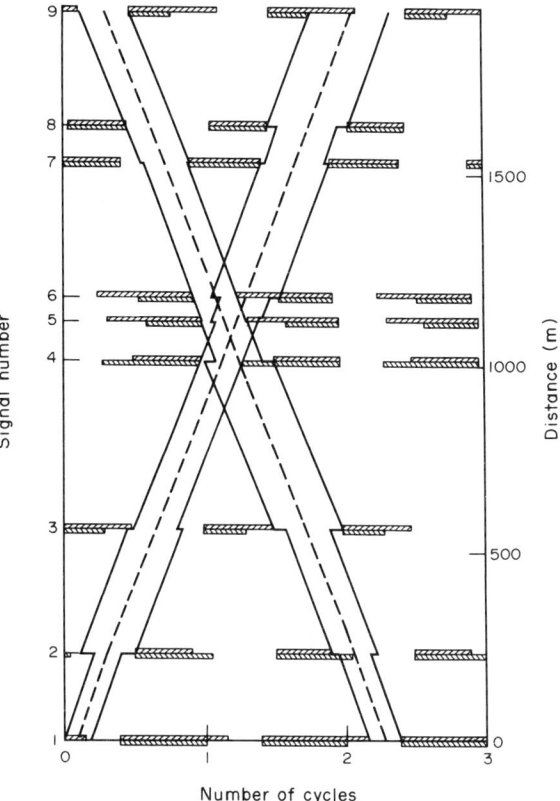

Figure 6
MULTIBAND time–space diagram: band weights equal to (platoon volume–capacity ratio)^4 and $C = 95$ s

objectives for each link. The objective function has the following form:

$$\text{maximize } B = \frac{1}{n-1} \sum_{i=1}^{n-1} (a_i b_i + \bar{a}_i \bar{b}_i) \quad (10)$$

where a_i ($\bar{a}_i$) are the link-specific weights in the two directions. The following weighting options are commonly used:

$$a_i = \left(\frac{V_i}{S_i}\right)^p, \quad \bar{a}_i = \left(\frac{\bar{V}_i}{\bar{S}_i}\right)^p$$

where V_i ($\bar{V}_i$) is the directional volume on section i, outbound (inbound) (either the total volume or the through volume can be used, the latter being called the platoon volume); S_i ($\bar{S}_i$) is the saturation flow on section i, outbound (inbound) (this is the capacity volume in vehicles per hour of green); and p is the exponential power—for example, $p = 0$ (unit coefficients), $p = 1$ (i.e., volume–capacity ratio), $p = 2$ (i.e., (volume–capacity ratio)^2), $p = 4$ (i.e., (volume–capacity ratio)^4).

Other weighting options can also be specified. This provides considerable flexibility to the user. The result is a multiband/multiweight maximization program which is summarized in BMP 3.

BMP 3

Given: splits, queue clearances
target ratios of bandwidths for each section; limits on: cycle time, link speeds, changes in speeds
allowed left-turn phase patterns

Find: cycle time, offsets, interferences
link-specific bandwidths $b_i, \bar{b}_i$
link progression speeds
left-turn phase patterns

To: maximize $B = (n-1)^{-1} \sum_{i=1}^{n-1} (a_i b_i + \bar{a}_i \bar{b}_i)$

Subject to: cycle-time constraint
bandwidth-ratio constraints
interference constraints
loop-integer constraints
speed and speed-change constraints

Similar to MAXBAND, MULTIBAND is also based

on mixed-integer linear programming optimization. Examples of MULTIBAND outputs are shown in Figs. 5 and 6 (for a multiphase arterial). The expanded decision capabilities of MULTIBAND can lead, in general, to considerable improvements in performance when compared with MAXBAND.

See also: Road Network Control; Road Traffic Control, Demand-Responsive; Road Traffic Control: TRANSYT and SCOOT

Bibliography

Cohen S L 1983 Concurrent use of MAXBAND and TRANSYT signal timing programs for arterial signal optimization. *Transp. Res. Rec.* **906**, 81–4
Cohen S L, Liu C C 1986 The bandwidth-constrained TRANSYT signal-optimization program. *Transp. Res. Rec.* **1057**, 1–7
Cohen S L, Mekemson J R 1985 Optimization of left-turn phase sequence on signalized arterials. *Transp. Res. Rec.* **1021**, 53–8
Federal Highway Administration 1985 Traffic Control Systems Handbook. Report FHWA-IP-85-11. US Department of Transportation, Washington, DC
Forschungsgesellschaft für Strassen-und Verkehrswesen 1981 Richtlinien für Lichtsignalanlagen. FGSV, Cologne, Germany
Gartner N H, Assmann S F, Lasaga F, Hou D L 1990 MULTIBAND: A variable-bandwidth arterial progression scheme. *Transp. Res. Rec.* (in press)
Gartner N H, Assmann S F, Lasaga F, Hou D L 1991 A multiband approach to arterial traffic signal optimization. *Transp. Res. B.* **25**, 55–74
Homburger W S (ed.) 1982 *Institute of Transportation Engineers Transportation and Traffic Engineering Handbook*, 2nd edn. Prentice-Hall, Englewood Cliffs, NJ
Little J D C 1966 The synchronization of traffic signals by mixed-integer linear programming. *Oper. Res.* **14**, 568–94
Little J D C, Kelson M D, Gartner N H 1981 MAXBAND: A program for setting signals on arteries and triangular networks. *Transp. Res. Rec.* **795**, 40–6
Liu C C 1988 Bandwidth-constrained delay optimization for signal systems. *ITE J.* **58** (12), 21–6
McShane W R, Roess R P 1990 *Traffic Engineering*. Prentice-Hall, Englewood Cliffs, NJ
Morgan J T, Little J D C 1964 Synchronizing traffic signals for maximal bandwidth. *Oper. Res.* **12**, 896–912
Skabardonis A, May A D 1985 Comparative analysis of computer models for arterial signal timing. *Transp. Res. Rec.* **1021**, 45–52

N. H. Gartner
[University of Lowell, Lowell, Massachusetts, USA]

Road Traffic Control Systems in Japan

Although area traffic control systems are already in widespread use in Japan, efforts are continuing to make them more sophisticated in order to deal with ever increasing traffic density. This article introduces a new type of area traffic control system.

In this system, linear information, such as the length of the traffic queues and the travel time, is collected in addition to point information, such as traffic volume at key intersections, allowing signal parameters including subarea configurations to be determined by the overall traffic situation. Signal control focuses on developing ways to increase the traffic capacity at key intersections, and the development of a split determination approach applicable across the spectrum from undersaturated to oversaturated traffic flow.

Progress is now being made in using light-emitting diode (LED)-based traffic information boards and roadside radio facilities as a way of providing traffic information services to the driver on roads leading into cities. The roadside radio system is completely automatic, from determining what information to broadcast to actual voice synthesis.

Efforts are also underway to improve traffic information services to deal with the worsening congestion on expressways, especially those in the cities.

1. History

The rapid economic development of Japan in the 1960s and the accompanying increase in demand for road transportation, led to levels of traffic congestion and traffic accidents so severe they came to be known as the "traffic wars." In an attempt to reduce the number of traffic fatalities, the National Police Agency developed a five-year plan in 1971 to introduce traffic safety systems on a nationwide basis, focusing on signal control systems in cities and on major trunk routes.

This five-year plan guided the introduction of area traffic control systems across the nation, starting with a system installed in the center of Tokyo in 1971. The number of systems in place as of March 1989 is as follows:

area control (74 systems) 38 298 signal intersections
coordinated   20 402 signal intersections
isolated   69 310 signal intersections

Now that the mass introduction of signal control systems is largely completed, the focus is turning to ways of increasing the sophistication of the installed systems, the development of more effective signal control approaches, the provision of enhanced traffic information services, and improved system operability. During this period, the Tokyo area traffic control system was expanded until it controlled approximately 6000 signal intersections, the largest system of its kind in the world.

Note that although traffic fatalities in Japan started decreasing in 1971, reaching one-half of the 1970 level by 1979, they are now increasing again.

Table 1
Concept of signal control

Traffic	Control aim	Control method	
		Split	Cycle length/offset
Low traffic	safe flow on streets (speed constraint)	split formation by inflow saturation degree	cycle length program selection
Near saturation	delay minimum (high green-time efficiency of critical intersections)	green-length control by right-turn actuation	offset program selection (corresponding to cycle length)
Oversaturation	flow maximum (maximizing of passing volume at critical intersections) preference control (control of queue length by predetermined preference)	maximum flow rate actuation preference control by travel time of all traffic at critical intersection	offset program selection (preference control of main-street traffic by restraint of incoming traffic from minor streets)

2. Control Methods

The severity of traffic congestion in large Japanese cities has led researchers to focus on how to improve traffic flow, using approaches such as signal control methods to increase the traffic processing capacity of major intersections and provision of traffic information services to drivers. Emphasis has also been placed on the use of offsets to improve traffic flow safety.

2.1 Gathering Traffic Data

Vehicle detectors for traffic signal control are installed at points about 150 m upstream from critical intersections, approximately corresponding to the end of the line of waiting vehicles that could pass through the intersection during one signal cycle. To measure traffic congestion, detectors are also placed 300 m, 500 m and 1000 m upstream from the intersection, this data being used for signal control.

A recently developed method for synchronizing the data obtained from each detector with the signal phase when gathering traffic data is helping to reduce its fluctuations.

Ultrasonic sensors are most widely used due to their ease of installation and maintenance.

2.2 Signal Control

Signal control is based on control objectives established for each of three sets of traffic conditions at critical intersections, as shown in Table 1. In low-traffic situations, offsets are used to discourage speeding and improve traffic safety. In near-saturated conditions, the goal is to reduce the amount of delay to a minimum and this is accomplished by determining the split appropriate to the traffic demand at the major intersection. In oversaturated conditions, the goal is to maximize traffic flow, so priority shifts to splits reflecting the nature and length of congestion on incoming routes.

(a) *Green split.* A key issue in dramatically reducing traffic congestion is how to control critical intersections that have reached near-saturation level. In the past, splits even at critical intersections were determined, as well as cycle lengths and offsets, through traffic-responsive program selection; several years ago, a method to calculate splits from traffic volume and the number of waiting vehicles was developed. This method, based on the principles of split determination, uses saturation degree, defined as traffic demand divided by saturation flow rate, as the key traffic measure. Traffic demand is computed as the sum total of vehicles expected to arrive plus those already waiting.

According to this method, the split of an intersection is represented as:

$$S_i = \frac{\rho_i}{\Sigma \rho_i} \quad (1)$$

The phase saturation level is given by formula:

$$\rho_i = \max(\rho_{i1}, \rho_{i2}) \quad (2)$$

The saturation level of each inflow route is given by formula:

$$\rho_{ij} = \frac{Q_{ij} + k_{ij} E_{ij}}{s_{ij}} \quad (3)$$

Where i is the phase number, j is the inflow route number, s is the saturation flow rate, E is the number of vehicles queued at each inflow route and k is the strategic flow rate of E ($0 \leq k \leq 1$).

Saturation flow rate can be measured by the sensor at the 150 m point when the intersection is congested.

This new approach greatly simplifies the specification of signal control parameters to the computer. This method can also be applied to oversaturated intersections.

To maximize the transit traffic volume at oversaturated intersections, it is insufficient to hold up traffic on major approaches (those with high saturated traffic flow rates). Unfortunately, this may have the effect of causing congestion on minor approaches, with lower saturated traffic flow rates. To solve this problem, splits are determined so that the ratio of travel times for a fixed interval obeys an objective function for each intersection, perhaps by controlling the flow rate on both approaches when congestion on minor approaches exceeds some value.

The travel time for a congested interval on an approach to an intersection is expressed by the following formula:

$$T = \sum_i L_i \frac{1}{H_i} \frac{1}{Q_i} = \sum_i L_i (K_m - aQ_i) \frac{1}{Q_i} \quad (4)$$

where L_i is a short interval distance where the vehicle detector is located, H_i is the average headway distance, Q_i is the short interval traffic volume and K_m and a are the traffic volume and density regression coefficients, respectively.

Yet another method uses a local signal controller with a microcomputer to monitor the timing of each vehicle to pass the intersection, automatically determining the point on each signal cycle at which ending the green time will maximize the flow rate.

(*b*) *Offset*. In the Tokyo area traffic control system begun in 1971, a method of minimizing delay and number of stops at each intersection on roads by measuring them and modifying the offset appropriately underwent trials. Eventually, however, it came to be believed that rather than varying offset in small amounts in order to improve traffic safety, it was better to establish a stable offset; offset is, thus, now determined via pattern selection based on the amount of traffic in both directions. The offset patterns are predetermined through simulations. In particular, offset can be adjusted to restrict speeding during low-volume traffic conditions at night.

(*c*) *Cycle length*. The goals of cycle-length control are to minimize the delay and number of stops throughout undersaturated subareas, while maximizing traffic processing capacity at oversaturated critical intersections. Delay and number of stops are governed by a combination of cycle length and offset. The approach is to select the cycle-length pattern for critical intersections based on the saturation level, and then to select one of three offset patterns to use in conjunction with the cycle length.

Subareas whose selected cycle lengths are close are grouped and controlled by the same cycle as a unit.

2.3 Traffic Information Services

(*a*) *Roadside traffic information services*. Information services based on film-type variable signs which can display any of several messages have been in use for many years. The next development was the variable message board which can display any dot pattern, whether text or image, transmitted from a central location. The latest trend is the replacement of incandescent lamps in these boards with LEDs. A combination of red and green LEDs can be used to produce orange displays, effective at catching the attention of drivers.

In order to provide more information than can be fitted onto a variable message board, roadside radio systems have been installed in cities. These systems broadcast traffic information using weak radio waves of 1620 kHz, from antennas installed along 1–2 km of roadside. Drivers need simply to turn on their car radios to obtain information about traffic congestion and its causes on the roads near them, and can then make informed decisions about what routes to use. The information is created completely automatically by computer, then converted into speech by voice synthesis.

(*b*) *Providing traffic information services directly to the vehicle*. Vehicle navigation systems are now a major focus of research around the world. In Japan too, a navigation system combining a technique known as the map-matching method with a self-contained location system has been developed. In 1988, an experiment was carried out to provide traffic information to the navigation system. This information service system dynamically sends information on congestion levels for each road segment, the availability of parking lots, and so on, to the vehicle using radio waves, and this information is then displayed on top of a map shown on a cathode-ray-tube display in the vehicle.

3. System Configuration

The Japanese police are organized into units for each of the 47 prefectures. The police are responsible for installing and managing traffic signals, with all the prefectural traffic control systems being networked around the headquarters traffic control center in charge of the prefectural capital. The conceptual system configuration of this system is shown in Fig. 1.

The computer system in the headquarters has three subsystems. Controlling all systems in the prefecture, including urban centers installed at locations remote from the headquarters and subcenters in small and medium-sized cities, is the operations management subsystem. The signal control subsystem takes care of signal control in the region where the control center is located. Finally, the traffic information subsystem manages various equipment providing traffic information to drivers.

The computer in the center and each local signal are connected in a 1:1 star configuration, with signal

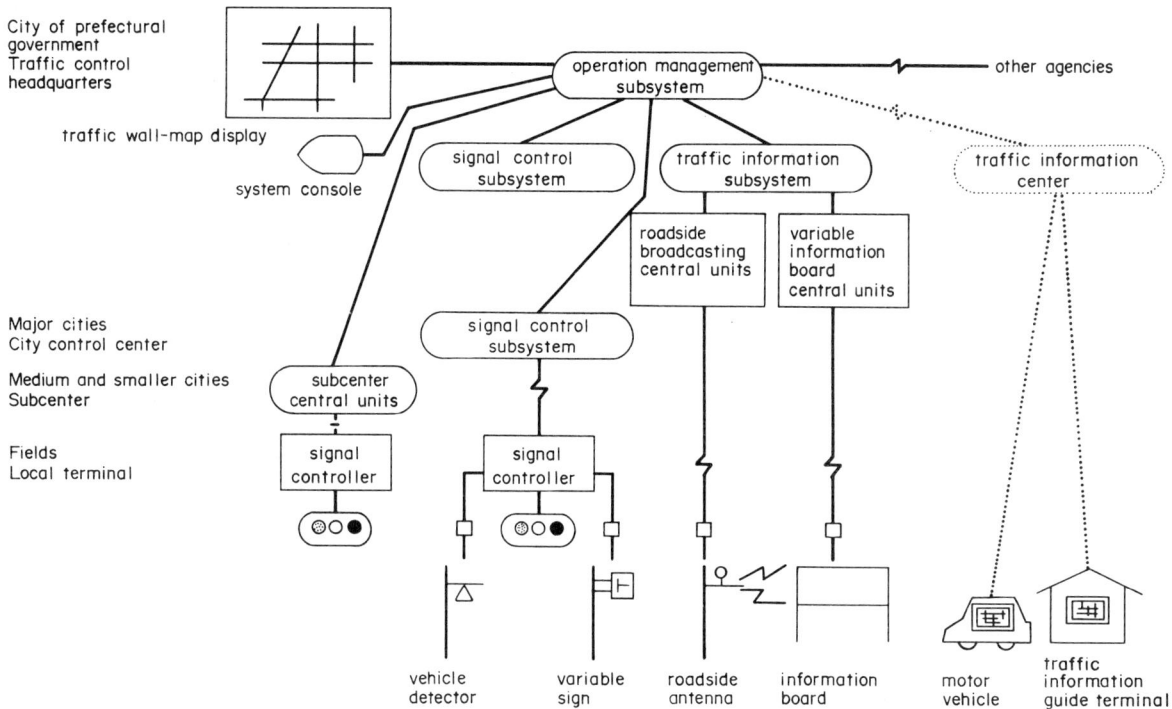

Figure 1
Conceptual system configuration of the area traffic control system

control commands sent down to the signal at 100 bits per second and the answer-back signal and detection signals from up to eight vehicle detectors sent back to the center at 600 bits per second.

The latest local signal controllers typically use a dedicated large-scale integration (LSI) for basic control, together with a microcomputer providing additional advanced capabilities. This type of traffic signal controller is highly reliable, with a mean time between failures (MTBF) of much more than 100 000 h.

4. Urban Expressway Traffic Control Systems

The construction of expressways has proceeded at a remarkable pace in recent years in Japan. Due to the extraordinarily high construction costs, these expressways are toll roads. However, this has not prevented heavy traffic demand on expressways in or near urban areas, resulting in frequent congestion.

Urban expressway traffic control systems, primarily designed to provide traffic information services to drivers, have been developed by the expressway administrators and been installed in major urban areas and their vicinity. The large-scale, sophisticated metropolitan expressway traffic control system, in use in Tokyo, is considered here.

The metropolitan expressways are a system of elevated toll roads connecting Tokyo to other expressways in the outlying regions with radial roads. The expressways total 200 km in length and congestion is severe, especially on the ring roads in the center of Tokyo. One of the approaches being taken to solve the problem focuses on traffic information services.

Vehicle detectors are installed in pairs a few meters apart to permit vehicle speed to be measured, with pairs located about every 300 m. Every minute, the detected information is analyzed to update the traffic information service. The existence of traffic congestion is determined for each 1 km road segment: if the average calculated speed drops below 20 km h^{-1}, the segment is considered "congested"; if between 20 km h^{-1} and below 40 km h^{-1} it is considered to be "slightly congested."

The resulting information is conveyed to drivers using variable message sign boards, roadside radios and the recently developed graphical information boards, which are shown in Fig. 2. These information boards display the road network, showing slightly congested sections in orange and congested sections in red. An accident is marked by a blinking red X mark, and a road closure by a solid red X mark. These boards are installed at key points leading into the center of Tokyo.

In its quest to offer more useful traffic information services and raise the efficiency of its road network, the metropolitan expressway system is also developing

Figure 2
Graphical information board

systems to forecast traffic conditions up to 1 h ahead, find optimal on-ramp closing algorithms, measure travel time and detect incidents.

Bibliography

Anon 1988 Research of signal control methods on main roads (in Japanese). Japan Traffic Management Technology Association, Tokyo, pp. 69–72

Anon 1989 *Traffic Control System of the Metropolitan Expressway*. Metropolitan Expressway Public Corporation, Tokyo

Furuscho T, Yoshikawa T 1985 Road-side radio system (in Japanese). *Traffic Eng.* **20**, 26–35

Ikenoue K, Usami T, Miyasako T 1986 Travel time prediction algorithm and signal operations at critical intersections for controlling travel time. *2nd IEE Int. Conf. Road Traffic Control*. Institution of Electrical Engineers, London

Koshi M 1968 One method of offset formation in area traffic control. *Seisan Kenkyo* **20**(3), 45–7

Koshi M 1988 State of the art and research needs of area traffic signal systems in Japan. *Transportation Research Board, 67th Annual Meeting*. TRB, Washington, DC

Mitoh K, Tenmoku K, Doi Y, Simizu O 1987 Development of a self-contained automobile navigation system (in Japanese). IEICE Technical Report SANE 87-47. Institute of Electronics, Information and Communication Engineers, Tokyo

Okamoto H, Tuzawa M 1989 Advanced mobile traffic information and communication system (AMTICS). *IEEE Int. Conf. VNIS '89*. Institution of Electrical and Electronic Engineers, New York

Suzuki M 1984 Area traffic control systems in Japan. *OECD Road Transport Research Program*. Organisation for Economic Co-operation and Development, Paris

Takada K, Tanaka Y 1989 Road/automobile communication system (RACS) and its economic effect. *IEEE Int. Conf. VNIS '89*. Institution of Electrical and Electronic Engineers, New York

Yamamoto T, Yamaoka S, Eikawa Y, Doi M 1988 Advanced local traffic signal controller for urban traffic control systems. *Sumitomo Electr. Tech. Rev.* **27**, 96–103

Yumoto N 1970 Multi-criterion area traffic control system with feedback features. *1st IFAC Int. Symp. Traffic Control*. International Federation of Automatic Control, Laxenburg, Austria

T. Kitamura
[Sumitomo Electric Industries, Tokyo, Japan]

Road Traffic Control: TRANSYT and SCOOT

A key aspect of the operation of road traffic control in urban areas is that of intersections, which have to safely and efficiently accommodate a variety of conflicting traffic movements. Traffic signals are now used extensively at busy intersections, particularly in urban areas, to safely separate conflicting movements in time and to minimize traffic delays. When signal-controlled junctions are closely spaced it is beneficial to coordinate the timings to allow efficient traffic progressions. This is usually achieved using a computer-based urban traffic control (UTC) system within which signal timings are calculated so as to optimize some measure of traffic performance, usually the minimization of journey time.

The program to optimize linked signal settings for UTC that is most common throughout the world is TRANSYT (Robertson 1969). TRANSYT operates on historic data, whereas the more recently developed SCOOT system (Hunt *et al.* 1981) operates by interpreting comprehensive detector information on line (see *Traffic Control Modes*).

1. TRANSYT

TRANSYT, the traffic network study tool, is a computer model to optimize the linking and timing of traffic signals in a network. Optimization is undertaken off line using historic data, and several traffic flow regimes may be simulated to produce a suite of predetermined programs for use in different traffic situations. Changes between programs may be made on a time-of-day basis, or from on-line measurements of traffic conditions, such as flow or queue length, taken at a few specific locations in the network.

The TRANSYT program was first released in 1969 (Robertson 1969) and, while the basic approach has remained the same, there have been several substantial revisions to enhance the facilities available to make it a more flexible tool for the traffic engineer (Vincent *et al.* 1980, Chard and Lines 1987).

1.1 Model Structure

The TRANSYT program produces a performance index (PI) from a traffic model of the network. The PI

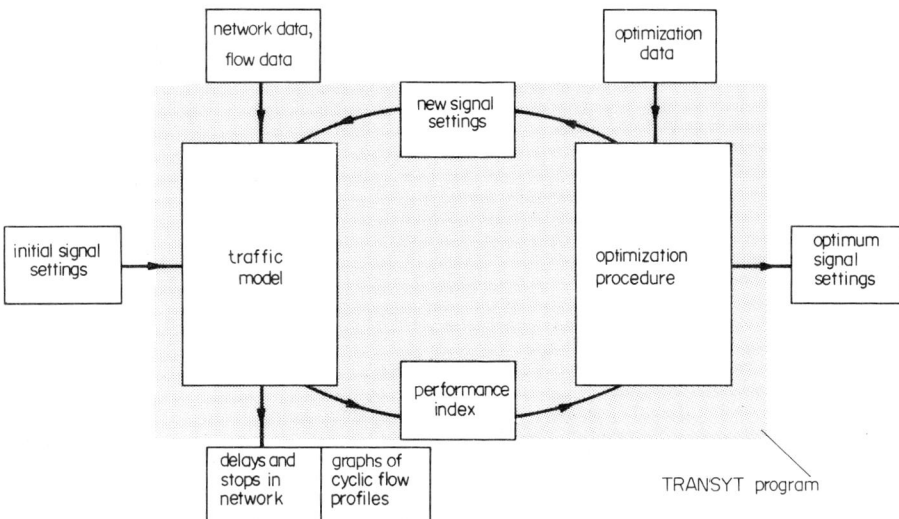

Figure 1
Structure of the TRANSYT program (after Robertson 1969)

is then subject to an optimization procedure to give new signal settings, which are evaluated by the traffic model to produce a new PI. The procedure is iterative until optimum signal settings are reached (see Fig. 1). Additional statistical and graphical output may be produced to describe traffic conditions, either globally or at specific locations in the network.

In order to be run, the program requires a network description, traffic flow estimates and optimization criteria and factors.

1.2 Initial Signal Settings

Input information on signal settings can be considered in two parts. The first relates to items such as stage structure and intergreen times which do not form part of the TRANSYT optimization process and the second includes cycle times, green splits and offsets which may be varied to optimize the predefined PI.

Up to seven stage changes may be specified by the user for each signal-controlled junction. Each change may either initiate or terminate a green signal, and "displacements" may be introduced to model starting and ending time lags. Displacements are necessary because of the simplistic linear time and distance assumptions adopted within TRANSYT and are akin to the starting and end lost times as defined by Webster and Cobbe (1966). Minimum green times are also set for safety and to ensure the discharge of a minimum queue.

The selection of an appropriate stage structure may be crucial to the efficient movement of traffic in a network, but is undertaken external to TRANSYT and, therefore, not subject to the optimization process. However, internal facilities in the program can consider double cycling or repetition of green phases at individual intersections.

Initial values for signal timings may be input either by the user, or estimated within the program on the assumption that critical conflicting links should have equal degrees of saturation.

1.3 Network and Traffic Flow Data

Road network and traffic flow data need to be determined for representation in the model.

Each road section between modelled intersections constitutes one or more links, a link being a movement that can be separately treated at the subsequent stop line, as illustrated in Fig. 2. Thus, in Fig. 2, the flow from nodes 1 and 2 would be considered as a single link, whereas that between nodes 3 and 1 would be identified as two parallel links with left-turning traffic having an exclusive lane and phase. Between two and five links can use a single shared stop line, with queued traffic discharging in the order of arrivals. The most common situation occurs when ahead and left turners are identified as one link, with opposed right turners on a separately marked lane forming an additional link with distinct queues. (This lane configuration is reversed when vehicles drive on the right.)

It is assumed, for the initial arrival profiles, that traffic flow is regular, the platooning of vehicles being the result of interruption of the arrivals by the signal control. In earlier versions of TRANSYT, it was necessary to identify "dummy" links to enhance the modelling process by producing better initial estimates of arrival patterns. This is now undertaken within the program.

Significant priority intersections and points of traffic generation such as car parks should be identifed as they may form part of the modelled network. Any bottlenecks that exist should also be input into the network

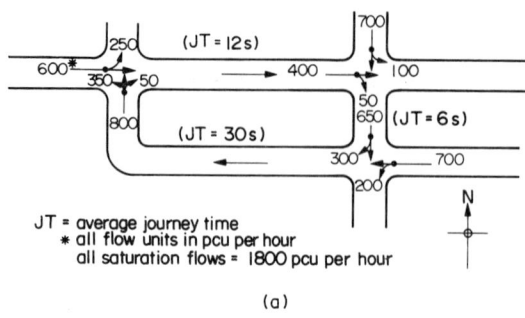

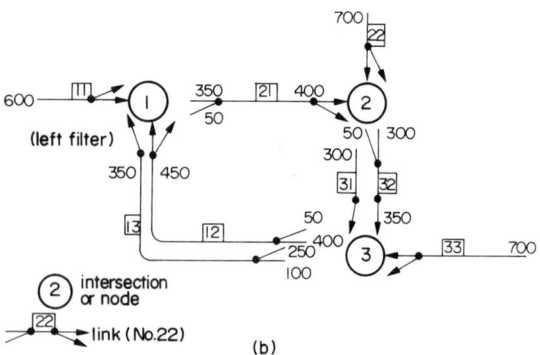

Figure 2
Simple triangular road network: (a) network and flows and (b) links and nodes (after Vincent et al. 1980)

description and are treated as intersections with 100% green time, but with a limited saturation flow that defines the bottleneck capacity. Traffic flow estimates are required for movements throughout the network. These can be given either as vehicles or as passenger car unit (pcu) equivalents. Flows are assigned to a link on the basis of historic data and no reassignment is undertaken within the TRANSYT model. However, a scaling factor may be introduced to vary flows by a fixed percentage to test the sensitivity of the signal settings produced by the model to changes in level of demand.

Saturation flows (i.e., the maximum rate of queue discharge) are required for each separately identified stop line. These can be measured or estimated from previously determined relationships.

The program can be run to simulate any period from 1 min to several hours. Queues and delays will build progressively on oversaturated approaches over the modelled period, and the period should, thus, be selected to represent actual conditions as closely as possible. Vertical queuing at the stop line is assumed within the model, but the control of queues to reduce the blocking of upstream intersections can be obtained by invoking a facility that introduces a severe penalty into the PI when a predetermined queue is reached.

Link journey times or cruise speeds need to be known in order to calculate platoon dispersion. These may be separately estimated for buses, to include a constant factor for their time at bus stops and for optimization to be based on appropriate weighting of the benefits to the different vehicle categories (Robertson and Vincent 1975).

Groups of adjacent intersections, or nodes, can be identified for which the offsets are altered together, rather than individually. This facility is used where a network is very large and must be treated as a series of subareas. When adjacent intersections are closely spaced or traffic considerations are such that a limited range of signal timings would be acceptable, a further facility can be used whereby the offsets and green times of an individual node can be optimized as well as the offsets for a group that includes that particular node. The selection of suitable node groups can make a substantial difference to the subsequent performance of a network and a range of alternatives should be identified for testing.

1.4 Optimization Data

TRANSYT can be run to assess the performance of a predetermined set of signal settings, in which case no optimization is undertaken. Optimization may otherwise be of offsets only, or of both offsets and green splits.

Optimization criteria are largely based on delays and stops, and the weighting to be applied to each must be specified. Monetary values may also be applied to these factors.

The sequence of node optimization and steps in the search process can be specified so as to minimize computing time and to generate "green wave" progressions for emergency vehicles.

1.5 Traffic Model

The traffic model section of the program models the operation of traffic movements at signal-controlled and priority intersections, and on the links connecting them. The model operates on a network described in terms of links and nodes (Fig. 2), the appropriate definition of which is critical to the successful optimization of the network.

A 1 s resolution is used for the timing of signals, whereas all other calculations within TRANSYT are undertaken on a time-unit or time-step basis. The time steps are an equal division of the common cycle time and typical values vary between 1 s and 3 s.

Priority intersections are modelled using an assumed entry capacity for the nonpriority arm, which is dependent on the level of priority flow in each time unit. For the model, levels of traffic flow are considered constant over specific modelled periods and the interruption of these flows at intersections leads to the creation of new discharge profiles which are then dispersed on each downstream link to form new arrival profiles at the downstream intersections. Each intersection is modelled in turn and "dummy" links are created between some nodes that are run prior to the actual links to give better initial profiles.

The relationship used to predict the dispersion of vehicles along a link is as follows:

$$q^1_{(k+t)} = Fq_k p + (1-F)q^1_{(k+t-1)}$$

where q^1_k is the flow in the step k of the downstream (in) profile, q_k is the flow in the step k of the upstream (out) profile, p is the proportion of the upstream (out) profile entering the link, t is $0.8 \times$ the mean cruise time over the distance for which the dispersion is being calculated, and F is the smoothing factor $1/(1+Kt)$.

The value of K in the smoothing factor affects the rate of dispersion and has a default value of 0.35. Measurements of dispersion taken at sites throughout the world have shown that this can vary substantially from about 0.2 to 0.6. An example of dispersion on a link is shown in Fig. 3.

Delays are estimated as pcu per unit time (i.e., the rate at which delays occur). The delay rate in pcu is thus a measure of the average number of pcu queuing throughout a cycle and is uniform. Network delay is a total of the delays at each stop line modelled, based on arrival and departure patterns. When a link is oversaturated (i.e., when the total number of vehicles arriving in a cycle exceeds the number able to depart) delays will increase during the period of oversaturation. Additionally, extra delay will occur because of the random nature of traffic not modelled in the platoon dispersion relationship.

Oversaturated and random delays are added to the uniform delays in TRANSYT using the following formula, which gives the random and oversaturated delay rate in pcu hours per hour:

$$\frac{T}{4}\left\{\left[(f-F)^2 + \frac{4f}{T}\right]^{0.5} + (f-F)\right\}$$

where f is the average arrival flow on the link (pcu per hour), F is the maximum flow able to discharge from the link (pcu per hour), and T is the duration of the flow conditions being modelled (hours). As saturation is approached, the nonuniform delay values are large and highly sensitive to small changes in demand or saturation, and particular care is required by the traffic engineer to ensure that the model is representing the actual situation.

The number of vehicle stops associated with the flow and delay conditions are estimated within the traffic model and are incorporated into estimations of fuel consumption. The traffic mix assumed for this calculation is 82% cars, 9% light commercial vehicles, 8% medium/heavy commercials and buses and 1% motor cycles.

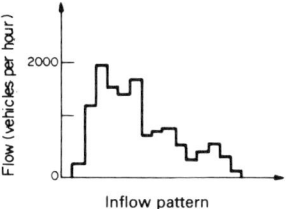

Inflow pattern

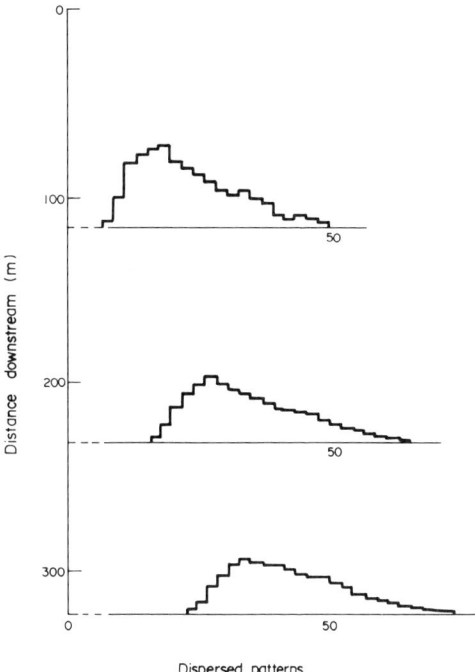

Dispersed patterns

Figure 3
Platoon dispersion

1.6 Performance Index

For each traffic flow pattern and link–node arrangement, a PI, the measure of the quality of the linking arrangement, is calculated as:

$$\text{PI} = \sum_{i=1}^{N}\left(Ww_i d_i + \frac{K}{100}k_i s_i\right)$$

where N is the number of links, W is the overall cost per average pcu hour of delay, K is the overall cost per 100 pcu stops, w_i is the delay weighting on link i, d_i is the delay on link i, k_i is the stop weighting on link i, and s_i is the number of stops on link i.

1.7 Data Output
The following data outputs are available:

(a) copy of data input,

(b) information to aid cycle-time selection,

(c) traffic model predictions for signal settings, and

(d) flow pattern graphs.

The traffic model prediction data is comprehensive and includes details of link performance, as well as overall network performance.

1.8 Applications
Results of the TRANSYT program have been the subject of substantial evaluation studies, both during its development and on a "before-and-after" basis. In general, studies have shown the benefits from linking signals using the program to be some 10–20% greater than with isolated vehicle-actuation systems depending, particularly, on signal spacing and levels of traffic flow. Savings have been shown in reduced journey times and vehicle operating costs. For example, experiments in Glasgow (Holroyd and Hillier 1979) showed that TRANSYT reduced journey times by some 16% on average, compared with the existing mixture of isolated and linked vehicle-actuated signals.

The flexibility of TRANSYT to meet the requirements of traffic engineers may be seen in its application to situations such as the signalization of large roundabouts.

2. SCOOT

SCOOT, the split, cycle and offset optimization technique, is a traffic-responsive UTC system developed in the UK for optimizing network traffic performance (Hunt *et al.* 1981). SCOOT was introduced into the UK in the early 1980s following a decade of research and development by the Transport and Road Research Laboratory (TRRL) in cooperation with the traffic signal companies Ferranti, GEC and Plessey. Development of SCOOT has been continued since then by TRRL and it is now installed, or is in the process of installation, in over 40 cities and large towns in the UK and overseas.

The main objective of SCOOT is to provide optimum up-to-the-minute network control, in contrast to the traditional systems such as TRANSYT which rely on fixed-time signal plans that are precalculated to suit average traffic conditions. A traffic-responsive system also removes the need to periodically prepare new fixed-time plans which, if not undertaken, can lead to an annual increase in delay estimated as some 4% (Bell and Bretherton 1986). The information obtained from a SCOOT system, such as traffic flows, delays and levels of congestion is also of considerable potential value.

2.1 General Principles
The flow of information in a SCOOT UTC system is illustrated in Fig. 4. Inductive-loop detectors measuring vehicle presence (occupancy) four times a second are placed at the upstream end of each link in the SCOOT network and transmit the occupancy data to the central SCOOT computer every second. In general, one loop is used for one or two lanes. The upstream location of loops was chosen as being good for measuring the cyclic flow profiles, which are needed to optimize signal coordination and for measuring and/or predicting queues and congestion.

The occupancy data is processed by the SCOOT software to provide measures of traffic demand and congestion on each link. This data, expressed in units specific to SCOOT (link profile units, or LPUs) is then analyzed within the SCOOT traffic model which contains optimizers for cycle time, green durations (splits) and offsets. The traffic model determines the signal timings that minimize a PI for each SCOOT region, based normally on a weighted sum of delays and stops. After optimization, these timings are then transmitted to the local controllers at each intersection, allowing the calculated timings to be implemented on street.

The SCOOT optimization process is "continuous," with signal timings being adjusted in frequent small increments to match the latest traffic situation. This philosophy ensures no large sudden changes in signal timings, which can occur during plan changes with other systems and give rise to increased delays (Bretherton 1979). It also means that predictions of average traffic behavior several minutes into the future are not required.

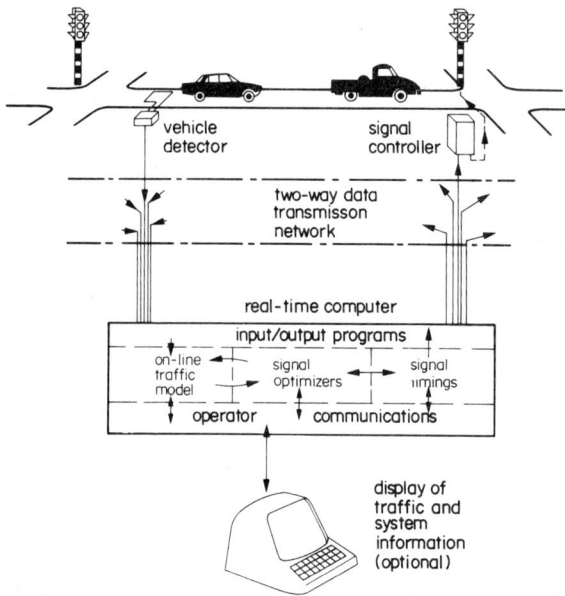

Figure 4
Information flow in a SCOOT UTC system (after Hunt *et al.* 1981)

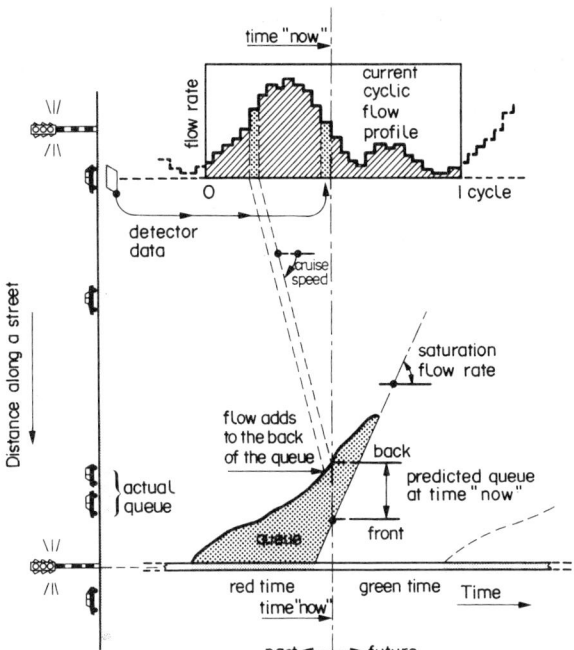

Figure 5
Principles of the SCOOT traffic model (after Hunt *et al.* 1981)

2.2 Traffic Model

The traffic model, which is fundamental to SCOOT, uses data that varies with time, such as the green and red times of the signals and vehicle-presence measurements, together with data that are preset for the area under control, such as detector locations, signal stage order and a variety of other parameters. The data from the detectors are stored in the SCOOT computer as cyclic flow profiles for each link. These reveal the variation in traffic demand during each cycle and are used during offset optimization to ensure good signal coordination. The same principles are used (off line) in TRANSYT.

The data described above are then used to predict traffic queues, delays and stops on each link. These predictions are undertaken by the SCOOT traffic model using the principles illustrated in Fig. 5, which shows a typical cyclic flow profile alongside a detector and a "time now" datum which moves to the right along the profile as time advances until the end of the cycle when the process is repeated. Vehicles (or LPUs) recorded at the detector are progressed along the link according to its cruise time (modified to take account of platoon dispersion) and are added to the back of any queue being modelled at the stop line. Alternatively, they proceed through the junction on green, undelayed. Queues grow during the red period, according to demand, and discharge during green at a rate known as the saturation occupancy which is determined during validation of the system (Sect. 2.4). Any queue remaining at the end of green is carried over to form part of the initial queue length at the start of the following green period. Figure 5 illustrates how the back of the queue and queue length are predicted at any time; the total delay during the cycle is equal to the dotted area within the "triangle." Other features of this modelling are that queues cannot exceed a maximum queue value for the link, dependent on the location of the detector, and that exit blocking by downstream links is modelled by assuming no saturation discharge when this occurs.

The green duration (split) optimizer in SCOOT functions a few seconds before each stage change and estimates whether the change should be advanced or retarded by a small amount. The criterion used in this decision is the minimization of the maximum degree of saturation on the approaches to that junction. The level of congestion can also be included as an optimization criterion. Temporary changes (of up to ± 4 s) can be made relative to permanent stage change positions in which adjustments of ± 1 s are allowed at each stage change. Each junction is treated by the split optimizer independently of other junctions.

The offset optimizer estimates every cycle whether or not to alter all scheduled stage change times at that junction. Any such change has the effect of altering the offsets between that and neighboring junctions. The criterion used in this decision is a comparison of the sum of the PIs on all adjacent streets for the scheduled offset with offsets that occur a few seconds earlier or later. The level of congestion can also be incorporated in the PI. The alteration (of -4 s, 0 s or $+4$ s) giving the minimum PI is then implemented.

The cycle time optimizer within SCOOT estimates every 2.5 min or, more usually, every 5 min, the optimum cycle time for the region. A SCOOT controlled network is typically divided into a number of regions, the junctions in each region operating to their own common cycle time to ensure good signal coordination. Additionally, SCOOT may operate some junctions at half the common cycle time (thus, double cycling) where this is beneficial. Cycle times are constrained to lie within a range between a minimum practical and a maximum cycle time, determined from operational and safety considerations. The criterion used in cycle-time optimization is that the most heavily loaded junction in the region should operate, if possible, at a maximum degree of saturation of about 90%. This target saturation is approached by allowing a change in cycle time of 4 s or 8 s at each cycle-time optimization. SCOOT can double cycle junctions where their maximum degree of saturation would be below 90% if they operate at a cycle time of half that for the region.

Various levels of optimization are provided, allowing either all optimizers, or subsets of them, to be operational simultaneously. This may be worthwhile where, for example, a junction with light side-road traffic does not require split optimization but would benefit from offset optimization.

The principles of operation and optimization described apply to cities with "conventional" network and traffic systems. However, SCOOT has also been developed to cope with unusual situations such as the socalled bicycle SCOOT which was designed to cope with the very high flows of bicycles in Beijing, China. Other cities have also adapted SCOOT to cope with their own circumstances.

2.3 System Configuration

SCOOT networks are designed on a hierarchical basis, the descending levels, in order, being area, region, node, link, detector and stage. Each hierarchical level has its own data requirements. For example, each node (junction) has a specified locational code, minimum and maximum cycle-time values, codes for single/double cycling requirements and some other specific data. Each link is referenced by its adjacent node(s) and the main downstream link is also specified for use when modelling exit blocking. Required data for links include validation parameters (see Sect. 2.5) and congestion importance factors (which are taken account of by the optimizers). There is also special data required for coping with filter links for turning traffic. Stage data includes a description of the stage order, minimum and maximum stage lengths and intergreen values. Optimum stage orders are predetermined and not calculated by SCOOT. It is important for optimal system performance that these, and other database values, are correctly specified.

Another important aspect of system configuration is detector siting; poor siting can lead to incorrect traffic information being received by SCOOT and inappropriate signal timings may then be implemented on street. The normal siting for detectors is one on every SCOOT link, located some 10–15 m downstream of the previous junction, with one detector monitoring up to two lanes of traffic. However, a range of site-specific circumstances may warrant alternative positioning of detectors to cope with, for example, bus stops, entrances, on-street parking and traffic "sinks" and "sources," where traffic may leave or enter the link.

2.4 Validation

Following installation of a SCOOT system and the setting up of the required fixed-parameters database, it is necessary to validate certain parameters in the model to reflect on-street traffic performance, before the system becomes operational.

Referring to Fig. 5, the cruise speed in the model has to be specified for each link by measuring the validation journey time. This is the time taken by an average vehicle travelling in a free-flowing platoon to travel from the point of detection to clearing the stop line. The average of a number of such readings is taken.

The saturation flow rate in Fig. 5 is specified in SCOOT as saturation occupancy; this is the maximum outflow rate of a queue from the stop line in LPUs per second. This is calculated by adjusting the saturation occupancy in the model until the predicted time for queues to discharge matches simultaneous measurements on street. Again, data averaged from a number of cycles are used.

The third important validation parameter is the maximum queue, which describes the largest number of vehicles that can normally be accommodated between the detector and the stop line. This parameter is calculated by finding the maximum queue value which makes the time for a full link to clear in the model equal to that occurring on street.

Carden et al. (1989) have shown that accurate validation of these three parameters—journey time, saturation occupancy and maximum queue—is important if accurate modelling is to be achieved, particularly in congested conditions.

Other parameters that require validation on each link include the default offset to which SCOOT reverts if the offset optimizer is inoperative, as well as green start and end "lags," which can vary between sites depending on the acceleration/deceleration characteristics of vehicles at the start and end of the green period.

2.5 SCOOT Information

During the course of its signal optimization, SCOOT measures and processes a variety of information on line that can be selectively output and used as required. The information can be at a variety of locational resolutions (e.g., detector, link, node or region) and at a variety of time resolutions, typically varying between 4 s and 5 m aggregations. Traffic and control information of particular use includes traffic flow, delay, number of stops, queue lengths, congestion, degree of saturation, spare capacity and signal timings. Carden et al. (1989) have shown that the information is sufficiently accurate to use for a variety of purposes (e.g., traffic management, transport planning) provided the system is well validated.

SCOOT information is normally requested by operator or timetable message and output to a line printer, although a new microcomputer-based database system for processing, storage and analysis is near completion. A number of graphical displays have also been developed which are useful in traffic monitoring and day-to-day running. For example, a display of congestion levels in central London is used by the police in New Scotland Yard.

Other system data of particular value are the detector fault information and a range of data enabling interrogation of the decision and the effects of the signal optimizers.

2.6 SCOOT Evaluation

The performance of SCOOT has been evaluated both by simulation and by direct on-street measurements. Simulations carried out by TRRL, using repeatable traffic conditions, have been aimed mainly at evaluating different control strategies within SCOOT.

On-street measurements, on the other hand, have been undertaken in a number of cities using before-and-after surveys to evaluate the benefits of SCOOT compared with one or more existing forms of control. The results of the main surveys undertaken are given in Table 1.

On average, it can be seen that SCOOT achieved delay savings of around 12% compared with up-to-date fixed-time plans and up to 40% savings in peak periods compared with networks operating with isolated, vehicle-actuated signals; against this letter base, savings were generally much lower in off-peak periods, probably because of the decreased benefits of coordination in these lower-flow conditions and the greater flexibility of isolated vehicle-actuated signals. Surveys have also been conducted in low-flow conditions (e.g., late evening and overnight) where SCOOT has sometimes been found to be marginally counter productive. Overall, the journey time savings achieved by SCOOT usually allow systems to pay for themselves in about a year.

The savings in journey time and delay with SCOOT are accompanied by a range of other second-order benefits. These include reductions in vehicle operating costs (particularly fuel consumption), vehicle noise and exhaust pollution. As savings apply equally to all vehicles, bus journey times are also reduced and operational efficiency can be increased. There are potential savings in accidents related to the reductions in journey times and stops, although these have not, as yet, been confirmed from observations. The dynamic nature of the system also has advantages in its ability to cope with abnormal demands brought about by special events or diversions (e.g., which may follow a traffic incident). Finally, dynamic traffic management is possible, such as interactive signing to divert motorists around a congested area or away from a full car park.

2.7 Traffic-Management Facilities

A range of situations can arise where traffic engineers may wish to override the "automatic" optimization of SCOOT to some extent, and have more control over its behavior. In particular, operational or traffic-management considerations may warrant the weighting or biasing of signal timings in favor of specific traffic streams. The current version of SCOOT (version 2.3) contains a number of facilities for this, particularly for the weighting of offsets and green splits and the fixing or biasing of offsets.

The split weighting facility allows any specified link to run at a higher degree of saturation, with all the spare green time being allocated to the opposing stages. This could be used to advantage to give more weight (green time) to a heavily trafficked main road at the expense of a minor side road; the dominant main road will then incur reduced delay due to the increased proportion of green time available and the better offsets that should be achieved. Split weighting is implemented by specifying a maximum degree of saturation for the weighted link and an index of the severity to which the link should be weighted.

Offset weighting in SCOOT is similar in concept to link weighting in TRANSYT; a multiplier is specified for the PI on a chosen link and SCOOT then adjusts the signal timings to reduce delay and stops on that link to achieve the new target PI. Offset weighting can be used to favor specific links or specific progressions; for example, high-cost arterial links could be favored (e.g., those containing high proportions of commercial vehicles and/or buses). However, offset weighting would be expected to produce some increase in overall vehicle delays, as the solution is no longer optimum.

Fixed offsets can be used to advantage in SCOOT where junctions are closely spaced. Certain movements can also be favored by giving, for example, preferential coordination to a turning movement rather than the main through movement. Particular offsets could also be biased to favor a chosen route while allowing the offset optimizer flexibility to determine the exact offset.

Operational improvements and some new traffic-management facilities are also available in the latest version of SCOOT (version 2.4). These include the automatic on-line calibration of SCOOT saturation occupancy, by interrogation of the detector(s) immediately downstream of the stop line, and the feedback of actual signal status into SCOOT to improve its modelling of junctions with demand-dependent stages. Traffic management improvements in version 2.4

Table 1
Results of surveys of traffic control systems

Location	Existing form of control	Journey time savings (min)		
		am peak	off peak	pm peak
Glasgow	fixed time (TRANSYT)	−1 (−2%)	7 (14%)	6 (10%)
Coventry, Foleshill	fixed time (TRANSYT)	5 (23%)	4 (33%)	8 (22%)
Coventry, Spon End	fixed time (TRANSYT)	2 (8%)	0 (0%)	2 (4%)
Worcester	fixed time (TRANSYT)	4 (7%)	3 (8%)	11 (19%)
London	fixed time (TRANSYT)	5	13	5
Southampton	isolated, vehicle actuated	18 (40%)	0 (1%)	26 (48%)
Worcester	isolated, vehicle actuated	18 (31%)	7 (18%)	13 (20%)

center on various methods for dealing with congestion, such as congestion offsets and "gating" (the restriction of flows entering the network to maintain smooth flow further downstream).

Bibliography

Bell M C, Bretherton R D 1986 Ageing of fixed-time traffic signal plans. *2nd Int. Conf. Road Traffic Control*. Institution of Electrical Engineers, London

Bretherton R D 1979 Five methods of changing fixed time traffic signal plans. Department of Transport, TRRL Report No. LR 879. Transport and Road Research Laboratory, Crowthorne, UK

Carden P J C, Hounsell N B, McDonald M, Bretherton R D 1989 SCOOT model accuracy. Department of Transport, TRRL Contract Report No. 153. Transport and Road Research Laboratory, Crowthorne, UK

Chard B M, Lines C J 1987 TRANSYT—The latest developments. *Traffic Eng. Control* **28** (718)

Holroyd J, Hillier J A 1979 The Glasgow experiment: Plident and after. Department of the Environment, TRRL Report No. LR 384. Transport and Road Research Laboratory, Crowthorne, UK

Hunt P B, Robertson D I, Bretherton R D, Winton R I 1981 SCOOT—A traffic responsive method of coordinating signals. Department of the Environment, Department of Transport, TRRL Report No. LR 1014. Transport and Road Research Laboratory, Crowthorne, UK

Robertson D I 1969 TRANSYT: A traffic network study tool. Road Research Laboratory, Ministry of Transport, TRRL Report No. LR 253. Transport and Road Research Laboratory, Crowthorne, UK

Robertson D I, Vincent R A 1975 Bus priority in a network of fixed time signals. Department of Transport, TRRL Report No. LR 666. Transport and Road Research Laboratory, Crowthorne, UK

Vincent R A, Mitchell A I, Robertson D I 1980 User guide to TRANSYT version 8. Department of Transport, TRRL Report No. LR 888. Transport and Road Research Laboratory, Crowthorne, UK

Webster F V, Cobbe B M 1966 Traffic Signals. Ministry of Transport, Road Research Technical Paper 56. Ministry of Transport, London

M. McDonald and N. B. Hounsell
[University of Southampton, Southampton, UK]

Road Traffic Monitoring Equipment

Road traffic monitoring has become more sophisticated with the advent of new hardware and equipment technologies. Increasing demand is being placed on traffic engineers to provide real-time data for surveillance and control, as well as for environmental monitoring. Recent developments in vehicle classification technology have increased the quantity and quality of data collected, which now include vehicle type, weight and speed. New systems have been developed to assist with traffic control and enforcement procedures, thus improving the overall safety and efficiency of the highway network. Vehicle identification systems are also beginning to be employed for applications such as road pricing, automatic toll collection and traffic control tasks. These types of systems are likely to become common on urban and interurban highways by early in the twenty-first century.

1. Vehicle Monitoring

Developments in microprocessor and traffic sensor technology have made several forms of automatic vehicle classification possible for traffic monitoring and control.

In the past, manual counts provided the only source of classified traffic flow figures, while speed distribution data were based upon radar or other semiautomated measurements. Axle load surveys were infrequent and often inherently biased. Data were based on labor-intensive static weighing at the roadside. Finally, turning traffic flows at intersections have been based on short-period manual counts as highway authorities responded to specific requests for data.

Current demands for classified traffic count data are already diverse. The long-term monitoring of trends provides a basis for statistical forecasts of future road usage. More localized classified counts are commonly required for highway scheme appraisal or traffic management purposes. Speed classification may be undertaken as part of the assessment of speed limits or to monitor the effectiveness of enforcement measures. Weight data may be required for enforcement appraisal or for the determination of standard-axle-equivalent traffic flows used in highway pavement design. To satisfy these increasing demands for data, highway authorities are developing traffic monitoring systems, utilizing new techniques founded on information technology. These permit different types of data to be collected more reliably and at lower costs.

Vehicle detection is a basic element in classification: it is the ability to sense and indicate the presence or passage of a vehicle. The first vehicle detectors were pneumatic systems, developed for vehicle actuation at traffic signals. Pneumatic detectors were later adapted for traffic counting using portable tubes and electro-mechanical counters. More recently, inductive loops have become the predominant means of vehicle detection for traffic control systems, except in Japan where overhead sonar has been more popular.

New developments in sensor technology have led to several transducers and other devices that may be regarded as alternatives to the pneumatic tube and inductive loop. Optical sensors take many forms ranging from simple light beams broken by vehicle passage to complex data acquisition systems using video image processing cameras to analyze maneuvers (see *Video Sensors*). Recording vehicles on film or video for subsequent analysis has proved popular in many research applications. Magnetic sensors that operate by detecting the residual magnetism induced in vehicles by their passage through the earth's magnetic field have also been developed. These types of sensor have very tightly defined zones of detection.

Axle detection requires different approaches from those used in overall vehicle detection. Recent developments have provided several alternatives to the pneumatic tube for temporary and permanent installations. Axle weighing presents more specific challenges, including uniformity of response along the sensor length and independence from temperature, speed and vehicle dynamics effects.

Contact sensors use the mass of the axle to close metallic contacts in or upon the pavement. A strip contact sensor known as a tape switch has been used for a variety of research applications, particularly in the USA, and has proved to be a viable method for temporary axle detection.

Capacitive sensors use dynamic tire forces to compress two or three parallel conducting plates, thereby changing their capacitance. The original capacitive sensor took the form of a portable steel and rubber mat fixed to the road surface. Capacitive strip sensors can be either portable or permanent, and can be manufactured at lower costs than the earlier mats.

Triboelectric sensors use the effect of charge generation with friction between certain materials. The sensor comprises a coaxial cable that exhibits the triboelectric effect when vibrated or flexed. It may be fixed directly to the road surface for short-term applications or can be located in a slot in the pavement surface to form a more permanent sensor.

Piezoelectric sensors operate by generating a charge proportionate to dynamic stress. The most common sensor comprises a 3 mm diameter coaxial cable insulated with a piezoelectric powder. It can be surface mounted or mounted in a slot in the road. More recently, poly(vinylidene fluoride) (PVDF) polymers have also come to be used for axle detection at permanent installations.

During the 1980s, there has been an increasing need for the development of comprehensive data collection systems, capable of giving much more information on real-time traffic conditions than simple vehicle flows or loop occupancies. First, a need is recognized for the classification of traffic counts by vehicle type. This not only increases the relevance of information on traffic control characteristics such as speed and headways, but also permits the allocation of costs to the various classes of road user. Second, the tendency towards the use of more and heavier vehicles has increased the attention paid to environmental standards and has emphasized the need for more sophisticated enforcement techniques. Third, driver information systems require data on traffic conditions, including volume, speed and headway, together with associated road configurations, to be extended to cover a wider range of locations and times.

Developments have taken place to accommodate a number of these requirements. A typical example is a system designed for the UK Department of Transport to monitor heavy vehicles within environmentally sensitive zones. The system has two main components: a vehicle detection element and a camera subsystem.

The vehicle detection element uses two piezoelectric axle sensors in each traffic lane to measure the number of axles and axle spacings for each vehicle. This enables classification within specified gross vehicle weight bands. Piezoelectric axle sensors are utilized for the vehicle monitoring system because they can also be used to measure axle loads. The numbers of axles, vehicle speeds and axle spacings are determined by software running in a roadside processor. A gap-seeking algorithm is used to determine where one vehicle ends and a new one starts, based on maximum axle spacings and an analysis of axle speed differences.

The camera subsystem is activated by a signal transmitted via a cable link from the vehicle detection subsystem whenever a suspected violator is identified. The photograph then allows the license plate of the vehicle to be read. A data unit is used to superimpose the date and time on all photographs. This information is particularly important for enforcement, as it provides the means for distinguishing between actual violators and overlimit vehicles entering the area for legitimate access purposes.

2. Environmental Monitoring

Environmental monitoring systems are increasingly viewed as integral parts of traffic monitoring and control systems for urban and interurban highways. Two principal areas are air pollution monitoring and the enforcement of heavy-vehicle limits. Heavy vehicles are particularly important components of the traffic stream in the context of environmental traffic controls.

Three air pollutants have received particular attention in the development of pollution monitoring hardware equipment. These are carbon monoxide (CO), nitrogen oxides (NO_x) and hydrocarbons (HC).

Flame ionization was developed for gas chromatography and is widely applied to monitoring HC. Hydrogen gas, combustion air and sample air are introduced at a steady rate. Organic compounds cause the flame to produce ionized carbon atoms, with a resultant ion flow between the flame and the collecting electrode.

Chemiluminescent instruments measure the reaction between an appropriate reagent and the compound to be monitored, using a photomultiplier tube for measurement. Nitric oxide is monitored by its reaction with ozone at reduced or near-atmospheric pressure.

The most commonly used instruments for CO measurement employ the principle of nondispersive infrared. These analyzers depend on the characteristic energy of absorption of the CO molecule at its absorption wavelength maximum and at a number of other specific lines.

Environmental monitoring and prediction also require hardware equipment for meteorological sampling of temperature, wind speed and humidity. These complex subsystems must then be integrated into an effective real-time system giving acceptable accuracies and running costs. Development and demonstration of suitable equipment is continuing in several countries and considerable progress is being made.

The second area of activity in environmental traffic monitoring is that of vehicle weight enforcement. A need is recognized throughout the world for the effective control of heavy vehicles and the enforcement of axle weight limits. These are necessary for environmental control, to reduce pavement damage and to enforce safety standards. Conventional procedures for the enforcement of these regulations involve slow-speed or static weighing at authorized locations.

For static weighing, vehicles stop with each axle on the scale, pulling forward several times to complete a weighing operation. Each time the vehicle is moved, redistribution of its load takes place within the suspension system, significantly reducing the accuracy of the final result. Although the scales themselves may be highly accurate, their method of utilization introduces errors such that a margin has to be allowed before any action is taken.

An alternative approach provides very long, segmental scales, with segment lengths matched to the axle spacings of typical heavy vehicles. Installation, maintenance and calibration of this equipment is very costly, while segmental configurations may be superseded by changing vehicle dimensions.

Some European countries adopt a different approach to enforcement weighing. Vehicles pass over a slow-speed scale at a constant, minimum speed in bottom gear. This allows an increase in the throughput of vehicles of around a factor of three, relative to static installations. With appropriate equipment and smooth approach aprons, accuracies similar to those of static weighing can be achieved.

Enforcement scales are installed at fixed locations adjacent to the highway and require vehicles to be selected from the traffic stream for weighing. Weigh-in-motion (WIM) systems which can weigh vehicles at normal highway speeds have recently been adopted as an aid in the enforcement process through weight screening on the highway.

The types of system available for weight screening can be categorized as in-pavement pit systems, bridge and culvert systems, capacitive systems and piezoelectric cable systems.

In-pavement systems require a pit to be constructed in the highway to contain the weighing sensors. There is considerable operating experience with these permanent systems, which have proved to be relatively reliable and generally capable of consistent, acceptably accurate measurements. The high installation and operating costs, however, preclude these systems from widespread enforcement screening applications.

Strain-gauged systems using bridges and culverts have also proved to be relatively expensive, because of the processing power needed to analyze the signals. They are restrictive in terms of location because of the need for a suitable structure to use as an instrument and in terms of operation because only one vehicle should be crossing the structure at any one time. Accuracies for gross weights are generally good, but are poor for individual axles.

Systems based on capacitive technology have been available for several years. Operating experience with capacitive weight mat systems has been favorable when they are used as portable systems for short durations. For long-term usage and under heavy traffic conditions, however, the reliability of the weigh mat is unproven.

Piezoelectric cable systems offer a lower-cost alternative to the pit systems as they are considerably easier and quicker to install, requiring only a narrow slot to be cut into the highway pavement. The relatively low system costs, coupled with the ease and permanence of the sensor installation, make this approach particularly suitable for heavy-vehicle screening.

The simplest methodology for screening is for the measured vehicle weights to be transmitted to a computer terminal in a control room. The operator can examine data for each vehicle and decide whether to select that vehicle for check weighing. The computer can assist in this selection process by automatically identifying vehicles that exceed specified thresholds. More complex strategies require automated forms of vehicle identification.

3. Vehicle Identification

There are a number of requirements for vehicle identification within traffic control systems. The most obvious of these is for enforcement of violations such as speed and weight. Identification can also provide journey time information for feedback through driver information systems. In addition, road pricing and automated toll collection systems require vehicle identification for billing and enforcement purposes.

Two basic methods of vehicle identification are license plate readers and automatic vehicle identification (AVI).

License plate readers constitute a specialized form of vehicle identification technology, capable of reading the characters contained within conventional vehicle license plates. These systems comprise three functional elements:

(a) an electronic imaging system;

(b) a digital image processor; and

(c) a computer, containing special license plate recognition algorithms.

The electronic imaging system typically utilizes a videocon or CCD camera. A strobe infrared light source is used in synchronization with the camera, producing a still frame of the license plate. The use of infrared light allows differentiation to be made between the reflective license plate and other surrounding non-reflective materials, without the risk of blinding drivers.

The video signal produced by the camera is input to the digital image processor, which stores the still frames in its image memory. If the software in the microprocessor-based controller locates a license plate within the digitized image, it processes the image further to read the plate. The license plate character recognition algorithms contained within the software must be adjusted to optimize character reading accuracy in different states or countries of application.

In the USA, a commercial automatic license plate reader has been developed. In field trials, the system was able to read approximately 65% of plates correctly. Some significant shortcomings were identified, including the inability to identify dirty, partially obscured or damaged plates. In addition, since license plate character fonts vary between states, the system was less successful in reading plates from outside the surrounding area.

AVI is the term used for techniques that uniquely identify vehicles as they pass specific points, without requiring any action by the driver or an observer. AVI is a rapidly developing technological area that is being applied to a number of traffic control applications.

AVI systems essentially comprise three functional elements: a vehicle-mounted transponder, a roadside reader unit with its associated antennas, and a system for the transmission, analysis and storage of data. At the simplest level, information that identifies the vehicle is encoded onto the transponder. This normally consists of a unique identification number, but can also include other coded data. As the vehicle passes the reader site, the transponder is triggered to transmit its data message, via a receiving antenna, to the roadside reader unit. Here, the data are checked for integrity before being transmitted to the computer system for processing and storage. More complex systems with two-way communications capabilities also allow data to be sent back from the roadside to the vehicle.

There are a number of approaches to automatic vehicle identification. The main technical options are optical, infrared, inductive-loop, radio-frequency and microwave systems.

Optical systems, which formed the basis of the earliest AVI technologies developed in the 1960s, typically utilize a tag in the form of a coded label mounted on the outside of the vehicle. The coded data message is held in a series of lines of varying width or color, similar to a bar code, and is extracted from the label by passing it through a laser scanner. As the light tracks across the label, varying amounts or colors of light are reflected back to the reader. These unique patterns of reflected light are analyzed automatically and the coded information is extracted.

Infrared systems were developed during the 1970s as a substitute for the earlier optical approaches, utilizing similar operating principles. They were found to share some of the problems of the optical systems, being similarly sensitive to environmental conditions. However, infrared AVI technology is being utilized for vehicle–roadside communication in the ALI-SCOUT electronic route guidance system, which is demonstrated in London and Berlin in the Autoguide and LISB projects.

Inductive-loop systems use conventional traffic detection and counting loops in the highway as antennas for relaying signals to or from transponders mounted on the underside of vehicles. Inductive transponders use a simple coil or ferrite rod antenna whose dimensions are a function of the communications carrier wavelength.

Radio-frequency (rf) and microwave transmissions are the basis of several current AVI systems. Systems may operate on a wide range of frequencies in the kilohertz, megahertz and gigahertz ranges. Transmission on frequencies of 915 MHz or 2.45 GHz forms the basis of several AVI systems available internationally, while higher-frequency communications are being investigated in European Commission research programs. The basis of the heavy-vehicle electronic license plate (HELP) system, currently being implemented in the USA, is formed by rf AVI technology.

Two-way microwave AVI equipment is the subject of significant research and development taking place in Europe. This aims to develop and test hardware suitable for use in various application areas in a future European integrated road transport environment, with interfaces to other in-vehicle technologies such as smart cards and on-board computers.

Surface acoustic wave (SAW) technology forms the basis of another recently developed RF AVI system. In a SAW AVI system, the vehicle-borne transponder contains an antenna for reception of the triggering signal and a SAW chip that has a series of notches etched on its surface to give the transponder its unique identity. The key characteristic of the chip is its ability to convert the electromagnetic wave of a triggering signal into a SAW.

Data message types for AVI systems can be fixed code, part-variable code or variable code. Fixed codes allow vehicle identities to be established, without allowing additional variable information to be passed between the vehicle and the roadside. They can be made to appear pseudovariable for certain purposes by means of lookup tables within the computer system of the reader. This can offer a higher degree of security than variable or part-variable identity codes, but has significant penalties in requiring more powerful communications and data processing capabilities.

Part-variable codes contain a fixed and a variable portion, which can pass a range of additional information to or from the vehicle. This results in savings in data processing and communications costs over the lookup table method. Real-time data, which originates while the vehicle is on the highway, can only be communicated in this way.

Wholly variable codes, as the name implies, contain no fixed code portion whatsoever. This leads to even greater flexibility in the range of information that may be transmitted. The tradeoff for this, however, may be reduced system security. Any variable part of a communications code is likely to be vulnerable to outside interference since, by definition, it is designed to be changed easily.

There are essentially two options relating to the communication capabilities of AVI systems: one-way and two-way communication.

One-way communication from vehicle to infrastructure reader unit provides for a basic identification and information transmission capability. At the simplest level, a fixed code is relayed, identifying the vehicle at a particular location and time. One-way communication from vehicle to infrastructure with part-variable codes implies that the variable part must be programmed from an on-board facility.

Two-way communication with fixed, part-variable or variable codes offers the potential for infrastructure-to-vehicle data flow, in addition to the vehicle-to-roadside communication previously outlined. This facility can be used to program an on-board computer or to write to variable code memory within the electronic tag for subsequent reading at another location.

At the simplest level of two-way communication, fixed codes can be passed from roadside to vehicle, essentially telling the vehicle where it is. These codes can be input to an on-board computer or smart card on the vehicle, keeping track of time, location and other information required. At the next level of two-way communication, part-variable or variable messages can be passed from a control center to drivers. The facility for passing messages to the driver can be of considerable value in fleet management and control.

The third and most sophisticated level of two-way communication allows writing to part-variable code memory in the transponder from the off-vehicle electronics, for subsequent reading at another location. This option offers many potential benefits, mainly in increased flexibility and a broader scope of applications. The distributed processing promised by this approach can lead to considerable savings in central processing and data communications requirements, though with the penalty of increased cost of the on-vehicle unit.

With regard to the source of power for the vehicle-borne transponder, there are three approaches. These are termed active, passive and semiactive.

Active systems use an electronic transponder which takes its power supply from the vehicle on which it is mounted. These systems may transmit a data message continuously or, more commonly, may be triggered by a signal from a suitable antenna.

Passive systems use an electronic tag which is energized by power transmitted from the in-pavement antenna. Passive tags are sealed units with no external power supply. When outside the field of the powering antenna, these tags are totally inactive. Passive systems are generally less vulnerable than active units to outside interference and damage, whether accidental or deliberate.

Semiactive systems have been developed more recently and use an internal battery to provide power to transmit the vehicle identification code when triggered by an antenna. These totally sealed units require no external power supply and, therefore, overcome the problems of the fully active system.

See also: Broadcasting Communications Systems; Detectors for Road Traffic; Mobile Communication

Bibliography

Catling I 1987 Automatic vehicle identification. *Information Technology Applications in Transport.* VNU, Utrecht, The Netherlands, Chap. 3

Davies P 1983 Automatic vehicle classification techniques: Lane, speed and basic type classification. *Traffic Eng. Control* **24**, 195–201

Davies P, Ayland N D 1985 New developments in weigh-in-motion. US Society of Automotive Engineers, Technical Paper No. 851454

Davies P, Salter D R and Bettison M 1982 Loop sensors for vehicle classification. *Traffic Eng. Control* **23**, 55–9

Gonzales R 1985 The use of computer vision in parking lot technology. *Parking* May–June

Head J R 1982 Detecting vehicles with inductive loops and microwaves. *Proc. Int. Conf. Road Traffic Signalling.* Institute of Electrical Engineers, London, pp. 111–14

Moore R C, Davies P, and Salter D R 1982 An automatic method to count and classify road vehicles. *Proc. Int. Conf. Road Traffic Signalling.* Institute of Electrical Engineers, London, pp. 119–22

Salter D R, Davies P 1984. Development and testing of a portable capacitive WIM system. *Transp. Res. Rec.* **997**, 61–9

Von Tomkewitsch R 1986 ALI-SCOUT—A universal guidance and information system for road traffic. *Proc. 2nd Int. Conf. Road Traffic Control.* Institute of Electrical Engineers, London, pp. 22–5

P. Davies
[Castle Rock Consultants, Nottingham, UK]

Route Guidance, Collective

Efforts have been made in recent years to increase the efficiency, speed, safety and quality of travel on both urban and interurban networks by the use of traffic engineering innovations. Usually the driver is required to rely on personal skill for estimation, prediction and judgement to perform driving tasks. In high-traffic-density situations, because of the complex decisions required, there is a higher probability that the driver will make an error of judgement. One possible method for providing a higher level of service on existing highways, however, appears to be a redistribution of demand. This requires some type of an information system that allows the driver to intelligently choose a suitable route from the alternatives available.

1. Definitions

Traffic information is information on speed, volume, minutes of delay and so on associated with a given route. A routing system that provides instructions to drivers based upon best route solutions will be referred to as a route guidance system. A static system routes drivers over historical "best" paths. Dynamic route guidance implies traffic-responsive real-time surveillance and control of traffic operations. It is conceived as the integration of the routing and the traffic control functions, and it is based on instantaneous information regarding conditions on the network.

The goal of route guidance systems is to balance the level of service on all major network links so as to minimize travel time and accident occurrences. Longer trip distance, slower trip speed, longer trip time, greater driver tension, increased accident potential and more air pollution production can all be attributed to less-than-optimal route guidance. Routing errors can result in the driver becoming temporarily lost, and improper route selection is usually attributed to unfamiliarity with available alternative routes and unfamiliarity with the immediate destination.

Route guidance systems can be distinguished into three types. All highway route markings (e.g., highway route numbers, mileage signs, street names, maps) are described as first generation and current active systems (e.g., commercial radio, changeable message signs, audio signs) as second generation. Both these systems furnish information to all drivers and, therefore, can be described as utilitarian or "one-way" collective systems. Communication is one way in the sense that information flows in one direction only—from the highway sign to the vehicle operator. There is no provision for the driver to make the destination known to the system. New concepts of route guidance based on individualized communication methods constitute third-generation systems. Such systems are based on the notion that the highway and the vehicle are tied together in a single entity (see *Route Guidance, Individual*).

2. Information Needs

The motorist beginning a journey usually has little or no knowledge of the operating conditions existing on any of the available routes between the origin and destination points. Generally, the driver's expectations of the conditions that are likely to prevail and of the alternatives available are founded solely on previous experience. The motorist's perceived notions of the traffic conditions on any given route can be the result of several information sources such as the news media or past driving experience. Trip planning requirements are, therefore, primarily served by human memory, previous transferable experience, and the ability of the driver to make accurate decisions regarding relative distances and durations associated with different routes. Other information regarding nonrecurring congestion or delays on various links is not usually available to the driver during the planning stages of trip making.

If the motorist selects a route that is heavily congested, the journey becomes personally taxing and serves to increase the inefficiency of the overall system. Better alternative routes may be available, but there is seldom sufficient information on which to base an intelligent choice. In congested areas, frequent changes in routing may be necessitated by various levels of traffic demand. The driver who is very familiar with the highway network has less difficulty in selecting alternate routes. However, even here the driver is placed in situations where memory may be taxed or decision process rate exceeded, or information regarding the state of congestion on alternate routes may simply not be available.

A driver appears to choose routes that provide significant time savings, even though a greater distance may have to be driven. A saving in travel time is defined as the time saved by travelling on route *i* instead of route *j* under normal conditions on both routes. A saving in travel time delay is defined as the time saved by diverting from route *i* to route *j* to avoid a delay on route *i*. Results from various studies imply that the driver values time directly and, hence, scales that variable. It is interesting to note that most studies substantiate the view that, in general, drivers are more receptive to a diversion to avoid a delay than to one that saves travel time. This is a very important factor to be considered when estimating the probability of diversion for drivers receiving information on traffic conditions.

Once a driver has begun a journey, congestion may be encountered which is caused either by the heavy demand placed on the route, or by the occurrence of an incident or an accident. (An incident may be anything that causes a breakdown in operation.) Under such circumstances some drivers will voluntarily divert to other routes to avoid the congested conditions. Others will simply remain on the preselected route. The provision of information at the appropriate point along the route can, and probably will, cause a voluntary diversion of a proportion of drivers. In turn, this will cause a shift in demand and a more rapid return to an efficient level of overall system operation. At this instance, system stability must be considered. In many highway corridors in which the demand is fairly close to capacity, shifts in demand may cause congestion on the best route, thereby ending its status as best route. The information system may then oscillate between the alternate routes in designating them as best route.

Collective route guidance systems furnish information aimed at all drivers; the information is necessarily incomplete and presented intermittently, and the system relies on the driver to integrate it with the desired destination, making decisions at each choice point. Drivers evaluate highways and alternate routes on a predominantly subjective basis. If a message does not conform to a driver's expectations or if it lacks good orientation, the result is hesitation and indecision at crucial decision points. Thus, accident potential is increased at decision points, and turbulence can be introduced into the traffic stream with resultant adverse effects on capacity and safety. A wrong decision means extra travel and driver frustration. Information conveyed by signs must be designed for the traffic stream at large. Consequently, with sign messages, it is inherently difficult to convey precise meanings according to individual driver's needs or destinations.

The information needs of the driver vary with the portion of the trip and the immediate maneuver to be accomplished. The environment in which individual trips are made is characterized by roadway segments, frequently composed of multiple parallel paths connected at junctions or level intersections. Driver actions are highly stochastic. In addition, the human response to various stimuli varies with the type of stimulus and the level of the stimulus (intensity and/or frequency) with respect to some threshold value. Therefore, a finite time is required to process information. The processing time is added to the transmission time to create a total reaction time to the various stimuli. Driving tasks can be thought of as responses to information-containing stimuli. A practical consideration is evolving from the question of where should information be presented to drivers. Analyzing the driver's tasks, two elements are most relevant for the determination of information lead distance (the exact location prior to a choice point where such information will have maximum utility for the driver). These elements are changing lanes to prepare for a maneuver and changing speed to perform the maneuver. At an intersection, the lead distance for the initial information can, thus, be determined as a function of the requirement to change lanes and the requirement for the exit speed from the choice point. These factors are influenced by other limitations such as vehicle characteristics, environmental considerations and other generic network characteristics. Therefore, the determination of information lead distances for any specific choice point should take these factors into account.

The driver's tasks must be developed around the concept of the unfamiliar driver. If the system can adequately guide the unfamiliar driver through the network, then all the drivers can use those portions of the displayed information that satisfy their particular needs. The worst-case condition must also be taken into account; that is, where the driver is in the worst possible lane in a situation of heavy traffic. Once these have been established, the time factors affecting these tasks can be developed.

In general, drivers with a greater familiarity of the road situation ahead would make choice decisions significantly faster and are expected to show a greater tendency to switch routes than drivers with less familiarity. These assumptions, nevertheless, have not been adequately proven by any research study. In the absence of available evidence, it cannot be concluded that drivers' route choice preferences are significantly influenced by their expectations of the situation, or by their intentions or even their trip purposes.

A desirable information format, whether audio or visual, would be one that is satisfactory to a wide range of drivers and yet will not discriminate against any particular subgroup. This capability of furnishing the desired information becomes a function of design.

There are times during the driving task when the driver's processing capacity is fully used or overloaded, as well as times when it is almost completely unused. Both situations present serious problems. For example, in low-signal-concentration areas, a driver's vigilance is usually low and drivers have been found to miss signals for no apparent reason.

Driving involves stimuli that are continuously changing; hence, it presents the driver with a flow of information. In high-signal-concentration areas, where many signals compete for the driver's attention, the flow of information exceeds the driver's processing capacity. When this capacity is overloaded, the motorist may miss signals because of not load shedding properly. In addition, there may be the problem of not having enough time to make a decision, leading to confusion or decisions made on the basis of incomplete information. Thus, the driving environment should not present information at a rate that exceeds the information processing capacity of drivers. For a given sampling rate of information, it can be hypothesized that a driver will adjust speed to prevent the uncertainty between samples from becoming too great (Gerlough and Huber 1975).

As shown previously, considerable attention should be paid with regard to the frequency and the detail of information that will be supplied to the drivers. Route guidance should exhibit, at any time, two properties: it must provide information in geographic and in hierarchical primacy, so that selective transmittal reduces the total information loading and handling requirements on the system. This is done by furnishing complete information about the network in the driver's vicinity and less information concerning the rest of the network. Concentrating on visual information techniques (e.g., changeable matrix signs which, according to the existing evidence, is the most preferred technique), the previous requirements relate to hardware density. There exists a direct relationship between hardware density and display complexity. Based on experimental results that viewing time for a sign often takes about 1 s, the feasible content of a sign message is limited to three or four short, or easily recognized words for normal highway speeds. Thus, in order to deploy changeable message signs displaying messages that are optimal for directing drivers through various combinations of decisions, considerable effort should be devoted to designing the content of each message display.

3. Route Diversion Strategies

Route diversion strategies can be categorized into various types. These strategies depend on the policies that are pursued by the traffic administrators. Successful network management is based on the ability of the system to affect drivers' decisions so as to achieve an optimum redistribution of traffic in response to existing conditions. The information transmitted to the driver, which is intended to cause a desired decision, must be in a form usable by the driver, and must be received at a time and location appropriate to the decision. The essential consideration in the design of diversion strategies is the choice of a performance criterion with which to perform the evaluation of alternate routes. It should be stressed that collective route guidance constitutes a voluntary choice scheme. Research has shown that these types of schemes are relatively successful if the provision of information to the drivers extends to information on the alternatives available to them and the level of service they offer; that is, a system based on voluntary choice has to give the driver a good reason for changing route.

The rest of this article will be focused on real-time systems. Mandatory choices will not be examined in detail as this option is characteristic of the dynamic route guidance of individual vehicles. Mandatory choice enforces the use of designated routes. At the beginning of a journey, the driver would be assigned a route and would be required to follow it. In contrast, voluntary choice schemes would include the provision of information to the driver that would enable the selection of a more appropriate route.

The problem with voluntary choice schemes is the degree of compliance. Driver acceptance and/or credibility is, therefore, a major concern in the decision to spend public funds for real-time information displays. A traffic information system must be able to influence the actions of drivers in such a way that they are encouraged to increase the level of efficiency of the transportation network.

Research in motorist communication techniques has shown that motorist compliance (i.e., the number of motorists who will divert) varies. Given adequate trip planning, all that is needed by drivers in the way of on-line guidance is information that relates the driver's trip plan to what is seen in transit; that is, by giving present location and direction of travel. Experience shows that displays that present speed information or descriptive terms are preferable to displays that present quantitative travel time or delay information. This would indicate that drivers do not evaluate alternate routes in terms of travel time and delay, but rather that they tend to think in terms of speed or some other reference parameter—perhaps comfort and convenience.

The factor that appears to dominate is learnt by direct experience and relates primarily to the negative characteristics of certain transportation facilities or particular routes. This has led various researchers to conclude that comfort or the lack of it is closely related to the stress arising while driving. This tension is caused by interferences that have purely negative effects. It seems reasonable to conclude that shifts in traffic to alternate facilities or routes are made in order to avoid the stress-inducing characteristics of certain facilities. Stated more generally, drivers make choices among routes in order to minimize the total stress to which they are subjected during driving. Since driving is a stressful and energy-consuming task, drivers have tolerance limits beyond which their subjective "cost" of driving becomes excessive. Thus, for the driver of a passenger car, the basis for scaling the benefits to be obtained from an expressway are neither economic nor time saving, but rather stress saving (Michaels 1966).

This does not necessarily mean that drivers do not end up taking the minimum-travel-time route, but merely that they do not evaluate the alternate specifically in terms of travel time. Because most traffic assignment procedures are based on some mininum-travel-time principle, this point is quite significant and suggests that further investigation into the philosophy of traffic assignment should be encouraged (see *Traffic Assignment*).

As evidenced by the previous discussion, research results seem to indicate that few people think they know what their accurate travel time might be to various critical points ahead, even when traffic conditions are good. In addition, most drivers seem to feel that they have a poor idea of their average travel speed. Travel time and speed, therefore, may not be so valuable to motorists as qualitative information in terms of location, length and degree of congestion. If drivers cannot relate information regarding the average speed attainable between choice points on the network, then they will have no use for this information. Thus, there is converging evidence that although travel time is a tool for the traffic engineer, it does not constitute a primary information need for the average motorist (Dudek and Messer 1971).

Consequently, if this is the case, motorists will not respond to (or comply with) route guidance messages to the desired degree, as they will think that there is no benefit in utilizing the information. This is an important factor that should be seriously considered when route guidance strategies are laid down. Route guidance schemes will not produce the desired operational results if drivers are being asked to do something that they object to.

The formulation of optimal (network-efficient) strategies must be based on the relative performance of each alternate route as defined with respect to the parameters that are of relevance to drivers. Short-lived incidents or a congestion that has just started building up may not be considered by motorists as serious events warranting a route diversion. Consequently, the optimum strategy may be to divert traffic at some time $t + \Delta t$, when the overall impact is beneficial. The best interests of the traffic control authority or the driver may not be served by furnishing all information in real time. When it is more desirable to delay the furnishing of information by some specific time interval Δt, the delayed information is in pseudo-real-time.

4. Traffic Control

In the preceding sections, it was hypothesized that the reliability and efficiency of traffic networks can be greatly enhanced by making the system dynamically responsive to individual user requirements. Success, in the form of system optimization, will be achieved if drivers' actions are predictable on a statistical basis. These actions will be predictable if they are rational.

The functions of the highway routing system are to provide course and routing information. These functions may be further delineated by specifying the operations involved in pretrip planning, route finding, choice point path selection, and terminal or destination recognition. Each of these tasks, presently demanded of the driver, becomes more important as the trip becomes longer, congestion more variable, and the perceptual and cognitive capabilities of the driver are lowered or stressed.

Collective route guidance systems provide information aimed at all drivers. The system relies on drivers to integrate it with their desired trip plans and to make their own decisions. If this information can serve to appropriately structure drivers' expectancies, it is possible to bridge the credibility gap that is built into the system. In order to link route guidance to the traffic control subsystem, it is essential to develop a relationship between a meaningful measure of driver efficiency and network efficiency. This goal can be achieved if a dynamic routing system with real-time feedback and predictive capabilities can be devised. This implies a network-wide traffic surveillance system that is capable of detecting traffic flow characteristics. These are subsequently input to a central control facility that utilizes a combination of optimum routing and traffic control algorithms to determine the best routing solutions and signal settings for the next time slice of network operations. Thus, network optimization is achieved by predicting the operation of the network in any control period based upon the information gathered during the past period. If the period can be made small enough (e.g., 10–15 min) so that conditions are not significantly altered between subseqeunt periods, this dynamic system can significantly improve both driver and network efficiency.

Historically, there have been a number of constraints on methods of providing reliable route guidance aids.

The transportation network information system is only a subsystem of the entire driver's information system and, therefore, is constrained by various elements of the larger system. These constraints may be loosely classified as either technological or socioeconomic. Research has demonstrated that the device most preferred by the driver for presenting information will not necessarily cause the proper actions on the part of the driver (Hoff 1971). It has been demonstrated that there must be a personal need associated with a traffic information system if the driver is to respond in the desired manner. The important point to recognise is the potential ability of a particular information system to significantly influence the redistribution of demand. Experience shows that a real-time information system can be made more appealing and useful to motorists by providing the information when it is needed, and in the desired form and mode (Dudek and Messer 1971).

Considerable benefits are expected if route guidance systems are extended to provide real-time information on parking availability (see *Parking Control Systems*). A considerable proportion of the traveller's time is occupied in the task of finding parking. Without implying that the terminal phase of travel is entirely constrained by routing information, parking availability is undoubtedly important in reducing both travel time and stress suffered in trying to find an available parking space near the destination.

See also: Freeway Network Control; Route Guidance, Individual

Bibliography

Armstrong B D 1977 The need for route guidance. *TRRL Suppl. Rep.* **330**, 1–20
Boyce D E 1988 Route guidance systems for improving urban travel and location choices. *Transp. Res. A* **22**(4), 275–81
Dudek C L, Messer C J 1971 Study of design considerations for real-time freeway information systems. *TRB Highw. Res. Rec.* **363**, 1–17
Gerlough D L, Huber M J 1975 Traffic flow theory. *TRB Spec. Rep.* **165**, 71–86
Hoff G C 1971 A comparison of selected traffic information devices. *TRB Highw. Res. Rec.* **366**, 116–29
Jeffery D J 1981 Ways and means for improving driver route guidance. *Transp. Road Res. Lab. (U.K.), TRRL Rep.* **1016**, 1–21
King G F, Lunenfeld H 1971 Development of information requirements and transmission techniques for highway users. TRB NCHRP Report No. 123
Michaels R M 1966 Attitudes of drivers towards alternative highways and their relation to route choice. *TRB Highw. Res. Rec.* **122**, 50–74

A. Stathopoulos
[National Technical University of Athens,
Athens, Greece]

Route Guidance, Individual

Individual traffic has been increasing for many years and shows no sign of slowing down. Cities are more and more concerned by the problems of congestion which yield a tremendous waste of time and money and a diminution of quality of life. Traffic engineers can fight congestion by:

(a) improving the capacity of the infrastructure, but that capacity can hardly be extended any further and, moreover, any improvement induces a new demand;

(b) using management schemes optimized through assignment procedures; and

(c) using adapted urban signal traffic control strategies—real-time traffic control algorithms such as SCOOT in the UK (Hunt et al. 1981) and PRODYN in France (Henry et al. 1983)—to react to the variation of traffic and ensure some protection against congestion.

The efficiency of these methods is limited because there is no control of routes and no control of demand. Attempts were made in Singapore and Hong Kong to control the demand by using road pricing. However, such systems, working in open loop, seem difficult to introduce because of the complex reactions of human beings at all levels (Borins 1988).

Another way to fight congestion would be to control the route of the driver: more and more systems intend to do this. They take advantage of information concerning the road network, and help the drivers to find the best routes. They can be classified into several categories. There are static systems, with no reactuated information. These make use of an on-board processor to compute the location of the vehicle and then give instructions to the driver. There are also dynamic systems, which can receive real-time information concerning the traffic conditions. This information is broadcast by outboard devices. Some of these systems just select, from among the collected information, that which is related to the destination of the vehicle. They give the driver a "map" and are information systems. In some cases, the on-board processor is able to compute a route according to this information. The system is a navigation system (the computed route is not available to the central processor, which therefore cannot use this information). Some dynamic systems can process the information inside the central computer, and find the optimal routes for every equipped vehicle and for every destination. They can propose the computed route to the on-board computer, and so to the driver. The latest systems are called dynamic route guidance (DRG) systems.

Route guidance has been considered for several years in Europe, the USA and Japan and has run into technological problems with vehicle positioning. Now, the technology makes it possible to realise DRG systems, and pilot systems such as RACS in Japan and ALI-SCOUT in Germany are in operation.

The aim of DRG systems is, for the user, to ensure guidance, to reduce stops and delays, to give real-time information on traffic, weather and so on, and, for the authorities, to optimize the road network traffic flow.

A DRG system is composed of on-board devices and outboard devices (roadside units and control centers) communicating on the one-way or two-way modes. The control algorithms actually implemented can be much improved depending on the hardware, on the architecture, on the type of collected data, and so on.

1. Existing Systems

Route guidance systems can be classified according to their communications characteristics:

(a) no communication—the vehicle has on-board equipment;

(b) one-way communciation; and

(c) two-way communication.

Of course, the early systems, such as NAVIGATOR, which belong to the first category, were not adapted to DRG because of the absence of communication.

All these systems use the same on-board equipment—an electronic navigation aid, a microcomputer, a display unit and a numeric keypad. A description of the road network is stored on a memory device; the driver enters the grid references for current position and required destination on the keypad. The position of the vehicle is sometimes superimposed on the map and the microcomputer determines the best route to follow. With the electronic navigation aid device, the current position of the car is always made known to the microcomputer. This electronic navigation aid is exclusively based on one or several of the following techniques.

(a) The dead-reckoning technique, which relies on distance and heading sensors fitted to the vehicle so that progress from a known initial position can be continually monitored and the position updated. Magnetic sensors are used to determine the heading, but can be affected by metal masses such as bridges, other vehicles, and so on. Dead-reckoning systems need to be reinitialized periodically with, for example, map-matching or beacons. Accuracies of about 2% distance travelled can be obtained.

(b) Trilateration techniques, which rely on the detection of radio transmissions from three or more fixed points. Loran-C or the global positioning system (GPS), a satellite navigation system, are examples of trilateration techniques.

(c) Beacon-based techniques where the vehicle uses the location information sent by the beacon to update its position. This technique is always used in conjunction with other techniques (e.g., to reinitialize dead-reckoning systems).

1.1 One-way Communication

(*a*) *Systems*. In one-way communication systems, the communication always takes place from the central computer to the vehicles via different infrastructures. The information can be broadcast in a determined area or transmitted through a beacon at precise geographical points.

If broadcasting in a zone, the system transmits, in each area, traffic information processed by the central computer and relating to that area. In the same area, all vehicles receive the same information at the same time. This information is either gathered by the police from traffic control and surveillance centers (e.g., the Japanese advanced mobile traffic information and communication system (AMTICS)) or by national or regional authority offices (e.g., Carminat). It concerns congested roads, weather conditions, traffic incidents or changes in the road network introduced, for example, by roadworks.

In AMTICS, the broadcasting system is a new radio data communications system which will be able to communicate between user centers and vehicles by radio broadcasting stations called teleterminals (from telecommunications terminal). Each terminal transmitter covers an area of approximately 3 km^2 and 40 terminals would be required to cover the whole City of Tokyo. So far, 11 vehicles and a bus have been equipped (see *Broadcasting Communication Systems*).

The Carminat project uses the radio data system (RDS) which works on FM broadcasting networks. As the FM range is small, one of the first applications of RDS was automatic frequency tuning. The principle is that a numeric subcarrier is added to the basic sound signal. The resulting numeric channel is organized into frames, some of which contain the frequencies of the surrounding transmitters broadcasting the same program. The receiver set scans these frequencies and chooses the one that gives the best broadcast in its zone. That principle can be expanded to other applications; for example, a type of frame can contain coded traffic information messages (traffic message channel).

If transmitting at precise geographical points, communication takes place at beacons and the vehicles can only receive information when they pass in an area of 25–50 m radius, as for the Japanese road automobile communication system (RACS). There is a beacon every 2–5 km in such a way that a vehicle whose speed is 100 km h^{-1} passes one beacon every 72 s. This defines a minimum radio zone structure placed intermittently. This system will be implanted on an area 350 km^2 in the south west of Tokyo center and uses beacons that will work at frequencies between 1 GHz and 3 GHz. This will be a two-way communication system but, as a first step, will be used as a one-way system.

(*b*) *Algorithms*. The algorithms used in one-way communication systems are shortest-path algorithms (see *Shortest-Path Algorithms*). To compute these shortest paths it is necessary to have an estimate of the travel times on the different links. This estimate itself depends on the precision of the real-time information given by the system. As an example, RDS cannot transmit a useful data flow of more than 100 bits per second, so in Carminat, which is based on RDS, the information must be compacted. It is proposed that three levels of information be sent to the vehicle:

(a) fluid traffic,

(b) heavy traffic, and

(c) congested traffic.

In the different one-way communication systems, the on-board computer makes the shortest-path computations a function of its estimate of the travel times on the different links. This estimate is based on the information sent by the system. Following the nature of these levels, it is obvious that, in Carminat, forecasting the travel times will be crude and that the route prescribed to the vehicle may be a good one but will probably not be the best one.

For one-way communication systems, the information sent to the vehicles could be, because they are easily obtained, the flows of vehicles on the different links and, in the case of fixed-time traffic control, the splits and cycle times. Therefore, to forecast the travel times, the on-board computer could use travel-time models as a function of flows. Many of these models exist, such as Wardrop's model:

$$TP_j(Q_j) = \tau_{0j} + \left(\frac{r_j^2}{2c_j}\right)\left(\frac{S_j}{S_j - Q_j}\right) \quad (1)$$

or Webster's model:

$$TP_j(Q_j) = \tau_{0j} + \left(\frac{r_j^2}{2c_j}\right)\left(\frac{S_j}{S_j - Q_j}\right) + \frac{Q_j}{2C_j(C_j - Q_j)} \quad (2)$$

where $TP_j(Q_j)$ is the travel time on the link j as a function of the flow Q_j of that link, τ_{0j} is the free travel time of the link j, c_j is the cycle of the signals, r_j is the red time, S_j is the saturation flow, and C_j is the capacity of the link j (see *Road Traffic: An Introduction*).

These models imply that the signal timings are constant. Thus, if traffic-responsive strategy is in operation, the model will be inadequate, and so will the guidance. In a first step, the best route determination will probably be made using fixed-time strategies.

A DRG system needs a prediction of the future travel times to process the best routes. This prediction will be done by estimating the future flows on links, and then applying the travel-time model (see *Prediction of Traffic Flow*).

It should be noted that the route computation problem for a vehicle which is at present time t at the origin O and which wants to go to the destination D can be expressed as:

$$\min_{n,\, l(2),\ldots,\, l(n-1)} t_n \qquad (3)$$

under the constraints:

$$\left.\begin{array}{l} l(1) = \mathrm{O} \\ l(i+1) \in S(l(i)), \quad i = 1, \ldots, n-1 \\ l(n) = \mathrm{D} \\ t_{i+1} = t_i + TP_{l(i)}(Q_{l(i)}(t_i)), \quad i = 1, \ldots, n-1 \\ t_1 = t \\ Q_{l(i)}(t_i) = Q^G_{l(i)}(t_i) + Q^{nG}_{l(i)}(t_i) \end{array}\right\} \quad (4)$$

where $S(l)$ is the set of downstream links of link l, and n is the number of links in a route between O and D (n is not fixed and will be determined solving the optimization problem). Assuming that:

$$TP_l(Q_l(t + \Delta t)) \geqslant TP_l(Q_l(t)) - \Delta t, \quad \forall t, \Delta t, l \quad (5)$$

which corresponds to the physical fact that a vehicle cannot overtake another, the preceding problem can be solved, for instance, using forward dynamic programming (FDP), or the Dijkstra algorithm (Dijkstra 1959). Otherwise, the problem can be solved by FDP using time and links as states. A solution would be to filter predicted flows so that Eqn. (5) is verified. Note that, if Eqn. (5) is true for Δt equal to the sampling time, it is true for any multiple of the sampling time; thus, the filtering can be done step by step.

1.2 Two-way Communication

(a) Systems. Even if it does not really perform route guidance, the American project Pathfinder is based on a two-way communication principle: the system transmits the state of congestion of the different links to the vehicles and the vehicles send their travel times to the system. Thus, it is easier for the system to forecast the new states of congestion. On the other hand, the drivers have to find their own paths to their destinations using the information they have received on their screens. The principle of communication which is used here is packet radio transmitting at 4800 bits per second on 419.975 MHz. This principle will probably be used in the future for route guidance; however, existing systems use beacons.

The control center computes the best routes to go from a point to another depending on the information given by the police and the different organizations and, overall, as a function of the travel times of the different links transmitted by the equipped drivers. If the system has been informed of the destination of the vehicle, it sends the vehicle the shortest path to reach its destination. An example of such a system was the comprehensive automobile traffic control project (CACS) (Yumoto *et al.* 1979). Otherwise, it sends the tree of the best routes to go from a fixed point (the beacon) to every other zone or beacon of the network; AUTOGUIDE and ALI-SCOUT work in this way.

For these three systems, the principle of communication on a fixed point uses broadcasting beacons for CACS and infrared beacons for AUTOGUIDE and ALI-SCOUT.

(b) Algorithms. An original feature of CACS, developed in 1973 and no longer operating, with computers of lower capacity than are used nowadays, was a processor called the traffic network simulator system, specially made for the determination of shortest paths. This processor was composed of a large number of chips, each of them representing an intersection. The chips were connected one to another to represent the network. Every 15 min, the processor was:

(i) forecasting the traffic flows for the future hour with the past travel times measured—using a model of travel time in function of flows, it was able to forecast these travel times and then, using the travel times sent by the equipped vehicles, it was able to adjust the parameters of the model; and

(ii) computing the best routes for each origin to each destination.

For AUTOGUIDE and ALI-SCOUT, the principle is different. The city is divided into zones; each zone is defined as a convex quadrangle or, in some cases, a triangle. The system computes the best routes for each origin to each destination zone. For each origin, the destination zones that can be reached using the same guidance vector define a destination area. Destination area boundaries and guidance vectors are transmitted to the vehicle. Forecasting the travel times is done every 15 min using the past and present measured travel times. The system uses a model of travel time as a function of time; and the algorithm used to find the shortest paths is Loubal's (Loubal 1971).

In summary, these two-way communication systems can directly use the past and present travel times measured by the equipped vehicles.

Future travel times are predicted for each link and each sampling time from past measured travel times. The prediction can use historical (depending on the type of day and on the hour) and recently measured data. The route computation problem, for a vehicle which is at the origin O at time t and which wants to go to the destination D can be expressed in the form of

Eqn. (3) under the constraints:

$$\left.\begin{array}{l} l(1) = O \\ l(i+1) \in S(l(i)), \quad i = 1, \ldots, n-1 \\ l(n) = D \\ t_{i+1} = t_i + TP_{l(i)}(t_i), \quad i = 1, \ldots, n-1 \\ t_1 = t \end{array}\right\} \quad (6)$$

where $TP_l(t)$ is the predicted average travel time for the link l at time t.

Once again, this problem can be solved using FDP. The assumption that a vehicle cannot overtake another can be written:

$$TP_l(t + \Delta t) \geq TP_l(t) - \Delta t, \quad \forall t, \Delta t, l \quad (7)$$

Predicted travel times can be filtered step by step, so that Eqn. (7) is verified; and then the Dijkstra algorithm can be used to solve this route computation problem.

(c) *Robustness of control.* Considering two routes A and B from an origin O to a destination D, route A may have a slightly shorter average travel time than route B, but an important variance, whereas B may have a small variance. In this case, route B may be more interesting because it is more reliable. It should be interesting to build a travel-time model taking into account standard deviation of measured travel time.

Only two-way communication systems are considered in Sect. 2.

2. Perspectives of the Development of Algorithms

2.1 With Signal Coordination

Theoretically, if the sampling time had been small, the preceding models would take into account signal coordination. In practice, as the sampling time is greater than the signal cycle, or the signal strategy is traffic responsive, or the travel-time measurements are not so frequent, the modelled travel times are not related to offsets. Thus, to take into account signal coordination, one way to proceed is to formulate the travel time of a link as a function of the travelled upstream link as well. The algorithm must then predict a travel-time value for each future sampling time, each link and each upstream link. The route computation problem can then be formulated as Eqn. (3) under the constraints:

$$\left.\begin{array}{l} l(1) = O \\ l(i+1) \in S(l(i)), \quad i = 1, \ldots, n-1 \\ l(n) = D \\ t_{i+1} = t_i + TP_{l(i)}(t_i, l(i-1)), \quad i = 2, \ldots, n-1 \\ t_1 = t \\ t_2 = t + TP_O(t) \end{array}\right\} \quad (8)$$

where $TP_l(t, m)$ is the travel time for the link l coming from the link m at time t, and $TP_l(t)$ is the travel time for the link l at time t.

The shortest-path resolution is the same as before but the number of travel times to be processed is higher.

Another way to proceed is to use a travel-time model depending on directional flows Q_{jk} where j is the considered link and k is an upstream link of j. To compute Q_{jk}, the origin–destination of guided vehicles and the traffic control policies must be known. The directional flows of the guided vehicles are known (control variables), the directional flows of the nonguided vehicles are not known and have to be estimated; this can be done, for example, by Kessaci's method (Kessaci et al. 1989) or by Cremer's method (Cremer and Keller 1987) (see *Origin–Destination Matrix: Dynamic Estimation*).

2.2 Queue-Dependent Model

The preceding models $TP(t)$ and $TP(Q)$ cannot model congestion as they do not consider queues. Thus, in order to study congestion, a model of travel time that depends on the state of the link (queues) is required. A queue is composed of guided and nonguided vehicles. Forecasting future numbers of guided vehicles in the queues depends on present queues, present routing and future vehicles entering the network and asking for guidance. In the same way, the prediction of the future number of nonguided vehicles in the queues relies on the present queues and on future nonguided vehicles entering the network. (Note that the assumption that routing does not affect nonguided flows is made.)

A prediction of the actual queues can be obtained with the help of magnetic loop sensors which measure flows, and with the help of equipped vehicles, which send their travel times.

It should be noted that, knowing the origin–destination demand for the guided vehicles, and making a projection of the nonguided demand into the future, queues and then travel times can be predicted. The route computation problem can be written as Eqn. (3) under the constraints:

$$\left.\begin{array}{l} l(1) = O \\ l(i+1) \in S(l(i)), \quad i = 1, \ldots, n-1 \\ l(n) = D \\ t_{i+1} = t_i + TP_{l(i)}(X_{l(i)}(t_i)), \quad i = 1, \ldots, n-1 \\ t_1 = t \\ X_{l(i)}(t_i) = X^G_{l(i)}(t_i) + X^{nG}_{l(i)}(t_i) \end{array}\right\} \quad (9)$$

where $X_j(t)$ is the number of vehicles in the queue of the link j at time t.

2.3 Global Criterion

When the penetration ratio grows, it is no longer possible to ignore the influence of routed vehicles on nonguided vehicles. It seems topical to consider community criterion.

Assuming that the queues are vertical (a vehicle travels the totality of a link with its free travel time and packs into a vertical queue) and that the vehicles having the same origin–destination are not necessarily grouped, the formulation can be:

$$\min_{Q_j^{OD}(t)} \sum_t \sum_{O,D,j \neq D} \left[X_j^{OD}(t) + \sum_{k=0}^{M_j} Q_j^{OD}(t-kT) \right] \quad (10)$$

where T is the sampling time and M_j is an integer s.t. $M_j T$ is the free travel time of the link j, under the constraints:

$$A_j^{OD}(t + M_j T) = Q_j^{OD}(t), \quad \forall j \neq D$$

$$A_D^{OD}(t) = 0$$

$$X_j^{OD}(t+T) = \frac{X_j^{OD}(t) + A_j^{OD}(t)}{\sum_{O,D}(X_j^{OD}(t) + A_j^{OD}(t))}$$

$$\times \max\left\{ \sum_{O,D}[X_j^{OD}(t) + A_j^{OD}(t)] - D_j(t), 0 \right\},$$

$$\forall O, D, j \neq D \quad (11)$$

$$\sum_{j \in U(n)} \frac{X_j^{OD}(t) + A_j^{OD}(t)}{\sum_{O,D,j \neq D}(X_j^{OD}(t) + A_j^{OD}(t))}$$

$$\times \min \sum_{O,D}[X_j^{OD}(t) + A_j^{OD}(t), D(t)]$$

$$= \sum_{j \in D(n)} Q_j^{OD}(t), \forall O, D, n \neq A(D)$$

$Q_0^{OD}(t)$ known, $\forall O, D$

where $A(D)$ is the upstream node of link D, $A_j^{OD}(t)$ is the flow entering the queue of the link j for the origin O and the destination D at time t, and $D(t)$ is the flow leaving the queue at time t. During the green phase, $D(t) = S(t)$ (the saturation flow), and during the red phase $D(t) = 0$.

The capacity constraint can be written:

$$\sum_{O,D} X_j^{OD}(t) + X_j^{nG}(t) \leq n_j, \quad \forall j \quad (12)$$

where n_j is the maximal number of vehicles that link j can contain.

Then, because of those constraints that are complex, this is a nonlinear programming problem. It can be solved using a gradient projection, by Newton's method or by an optimal control approach (Charbonnier et al. 1991).

In this formulation, small sampling time and known traffic signal strategies are required to consider signals coordination. This is difficult to implement.

It should be remarked on that the global criterion has been presented here with a queuing model. If the models were time or flow dependent, the problem would be solved in the same way by linear or nonlinear programming, respectively.

3. The Future

Up to now, there are few DRG systems. It is intended that more are developed, but the penetration rate that could be obtained when guided and unguided vehicles have nearly the same gains will be highly dependent on the level of information exchange and on the credibility of that information.

The efficiency of these systems is related to the credibility of the information that has to be accepted by users. Moreover, users could be upset by an authoritative system and then may react by not obeying the "controls," so making them less effective.

DRG systems alone will not resolve the congestion problem if too many cars use the infrastructure; they are a step forward towards complete demand-management schemes and, thus, more complex management controls.

The main difficulty lies in the modelling of the driver and the user behavior, and also in the size of the problem.

See also: In-Vehicle Equipment for Future Traffic Control Systems; Route Guidance, Collective

Bibliography

Babsky J, Cohen H, Mauge J, Sarignac A 1988 Carminat—Programme Eureka (EU 55)—Réalisation d'un système complete d'information, de gestion et de navigation à vocation européenne pour les véhicules routiers. In: Milne M, Knoll E (eds.) *Proc. Int. Road and Traffic Conf. Road and Traffic 2000*, Vol 1. Road and Transportation Research Association, Cologne, Germany, pp. 147–50

Borins S F 1988 Electronic road pricing: An idea whose time may never come. *Transp. Res.* **22**, 37–44

Charbonnier C, Farges J L, Henry J J 1991 Models and strategies for dynamic route guidance: Optimal control approach. *Proc. DRIVE Conf. Advanced Telematics in Road Transport*. Elsevier, Brussels, pp. 106–12

Cremer M, Keller H 1987 A new class of dyanmic methods for the identification of origin–destination flow. *Transp. Res. B* **21**, 117–32

Dijkstra E W 1959 A note on two problems in connection with graphs. *Numer. Math.* **1**, 269–71

French R L 1989 Assessment of technologies, institutional barriers, and cooperative programs on dynamic route guidance in the USA. In: Henry J J, Perrin J P (eds.)

Proc. 6th IFAC–IFIP–IFORS Int. Conf. Control Computer Communication in Transportation Systems. Pergamon, Oxford, pp. 403–21

Henry J J, Farges J L, Tuffal J 1983 The PRODYN real time traffic algorithm. In: Klamt D, Lauber R (eds.) *Proc. 4th IFAC–IFIP–IFORS Int. Conf. Control in Transportation Systems.* Pergamon, Oxford, pp. 307–11

Hoffman G 1989 LISB an individual route guidance and information system in Berlin. In: Henry J J, Perrin J P (eds.) *Proc. 6th IFAC–IFIP–IFORS Int. Conf. Control Computer Communication in Transportation Systems.* Pergamon, Oxford, pp. 399–402

Hunt P B, Robertson D I, Bretherton R D, Winton R I 1981 SCOOT: A traffic responsive method of coordination signals. *Transp. Road Res. Lab. (U.K.), TRRL Report* **1014**

Jeffery D J, Smith J C 1989 Driver information systems. TRRL Note 240489. Transport and Road Research Laboratory, Crowthorne, UK

Kawashima H, Ishii Y, Fukui R 1989 Discrete minimal radio zone communication system in RACS project and the performance evaluation. In: Henry J J, Perrin J P (eds.) *Proc. 6th IFAC–IFIP–IFORS Int. Conf. Control Computer Communication in Transportation Systems.* Pergamon, Oxford, pp. 29–36

Kessaci A, Farges J L, Henry J J 1989 On-line estimation of turning movements and saturation flows in PRODYN. In: Henry J J, Perrin J P (eds.) *Proc. 6th IFAC–IFIP–IFORS Int. Conf. Control Computer Communication in Transportation Systems.* Pergamon, Oxford, pp. 387–93

Loubal P S 1971 Persönliche Mitteilung über iterationsfreie Routensuche. Munich

Mammano F, Sumner R 1989 PATHFINDER System Design. In: Reekie D H M, Case E R, Tsai J (eds.) *Proc. 1st IEEE Int. Conf. Vehicle Navigation and Information Systems.* Institute of Electrical and Electronics Engineers, New York, pp. 484–8

Okuaki T, Yamazaki T, Iwaizumi K 1986 Trial model of data base for use of road traffic information service. In: Huddart K W (ed.) *Proc. 2nd Int. Conf. Road Traffic Control.* Institution of Electrical Engineers, London, pp. 18–21

Poignet A 1988 Application du système RDS à l'amélioration du déplacement en automobile. In: Milne M, Knoll E (eds.) *Proc. Int. Road and Traffic Conf. Road and Traffic 2000*, Vol 1. Road and Transportation Research Association, Cologne, Germany, pp. 181–6

Russam K, Jeffrey D J 1986 Route guidance and driver information systems—An overview. In: Huddart K W (ed.) *Proc. 2nd Conf. Road Traffic Control.* Institution of Electrical Engineers, London, pp. 1–5

Shibata M 1989 Development of a road/automobile communication system. *Transp. Res. A* **23**, 63–71

Tsugawa M, Okamoto H 1989 Advanced mobile traffic and communication system—AMTICS. In: Reekie D H M, Case E R, Tsai J (eds.) *Proc. 1st IEEE Int. Conf. Vehicle Navigation and Information System.* Institute of Electrical and Electronics Engineers, New York, pp. 475–83

Tsugawa S, Tabei S 1986 Route guidance system for automobile drivers by speech synthesis. In: Genser R, Etschmaier M M, Hasagawa T, Strobel H (eds.) *Proc. 5th IFAC–IFIP–IFORS Int. Conf. Control in Transportation Systems.* ÖPWZ, Vienna, pp. 215–20

Tsuji H, Takahashi R, Kawashima H, Yamamoto Y 1985 A stochastic approach for estimating the effectiveness of a route guidance system and its related parameters. *Transp. Sci.* **19**, 333–51

Yumoto N, Ihara H, Tabe T, Namiwada M 1979 Outline of the comprehensive automobile traffic control pilot test system. *Trans. Res. Rec.* **737**, 113–21

J. J. Henry, C. Charbonnier and J. L. Farges
[ONERA, Toulouse, France]

Safety of Road Traffic

Road safety is one of the major topics covered by traffic and transportation engineers in close collaboration with other disciplines. It deals with the systematic recording, storage and analysis of accident data and the use of the results in the study and research of single accidents, as well as of groups of accidents, in order to define causes, identify hazardous locations and propose remedies. The final aim of road safety studies and research is to reduce the number and severity of road accidents which are one of the major causes of deaths, resulting in an annual toll of some 300 000 killed and 8×10^6 injured persons worldwide.

1. The Problem

The scale of the road safety problem is well illustrated by just a few estimates from global accident statistics.

(a) Road accident fatalities every year amount to some 300 000 world-wide and 55 000 in the EC countries.

(b) In 1987, 46 330 deaths were caused by road accidents in the USA. This figure amounts to 94% of the 49 306 deaths caused by all transportation modes and is about equal to the number of Americans killed in the Vietnam War.

(c) The number of persons injured every year amounts to some 8×10^6 worldwide, 3×10^6 in the USA and 1.7×10^6 in the EC countries. Of these EC accidents, 150 000 result in permanent handicaps.

(d) The financial cost of road accidents in the EC is estimated at about US$$35 \times 10^{12}$ per year, while social costs in terms of human misery and suffering cannot be measured.

The occurrence of a road accident indicates a failure by the road user, the road and its environment, the vehicle, or any combination of the three. The probability of the occurrence of such a failure is directly related to the amount of travel.

In spite of the measures being taken and the improvements achieved in relation to each one of these three accident-contributing factors, the absolute numbers of accidents and their effects—fatalities, injuries, damages—are fairly stable or even increasing. This is the result of the fact that the amount of road travel, expressed in vehicle kilometers, is increasing at

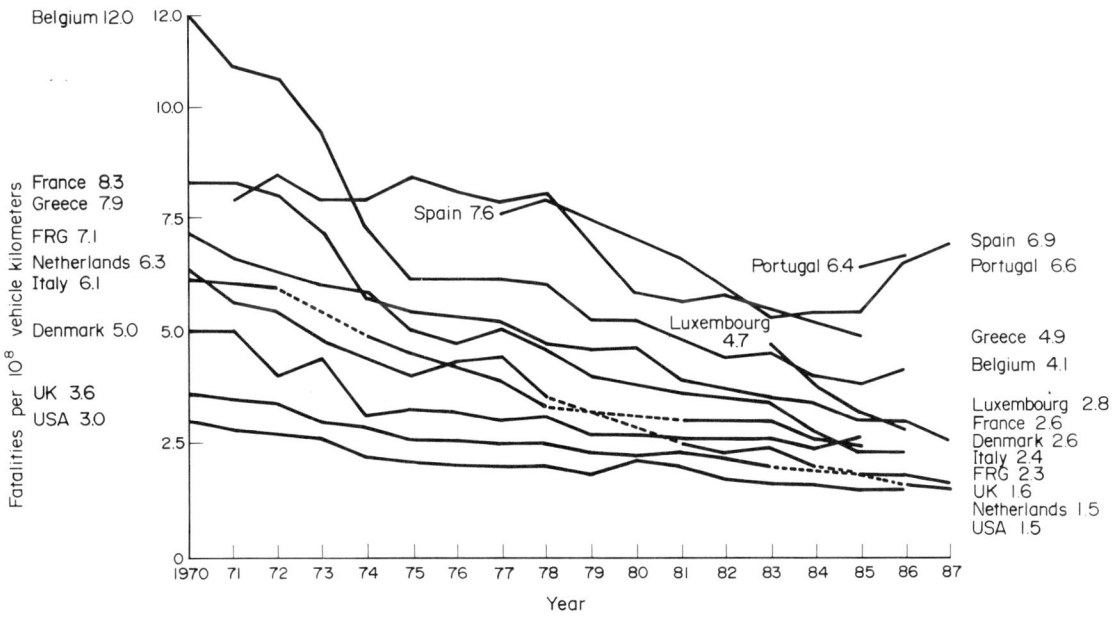

Figure 1
Evolution of fatality rates in the EC and the USA (prepared from the annual road statistics of the International Road Federation)

a high rate, while the accident rates, expressed in accidents per vehicle kilometer, are decreasing at a lower rate or remain practically constant, especially in the EC, the USA and other developed countries where extensive road safety programs have already been implemented. As illustrated in Fig. 1, fatality rates (number of fatalities per 10^8 vehicle kilometers) gradually decreased in the EC countries and the USA between the early 1970s and mid-1980s, but remained practically stable during the late 1980s. Accident rates in developing countries are much higher but are decreasing faster.

2. Causes

As mentioned, road accidents are attributed to three major factors:

(a) the human factor (i.e., the road user: driver, passenger or pedestrian),

(b) the conditions of the road and its environment, and

(c) the conditions of the vehicle.

In many cases, two or all three of these factors contribute to an accident. The complexity of the situation and the lack of proper accident reporting and analysis do not always allow the contribution of each one of the factors to be attributed. However, several studies on indepth accident analysis clearly point to the human factor, alone or combined with the other two factors, as the main cause of road accidents. The magnitude of the contribution of the three factors is indicated by a study conducted in the UK in the period 1970–1974 (Sabey 1980) showing that

(a) road users were responsible in 95% of the accidents examined (65% alone, 24% with road conditions, 4.5% with vehicle conditions, and 1.25% with road and vehicle conditions),

(b) road and environmental conditions were responsible in 28% of the cases examined (2.5% alone and 0.25% with vehicle conditions), and

(c) vehicle conditions were responsible in only 8.5% of the cases (2.5% alone).

2.1 The Road User

The accidents attributed to the road user are mainly caused by violations of traffic rules and regulations such as improper overtaking, turning or stopping, reckless driving and driving under the influence of alcohol, non-observance of traffic signals and signs by drivers and pedestrians, and careless road crossing by pedestrians.

Road user accidents are reduced through proper education, and more control and enforcement regarding the application of traffic rules, as well as through better information being provided from the road, the road environment, other road users and the vehicle.

2.2 Road and Environmental Conditions

The accidents attributed to the road and environmental conditions are caused by

(a) poor geometric characteristics such as narrow lanes and shoulders, lack or reduced width of medians, small radii of horizontal and vertical alignment resulting in poor visibility, and poor intersection layout;

(b) poor construction standards, mainly reduced skid resistance of pavements and poor drainage;

(c) improper design, location and construction of roadside features such as poles, guardrails, trees, curbs, embankments and ditches;

(d) poor organization of traffic flow such as lack of proper signing, poor control of accesses (entrances, exits, driveways) and on-street parking;

(e) complete lack of or insufficient road lighting;

(f) poor management of work zones; and

(g) adverse environmental conditions, mainly fog, rain, wet surfaces, snow and ice, but also dust, smoke and wind.

A more detailed presentation is given by the Federal Highway Administration (1982).

Reducing the effect on road safety of road and environmental conditions requires the proper design, construction, operation and maintenance of new facilities, on the basis of accident experience from similar existing facilities. Concerning existing facilities, this requires identifying through proper statistical analysis the hazardous locations, determining the road deficiencies that most probably caused the accidents at these locations and implementing the proper countermeasures dictated by experience from similar cases.

2.3 Vehicle Conditions

Some accidents are attributed to mechanical or other failures caused by the poor maintenance and age of vehicles, as well as to oversize loads carried by trucks. Better maintenance and periodic checks of vehicles, as well as an effort to reduce the average age of the vehicular fleet reduces the number of these accidents. In addition, research and production of safer vehicles contributes to the reduction of the number of road accidents and of their severity.

3. Data Collection and Storage

The proper collection and storage of road accident data in databases allowing for easy and systematic information retrieval is a prerequisite for correct accident analysis and use. A road accident information system, which in most cases constitutes part of a broader road information system, provides useful information for a wide range of public services and private organizations and individuals, such as the police, insurance companies, courts, legislative services, car industries, statistical services, education and research institutions, hospitals, and traffic and highway engineers.

3.1 Accident Reporting

In each country, standard accident record forms exist which are filled in by the police and constitute the basic source of accident information. Motor vehicle laws in certain countries also require drivers to submit a report for any motor vehicle accident that results in death, injury or property damage of more than a specified value.

The extent of data collected for each accident differs from country to country. For example, the number of questions included in the accident record forms of the 12 EC countries varies from 348 in France to 82 in Portugal.

The major problem in accident data collection is the number of traffic accidents that are not reported to the police and are, therefore, not considered in any safety analysis. Furthermore, in certain countries, due to lack of sufficient staffing and to fiscal restraints, police policies do not require the responding officer to prepare a report unless someone is injured or a vehicle has suffered severe damage (e.g., needing to be towed).

Other problems are unclear definition of the accident location and insufficient or inaccurate reporting.

To improve the accuracy and efficiency of accident location, various reference methods are used in a computerized accident data system. These are discussed in detail by the Transportation Research Board (1974) under the following three categories:

(a) sign-oriented methods, using physical milepost signs or other field reference signs;

(b) document-oriented methods, involving the referencing of office maps and/or files; and

(c) other methods such as coordinate methods using a unique set of *x* and *y* coordinates, and referencing to road features (e.g., bridges, utility poles, railroad crossings).

3.2 Data Storage and Retrieval

Traffic accident reports filled in by the police, sometimes supplemented by driver reports, are coded and keyed into the computer accident file. They are usually combined with other standard files maintained for highway systems and/or highway users such as

(a) the road file, consisting of physical information on roads and on their environment;

(b) the traffic file, containing information on traffic volumes (this may be part of the road file);

(c) the driver licensing file, containing information on individual drivers; and

(d) the motor vehicle registration file, containing information on vehicle types, age, horsepower and so on.

The merging or interfacing of the accident file with these files is necessary for various reasons. For example, it is necessary to merge

(a) with traffic files to compute accident rates for specific locations or facilities,

(b) with road files to assess the effect of various road elements on accidents, and

(c) with driver licensing or motor vehicle registration files to analyze the effect of a specific category of drivers or vehicles on accidents.

All of the computerized data files are used in accident analysis to provide

(a) periodic (monthly, quarterly or annual) lists of all accidents by location;

(b) periodic lists of high-occurrence accident spots and sections, and detailed summaries of the accidents that occurred at these locations;

(c) special report summaries and accident statistics for public distribution (usually once per year); and

(d) special information for research studies.

4. Accident Analysis

Accident analysis is carried out for the following purposes:

(a) to identify causes and assign responsibilities for each accident,

(b) to identify hazardous locations and to study and program the necessary improvements at these locations,

(c) to evaluate road safety improvements, and

(d) to study the effects of the three main factors (i.e., the road user, the road/environment and the vehicle) on traffic accidents and to define the necessary general policies and specific improvements for each case.

The analysis begins with the investigation of a single accident and continues as a broader analysis of all the accidents occurring during a specific time frame at a single location or in a broader environment, such as a single road or a road network, at an urban, regional, national or international level.

4.1 Study of a Single Accident

A first analysis of an accident may be carried out on the basis of the standard report prepared by the police for each accident. However, an in-depth analysis can only be performed by a multidisciplinary team visiting the scene of the accident as soon as possible after the accident occurrence. This is, however, possible only for a selected number of accidents, within the framework of programmed research. Examples of such in-depth accident studies are the REAGIR program initiated in France, the special "at the scene" file in the UK describing over 2000 accidents that have been attended and investigated in depth by specialist staff of the Transport and Road Research Laboratory (TRRL), and the *Adelaide In-Depth Accident Study 1975–1979* (Adelaide University Road Accident Research Unit 1979). In all other cases, the standard accident reports are used, supplemented in some cases by a detailed examination of the accident locations so as to study the actual conditions from the point of view of geometry, environment and traffic behavior.

The analysis of a single accident is carried out either through simple manual computations based on the principles of momentum and energy, or through computer programs where accidents are reconstructed with the aid of more sophisticated procedures. In the former, simple equations are used to estimate the speeds of the vehicles prior to collision. For example, in the simple case where a vehicle stops after a continuous uniform braking over a length of s meters, measured by the skid marks, the speed v before braking is computed by

$$v = 2sfg \tag{1}$$

or, because $g = 9.81 \text{ m s}^{-2}$,

$$v = 4.43sf \text{ in } \text{m s}^{-1} \tag{2}$$

or

$$v = 15.95sf \text{ in } \text{km h}^{-1} \tag{3}$$

where f is the drag or deceleration factor, which is equal to the friction coefficient plus or minus the slope (tangent of angle) in the direction of movement. The coefficient of friction may be estimated according to the pavement conditions or, better, computed through Eqn.(1), using the same or a similar vehicle braking at the same location and measuring s at various speeds.

Such simple methods, usually used by police in cases such as the one mentioned previously, are described in relevant textbooks such as the *Transportation and Traffic Engineering Handbook* (Institute of Traffic Engineers 1976).

More complicated methods, providing simulations of collosions and vehicle trajectories to varying levels of complexity, sophistication and ease of use, are applied through various computer programs established during the 1980s.

Three of the older and more widely known of these programs are CRASH (Calspan reconstruction of accident speeds on the highway), SMAC (simulation model of automobile collision) and HVSOM (highway vehicle object simulation model). All three programs were developed by the Cornell Aeronautical Laboratories (later Calspan Corporation) in a series of research projects sponsored by the US Department of Transportation.

A fourth program, the EES-ARM (equivalent energy speed-accident reconstruction program), has been widely used in Europe for speed reconstruction in automobile crashes. It automates the methods usable for manual calculations based on the principles of momentum and energy. These and other programs are examined by Ronald et al. (1986).

4.2 Identification of Hazardous Locations

A hazardous location, sometimes called a blackspot, is a location where more road accidents occur than those expected in similar locations, within a specified degree of certainty. The identification of hazardous locations is a very difficult task because, as already mentioned, accidents caused by deficiencies in the road and its environment, with or without the involvement of human error or vehicle failure, account for about 25% of all accidents. All other accidents, which are caused by human error on its own or combined with a vehicle failure, do not depend on road and environmental conditions, thus interfering with the analysis for the identification of hazardous locations. Furthermore, the consideration of accident severity and type as well as the need to compare "similar" locations increase the difficulties of the analysis.

Hazardous locations are usually identified through numerical and statistical methods. The former are simple techniques not taking into account the random variation in the number of accidents while, in the latter, similar road sections are compared and locations are identified where significantly more accidents occur than the number expected with a certain probability.

In all these methods, either the number of accidents or the number of involvements is considered (i.e., the number of vehicles or pedestrians involved in an accident). Efforts are made to weight accidents according to their severity or their cost. In the first case, three basic categories of accidents are normally used: accidents with fatalities, accidents with injuries (sometimes classified into serious and light) and damage-only accidents. Various arbitrary weighting factors have been proposed for these three categories of accidents (e.g., 12:3:1). Weighting according to cost is also questionable considering, among other reasons, the difficulties in assessing the cost of a fatality. Thus, a mathematical weighting of accidents is usually avoided. Instead, various categories of accidents are considered separately in the analysis and judgement is used to reach overall conclusions with the aid of various indexes such as the severity index.

The accident severity index is defined as the ratio of the number of accidents with casualties (fatalities plus injuries) to the total number of accidents. The severity of a certain group of accidents is increasing as this index approaches 1.0. For example, it was found that the severity index for accidents involving pedestrians approaches 1.0, for head-on collisions it is about 0.5, for rear-end collisions it is below 0.3 and for parking it is below 0.2.

Separate analysis should be carried out for urban and rural roads, and for the differing elements of each of these two basic categories. Road sections are usually classified according to the type of road (e.g., freeways, other four-lane divided, four-lane undivided and two-lane roads). Road intersections are usually classified according to type and control (e.g., no control, yield signs, stop signs, signalization, interchange).

A proper exposure index should be defined to take into account the differing traffic flows served by an examined road element which create different probabilities for the occurrence of an accident. Two such indexes that are often used are vehicle kilometers for road sections and number of vehicles entering an intersection for all types of intersections. In the latter, pedestrians crossing the intersection should also be considered where their number is significant. For road sections the accident rate is given by

$$R_s = 10^6 A/TVL \quad (4)$$

measured in accidents per 10^6 vehicle kilometers and for road intersections it is given by

$$R_j = 10^6 AT \sum_{i=1}^{n} V_i \quad (5)$$

measured in accidents per 10^6 vehicles, where T is the period examined in days (usually one or more years), A is the number of accidents or involvements recorded during T, V is the average daily volume of traffic in the examined section during T, L is the length of the examined section in kilometers (usually 0.2–0.5 km for urban and 0.5–1.0 km for rural roads), V_i is the average daily volume of traffic for leg i of the intersection and n is the number of legs of the intersection.

Such accident rates imply that traffic accidents increase in proportion to traffic volumes, which is an oversimplified assumption. Frantzeskakis and Damlanos (1987) suggest that the volume-to-capacity ratio V/C, which is one of the parameters affecting the level of service, may offer a better measure of exposure than taffic volume *per se*. Analyzing accidents on two interurban four-lane highways in Greece (see Fig. 2), they found that the accident rates remain almost constant up to a V/C of 0.65 (i.e., for levels of service A, B

and C) and increase considerably at levels of service D, E and F, more than doubling at level F when $V/C > 1.0$.

Numerical methods identify probable hazardous locations on the basis of simple relations, for example, when accident indexes are higher than the average value for all examined locations or higher than the double of the average value, depending on the resources available for a detailed examination of all hazardous locations and implementation of the appropriate countermeasures.

Statistical methods comprise more sophisticated procedures such as the quality control method, similar to the method used in industry to check products through sampling. A critical value R_c of the accident rate is computed for each location examined (road section or intersection), based on the average rate for all locations. Locations where the actual accident rate is higher than R_c are considered—within the confidence level used—out of control (i.e., hazardous). This means that the high number of accidents recorded there cannot be attributed to chance alone, but also to some unfavorable characteristics.

Values of R_c are computed from

$$R_c = R_a + K(R_a/M)^{1/2} + 1/2M \quad (6)$$

where R_a is the average accident rate for all locations examined in accidents per 10^6 vehicle kilometers, M is 10^6 vehicle kilometers for the examined road section or 10^6 vehicles for the examined intersection, and K is the probability constant, defining the confidence level. Usually values of K of 1.28, 1.64 and 2.32 are used, corresponding to confidence levels of 0.10, 0.05 and 0.01 or to probabilities of 90%, 95% and 99%.

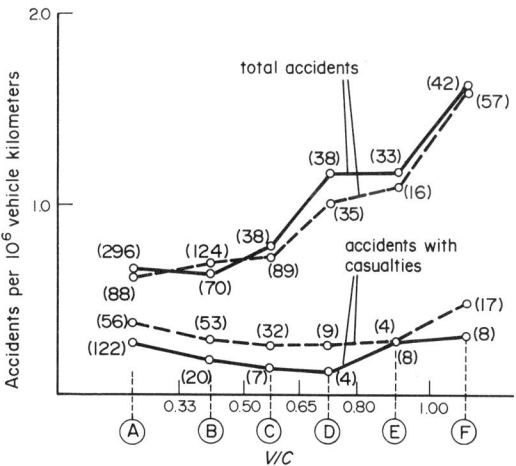

Figure 2
Accident rates vs V/C ratio and level of service (A–F): —— Athens–Corinth national highway; ——— Athens–Salonica national highway; number of accidents in brackets

Safety of Road Traffic

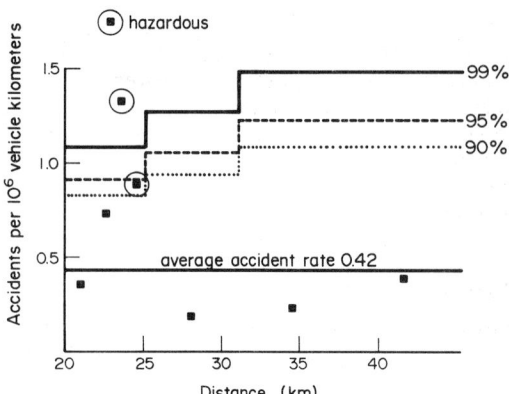

Figure 3
Quality control diagram for identifying hazardous locations

Figure 3 illustrates a quality control diagram for a 25 km road section. The analysis shows that at a distance of 24 km the road is hazardous with a 95% accident probability, while the location at a distance of 25 km is also indicated as hazardous if a 90% probability is examined.

4.3 Traffic Conflict Technique

Studies conducted in the USA, Canada, the UK and Sweden have developed a promising technique to evaluate the safety performance of a road element, especially an intersection, through observation of "conflicts" between vehicles. A traffic conflict is defined as an event involving two or more road users in which one user performs some nontypical or unusual action, such as a change of direction or speed, that places another user in jeopardy of a collision unless an evasive maneuver is undertaken.

The traffic conflict technique was developed to cope with the problems of unreliable accident records and of the time required to wait for adequate sample sizes of real accidents. Although traffic conflicts have not yet been statistically related to traffic accidents, relevant research suggests that such a relationship may exist (Glauz and Migletz 1980).

4.4 Analysis of Causes and Determination of Remedial Measures

Analysis of the causes of accidents at a certain location is usually aided by collision and condition diagrams supplemented, if possible, by an on the spot examination of road and traffic conditions.

A collision diagram (see Fig. 4) is a schematic drawing of the location examined showing, by conventional symbols for each accident, the direction and maneuvers of vehicles and pedestrians, as well as information on the time and date that the accident occurred, the weather and any special conditions (e.g., slippery surface). It is usually supplemented by an accident summary table showing the number of accidents by type and severity.

The condition diagram is a drawing usually at a scale of between 1:100 and 1:250 showing existing physical conditions with an emphasis on those related to traffic accidents (e.g., curbs, poles, view obstructions).

Accident causes and patterns, examined in relation to the existing conditions obtained from the diagrams and the site visit, suggest possible remedies for specific types of accidents or accidents in general.

5. Evaluating Road Safety Improvements

Road safety improvements are evaluated both before and after implementation. Before implementation, the necessary road improvements, which are identified through the procedures described in Sects. 4.2, 4.3 and 4.4, are evaluated to assure that there is a high probability of reducing the number and/or the severity of accidents to an extent resulting in economic benefits. After implementation, each improvement is evaluated in order to compare actual with predicted economic

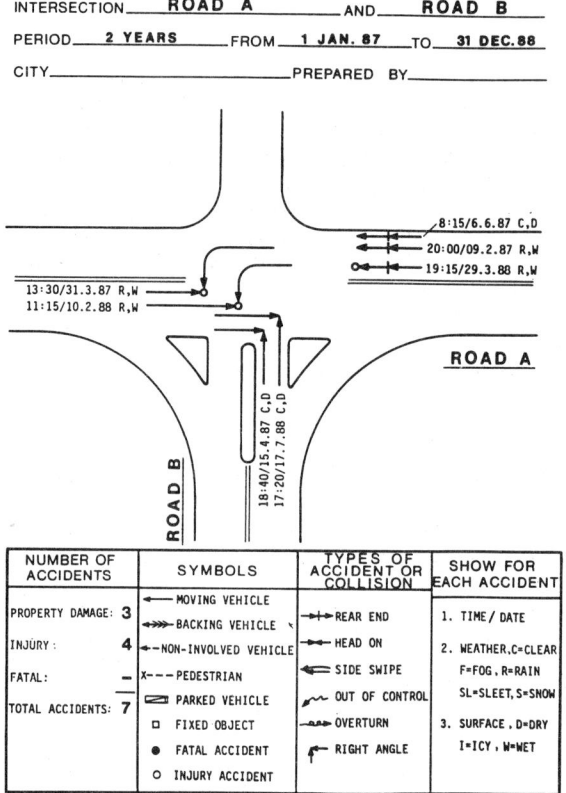

Figure 4
Typical collision diagram

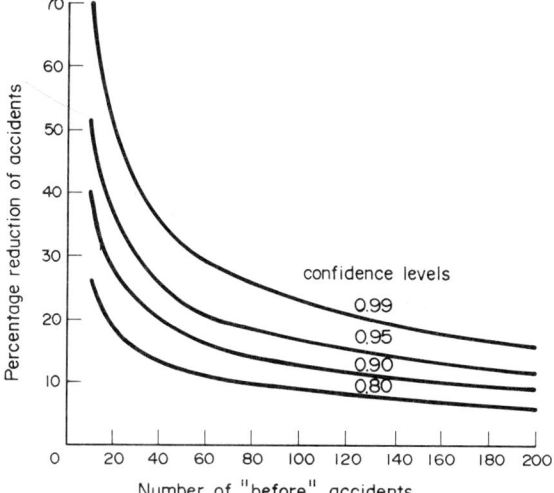

Figure 5
Poisson test for significance in "before" and "after" studies

and other benefits. Furthermore, postimplementation evaluation provides an objective basis for forecasting future improvements.

5.1 Preimplementation Evaluation

Preimplementation evaluation usually consists of a typical cost–benefit analysis where the construction, maintenance and operational costs of each improvement are compared to the benefits resulting from the improvement through accident reduction, as well as through any other probable operational improvement. Because safety improvements are, as a rule, inexpensive while benefits from the reduction of accidents are high, cost–benefit ratios or internal rates of return resulting from safety improvements are usually very high. Thus, road safety improvements are, in most cases, economically justified and have a high priority in the programming of road projects.

5.2 Postimplementation Evaluation

Postimplementation evaluation determines whether the predicted results were achieved. A "before" and "after" study is conducted where the measured reduction of accidents is tested to check whether it is statistically significant or whether it could have occurred by chance.

A simple method treats the before accidents at a location as a known Poisson mean and indicates the percentage change in the after accidents that must be observed to be judged statistically significant at a certain confidence level (see Fig. 5).

Improvements to this simple method, in order to avoid the phenomenon of regression to the mean, have been proposed by Weed (1986) by properly treating the before data as a random variable.

A by-product of the postimplementation evaluation is that it provides useful information to refine methods of predicting the results of safety improvements. In this respect, databases of actual results of road safety improvements provide a valuable source of information which is continuously updated by the input of more data resulting from the application of new improvements.

6. Research

Extensive research on road safety is being carried out by universities, industry and various research bodies such as the Transportation Research Board in the USA, and the Transport and Road Research Laboratory in the UK. The results can be found in the special publications of all these bodies as well as in articles in traffic engineering and safety journals such as the *ITE Journal* of the Institute of Transportation Engineers in the USA and *Traffic Engineering and Control* in the UK.

An extensive research effort aimed at improving road safety and transport efficiency, as well as reducing environmental pollution, through the application of road transport informatics both to vehicles and to the infrastructure, has started in the EC under the DRIVE program (dedicated road infrastructure for vehicle safety in Europe) (see *DRIVE*).

See also: Safety of Road Traffic: Intervention and Evaluation

Bibliography

Adelaide University Road Accident Research Unit 1979 *Adelaide In-Depth Accident Study 1975–1979*, Parts 1–10. University of Adelaide, Adelaide, Australia

Babkov V F 1975 *Road Conditions and Traffic Safety* Mir Publishers, Moscow

Federal Highway Administration 1982 *Synthesis of Safety Research Related to Traffic Control and Roadway Elements*, Vols. 1, 2. FHWA, US Department of Transportation, Washington, DC

Frantzeskakis J, Damianos I 1987 Volume to capacity ratio and traffic accidents on interurban four-lane highways in Greece. *Transp. Res. Rec.* **1112**, 29–38

Glauz D W, Migletz D J 1980 Application of traffic conflict analysis at intersections. National Cooperative Highway Research Program Report No. 219. Transporation Research Board, Washington, DC

Institute of Traffic Engineers 1976 *Transportation and Traffic Engineering Handbook*. Prentice-Hall, London

Organisation for Economic Co-operation and Development Road Research Group 1976 *Hazardous Road Locations: Identification and Countermeasures*. OECD, Paris

Ronald L W, Charles Y W, Thomas R P 1986 An overview of selected computer programs for automotive accident reconstruction. *Transp. Res. Rec.* **1068**, 18–33

Sabey B 1980 Road safety and value for money. Transport and Road Research Laboratory Supplementary Report No. 581. TRRL, Crowthorne, UK

Transportation Research Board 1974 Highway Location reference methods. NCHRP Synthesis 21. Transportation Research Board, Washington, DC

Transportation Research Board 1975 Methods for evaluating highway safety improvements. NCHRP Report No. 162. Transportation Research Board, Washington, DC

Weed R W 1986 Revised decision criteria for before and after analyses. *Transp. Res. Rec.* **1068**, 8–17

Zeeger V C 1982 Highway accident analysis systems. NCHRP Synthesis 91. Transportation Research Board, Washington, DC

J. M. Frantzeskakis
[National Technical University of Athens,
Athens, Greece]

Safety of Road Traffic: Intervention and Evaluation

An efficient road transportation system is beneficial to economic development and social progress, both in industrialized and in most nonindustrialized countries. Whether its importance is measured in terms of the proportion of gross national product (GNP) devoted to roads or the proportion of passenger kilometers and tonne kilometers of freight carried, the "road mode" is preeminent in virtually every country.

Roads do, however, have several disadvantages, of which by far the most important is damage arising from collisions between vehicles. This damage, in the form of injury and death experienced by road users and the destruction of property, imposes substantial penalties on society, as well as on the users of the system. The loss of life in the road transportation network is an important component in mortality totals and, in highly motorized nations, traffic may account for as much as 3% of deaths.

In monetary terms, this negative effect of road transportation is even more significant. The reason for this is that the age distribution of victims injured or killed is, especially in highly motorized countries, skewed towards the younger ages, with consequent loss of much productive life. This is in sharp contrast to deaths caused by the major diseases of the industrialized world, which typically affect the elderly. As a result, according to Hartunian *et al.* (1981), traffic injury in the USA in 1975 was one of the most costly health problems, surpassing strokes and heart disease, and approaching cancer (see also Baker *et al.* 1984 p. 13). Although the comparable figures for less motorized societies may be smaller, owing to a greater incidence of infectious diseases, the cost is still substantial.

There can be no doubt that reducing traffic injury is also an intractable problem, as shown by only moderate reductions in per capita fatality rates in the century since the invention of the motor vehicle. While it is true that the fatality rate per unit of travel has been steadily decreasing in every jurisdiction for which significant data exist, a corresponding increase in travel has nearly compensated for this increased safety. Some data from the USA that illustrate these trends is shown in Table 1.

It is not easy to identify exactly the causes of these trends. It would seem obvious that a multitude of factors must have contributed:

(a) better roads, built to facilitate travel;

(b) more modern vehicles, which are incidentally safer;

(c) more skilled and responsible drivers produced, in part, by educational programs but, most importantly, by experience.

All of these causes can be aggregated under the general heading "maturation of the road transportation system." The contrast between a mature traffic system in an industrialized, motorized society and the chaotic road use in a developing nation is striking.

In addition to these obvious factors, there are others that are less well known such as increased urbanization (with lower travelling speeds); an aging population (with less risky behavior); better rescue, treatment and rehabilitation services; as well as the decreasing use of alcohol.

These gross tendencies, which favor increased safety and which are mostly not subject to deliberate control, are frequently reinforced by "safety" programs, designed specifically to suppress road transportation risks. This article outlines how such programs are (and should be) designed, implemented and evaluated. (For more detail, see Arthur D Little Incorporated (1970), Klein and Waller (1970), Trinca *et al.* (1988).)

The treatment given in this article deals mainly with industrialized countries, by assuming a sophisticated road system and relatively stable levels of funding. The situation in developing countries is often quite different and it can be a serious mistake to attempt to transfer undiluted road safety programs that have proved to be successful in industrialized countries. Local conditions must be taken into account much more, for these show greater variability in developing countries: in the road network, in the funding available, in the relative importance of traffic hazards vis-à-vis disease, in social structure and in transport demand. A discussion of safety measures for newly or partially motorized societies must take into consideration not only general

Table 1
Selected US traffic death rates

	1932	1952	1982
Traffic deaths per 10^8 vehicle kilometers of travel	9.4	4.6	1.82
Traffic deaths per 10^5 resident population	23.6	24.3	19.90

Source: National Safety Council (1987)

principles, as discussed in this article, but, more importantly, the local needs and constraints. (For a more complete description of the problems in developing countries, see Haight (1980, 1983).)

1. Countermeasures

The word countermeasure is used rather loosely to describe any category in which a safety program might be classified. There are countermeasures relating to the behavior of road users: speed control, seat belt use, sobriety, obedience to traffic control devices and so on. Other countermeasures affect the design and performance of vehicles, notably the US federal motor vehicle safety standards, but also corresponding European, Japanese and Australian standards. Still other types of countermeasures deal with the road system: its geometric design, presence or absence of roadside obstacles, treatment of dangerous locations (black spots), drainage, surfaces and many others. Finally, changes in laws and regulations that affect traffic and, hence, safety can be considered as a type of countermeasure.

There is no economic way—and certainly no practical way—to compare the desirability of disparate types of countermeasures. Some attempts at cost–benefit analysis have become bogged down in questions of, for example, who the cost should apply to and of what value can be placed on a life or an injury.

For this reason, the choice of a particular countermeasure area is invariably a political choice, influenced to some degree by public opinion which, in turn, is often manipulated by publicity. There have been periods of great concern for the safety of vehicles, stimulated by the work of Ralph Nader. At other times, the public has been especially concerned by heavy-truck risks, by the failure to wear seat belts, by roadside obstacles ("booby traps") or by youthful (or, more recently, elderly) drivers. Activist groups such as Mothers Against Drunk Driving (MADD) have been highly successful in promoting awareness of the dangers of drunk driving.

All of these areas may be worthy of attention, but a decision regarding which area is worthy of action at a given time and place cannot be based solely on rational factors. It is not the safety analyst or scientist who is called upon to decide on the relative merits of, for example, a belt-wearing law vis-à-vis reconstructing a dangerous segment of highway or installation of traffic signals vis-à-vis raising the drinking age. These are necessarily, and properly, political questions, which involve not only safety, but numerous other social values.

In making the political choice, the immediate context is, of course, the road transportation system itself. Government must provide not only for safety, but for mobility, and must do so at a cost commensurate with the perceived relative importance of safety and mobility. Some countermeasures make the road network more efficient (e.g., the construction of limited access roads), some make it less efficient (e.g., stop signs or sobriety checkpoints), while others are virtually transport neutral (e.g., the installation of seat belts). There can also be countermeasures that affect broader social issues, especially those relating to public health. Emergency medical services promote public health in addition to their effects on traffic injury; alcohol countermeasures may also help to reduce alcohol-induced diseases. Pedestrian crossings and lane markings contribute to safety but, at the same time, indicate how limited space is to be shared by different groups of road users. Police enforcement of traffic regulations is an example of a primary transportation management function, with safety as a particularly important aspect.

All these factors, plus many others, will go into a political decision to designate (and perhaps fund) a particular traffic safety countermeasure. Unbiased information on the cost and effectiveness of earlier programs can provide some assistance, but not as much guidance as is often supposed.

The fundamental political problem is to balance safety, mobility and economy in an optimal manner. There are well-known examples where two of these desirable quantities are optimized at the expense of the third, but there is no known way to make an optimal choice of all three. To find effective countermeasures is easy; to find appropriate countermeasures is difficult.

2. Interventions

When a countermeasure has been chosen, it is necessary to design an intervention with which to implement the countermeasure. The fact that a given countermeasure can provide the basis for many different interventions can be shown by the varying legal and administrative requirements of the 50 states of the USA with respect to driver licensing, vehicle registration, heavy-truck limitations, road standards, signing and so on. Even in areas where some uniformity does exist, such as the (former) US nationwide 55 mph (89 km h^{-1}) speed limit, there can be important variations in the level of enforcement, the amount and collection methods of fines, speed tolerance limits, public information campaigns and so on.

There is a tendency for the general public to believe that countermeasures of any nature are necessarily—or, at least, optimally—addressed by use of the criminal justice system. Passing new laws or more rigorous enforcement of old ones has a particular appeal to concerned citizens. Thus, the middle-level authorities responsible for the design of an intervention are sometimes constrained by public opinion to almost the same extent as the politicians. An excellent example is provided by the approaches that have been suggested to the problem of drunk driving. Measures that would assist intoxicated persons to avoid the risk of damage to themselves or others (e.g., free taxi services, or increased visibility of road signs and markings) lack the public appeal of measures containing a punitive component, whatever the relative safety consequences may be.

For an efficient choice among many possible interventions, it is useful to specify as clearly as possible the goal of the intervention. In some circumstances, the goal may be given by the political authorities. If so, it is worth negotiating a precise parameter-based definition of the goal. If the goal is vague or entirely missing, it may be necessary to design the goal simultaneously with the formulation of the intervention. It is important, in either case, that the goal is stated as exactly as possible.

As an example, it will be supposed that a new law concerning seat belt wearing (requiring front seat occupants to be belted) has been passed and a choice is to be made as to the division of funding for enforcement, school education, prosecution and public relations. There might be several goals under discussion:

(a) to reduce traffic fatalities;

(b) to reduce traffic fatalities by 10%;

(c) to reduce traffic fatalities among front seat occupants; and

(d) to reduce traffic fatalities among front seat occupants by 10%.

These goals, although plausible, are hardly satisfactory and will not hold up to close scrutiny. In the first place, there is no specified time limit. Is this reduction supposed to be accomplished in a week, a month or a year? Is it then supposed to stay in effect for another week, another month or forever?

In the second place, fatalities are extremely rare and so their total number is subject to large statistical fluctuations, except in populations of enormous size. Even if the sample size is statistically adequate, fatalities are determined by many more important factors than the intervention being proposed.

A more modest set of goals would involve the specific intervention, that is, seat belt wearing. Examples of this type of goal would be

(e) that after one month, a survey would show 50% of front seat occupants to be wearing seat belts;

(f) that after one month, a survey would show that 50% of adults interviewed stated that they "usually" wore seat belts in the front seat, and

(g) that in police-reported crashes, 50% of seat belts were stretched, showing that they had been fastened at the moment of collision.

Goal (e) has some problems. It is difficult to observe belt wearing at night, and studies have shown that individuals likely to flout belt-wearing laws are more likely to drive at night or that people are more likely to flout the law in the dark. Goal (f) has more severe problems, involving as it does veracity. Goal (g) involves a particular, and probably not entirely typical, subgroup of the population, as well as sample-size difficulties. There are also subjective judgements required on the part of the police at the scene of the crash.

If the intent is to improve the statistics of traffic safety, it would be sensible to concentrate on the groups most at risk (e.g., young, working-class males). If, however, it is to improve the statistics of beltwearing, a quicker payoff would be probable with the emphasis on the drivers most likely to respond to legal requirements and social pressure.

The examples given are illustrative and should not be take to mean that the goal of an intervention must invariably be an exact value; obviously, a frequent goal is simply to improve the value of one or more parameters. It is, however, important to express the goal in parametric terms, rather than in language that can be interpreted *post facto* in a variety of ways. This pre-assigned numerical criterion is called by the US Federal Highway Administration the measure of effectiveness (MOE).

In connection with this, it is also necessary to mention a goal that is often implicit, but seldom mentioned; making the decisions—both political and administrative—appear justified by the outcome (i.e., making the intervention a "success," whatever the outcome). This might be called a "shadow" goal. Shadow goals are relatively easy to build into an intervention. For example, an effective publicity campaign designed to increase compliance with a speed limit followed by a public opinion survey is likely to show "increased public awareness" of the need for speed control. There is, of course, nothing wrong with increasing public awareness, if this was the stated goal, but its use to mask failure to reach some other goal (in this example, compliance with a speed limit) is an all-too-frequent flaw in traffic safety interventions. The use of an MOE (in this example, a measure of compliance) can help to prevent the waste of money often associated with the adoption of shadow goals.

One further step is needed before the intervention design can be considered complete: a specification of the means for accomplishing the stated goal in terms sufficiently precise to specify the agency that is to be responsible for the work ("the lead agency"). There are many types of public departments involved with aspects of traffic safety: education departments, transportation departments, police departments, justice departments, road construction departments, public health departments and, in some countries, even military departments. Although many agencies can be involved in an intervention, there should be one with clear responsibility for budgeting, staffing, implementation, monitoring and evaluation.

When the budget has been established, a specific department identified and a clearly specified goal (expressed as an MOE) defined, the intervention planning is reasonably complete.

3. Programs

It is difficult to make a clear-cut distinction between interventions and the programs intended to carry them out. For convenience, it is assumed here that program design begins at the point where the budget, the goal (MOE) and the agency are known.

For some interventions, the program design is relatively simple. For example, if a dangerous intersection is to be improved by the installation of traffic lights, the program design is essentialy engineering design. There are questions of cost, supplier, contractor and so on, but these are straightforward and hardly deserve further comment. The principal decision—that traffic lights will improve safety—has already been made at an earlier stage.

Other situations may be more complex. Suppose, for example, that a police agency is given an increased budget with the intention of increasing observed seat belt wearing by means of ticketing drivers not wearing belts. The following are examples of important decisions involved in the program design.

(a) Should extra patrols be established by payment of overtime or should officers be given relief from other duties?

(b) Should more patrols be assigned to rural areas where belt wearing is not normal behavior or to middle-class urban areas which may be quick to change?

(c) How much (if any) of the budget should be contracted out to advertising aimed at informing the public of the new program? It might be unrealistic to expect severe enforcement of a belt-wearing law if the public believed that all traffic fatalities were the consequece of speeding or of drunk driving. This might also be true if a spectacular crime of some entirely different nature led to an emphasis by police on that type of enforcement.

(d) Is the cooperation of the judiciary likely or will tickets for not wearing a seat belt be regarded as a waste of court time?

(e) Will officers issuing tickets need a short course in the legal and psychological aspects of ticketing a new offense? How can officers be persuaded of the usefulness of this kind of work?

These questions suggest some criteria for the design of programs. First, it must be considered whether it is feasible to expect the cooperation and support of those responsible for carrying out the program. Instruction—even in writing—to subordinates may not be automatically translated into full and enthusiastic cooperation. Officials do not always comply with instructions from a central authority to the best of their ability, especially if they perceive that the instructions go against local sentiment or lead to their agency being viewed unfavorably by the public, or if they believe that they have more important tasks at hand. The lack of enforcement of pedestrian priority at crosswalks in the Eastern USA provides an interesting example.

A second factor to be considered is the stated goal of the intervention. It is clear from the examples given in Sect. 2 that similar goals, expressed in differing MOE, can call for quite different programs.

A third factor is the extent of control over other agencies that must be involved. Without direct job assignment and job rating, the problem can be greatly exacerbated, especially if traffic safety is a relatively minor interest of the "cooperating" agency. A familiar example is driver education. These programs are often funded by a transportation agency, but administered by an education department, which may have entirely different goals and procedures, and which may be unwilling to be subjected to direction or scrutiny. Furthermore, the education department often passes the responsibility (and funds) on to local schools, where the expertise, motivation and facilities vary widely.

A fourth consideration is record keeping. There are few, if any, programs that are immune from some form of accountability. Eventually, evidence may be needed that the program addressed the goals and spent the money in a manner designed to achieve those goals. In many cases, it is also demanded that an estimate be made of the extent to which the goals have been achieved. This involves monitoring and data collection during the lifetime of the program and leads into the important subject of program evaluation.

4. Statistical Design and Data Collection

As the countermeasure is developed into an intervention and then into a program design, the questions of monitoring (while the program is running) and evaluation (after the program is complete) gradually come to the fore. Even with a specific MOE, there are many statistical tests that can be chosen to see if the goal has been achieved and each of these tests carries its own requirement for data collection and analysis.

The design and analysis of statistical experiments is beyond the scope of this article and evaluation will be discussed in Sect. 7. What is important to emphasize at this stage is that data gathering should not be left to the conclusion of the intervention, but should be built into the program structure from the beginning. In fact, if the statistical test is based on trends or time series analysis, data acquisition must extend to the period before the program begins, often for a considerable time in advance of implementation.

It is also important to realize that any statistical test

involves a "significance level" and that there is a trade-off between the level needed and sample size. The choice of these values should concern the program management as the intervention is being designed. By delaying specification of monitoring and testing procedures, the program is laid open to the suspicion of "fishing" for favorable conclusions. It is precisely to avoid the possibility of fishing that MOE are specified in the basic design of the intervention.

5. Obtaining Consensus

5.1 Public Support

The idea of countermeasures usually attracts considerable public support; the programs that implement the countermeasures are often less enthusiastically received. An intervention normally affects road users in some way; of course, it also affects the conventional routines of the agencies involved. Some programs may be perceived as a threat to the interests of various commercial and public interest groups, especially those with a single-issue agenda not covered by the programs in question.

It is difficult to prescribe any universal method for forestalling disruption of an intervention by affected (or concerned) groups and organizations. Some would recommend a preliminary softening up of opinion through news releases and public hearings. Others would try to build a consensus by political means.

5.2 Pilot Projects

One method that can given an early warning of possible trouble is the use of a pilot study; that is, a small-scale experimental project before full implementation is undertaken. In this way, it is possible not only to assess any potential threats to the project, but also to evaluate the methods of training, data collection and so on.

Obviously, not all programs are amenable to a pilot project. For example, it would probably not be feasible to test a change in the law specifying the maximum blood alcohol level by applying it to one county only or on one week only.

6. Evaluation

There is considerable literature on the evaluation of social programs in general (e.g., Majone and Quade 1980, Caro 1982, and especially Bickner 1980, Hatry 1980 and Rossi 1982) and also several studies dealing more specifically with traffic safety program evaluation (e.g., Griffin *et al*. 1975, Tarrants and Veigel 1978, Council *et al*. 1980, Organisation for Economic Co-operation and Development 1981, 1984). In this article, it is only possible to mention the main principles discussed in these works, with a few examples.

First, it is necessary at an early stage to consider whether a project does indeed require evaluation. Just as there are some projects that deserve a full evaluation, there are some that do not. Evaluation is usually regarded as an aid to decision making and is normally performed only in circumstances where some important decisions are envisaged. Being a difficult and costly task, it should not be undertaken for purely frivolous reasons, such as justifying decisions already made.

In deciding whether to evaluate, not only must the importance of future decisions be considered, but also the cost and, especially, the chance that the evaluation will prove to be sufficiently conclusive one way or the other. It is obviously fruitless to search for the effect of inconsequential programs or to depend on a statistical procedure requiring larger samples than can reasonably be obtained.

If an evaluation is warranted, it may be difficult or impossible to persuade the sponsoring agency, with its political agenda (and responsiveness to public opinion), that the money allotted to "save lives" (i.e., the intervention) should, in part, be diverted to what may appear to be "satisfying curiosity" (i.e., to assessing the effectiveness of the same intervention). In proposing the countermeasure in the first place, it is usually implied that "we know it will work, otherwise why waste money on it?" and the suggestion that data will be needed to establish this "obvious" fact may appear to be a poor use of resources. Experience in the traffic safety field has shown, however, that money is more likely to be wasted through programs that are superficially plausible but that, on close examination, turn out not to have accomplished their purpose.

Second, it is universally agreed that the evaluation team should be independent of the implementation team and that those responsible for the evaluation should have no stake (financial, professional or emotional) in the outcome. It is seldom that traffic safety evaluations have the benefit of this degree of objectivity.

Third, before assessing the effect of the program on the MOE, it is necessary to see that the program has indeed been carried out. This is often called administrative evaluation in contrast with the observation of MOE changes, which is called impact evaluation. In most cases, the administrative evaluation would coincide with the monitoring of the project during its lifetime.

When the probable consequences of a program appear to be several steps removed from the safety payoff, it is frequently better to settle for a well-designed administrative evaluation. For example, if it is proposed to establish a computerized driver license file, with the intention of not only rationalizing office work but also of monitoring hazardous drivers, the consequences for traffic safety, although real, would be virtually impossible to document. The administrative evaluation, however, would be a relatively simple and thoroughly worthwhile exercise.

Fourth, it is necessary to demonstrate a connection between the program and the MOE changes, to be sure that the changes were not due to some factor other than the program being evaluated. One way to test a connection is by the common statistical procedure of using a control group and monitoring changes in the corresponding MOE. Unfortunately, control groups are frequently difficult to define or impossible to monitor in the traffic context.

Another approach would be to search for other uncontrolled variables that could have affected the outcome: traffic, climate, economics and so on. Suppose there is an unforeseen spell of bad weather leading to skidding and consequent deaths; is a belt-wearing program to be pronounced a failure? Conversely, suppose some extraneous circumstance, such as road construction, reduces traffic or slows it down, leading to fewer high-speed collisions; are the consequences to be attributed to the belt-wearing intervention? Recessions are well known for improving most measures of traffic safety so that, without due care, a program conducted during such a period would be more likely to be judged successful than the same program during a period of expansion in the economy.

An ideal evaluation would refer to the costs and benefits of the program. However, there are pitfalls here, not only in the difficulty in measuring both costs and benefits, but also in the risk of suboptimization; that is, in ignoring larger social contexts (Bickner 1980). An interesting example is driver education, where the benefits are usually underestimated. Viewed solely as a safety measure, these programs are typically found to be not worth their cost. However, viewed as a necessary prerequisite to introducing new cohorts of drivers to the road transportation system, a wider benefit can be perceived. An example in which costs are underestimated is provided by some types of traffic control. A stop sign is costed, in one official handbook, on the basis of manufacture and installation costs, without regard to the (much greater) costs incurred by vehicles stopping and then starting up in obedience to the sign.

In insisting on the definition of an MOE at a very early stage of an intervention, a false impression may be given that faithful observation of MOE changes can remove judgemental factors from the evaluation. The fallacy of such a view is apparent when consideration is given to all the subjective choices—of the MOE, of the statistical test, of the level of significance—that go into the design.

In fact, complete reliance on numerical changes is as unreliable as complete reliance on personal judgement. Evaluation of a program is analogous to medical diagnosis; the laboratory tests are important and should not be ignored, but the final conclusion must be in the hands of a person (or evaluation team) capable of objective analysis, taking into consideration all relevant factors, including the MOE changes. This "diagnostic" evaluation will, of course, depend on the skill (Majone (1980) uses the expression "craft knowledge") with which the evaluation is planned and carried out.

An appropriate slogan for evaluation is given by Enthoven (quoted by Lynn (1980)): "It is better to be roughly right than exactly wrong."

See also: Evaluation of Traffic Control Systems; Safety of Road Traffic; Social Issues of Transportation

Bibliography

Arthur D Little Incorporated 1970 *The State of the Art of Traffic Safety*. Praeger, New York
Baker S P, O'Neill B, Karpf R S 1984 *The Injury Fact Book*. Lexington Books, Lexington, MA
Bickner R E 1980 Pitfalls in the analysis of costs. In: Majone and Quade 1980, pp. 57–69
Caro F G (ed.) 1982 *Readings in Evaluation Research*. Russell Sage Foundation, New York
Council F M, Reinfurt D W, Campbell B J, Roediger F L, Carroll C L, Dutt A K, Dunham J R 1980 *Accident Research Manual*, No. FHWA/RD-80/016. Federal Highway Administration, Washington, DC
Griffin L I, Powers B, Mullen C 1975 *Impediments to the Evaluation of Highway Safety Programs*. Highway Safety Research Center, University of North Carolina, Chapel Hill, NC
Haight F A 1980 Traffic safety in developing countries. *J. Saf. Res.* **12**, 50–8
Haight F A 1983 Traffic safety in developing countries II. *J. Saf. Res.* **14**, 1–12
Hartunian N S, Smart C N, Thompson M S 1981 *The Incidence and Economic Costs of Major Health Impairments*. Lexington Books, New York
Hatry H P 1980 Pitfalls of evaluation. In: Majone and Quade 1980, pp. 159–78
Klein, D, Waller J A 1970 *Causation, Culpability and Deterrence in Highway Crashes*, Department of Transportation Automobile Insurance and Compensation Study. US Government Printing Office, Washington, DC
Lynn L E Jr 1980 The user's perspective. In: Majone and Quade 1980, pp. 89–115
Majone G 1980 An anatomy of pitfalls. In: Majone and Quade 1980, pp. 7–22
Majone G, Quade E S (eds.) 1980 *Pitfalls of Analysis*, International Series on Applied Systems Analysis, Vol. 8. Wiley, New York
National Safety Council 1987 *Accident Facts*. National Safety Council, Chicago, IL
Organisation for Economic Co-operation and Development 1981 *Methods for Evaluating Road Safety Measures*. OECD, Paris
Organisation for Economic Co-operation and Development 1984 *Costs and Benefits of Road Safety Measures, European Conf. Ministers of Transport*. OECD, Paris
Quade E S 1980 Pitfalls in formulation and modeling. In: Majone and Quade 1980, pp. 23–43
Rossi P H 1982 Boobytraps and pitfalls in the evaluation of social action problems. In Caro 1980, pp. 239–48
Tarrants W E, Veigel C H (eds.) 1978 *The Evaluation of Highway Safety Programs*. US Department of Transportation, National Highway Traffic Safety Administration, Washington, DC
Trinca G W, Johnston I R, Campbell B J, Haight F A, Knight P R, Mackay G M, McLean A J, Petrucelli E

1988 *Reducing Traffic Injury—A Global Challenge*. Royal Australasian College of Surgeons, Melbourne, Australia

F. A. Haight
[University of California, Irvine, California, USA]

Ship Automation and Control

Ship control is an imprecise term, widely used in marine circles, whose meaning largely depends on who is using it. For example, it may refer to the control of:

(a) the internal systems of a vessel,

(b) the motions of a vessel, or

(c) a vessel as a whole, on a voyage or in company with others.

It is necessary, therefore, to establish that this article is concerned primarily with areas (a) and (c).

Both these areas involve the participation of people. Hence, the main theme explored here is the developing relationship between people and machines in the operation and control of merchant ships.

1. Historical Background

Shipping is an old industry. Many of the practices it still employs have evolved over many centuries—through all the days of sail until after the introduction of mechanical propulsion during the industrial revolution. Tradition, therefore, greatly influences the way in which ships are operated.

This is evident in several ways, but most significantly in the relationship between crews and the vessels in which they serve. The organization of the crew, the tasks they are required to perform, even the language they use still retain recognizable links with the past.

Sailing ships could be regarded as independent, totally self-sustaining units. Their crews were responsible for everything, from the safety of navigation to the commercial success of the voyage, from all repairs and maintenance to the handling and care of the cargo. Mechanical propulsion, which arrived with the industrial revolution, did little to change this and the attitudes of many involved in the shipping industry, even today, have been formed around this view.

But it is a distorted view. Global communications—telegraph, radio and now satellites—have seen to that. Ships are now elements in wider transportation systems; their crews provide only a part of the required labor input. How has this come about?

It is no coincidence that shipping markets became established when communications began to improve. Their effect was to remove from the shipmaster responsibility for the commercial success of the voyage.

As the twentieth century has progressed and communications have improved still further, responsibility for more and more of the day-to-day affairs of a ship has passed to managers working in offices ashore, to the extent that in many ships, the crew (i.e., those people continuously engaged on board) are concerned with nothing other than ensuring a safe passage from one port to the next.

Crew numbers have been reduced. A typical ship in the 1950s probably had a crew of over 40, whereas today under 20 is common. Hence, it has become less and less practicable for the crew to undertake all the tasks that have hitherto been associated with shipboard life, and the traditional seafaring skills have lost most of their value.

Such changes call into question the conventional wisdom that has served the maritime community for so long. What, for instance, are the qualities required by a modern seafarer? Opinions span the entire range of human capabilities. Alternatively, what are the implications of further reducing crew numbers by increasing shipboard automation? Smaller permanent crews may simply lead to the casual engagement of more people whenever or wherever they may be needed.

These and other questions in a similar vein have been taxing the international shipping industry and some of them are considered here.

The shipping industry was once at the forefront of technological advance. That was in the nineteenth century. It is no longer so. The main thrust of technological development is taking place elsewhere—in electronics, information technology, biotechnology, and so on.

It is up to the shipping industry, therefore, to recognize, select and adapt for its own purposes, such non-marine advances as may be useful to it. Sometimes the industry has been accused of being slow to do this, perhaps because of the inertia associated with so many long-established maritime practices (see *Ship Stabilization: History*).

2. Shipboard Automation

2.1 Beginnings

One of the first questions to ask when considering the introduction of any form of automation is "What are the benefits?" This is often a difficult question to answer.

However, when starting out from a point where operation is totally manual, saving manpower may be at least part of the answer. This was arguably the case with the introduction of the automatic pilot, the first experiment with automation in the shipping industry which took place during the 1920s and 1930s.

Steering is an obvious candidate for automation because it is:

(a) boring,

(b) continuous, and

(c) labor intensive.

The introduction of the autopilot was made possible by the invention of the marine gyrocompass which provided a heading reference that could be transferred electromechanically to a variety of remote devices. It was, moreover, extremely successful insofar as it released crew members from the drudgery of steering and generally achieved more accurate course keeping than the human helmsmen it replaced.

Early autopilots were very simple, often little more than a proportional controller to regulate the telemotor valve and, hence, the steering gear. Minor adjustment for the effects of yawing could be made.

Autopilots were only used in open waters. They were, nonetheless, effective in reducing the number of crew by as many as three in any ship so fitted. Most ocean-going ships had one by the end of the 1950s.

The first attempts at engine-room automation were made during the 1960s. Here, the problem was to devise means for monitoring the operation of the main engine and auxiliary plant. Ever since the coming of steam it had been standard practice for the watchkeeping engineers to tour the engine room in order to inspect the machinery, to record temperatures and pressures and other relevant parameters, and to make manual adjustments as required. Local indicators and controls were normally all that was available. Thus the task of overseeing the operation of the many different items—main engine, boilers, pumps, generators, and so on—in the machinery space of a large ship was a major one, time consuming and often arduous.

The invention of marine data loggers capable of polling numbers of remote sensors made it possible to concentrate the machinery surveillance task in one place within the engine room. Thus, in 1961, the first vessel to have a centralized machinery control room entered service.

Once the information that describes the performance of the plant has been brought to one place, it is clearly sensible to make adjustments to the plant from the same place. Hence, remote operation of all the machinery from the control room became increasingly common in vessels built during the 1960s.

In some cases, the main engine controls, even the control room itself, were located on the bridge, bringing together everything associated with conning and navigating the vessel in a manner reminiscent, conceptually at least, of sailing-ship practice.

Initially, remote monitoring of the machinery from a central control room still involved watch-keeping engineers. It is true that they were no longer required to undertake their routine perambulations around the machinery spaces, but there were no significant operating economies or other obvious benefits to the shipowner arising from this development. Nevertheless, it was a necessary step along the path leading to such benefits.

Clearly, once remote monitoring systems were in place it was only a matter of refinement to incorporate alarms, automatic shut down and other safety arrangements to allow the plant to be left unattended. Such arrangements have led directly to the unmanned machinery spaces that are commonplace in merchant ships and that have inevitably brought about further reductions in the number of crew engaged on board.

2.2 The Labor Question

Since the late 1950s, shipping companies, especially those of Europe and Japan, have given very high priority to reducing crew numbers. There are several reasons for this. The most important is that many operators see labor costs as the most significant determinant of their ability to compete internationally.

It is unnecessary to rehearse all the arguments for or against this view here, but substantial reductions in numbers have undoubtedly been achieved. They have often been gained at the cost of sacrificing long-established shipboard practices (see *Social Issues of Ship Automation*).

From the time that ships were first mechanically propelled their crews have been divided into three departments:

(a) deck,

(b) engine, and

(c) catering.

Personnel within each department have normally been ranked into three categories:

(a) officers,

(b) petty officers, and

(c) ratings,

with the master, rising from the ranks of deck officers, in overall command.

The functions that the crew carry out on board have mainly fallen into four areas:

(a) navigation,

(b) cargo,

(c) maintenance, and

(d) life support.

In general, duties in the areas (a)–(c) have been shared between members of both engine and deck departments.

Thus, the safe navigation of the vessel from one port to the next has been the deck officers' responsibility, while providing the force that moves the vessel has been the engineer officers' responsibility.

Similarly, loading and stowing the cargo has traditionally been a deck department function, while looking after the cargo on voyage (e.g., refrigeration and environmental control) has been an engine department function.

Also, as a final illustration, the deck department has always been responsible for the upkeep of the hull and safety equipment, whereas the engine department has been responsible for maintaining the mechanical plant.

As time has passed, several changes have taken place:

(a) the level of technology employed on board has risen, increasing the job content of the engineers at the expense of the deck personnel;

(b) the number of people on board has diminished, thereby threatening the traditional organizational structure, and even rendering it untenable; and

(c) the labor input into the transportation system of which the ship is a part has become more diffuse, with a decreasing proportion of the total being supplied by the crew.

As a consequence of these changes, it has become necessary to review the role of the human crew and to consider how best to arrange the shipboard systems so as to ensure their safe and efficient operation. There are several important considerations here, of which the following are examples.

(*a*) *Safety of navigation*. Ships operate night and day. In the past a watch keeper has been required to be awake and on duty the whole time. Is there still a requirement for a human watch keeper when a ship is operating with unmanned machinery spaces, autopilot and automatic collision avoidance aids? Will it always be so?

(*b*) *Security*. A large, slow moving ship with a small crew is not easily secured against illegal boarding. Hence, piracy is once more becoming a serious problem at sea.

(*c*) *Maintenance*. No matter how well engineered a ship may be, breakdowns will occur. How can a small crew deal with such eventualities? Routine maintenance and even low-grade tasks such as hold cleaning cannot be undertaken except with the help of temporary labor. There is an obvious parallel here with the operation of aircraft.

(*d*) *Crew training*. It has been argued that the higher the level of technology in a ship the more highly trained and qualified the crew need to be. It has also been argued that a high level of technology is a substitute for a human involvement in the operation of a ship. The true position is probably somewhere between these two extremes. Even so, the specification of shipboard systems must be matched to the capabilities of the crew.

2.3 Use of Computers

From the earliest days, shipboard automation has employed mainly electrical and mechanical (including hydraulic and pneumatic) devices such as actuators and comparators installed in fairly simple and robust feedback control loops. This continues to be the case.

However, as time has gone by, computers have begun to play an increasingly prominent role in the operation of merchant ships.

The history of the use of shipboard computers begins in the 1960s, but only in the 1980s has their impact been really significant in merchant ships. In some respects, this experience parallels the general development of computers.

During the 1960s, it was widely thought that a single central computer might be employed to control many routine operational monitoring and control functions. Such ideas were incorporated to a greater or lesser extent into experimental shipboard installations, for example those in *Sea Sovereign, Queen Elizabeth 2* and one or two other noteworthy vessels.

At that time, the greater part of the shipping industry was slow to follow this lead. There were several reasons for this, including:

(a) high initial cost,

(b) inadequate or ill-defined benefits, and

(c) low hardware reliability in the shipboard environment.

During the 1970s, progress towards the introduction of general purpose computers remained slow. Neither the rapidly falling cost of hardware nor the increasing ratio of processing power to its physical size had much impact on this. The reason was, primarily, that few ship operators were able to see what the benefits of using computers at sea might be.

This is an over simplification of course. Many ship operators had experience of using computers in their general administration ashore. They were also becoming increasingly accustomed to radar and other instruments having dedicated digital circuitry. However, as has already been mentioned, much had been achieved by employing noncomputerized techniques. Crew numbers had been substantially reduced. Where then was the point of doing more? With hindsight this is an easy question to answer.

Computers are tools. Like any tools, they facilitate operations, the creation of artifacts or the achievement of results that would not be possible without them. This applies as much to computers as to the contents of a carpenter's chest.

Thus, at sea, computers have brought benefits by making possible new operational practices that could not otherwise be contemplated. A few examples can illustrate this.

(*a*) *Cargo planning*. It has always been necessary to ensure that a ship is properly loaded, that it has adequate stability and that its cargo is satisfactorily arranged within the holds. Using a computer in this process provides opportunities and advantages not otherwise available including:

(i) allowing practical consideration of alternative arrangements of cargo and other weights on board;

(ii) allowing practical consideration of damaged stability and, hence, the possibility of reducing margins;

(iii) allowing better use of ship space; and

(iv) allowing direct links to port/terminal information systems (see *Container Terminal Management*).

(*b*) *Performance monitoring.* It has long been the practice on board ships to collect and record performance data—speed, fuel consumption, temperatures, pressures, and so on. Much of this data has been of little practical value, largely because of the unavailability of proper analytical tools. Thus, using a computer can provide opportunities and advantages not otherwise available including:

(i) allowing more accurate performance monitoring and, hence, earlier remedial action;

(ii) allowing direct and easy comparison of the performance of similar ships;

(iii) allowing the development and use of more efficient energy policies; and

(iv) allowing direct links with commercial information systems employed in voyage planning/estimating.

(*c*) *Spares ordering/inventory control.* It is difficult to determine the correct level of spare parts that should be carried on board a ship. Even so, it is an area of operation where opportunities exist for reducing both inventory costs and transport costs provided that efficient methods for monitoring and control are available. Shipping companies are increasingly turning to computer-based systems for assistance with this. Such systems offer possibilities for:

(i) allowing closer integration of spares ordering (on board) and purchasing (ashore);

(ii) allowing direct and easy comparison of the spares usage on similar ships; and

(iii) allowing the creation of better service histories and, hence, providing early warning of problems arising in plant consuming more spares than expected.

These are three of several possible examples of the way in which general purpose computers are being employed in the operation of ships. It is clear that they are applications which are not so much aimed at reducing the need for labor on board as ones which extend the capabilities of the labor that is available.

Moreover, they are applications that greatly strengthen the link between ship and shore and that, therefore, reduce the isolation and independence of each ship, weakening still further the foundation on which many shipping practices have been traditionally based.

It would be wrong to conclude from this that there is no direct role for computers in shipboard automation, however.

2.4 Expert Systems

A reduction in the stock of experience available on board is an inevitable consequence of smaller crews. With a crew of 40 or more, most of whom were lifelong seafarers, there was always a fund of practical experience that could be brought to bear on any problem that might arise—breakdown, navigational hazard, medical emergency, and so on. This was especially valuable when far out at sea, beyond the range of immediate external assistance. However, the situation is different today.

Small crews, consisting of men and women who may have no long-term commitment to seafaring, who may come from countries that have only recently become active in the maritime field, cannot count on such experience being available. There is thus a need to supplement the experience of those remaining on board.

Training provides the conventional means of doing this—the training/recruitment dilemma that has confronted the shipping industry has already been mentioned in Sect. 2.2.

Detailed and specific guidance in the form of operational manuals provides another method which may be very useful in some circumstances although it, too, has training implications.

However, a less conventional approach to the problem of supplementing the practical experience of individuals is to employ expert systems. This is becoming practicable, as techniques and software are developed and refined, now that computer hardware of sufficient power is cheap and widely available.

As yet, shipping has been slow to follow the oil industry, the medical profession and other pioneering users of expert systems. However, this does not mean that it has less to gain.

Expert systems—intelligent knowledge-based systems (IKBSs)—offer a variety of benefits which, in a marine context, might accrue through their application in areas such as the following.

(*a*) *Machinery breakdowns.* Fault diagnosis is a well-established and obvious application for expert systems.

(*b*) *Navigation—traffic operations.* A small scale application directly linked to radar/electronic watch-keeping equipment.

(*c*) *Navigation—pilotage.* Ships are often directed to ports where none of the watch-keeping officers has previously been.

(*d*) *Company procedures.*

The question of whether such systems should be installed in particular vessels arises.

In some cases the answer may well be affirmative. This is certainly so where the system is an adjunct to another item of shipboard equipment—a radar for example, where the expert system may be little more than an extension of the techniques already employed in intelligent plotting aids (ARPAs) which have been employed for collision avoidance throughout the 1980s and are now mandatory for many vessels.

However, in other cases the answer is almost certainly negative. Fault diagnosis may be one such area

439

because machinery manufacturers themselves are probably best placed to provide information and advice that is based on operating experience with their own products. Thus, expert systems installed ashore, accessible to ships via data link as required is a more practicable approach here.

2.5 Robots

A feature of the evolution of shipboard automation has been that many low-grade tasks have been untouched by it. Many of these tasks have little skill content and, thus, there has been little incentive to devise alternative means of doing them. A small crew is often forced to ignore them.

In the past it was usual for the crew to be employed in routine maintenance and cleaning such as:

(a) chipping rust,

(b) painting, and

(c) cleaning holds,

as well as many other similar jobs.

Often, but by no means always, such jobs provided a way of keeping the crew occupied on long passages.

Today, such work is rarely done at sea because:

(a) crews are too small,

(b) ships are too big, and

(c) ships are too fast.

It has, therefore, become increasingly necessary for such work to be undertaken by specially contracted labor.

However, much of this work is really quite straightforward and could be done by machines that employ the manipulators and other similar techniques developed for intelligent manufacturing systems (i.e., by robots).

An advantage of using robots on board a ship is that they can be set to work continuously at sea. Time on passage can, thus, be used productively as it was in the days when ships were slower and sailors more numerous. A variety of shipboard operations are amenable to this approach.

(a) *Hold/tank cleaning.* Mechanical cleaning/washing systems are installed in many ships, especially tankers, but most are not "intelligent" or even automatic—they are not fitted with sensors that would allow them to adjust their own operation to suit the task in which they are engaged.

(b) *Deck maintenance—painting.* This is an obvious marine adaptation of an established application of robots in manufacturing industry.

(c) *Automated machinery maintenance.* Plant having the ability to monitor its own performance and diagnose its own faults is becoming more common. By incorporating manipulators, self-maintenance and repair become potentially possible. This would clearly be a great advantage in a ship where labor is limited.

Even though labor levels raise serious problems, it is unrealistic to assume that robots will supply solutions that can be universally applied in the immediate future. It is clear, however, that there are marine applications of robotics and no doubt the time will come when they are widely recognized and accepted. Simply observe the other changes and developments that have been adopted by the shipping industry.

3. Vessel Routing: *Operation on Passage*

The command and control of a ship at sea is the responsibility of the master acting without external interference. This has been the position throughout history and has only begun to change during the twentieth century.

One of the principal factors influencing this change has been the introduction and development of marine communications that have allowed many of the master's traditional functions, such as

(a) responsibility for commercial aspects of the voyage,

(b) responsibility for financial aspects of the voyage,

(c) responsibility for planning the voyage, and

(d) responsibility for the upkeep of the fabric of the vessel,

to be transferred to centers ashore—usually the company office.

It can be argued that a logical extension of this process is to transfer the monitoring and control of shipboard systems to centers ashore as well.

There are significant technical problems attached to this, but none that can be regarded as insurmountable within a reasonable period of time, given sufficient commitment to overcome them. However, the important question is whether this is really serving the best interests of ship operators.

Is it possible, for example, to ensure that ships wholly operated remotely would be, or could be made:

(a) safe, and

(b) secure,

in all circumstances? There must be some doubt. It is not a matter that can be resolved by reference to technology alone.

Crewless ships are technically feasible. The concept of crewed and crewless vessels operating in convoy has received some attention and provides one possible model for the future. However, such a future is viewed with little enthusiasm by many who are now engaged in maritime affairs.

Notwithstanding the diminution of the master's role, it is the master who is still responsible for the safe navigation of his ship at sea. For a variety of historical and legal reasons, traffic control, in the sense in which that term is understood in aviation, does not generally exist in the marine world.

Nevertheless, since the mid-1960s, the flow of marine traffic has become increasingly orderly through the introduction of mandatory routing and traffic separation schemes in areas of convergence or high density (see *Vessel Traffic Services: Management from Shore*).

The first traffic separation scheme was established in the Dover Strait and, today, many sea-lanes have been so designated.

In contrast to air traffic control, marine traffic control is conducted in two-dimensional space and involves vehicles that are slow moving. Moreover, collisions are avoided by

(a) following defined routes, and

(b) employing maneuvering rules,

without the intervention of an external ground or shore controller. However, in some areas where traffic routing exists, radar monitoring stations and systems of vessel traffic services (VTS) have been set up which can advise vessels about the general traffic conditions that may exist throughout the area.

4. Conclusion

The nature of any system depends to a large extent on where the boundaries are drawn around it. Many discussions of ship control concentrate on very narrow aspects of the ship system, including, for example,

(a) motions (e.g., roll stabilization), and

(b) direction (e.g., steering).

Sometimes, however, it is useful to take a wider view of ship control than is implied by such considerations.

After all, a ship often exists simply as an element within a much wider transportation system. The traditional viewpoint must, therefore, be adjusted to recognize that control in the sense of management has become an important, perhaps even the most important, aspect of ship control.

See also: Navigation Control of Ships; Ship Stabilization: History; Social Issues of Ship Automation

J. King
[University of Wales, Cardiff, UK]

Ship Dynamics: Modelling

The growth of traffic and of the size of certain types of ships has led many countries to develop simulation methods aimed at the training of mariners, on one hand, and at the optimal design of harbor accesses, on the other. These methods are based on mathematical models, reproducing as accurately as possible the behavior of the ship under the influence of the numerous external factors to which it is submitted.

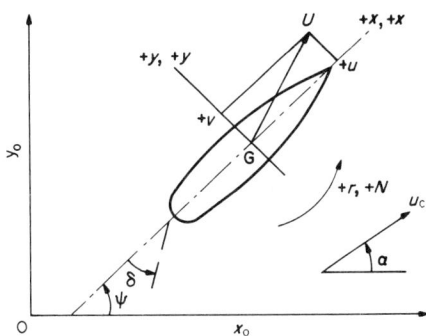

Figure 1
Coordinate system for ship

1. Structure and Elaboration of Models

1.1 Equations of Motion

Only the movements of the ship in the horizontal plane will be considered. Thus, Newton's laws of dynamics may be written, using a system of coordinates whose origin coincides with the center of gravity of the ship G (see Fig. 1), as

$$\left. \begin{array}{l} m(\dot{u}_0 - v_0 r) = X_h + X_p + X_r + X_e \\ m(\dot{v}_0 + u_0 r) = Y_h + Y_p + Y_r + Y_e \\ I_z \dot{r} = N_h + N_p + N_r + N_e \end{array} \right\} \quad (1)$$

where u_0 and v_0 are the longitudinal and the lateral velocities with respect to the bottom, r is the yaw rate, $\dot{u}_0, \dot{v}_0, \dot{r}$ are the derivatives of these quantities, and m and I_z are the mass and the moment of inertia about a vertical axis trough G, respectively. X and Y are the force components and N is the moment acting on the ship where the suffixes h, p, r and e denote the components of hull, propeller, rudder and external force, respectively. The external forces and moment contain the effect of the wind, depending on the direction with respect to the heading, the direction and period of the sea waves, the effect of the currents, the bank effect and so on.

X_h, Y_h and N_h are functions depending on the velocity components u and v of G with respect to the mean value u_c of the current in the vicinity of the ship, on r and on the derivatives of these quantities, and may be expressed (Hirano *et al.* 1987) by

$$\left. \begin{array}{l} X_h = X_{\dot{u}}\dot{u} + X_{vr}vr + X_{uu}u^2 \\ Y_h = Y_{\dot{v}}\dot{v} + Y_{\dot{r}}\dot{r} + Y_v Uv + Y_r Ur \\ \quad + Y_{v|v|}v|v| + Y_{v|r|}v|r| + Y_{r|r|}\dfrac{u}{U}r|r| \\ N_h = N_{\dot{r}}\dot{r} + N_{\dot{v}}\dot{v} + N_v uv + N_r Ur \\ \quad + N_{wr}v^2 r + N_{vrr}\dfrac{uvr^2}{U^2} + N_{r|r|}r|r| \end{array} \right\} \quad (2)$$

where $U = (u^2 + v^2)^{1/2}$ and the hydrodynamic coefficients $X_{\dot{u}}, \ldots, N_{rr}$ are functions depending on the following hull characteristics: length L, breadth B, draft T, block coefficient C_B = displaced volume/LBT, and the water depth H.

By taking into account these expressions of the forces Eqn. (1) may be put in the form

$$(m - X_{\dot{u}})\dot{u} = f_1(u, v, r, \delta)$$
$$(m - Y_{\dot{v}})\dot{v} = f_2(u, v, r, \delta) \quad (3)$$
$$(I_z - N_{\dot{r}})\dot{r} - N_{\dot{v}}\dot{v} = f_3(u, v, r, \delta)$$

where f_1, f_2 and f_3 are functions depending on u, v, r and the rudder angle δ. The following form of f_1 corresponds to a 3rd-order Taylor expansion about the straight-ahead motion (corresponding to the constant X_* (Abkowitz 1980)) where Δu is the change in forward speed around the constant value considered:

$$\begin{aligned}f_1(u, v, r, \delta) =\ & X_* + X_u \Delta u + X_{uu} \Delta u^2 + X_{uuu} \Delta u^3 \\ & + X_{vv} v^2 + (X_{rr} + mx_G) r^2 + X_{\delta\delta} \delta^2 \\ & + X_{vvu} v^2 \Delta u + X_{rru} r^2 \Delta u + X_{\delta\delta u} \delta^2 \Delta u \\ & + (X_{vr} + m) vr + X_{v\delta} v\delta + X_{r\delta} r\delta \\ & + X_{vru} vr \Delta u + X_{v\delta u} v\delta \Delta u \\ & + X_{r\delta u} r\delta \Delta u \end{aligned} \quad (4)$$

The corresponding expressions of f_2 and f_3 are of the same type and complexity. By integrating Eqn. (3) (a computer will treat the discrete version of the system) for a given rudder angle control, a simulated ship trajectory is obtained. Thus, modelling the ship dynamics means finding the most adequate form of f_1, f_2 and f_3 together with the method of estimating the hydrodynamic coefficients as a compromise between the degree of accuracy actually needed and the experimental data that are available.

1.2 Experimental Estimation of the Coefficients

Most of the coefficients cannot be calculated by hydrodynamic theory to acceptable accuracy so that model tests must be heavily relied upon for estimating the magnitudes of the coefficients. The obtained values are tested and possibly modified after comparison with results of full-scale trials, made mostly during delivery tests.

(a) Model tests. These tests make it possible to fix the experimental conditions and to eliminate the environmental effect. The most classical are constrained scaled-model tests made in a towing tank, using a special apparatus such as a rotating arm or the planar motion mechanism (PMM) where the model is forced into a specific motion (see Fig. 2) and the forces acting on it are measured.

Since the PMM and the rotating-arm facility are quite expensive, a new tendency is to practise free-steering experiments, where the rudder is changed and the motion observed. In order to estimate the hydrodynamic coefficients from these experiments, system identification techniques are used (see Sect. 2.2).

The drawback of using model tests is that results may be suspect to the scale effect because of the large difference in Reynolds number ($Re = VLv^{-1}$, where V is the speed of the ship, L is the length of the ship and v is the kinematic water viscosity) between ship and model. These effects are amplified by ships operating in shallow water. In order to limit the scale effect, large-scale models the length of which may reach 30 m are sometimes used.

(b) Full-scale tests. These are mainly delivery tests, consisting of zigzag maneuvers, turning circles and stopping maneuvers. Ideally, free-scale tests should be made in deep water and in the absence of wind, currents and waves. These conditions are difficult to find simultaneously. Because of the difficulty in containing the environmental conditions, extended full-scale tests, aimed at a complete estimation of the ship coefficients, are very scarce. The most important experience of this type was carried out with the very large crude carrier (VLCC) *Essor Osaka* in the Gulf of Mexico (Crane 1979). Generally, delivery tests are used to validate mathematical models established with the aid of scale models (see Fig. 3).

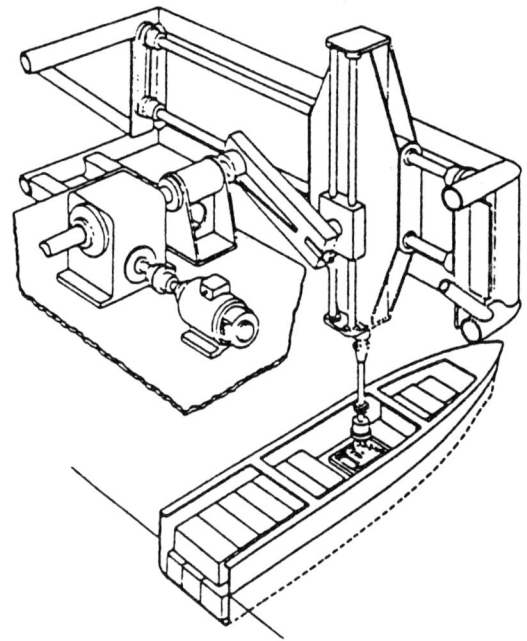

Figure 2
Planar motion mechanism

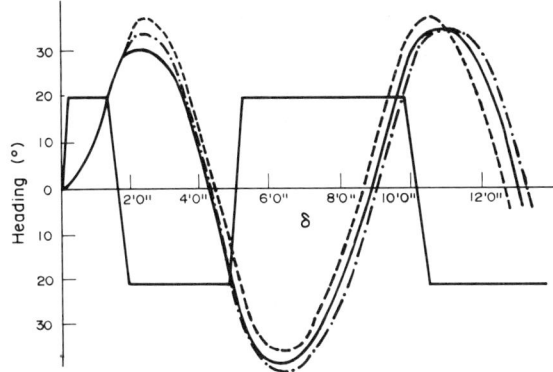

Figure 3
Comparison of zigzag test results: —·—, ship; ——, PMM; ———, scale model

2. The Linear Model

In the linear model, which makes it possible to use all the classical concepts of automatic control (Clarke *et al.* 1983), the hydrodynamic forces X and Y and moment N in Eqn. (1) are considered as small perturbations about a steady ahead speed u_0. It is supposed that these forces and moment depend only on the instantaneous value of the state vector $(u, v, r, \dot{u}, \dot{v}, \dot{r}, \delta)$ and it is usual to express these quantities by using a dimension-free system called prime where the units are L for the length, u_0 for the speed, L/u_0 for the time and $\rho L^3/2$ for the mass, with ρ being the density of water. Thus, Eqn. (1) becomes

$$\left.\begin{array}{l} m'(\dot{u}' - r'v') = X'(u', v', r', \dot{u}', \dot{v}', \dot{r}', \delta) \\ m'(\dot{v}' + u'r') = Y'(u', v', r', \dot{u}', \dot{v}', \dot{r}', \delta) \\ I'_z \dot{r}' = N'(u', v', r', \dot{u}', \dot{v}', \dot{r}', \delta) \end{array}\right\} \quad (5)$$

By retaining only the linear part of the Taylor expansion of the right-hand terms in Eqn. (5) around $u' = u'_0(=1)$, $v' = 0$, $r' = 0$, and $\delta = 0$ gives

$$m'\dot{u}' = X'_u \Delta u' \quad (6a)$$

$$m'(\dot{v}' + r') = Y'_{\dot{v}}\dot{v}' + Y'_{\dot{r}}\dot{r}' + Y'_v v' + Y'_r r' + Y'_\delta \delta \quad (6b)$$

$$I_z \dot{r}' = N'_{\dot{r}}\dot{r}' + N'_{\dot{v}}\dot{v}' + N'_r r' + N'_v v' + N'_\delta \delta \quad (6c)$$

where

$$Y'_{\dot{v}} = \frac{\partial Y'}{\partial \dot{v}'}, \quad Y'_v = \frac{\partial Y'}{\partial v'}, \quad \text{etc.} \quad (7)$$

and $\Delta u'$ is the variation in longitudinal speed.

Equation (6a), which describes the longitudinal movement of the ship, is decoupled from Eqns. (6b) and (6c) and will not be taken into further consideration. It may be shown that Eqns. (6b) and (6c) are equivalent to the second-order system of decoupled equations

$$T'_1 T'_2 \ddot{r}' + (T'_1 + T'_2)\dot{r}' + r' = K'\delta + K'T'_3 \dot{\delta} \quad (8a)$$

$$T'_1 T'_2 \ddot{v} + (T'_1 + T'_2)\dot{v}' + v' = K'_v \delta + K'_v T'_4 \dot{\delta} \quad (8b)$$

where the coefficients $T'_1, \ldots, K'_v$ are functions depending on the coefficients of Eqn. (6). In the following, the main interest will be in Eqn. (8a) which is used in practice to describe, in terms of yaw rate, a great many of the situations encountered in ship maneuvering.

2.1 Maneuvering Criteria

(a) Dynamic stability. The solution of the homogeneous system associated with Eqns. (8a) and (8b) is of the form

$$\begin{bmatrix} v' \\ r' \end{bmatrix} = \begin{bmatrix} v_1 \\ r_1 \end{bmatrix} \exp(-t'/T'_1) + \begin{bmatrix} v_2 \\ r_2 \end{bmatrix} \exp(-t'/T'_2) \quad (9)$$

where v_1, v_2, r_1 and r_2 are constants of integration that depend on the initial conditions.

The ship will join a straight course with constant speed after a small course perturbation if the time constants T'_1 and T'_2 are positive and this case is called dynamically stable. Since T'_1 and T'_2 may be expressed as functions depending on the coefficients of Eqn. (6), it can be shown that this stability condition is equivalent to the inequality

$$N'_r(Y'_r - m')^{-1} > N'_v/Y'_v \quad (10)$$

which indicates that the center of pressure in pure yaw should be ahead of the center of pressure in pure sway.

(b) Turning ability. It is current, in the analysis of trial maneuvers, to use a simpler model than Eqn. (8a) (Nomoto *et al.* 1957):

$$T'\dot{r}' + r' = K'\delta' \quad (11)$$

where

$$T' = T'_1 + T'_2 - T'_3 \quad (12)$$

having the complementary function

$$r' = r_1 \exp(-t'/T') \quad (13)$$

Thus, Eqn. (13) approximates to the component in r of the second term of Eqn. (9) when T'_1 is much greater than T'_2 and T'_3, as often occurs in practice.

By solving Eqn. (11) in the particular case of the beginning of a zigzag maneuver (see Fig. 4), it may be seen that K' and T' are solutions of the system:

$$1/K' = t' - T' + T' \exp(-t'/T') \quad (14)$$

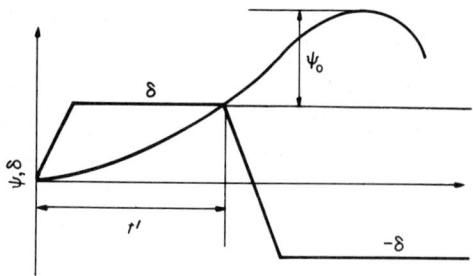

Figure 4
Heading change and rudder angle vs time, for the beginning of a zigzag maneuver

$$\frac{\psi_0}{\delta} = K'[T' - T' \exp(-t'/T') - T' \log(2 - \exp(-t'/T'))] \quad (15)$$

where t' (in ship lengths) is the time after the heading reaches the rudder angle value corresponding to the trial (generally $\delta = \pm 10°$) and the overshoot angle ψ_0 may easily be deduced from experimental data. It is considered that $t' = 1.75$ is a limit for all ships whereas ψ_0/δ has a large range of values and depends closely on the hull shape.

Norrbin (1965) introduced the turning index

$$P = \psi(1)/\delta \quad (16)$$

defined as the heading change per unit rudder angle in one ship length travelled (see Fig. 5).

By solving Eqn. (11) for the corresponding function

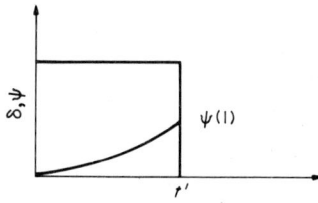

Figure 5
Heading change and rudder angle vs time, in travelling one ship length

with δ = constant gives

$$P = K'(1 - T' + T' \exp(-1/T')) \quad (17)$$

which reduces to

$$P \approx \frac{1}{2}\frac{K'}{T'} \quad (18)$$

when T' is large. In practice, $P = 0.3$ (which is equivalent to a 10° heading change in one ship length, when the rudder is placed in excess of 30°) is considered as an acceptable turning index for all ships.

(c) *Manual steering ability*. Figure 6 illustrates the block diagram of the control loop of the ship, in which, if the Laplace transform of $x(t)$ is denoted by

$$\mathscr{L}x(s) = \int_0^\infty \exp(st) x(t) \, dt, \quad s \in \mathbb{C} \quad (19)$$

the components are $S(s) = -\mathscr{L}r'(s)/\mathscr{L}\delta(s)$, the transfer function of the ship represented by Eqn. (8a) relating the rate of turn to the rudder angle, $E(s)$, the transfer function of the steering engine relating the actual rudder angle to that demanded, and $H(s)$, the transfer function of the helmsman relating the demanded rudder angle to the heading error.

It has been shown by Nomoto (1977) that $H(s)$ may be represented by a lead–lag compensation network of the form

$$H(s) = \frac{1 + \alpha T s}{1 + T s} \quad (20)$$

for moderate frequencies ($s = if, f < 0.8$ rad s^{-1}). Thus, the transfer function of the system in closed loop is

$$R(s) = \frac{H(s)E(s)S(s)1/s}{1 - H(s)E(s)S(s)1/s} \quad (21)$$

It may be seen by using the Nyquist stability criterion that, for an appropriate choice of α and T in Eqn. (20), the system in closed loop may become stable (the response at an impulsion vanishes at infinity), even if it is not stable in the sense of Sect. 2.1. Existing course stabilizing systems are based on this property.

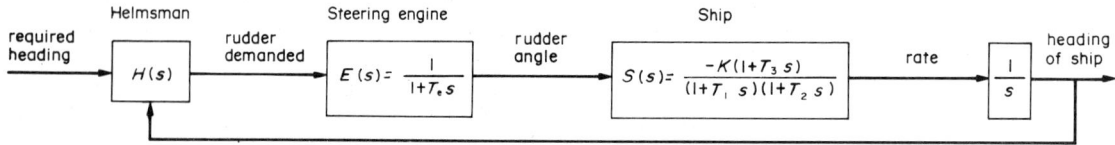

Figure 6
Steering control loop of ship

2.2 Estimation of Coefficients

(a) *Slender-body theory*. This theory, originally developed for aircraft, aims at obtaining the coefficients of Eqn. (6) as functions depending on the main geometrical parameters of the hull. In the case of ships, the application of the slender-body theory does not give very accurate results, but it does lead to good predictions of the algebraic form of the coefficients. By using multiple regression techniques, it is possible to fit such forms to a large amount of data from constrained tests, in order to obtain expressions of the form (Inoue *et al.* 1981):

$$-Y'_v/\pi(T/L)^2 = 1 + 0.16 C_B B/T - 5.1(B/L)^2 \quad (22)$$

(b) *Free-steering experiments*. Since the linear model in Eqn. (6) can be put into the standard form of a stochastic differential equation

$$dX = AX\,dt + Bu\,dt + dW(t) \quad (23)$$

and the measurements at discrete instants t_k can be put into the form

$$Y(t_k) = CX(t_k) + Du(t_k) + c(t_k), \quad k = 1, 2, \ldots \quad (24)$$

where $\mathbf{A}$ and $\mathbf{B}$ are matrices the coefficients of which are related to the coefficients of Eqn. (6) $u = \delta$, $\mathbf{C}$ and $\mathbf{D}$ are known matrices, and $c(t_k)$ and $W(t)$ are known white noise, it is possible to use classical system identification techniques in order to estimate the coefficients of $\mathbf{A}$ and $\mathbf{B}$, from which estimations of the coefficients of Eqn. (6) may be deduced. Flobakk (1983) used these techniques to analyze free-steering experiments in which a scale model, under remote control, executes modified zigzag maneuvers in order to hold the limits of the towing tank, and the position and heading angle are measured.

Such methods were also applied by Astrom and Kallstrom (1976) for experiments with full-scale ships, and by Abkowitz (1980) in order to identify the coefficients of the VLCC *Esso Osaka* (see Sect. 1.2).

See also: Ship Automation and Control

Bibliography

Abkowitz M A 1980 Measurement of hydrodynamic characteristics from ship manoeuvring trials by system identification. *Trans. Soc. Nav. Archit. Mar. Eng.* **88**

Astrom K J, Kallstrom C G 1976 Identification of ship steering dynamics. *Automatica* **12**

Clarke D, Gelding P, Hine G 1983 The application of manoeuvring criteria in hull design using linear theory. *Trans. R. Inst. Nav. Archit.* **125**

Crane C L 1979 Manoeuvring trials of 278 000 DWT in shallow and deep waters. *Trans. Soc. Nav. Archit. Mar. Eng.* **87**

Flobakk T 1983 Parameter estimation of ship manoeuvring equations. *Automatica* **4**

Hirano M, Takashina J, Moriya S 1987 A practical prediction method of ship manoeuvring motion and application. *Trans. R. Inst. Nav. Archit.* **129**

Inoue S, Hirano M, Kijima K 1981 Hydrodynamic derivatives of ship manoeuvring. *Int. Shipbuild. Prog.* **28** (321)

Nomoto K 1977 Some aspects in simulation studies in ship handling. *Trans. Int. Symp. Practical Design in Shipbuilding*, Tokyo

Nomoto K, Tagushi T, Honda K, Hirano S 1957 On the steering qualities of ships. *Int. Shipbuild. Prog.* **4** (35)

Norrbin N H 1965 Zig-zag test technique and analysis with preliminary statistical results. SSPA Allman Report No. 12

G. Haiman
[Université de Lille, Lille, France]

Ship Positioning: Adaptive Control

Ship positioning, or dynamic positioning, describes systems that are primarily applied for the automatic position control of drilling vessels, platforms and support vessels in the offshore oil industry. Such systems have been manufactured since the 1960s. In the beginning, these systems were designed using conventional control principles (Morgan 1978). Around 1974, a new model-based control concept was introduced in which mathematical modelling of vessel movement, and state and parameter estimation was employed in the form of extended Kalman filters and multivariable control using linear quadratic Gaussian (LQG) theory (Balchen *et al.* 1976). In dynamic positioning systems, vessels are moved by forces exerted by a variety of devices such as controllable thrusters (propellers) or even winches pulling anchor chains. A number of different systems are available for measuring the motion of a vessel (position-reference systems) such as hydroacoustic transponder systems, taut wire systems, radio transponder systems and satellite systems.

A good system for dynamic positioning should keep a vessel within specified position and heading limits with minimal fuel consumption and wear on the propulsion equipment. Furthermore, the system should tolerate transient failures in the measurement and propulsion systems.

Figure 1 shows a typical vessel with dynamic positioning and Fig. 2 presents a block diagram indicating the main elements in the control structure.

1. Modelling Vessel Motion

The behavior of the vessel is modelled as a combination of low-frequency (LF) drift and high-frequency (HF) oscillation. The slow motion is caused by the thrusters and the forces from the wind and water currents. Since the wind can be measured fairly precisely, only the current forces need to be estimated. It is assumed that the current is relatively constant both in magnitude and

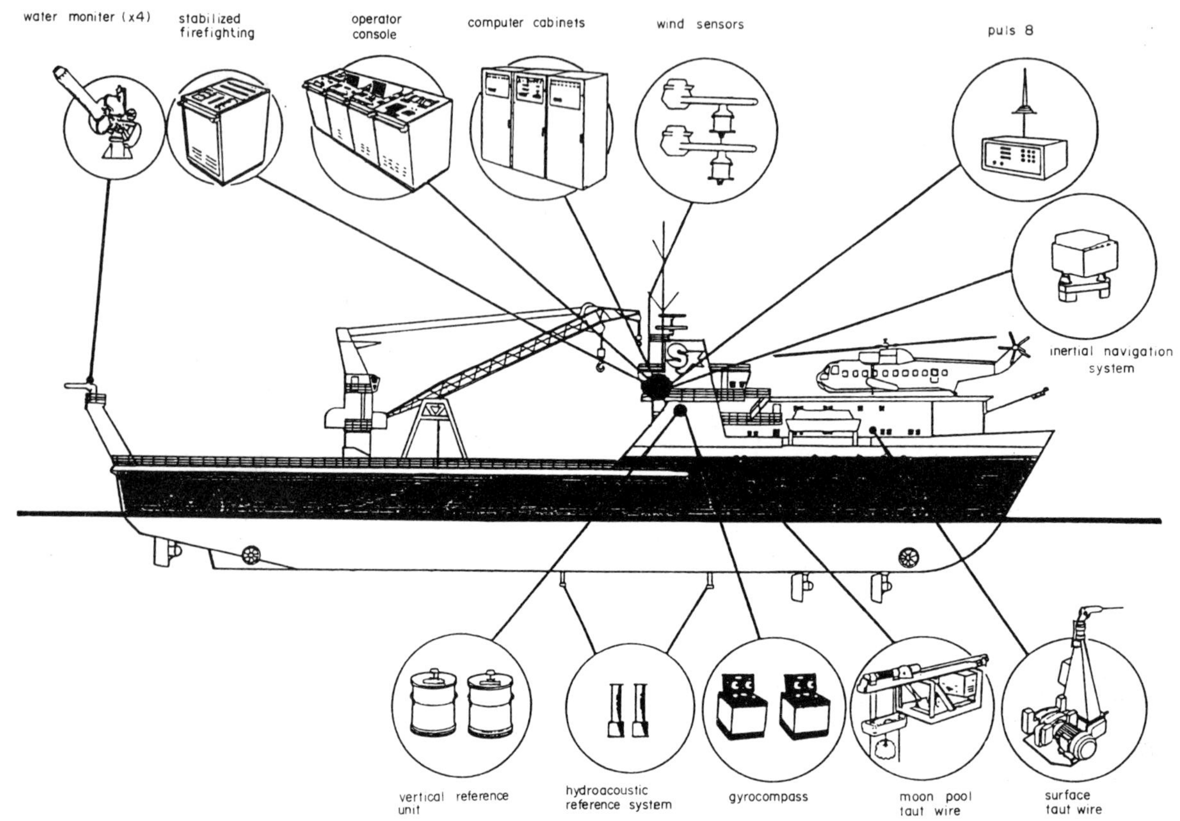

Ship Positioning: Adaptive Control

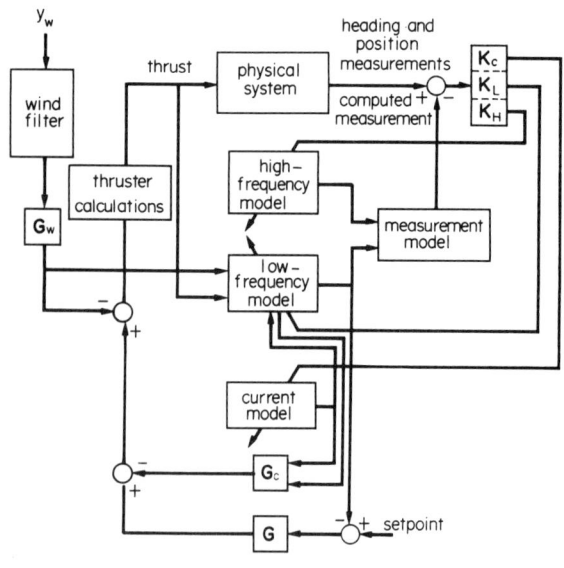

Figure 1
Seabex One, a support vessel for offshore operations equipped with dynamic positioning system (courtesy Simrad-Albatross, Norway)

Figure 2
Control system structure

direction, leading to a model containing only two integrators. Since it is not possible to counteract the oscillatory movement due to waves, the control action of the thrusters is derived from the state of the LF model, the state of the current model and a feedforward from the wind measurements.

The most significant feature of the structure of Fig. 2 is that the vessel motion is the sum of the LF and HF motions. As can be seen from Fig. 2, the Kalman filter structure contains three gain matrices $\mathbf{K}_C$, $\mathbf{K}_L$ and $\mathbf{K}_H$ for the updating of the current state, the LF state and the HF state, respectively. Since its introduction in 1974 (Balchen *et al.* 1976), this structure has been employed in a number of subsequent investigations (Balchen *et al.* 1980a, b, Grimble *et al.* 1980a, b, Fung and Grimble 1983, Saelid *et al.* 1983). A good approximation to the LF behavior of a vessel is given by the following model:

$$\dot{x}_L = u_L \tag{1}$$

$$\dot{u}_L = \frac{1}{m_x}[m_y(v_L - v_c)r_L + m_x v_c r_L$$
$$- d_x|u_L - u_c|(u_L - u_c) + F_{wx} + F_{Tx}] + \eta_{Lx} \tag{2}$$

$$\dot{y}_L = v_L \tag{3}$$

$$\dot{v}_L = \frac{1}{m_y}[m_y u_c r_L - m_x(u_L - u_c)r_L$$
$$- d_y|v_L - v_c|(v_L - v_c) + F_{wy} + F_{Ty}] + \eta_{Ly} \tag{4}$$

$$\dot{\psi}_L = r_L \tag{5}$$

$$\dot{r}_L = \frac{1}{m_\psi}[-(m_y - m_x)(u_L - u_c)(v_L - v_c)$$
$$- d_\psi|r_L|r_L + M_c + M_w + M_T] + \eta_{L\psi} \tag{6}$$

where x_L and u_L are the position and velocity, respectively, in the surge direction (LF part), y_L and v_L are the position and velocity, respectively, in the sway direction (LF part), ψ_L and r_L are the heading and heading rate, respectively (LF part of yaw), u_c and v_c are the current velocities in the surge and sway directions, respectively, F_{wx} and F_{wy} are the wind forces in the surge and sway directions, respectively, F_{Tx} and F_{Ty} are the thrust forces in the surge and sway directions, respectively, M_w and M_T are the moments from wind and thrust, respectively, and M_c is the residual current moment. The terms η_{Lx}, η_{Ly} and $\eta_{L\psi}$ are assumed to be zero-mean Gaussian white-noise processes, d_x, d_y and d_ψ are viscous drag coefficients and m_x, m_y and m_ψ are inertial coefficients that are assumed to be constant. The residual current-generated moment is included to account for errors in Eqn. (6).

A number of different model structures have been used in the literature to represent the HF behavior of a vessel. In the early versions of a Norwegian system (Balchen *et al*. 1976, 1980a, b) the model consisted of three harmonic oscillators without damping (two state variables in each) in which the frequencies were updated using a simplified extended Kalman filter. This structure has been improved introducing a damping term in the oscillators and has employed a more robust parameter-estimation algorithm. In a British system (Grimble *et al*. 1980a) the HF behavior has been represented by three fourth-order decoupled sub-models with five fixed parameters in each.

Using a notation similar to that in Eqns. (1)–(6), an HF model would have the form

$$\dot{x}_H = u_H \tag{7}$$
$$\dot{u}_H = -\theta_x x_H - \zeta_u u_H + \eta_{Hx} \tag{8}$$
$$\dot{y}_H = v_H \tag{9}$$
$$\dot{v}_H = -\theta_y y_H - \zeta_v v_H + \eta_{Hy} \tag{10}$$
$$\dot{\psi}_H = r_H \tag{11}$$
$$\dot{r}_H = -\theta_\psi \psi_H - \zeta_\psi r_H + \eta_{H\psi} \tag{12}$$

The η factors are zero-mean white-noise processes, the ζ factors are damping factors that can be chosen *a priori* to fit ocean wave spectra and the θ factors can be interpreted as the square of the oscillation frequencies in each mode (Saelid *et al*. 1983).

2. Need for Adaptivity

The dynamic models of vessel behavior given by Eqns. (1)–(12) contain a number of parameters that have to be determined. Most of them can be calculated and estimated from vessel construction data using knowledge available in the literature on ship hydrodynamics. The state variables of both the LF and HF models are updated through the Kalman filter gain matrices $\mathbf{K}_L$ and $\mathbf{K}_H$ which can be determined *a priori*. The most important unknown parameters are those associated with the frequencies and damping factors of the HF model. The damping factors may be estimated *a priori* based on known spectral distributions of ocean waves, leaving the frequencies ($\theta_x^{1/2}$, $\theta_y^{1/2}$, $\theta_\psi^{1/2}$) to be determined by an automatic parameter-estimation technique. Balchen *et al*. (1976) did this by means of an extended (augmented) Kalman filter where the three frequencies were regarded as augmented states. Later (Balchen *et al*. 1980a, b), a technique based on sensitivity multipliers was employed, this being quite similar to the

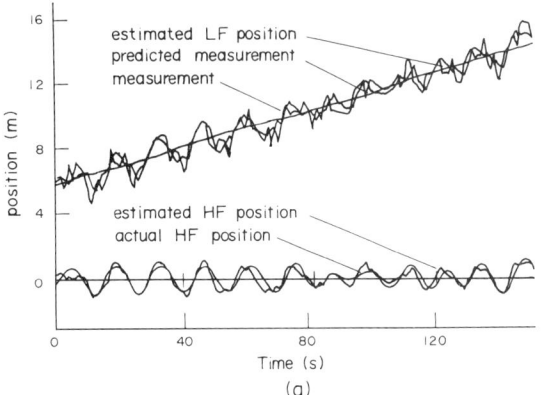

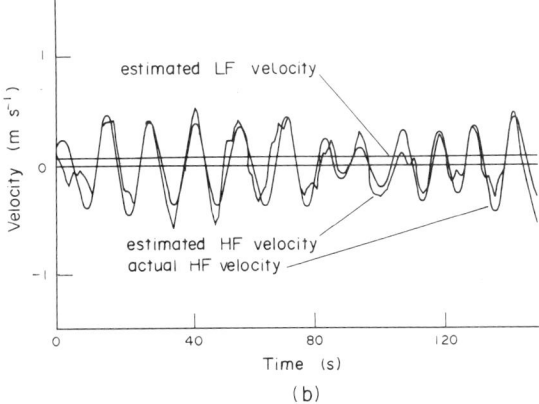

Figure 3
Simulated and estimated position components (a) and velocity components (b)

prediction error technique. This technique is quite satisfactory, but is rather less robust than that described by Saelid *et al.* (1983) which employs more terms in the sensitivity multiplier equations and takes the interaction between the LF and HF models into account. Fung and Grimble (1983) have successfully demonstrated how to use one of the standard self-tuning filtering algorithms for adapting a second-order HF model to a simulated fourth-order wave motion. This indicates the adequacy of the assumption of the second-order HF model made by Balchen *et al.* (1976).

Figure 3 presents examples of simulations of an actual dynamic positioning system (Balchen *et al.* 1980a) where it is clearly seen how the HF estimates of velocity and position are close to the actual behavior and how the oscillatory motion is removed from the LF position estimates. Figure 4 shows the estimated surge frequency of HF motion.

3. Control Structure

As can be seen from Fig. 2 the thrusters (or other control devices) are controlled by feedback matrices (G, G_w) from the states of the LF model and the current model respectively. The wind measurement acts as a feedforward through matrix G_w after a certain filtering action. An important block in Fig. 2 is the calculations that are necessary to determine the actual thruster signals based on the specific geometries of the thrusters in the particular vessel. The elements in the feedback matrix G may be calculated using standard LQ theory on the basis of a quadratic performance criterion in which the states and the control actions are quadratically weighted. The feedback matrix G_C may be derived from a specificiation of the eigenvalues of the resulting dynamic systems (Balchen *et al.* 1980a).

4. Operational Experience

The model-based control concept has several advantages over the classical proportional-plus-integral-plus-derivative (PID) control strategy, such as improved noise suppression, more precise dynamic behavior and improved reliability through prediction and error detection. However, to achieve these advantages, a substantial amount of computational work has to be carried out prior to installation and testing. Vessel displacement and added water mass, drag and lift coefficients, wind force parameters, thruster parameters and filter and control system parameters all have to be computed.

One major advantage of the model-based concept is that many inherently different measurement systems can be integrated through the Kalman filter procedure. This makes the total system insensitive to sensor failure (high integrity) and provides a direct means of detecting sensor malfunction.

A large number of dynamic positioning systems have been installed since about 1980 on a worldwide basis and most of them have used the control concepts described in this article.

See also: Ship Automation and Control; Ship Steering: Model-Reference Adaptive Control

Bibliography

Balchen J G, Jenssen N A, Mathisen E, Saelid S 1980a A dynamic positioning system based on Kalman filtering and optimal control. *Modeling, Identification and Control* **1**, 135–63

Balchen J G, Jenssen N A, Saelid S 1976 Dynamic positioning using Kalman filtering and optimal control theory. *IFAC/IFIP Symp. Automation in Offshore Oil Field Operation*. North-Holland, Amsterdam, pp. 183–8

Balchen J G, Jenssen N A, Saelid S 1980b Dynamic positioning of floating vessels based on Kalman filtering and optimal control. *Proc. 19th IEEE Conf. Decision and Control*. Institute of Electrical and Electronics Engineers, New York, pp. 852–64

Fung P T K, Grimble M J 1983 Dynamic ship positioning using a self-tuning Kalman filter. *IEEE Trans. Autom. Control* **28**, 339–50

Grimble M J, Patton R J, Wise D A 1980a The design of dynamic ship positioning control systems using stochastic optimal control theory. *Optim. Control Appl. Meth.* **1**, 167–202

Grimble M J, Patton R J, Wise D A 1980b The use of Kalman filtering techniques in dynamic ship positioning systems. *Proc. Inst. Electr. Eng. D* **127**, 93–102

Morgan J M 1978 *Dynamic positioning of offshore vessels*. Petroleum, Tulsa, OK

Saelid S, Jenssen N A, Balchen J G 1983 Design and analysis of a dynamic positioning system based on Kalman filtering and optimal control. *IEEE Trans. Autom. Control* **28**, 331–9

J. G. Balchen
[University of Trondheim, Trondheim, Norway]

Ship Rudder Roll Stabilization

In addition to control of the heading, reduction of the roll motions is also desired on some ships. One solution is rudder roll stabilization (RRS) where the rudder

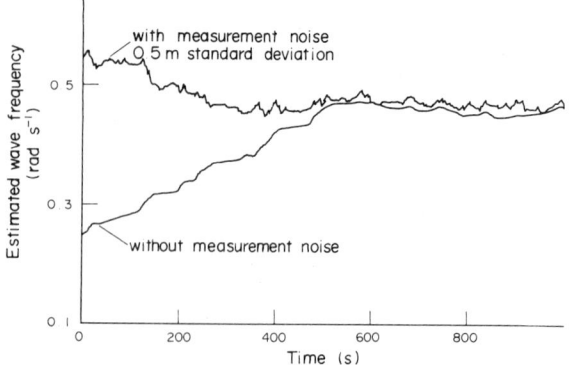

Figure 4
Estimated surge frequency of HF motion

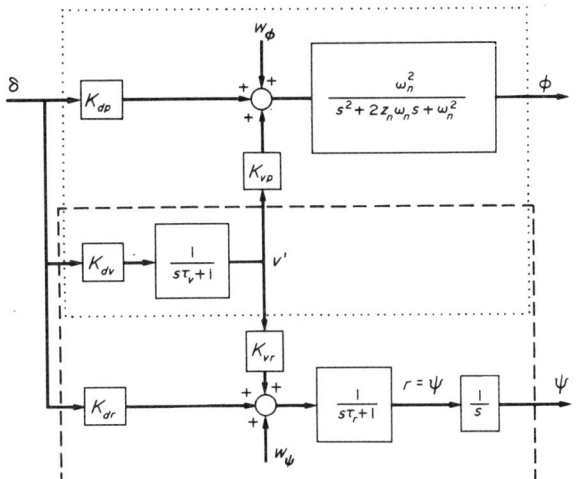

Figure 1
Simplified dynamics between rudder and yaw and roll

alone is used for controlling the heading as well as reducing the roll. This can be achieved by using low-frequency rudder motions for control of the heading and high-frequency rudder motions for roll reduction.

The idea of RRS has been investigated; however, the attempts did not result in successful applications, possibly because appropriate control algorithms were not then available. The first successful full-scale trials were reported by Baitis (1980) who used the rudder for automatic roll stabilization, while the heading remained under manual control. A system that simultaneously controls both the heading and the roll of a ship is described in this article.

1. Mathematical Models

1.1 Ship Dynamics

Models that describe the transfer from rudder angle to heading, and from rudder angle to roll, can be derived from the hydrodynamical models used by shipbuilding engineers. The model used in this article is shown in Fig. 1, where δ is the rudder angle, ϕ is the roll angle, ψ is the heading or yaw angle, v' is the sway velocity caused by the rudder, w_ϕ and w_ψ are white noise with nonzero mean, $H_{w\phi}$ describes the influence of the disturbances on the roll moment, and $H_{w\psi}$ the influence of the disturbances on the yaw moment.

The parameters of this model were found from a series of full-scale modelling trials and are dependent on, for example, the design and speed of the ship. A relation between these parameters and the hydrodynamical models can also be found.

1.2 Disturbances

The disturbances acting on a ship are due to the wind, waves and current, although when the latter is steady, uniform and horizontal, it does not play a role in the control system considered in this article.

Wind can be modelled as a stochastic signal having a nonzero mean, with only the mean value of the wind disturbance begin taken into account. Stochastic variations can be added as a white-noise signal. The nonzero mean results in a constant roll angle as well as a stationary heading error. As the constant roll angle cannot adequately be compensated for by the RRS system, the mean value of the measured roll angle is suppressed by an appropriate high-pass filter. Variations in roll angle and heading are mainly caused by the waves.

Waves can be described by means of a frequency spectrum; for example, the Bretschneider spectrum (Bhattacharyya 1978). This spectrum can be simulated either by a summation of a series of sinusoidal signals with appropriate amplitudes, or by using a coloring filter driven by white noise. The following filter gives a good approximation:

$$H = \frac{Ks}{s^2 + 2z\omega_f s + \omega_f^2} \quad (1)$$

where ω_f is the dominating wave frequency and z is the damping ratio. The disturbances can be added to the model of the ship dynamics by means of the filters $H_{w\phi}$ and $H_{w\phi}$ as indicated in Fig. 1.

1.3 Steering Machines

For the purpose of designing a controller and for simulation of the system, the steering machine is sufficiently accurately described by the block diagram of Fig. 2. The rudder angle is either limited by the mechanical constraints of the steering machine (in general it is always smaller than 35°) or it is intentionally set at a lower value. The maximum rudder speed is determined by the maximum capacity of the hydraulic pumps.

2. Controller Design

Controller design proceeds through two stages. First, a controller without a steering machine is designed for the system. The controller is then modified to deal with the nonlinear dynamics of the steering machine.

2.1 Linearized Systems

Let the process be described by the model shown in Fig. 1. A state-feedback controller for this process

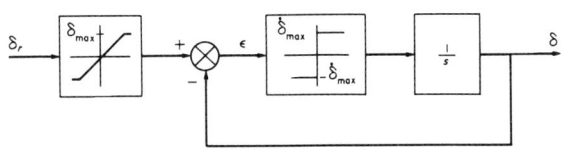

Figure 2
The steering machine

requires that angle ψ, $d\psi/dt$, ϕ, $d\phi/dt$ and v' be available to the controller. Angles ψ and ϕ can be measured with gyroscopes, and their derivatives can be measured with rate gyroscopes or may be obtained from a state estimator. In general, the signal v' can only be obtained from a state estimator. The system can be described by the following state-space equations:

$$\dot{x} = Ax + Bu + Dw \quad (2)$$

where

$$x^T = (\phi, \dot{\phi}, v', \dot{\psi}, \psi) \text{ and } u = \delta$$

A and B are described by

$$A = \begin{pmatrix} 0 & 1 & 0 & 0 & 0 \\ -\omega_n^2 & -2z_n\omega_n & \omega_n^2 k_{vp} & 0 & 0 \\ 0 & 0 & -1/\tau_v & 0 & 0 \\ 0 & 0 & k_{vr}/\tau_r & -1/\tau_r & 0 \\ 0 & 0 & 0 & 1 & 0 \end{pmatrix}$$

$$B = \begin{pmatrix} 0 \\ \omega_n^2 k_{dp} \\ k_{dv}/\tau_v \\ k_{dr}/\tau_r \\ 0 \end{pmatrix} \quad (3)$$

and

$$D = \begin{pmatrix} 0 & 0 & 0 & H_{wr}/\tau_r & 0 \\ 0 & H_{wp}\omega_n^2 & 0 & 0 & 0 \end{pmatrix} \quad (4)$$

where H_{wr} and H_{wp} are gains determined by the influence that the waves have on the rate of turn and roll. Application of the linear quadratic Gaussian (LQG) method requires that a quadratic criterion be defined:

$$J = \lim_{T \to \infty} \frac{1}{T} \int_0^T (x^T Q x + u^T R u) \, dt \quad (5)$$

where Q is a (semi-) positive-definite weighting matrix and R is a positive-definite weighting matrix.

A problem (discussed in more detail in Sect. 2.3) is the selection of weighting factors in this criterion. Feedback gains can be found by means of a computer program that solves the matrix Ricatti equations. A model-reference adaptive state estimator is used to suppress high-frequency components in the heading and rate-of-turn signals. The low-frequency components of the roll angel are suppressed by means of an adaptive high-pass filter. Large roll reductions can be obtained with this system although the required rudder angles and rudder speeds are too large to be realistic. It is essential, therefore, that the nonlinearities of the steering machine be taken into account.

2.2 Nonlinear Systems

The nonlinear model of the steering machine has been given in Fig. 2. The maximum rudder angle directly limits the roll-reduction ability of the system, and the limited rudder speed reduces the amplitude of the controller output and introduces phase lag. This phase lag is not only a function of the frequency, but also of the amplitude of the controller signal. Even for small phase lags, the performance of the system rapidly deteriorates; it is therefore essential that phase lag be prevented. In addition to redesigning the steering machine to ensure higher rudder speeds, the controller must prevent the steering machine from saturating. Three solutions are discussed in this article—gain scheduling, the introduction of automatic gain control and the introduction of an adaptive criterion.

(a) Gain scheduling. The control system (including the steering machine) has been simulated using the simulation package PSI which enables optimization of a system by means of a hill-climbing procedure. Its use is neither restricted to linear systems nor to quadratic criteria, which makes it possible to use more appropriate criteria and to take into account the nonlinear steering-machine dynamics.

These experiments enable determination of the values of the maximum rudder speed necessary for realizing the required roll reduction. Because of the nonlinear nature of the problem, it is not possible to find one set of controller parameters for all situations; however, the method may be used to determine a gain-scheduling table containing the controller gains as a function of the amplitude and dominating frequency of the disturbances. This table can be used for manual adjustment of the controller during the experiments or, when estimates of the amplitude and frequency of the waves are available, for automatic gain scheduling. The problem remaining is to measure or estimate the amplitude and frequency of the disturbances during normal operation. A Kalman-filter type of observer designed for this purpose gave good results in simulations but did not perform satisfactorily during the full-scale trials.

The controller described previously has been tested during extensive computer simulations. In addition to simulations with the model according to Fig. 1, a series of simulations has been carried out with a more extensive model, based on a hydrodynamical modelling approach. The main purpose of these experiments was to determine the required rudder speed for an RRS system as well as to carry out a sensitivity analysis for variations in controller gains. It could be concluded that for the naval ship simulated during the experiments, a rudder speed of $15°\,\text{s}^{-1}$ would be appropriate. In addition, the sensitivty analysis showed that it is

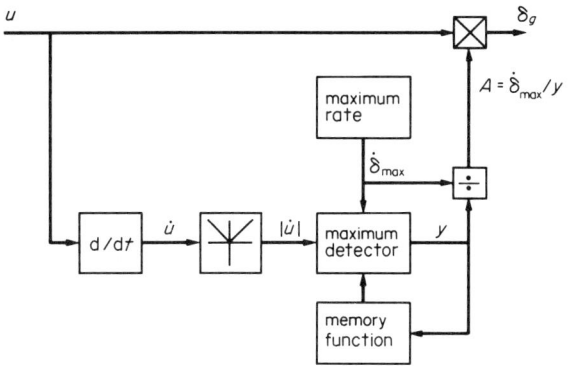

Figure 3
The AGC

important that the controller gains selected are not too large.

(*b*) *Introduction of automatic gain control.* The method described previously in Sect. 2.2 gives the best controller for each situation and for an arbitrary criterion. A disadvantage is that generation of the gain-scheduling table necessitates several computations for each particular situation. In addition, it does not guarantee the prevention of saturation of rudder speed.

This problem can be solved by a mechanism that reduces the output of the controller automatically and instantaneously, as soon as the rate of change of the controller output is so large that this would cause saturation in the steering machine. When there is no further risk of saturation, the gain is gradually increased until the standard value of unity is again reached. The result of application of such a mechanism is that the rudder-speed limiter is removed from the control loop, and therefore its phase lag is not able to deteriorate the performance of the system further. This patented mechanism (the automatic gain controller (AGC)) has proven to be a robust and simple algorithm. The AGC can be explained with the aid of Fig. 3 where u is the controller output, δ is the setpoint of the rudder, $\dot{\delta}_{max}$ is the maximum rudder rate, y is the maximum of three signals (the maximum rudder rate, the absolute value of the derivative of u and the output of a memory function) and A is the gain needed to adjust the controller output $(0 < A \leq 1) = \dot{\delta}_{max} y^{-1}$.

When the absolute value of du/dt is larger than $\dot{\delta}_{max}$, A is instantaneously decreased, resulting in a decrease in the desired rudder angle δ_g and its derivative. When the absolute value of du/dt is no longer too large, the memory function takes care of a slow increase of A. This memory function is the major reason that the phase lag introduced by the steering machine is reduced to a minimum.

The performance of the AGC can be judged from Fig. 4. A sinusoidal signal with increasing amplitude forms the input u. Without the AGC, the rudder angle δ_w shows the typical triangular shape caused by the rate

limiter. With the AGC, δ_w remains a sinusoidal signal with a constant maximum amplitude. The smaller phase lag when the AGC is applied is clearly visible in this figure.

The controller, extended with the AGC, has been tested in several series of full-scale trials. The parameters of the state feedback controller were adjusted manually, based on an off-line optimization procedure (hill climbing). The AGC mechanism appeared to contribute a great deal to the success of these trials. Because the ship used had a rudder speed of only $7°\,s^{-1}$, the achievable roll reduction was limited. A typical example of these trials is given in Fig. 5 where the RRS autopilot is compared with an adaptive stabilization autopilot (ASA) and with the standard autopilot. The roll angle ϕ, the heading error ψ and the actual and desired rudder angle, δ_w and δ_g, are shown. Even with this "slow" rudder, the roll reduction is clearly visible, while the variance of the heading error does not increase when rudder roll reduction is applied. By comparing the results of the full-scale trials with those of the simulations, it may be concluded that the roll reduction with a rudder speed of $15°\,s^{-1}$ will be at least as good as the reduction that can be obtained with the present fine-stabilizer system.

2.3 Adaptive LQG Method

(*a*) *Introduction of an adaptive criterion.* Although the AGC is able to solve the problem of the limited rudder speed in a robust way, it does not realize an optimum controller. Its effect on the controller can be expressed as a reduction of all the feedback gains simultaneously and with the same rate (which is not necessarily an optimum solution). Since optimization of the controller gains with the aid of a gain-scheduling mechanism did not give good results in practice, another mechanism had to be sought.

In this section, the idea of an adaptive criterion combined with the LQG approach (adaptive LQG (ALQG)) is introduced. This enables the definition of criteria that are more appropriate for a particular problem than the otherwise necessary, quadratic criteria.

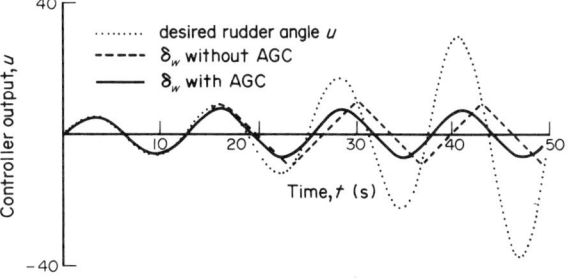

Figure 4
Influence of the AGC with increasing controller output u

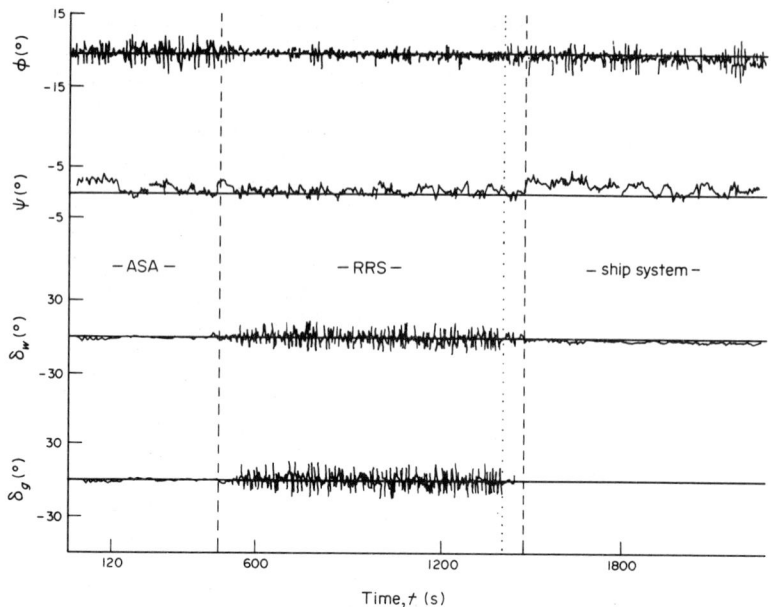

Figure 5
RRS during full-scale trials

Let a process, described by the following state-space equations, be given by:

$$\left.\begin{aligned}\dot{x} &= Ax + Bu + Dw \\ y &= Cx\end{aligned}\right\} \quad (6)$$

Without loss of generality for the method mentioned below it is assumed that w denotes white noise with a zero mean. If the process is time invariant, the optimal controller, with respect to the criterion defined in Eqn. (5), can be calculated off line (e.g., Kwakernaak and Sivan 1972):

$$u = -Kx \quad (7)$$

where

$$\left.\begin{aligned}K &= R^{-1}B^TP \\ 0 &= A^TP + PA + C^TQC - PBK\end{aligned}\right\} \quad (8)$$

When the parameters of the process (A, B, C and D), or the weighting factors (Q and R) change, new optimal controller gains have to be computed. A robust real-time method has been proposed to calculate the optimal controller. It is based on the translation of Eqns. (8) to the nonlinear innovation process given by Eqn. (9) which has as inputs u_m the weighting factors of Eqn. (5). This method can be used to compute the controller gains when the process parameters or the weighting factors in Eqn. (5) vary slowly.

$$\left.\begin{aligned}\dot{x}_m &= A_m x_m + B_m u_m \\ y_m &= C_m x_m\end{aligned}\right\} \quad (9)$$

where

$$A^TP + PA - PBK \leftrightarrow A_m x_m$$
$$C^TQC \leftrightarrow B_m u_m$$
$$R^{-1}B^TP \leftrightarrow C_m x_m$$

On-line simulation by means of numerical integration yields the optimal controller gains K, as outputs (y_m) of the innovation process.

When the process parameters are known by on-line parameter identification or by gain scheduling (e.g., as a function of the speed of the ship), the proposed mechanism takes care of the adaptive controller adjustment. In addition, changing the weighting factors of the criterion (e.g., if the steering machine is saturating) will gradually result in another "optimal" controller. By multiplying each element of dx_m/dt with a scaling factor l_i, the rate of convergence of this innovation process (and thus the speed of adaptation) can be controlled.

(b) *Adaptation of the criterion.* The term optimal in relation to the LQG method is more an indication of the method than a guarantee of optimum performance. This is even more true when an adaptive criterion is used. Apparently, there is a criterion behind the quadratic criterion that really defines the optimum performance. Suitable adjustment mechanisms exist for various types of nonlinear elements, such as dead bands, limiters and rate limiters. Rate limiters are most relevant for RRS. In practice, however, it is not possible to solve this problem with a single linear controller: a controller that gives satisfactory results for small roll angles may give no roll reduction when the roll angles are large in rough weather. Furthermore, the operational requirements may change; operators of ships may want to have as much roll reduction as possible, even if that introduces larger heading deviations, or they may be satisfied if the heading error and roll angle stay below a certain limit. This indicates that, in practice, it is not possible to define one criterion that covers all conditions: the criterion has to change with the conditions. In addition, it should be possible for the operator to change the criterion easily, based on the operational demands.

The desired performance of the RRS system can be defined as a series of demands.

(i) The roll angle must not exceed a certain value set by the operator of the ship.

(ii) The demanded rudder speed must not be larger than the limitation posed by the steering machine.

(iii) Under some conditions, roll stabilization by the rudder might increase the heading deviations. If these deviations reach a certain limit (set by the operator), more weight should be given to a good course-keeping performance.

(iv) If the roll remains below a certain limit (set by the operator), less weight should be given to roll reduction in order to reduce the wear and tear of the steering machine.

(v) The controller design, indicated in Sect. 2.2, will result in a stable system. However, due to nonlinear and unmodelled dynamics, problems may occur. Therefore, to avoid stability problems, the controller parameters are not allowed to become too large.

(vi) The adjustment of the controller parameters should be slow enough to follow only weather changes.

For given disturbance conditions, sufficient knowledge is available (whether *a priori* or from measurements) to derive a proper criterion. Only if the disturbance conditions, or the desired performance, change is criterion adjustment necessary.

If a ship is considered with the rudder as its only actuator, the criterion of Eqn. (5) may be rewritten as

$$J = \lambda(q_\phi J_\phi + J_\psi) \qquad (10)$$

where

$$J_\phi = \sum_{i=k}^{3} q_k E[y_k \cdot y_k] + E[\delta_\phi \delta_\phi] \qquad (11)$$

described the influence of the roll motions on the criterion while J_ψ is selected to be similar to the course-keeping criterion. The term δ_ϕ indicates the components of the rudder angle needed for roll reduction. Further simplification is obtained by choosing fixed values for the weighting parameters q_k. Thus, it remains necessary only to choose the weighting parameter q_ϕ depending on the weather conditions.

The above-mentioned demands can easily be translated into a rate of change Δ_q of the weighting parameter q_ϕ. For two of the demands, this is illustrated in Fig. 6 where Δ_{q_i} is the rate of change of weighting parameter q_i, $\dot\delta$ is the demanded rudder speed, $\dot\delta_{max}$ is the maximum rudder speed, σ_ϕ^2 is the variance of the roll angle ϕ, and $\sigma_{\phi_g}^2$ is the allowed variance of the roll angle. The resulting rate of change Δ_{q_ϕ} of parameter q_ϕ incorporating the demands mentioned above, is chosen to be:

$$\Delta_{q_\phi} = \Delta_{q_1} + \Delta_{q_2} + \Delta_{q_3} + \cdots \qquad (12)$$

The weighting parameter q_ϕ will be adjusted according to:

$$q_\phi = q_0 + \alpha \int \Delta_q \, \mathrm{d}t \qquad (13)$$

where α is a parameter that is introduced to determine the speed of the adaptation.

The weighting parameters q_ϕ and q_k are used as the input variables of the innovation process mentioned in Sect. 2.3(*b*). The outputs of this process approach the desired controller parameters. If the weather conditions change slowly compared with the convergence speed of the innovation process, the resulting controller will be optimal with respect to the demands stated above.

Experiments with an extensive hydrodynamical model indicate that the performance of the adaptive controller is close to the performance of the optimally manually adjusted controller tested at sea. The latter, however, requires careful tuning, while the adaptive controller requires no manual adjustments to compensate for a changing environment.

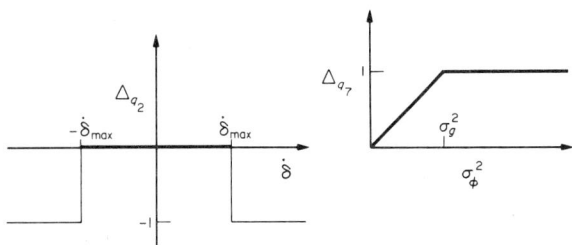

Figure 6
Controller design demands

3. Conclusions

Linear control techniques are no longer applicable when the saturation type of nonlinearities are dominating the behavior of the process. The AGC algorithm, which prevents the rate of change of the actuator input from becoming too large, was therefore developed. Full-scale experiments with this algorithm have demonstrated its usefulness and robustness. However, because it reduces all controller gains simultaneously, the resulting controller will not be an optimal controller and should only be applied as a safety mechanism.

The adaptive adjustment of the weighting factors of the criterion in combination with the on-line calculation of the optimal controller solves this problem. The adaptation mechanism is based on a series of simple rules that translate the operational demands into the weighting factors themselves. Simulation results have demonstrated that this method is robust against variations in the characteristics of the disturbances and process parameters, including variations in nonlinearity.

During the experiments with the ALQG method, it was assumed that the parameters of the process were known (the influence of variations in the ship speed on these parameters was taken into account by a gain-scheduling table). Large variations in these parameters were made in order to determine the sensitivity for these variations. Although no serious problems were encountered, the addition of an on-line parameter estimator may improve the performance. The first application of this system was in a series of frigates of the Royal Netherlands Navy.

See also: Ship Automation and Control; Ship Positioning: Adaptive Control; Ship Steering: Model-Reference Adaptive Control

Bibliography

Baitis A E 1980 *The Development and Evaluation of a Rudder Roll Stabilisation System for the WHEC Hamilton Class.* DTNSRDC Report, Bethesda, MD
Bhattacharyya R 1978 *Dynamics of Marine Vehicles.* Wiley, New York
Källström, C G 1981 Control of yaw and roll by a rudder/fin stabilisation system. *6th Ship Control Systems Symp.*, Ottawa
Klugt P G M van der 1987 Rudder roll stabilisation, Ph.D. Thesis. Delft University of Technology, Delft, The Netherlands
Kwakernaak H, Sivan R 1972 *Linear Optimal Control Systems.* Wiley, New York
van Amerongen J 1984 Adaptive steering of ships—A model reference approach. *Automatica* **20**(4), 3–14
van Amerongen J, van Nauta Lemke H R, Veen J C T van der 1977 An autopilot for ships designed with fuzzy sets. *IFAC/IFIP Symp. Digital Computer Applications to Process Control, The Hague.* Pergamon, Oxford

J. van Amerongen
[University of Twente, Enschede, The Netherlands]

Ship Stabilization: History

With the change from sailing to steam-powered vessels, the problem of rolling of ships became worse; the large head of sail and the deep keel of sailing vessels had provided large forces to resist rolling motion and the sails had provided considerable damping to rolling, but these forces were absent from steamers. The control of rolling is important in both passenger and naval ships. For the navies of the world, not only is the comfort of the sailors a consideration, but the method used to aim and fire the guns is also affected by rolling. In the latter part of the nineteenth century, most navies had adopted the practice of elevating the guns to the approximate position and then firing them when the roll of the ship brought them on target. Large angles of roll reduced the rate at which the guns could be fired.

The methods used to control or reduce rolling fall into four groups: gyroscopic stabilizers, antirolling tanks, moving weights and fins. In each case the development has passed through two different stages, employing first a passive principle of action and then an active one. In systems based on the passive principle, no special actuator is used; the action depends on some force or forces inherent in the system. The drawback of all passive methods is that the coupling between the ship and the stabilizing agent is weak; hence, the ship has to roll to large angles to develop an appreciable stabilizing moment. As a consequence, passive methods only reduce the rolling motion; they do not eliminate it. Table 1 summarizes the early development.

1. Gyroscopic Methods

In 1904, Ernst Otto Schlick suggested placing a large gyroscope in the hull of a boat. The action of rolling causes the gyroscope to precess and generate a torque opposing the roll; the force of the precession couple was to be absorbed by a brake. The system was installed in a torpedo boat of the German Navy, the *See-Baer*, and trials took place in 1906. The trials were observed for the British Admiralty by Sir William White, who reported favorably, and the patent rights were taken up by the British firm of Swan, Hunter and Wigham Richardson, which even purchased the *See-Baer* from the German government for test purposes.

This work, as well as the work of Anschütz–Kaempfe on the gyrocompass, was reported in a series of articles on the use of the gyroscope, published in *Scientific American* in 1907. The articles attracted much attention; one of the people who became interested was Elmer Sperry who, in May of 1908, filed what proved to be the basic patent for the "active" gyrostabilizer. In the patent application Sperry wrote

"It has been known ... that gyroscopes mounted in swinging frames in a plane transverse of a ship would reduce the rolling to some extent ... such apparatus was sluggish, did not act and hence was only partially suc-

Table 1
Chronological list of stabilization installations to 1950

Date (c.)	Designer	Ship	Stabilizer	Control type	Control method
1870		almost all	bilge keels	passive	none
1880	Bridge (UK)	*Colussus, Edinburgh*	tanks (slosh)	passive	?
1883	Watts (UK)	*Inflexible*	tanks (slosh)	passive	tuning and damping
1891	Thorneycroft (UK)	*Cecile*	weight	active	position of weight proportional to effective wave-slope and roll angle
1906	Schlick (Germany)	*See-bar* and others	gyro	passive	tuning and damping
1909	Cremieux (France)	channel steamer	weight	passive	tuning and damping
1910	Frahm (Germany)	*Ypiranga, Europa* and many others	tanks (U-tube)	passive	tuning and damping
1912	Frahm (Germany)	*Deutschland, Hamburg* and many others	tanks (sea-ducted)	passive	damping
1915	Sperry (USA)	*Worden, Conte de Savoia* and many others	gyro	active	precessional velocity proportional to roll velocity
1924	Fieux (France)	French destroyer	gyro	passive	tuning and damping
1925	Motora (Japan)	*Mutsu Maru* and others	fins (variable angle)	active	manual control and ?
1929	Hort (Germany)	*Fuchs, Rossarol*	weight	active	velocity of weight proportional to roll angle
1930	Hort (Germany)	*Konige Louise*	tanks (sea ducted)	active	velocity of fluid proportional to roll angle
1933	Kefeli (Italy)	*Aviso Etourdi*	fins (variable area)	active	area of fins proportional to roll velocity
1935	Hort (Germany)	*Prins Eugen* and others	tanks (U-tube)	active	velocity of fluid proportional to roll angle
1936	Denny-Brown (UK)	*Isle of Sark, Queen Elizabeth* and many others	fins (variable angle)	active	angle of fins proportional to roll velocity
1938	Dutch Engineers (The Netherlands)	?	fins (hydrfoil keels)	passive	none
1939	Minorsky (USA)	*Hamilton*	tanks (U-tube)	active	velocity of fluid proportional to roll acceleration
1949	US Navy (USA)	*Peregrine*	tanks (U-tube)	active	velocity of fluid proportional to roll acceleration

Source: Chadwick (1955)

cessful in dampening the roling.... My invention becomes active promptly on incipient rolling."

Sperry's first proposal was to use a pendulum to sense the rolling motion, the signal from this being used to cause a large gyroscope to precess, thus generating a large torque opposing the rolling torque. The pendulum was eventually replaced by a small "control" gyroscope as shown in Fig. 1. As can be inferred from the figure, the system was an on–off controller, the control gyroscope detecting the direction of the roll (Hughes 1971, Bennett 1979).

Initially, when applied to small ships, the gyroscopic method showed great promise, but it became impractical on large ships. Installation costs were high, considerable space in the center of the ship was required and the maintenance requirements were excessive. In the 1930s, the method was used in one large steamship, the *Conte di Savoia* (41 000 t). Three gyroscopes, each 4 m in diameter, running at 900 rpm and with a total weight of 650 t, were used; after this the method seems to have been abandoned.

Elmer Sperry was not the first person to use a gyroscope as an instrument to detect the rolling motion of a ship; in 1874, Henry Bessemer installed a system on board a steamship designed to keep the saloon level. This mechanism used a gyroscope to detect the motion and a hydraulic control system to move the saloon. Bessemer hoped to use the gyroscope to detect both rolling and pitching motion, but he mounted the gyroscope in such a fashion that it detected only rolling motion and that only through the pendulum action. He also had difficulty with the operation of the hydraulic servosystem and the attempt was abandoned.

2. Antirolling Tanks

The earliest of the passive methods was developed by Sir Philip Watts in 1875. Details of the system were published in 1883 and 1885. It used water located in connected tanks; rolling of the ship caused the water to move from tank to tank and air displaced from one tank passed through a restriction into the outer tank. The tank connections were adjusted so that the water motion resonated in antiphase with the rolling motion of the ship. The principle was based on work by William Froude published in the *Transactions of the Institute of Naval Architects* in 1861. Watts' work was discontinued because of the noise of the system due to the churning of the water.

Some years later, the method was revived, and the shape and method of interconnection of the tanks were improved by H. Frahm, a German naval engineer at the Blohm and Voss yards in Hamburg. The results of trials of the improved system were published (Frahm 1911). The principle of the system is shown in Fig. 2. Tanks on each side of the ship were partly filled with water and connected at the bottom; they were sealed and further connected at the top through a restriction. This restriction enabled the resonant frequency of the tank system to be adjusted.

Frahm tanks were installed in a number of ships, mainly German, and met with moderate success. It was found that, for ships with low stability (small metacentric height) operating in seas with a regular wave pattern, good results were obtained with roll quenching of approximately 50%. However, in choppy seas with no regular wave pattern, the results were poor, frequently with no observed quenching of the roll. Attempts to use Frahm tanks on ships with good stability (high metacentric height) were complete failures and the method was abandoned immediately after World War I.

Minorsky (1941) showed that Frahm and Watts were correct in assuming that the tanks depended on the

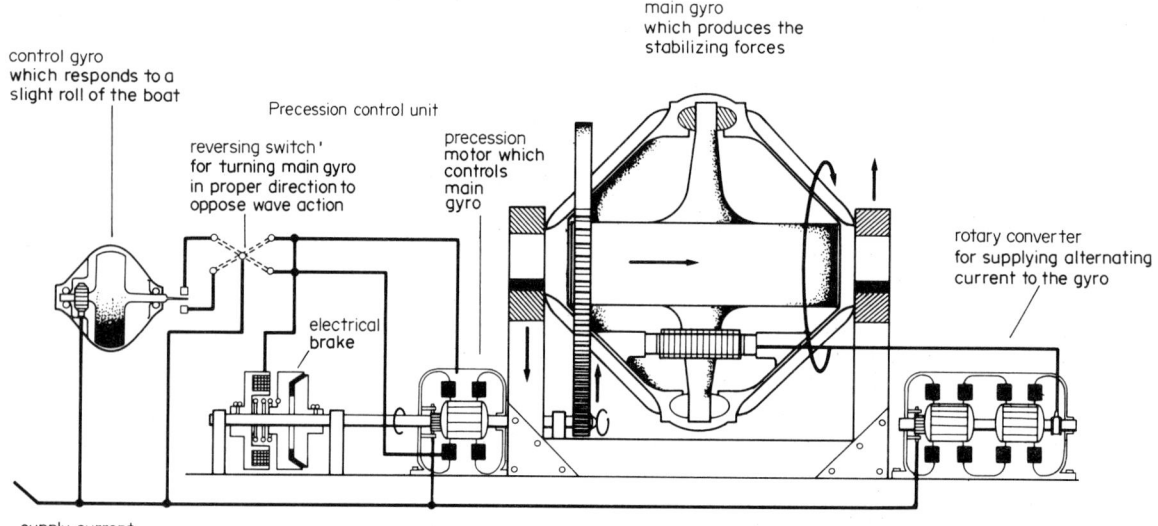

Figure 1
Sperry ship stabilizer (after Hughes 1971)

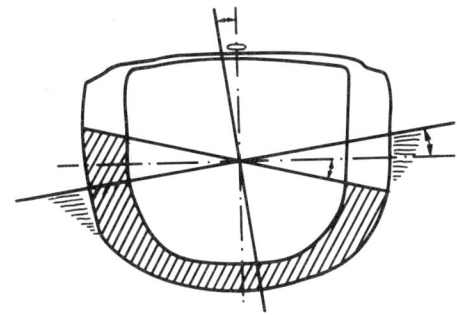

Figure 2
Antirolling tanks

estabishment of a resonant system; the weakness was that in a confused seaway the proper rhythm in the tanks could not be established. He proposed a method in which the oscillating movement of the water in the tanks was established and maintained by the use of an external source of power (e.g., compressed air, variable-pitch pumps), the motion being maintained substantially in phase with the external disturbing motion. This involved a control system that kept the excess ballast in one of the tanks proportional to the angular rate of roll.

The first work on active tanks had been carried out in Germany by Hort and Rellstab in the 1930s; they used compressed air to move the water between the tanks. Minorsky, in addition to carrying out theoretical and model studies, supervised the practical trials of his activated-tanks scheme. He considered that the use of a pump system had many advantages over the use of compressed air and noted that the later German systems used pumps.

Minorsky's scheme involved the use of a variable-delivery pump, with the blades of the pump adjusted by means of a two-phase ac motor. Since the required control was that the level in one of the ballast tanks should be proportional to the rate of roll and since the pump controlled the rate of supply of water to the tank, he decided to use an accelerometer to measure the rate of change of rate of roll and to use this signal to control the pump. As he had done in his work on automatic steering of ships, Minorsky also introduced into the control equation two higher-order derivative terms so that the control signal obtained was

$$e = m\,d^2\theta/dt^2 + n\,d^3\theta/dt^3 + p\,d^4\theta/dt^4 - k\alpha \quad (1)$$

where θ is the angle of roll and α is the angle of the pump vanes. The system met practical difficulties in that violent water hammer occurred in the blade-shifting mechanism and the vibrations produced were picked up by the accelerometer thus causing unstable oscillation of the system. In 1940, before modifications could be carried out, the ship used for the tests was called into active service and further work was abandoned.

This work was carried out during the period 1934–1940, when Minorsky was working for the US Navy. Between 1940 and 1946, Minorsky was special consultant to the Director of the David Taylor Model Basin. After the war, he went to Stanford University where he continued work on the active-tank method.

Up to the 1970s, the US Navy continued to use antiroll tanks for ships that operated at medium-to-high speeds. This was despite evidence of dissatisfaction with the performance from crews (Cox and Lloyd 1977) and performance inferior to that of ships of other navies which were fitted with antiroll fins (Kehoe 1973).

3. Moving Weights

At the beginning of the twentieth century, stabilization schemes using moving weights were frequently proposed as an alternative to the passive-tank system, largely on the grounds that by using ballast of a higher density, less space would be occupied. Few systems were tried, possibly because of the inherent danger of a large mass of metal moving about inside a ship. One scheme was that of Cremieu, who reported in 1910 the results of a trial of his system. This consisted of a carriage, weighing about 10 t, mounted inside the ship on a curved track with the center of curvature of the track 7 m vertically above the metacenter of the ship.

The track and carriage were all encased in a steel enclosure which was filled with a mixture of glycerine and water; a clearance of 4 cm was left between the carriage and the casing. The purpose of the liquid was to dissipate a proportion of the kinetic energy and the clearance was calculated to obtain a particular frequency of oscillation. Unfortunately for Cremieu (who was reported to be a bad sailor), he miscalculated the clearance required; on the trial run in heavy seas, the carriage settled to one side of the ship or the other causing a 15° list, and periodic shocks were reported as the wave motion moved it to the other side (Frahm 1911).

An active moving-weight scheme was tested by Sir John Thorneycroft in 1892 and in 1929 C. P. Norden was granted a US patent for a similar scheme.

4. Fins

The use of projecting fins, either in the form of keels or fins attached to the side of the hull, can increase the damping of rolling motion of a ship. However, the size of the fins necessary to have a significant effect on the rolling motion is such as to make passive fins impractical. There were some early attempts to use movable fins, most notably by Wilson in 1890 in the UK and by S. Motora in Japan at the beginning of the twentieth century. The major development, however, was between the two world wars, when Brown Brothers and Co., William Denny and Bros. and the Admiralty

Research Laboratory in the UK collaborated on the design and testing of an active fin system.

The active fin system works on the principle of the aircraft wing; there is a forward part of the fin which is fixed and a rear part which can be moved and acts in a similar fashion to an aileron. By making use of the hydrodynamic forces generated by the movement of the fin through the water, the size of the fin can be kept small. The disadvantage of this system is that there is almost no stabilizing effect when the ship is stationary or moving at a slow speed. It has been found that the active fin system works best on small high-speed ships and it has become widely used.

The Denny–Brown system developed during the 1930s first used on–off control, based on the use of a gyroscope, to detect the direction of roll. It was later modified to use both a position gyroscope and a rate gyroscope, the two signals being combined and coupled to the Admiralty Research Laboratory (ARL) oil-hydraulic unit used to operate the fins (Allan 1945). An example of the behavior of the system is shown in Fig. 3.

Chadwick (1955) reviewed the history of the development of stabilizers and presented an analysis based on control systems theory. Attention then turned to detailed analysis of the performance of active fins, to the development of design methods and to methods of predicting the performance under different sea conditions. Investigations into the shape of fins, the relationships between keel shape and fin positioning, cavitation and controller design were carried out. Adjustable controllers such as the Muirhead Multa which gave a range of switchable settings for the fin controller parameters were introduced. The Admiralty Engineering Laboratory, UK, introduced an adaptive system in which the controller gain varied inversely with the ship speed.

5. Development

The pattern of development of stabilization systems is typical of many control systems during the first part of the twentieth century: initially, attempts were made to use inherent characteristics to obtain control, then on–off controllers were applied and, finally, continuous control was used. It also shows the importance of practice: Minorsky was one of the few people who understood the dynamics of both the ship and the control systems, yet the successful schemes were the

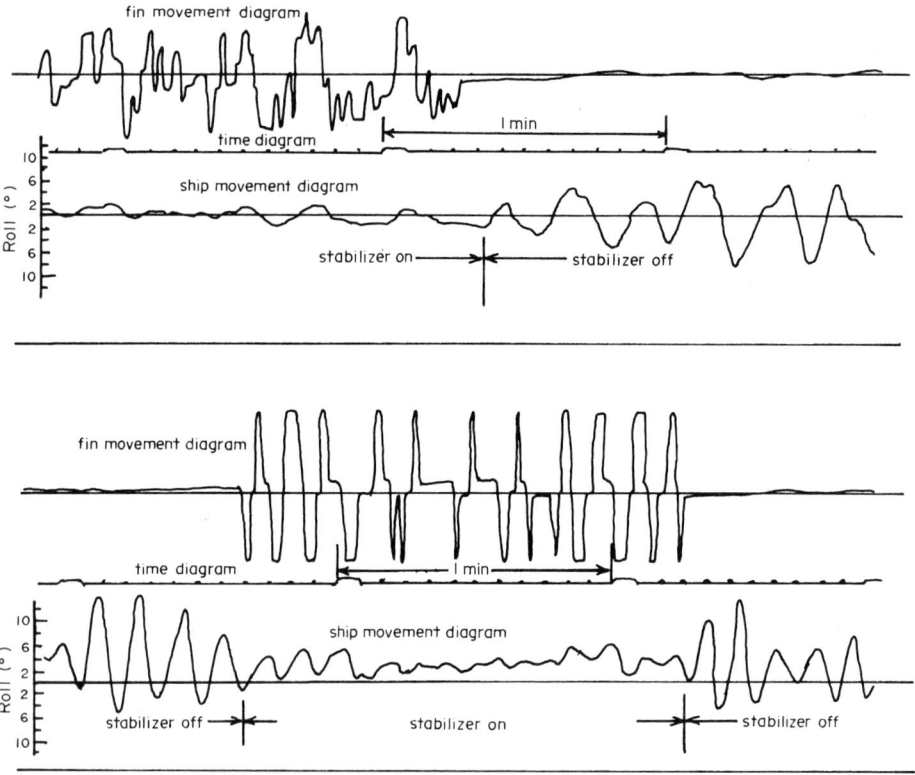

Figure 3
Stabilization by the use of active fin system (after Allan 1945)

Denny–Brown and Sperry systems, largely because their control systems worked. In both cases, "anticipation" was included in the controller on the basis not of theoretical arguments, but of an intuitive understanding.

Gradually the position changed. Chadwick recast the problem in systems terms and began using simulation techniques. During the 1960s and 1970s, staff at the Admiralty Experiment Works, Gosport, UK, carried out intensive investigations into hydrodynamic action. They developed design methods concerned with predicting unstabilized roll motion, bilge keel size effects antiroll fin controller gain effects, antiroll fin size effects, inception of leading-edge cavitation on antiroll fins and open-loop stability of ship–fin systems.

In the late 1970s, the staff of the David W. Taylor Naval Ship Research and Development Center, USA, carried out a similar program to upgrade methods of design and performance assessment of antiroll tanks.

See also: Ship Automation and Control; Ship Rudder Roll Stabilization; Social Issues of Ship Automation

Bibliography

Allan J F 1945 The stabilization of ships by activated fins. *Trans. Inst. Nav. Archit.* **87**, 123–59
Bennett S 1979 *A History of Control Engineering 1800–1930*. Peregrinus, Stevenage, UK
Chadwick J H 1955 On the stabilization of roll. *Trans. Soc. Nav. Archit. Mar. Eng.* **63**, 234–80
Cox G G, Lloyd A R 1977 Hydrodynamic design basis for navy ship roll motion stabilization. *Trans. Soc. Nav. Archit. Mar. Eng.* **85**, 51–93
Flügge-Lotz I 1971 Memorial to N Minorsky. *IEEE Trans. Autom. Control* **16**, 289–91
Frahm H 1911 Results of trials of the anti-rolling tanks at sea. *Trans. Inst. Nav. Archit.* **53**, 183–216
Hughes T P 1971 *Elmer Sperry: Inventor and Engineer*. Johns Hopkins, Baltimore, MD
Kehoe J W Jr 1973 Destroyer sea keeping, US and USSR. *Nav. Eng. J.* **85**(6), 13–23
Minorsky N 1941 Note on the angular motion of ships. *Trans. ASME Ser. A* **63**, 111–20
Minorsky N 1947 Experiments with activated tanks. *Trans. ASME Ser. A* **69**, 735–47

S. Bennett
[University of Sheffield, Sheffield, UK]

Ship Steering: Model-Reference Adaptive Control

The application of model-reference adaptive control (MRAS) to the automatic steering of ships is of interest because it simplifies controller adjustment and decreases fuel costs. Algorithms can be derived for direct adaptation of the controller gains applicable after set-point changes, as well as for identification and adaptive state estimation to be used when the input is constant. These algorithms are based on simple mathematical models of the process and disturbances and on criteria for optimal steering.

A simple model that describes the transfer between rudder angle δ and rate of turn $\dot{\psi}$ is

$$\tau^*\left(\frac{L}{U}\right)\ddot{\psi} + \dot{\psi} = K^*\left(\frac{U}{L}\right)\delta \quad (1)$$

where $\dot{\psi} = d\psi/dt$, ψ is the heading of the ship, K^* and τ^* are constants (with respect to speed variations) in the region of 0.5–2.0, U is the speed of the ship and L is its length.

The rudder is actuated by a hydraulic steering machine that has nonlinear dynamics: the rudder angle and the rudder speed are limited. Compared with the limited rudder speed, other lags of the steering machine are negligible and may be disregarded for the controller design.

Disturbances that play a role in heading control are wind and waves. Wind can be taken account of in the model of Eqn. (1) by adding the slowly varying moment of the wind K_w to the moment excited by the rudder. The moments caused by the waves will be considered as (high-frequency) noise, added to the desired movements caused by the rudder.

1. Steering Criteria

During course changing, the "optimal" performance can be defined as a step response in the time domain. Three phases may be distinguished: start of the turn, stationary turning and end of the turn. The turn should have a quick start and should stop without overshoot of the heading. From the user's point of view, only the stationary rate of turn should be adjustable; all the conventional settings should be automatically adjusted.

During course keeping, the controller should be adjustable between maximum accuracy and maximum fuel economy. Maximum accuracy can be obtained by selecting a high-gain controller. A commonly used criterion for the minimization of fuel cost is

$$J = \int_0^T (\varepsilon^2 + \lambda \delta^2)\, dt \quad (2)$$

where ε is the heading error and λ is a weighting factor.

If the steering-machine dynamics are neglected, a state feedback controller for the process of Eqn. (1) is described by

$$\delta = K_p\varepsilon - K_d\dot{\psi} + K_i \quad (3)$$

K_i has been added to compensate for the moment of the wind K_w. If the sum of K_i and K_w is assumed to be zero, the optimal feedback gains can be straightforwardly computed:

$$K_p = \lambda^{-1/2} \quad (4)$$

$$K_d = \frac{L}{K^* U}\left[\left(1 + 2\frac{K^*\tau^*}{\lambda^{1/2}}\right)^{1/2} - 1\right] \quad (5)$$

K_i can be separately computed by simply taking the mean of the rudder angle necessary for compensation of the disturbance moment.

In the literature there is no consensus about the value of λ. Van Amerongen (1984) suggests the following extremes: $\lambda = 0.1$ for accurate steering, smooth conditions and large ships, and $\lambda = 4$ for small ships in high-sea conditions. Optimal course keeping is obtained by applying the controller gains of Eqns. (4) and (5) and by designing a low-pass filter which suppresses useless high-frequency rudder motions caused by the waves.

2. Course-Changing Controller

During course changing, the optimal performance is defined as a step response with constant slope (rate-of-turn control). Such a response is obtained when the heading control system itself is preceded by a series model which modifies the heading reference. The series model is a second-order system with an adjustable maximum rate of turn. The desired response will be followed when high controller gains are selected. In practice, the controller gains should be carefully turned to their maximum allowable values, for reasons of stability. To compensate for variations in the speed of the ship, Eqn. (5) may be used. Other parameter variations can be dealt with by applying a second, parallel reference model.

A straightforward design of MRAS based on stability theory requires that the process model and the parallel reference model be linear and of the same order and structure. For nonlinear process dynamics, a stable MRAS can also be designed, but the proposed algorithms are too sensitive to small structural differences between the process model and the reference model. It can be shown that the series model can be modified in such a way that it generates only inputs for the course control system that do not saturate the steering machine, which forms the dominant nonlinearity of the process. The process is then linearized and the design of a stable adaptive controller becomes straightforward (Landau 1979). This yields the adjustment laws for the controller gains:

$$dK_p/dt = \beta(p_{12}e + p_{22}\dot{e})\varepsilon \quad (6)$$

$$dK_d/dt = -\alpha(p_{12}e + p_{22}\dot{e})\dot{\psi} \quad (7)$$

$$dK_i/dt = \gamma(p_{12}e + p_{22}\dot{e}) \quad (8)$$

where e is defined as $e = \psi_m - \psi$ and $\dot{e} = \dot{\psi}_m - \dot{\psi}$; α, β and γ are "arbitrary" positive constants and p_{12} and p_{22} are elements of the matrix **P**. **P** can be solved from

$$\mathbf{A}_m^T \mathbf{P} + \mathbf{P}\mathbf{A}_m = -\mathbf{Q} \quad (9)$$

where **Q** is an arbitrary positive-definite matrix and **A** is the system matrix of the reference model. By computing K_i during course changing in an adaptive manner according to Eqn. (8), it is not necessary to stop the integration during course changing, which is common practice in conventional autopilots. To prevent the controller gains from drifting in the presence of noise, the concept of decreasing adaptive gains can be applied and the adaptation can be switched off entirely for a certain period of time after a set point change.

3. Course-Keeping Controller

Optimal course keeping is achieved by optimizing the criterion of Eqn. (2) (which requires that the process parameters be known) and by filtering noisy signals. For identification and state estimation, MRAS can also be applied. A simple first-order adjustable mode is placed parallel with the transfer between δ and $\dot{\psi}$:

$$\tau_m \ddot{\psi}_m + \dot{\psi}_m = K_m(\delta - K_{i,m}) \quad (10)$$

where $K_{i,m}$ is the rudder offset. Defining $e = \dot{\psi}_m - \dot{\psi}$ yields the simple adjustment laws

$$d(K_m/\tau_m)/dt = -\beta e(\delta - K_{i,m}) \quad (11)$$

$$d(1/\tau_m)/dt = \alpha e \dot{\psi}_m \quad (12)$$

$$d(K_{i,m})/dt = -\gamma e \quad (13)$$

Besides unbiased estimates of the process parameters, the adjustable model also produces a noise-free estimate of the actual rate-of-turn signal, thus simultaneously solving the filtering problem. However, this filter is not the best possible. When the noise level is low, it is not necessary to rely on the output of the adjustable model alone. The prediction may be updated, based on the measurements. So as not to influence the identification process, a second adjustable model can be introduced whose parameters are adjusted simultaneously with the first model (see Fig. 1). The output of the second model is updated with the measurements.

The weighting between prediction and measurements is determined by the relation between the low-frequency components of the error signal, which should not be filtered, and the high-frequency components, which should be suppressed (van Amerongen 1984). This filter does not distinguish between system noise and observation noise (as a Kalman filter does), but between the low-frequency and high-frequency components of the system noise. Observation noise is assumed to be completely absent.

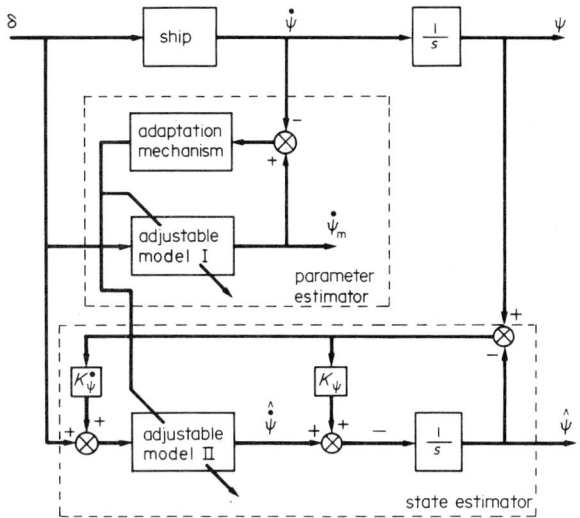

Figure 1
Block diagram of two-model MRAS controller for course keeping

4. Test Results

The algorithms given previously are continuous-time algorithms, which can easily be combined with the proposed gain scheduling. However, practical realization is made more robust and simple by using a digital computer. Full-scale tests on various ships as well as experiments with a scale model have demonstrated the practical usefulness of the autopilot designed. The advanced filter algorithm, combining MRAS with the concepts of Kalman filtering, provides a major contribution to improved fuel economy. During full-scale trials, the increase in speed has been shown to be 0.5–1.5%. This leads to fuel savings of 1–3%. During model tests, in which the speed of the ship could be corrected for variations in the thrust, increases in speed of up to 5% were demonstrated.

See also: Marine Propulsion Plants: Control; Navigation Control of Ships; Ship Automation and Control; Ship Positioning: Adaptive Control

Bibliography

Landau I D 1979 *Adaptive Control—The Model Reference Approach*. Dekker, New York
van Amerongen J 1984 Adaptive steering of ships—A model reference approach. *Automatica* **20**, 3–14
van Amerongen J, Udink ten Cate A J 1975 Model reference adaptive autopilots for ships. *Automatica* **11**, 441–9

J. van Amerongen
[University of Twente, Enschede,
The Netherlands]

Shortest-Path Algorithms

Shortest-path algorithms are classical problems of combinatorial optimization and have widespread applications. To get a feel for these problems, some typical situations that can be handled by shortest-path algorithms will first be described.

EXAMPLE 1
Consider a network of freeways. Every freeway entry, exit or crossing is called a node and the freeway between two adjacent nodes is modelled as an arc. A weighting can be attached to every arc; for example, the average time needed to pass this arc. The obvious question to ask is "what is the fastest way from entry (node) A to exit (node) B?" Considering the weight of an arc to be the length of the corresponding freeway section, either the shortest way from A to B or all shortest paths between any pair (A, B) of different nodes in the network can be determined.

Questions concerning shortest or fastest ways are not the only ones that can be dealt with by shortest-path problems. Many other situations can also be modelled in this way.

EXAMPLE 2
Consider a communications network. The arcs represent the communication links and the weight of an arc is a measure for the probability of no-breakdown. The most reliable path from a given start node A to a given destination can be found by looking for a path such that the product of weights along this path is as large as possible.

EXAMPLE 3
A tourist travelling to North America has an air ticket that is valid for five arbitrary links on a certain airline. Which airports can be reached when the tourist decides to travel only by plane and when can the tourist return home from any desired place?

EXAMPLE 4
A heavy transport should bring a turbine to a new power station in the mountains. This turbine is not only very heavy so that it cannot pass certain bridges, but is has also a huge size so that narrow passages must be avoided. On which route can this turbine be transported?

Example 4 is a so-called bottleneck problem, where a path from A to B is sought such that the minimum arc weight along the path is as large as possible.

The following sections will describe how these situations can be modelled by weighted (directed) graphs. Furthermore, some basic questions related to shortest paths will be formulated.

For more extensive surveys on shortest-path problems, the reader is referred to Gallo and Pallottino (1988) as well as to Ahuja *et al.* (1989). Gallo and Pallottino also give FORTRAN codes for several

Shortest-Path Algorithms

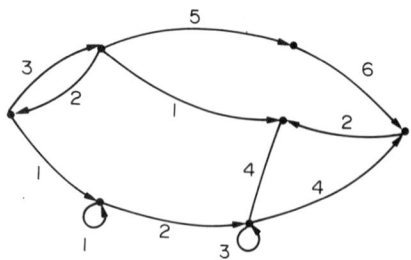

Figure 1
A digraph with two loops

shortest-path algorithms as well as computational comparisons. Ahuja et al. survey shortest-path problems up to 1988 and stress algorithms of good (theoretical) complexity.

1. Formulation of Shortest-Path Problems

A directed graph or digraph $G = (N, A)$ consists of a set N of nodes and a set A of arcs. Every arc (a, b) is a pair of nodes, the node a is called the tail and b is called the head of the arc. An arc (a, a) is called a loop. If weights are attached to the arcs, a weighted digraph $G = (N, A, w)$ is obtained (see Fig. 1).

The concept of an undirected graph is very similar. In an undirected graph, the arcs are not directed and just express the fact that nodes a and b are adjacent or connected by an edge. Obviously, an undirected edge can be replaced by two directed arcs (see Fig. 2). Think, for example, of the two lanes of a highway. Thus, the following deals only with digraphs.

A path in the digraph $G = (N, A)$ is a sequence of arcs $(a_1, a_2, \ldots, a_n)$ such that the head of a_i equals the tail of a_{i+1} for $i = 1, 2, \ldots, n-1$. Therefore, a path is also uniquely determined by the sequence of its nodes. A path is termed simple if no node is repeated. Furthermore, a path is termed a cycle if the tail of a_1 coincides with the head of a_n.

LEMMA 1. *If there is a path from node i to node j there is also a simple path from i to j* (see Fig. 3).

Now consider paths in weighted digraphs $G = (N, A, w)$. Let $p = (a_1, a_2, \ldots, a_n)$ be a path. As length $w(p)$ of p define

$$w(p) = w(a_1) + w(a_2) + \cdots + w(a_n)$$

Different definitions for the length of a path are possible and will be discussed later.

Figure 2
Undirected edge and corresponding directed arcs

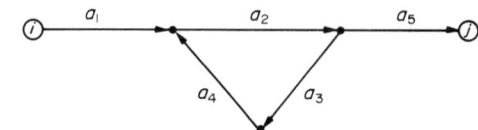

Figure 3
Shortcutting of a path: the path $(a_1, a_2, a_3, a_4, a_2, a_5)$ can be shortcut to the simple path (a_1, a_2, a_5)

The shortest-path problem can now be stated as follows.

SHORTEST PATH PROBLEM. *Let s be a given start node and t be a given terminal node. Find a path p from s to t with minimum length $w(p)$.*

Consider the weighted digraph shown in Fig. 4. Starting from S_1 with the aim of computing the length of a shortest path from S_1 to t, a unique solution (S_1, a, b, t) with length $5 + (-2) + 6 = 9$ is obtained.

Should the same thing be attempted from S_2 to t, a length for the shortest simple path of $2 + 3 + 4 + 6 = 15$ is obtained. However, if not restricted to simple paths, the cycle (c, d, e, c), which has weight -4 (a negative cycle), can be used as often as desired. Thus, the length is found to be

$$2 + k(3 - 5 - 2) + 4 + 6$$

which tends to minus infinity as k tends towards infinity. In this case, the shortest-path problem has no solution, thus Proposition 1 can be stated as follows.

PROPOSITION 1. *There is a shortest path from s to t if and only if no path from s to t contains a negative cycle.*

COROLLARY. *If there is a shortest path from s to t, then there is also a simple shortest path from s to t.*

The corollary follows easily from the fact that, by shortcutting, only the length of the path will be reduced since there is no negative cycle.

There are several variants of shortest-path problems. With respect to the problem structure the following types can be distinguished.

(P1)   *Single-pair problem.* Find the shortest path from source s to sink t.

(P2)   *Single-source problem.* Find the shortest path from source s to all other nodes.

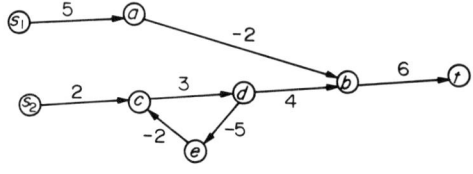

Figure 4
Example for shortest paths

(P3) *Single-sink problem.* Find the shortest paths from every node to sink t.

(P4) *All-pairs problem.* Find the shortest paths between all pairs of nodes.

As already mentioned, different objective functions can also be treated, important examples of which are the following.

(OF1) *Classical shortest-path problem.* Minimize the sum of weights.

(OF2) *Connectivity problem.* Determine whether there is a path from s to t or not.

(OF3) *Longest path problem.* Maximize the sum of weights along a path.

(OF4) *Most reliable path problem.* Maximize the product of weights along a path.

(OF5) *Bottleneck problem.* Find a path such that the maximum weight is as small as possible or the minimum weight is as large as possible.

All problems (OF1)–(OF5) differ only in the definition of the length of a path and in the used order relation (i.e., which path is considered to be shorter). Problem (OF1) refers to the definitions made so far. Problem (OF2) can be modelled with weights 0 and 1; if nodes a and b are joined by an arc in A, this arc (a, b) is given weight 0, otherwise the arc (a, b) is added and weight 1 attached to it. The length of a path p from s to t is again defined as sum of all arc weights. Clearly, if and only if $w(p) = 0$ can t be reached from s. Problem (OF4) uses as weights real numbers between 0 and 1. Here, the length of a path is defined as the product of the arc weights. To minimize the maximum weight (OF5), the weight of a path is defined as the maximum of its arc weights. Finally, when considering (OF3), (OF4) or (OF5) where the minimum weight should be as large as possible, "minimize" is replaced by "maximize" and the desired results are obtained.

2. Dijkstra's Algorithm

One of the fastest algorithms for computing shortest paths is due to Dijkstra (1959). This algorithm determines a shortest path from a single source to one or all terminal nodes under the assumption that all arc weights are nonnegative. In the general setting, this means that for any two weights w_1 and w_2 the composition (classically the sum) of w_1 with w_2 is not shorter than w_1. This implies that the path length increases if an arc is added to the path.

The underlying idea of Dijkstra's algorithm is the following. Let X be a proper subset of the nodes which contains the source. Assume that the shortest distance $d(x)$ is already known from the source to all other nodes x in X. For any node $y \notin X$ define $d(y)$ as the shortest length of those paths that, apart from y, use only nodes in the set X. Now determine $d(\bar{y}) = \min$ $\{d(y) | y \notin X\}$ and add $\bar{y}$ to the set X. The shortest distance of $\bar{y}$ from s is now $d(\bar{y})$. This follows from the fact that every path from s to $\bar{y}$ uses at least one arc (x', y') with $x' \in X$, $y' \notin X$. But then $d(y') \geq d(\bar{y})$. Since there are no arcs with negative weight, the path from s to $\bar{y}$ via y' is at least as long as the direct path from s via nodes in X to $\bar{y}$.

To find a minimum path from s to t, start with $X = \{s\}$ and determine in each step recursively the labels $d(y)$, $y \notin X$. At the beginning, set $d(y) = w(s, y)$, or, if there is no arc (s, y), equal to infinity. As soon as t is added to X a shortest path from s to t has been found. Obviously, this process can be continued until all nodes x belong to X. In this case, a solution of the single-source problem (P2) is obtained. Similarly, interchanging the roles of the source and sinks gives a solution to the single-sink problem (P3).

What remains to be shown is the update of the labels $d(y)$, $y \notin X$. Assume that $d(\bar{y}) = \min \{d(y) | y \notin X\}$. For any $y \notin X$, $y \neq \bar{y}$ two values, must now be compared: $d(y)$, the length of a shortest path to y via node in X, and $d(\bar{y}) + w(\bar{y}, y)$, the length of the shortest path from s to $\bar{y}$ and along the arc $(\bar{y}, y)$ to y. Thus, for all $y \notin X$, $y \neq \bar{y}$: if $d(y) > d(\bar{y}) + w(\bar{y}, y)$, then $d(y)$ has to be replaced by $d(\bar{y}) + w(\bar{y}, y)$.

Since not only the shortest path length, but also the corresponding path is of interest, for any $x \neq s$ a label $p(x)$ is stored which finally will be the predecessor on a shortest path from s to x. If $d(y) > d(\bar{y}) + w(\bar{y}, y)$, then the predecessor $p(y)$ changes to $\bar{y}$. Summarizing, the following algorithm is obtained.

DIJKSTRA'S ALGORITHM
Start:

For $x \in N \setminus \{s\}$: $d(x) := w(s, x)$
$\qquad p(x) := s$
$\qquad X := \{s\}$

Recursion:

Determine $d(\bar{y}) := \min \{d(x) | x \notin X\}$
$\qquad X := X \cup \{\bar{y}\}$
If $t \in X$ then STOP
else for all $y \notin X$
$\qquad$ if $d(y) > d(\bar{y}) + w(\bar{y}, y)$ then $d(y) := d(\bar{y}) + w(\bar{y}, y)$
$\qquad\qquad p(y) := \bar{y}$

Example 5 shall illustrate this method (cf. Fig. 5).

EXAMPLE 5
Determine the shortest paths from s to all other nodes in Fig. 5.

The initial position is given in Table 1, with $X = \{s\}$. The minimum d value is 3 attained for node 4. Thus $X = \{s, 4\}$, $p(4) = s$ and Table 2 is obtained. The minimum d value is now attained for $x = 1$, thus $X = \{s, 4, 1\}$, $p(1) = s$ and an update of Table 2 yields Table 3. Now, min $d(x) = 5$ for $x = 2$ and $X = \{s, 4, 1, 2\}$ with $p(2) = 1$, giving Table 4. The next step yields min $d(x) = 5$ for $x = 5$. Thus $p(5) = 2$ and $X = \{s, 4, 1, 2, 5\}$, giving Table

5. The minimum d value is now 6 attained for node $x = 6$. Thus $X = \{s, 4, 1, 2, 5, 6\}$ and $p(6) = 5$, giving Table 6. Thus, $\min d(x) = 8$ for $x = 3$, $p(3) = 2$ and, finally, $d(7) = 9$ with $p(7) = 4$. The problem has been solved. The solution can be represented by the so-called shortest-path tree (see Fig. 6), taking into account that for every node $x \in N \setminus \{s\}$ the predecessor $p(x)$ on a shortest path from the source s to node x is known. This tree consists of all arcs $(p(x), x)$ for all $N \setminus \{s\}$.

The amount of work needed for Dijkstra's algorithm is $O(|N|^2)$, since at most $n - 1 = |N| - 1$ recursion steps must be performed and every recursion step needs $O(|N|)$ arithmetic operations.

For graphs with $O(|N| \log |N|)$ arcs only, Dijkstra's algorithm can be speeded up by using advanced data structures (e.g., Tarjan 1983, Fredman and Tarjan 1984).

Dijkstra's algorithm can easily be modified to solve maximum reliability problems with objective function (OF4) or bottleneck problems with objective function (OF5). The essential change necessary concerns the update of d labels.

For bottleneck problems of the form minimize the maximum distance, use

$$\text{if } d(y) > w(\bar{y}, y) \text{ then } d(y) := \max (d(\bar{y}), w(\bar{y}, y))$$

To solve bottleneck problems of the form maximize the minimum distance or the maximum reliability problem, the order must be reversed. Thus, the recursion step of the algorithm becomes

Recursion:

Determine $d(\bar{y}) := \max \{d(x) | x \notin X\}$
$\qquad X := X \cup \{\bar{y}\}$
If $t \in X$, then STOP
Else for all $y \notin X$
$\quad$ if $d(y) < w(\bar{y}, y)$ then $d(y) := \min \{d(\bar{y}), w(\bar{y}, y)\}$
$\qquad p(y) := \bar{y}$

For reliability problems, use

$$\text{if } d(y) < d(\bar{y}) w(\bar{y}, y), \text{ then } d(y) := d(\bar{y}) w(\bar{y}, y)$$

Table 1
Step 1 of Dijkstra's algorithm for Fig. 5

Node $x \in X$	1	2	3	4	5	6	7
$d(x)$	4	∞	∞	3	∞	∞	∞
$p(x)$	s	s	s	s	s	s	s

Table 2
Step 2 of Dijkstra's algorithm for Fig. 5

Node $x \in X$	1	2	3	5	6	7
$d(x)$	4	7	∞	6	∞	9
$p(x)$	s	4	s	4	s	4

Table 3
Step 3 of Dijkstra's algorithm for Fig. 5

Node $x \in X$	2	3	5	6	7
$d(x)$	5	∞	6	∞	9
$p(x)$	1	s	4	s	4

Table 4
Step 4 of Dijkstra's algorithm for Fig. 5

Node $x \in X$	3	5	6	7
$d(x)$	8	5	14	9
$p(x)$	2	2	2	4

Table 5
Step 5 of Dijkstra's algorithm for Fig. 5

Node $x \in X$	3	6	7
$d(x)$	8	6	9
$p(x)$	2	5	4

Table 6
Step 6 of Dijkstra's algorithm for Fig. 5

Node $x \in X$	3	7
$d(x)$	8	9
$p(x)$	2	4

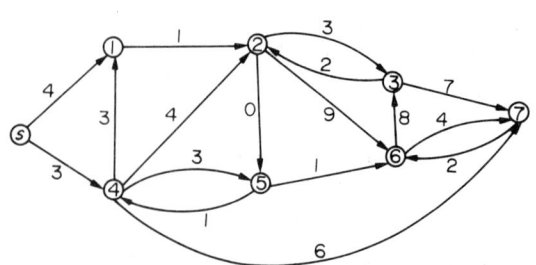

Figure 5
Example for Dijkstra's algorithm

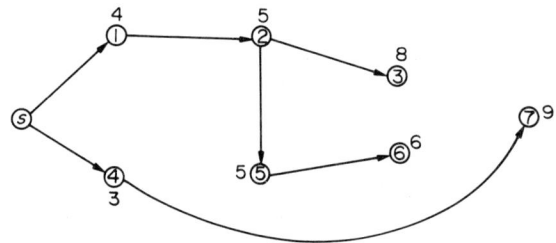

Figure 6
Shortest paths from s to all other nodes: the numbers attached to the nodes represent their shortest distance from the source s

In the derivation of Dijkstra's algorithm, use was made of the fact that all arc-weights are nonnegative. Dijkstra's algorithm does not yield correct answers in the presence of negative arc weights. In the next section another method is described that determines shortest paths, even if some arcs have negative length.

3. Negative Arc Weights

At a first glance it might appear strange to allow negative arc weights. However, consider the problem of finding a longest path in a digraph without cycles (acyclic digraph) and with positive arc lengths. This problem arises, for example, in connection with determining a critical path in a project network—the arcs of the project network correspond to activities that should be carried out in the order described by the underlying digraph. Every activity has a certain duration, the length of the corresponding arc in the digraph. The longest path (the critical path) determines the minimal time needed to carry out all activities and therefore provides a lower bound for the duration of the project.

The problem of finding a longest path can be reduced to the problem of finding a shortest path in the same digraph by taking all arc weights negative. Thus a shortest path in an acyclic digraph with negative weights has to be found. According to Proposition 1, there exists an optimal solution, since the digraph does not contain a cycle. However, Dijkstra's algorithm fails in such a situation. The following method due to Bellman (1958) and Moore (1959) does solve shortest-path problems in the presence of negative arc lengths. Its complexity $O(|A||N|)$ or $O(|N|^3)$, however, is worse than that of Dijkstra's algorithm.

Let $G = (N, A, w)$ be a weighted digraph where the arc lengths $w(x, y)$ may be positive, zero or negative. Denote by $d(x)$ the length of the shortest path from source s to x found so far, and by $p(x)$ the immediate predecessor of x on such a path. All possible paths in the sequence are investigated according to their number of arcs. Thus, the first pass deals with the arcs leaving the source.

BELLMAN–MOORE ALGORITHM
Start (Pass 1):

For all $(s, x) \in A$: $d(x) := w(s, x)$
$\qquad p(x) := s$
$\qquad$ store node x in Q_1
If there is no arc from s to x define
$\qquad d(x) := \infty$
$\qquad p(x) := s$

Now continue with paths using two arcs. So, the nodes that can be reached via two arcs from the source must be found. Obviously, only the successors of nodes in Q_1 need be considered. This leads to the following general recursion.

Recursion (Pass $i + 1$):

For all $x \in Q_i$ do
$\qquad$ for any $(x, y) \in A$:
$\qquad\qquad$ if $d(y) > d(x) + w(x, y)$ then $d(y) := d(x) + w(x, y)$
$\qquad\qquad\qquad p(y) := x$
$\qquad\qquad\qquad$ if y is not in Q_{i+1} store
$\qquad\qquad\qquad$ it there

So in pass i the shortest paths are obtained from s to node x (with $d(x) < \infty$) using, at most, i arcs. Since only simple paths with at most $|N| - 1$ arcs need be considered, the procedure will be finished after at most $|N| - 1$ passes. The computational amount for every pass is at most $O(|A|)$. This yields an overall complexity of $O(|N||A|)$. Example 6 shall illustrate this approach (see Fig. 7).

EXAMPLE 6
Find a shortest path from the source s to the sink t in the digraph shown by Fig. 7.

The initial position is given by Table 7 and $Q_1 = \{1, t\}$. Step 2 yields Table 8 and $Q_2 = \{2, 4\}$. Step 3 yields Table 9 and $Q_3 = \{3, t\}$. The next iteration yields the shortest paths using at most four arcs, shown in Table 10 and $Q_4 = \{4\}$. Now the fifth and last pass gives Table 11.

Table 11 shows that the shortest path from s to t has length 3. Starting from t, iteratively, to the predecessors, the following nodes are obtained:

$$t, p(t) = 4, \quad p(4) = 3, \quad p(3) = 2, \quad p(2) = 1, \quad p(1) = s$$

Thus, the optimal path with length 3 is $(s, 1, 2, 3, 4, t)$.

Applying Dijkstra's algorithm to this example gives the path $(s, 1, 4, t)$ with length 5, which obviously is not optimal.

4. Matrix Algorithm

Up to now only the single-source problem has been considered. If the solution to the all-pairs problem is being sought, it would be possible to solve single-pair problems for any node as source node, but this would blow up the complexity. Therefore, an algorithm due to Floyd (1962) is derived in this section that solves the all-pairs problems in $O(|N|^3)$ steps.

Assume that the underlying weighted digraph (N, A, w) has n nodes and does not contain negative cycles. In this case, the shortest path between any two nodes is attained by a simple path using at most $n - 2$ intermediate nodes. The length of the shortest paths between all pairs of nodes is stored in an $n \times n$ matrix $\mathbf{D} = (d_{ik})$ where d_{ik} is the length of a shortest path between node i and node k, or equal to infinity if k cannot be reached from i. Since the construction of the shortest paths is of interest a predecessor matrix $\mathbf{P} = (p_{ik})$ is also introduced. The meaning of $\mathbf{P}$ will be explained in the algorithm.

Begin with the matrices $\mathbf{P} = (p_{ik}) = \mathbf{0}$ and $\mathbf{D}(0) = (w_{ik})$, where w_{ik} is the length of the arc (i, k) or infinity if this arc is not present. Additionally, define $w_{ii} = 0$ for

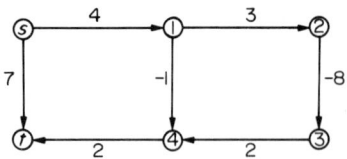

Figure 7
Example for the Bellman–Moore algorithm

Table 7
Pass 1 of the Bellman–Moore algorithm for Fig. 7

Node $x \in X$	1	2	3	4	t
$d(x)$	4	∞	∞	∞	7
$p(x)$	s	s	s	s	s

Table 8
Pass 2 of the Bellman–Moore algorithm for Fig. 7

Node $x \in X$	1	2	3	4	t
$d(x)$	4	7	∞	3	7
$p(x)$	s	1	s	1	s

Table 9
Pass 3 of the Bellman–Moore algorithm for Fig. 7

Node $x \in X$	1	2	3	4	t
$d(x)$	4	7	-1	3	5
$p(x)$	s	1	2	1	4

Table 10
Pass 4 of the Bellman–Moore algorithm for Fig. 7

Node $x \in X$	1	2	3	4	t
$d(x)$	4	7	-1	1	5
$p(x)$	s	1	2	3	4

Table 11
Pass 5 of the Bellman–Moore algorithm for Fig. 7

Node $x \in X$	1	2	3	4	t
$d(x)$	4	7	-1	1	5
$p(x)$	s	1	2	3	4

all $i = 1, 2, \ldots, n$. Next, compute $\mathbf{D}(1) = (d_{ik}(1))$. Here, $d_{ik}(1)$ is the minimum of the arc length w_{ik} and of the length of the path from i to node 1 and from there to node k (cf. Fig. 8). If this minimum is attained for

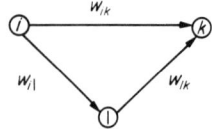

Figure 8
Recursion in the matrix algorithm

the path via node 1, store $p_{ik} = 1$. Thus

If $w_{ik} \leq w_{i1} + w_{1k}$, then $d_{ik}(1) := w_{ik}$
else $d_{ik}(1) := w_{i1} + w_{1k}$
$\quad p_{ik} := 1$

Now continue, inserting the nodes $2, 3, \ldots, l$. Matrix $\mathbf{D}(l) = (d_{ik}(l))$ has the following meaning: $d_{ik}(l)$ is the length of the shortest path from i to k using only nodes from $\{1, 2, \ldots, l\}$ as intermediate nodes. Inserting node $l + 1$ the following recursion is obtained:

If $d_{ik}(l) \leq d_{i,l+1}(l) + d_{l+1,k}(l)$, then $d_{ik}(l+1) := d_{ik}(l)$
else $d_{ik}(l+1) := d_{i,l+1}(l) + d_{l+1,k}(l)$
$\quad p_{ik} := l + 1$

The process stops after finishing the step for $l = |N| - 1$. Thus, n iterations have to be performed. The amount of work for one iteration l to $l + 1$ is $O(|N|^2)$. This yields a total complexity of $O(|N|^3)$.

The element $d_{ik}(n)$ is the length of a shortest path between node i and node k. To find this path, matrix $\mathbf{P}$ is used. The element $p_{ik} = l$ indicates that the shortest path from i to k uses node l as intermediate node. Next, turn to p_{il} and p_{lk}. Element p_{il} indicates an intermediate node on the shortest path from i to l, and p_{lk} indicates a node on the path from l to k. By continuing until $p_{rs} = 0$ the whole path can be constructed. Element $p_{rs} = 0$ indicates that there is no intermediate node between nodes r and s, and r and s are directly linked by an arc. Summarizing, the following algorithm is obtained.

FLOYD'S ALGORITHM
Initialization:
For $i := 1$ to n do
$\quad$ For $k := 1$ to n do
$\quad\quad d_{ik}(0) := w_{ik}$
$\quad\quad p_{ik} := 0$

Recursion:
For $l := 0$ to $n - 1$ do
$\quad$ For $i := 1$ to n do
$\quad\quad$ For $k := 1$ to n do
$\quad\quad\quad$ If $d_{ik}(l) \leq d_{i,l+1}(l) + d_{l+1,k}(l)$,
$\quad\quad\quad$ then $d_{ik}(l+1) := d_{ik}(l)$
$\quad\quad\quad$ else $d_{ik}(l+1) := d_{i,l+1}(l) + d_{l+1,k}(l)$
$\quad\quad\quad\quad p_{ik} := l + 1$

The following example shall illustrate this approach (see Fig. 9).

$$\mathbf{D}(0) = \begin{pmatrix} 0 & 4 & 2 & 8 & \infty \\ 7 & 0 & \infty & \infty & \infty \\ \infty & 1 & 0 & 2 & 7 \\ 3 & \infty & 1 & 0 & 9 \\ \infty & 8 & 3 & 1 & 0 \end{pmatrix}, \quad \mathbf{P} = \begin{pmatrix} 0 & 0 & 0 & 0 & 0 \\ 0 & 0 & 0 & 0 & 0 \\ 0 & 0 & 0 & 0 & 0 \\ 0 & 0 & 0 & 0 & 0 \\ 0 & 0 & 0 & 0 & 0 \end{pmatrix}$$

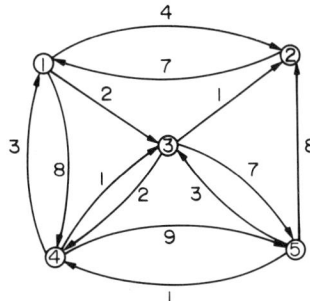

Figure 9
Example for Floyd's algorithm

In the first iteration, insert node 1 and compare the direct arc lengths with the paths via node 1. The following improvements are obtained:

$$d_{23}(1) = 9, \quad d_{24}(1) = 15, \quad d_{42}(1) = 7$$

This leads to

$$\mathbf{D}(1) = \begin{pmatrix} 0 & 4 & 2 & 8 & \infty \\ 7 & 0 & 9 & 15 & \infty \\ \infty & 1 & 0 & 2 & 7 \\ 3 & 7 & 1 & 0 & 9 \\ \infty & 8 & 3 & 1 & 0 \end{pmatrix}, \quad \mathbf{P} = \begin{pmatrix} 0 & 0 & 0 & 0 & 0 \\ 0 & 0 & 1 & 1 & 0 \\ 0 & 0 & 0 & 0 & 0 \\ 0 & 1 & 0 & 0 & 0 \\ 0 & 0 & 0 & 0 & 0 \end{pmatrix}$$

Next, compare the shortest paths of $\mathbf{D}(1)$ with those also using node 2 as intermediate node. The following improvements are obtained.

$$d_{31}(2) = 8, \quad d_{51}(2) = 15$$

This leads to

$$\mathbf{D}(2) = \begin{pmatrix} 0 & 4 & 2 & 8 & \infty \\ 7 & 0 & 9 & 15 & \infty \\ 8 & 1 & 0 & 2 & 7 \\ 3 & 7 & 1 & 0 & 9 \\ 15 & 8 & 3 & 1 & 0 \end{pmatrix}, \quad \mathbf{P} = \begin{pmatrix} 0 & 0 & 0 & 0 & 0 \\ 0 & 0 & 1 & 1 & 0 \\ 2 & 0 & 0 & 0 & 0 \\ 0 & 1 & 0 & 0 & 0 \\ 2 & 0 & 0 & 0 & 0 \end{pmatrix}$$

Now insert node 3. The following improvements are obtained.

$$d_{12}(3) = 3, \quad d_{14}(3) = 4, \quad d_{15}(3) = 9$$
$$d_{24}(3) = 11, \quad d_{25}(3) = 16$$
$$d_{42}(3) = 2, \quad d_{45}(3) = 8$$
$$d_{51}(3) = 11, \quad d_{52}(3) = 4$$

This gives

$$\mathbf{D}(3) = \begin{pmatrix} 0 & 3 & 2 & 4 & 9 \\ 7 & 0 & 9 & 11 & 16 \\ 8 & 1 & 0 & 2 & 7 \\ 3 & 2 & 1 & 0 & 8 \\ 11 & 4 & 3 & 1 & 0 \end{pmatrix}, \quad \mathbf{P} = \begin{pmatrix} 0 & 3 & 0 & 3 & 3 \\ 0 & 0 & 1 & 3 & 3 \\ 2 & 0 & 0 & 0 & 0 \\ 0 & 3 & 0 & 0 & 3 \\ 3 & 3 & 0 & 0 & 0 \end{pmatrix}$$

Next, insert node 4 which leads to the following improvements:

$$d_{31}(4) = 5$$
$$d_{51}(4) = 4, \quad d_{52}(4) = 3, \quad d_{53}(4) = 2$$

This gives

$$\mathbf{D}(4) = \begin{pmatrix} 0 & 3 & 2 & 4 & 9 \\ 7 & 0 & 9 & 11 & 16 \\ 5 & 1 & 0 & 2 & 7 \\ 3 & 2 & 1 & 0 & 8 \\ 4 & 3 & 2 & 1 & 0 \end{pmatrix}, \quad \mathbf{P} = \begin{pmatrix} 0 & 3 & 0 & 3 & 3 \\ 0 & 0 & 1 & 3 & 3 \\ 4 & 0 & 0 & 0 & 0 \\ 0 & 3 & 0 & 0 & 3 \\ 4 & 4 & 4 & 0 & 0 \end{pmatrix}$$

In the last step, inserting node 5 produces no further improvement. Thus, $\mathbf{D} = \mathbf{D}(4)$ is the matrix of the shortest path lengths.

In order to construct now the shortest path, say from node 5 to node 2, matrix $\mathbf{P}$ is used. Since $p(5, 2) = 4$, node 4 is an intermediate node on this path. Now $p(5, 4) = 0$ which says to go directly from node 5 to node 4. Furthermore, $p(4, 2) = 3$ says that from node 4 to node 2, pass node 3. Since $p(4, 3) = p(3, 2) = 0$ the shortest path is found to be $(5, 4, 3, 2)$.

There are also other algorithms for the all pairs problem that have a structure very similar to matrix multiplication. This can be exploited in parallel algorithms for these problems. The interested reader is referred to Rote (1985).

See also: Queuing Theory Applications; Route Guidance, Individual

Bibliography

Ahuja R K, Magnanti T L, Orlin J B 1989 Network flows. In: Nemhauser G L, Rinnooy Kan A H G, Todd M J (eds.) *Optimization*, Handbooks in Operations Research and Management Science, Vol. 1. North-Holland, Amsterdam, pp. 211–369
Bellman R E 1958 On a routing problem. *Q. Appl. Math.* **16**, 87–90
Dijkstra E W 1959 A note on two problems in connection with graphs. *Numer. Math.* **1**, 269–71
Floyd R W 1962 Algorithm 97 shortest path. *Commun. ACM* **5**, 345
Fredman M L, Tarjan R E 1984 Fibonacci heaps and their use in improved network optimization algorithms. *Proc. 25th Annual IEEE Symp. Foundations of Computing FOCS-1984*. Institute of Electrical and Electronics Engineers, New York, pp. 338–46

Gallo G, Pallottino S 1988 Shortest path algorithms. In: Simeone B, Toth P, Gallo G, Pallottino S (eds.) *FORTRAN Codes for Network Optimization*, Annals of Operations Research, Vol. 13. Baltzer, Basel, Switzerland, pp. 3–79

Moore E F 1959 the shortest path through a maze. *Proc. Int. Symp. Theory of Switching*, pt. II, April 2–5, 1957, The Annals of the Computation Laboratory of Harvard University, vol. 30. Harvard University Press, Cambridge, MA

Rote G 1985 A systolic array algorithm for the algebraic path problem. *Computing* **34**, 191–219

Tarjan R E 1983 *Data Structures and Network Algorithms*. SIAM, Philadelphia, PA

R. E. Burkard
[Technische Universität Graz, Graz, Austria]

Signal Control and Traffic Assignment

Control strategies for traffic signal installations have been well researched over the years. Since Webster's first experiments in the late 1950s, numerous strategies have been proposed and developed. Initially, these strategies were mainly concerned with the optimization of green times at isolated intersections. With progress in technology, an approximate method for area-wide optimization of green splits, cycle time and offsets was realized with TRANSYT; later developments, such as SCOOT, even allow on-line control, whereby signal settings can be altered directly whenever changing traffic conditions so require (see *Road Traffic Control: TRANSYT and SCOOT*). Much of the work in this area has a practical purpose, but theoretical work has also been carried out.

However, traffic assignment or route choice has mainly attracted academic attention. Since Wardrop stated his two equilibrium principles, a considerable amount of work has been carried out on the development of solution algorithms to the capacity restraint assignment problem. A thorough understanding of this problem has been achieved and equilibrium assignment has increasingly become the standard in transport network modelling (see *Traffic Assignment*).

As a rule, signal control optimization has been concerned with static traffic conditions. Flows are assumed to be fixed and signal timings are changed to best accommodate these flows. Even modern on-line control accepts the current flow pattern as an externally fixed input.

Traffic assignment has also mainly been applied to fixed conditions. Network characteristics, such as capacities, free flow speeds and signal settings, are considered to be given and the flow is distributed over the network in adjustment to these conditions. This is termed a steady-state model.

It should be recognized, however, that there is a mutual influence between signal settings and traffic assignment. In an urban network, traffic signals impose a considerable delay on traffic passing the junction and they may well be the main determinants in drivers' route choices. The delays at signal-controlled junctions differ for the various movements, according to, for example, the green time for that particular movement and the turning volume. A change in green times will alter the delay experienced by the various turning movements and, as a result, the route choice of the drivers passing the junction might well change. However, the optimality of signal settings depends on the flows passing through the junction. If a change in routes takes place, hitherto optimal signal timings may no longer be the best setting and alterations may be required. This article will be mainly concerned with green splits; cycle times and offsets are of a different nature and will be discussed in Sect. 4.

1. Statement of the Problem

Allsop (1974) was probably the first to formally state the interaction between signal control and traffic assignment, by realizing that the link cost functions in a signalized network are dependent on the actual settings and flows. This can be expressed by

$$c = c(f, g) \quad (1)$$

where f is the flow pattern, g is the set of green times and c is the link cost vector.

In user-equilibrium conditions, as first stated by Wardrop (1952), drivers will choose their routes in such a way that no one can reduce personal travel costs by unilaterally switching routes. In game theory, this is known as a Nash noncooperative equilibrium and it represents what must be expected to happen in a real-life network. A user-equilibrium flow f satisfies

$$f \in E(c) = E(c(f, g)) \quad (2)$$

where $E(c)$ is the set of equilibrium flows when the link cost vector is c. The flow pattern depends on the link cost functions, which themselves depend on the signal settings. A signal control policy determines optimum settings based on link flows and costs (however defined):

$$g = G(f) \quad (3)$$

where f depends on the link costs c as expressed in Eqn. (2).

A user equilibrium will usually not be the best possible distribution of flows over the network from a system cost point of view. In a system, optimum drivers (cooperatively) choose their routes so as to minimize the total travel costs in the network. However, in such conditions, some drivers will be able to improve their own travel costs by unilaterally switching routes, but by doing so they will influence other drivers more negatively. Such a system optimal state is not likely to appear under free choice conditions in real life.

Maher and Akcelik (1977) realized the potential of signal setting changes to push a user-equilibrium travel pattern closer to a system optimal travel pattern. In their so-called route control work, user-equilibrium and system optimum flow patterns were compared, and measures such as changes in signal settings were proposed to make the two patterns more similar while maintaining user equilibrium. No systematic attempt to find the optimal settings was made, however.

2. The Combined Problem as a Network Design Problem

The network design problem (NDP) is that of choosing optimum network additions or improvements (i.e., investments), so that some measure of total cost is minimized. Usually, a budget constraint applies and the drivers in the network are expected to follow a user equilibrium, so that they adapt their route choices to the changed conditions (an overview is given by Magnanti and Wong (1984)). It is evident that, in the case of signal control, the investment variables are the green splits at each junction, while the budget consists of the total green time per cycle. The NDP for traffic signal control can thus be stated as

$$\min_{g} f \cdot c(f, g) \qquad (4)$$

where the link flows f satisfy the user equilibrium condition

$$\min_{f} \sum_{i} \int_{0}^{f_i} c_i(x, g_i) \, dx \qquad (5)$$

subject to nonnegativity constraints and conservation of flow. For a more complete discussion of the calculation of a user equilibrium see *Traffic Assignment*. For simplicity in this formulation, separable cost functions c_i are assumed, where costs of traversing a link depend solely on the flow and green time on that particular link. This NDP is known as a leader–follower or Stackelberg game, where a leader (a traffic engineer) decides upon an optimum strategy for signal control, knowing how the follower (the drivers) will respond in their route choice.

Thus, the network design problem attempts to determine an optimal set of signal timings and related link flows in a systematic way. A number of authors has been engaged in this field.

The main problem with the NDP is that it is ill-conditioned: it is generally nonconvex and nondifferentiable. Therefore, although the statement of the optimal signal control problem under equilibrium conditions as a network design problem is straightforward, exact solutions are very hard to find, particularly for larger networks. Several local optima may exist.

An equilibrium link flow pattern can be expressed as a set of constraints. Tan *et al.* (1979) used this observation to eliminate the optimization given in Eqn. (5) from the NDP, so that Eqn. (4) must be solved under an increased number of constraints, some of these nonlinear. They called this a hybrid optimization formulation and suggested an augmented Lagrangian method for its solution, but the path enumeration this requires makes this algorithm inappropriate for networks of even reasonable size.

Fisk (1984) approached the NDP problem from game theory. She stated it as a max–min problem and suggested a penalty approach for its approximate solution. Marcotte (1983) suggested a constraint relaxation technique. Unfortunately, both solution methods again suffer from nonconvexity under realistic assumptions.

The proposed solution algorithm of Sheffi and Powell (1983) to the optimal signal control problem is a feasible descent procedure which, because of the nonconvexity, is bound to find only a local optimum. To determine a descent direction, a gradient needs to be determined for every link with respect to all stages in the network. Because of nondifferentiability this is done numerically, but this requires in each iteration a number of equilibrium assignments equal to the overall number of stages in the network, which will be prohibitive for real-sized networks. The solution methods that Abdulaal and LeBlanc (1979) suggest suffer from the same drawbacks.

A bilevel linear programming formulation for the NDP is suggested by LeBlanc, Boyce and others (LeBlanc and Boyce 1986, Ben-Ayed *et al.* 1988). As the name indicates, the NDP is split into an upper level optimization and a lower level optimization; at the upper level the total system costs are minimized by changing green times, while at the lower level the user equilibrium is pursued by minimizing cumulative user travel costs, as in Eqn. (5). To simplify solution, a piecewise linearization of the link cost functions is proposed. Unfortunately, no convergent solution algorithm for the bilevel linear program is known.

Finally, some authors have suggested omitting the user-equilibrium constraints as given in Eqn. (5) (Dantzig *et al.* 1979, Marcotte 1981). The resulting problem is generally convex and, at its solution point, the flow pattern will be a system optimum; computation is straightforward. As the original NDP has the additional constraint of user-equilibrium flows, this program will provide a lower bound to the optimum signal control problem under equilibrium conditions. The appropriateness of this approximation relies heavily on the closeness of the user-equilibrium and system optimum flow patterns. Though some authors (e.g., Sheffi and Powell 1983) argue that these patterns should be very similar under low and high congestion (though not necessarily in the intermediate area), this depends heavily on the assumptions for the link cost functions. Others, however, predict a divergence of the two patterns when congestion increases. Dantzig *et al.* (1979)

suggest solving the unconstrained problem after which an equilibrium reassignment should be carried out with fixed green times. These two solutions would give an upper and a lower bound to the actual NDP.

Summarizing, although the literature is very rich in formal statements of the optimum signal control problem when drivers follow a user-equilibrium route pattern, no convergent solution algorithms to this NDP are known. Essentially, the problem is very ill behaved, with many possible local optima.

All of the previously mentioned approaches suffer from at least one of the following shortcomings.

(a) The solution (if any) is only roughly approximated. Because of the ill conditioning of the problem, existing convergent methods cannot be applied. Therefore, heuristics have to be used instead, which approximate the exact solution to some extent; sometimes only a local solution is found.

(b) The networks investigated are very small. The only possible application of rigorous solution methods, which guarantee convergence, are small example networks, so that the number of variables is limited and the amount of interaction is tractable.

(c) The assumptions made are too restrictive. In order to enforce a better behavior on the NDP, strong assumptions are needed, particularly with respect to the delay formula and network structure. This severely undermines the real-life applicability of these methods.

It is clear that the network-wide optimization of assignment and signal control remains unsolved, and is unlikely to be solved in the near future. This is mainly due to the complicated interactions between junctions and the subsequent influence of signal changes on equilibrium flows throughout the network. However, the optimization of traffic signals at isolated intersections is relatively easy to solve. In addition, extensive origin–destination data are not required, as only local data are needed in such calculations.

While an attempt to solve the optimal signal control problem via an NDP approach will be more appropriate for nonexistent, future networks, other tactics are more suited for currently existing situations.

3. Mutually Consistent Signal Control and Assignment

Many of the authors mentioned in Sect. 2 suggested solving the NDP via an iterative optimization assignment approach (e.g., Allsop 1974, Sheffi and Powell 1983). In this method, the two subproblems in Eqns. (4) and (5) are alternately solved, keeping the flows fixed when optimizing the green times and, subsequently, keeping the green times fixed when calculating user-equilibrium flows. This method works as shown in Fig. 1.

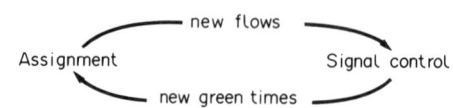

Figure 1
The iterative optimization assignment method

The method accurately represents real life when a traffic authority regularly updates its signal timings according to changed traffic conditions; traffic-responsive policies also fall under this category. Note that in such a case, traffic control merely follows the traffic assignment (see *Traffic Control Modes*).

Early computational work with the iterative optimization assignment method was carried out by Allsop and Charlesworth (1977). Using the signal optimization program TRANSYT in interaction with an equilibrium assignment model, they found that, dependent on the initial state, several mutually consistent points existed where the flows are in equilibrium and the signals optimal for those flows. Green splits and cycle times varied between such solutions, and they concluded that these gave scope for choosing the most advantageous ones. However, it is now known that such an approach will not generally find an optimum and it might well lead to deteriorating network conditions (e.g., Tan et al. 1979, Dickson 1981, Marcotte 1981).

This apparent paradox is caused by the phenomenon that, when using such appealing delay-minimizing policies in interaction with assignment, link costs may actually decrease with increasing flow. A further explanation of this is given by Heydecker (1983).

Smith (1981) derived properties of a signal control policy that guarantee the existence of a traffic equilibrium consistent with the policy. He also showed that most natural-looking, standard control policies that in some way try to optimize the signals for current conditions do not satisfy these conditions.

Based on these properties, Smith introduced a signal control policy that is guaranteed to find a feasible solution to the combined assignment–signal control problem. Under natural, but rather severe, conditions upon the assumed delay function, this solution is unique.

The practical use of this is that the iterative optimization assignment procedure can be used to determine so-called mutually consistent points for appropriate signal control policies in interaction with traffic assignment. In this way, the combined problem falls apart into two separate, well-researched problems, for which efficient solution methods are known. Properties of such mutually consistent points can be investigated in order to determine the merits of these policies. In addition, and most importantly, this method allows the investigation of the optimal control problem under realistic assumptions. The complexity of the NDP enforces limits on the assumptions with respect to, for example, network topography and junction delays, which may make the model questionable in real-life applications. The iterative optimization assignment procedure, however, can handle realistic delay formulae, complicated junction layouts (such as shared lanes) and sophisticated assignment models (such as SATURN (Van Vliet 1982)). It is better to approximately solve a realistic model, than to exactly solve an abstract mathematical model that bears no resemblance to the actual world.

Computer tests using the iterative optimization assignment procedure have been carried out to compare the performance of Smith's new policy with existing ones. The tests, using both artificial networks and networks drawn from real life, reconfirmed the findings of previous researchers that, for policies that try to minimize delays explicitly or approximately, no unique mutual equilibrium exists (Smith *et al.* 1987, van Vuren *et al.* 1988, van Vuren *et al.* 1990).

At low flow levels, these standard policies generally lead to optimal or near-optimal solutions. This is not surprising as in this case interactions between the two subproblems will be small, so that they can be regarded as separate problems. A delay-minimizing control strategy will, therefore, perform well in such conditions.

However, under highly congested conditions, interactions cannot be neglected and these policies can then unnecessarily increase network travel times by shifting green time and flows to inefficient routes with high marginal costs.

Smith's new policy is developed with a redistribution of traffic in mind. The existence of a consistent equilibrium is guaranteed; the policy is capacity maximizing, as explained by Smith (1979). Whereas the conventional policies attempt to minimize delay for the current flow pattern (so that an eventual traffic redistribution is an often undesirable side effect), Smith's policy weighs experienced delays with the saturation flows of the relevant approaches. More green time is awarded to high-capacity roads and so traffic is enticed to reroute to these roads.

The computer tests carried out show that this policy works particularly well when travel demand is high. Under such circumstances, a redistribution of traffic to high-capacity roads is very desirable and the average performance of Smith's policy over a range of networks is some 8% better than the traditional policies (measured in total network travel time at the mutually consistent point; cf. the objective in Eqn. (4)).

Although, with relatively simple delay assumptions, Smith's policy leads to a unique mutually consistent point, tests under more realistic assumptions show a disappointing lack of stability. The necessary conditions for stability in Smith (1981) are not always satisfied in real life and more work is needed in this area.

As stated previously, the search for a mutually consistent point through an iterative optimization assignment approach is a representation of what happens in reality when signal settings are regularly updated or is a representation of traffic-responsive control. As such, it can only provide an upper bound for the true solution of the NDP (as defined in Eqns. (4) and (5)). The value of this approach can, therefore, only be tested by comparing the performance of control policies at a point of mutual consistency with assignment, with the system optimum (which provides a lower bound to the problem) or, where possible, with the optimal NDP solution. Limited tests have been carried out in this respect and results suggest that, particularly in regions of intermediate congestion, differences between the lower (system optimal) bound and mutually consistent points for various control policies can be substantial. There is clearly scope for the development of more efficient control policies that interact well with assignment at all levels of congestion.

A unifying framework for such developments was first set up by Smith (1981); this work was extended by Smith *et al.* (1987). At the basis of this approach lies the observation that a signal control policy can be expressed in similar terms to a user-equilibrium flow pattern. There are many efficient solution algorithms known for the solution of the user-equilibrium assignment problem that could be used to solve the signal control subproblem or, indeed, the combined problem. In a user equilibrium: more costly routes carry no flow. A similar principle can be applied to the allocation of green time to stages: less-pressurized stages receive no green time. The essence of this expression is in the definition of the pressures on stages. In the same way as the cost C_r of route r consists of a summation of link costs c_i, the pressure P_r on stage r is defined as the summation of pressures p_i on movements that have right of way during that stage. Thus,

$$p_i = p_i(f_i, g_i) \qquad (6)$$

and

$$P_r = \sum_{\substack{\text{link } i \\ \text{belongs to stage } r}} p_i \qquad (7)$$

For a delay-minimizing policy, pressure p_i is given by

$$p_i = -f_i \frac{\partial d_i}{\partial g_i} \qquad (8)$$

For Smith's new policy, pressure p_i is given by

$$p_i = s_i d_i(f_i, g_i) \qquad (9)$$

where $d_i(f_i, g_i)$ is the delay experienced by a movement at the traffic signals and s_i is the saturation flow for that movement. The existence, uniqueness and stability of a mutually consistent point for the combined signal control–assignment process depends on the precise expression for p_i. In the same way, the suitability of efficient solution algorithms, such as feasible descent methods, is influenced.

Research into the development of practical pressure definitions, in terms of both resulting network travel times at mutual consistency and stability, uniqueness and applicability of existing solution algorithms, is ongoing.

4. Cycle Times and Offsets

Whereas the mutual influence of green splits and link flows is well researched, much less is known about the redistributional effect of changes in cycle time and/or offsets in a network. Changes in green times can be used to favor particular movements, thus rerouting traffic in a systematic and predetermined way. The relationships between delays per movement and cycle time and offsets is much less clearcut. From Webster (1958), it is known that optimum cycle times per junction can be determined, while Gartner et al. (1976) determine an optimum network-wide cycle time. Starting from here, it might be possible to determine cycle times in a way that favors particular junctions, but no research has been published that reports on redistributional effects, accidental or intended.

A slightly different argument can be set up for offsets. The optimality of signal offsets depends largely on the junction distances and much less on traffic volumes, unlike green times. Therefore, it can be expected that a mutual consistency between offsets and assignment can be established with relative ease. For, although an improvement in the relative offsets in a signal-controlled network may cause an initial redistribution of traffic, the relations between junctions stay structurally the same, so that further offset changes and/or reassignments will be of negligible proportions. Preliminary results by Heydecker et al. (1990) support this. The questions of whether and how cycle time and offsets can be used to achieve an optimum mutual consistency between signal control and assignment remain open.

5. Future Developments

The relevance of an understanding of the combined signal control–assignment problem is particularly evident with the pending introduction of electronic route guidance systems (see *Route Guidance, Individual*). For an optimal performance of such systems, efficient signal settings are indispensable. There are, however, two important differences from the current situation. First, under route guidance there is, at the least, partial control over route choice by drivers. This can be exercised to simplify the determination of advantageous green time/flow patterns. Second, under electronic route guidance, time, in addition to space, will play a crucial role. A static assignment and fixed signal timings over an extended period will not suffice. Mutual interdependencies then exist over time as well as over space, so that changes in green time will affect the flow patterns in later time periods and vice versa. The extra complexity this introduces makes the solution of the combined problem even less tractable (see *Traffic Assignment, Dynamic*).

See also: Traffic Assignment; Traffic Assignment, Dynamic

Bibliography

Abdulaal M, LeBlanc L J 1979 Continuous equilibrium network design models. *Transp. Res. B* **13**, 19–32

Allsop R E 1974 Some possibilities for using traffic control to influence trip distribution and route choice. *Proc. 6th Int. Symp. Transportation and Traffic Theory.* Elsevier, Amsterdam, pp. 345–74

Allsop R E, Charlesworth J A 1977 Traffic in a signal-controlled road network: An example of different signal timings inducing different routings. *Traffic Eng. Control* **18**, 262–4

Ben-Ayed O, Boyce D E, Blair C E 1988 A general bilevel linear programming formulation of the network design problem. *Transp. Res. B* **22**, 311–18

Dantzig G B, Harvey R P, Lansdowne Z F, Robinson D W, Maier S F 1979 Formulating and solving the network design problem by decomposition. *Transp. Res. B* **13**, 5–17

Dickson T J 1981 A note on traffic assignment and signal timing in a signal-controlled road network. *Transp. Res. B* **15**, 267–71

Fisk C S 1984 Game theory and transportation systems modelling. *Transp. Res. B* **18**, 301–13

Gartner N H, Little J D C, Gabbay H 1976 Simultaneous optimization of offsets, splits and cycle time. *Transp. Res. Rec.* **596**, 6–15

Heydecker B G 1983 Some consequences of detailed junction modeling in road traffic assignment. *Transp. Sci.* **17**, 263–81

Heydecker B G, van Vuren T, Van Vliet D 1990 Optimal signal offsets for traffic assignment networks. In: Yagar S, Rowe S E (eds.) *Traffic Control Methods.* Engineering Foundation, New York, pp. 295–305

LeBlanc L J, Boyce D E 1986 A bilevel programming algorithm for exact solution of the network design problem with user-optimal flows. *Transp. Res. B* **20**, 259–65

Magnanti T, Wong R T 1984 Network design and transportation planning: Models and algorithms. *Transp. Sci.* **18**, 1–55

Maher M J, Akcelik R 1977 Route control—Simulation experiments. *Transp. Res.* **11**, 25–31

Marcotte P 1981 An analysis of heuristics for the continuous network design problem. *Proc. 8th Int. Symp. Transportation and Traffic Theory.* Toronto University Press, Toronto, pp. 27–34

Marcotte P 1983 Network optimization with continuous control parameters. *Transp. Sci.* **17**, 181–97

Sheffi Y, Powell W B 1983 Optimal signal settings over transportation networks. *J. Transp. Eng.* **109**, 824–39

Smith M J 1979 A local traffic control policy which automatically maximises the overall travel capacity of an urban road network. *Traffic Eng. Control* **21**, 298–302

Smith M J 1981 Properties of a traffic control policy which ensure the existence of a traffic equilibrium consistent with the policy. *Transp. Res. B* **15**, 453–62

Smith M J, van Vuren T, Heydecker B G, Van Vliet D 1987 The interaction between signal control policies and route choice. *Proc. 10th Int. Symp. Transportation and Traffic Theory.* Elsevier, New York, pp. 319–38

Tan H, Gershwin S B, Athans M 1979 Hybrid optimization in urban traffic networks. Massachusetts Institute of Technology Final Report No. US-DOT-TSC-RSPA-79-7

Van Vliet D 1982 SATURN—A modern assignment model. *Traffic Eng. Control* **23**, 578–81

van Vuren T, Smith M J, Van Vliet D 1988 The interaction between signal setting optimization and re-assignment: Background and preliminary results. *Transp. Res. Rec.* **1142**, 16–21

van Vuren T, Smith M J, Van Vliet D 1990 Mutually consistent signal control and traffic assignment. In: Yagar S, Rowe S E (eds.) *Traffic Control Methods.* Engineering Foundation, New York, pp. 315–27

Wardrop J G 1952 Some theoretical aspects of road traffic research. *Proc. Inst. Civ. Eng. II* **1**, 325–78

Webster F V 1958 *Traffic Signal Settings*, Road Research Technical Paper, Vol. 39. Her Majesty's Stationery Office, London

T. van Vuren
[Frank Graham Consulting Engineers,
Worcester, UK]

Signal Control at Individual Junctions: Phase-Based Approach

The traffic signal setting for an individual junction can be evaluated by mathematical programming techniques. The various methods can be grouped into two classes.

In the first class, the composition and the sequence of the stages are assumed to be fixed initially and the green times for the stages optimizing a given performance index are calculated. This is called a stage-based approach.

In the second class, the optimal timing (green time for each stream and green scheduling) can be obtained directly using knowledge of the incompatibilities among the streams, rather than through the intermediary of stages. This is called a phase-based approach.

1. Definitions

A junction is defined as isolated if there are no "interactions" between it and the other surrounding junctions. In general, this means that the distances between junctions are long enough to eliminate the platoon effect generated by the traffic signals on the vehicle flow (see *Road Traffic: An Introduction*; *Road Traffic Control: Progression Methods*; *Road Traffic Control: TRANSYT and SCOOT*).

Interacting junctions can be treated as isolated only if the control objective can be assumed to be independent of the characteristics of the arrivals (capacity factor maximization, cycle time minimization). This type of junction is termed an individual (or single) junction (see *Signal Control at Individual Junctions: Stage-Based Approach*).

A junction can be described, schematically, through a set of approaches and a common crossing area. An approach is a part of a road leading to the junction such that the traffic in it has right of way simultaneously and a vehicle joining the back of the queue can expect to pass the signal at roughly the same time, whichever lane it chooses.

The traffic at a junction is usually divided into streams. A stream is formed by all the users who cross the junction from the same approach; it is the smallest portion of the traffic that is generally distinguished in the analysis. A stream can be formed by a set of maneuvers that share the same approach.

When considering the control system, users are divided into groups, which are sets of streams that receive identical signals from the controller. The streams in a group may, therefore, use different approaches. The group represents the smallest unit that is often considered in signal control problems at individual junctions. In many practical cases, each group is formed by one stream.

It is assumed, in general, that the vehicle departures occur at a time interval. Therefore, an effective green time g, during which the vehicles can leave the stop line, and an effective red time r, during which no departure occurs, are considered (see *Signal Control at Individual Junctions: Stage-Based Approach*).

Let $s(t)$ be the flow leaving the stop line during the green and amber periods. A saturation flow s is defined, for each stream, as the average flow that can cross the junction in the unitary time when a queue is present at the approach.

Let G, R and A be the displayed green, red and amber times for a stream, respectively. According to Webster (1958) and Webster and Cobbe (1966), it is possible to define

$$g = \frac{1}{s}\int_0^{G+A} s(t)\,dt \qquad (1)$$

$$r = G + R + A - g \qquad (2)$$

In addition, the lost time is defined as

$$l = G + A - g \qquad (3)$$

The cycle time c is the minimum time in which a complete succession of signals occurs. The cycle time is, obviously, equal to the sum of the effective green and the effective red times for each group.

Let $q(t)$ be the flow reaching the stop line. An approach is defined as undersaturated if

$$\int_0^c q(t)\,dt \le \int_0^{G+A} s(t)\,dt \qquad (4)$$

If the arrival rate is assumed uniform with a value q and the discharge of the queue is supposed to be at an average flow equal to s, the undersaturation condition becomes

$$qc \leq gs$$

An undersaturated junction is a junction for which all the approaches are undersaturated.

The cycle can be subdivided into stages, which are elementary periods during which particular sets of groups receive green simultaneously (Transportation Research Board 1985).

The order in which n groups receive right of way in m stages is generally specified by a Boolean matrix $\mathbf{T}(m, n)$, called the stage matrix, such that $t_{ij} = 1$ if group j has right of way in stage i and $t_{ij} = 0$ otherwise.

Two groups that can safely cross the junction at the same time are called compatible; otherwise they are called incompatible or conflicting. The compatibility between two groups is defined taking into account the layout of the junction, the arrival flows and the characteristics of the streams.

The clearance time for two incompatible groups is the minimum time between the end of the amber time of one and the beginning of the green time of the other.

The compatibilities among n groups can be expressed by a Boolean matrix $\mathbf{A}(n, n)$, called the compatibility matrix, in which $a_{ij} = 1$ if groups i and j are compatible, and $a_{ij} = 0$ otherwise. A compatibility graph can be defined from matrix $\mathbf{A}$. It is an undirected graph with n nodes (one node for each group) and one edge ij for each pair of compatible groups i and j.

The edges corresponding to a set of groups that are mutually compatible determine a clique on the compatibility graph (i.e., a complete subgraph). For this reason, a set of groups that are mutually compatible is called a clique. Usually, the set of groups having the right to cross the junction during a stage forms a maximal clique. Any ordered set of cliques respecting various imposed constraints is a possible stage sequence or, simply, a sequence.

In a similar way, a matrix and a graph of incompatibility, as well as cliques of incompatibility, can be defined (i.e., conflict matrix, conflict graph, conflict cliques).

As an example, Fig. 1 shows the layout of a three-arm junction and the corresponding conflict matrix. The conflict matrix has been obtained assuming that each stream is formed by just one maneuver and that each pair of merging or crossing streams is incompatible. Figure 2 shows the conflict graph and cliques for the example junction of Fig. 1. Figure 3 shows the compatibility graph and cliques, and corresponding stages for the same junction.

Two dependent variables are of great importance in defining the performance of a single junction. First, the capacity factor μ is, for an assigned signal setting, the maximum common multiplier for all the rates of flow

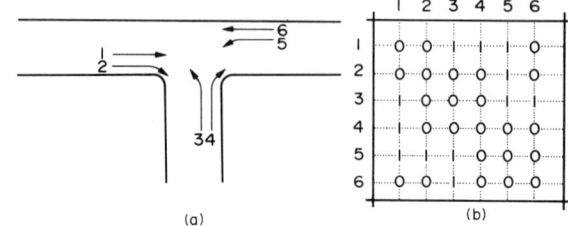

Figure 1
Three-arm junction: (a) layout and (b) corresponding conflict matrix

arriving at a junction for which the undersaturation still holds. Second, the delay is the difference between the travel time of a vehicle crossing the signalized junction and the time that vehicle would take if no other traffic were present and the stream in which it travelled had constant right of way.

Corresponding to these variables, the usual objectives of optimization are maximization of the capacity factor of the junction and minimization of the total rate of delay.

A third objective corresponds to the evaluation of the critical cycle time (i.e., the minimum cycle time for which $\mu = 1$). This is the least cycle time that gives a useful operating condition for the junction.

Maximization of μ is a generally chosen objective for heavily congested junctions. A delay minimization control scheme can be adopted if μ is satisfactory. The critical cycle time is useful for the calculation of the master cycle in signal coordination (see *Road Network Control*).

The signal setting for an individual junction is fully defined by determining

(a) the green timing (or green split) (i.e., the green time assigned to each group),

(b) the green scheduling (i.e., the succession of green periods of groups), and

(c) the cycle time.

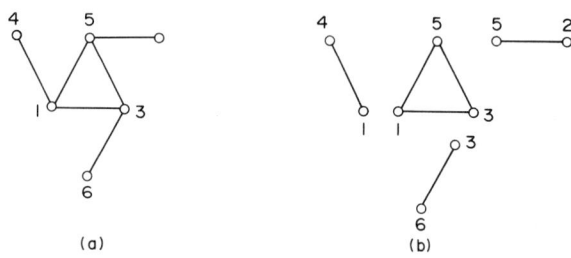

Figure 2
(a) Conflict graph and (b) conflict cliques for junction in Fig. 1

Signal Control at Individual Junctions: Phase-Based Approach

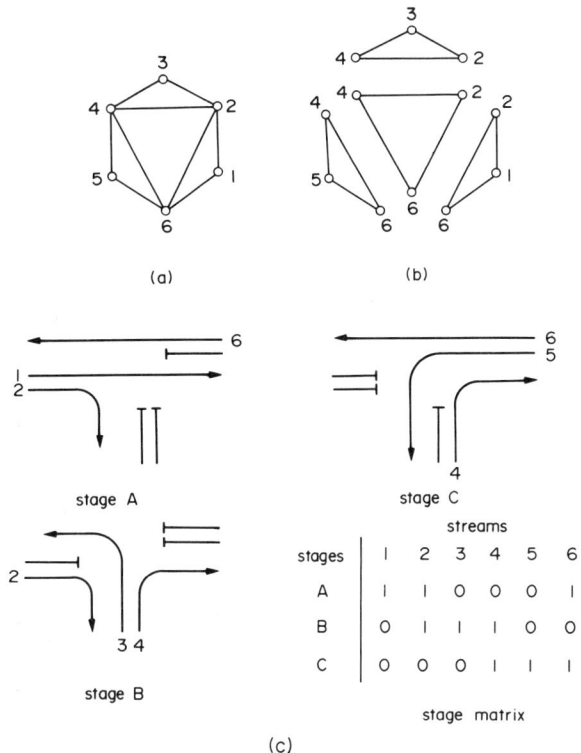

Figure 3
(a) Compatibility graph, (b) compatibility cliques, and (c) stage matrix and corresponding stages for junction in Fig. 1

2. Mathematical Programming Methods

The evaluation by mathematical programming techniques of optimal values for control variables (green timing, green scheduling and cycle time) is carried out by various methods.

According to the general statements given in Sect. 1, mathematical programming methodologies for single-junction control assume as possible objectives of optimization the maximization of μ (linear), the minimization of the total rate of delay (nonlinear but convex, by introducing the reciprocal of the cycle time as a variable, or linear, if using piecewise linearization of delay functions) and the minimization of the cycle time (linear).

Mathematical-programming-based methodologies for single-junction control can be grouped into two classes of different complexity.

In the first class, the flows and saturation flows of the streams being known, the stage matrix is initially fixed and the optimal green times for the stages calculated. This is the stage-based approach (see *Signal Control at Individual Junctions: Stage-Based Approach*).

The methods for calculating the optimal green times for the stages use sets of constraints including

(a) undersaturation constraints (for each group);
(b) minimum-green constraints (for each stage);
(c) maximum-red constraints (for each stage);
(d) congruence constraints between cycle times and green times of stages, and
(e) constraints on minimum and maximum values of the cycle time.

All constraints are linear and variables of optimization are real.

The stage-based approach has the disadvantage that it requires the user to specify in advance the sequence of stages and the duration and details of the transition periods. Furthermore, no direct method is available within these methods to constrain the duration of the effective red and green periods for the groups themselves.

In the second class, the flows and saturation flows are still assumed known and the optimal timing is evaluated from the knowledge of the crossing compatibilities of the different streams.

It is possible to divide the methods belonging to the second class into two subclasses (Allsop 1983).

The first subclass is based on the analysis of possible sequences of compatibility cliques and can be regarded as a natural evolution of the stage-based approach. Any clique is a possible stage and any sequence of cliques respecting some imposed constraints is a possible stage succession. The optimization problem consists of calculating the optimal clique sequence, the green time for each clique and the cycle time (Stoffers 1968, Zuzarte Tully 1977, Zuzarte Tully and Murchland 1978, Cantarella and Improta 1981).

In Stoffer's approach, assuming as known the compatibility matrix, the computation of optimal values for control variables is divided into four steps:

(a) determination of all the maximal compatibility cliques, which are cliques with the requirements that
 (i) all groups in a clique must be compatible, and
 (ii) a clique should contain as many compatible groups as possible;
(b) determination of all the sequences of these cliques with the requirements that
 (i) each group must get green in at least one clique,
 (ii) each group must have a single green interval (i.e., the green cliques for any particular group must be consecutive when the sequence of cliques is considered to be circular), and
 (iii) each sequence must be maximal (i.e., no further cliques can be inserted into it);
(c) assuming each sequence as a possible stage matrix and computation of the optimal green timing; and
(d) choice of the best sequence.

Zuzarte Tully and Murchland (1978) heuristically modified Stoffer's procedure to reduce computation time. In particular, steps (a) and (b) have been reduced. If a great number of sequences occurs, they reduce the dimension of the problem by discarding all the sequences with a high value of the critical cycle time (calculated by a fast procedure).

For all the remaining sequences, they compute the green timing using a stage-based model, by iterating two steps:

(a) computation of green times and determination of the cliques with zero green time; and

(b) generation of subsequences without these cliques.

This procedure is repeated until all the generated subsequences are formed by cliques with nonzero green times.

The best value of the performance index used in the green-timing step indicates the best (but not necessarily optimal) solution.

The methods belonging to this subclass provide only a partial solution to the problem of determining a suitable sequence. They generate all maximal sequences of stages that give a single green interval for each group in each cycle. The original intention was that any superfluous stages would be removed when suitable stage durations were calculated. However, no automatic procedure has been developed to do this. Instead, it requires a considerable amount of manual intervention to reformulate the constraints when cliques (with zero green time) and the corresponding transition periods are removed (Heydecker and Dudgeon 1987). Moreover, the need to eliminate cliques with zero green time does not allow the inclusion of minimum-green constraints in the model.

An end point and the green time for each stream are used in the second subclass to determine the optimal cycle time, green timing and green scheduling. This is called a phase-based approach or, better still, a stream-based or group-based approach.

In this alternative approach, the durations of the signal indications for the groups are considered directly rather than through the intermediary of stages. This allows for constraints such as minimum green times and group-to-group clearance times to be expressed directly. Furthermore, the stage sequence and structure of the transition periods need not be specified in advance of optimization.

Two methods belonging to this approach have been proposed in recent years. Both methods are based on models that include

(a) undersaturation constraints (for each stream),

(b) minimum-green constraints (for each group),

(c) maximum-red constraints (for each group),

(d) constraints on minimum and maximum values for the cycle time, and

(e) incompatibility crossing constraints (for each pair of incompatibility groups).

In the first method (Gallivan and Heydecker 1983, Heydecker and Dudgeon 1987, Moller 1987), the green timing and scheduling are represented in the complex plane by two variables for each group, one representing the start and the other the duration of the green interval. That is, the cycle time can be represented by a circumference and the green times of the groups by circular arcs duly located.

Using this representation, a linearly constrained mathematical programming model with real variables can be formulated if the following are specified:

(a) for each group, the set of groups that have right of way after it; and

(b) the set of groups that start the sequences.

In this way, a possible order of conflict groups is initially fixed.

This method is less constrained than stage-based methods. Moreover, the exogenous specification of the indicated parameters enables some control on the form of the solution. However, the determination of the optimal solution requires the complete enumeration of all the possible orders of incompatible groups.

The second method (also called single-junction control system design) is based on a mathematical programming model in which control variables (cycle time, green timing and scheduling) are calculated simultaneously.

This model is formulated representing the control variables on a time axis, measured in time units (Improta and Cantarella 1982, 1984) or defined as a proportion of the cycle time (Cantarella and Improta 1988).

All the constraints are linear and the nonoverlapping condition among incompatible groups is modelled using binary variables. An alternative formulation using only real variables requires nonlinear constraints (Cantarella and Improta 1983).

The model can be solved by usual mathematical programming techniques. A problem-oriented algorithm can be adopted for capacity factor maximization and critical cycle time evaluation (Cantarella and Improta 1988).

3. The Phase-Based Approach

To introduce the signal control phase-based approach in a simple way, the model used by Improta and Cantarella (1982, 1984), and Cantarella and Improta (1988) will be referred to. For clearness of exposition, a situation in which each group is formed by just one stream and all the clearance times are zero will be considered.

3.1 Variables and Parameters

Let f be the junction capacity factor, $z = 1/c$, the reciprocal of cycle time, q_k the arrival rate of stream k, s_k the saturation flow of stream k and $y_k = q_k/s_k$, the flow

Signal Control at Individual Junctions: Phase-Based Approach

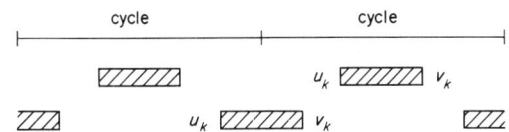

Figure 4
Reference interval for model variables

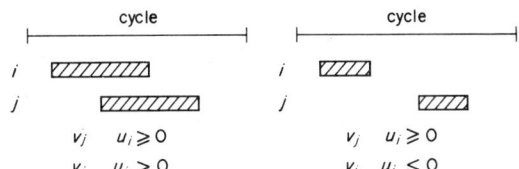

Figure 5
Incompatibility constraints

ratio of stream k, w_{ij} a binary variable associated with a pair of i, j incompatible streams ($w_{ij} = 0$ if the green of i precedes the green of j in a sense to be defined, $w_{ij} = 1$ otherwise), and $l_k \geq 0$, the lost time of stream k (part of the green–amber interval unused because of the starting and stopping transients). For each stream k, the following are defined as a proportion of the cycle time: u_k the starting time of green for stream k, v_k the ending time of amber for stream k, g_k the effective green of stream k and r_k the effective red of stream k. Thus,

$$g_k + r_k = 1 \qquad (5)$$

$$v_k - u_k = g_k + l_k \qquad (6)$$

3.2 Constraints

It is necessary that each stream k has enough green time to enable all the traffic arriving in its stream to cross the junction (undersaturation condition):

$$g_k \geq f y_k \qquad (7)$$

In addition, since the green and amber period of stream k cannot be greater than the cycle time, then

$$g_k + l_k \leq 1 \qquad (8)$$

It is assumed, for simplicity, that each stream has just one green period in the cycle time. In order to position the green and amber period of a stream k inside the cycle, a reference interval $(-1, +1)$ on the time axis will be considered. In this interval, each stream k has, obviously, two green and amber periods, at least one of which is represented by one segment (see Fig. 4). Hence,

$$v_k \geq 0, \qquad v_k \leq 1 \qquad (9)$$

$$u_k \geq -1, \qquad u_k \leq 1 \qquad (10)$$

$$v_k - u_k \geq 0 \qquad (11)$$

The relative positions of the green and amber periods of two conflict streams i and j are defined by the quantities $v_i - u_j$ and $v_j - u_i$ (see Fig. 5) which give

$$v_i - u_j \geq -1 \qquad (12)$$

$$v_j - u_i \geq -1 \qquad (13)$$

$$v_i - u_j \leq 1 \qquad (14)$$

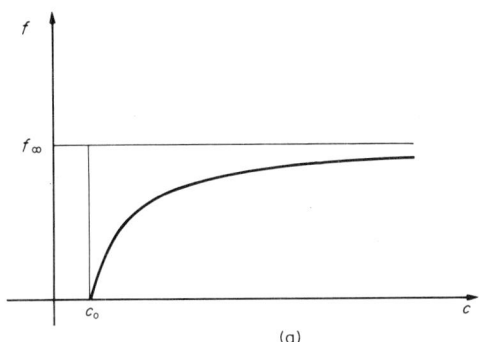

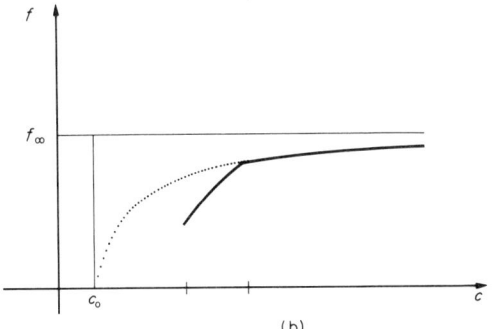

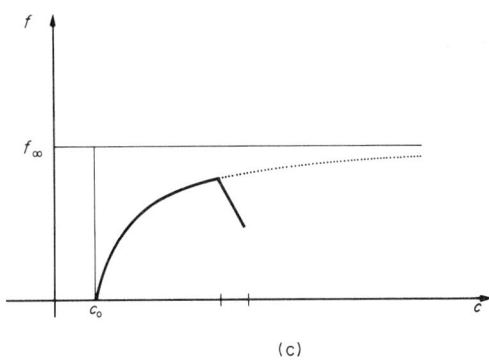

Figure 6
Capacity factor vs cycle time: (a) without minimum-green and maximum-red constraints, (b) with minimum-green constraints and (c) with maximum-red constraints (dotted lines indicate correspondence without any constraints)

$$v_j - u_i \leq 1 \quad (15)$$

It is easy to verify that, for two conflict streams i and j, the quantities $v_j - u_i$ and $v_i - u_j$ must have opposite signs, in order to avoid overlapping. This condition can be expressed by

$$v_i - u_j \leq w_{ij} \quad (16)$$

$$v_j - u_i \leq 1 - w_{ij} \quad (17)$$

$$w_{ij} = 0 \text{ or } 1 \quad (18)$$

Equations (5–18) are not all independent; that is, Eqns. (8) and (10–15) can be omitted and variable u_k can be drawn from Eqn. (6) and substituted in the other constraints. Thus, g_k, v_k, w_{ij}, z and f are the decisional variables. Equation (5) allows r_k to be computed from g_k.

The incompatibility crossing constraints are at the heart of the phase-based approach for signal control at individual junctions.

It is very easy to complete the optimization model by adding some maximum-red, minimum-green and cycle time constraints and a usual objective function.

It is interesting to note that the method proposed by Heydecker and Dudgeon (1987) can be regarded as a particular case of the introduced model, in which a set of values for w_{ij} is initially fixed.

As previously introduced, a problem-oriented algorithm can be adopted for capacity factor maximization and critical cycle time evaluation. This algorithm is based on the calculation of the conflict cliques. The knowledge of these cliques generally allows the calculation of the green times for the streams and of the capacity factor–cycle time relationship. Figure 6 shows some examples of this correspondence.

See also: Road Traffic: An Introduction; Signal Control at Individual Junctions: Stage-Based Approach

Bibliography

Allsop R E 1983 Optimization of timings of traffic signals. *Proc. 1983 AIRO Conf.* Guida Editori, Naples, pp. 103–20

Cantarella G E, Improta G 1981 Una metodologia per il progetto globale di intersezioni semaforizzate. *Atti Delle Giornate di Lavoro AIRO 1981*. AIRO, Genoa, pp. 33–48

Cantarella G E, Improta G 1983 A non-linear model for control system design of an individual signalized junction. *Proc. 1983 AIRO Conf.* Guida Editori, Naples, pp. 709–22

Cantarella G E, Improta G 1988 Capacity factor and cycle time optimization: A graph theory approach. *Transp. Res. B* **22**, 1–23

Gallivan S, Heydecker B G 1983 Optimising the control performance of traffic signals at a single junction. *15th Annual Conf. Universities Transport Study Group*. Imperial College London

Heydecker B G, Dudgeon I W 1987 Calculation of signal settings to minimize delay at a junction. In: Gartner N H, Wilson N H M (eds.) *On Transportation and Traffic Theory*. Elsevier, Amsterdam, pp. 159–78

Improta G, Cantarella G E 1982 Signalized junction control system design. *EURO–TIMS XXV*

Improta G, Cantarella G E 1984 Control system design for an individual signalized junction. *Transp. Res. B* **18**, 147–67

Moller K 1987 Calculation of optimum fixed-time signal programs. In: Gartner N H, Wilson N H M (eds.) *On Transportation and Traffic Theory*. Elsevier, Amsterdam, pp. 179–98

Stoffers K E 1968 Scheduling of traffic lights—A new approach. *Transp. Res.* **2**, 199–234

Transportation Research Board 1985 *Highway Capacity Manual*. Special Report No. 209. Transportation Research Board, National Research Council, Washington, DC

Webster F V 1958 *Traffic Signal Setting*, Road Research Technical Paper, Vol. 39. Her Majesty's Stationery Office, London

Webster F V, Cobbe B M 1966 *Traffic Signals*, Road Research Technical Paper, Vol. 56. Her Majesty's Stationery Office, London

Zuzarte Tully I M 1977 Synthesis of sequences for traffic signal controllers using techniques of the theory of graphs. Ph.D. thesis, University of Oxford

Zuzarte Tully I M, Murchland J D 1978 Calculation and use of the critical cycle time for a single traffic controller. *Proc. PTRC Summer Annual Meet.* PTRC, Brighton, UK, pp. 96–112

G. Improta
[Università di Napoli, Naples, Italy]

Signal Control at Individual Junctions: Stage-Based Approach

Traffic signals are used to resolve conflicts between movements of vehicles and pedestrians at road junctions. Their provision and operation involves deciding for each junction

(a) which movements are to be provided for and which of these are to be separately signalled,

(b) what the physical layout of the junction should be,

(c) in what order the movements should proceed, and

(d) for how long each movement should proceed.

The overall optimization of the design and operation of a junction can be seen as identifying various practicable alternatives for (b) in the light of (a), determining (c) and (d) so as to optimize some criterion of performance for each of the alternatives, and then choosing between the alternatives in the light of the resulting optimal values of the criterion. The usual criteria of performance are traffic capacity, in a sense to be defined, and average delay per vehicle.

There are two main types of operation of traffic signals. In fixed-time operation, different sets of movements are allowed to proceed in a prescribed cyclic order for predetermined lengths of time regardless of the arrival of vehicles or pedestrians. Alternatively, in vehicle-actuated operation, the changing of the signals may be influenced by vehicles being detected or pedestrians pressing buttons. This article refers mainly to fixed-time operation, but the resulting signal timings are also relevant to vehicle-actuated operation, because many forms of such operation require timings to be specified to act as constraints on the changing of the signals when traffic is heavy.

1. Traffic Streams

A stream of traffic at a signal-controlled junction comprises either one lane or several adjacent lanes of traffic that behave as a single queue, independent of the behavior of traffic in any other adjacent lane. Thus, all vehicles in a stream are subject to the same signal indication and, given their intended direction at the junction, none of them may choose a lane used by another stream. Conversely, if the stream uses more than one lane, an appreciable proportion of its vehicles must have a free choice between these lanes. A stream is the smallest subdivision of traffic that need be considered in calculating signal timings. The pattern of traffic for which timings are to be calculated is specified by the arrival rate of traffic in each steam together with, for each stream from which traffic has a choice of exits from the junction, the proportion taking each exit. These quantities are averages over the period for which timings are to be calculated.

1.1 Model of Signal Control of a Stream

Traffic engineers (Webster and Cobbe 1966) have shown that, for each stream, signal control can usually be regarded as dividing all time into alternate periods called effective red and effective green. In the effective red period, traffic in the stream is modelled as not passing the signal. In the effective green period, it is modelled as passing the signal at a uniform rate called the saturation flow, if there is a queue in the stream, or as passing as it arrives, if there is no queue. The effective green period is related to, but is not the same as, the period for which the signals controlling the stream display green. The relationship between the end points of the effective green period and those of the displayed green is assumed to be independent of the duration of the latter, which therefore determines the duration of the former. The relationship does, however, depend on the behavior of traffic in the stream and may, therefore, differ among streams receiving the same signal indications.

The relationship and the saturation flow can be estimated for an existing signal-controlled traffic stream by making observations during green periods when the

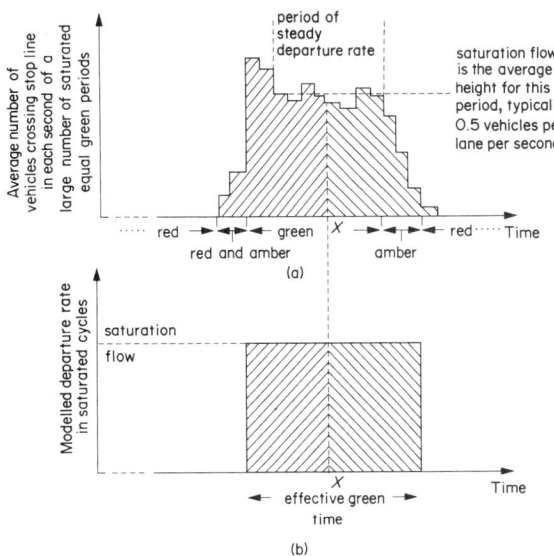

Figure 1
Model of the effect of fixed-time signal control upon a vehicular traffic stream, showing how saturation flow and effective green time can be estimated from (a) observations followed by (b) use of a model

queue has still not cleared by the time the signal changes back to red. These are called saturated green periods. If the numbers of vehicles passing the signal in each second during which traffic moves in such periods of equal duration are counted and the observations superimposed on the same time axis, then a diagram such as Fig. 1a results, from which the saturation flow is estimated as shown. In Fig. 1b, the irregularly shaped area is simplified to a rectangle, keeping the point X on the time axis fixed, by taking the height of the rectangle as the saturation flow and choosing the end points to make the shaded areas to the left of X in Fig. 1a and Fig. 1b equal and, similarly, the areas to the right of X. The width of the resulting rectangle is the effective green time and its end points are well defined relative to the time X. The saturation flow and the relationship between the end points of the effective and displayed green periods can also be estimated from the layout of the parts of the carriageway used by the stream. In either case, they are assumed to be independent of the duration of green, so that the effect of lengthening or shortening the displayed green is represented by adding or removing time at X. The equivalence of X in the model and in reality enables results of calculations made using the model to be translated into timings of the signal display.

In measuring or estimating the saturation flow for a vehicular traffic stream, it is necessary to take account of the fact that vehicles of different types take different lengths of time to pass the signal. Larger and heavier vehicles take longer than smaller and lighter ones.

Even for vehicles of the same type, those making a turning maneuver take longer than those going straight ahead. These differences are allowed for in streams in which all vehicles make the same maneuver by estimating for each type of vehicle how many queuing passenger cars would, on average, pass the signal in the time taken by one vehicle of the type concerned. This number is called the passenger car unit (pcu) value of the vehicle type. In streams from which vehicles make two or more different maneuvers, the procedure is the same except that the basic unit is taken as the passenger car going straight ahead, and the numbers for other vehicle types or maneuvers are given in through car unit (tcu) values. These values and their estimation are discussed further by Kimber et al. (1985).

Saturation flows are often expressed in pcu per unit time or tcu per unit time but, when estimating delay and quantities related to it, the unit of measurement should be vehicles per unit time, because the distribution of headways between vehicles influences the delay. It is useful to note that if a traffic stream contains K types of vehicles and if for $k = 1, 2, \ldots, K$ a fraction f_k of vehicles in the stream are of type k and have a pcu or tcu value of u_k, then the saturation flow in vehicles per unit time is

$$\frac{\text{saturation flow in pcu or tcu per unit time}}{\sum_{k=1}^{K} f_k u_k}$$

In terms of notation and terminology for a stream, c is the cycle time, Λ is the proportion of the cycle effectively green, s is the saturation flow in vehicles per unit time, q is the arrival rate in vehicles per unit time, x is the degree of saturation $= q/\Lambda s$, and p is the maximum acceptable degree of saturation. All these quantities are also relevant to pedestrian streams with flows in pedestrians per unit time but, at most junctions in developed countries, the degrees of saturation for pedestrian streams are small. Calculations of capacity and delay are relevant to pedestrian streams in so far as the pedestrians obey the signals and do not divert to other crossing places when they encounter signals at red.

1.2 Indicators of Performance

The traffic capacity of a stream with given signal timings is the average flow of traffic per unit time when every green time is saturated. In the notation this is Λs and if traffic arrived at this rate the degree of saturation would be unity. For practical and theoretical reasons, it is wise to design for operation at lower degrees of saturation; p is therefore usually taken as 0.9 or less and an aim is to keep $x \leq p$, if possible. The stream is said to have a practical capacity of $p \Lambda s$. In terms of queuing theory, x is the traffic intensity of the queue of traffic in the stream.

The delay caused by signal control to a vehicle in a stream is the amount by which the time the vehicle takes to negotiate the junction exceeds the time it would have taken if the signal controlling the stream were always green. For purposes of calculation, this delay may be regarded as occurring at the stop line. The delay is estimated by analyzing the process of queuing at the stop line as implied by this model and arises from three causes: randomness in the arrival of approaching vehicles in the stream, overload occurring when the average arrival rate exceeds capacity for several minutes or longer (usually in peak periods), and alternation of red and green, which causes delay even to traffic arriving perfectly uniformly at a rate less than capacity. Delay arising from the first two causes is called random delay (short for random and overload delay) and additional delay arising from the third cause is called uniform delay. They are estimated separately.

Random delay is reflected in queuing continuing until the latter part of the green period and the occurrence of saturated cycles. It depends strongly on the degree of saturation. The capacity changes relatively little over time, unless the signal timings are altered. The arrival rate (averaged over several minutes) and, hence, the degree of saturation can, however, change quite rapidly, especially during peak periods, and it is important to take such changes into account in estimating delay. This can be carried out by dividing the period over which delay is to be estimated into intervals of, for example, 10 min or 15 min during which the average arrival rate can be regarded as constant.

Various approximate time-dependent expressions have been developed for the expectation of the number of vehicles queuing at the end of green at the end of such a period and the average random delay per unit time D_r over the period (Kimber and Hollis 1979, Burrow 1987) in terms of c, Λ, q, s, and for the duration of the period and the number of vehicles queuing at the end of green at the start of the period. These expressions all assume that the headways between arriving vehicles have an approximately exponential distribution, which is appropriate for junctions that are some distance from other signal control, but not for closely spaced signal-controlled junctions on arterial roads or in networks. The uniform delay per unit time D_u can be estimated simply in terms of c, Λ, q and s from the model in Sect. 1.1. Delay per unit time has the dimension vehicles and represents the average excess of the number of vehicles in the stream in the neighborhood of the junction over the number that would be present if the signal were always green. Examples of graphs of D_r and D_u as functions of q and Λ are shown in Fig. 2. For streams having $x < 1$, a range of approximate steady-state expressions for $D_r + D_u$ have been derived and shown to be substantially equivalent for practical purposes (Allsop 1972a, Hutchinson 1972), but their use is not recommended when x is close to unity because of the long time then taken to reach the steady state. Special expressions for delay have to be used for streams of turning traffic that turn through gaps in an opposing stream (Allsop 1977) or that receive more than one green period in the cycle.

2. The Signal Cycle

The order in which different streams have green is determined entirely by the signals and, in specifying this order, it is therefore unnecessary to distinguish between traffic streams that always receive identical signal displays. Such a set of streams is called a group. Two groups are said to be compatible if every stream in one group can safely have green at the same time as every stream in the other. If two groups are incompatible then, for safety reasons, a certain period called the clearance time must elapse between the end of displayed green for one and the start of displayed green for the other.

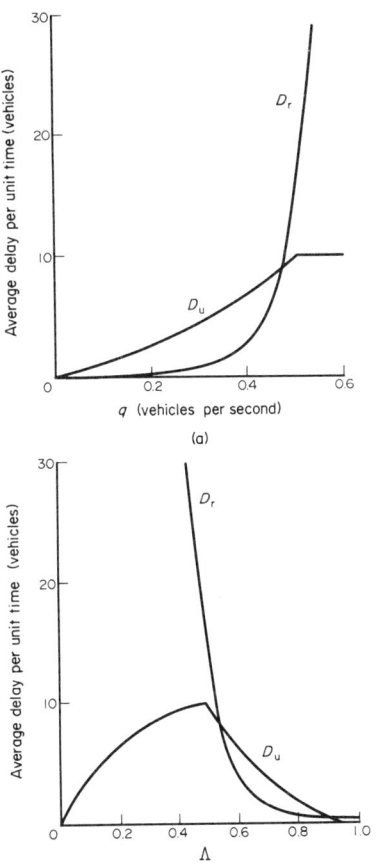

Figure 2
D_r and D_u, averaged over a 10 min period in which average arrival rate is steady, saturation flow is 1 vehicle per second, cycle time is 80 s, effective green time is constant and initial queue at end of green is 10 vehicles, vs (a) q and (b) Λ

In fixed-time operation, specified sets of mutually compatible groups receive green in turn for specified times in each cycle. The composition of the sets of groups and their order in the cycle together form what is called a sequence. A part of the cycle in which a particular set of groups has green and the signals are not in the process of changing is called a stage. So that traffic is not denied green unnecessarily, the groups that have green in a stage usually form a maximal set of mutually compatible groups. This means that at each change of stage, one or more groups loses green in order that one or more others may gain it and the losers are incompatible with the gainers. The relevant clearance times must, therefore, be respected at each change of stage, and an arrangement of ends of green for losers and starts of green for gainers that respects all the clearance times is called an interstage structure. If the number of groups losing green and the number gaining it both exceed one, there will, in general, be more than one structure providing the required clearance times, each with its own implications for the overall signal timings.

For any given sequence and interstage structures, the duration of each stage can be defined as the intersection of the displayed greens of the groups having green in the stage. The time between one stage and the next is then called the interstage time, which is determined by the clearance times and the interstage structure. The stage durations are then sufficient to determine the cycle time, the beginning and end of displayed green for each group and, hence, of the effective green for each stream. This is the basis of the stage-based approach to calculation of signal timings. In this approach, it is convenient to define the effective green time for a stage as the intersection of the effective green times for the streams that have green in the stage.

In many cases, the choice of sequence is restricted to those in which each group has only one green period per cycle and the calculation of timings is discussed here in terms of such sequences. Some procedures mentioned extend, however, to sequences in which designated groups may have two separate green periods per cycle, as is sometimes useful.

3. Practical Constraints on Signal Timings

For various practical reasons, some or all of the following constraints may apply to the signal timings.

(a) The cycle time should lie in a certain range or take a specified value.

(b) Clearance times between incompatible groups should be respected.

(c) Green times or red times for particular groups should lie in certain ranges.

(d) The durations of particular stages should lie in certain ranges.

In addition, as discussed in Sect. 1.2, it is desirable that each stream should have enough green to enable all the traffic arriving in the stream to pass the signal in the long run, with a small margin to allow for variability. If this is practicable, it is achieved by requiring that

(e) $x \leq p$ for each stream.

If queuing space is limited on roads leading to the junction, it may be desirable that

(f) the queue lengths in the various streams should not exceed certain maxima.

Constraints (e) and (f) are explicitly dependent upon the arrival rates, whereas (a–d) can be specified independently of the pattern of traffic.

4. Optimization of Timings

Subject to constraints, the signal timings can be optimized with respect to any of several criteria. To determine what common multiple of the specified arrival rates can be accommodated by the junction, and with what timings, the problem is to

$$\text{maximize the common multiple of the arrival rates} \qquad (1)$$

In problem (1), the constraint (e) is applied to the common multiple of the specified arrival rates. Solving this problem maximizes the traffic capacity subject to proportional changes of flow in all streams.

For any given multiple less than the maximum value, the steady-state delay per unit time to traffic in each stream can be estimated and, for any given multiple, the average delay per unit time over a specified period can be estimated. In either case, it is useful to find timings to

$$\text{minimize the delay per unit time to traffic at the junction} \qquad (2)$$

The delay per unit time at the junction is obtained by summing the values for the various streams.

Again, for any given multiple less than the maximum value, it is possible to

$$\text{minimize the cycle time} \qquad (3)$$

This objective arises in the coordination of signals at different junctions.

Problems (1) and (3) are relevant not only to isolated junctions, but also to closely spaced junctions on arterial roads or in networks. However, problem (2), because it involves use of an expression for delay per unit time, is relevant only to isolated junctions. For a discussion of the calculation of timings to reduce delay in the coordinated operation of signals at closely spaced junctions see *Road Network Control*. The control variables are a mixture of discrete variables, determining the sequence and the interstage structures, and continuous variables, determining the durations of the stages.

In the stage-based approach, these problems can be formulated (Allsop 1971, 1972b) for any given sequence of m stages and corresponding interstage structures in terms of the following control variables: λ_0 is the proportion of the cycle not effectively green for any stage, λ_i is the proportion of the cycle effectively green for stage i ($i = 1, 2, \ldots, m$) and μ is the common multiplier applied to all arrival rates. Since $\lambda_0 c$ is of known duration, λ_0 determines the cycle time which, together with the λ_i, determines the whole set of timings. Problem (1) is linear in the λ_i and μ, and, when one of the steady-state approximations (Webster and Cobbe 1966) for $D_r + D_u$ in each stream is used, problem (2) is convex in the λ_i for given μ less than the maximum. Solutions to these problems incorporating many of the constraints (a–e) have been implemented in readily usable computer programs (Allsop 1976, Burrow 1987). Problem (3) is also linear in the λ_i and is soluble for any given μ less than the maximum. The solution of problem (2), formulated in this way, has been extended to include constraint (f) and the estimation of delay by a time-dependent expression (Reljic 1988). This permits delay minimization subject to queue-length constraints over a period during which arrival rates exceed capacity and the nonconvexity implicit in Fig. 2 is avoided by neglecting the change in gradient of the uniform delay curve at $x = 1$. Another computer program, less rigorous in optimization but providing many useful features, has been developed by Akcelik (1981) on the basis of a comprehensive overview of the design of signal-controlled junctions.

5. Limitations

The principal limitation of the stage-based approach lies in the fact that optimization of the stage durations presupposes that the sequence of stages and the interstage structures have been specified. At the simplest junctions or where the traffic engineer wishes to consider only a limited range of possibilities, there is little difficulty in enumerating the sequences and interstage structures that are to be considered, and optimizing signal timings for each of them. Even in such cases, however, this process can be tiresome and, at many junctions, the number of possibilities is too great for manual enumeration to be practicable. Techniques of graph theory have been used (Tully 1977) to generate, from information about the compatibilty or otherwise of the groups of streams at any given junction, a set of sequences that contains all relevant sequences as subsequences. However, the problem of enumerating the interstage structures, as well as of identifying those subsequences and associated interstage structures that yield good optimal timings and meet various constraints on sequencing that arise from signal control practice in different countries, remains formidable at all but relatively simple junctions.

A promising way forward is to transfer the specification of interstage structures, subject to the required clearance times, to the continuous part of the optimization; that is, to the calculation of timings, thus leaving the identification of the sequence as the only discrete choice. This can be achieved by taking the cycle time (in fact, its reciprocal) and the starting point and duration of displayed green for each group of streams, expressed as proportions of the cycle, as the continuous control variables. Such a choice of control variables forms the starting point for the alternative phase-based approach to signal control and associated calculations, so-called because traffic engineers refer to the timed succession of signal indications displayed during the cycle to a particular group of streams as a phase (see *Signal Control at Individual Junctions: Phase-Based Approach*). This is already leading to user-oriented implementations of solutions to the resulting wider optimization problems.

See also: Road Traffic: An Introduction; Signal Control at Individual Junctions: Phase-Based Approach

Bibliography

Akcelik R 1981 Traffic signals: Capacity and timing analysis. Australian Road Research Board Report No. ARR 123. ARRB, Nunawading, Australia

Allsop R E 1971 Delay-minimising settings for fixed time traffic signals at a single road junction. *J. Inst. Math. Its Appl.* **8**, 164–85

Allsop R E 1972a Delay at a fixed time traffic signal I. Theoretical analysis. *Transp. Sci.* **6**, 260–85

Allsop R E 1972b Estimating the traffic capacity of a signalized road junction. *Transp. Res.* **6**, 245–55

Allsop R E 1976 SIGCAP: A computer program for assessing the capacity of signal-controlled road junctions. *Traffic Eng. Control* **17**, 338–41

Allsop R E 1977 Treatment of opposed turning movements in traffic signal calculations. *Transp. Res.* **11**, 405–11

Burrow I J 1987 OSCADY: A computer program to model capacities, queues and delays at isolated traffic signal junctions. Transport and Road Research Laboratory Report No. 105. TRRL, Crowthorne, UK

Hutchinson T P 1972 Delay at a fixed time traffic signal II. Numerical comparisons of some theoretical expressions. *Transp. Sci.* **6**, 286–305

Kimber R M, Hollis E M 1979 Traffic queues and delays at road junctions. Transport and Road Research Laboratory Report No. LR 909. TRRL, Crowthorne, UK

Kimber R M, McDonald M, Hounsell N 1985 Passenger car units in saturation flows: Concept, definition, derivation. *Transp. Res. B* **19**, 39–62

Reljic S 1988 TRAFSIG: A computer program for signal settings at an isolated under- or oversaturated fixed-time controlled intersection. *Traffic Eng. Control* **29**, 562–6

Tully I M Z 1977 Synthesis of sequences of traffic signal controllers using techniques of the theory of graphs. Ph.D. thesis, University of Oxford

Webster F V, Cobbe B M 1966 *Traffic Signals*, Road Research Technical Paper No. 56. Her Majesty's Stationery Office, London

R. E. Allsop
[University College London, London, UK]

Simulation of Urban Traffic: Software Environments

Computer simulation has always been considered as a tool especially well suited for answering what . . . if questions, and this assertion has been regarded as especially true in the field of traffic engineering. Since the advent of electronic computers in early 1950s (Blum 1964, 1970), traffic simulation was thought particularly useful for traffic studies and, since then, a wide variety of specific or general purpose traffic simulators has been built.

Simulation modelling is the activity of building simulation models that properly represent the features of the system object of study, so a good traffic simulator should be based on a model of the traffic system under study that adequately represents its entities, attributes and relationships; that is, characteristics of the traffic network such as links, link capacities, intersections, turning movements and traffic flows (Wattleworth 1976, Kaltenbach 1984, Improta *et al.* 1986), traffic flow models such as microscopic leader–follower models (Herman *et al.* 1959) and macroscopic fluid models and platoon dispersion models (Michalopoulos 1986), traffic control plans, and so on (see *Simulation Programs, Macroscopic*).

Summarizing, in order to give proper answers to what . . . if questions in traffic engineering, a traffic simulator must satisfy, among others, the following functions.

(a) It must be a tool to analyze and obtain a better understanding of traffic behavior, namely for urban traffic studies.

(b) It must be built in such a way so as to allow a flexible design of simulation experiments for testing different modelling hypotheses about traffic behavior, and so on.

(c) It must provide the traffic engineer assistance with the design and computation of traffic control plans.

(d) The analysis of traffic behavior under different modelling hypothesis should serve as a basis for the process of defining control strategies leading to improved control plans.

(e) Usually, control plans computed from a heuristic or a mathematical model must be tuned before being applied (Gershwin *et al.* 1978, Hall *et al.*

1980). A traffic simulator should serve as basis of a tuning process using its simulation experiment design capabilities to reduce the tuning times substantially.

(f) Control strategies lead in certain circumstances to alternative control plans that must be evaluated before a decision is made. A traffic simulator should work as a decision support system for traffic engineering purposes and that requires that it incorporates the function of an easy evaluation of alternative control plans.

(g) Traffic control should be understood in many cases as one of the main functions of a more complex system—a traffic management system (Gartner *et al.* 1980, Hall *et al.* 1980). Traffic management systems propose traffic management schemes involving network operation conditions (e.g., tidal flows responsive systems) and traffic control strategies. Such traffic management schemes should be tested before implementing. Traffic simulators must be an easy-to-use testing tool.

Unfortunately, most of the traffic simulators built to date fail to cover these functions properly. Users of traffic simulation systems report the difficulties found in practice. Problems appear, typically, in the following areas:

(a) generation of the data required for the simulator;
(b) building of the simulation model; and
(c) design and performance of the simulation experiments, including the analysis and interpretation of the results provided by the simulation run.

These flaws are common to all traffic simulation systems but are especially true of microscopic simulators. To overcome these problems, different approaches have been proposed and developed. The first approach, perhaps the most common and widely used, consists of building *ad hoc* interfaces for one already existing traffic simulation system; other approaches bring to the traffic simulation domain the current trends in simulation theory and practice (see *Simulation Programs, Microscopic*).

Examples of the first approach are the last versions of NETSIM, NETWIM/ICG (Chin and Eiger 1982) and TRAF-NETSIM version 2.00 (Rathi and Santiago 1988).

Chin and Eiger group the above deficiencies into three groups: data input, data debugging and output analysis. With regard to the three problem elements associated with the use of the NETSIM program, a set of three interactive computer graphics (ICG) interfaces have been added to aid in reducing or even eliminating many of these difficulties. The three ICG programs are PRE-NETSIM (PRENET), NETSIM DISPLAY (NETDIS) and NETSIM POST-DISPLAY (POSDIS). Figure 1 depicts the overall framework of NETSIM/ICG.

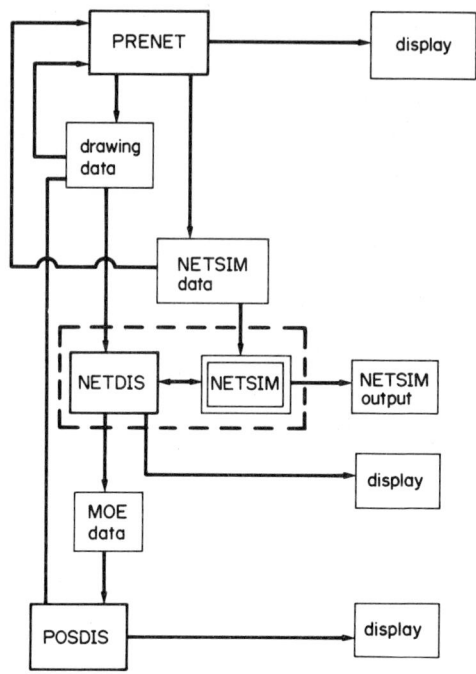

Figure 1
Overall framework of NETSIM/ICG

The PRENET program provides the capability for interactive data input and modification, data display, and preparation of the input data files for both NETSIM and NETDIS.

The NETDIS program runs in conjunction with NETSIM and provides real-time display of link-specific measures of effectiveness (MOES) generated by NETSIM and, in addition, prepares a display file for POSDIS.

The POSDIS program provides passive displays of user-selected link-specific MOES.

Rathi and Santiago (1988) report the enhancement of input–output capabilities through the addition to NETSIM of an input processor providing an echo print of the user input file and tables containing the user input information in logical groupings for link information, turning movement data, fixed-time and actuated control data, entry link traffic volume data, and so on, and the output processor.

The following are other enhancements of NETSIM, based on the same approach.

(a) *GTRAF* (Andrews *et al.* 1987), which is a mocrocomputer-based interactive graphics program that provides users with the highly effective capability of reviewing the input and analyzing the results generated by the NETSIM simulation program.

(b) The integrated traffic data system (ITDS) (Santiago 1985) which is an integrated microcomputer-based database system designed to store, maintain and update traffic network information in a centralized database that can be automatically queried to create input data files for traffic simulation and network timing optimization models included in TRAF.

(c) The NETSIM input data editor (NEDIT) (Santiago 1985) which is a microcomputer-based program that can be used to modify and create input data sets for the NETSIM models. It is a menu-driven system that assists the user in entering the input data but runs separately.

The other approach mentioned proposes to imbed the model-building process into a simulation environment, in other words to apply the concept of software architecture for simulation environments to traffic simulation. This means integrating the traffic simulator into a software system which can be understood as a computer-assisted modelling tool for traffic simulation.

1. Artificial Intelligence Approach

To accomplish the functions defined previously, a traffic simulator must be a user-friendly computer system, which means that, to solve the simulation data input collection problem, the traffic simulation systems must include components such as an automatic data input generation subsystem and an interactive data handling subsystem.

To overcome the difficulties in building traffic simulation models, the traffic simulation system must include model-building facilities. Among them is proposed the one known as automatic simulator generator; that is, "an interactive software tool that translates the logic of a model described with a relatively general symbolism to a language code as a computer executable program" (Mathewson 1984). This means developing a system that will allow a simulation model to be built by selecting appropriate modules from an available library and linking them together.

To facilitate the design, performance and interpretacion of simulation experiments, the traffic simulation system must include facilities to design and run simulation experiments interactively. However, a simulation system with all the capabilities just described is what is known as simulation intelligent front end—a kind of expert system interfacing the user that generates the simulation program code as a result of the dialog with the user, runs the simulation, and helps the user to analyze the simulation results interactively. Consequently, the purpose is to design and build a simulation intelligent front end for traffic simulation, specifically for microscopic urban traffic simulation.

Once the problem has been stated in such terms, the question becomes what is the best approach to design and build this system? Among the current trends to combine computer simulation and artificial intelligence tools, three areas of design are of particular interest: advanced simulation methodologies (Oren and Ziegler 1979); simulator generators (Davies 1979, Mathewson 1984, Haddock 1987); and knowledge-based simulation systems (Ramana *et al.* 1986).

1.1 Advanced Simulation Methodologies

Henriksen (1983) proposes an approach to implement the main concepts of advanced simulation methodology, which he calls the software architecture for simulation environment, whose conceptual structure is shown in Fig. 2.

The starting point for Henriksen is the methodological chain model design, program design, model programming and the links between model programming and model input specification of experimental design on one side and output analysis module on the other side. The software architecture he proposes has, as components, the following.

(a) *Syntax-directed editors*. These are "aware" of the syntactic structure of a programming language.

(b) *Model-preparation facilities*. These are software components which operate on the objects and comprise the following:
  (i) *Model editor*. This produces an internal representation of a model from input specifications and by manipulating previously entered specifications.
  (ii) *Input preparation subsystem*. This is used for transforming raw input data into forms usable in a model. It includes the capability of fitting distributions to data.
  (iii) *Model design languages*. This is used to specify the components and rules of operation of the system to be modelled.
  (iv) *Knowledge base*. This is the central repository of all information about a model.

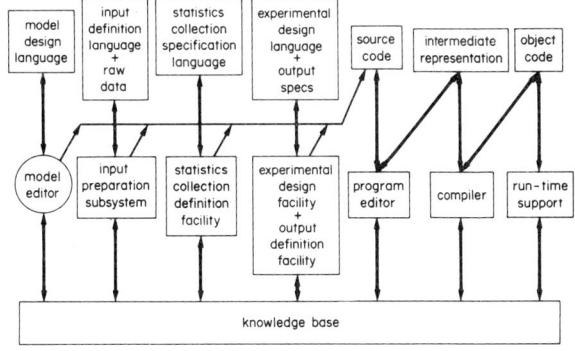

Figure 2
Henriksen's software architecture for the simulation environment

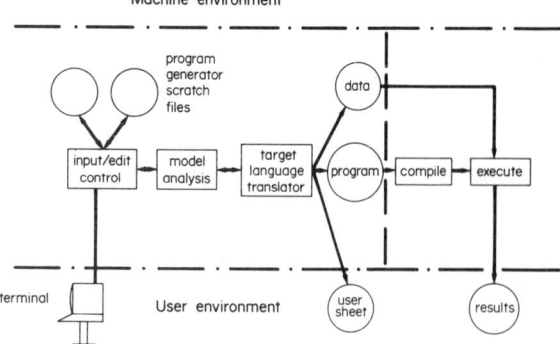

Figure 3
Structure of the program generator

1.2 Simulator Generators

Mathewson (1984) defines a program generator as an interactive software tool that translates the logic of a model described in a relatively general symbolism into the code of a simulation language and so enables a computer to mimic model behavior.

This facility gives the user the benefits of a simple symbolic input (especially valuable for use by those not conversant with programming practices) combined with the power of a high-level language in which to develop models of complex systems.

The development of this concept creates a tool to integrate all aspects of the use of simulation in problem solving. The DRAFT system, developed by Mathewson, represents a practical implementation of this concept.

"DRAFT is a family of program generators whose structure is formed by a group of modular units linked as shown in Fig. 3. The input/editor module accepts the model description, identifies minor semantic errors (e.g. the use of reserved variable names), and lets the user correct them. It also creates a backup copy of the input on a temporary file, which later can be accessed for correction or stored as a compact copy of the model. The analysis module checks the input for errors, which may often be corrected on-line, and prepares a coded file of the entity interaction within the model. This file forms a general input to the program writer selected by the user. It is only after this stage that the model description is mapped into the particular structure—event, activity, process—associated with the target language. The code is then produced."

To describe the logic of a model in a relatively general symbolism, Mathewson proposes the specification of a model as an entity cycle diagram, given that life cycles of the individual entities taken together form a complete specification of the model. With the assumption of a number of simple queuing rules, entity cycles can be assembled in a network diagram. The specification of a model as a network allows the user to have access to other more general network tools.

Finally, SIMSCRIPT II.5 is proposed by Mathewson as the target language in his experimental version of DRAFT.

1.3 Knowledge-Based Simulation Systems

Knowledge-based simulation (KBS) systems (Ramana et al. 1986) are object-oriented modelling systems that contain attribute and behavioral descriptions and provide interactive access and display. The model is the kernel of the system which can be understood as an interpreter that accesses the model and provides simulation, model checking and data analysis.

A KBS model is a collection of schema representation language schemata that represents physical and abstract system entities. The schema is the basic unit that represents objects, processes, ideas and so forth. Schemata may form networks. Each slot in a schema may act as a relation tying the schema to others. The schema may inherit slots and their values along these relations.

Model creation in KBS is simply the creation of schemata that represent the entities of the model, including specifications for their event behavior and interconnections with other schemata.

The KBS kernel interprets the model once it has been defined and uses a discrete-event simulation approach.

Some of the facilities that should be included in the simulation environment represented by a KBS are (Ramana et al. 1986):

(a) different programming paradigms such as object-oriented, logic, data-oriented and rule-based programming;

(b) behavioral representation of system entities through an object-oriented (frame-based) knowledge representation that lets entities be altered without altering the simulation model interpreter;

(c) alternate command interfaces such as text commands, natural language, and graphics;

(d) model validation to save time otherwise spent verifying model consistency and completeness;

(e) interactive access to model building and simulation through a command interface using windows and graphics;

(f) color graphics to depict models statically and dynamically;

(g) selective report generation to schematically design reports about the simulations goals; and

(h) simulations integrated with expert systems to examine the performance of a scenario and suggest model modifications.

Figure 4 shows the ideal conceptual structure of a KBS system.

The three approaches described thus far in Sect. 1 represent three independent attempts to go beyond the

Simulation of Urban Traffic: Software Environments

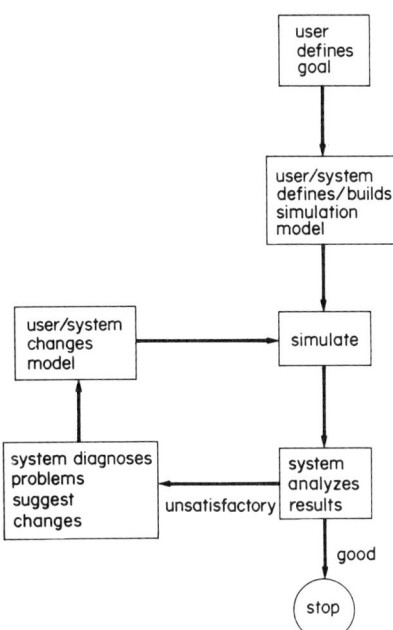

Figure 4
Conceptual structure of a KBS system

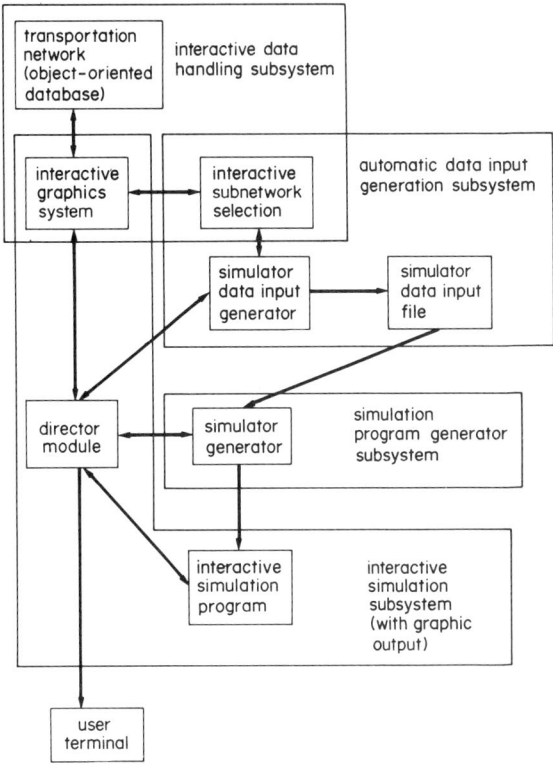

Figure 5
Conceptual structure of the computer-assisted modelling system for traffic simulation

classical simulation concepts, creating new hybrid systems through a combination of simulation and artificial intelligence concepts. All three have been conceived as general purpose approaches and careful analysis identifies the many common aspects that all three share. From these considerations, the following approach is proposed.

First, use the concepts common to the three approaches described above for designing a simulation environment that imbeds a traffic simulator.

Second, instead of developing a general purpose approach, design the components of such a simulation environment as special purpose components exploiting the features and properties of traffic systems.

That means integrating the traffic simulator into a software system that can be understood as a computer-assisted modelling system for traffic simulation. According to the proposed approach, such a software system consists of three main structural components: a knowledge database object oriented and an intelligent front end with two subsystems; an automatic data-generation subsystem; and a simulation program generator subsystem. These three structural components are linked together and run in an interactive environment. The block diagram in Fig. 5 shows the conceptual structure of the proposed system.

The knowledge database (KDB) contains the information relative to the objects describing the traffic network—links, nodes, turning movements, link capacities, traffic flows, traffic control parameters, and so on. As far as the system (the traffic network) to be represented through the data, facts and rules contained in the KDB, admits a geometric representation, it is natural to conceive that the editor that allows the user the creation and updating or modifying the representation of such a system be an editor with graphics capabilities, allowing graphics displays either of the whole network of subnetworks accessed through the windowing capabilities.

In implementing a KDB like the one proposed, the knowledge representation to be used is very important given that one of the main functions of the KDB is to provide the simulator generator with the information required for the model-building process. Taking into account the concepts introduced in the description of the KBS systems, the frame structures used to represent entities and attributes can be organized to serve as schema descriptions useful for the KBS model-building process.

The automatic data input generation subsystem has two main components. The first component is, in fact, a special function of the graphic network editor which allows the user to select the subnetwork of the traffic network to be studied by simulation by means of the windowing facility of the graphic system. The second component accesses the KDB and creates a special file

487

containing the schemata and the related information corresponding to the selected subnetwork.

The simulation program generator subsystem builds the simulation model corresponding to the selected subnetwork, analyzing the file created by the automatic data generation subsystem, linking together the schemata corresponding to the entities of the selected subnetwork according to the complementary information and checking the resulting model for completeness and consistency according to the operating rules.

The resulting simulation model may be understood as a set of data structures especially linked. The simulation program is built by linking such a set of data structures to a main program that represents the logic of a general microscopic traffic simulator (see Sect. 2).

Simulation environments designed and implemented following the proposed approach are a solution to the traffic simulation problems listed above. The kind of KDBs suggested, together with the graphic network editor and the data generation subsystem, represent a solution to the data management problems appearing in traffic simulation studies; an intelligent front end with a simulator generator constituting a model-building facility (with the described features) is a feasible solution to the problem of building traffic simulation models for experts in traffic engineering, but without specific training in computer programming and simulation languages.

Finally, this simulation environment for traffic simulation allows an interactive run of the simulation programs, including dynamic model-manipulation capabilities, and provides the user with an experimental framework for designing, running and interpreting the simulation results; it solves the problem arising, in practice, in performing traffic simulation studies, mentioned previously.

2. System Implementation

The design and implementation of a traffic simulation environment of the characteristics described was the objective of a project developed for the Transportation and Traffic Department of the Municipality of Barcelona by the Technical University of Catalonia, that led to an experimental prototype (Barecló 1988).

The main information required to build a microscopic traffic simulator is the following.

(a) *Link network data.* Coordinates of link ends, number of lanes, number of reserved lanes (e.g., bus lanes), lane width, lane capacity, link flows and turning movements allowed for each lane.

(b) *Traffic control data.* Traffic lights acting on each link and intersection and traffic control parameters of each traffic light—cycle, split and offset.

Given that a substantial amount of this information is also the information required for other types of transportation studies, the prototype network database was organized as two linked databases, one containing the information about links and intersections in the network, and the other one including the traffic control information and other complementary data.

The first database is just the one containing the implementation common to other transportation studies such as transportation planning studies. The network database system from the EMME/2 system for transportation planning, developed at the University of Montreal (Babin *et al.* 1982), has been used as first database, and the network editor and graphics facilities of that system as part of the equivalent facilities in the prototype.

Using the windowing facility of the graphics system, the user can select, on the graphics display of the transportation network, the subnetwork chosen to be simulated. Figure 6 displays an example of the process of selecting a subnetwork containing information on traffic flows, from the Barcelona base network, through the interactive definition of the corresponding window in the graphics workspace.

An additional EMME/2 facility has been developed to access the information stored in the network database, relative to the subnetwork defined through the windowing process. This facility, called dump window facility, dumps out the required information creating an auxiliary file. This is the first step in the practical implementation of the automatic data input generation subsystem. Once the auxiliary file has been created, the second step is performed by a computer program that accesses the second database, linked to the first, to collect the traffic control data and complementary information corresponding to the subnetwork selected in the first step. Once all the information required by the simulator has been collected, it is structured in a frame way to build the schemata that will constitute the model.

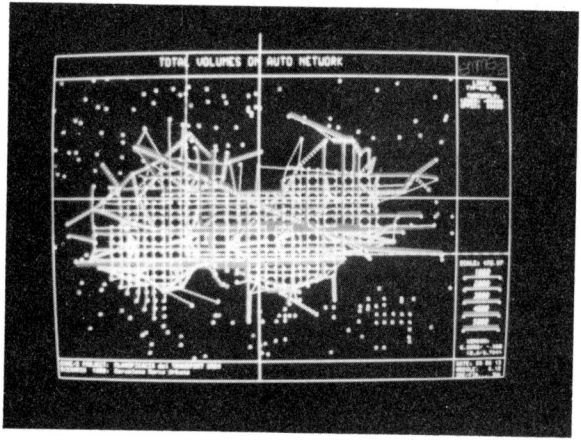

Figure 6
Interactive subnetwork selection using the windowing facilities

Simulation of Urban Traffic: Software Environments

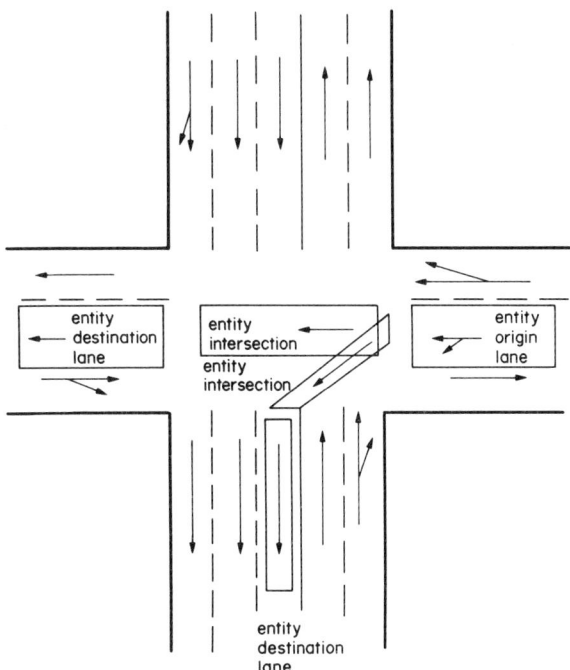

Figure 7
Schema intersection entity

Schemata, described as frames, correspond to the entities of the simulated system. There are two main classes of entities—lane entities and intersection entities. Lane entities can be origin entities or destination entities depending on the flow direction. An intersection entity is, in fact, a set of entities that connect lane origin entities to lane destination entities (see Fig. 7).

The simulation model building is performed in two steps, such as has been described in Sect. 1. The first step corresponds to the KBS model-building logic. The simulation model is the set of data structures that results from linking together the schemata for the entities following the subnetwork operating conditions (e.g., directions, turning movements, origin–destination relationships between links). The resulting data structures continue a description of the subnetwork being simulated.

The simulation program is the result of linking the data structures with a simulation module whose structure is shown in Fig. 8.

The simulator updating is performed through an activity scanning process at fixed time intervals. This process computes the state of the traffic lights according to the traffic control plan given as input, the individual vehicle data (as corresponds to a microscopic simulation—position (coordinates), speed, and acceleration), generate the new arrivals to the system as a function of the traffic flow model (e.g., a headway distribution of given parameters) and refreshes the graphics display showing the state of the system.

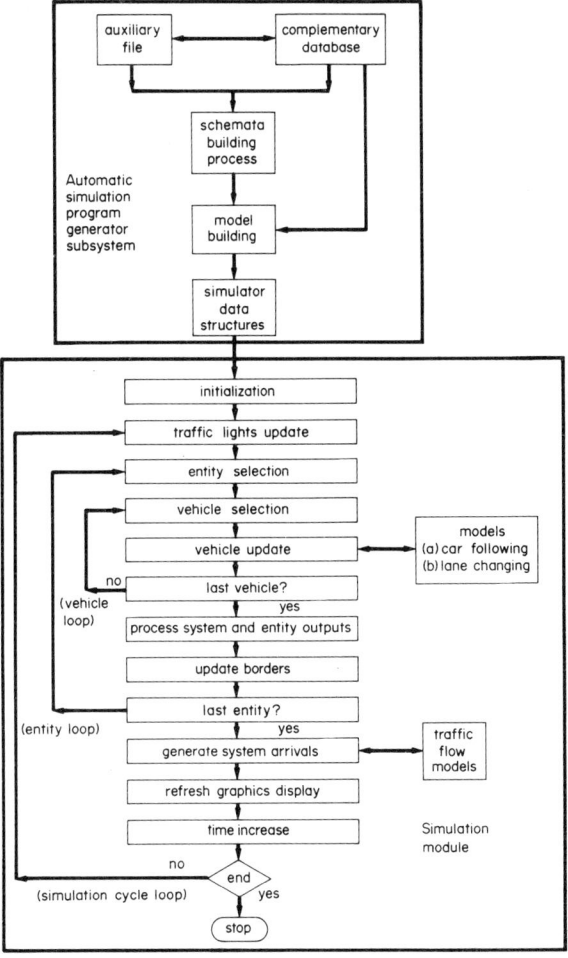

Figure 8
Structure of the simulation module

The individual vehicle data updating is done entity by entity (vehicles belong to entities and can move from one entity to another). The updating process is governed by two models:

(a) a lane-changing model that works only when it is desirable and feasible among the entities (lanes) belonging to the same link; and

(b) a car-following model (or leader–follower model) between vehicles from the entity (lane) when no lane changes occur (see *Car-Following Models*).

The simulation module has been programmed in a modular way that allows the user to modify easily all the underlying submodels—traffic flow models, lane-change models, car-following models, and so on.

The simulation runs in an interactive way that can be interrupted by the user at any time; a dialog subsystem then allows changes to be made in the model parameters. Figure 9 shows an example of a screen from the

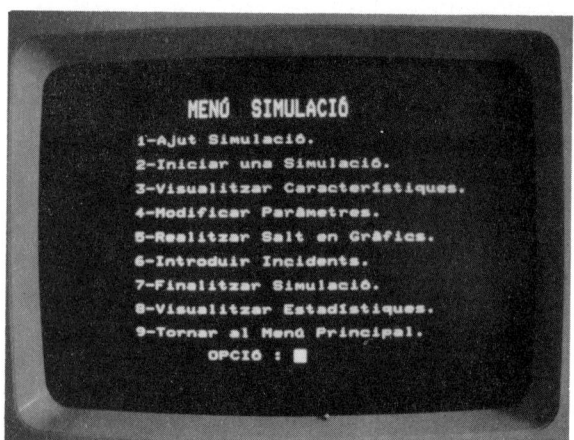

Figure 9
Example of menu-driven dialog

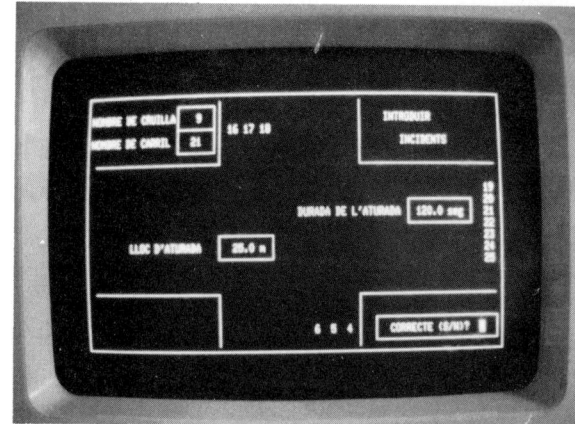

Figure 11
Example of menu-driven dialog for incident definition

menu-driven dialog. This is the basis of the simulation experiment design support system.

Animated graphics displays of the state of the system, such as the ones displayed in Fig. 10, provide a visual interpretation of the simulation results. Queues and bottlenecks are easily identified and evaluated. However, the dialog subsystem also allows numerical estimates of queue lengths, average delays, and so on to be obtained as desired.

Special purpose screens of the menu-driven dialog, such as the one shown in Fig. 11 displaying the structure of an intersection, allow the user to define special events, such as lane blocking, to analyze the impact on the system behavior and performance.

3. Future Trends

The experience obtained by mid-1989 has definitely been positive. A new version is being worked on that will include an improved KDB system and two new features—a traffic control plan module and an evaluation and diagnosis module.

While, in the present version, traffic control plans are defined as data input, in the future version, the traffic control plan module will be capable of computing and testing demand-responsive control plans.

The evaluation and diagnosis module will be a further step in the incorporation of artificial intelligence components. This module is being designed as a hybrid expert system that will perform both symbolic and algorithmic computations. It will complete the KBS structure including automatic results interpretation capabilities.

This development plan is now part of a project of the DRIVE Programme of the EC (see *DRIVE*).

See also: Road Traffic: An Introduction; Simulation Programs, Microscopic

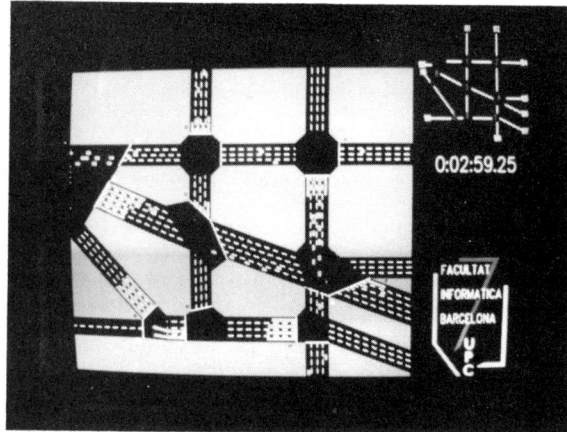

Figure 10
Animated graphic display of the subnetwork being simulated

Bibliography

Andrews B, Liberman E, Santiago A 1987 GTRAF: The NETSIM graphics systems. *Proc. North American Conf. Microcomputers in Transportation*. American Society of Civil Engineers, New York, pp. 103–13

Babin A, Florian M, James-Lefevre L, Spiess H 1982 EMME/2. Une méthode interactive-graphique pour la planification du transport urbain mutimodel. Université de Montréal, Centre de Recherche Sur les Transports Publication No. 204-F

Barceló J 1988 Inteligencia Artificial y Simulación: Aplicaciones a la Simulación de Tráfico Urbano. *Automática e Instrumentación* **181**, 239–45

Blum A M 1964 A vehicle traffic simulator. *IBM Syst. J.* **3**(1), 41–50

Blum A M 1970 general purpose digital traffic simulator. *Simulation* **14**(1), 9–25

Chin S-M, Eiger A 1982 Network simulation interactive computer graphics program. *Transp. Res. Rec.* **835**

Davies N R 1979 Interactive simulation program generation. In: Zeigler B P, Elzas M S, Klir G J, Oren T I (eds.) *Methodology in Systems Modelling and Simulation*. North-Holland, Amsterdam

Gartner N H, Gershwin S B, Little J D C 1980 Pilot study of computer-based urban traffic management. *Transp. Res. B* **14**, 203–17

Gershwin S B, Little J D C, Gartner N 1978 Computer-assisted traffic engineering using assignment, optimal signal setting and modal split. US Department of Transportation, Report No. DOT-TSC-RSPA-78-10

Haddock J 1987 An expert system framework based on a simulator generator. *Simulation* **47**(2)

Hall M D, Van Vliet D, Willumsen L G 1980 Saturn: A simulation-assignment model for the evaluation of traffic management schemes. *Traffic Eng. Control* **21**(4), 168–76

Henriksen J O 1983 The integrated simulation environment (simulation software of the 1990s). *Oper. Res.* **31**, 1053–73

Herman R *et al.* 1959 Traffic dynamics: Analysis of stability in car following. *Oper. Res.* **7**, 86–106

Improta G, Allsop R E, Heydecker B G 1986 Network models for traffic management. *Int. Sem. Management and Planning of Urban Transport Systems from Theory to Practice*. Université de Montréal, Centre de Recherche Sur les Transports, Montreal, Canada

Kaltenbach M 1984 Network flow modelling for traffic control. Université de Montréal, Centre de Recherche Sur les Transports, Publication No. 348

Mathewson S C 1984 The application of program generator software and its extensions to discrete event simulation modelling. *IEE Trans.* **16**(1), 3–17

Michalopoulos P G 1986 Macroscopic models of traffic flow for simulation and control. *Proc. Symp. Traffic Control Systems*. Institut Catalá per al Desenvolupament del Transport Generalitat de Catalunya, Barcelona, Spain

Oren T I, Zeigler B P 1979 Concepts for advanced simulation methodologies. *Simulation*, **32**(3), 69–82

Ramana Reddy Y V, Fox M S, Nizwar Husain 1986 The knowledge based simulation system. *IEEE Software*, **3**(2), 26–37

Rathi A K, Santiago A J 1988 The new NETSIM: TRAF-NETSIM version 2.00 simulation program. *Prep. 68th Transportation Research Board Annual Meeting*. Washington, DC

Santiago A J 1985 ITDS: A database driven interface to traffic models using a microcomputer. *Proc. National Conf. Microcomputers in Urban Transportation*, American Society of Civil Engineers, New York, pp. 383–92

Wattleworth J A 1976 Traffic flow theory. In: Baerworld J E (ed.), *Transportation and Traffic Engineering Handbook*. Prentice-Hall, Englewood Cliffs, NJ

J. Barceló
[Universitat Politecnica de Catalunya,
Barcelona, Spain]

Simulation Programs, Macroscopic

The macroscopic simulation approach is to treat platoons of vehicles rather than individual vehicles. Vehicular traffic macroscopic simulation programs range from special purpose programs directed towards studying the impact of trucks on the traffic flow to general purpose programs that include most known variables of importance. Macroscopic simulation has attracted considerable interest since the 1960s. It began with the simulation of vehicles approaching and departing from isolated signal-controlled intersections. Since the late 1960s, simulation has been applied to freeway traffic and related features.

A careful examination of the existing models indicates that there was a lack of coordination in the development of models. There were no standards for the models and no application guidelines, which makes it difficult for users to determine which model to select for their needs. Because of the lack of a universally accepted traffic flow theory and varying operational characteristics, each model was developed largely through intuition. On the contrary, considerable knowledge of traffic flow phenomena has been obtained through simulation. The following sections briefly present the major macroscopic simulation models.

1. SIMAUT

The SIMAUT model is based on the hydrodynamic theory of traffic flow (shock-wave propagation (see *Kinematic Wave Theory*)) as it was proposed by Morin (1985a). The model continuously monitors the propagation and collisions of shock waves and displays a photograph of the different dynamic cells along the freeway in terms of average speed, flow and density at the end of each time slice; SIMAUT simulates both the freeway and the on- and off-ramps.

Apart from the physical description of the freeway, the following parameters are to be calibrated:

(a) flow-density relationship characteristics—free speed, capacity, capacity cut-down when passing to congested flow (hysteresis), density at capacity and jam density (maximum density);

(b) time-slice length (e.g., 2 min for on-line use, 15 min up to 30 min for off-line use);

(c) smoothing parameters (e.g., moving average, exponential smoothing);

(d) mean vehicle length, in order to convert occupancy rate into density;

(e) occupancy rate threshold between free flow and congestion (used for detecting real congestion in certain locations and, thus, specifying current measured traffic volumes); and

(f) speed thresholds between free flow, precongestion and saturation for display features.

The model has been validated on the A1 autoroute (freeway) north of Paris, which is one of the most heavily trafficked freeways in France, with many disturbing effects such as weaving stretches, congested exits, lane-number variations and a high percentage of trucks. These enhancements have made it possible to simulate the effect of queues backing up from the freeway (effects on exits and entrances) and from exits (effects on the main stream).

SIMAUT is quite easily implementable on any site and probably able to handle most freeway patterns.

The input data are

(a) traffic volumes at on-ramps;

(b) traffic volumes and occupancy rates at off-ramps and on the carriageway immediately downstream of the off-ramps;

(c) traffic volumes and occupancy rates at usual bottlenecks on the carriageway;

(d) traffic volumes entering the on-ramps from the surface street network and nominal metering rates of the on-ramps;

(e) traffic volumes exiting the off-ramps to the surface street network; and

(f) forecasts of entering demands (main line and on-ramps), forecasts of operational capacities and forecasts of exit percentages at off-ramps.

The output data are

(a) volume, speed and density in each homogeneous traffic cell (between two following shock waves) along the freeway;

(b) volume, speed and occupancy rate at any location specified in advance (detector simulation) on the freeway;

(c) queue lengths on the freeway and on the on- and off-ramps;

(d) travel times on the freeway and delays at ramps; and

(e) total number of vehicles observed, total travel times, total travel distances, mean speed and total and mean fuel consumption, and so on for each time slice and globally at the end of the period.

For on-line use, the layout of the freeway on the screen exhibits the current status of traffic flow in terms of queue lengths, travel times and delays, and this display is updated every 2 min. At the end of a period, it is possible to obtain diagrams showing the evolution of queuing and travel times during that period.

SIMAUT has been used for off-line evaluation of control strategies. For example, on the A6 autoroute, for the assessment of ramp control during weekend back rushes to Paris, and its results have been found to be in good accordance with the results of a survey (travel time, volume counts).

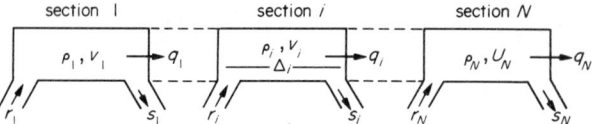

Figure 1
A freeway axis subdivided into sections

2. META

META (modèle d'écoulement du traffic sur autoroutes), a freeway traffic model developed by Papageorgiou et al. (1989), is based on a geometrical subdivision of a given freeway axis into several sections, the length of which may be chosen up to 1000 m. Typical section lengths are around 500 m. For each freeway section, META describes the time evolution of three traffic variables: traffic density, mean speed and traffic volume (see Fig. 1) (see *Freeway Traffic Modelling*).

The macroscopic traffic flow model META consists of four equations applying to each section of a given freeway axis.

The input data are the geometric characteristics of the freeway sections (length, number of lanes, existence of ramps):

(a) traffic volume $q_0(k)$ at the entry of the freeway axis;

(b) mean speed $v_0(k)$ at the entry of the freeway axis;

(c) traffic density (or occupancy rate) $\rho_{N+1}(k)$ at the exits of the freeway axis; and

(d) all on-ramp volumes $r_i(k)$ and off-ramp volumes $s_i(k)$.

As output data, the model provides the time evolution of density, mean speed and traffic volume for each freeway section.

Previous versions of the model were validated on a Los Angeles freeway and a German Autobahn (freeway). The present form of META was validated on the Boulevard Périphérique in Paris and the French A1 autoroute.

High accuracy of results was reported from all these studies. In particular, the model reproduces complicated real traffic phenomena such as discontinuities of volume-density diagrams, hysteresis phenomena and sudden speed breakdown in the case of congestion.

META can be used:

(a) as a simulation tool, with input data provided by the user (e.g., for testing efficacy of coordinated on-ramp metering strategies with hypothetical or historical input data);

(b) for on-line description of traffic conditions between two distant (e.g., 2–10 km or more) detector stations (the input data is provided by the two detector stations and on-ramp and off-ramp detectors, and full information about the internal traffic state can be furnished by META in real time);

(c) for on-line prevision of traffic flow based on prevision of the input data;

(d) for modelling of freeway networks after suitable modifications; and

(e) for systematic design of control strategies (state-space form).

3. Other Models

3.1 CONTRAM

In the UK, the Transport and Road Research Laboratory (TRRL) has recently begun studying queues and delays on freeway networks using a version of a model known as CONTRAM. This model was originally developed at the TRRL (Leonard and Tough 1979) for use in the assessment of urban traffic management schemes, but new developments by Kimber and Daly (1986) have extended the scope of the model to freeways and other high-speed road applications. It is unlikely that CONTRAM could sensibly be adapted for use on line, but off line it does offer a method of predicting the effects of traffic control strategies on freeway networks, in terms of traffic delay and other parameters.

3.2 DYNEMO

The DYNEMO model developed at the Karlsrühe Institut für Verkehrswesen (Schwerdtfeger 1984) is a micro/macroscopic model for freeway networks allowing for traffic management schemes, such as ramp metering and speed limits. This model claims to combine the computational simplicity of a macroscopic model with the individual vehicle modelling available from a microscopic model.

3.3 Arizona Transportation and Traffic Institute Traffic Simulation Model

The Arizona Transportation and Traffic Institute traffic simulation model was developed by Richard *et al.* (1965) to simulate freeway traffic that may be used to establish freeway interchange design criteria.

The freeway geometry is restricted to three through lanes, one ramp and an acceleration lane or an auxiliary lane. However, with minor program changes, one and two through-lane systems can be simulated. Ramps are restricted to direct connections and loop connections. Freeway slope is handled in this study by changing both the operating speeds and the vehicle acceleration and deceleration rates. Simple logic for vehicle distribution among lanes, car following, and lane changing is provided.

In preparing simulation runs, freeway volume and ramp volume are specified. There are three alternatives in choosing vehicle processing time, starting from 1.5 s with increments of 1.5 s. Vehicles are generated from a binary decision rule within the review period according to the input volume. This gives, essentially, a negative exponential distribution. Desired speeds are generated from a normal distribution with modification for trucks and slopes.

This model is probably flexible enough to allow some simple additional capabilities. Since the overall logic is very simple, it is doubtful that the model realistically represents traffic flow in any detail.

3.4 Midwest Research Institute Freeway Simulation Model

The purpose of the Midwest Research Institute freeway simulation model, developed by Kobett (1968), is to assist in the design of interchanges by providing a method for assessing the efforts of design variables on traffic capacity, safety and level of service. Special emphasis is therefore placed on traffic flow in the vicinity of entrance and exit ramps.

The freeway section can be up to 25 km long, with between two and four through lanes and up to six right-hand and six left-hand ramps (a maximum possible total of 12 ramps). The ramps can be any combination of on- and off-ramps located arbitrarily along the freeway section. Any or all of the on-ramps can be equipped with traffic signals and, therefore, the mode is capable of testing ramp-control strategies. The simulation vehicles are designated by driver type, vehicle type, desired speed and cooperator, which indicates that the vehicle will oblige would-be lane changers by trying to provide a usable gap in front. The volumes are specified for each lane and ramp.

3.5 Midwest Research Institute Mountainous-Terrain Model

The Midwest Research Institute mountainous-terrain model was developed to study traffic characteristics on four-lane divided highways in mountainous terrain. The geometric configuration of the model allows simulation of a freeway section up to 40 km long with the two lanes and an intermittent right climbing lane. There is no provision for on- or off-ramps. The slope and the front and rear sight distances are defined for the entire section. Different vehicle characteristics are also defined and curve-limited or downslope-limited maximum speeds may be specified within certain zones.

Most simulation dynamics are the same as those of the previous model except where the desired speeds and acceleration capabilities are functions of slopes and horizontal curvature. Validation was performed at two levels—microscopic and macroscopic. The microscopic level includes vehicle performance characteristics, car-following behavior, and gap acceptance in lane changing. The macroscopic level considers gross flow characteristics such as flow to lanes, lane-change frequencies, spot-speed distributions, time headways and overall travel speeds.

3.6 Northwestern University Lane-Changing Model

A detailed examination of freeway lane-changing behavior was the motivation for Worrall and Bullen developing a lane-changing model at the Northwestern University. For this reason, the freeway geometry is limited to a four-lane straight section without ramps. The simulated freeway length can be up to a few kilometers.

The car-following logic is fairly complex. The lane-changing logic is based on a lane-changing desire flag for each vehicle and the available gap. Vehicles are generated from a shifted negative-exponential distribution with desired speeds chosen from a normal distribution. The model produces output showing lane frequencies, lane-change delays, vehicle redistribution, and so on, but does not accept a mix of vehicle types. The computer running efficiency of the model is relatively high, the computer-time to simulated real-time ratios ranging from 1:4 to 1:20 for two-, three- and four-lane situations with volumes ranging from 600 to 1800 vehicles per lane per hour. The flexibility of the model is fairly high, although the relative ease with which the model can be recalibrated is not. Lane-changing frequency and speed-volume outputs of the model match favorably with field data collected from various Chicago freeways.

3.7 Sinha Freeway Simulation Model

The freeway simulation model developed by Sinha (1969) is a general purpose simulation model for use as a tool in the analysis of freeway phenomena. The model has a capacity for the simulation of five lanes, four on-ramps and six off-ramps. The ramps may be located on either the right-hand or left-hand side of the freeway. The model can simulate up to 6 km in length.

The car-following and lane-changing logic is fairly complex. The gap-acceptance logic is similar to that of the Texas Transportation Institute model (see Sect. 3.9). Only two types of vehicles are assumed. Freeway main-line traffic was generated from a shifted exponential distribution, while ramp vehicles were generated from a hyper-Erlaug distribution. Desired speeds were generated from a normal distribution.

Because the model was developed as a general-purpose tool for analyzing traffic operating characteristics, the computer output provides detailed information, at each of the several control points (points of interest), on the following distributions:

(a) headways in each lane,

(b) speeds in each lane,

(c) traffic volumes in each lane,

(d) exiting, entering and through vehicles in each lane, and

(e) exiting, entering and through-vehicle speeds in each lane.

3.8 Connecticut Department of Transportation Expressway Simulation Model

The Connecticut Department of Transportation expressway simulation model was developed by Leland (1970) and is similar to the model discussed in Sect. 3.3; it was also developed for the purpose of investigating, evaluating and solving freeway design problems. It allows an 8 km seven-lane section with ten on-ramps and ten off-ramps. On-ramps are restricted to direct connections. The model can handle 1000 vehicles in the system during any given second.

Driver characteristics include the assignment of acceptance gaps to individual vehicles, desired speeds from a near-normal distribution based on field data, the generating of vehicles similar to that of SIMAUT (see Sect. 1) and the acceleration and deceleration capability as linear functions of speeds. The car-following and lane-changing logic is very simple as compared with most of the other models.

3.9 Texas Transportation Institute Freeway-Merging Model

The Texas Transportation Institute freeway-merging model was developed by Buhr et al. (1968) for the purpose of simulating traffic operations under different modes of on-ramp control. The number of off-ramps is limited to two, while the number of entrance ramps plus freeway lanes is limited to six, with a maximum freeway length of 2 km.

During simulation, road sections must be preloaded. New vehicles are generated from a Poisson distribution. Each vehicle is assigned a number of characteristics such as length, current speed, desired speed, and distance from the zero reference point (beginning of the simulation section). The desired speed is generated from a normal distribution, but if the generated speed is higher than the designated maximum speed, it is reduced to the maximum speed. The simulation program consists of one monitor routine and 16 subroutines. Each subroutine is completely modular so that any logic changes in any subroutine will not affect the remainder of the program. The various ramp-control modes the model can handle include: no control, fixed-time metering and demand–capacity and gap-acceptance control. The computer scan time is 1 s. Besides rather simple car-following and lane-changing logic, the model provides extensive ramp-merging logic for the purpose of testing the various ramp-control strategies.

3.10 System Development Corporation Diamond Interchange Model

The System Development Corporation diamond interchange model is one of only two models discussed in this article with both microscopic and macroscopic features built into the model. The model, developed by Nemeczky and Widdice (1970) as a tool to aid in the design and operation of signalized diamond interchanges, contains both freeway and signalized arterial submodels, the freeway section being limited to two on- and off-ramps.

The program logic is fairly complicated in terms of a macroscopic model. Vehicles are generated from a truncated exponential distribution to eliminate the possibility of unusually large time headways, and five driver types are allowed.

Since the model was developed to evaluate selected operational or design alternatives, the following output were provided:

(a) average travel time through the system,
(b) average travel time through each individual model region,
(c) average speed through the system,
(d) average speed through each individual model region,
(e) average delay through the system,
(f) average delay through each individual region,
(g) number of stops, and
(h) acceleration noise.

The model allows inspection of changes of a diamond interchange geometry (full, split or partial diamond, changes in through and turning lanes and pockets, etc.) as well as the change in signal control parameters. Model validation was conducted by comparing the simulation outputs of the number of cars through the system and through each section of the model by origin–destination and the travel times by origin–destination with data collected at the Cold Water Canyon diamond interchange of the Ventura freeway in Los Angeles. The Wilcoxon signed-rank tests indicated the model was valid at the 5% level of significance.

3.11 System Development Corporation Freeway Simulation Model

A series of general purpose freeway simulation models were developed by Warnstmis (1978) in the order of increasing complexity. A unique feature of these models is that multilane highways were modelled by circular tracks. This creation generated many advantages in the simulation. The model is considered very general, so that any reasonable freeway configuration (including freeway interchanges) can be modelled. The network size is limited primarily by the number of cars that can be handled. For a 65K core computer the number is 2500 cars. The model provides extra capabilities such as the generating of position–time plots, and its structure allows direct simulation of sensors, controls and control algorithms. The logic is not complicated in terms of the capabilities it provides.

The position–time plots could be viewed as a computer-generated movie, so that it is easy to bring out the turbulent aspects of the overall flow or so that the behavior of individual vehicles can be looked at to determine the realism of the simulation logic.

3.12 Mikhalkin Freeway Simulation Model

The freeway simulation model developed by Mikhalkin (1975) provides a means for systematic experimentation with the capability of controlling factors of the driver vehicle-roadway system that usually cannot be controlled in real traffic flow. The simulated roadway is a straight, level freeway with no ramps, up to four lanes wide and 6km long and with sensors.

The car-following logic is based on the nonlinear car-following rule of Gazis et al. 1961 and the lane-changing logic is based on that of the Northwestern University model (see Sect. 3.6). Detectors were simulated in the model to provide volume and occupancy measurements. The sensor sampling rate was $15\,s^{-1}$. Procedures for estimating the roadway local density and space mean speed from detector measurements were developed with a high degree of accuracy. Vehicles were generated randomly, but no specific speeds and vehicle lengths were obtained from truncated normal distributions whose parameters were input data. Because of the various traffic parameter estimating algorithms implemented in the model that use detector-measured data, this model is extremely useful for freeway surveillance and incident-detection purposes.

3.13 Georgia Model

The Georgia model, developed by Wildermuth (1979), was primarily concerned with the assessment of truck effects on freeway flow characteristics. Model development was based on an extensive evaluation of the Midwest Research Institute, Northwestern University and Texas Transportation Institute models (see Sects. 3.4, 3.6 and 3.9). Successful components from the earlier models were adapted and modified so that trucks could be properly introduced as a distinct element into the traffic flow simulation.

In the Georgia model, each vehicle is associated with a vector containing 12 specific characteristics such as the desired speed, current speed and vehicle type. Vehicles are generated from a shifted exponential distribution. The desired speed is generated from a normal distribution, with the mean and standard deviation as separate input variables for each lane. Simulation starts with a preloaded condition without requiring a warm-up time to achieve a stable flow.

Model validation was done in terms of comparing the generation of different vehicle types, headway distributions, lane volumes, speed distributions and lane-changing frequencies from the simulation runs to those of the real data, and good results were shown.

3.14 SCOT Corridor Model

The SCOT (simulation of corridor traffic) model was originally conceived by Wicks (1972), as a concatenation of two existing models—the urban traffic control system (UTCS-1) simulation model of urban traffic (developed in 1975) and the DAFT simulation model of freeway traffic. The SCOT model is also a dual microscope (the UTCS-1) and macroscopic (the dynamic analysis of freeway corridor traffic (DAFT)) model, like the System Development Corporation model (see Sect. 3.10). The SCOT model treats vehicles microscopically on the arterial street system (including ramps) and macroscopically (as platoons) on the freeway.

Any arbitrary freeway and surface street network containing up to 200 intersections can be represented and traffic flow is described by specifying either origin–destination volumes along the peripheral entry links or turning movements at each node. The objective of developing this model was to use it as a medium for assisting in defining the surveillance and control requirement for both existing and planned freeway corridors. The crux of the freeway component of the SCOT model is the speed–density formulation resulting from a general form of the noninteger car-following rule of Gazis *et al.* (1961). The unknown parameters of the speed–density equation have to be determined experimentally before the simulation run.

3.15 Priority Lane Model

The priority land model was directed toward evaluating traffic operations on freeways with priority lanes such as those allocated for buses or vehicles containing a required minimum number of passengers. The model geometry allows a maximum of 50 freeway subsections, with not more than one ramp in each subsection.

The model logic by Blaukenborn and May (1972) is sophisticated and efficient. It is essentially macroscopic, with input data provided for each 15 min. The computer program has a modular structure and additional capacity can easily be obtained by modifying or including the appropriate subroutines. Future changes can be made with minimum effort to match model results with empirical data. Running instructions and input data formats are provided for using the model. The model idealizes physical queues, and this may obscure some of the effects being studied. Furthermore, subsection capacities and demand are assumed to remain constant over 15 min time intervals. The study section is limited to 15 km and no off-ramp queuing calculations are attempted if off-ramp demand exceeds ramp capacity. Otherwise the model affords sufficient realism for representing traffic flow on freeways with any kind of priority lanes or reversible lanes and for ramp-control schemes for priority vehicles.

3.16 Aggregate Variable Models

In the aggregate variable models developed by Payne (1972), a freeway is partitioned into sections and the freeway traffic is described by a set of dynamic equations in terms of the aggregate variables of flow rate, section density and section speed. The purpose of the models is to study the problem of developing ramp-control strategies with a high simulated real-time to computer-time ratio. The model does not distinguish flow by lanes, and the traffic flow is described as an extension of the simple continuum models. However, it seems that the model produces good results for ramp-control purposes.

See also: Freeway Traffic Modelling; Prediction of Traffic Flow; Simulation Programs, Microscopic

Bibliography

Babcock P S 1984 Improved dynamics and performance for the FRECON free-way simulation model. Technical Document UCB-ITS-TD-84-1. Institute of Transportation Studies, University of California, CA

Blaukenborn R C, May A D 1972 FREQZ: A revision of the "FREQ" freeway model. Special Report. Institute of Transportation and Traffic Engineering, University of California, Berkeley, CA

Buhr J H, Meserole T C, Drew D R 1968 A digital simulation program of a section of freeway with entrance and exit ramps. *Highw. Res. Rec.* **230**, 15–31

Cremer M, May A D 1986 An extended traffic flow model for inner urban freeways. *Proc. 5th IFAC/IFIP/IFORS Symp. Control in Transportation Systems*. Pergamon, Oxford pp. 29–37

Cremer M, Papageorgiou M 1981 Parameter identification for a traffic flow model. *Automatica* **17**, 837–43

Gabard J F 1988 A melioration du programme de simulation de trafic autoroutier—SIMAUT. INRETS/DERA Report No. 2/7507. INRETS/PART, Arcueil, France

Gazis D, Herman R, Rothery R 1961 Non-linear follow-the-leader models of traffic flow. *Oper. Res.* **9**, 545–67

Hadj-Salem H, Davee M M, Blosseville J M, Papageorgiou M 1988 ALINEA: Un outil de regulation d'access isolé sur autoroute. INRETS Report No. 80, INRETS, Arcueil, France

Isaksen L 1971 Suboptimal control of large scale systems with application to freeway traffic regulation. Ph.D. dissertation, University of Southern California

Kimber R M, Daly P N 1986 Time-dependent queuing at road junctions. Observation and prediction. *Transp. Res. B* **20**(3), 187–203

Kobett D R 1968 A digital simulation model of freeway traffic, vol. I: Model description. Final Report, Midwest Research Institute, Contract No. CPR-11-3661. Federal Highway Administration, Washington, DC

Kobett D R, Levy S L Study of expressway traffic flow through digital simulation. Final Report, Midwest Research Institute, Contract No. CPR-11-0963. Federal Highway Administration, Washington, DC

Leland S D 1970 A general traffic flow simulation model for freeway operation. Final Report, Connecticut Department of Transportation, Wethersfield, CT

Leonard D R, Tough J B 1979 Validation work on CONTRAM—A model for use in the design of traffic management schemes. *Planning and Transport Research and Computation Summer Annual Meeting*. Transport and Road Research Laboratory, Crowthorne, UK

Lieberman E 1970 Dynamic analysis of freeway corridor traffic (DAFT). General Applied Science Laboratories, Inc. TR-744

Man A D, Minister R D, Lew L P, Ovaicki K 1976 A computer simulation model for evaluating priority operations on freeways. Special Report, University of California, Berkeley, Contract No. FH-11-7704. Federal Highway Administration, Washington, DC

Masher D P, Ross D W, Wong P J, Tuan P L, Zeidler H M, Petracek S 1975 Guidelines for design and operation of ramp control systems. Report of Stanford Research Institute, Menid Park, CA

Mikhalkin B 1975 Estimation of Roadway, Behavior using occupancy detectors. Ph.D. dissertation, University of Southern California

Morin J M 1985a SIMAUT: Un programme de simulation du trafic autoroutier. INRETS Report No.4. INRETS, Arcueil, France

Morin J M 1985b SIRTAKI, an aid to real time decision-making for motorway access control. Planning and Transport Research and Computation, Brighton, UK

Morin J M, Pierrlic J C 1988 La regulation du trafic dans un ressau maiki autoroutier urbain et periurbain. INRETS Research Report No. 63. INRETS, Arcueil, France

Nemeczky J A, Widdice B D 1970 Development and validation of a digital simulation model for research in geometric design and control of diamond interchanges. *Proc. Summer Simulation Conf., Vol. 2*

Papageorgiou M 1988 Modelling and real-time control of traffic flow on the southern part of Boulevard Périphérique in Paris. INRETS Internal Report. INRETS/DART, Arcueil, France

Papageorgiou M, Blosseville J M, Hadj-Salem H 1989 Macroscopic modelling of traffic flow on the Boulevard Périphérique in Paris. *Transp. Res. B* **23**, 29–47

Papageorgiou M, Blosseville J M, Hadj-Salem H 1990 Modelling and real-time control of traffic flow on the southern part of the Boulevard Périphérique in Paris. Part I: Modelling. *Transp. Res A* **24**, 345–59

Papageorgiou M, Blosseville J M, Hadj-Salem H 1990 Modelling and real-time control of traffic flow on the southern part of Boulevard Périphérique in Paris. Part II: Coordinated on-ramp metering. *Transp. Res. A* **24**, 361–70

Payne H J 1972 Aggregate variable models of freeway traffic. NSF grant GK-24520. University of Southern California, Los Angeles, CA

Peat, Marwick, Mitchell and Co, General Applied Science Laboratories Inc 1975 Network flow simulation for urban traffic control system. Federal Highway Administration Report No. FH-11-7462. FHWA, Washington, DC

Richard R M, Baker R L, Seldon W P 1965 Simulation of traffic flow as a basis for interchange design. Final Report, Arizona Transportation and Traffic Institute, Phoenix, AZ

Schwerdtfeger Th 1984 DYNEMO: A model for the simulation of traffic flow in motorway networks. *9th Int. Symp. Transportation and Traffic Theory*. VNU, Utrecht

Sinha K C 1970 The development of a digital simulator for the analysis of freeway traffic phenomena. PhD dissertation, University of Connecticut

Sinha K C, Dawson R F 1969 The development and validation of a freeway traffic simulator. *Highw. Res. Rec.* **308**, 34–47

St John A D, Kobett Dr R, Sommerville D, Colanz W D 1970 Traffic simulation for the design of uniform service roads in mountainous terrain, vol. 1–4. Final Report, Midwest Research Institute, Contract No. CPR-11-5093. Federal Highway Administration, Washington, DC

Warnstmis P 1978 Simulation of traffic flows through interchange systems of arbitary configuration. System Development Corporation, Contract No. FH-11-7628. Federal Highway Administration, Washington, DC

Wicks D 1972 Traffic flow simulation study: The SCOT model, vol. 1–4. Final Report, General Applied Science Laboratories Inc, Contract No. DOT-TSC-161. Department of Transportation, Washington, DC

Wildermuth B R 1979 Effect of lane placement of truck traffic freeway flow characteristics. Final Report, Wilbur Smith and Associates, Project 6909. Georgia Highway Department, Atlanta, GA

Worral R D, Bullen A G R Lane-Changing on Multi-Lane Highways. Final Report, Northwestern University, Contract No. CPR-11-5228. Federal Highway Administration, Washington, DC

J. Chrisoulakis
[Athens, Greece]

Simulation Programs, Microscopic

Digital computer programs to simulate traffic flow have been developed from 1950 onwards. The great appeal of the simulation approach is that this technique offers the user an opportunity to evaluate alternative strategies before implementing them in the field. Thus, the optimal strategy may be identified prior to the commitment of substantial funds for the implementation of large systems.

Simulation models may be classified as microscopic or macroscopic. The term microscopic, as used in this article, denotes a model that simulates the movement of individual vehicles. Macroscopic denotes a grouping of vehicles and the application of flow relationships to determine successive traffic states.

Gibson and Ross (1977), Lieberman (1979), Transportation Research Board (1982) and May (1988) have reviewed various computer traffic simulation programs. Some of the simulation programs have been examined with other traffic management and control systems (Byrne *et al.* 1982, Van Aerde *et al.* 1988). This article concentrates on microscopic traffic simulation programs that are either continuously maintained, supported and enhanced and, therefore, well received in the traffic engineering community, or newly developed but not covered by the aforementioned review papers.

1. Models Developed in the USA

The US Department of Transportation (US DOT), Federal Highway Administration (FHWA) plays a leading role in sponsoring and cosponsoring the development of traffic simulation programs. This section reviews models that simulate isolated intersection, two-lane, two-way rural road, urban surface street network and freeway traffic flow conditions.

1.1 TEXAS

With the growing complexity of intersection design and the concern for improved planning and design of the highway and street system, the need for a relatively inexpensive method for detailed study of the variety of intersections and control techniques has become evident. To meet this need, the University of Texas has developed the TEXAS (traffic experimental and analytical simulation) simulation model for the Texas State Department of Highways and Public Transportation to perform microscopic simulation of isolated intersections. The Texas State Department of Highways and Public Transportation maintains the model.

The model will simulate any intersection from two uncontrolled one-way streets to complex intersections with multiphase control, and/or multilane movements. Traffic control may be none, priority movement (e.g., stop or yield sign controlled) or signalized.

Statistics about delays and queue lengths are gathered by the model for evaluating the performance of traffic at the intersection. Delay statistics include the average total delay and the average stop delay incurred by each vehicle. Each delay is summarized by left-turn, right-turn and straight movements. Queue length statistics include average queue length and maximum queue length. The model incorporates models to predict the instantaneous vehicle motions and fuel flow for both light-duty and heavy-duty vehicles.

A microcomputer version with added user-friendly interactive capability has been developed (Lee and Machemehl 1988). Special features include animated graphics that show drawn-to-scale, color-coded vehicle types moving through the geometry of the intersection. This allows viewing the overall performance of a design while providing an excellent means of communicating these patterns to officials and the general public.

1.2 ROADSIM

ROADSIM is a microscopic traffic simulation model for two-lane, two-way rural roads that was developed for the FHWA to replicate traffic operations observed on existing two-lane rural roads. The model can be used to determine the traffic operational effects of geometric changes, volume, and/or traffic composition variations. Simulation of rural traffic on two-lane roads developed at a slower pace compared with the urban network and freeway microscopic simulation models. This is because the two-lane traffic flow is complicated by platoon formation and passing decisions and, therefore, is not easily modelled. Also, the low volumes of traffic on rural roads usually do not make simulation cost-effective.

The model has the ability to simulate traffic operation on two-lane, two-way rural highways with or without climbing lanes. Input consists of horizontal alignment, passing, no-passing and climbing lane sections, sight distance, crawl zones, traffic volumes, desired speed, and traffic composition. In addition to overall statistics, the output also consists of link-specific and direction-specific measures of effectiveness (MOEs) by vehicle category (e.g., distance travelled, travel time, standard deviation of travel time, delay, standard deviation of delay, speed, standard deviation of speed, and passes attempted, completed and aborted) and link-specific information (e.g., headway, speed and platoon-size distributions).

The model has been evaluated (Morales and Paniati 1985). Field data was collected on a two-lane rural road in Loudoun County, Virginia and coded as input for the model. Statistical analyses performed to compare the MOEs observed in the field with those obtained from the simulation showed that simulation results for ROADSIM compared favorably with those observed in the field. Results support its potential usefulness to the traffic engineering community. A microcomputer version of ROADSIM has become available.

1.3 NETSIM

The urban traffic network contains a mix of geometric, control and traffic management strategies and the traffic engineer needs to analyze potential designs more rigorously or consider the stochastic variations of traffic flow. Microscopic traffic simulation is the logical choice as a tool to deal with complex urban network traffic. One of the first successful large-scale network simulation models was the urban traffic control system (UTCS) developed by Peat, Marwick, Mitchell and Co (1971) for the US DOT, FHWA. The model was extended by KLD and Associates (and others) for FHWA and renamed as network simulation (NETSIM) (Federal Highway Administration 1980). The model was later included in the TRAF family models (TRAF-NETSIM).

NETSIM can evalute any configuration of an urban network, including normal forms of traffic control at individual intersections. Control features include stop and yield signs, turn controls, parking controls, fixed-time signals, vehicle-actuated signals, and real-time traffic control and surveillance systems. The modular format enables analysis of extremely flexible design configurations and strategies. These include incorporating detailed treatment of car-following behavior, network geometry, slopes, bus traffic, queue formation, intersection discharge, intralink friction and midblock blockages, and pedestrian–vehicular conflicts.

A summary of important statistics of MOEs is given at the end of simulation intervals. The cumulative performance on each link and the entire network are generated. The user may also request this report at any time in the simulation interval.

A summary of fuel consumption and vehicle emissions for each link and the network as a whole is obtained for each run. A variety of optional outputs, such as origin–destination pattern, type and locations of detectors, all "rare" events, and bus performance, can be requested for detailed analysis.

NETSIM has been converted to run on microcomputers and other related microcomputer programs have been developed to assist the user in preparing input data and analyzing results. These include NETSIM editor (NEDIT), integrated traffic data system (ITDS) as part of the traffic software integrated system (TSIS) and graphics for TRAF (GTRAF). GTRAF provides graphic display of both input and output data. For example, details about intersection geometries, highlighting of potential problem areas or "hot spots," and animation of simulated traffic flow can be displayed.

1.4 INTRAS

During the 1970s, attention was focused on the need to develop effective freeway incident detection techniques. It is believed that one of the important problems in urban freeway traffic operations is the detection of stopped vehicles and the necessary steps required to remove the stoppage. Flow-disruptive incidents are a substantial cause of congestion and are considered by some to be a greater problem than recurring daily congestion.

The INTRAS (integrated traffic simulation) model (Wicks and Lieberman 1980) is an adaptation and extension of the simulation logic of the UTCS model for evaluating traffic operations in a freeway corridor. The model was developed with the primary objective being the investigation of freeway incident detection and control strategies. However, the available user options make INTRAS a potentially versatile analysis tool.

The model considers a corridor system consisting of a freeway, frontage road and associated ramps. The simulation process is microscopic and stochastic, and allows treatment of a variety of geometric features, intersection control, demand characteristics and freeway surveillance and control strategies. In addition to a comprehensive set of flow characteristics, INTRAS can also provide vehicle trajectory plots, contour maps of selected MOEs, analyses of fuel consumption and emission, and statistical analyses and comparisons of output data sets. Simulation of detector (single loop, double loop, radar) surveillance systems for freeway with off-line processing (incident detection, MOE estimates) is included.

Major enhancement of INTRAS has been carried out by JFT and Associates, under the sponsorship of the FHWA. Only the freeway portion of the INTRAS has been extracted and reprogrammed according to structured design techniques. Reportedly, the execution speeds of the model have been improved 4–6 times. Further enhancements to make the model more user friendly and appropriate for a wider range of applications have been carried out. The revised model is called FRESIM (freeway simulation) and is incorporated into the TRAF family of programs.

2. Canadian Model

Transporation and traffic researchers and engineers in Canada work closely with their counterparts in the USA. However, due to the increasing needs to manage congestion traffic around major urban transportation corridors, advanced models with the capability to simulate the dynamics of delay and route choice have been developed. The model INTEGRATION will be described in this section.

The growing traffic congestion problem has increased both the scope and scale of traffic management tasks, while simultaneously constraining the range of alternative solutions that are available to alleviate such problems. Clearly, there is a need for simultaneously modelling freeways and urban streets, traffic signals, real- and fixed-time control, queuing, traffic assignment, traffic diversion and congestion management within a single simulation/optimization model.

The design of a model that can deal specifically with integrated networks was formulated (Van Aerde *et al.* 1988). Under a sponsorship of the Ministry of Transportation of Ontario, INTEGRATION has been developed at the Queen's University in Kingston, Ontario (Van Aerde *et al.* 1988).

INTEGRATION consists of a discrete simulation that traces the path of each vehicle throughout the network. The links that a vehicle utilizes are selected in accordance with its estimate of the "best" route and, along its path, the route of each vehicle is further adjusted in view of any changes in the prevailing traffic congestion and traffic controls. In addition, simulation of traffic incidents and traffic demands is also included.

During the simulation, the model provides a real-time graphical illustration of the performance of the traffic network, which indicates the amount of traffic and queuing on each network link, as well as the status of traffic signals. At the conclusion of the simulation run, the model produces two types of summary reports. The first provides user-oriented statistics on the trips between origin–destination pairs, while the second provides system-oriented statistics on the operation of each network link.

This model is suitable for incident management plan development, evaluation of fixed-time and real-time signal control strategies and the assessment of effective alternative driver-information systems.

3. Models Developed in the UK

Keeping close contacts with the traffic and transportation professionals in North America, experts have made parallel efforts in developing computer simulation models in the UK. For example, the TRANSYT model—a network signal timing optimization with macroscopic simulation developed at the Transport and Road Research Laboratory (TRRL)—has been used in all kinds of network traffic analyses worldwide. However, another microscopic model, TRAFFICQ, which is well maintained, supported, enhanced and utilized by traffic engineers in a number of countries, will be reviewed in this section.

Some models developed at the University of Bradford are worth mentioning. Although wide usage of these models for purposes other than research is not found, Salter and his associates have been continuously developing isolated intersection simulation models for studying junction-traffic-related problems since around 1970.

3.1 TRAFFICQ

TRAFFICQ is a computer simulation model designed to aid the evaluation of alternative traffic management plans for networks. The plans may incorporate complex traffic management techniques (e.g., computer-controlled traffic signals, pedestrian management and complicated one-way systems). Each vehicle and pedestrian is modelled as a single entity and the time-varying and random characteristics of traffic flow are integrated into the modelling.

TRAFFICQ can easily model most commonly encountered traffic management systems. Additionally, users may add their own modelling to the framework already provided by TRAFFICQ. This is done through a facility that gives users control over such matters as signal setting, vehicle access to the network and vehicle discharge at intersections.

The model provides an output in several different ways. Distributions are used where the best- and worst-case information might be as important as average values. This format is used to present information on queues, travel times and pedestrian delays. Link-specific information, such as travel times, traffic flows, and fuel consumption, is presented in a link-by-link format. TRAFFICQ also provides various measures of economic benefits.

MVA Systematica (1989) is responsible for maintaining and distributing TRAFFICQ on behalf of the UK Department of Transport. The current version is in FORTRAN and it is available on a wide range of mainframes, minicomputers and microcomputers.

3.2 Models Developed at the University of Bradford

(*a*) *SIGART*. The use of roundabouts to control vehicle conflicts is common in the UK. The increased use of area-wide control of signals and the difficulty experienced by roundabouts when dealing with unbalanced peak flows has led to the development of signal-controlled roundabouts. To extend knowledge of the operation of this type of junction, a simulation model, SIGART, has been developed (Salter and Okezue 1988).

(*b*) *SEGOPT*. This model simulates four straight-ahead movements in a four-leg intersection that is controlled by a two-phase system with the fixed cycle time divided into effective green and red. The program can be extended to approaches with more than one lane if drivers have freedom of lane choice as they approach the intersection. The drivers select the lane with minimum queue. The saturation flows for each approach lane are identical. At the completion of the run, the model generates delay to vehicles for each leg, the queue for each leg and total intersection delay.

(*c*) *JUNOPT*. This model is an extension of SEGOPT in that it provides for an exclusive right-turn lane on each approach with vehicles making the right-turn movement through gaps in the opposing flow.

(*d*) *EARCUT*. This model is similar to JUNOPT in that it simulates a four-leg intersection controlled by fixed-time, two-phase signals with an exclusive right-turn lane on each approach where right-turn movements are made by means of an early cut-off or late-start facility.

(*e*) *THREPHS*. This model is a further development of the previously described programs. It simulates a four-leg intersection controlled by fixed-time signals with an exclusive right-turn movement to take place.

(*f*) *CYPED 1, 2, and 3*. These programs allow the simulation of traffic flow at four-leg intersections controlled by fixed-time signals where provisions are made for pedestrian movements. CYPED 1, 2, and 3 are adaptations of JUNOPT, EARCUT, and THREPHS, respectively. (Salter and Okezue 1988.)

4. Models Developed in Germany

Based on the research of published reports and papers from technical journals in English, some microscopic models as well as two mixed macroscopic and microscopic models have been identified. These models have been developed at two universities.

4.1 SIMNET

According to Hoffmann (1988), SIMNET is a stochastic simulation program that was developed at the Berlin Technical University and tested in various case studies. With the aid of SIMNET, it is possible to assess, on the basis of a simplified movement of individual vehicles in connected street networks, all common control procedures in urban areas in integrated or single use with respect to their effect in terms of traffic engineering improvements.

An extension of this model has been completed that allows the additional microscopic level of car-following theory for selected network elements and integration at the macroscopic level of traffic assignment. The traffic engineer, thus, has a comprehensive tool to assess integrated traffic control measures and judge the advantages of the proposed traffic management strategies for an urban street network.

4.2 Models Developed at the Karlsrühe Institut für Verkehrwesen

There is a group of models which have evolved from their predecessors SIM and SIM-2/S that simulate one-lane one-way, multilane one-way, and multilane two-way traffic flows. However, the recently developed DYNEMO is a fresh-start endeavor.

(*a*) *DYNEMO*. This has been developed as a tool for the development, evaluation and optimization of traffic control systems for urban networks. For off-line applications, a traffic control model can be easily included with the simulation model.

DYNEMO is a mixed microscopic and macroscopic simulation model that uses both knowledge about general speed-flow relationships and the desired speeds to calculate the actual speeds of vehicles. The mean speeds on neighboring sections of the system are taken into account to avoid discontinuities in the speed profiles.

The information available from the simulation model DYNEMO includes

(i) patterns over times of traffic volume and mean speed at any point in the network;

(ii) travel times of individual vehicles or travel time distribution on any route in the network; and

(iii) fuel consumption of vehicles.

(*b*) *Other models.* INTAC is a microscopic simulation model for traffic in a single lane and is suitable for studying the traffic flow in a two-way road with no passing. The model is also useful for simulating traffic flow on a one-way street to determine stability criteria for platoons described in car-following equations. The model can also be used to study collision avoidance systems. The model, then, is expanded to simulate two-lane traffic (INTAC-2) and three-lane traffic to study lane-change regulations (SIM-2,3/F). The model is further modified (MISSIS) to simulate uphill two-lane roadways with passing situations for fast-following vehicles.

SIMLA programs are also derived from INTAC but are microscopic simulation models for rural roads. The programs can handle both straight flat and curved inclined rural roads.

MISSION is a programme environment for the simulation of general traffic, buses and trams within urban street networks. This environment also includes an interface for traffic control systems and public transport vehicle location and control systems.

MISSION was employed to study the coupling of bus and light-rail transit guidance and control systems with traffic signal control systems and its effects on the performance of the traffic system.

The output information produced by SIMLA/2 and MISSION can be evaluated by other programs with respect to noise level, emissions and fuel consumption (Leutzbach and Wiedemann 1986).

5. Models Developed in The Netherlands

Dutch traffic engineers have developed a programming language, traffic control language (TRAFCOL version 1976), to describe the control program. TRAFCOL-76 is an event-oriented language. FLEXSYT provides compiler and control program evaluation facilities as well as traffic simulation capabilities. In fact, FLEXSYT is the tool that makes TRAFCOL-76 a powerful and living language with almost no limitations as to the type of control program or the specialties in the design of the traffic engineer.

FLEXSYT (flexible network and traffic control simulation study tool) (Middleham 1986) is a series of computer programs that can be used effectively in the study and design of traffic control programs and networks. This series of programs can accommodate a variety of traffic control strategies. Practically all controlled and uncontrolled intersections in roadway networks can be modelled. Controlled intersection, fixed-time, traffic-responsive controlled and arterial coordinated situations can be accommodated. The program is developed in FORTRAN and can be easily ported to a variety of computers.

With the aid of a number of pseudo-random-number generators, the actual simulator simulates both the traffic process and the control process and their mutual influence by means of a number of models; simulation models, platoon diffusion models, route choice models and interaction models. All individual vehicles of the traffic process are simulated. For this reason, it is possible to study the behavior and handling of individual vehicles in the network. The detectors can also react to individual vehicles and the traffic control programs are not subject to any restrictions regarding their structure and operation.

The results of FLEXSYT are generated by statistics programs. Some of the results are printed in the forms of histograms, tables and other statistics reviews. Signal control statistics such as cycle length and red and green times, as well as traffic flow performance information such as queues, stops, delay, irregularity and load ratios are generated.

6. French Model

Only limited information has been found regarding the simulation developed by the Institute National de Recherche sur les Transports et leur Securité (INRETS). Several macroscopic models (SSMT, SIMAUT and PHEDRE) and one microscopic model have been identified.

The microscopic model is SIMIR (simulation microdcopique de petits reseaux routiers). According to the Transportation Research Board (1986), this is a microscopic simulation model for small road networks. A microcomputer version has been developed. In addition, capability to graphically represent the simulated results has been incorporated.

See also: Car-Following Models; Simulation Programs, Macroscopic

Bibliography

Byrne A S, Courage K G, Culpepper T H, Wallace C E 1982 handbook of computer models for traffic operation analysis. Technical appendix: Summary of models and

references. Technology Sharing Report FHWA-TS--82-214. Federal Highway Administration, Washington, DC
Byrne A S, de Laski A B, Courage K G, Wallace C E 1982 Handbook of computer models for traffic operation analysis. Technology Sharing Report FHWA-TS-82-213. Federal Highway Administration, Washington, DC
Federal Highway Administration 1980 Traffic network analysis with NETSIM: A user guide. Implementation Package FHWA-IP-80-3. FHWA, Washington, DC
Gibson D, Ross P 1977 Simulation of traffic street networks. *Transp. Eng.* December, 19–27
Hoffmann G 1988 State and future development of traffic signal control in the Federal Republic of Germany. *67th TRB Annual Meeting*. Transportation Research Board, Washington, DC
Lee C E, Machemehl R B 1988 The TEXAS model for intersection traffic—A user-friendly microcomputer version with animated graphics screen display. *Transp. Res. Rec.* **1142**, 1–5
Leutzbach W, Wiedemann R 1986 Development and applications of traffic simulation models at the Karlsrühe Institute für Verkehrwesen. *Traffic Eng. Control* **28**(5), 270–8
Lieberman E B 1979 Traffic simulation: Past, present, and potential. *Proc. Int. Symp. Traffic Control Systems*. Report No. UCB-ITS-P-79-2. Federal Highway Administration, Washington, DC, pp. 242–61
May A D 1988 Freeway Simulation Models Revisited. *Transp. Res. Rec.* **1132**, 94–9
Middelham F 1986 *Manual for the Use of FLEXSYT-I with FLEXCOL-76*. Dutch Ministry of Transport, The Hague, Netherlands
Morales J M, Paniati J F 1985 Two-lane traffic simulation: A field evaluation of ROADSIM. *Public Road* **49**(3)
MVA Systematica 1989 *TRAFFICQ: A Simulation Model for Detailed Evaluation of Vehicle and Pedestrian Activity in Complex Road Networks*. Surrey
Peat, Marwick, Mitchell and Co, General Applied Science Laboratories Inc 1971 Network flow simulation for urban traffic control system. Final Report, Contract No. DOT-FH-11-7462. Federal Highway Administration, Washington, DC
Salter J, Okezue O G 1988 Simulation of trafffic flow at signal-controlled roundabouts. *Traffic Eng. Control* **30**(3), 142–7
Salter J, Tadayon M 1987 Optimization of signal cycle time by a computer simulation. *Traffic Eng. Control* **29**(5), 290–3
Transportation Research Board 1982 The application of traffic simulation models. Special Report No. 194. TRB, Washington, DC
Transportation Research Board 1986 Traffic Signal Systems Committee Newsletter, TRB Committee A3A18. TRB, Washington, DC
Van Aerde M, Voss J 1988 *Integration—1. User's Manual*. Ontario, Ministry of Transportation
Van Aerde M, Yagar S 1988 Dynamic integrated freeway/traffic signal networks: A routing-based modelling approach. *Transp. Res. A* **22**(6), 445–53
Van Aerde M, Yagar S, Ugge A, Case E R 1988 A review of candidate freeway-arterial corridor traffic models. *Transp. Res. Rec.* **1132**, 53–65
Wicks D A, Lieberman E B 1980 Development and testing of INTRAS, a microscopic freeway simulation model, vol. 1. Program design, parameter calibration and freeway dynamics component development. Final Report. No. FHWA/RD-80/106. Federal Highway Administration, Washington, DC

S-M. Chin
[Oak Ridge National Laboratory, Oak Ridge, Tennessee, USA]

Social Issues of Ship Automation

Since ancient times, people have been fascinated by the sea as a challenge to their control over the elements and as a chance to discover the unknown. Throughout the centuries, the evolution of navigation has allowed humanity to attain a certain level of domination over the sea and has, thus, enabled trading across the oceans. The most recent innovations are the transformation from sail to steam propulsion and the ability to obtain position information, which allows analysis of the environment in poor visibility weather conditions, in deep-sea areas or close to the coastline (see *Ship Stabilization: History*). Consideration is now being given to the application of advanced technologies to ship management, operation and propulsion. The major goals of such developments are the reduction of crew size and the integration of the whole transportion process, from production of goods to overseas distribution. These are the only two developments occurring on a worldwide level and they are evolving quickly in parallel, rather than in collaboration, with the shipping industry. This industry is taking advantage of the advances by on-shore industries in automation and robotics. Irreversible developments in automation in the marine industry are leading to economic and social changes.

1. The Shipping Crisis

In the 1980s, the worst ever maritime economic crisis spread across the world. The social consequences were drastic and the effects have still not been overcome. The emergence of nontraditional maritime trading countries caused a panic in the conservative world of the shipping business.

The lack of balance in ship operation costs in individual countries led to an overtonnage of the world fleet. The largest part of the world tonnage was owned by the richer countries; the arrival of newcomers, such as developing countries and the "five dragoons," on a more-or-less balanced market reduced freight rates. With the plentiful and cheap labor of these countries, as well as their deliberately low-profit or non-profit policies, the market was severely depreciated. Social protection of European seafarers and government economic protective measures prevented shipowners from reacting appropriately to the market. To compete on level terms, deflaged convenience registers granted similar possibilities to shipowners; the consequences were low-cost-labor recruitment and low levels of maintenance of the vessels.

Nevertheless, the positive answer to its problem was to do what other industries had already done: reduce labor by increasing the level of technology applied to the ship. Thus, this crisis generated contradictory reactions from the traditional maritime countries, mainly Western European, for reducing costs:

(a) restraint on expenses (e.g., by reducing wages and maintenance costs) as a short-term measure, and

(b) expansion towards new technologies to increase the productivity of the ship (i.e., larger ships for reduced crews) as a long-term policy.

2. Cost Cutting Policy

Cost cutting became the key theme for all shipowners in the developed world. Cost cutting by reducing fuel consumption is a technological matter and progress has been made on low- and medium-speed engines (see *Fuel Conservation: Sea Transport*).

The cost cutting of crews has not been so easy to achieve because of strong union protection, social welfare commitments and national legislation. Negotiations between shipowners, unions and governments gave no return advantages when shipping and ship owning were profitable businesses.

Currently, maritime economic indicators seem to have stabilized. The situation was so bad that experts remain reluctant to carry out any optimistic analysis and many ship management companies remain weak (as a result of the bankruptcy of the US lines); nevertheless, the sea transport industry looks to have slowly overcome its difficulties. The average age of the world fleet is old and the shipbuilding market, with the shutdown and reduction in capacity of yards, is encouraging for the overall situation.

Orders for highly automated tankers and container vessels are received by advanced yards. Besides the automated Japanese vessels, the German ships operate in the Far East and these yards have orders for fully automated ships with a capacity of up to 4000 TEU containers. Competition is so fierce in this field of activity that overcapacity could occur in the near future.

Some ship management operators consider that savings on ship operations and crews are not endless. Overheads cannot only be reduced by high-technology vessels. The essential function of a ship is to safely transport goods from one place to another to the shipper's satisfaction; optimizing management of a flow of containers is a modern way to reduce running costs. An increase of 20% on the turnround of a container, which is not unrealistic, means a reduction of 20% on the investment for containers. This is worthy of consideration because container investment represents nearly 60% of the investment for a container vessel.

The "door-to-door" concept of transport is now being used by ship operators; transport is an entity in itself and is no longer considered transport by sea with minor operations at both ends. The fully developed network of communication, if managed by a ship operator, between origin, ship and destination is making the operator a real partner of the shipper. In the resolution of the "tense flow" of the merchandise, the vessel is an "obliged passage" which ship operators have to take advantage of.

3. High-Technology Ships

Ships can be rationalized, efficient, futuristic or intelligent; the common denominator of the research undertaken in Japan, Norway, Germany and the UK is the resolution of the tasks imposed by the operation of the ship. The functions to resolve the tasks are centralized in integrated systems on the bridge, for easy access to all information simultaneously.

Automation is a possible way of giving an order that will generate a function at a certain distance, the follow-up of the processing, control and modification (if necessary, within homogeneous entirety, where the role and task of this particular function is taken into consideration) and the self-integrating operating parameters.

The Japanese approach started in the late 1970s. Their long-term objective was reduced-crew vessels. The processing used to cope with this target was a scientific approach:

(a) determination of the aim: crew reduction on board;

(b) existing means: state of development of shipbuilding;

(c) means to approach the aim: technologies to be installed on board, adjustment to vessel and crew, work assignment of crew members, operation of the ship for a time, follow-up of operation of the ship, report of operation; and

(d) criticism of the experience in itself and in comparison to the aim.

The project then starts a new phase of development and repeats the steps.

The technological level of development is not, in itself, an end that renders solutions, possible or not; it is considered as a means at a certain level of development in knowledge. This concept finds a place within an organization prepared to leave room for new devices and systems on board.

To allow a step by step crew reduction on ships and considering the level of development of automation, several solutions have to be found to ship problems before a computerized vessel can be spoken of, such as preventative maintenance, inventory systems, remote control of the engine, robotic devices, fault-tolerant and "event-driven" computer systems, automatic mooring, and an Immarsat radio communications system, which makes the ship an office of the headquarters. Engine breakdown in deep-sea areas seems to be the only matter that could stay unscheduled and outlined by an alternative propulsion system.

4. The Integrated Bridge Concept

On an integrated bridge, the environment of the operation center of the ship is redesigned for the purpose of single-person bridge operation. Organization of working and leisure times are changing. Trial tests before sailing are increased and routine control operations systematized. The internal communications system is redesigned and complex.

On an integrated bridge, different functions are located at different places on the bridge and complete control of the ship is carried out from here. Navigation is controlled from the chart table and there is the capability for automatic navigation and position determination, meteorological routing, electronic chart systems and planning of navigation, weather condition evaluation and simulation. Expert systems and artificial intelligence aids assist the captain in decision making; this means computerized experience from similar situations is at the officer's disposal. There are consoles for the following functions:

(a) navigation, maneuvering, mooring and deck operations;
(b) engine control;
(c) safety control;
(d) external and internal communications board; and
(e) a black box.

The purpose of such an organization at the highest level of the ship is to place the control of every function of the ship and its environment under the authority of one person: the watch officer.

5. Ship–Shore Policy

The aim of reducing the number of crew implies a different sharing of duties between ship and shore: the crew runs the ship, the shore maintains it.

This evolution has continued and the relationship between vessel and shore has gradually increased because of the increase in the number and size of ships worldwide, since the late 1950s. The cautious control of vessels came about through the International Maritime Organization Conventions and Recommendations, and the Vessel Traffic Services (see *Vessel Traffic Services: Management from Shore*).

Considered as a "branch office aboard," the management of a ship can be considered as part of a globality, or as an independent unit or a satellite of an organization. In the first case, ships will provide information by radio communication for the management and administration of the ship. The second possibility could render the ship independent, making it a fully autonomous enterprise managing its own business; this solution has proved to be very motivating for the crew. New facilities for communication between ship and shore act in favor of trading cooperations between shipowners (e.g., trade coordination, catering, bunkers). Development of communications allows automatic transmission of prerequisite information to shore parties.

6. Social and Educational Impact

For centuries, the ship has been an autonomous social entity which was managed to navigate and trade with a minimum of interaction between ship and shore. Because of the hazardous nature of sea ventures, the organization of life on board was structured and rigid under the strong legal and individual authority of the captain. The skills of the crew were varied to cope with every aspect of possible encounters with uncircumventable hazards. The ship was a community organized to care for its own survival, independently of any rescue. The seafaring world was a closed one, with proper traditions and laws for living and trading. The running of the ship was assumed by the crew.

Plans for reduced-crew vessels have been designed. These vessels are highly automatized. From 40 crew members in the late 1960s, owners are now operating ships of 24, 17, 14 and, even, 12 crew members. Task studies from Norway and Japan indicate that these numbers could be reduced to 6; option zero is considered as theoretically possible. These expectations are of consequence to several social aspects. Some studies pay attention to labor; in future, ships will be operated by carefully selected and highly qualified officers only.

Considering the future of the automated ship, recruitment and motivation, education and training, and career planning must be studied in parallel. Without the proper trained officer, the automated ship is a danger to navigation. The description of the operation of this vessel clearly shows that it is not so different from that of an aircraft. Besides the time spent on board, the ship officer's work should be similar to that of an aircraft pilot. The career prospects of the merchant marine officer will have to be considered as for any other career and recruitment will have to be enlarged to nontraditional resources of personnel, to enlarge selection possibilities. To run these different ships, extra knowledge and peculiar human qualities will be required from the officers. Their selection will be at a higher level and their individual qualities will be tested in order to determine the capability of the cadet to adapt to the reduced-crew environment. The romantic side of navigation is left to sailing boats. Intelligent, pragmatic, determined and individualistic personalities are needed to cope with the requirements of this new profession. The attractiveness of this career to young people will have to be reinforced by social measures and motivating salaries, together with a clear and straightforward definition of the profession and the high level of responsibility and devotion involved.

The establishment of a program of education and training in maritime academies should be the result of close cooperation among the shipping partners and the boards of education. The requirements of the profession will have to be redefined on the basis of the objective of a dual purpose, engine and deck, education. Only one speciality officer is necessary on the vessel. The means to achieve these goals will be incorporated into the curriculum: the use of simulation, training in workshops, research programs, passage planning and so on. The qualifications received will have to be recognized as equivalent to university diplomas. Knowledge of the English language will have to be of the highest level. The standard of maritime education will have to be equivalent in all countries providing this education.

7. Future Developments

The effective cost of automation is confidential. Relevant studies provide information by comparing the operations of ships having large crews and little automation, and those having small crews and a high level of automation. Comparisons are said to be in favor of the latter.

A systematic cost cutting policy together with a ship management profit mentality certainly leads to a lower standard of vessel and of crew. This consideration, which is valuable for a part of the world fleet, is a real danger to the environment of the seas and the coasts, to safe navigation and to the quality of sea transport. Fortunately, this trend does not appear to be the major one. Reliable and numerous studies on high-level-technology ships, as well as the number of orders for these vessels, imply that an important part of future world tonnage will be of a sophisticated design together with a reduced crew.

See also: Marine Fleet Planning and Scheduling

Bibliography

Graf H 1988 The ship of the future project. *Ocean Voice*. IMMARSAT, UK
Leaf J J 1988 Branch officer aboard. *Ocean Voice*. IMMARSAT, UK
Motozuna K, Yoshida O 1988 Advanced ship operation and shipboard administration system. *Conf. Society of Naval Architects and Marine Engineers*. SNAME, Jersey City, NJ
Vieljeux T 1989 La technologie sauvera-t-elle les activités maritimes françaises? *Conf. Academie de Marine*. Academie de Marine, Paris

L. Courcoux
[Inspection Generale Enseignement Maritime, Paris, France]

Social Issues of Transportation

Since ancient times, cultural and economic life has meant the development of transportation for people and goods. It is also necessary for political purposes: the road network of the Roman Empire is the most perfect example of a transportation system in Mediterranean history. The development of railroads in the late nineteenth century changed the geographic space–time scale; since World War II, the increasing importance of the car has completely changed the mobility of people and goods, and designed new patterns and new structures in geographic space. Air transport has had the same effect on a worldwide scale.

After being a necessity, transportation and mobility became an ideal, an important part of our life-style and, in some cases, a symbol of freedom.

The car has modified rural and urban patterns, and given new opportunities for the development of the employment market and for an increase in leisure. The creation of the EC after World War II, with the intention of forming a common market (1992) created a common policy for transportation; this has not been as efficient as the common policy for agriculture. This common policy was foreseen as a necessity for the creation of a new political, social, cultural and economic entity.

1. The Importance of Transportation in Daily Life

There are 320×10^6 inhabitants in the EC. They spend 14% of their budget on transportation needs (car equipment, maintenance and use, train, urban mass transit, air transport). This percentage is increasing slightly, mostly because of the increase in car equipment and use. In 1987, private consumption in France amounted to US$$650 \times 10^{12}$, of which US$$84.5 \times 10^{12}$ was for transportation needs (13%) and US$$74 \times 10^{12}$ for car expenditures. Transportation is the third largest percentage of expenditure after food and housing.

Many studies have shown that people spend 20 min for each home–work trip in Europe. The improvement in the infrastructure and mass transit services has allowed people to increase the distance between home and work: 30% of the active population in the EC are working at a distance of more than 10 km from home.

2. Car Transport

In interurban trips, at least 85% of trips are by car. The modal split depends on the distance of the journey. The use of rail and air transport increases with the distance and the duration of the journey. The split also depends on the purpose of the trip. Table 1 shows some results for long-distance trips (over 100 km) in France.

In urban areas, car usage depends on the size of the conurbation. For example, the modal split is 25% in the

Table 1
Transportation modal split in France

	For personal motivation (%)	For professional motivation (%)
Car	81	68
Train	12	18
Air	2	10
Bus	4	2
Others	1	2

Source: INSEE 1981–1982

Paris area and 59% in other French towns (see Table 2). The modal split is a function of the efficiency of the mass transit system and of the general policy for land use, parking, traffic control and general infrastructure capacity.

The trend is towards car use both in rural and urban areas; in spite of energy and pollution problems, car use has managed to expand, even during high fuel prices. In France, 75% of households possess a car; nearly 25% of households have more than one. In Europe, it is expected that, by around the year 2000, there will be nearly one car for each person who is old enough to have a licence. The increase, in recent years, in traffic volume has been very rapid on main roads (5% per year in France).

The car is ommipresent and there is no doubt that its future is bright. The research and development program in Europe (Prometheus–DRIVE (see *DRIVE*)) is allowing the challenge of increasing traffic volume and safety problems to be met. There seems to be no medium-term problem with energy and so the main problems may be solved with the possible exception of the biggest conurbations in industrialized countries, where traffic congestion and pollution from traffic are serious and difficult issues, especially in surburbs. Road safety is also an important issue: how long can more than 45 000 fatalities a year in the EC alone be tolerated?

Urban mass transit and rail transport represent an opportunity to cope with these issues by offering an alternative to car use. To make this alternative a real one, it is necessary to offer very effective mass transit services to attact passengers, including those who can afford the cost of car use. It is not only a question of low tariffs, but also of the quality of services (frequency, speed, comfort). The growth of traffic using the new French high-speed train (TGV) gives an illustration of this scheme for interurban traffic (see *High-Speed Railroad Networks in Europe*). A good comprehensive system of transportation cannot depend on only one mode: this does not offer sufficient choice to cope with the different needs or wishes of the population.

3. Speed of Transportation

The history of transportation has been a battle to increase the speed of travel: after the horse (20 km h^{-1}) came the railroad steam engine (140 km h^{-1}) followed by the railroad electric engine (now 300 km h^{-1}) and air travel ($>600 \text{ km h}^{-1}$). Each progressive step has profoundly modified the relationships between transportation and space, transportation and society and, possibly, transportation and civilization. Tourism is a good example of the relationship between these different concepts. Middle-class people may now spend a week or two in any country in the world. Chartered plane fares are continually getting cheaper and the whole world is now a tourist market.

4. Travelling Time

There is a paradox concerning the value of travelling time. From traveller behavior, a value can be derived that is a part of the work time value; that is, people behave so as to minimize the travelling time, accepting the payment of a higher tariff (as shown on the French toll freeway). However, each time there is an improvement in the duration of travelling time, many people move house, so that there is no general improvement in the average time spent commuting from home to work. New travel opportunities are opening up new choices such as a home instead of an apartment or a bigger house at the same price. The paradox is that the reduction in travelling time is generally used to increase the distance of travel.

Beyond this, travelling time may also be used for leisure activities, such as relaxing on a train or an airplane, or reading without being disturbed. Travelling is also an opportunity to work: the train offers various possibilities such as the use of a telephone or a typewriter. In the future, it will be the same in a car and some rent-a-car firms offer an "office on wheels," with a telephone, telex and computer. Travel can also be a cultural opportunity; some airports and railway stations put on exhibitions. Some trains display the history and geography of the region and even present local or cultural shows or movies. Moreover, the car offers other pleasures such as that of driving for its own sake or for the sensation of speed beyond the speed limit. Transportation is not only utilitarian, it is also a part of our daily life-style. The choice of a new car is as much affective as guided by its use or its economy.

5. The Goods Transportation System

The goods transportation system is different to the passenger transportation system: it is highly professional. The big markets are now worldwide and the transportation system, with logistics, is an increasing part of the commercialization of goods. Transportation

Table 2
Work trips in France

	Average time (min)	Average distance (km)	Return home at noon (%)	Modal split (%)			
				By foot	Two wheels	Mass transit	Car
Paris	31	11	17	15	5	55	25
Others	18	10	50	15	9	17	59
Total	21	10.5	42	15	8	24	53

Source: CREDOC 1983–1986

systems have to offer good services in very diverse conditions: high-value products in small quantities, such as drugs, or a complete plan. The transportation system is often a complex one, using different modes, such as maritime, air and land, and needing a close connection and perfect coordination between the modes. The goods transportation economy is quite different from that of passenger transport. However, the needs of a transportation system are even more important in the goods market. The percentage of product price made up of transportation costs varies up to a maximum of 15%, depending on the value of the product. It is obvious that even a small difference in the price or the quality of the transportation may have important effects on the competition. Transportation services are a strategic part of the system of production and commercialization in industry. This is why the logistics and storage policy of a firm are so connected with their transportation policy.

It is not possible to conceive of a world without cars or trucks as they have such a strong interaction with our daily way of life. The mobility of people and goods is taken for granted and there is no reason to imagine a fundamental change until the end of cheap oil energy. At some time in the twenty-first century, another cheap source of energy will succeed in replacing oil. Society is so tied to the transportation system that a regression of the characteristics of the system appears to be impossible.

See also: Social Issues of Ship Automation

Bibliography

Dobias G 1989 Les transports interrégionaux de voyageurs. Les Presses des Ponts et Chaussées, Arcueil, France
Orfeuil J P 1989 Un milliard de déplacements par semaine—INRETS. La Documentation Française, Arcueil, France

G. Dobias
[INRETS, Arcueil, France]

Speed Limitation on Freeways: Traffic-Responsive Strategies

The driver adjusts speed to the road alignment, the weather conditions, the traffic situation and the individual driver-vehicle characteristics. Driver behavior can be improved through variable information reflecting the current situation with regard to the road section capacity (maximum flow, mean speed) and traffic safety. An appropriate way to achieve this is flexible speed limits.

1. Basic Principles

At very low traffic volumes (i.e., if the vehicles do not influence each other), the road geometry (e.g., "curviness"), the road surface condition (e.g., dry, wet, slippery), the range of vision, specific vehicle features (e.g., maximum speed, maximum deceleration or acceleration) and the individual driver characteristics (e.g., aggressiveness, idea of comfortable driving, compliance with permitted maximum speed) determine the desired speed or the "free" speed (see *Flow Variables*). This, therefore, varies from driver to driver and according to vehicle features, even if the other features remain constant. As traffic volume (vehicles per time unit) and traffic density (vehicles per distance unit) increase, the probability will grow that a vehicle will approach another vehicle travelling at a slower speed that cannot be overtaken, at least not without delay. The approaching vehicle therefore has to reduce its speed, at least temporarily. The degree of deceleration (and acceleration) and, therefore, distance behavior with regard to a preceding vehicle, as well as lane-changing behavior depending on position and speed of neighboring vehicles, affect speed distribution.

As speeds are partially adjusted with increasing traffic volume, the mean value and the variance of the speed distribution decrease. The traffic volume may further increase to a maximum (capacity, maximum flow), entailing a growing probability of an unstable traffic condition, characterized by disturbances (high speed reductions). This situation is intensified more than eased by the following vehicles. Compared with the state of "capacity," traffic volume and mean speed decrease, whereas traffic density increases. The variance in traffic conditions in the unstable range is very large, particularly when shorter time intervals are considered (~1–5 min).

If a variable speed is indicated which is adapted to the current situation (as maximum or recommended speed), drivers change their behavior in two respects: they approximately adjust their desired speed to the indicated speed and they relax their aggressive driving behavior by adapting their distance behavior. This results in lower mean speed differences between vehicles following each other, as well as in lower mean accelerations and decelerations. The altered behavior is due to the informative contents of the indication, which is interpreted as follows: it is no use wanting to overtake under any circumstances because, in the road sections further downstream, the traffic density will also not allow higher speeds. A precondition for this kind of interpretation is confidence in the traffic sign, which must be confirmed again and again by experience.

Figures 1 and 2 show the direct effects of speed limits on traffic flow: at lower or mean traffic volumes, the mean speed is lower due to the reduction effect whereas, at higher speeds, an increase is detected due to the

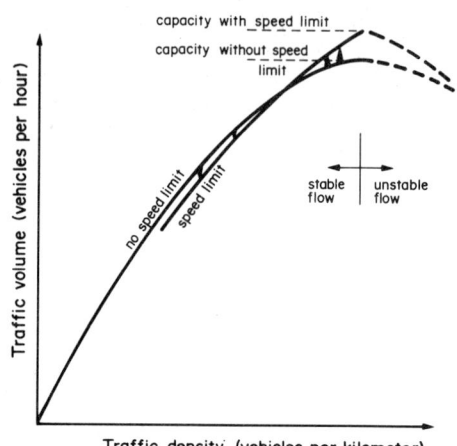

Figure 1
Changes of traffic state by adaptive speed limits

stabilizing effect. Thus, both capacity and speed rise by about 5% to 10% at the same time. The variation coefficient of speed is reduced by 30% to 50%. The success of speed control can be actually achieved or

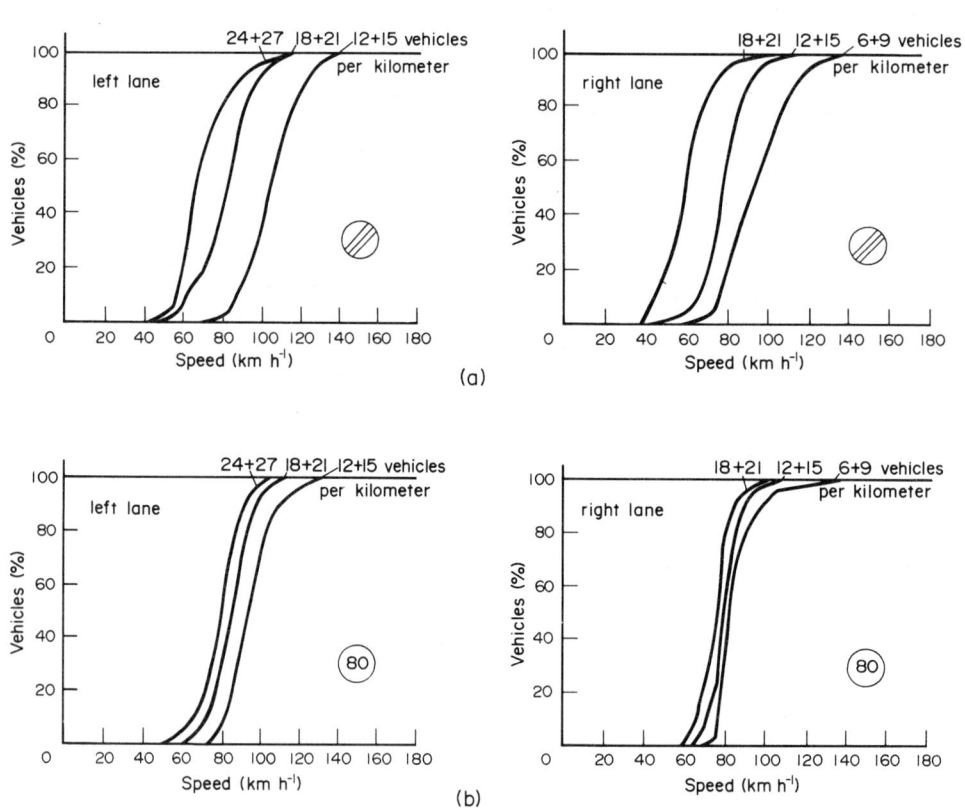

Figure 2
Distribution of speeds: (a) without speed limit and (b) with speed limit

supported by direct explanations on the display panel (e.g., congestion, fog) or by indirect information via the mass media (e.g., newspapers, television).

It is assumed that speed is indicated by locally fixed variable message signs. New technology using in-vehicle display devices will not be considered in this article.

2. Control Strategies

2.1 Stabilization of Traffic Flow at High Traffic Loads

Due to their attractiveness, freeways frequently have very high traffic loads. Therefore, there is the risk of overload and transition to an unstable traffic condition. As a recommendation, the speed limit aims at the mean value of the currently measured speed. This means that it does not aim at a lower average speed, but at a smoother traffic flow. The risk of disturbances and traffic breakdowns with resulting rear-end collisions is thus reduced.

2.2 Reduction of Very High Speeds at Low Traffic Volume

If there is no general speed limit, low traffic loads make it possible to drive at the highest speed, as far as the alignment of the freeway allows. If drivers falsely estimate the maximum safe speed, they run the risk of leaving the road and having an accident. In such cases, the accident consequences are usually very severe due to the high kinetic energy involved. If accident sites where excessive speed has often been the cause of accidents are known, the correct measure is to indicate a maximum speed. It can be further reduced on the basis of current road and traffic conditions.

2.3 Restabilization of Unstable Traffic Flow

Once congestion has occurred on a road section due to overloading, the aim is to disperse it as soon as possible, since congestion always implies loss of capacity and safety, as well as an increase in fuel consumption and pollutant emission. A speed limit in the sense of a recommendation just allowing stable-flowing traffic (~ 60 km h^{-1}) can cause a speeding up of the traffic flow and restabilization. The display must make drivers realize that the road section is overloaded not just as far as their range of vision, but over a much longer distance. They must also realize that aggressive driving will not help the situation. Here, stabilization is also aimed at by harmonizing speed, but at a higher level.

2.4 Safety Measures at Approaches to Congestions

In the approaching area of a congestion, speeds are often reduced very abruptly because the drivers recognize the situation too late. In such situations, rear-end collisions frequently occur. In order to make safe the end of a congestion, speed must be reduced. This is done by a gradual reduction in several steps. Such a system requires display panels at short intervals (every few hundred meters), since otherwise they might be considered inadequate and not accepted. Furthermore, the current position of the congestion must be detected fairly exactly. For these two reasons, the technical effort per distance unit (section length) is relatively high, so that such systems can only be used in bottleneck areas known to be critical.

2.5 Speed Reduction in Severe Weather Conditions

The accident rate (number of accidents related to mileage) not only depends on traffic volume but also, to a large extent, on weather conditions. Drivers frequently underestimate the danger of darkness, road surface wetness, slipperiness, fog or combined weather conditions. Therefore, systems equipped with appropriate sensors to detect conditions and then indicate the adequate speed based on these measurements are required to increase safety.

2.6 Speed Reduction to Prevent the Exceeding of Pollution Threshold Values

As speed rises, the emission of most pollutants, particularly nitrogen oxides, increases considerably. In densely populated areas, it can be helpful to reduce speed on the heavily loaded road sections in adverse weather conditions (inverted atmospheric thermal layers) to prevent pollution threshold values from being exceeded. Speed limits can be used preventively when weather forecasts predict adverse conditions.

2.7 Merging Aids at Access Ramps

At access points to freeways, the sidestream flowing in by the ramp cannot necessarily fill the gaps in the main traffic stream to capacity. This is possible only if the gaps in the main stream are concentrated on the outer, right lane and if the speeds on that lane and on the access ramp (acceleration lane) are adjusted to each other. This can be achieved by a lane-related speed limit in the area of the access point: the speeds on the outer, right lane are reduced to the speed limits on the access ramp. The other lanes on the main carriageway have higher speed limits in order to make it attractive to change to the lane to the left if there are still gaps.

2.8 Integration of Several Measures

Speed limits on a road section are often combined with other supportive and complementary measures that can also be indicated by variable message signs:

(a) congestion warning,

(b) warning by bad road surface or adverse weather conditions,

(c) overtaking prohibitions for trucks, and

(d) lane signalization.

Further integration is possible if the section-related measures are included in a network-related alternative route guidance by means of which substreams are diverted to other network sections.

Speed Limitation on Freeways: Traffic-Responsive Strategies

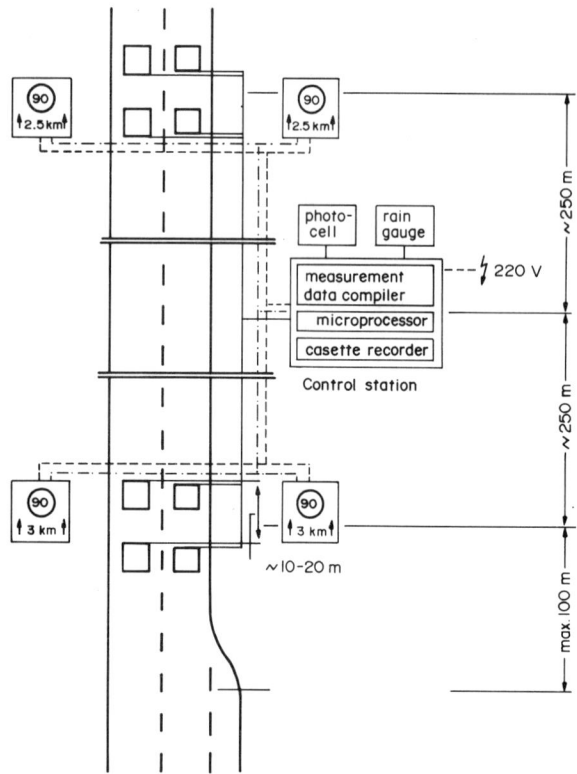

Figure 3
Decentralized speed control: ---, power supply; ——, detector data transmission; —·—, actuation of variable signs

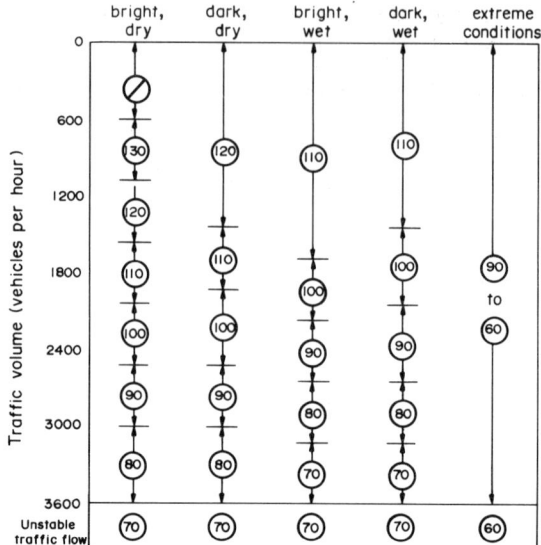

Figure 4
Switching plan

3. System Architecture

A system architecture that allows the practical use of the control strategies mentioned in Sect. 2 consists basically of three subsystems with the following functions (see *Traffic Control Systems: Architecture*):

(a) measurement data acquisition,
(b) data processing and decision on measures,
(c) communication of measures to the driver.

They are complemented by the necessary data links, as well as the power supply for all system components.

Traffic and weather condition detectors are used for measurement data acquisition. Often, inductive loops laid in the road surface record the number of private cars and trucks, as well as their speeds, separately for each lane. The distances between the measurement points must not be greater than 1–2 km. Weather conditions are determined by rain meters, photoelectric cells, visibility measurement devices, ice meters and so on. Their selection and specifications, however, are dependent on the requirements of the control strategy.

At the measurement points, the data are automatically checked for plausibility, precondensed and transmitted to a control center. There, a speed indication is assigned to the measured values on the basis of a decision logic and this is transmitted via a data link to an outstation (control device). These control devices switch the variable message signs to the newly determined value, if necessary.

The task of the control center can also be fulfilled decentrally (i.e., directly by the outstation). Figure 3 shows an example of such a decentral speed control with two display points located in the area of an access point. Such a solution is not cost intensive since a control center is not necessary. The disadvantage is, however, the missing central system control. In addition, the control states of several outstations following each other cannot be matched. This is possible only indirectly through a good adjustment of the measurement data acquisition.

As an example, Fig. 4 shows a simple control plan for a speed limit depending on traffic and weather conditions.

Most speed control systems using variable message signs are situated in Germany, but variable speed limits are also in operation in some other countries (e.g., the UK, The Netherlands, France, the USA).

See also: Freeway Capacity: Reliability and Control; Freeway Control: An Overview; Freeway Network Control

Bibliography

Kühne R 1989 Freeway control and incident detection using a stochastic continuum theory of traffic flow. *Proc.*

1st Int. Conf. Applications of Advanced Technologies in Transporation Engineering

Zackor H 1972 Beurteilung verkehrsabhängiger Geschwindigkeitsbeschränkungen auf Autobahnen. *Strassenbau Strassenverkehrstech.* **128**

Zackor H 1979 Self-sufficient control of speed on freeways. *Proc. Int. Symp. Traffic Control Systems*, Vol. 2A

Zackor H, Kühne R, Balz W 1989 Untersuchungen des Verkehrsablaufs im Bereich der Leistungsfähigkeit und bei instabilem Fluss. *Forsch. Strassenbau Strassenverkehrstech.* **524**

H. Zackor
[Steierwald Schönharting und Partner,
Stuttgart, Germany]

Traffic Assignment

A transportation system basically consists of two elements: transportation supply and travel demand. The transportation supply is the set of facilities and means available to the users of the transportation network. The travel demand is expressed by the number of users using the network for a given reason, with a transportation means, at a given time of day. The interaction between transportation supply and travel demand produces a flow pattern on network links.

A representation of transportation supply and a model of travel demand are necessary to compute the network flow pattern. Moreover, a model has to be built to simulate the interaction between the two elements. This model is known as the traffic assignment model (Friesz 1985, Sheffi 1985).

In problems of traffic management and control, knowing the network flow pattern allows the congestion level on each link to be determined by comparing link capacity with the corresponding flow. Knowledge of the flow pattern is also very important in network design problems, when changes to the network are needed to improve existing links, to insert new links or to enhance the performance of the transportation system.

1. Representation of a Transportation System

1.1 Transportation Supply

The transportation supply is the set of facilities and means available for the users' movements. In the case of traffic assignment it can be identified with the road network, generally represented by a directed graph where arcs correspond to network links and/or to junction movements.

A directed graph is a pair of sets $G = (N, A)$, where N is a set of n nodes and A is a set of m directed arcs, each one defined as an ordered pair of nodes (i, j). The density of a graph is the m/n ratio. A path is a finite sequence of arcs such that the origin node of an arc coincides with the destination node of the previous arc. In traffic assignment problems elementary paths are considered, where a node can be traversed at most once.

In terms of computer data structure, a graph can be represented with matrices or arrays. The choice depends on graph topology, problem characteristics and the algorithm used to solve the problem. Traffic networks are sparse (i.e., they have a low density). For this reason, it is advisable to adopt an array for data structure. This structure is based on lists that contain the set of successors for each node i of the graph; that is, the destination nodes of the arcs issuing from node i (forward star). The memory requirement is a linear function of n and m.

A graph where numerical functions are associated with arcs is generally indicated as a network. For a traffic network, these functions express travel costs c_{ij} on link ij or the delay cost at a junction approach. It can be the time spent to travel along the link or even a more general cost, such as the sum of vehicle operation, time and other costs, measured with the same unit. Thus, travel cost becomes a measure of transportation supply.

Hence, a network is a triplet of sets $N' = (N, A, C)$, where N is the set of nodes, A is the set of links and C is the set of link costs. The travel cost of a link is assumed to be a function of flow f_{ij}, which represents the number of users moving on the link in a time unit; that is, $c_{ij} = c_{ij}(f_{ij})$ (see Fig. 1). In this case, the network is said to be congested. Each point of the function represents the mean value of the users' cost at that flow value. On the other hand, if the link cost is assumed to be constant and not dependent on flow, the network is said to be not congested.

The functions most frequently used in practical application are derived from theoretical considerations and experimental observation. They contain some parameters that should be estimated from observed data. For example, travel time $t_{ij}(f_{ij})$ on a road link can be expressed by Davidson's function:

$$t_{ij}(f_{ij}) = t_0\left(1 + \frac{\sigma f_{ij}}{Q_{ij} - f_{ij}}\right)$$

where t_0 is travel time without congestion, Q_{ij} is link capacity and σ is a calibration parameter. Similar functions can be defined for delay on a signalized or unsignalized junction. Useful references on this subject are Branston (1976), Boyce *et al.* (1981) and Taylor (1984).

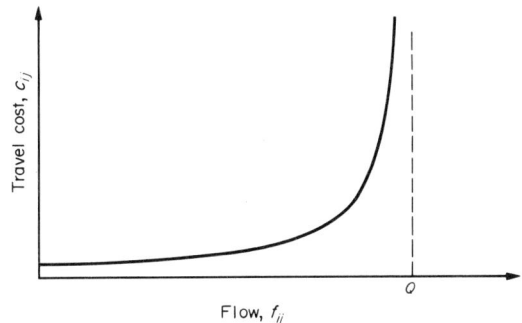

Figure 1
Cost-flow function on a road link

The application of models and methods for traffic assignment requires the definition of the level of detail necessary to represent the study area. It is impossible (and useless) to represent all the links of the road network and it is impossible to define travel demand between all origin–destination (OD) points. For this purpose, the study area is partitioned into traffic zones. For an urban network, the size of a zone can vary from a whole neighborhood to a set of buildings and, therefore, the number of zones can vary from a few units to several thousands. Each zone is represented by a centroid node assumed as the gravity center of the zone.

Fictitious links (centroid connectors) are used to connect the centroids to the road network, associating them to travel time from the centroid to a real node. Therefore, the network representation of the urban area will include both real nodes (junctions of the road network) and centroids (origin and destination of flows).

The level of detail to be adopted for the road network depends on the level of detail adopted for area zoning. All the main arteries and principal roads should be represented on the graph, while the same does not apply to secondary and minor roads.

1.2 Travel Demand

Travel demand represents the users' need to travel. It can be classified by reason (work, study, business, leisure, etc.), means (private car, public transport, on foot) and time of day. It is numerically expressed by the number of users D_{OD} of an OD pair of nodes. The number of OD pairs depends on the adopted zoning. In general, it is a decreasing function of the travel cost C_{OD} between nodes O and D (elastic demand); thus, $D_{OD} = D_{OD}(C_{OD})$. Its shape depends on the socio-economic characteristics of the urban area. Travel demand can also be assumed to be independent of cost (fixed demand). In this case, the D_{OD} values form the so-called matrix of transportation demand.

Origin and destination of an OD pair r are connected by several, sometimes many, paths. The definition of efficient path allows their number to be reduced. A path can be said to be efficient if a user, going from origin to destination, gets away from his origin and/or approaches his destination. Travel demand D_r spreads on the paths available for each pair r. The flow on each path k is denoted F_k. Hence, a flow conservation equation has to be satisfied for each pair r:

$$\sum_{k \in K_r} F_k = D_r, \quad \forall r \qquad (1)$$

where K_r is the set of all available paths for OD pair r.

The relation between links and paths is expressed by the link–path incidence matrix **A**, where each row is associated to a link h and each column to a path k. The elements a_{hk} of a row h take a value of 1 in the columns corresponding to the paths containing link h, and a value of 0 in the other cases.

Path cost C_k is the sum of the costs of the links that compose it:

$$C_k = \sum_h a_{hk} c_h, \quad \forall k \qquad (2)$$

The flow on link h is the sum of the flows on the paths containing that link:

$$f_h = \sum_k a_{hk} F_k, \quad \forall k \qquad (3)$$

In general, travel demand is not uniform during the day. So a transportation system is analyzed at a specified time of day, such as the morning peak hour or the evening peak hour. During this time, travel demand is assumed stable and a steady-state analysis is developed.

1.3 Demand–Supply Interaction

Travel demand depends on travel costs. On the other hand, costs on links and paths depend on flow pattern. Therefore, a mutual interaction between demand and costs (as a measure of supply) is generated. This interaction has two simultaneous effects, definition of demand value D_{OD}, a function of the travel cost, for each OD pair, and a flow pattern resulting from the distribution of demand among available paths. If demand is assumed to be fixed, the interaction produces only the latter effect.

A mutual interaction between flows and costs is generated. Through the stages of this interaction the system reaches a state of equilibrium, where the flow pattern produces a set of costs which induce a choice of paths which generates the same flow pattern. In order to determine the flow pattern, it is necessary to define a criterion for users' behavior. In general the following assumptions are made:

(a) each user has a full knowledge of the networks (links, available paths and related costs); and

(b) all users follow the same behavior rules.

The two most commonly used equilibria—deterministic and stochastic—are defined in relation to the behavior of two different users. A different approach, based on the theory of stochastic processes, is described in the article *Traffic Assignment, Stochastic*.

2. Deterministic Equilibrium

In the deterministic equilibrium model, each user is assumed to be a perfect economic subject: the user knows the network and has a correct perception of actual costs.

2.1 Deterministic Behavior of Path Choice (Wardrop's Principles)

If users are autonomous, it can be assumed that they behave rationally, choosing a path that minimizes travel costs from origin to destination. An equilibrium condition is reached (if any) when no user can reduce cost by selecting a different path. In this condition, all used paths of an OD pair have the same cost. This kind of equilibrium is formally expressed by Wardrop's first principle, referred to a single OD pair (Wardrop 1952).

"The journey times on all routes actually used are equal and less than those which would be experienced by a single vehicle on any unused route."

As a consequence of this principle, the paths actually used are the shortest paths at the existing congestion level. This flow pattern is called descriptive because it can be assumed to describe the real phenomenon of a transportation network. Therefore, Wardrop's first principle is used to build a traffic model for a transportation system with autonomous users.

If external interventions aimed at minimizing total users' costs do exist, then users cannot be assumed to be autonomous. In this case, another behavior criterion has to be defined to reach the optimal flow pattern of the system. This condition is expressed by Wardrop's second principle (Wardrop 1952).

"The average journey time is minimum."

As a consequence of this principle, not all the users of an OD pair use a shortest path. The resulting flow pattern is such that marginal costs on all paths actually used for an OD pair are equal and less than those on any unused path.

Such a user behavior is not plausible and, therefore, the related flow pattern is called normative, to indicate that it corresponds to a norm imposed by an agency. It can be used profitably to build a model for a public transportation system (people or freight).

2.2 Mathematical Formulation

If k is a path for OD pair r, C_k the cost on this path, C_r the minimum cost on paths actually used for r, Wardrop's first principle can be expressed by the following relations:

$$\left. \begin{array}{l} C_k = C_r, \quad \text{if } F_k > 0 \\ C_k \geq C_r, \quad \text{if } F_k = 0 \end{array} \right\} \forall k, r \quad (4)$$

The deterministic equilibrium problem can be formulated using the conditions of Eqn. (4), and the network constraints of Eqns. (1) and (3) (flow conservation and link flow–path flow relation), as well as nonnegativity conditions:

$$\sum_k F_k = D_r$$

$$f_h = \sum_k a_{hk} F_k, \quad \forall h$$

$$f_h \geq 0, \quad F_k \geq 0, \quad \forall h$$

where $D_r = D_r(C_r)$ and $C_k = \Sigma_h a_{hk} c_h$. It is also assumed that $c_h = c_h(f_h)$ is a separable cost-flow function. This hypothesis will be relaxed in the following.

It can be demonstrated that the solution to this equilibrium problem can be obtained by solving a mathematical programming model, if the previous model relations of Eqn. (4) are replaced by the following objective function (Beckmann et al. 1956, Dafermos and Sparrow 1968), usually defined as the integral cost function:

$$\min z(\mathbf{f}) = \sum_h \int_0^{f_h} c_h(x) \, dx - \sum_r \int_0^{D_{OD}} C_r(x) \, dx \quad (5)$$

where $C_r = C_r(D_r)$ is the inverse function of the demand function. In the case of fixed demand, the objective function becomes:

$$\min z(\mathbf{f}) = \sum_h \int_0^{f_h} c_h(x) \, dx \quad (6)$$

The Kuhn–Tucker optimality conditions of this optimization model are coincident with Eqn. (4) expressing Wardrop's first principle.

If Wardrop's second principle is adopted, the objective function has to express total cost minimization:

$$\min z(\mathbf{f}) = \sum_h c_h(f_h) f_h \quad (7)$$

If k is a path for the OD pair r, C'_k the marginal cost on this path and C'_r the marginal cost on paths actually used for pair r, Wardrop's second principle can be expressed by the following relations:

$$\left. \begin{array}{l} C'_k = C'_r \quad \text{if } F_k > 0 \\ C'_k \geq C'_r \quad \text{if } F_k = 0 \end{array} \right\} \forall k, r \quad (8)$$

The mathematical formulation presented is defined as a link–path formulation because it adopts flows on links and paths as variables. Another formulation of the constraints is defined as a link–node formulation because the only variables are flows on links. Denoting link h by ij, and the flow on link ij originating at O by $f_{O,ij}$, the set of constraints is formed by flow conservation equations at nodes and flow additivity equations on links

$$\sum_l f_{O,il} - \sum_i f_{O,ij} = \begin{cases} -D_{Ol}, & l \text{ destination} \\ \sum_D D_{iO}, & l \equiv O \\ 0 & \text{otherwise} \end{cases} \quad ((\forall O) \, \forall i)$$

$$\sum_O f_{O,lj} = f_{lj}, \quad \forall ij, f_{ij} \geq 0$$

Traffic Assignment

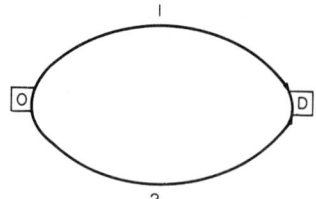

Figure 2
A simple example network

It can be shown that the equilibrium solution:

(a) exists if $D_{OD}(C_{OD})$ is a continuous and positive function or if it is constant and $c_h(f_h)$ is an nondecreasing function; and

(b) is unique when $c_h(f_h)$ is an increasing function.

In the described model, the cost on link h has been assumed to be a function of flow f_h alone on the same link (separable costs). As a rule, it can also be a function of flows on other links, and then $c_h = c_h(f)$ (nonseparable costs). Three cases are possible in relation to the structure of the Jacobian matrix of cost functions $\mathbf{J} = (\partial c_h/\partial f_k)$.

(a) Separable costs: $\partial c_h/\partial f_k = 0$, $\forall h \neq k$ ($\mathbf{J}$ diagonal).

(b) Nonseparable costs: $\partial c_h/\partial f_k = \partial c_k/\partial f_h$, $\forall h \neq k$ ($\mathbf{J}$ symmetric).

(c) Nonseperable costs: $\exists h \neq k$: $\partial C_h/\partial f_k \neq \partial c_k/\partial f_h$ ($\mathbf{J}$ asymmetric).

With this classification, the above-described model is called a model of equilibrium with separable costs. It can be extended to the symmetric Jacobian case, thus yielding the so-called symmetric equilibrium model (Dafermos 1980).

In the more general case of an asymmetric Jacobian matrix, a different formulation of the problem, called asymmetric equilibrium, is needed (Hearn et al. 1984). Asymmetric equilibrium can also be formulated as a variational inequality model (Dafermos 1980) or as a nonlinear complimentarity model (Aashtiani 1979).

2.3 Graphical Representation of the Deterministic Equilibrium

The equilibrium problem can be graphically represented through a simple network (see Fig. 2) by a single OD pair and two paths, 1 and 2, each one with a single link. Separable costs are assumed.

If D is the travel demand between O and D, it is distributed among available paths in such a way that $f_1 + f_2 = D$. If Wardrop's first principle is assumed, the equilibrium condition is reached for $C_1 = C_2$; that is $c_1(f_1) = c_2(f_2)$. If Wardrop's second principle is assumed, the optimal flow pattern of the system is reached for $C_1' = C_2'$, that is, $c_1'(f_1) = c_2'(f_2)$. Figure 3 reports the functions of actual and marginal cost on opposed diagrams, in such a way that distance between vertical axes equals demand D. Point E on the horizontal axis identifies the users' equilibrium solution (minimum integral cost). Point O identifies the optimal solution of the system (minimum total cost).

2.4 Model Solution

(a) *Constant costs.* If link costs are assumed to be constant (noncongested networks), that is independent of the flow, the model strongly simplifies. In this case, both the minimum cost for each OD pair and the corresponding value of demand are univocally determined. Hence, only the fixed demand case is considered in the following.

The objective functions relative to user equilibrium and system optimum models coincide; that is $z(\mathbf{f}) = \Sigma_h c_h f_h$. Then, since $f_h = \Sigma_O f_{O,h}$, it turns out that $z(\mathbf{f}) = \Sigma_h c_h \Sigma_O f_{O,h} = \Sigma_O \Sigma_h c_h f_{O,h}$. The whole model can be written in the following form:

$$\min \sum_O \sum_h c_h f_{O,h}$$

$$\sum_l f_{O,il} - \sum_i f_{O,ij} = \begin{cases} -D_{Ol}, & l \text{ destination} \\ \sum_D D_{lD}, & l \equiv O \\ 0 & \text{otherwise} \end{cases}$$

$$((\forall i)\, \forall O)$$

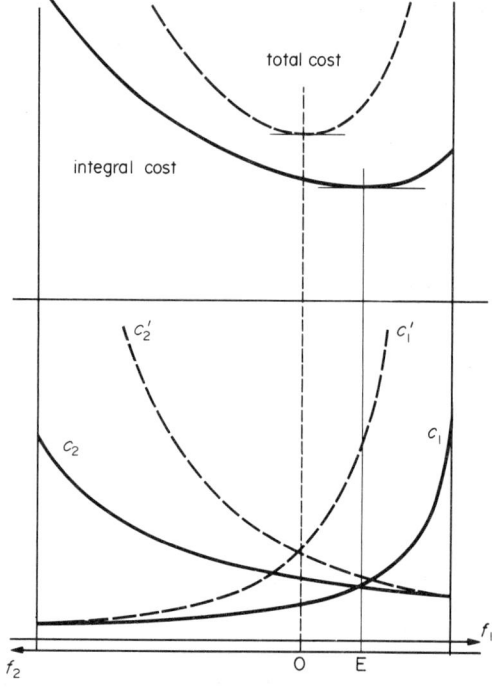

Figure 3
Graphical representation of equilibrium

The set of variables $\{f_O, h\}$ can be partitioned into subsets, one for each origin O. The objective function can be expressed as the sum of components relative to the different OD pairs, and therefore:

$$\min \sum_O \sum_h c_h f_{O,h} = \sum_O \min \sum_h c_h f_{O,h}$$

Moreover, each group of constraints is relative to a subset of variables and there are no constraints with variables belonging to different subsets. Hence, the problem is reduced to a set of linear programming problems, each of them relative to one origin.

Each subproblem can be considered as the minimization of the total travel time between the origin node O and all the destination nodes, having assumed that link costs are constant. Therefore, each demand D_{OD} can be entirely assigned to the links belonging to the shortest path between O and D, while no flow is assigned to the other links of the network. Thus, all the constraints of the problem are certainly satisfied. This technique is called "all-or-nothing" assignment. The algorithms for the solution of the shortest-path problem are treated in great detail by Gallo and Pallottino (1986) and in the article *Shortest-Path Algorithms*.

(b) Variable separable costs with fixed demand. If link costs vary according to link flows, until the 1970s traffic assignment was achieved by means of heuristic loading techniques, based on the all-or-nothing assignment. These techniques are easy to implement but they do not always converge to the solution of the problem. In the most frequently used technique, called incremental assignment, the travel demand is assigned to the network by successive increments through constant cost assignment. After each assignment, link costs are updated according to the current flow pattern (Martin and Manheim 1965).

If demand is assumed to be fixed and link costs separable, the model can be solved by a network implementation of convex programming techniques (Nguyen 1976). Generally, the network implementation of the Frank–Wolfe algorithm is applied. A particularly easy implementation makes it suitable to network problems.

Starting from an initial solution f^0, it is a two-phase procedure that involves:

(i) determination of the descent direction (if any) of the objective function; and

(ii) determination of step length.

In particular, the descent direction d^k is expressed at the generic iteration k as the difference between the current solution relative to the previous iteration f^{k-1} and the unknown solution y^k. This solution, and hence the descent direction, can be obtained by solving a linear program formed by the equilibrium constraints and a linear objective function obtained with a linear approximation of the original objective function z at point f^{k-1}. Denoting the gradient of the objective function at point f^{k-1} by $q^{k-1} = \nabla(z(f^{k-1}))$, the linear program has the objective function $q^{k-1}(y - f^{k-1}) + z(f^{k-1})$ with the unknown y. By dropping the constant terms it becomes $q^{k-1} \cdot y$.

In the case of user equilibrium, with objective function given by Eqn. (6), the gradient coincides with the link costs computed according to the current solution $q_{k-1} = c(f^{k-1})$. Similarly, in the case of system optimum, with the objective function given by Eqn. (7), it turns out that $q_1^{k-1} = c_1(f_1^{k-1}) + c_1'(f_1^{k-1})f_1^{k-1}$.

In both cases, the model obtained after the update of the gradient is a constant cost assignment one and it is solved by an all-or-nothing assignment. Let y^k be the solution of this problem. The step length along direction $d^k = y^k - f^{k-1}$ can be obtained by solving the following one-dimensional optimization problem with the golden search or the bisection algorithms:

$$\min_{\mu \in [0,1]} z(f^{k-1} + \mu d^k)$$

Let μ^k be the solution of this problem. The new current solution is given by $f^k = f^{k-1} + \mu^k d^k = (1 - \mu^k)f^{k-1} + \mu^k y^k$. From this solution, the gradient is updated and then the search of a descent direction is repeated, and so on until convergence.

The convergence of the Frank–Wolfe algorithm can be accelerated with some modifications of the direction computation. Moreover, the algorithm can be modified by eliminating the one-dimensional optimization and computing the new current solution with a convex combination of the all-or-nothing solution y^k and the previous current solution f^{k-1}. The algorithm can be shown to converge if a combination parameter is adopted that decreases with the iteration number. In general, the value $\mu^k = 1/k$ is adopted.

(c) Variable nonseparable costs and/or elastic demand. The case of elastic demand can be solved in two ways—by applying the algorithm on a modified network, or by modifying the algorithm itself.

In the first case, let D'_{OD} be an upper bound to the demand value and $C_{OD} = C_{OD}(D_{OD})$ the inverse function of the demand function. A fictitious link is introduced for each OD pair with a cost function $c(f) = -C_{OD}(D_{OD})$ and a traffic assignment of the fixed demand D'_{OD} is performed. If f'_{OD} is the resulting flow on fictitious link O–D, the real demand value for the OD pair is $D_{OD} = D'_{OD} - f'_{OD}$.

In the second case, the pair (f, D) formed by the link flow vector f and the demand vector D is assumed to be the current solution. In the all-or-nothing assignment for a given origin O, the demand value $Y_{Oi} = D_{Oi}(C_{Oi})$ to destination i is computed as a function of the shortest path cost C_{Oi}, for each node i. Then the flow pattern y and the new direction $(y - f, y - D)$ are obtained. The algorithm continues with the one-dimensional optimization to compute step length.

The asymmetric equilibrium model can be solved using several approaches (Fisk and Nguyen 1982, Hearn et al. 1984, Nagurney 1986). The more consolidated algorithm is the so-called diagonalization algorithm which is simple to implement and has reduced memory requirements (Sheffi 1985). The convergence is guaranteed under some hypotheses on the cost Jacobian. In this algorithm, let $e(i)$ be the current solution. A separable-cost equilibrium problem is built, with the separable cost functions $c_h = c_h^{(i)}(f_h) = c_h(f_1 = e_1^{(i)}, \ldots, f_h, \ldots, f_m - e_m^{(i)})$. This is equivalent to diagonalizing the Jacobian of the cost functions and, thus, the new problem can be solved with the previously described algorithm. The solution obtained is used to update the current solution, by performing a convex combination of the two solutions. The algorithm iterates until a convergence criterion is reached.

A simplified version of this algorithm perfoms an all-or-nothing assignment instead of the equilibrium assignment at each iteration. This version requires a shorter computation time. It coincides with a Frank–Wolfe algorithm where at each iteration the gradient is updated using the nonseparable cost functions.

3. Stochastic Equilibrium

In stochastic equilibrium models, costs perceived by users are considered different from actual costs. The perceived cost is modelled as a random variable distributed among users. This way, each user perceives a different cost (Daganzo and Sheffi 1977).

3.1 Stochastic Behavior of Path Choice

The stochastic path choice behavior can be represented by a discrete choice model, based on the random utility theory. In fact, this kind of choice model can be applied when decision makers are allowed to choose among a discrete set of alternatives. A utility value, modelled as random variable, is associated with each alternative. In the case of traffic assignment, the users relative to an OD pair are considered as decision makers and the set of alternatives corresponds to the set of available efficient paths.

Therefore, a user associates to each path k a perceived cost U_k which is considered as a random variable, the sum of a deterministic term C_k and a random residual ε_k, expressing the perception error. Then $U_k = C_k + \varepsilon_k$, with expected values $E(C_k) = C_k$, $E(\varepsilon_k) = 0$, and variances $\text{var}(C_k) = 0$, $\text{var}(\varepsilon_k) = \sigma_k^2$. Hence, $E(U_k) = C_k$, $\text{var}(U_k) = \sigma_k^2$.

The probability that a user of OD pair r chooses the path k belonging to the set K_r of all available paths is equal to the probability that the perceived cost on path k is less than the perceived cost on the other paths connecting the pair r.

Assuming homogenous and independent users, C_k is the actual path k cost and U_k is the average perceived cost. The random residual ε_k also takes into account cost dispersion among users. Under these assumptions, p_k is the expected value of the random variable representing the fraction of users choosing path k. The explicit expression of p_k depends on the random distribution of residuals.

If random residuals are assumed to be distributed according to the Weibull–Gumbel (WG) distribution with parameter Θ and zero mean, that is $\varepsilon_k \sim \text{WG}(\Theta)$, $E(\varepsilon_k) = 0$, $\text{var}(\varepsilon_k) = \pi^2/6\Theta^2$, the perceived cost is a WG random variable, since it is the sum of a deterministic variable and of a WG variable, $U_k = C_k + \varepsilon_k$. Moreover, because the minimum of WG variables is a WG variable, for a user of OD pair r it results that:

$$U_{\min}^r = \min_{k \in K_r}(U_k) \sim \text{WG}$$

$$E(U_{\min}^r) = (1/\Theta) \sum_k \exp(-\Theta C_k) = Y/\Theta$$

$$\text{var}(U_{\min}^r) = \pi^2/6\Theta^2$$

where $Y = \ln[\Sigma_k \exp(-\Theta C_k)]$ is called logsum variable, Θ is a calibration parameter inversely proportional to standard deviation σ and common to all paths connecting OD pair r. Under these assumptions the choice probabilities are given by the relations:

$$p_k = \exp(-\Theta C_k)/\exp(Y), \quad \forall k \in K_r, \forall r$$

The choice model thus defined is called logit.

In the case of the network represented in Fig. 2, these relations give $c_1(f_1) + (1/\Theta) \ln(f_1) = c_2(f_2) + (1/\Theta) \ln(f_2)$ which coincides with the deterministic equilibrium condition if $\sigma = 0$ (i.e., $1/\Theta = 0$). As the perception variance increases, the stochastic equilibrium point gets farther from the deterministic one and it tends towards a completely random perception (equiprobable alternatives).

The main limiting hypotheses of the logit model are:

(a) common variance for all paths connecting an OD pair and the same perception error for both short and long paths; and

(b) null covariance between two different paths—thus, two paths are always perveived as two completely different alternatives, even though they have a common part.

The probit choice model allows these limiting hypotheses to be removed. Random residuals are assumed to form a random vector ε distributed according to a multivariate normal law (MVN), with zero mean and covariance matrix Σ, that is $\varepsilon \sim \text{MVN}(\mathbf{0}, \Sigma)$ with $E(\varepsilon_k) = 0$, $\text{var}(\varepsilon_k) = \sigma_k^2$, $\text{cov}(\varepsilon_i, \varepsilon_j) = \sigma_{ij}$. In particular, it is assumed $\sigma_k^2 = \tau C_k$ and $\sigma_{ij} = \tau C_{ij}$, where C_{ij} is the cost relative to the part common to paths i and j, and τ is a calibration parameter equal for all paths connecting an OD pair.

The probit model cannot be solved in a closed form, since the normal distribution function is not known in a closed form. It can only be roughly solved with a Monte Carlo simulation technique or, if the number of alternatives is small, with the Clark method (Sheffi 1985).

3.2 Mathematical Formulation

The stochastic equilibrium can be modelled through a mathematical programming scheme, assuming separable costs and fixed demand. For this purpose, the satisfaction $S_r = E(U_{r,\min})$ of the users of OD pair r is defined as the expected value of the minimum cost on the various paths connecting pair r. This value can be adopted as a system effectiveness parameter. It can be proved that the partial derivatives of satisfaction are proportional to the choice probabilities: $\partial S_r / \partial C_k = p_k C_k$ if $k \in K_r$, and equals zero otherwise.

The equilibrium flow pattern can be proved to be the soluton of the following unconstrained optimization model:

$$\min S \cdot d + c \cdot f - \sum_l \int_0^{f_l} c_l(x)\,dx \qquad (9)$$

where S is the vector of the satisfaction values of the users of different OD pairs. Generally, it results $S = S(C) = S(\mathbf{A}^T c)$ and $c = c(f)$. The problem is unconstrained because the introduction of network constraints is not necessary, since it can be proved that the optimal solution of the problem given by Eqn. (9) certainly meets these constraints.

Denoting by $\mathbf{P}$ the probability matrix the equilibrium path flow vector is given by:

$$F^* = \mathbf{P}[\mathbf{A}^T c(f^*)]\,d \qquad (10)$$

with expected value $E(F) = F^*$ and covariance matrix Σ_F such that:

$$\text{var}(F_k) = \sigma_k^2 = d_r p_k (1 - p_k) \quad k \in K_r$$

$$\text{cov}(F_k, F_h) = \sigma_{Kh} = \begin{cases} -d_r p_r p_n & \text{if } k, h \in K_r \\ 0 & \text{otherwise} \end{cases}$$

The corresponding equilibrium link flow vector is given by

$$f^* = \mathbf{A}F$$

with

$$E(f) = f^*$$
$$\Sigma_f = \mathbf{A}^T \Sigma_F \mathbf{A}$$

3.3 Solution of the Model

(a) *Constant costs.* In the case of flow-independent costs, from a theoretical point of view, the problem can be solved by computing the choice probability matrix $\mathbf{P}$, function of path costs $C = \mathbf{A}c$ and then the path flows with relation $F = \mathbf{P}d$. Finally, link flows are determined from path flows through $f = \mathbf{A}F$.

This approach requires prior explicit enumeration of all the efficient paths connecting each OD pair. This computation is very demanding for running time and core memory and can be avoided by using methods that allow an implicit enumeration of efficient paths to solve the choice models described above. Dial's algorithm can be adopted for the logit model and a Monte Carlo simulation algorithm for the probit model.

Dial's algorithm successively considers all the origin nodes. For origin O, having computed the costs of the shortest paths C_{Oi} from O to each node i, a path between O and destination D is said to be efficient if it is formed by links ij such that $C_{Oi} < C_{Oj}$.

Likelihood is computed for each link:

$$l_{ij} = \begin{cases} \exp[\Theta(C_{Oj} - C_{Oi} - c_{ij})] & \text{if } C_{Oi} < C_{Oj} \\ 0 & \text{otherwise} \end{cases}$$

Then, starting from origin O, all nodes i are examined, following the C_{Oi} increasing order. For each link ij exiting from node i, the weight w_{ij} is computed, with relations:

$$w_{ij} = \begin{cases} l_{ij} & \text{if } i = O \\ l_{ij} \Sigma_{ki} w_{ki} & \text{otherwise} \end{cases}$$

Finally, all nodes j are examined following the C_{Oj} decreasing order, and quantity $d_{Oj} W_{ij} / \Sigma_{kj} W_{kj}$ is added to the flow of link ij.

The Monte Carlo algorithm for the probit model determines at each iteration a perceived cost for each link ij as a pseudorandom realization of a Gaussian random variable with mean c_{ij} and variance τc_{ij}. Then, the all-or-nothing assignment is performed and the resulting link flows are averaged with the current average link flows. The procedure ends when the difference between two successive solutions is less than a given tolerance threshold.

(b) *Variable costs.* In the case of congested networks, the stochastic equilibrium flow pattern can be determined by adopting a path choice model and then by solving the unconstrained optimization model of Eqn. (9) Starting from an initial solution, this model can be solved by choosing a descent direction of the objective function from the current point and by moving along this direction. From this new point, the procedure is repeated until no descent direction exists or until a convergence criterion is met.

It can be proved that $\mathbf{AP}[\mathbf{A}^T c(f)]d - F$ is a descent direction (if any), obtained by duly deflecting the gradient of the objective function. This direction can be easily determined by computing link costs defined by the current solution and then by performing a stochastic assignment with constant costs. Moreover, it can be proved that the procedure converges if the moving step decreases with the number of iterations.

Hence, a procedure similar to the one adopted for deterministic equilibrium is obtained, as follows.

ALGORITHM
Choose an initial point $f^{(O)} = f_O$; $k = 0$
Repeat

$$k = k + 1$$

Compute link costs corresponding to the current solution

$$c^{(k)} = c(f^{(k-1)})$$

Perform a constant cost stochastic assignment

$$y^{(k)} = \mathbf{AP}(\mathbf{A}^T c^{(k)})d$$

Compute the direction of movement

$$y^{(k)} - f^{(k-1)}$$

Update the current solution

$$f^{(k)} = [(k-1)f^{(k-1)} + y^{(k)}]/k$$

Until the direction of movement is not descent.

The initial solution f_O can be found by using a constant cost stochastic assignment with zero-flow costs.

In the case of elastic demand and/or nonseparable costs the solution can be obtained with algorithms similar to those adopted to solve the deterministic equilibrium model.

4. Future Developments

Stochastic equilibrium is the most satisfying model, but its solution requires a greater computational effort than the other models. However, for heavily loaded networks the deterministic equilibrium assignment can be adopted since it is very close to the stochastic equilibrium assignment as it can be proved also from a theoretical viewpoint. Developments of models for traffic assignment are the recently proposed dynamic assignment models, which try to simulate the day-to-day adjustment process, not necessarily seeking an equilibrium state of the system (see *Traffic Assignment, Dynamic*; *Traffic Assignment, Stochastic*).

See also: Road Networks: Dynamic Equilibrium Models; Signal Control and Traffic Assignment; Traffic Assignment, Dynamic; Traffic Assignment, Stochastic

Bibliography

Aashtiani H Z 1979 The multimodal traffic assignment problem. Ph.D. dissertation, Massachusetts Institute of Technology

Beckmann M J, McGuire C B, Winston C B 1956 *Studies in the Economics of Transportation*. Yale University Press, New Haven, CT

Boyce D E, Janson B N, Eash R W 1981 The effect on equilibrium trip assignment of different link congestion functions. *Transp. Res. A* **15**, 223–32

Branston D 1976 Link capacity functions: A review. *Transp. Res. B* **10**, 223–36

Dafermos S C 1980 Traffic equilibrium and variational inequalities. *Transp. Sci.* **14**, 42–54

Dafermos S C, Sparrow F T 1968 The traffic assignment problem for a general network. *J. Res. Natl. Bur. Stand., Sect. B* **73**(2), 91–117

Daganzo C F, Sheffi Y 1977 On stochastic models of traffic assignment. *Transp. Sci.* **11**, 253–74

Fisk C S, and Nguyen S 1982 Solution algorithm for network equilibrium models with asymmetric user costs. *Transp. Sci.* **16**, 361–81

Friesz T L 1985 Network equilibrium design and aggregation: Key developments and research opportunities. *Transp. Res. A* **19**, 413–27

Gallo G, and Pallottino S 1986 *Shortest Path Methods: A Unifying Approach*. Mathematical Programming Studies, Vol 26. North-Holland, Amsterdam, pp. 38–64

Hearn D W, Lawphongpanich S, Nguyen S 1984 Convex programming formulations of the asymmetric traffic assignment problem. *Transp. Res. B* **18**, 357–65

Martin B V, Manheim M L 1965 A research program for comparison of traffic assignment techniques. *Highw. Res. Rec.* **88**

Nagurney A 1986 Computational comparisons of algorithms for general asymmetric traffic equilibrium problems with fixed and elastic demands. *Transp. Res. B* **20**, 78–84

Nguyen S 1976 A unified approach to equilibrium methods for traffic assignment. In: Florian M (ed.) *Traffic Equilibrium Methods*. Springer, Berlin, pp. 148–82

Sheffi Y 1985 *Urban Transportation Networks*. Prentice-Hall, Englewood Cliffs, NJ

Taylor M A P 1984 A note on using Davidson's function in equilibrium assignment. *Transp. Res. B* **18**, 181–99

Wardrop J G 1952 Some theoretical aspects of road traffic research. *Proc. Inst. Civ. Eng.* **2**(1), 325–78

G. E. Cantarella
[Università di Napoli, Naples, Italy]

A. Sforza
[Università di Salerno, Fisciano, Italy]

Traffic Assignment, Dynamic

The problem of dynamic traffic assignment is to predict the temporal evolution of the traffic flow pattern on a congested transportation network where travel demands and travel costs vary over time and space. Its solution may be represented by trajectories of arc flow rates or densities over a specified time period. Depending on the behavioral assumption of individual routing decisions, the dynamic traffic assignment problem can be divided into two classes:

(a) dynamic user optimal traffic assignment, wherein network users attempt to minimize individual travel costs; and

(b) dynamic system optimal traffic assignment, wherein network users cooperate in minimizing the total transportation cost.

In contrast, the problem of static traffic assignment is to predict the traffic flow pattern under the following steady-state assumptions: travel demands and costs are time-invariant; and inflow rate is equal to outflow rate on each arc at each instant. The solution of a static assignment problem can be either arc flows or path flows measured in the number of vehicles per period of analysis. It is important to note that any static traffic assignment model cannot consider the dynamic changes in traffic congestion level that are often observed on urban transportation networks during peak hours.

1. Literature Review

There have been three different approaches to the dynamic traffic assignment problem:

(a) computer simulation approach,

(b) mathematical programming approach, and

(c) control theoretic approach.

Yagar (1970, 1971) presented the first computer simulation model to emulate the traffic assignment according to Wardrop's first principle (Wardrop 1952) where each network user minimizes his travel cost, while taking into account both time-varying demand and queue evolution. Time-varying demands were represented by dividing the peak period into time slices. The demand was assumed to remain constant within each time slice. The demand that was queued in the previous time slice is added to the demand of the present time slice to obtain the total demand to be served in the present time slice. Yagar (1970) also presented a heuristic solution algorithm for the dynamic system optimal traffic assignment problem.

The first mathematical programming approach to the dynamic system optimal traffic assignment problem was made by Merchant (1974) and Merchant and Nemhauser (1978a, b). A macroscopic model for minimizing total transportation cost was formulated as a discrete-time, nonlinear, nonconvex mathematical programming problem. To consider the physical process of congestion explicitly, the arc exit functions were assumed to be nondecreasing and concave, and to represent the disutility of congestion, the arc cost functions were assumed to be nondecreasing and convex. The Kuhn–Tucker optimality conditions for the model were interpreted as a dynamic generalization of Wardrop's second principle which requires equalization of certain marginal costs for all the paths that are being used. The behavior of the dynamic model was examined under the steady-state assumptions and it was proven to be a proper extension of a conventional static system optimal traffic assignment model in a dynamic setting. Although the model did not have a convex constraint set, it was solved for a global optimum using a one-pass simplex algorithm without requiring branch-and-bound. It was also shown that the piecewise linear program has a staircase structure so that decomposition techniques or compactification methods can be applied. It should be noted that the model was restricted to a network with multiple origins and single destination.

Ho (1980) resolved the algorithmic question of implementing the Merchant–Nemhauser (MN) model. It was shown that a global optimum exists for a piecewise linear version of the MN model. A sufficient condition for optimality was derived which implies that a global optimum can be obtained by successively optimizing at most $N+1$ objective functions for the linear program, where N is the number of time periods in the planning horizon.

Carey (1986) showed that the MN model in fact satisfies a linear independence constraint qualification, which establishes the validity of the optimality analysis presented by Merchant and Nemhauser (1978b). Originally, their optimality analysis was based on an assumption that the Kuhn–Tucker conditions hold at an optimum and that their model, in turn, satisfies a constraint qualification. The MN model was reformulated by Carey (1987) as a well-behaved convex nonlinear program. The new formulation could then have analytical, computational and interpretational advantages in comparison with the nonconvex MN model. In particular, the Kuhn–Tucker conditions for the new formulation became both necessary and sufficient to ensure a global optimum. The new formulation also allowed an optimal flow control pattern to be determined so that the total network cost could be improved by reducing the arc outflow rates below their natural capacity levels.

In contrast with the aforementioned computer simulation and mathematical programming approaches, Wie (1988) and Friesz et al. (1989) provided a new insight into the problem of dynamic traffic assignment with application of optimal control theory. First, the MN model was reformulated as a continuous-time optimal control problem. The optimality conditions were derived using the Pontryagin maximum principle. The existence of singular controls was considered and singular controls were tested for optimality using the generalized Legendre–Clebsch condition. Second, for the first time, a dynamic user optimal traffic assignment model was formulated as an equivalent continuous-time optimal control problem. Its optimality conditions were interpreted as a dynamic generalization of Wardrop's first principle which requires equilibration of certain instantaneous measures of unit travel costs for all the paths that are being used at each instant for a given origin–destination pair. Third, various extensions of a dynamic user optimal traffic assignment model were presented by Wie (1988), including multiple destinations, time-varying elastic travel demands, and penalties for late arrivals.

2. Mathematical Formulations

The mathematical models for the dynamic traffic assignment problems that have been developed will briefly be reviewed. Both discrete- and continuous-time models for the dynamic system optimal traffic assignment problem will first be presented. A continuous-time model for the dynamic user optimal traffic assignment problem will then be presented. For simplicity, the review will be restricted to the dynamic models for a network with many origins and single destination.

2.1 Dynamic System Optimal Traffic Assignment

Consider the traffic network represented by a directed graph $G(N, A)$, where N is the set of nodes and A is the set of arcs. Assume that the cardinality of the set N is n and that nodes $1, 2, \ldots, n-1$ are origins while node n is the only destination. The set of all origins will be denoted by M. In general, the index a will be used to denote an arc, the index k to denote a node and the index p to denote a path. The fixed planning horizon is divided into equal time intervals of suitably small length and the individual periods are denoted by $\{i \mid i = 0, 1, \ldots, I\}$.

The dynamic evolution of the state of each arc is described by nonlinear difference equations:

$$x_a(i+1) - x_a(i) = u_a(i) - g_a[x_a(i)], \quad \forall a \in A, i = 0, \ldots, I-1 \quad (1)$$

where the state variable $x_a(i)$ is the number of vehicles on the arc a at the beginning of the ith time period; the decision variable $u_a(i)$ is the number of vehicles that is admitted onto the arc a during the ith time period; and $g_a[x_a(i)]$ is the number of vehicles that leaves the arc a during the ith time period. In order to depict the physical phenomenon of congestion on each arc, the exit function $g_a[x_a(i)]$ is assumed to be nonnegative, nondecreasing, continuously differentiable, and concave for all $x_a(i) \geq 0$ with the additional restriction that $g_a(0) = 0$ for all $a \in A$. In addition, it is assumed that the number of vehicles on each arc is a known nonnegative constant at the beginning of the initial time period.

The flow conservation constraints are stated for each node $k \in M$:

$$S_k(i) + \sum_{a \in B(k)} g_a[x_a(i)] = \sum_{a \in A(k)} u_a(i),$$
$$\forall k \in M, i = 0, \ldots, I \quad (2)$$

where $S_k(i)$ is the number of vehicles generated at node k for the ith time period; $A(k)$ is the set of arcs whose tail node is k; and $B(k)$ is the set of arcs whose head node is k. It is assumed that the external input $S_k(i)$ is a known nonnegative constant for each time period.

Let $C_a[x_a(i)]$ denote the total cost incurred on the arc a during the ith time period when the number of vehicles at the beginning of the ith time period is $x_a(i)$. It is assumed that the cost functions $C_a[x_a(i)]$ are nonnegative, nondecreasing, differentiable and convex for all $x_a(i) \geq 0$. A discrete-time, nonlinear, nonconvex mathematical program for the dynamic system optimal traffic assignment problem that was formulated and analyzed by Merchant (1974) and Merchant and Nemhauser (1978a, b) can now be stated as follows:

$$\text{minimize} \sum_{i=1}^{I} \sum_{a \in A} C_a[x_a(i)] \quad (3)$$

subject to

$$x_a(i+1) - x_a(i) = u_a(i) - g_a[x_a(i)],$$
$$\forall a \in A, i = 0, \ldots, I-1$$

$$S_k(i) + \sum_{a \in B(k)} g_a[x_a(i)] = \sum_{a \in A(k)} u_a(i),$$
$$\forall k \in M, i = 0, \ldots, I$$

$$x_a(0) = R_a \geq 0, \quad \forall a \in A$$
$$u_a(i) \geq 0, \quad \forall a \in A, i = 0, \ldots, I$$
$$x_a(i) \geq 0, \quad \forall a \in A, i = 0, \ldots, I$$

The objective of the problem in Eqn. 3 is to minimize the sum of the total costs incurred on all the arcs over all the time periods. The Kuhn–Tucker conditions for the problem were derived and given economic interpretations by Merchant and Nemhauser (1978b).

The discrete-time model of Eqn. (3) can now be transformed into a continuous-time optimal control problem. Let $[0, T]$ denote a fixed planning horizon of length T. The dynamic evolution of the state of each arc is described by first-order nonlinear differential equations:

$$\frac{dx_a(t)}{dt} = \dot{x}_a(t) = u_a(t) - g_a[x_a(t)], \quad \forall a \in A, \forall t \in [0, T] \quad (4)$$

where $x_a(t)$ is the state variable, denoting the traffic volume on the arc a at time t; $u_a(t)$ is the control variable, denoting the flow entering the arc a at time t; and $g_a[x_a(t)]$ is the flow exiting the arc a at time t. The flow conservation constraints are stated for each node $k \in M$:

$$S_k(t) + \sum_{a \in B(k)} g_a[x_a(t)] = \sum_{a \in A(k)} u_a(t), \quad \forall k \in M, \forall t \in [0, T] \quad (5)$$

It is assumed that $S_k(t)$ is a known, nonnegative and continuous function of time $t \in [0, T]$. It is also assumed that the traffic volume on each arc $a \in A$ is a known nonnegative constant at $t = 0$.

$$x_a(0) = x_a^0 \geq 0, \quad \forall a \in A \quad (6)$$

Both the control and state variables are required to be nonnegative:

$$u_a(t) \geq 0, \quad \forall a \in A, \forall t \in [0, T] \quad (7)$$

$$x_a(t) \geq 0, \quad \forall a \in A, \forall t \in [0, T] \quad (8)$$

Since the assumption $g_a(0) = 0$ ensures that the state variables are always nonnegative, Eqn. (8) does not have to be considered in an explicit manner. Define $x(t) \equiv (\ldots, x_a(t), \ldots)$ and $u_a(t) \equiv (\ldots, u_a(t), \ldots)$. In the sequel, the set Ω of feasible solutions is used

$$\Omega \equiv \{[x(t), u(t)]: \text{Eqns. (4)-(7) are satisfied}\} \quad (9)$$

to effect some economy of notation.

Let $C_a[x_a(t)]$ denote the instantaneous total cost rate on arc a when the number of vehicles travelling on the arc a is $x_a(t)$ at time t. It is assumed that the total cost rate function $C_a[x_a(t)]$ is nonnegative, nondecreasing, differentiable and convex for all $x_a(t) \geq 0$, $a \in A$ and $t \in [0, T]$. Wie (1988) and Friesz et al. (1989) formulated a continuous-time optimal control problem for dynamic system optimal traffic assignment as follows:

$$\text{minimize } J_1 = \sum_{a \in A} \int_0^T C_a[x_a(t)] \, dt \quad (10)$$

subject to

$$[x(t), u(t)] \in \Omega$$

The performance index J_1 has an economic interpretation as the total transportation cost incurred during the fixed time interval $[0, T]$. The optimal control problem given by Eqn. (10) is to find the optimal trajectories of both the state and control variables that minimize the performance index J_l while satisfying all the associated constraints. An optimal solution of the problem is said to be a time-varying traffic flow pattern that minimizes the total transportation cost.

Let $A(p)$ denote the set of arcs contained on a path p. Let $\lambda_a(t)$ denote the costate variable associated with the state Eqn. (4) for the arc a at time t. Wie (1988) and Friesz et al. (1989) showed that the control problem given by Eqn. (10) has an optimal solution that equilibrates the instantaneous marginal cost

$$\Phi_p(t) = \sum_{a \ni A(p)} \frac{\dfrac{dC_a[x_a(t)]}{dx_a(t)} + \dfrac{d\lambda_a(t)}{dt}}{\dfrac{dg_a[x_a(t)]}{dx_a(t)}} \quad (11)$$

for all the paths that are being used at each instant for a given origin–destination pair. It was also proven that the following theorem holds.

THEOREM 1. *If $u_a(t) > 0$ for all $a \in A(p)$ at some time $t \in [0, T]$, then $\Phi_p(t) = \inf\{\Phi_q(t): \forall q \in P_{kn}\}$ for a solution of the control problem given by Eqn. (10).*

Evidently, this theorem establishes that, for each origin–destination pair, the marginal cost given by Eqn. (11) is instantaneously equilibrated for all the paths whose comprising arc has positive inflow rate. The same economic interpretation of the Kuhn–Tucker conditions for the problem of Eqn. (3) was given by Merchant and Nemhauser (1978b) as a dynamic generalization of Wardrop's second principle.

2.2 Dynamic User Optimal Traffic Assignment

Consider a special case where all network users have complete information on the current state of the network at each instant in time and they attempt to minimize individual travel costs by continuously revising their current route choices. It is assumed that no driver has information as to how travel costs for further downstream arcs may change by the time of arrival at those arcs. However, as drivers move downstream, they are free to revise their route choices at any en route intersection nodes if their current routes are no longer optimal on the basis of updated traffic information. Such an instantaneous adaptive routing strategy appears to be quite appropriate on the network where travel demands significantly fluctuate from day to day and where arc capacities unexpectedly vary from day to day due to traffic accidents, weather conditions or road construction. In such a case, network users may try to optimize their routing decisions each day on the basis of continually updated traffic information.

Let $D_a[x_a(t)]$ denote the instantaneous expected unit travel cost on the downstream arc a that can be measured as a function of the traffic volume on the arc a at time t. It is assumed that the expected unit cost function $D_a[x_a(t)]$ is nonnegative, nondecreasing, continuously differentiable and convex for all $x_a(t) \geq 0$, $a \in A$ and

$t \in [0, T]$. An equivalent continuous-time optimal control problem for dynamic user optimal traffic assignment can now be stated that was formulated by Wie (1988) and Friesz et al. (1989):

$$\text{minimize } J_2 = \sum_{a \in A} \int_0^T \int_0^{x_a(t)} [D_a(\omega) \, dg_a(\omega)/d\omega] \, d\omega \, dt \quad (12)$$

subject to

$$[x(t), u(t)] \in \Omega$$

where Ω is the set of feasible solutions defined in Eqn. (9) and ω is a dummy variable of integration. The performance index J_2 does not have any intuitive economic interpretation, whereas the performance index J_1 has an economic interpretation as the total transportation cost. In fact, the derivation of J_2 is analogous to that of the objective function in a mathematical programming formulation for the static user equilibrium traffic assignment problem (Beckmann et al. 1956).

Wie (1988) and Friesz et al. (1989) showed that when J_2 achieves its minimum value the optimal trajectories of both the state and control variables represent a time-varying traffic flow pattern consistent with the following dynamic generalization of Wardrop's first principle:

"If the instantaneous expected unit travel costs for all the paths that are being used are identical and equal to the minimum instantaneous expected unit path cost at each instant for each origin–destination pair, the corresponding time-varying flow pattern is said to be user optimized."

It was shown that the control problem defined by Eqn. (12) has an optimal solution that equilibrates the instantaneous expected unit travel cost

$$\Psi_p(t) = \sum_{a \in A(p)} \left\{ c_a[x_a(t)] + \frac{\frac{d\lambda_a(t)}{dt}}{\frac{dg_a[x_a(t)]}{dx_a(t)}} \right\} \quad (13)$$

for all the paths that are being used at each instant for a given origin–destination pair. It was also proven that the following theorem holds.

THEOREM 2. *If $u_a(t) > 0$ for all $a \in A(p)$ at some time $t \in [0, T]$, then $\Psi_p(t) = \inf\{\Psi_q(t): \forall q \in P_{kn}\}$ for a solution of the control problem defined by Eqn. (12).*

Evidently, this theorem establishes that, at each instant in time, for each origin–destination pair, the expected unit travel cost $\Psi_p(t)$ is instantaneously equilibrated for all the paths whose comprising arc has positive inflow rate. It is important to note that a more appropriate dynamic generalization of Wardrop's first principle is one in which all network users with the same origin, destination and desired arrival time experience the same travel cost to the destination regardless of route and departure time selected.

See also: Traffic Assignment; Traffic Assignment, Stochastic

Bibliography

Beckmann M, McGuire C, Winston C 1956 *Studies in the Economics of Transportation*. Yale University Press, New Haven, CT
Carey M 1986 A constraint qualification for a dynamic traffic assignment model. *Transp. Sci.* **20**, 55–8
Carey M 1987 Optimal time-varying flows on congested networks. *Oper. Res.* **35**, 58–69
Friesz T L, Luque F J, Tobin R L, Wie B W 1989 Dynamic network traffic assignment considered as a continuous time optimal control problem. *Oper. Res.* **37**, 893–901
Ho J K 1980 A successive linear optimization approach to the dynamic traffic assignment problem. *Transp. Sci.* **14**, 295–305
Merchant D K 1974 *A study of dynamic traffic assignment and control*. Ph.D. dissertation, Cornell University
Merchant D K, Nemhauser G L 1978a A model and an algorithm for the dynamic traffic assignment problems. *Transp. Sci.* **12**, 183–99
Merchant D K, Nemhauser G L 1978b Optimality conditions for a dynamic traffic assignment model. *Transp. Sci.* **12**, 200–7
Wardrop J G 1952 Some theoretical aspects of road traffic research. *Proc. Inst. Civ. Eng.* **2**(1), 235–378
Wie B W 1988 *Dynamic models of network traffic assignment: A control theoretic approach*. Ph.D. dissertation, University of Pennsylvania
Yagar S 1970 *Analysis of the peak period travel in a freeway-arterial corridor*. Ph.D. dissertation, University of California, Berkeley
Yagar S 1971 Dynamic traffic assignment by individual path minimization and queueing. *Transp. Res.* **5**, 179–96

B. W. Wie
[University of Hawaii, Honolulu, Hawaii, USA]

Traffic Assignment, Stochastic

Assignment, or demand–supply interaction models simulate flows on the arcs of a transportation network.

Traditional assignment models assume that travel demand is (approximately) constant over subperiods of a reference period (e.g., the morning peak period) long enough to allow the system to reach a stationary flow pattern. In particular equilibrium models, a "fixed-point" configuration of the system is looked for; that is, a flow pattern giving rise to generalized travel costs which, in turn, induce users' choices reproducing the same flow pattern (see *Traffic Assignment*).

Equilibrium assignment models simulate congestion phenomena on transportation networks in a "static" framework, failing to predict how congestion builds up and vanishes in different parts of the network at different times and how the flow pattern changes from one day to another adjusting to actual or anticipated changes in the supply.

Increasing congestion and oversaturation phenomena in urban and metropolitan areas, as well as the development of new information-based control systems, require dynamic assignment models able to simulate time-dependent flows explicitly.

Two dimensions of dynamics are possible: an interperiodal, or day-to-day, dynamics in which the system can take on different states in different period (or "days") and an intraperiodal, or within-day, dynamics in which departure times and travel demand are not uniformly distributed over the reference period.

Many models have been proposed dealing with one or both of the above temporal dimensions; for a review of dynamic assignment models proposed up to the late 1980s see Ben Akiva and De Palma (1987) and also *Traffic Assignment, Dynamic*.

Most of the proposed models are dynamic equilibrium models insofar as they define a self-reproducing state of the system in which no users can reduce their generalized costs (increase utility) by unilaterally changing paths and/or departure times.

Most of these models, however, refer to "elementary" demand and/or supply configurations and cannot be applied to general networks.

Assignment models can also be classified according to the type of users' behavior models adopted. Deterministic models assume that users are endowed with a perfect knowledge of network attributes and choose to minimize their generalized costs.

Stochastic models are based on probabilistic choice models allowing perception errors, limited information, habit, and so on. These models are, by their nature, more suitable to capture explicitly users' behavioral reactions. In the following, two types of stochastic dynamic models will be described, the first dealing with the day-to-day and the second with the within-day temporal dimension.

1. Day-to-Day Dynamic Stochastic Model

The day-to-day assignment model simulates the system evolution over time under the assumption of demand uniformly distributed over the reference period. The model follows a nonequilibrium approach; for an examination of uniform demand equilibrium models, see *Traffic Assignment*.

1.1 Definitions and Notation

The transportation system is assumed represented by a network made up of a set of nodes, a set of directed links and a set of generalized transportation cost functions for each link.

The generic origin–destination (OD) pair of centroid nodes will be indicated by (o_r, d_r), the number of OD pairs by n_r, the generic path of the network by k, the number of paths by n_k and the finite set of indices relative to acyclic paths connecting the OD pair r by K_r. Let d denote the OD demand vector whose components d_r are the average number of homogeneous travellers moving between each OD pair r. Let F denote the path-flow vector whose components F_k are the number of trips using path k in the reference period and f denote the link flow vector with components f_l expressing the users' flow on link l. All vectors are column vectors.

Demand and path flows are linked by the demand conservation equation requiring that the whole OD demand is distributed among paths connecting each OD pair:

$$d_r = \sum_{k \in K_r} F_k \qquad (1)$$

Link and path flows are linked through the link–path incidence matrix $\mathbf{A}$ whose elements a_{lk} are equal to one if link l belongs to path k, and zero otherwise:

$$f_l = \sum_k a_{lk} F_k \qquad (2a)$$

or

$$f = \mathbf{A}F \qquad (2b)$$

The travel cost on path k, C_k can be obtained as the sum of costs on component links c_l as:

$$C_k = \sum_l a_{lk} c_l \qquad (3a)$$

or

$$C = \mathbf{A}^T c \qquad (3b)$$

Because of congestion, the generalized travel cost (time, money, discomfort, etc.) of a link is an increasing function of the flow on that link and, possibly, on other links. This dependence is expressed through the link cost functions $c_l = c_l(f)$ whose analytical form depends on the particular type of link they refer to (highway stretches, urban roads, signalized junctions, toll barriers, etc.). For a more detailed description of transportation network models, see Sheffi (1985).

1.2 Choice Models

Typical assignment models for within-day constant demand simulate users' path choices. Other choice dimensions, such as trip frequency, destination and mode, can be included in the same framework by network expansion techniques (see *Transportation Modelling*) (Sheffi 1985).

Stochastic assignmnet models assume that travellers choose their path following a random-utility model.

Each user associates a perceived generalized cost C_k^l to each path k connecting his OD pair and chooses to minimize this cost. Because of a number of causes, such as random fluctuation of travel costs, omitted and/or inexact evaluation of some attributes making up the generalized cost, limited information, variation of tastes across the population and, for the same user, it is assumed, over time, that perceived path costs are random variables with mean value given by average predicted transportation cost $\tilde{C}_k$:

$$C_k^l = \tilde{C}_k + \varepsilon_k \qquad (4)$$

The variable $\tilde{C}_k$ will be better defined in Sect. 1.3.

Under the above assumptions, the probability $p_r(k)$ of choosing path k connecting OD pair r can be expressed as:

$$p_r(k) = P\{\tilde{C}_k - \tilde{C}_h < \varepsilon_h - \varepsilon_k\}, \quad \forall h \neq k; h, k \in K_r \quad (5)$$

The functional form of path choice probabilities depends on the joint probability law assumed for random terms ε_k.

If error terms are assumed to be independent, identical Weibull variates $W(0, \sigma)$ with zero mean and standard deviation σ, Eqn. (5) reduces to the well-known multinomial logit model:

$$p_r(k) = \frac{\exp(-\delta C_k)}{\sum_{h \in K_r} \exp(-\delta C_h)} \qquad (6)$$

where the parameter δ is inversely related to the standard deviation σ of the residuals through the expression:

$$\delta = \frac{\pi}{\sigma\sqrt{6}}$$

If residuals are assumed to follow a joint multivariate normal (MVN) distribution with moments:

$$E(\varepsilon_k) = 0$$
$$\text{var}(\varepsilon_k) = \delta C_k \qquad (7)$$
$$\text{cov}(\varepsilon_k \varepsilon_h) = \begin{cases} \delta C_{kh} & \text{if } h, k \in K_r \\ 0 & \text{otherwise} \end{cases}$$

where C_{kh} is the cost of the overlapping parts of paths k and h, then path choice probabilities are given by the probit model:

$$p_r(k) = \int_{C_h^l \in S_k} \Phi(x/C, \Sigma) \, dx \qquad (8)$$

where S_k is the set of values $C_h^l > C_k^l$ for any value of C_k^l and $\Phi(.)$ is the MVN probability density function.

The probit model rests on theoretically sounder assumptions insofar as different error variances for paths of different cost and a covariance structure proportional to the extent of overlapping between each pair of alternative paths are allowed.

On the other hand, computing path choice probabilities with a probit model is a more demanding task as no closed form exists for Eqn. (8).

A more detailed exposition of random-utility path choice models with their relative limits has been given by Sheffi (1985).

1.3 Stochastic Process Model

The model simulates the evolution of the transportation system over successive reference periods or "days" as a stochastic process.

The state occupied by the system on the day t is defined by the vector of path flows F^t occurring that day.

The number of feasible states the system can take on a given day is finite because such is the number of path flow vectors with nonnegative integer components respecting demand conservation Eqn. (1).

Path flows occurring on day t are the result of path choices made by users that day; it is assumed that these choices are not predictable *a priori* (i.e., they are random variables) and the probability $p_r(k)^t$ of observing the choice of path k on day t is expressed by the random-utility model of Eqn. (5). To increase the system inertia it can be assumed that only a fraction of users may (but not necessarily must) change path choices each day. Travellers moving on a given day do not know the actual cost they are going to face on each path but base their choices on cost predictions which, in turn, depend on costs incurred on previous days.

Average values of predicted cost on path k can be expressed as:

$$\tilde{C}_k^t = \theta[C_k^{t-1}(F^{t-1}), \ldots, C_k^{t-m}(F^{t-m})] \qquad (9)$$

where $\theta[.]$ represents the learning and forecasting mechanism followed by the average traveller; the most used specification is a moving average:

$$\bar{C}_k^t = \sum_{i=1}^{m} w_i C_k^{t-i} \qquad (10)$$

with weights w_i, not necessarily equal, summing up to unity.

In conclusion, the state adopted by the system on day t is a discrete multivariate random variable whose probability law depends on costs and, through congestion, on the states that occurred on previous days.

It can be shown that the resulting stochastic process admits a unique stationary distribution of probabilities of occupying any feasible state and it is ergodic if the following sufficient conditions hold:

(a) users base their choices on day t upon system states that occurred on a finite number (m) of previous days

$$p_r(k)^t = p_r(k)^t [\boldsymbol{F}^{t-1}, \ldots, \boldsymbol{F}^{t-m}] \qquad (11)$$

(b) travellers' choice mechanisms are time homogeneous (i.e., invariant with respect to a time translation)

$$p_r(k)^t [\boldsymbol{F}^{t-1} = \boldsymbol{F}_p, \ldots, \boldsymbol{F}^{t-m} = \boldsymbol{F}_q]$$
$$= p_r(k)^h [\boldsymbol{F}^{h-1} = \boldsymbol{F}_p, \ldots, \boldsymbol{F}^{h-m} = \boldsymbol{F}_q] \qquad (12)$$

(c) choice probabilities are strictly positive for each feasible path

$$p_r(k)^t > 0, \quad \forall k \in K_r \qquad (13)$$

Under the above conditions, the stochastic process results as an m-dependent homogeneous Markov chain. The expected vector $\boldsymbol{F}^+$ and the variance–convariance matrix $\boldsymbol{\Sigma}_F$ of the path flow vectorial random variable can be formally expressed as:

$$\boldsymbol{F}^+ = \sum_i \pi_i \boldsymbol{F}_i \qquad (14)$$

$$\boldsymbol{\Sigma}_F = \sum_i \pi_i (\boldsymbol{F}_i - \boldsymbol{F}^+)(\boldsymbol{F}_i - \boldsymbol{F}^+)^T \qquad (15)$$

where π_i represents the unique steady-state probability of finding the system in any feasible state $\boldsymbol{F}_i$.

Average link flows $\boldsymbol{f}^+$ and the relative variance–covariance matrix can be obtained from Eqn. (2) as

$$\boldsymbol{f}^+ = \boldsymbol{A}\boldsymbol{F}^+ \qquad (16)$$

$$\boldsymbol{\Sigma}_f = \boldsymbol{A}\boldsymbol{\Sigma}_F \boldsymbol{A}^T \qquad (17)$$

Proofs and further details on the model can be found in Cascetta (1989).

The algorithm for computing average link flows and, eventually, higher order moments, is based on the ergodicity of the stochastic process. A realization of the process is simulated to estimate the relevant link flow moments. For each day t of the simulation period, the following operations are carried out.

ALGORITHM 1.
Step 1: Compute average predicted costs by using the last m iteration link flows via Eqn. (10):

$$\bar{C}_k^t = \sum_{i=1}^{m} w_i C_k(f^{t-i})$$

Step 2: Simulate travellers' path choices via a random-utility model and compute resulting link flows f^t.
Step 3: Compute the mean and jth-order moments for each link l by averaging over all iterations

$$f_1^{+t} = \frac{1}{t} f_1^t + \frac{t-1}{t} f_1^{+t-1}$$

A sequence of days is generated until a test on the accuracy of moments estimates is satisfied.

Different methods can be used to obtain link flows in step 2; the computationally less demanding method consists of computing link flows via a logit or probit network loading algorithm (Dial or Monte Carlo) (see *Traffic Assignment*). The resulting sequence in this case is a pseudorealization of the process.

A number of initial iterations have to be discarded in order to eliminate the initial-point dependence.

2. Within-Day Dynamic Stochastic Models

Two approaches to the simulation of within-day dynamics can be followed. The equilibrium approach looks for a self-reproducing configuration of the system while nonequilibrium models simulate both its day-to-day and within-day temporal evolution.

2.1 Definitions and Notation

Variables defining the demand–supply system introduced in Sect. 1.1 must be modified to take into account the within-day temporal dimension.

This can be done by dividing the day into n_h subperiods or intervals; the alternative continuous-time approach is only of theoretical interest and will be referred to whenever required. Let d_{rh} denote the average number of homogeneous travellers moving between OD pair r and starting their trip in interval h, d_h the $n_r \times 1$ vector of OD flows leaving in interval h, F_{hk} the flow on path k leaving in interval h, $\boldsymbol{F}_h$ the vector of path flows leaving in interval h and F the vector of path flows in all intervals obtained by ordering vectors F_h with increasing period indices. In the following, the set of users following path k and leaving in interval p will be also referred to as "group" (k, p).

Analogously, f_{lh}, f_h and f denote the time space average link flow in interval h on link l, and the interval h and the whole day link flow vectors, respectively.

Characteristic flow for a link of length L_l can then be formally expressed as:

$$f_{lh} = \frac{1}{TL_l} \int_0^T \int_0^{L_l} \sigma_l(t, s) \, ds \, dt$$

where T is the duration of interval h and $\sigma_l(t, s)$ is the instantaneous flow rate at section s and time t.

Demand and path flows are related through the following equation:

$$d_{rh} = \sum_{k \in K_r} F_{kh} \qquad (18)$$

C_{hk} and W_{hk} denote the generalized transportation cost and travel time spent on path k by travellers leaving in interval h, while c_{hl} and w_{hl} are the average generalized cost and travel time on link l during interval h, respectively.

Because of congestion, travel time and generalized cost in interval h usually depend on link flows in the same interval and, for oversaturated links, in previous intervals. The dependence is expressed through travel time and cost functions $w_{lk}(f)$ and $c_{lh}(f)$.

In within-day dynamic models, link and path flows are no longer linearly related through Eqn. (2).

In fact, link flow f_{lh} is made up by path flows leaving in that interval and/or in previous intervals and covering (totally or partially) link l during interval h. This relationship can be formally expressed as

$$f_{lh} = \sum_{kp} \alpha_{lh}^{kp}(f^h, f^{h-1}, \ldots) F_{kp} \qquad (19)$$

The crossing fraction $\alpha_{lk}^{kp} \in [0, 1]$ is the fraction of link l covered by group (k, p) during interval h; this fraction depends on the time spent by each group (k, p) on links preceeding l on path k and, because of congestion, on flows on those links.

The actual functional form of fractions $\alpha(.)$ depends on the link crossing model adopted; an example will be given in Sect. 2.4. Generalized cost for path k leaving in interval p can be obtained from those of component links as:

$$C_{pk} = \sum_{l \in K_r} \sum_{h=1}^{nh} \alpha_{lh}^{kp} C_{lh}$$

and analogously for travel time.

2.2 Choice Models

In within-day dynamic models, the distribution of demand among different departure times must be simulated in addition to path choice.

One possibility is to assume the demand distribution as known and fixed at the price of failing to simulate demand elasticity on this dimension. Alternatively, departure time choice can be simulated with a random-utility model to be combined with the path choice model. Each user associates a perceived utility to an interval–path pair (h, k); this is a random variable U_{hk} with average value V_{hk}:

$$U_{hk} = V_{hk} + \varepsilon_{hk} \qquad (20)$$

The expected utility can be expressed as:

$$V_{hk} = -\max\{\beta[(a_r - \delta_1) - \tilde{B}_{hk}], 0\} \\ - \max\{\mu[\tilde{B}_{hk} - (a_r + \delta_2)], 0\} - \alpha \tilde{C}_{hk} \qquad (21)$$

where $\tilde{C}_{hk}$ is the average predicted transportation cost along path k leaving in interval h; $\tilde{W}_{hk}$ is the average predicted travel time leaving in interval h and following path k; $\tilde{B}_{hk} = (h - 1)T + \tilde{W}_{hk}$ is the average predicted arrival time starting in interval h and following path k, a_r is the desired arrival time, $-\delta_1$ and δ_2 are the extremes of a tolerance interval around the desired arrival time; β and μ are coefficients allowing early or late arrivals to be weighted differently and α is a reciprocal substitution coefficient.

The above model assumes that users have a tolerance interval $(-\delta_1, \delta_2)$ around the desired arrival time a_r and early or late arrivals cause a disutility proportional to the anticipation or the delay.

For further details on this specificiation, see Small (1982); for other possible specifications of departure time choice models, see Alfa (1986).

The probability of choosing departure interval h and path k can be formally expressed as

$$p_r(hk) = P\{V_{hk} - V_{pq} < \varepsilon_{pq} - \varepsilon_{hk}\} \\ \forall p \neq h, q \neq k, q, k \in K_r \qquad (22)$$

The functional form of $p_r(hk)$ depends on the probability function assumed for residuals ε_{hk}, the most popular departure choice model being the multinomial logit:

$$p_r(hk) = \exp \frac{V_{hk}}{\sum_{pq} \exp V_{pq}} \qquad (23)$$

A continuous departure time version of the above model has been developed by Ben Akiva et al. (1986).

2.3 An Equilibrium Model

Average path flows can be obtained by multiplying total OD demand by departure time and path choice probabilities which depend on travel costs and times and, because of congestion, on the vectors of link flows:

$$F_{hk} = d_r p_r(hk)[W(f), C(f)] \quad (24)$$

The equilibrium approach consists of finding path flows F^*_{hk} and link flows f^*_{lh} that are mutually consistent; that is, satisfying simultaneously the two sets of nonlinear equations, Eqns. (24) and (19).

In that configuration, no users can reduce their perceived disutility by leaving earlier or later and/or by following different paths. At the present state of the research, no general statement can be made on the existence and uniqueness of flow vectors comparable to those for the static stochastic users equilibrium model (see *Traffic Assignment*); the above properties largely depend on the structure of functions $a(.)$, $c(.)$ and $w(.)$.

A continuous-time version of the eqilibrium model described has been developed by Ben Akiva et al. (1986).

2.4 Stochastic Process Model

The system evolution on both day-to-day and within-day dimensions is simulated as a stochastic process extending the model described in Sect. 1.3.

The system state on day t is similarly defined by the vector of path flows F^t occurring that day; the number of feasible states with integer components satisfying demand conservation Eqn. (18) is finite.

Path flows occurring on day t are the result of departure time and path choices made by users which are random variables with probabilities given by the random-utility model described in Sect. 2.2. It is also assumed that travellers base their choices on predicted travel costs and times whose average values depend on times and costs incurred in m previous days through functions representing their learning and forecasting mechanism:

$$\bar{C}^t_{hk} = \theta_C[C^{t-1}_{hk}(F^{t-1}), \ldots, C^{t-m}_{hk}(F^{t-m})]$$
$$\bar{W}^t_{hk} = \theta_W[W^{t-1}_{hk}(F^{t-1}), \ldots, W^{t-m}_{hk}(F^{t-m})]$$

with a moving average being the most popular specification.

Also, in this case only, a fraction of users can be allowed to reconsider their choices on each day.

In analogy with the model described in Sect. 1.3, the stochastic process simulating the system evolution in path-flow space is ergodic with a unique stationary distribution of probability for each feasible state if the travellers' departure time and path choice mechanism are stable over time and each interval/path alternative has a strictly positive probability of being chosen in the generic day t.

Average values and moments can be obtained for path and link flows even though, for the latter, linear expressions corresponding to Eqns. (16) and (17) are no longer valid.

Demonstrations and further details on the doubly dynamic stochastic process model can be found in Cascetta and Cantarella (1989).

The algorithm proposed for computing equilibrium link flows f^*_{lh} simulates flows resulting from users' departure time and path choices over a sequence of days until convergence is obtained (i.e., flows remain equal over time). Each day, only a fraction R of users is allowed to change their travel choices with respect to the previous day. For each day t, the following operations are carried out.

ALGORITHM 2.

Step 1: Compute travel costs and times by using previous day link flows f^{t-1}_{lh}.

Step 2: Compute path flows combining the fraction $(1-R)$ of previous day path flows with flows resulting from the choices of the users' fraction R:

$$F^t_{hk} = (1-R)F^{t-1}_{hk} + Rd_r p_r(hk)[C^{t-1}, W^{t-1}]$$

Step 3: Simulate link flows f^t_{lh} by assigning path flows F^t_{hk} to the network.

The algorithm stops when flows computed in successive days are equal within previously fixed tolerance limits.

The algorithm for the doubly dynamic model simulates a realization of the stochastic process on which average flows are computed. For each day t of the sequence, the following operations are carried out.

ALGORITHM 3.

Step 1: Compute average predicted costs and times by using the last m interaction link flows:

$$\bar{C}^+_{hk} = \sum_{i=1}^{m} w_i C_{hk}(f^{t-1})$$
$$\bar{W}^t_{hk} = \sum_{i=1}^{m} w_i W_{hk}(f^{t-1})$$

Step 2: Compute path flows via a random-utility model of departure time and path choice:

$$F^+_{hk} = d_r p_r(hk)[\bar{C}^{t-1}, \bar{W}^{t-1}]$$

Step 3: Simulate link flows f_{lh} by assigning path flows F^t_{hk} to the network.

Step 4: Compute the mean and the jth order moments for each link l and period h by averaging over all iterations:

$$f^{+t}_{lh} = \frac{1}{t}f^t_{lh} + \frac{1-t}{t}f^{+t-1}_{lh}$$

A sequence of days is simulated until a properly defined test on the accuracy of relevant moments esimtates is met.

A number of initial iterations has to be discarded in order to obtain a stationary realization of the stochastic process.

In spite of their apparent similarity, Algorithms 2 and 3 differ substantially. Algorithm 2 generates a number of flows until a self-reproducing pattern is reached, while Algorithm 3 produces a sequence of flows in order to estimate moments on it.

In both algorithms, link flows for each interval have to be computed from path flows. Algorithms of different complexity and levels of detail can be used for this purpose, from microsimulation algorithms tracking single cars (or group of cars) on the network to macrosimulation models extending usual shortest-path algorithms to cope with time-varying demand.

An intermediate approach can now be described based on two simplifying hypotheses.

The grouping hypothesis assumes that all members of a group (h, k) experience the same travel times and costs and cover the same length as the group leader; the equal running time hypothesis assumes that all users travelling on each arc experience the same average speed.

Algorithm 4 computes link flows by solving the fixed-point problem described in Sect. 2.1.

ALGORITHM 4.

Step 1: Compute travel time w_{ih} for each link and each period by using link flows resulting from the previous iteration:

$$w^i = w(f^{i-1})$$

Step 2: Compute link flows assigning path flows:

$$f_{lh} = \sum_{kp} \alpha_{lh}^{kp} F_{kp}$$

Crossing fractions α_{lh}^{kp} are computed recursively by using travel times w^i:

$$\alpha_{lh}^{kp} = \min\left\{1 - \sum_{m=1}^{h-1} \alpha_{lm}^{kp}, \left(T - \sum_{j \in K_{lk}} \alpha_{jh}^{kp} w_{jh}^1\right) \Big/ w_{lh}^i\right\}$$

where K_{lk} is the set of links preceding l on path k.

Step 3: Convergence criterion to be applied to flows two iterations apart.

Further details on Algorithm 4 are given by Cascetta and Cantarella (1989).

3. Further Developments

Stochastic dynamic assignment is a recent and open research field so that consistent developments are underway or are foreseeable in the near future.

Models simulating users' behavior in choosing and changing departure time and path in alternative and more sophisticated ways are currently being studied (Mahmassani and Chang 1986).

Efficient algorithms able to deal explicitly with oversaturation at bottlenecks and different informative systems on general, large-scale networks are also under development.

See also: Queuing Theory Applications; Road Networks: Dynamic Equilibrium Models; Traffic Assignment

Bibliography

Alfa A S 1986 A review of models for the temporal distribution of peak traffic demand. *Transp. Res. B* **20**, 477–99

Ben Akiva M, De Palma A, Kanaroglou P 1986 Dynamic model of peak period traffic congestion with elastic arrival rates. *Transp. Sci.* **20**, 164–81

Ben Akiva M, De Palma A 1987 Dynamical models of transportation networks. *Acts of PTRC Summer Annual Meeting*. PTRC, Brighton, UK

Cascetta E 1989 A stochastic process approach to the analysis of temporal dynamics in transportation networks. *Transp. Res. B* **23**, 1–17

Cascetta E, Cantarella G 1989 A doubly dynamic assignment model. *Acts of PTRC Summer Annual Meeting*. PTRC, Brighton, UK

Mahmassami H S, Chang G L 1986 Experiments with departure time choice dynamics of urban commuters. *Transp. Res. B* **20**, 297–320

Small K 1982 The scheduling of consumer activities: Work trips. *Am. Econ. Rev.* **72**, 467–79

Sheffi Y 1985 *Urban Transportation Networks*. Prentice-Hall, Englewood Cliffs, NJ

E. Cascetta
[Dipartimento di Ingegneria dei Trasporti,
Naples, Italy]

Traffic Control Modes

Control engineering began evolving into a self-reliant scientific discipline in the 1940s. Currently, control engineering comprises a wealth of theoretical methods and tools that have proved to be irreplaceable for a series of practical applications. Traffic engineering is concerned with the design of traffic systems, with improving their safety and efficiency, and with reducing the environmental degradation they caused. Obviously, there is a field of common interest for these disciplines in the area of traffic control. Undoubtedly, the goals of traffic control are best statisfied if relevant knowledge from both disciplines merges into a common research field.

Consider a basic control scheme as depicted in Fig. 1. The process under control (e.g., a road network) is affected by controllable inputs (e.g., traffic lights) and noncontrollable disturbances. Disturbances are classified as nonpredictable (e.g., incidents) or predictable (e.g., demand, origin–destination rates). The control

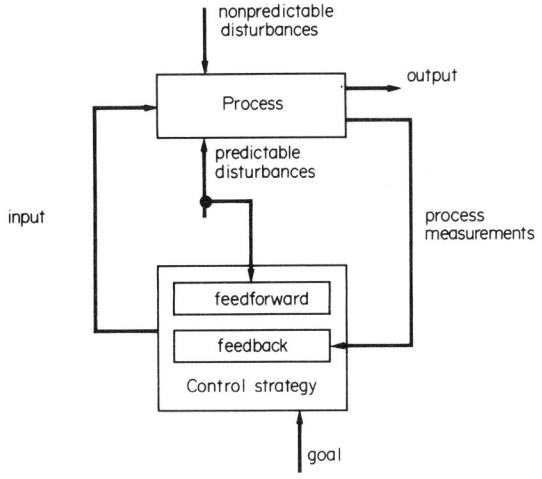

Figure 1
A basic control scheme

problem is to appropriately select the controllable inputs from an admissible control region so as to achieve a desired process output (e.g., minimum travel time).

The control problem may be solved automatically (i.e., with reduced or no human intervention) by a control strategy. In some cases, measurements of process quantities (e.g., traffic volumes) or of disturbances may be available. If the task of the control strategy is performed continuously during process operation, using current measurements, this is a real-time (or, in traffic engineering, a traffic-responsive) control system. Real-time control systems may be feedback (closed loop), if they make use of current measurements of process variables, or feedforward (open loop), if they make use of current disturbance measurements. Mixed forms are possible. Alternatively, if control input specification is performed before operation, perhaps using historical data, this is a fixed (or, in traffic engineering, a fixed-time or time-of-day) control system. Furthermore, calculations performed during process operation (usually in the context of a real-time system) are called on line, whereas calculations connected to the design of a control strategy and needed before the actual implementation of the control system are called off line.

M. Papageorgiou
[Technische Universität München, Munich,
Germany]

Traffic Control Systems: Architecture

Traffic engineers have been developing and evaluating methods of signal control since the mid-1960s. Throughout this period, many diverse attempts have been made to ease congestion, decrease travel times and reduce the impact of vehicular traffic on the environment. However, despite considerable advances and successes, the effects have been short lived. After a time, following the implementation of a new urban traffic control system, congestion reappears more severely. This happens because, with increasing freedom of movement, more people are encouraged to use cars, thus taking up the spare capacity the control system has been designed to give.

Up to now, signal control policy has sought to minimize vehicle delay and reduce stops so as to cope with the traffic moving along the routes that drivers have chosen to take to reach their destinations. Such a policy performs well, provided that there is adequate capacity at junctions and along roads. This is no longer the case in most cities during peak periods. Therefore, serious consideration needs to be given to a change in signal control policy to one that seeks to maximize the use of the road space in urban areas by diverting traffic along routes that drivers would not choose to take.

The hardware (signal control and monitoring systems), software (both for design and optimization) and technology (in the form of route guidance systems) are available, but these resources are exploited and used independently of one another. This article proposes a design for a future generation of strategic signal control that will integrate these resources. This will ease the current serious congestion problems and provide the basis of the traffic control systems of the future.

1. Background

Traffic signal control systems implemented in a town or city and the surrounding urban area are of two types, either isolated or coordinated. The choice of signal control employed depends on the arrival patterns of vehicles at the junction. Isolated junction control is employed when vehicle arrivals at all stop lines are either random or uniform. However, when arrivals are from upstream signal-controlled junctions in close proximity, they tend to be platooned. In these cases, the signal timings of consecutive junctions should be linked or coordinated. This means that the green starts of adjacent links are offset by an amount equal to the journey time for the platoon of traffic travelling between the junctions. When signal timings of groups of junctions need to be linked, this is achieved either by a green-wave or area-wide policy. The former control policy seeks to enhance vehicle movements along pre-defined routes while increasing the delay to traffic on side roads. The latter, however, gives equal priority to all traffic movements in the network.

1.1 Types of Control System

Whether isolated or coordinated, traffic control systems are operated in three different modes, either vehicle actuated, fixed time or demand responsive (see *Traffic Control Modes*). Most systems operate in the vehicle-actuated mode when traffic flows are low, particularly during nighttime periods. Usually, green is assigned to the "main" road traffic stream until a demand for minor road green is triggered by a vehicle sensed by loop detectors sited close to the stop line. As flows increase, the demand for minor road green can occur every cycle, eventually leading to the signal control assuming a fixed-time mode. The durations of greens for each stage at the various junctions in the network, whether operating as isolated junctions or as coordinated groups, are usually defined by signal optimization programs. Examples of isolated junction software include OSCADY (Department of Transport 1987), LINSIG (Simmonite 1985) and SIGCAP (Allsop 1975). For operating green-wave maximum-bandwidth methods, the work of Morgan and Little (1964), and Lear and Bell (1988) is often employed, while the area-wide philosophy is modelled with programs such as the **traffic network study tool** (TRANSYT) (Robertson 1969, Vincent *et al.* 1980).

1.2 Choice of Control System

Fixed-time control works well in traffic environments where flows change dramatically at fixed times of the day such as morning, off-peak and evening peak periods. For each of these periods, a different fixed-time plan is in operation. The flows during these plan periods should typically vary by less than about 10–15% (Bell and Bretherton 1986) and the transition between signal plans should be fairly short, occurring at a similar time each day. Provided signal plan changes occur during off-peak flow levels (Bell and Gault 1982) and adequate attention is paid to cater for large seasonal and yearly changes in flow patterns by regular updating of fixed-time plans (Bell and Bretherton 1986), significant benefits can be achieved.

However, in some urban areas, traffic patterns continuously vary, with little consistency of traffic movements from day to day, week to week and throughout the year. This traffic environment is best controlled by demand-responsive systems. These vary widely. An isolated junction control example is MOVA (Vincent and Young 1986). Examples for area control are SCATS (Lowrie 1982, Luk *et al.* 1982, Sims and Finlay 1984), which continuously updates green splits but operates offsets according to plan selection by time of day based on historic data, and split cycle offset optimization techniques (SCOOT) (Hunt *et al.* 1981), which continuously updates signal timings on a cycle-by-cycle basis and responds well to transients in traffic demand. While demand-responsive systems are more expensive both to install and operate compared with fixed-time systems, when operated in the correct environment they usually achieve benefits that compensate for their capital expenditure within the first few years of operation.

1.3 Typical Control Configuration

A typical city with its surrounding urban area will be divided into subareas within which one or other of the three different control systems will operate. They are designed to manage traffic in undersaturated or nearly saturated conditions. When overload at junctions occurs or short links become blocked by queuing vehicles for periods of more than about 10 min, the current signal control systems are unable to cope. Congestion occurs and builds up rapidly to form "congestion trees." These often lead to a network traffic jam. The signal timings are no longer relevant to the traffic conditions and, to some extent, can hinder the dispersal of traffic from the area as drivers seek *ad hoc* alternative routes. At this stage, a more strategic approach to controlling the traffic across the entire city and its surrounding areas is needed. Such strategic traffic signal control needs to embrace all subareas and all control system types.

As drivers seek alternative routes shifts in traffic flows create the need for signal coordination along routes with junctions that normally operate in isolation. This is because, while distances between junctions remain the same, the flows travelling between them increase sufficiently to give platooned arrivals at junctions. During congestion, the boundaries between subareas of adjacent fixed-time control are no longer valid. In fact, controlling traffic redistribution following recurrent congestion can be enhanced by setting up green waves on the alternative routes to accelerate the dispersal of traffic (Lear and Bell 1988). Further steps can be taken to enhance congestion control by identifying methods that maximize the use of spare capacity in a network and by implementing a system of on-street or in-car route guidance to advise or instruct drivers to take the particular alternative routes.

2. Integrated Traffic Control

In order to make progress in managing and controlling congestion, it is necessary to integrate signal control systems across the whole city and its surrounding area. The architecture of such a system embraces all types of signal hardware and software as well as traffic monitoring systems. It is proposed that the various signal control and traffic monitoring systems form the building blocks or foundations for an advanced strategic signal control system.

This article proposes an architecture through which isolated and subarea traffic control, which conventionally work independently of each other, can evolve into an advanced system that achieves strategic planning for more efficient management and control of traffic over a whole city and the wider urban area.

The main functions of such a system will be performed at a higher level, leaving conventional control systems to perform tasks uninterrupted until a particular traffic congestion problem and solution has been

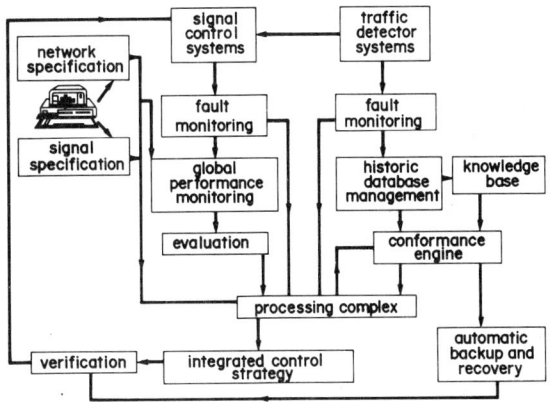

Figure 1
The main functions of strategic traffic control

identified. It would be at this point that a strategic control system will override all signal control timings. By continuous monitoring and evaluation by the strategic control, the point in time when there is no further need for strategic control would be identified and the control would be returned to the conventional systems.

With reference to Fig. 1, the main functions of the strategic control system are

(a) to interface with the conventional signal control and traffic monitoring systems irrespective of type or manufacture (there will be an on-line and off-line component and a means of transferring fault information to the strategic control system);

(b) to provide a user-friendly component through which network and signal specifications can be entered into the strategic control system;

(c) to establish global performance monitoring and evaluation;

(d) to have database management capabilities to develop a historic picture of events and build up a knowledge base;

(e) to interrogate the system to identify problems and their causes, and derive remedial strategies (this will be achieved by the processing complex); and

(f) to provide complete systems support to ensure automatic backup and recovery (a conformance engine will keep data in step and verify all control strategies before implementation to ensure safe on-street control).

3. Signal and Detector Systems

For the cost, the most reliable and frequently used traffic detector employed in current control systems is the loop detector. Obviously, demand-responsive control systems rely on on-line data retrieved from loop detectors sited on most links. With the advent of microelectronics and the development of microprocessor controllers, both isolated and linked signals have gained automatic traffic and fault monitoring capabilities. In addition, permanent remote monitoring systems are becoming a common feature throughout European cities. Typically, several hundreds of detectors can be sited throughout the city and its surrounding area providing a wealth of information which is mainly used for traffic planning and management purposes. This traffic monitoring provides a local authority with a wealth of information that has potential uses for strategic signal control. More advanced traffic control centers have data collection capabilities from systems other than loop detectors. These include closed circuit televisions, ramp metering, queue detectors and speed sensors. For strategic signal control, integration of all types of data monitoring systems, irrespective of source, is a fundamental requirement (see *Detectors for Road Traffic*).

4. Fault Monitoring

Fault monitoring occurs at the street level and relates to the individual control system hardware. These systems automatically alert traffic and maintenance engineers of hardware faults. What is required by the strategic control system is knowledge of when and where faults occur and for how long they last. This will trigger the need for a pseudotraffic pattern, based on historic data, to be substituted.

5. Network and Signal Specifications

A user-friendly interface enables data relating to the network geometry (e.g., number of nodes, links and lanes, road lengths and widths, the origin–destination matrix of traffic) and to the signal control (e.g., number of stages, phases, minimum greens, lost time) to be directly input by hand. However, given that the individual control systems, whether isolated or coordinated, have usually been derived by the application of standard software, the option to read and write relevant data from the standard package input files to the strategic control system will be essential.

6. Performance Monitoring and Evaluation

A software component will take network-, signal- and traffic-related data to derive performance measures for the network. Such performance measures will quantify the benefits of coordination, the reduction in delay, stops, queues and congestion, and so on. These parameters will then feed into the evaluation process. This will compare the different strategies operated and produce a printed record of the performance of the system for given control scenarios. This information is valuable for further development of the strategic control system and will be used by the processing complex (see *Evaluation of Traffic Control Systems*).

7. Database Management System

The main input to the database management system is traffic monitoring. The database management system will perform simple mathematical manipulations to develop tabulated reports and build up a historic picture of events. In general, this historic database evolves off line and provides a reference framework for the central processing complex of the strategic control system. One of its tasks is to pass data to the processing complex to replace dirty or missing data.

8. Knowledge Base System

Historic data feeds directly into the knowledge base software. This uses mathematical relationships and statistical analysis of the stored data and develops system measures such as congestion, spare capacity, queue lengths and flow distributions. These form the basic data for building up a knowledge base. By applying thresholds to these system parameters, the events or "data of substance" is isolated from the wealth of raw data stored in the historic database. The algorithms of the knowledge base system are likely to highlight problems of recurrent congestion and their associated congestion trees, changes in traffic route patterns, definition of spare system capacity and so on. The frequency of occurrence of these recurrent congestion incidents, where and when they take place, and for how long they persist is an example of the information built up by the knowledge base component of the system.

9. Conformance Engine

The conformance engine takes reporting tables from the historic database and system measures and indicators from the knowledge base and performs cross-checking to make sure that the two are kept in step. The conformance engine outputs both the knowledge base and the historic data to a backup system, making regular updates, during quiet system periods, to meet the needs of automatic backup and recovery.

10. Automatic Backup and Recovery

All computer control systems are open to failure or system breakdown, whether caused by a hardware or software fault, or system overload. Therefore, it is necessary to maintain a volume backup which will establish corrective action in the event of system failure. In the framework of the strategic control system discussed here, the corrective action in the event of failure could be to pass back control to the conventional street level systems. Before this can be done, however, conventional control strategies must be available to implement on the streets. These must be ratified before implementation by the signal verification component of the software (see *Maintenance and Reliability of Traffic Control Systems*).

11. Verification Software

The verification component of the software is designed to ensure that the integrity of the control strategies to be implemented are consistent with signal constraints and remove the risk of accident. In addition, the verification software secures the control system against accidental system override and codes elements to ensure that operators have legal access.

12. Processing Complex

The processing complex is the highest level of the strategic control system. It works on line receiving processed traffic monitoring data from the streets. It is kept informed of the signal control currently operating and is backed up by the information banks (the historic data and knowledge base) through the conformance engine. It needs the basic network and signal control specifications for simulation.

The processing complex as outlined in Fig. 2 mainly consists of five functions.

12.1 Volatile Interface

The volatile interface provides the channel for on-line processed raw data and off-line knowledge base and historic data to pass from the streets to the main central processor and vice versa. The raw traffic monitoring data feed continuously through the volatile interface. This selects the relevant knowledge base and historic data which are merged and validated by the conformance engine. These three data systems are then compared by the interrogation element of the processing complex.

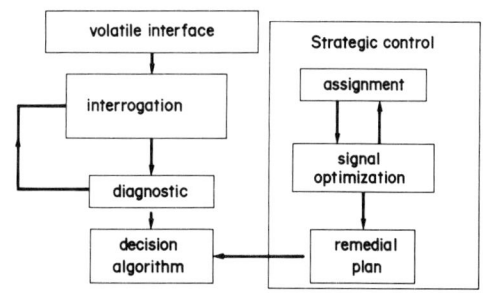

Figure 2
The components of a central processing complex

12.2 Interrogation

A comparison of the current situation with that of the knowledge base will establish the likelihood of an incident developing. This sets up a warning to the system that an event is to take place and gives some indication of the probability that system intervention will be required by assigning a severity code. At this stage, a problem is known to be occurring but its cause is unknown. When this position is reached, the information will be passed to the diagnostic component.

12.3 Diagnostic Algorithms

A series of algorithms will be employed, which compare the description of the event identified by the interrogation software with the historical database, to derive a cause for the problem that is about to occur. It is likely that more than one cause can be assigned to the problem. The diagnostic algorithms will evaluate the problem calling for information via the interrogation and volatile components of the central processing complex. This iterative process will eventually define a list of possible causes of the problem with some measure of statistical confidence. This information forms the basic input to the decision-making software.

12.4 Strategic Control Definition

The strategic planning of infrastructure, including land use, building of roads and traffic management, has historically developed as a process independently of traffic signal control. Planners use transportation modelling and traffic assignment techniques to assess the impact of changes in the infrastructure. Traffic engineers, however, have concentrated on developing software for signal design to control traffic mainly in undersaturated traffic conditions.

It is essential, within the framework of the strategic control system proposed here, to integrate the modelling techniques of assignment with signal opimization (see *Signal Control and Traffic Assignment*). This is because, in designing a system that optimizes the use of road space, assignment techniques must be employed to predict the redistribution of traffic. This is achieved by assigning appropriate costs of travel, based on measured levels of congestion, and weighting factors, to respect environmentally sensitive areas, giving due respect to capacity restraint. The resulting flows then need to be transferred to a signal optimization program to define the appropriate signal control strategy. This may simply be modified isolated junction control, but it will most likely involve signal coordination over a much wider area crossing and including several parts of different subareas. The signal plans will use gating and metering techniques, and seek to reduce the degree of saturation at junctions as far as possible and to avoid queue back. It may be that the area-wide control philosophy needs to be modified by green waves to enhance movement on particular routes (Lear and Bell 1988).

This off-line simulation capability will form part of the main processing complex. It will

(a) require an origin–destination matrix, the total network geometry and signal specifications;

(b) use the historic database for calibration and validation of the predicted flows at particular times of the day; and

(c) model the problems identified by the knowledge base to predict the capacity-restrained redistributed flows.

It is worth pointing out that the simulated flows may be different from the redistributed flows actually observed on street through traffic monitoring. This is because drivers chose an alternative route with incomplete information. Often by *ad hoc* rerouting, drivers merely transfer the problems of congestion from one part of the network to another. Successful simulation with route assignment that respects signal control constraints is expected to result in a redistribution of flows that can be accommodated by the network without overload.

The final model flow predictions are fed into the signal optimization model to produce remedial strategic signal plans. These will be derived off line for each of the problems identified by the knowledge base. The conformance engine will make sure that the plan strategy is kept in step with the problem it is designed to solve.

12.5 Decision Algorithm

Decision algorithm software will analyze the set of problems, taking note of the level of confidence assigned to its correctness. Consideration will be given to an appropriate remedial strategy, along with information from the performance and evaluation software records of its use on previous occasions. The end result of this component is the definition of a control strategy to be recommended for implementation.

13. Verification

The signal control recommendations after verification and validation will override the signal timings of the conventional control. The loop is then closed and the whole process is repeated. The performance of the strategic signal control is monitored and the effect it has on the historic data and knowledge base is evaluated. The algorithms that form the basis of the learning and decision-making processes can in turn be developed further as the strategic control system evolves.

14. Exceptional Condition Analysis

It is necessary to develop software that continuously monitors all stages of the integrated control system. Its main function would be to use a series of step-by-step tests which will ensure that the system operator is alerted when any component, whether hardware or software, is malfunctioning. The exceptional condition analysis will house sets of routines that automatically achieve functional recovery of each component of the strategic control system while ensuring safe control of traffic through the conventional systems.

15. Future Developments

Much research and development is needed to realize the system integration proposed in this article. Active research at the Nottingham University Transport Research Group, funded by the UK Science and Engineering Research Council and the European Community, is contributing to the development of such a traffic control system. However, for the proposed strategic control system to be successful in tackling congestion problems in the future, it will be essential to exercise traffic restraint policies in parallel. This is necessary to prevent more vehicles than at present using the urban roads at peak times. If this is ignored, very serious levels of congestion will result. Cities have a finite capacity in terms of time and space: the proposed strategic control will have made maximum use of both at peak periods and there is, therefore, unlikely to be room for any more.

Acknowledgement

The author wishes to thank her husband Dr Ian Bell for useful discussion without which this paper could not have been written.

See also: Road Network Control; Road Traffic Monitoring Equipment

Bibliography

Allsop R E 1975 Computer program SIGCAP for assessing the traffic capacities of a signal-controlled road junction. Description and manual for users. Transport Operations Research Group Research Report No. 11
Bell M C, Bretherton R D 1986 Ageing of fixed-time traffic signal plans. *Proc. IEE, 2nd Int. Conf. Road Traffic Control.* Institution of Electrical Engineers, London, pp. 53–8
Bell M C, Gault H E 1982 An empirical study of plan change algorithms for area traffic control systems. *Proc. IEE Conf. Road Traffic Signalling.* Institution of Electrical Engineers, London, pp. 77–80
Department of Transport 1987 OSCADY: Capacities, queues and delays at signal controlled junctions. Report No. HCSL/R/42. DOT, London
Hunt P B, Robertson D I, Bretherton R D, Winton R I 1981 SCOOT—A traffic responsive method of co-ordinating signals. Transport and Road Research Laboratory Report No. LR 1014. TRRL. Crowthorne, UK
Lear D W, Bell M C 1988 An investigation of the differences between British and German traffic control policies. Nottingham University Transport Research Group, Nottingham, UK
Lowrie P R 1982 The Sydney co-ordinated adaptive traffic system. Principles, methodology, algorithms. *Proc. IEE Int. Conf. Road Traffic Signalling.* Institution of Electrical Engineers, London, pp. 67–70
Luk J Y K, Sims A G, Lowrie P R 1982 SCATS—Application and field comparison with a TRANSYT optimised fixed-time system. *Proc. IEE Int. Conf. Road Traffic Signalling,* Institution of Electrical Engineers, London, pp 71–4
Morgan J T, Little J D C 1964 Synchronizing traffic signals for maximal bandwidth. *Oper. Res.* **12**, 896–912
Robertson D I 1969 TRANSYT A **TRA**ffic Network Stud**Y** Tool. Transport and Road Research Laboratory, Report No. LR253. TRRL, Crowthorne, UK
Simmonite B F 1985 LINSIG—A computer programme for traffic signal design. *Traffic Eng. Control* **26**(6), 310–15
Sims A G, Finlay A B 1984 SCATS—Splits and offsets simplified SAS. *Aust. Road Res. Board. Proc.* **12**(4), 17–33
Vincent R A, Mitchell A I, Robertson D I 1980 User guide to TRANSYT, version 8. Transport and Road Research Laboratory Report No. 888. TRRL, Crowthorne, UK
Vincent R A, Young C P 1986 Self optimising traffic signal control using microprocessors—The TRRL MOVA strategy for isolated intersections. *Traffic Eng. Control* **27**(7/8), 385–7

M. C. Bell
[University of Nottingham, Nottingham, UK]

Traffic Control Systems: Trends

Traffic control systems have evolved during the twentieth century in response to the growth in automobile travel. Public attitudes towards traffic controls have been generally supportive. In general, drivers want a safe, orderly, deterministic and efficient traffic environment, while wishing that this could be achieved with as much individual freedom of movement as possible. However, traffic systems must be organized and controlled carefully in order to operate efficiently, and this must necessarily limit individuals' freedom of movement.

Efficiency in traffic is generally measured in terms of speed, safety and cost. Clearly, there have been major improvements on all levels since the early days of the automobile. However, automotive traffic has been a victim of its own success, with the growth caused by its popularity threatening to reverse the gains in travel speed, at a time when society has become dependent on this speed. Thus, the concerns for future mobility may necessarily see a trade-off of some of our autonomous freedoms for reasonable trip speeds.

Since critical urban traffic control is accomplished by signal systems, some emphasis is given to this area in this article.

1. Past Developments

Based on the basic axiom that two vehicles should not share the same space and time, some rules were adopted in the interest of safety. First, drivers were

Table 1
Chronology of significant events in traffic management and control

Past
    drive on right (or left) side
    right of way at intersections
        absolute (YIELD)
        alternating (all-way STOP)
        bulk service in green period
            traffic police
            timing isolated signals
    Signal systems
        one-way progression
        arterial systems
        signal networks

Present and short term
    transportation systems management (TSM)
    automatic incident detection
    real-time control

Inevitable medium-range future
    corridor management and integrated networks
    in-vehicle route guidance
    demand management

Not yet committed long term
    trip allocation
    automatic vehicle control

required to drive on the right-hand side in some countries and on the left in others. This initial piece of organization was really for traffic capacity. Next, collision avoidance at intersections required that conflicting movements did not have simultaneous right of way. The proliferation of traffic signals in the interests of safety and capacity led to concerns for progressive movement through these signals. This led to simple progression on one-way streets, arterial systems and, finally, optimization of closed networks. As early as 1917, six intersections were controlled manually as a single system (Sessions 1971) and, in 1928, an automated flexible-progressive pretimed system was introduced.

By the 1960s, congested freeways led to the metering or closure of key on-ramps in order to strategically divert trips to less congested parts of a corridor. This required analysis of whole corridors, where complete control plans involving series of ramps and traffic signals could be simulated to predict overall impacts in terms of flows, queues and trip times. This suggested the need for integrated control of freeways and signalized arterials (Yagar 1985), and models to evaluate them (Yagar 1975).

Table 1 orders the significant developments in the evolution of traffic "control" to the present. It then continues with the modern emphasis on traffic "management" rather than control to measures that are being developed for likely implementation in the 1990s. It ultimately returns to the more severe "controls" that may have to be implemented in the twenty-first century to enforce future traffic and demand management strategies.

2. Traffic Signal Systems to the Present

The key element in urban traffic systems to date has been the system that links traffic signal timings at the different intersections. The development of computer-aided traffic signal control systems has evolved through a number of different methods for combining intersections. These are known as generations. Of these, the 1st, 1.5th and 3rd generations are currently quite common, while the 2nd never became established, for reasons best described by Euler (1988). Euler summarized these generations of control and discussed some desirable characteristics that a 4th generation should have.

2.1 1st Generation

Signal timing plans are developed off line (see *Traffic Control Modes*) using manual or computerized techniques. These predetermined timing plans are implemented by the system according to the time of day. Any variation from this, such as an early start for a peak period plan, requires operator intervention to call the plan in at the desired time.

2.2 1.5th Generation

In 1.5th generation systems, a library of timing plans is calculated/updated automatically, based on detected traffic volumes, using an off-line model such as TRANSYT (Robertson 1969). The system selects the one it thinks is best suited to traffic conditions at the time, usually based on current measurements of volume and occupancy. These plans offer considerable flexibility; for example, the Los Angeles automatic traffic surveillance and control (ATSAC) system (Rowe 1988) supplements its 1.5th generation plans with critical intersection control to adjust phase times on a cycle-by-cycle basis.

2.3 2nd Generation

In this aborted system, timing plans were to have been developed and used on line by the system. In general, 2nd generation traffic signal systems have not been as effective as anticipated. Some of the reasons for this will be given in an attempt to put the essence and problems of traffic signal systems into some perspective.

(*a*) *Too many timing plan changes.* By their nature, 2nd generation systems attempt to change timing plans on a periodic basis. In many systems, timing plans were changed too often. The benefits to be gained by implementing a better timing plan were often offset by the negative effects of making the transition from one plan to another.

(*b*) *Too reliant on prediction.* Second generation systems attempt to develop timing plans that best accommodate expected traffic demands and patterns. These are predicted based on recently detected traffic volumes and occupancies. In some cases, the predictions were heavily weighted to measurements made during the most recent signal cycle. This did not work very well because traffic demand is quite variable from cycle to cycle. In other cases, the predictions used detector data that were smoothed over, for example, a 15 min period. This was often not responsive enough to predict sudden increases in demand in a timely fashion.

(*c*) *Narrow objectives*. A number of the systems tested were designed to move traffic along major arterials. As a result, traffic on the side streets often incurred large delays. Traffic performance was not necessarily optimized from a total system point of view.

(*d*) *Inflexibility*. Though responsive, these systems were very structured and unable to deal effectively with unanticipated events.

2.4 3rd Generation

With a 3rd generation system, the signal timings are free to evolve in response to detected traffic volumes and queues. While this leaves analysts and researchers unconstrained, it also leaves them in a virtual vacuum from which to synthesize an evolving dynamic plan for a very complex multivariate stochastic problem. While many struggled with the enormity of this totally unconstrained problem, two working real-time systems were developed by reintroducing the partial constraint of a system or subsystem cycle that is updated on a continuing basis.

In the Australian SCATS system (Sims 1979), small subareas of up to ten intersections share a common cycle length which can be altered by up to 6 s once per cycle. It can also strategically combine certain subareas for varying lengths of time in order to improve on overall network performance. It draws its data from stop line detectors.

In the British SCOOT system (Hunt *et al.* 1981), the network shares a common cycle time which is also periodically adjusted in small steps. SCOOT employs detectors located just downstream of the upstream intersection to measure cyclic flow profiles in real time. It uses these to predict arrival profiles at the downstream intersection and to adjust phase times and offsets by up to 4 s, again once per cycle.

While these 3rd generation systems have met expectations better than 2nd generation systems, a number of shortcomings remain.

(a) Traffic signal control is not yet integrated with freeway control (i.e., ramp metering). Signal timing and ramp metering rates are obviously related and should be considered together in order to achieve total system optimization.

(b) The current systems are relatively slow to respond to sudden changes in traffic flows caused, for example, by an indicent or a work shift change in a factory. This is because the systems have been designed to implement small changes over time in order to overcome the problem of frequent transitioning.

(c) They rely on data from detectors, which are not yet reliable. This should improve as inductance loops are made obsolete by a system such as image processing which should ultimately be more reliable, as well as providing more key information (e.g., queue lengths).

(d) They cannot deal with specific origin–destination (OD) trip demands, but respond only to flows measured on the roadway.

(e) They still require extensive operator involvement in their design, calibration and operation.

3. Complementary Developments

Past developments in signal systems, as described in Sect. 2, attempt to serve given vehicle volume demands as efficiently as possible. Efficient overall management should combine various types of supply and demand management. Systems of signalized networks have been complemented by the four key strategic developments that will be given in this section which increasingly affect the traffic that is controlled by signals. These four are certainly not all of the complementary developments that have impacted on traffic signal systems. However, they serve to indicate that signal systems are only a subsystem in urban traffic control, though perhaps the most important one.

3.1 Mass Transit

Transit systems preceded the automobile congestion that characterized our modern traffic control systems. Still, traffic control systems tended to grow without concern for transit, which usually either had its own exclusive right of way or received little consideration in the setting of traffic signal timings. Increasing concern for the movement of people rather than vehicles has led to greater consideration for transit in the setting of traffic signals, with light rail transit (LRT) and even buses often preempting traffic signals (Philipps 1988). Integrating traffic and transit controls can only help, as illustrated by Teply (1985).

3.2 Urban Freeway Systems

The freeways that joined and passed through urban centers were found to provide efficient intraurban travel and even took some load off the signalized networks. However, they also increased urban travel and caused problems at signals near freeway–arterial interchanges, where they generated large volumes. Since the mid-1960s, this has been exacerbated by traffic volumes being metered onto signalized streets as freeways have become congested, thus shifting freeway queues onto the surface streets which are often more congested than the freeways. Integrating freeway and signal controls can only help (see *On-Ramp Control: Realization Principles*).

3.3. Transportation Systems Management

As land and money for expansion of urban road systems became increasingly scarce in the 1970s, a concern for the efficient use of transportation infrastructure and expenditure grew. The objective of moving vehicles gave way to that of moving people. This led to formal management of demand as well as supply, under the umbrella of transportation systems management (TSM). Strategies such as flextime, for spreading the peak period and reducing the maximum demand intensity, were considered in conjunction with reversible lanes. Virtually any strategy that would provide cost-effective marginal improvements to people movement with short lead times and low capital costs was considered as TSM. This was an effective fine tuning of the urban transportation system and saw a wide ranging variety of improvements, such as those described by Urbitran Associates Inc. (1982).

3.4 Automatic Incident Detection

Just as the 3rd generation signal systems attempt to respond to changes in traffic demand, by using detectors to provide real-time information, algorithms have been developed to detect incidents on freeways based on the continuity of counts over time and space. This allows officials to clear incidents faster and divert demands around the incident. Thus, incident detection is used for both supply and demand management (see *Incident Detection*).

4. Designing for the Future

4.1 Objectives

In what he called brainstorming for the 4th generation of traffic control, Euler (1988) described a "wish list" for the future:

(a) minimize total travel cost (delay, vehicle operating cost) to vehicles, people and goods;
(b) maximize safety;
(c) minimize negative environmental impacts (air, noise);
(d) protect neighborhoods;
(e) minimize travel time variability;
(f) maximize equity;
(g) minimize wasted travel;
(h) minimize unnecessary travel; and
(i) automate control decisions to the extent possible.

4.2 Features

To provide the objectives of Sect. 4.1, a system should have the following features, as indicated by Euler:

(a) integrations of traffic signal control systems with other control systems (e.g., freeway surveillance and control systems, transit preference schemes, LRT control systems, reversible lane operation)—this means more than just sharing information, it means coordinated control for all of these systems;

(b) prompt detection of and response to events—changes to signal timing, ramp metering or high-occupancy vehicle (HOV) lane operation when events so dictate;

(c) ability to predict and respond to OD information to accommodate priority paths or detours;

(d) integration with vehicle location/identification/classification systems—this would have application in control of transit or goods vehicle movement and also in congestion pricing schemes;

(e) inclusion of artificial intelligence/expert system features—these concepts could be applied in many areas, for example, in automating control decisions (perhaps also including self-learning features) or in hardware diagnostics (see *Expert Systems Approach to Road Traffic Control*);

(f) accommodation of demand control and congestion pricing—methods to encourage ride sharing and transit usage, provision of information for pretrip planning and, ultimately, congestion pricing through use of automatic vehicle identification and location systems;

(g) flexibility to accommodate different control objectives in different parts of an urban area or during different time periods;

(h) real-time communications with motorists—provide timely information for trip planning purposes and for navigation in conjunction with an in-vehicle device (a vehicle-to-system link should also be included for purposes of monitoring traffic flows and collecting OD information);

(i) inclusion of visual surveillance—cable television (CATV) for detection/confirmation of incidents on freeways and arterial streets, use of video scanning techniques for improved detection capability and, perhaps, direct measurement of vehicle queues (see *Video Sensors*);

(j) variable speed control—determine appropriate speeds during periods of congestion and inclement weather, advise motorists of that speed and adjust control parameters based on that speed (see *Speed Limitation on Freeways: Traffic-Responsive Strategies*);

(k) provisions for headway control—automated highways and vehicle safety technology such as crash avoidance systems hold promise for reducing headways and accidents, and increasing capacity.

4.3 Impediments

Euler forewarned of the following impediments that must be dealt with in designing traffic systems of the future:

(a) cost and sources of funding—there are questions of who should provide the funds to do the research and development work, design and install the systems, and then maintain them, and of whether the private sector will perceive a profit potential;

(b) political problems and institutional frictions;

(c) consensus building;

(d) commitment;

(e) public acceptability—it is a question of whether the public will accept control decisions and traffic advisories from a machine, the "intrusion" of automatic vehicle identification and location, and the notion of congestion pricing;

(f) liability;

(g) investments in present systems;

(h) complexity of systems and shortage and lack of qualified people in the traffic engineering profession;

(i) software development—it is a question of whether effective software can be developed to predict ODs, assign traffic, simulate traffic flow and make good decisions on control parameters;

(j) the stifling of creativity of talented innovators.

4.4 Assets

Euler also listed a number of assets, some of which are being brought about by necessity. It is interesting to note that although traffic was a victim of its own growth success, it may, however, benefit from current congestion levels. Only when the public and political representatives perceive the effects of congestion will they allow traffic control personnel to plan for the control measures that may well be required in the future. The assets listed by Euler are

(a) high levels of congestion—congestion is already a paramount problem in Australia, Europe, Japan and the USA causing the public's frustration to grow and productivity to be lost (as mentioned, it is unfortunate that traffic personnel must see this as an asset);

(b) public demand—the public's frustration is being translated into action through pressure on political leaders and transportation is now often the primary issue in local elections;

(c) the economic cost of congestion;

(d) growing awareness of the potential of traffic management;

(e) technology push—recent advances in computer technology permit more real-time processing by small computers and will thereby enable development of better real-time control systems and in-vehicle route guidance systems;

(f) private sector interest—automobile manufacturers worldwide are becoming interested, as congestion is a detriment to the sale of new vehicles;

(g) international competition will bring a sense of urgency to research and development within individual nations, but there is also great potential for international cooperation in research and development (the European Prometheus project is an example of such cooperation as well as being an example of cooperation between the public and private sectors);

(h) the spread of congestion—urban congestion is no longer a problem except in central business district (CBD) areas;

(i) demographic changes, such as the increase of women in the workforce, the increase in the number of single-person households, the general aging of the population and the accompanying increase in discretionary travel;

(j) fuel and environmental costs; and

(k) the opportunity to improve safety.

5. The Near Future

The 1990s should see some prototype applications of developments that combine systems analysis, computer power and institutional cooperation.

5.1 Integrated Corridor Management

Techniques for development and preevaluation of schemes for integrated management of freeway–arterial corrridors have been available since the mid-1970s (Yagar 1975) and institutional barriers to integrated management of systems under different jurisdictions are falling. There are now some instances of partial integration, and the 1990s should see major applications and moves towards fully integrated management of traffic signals on surface arterials and freeway ramps.

5.2 In-Vehicle Route Guidance Systems

Prototype systems for dynamically guiding drivers to their individual destinations using in-vehicle audio and visual information are already planned for London (Department of Transport 1986), Tokyo and Berlin (von Tomewitsch 1986). These have provision for updating their databases via two-way communication with vehicles. The 1990s should see some measurable savings in trip times and urban demand in these and other major cities where in-vehicle route guidance systems are employed (see *Route Guidance, Individual*).

5.3 Serious Demand Management

Historically, peak period demands have been managed around life-styles rather than transportation. Thus, trips tended to converge and cause major peak demands on transportation systems. Since the 1970s, there has necessarily been some reversal towards staggered working hours and flextime in order to spread the peak travel demands. This has been complemented by differential peak and off-peak transit fares to spread demand, and peak period parking and stopping restrictions to increase capacity. The 1990s should see increasingly selective restrictions on vehicle access to busy urban areas as a way of reducing discretionary travel and increasing vehicle occupancy. This could be just the beginning of serious demand management in larger cities.

6. The Twenty-First Century

The technology exists for combining supply and demand management to produce a more productive urban transportation system, albeit one that restricts freedom of movement. It requires only the political and public will to change to a system that can increase vehicle throughput by automatic vehicle control, and reduce the demand on the system by allocating routes and even trip times. When and if such a system will become more palatable than the queuing and congestion that result from free access to a system will depend on local conditions.

Clearly, a system cannot serve beyond its capacity. Some systems are already dangerously close to capacity and suffer an increasing lack of mobility due to congestion. However, combining automatic vehicle control with trip allocation would effectively reduce travellers to the equivalent of packaged goods or transit passengers being shipped in their own vehicles. Would not an expanded transit system as is currently known be more acceptable than such a controlled system?

7. Problems

As long as the supply of urban transportation could increase at a greater rate than the demand for it, the quality improved. However, the quality level in most major urban areas has peaked and is on an increasing downward trend. As long as demand continues to increase, the quality of service will deteriorate further. It is only possible to try to minimize the adverse affects by appropriate integrated planning and management of the supply and demand. Help can be given to the decision makers by providing likely scenarios that would unfold in response to alternative political and philosophical choices.

These are exciting times for professionals involved in traffic control. The enormous problems related to congestion and urban movement will necessarily provide increased visibility and opportunity to contribute. However, the past frustrations of being ignored while congestion got irreversibly out of hand will be replaced by the frustration of not being able to return to the quality of urban movement that has been known.

See also: Transportation Systems: Trends

Bibliography

Department of Transport 1986 AUTOGUIDE: A better way to go? Central Office of Information, London
Euler G 1988 Issues in real time control of traffic—Workshop report. *Management and Control of Urban Traffic Systems.* US Engineering Foundation Press, New York, pp. 53–63
Hunt P G, Robertson D I, Bretherton R D, Winton R I 1981 SCOOT—A traffic responsive method of coordinating signals. Transport and Road Research Laboratory No. 1014. TRRL, Crowthorne, UK
Philipps P 1988 Transit companies sharing in computerized traffic control for bus priority. *Management and Control of Urban Traffic Systems.* US Engineering Foundation Press, New York, pp. 29–36
Robertson D I 1969 TRANSYT methods for area traffic control. *Traffic Eng. Control* October, 276–81
Rowe E 1988 The Los Angeles ATSAC system. *Management and Control of Urban Traffic Systems.* US Engineering Foundation Press, New York, pp. 9–17
Sessions G M 1971 *Traffic Devices—Historical Aspects Thereof.* Institute of Transportation Engineers, Washington, DC
Sims A G 1979 The Sydney co-ordinated adaptive traffic system. *Research Directions in Computer Control of Urban Traffic Systems.* American Society of Civil Engineers, New York, pp. 12–27
Teply S 1985 Interaction of two control systems: The Edmonton LTR example. *Traffic Monitoring and Control Systems.* Institute of Transportation Engineers, Washington, DC, pp. 43–61
Urbitran Associates Inc. 1982 Transportation systems management: Implementation and impacts. Urban Mass Transit Administration Report No. DOT-1-82-59. UMTA, Washington, DC
von Tomewitsch R 1986 ALI-SCOUT—A universal guidance and information system for road traffic. *2nd IEE Conf. Road Traffic Control.* Institution of Electrical Engineers, London
Yagar S 1975 CORQ—A model for predicting flows and queues in a road corridor. *Transp. Res. Rec.* **533**, 77–87
Yagar S 1985 The future for integrated control. *Traffic Monitoring and Control Systems.* Institute of Transportation Engineers, Washington, DC, pp. 85–8

S. Yagar
[University of Waterloo, Waterloo, Canada]

Traffic Management Systems

Usually, traffic control techniques have had the aim of coping with increasing traffic demand trying to maximize the capacity of the network by means of the

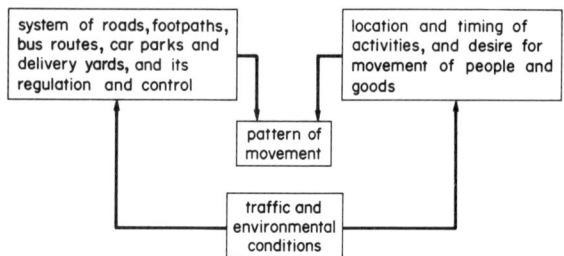

Figure 1
Interactions that form the context for traffic management and control (after Improta *et al.* 1986)

control over the time. Nevertheless, congestion is nowadays one of the main problems of urban areas all over the world, showing that, in spite of the advances in traffic control, control techniques alone are not enough to solve the problem.

Reduction of congestion is a main objective not only for ensuring shorter and more reliable travel times, but also in reducing its indirect consequences on pollution and fuel consumption, thus improving the quality of life. These aspects have also to be seen as major needs for road transport operations and drivers in general, although the fact that convenience to individuals is not necessarily also convenience to the community will often lead to conflicting strategies.

Here is one of the keys to the problem. In deciding what should be done in addition to traffic control measures, traffic engineers agree to a suite of complementary actions and strategies that puts the emphasis on those techniques that tend to ensure a better use of the available capacity, or, according to the Organisation for Economic Co-operation and Development (1987) "what still can be done on existing urban networks and facilities to alleviate the imbalances in the use of available capacities through intelligent rerouting strategies." Traffic management systems and facilities have these objectives, integrating not only control over time, but also control over space while guaranteeing the best trade-off between the individual and the community interests.

Improta *et al.* (1986) formulate the problem in a more precise way.

"The pattern of traffic in an urban area is the result of interaction between people's wish to travel or move goods in the area and the available road system, including the regulations governing its use and any control system that is in operation. The pattern of traffic itself gives rise to traffic and environmental conditions which may themselves both influence the location and timing of activities, and hence the demand for movements, and give rise to ideas and pressures for changes in the road system and its managements and control. These interactions are summarised in figure 1."

These interactions lead in current practice to imbalances in the use of available network capacity due to stochastic variations in traffic demand, incidents and so on and, meanwhile, routes not overcrowded are not used by drivers due to lack of information and control actions. Therefore, a traffic management system which, as has been pointed out above, integrates control over time and over space, needs:

(a) practical methods of measuring the degree of change in network traffic flows resulting from system modifications;

(b) real-time identification of imbalance situations in the use of available capacity;

(c) definition and assessment of adequate strategies; and

(d) implementation of real-time management decisions and control measures.

Most of these needs were recognized at early stages of the development of traffic management systems; however, the lack of suitable tools to handle them did not allow the implementation of efficient solutions, although the early attempts made possible the identification of some of the key issues.

Gartner (1977) reported that many traffic management schemes indicate that it may be desirable, for one reason or another, to reduce the amount of traffic using parts of the road network and that most of the operational improvements experienced so far tended, in practice, to reinforce the existing traffic patterns rather than to explore alternatives to them, underlining that this is most evident in calculating and updating signal control settings, which are based on the existing or projected pattern of traffic. The result is that signal controls are regularly set to favor existing heavy flows of traffic over light flows. This is particularly evident when arterial progression schemes rather than area-wide optimization schemes are being implemented. Gartner concluded "thus the choices that created existing traffic flow patterns receive constant reinforcement," in accordance with the results of Allsop (1975) and Berg and Tarnoff (1976).

In investigating the relationships between traffic control and route choice, Gartner pointed out that, traditionally, traffic engineers have chosen signal settings that minimize travel costs given a fixed traffic flow pattern. In doing this, it is assumed that all route choices are predetermined and result in constant flows on each link, irrespective of the controls imposed on that link and, hence, irrespective of the level of service offered by the link. This assumption is, in most cases incorrect. As shown in Fig. 2, there exists an interdependence between traffic controls and link volumes; one determines the other and vice versa. Models must, therefore, be developed that incorporate both traffic control and route choice (traffic assignment) and, thus, provide tools for establishing mutually consistent signal settings and traffic flow patterns (see *Traffic Assignment*).

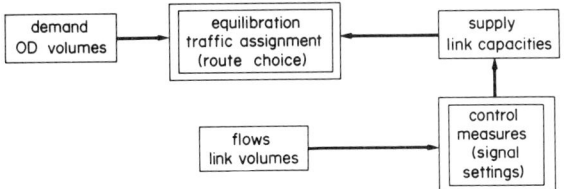

Figure 2
Interdependence of link volumes and traffic controls

Figure 2 represents a step ahead towards a better understanding of the interactions at work in the context of traffic management. Gartner *et al.* (1980) address these interactions, considering the traffic management problem as representing the interactions between the urban traffic manager and the individual traveller and their (sometimes differing) perceptions of the performance of, or the supply provided by, the transportation system.

Gartner *et al.* (1980) model these interactions in a transportation network by means of a schematic illustration that has served as basis for a more elaborate model of the interactions between public policy and user preferences, formulated by the Organisation for Economic Co-operation and Development (1987). This model, displayed in Fig. 3, represents the interactions between the physical transportation system and the socioeconomic activity system, via the equilibrium process, to produce a set of flows on the links of the network.

The operation of this model of a transportation system management (TSM) can be explained as follows (Gartner *et al.* 1980). In the system loop, the traffic manager assesses the system performance according to his measures of effectiveness and intervenes in the physical transportation system to achieve the best trade-off between the individual's interests and the interests of the community. The travellers, on the other hand, perceive the flows according to their own values, which may be different from those of the traffic manager, and propagate an adjustment in the travel demand pattern via the user loop. While the traffic manager can intervene fairly quickly to further his objectives, it is assumed that the reactions of the trip makers have a longer time constant. Therefore, in a short-range analysis, the demand pattern may be regarded as fixed.

From these considerations, the traffic management problem can be set in the following context: given (fixed) demands for travel in an urban area and a (fixed) supply of transportation facilities, the traffic manager must consider a variety of management strategies to induce a traffic flow pattern that will meet, in an optimal way, the overall objectives of the community.

Traffic equilibrium in an urban road network can thus be considered as the result both of control and management policies taken by the traffic engineer and of choices made by individual drivers according to their perceptions on the network performance.

Many examples of this interplay between the not always consistent user and community interests may be given. For instance, the Organisation for Economic Co-operation and Development reports that, as a matter of fact, it is possible to prove that to minimize the total time spent in a freeway corridor (community optimum), for example, it can be advantageous in some cases to discourage some drivers from entering the freeway at the beginning of the peak period, instead making use of the surface streets (by restricting access), in order to delay as long as possible the beginning of congestion on the freeway. The total time saving for the other drivers is greater than the total delay experienced by diverted drivers. In this case, community and user interests are in competition. However, in most cases, any control action that attempts to establish a time equilibrium between competing routes globally decreases the total time spent in the whole system, although it does not, in general, minimize it (see *Corridor Control Systems*).

Given the above formulation of the traffic management problem, and its related consequences, the decision variables for this problem are the set of management variables under control of the traffic manager. A comprehensive list of management actions to be considered has been given by the US Department of Transportation (1975). In the area of traffic signalization it lists

(a) cycle time,

(b) green-time splits,

(c) offsets (throughbands),

(d) signal phasing,

(e) phase sequencing, and

(f) variable message signs (off-vehicle route guidance).

In the area of traffic operations improvements it lists

(a) channelization of traffic,

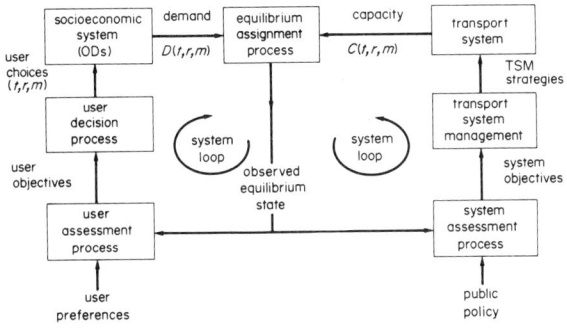

Figure 3
Transport system interactions

(b) one-way streets,

(c) metering access to freeways, and

(d) reversible (tidal) traffic lanes.

In the area of preferential treatment it lists

(a) reserved or preferential lanes on freeways and city streets,

(b) exclusive lanes to by-pass congested points,

(c) conversion of selected downtown streets to exclusive bus use,

(d) exclusive access ramps to freeways,

(e) bus preemption of traffic signals,

(f) special turning lanes,

(g) truck routes, and

(h) parking control.

New technological developments allow some other possibilities to be added to these lists, such as vehicle guidance (in-vehicle guidance systems) and fleet management.

Dynamic management systems enable the traffic manager to act on the traffic system by means of the management actions with the purpose of acting on traffic supply or on traffic demand, or both, in an attempt to ensure that traffic flows resulting from travel requirements meet traffic conditions established by the set management actions; that is, the traffic management scheme. Figure 4 shows the way in which these actions operate.

According to the Organisation for Economic Co-operation and Development (1987), systems providing dynamic management facilities fall into three distinct categories, depending on whether they act mainly on supply, demand or both.

Systems that affect demand (loop 1 in Fig. 4) are driver information systems. Their main function is either of the following.

(a) To provide drivers with trip-planning advice: drivers' routes, mode of travel and departure times may be influenced if they can be given advance warning of current and anticipated conditions on the road network.

(b) To provide drivers with route-following advice: the actual routes followed by drivers may be influenced to avoid congestion sites as well as to reduce waste caused through inefficient route choice.

Systems that affect supply (loop 2 in Fig. 4) are the direct traffic control systems. Their main function is either of the following.

(c) To provide flow-control measures: the numbers and speeds of drivers entering congested areas may be influenced by introducing impedances (e.g., traffic signals or warning signs) at selected points in the network or by giving warnings (e.g., area broadcasts) in selected areas of the network.

(d) To provide traffic management information: improved systems for collecting and collating information may reduce both the time and resources needed to deal with traffic incidents and can enable systems to respond automatically to changing traffic conditions.

A further method of affecting both supply and demand is the following.

(e) Through public transport and fleet management systems, public transport, taxi and emergency service vehicles may be organized and deployed more efficiently by a controller if that controller can locate and communicate with the drivers while they are on the move, and give them priority at traffic signals; this may, in turn, affect demand through influencing mode choice or supply by facilitating the early alleviation of incidents and hazards.

1. Driver Information Systems

The nature and functions of driver information systems depend on whether their primary function is to help the driver prepare for a trip (trip-planning function), or to help the driver while on the move (route-following function); among this last class, the expected role of in-vehicle route guidance aids with radio-broadcast updating will be noted (see *Route Guidance, Collective*).

Research on so-called autarkic (i.e., self-contained) route guidance systems is actively being undertaken by

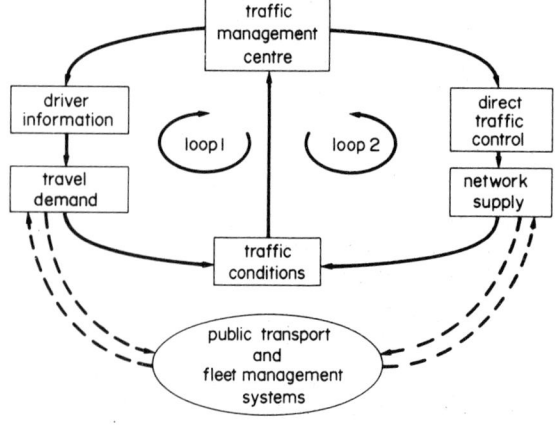

Figure 4
Dynamic traffic management overview (after Organisation for Economic Co-operation and Development 1987)

numerous motor and electronics manufacturing companies in several countries. These devices rely on an electronic "map" carried in the vehicle and some form of navigation system to enable drivers to keep track of their position in the network. Some systems, (e.g., ETAK in the USA), display a map of the surrounding road network with the driver's position superimposed, while others (e.g., EVA in Germany or CARIN in The Netherlands) offer route guidance by deducing the best routes and giving drivers turn instructions as they proceed on their journey. In these systems the electronic maps are historic but research is being carried out on the possibility of updating the maps using digital radio broadcasts. The systems would then be useful for trip planning as well as route following and, if enough vehicles became equipped, for flow control. This is one of the objectives of the Socrates project of the DRIVE programme.

2. Direct Traffic Control Systems

Direct traffic control systems control traffic by setting variable impedances in the network. They essentially fulfill a flow control function, although systems that incorporate an automatic data collection facility, also provide traffic management information.

2.1 Variable Message Sign and Route Control Systems

Variable message signs are used at the roadside in most countries to set speed limits and to advise drivers of conditions on the road ahead or to advise them on the best route to follow to reach their destination.

Data collection systems may also be used so that the signs can be automatically varied in real time to take account of changing traffic conditions. These systems, therefore, provide route-following, flow control, and traffic management information functions. The amount of information that can be given is very limited, however, and destinations can only be described in very broad terms. The route control aspect of these systems is, therefore, generally reserved for known congestion sites (e.g., bridges and tunnels) where a limited number of clearly identifiable alternative routes are available.

New forms of variable message signs are currently under investigation by some projects of the DRIVE programme.

2.2 Historic Urban Traffic Control, Tidal-Flow and Ramp-Metering Systems

Historic urban traffic control, tidal-flow and ramp-metering systems are all traffic signal control systems. The urban traffic control (UTC) systems control and coordinate traffic signals on junction approaches in an attempt to minimize overall delays experienced by crossing flows of traffic in urban networks. Examples include most UTC systems currently in use which employ programs such as TRANSYT to determine signal timing plans (see *Road Traffic Control: TRANSYT and SCOOT*).

Tidal-flow systems control special traffic signals over the individual lanes of a carriageway so that the number of lanes assigned to each direction can be varied to accommodate the direction experiencing the greatest demand. Ramp-metering systems employ traffic signals and sometimes variable message signs at on-ramps to control the rate at which vehicles can enter a freeway (see *On-Ramp Control: Realization Principles*).

All these systems are not strictly dynamic traffic control systems. They use fixed time plans based on historic data to control a system of traffic signals. However, because the plans may be changed as often as every 15 min during the peak hours, they exhibit a degree of "dynamicness" and are, therefore, included for completeness. The historic data must be updated every year or so, so that some traffic management information is obtained, but these systems essentially provide only flow control.

2.3 Adaptive UTC, Tidal-Flow and Ramp-Metering Systems

Adaptive UTC, tidal-flow and ramp-metering systems represent a more recent development in traffic control signalling systems and are truly dynamic in that they collect their own traffic data in order to respond to changing conditions in real time. They therefore provide flow control and traffic management information. Examples include: UTC systems based on SCOOT from the UK, PRODYN from France and SCATS from Australia; tidal-flow systems such as those operating within the SCATS system; and most ramp-metering systems such as those used on Highway 401 in Canada and in the Northern Long Island Corridor in the USA.

2.4 Automatic Route Guidance Systems

No examples of automatic route guidance systems exist except as experiments (e.g., ALI from Germany, CACS from Japan and AUTOGUIDE from the UK). However, they have the potential to significantly increase the efficiency of network usage and, hence, to provide the next generation of traffic control tools. They require an extensive roadside infrastructure and units in vehicles which together enable drivers to receive guidance at junctions to direct them over a minimum path route to their individual destinations. At the same time, data on traffic movements is collected. Equipped vehicles can then be guided in real time and in such a way as to distribute traffic more evenly throughout the network and to alleviate congestion at the sites of incidents. These systems therefore provide route-following, flow control and traffic management information functions (see *Route Guidance, Individual*).

2.5 Road Pricing and Automatic Vehicle Location Systems

Electronic road pricing (ERP) systems also require an extensive roadside infrastructure and vehicle units. An example is the experimental ERP System in Hong Kong in which vehicles are both located and charged as they cross toll points distributed in the network. For these systems, however, the fitting of vehicle units is mandatory and an exceptionally high integrity road-vehicle communication link is required to ensure that vehicles are not mistakenly identified or wrongly charged. These systems provide flow control and traffic management information functions.

3. Assessment of Traffic Management Schemes

The implementation of traffic management schemes is one of the most flexible and effective ways of promoting a better use of the transport facilities available in an urban area. As Hall *et al.* 1980 pointed out in their introduction to the paper on the SATURN model: "currently the design of traffic management schemes relies on a mixture of intuition, past experience, established traffic engineering practices and 'before and after' assessments." Obviously, this practice becomes insufficient when management schemes involve more complex actions such as the ones just described.

There is a need for more powerful modelling techniques suitable for the assessment of such management schemes. SATURN and CONTRAM provide a first answer to that need.

3.1 The SATURN Model

SATURN (simulation and assignment of traffic to urban road networks) is a computer model developed at the Institute for Transport Studies, University of Leeds, for the analysis and evaluation of traffic management schemes over relatively localized networks (typically of the order of 100 to 150 intersections). It is primarily intended to be used as, in effect, a highly sophisticated traffic assignment model in which the sophistication is primarily due to a highly detailed simulation of delays at intersections. It is most suitable for the analysis of movement-based control systems such as one-way streets, changes to junction controls, bus-only streets, and so on.

SATURN seeks to answer the question of how traffic flows in a network will change in response to a traffic management scheme. Being concerned primarily with routes taken by drivers, it is most naturally suited to evaluate schemes that affect traffic movement, such as the introduction of one-way streets, banned turns, cell schemes, bus-only lanes, pedestrianization schemes, changes in junction design, and so on.

Other attractive features of SATURN include the wide range of junction types modelled, the relative simplicity of its data input, the inclusion of performance indexes such as the number of vehicle stops, and the inclusion of a technique for updating trip matrices to match observed counts.

SATURN may be thought of as a sophisticated traffic assignment model. In common with virtually all such models, it assumes that the underlying traffic pattern represented by an origin–destination (OD) trip matrix is fixed (although changes in modal split and/or distribution may be incorporated externally using conventional techniques). The model incorporates two phases: a detailed simulation model of delays at intersections coupled with an assignment phase which determines the routes taken by the above-mentioned OD trips.

(*a*) *The simulation model.* The primary objective of the simulation is to determine junction delays resulting from a given pattern of traffic. To do this, two fundamental assumptions are made:

(i) that the pattern of traffic flows is constant over time periods of the order of 30 min; and

(ii) for simulation purposes, that a cyclical behavior is imposed on the flows by traffic signals operating with a common cycle time (typically in the range 60–120 s).

The first of these assumptions is shared by virtually all models in this area and enables us to rely on a fixed trip matrix. It also restricts the investigation to the average behavior of the system within the given time period.

The second assumption allows the simulation to concentrate on a single cycle, making a detailed simulation possible while avoiding many of the overheads normally associated with simulation. Within each cycle, traffic is represented by semicontinuous "flow profiles" as opposed to modelling individual vehicles or packets of vehicles.

These features, together with the use of platoon dispersion of traffic along links, represent techniques used successfully in the signal optimization program TRANSYT.

(*b*) *The simulation structure.* The cyclic flow profile (CFP), the flow of traffic past a certain point as a function of time over a single cycle (the cycle length being that of the traffic signals in the network), is the main building block for the simulation. The model can be seen as a collection of specialized routines, each designed to modify the CFPs according to prevailing conditions. To achieve this, CFPs are attached to individual turning movements and provide the basis for the detailed analysis of delays.

(*c*) *The modelling of junctions.* Since junction delays are one of the major determinants of urban travel times, the accurate modelling of such delays assumes considerable importance in SATURN. The following junction types are modelled separately—traffic signals, pelican crossings, priority junctions (including merges) and roundabouts.

In contrast to the majority of assignment models, delays at intersections differ for different turning movements. The careful calculation of turn-dependent patterns is clearly important and involves much of the computational effort. Purpose-built routines are employed for each type of turning movement at an intersection. Thus, for example, there are separate routines that specifically calculate the flows conflicting with a left turn, a right turn or a straight-on movement at a stop line of a priority junction.

Other innovations in the simulation include the adoption of a gap-acceptance formulation in the calculation of merging and crossing capacities and the classification of junction approaches according to lane markings. The first means that the level of merging flow varies throughout the cycle in response to the instantaneous strength of opposing flows; the second enables us to calculate the distribution of turning flows between the available lanes and so determine interactions between distinct movements with greater confidence.

By using these techniques in combination with the information concerning turn and link capacities the following can be accounted for:

(i) delays caused by opposing flows (e.g., to right-turning traffic);

(ii) delays due to vehicles on the same link (e.g., the effect on straight-ahead traffic of impeded right-turners in a shared lane);

(iii) the shape of arriving platoons;

(iv) the effect of phase structure and offsets at traffic signals; and

(v) individual lane capacities.

(d) *The assignment model.* The objective of the assignment phase is to select, for each element in the trip matrix, minimum time routes through the network, bearing in mind the relationships between travel time and flows. The model uses an equilibrium technique that optimally combines a succession of all-or-nothing assignments such that the ultimate flow pattern satisfies the above criterion (commonly referred to as Wardrop equilibrium). While the concept of fastest routes only is open to objections in large networks, where obvious exceptions are frequently observed, it is probably much more valid for small networks where the potential routes open to each OD pair are relatively distinct.

The model assumes:

(i) that the travel time of each link is fixed independent of flow; and

(ii) that the delay to each turning movement at an intersection is a function of that turning volume.

The simulation model is used to model the flow-delay curves by calculating the delays for each turning movement at zero flow, current flow and capacity with all other flows (i.e., opposing traffic) fixed.

(e) *The complete model.* The complete model is based on an iterative loop between the assignment and simulation phases (see Fig. 5). Thus, the simulation determines flow-delay curves based on a green set of turning

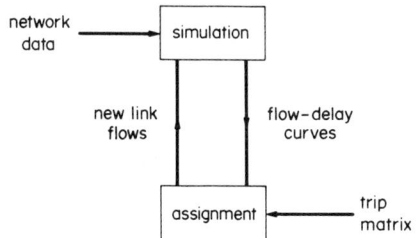

Figure 5
The simulation and assignment phases of SATURN

movements and feeds them to the assignment. The assignment, in turn, uses these curves to determine route choice and, hence, updated turning movements. These iterations continue until the turning movements reach reasonably stable values.

The reason for this "outer" iteration is essentially that the delay curves used by the assignment are based on fixed values of opposing traffic at each intersection. However, as each assignment gives different opposing flows, so too does the simulation give different delay curves and, in turn, different assignments. Thus, the dependence of delay, on flow V is incorporated within the assignment phase, whereas the wider dependence on opposing flows is catered for by the outer loop.

While ultimate convergence to stable values is difficult, if not impossible, to establish theoretically, tests on realistic networks have all shown that reasonably stable values are reached after 4–7 iterations.

3.2 The CONTRAM Model

CONTRAM 5 is a traffic assignment model intended for use in the design of traffic management schemes. It models the build up and decay with time of peak period congestion, and its effect on the routes and journey times of traffic, and includes the effects of priority junctions, signal control and signal coordination.

The time-dependent assignment by CONTRAM of time-varying traffic demands ensures that where and when queues and delays build up is modelled realistically.

The CONTRAM assignment method assumes that individual drivers seek to minimize either their total journey time or total journey cost on the basis of their expectation of the queues and delays that they will meet on their journey.

The principal features of CONTRAM are that it:

(a) models signal-controlled junctions, give-way junctions (including roundabouts) and bottlenecks at merges;

(b) represents variation in traffic conditions with time, particularly where demands temporarily exceed capacity (e.g., during peak periods);

(c) allows for blocking-back effects where a queue from one junction fills a street and restricts the capacity at an upstream junction;

547

(d) subdivides results into three classes of vehicle—usually cars, buses and lorries (heavy-goods vehicles);

(e) bans selected traffic movements—prohibiting access to selected streets for any or all of the specified vehicle classes;

(f) includes a fixed route option (e.g., for modelling bus services);

(g) incorporates three methods for setting signals;

(h) estimates fuel consumption for each class of vehicle; and

(i) calculates average linear speeds for selected OD movements as a measure of fairness.

(*a*) *The simulation model.* CONTRAM, which stands for continuous traffic assignment model (Leonard and Gower 1982), is a traffic assignment model that can deal with time-varying traffic conditions in a network. The network of streets in a town is represented by a series of unidirectional links and junctions. Traffic demands enter the network at origins and leave at destinations. Origins and destinations may occur within the network as well as on the periphery. Time variation is modelled by dividing the simulation period into a number of consecutive time intervals of, typically, 15 min duration. The demands for each OD movement are specified as a flow rate (vehicles per hour) for each time interval.

Vehicles from each OD pair are grouped together to form "packets." The packets from any OD pair enter the network equally spaced in each time interval. The vehicles in a packet are assumed to remain together, for computational purposes, as they travel through the network; each packet is assigned to its minimum journey time route. The journey time for a packet along each link of its route consists of two parts—a "cruise" time (i.e., a free-running time), which corresponds to an average unimpeded travel time, and a delay time, which is dependent on the level of flow on a link and on the method of junction control. Three types of junction are modelled: signal controlled, give way and uncontrolled. Uncontrolled junctions arise where roads merge or diverge with no one traffic stream having priority, and are the controlling links at give-way junctions. Roundabouts are modelled by considering each approach road as a give-way link and the road around the island as a series of uncontrolled links.

Delay calculations are based on an estimate of the average queue on each link at the end of each time interval. The estimate of queue depends on the queue at the end of the previous time interval (the initial queue), the number of arrivals at the stop line in the interval, the maximum rate at which vehicles can leave the link and the duration of the time interval. The theory of time-varying queues (Kimber and Hollis 1979) is used to take account of random variations in vehicle arrival patterns and the type of junctions (e.g., signal controlled, give way, etc.). It is assumed that the queue varies linearly between the queue values at the start and end of the interval. The delay for an individual packet is calculated from the length of the queue encountered by the packet at the time that it reaches the stop line (stop line arrival time equals entry time to link plus cruise time on link). The delay is then the time taken for the queue encountered by the packet to be discharged. The rate of discharge is the maximum throughput capacity rate for that link (i.e., allowing for the type of junction control).

CONTRAM uses an iterative procedure to predict the patterns of routes, flows, queues and delays on a network that result when drivers are familiar with network conditions. An iteration of the model consists of assigning each packet of vehicles to its minimum journey time route through the network using a tree-building algorithm. This produces patterns of traffic on the network which are used in the next iteration when each packet is reassigned to its new quickest route. Before each packet is assigned the flow corresponding to that packet is removed from each link of the route determined in the previous iteration. Thus, a packet does not incur delays due to itself when being assigned to a new route. As the reassignment of packets is made in the time order in which packets enter the network, the delays calculated for a given packet will be determined by flows due to packets that have entered the network prior to that packet, during the current iteration, and by flows due to subsequent packets assigned to routes in the previous iteration. Several iterations are required before stable traffic patterns are obtained. Full convergence of the process is attained when all vehicles are assigned to their same routes during successive iterations. This iterative process can be considered as modelling the gradual familiarization of drivers to network and traffic conditions.

(*b*) *Signal control.* Specifies cycle time, order and length of green stages, and offset of stage 1 green in a coordinated system. Vehicle actuation is not simulated but isolated signals can be optimized using the equal saturation method.

CONTRAM models the growth and decay of queues from time interval to time interval including the effects of temporary oversaturation, such as occurs during peak periods, resulting in the growth and decay of queues. Allowance is made for blocking back effects at junctions when vehicles queue back along the full length of links, blocking upstream junctions, and thereby restricting the flow of vehicles from feeding these junctions.

The banned-vehicle facility can be used to prevent selected vehicle classes from using specified links in a network. This provides an easy method for examining, for example, the pedestrianization of links or restricted access for heavy-goods vehicles. The modelling allows for up to three classes of vehicle—C, B and L, where C represents cars and B and L usually represent buses and lorries. Bus routes are modelled using a fixed-route option in which the routes for specified OD movements are prescribed even though they may not correspond to the minimum journey time routes for those movements.

Within a given memory size, different network parameters can be traded off (e.g., reducing the number of time slices allows more links to be accommodated). On a 640K 386 PC with typical network configurations and time-varying traffic demands, 300–400 links can be handled (run time 2–3 h). Dynamic allocation of extension of memory is supported on machines with fixed memory (e.g., CDC Cyber and IBM PCs).

3.3 Other Assignment-Based Alternatives

Further developments in transportation planning systems based on traffic assignment approaches such as EMME/2 (Babin *et al.* 1982, INRO Consultants 1989), enable them to be used for the assessment of traffic management schemes.

Traffic assignment models can be viewed as static network models that reproduce average flow behavior in a given time period and typically model a large metropolitan or regional scenario. This modelling approach has the following features:

(a) a macroscopic point of view providing a good description of the road network topology, but not very detailed for other objectives (e.g., traffic control purposes);

(b) demand behavior described by a static average OD matrix for the time period under study; and

(c) used for static assignment analysis for *tactical planning purposes*, which search for equilibrium solutions based on Wardrop's principle as fundamental modelling hypothesis.

Traffic management schemes usually concern only smaller urban scenarios within the large metropolitan or regional scenario and must take into account dynamic aspects (i.e., time-varying flows over short periods of time) that usually require a microscopic analysis approach.

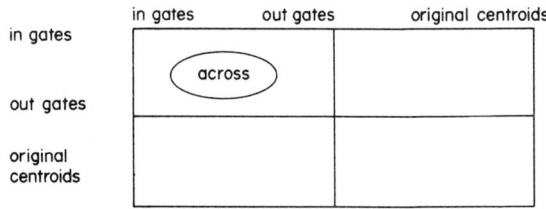

Figure 7
The traversal matrix (across denotes traffic that goes across the subarea from in gates to out gates)

Microscopic models have an excellent and very detailed network description, including traffic control aspects, and are used for dynamic analysis for operational planning purposes, but they require, as input, a time-varying demand behavior description.

To use a traffic assignment model such as EMME/2 for traffic management purposes, both levels of modelling (macroscopic and microscopic) must be linked in some way. Enhancements in such systems supply the tools enabling the user to link macroscopic and microscopic models.

Assuming that interest is focused on a specific subarea within a large region, as in the example displayed in Fig. 6, the following can be done:

(a) static equilibrium traffic assignment at the regional level that computes the paths used between the OD pairs;

(b) tag in (out) links in the selected subarea, which can be labelled and converted into centroids for the subarea;

(c) completion of the traversal matrix (the OD matrix for the subarea) (see Fig. 7); and

(d) assuming that the zone to be assessed by the traffic management schemes is a zone equipped with detectors that supply measurements of flows in the

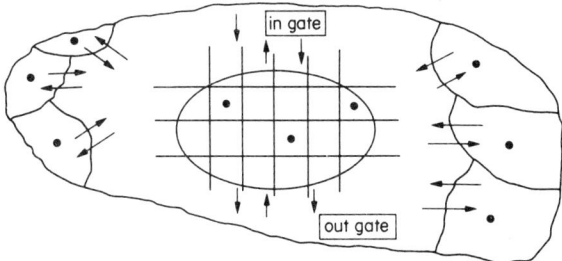

Figure 6
Examination of a specific subarea within a large region

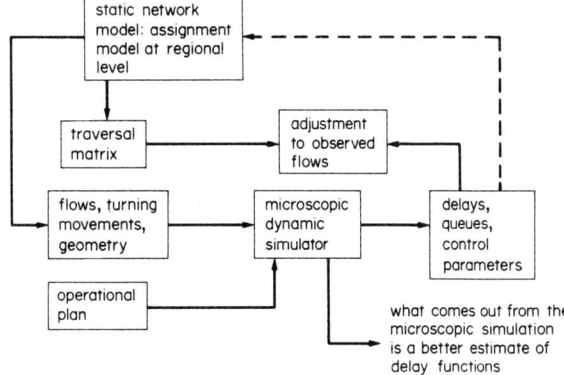

Figure 8
Iteration procedure for estimating cost functions

zone links, the traversal matrix, combined with the observed flows can be used to compute an adjusted matrix which provides the basis for the estimation of a time-sliced OD matrix (Spiess 1987, Cascetta and Nguyen 1989).

When using this model, cost functions can be replaced by estimates made by microscopic models (e.g., TRANSYT) if traffic control aspects are taken into account, iterating in the way displayed by Fig. 8.

Bibliography

Allsop R E 1975 Use of traffic signals to influence the amount and routing of traffic. *Conf. Getting the Most from Our Transport Facilities—The Role of Traffic Engineering.* Transport and Road Research Laboratory, Crowthorne, UK

Babin A, Florian M, James-Lefevbre L, Spiess H 1982 EMME/2: Une métode interactive-graphique pour le planification du transport urbain multimodel. Centre de Recherche sur les Transports Publication No. 204-F. University of Montreal, Montreal, Canada

Berg W D, Tarnoff P J 1976 Research in urban traffic management. *Transp. Res. Rec.* **603**, 55–6

Cascetta E, Nguyen S 1989 A unified framework for estimating or updating origin–destination matrices from traffic counts. *Transp. Res. B* **22**, 437–55

Florian M 1986 Nonlinear cost network models in transportation analysis. *Math. Program. Study* **26**, 167–96

Gartner N H 1977 Influence of control measures on traffic equilibrium in an urban highway network. *Transp. Res. Rec.* **644**, 125–9

Gartner N H, Gershwin S B, Little J D C 1980 Pilot study of computer-based urban traffic management. *Transp. Res. B* **14**, 203–17

Hall M D, Van Vliet D, Willumsen L G 1980 SATURN: A simulation–assignment model for the evaluation of traffic management schemes. *Traffic Eng. Control* **21**(4), 168–76

Improta G, Allsop R E, Heydecker B G 1986 Network models for traffic management. Centre de Recherches sur les Transports Seminar on Transportation Systems. University of Montreal, Montreal, Canada

INRO Consultants 1989 *EMME/2: Release 4.0, User's Manual.* INRO

Kimber R M, Hollis E M 1979 Traffic queues and delays at road junctions. Transport and Road Research Laboratory Report No. LR 909. TRRL, Crowthorne, UK

Leonard D R, Gower P 1982 User guide to CONTRAM version 4. Transport and Road Research Laboratory Supplementary Report No. 735. TRRL, Crowthorne, UK

Matzoros T, Van Vliet D, Randle J, Weston B 1988 A validation of the SATURN and ME2 models using before-and-after survey data from Manchester. *Traffic Eng. Control* **29**(12)

Morin J M 1987 Traffic corridor control: General principles. In: Singh M G (ed.) *Systems and Control Encyclopedia*, Vol. 7. Pergamon, Oxford, pp. 4910–13

Morin J M 1987 Traffic corridor control: Ramp assignment. In: Singh M G (ed.) *Systems and Control Encyclopedia*, Vol. 7. Pergamon, Oxford, pp. 4913–15

Morin J M 1987 Traffic corridor control: Ramp metering. In: Singh M G (ed.) *Systems and Control Encyclopedia*, Vol. 7. Pergamon, Oxford, pp. 4915–17

Organisation for Economic Co-operation and Development 1987 *Dynamic Traffic Management in Urban and Suburban Road Systems.* OECD, Paris

Spiess H 1987 A maximum likelihood model for estimating origin–destination matrices. *Transp. Res. B* **21**, 395–412

US Department of Transportation 1975 Transportation improvement program. Federal Register 40 (181), 42976–84

Van Vliet D 1982 SATURN: A modern assignment model. *Traffic Eng. Control* **23**(12), 578–81

Van Zuylen J H, Willumsen L G 1980 The most likely trip matrix estimated from traffic counts. *Transp. Res. B* **14**, 281–93

Webster F W, Cobbe B M 1966 Traffic signals. Road Research Technical Paper 56. Her Majesty's Stationery Office, London

J. Barceló
[Universitat Politècnica de Catalunya,
Barcelona, Spain]

Traffic Office Management

Organizations responsible for traffic and transportation engineering suffer from the relative weakness of the relevant professions and the strength of lay experience of the practical outcome. Administrators and decision makers believe themselves to have a full understanding of the subject and issues involved, notwithstanding their lack of professional qualifications or background information. This Encyclopedia demonstrates the wide range of subjects that have to be covered and the wealth of experience that is available in each subject.

1. Range of Organizations

Traffic offices have perforce to be provided by local, regional and national governments. The resources available to these different authorities vary widely. The greater these resources are, the more varied are the options for organization. If the resources are very limited, as in small authorities, only limited work can be carried out. Some large authorities, particularly in developing countries, suffer from inadequate resources, in which case the quality of work suffers or much of it has to be contracted out.

Consultants may provide traffic offices, particularly for major developing country projects. When operating in their own countries, the continuity of workload is usually insufficient to support the full range of skills required.

1.1 Advantages of Scale

Table 1 indicates the variation of team strength for traffic management. Traffic engineers are omitted from the table if they are in other teams such as transportation planning, highway planning or highway design.

Table 1
Numerical strengths of some example traffic offices

Category of office and function	UK metropolitan authority	UK local authority	Local traffic management consultancy	Foreign work consultancy
Traffic signals				
maintenance	(30)[a]		20	2
fault inspection	5			
fault reporting	8			
signal design service	20			
signal plan production	10			
computer software	5			
computer hardware	10	25		
Special interest groups				
transit priority	30		20	2
cycling facilities	15			
pedestrian measures	8			
truck controls	8			
minor schemes	8			
parking provision	10	20		
traffic order making	20	1		
Road accident statistics				
collection (by police)	(50)		20	1
distribution to users	5			
analysis	5			
remedial scheme design	10			
road safety publicity	1			
education	10			
Surveys (permanent staff)				
land survey	(10)		(5)	
traffic scheme design	20		50	
traffic monitoring	10		20	

a Items in brackets are often carried out by other organizations or under contract

In addition, traffic departments of several thousand staff such as that of New York City are not included.

It is immediately apparent that, in smaller authorities, each staff member has to deal with a wide range of subjects and has little opportunity to become expert in any of them. For example, in a team of only six, it is unlikely that knowledge of traffic signals will include the full range of different equipment available, the relative benefits of local and central control, the design programs available to set the timings or, even, the ability to fully specify programming of the on-street equipment. To achieve such expertise, a traffic team should number at least 100 people.

To offset problems of inadequate resources, traffic managers in small organizations will probably have a more direct knowledge of the sites and of the requirements of the local populace.

1.2 Geographical or Functional Organization

A choice has to be made between two distinct approaches. If an organization is large enough, it is possible for it to be divided on a functional basis. Teams, each of at least five people, can specialize. They can build up knowledge, standards and programs of work to optimize their performance. In some cases, this substantially improves productivity; for example, the writing of traffic orders can be several times quicker if the staff have already written many previous orders and an experienced team can use computer processing of model orders to further improve its speed and accuracy. Alternatively, the superior performance may be more technical, such as the ability of a traffic signal team to design a complex intersection or of a signal maintenance team to discover frequently occurring fault categories.

Some organizations are split geographically. In spite of the advantages of teams knowing their own area, this is usually a mistake and represents lost opportunities.

Very large organizations can afford the luxury of both geographical teams and specialist functional teams working side by side; care is needed by management to ensure harmonious working. A common arrangement with large specialist teams is to divide the workload on a geographical basis as well; sometimes the most specialized skills are kept in a central part of the team and made available to all the geographical subdivisions.

1.3 Relationships with Other Organizations

It is rare for a traffic office to be totally self-sufficient. Normally, a part of the task is undertaken by other organizations, including different levels of local government, the police and independent contractors.

The police have traditionally been responsible for traffic management. This still applies in developing countries and where other professional organizations are poorly developed. The increasing sophistication of traffic management tools and the requirements for public consultation and forward planning are not entirely consistent with normal police activities; these are normally in terms of intervention at the time of an event or when undisputed authority is required. In developed countries, an increasing range of traffic conditions is handled by correct traffic engineering design and automatic equipment. This is not usually organized by the police.

Private contractors often carry out work as an alternative to the local authority. This has the advantage of allowing the resources to be redeployed once a project is complete. Public authorities have more difficulty in reducing size to match a reduced workload or in removing ineffective staff. Contracting the work out can ease these problems. It also provides a contractual interface at which performance can be measured and commercial incentives applied to improve it.

2. Functional Requirements

The boundary between a traffic office and other offices is imprecise. Table 2 lists some of the functions that have to be carried out. In small organizations, these are all in the traffic office. In larger centers, they may be separate organizations.

2.1 Traffic Surveys

The traffic survey function is frequently contracted out to specialists able to mobilize suitable staff or to use sophisticated tools.

Many of the staff may be part time, held on a register by the survey office and deployed when surveys are required near their homes. Similarly, students may be used. Such arrangements make efficient use of staff time. It is, however, important that full training is given and that the results are thoroughly checked. The permanent staff of the survey office have the skills and experience to effectively control the part-time staff.

The sophisticated tools include hand-held computers for data collection. These ensure that the data format is correct and facilitate transfer to a computer for further analysis. The analysis computers will perform statistical checks to identify bad or missing data. If the survey is part of a comparative study (such as before and after traffic management changes are made), the survey service can include statistical analysis of the changes.

Some surveys may still be carried out by the traffic management designers. Their knowledge of the situation can enhance the data quality and they are likely to notice other events through having to stand watching the traffic for long periods.

The objective of ensuring that traffic engineers have first-hand knowledge of traffic situations is often helped

Table 2
The necessary functions and facilities

Office function	Method
Transportation planning	models
Surveys	household interviews? origin–destination from interviews on ground computing aids automatic traffic counters speed measuring by radar or twin tube video recording
Accident statistics	data collection network data input computer system design analysis remedial scheme design
Land surveys	data collection tools drawing or plotting site visits for schemes
Traffic order making	consultation regular meetings reaction to press and public
Traffic engineering	design
Traffic signal design	signal laboratory holding of stocks installation arrangements
Central control computer	remote monitoring signal database signal calculation facilities
Signal maintenance	fault control center signals inspectorate contractual control
Performance indicators	traffic delays accidents pedestrian satisfaction
Signal system	time controlled intersections off control detectors lamps
Organizational targets	schemes in a given time budget control elapsed time to design decision to completion time projects awaiting attention
Research component	

by providing them with cars and allocating them specific routes to work. Care is needed to ensure that an adequate spread of times and routes is consistent with the engineers' travel needs.

2.2 Central Control of Traffic Signals

As an on-line task, the central control of traffic signals makes special demands on an organization. Since a response to special situations may be needed whenever there is significant traffic, there is some advantage in having extended staffing hours, certainly covering the morning and evening traffic peaks. Surprisingly, this need is not as great as it might be. First, many modern control systems respond well to changes in traffic levels and there is little else required for human interaction. Second, there is often a need for the police to have on-line information and to be available at all times, so that no additional cover is required.

2.3 Performance Monitoring

Since traffic management is provided by a public service organization, there is no explicit commercial control over performance or value for money. It is useful for such controls to be introduced within the organization.

The monitoring of traffic management schemes is normal practice in responsible organizations. On a short-term basis, traffic flow and delays are measured before and after changes are made. Accidents can also be assessed once sufficient time has elapsed for the statistics to be meaningful.

The monitoring of the performance of the organization is less common, even though it is essential to its efficiency. Sets of targets should be developed. An example is the number of schemes of a particular type, such as bus priority traffic management measures, which are introduced each year. Another is the time it takes for a scheme to be introduced once it is duly approved. For centrally controlled traffic signals, there should be a target maximum number of signals not under central control at any time; otherwise, numerous partially valid excuses will allow excessive numbers of signals to perform poorly.

Table 3
Techniques in traffic management

Objectives	Techniques
Relating to land use	identifying land uses identifying appropriate traffic identifying road purposes
Control of accesses	side roads to main roads environmental protection minor road closures banned movements road pricing systems

Traffic efficiency	one-way systems
	banned turns
	barriers
	    central
	    kerbside
	    pedestrian control
	    vehicle control

Intersection types	no control give way four-way stop rotaries, various sizes traffic signals
Speed control	speed limits road humps and rumble strips character of road
Enforcement	self-enforcing designs moving vehicle offences—police stationary vehicle offences
Parking control methods	kerbside markings disks or cards, paid or free parking meters
Facilities for special road users	pedestrians buses cyclists heavy-goods vehicles

Road safety	identifying blackspot features
	applying successful techniques widely
	    antiskid surfacing
	    improved street lighting
	    changed signal clearance time

3. Traffic Management Techniques

The traffic office has to be skilled in a range of techniques that can be applied on the road. Some of these are listed in Table 3. Clearly, the techniques to be deployed relate to the functions of the office, but there is no one-to-one relationship. For example, roundabouts and traffic-signalled intersections may be alternatives selected as part of area-wide schemes to improve traffic flow. Alternatively, they may be selected by a specialist road safety team as a result of its accident analysis work. They may also be selected by a bus priority team having analyzed relative delays to buses and other traffic.

4. Procedures in a Scheme

In view of the range of people knowledgeable and interested in traffic management, it is important that the work of the traffic office should be generally acceptable. Carefully thought out procedures are essential. They should be negotiated with other interacting teams. They should then be followed accurately and the office should be expert in doing this as quickly as possible. Speeding up scheme implementation should be achieved by carrying out the procedures promptly rather than by leaving out significant steps.

Table 4 lists, in sequence, the steps that normally need to be taken to introduce a traffic management scheme. It is instructive to examine the consultation stages which are of different types.

4.1 Public Consultation

The agreement of the public is so important that steps should be taken to secure it. However, the public expects the traffic office to be competent and to recommend suitable schemes. It is, therefore, difficult to decide whether consultation should be on the basis of very broad principles so that members of the public can readily suggest their own ideas, on the basis of a well-designed scheme with few remaining choices or on some intermediate range of options. The broader-based consultation is appropriate at an early stage of design. For large schemes, it may be advisable to consult the public more than once.

The public is normally represented by councillors. Engineers should consult the councillors on the scheme more frequently. This is normally provided for if approvals are to be given for work to be carried out on scheme preparation or, eventually, on implementing an agreed scheme. The problem of the degree of detail to provide is usually overcome by presenting the options and, later, the fully designed schemes to the councillors on a number of separate occasions.

4.2 Professional Consultation

The initiative for producing a scheme is normally entrusted to a part of the office. For a geographical organization, the allocation is obvious. For a functional organization, more care is needed in allocating responsibility. Thus, changing the signal layout of an intersection may be the result of area traffic management or it may be part of a program run by the traffic signals team to introduce a better form of signalling, perhaps taking advantage of the much greater flexibility of modern microprocessor-based traffic signal controllers. Either way, it is clear that other teams have an interest. In particular, it is now considered good practice for the road safety team to audit all designs several times during their development. The procedure has to provide for these different professional teams to provide their inputs at prescribed times. For it to be provided promptly, there may be a need for advance warning to be given that a scheme has reached the appropriate consultative stage and will shortly require comment.

In order to control workloads of different teams smoothly, there is some merit in having alternative procedures. For example, if a traffic management scheme contains new signalized intersections, the signals team would ideally need to carry out full calculations to ensure that it was possible to signalize the intersection and carry the estimated traffic within the space envisaged. At times of high workload, it may be advisable to adopt an alternative procedure and promptly provide cost and space estimates on the basis of general experience (which should be properly codified) so that the elected representatives can decide whether further work should proceed. If such a shortened procedure is adopted, it is doubly important that this fact is recorded on the documents so that adequate time is allowed for the consequential extra work which will be required later in the process.

5. Efficient Operation

All too often, traffic offices grow haphazardly as different projects are directed to them. This is not efficient and it is important to set up the best organization that size permits, preferably on a functional basis. It is also important to preplan all the necessary procedures and

Table 4
Progress of a scheme

request to make changes
studies and surveys to assess the range of problems
broad concepts of solutions
identify consequences
consider costs and benefits
ranking of schemes in merit order
report to decision makers for consultation

consultation by
   local people
   elected representatives
   police
   private contractors
   advisory bodies—try to avoid
report for decision

further surveys
detailed design
on-site experiments
consultation with cooperating organizations
   internal specialist design groups
      traffic signals
      traffic signs
      road safety audit
design adjustments
report to decision makers for detailed approval

place orders on contractors and agents
traffic order making process
contract management
   rate of progress
   sequence of activities
   temporary traffic control
   local information and cooperation
inspection of finished works
rectification of design errors
traffic measurements for monitoring studies
commissioning
assessment
   experience in actual use
   formal performance monitoring
report to decision makers on performance

   modification design, approvals and implementation

interactions with other organizations and with the public.

See also: Transportation Management: Systems Engineering Approach

K. W. Huddart
[Esher, UK]

Transportation Management: Systems Engineering Approach

Management is the use and creation of resources (e.g., finance, personnel, tools, energy, information, authority) for action and control to achieve objectives such as profit, fulfillment of missions, realization of projects and solving of problems, and to keep conditions as desired.

Environmental pollution and the impact of transportation on society and national economy require a holistic approach. The demand by customers, for example, concerning logistics, and the progress of new technologies, especially in new information and communications systems, are compelling the coordination and integration of different modes of transport. A systems engineering approach is a necessity.

A systems engineering approach means that the methods, tools, knowledge and skill available have to be used for grasping and treating the causes of effects instead of just the symptoms. The investigation has to take into account functions, structures, different elements (e.g., human, devices) of the target system, and the influence of and to other external systems, conditions or environments, in general, over the time range of importance. The dynamic and self-organizing behavior of systems cannot be neglected. Often in the past, the impact of feedback was not recognized and this led to solutions that made the situation worse instead of better. Of course, only some aspects of interest can be grasped or covered by a model of the mental or any other type. A systems engineering approach, therefore, needs multistep decision making, depending on learning from feedback.

1. Transportation Systems

Transportation systems

(a) cover the world;

(b) have been used for generations;

(c) tie up huge amounts of investment; and

(d) fix other systems such as production systems, military systems and societies.

The network structure of the infrastructure for transportation covers vast spaces and makes it almost prohibitive, due to limitations concerning space (e.g., capacity, environmental conditions) or medium (i.e., channels of radio frequencies), to install additional networks of the same kind. Furthermore, the sources of demand for transport are fixed to given locations with little flexibility for change.

Technical and economic reasons, as well as historical development, limit the attachment of income to effort for services, which cannot be measured and stored like products.

Numerous and embracing functions, structures and subsystems with vast extensions cause complex control relations and the need for extensive communications.

Operation can cause danger to life, limb and property, not only to users of the system, but also to parties unconcerned with the system, as well as to the environment. From the very beginning, safety was one of the main objectives in transportation.

The risk of misinterpreting the system dynamics and behavior increases with the time span of the mission or the life of the system. In addition, the efficiency of control mechanisms may change because of human adaptability and learning ability, by psychological facts given, and because of feedback with long time delays (i.e., in society, culture) and the self-organizing effects of natural processes.

The limitations of human behavior and rationality (e.g., alcoholism, short range of political decisions) have to be considered.

All these reasons make the management of transportation systems different from that of production plants, trading companies, the military, social welfare and so on.

The phases that can be distinguished during the time range of managing a transportation system are given in Table 1.

2. Objectives

Management is only sensible with objectives, which may, for example, be from intuition or given by an order. Because of the time span and the complex distribution of decision makers in transportation (e.g., governments, military, commerce, municipalities, regional planning groups, road administration, forwarding companies, transportation enterprises, political parties, labor unions, insurance companies, driver clubs, private drivers), the evaluation of objectives needs special consideration.

Attention must be paid to the change of objectives. The purpose of consequential or secondary objectives is that the primary objectives can be reached or kept. They are attached to the subsystems for control, monitoring, and so on, and will be changed if the subsystem or the strategy alters. However, depending on time or on the state of fulfillment of the setting of all goals, primary goals may also change themselves or vary their importance. Social systems have, besides impartial or rational objectives, subjective ones (Huse and

555

Table 1
Management phases

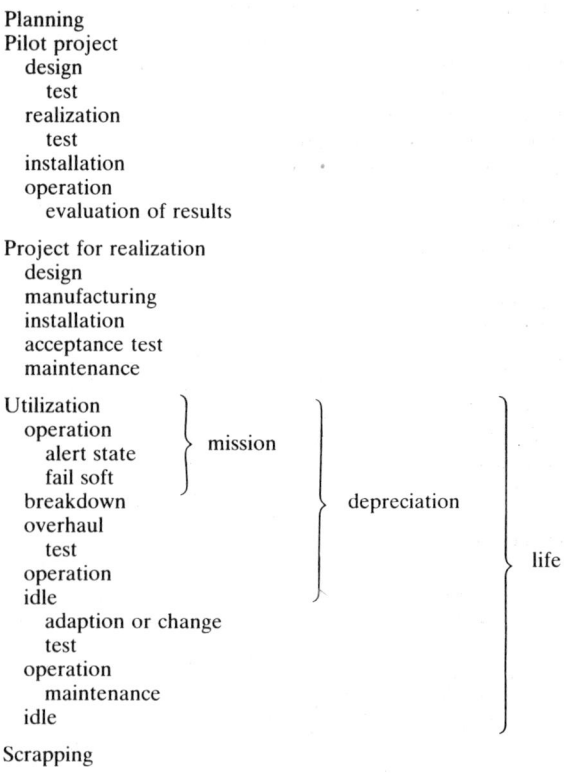

Planning
Pilot project
  design
    test
  realization
    test
  installation
  operation
    evaluation of results
Project for realization
  design
  manufacturing
  installation
  acceptance test
  maintenance
Utilization
  operation
    alert state
    fail soft
    breakdown
  overhaul
    test
  operation
  idle
    adaption or change
    test
  operation
    maintenance
  idle
Scrapping

Bowidtch 1973) attached to individuals or groups, which influence the behavior of the system. In social systems, the main problem consists of reducing the conflict situation between the different goal settings. Even if an individual intends to strive for a rational goal, this effort may be for only a short time if the surroundings and human nature are not compatible. The instinct of self-preservation has to be understood in a spiritual as well as a physical sense. Self-interest may be hidden by altruistic decisions, depending on long-term or short-term strategies of the individual or group, by limitations concerning strategies available or by ignorance of possibilities in a given situation. An asymmetry may exist between the motive for achieving a goal and the attempt to reach it. The dissatisfaction, influencing the forces to change a situation or objectives, is dependent on consciousness of deviation of goal settings and consciousness of interest for change.

The distinction of long-range (fundamental, strategic), medium-range (guiding, tactical) and short-range (operational) objectives brings a stabilizing effect to their application. Functional requirements should be attached to the long-range objectives, and structural conditions (infrastructure given, location in consideration) and time-dependent aspects (conditions for construction phases, war or catastrophe, normal operation) should be covered by the settings for the operational goals. Fundamental objectives are less prone to conflicts because of soft (fuzzy) information. The conflicts arise in the detail. However, at this level, the complexity will be reduced by limited range of time, space or functional variety. Moreover, such conflicts can be reduced in many cases by the objectives with consensus at the higher levels.

A structure of objectives (as used in the Austrian general conception of transportation), without the operational objectives, will be used as an example.

(a) *Fundamental objectives.* These are policy of state (e.g., aspects of constitution), society (e.g., aspects of human rights), and economy.

(b) *Guiding objectives from other fields.* These are regional planning, economy (e.g., industry, commerce, farming, tourism), international relations, national defense, environment and health, social and educational, safety and security, energy and resources, and public housekeeping.

(c) *Guiding objectives from the field of transportation.* These are general objectives (demand should be satisfied, holistic point of view, possible substitution among different modes of transportation should be considered concerning specific situations, compatibility, flexibility and adaptability of systems should be considered, etc.), special guiding objectives such as accomplishment of main task for transportation (e.g., rapidity, dependability, comfort), avoidance of damage, disadvantage, disturbance and injury, rational use of resources, high economy and efficiency, and optimal receipts.

In case of developing the measures or in conflict situations, more than one level of objectives has to be taken into consideration. There is movement from top to bottom and vice versa. The importance of single objectives may not only change with time, but may also be changed by this decision-making process, and new goals may be created in lower levels. More than one general main objective (as in the Swiss general conception of transportation: the greatest contribution possible of the transportation system to the quality of life) is given in the highest level (fundamental). The ranking in one level depends on the situation.

The following example shows the development of measures for the case of access to platforms in stations for public transport.

(a) *Fundamental objectives* (only excerpts from the relevant one). These are that living conditions should be equal for the population, safety for the public should be high, accessibility to the transportation system should be given and so on.

(b) *Guiding objectives from other fields.* These are that physical and mental effort should be considered, disabled people should be considered, high dependability or reliability should be gained, use of energy and resources should be rational and so on.

(c) *Guiding objectives from the field of transportation.* These are quality, disadvantages for user, energy and resources, and so on.

The operational objective was developed: accessibility for the disabled in normal and abnormal situations. This leads to the operational objective for the range in consideration: to prefer ramps instead of escalators. Thus, this leads to a measure in the range of the planning level for new stations for commuter transport: a new regulation for the construction of station buildings considering the operational objective.

It is a necessity for systems with self-organizing behavior to state objectives, which should support the achievement of the goal settings, and to have objectives, which are dealing with adaptability of the system and with early perceptibility of change of objectives. Objectives for achieving goals and smoothing adaption may be

(a) control of results,

(b) responsibility and guaranteeing consequences,

(c) possibility of comparisons,

(d) observation and access to development in the world,

(e) motivation,

(f) cooperative competition (optimization by evolutionary processes),

(g) freedom for correction and control actions (i.e., by limiting commitments),

(h) robust and practical solutions,

(i) early conflict solving, and

(j) early corrections of shortcomings.

3. Decision Making

In control theory, the criteria function is usually developed by weighting the conflicting goals (e.g., time optimality, smallest deviation from state wanted, and minimum of energy needed). The distinction, if all have the same importance or all are very important, as well as the information on individual preferences or opinion, are lost by using weights and an allover criteria function. The change of objectives, in case of a multistep decision process, creates problems in regard to recognizing which changes of weights are suitable or which effects would be entailed. The objectives of decision makers or actuators depend, for example, on states and information available. Moreover, it has to be recognized that the information available is given in a fuzzy way.

Fuzzy information cannot be improved in all cases by introducing probability functions; rather, a feedback loop has to be provided for handling possibilities. If any information is missing, then an investigation must be carried out to see if this information has any influence or can be neglected. Sensitivity analysis may be applied. However, if the influence of this missing information should not be neglected, then, for example, test or pilot projects have to be installed, or simulation has to be applied. It is essential to have the ability to learn. Ignorance should not be hidden by probabilities that are only assumed. Dynamic planning is required for dynamic processes. This means that changes must also be planned for and, importantly, a learning feedback has to be provided.

Decision making in transportation is a multistep process for these reasons. However, this is also due to the amount of data that has to be processed and the time limits given (especially during the operational phase). The decision-making process follows according to a hierarchical structure with long-range (strategic) planning at the top. Very simplified models with regard to the boundary conditions and actual parameters of influence are used at this level, and both forecasted and predicted data are relied on. At the short-range (operational) level, the variety of real states of the system has to be considered. However, the freedom in the solution domain will have already been reduced by the investigation of a solution grid during the planning stage (long and medium range). The output of an upper level is an input, boundary condition or optimization objective for a lower level. Of course, feedback exists, but with increasing distance between the levels, the time lag also increases. The state of other subsystems is regarded as quasistatic and is represented, for example, by boundary conditions. It is assumed that the dynamics or changes as well as the objectives of another subsystem are known *a priori*.

Decision making comprises

(a) information gathering, with the problems or what, where, when, and how to collect and use;

(b) investigation of alternatives for actions of events;

(c) evaluation of outcome considering objectives or criteria;

(d) selection of results from decision-making processes with the measures, which are essential for achieving the objectives given; and

(e) documentation of the decision-making process for learning and explanation.

A special problem consists in recognizing "weak signals," which can sometimes become very important.

Optimization has to be carried out by evolutionary processes in case of complex long-range problems. In general, the theoretical best value of a criteria function does not yield the best results in the long run. The best strategy will be to act between the extrema. This supports not becoming endangered because of saturation of a market or by slight changes of assumed conditions. If the timelag of the feedback needed for optimization is too long, then a feedforward strategy has to be used and competent groups, such as research institutes, authorities, industries, and user or consumer groups, who are struck by the problem, by planning or by measures, have to be involved in the decision-making process.

Approaches are under development (e.g., Fandel et al. 1986) that consider these aspects of decision making (e.g., multiobjectives, committee decisions).

4. Measures

It is a precondition that the time schedule for realization of the actions intended, the responsibilities, the structure of control and the covering of effort needed for a measure are fixed. Otherwise, it would be an objective concerning the action strived for. Measures can be categorized according to the increase of effort for realization:

(a) information, conviction, education and training;
(b) arrangements, agreements and treaties;
(c) funding;
(d) legislation and regulations;
(e) organizational solutions;
(f) structural planning and design; and
(g) technical solutions.

It should be recognized that a time lag of more than 100 years may occur between design and realization of the infrastructure in transportation (e.g., Suez Canal, Channel Tunnel).

The stability and efficiency of measures are mainly influenced by the organizational conditions given.

5. Organization

A suitable organization is a precondition for achieving objectives by ensuring an optimal information flow, effective information processing, the piling up of knowledge, control, motivation and correction, and therefore allowing optimal decision making and management. The qualified person at the proper place, able to find ingenious solutions, will only happen in efficient organizations. However, a correct organization does not mean a frozen one. Problems should not be administered.

Centralization has a positive influence on increasing the coordination of actions and on reducing trends towards autonomy. The central allocation of power has advantages when quick actions are important. Surveillance should be centralized (e.g., audit office). The disadvantages are

(a) vulnerability;
(b) problems concerning variations of spatial or functional particularities;
(c) if not only centralized but also activity elements (information center) are concentrated, then problems will arise with shutdown during mission time, or concerning adaptation or changes; and
(d) limits concerning transmission speed of information in case of operational control.

The gathering and acceptance of information will be decentralized in large systems, because the feedback distance for correction or inquiries is shorter and the transmission capacity to centralized processing can be smaller, for example, because of filtering or comprehending of information by preprocessing before transmission. This also reduces time racing problems. Decentralized systems allow the extension of a system step by step and the gain of profit at early stages. It is also possible to make changes step by step. The disadvantages are

(a) coordination or synchronization problems, and
(b) the increase of activities for separation or particularization in case of social systems.

The information flow between subsystems should be minimal in view of mutual influence on dynamics and according frequency and periodicity. Decomposition will be found, in general, in accordance with location and functions.

A monopoly is applied reasonably in international standardization and in interregional long-term use of limited resources (e.g., radio channels), if participation and consensus are considered for the decision-making process. If the control functions are active and are not hampered by limitations of social systems and psychological facts, the long-term results are theoretically the same, independent of whether enterprises are owned by state (controlled by government), by cooperative society or are private (controlled by the market). Usually, market forces are highly reliable at being effective. However, if the feedback of the market has a time lag that takes too long or if the elimination of ill-managed systems is not tolerable, then a feedforward strategy has to be installed, which allows cooperative decision making of participants with responsibility, controlled by consequences. However, even in such systems, the possibility of an evolution of optimal solutions should be improved by cooperative competition of subsystems. Cooperative competition means achieving common objectives by measures that can be selected by the competitors according to their own decisions. In addition, quality circles (e.g., Gudnason and Kofoed 1987) may improve systems efficiency and may even facilitate the recognition of weak signals.

Considering the safety requirements and the conditions for competitions in transportation (e.g., characteristics of infrastructure, limited capacity of space, empty car problem), deregulation is rational as long as

(a) qualification of persons acting in transportation is given and quality of service for customers is controlled,

(b) conditions for competitors in the same market are the same, and

(c) nondiscriminating objectives of higher levels (government policy) are given attention.

The control of qualification in transportation is a necessity as it is, for example, for public health, because of the possibility of endangering other systems and of severely violating higher objectives.

Functional-oriented units have evolved in transportation and this structure has also proved to be suitable for electronic data processing systems in transportation in the long run. Pyramid-shaped hierarchical systems (all information is concentrated in comprehensive form in a single organizational element (i.e., the director) for making decisions in its own responsibility) are reasonable only for subsystems. At present, the problem with managing transportation is that transportation systems are organized mainly according to modes of transport and these do not take into account the allover functions such as freight transport or the demand of passengers. This is due to historical development and to the state of technologies in the past (see *IFAC Working Group on Transportation Systems*).

The trend will be to have organizations similar to those in air transport. This development will also be supported by the further progress of technologies. Mobile units and services will be operated by companies in competition (also in railroads) in the case of commercial transport. The infrastructure will be controlled by regional governments or states. The traffic will be controlled by neutral organizations similar to air traffic control or urban traffic control. International organizations are responsible for transnational planning and standardization. Organizations with matrix structures are used in the case of planning and developing complex projects. This means, during a given phase of a problem, experts are taking part in groups that are established according to different problem areas and that may also have areas with joint interest. An example of this approach on a large scale is the trade electronic data interchange system (TEDIS) of the European Community (see *DRIVE*), whose purpose is to develop electronic data interchange for administration, commerce and transport.

Safety and dependability are main aspects in transportation that are influenced to a great extent by the organizational conditions (see *Safety of Road Traffic*). This is not only in respect to operational and maintenance phases, but also concerns planning, certification and the feedback of operational data for learning. In most countries, the main interest in the safety of sea and road transport is shown by insurance companies. Authorities of only a few countries are covering the full control circle for safety and are comparing other systems for evaluating improvements. Of course, road transportation is difficult to tackle due to the large number of private cars in comparison to commercial vehicles. According to recent trends, more countries have separate ministers for particular transport modes (Beckenham 1988). However, this development must be compared against the demand for smoothing coordinated planning of transportation systems for improving performance and efficiency.

At present, on an international scale, the organizations in the field of aviation are very effective at achieving high safety and efficiency, and are able to apply new technologies smoothly and rationally for the advantage of customers.

6. Control

Control is the attempt to use actions to obtain the desired state of a system. The control input influencing the system states is framed according to some criteria function. The law of requisite variety states that the capacity of a device as a regulator cannot exceed its capacity as a channel of communication. In control theory, optimality is usually considered by the variance around the mean value of the fluctuating variable, which is an interval (metric) measure. However, a good regulator in complex organizations should also minimize the variability of the output.

An error-controlled regulator cannot keep the output constant. A cause-controlled regulator is able to perfectly regulate the output, because the regulator reacts directly to the disturbances that affect the system. The success of the cause-controlled regulator depends on the adequacy and validity of the system model in the regulator. This can be gained by also installing a feedback loop, eventually with learning ability.

On-line error-controlled regulators smooth bad states, which prevents the investigation of real weak aspects in the system. Therefore, such control is found in subsystems, but not in systems with self-organizing behavior and long-term cause–effect relations. In large organizations, decisions for control actions are based on feedforward information. Feedback information is used for spot checks or random tests of subsystems (see *Traffic Control Modes*).

In systems with biological elements, it is necessary to examine not only immediate actions, but also those depending on the past, such as actions belonging to the memory or historical events. The changes that the biological system has undergone must be considered.

In general systems, irreversible processes can occur. Attention has to be given to keeping freedom (e.g., availability of resources, reduced stress of time, limitation of commitment) for actions.

Control needs information on the state of a system and on achievement of the objectives. The achievement of aims has to be made observable by goal settings and has to be made measurable by attaching values or classes to the objectives according to the grade of deviation from the goals.

In many cases, the effort in transportation is paid by fiscal money transfer or lump sums and the income is not attached to the performance. In the past, technical solutions were not available for a reasonably direct billing of the expenses, especially in the case of the road system, which is used by different classes of transport performance (e.g., heavy lorry, sports car, pedestrian or transit traffic, distribution traffic). It seems that the expenses could be kept low for administration. Moreover, the models are still under development that are needed for the correct attachment of expenses to the user in a road system that has been developed over centuries. However, even a coarse attachment of financial input to output would allow the improvement of efficiency because of the psychological facts and because of the possibility of pinpointing the weak parts of the system for elimination, instead of making the expenses unbearable for the public budget or bringing down the whole system.

Research is being carried out to develop identification methods for recognizing the state of complex economical systems and for finding metrics suitable for measuring the efficiency of transportation. Productivity values and even classical financial balance sheets are not reasonable for judging transportation systems.

In addition, methods for analyzing the stability of such self-organizing and teleogenic (generating their own goals) systems are demanded. The ability of a system to compensate for all defined variations of disturbances or environmental situations, and to keep or reach a state that is aimed for, with an equilibrium over a given time interval, is called stability.

See also: Evaluation of Traffic Control Systems; High-Speed Railroad: Systems Approach; Traffic Office Management

Bibliography

Beckenham A F 1988 Transport and nation-state governments: A global review. *Transp. Rev.* **8**, 267–79
Fandel G, Grauer M, Kurzhanski A, Wierzbicki A P (eds.) 1986 *Large-Scale Modelling and Interactive Decision Analysis.* Springer, Berlin
Gudnason C H, Kofoed C A 1987 *Quality and Quality Management.* Elsevier, Amsterdam
Huse E F, Bowidtch J L 1973 *Behavior in Organizations.* Addison-Wesley, Reading, MA

R. Genser
[Austrian Federal Railways,
Vienna, Austria]

Transportation Modelling

Urban and regional transportation systems are among the most important resources of any metropolitan area. The quality and spatial distribution of transit services help to shape future regional growth and settlement patterns, affecting both the economic health of the region and the quality of life within the region in the years to come. For these reasons, transportation planning is a fundamental public concern.

Effective transportation planning recognizes relationships between land use and transportation. Current land-use patterns largely determine existing traffic patterns and the immediate need for improvements in the transportation network. The existing transportation network, together with planned changes and augmentation of that network, combine to determine future land-use patterns. Transportation planning seeks to provide an evolving network that best serves current transportation needs, while promoting employment growth and desirable patterns of land-use development.

To achieve these ends, transportation planning requires the means:

(a) to forecast future land-use patterns based on the proposed evolution of the transportation network; and

(b) to forecast future traffic patterns based on the evolution of land uses.

The first need is addressed by land-use forecasting and spatial-allocation models. The second need is met by an integrated set of models, collectively called the urban transportation planning modelling system (UTPS).

This article describes each of the four major components of the UTPS. These are:

(a) a trip-generation model, which forecasts the number of daily trip productions and trip atttractions for each zone of the study area, according to trip purpose;

(b) a trip-distribution model, which forecasts the travel flow between each pair of zones;

(c) a modal-choice model, which forecasts the apportionment of trips among transportation modes for each pair of zones; and

(d) an assignment model, which forecasts the routes traversed between each pair of zones, according to transportation mode.

1. Traffic Zones

In order to formulate land-use and transportation models, it is first necessary to divide the area to be modelled into a number of smaller spatial units. These units are called traffic or analysis zones. Boundaries are usually chosen to coincide with physical features such as streets, railroad tracks, rivers, escarpments or a shoreline. A common practice is to define zones that are subdivisions of census enumeration tracts. While census tracts in most cases are too large in area to permit meaningful analysis, this practice allows the use of census data in checking zonal aggregates.

The number and size of zones represents a compromise between precision and tractability. Ideally, each zone should be sufficiently small in area to embrace a single type of land use. The availability of data and the costs of analysis, however, often demand zones that are larger than this ideal. The number of analysis zones, typically, ranges between 10 and 1000, with the size of zones ranging from a few city blocks to several square kilometers.

With sufficiently small zones, the location of each zone can be defined by a single point in space, called the zone centroid. The centroid may be the geographic center of the zone, or may be skewed to reflect some known distribution of activity within the zone. All distances, travel times or other accessibility indexes are measured between zone centroids.

2. Trip Generation

A trip-generation model links land use to travel demand. The purpose of the model is to determine T_i, the number of trip ends per day (trips that start or finish) in each of the n traffic zones, $i = 1, 2, \ldots, n$. As used in the UTPS, trip ends may be defined in one of two ways, depending on the assumptions used:

$$T_i = O_i + D_i, \quad i = 1, 2, \ldots, n$$

where O_i is the number of trip origins and D_i is the number of trip destinations in zone i, or

$$T_i = P_i + A_i, \quad i = 1, 2, \ldots, n$$

where P_i is the number of trip productions and A_i is the number of trip attractions in zone i. The principal difference between these two definitions is that trip origins are estimated based on travel that begins at private residences, whereas trip productions are defined as the "home end" of trips, regardless of where travel actually begins. Productions and attractions usually are estimated using trip-generation methods and subsequently converted into origins and destinations. This is the case illustrated here.

To estimate the number of trip ends in each zone, trip productions and trip attractions, typically, are disaggregated according to the purpose of the trip. Thus

$$P_i = \sum_{k=1}^{m} P_i^k, \quad i = 1, 2, \ldots, n$$

and

$$A_i = \sum_{k=1}^{m} A_i^k, \quad i = 1, 2, \ldots, n$$

where P_i and A_i are the total number of trip productions and attractions per day in zone i, respectively; P_i^k and A_i^k are the number of trip productions and trip attractions per day in zone i for purpose k, respectively; m is the number of trip-purpose categories; and n is the number of analysis zones.

The number of productions and attractions for each zone for each trip purpose is a function of land use within the zone, including the demographic and employment characteristics of the zone. For example, the amount of home-based travel generated is estimated as a function of the number of households living in the zone, together with the characteristics of these households, such as income and car ownership. Similarly, at the non-home end, a trip rate or attraction rate is determined, which reflects the number of trip attractions per employee in retail, commercial and industrial activities, the floor space devoted to these activities, and so on.

Trip productions usually are estimated using regression models or cross-classification techniques, sometimes called category analysis. Trip attractions, typically, are estimated using trip attraction rates per different land-use categories. For example, the number of trip productions for zone i for purpose k might be represented as

$$P_i^k = f_i^k(e_i^1, e_i^2, \ldots, e_i^q)$$

where e_i^l is the value of the lth predictor variable for the zone. The function f_i^k is arbitrary and is chosen to provide good fit with traffic survey data. The total number of trip productions at the home end is used as a control on the total number of trip attractions, since productions and attractions must balance for the region as a whole.

A calibrated trip-generation model can be used to forecast the pattern of future trip ends, based on the future values of the predictor variables for each zone. Forecasts for the predictor variables are derived from land-use forecasting models. Two major assumptions of the trip-generation model are apparent: the relationships between trip ends and predictor variables determined from current survey data are relatively stable over time, and the effects of future trip-end patterns on future land-use patterns can be captured by the land-use model *a priori*. Both of these assumptions improve with decreasing forecasting horizons.

3. Trip Distribution

Trip distribution is the process whereby trip productions or origins are assigned specific destinations. The results of the trip distribution model are stated in term of the origin–destination flows T_{ij}. Since each trip having an origin must also have a destination,

$$T_i = \sum_{j=1}^{n} (T_{ij} + T_{ji}), \quad i = 1, 2, \ldots, n$$

where T_i is the number of trip ends in zone i for all purposes k, T_{ij} is the number of trips originating in zone i and ending in zone j for all purposes, and T_{ji} is the number of trips originating in zone j and ending in zone i for all purposes k. If trips are assumed to be symmetric during a 24 h period, as is often the case, then $T_{ij} = T_{ji}$ and the trips originating and terminating in any zone are equal in number. Intrazonal trips are indicated by T_{ii} and are not excluded from the distribution.

The most widely applied trip-distribution model is the gravity model, which is similar in form and concept to the gravity model used in many land-use forecasting models. In terms of trip productions and trip attractions, the defining equation of the gravity model is

$$T_{ij} = \frac{P_i A_j F_{ij}}{\sum_{l=1}^{n} A_j F_{ij}}$$

In this equation, F_{ij} is a friction factor or spatial impedance which expresses the effect on traffic patterns of the spatial separation between zone i and zone j. The sum in the denominator ensures that $T_i = P_i + A_i$ for all i and $j = 1, \ldots n$. Only if it is assumed that the friction factor is symmetric, however, will the number of trips terminating in zone i equal the number originating in zone i; that is, only if $F_{ij} = F_{ji}$ will it be true (in general) that $T_{ij} = T_{ji}$. Symmetry of the F_{ij} can be achieved, if necessary, by assuming that the friction factors are based on the spatial impedance of the round trip between the corresponding zones.

A gravity model is calibrated by determining an appropriate set of friction factors F_{ij}. Although spatial impedance depends on a range of considerations, including the time, cost and general quality of a trip, F_{ij} is usually considered to be a single-valued function of the travel time between zone i and zone j. The basic relationship between travel time and spatial impedance is determined by regression analysis on a table of current travel times between each origin and destination pair, cross-plotted against a table of actual trip frequencies (called a trip table) between each corresponding pair. Both tables are obtained from traffic survey data. The friction factors determined from the regression may then be adjusted or corrected to account for other factors influencing spatial impedance, such as expensive tolls or more subtle psychological barriers to travel. Of course, friction factors based on unidirectional travel times will yield symmetric friction factors only if the travel times are themselves symmetric. In this case, iterative adjustments of gravity model results will be required in order to make the number of trips terminating in a given zone equal to the number of trips originating in that zone.

Two other models have also been used to forecast trip distributions. These are the growth-factor model and the intervening-opportunities model. The growth-factor model distributes trips among zones based on the so-called growth factors for each zone, where the growth factor is defined as the ratio of forecast future trips originating in a given zone (from the trip-generation model) to actual current trips originating in that zone (from the traffic survey data). The average-growth-factor formulation assumes that the growth in trips T_{ij} is proportional to the average of the growth factors for zones i and j, whereas the Fratar formulation assumes that such growth is proportional to the product of the respective growth factors. In either version, however, growth-factor models are not capable of adjusting forecast trip distributions in response to changes in the transportation network. These include planned changes in the network (such as the construction of new roads or bridges) or natural changes (such as the increased spatial impedance between zones resulting from the growth in the number of trips between zones). Because of this limitation, growth-factor models are rarely used in current practice.

The intervening-opportunities model is similar in form and concept to the model of the same name used in land-use forecasting. The assumption embodied in this model is that the probability that a trip originating in zone i will terminate in zone j is proportional to the number of trip destinations or trip attractions "intervening" between the two zones. The intervening-opportunities model was used for trip distribution in the Chicago area transportation study, but has found little practical application beyond this instance. The principal disadvantages of the intervening-opportunities approach involve difficulties in model calibration, as well as practical problems in accounting for the mutual ordering of analysis zones.

4. Modal Choice

A modal-choice model forecasts the apportionment of trips among alternative transportation modes, such as walking, private automobile, taxi, subway or train. The total number of trips between zone i and zone j may be viewed as a market or demand for transportation. If it is possible to make the trip from i to j by more than one mode of transportation, the various alternative modes compete for a share of the market. Determining modal choice, therefore, is fundamentally a matter of determining human preference and choice when faced with competing alternatives. Many modal-choice models are related to models of human decision making from economics and decision analysis.

An individual's preference for a given transportation mode for a given trip is thought to be influenced by two types of factors. First, there are those factors such as transit time, cost, comfort, convenience and status value which characterize the utility of the mode itself. Second, there are those factors such as age, income, social status and automobile ownership which characterize the individual making the choice. Modal-choice models seek to determine a subset of these mode attributes and personal attributes on which reliable forecasts of the traveller's choice of mode can be based.

In practical applications, the utility of each mode U_i ($i = 1, 2, \ldots, m$) is assumed to be a function of travel times and costs, differentiated by socioeconomic class. A logit model is then used to describe the choice curves for different socioeconomic groups, where the probability of choosing alternative (mode) i is given by

$$P_i = \frac{\exp(U_i)}{\sum_{j=1}^{m} \exp(U_j)}$$

Maximum-likelihood estimation can be used to calibrate the coefficients of the linear utility functions U_i. For example, for an alternative set consisting of three modes—autopassenger (1), autodriver (2) and public transit passenger (3)—the following utility functions were calibrated for each mode:

$U_1 = 0.0$

$U_2 = -1.4809 + 1.9500 HI$

$U_3 = 1.1636 + 0.0916 DX + 0.0563 DL + 0.0106 DC$

where HI is a transformed household income variable, DX is the difference in excess time of public transit over private automobile, DL is the difference in line time and DC is the difference in travel cost.

5. Assignment

An assignment model serves to forecast the specific route or routes taken by travellers between each pair of origin and destination zones. In this sense, trips T_{ij} are "assigned" to one or more links of the transportation network that connects zone i with zone j. Assignments, typically, are made on the basis of minimum travel time or travel cost. Trips using each alternative transportation mode are assigned separately, using the appropriate transportation network (see *Traffic Assignment*; *Traffic Assignment, Stochastic*).

Traffic assignment requires a representation of the transportation network, which usually takes the form of a directed graph or digraph. The network digraph comprises a set of links or branches, connecting a set of nodes. Nodes represent trip sources (origins), sinks (destinations) and intermediaries (intersections or transfer points). Traffic is assumed to enter the network only at the specified source nodes and to leave the network only at the specified sink nodes.

Links represent segments of the transportation network that are traversed in going from one node to the next. Each link is directed such that traffic flows along the link in one direction only. In addition, each link is assigned a generalized cost, most often taken to be the transit time between the connected nodes in the assigned link direction. This cost may be a constant, assuming that the corresponding segment of the transportation network operates below its capacity to handle traffic. Such a network is said to be fixed cost. Alternatively, where congestion is a significant factor in route selection, the link cost may be a function of the total number of trips assigned to the link. Such a network is said to be capacity constrained.

Traffic assignment amounts to determining the route or path through the network graph for each trip in the trip table such that the total transportation cost for all trips is minimized. For an "all-or-nothing" assignment in a fixed-cost network, all of the traffic from one zone to another is assigned to the single "shortest" or minimum-cost route between the origin and destination nodes. The cost for any route is simply the sum of the costs for all the links along that route. The total cost between each origin and destination pair is then the product of cost of the shortest route connecting the pair and number of trips assigned. Total network cost is the sum of the costs for each pair.

Capacity-constrained assignment is more difficult, principally because the assignments for each origin and destination pair cannot be determined independently. For this reason, capacity-constrained traffic assignment leads to some interesting network optimization problems, which frequently can be framed as mathematical programming problems. A number of computer packages provide capacity-constrained optimization algorithms for updating link travel times.

See also: Simulation Programs, Macroscopic; Traffic Assignment; Traffic Assignment, Stochastic

Bibliography

Dickey J W 1983 *Metropolitan Transportation Planning*, 2nd edn. McGraw-Hill, New York

Federal Highway Administration 1977 *Plan Pac/Back Pac General Information Manual*. US GPO, Washington, DC

Kanapini A 1983 *Transportation Demand Analyses*. McGraw Hall Bood, New York

Manheim M L 1979 *Fundamentals of Transportation Systems Analysis*, Vol. 1. MIT Press, Cambridge, MA

Sheffi Y 1985 *Urban Transportation Networks: Equilibrium Analysis with Mathematical Programming Methods*. Prentice-Hall, Englewood Cliffs, NJ

Stopher P R, Meyburg A H 1975 *Urban Transportation Modeling and Planning*. Lexington Books, Lexington, MA

Transportation Research Board 1982 New approaches to understanding travel behavior. NCHRP Report No. 250. TRB, Washington, DC

Transportation Research Board 1982 Applications of disaggregate travel demand models. NCHRP Report No. 253. TRB, Washington, DC

K. P. White Jr. and M. J. Demetsky
[University of Virginia, Charlottesville, Virginia, USA]

Transportation Planning: Activity-Based Approach

The role of travel is as the means of enabling people to engage in spatially and temporally distinct activities. Activities and travel are, therefore, interlinked components in the overall pattern of daily behavior. It follows that the characteristics of a single trip such as its timing, destination, mode and route can only be fully understood in relation to the activities in which the traveller participates during the day. The "activity approach" (AA) acknowledges this insight as the starting point for its analysis. The main questions addressed by the approach are the description and modelling of the daily and weekly activity and travel patterns as a function of the socioeconomic situation of the travellers and of the characteristics of the transport system.

1. Characteristics of the Approach

1.1 Definition

The AA to transport modelling originated during the 1970s as a response to the widely perceived shortcomings of the prevailing four-stage, trip-based transport modelling framework. The AA challenged the fundamental assumptions of this framework by focusing on the role of travel as a means of enabling people to engage in spatially and temporally distinct activities. It argued that observed patterns of trip making should be seen as a consequence of an individual's desire to participate in activities and that the analysis of travel behavior should, therefore, be based on an understanding of the linked sequence of activities in which people engage during a day. This formulation was heavily influenced by the work of geographers such as Hägerstrand (1970) and Chapin (1974) and by work carried out on the use of time budgets during the 1960s (Szalai 1972).

Thus, within the AA, the basic focus is on the activity pattern, and the associated, contingent travel behavior to which it gives rise. The activity patterns of individuals are considered to be formed on the basis of their role within the household, as a result of the joint allocation and scheduling of tasks (e.g., working, shopping, child care) and resources (e.g., cars) and taking account of the short run time–space constraints of the current situation. Although time–space constraints have an important effect in the short term, over the longer term the AA acknowledges that they are open to conscious change by the household (e.g., by means of residential relocation, vehicle acquisition, etc.). Activity patterns are also influenced by the activity space of the household which is formed through the personal experiences of its members and of their information gathering exercises.

1.2 Frameworks of Analysis: Time–Space Geography

The time–space path of an individual provides the natural representation of travel behavior for the AA. The three-dimensional view (see Fig. 1) and its two-dimensional simplification were introduced by Hägerstrand (1970) and the Lund School of Time–Space Geography. Figure 2 shows examples of two-dimensional space–time paths and of the three constraints that shape them. Hägerstrand classified these constraints into three categories. Capability constraints reflect those constraints that are imposed on individuals because of the limits of their physical abilities or because of the limits of the available technology. The first group of limits is typified by the need for regular sleep and food, that create the basic rhythms of daily life. The second may be related to, for example, the maximum speed or maximum carrying capacity provided by a vehicle. The capability constraints define a time–space prism of locations and times which the traveller can reach. Coupling constraints are created by both the rules that govern the time–space accessibility of the environment and the agreements into which a person enters for the purpose of meeting with other persons. For example, a bank or post office is only open for certain times; a bus or train arrives only at certain times at certain places; and a family meets at certain times and places for a joint lunch or shopping journey. Coupling constraints bundle the time–space paths of many people in one location. Authority constraints describe the temporal and spatial authority individuals have over a specific domain, such as their own homes, their places in a cinema or their places in a queue.

1.3 Frameworks of Analysis: Activity Choice

The time–space paths do not identify the reason why individuals choose to participate in a given activity at given time and location. The economic and econometric modelling of time used in the tradition of Becker (1965) and his colleagues at the American National Bureau of Economic Research provide a utility-maximization framework for this analysis. Winston (1987) extends this activity choice framework by considering the problems of goal- and process-related utility and by including activity-duration-related effects in the formulation.

1.4 Dynamic Frameworks of Analysis

Time–space geography identifies the constraints on activity behavior; the activity choice framework identifies the choices open to the traveller; however, neither enables the interactions between the activities over time to be identified. The interactions between the plan of a traveller, the execution of this plan and the need to modify the plan in the light of unexpected changes in circumstances are captured in a dynamic framework of daily behavior (e.g., Hayes-Roth and Hayes-Roth 1979, Root and Recker 1983). The dynamic framework has been expanded to cover longer-term decision making (see Fig. 3), which feeds back to daily behavior through changes in the experience (mental maps) of travellers and through changes in the longer-term constraints (e.g., home location, work location, car ownership).

2. Describing Activity Patterns and Traveller Groups

2.1 Activity Patterns

Activity patterns are generally described in terms of a number of dimensions including: purpose, location, mode used to reach the location, sequence of activity, number of other persons participating, timing, duration and importance to name only the most relevant ones. The most frequently used descriptors are participation rates (activities/period), activity time budgets (duration/period) and the sequence of activities (e.g., home to work to shop to home). For an overview of the empirical studies see Hanson (1979), Golob and Golob (1982), Damm (1983), Kitamura (1988) and Jones *et al.* (1990). Techniques to describe the distribution of activity patterns in a population were developed by Küchler (1985) and Mazurkiewicz (1985). Advanced pattern recognition techniques have been applied to the analysis of daily and weekly activity patterns by a number of researchers including Recker *et al.* (1983), Hanson and Huff (1986) and Pas (1988). The measurement of the variability of activity behavior and its implications for understanding travel behavior have been discussed by Herz (1983), Hanson and Huff (1988) and Jones and Clarke (1988).

2.2 Traveller Groups

The work of the Lund School identified the importance of constraints on the nature of activity participation. This finding has encouraged researchers to attempt to classify persons and households on the basis of similarities in their activity patterns and constraints. Two main approaches have been used. The first takes an *a priori* classification of households or persons (e.g., based on socioeconomic criteria) and examines the similarities in the activity patterns carried out by members of each class. The second approach classifies units according to the similarities in their activity patterns and then examines how these similarities relate to underlying socioeconomic, demographic and other factors.

An example of the first approach is the study of Jones *et al.* (1983). This study showed the fundamental importance of the life-cycle status of a household on its travel behavior. Eight life-cycle stages were identified dependent on the presence or absence of children in the age categories: preschool children, young school children or older school children. Jones *et al.* derived typical patterns of time use from theoretical considerations, which they were able to verify empirically. Figure 4 compares the prototypical patterns for group B (families with preschool children) with the measured patterns for households in Banbury belonging to group B.

In Germany, a similar set of studies was performed by Kutter (1972), Holzapfel (1980) and Schmiedel (1984), which used the socioeconomic characteristics of individuals, and not the life-cycle status of the household as an *a priori* classification. Kutter and Schmiedel showed that a small number of classes is sufficient to

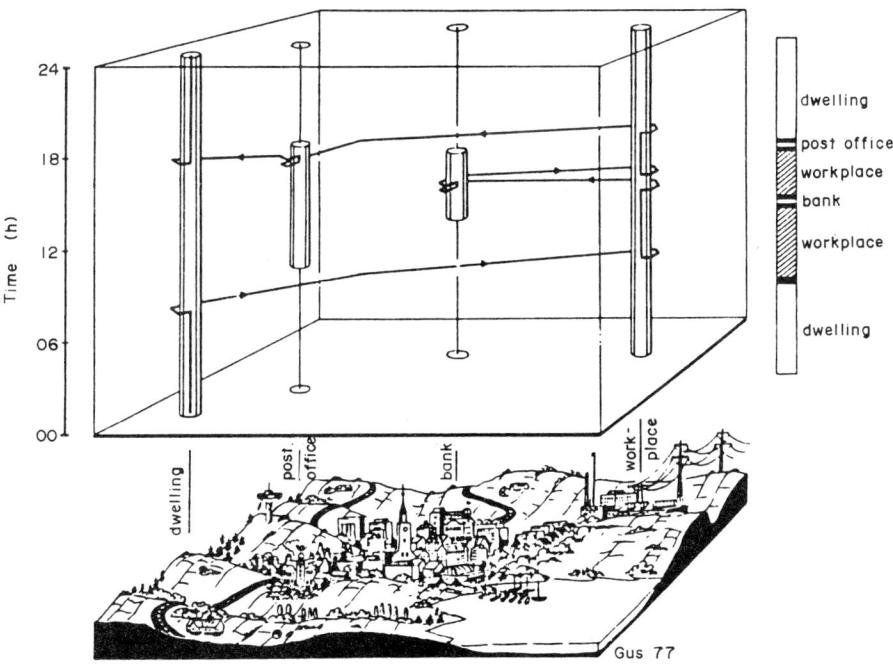

Figure 1
Three-dimensional view of a time–space path (courtesy of B. Lenntorp)

group the population into broadly similar, or homogeneous groups. The term homogeneous group does not imply that the members are homogeneous individually, but that the groups are, on average, homogeneous relative to each other.

Two American studies are examples of the second strand of research in this field. Root and Recker (1983) and Pas (1988) employ their pattern recognition techniques to classify the activity patterns of their samples. These reduced descriptions are then analyzed with clustering techniques to establish behaviorally similar groups of respondents.

3. Household and Person Scheduling

The AA has developed numerous approaches to predict where, when and why persons and households participate in activities. The simplest models were based on Markov chains and have been developed to incorporate the history dependency obvious in daily behavior (e.g., Marble 1964, Burnett 1978, Lerman 1979, Kitamura and Lam 1981). Choice models of activity participation during the whole day, based on simplified descriptions of location and time, were estimated (e.g., van der Hoorn 1979, Damm and Lerman 1981, Kitamura and Kermanshah 1984, Hirsh *et al.* 1986). Mentz (1984) estimated a series of sequential models of activity participation incorporating the generalized costs of the whole household. The allocation of resources and tasks within the household has been analyzed (e.g., Pas 1987, Koppelman and Townsend 1988).

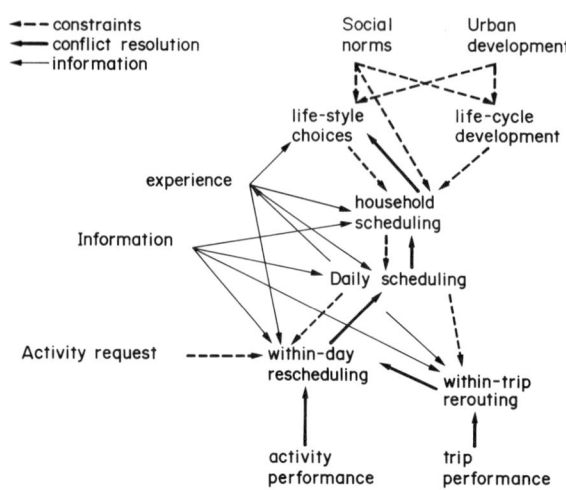

Figure 3
Dynamic framework of activity behavior (after EUROTOPP 1989)

A number of approaches to the question of scheduling have been proposed. Lenntorp's PESASP model (1978) generates all possible time–space paths for a given activity program and given time–space environment. The STARCHILD model developed by Recker *et al.* (1986) expands this approach by first generating all possible activity patterns, then classifying them and then using a choice modelling approach to select the optimal pattern. Both PESASP and STARCHILD are concerned with the synthesis of activity patterns. In contrast, Clarke's model CARLA (Jones *et al.* 1983) calculates the possible modifications to existing activity patterns given certain changes in the time–space environment, such as a change in school hours or the introduction of flexible working hours.

Gärling *et al.* (1989) propose a dynamic model of household scheduling, where the household members continuously update their plans in reaction to the mental or real execution of their plans.

4. Using the Approach

4.1 Improving Data Collection

The need for more accurate information about the activity participation of persons and households has led to numerous improvements in survey instruments, sampling techniques and in understanding the response biases involved in such surveys (Ampt *et al.* 1985).

Survey instruments developed to collect activity data include the household activity travel simulator (HATS) (Jones *et al.* 1983) and the situational approach (Brög and Erl 1983). The HATS interview process starts with recording the activity pattern of a person on a map and on a diary. This map and diary are the basis for observing how the interviewee's schedule is changed in

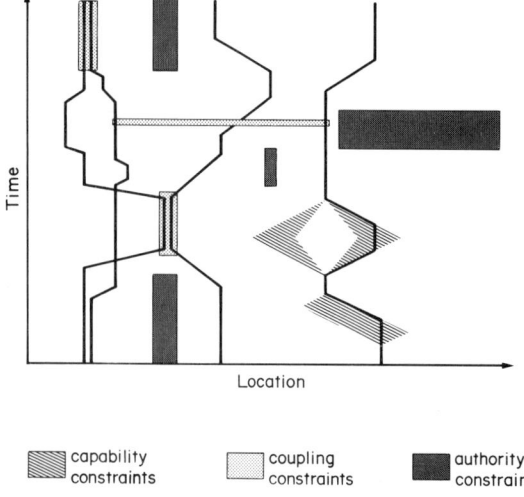

Figure 2
Two-dimensional view of the coupling, capability and authority constraints

reaction to a stimulus. The interviewee thinks aloud and the other household members are invited to comment and discuss. The HATS process has been transferred onto the computer (Jones *et al.* 1989) for increased accuracy of recording and for better customization of the stimuli. The situational approach is an attempt to identify the constraints in the decision making process of an individual. The constraints can be external or internal to the interviewee (physical, emotional or cognitive).

4.2 Simulating Activity Chains

German activity chain simulation models (Kutter 1984, Axhausen and Herz 1989, Zumkeller 1989) have used microsimulation techniques to develop operational models based on the AA. A particular feature of these models is that they attempt to use the implicit constraints within an activity chain in order to reduce the number of degrees of freedom within the modelling

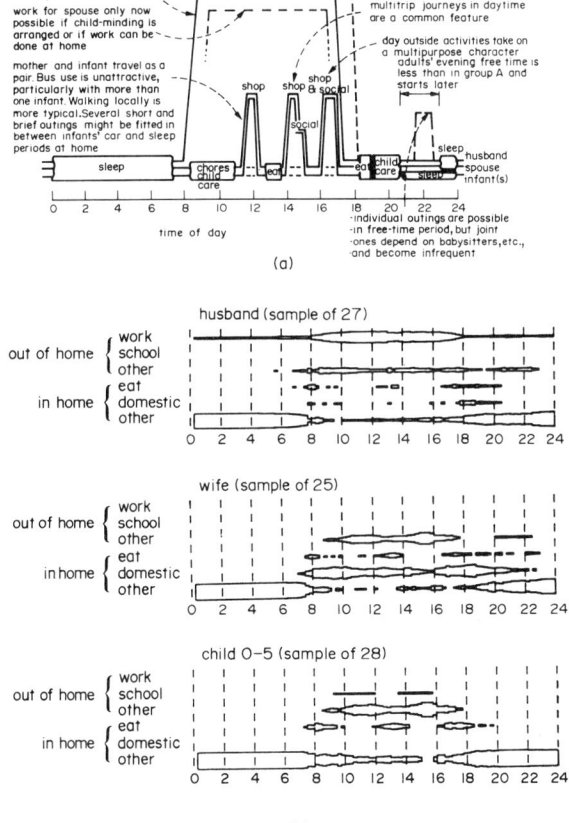

Figure 4
Comparing prototypical and empirical activity patterns: (a) group B, prototypical pattern; and (b) group B, measured activity pattern (after Jones *et al.* 1983)

process. For example, if a car has to be used throughout a journey, no further mode choice is necessary. Models of this type have been successfully applied to urban and interurban transportation planning problems.

5. Critique and Future Developments

The AA has been successful in four main areas. First, it has led to an increased understanding of human travel behavior and led to a greater appreciation of the true complexity of travel phenomena. Second, it has helped substantially to improve the methods and technologies for the collection of behavioral travel data. Third, it has called into question many of the underlying assumptions of conventional transport modelling that had, hitherto, been uncritically accepted. Finally, it has forced the pace of development of conventional transport modelling techniques. This has been most apparent in the addition of a large number of activity-related variables into choice modelling, in the development of models of departure-time choice for the morning commuter in the assignment process, in the introduction of dynamic frameworks and in the inclusion of journey-based models.

However, for all its success there are a number of important problems that the AA has still to address (EUROTOPP 1989).

(a) *Measurement issues.* There is still debate as to which activity and travel quantities should be measured, what techniques and protocols should be used and the relationship between measurement practice and the detection of variation and change in behavior.

(b) *Variability of behavior.* The increasing use and availability of multiday or multiperiod data requires better methods to distinguish between change in behavior and variability in behavior. Such methods are a precondition to modelling change, to establish sampling procedures and data-collection exercises.

(c) *Complexity.* A number of authors have recently warned that the AA risks being overwhelmed by the complexity of the behaviors it addresses (Mahmassani 1988, Pas 1990). These observations point to the tendency of much work in the AA to proceed inductively and not deductively from a stated hypothesis or theory. The inductive approach tends to describe and discover the complexities of human behavior, without providing general and transferable theoretical results.

(d) *Predictive ability.* The AA has not succeeded in producing a widely accepted replacement for conventional transport modelling, the German activity-chain modelling not withstanding. This failure has led to a situation where the credibility of traditional approaches is undermined without providing a credible replacement, leaving the nonacademic users of transportation planning models in a very uneasy position.

(e) *Demand–supply interactions.* The literature of the AA has mostly neglected the interactions between activity demand and supply, both in the short term on the transport networks and in the long term in the development of the urban structure and the preferences of the travellers.

(f) *Aggregation issues.* The focus of the AA on the individual or the household made it unnecessary to look at the question of how to derive aggregate results. Sample enumeration extensively used in the choice modelling area should be transferred into this area.

These points of criticism should not obscure the fact that the challenges facing transport planning require more complex and more comprehensive modelling approaches than before. The introduction of road transport informatics (RTI) weakens three central assumptions of the trip-based methodologies: the assumption of a perfectly informed driver, the assumption of an equilibrium as anything other than a theoretical construct and the assumption that it is feasible to separate trips from their journey or activity context. The multitude of interventions now possible during the trip and during the activities makes it necessary to adopt an approach that acknowledges this complexity. Models based on the concepts of the AA should be able to respond to this challenge and to provide support to the decision makers.

Bibliography

Ampt E S, Richardson A J, Brög W 1985 *New Survey Methods in Transport.* VNU, Utrecht, The Netherlands

Axhausen K W, Herz R 1989 Simulating activity chains: German approach. *J. Transp. Eng.* **115**, 316–25

Becker G S 1965 A theory of the allocation of time. *Econ. J.* **75**, 493–517

Brög W, Erl E 1983 Application of a model of individual behavior (situational approach) to explain household activity patterns in an urban area and to forecast behavioral changes. In: Carpenter and Jones 1983, pp. 350–70

Burnett K P 1978 Markovian models of movement within urban spatial structures. *Geogr. Analyses* **10**, 142–53

Carpenter S, Jones P M (eds.) 1983 *Recent Advances in Travel Demand Analysis.* Gower, Aldershot, UK

Chapin F S 1974 *Human Activity Patterns in the City.* Wiley, New York

Damm D 1983 Theory and empirical results: A comparison of recent activity-based research. In: Carpenter and Jones P M 1983, pp. 3–33

Damm D, Lerman S R 1981 A theory of activity scheduling behaviour. *Environ. Plann.* **13A**, 703–18

EUROTOPP 1989 Potential of Using Activity-Based Approaches. Transport Studies Unit Report to DRIVE. Oxford University, Oxford

Gärling T, Brännäs K, Garvill J, Golledge R G, Copal S, Holm E, Lindberg E 1989 Household activity scheduling. *Contemporary Developments in Transport Modelling.* Selected Proc. 5th World Conf. Transport Research, Vol. 4. Western Periodicals, Ventura, CA

Golob J M, Golob T F 1982 Classification of approaches to travel-behaviour analysis. *Transp. Res. Board Spec. Rep.* **201**, 83–106

Hägerstrand T 1970 What about people in Regional Science? *Pap. Reg. Sci. Assoc.* **24**, 7–21

Hanson S 1979 Urban travel linkages: A review. In: Hensher D A, Stopher P R (eds.) *Behavioral Travel Modelling.* Croom Helm, London, pp. 81–100

Hanson S, Huff J O 1986 Classification issues in the analysis of complex travel behaviour. *Transportation* **13**, 271–93

Hanson S, Huff J O 1988 Systematic variability in repetitious travel. *Transportation* **15**, 111–35

Hayes-Roth B, Hayes-Roth F 1979 A cognitive model of planning. *Cognit. Sci.* **3**, 275–310

Herz R 1983 Stability, variability and flexibility in everyday behaviour. In: Carpenter and Jones 1983, pp. 385–400

Hirsh M, Prashker J N, Ben-Akiva M 1986 Dynamic model of weekly activity pattern. *Transp. Sci.* **20**, 24–36

Holzapfel H 1980 Verkehrsbeziehungen in Städten. *Schriftenr. Inst. Verkehrsplanung Verkehrswegebau* **5**, Technische Universität Berlin

Jones P M, Bradley M, Ampt E 1989 Forecasting household response to policy measures using computerized, activity-based stated preference techniques. In: International Association for Travel Behavior (ed.) *Travel Behaviour Research.* Avebury, Aldershot, UK, pp. 41–63

Jones P M, Clarke M I 1988 The significance and measurement of variability in travel behaviour: A discussion paper. *Transportation* **15**, 65–87

Jones P M, Dix M C, Clarke M I, Heggie I G 1983 *Understanding Travel Behaviour.* Gower, Aldershot, UK

Jones P M, Koppelman F S, Orfeuil J P 1990 Activity analysis: State-of-the-art and future directions. In: Jones P M (ed.) *New Directions in Dynamic and Activity-Based Approaches.* Avebury, Aldershot, UK, pp. 34–55

Kitamura R 1988 An evaluation of activity-based travel analysis. *Transportation* **15**, 9–34

Kitamura R, Kermanshah M 1984 A sequential model of interdependent activity and destination choice. *Transp. Res. Rec.* **987**, 81–9

Kitamura R, Lam T N 1981 A time dependent Markov renewal model of trip chaining. In: Hurdle V F, Hauer E, Stewart G N (eds.) *Proc. 8th Int. Symp. Transportation and Traffic Theory.* University of Toronto Press, Toronto, pp. 376–402

Koppelman F S, Townsend T A 1988 Task allocation among household members: Theory and analysis. *Prep. 5th Int. Conf. Travel Behaviour.* INRETS, Arcueil, France

Küchler R 1985 Wegekettenorientierte Verkehrsberechnungsmodelle. Dissertation, TH Darmstadt

Kutter E 1972 Demographische Determinanten des städtischen Personenverkehrs. *Veröff. Inst. Stadtbauwesen* **9**, Technische Universität Braunschweig

Kutter E 1984 *Integrierte Berechnung Städtischen*

Personenverkehrs. Dokumentation der Entwicklung eines Verkehrsberechnungsmodells für die Verkehrsentwicklungsplanung Berlin West. Arbeitsberichte zur Integrierten Verkehrsplanung, Fachgebiet Integrierte Verkehrsplanung, Technische Universität Berlin

Lenntorp B 1978 *Paths in Space–Time Environments: A Time Geographic Study of the Movement Possibilities of Individuals*. Liber, Lund, Sweden

Lerman S R 1979 The use of disaggregate choice models in semi-Markov process models of trip chaining behaviour. *Transp. Sci.* **13**, 273–91

Mahmassani H S 1988 Some comments on activity-based approaches to the analysis and prediction of travel behaviour. *Transportation* **15**, 35–40

Marble D F 1964 A simple Markovian model of trip structures in a metropolitan region. *Pap. Reg. Sci. Assoc. West. Sect.* 150–6

Mazurkiewicz L 1985 A statistical model of a multitrip spatial-interaction pattern. *Environ. Plann.* **17**A, 1533–9

Mentz H J 1984 Analyse von Verkehrsverhalten im Haushaltskontext. *Schriftenr. Inst. Verkehrsplanung Verkehrswegebau* **11**, Technische Universität Berlin

Pas E I 1987 Vehicle usage patterns in two-car households: Identification and analysis of homogeneous market segments. *Prep. 5th Int. Conf. Travel Behaviour*. INRETS, Arcueil, France

Pas E I 1988 Weekly travel-activity patterns. *Transportation* **15**, 89–110

Pas E I 1990 Is travel demand analysis and modelling in the doldrums? In: Jones P M (ed.) *New Directions in Dynamic and Activity-Based Approaches*. Avebury, Aldershot, UK, pp. 3–33

Recker W W, McNally M G, Root C S 1983 Application of pattern recognition theory to activity pattern analysis. In: Carpenter and Jones 1983, pp. 434–49

Recker W W, McNally M G, Root G S 1986 A model of complex travel behaviour. *Transp. Res. A* **20**, 307–30

Root G S, Recker W W 1983 Towards a dynamic model of individual activity pattern formulation. In: Carpenter and Jones 1983, pp. 371–82

Schmiedel R 1984 Bestimmung verhaltensähnlicher Personenkreise für die Verkehrsplanung. *Schriftenr. Inst. Städtebau Landesplanung* **18**, Universität Karlsruhe

Szalai A (ed.) 1972 *The Use of Time*. Mouton, The Hague, The Netherlands

van der Hoorn T 1979 Travel behaviour and the total activity pattern. *Transportation* **8**, 309–28

Winston G C 1987 Activity choice: A new approach to economic behaviour. *J. Econ. Behav. Organ.* **8**, 567–85

Zumkeller D 1989 Ein sozialökologisches Verkehrsmodell zur Simulation von Massnahmewirkungen. *Veröff. Inst. Stadtbauwesen* **46**, Technische Universität Braunschweig

K. W. Axhausen and J. W. Polak
[University of Oxford, Oxford, UK]

Transportation Planning: Microscopic Approach

Transportation plays a growing role in modern societies. The transportation system itself consists of a large number of components and has to face a number of different, sometimes opposing, goals and purposes. Therefore, a well-designed transportation system (including network, traffic management and user behavior) is crucial. The process of designing (and running) such a system cannot be carried out properly without suitable planning models and tools.

Transportation planning processes can be based on several main approaches. In recent years, the focus of attention has shifted towards activity-based and dynamic methods. To model more detailed processes, "microscopic" approaches are gaining popularity, usually implemented as microsimulation models. Since about 1970, transportation planning methodology has increasingly focused its efforts on the individual and on individual decision making.

1. General Considerations

Microscopic transportation planning models concentrate on an individual decision-making unit such as single vehicles, individual persons or households. The travel behavior of these individuals is then modelled within the approaches. Therefore, a more comprehensive definition could speak of "microanalytical models of human behavior in transportation, using *simulation* and/or mathematical tools."

Assumptions on human behavior are fundamental to models of this type. Any assumptions on travel behavior are based on rules of how people experience and memorize the transportation system, how they make, change and realize their decisions and how all their decisions interact and influence each other. It is suggested that approaches explicitly dealing with cause–effect relations at the individual level are superior to aggregate methods when dealing with changes in individual decision-making or individual travel behavior. This is due to the fact that these methods are able to:

(a) represent interactions among several individuals more closely;

(b) consider dynamic changes in behavior; and

(c) address many different points at the same time.

Thus, all approaches described in this article are characterized by the following common key features:

(a) reference and decision unit are at the level of individuals;

(b) rules determining the behavior of the individuals are explicitly incorporated into the model;

(c) conditions, restrictions and interactions influencing the behavior of the individuals are incorporated into the model; and

(d) simulation approaches are employed in order to model the behavior of many different individuals.

In Sect. 2, definitions of key expressions and important terms are given. Selected examples of this type of models are described in Sect. 3. Results and conclusions are discussed in Sect. 4.

2. Definitions

In order to describe microscopic simulation approaches in transportation, some common key words and terms are defined (the definitions are primarily for usage within this article).

(a) *Trip*. This is the fundamental unit of travel. It is defined as the journey between two locations (e.g., from home to work place).

(b) *Tour/journey*. A tour is a sequence of trips that are "chained" and begin and end at home. Therefore, a tour contains at least two trips (e.g., from home to the work place and back home).

(c) *Journey chain*. This stands for the daily program of tours. A journey chain contains all tours made by a person during one day. In the minimal case, a journey chain includes one tour and thus two trips.

(d) *Activity*. This expression defines the business carried out at the destination of a trip. All trips are leading to an activity except where the trip itself is the activity (e.g., sight-seeing by bus). However, it does not follow that every activity implies a trip. At one destination, several activities can be carried out.

(e) *Activity chain*. This is the sequence of activities within one day. It is possible that all activities of a certain activity chain are carried out at home and thus no trips are made.

(f) *Reference Unit*. This is the subject of the simulation. In the model descriptions the term individual is synonymous with the term reference unit.

3. Model Descriptions

Microscopic simulation to study human travel behavior has been known since the 1960s. For partial reviews see Ruppert (1976, 1986), Clarke *et al.* (1980) or Golob and Golob (1982). Within transportation modelling, the following five groups of models can be distinguished.

(*a*) *Models of demographic change/housing choice*. The oldest model known by the authors concentrates on the demographic development of the household and as an application on household expenditure over time (Orcutt *et al.* 1961). Mackett (1988) combined a demographic model with mode choice and residential location choice. Wegener (1979) or Miller *et al.* (1987) focused on the housing market and the search for housing.

(*b*) *Models of time–space prisms*. Based on Hägerstrand's (1970) time-geographic approach, Lenntorp (1978) developed a model to generate the possible set of activity chains for individuals given the physical and social constraints on their travel behavior.

(*c*) *Models of trip chaining/activity generation*. A number of approaches concentrate on the simulation of the chaining and/or the generation of activity chains (e.g., Mentz 1984, Recker *et al.* 1986).

(*d*) *Models of activity chain performance*. These models were mainly developed in Germany. They assume activity chains as given and simulate the individual performing these chains: (Poeck and Zumkeller 1976, Sparmann 1980, Schmiedel 1984, Axhausen 1989). The approaches of Kutter (1984) or Küchler (1985) transformed this form of microscopic simulation into an aggregate approach. Additionally, Sonntag (1985) employed this idea for modelling urban freight traffic. Changes in the activity chain and its elements can be described with or in the models of Brög (1982) and Zumkeller (1989).

(*e*) *Models of behavioral change*. A number of models were developed for individual aspects of travel behavior. The potential of organized car sharing was explored by Bonsall (1980). Southworth (1985) used simulation to derive behaviorally sound accessibility measures. Route choice in the underground network of Tokyo was described using a combination of artificial intelligence (AI) and simulation techniques by Shimazaki and Matsumoto (1989).

Markov-chain-based models have an obvious similarity, but as they have analytical solutions, they will not be covered here. An example is Kitamura (1983).

To demonstrate the range of microscopic simulation a number of well-known, important or representative approaches will be described in more detail. The models will be presented in roughly chronological order.

3.1 PESASP, the Time–Geography Model

Time–geography as developed by Hägerstrand (1970) and the Lund School (Parkes and Thrift 1980) describes travel behavior as being severely restricted by physical and social constraints. Hägerstrand distinguishes between coupling, capability and authority constraints. Their effects become obvious when the "path" of an individual through time and space is marked in a three-dimensional map, where the surface of the planning area represents the first two coordinates (x, y) in the normal way while the third coordinate (height z) represents time. Within this three-dimensional map the time–space paths of the individuals are contained (through these constraints) in time–space prisms. Lenntorp's (1978) PESASP uses these constraints to generate all possible time–space paths in an effort to understand the effects of tightening or relaxing the constraints.

Input data for PESASP are:

(a) daily activity programs of individuals;

(b) possible destinations—locations where activities can be carried out (physical space, plus time conditions); and

(c) transport network, usually expressed in links and nodes.

As an extension to PESASP, a situational approach by Brög (1982) called SINDIVITAL can be mentioned. Certain constraints on individual behavior are defined and a three-stage method is used to forecast behavior.

The model technique within PESASP is deterministic simulation: fixed activity programs are matched to destinations and other constraints; permutations of activity programs are generated. The behavioral framework is only included in observed activity programs. No dynamics are provided; the activity program is fixed as observed. The level of geographical detail is open to the user. The model can be used to identify transportation constraints and individual trip realization, applications include smaller Swedish cities (e.g., Karlstad with c. 75 000 inhabitants) and Tepoztlan, Mexico, as well as Newcastle, Australia.

3.2 Dortmuder Housing Market Model

The Dortmunder housing market simulation model (Wegener 1979) describes the relationship between economic change, locational choice, mobility and land use in urban areas. It defines the factors influencing the housing market in one region.

The core of Wegener's approach consists of two submodels: an aging model and a model of the housing market.

The aging submodel simulates changes in the households and in the housing stock in the model region over time; it encompasses household status like birth, aging, death, marriage, divorce, and so on, as well as deterioration, rehabilitation and demolition of houses. The submodel is realized by Markov chains using dynamic transition rates.

The second submodel simulates the migration decisions of households. They depend on the actual situation of households (aging model) and decisions taken by landlords. The first submodel provides the frame in which the second model operates.

The input data for Wegener's model is based on traffic zones. The data requirements are comparatively high as the model incorporates more than 300 variables on population, employment, housing stock, public amenities, transportation infrastructure and land use as well as accessibility measures for a number of trip purposes (work, education, shopping, etc.). The model technique is multistage simulation. Long-term housing market decisions are simulated. The behavioral context is mainly the individual's position in the life cycle. External factors such as the actual housing market and landlords' decisions influence behavior. The situation of households and dwellings is dynamic: changes over time are modelled using a fairly long pulse time (about one year). The reference unit is one household. The model was applied to the Dortmund region with about 2.4 million inhabitants (1978).

3.3 The Models of Zumkeller and Sparmann, Including Extensions

The models of Zumkeller, described for example by Zumkeller (1989), although developed earlier, and Sparmann (1980) illustrate the early German activity chain models. The activity chain is used as a fixed frame. The population is divided into behaviorally homogeneous groups.

Four types of input data are needed:

(a) structure data (number of houses, working places, and so on, for each zone),

(b) population (share of each of the "homogeneous" groups in each zone),

(c) activity chains (distribution of activity chain types for each homogeneous group), and

(d) transport infrastructure (level of service, accessibility, distances).

The simulation process starts with the selection of an individual activity pattern. Based on the activity chain, destinations and modes are chosen. The probability distributions for these decisions are based on empirical results.

The model technique is microscopic Monte Carlo simulation of daily travel behavior. Individual behavior is determined by socioeconomic status and by exogeneous variables, such as level of service. Reference units are activity chains as realized by individuals. Applications include a number of cities (e.g., Konstanz and Pforzheim as well as Nürnberg).

An extension called ORIENT/RV (Axhausen 1989) includes an explicit time axis to model traffic flow movement through the urban network. Supply and demand interactions while travelling on the road are included as well as parking search processes. These fixed activity chains are divided into a sequence of discrete events, including a separation of activities in primary and secondary activities, nested logit models of destination and mode choice, explicit choice of parking type and location, and a more complex description of exchangeable modes.

Model techniques include deterministic and probabilistic rules of travel behavior, added to a behavior-oriented simulation approach. Certain aspects of dynamics are provided, but there are no interactions across days or based on experience. A representation of a network is included, but the zonal system is limited in detail due to the necessary data. Applicability is extended due to the incorporated time axis, a network, and demand–supply interactions. Data requirements are somewhat higher. The model might be useful for combinations with traffic flow models because a travel demand diversion is provided and parking traffic is explicitly addressed.

3.4 Car Sharing

The car sharing method (Bonsall 1980) explores the potential for intrahousehold car sharing. In the first stage, the model identifies those willing to participate in a car sharing scheme. In the second stage, the potential car sharers are matched and, in the third stage, a complex utility-maximization process identifies the best partners.

The model technique employs empirical rules and utility maximizing approaches. Behavior is based on utility maximizaton within given constraints of acceptability (timing, extra distances, risks, etc.). No dynamics are incorporated. The reference unit is the individual driver. Applications of the model require a detailed description of the decision framework of the travellers and stated-preference interviews for establishing the utility functions. Therefore, applicability is somewhat expensive due to the data requirements, but the model was used for a car sharing scheme in Yorkshire.

3.5 A Microsimulation of Household Interactions

The microsimulation of household interactions model (Mentz 1984) tries to capture the effects of household interactions on the decision to perform an "out-of-home" activity by constructing decision trees for the daily activity chains. Logit-type decision models are used to calculate probabilities. The utility functions include time and money costs of the previous activities for the individual and the entire household.

The model technique is simulation to choose between alternatives; choice probabilities are calculated with logit-models. The behavioral framework is based on sequential utility-maximizing choice processes of activity participation. Household constraints are included in the utility functions of individuals as generalized costs. There is a distinction between fixed and discretionary activities. The probabilities of the discretionary activities are conditional on the performance of fixed activities. Both individual and household constraints are included. The estimation of a large number of complex logit models requires extensive and expensive travel surveys, but the model was calibrated with and applied to a travel survey from Munich (Mentz 1984).

3.6 STARCHILD

STARCHILD (Recker *et al.* 1986) stands for simulation of travel/activity responses to complex household interactive logistic decision. It is an ambitious attempt to model activity generation within the larger context of the household. PESASP (see Sect. 3.1) and CARLA (Jones *et al.* 1983) are similar models, but are not so broad in scope.

The model itself is split into five parts dedicated to different tasks within the overall activity program generation.

(a) *Generation of household activity programs* (TROOPER). This generates the activity programs for the household member. An initial activity chain including timing, locational and coupling constraints is generated for each household member.

(b) *Establishment of all feasible activity patterns* (SNOOPER). This uses the initial activity pattern to generate the set of possible patterns respecting the inherent constraints.

(c) *Establishment of relevant feasible activity patterns* (GROOPER). This uses mathematical pattern recognition techniques to reduce the potentially large set of possible patterns to a set of relevant and distinguishable patterns.

(d) *Exclusion of inferior activity patterns* (SMOOPER). This employs multidimensional programming techniques to exclude inferior activity patterns according to user-specified criteria.

(e) *Choice of activity patterns* (CHOOSER). This calculates the utilities of the remaining set of activity patterns. Simulation could be used at this point to perform the choice. The utility function incorporates travel time elements, time spent at home and risk measures for the ability to participate in unplanned activities and/or not to participate in important activities.

The model techniques include a wide range of sophisticated modelling techniques, from simulation to complex mathematical programming and constrained combinatoric algorithms. Utility maximization is used for the final choice between activity patterns. Behavior is based on the time–geographic approach, combined with techniques of constraint optimization to establish the preferred activity patterns. The model is not dynamic, but it would be possible to establish feedbacks over time for time–space constraints The model requires a detailed geographic description of the action and information space of the population.

Reference units are households and individual members. Applications in real-world situations are not known to the authors. Initial tests were carried out with a sample of about 100 individuals. STARCHILD has to be scaled down for use in a planning model. The overall concept is appealing, but there are some limitations in the adopted time frame of one-day activity patterns.

3.7 Housing Search

A dynamic simulation model of the housing market (Miller *et al.* 1987) generates households and their characteristics including, for example, size, household income and home and work locations. The search process and its duration is modelled as a bifurcation surface, triggered under certain utility or opportunity constraints.

The model technique is simulation for updating transition probabilities which are empirically derived. The search process is described as a bifurcation surface, indicating a sudden increase in search willingness, slowly receding if the search is unsuccessful. Individual behavior is represented based on utility maximization. The model is dynamic in its description of the population, the housing market and the search process (smallest time frame usually one month). The level of geographic detail is set by the availability of zonal information on population and housing market. The reference unit is the household with some information "collected" through single members. Applications include the Toronto metropolitan area, but the large data requirements limit its practicability. The original form of the model is based on data from the 1971 Canadian census indicating the scope and amount of necessary data.

3.8 MASTER

MASTER (Mackett 1988) offers an approach to the simulation of transport demand, employment and residential choice based on the household. All important demographic processes necessary to forecast household development are included. Immigration and emigration are implemented based on assignments of available jobs to individual households. The household car allocation process is implemented as the basis of a rule oriented approach to the modal choice. A feedback is established between the different decision levels. For several choices a kind of learning process is employed: out of a possible set of alternatives a choice constrained by specific requirements can be made. The choice set can be reexamined (under changed requirements).

The model techniques are rule-based decisions with probability distributions empirically derived. The household interactions are the focus of the approach and constrain the behavior of the household members. The model is dynamic in its description of the households and the environment. No memory of earlier decisions is provided. Model geography is zonal based. Resolution is limited by the need for secondary statistics. The reference unit is the household or an individual constrained by household resources or decisions. The decision structures of the model are somewhat complex. The model is suitable for simulating household processes, because of the explicit time-axis, even for long-term household simulation. Applications of MASTER include tests with a data set of the city of Leeds, based on 1740 households representing 4800 inhabitants in 28 zones. The simulated time period was 20 years (1971–1991), which allowed a comparison of the 1981 simulation results with the 1981 census (giving good results).

4. Recommendations

Microscopic transport planning models have been used in a wide area of practical applications for some time. Depending on the actual problem, the approach is considered to be useful, feasible and successful.

Several aspects promoted this development. One key factor is the improvement in computer capabilities which enabled extensive simulation of individuals. A second important factor is the type of transportation problem itself which has to be addressed by the models.

In the years following World War II, transportation planning tried to satisfy growing demand by repairing and improving networks (often using macroscopic tools), but current applications require more sensitive (microscopic) models in order to model the effects of regulations and changes of individual behavior.

Therefore, microsopic models are needed in order to provide a finer resolution and a higher degree of realism. Microscopic models concentrate especially in the forecasting of temporal and spatial changes in traffic demand (time–geography) and in the modelling of choices and decisions (e.g., in the modal-split area).

When applying microscopic models, it should be realized that it is absolutely necessary to use an adequate data sample for the simulation. Contrary to macroscopic approaches, the input data has to provide for both number and type of individual trips as realized under *status quo* conditions as well as for reasons and conditions under which the individual choices are made. A high level of accuracy and detail is required for the modelling of the choice set and the surrounding restrictions of individuals. Here, the main problems for usage of microscopic simulation models can be found:

(a) data requirements may be higher than for extrapolations of macroscopic models;

(b) various relationships have to be observed and measured; and

(c) complex human decision processes have to be modelled.

These main problems define the area for future research and future improvements in the modelling process. The resulting approaches might help to answer some questions when studying social behavior and complex interactions between political decisions and individual reactions. Those interactions, restrictions and reactions can only be modelled by using at least parts of microscopic simulation approaches.

Acknowledgement

The work described in this article is part of a research effort carried out by the EUROTOPP consortium, composed of Transport Studies Unit (Oxford University), Bureau Goudappel Coffeng (Deventer), CETE Méditeranée (Aix-en-Provence), INOVAPLAN (Munich), Institut für Verkehrswesen (Universität Karlsruhe), Institute for Social and Behavioural Research (Universiteit Nijmegen), Robotiker (Mungia) and Syseca Temps Réel (St. Cloud) under the DRIVE programme of the European Commission.

See also: Transportation Modelling; Transportation Planning: Activity-Based Approach

Bibliography

Axhausen K W 1989 Eine ereignisorientierte Simulation von Aktivitätenketten zur Parkstandswahl. *Schriftenr. Inst. Verkehrswesen* **40**. Univerität Karlsruhe, Karlsruhe, Germany

Bonsall P W 1980 Microsimulation of organized car sharing: Description of the models and their calibration. *Trans. Res. Rec.* **767**, 12–21

Brög W 1982 *Individualverhaltensmodell des Personenfernverkehrs auf der Basis des Situationsansatzes: Ein situationsbestimmter Individualverhaltensalgorithmus (SINDIVITAL)*. Socialdata and Deutsche Versuchsanstalta für Luft- und Raumfahr, im Auftrag des Bundesministeriums für Verkehr, Bonn

Clarke M C, Keys P, Williams H C W L 1980 Microanalysis and simulation of socio-economic systems: Progress and prospects. In: Bennett R J, Wrigley N (eds.) *Quantitative Geography in Britain: Retrospect and Prospect*. Routledge and Kegan Paul, London

Golob J M, Golob T F 1982 Classification of approaches to travel—Behavior analysis. Transportation Research Board Special Report No. 201. TRB, Washington, DC, pp. 83–107

Hägerstrand T 1970 What about people in regional science? *Pap. Reg. Sci. Assoc.* **24**, 7–21

Jones P M, Dix M C, Clarke I G, Heggie I G 1983 *Understanding Travel Behaviour*. Oxford Studies in Transport, Oxford

Kitamura R 1983 Sequential, history-dependent approach to trip-chaining behavior. *Transp. Res. Rec.* **944**, 13–21

Küchler R 1985 *Wegekettenorientierte Verkehrsberechnungsmodelle*. Dissertation, TH Darmstadt

Kutter E 1984 *Integrierte Berechnung Städtischen Personenverkehrs*. Dokumentation der Entwicklung eines Verkehrsberechnungsmodells für die Verkehrsentwicklungsplanung Berlin (West), Arbeitsberichte zur Integrierten Verkehrsplanung, Fachgebiet Integrierte Verkehrsplanung, TU Berlin, Berlin

Lenntorp B 1978 *Paths in space-time environments: A time geographic study of the movement possibilities of individuals*. Liber, Lund, Sweden

Mackett R 1988 *Exploratory analysis of long term travel demand using micro-analytical simulation*. Prep. Oxford Conf. New Directions in Dynamic Activity-Based Approaches, Oxford

Mentz H J 1984 Analyse von Verkehrsverhalten im Haushaltskontext. *Schriftenr. Inst. Verkehrsplanung und Verkehrswegebau* **11**

Miller E. J., Noehammer P J, Ross D R 1987 A microsimulation model of residential mobility. In: Young W (ed) *Proc. Int. Symp. Transportation, Communication and Urban Form*, Vol. 2. Monash University, Clayton, Australia

Orcutt G, Greenberger M, Rivlin A, Korbel J 1961 *Microanalysis of Socio-Economic Systems: A Simulation Study*. Harper and Row, New York

Parkes D, Thrift N 1980 Time–geography: The Lund approach. *Times, Spaces, and Places—A Chronogeographic Perspective*. Wiley, New York, Chap. 6

Poeck M, Zumkeller D 1976 Die Anwendung einer massnahmenempfindlichen Prognosemethode am Beispiel des Grossraums Nürnberg. DVWG Workshop Policy Sensitive Models. Deutsche Verkehrswissenschaftliche Gesellschaft, Bergisch Gladbach, Germany

Recker W W, McNally M G, Root G S 1986 A model of complex travel behaviour. *Transp. Res. A* **20**, 307–30

Ruppert E 1976 *Modelle räumlichen Verhaltens*. Dortmunder Beiträge zur Raumplanung, XX. Universität Dortmund, Dortmund, Germany

Ruppert E 1986 *Simulation räumlicher Interaktion*. Dortmunder Beiträge zur Raumplanung, XX. Universität Dortmund, Dortmund, Germany

Schmiedel R 1984 Bestimmung verhaltensähnlicher Personenkreise für die Verkehrsplanung. *Schriftenr. Inst. Städtebau und Landesplanung* **18**. Universität Karlsruhe, Karlsruhe, Germany

Shimazaki T, Matsumoto Y 1989 Simulation of personal route choice mechanism. *Prep. Int. Conf. Dynamic Travel Behaviour Analysis*, Kyoto, Japan

Sonntag H 1985 A computer model for urban commercial traffic-analysis (concept/application). *Transp. Policy Decision Making* **3**, 171–80

Southworth F 1985 Multi-destination, multi-purpose trip chaining and its implications for locational accessibility: A simulation approach. *Pap. Reg. Sci. Assoc.* **57**, 107–23

Sparmann U 1980 ORIENT—Ein verhaltensorientiertes Simultationsmodell zur Verkehrsprognose. *Schriftenr. Inst. Verkehrswesen* **20**. Universität Karlsruhe, Karlsruhe, Germany

Wegener M 1979 Das Dortmunder Wohnungsmarktmodell. *Schriftenr. Bundesm. Raumordnung, Bauwesen und Städtebau* 07.011. Bonn, pp. 79–100.

Zumkeller D 1989 Ein sozialökologisches Verkehrsmodell zur Simulation von Massnahmewirkungen. *Veröff. Inst. Stadtbauwesen* **46**

U. J. Becker, R. Schneider and R. Schwartzmann
[Universität Karlsruhe, Karlsruhe, Germany]

Transportation Systems: Trends

The topic of transportation systems analysis and control represents a contemporary synthesis of theories and methods that have been used for two different levels of transportation study. Transportation analysis tools are typically employed to test and evaluate alternative transportation plans, while transportation control techniques are used to develop and evaluate traffic control plans. The body of knowledge encompassed in transportation systems analysis began with the application of systems analysis methods to urban transportation studies (Manheim 1979).

For a description of the basic tools of transportation systems analysis see *Transportation Modelling*. Manheim provided a collection of the theories that underlie the rationalization of these models used for transportation systems analysis. These models operate at a macroscopic level when compared with the analytical methods used for operations control analysis. They derive activity and travel patterns from land uses and are employed to identify where improved capacities are required in a transportation network to maintain a desired level of service. There are also techniques available to forecast land use and development trends from transportation investments (Hutchinson 1974).

Transportation systems control studies focus on the application of traffic control strategies to maintain efficient movement of traffic on existing networks. The analytical methods used derive from the classic traffic flow theories that developed between the 1930s and the 1970s, and were applied to explain and analyze traffic flow phenomena (Huber 1982).

This article describes the trend towards integration of planning analysis and control modelling systems by illustrating recent examples of methodologies used to address classes of problems that require the combination of the two levels of analysis. These applications illustrate expanded and creative uses of available methods that are closer to the realm of research than practice. In addition, trends in using expert systems to enhance the selection of appropriate models and their proper use are described.

1. Trends and Needs in Modelling

Since the early 1970s, the US Department of Transportation and other governmental transportation agencies throughout the industrialized world have sponsored the development and maintenance of computerized models for traffic and transportation engineering and planning. Limitations in the scope of specific models, such as the insensitivity of demand to changes in traffic controls and lack of real-time updates of the system parameters, lessen the effectiveness of the evaluation process of transportation system management strategies. Further limitations with these computer models are associated with the inability of professionals to select an appropriate model and to interpret the model results.

Efforts are underway to formally integrate traffic simulation models into complete systems. The result would provide traffic engineers with the ability to simulate traffic over large areas with any combination of roadway facilities at a reasonable computing cost.

It is hoped that the use of intelligent vehicle–highway systems (IVHSs) will be the solution to congestion problems (US Department of Transportation 1989). An IVHS incorporates advanced communications technology, computers and electronics. It permits highway operators to dynamically monitor traffic and communicate optimum routing directions and modify signalization schemes to increase effective highway capacity (US Department of Transportation 1989). Extensive analysis, testing and demonstration are needed to identify efficient and cost-effective technologies and packages of technologies. The analysis task will require some models and analytical strategies that derive from current tools and others that will be developed in time (see *In-Vehicle Equipment for Future Traffic Control Systems*).

A further area for enhancing the capabilities of analysis/control models is the application of expert systems (Faghri 1990). Specific developments from using expert systems include model-building guidance, knowledge-based simulation models, and integration of expert systems and simulation tools. An expert system that contains knowledge about the system and model can guide a simulation analyst through the different stages of a simulation study. Likewise, an expert system can assist the user in selecting the appropriate model for a problem and then show how to use it properly (see *Expert Systems Approach to Rail Traffic Control*; *Expert Systems Approach to Road Traffic Control*).

Finally, improvements in transportation systems analysis and control studies will probably result if new technology is employed for data collection and management. Included here are developments such as video recording and image processing; comprehensive inventory, traffic and accident information systems; weigh-in-motion and automated vehicle identification. Automated information processing methods are needed to make the system data acceptable to the analysis and control process (see *Video Sensors*).

2. Software Systems

Traditionally, separate simulation models were developed for various types of traffic facilities (e.g., freeways, urban streets), in order to represent the unique characteristics of each facility. In the 1970s, the idea was conceived for a single integrated simulation system to incorporate available simulation techniques for all traffic environments, thus providing engineers with a single software system that could simulate traffic over large areas with any combination of roadway facilities at a reasonable computing cost. This concept set the trend for future developments in systems analysis and traffic control modelling.

For example, a demonstration system was proposed for the Federal Highway Administration (FHWA) called TRAF—a traffic simulation software system. The TRAF simulation software consists of six models: four simulation models, an equilibrium assignment model and an intersection capacity model. It has been described by Yedlin and Lieberman (1988) as follows.

"The TRAF system integrates the individual component models. The user can simulate traffic on networks comprised of different facility types with a single computer run. Also, different sections of the network may be analyzed at different levels of detail. This ability to analyze integrated highway networks of any configuration, exhibiting any combination of facility types, constitutes a powerful analysis tool."

Logic developed in each TRAF tool permits it to receive and discharge traffic to/from adjacent sub-networks. The key interface in the TRAF system for the analysis control problem is the equilibrium traffic assignment model which uses origin–destination (OD) data and translates this information into a form suitable for simulation models. Link travel times are related to link capacities through the TRAFLO capacity model.

The TRAF model provides the planner with information on traffic control plans that did not previously enter into the analysis. At the same time, it expands the ability of the traffic engineer to look at system-wide congestion. This model system is a framework for considering two priority problems; traffic management during reconstruction and management of traffic congestion.

3. Applications

The trends toward a highly sophisticated real-time traffic analysis/control methodology are cited here as they currently exist—as isolated examples that illustrate an improvization of available tools. In time, all of these ideas will become intertwined into complete computer modelling packages that include the concepts of

(a) real-time updates of network measures and traffic controls, and

(b) sensitivity of demand to the contol strategy and real-time diversion of traffic.

These developments which extend beyond the TRAF-type system are introduced through examples of the purpose for which they were devised.

3.1 Managing Traffic Congestion in Combined Freeway and Traffic Signal Networks

While the majority of traffic control tools were developed to deal primarily with undersaturated conditions, traffic congestion has become a serious problem in populated urban and suburban areas. The problem of traffic congestion has increased both the scope and scale of the control problem, while reducing the range of alternative route choices available. This occurs because, as traffic diverts from a congested freeway, parallel arterials also become congested. The loss of diversion strategies results in large freeway queues, while on parallel arterials more advanced signal optimization/coordination algorithms are required to deal with oversaturated traffic signals. Then, as congestion spreads, congestion management becomes a complex multidimensional congested network problem, as opposed to a congested corridor problem (see *Corridor Control Systems*).

This is an example of a complex transportation systems analysis and control problem because none of the traffic models currently in use has been designed to deal with as broad a problem. What is lacking is one comprehensive model that deals with all aspects of congested signal/street networks. Furthermore, to reflect the effect of advanced communications technology, the model should be sensitive to behavior generated by on-board driver information systems.

Although the need for integrated network and control models has been established, prototypes will take time to develop. The primary reason for this is the need for models to be sensitive to real-time traffic flows so as to produce dynamic signal control/diversion plans and communicate them to motorists.

An early approach to this complex transportation problem was a model called INTEGRATION, which is being tested and evaluated at Queen's University in Kingston, Canada, under the sponsorship of the Ministry of Transportation of Ontario (Van Aerde *et al.* 1989b):

"INTEGRATION-1 consists of a discrete simulation that traces the path of each vehicle throughout the network. The links that a vehicle utilizes are selected in accordance with its estimate of the 'best route' from origin to destination, and, along its path, each vehicle's route is further adjusted in view of any changes in the prevailing traffic congestion and traffic controls."

The application of this model, thus, considers modelling traffic flow and controls as they react to incidents and congestion. Incidents are modelled as reductions in capacity of the number of lanes available.

INTEGRATION-1 reflects the most important attributes of congestion through its explicit account of queue growth/delay, while maintaining a dynamic equilibrium traffic assignment. The measures of queue size and delay are used to develop direct modelling of queue spillback from upstream links, continuous modelling of traffic signal progression and automatic delay of downstream link arrivals if they are held up at an upstream bottleneck. In addition, as the travel time difference between the shorter (congested) route and a longer (less congested) route changes, new arrivals will automatically redistribute themselves to avoid the congested link or area.

This model was applied to a series of incident scenarios and a range of incident response plans for each case. The study was for the Burlington Skyway FTMS in Ontario (Van Aerde *et al.* 1989a). The response plans were defined by real-time control on signal timings, real-time diversion of traffic and combinations of the two.

3.2 Reconstruction Project Travel Impact Evaluation Process

A problem similar in some respect but different in others to the incident/congestion situation as described in Sect. 3.1 is traffic management for reconstruction projects. The flow chart shown in Fig. 1 illustrates the impact analysis process. The outputs from the process are measures of effectiveness (MOEs) for selecting between traffic handling options and for finalizing a traffic management plan.

If the capacity through the construction zone is inadequate, the corridor-wide travel volumes and congestion should be compared to determine whether the available capacity on alternative routes and modes in the corridor compensate for the reductions in capacity on the highway being reconstructed. If the total corridor-wide capacity appears to be adequate, then a traffic control plane for the reconstruction zone and a public information program may be sufficient to maintain acceptable flow throughout the corridor. If, however, the total corridor-wide capacity appears to be inadequate, then it may be necessary to review the traffic management plan.

A corridor traffic management plan includes a basic traffic handling option for the highway being reconstructed and strategies to mitigate the travel impacts throughout the affected corridor. Impact mitigation strategies include techniques to increase the capacity of the reconstruction zone, and improvements of the

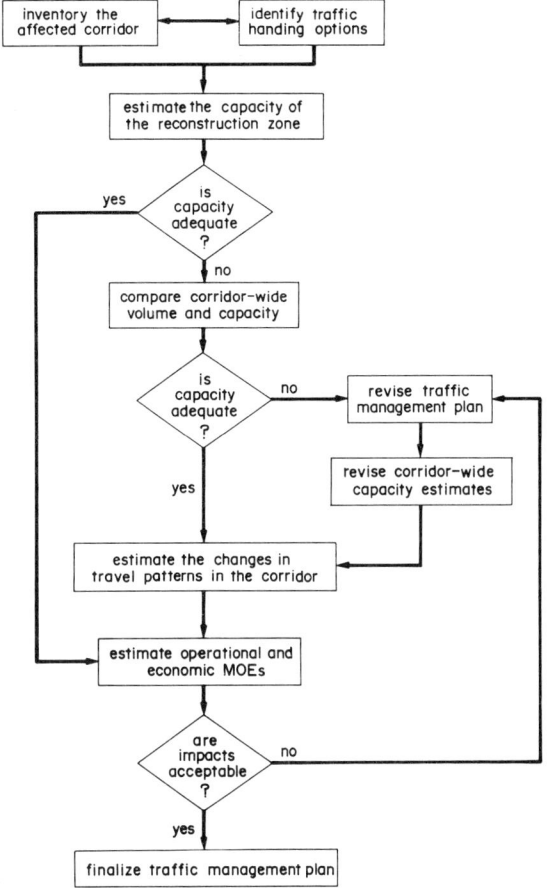

Figure 1
Flow chart of reconstruction project travel impact evaluation process (after Krammes *et al.* 1989)

transportation systems management (TSM) type on alternative routes and modes.

In selecting a traffic handling option, trade-offs must be considered between savings in reconstruction costs and increases in traffic management and road user costs. Generally, as more roadway space is allocated to the reconstruction activity, reconstruction costs decrease, but traffic management and road user costs increase.

Highway capacity analysis procedures can be used to estimate the changes in capacity associated with the impact mitigation strategies. However, further analysis with supplementary techniques is usually required, such as a transportation systems analysis of how current users of the highway being reconstructed will respond to a particular traffic management plan and what secondary impacts there will be on current users of alternative routes and modes in the corridor. This is currently the most difficult step in the evaluation process because information on how motorists respond to reconstruction projects is extremely limited. A broad database is needed to replicate the motorist responses to impacts and traffic plans (Krammes *et al.* 1989).

A study of experiences from five projects suggested that motorists respond to reconstruction projects in one of five ways (Krammes *et al.* 1989):

(a) cancel trips in the corridor,

(b) use an alternative route,

(c) travel at a different time of day,

(d) use another mode, or

(e) continue normal travel patterns.

Such are the types of diversion that are also required to be captured over a long time by planning models when the impact of a new facility is being considered. The problem addressed here focuses on short-term, temporary changes (see *Transportation Modelling*).

It follows that motorists' responses can be measured as changes in

(a) trip generation rates,

(b) trip distribution patterns,

(c) mode split and/or auto-occupancy rates, and

(d) traffic assignments among routes in the corridor.

The analysis here must go further than traffic assignment, as was the case in the previous example. Accordingly, congestion must also be addressed, which places the study at the microscopic level of traffic simulation models.

Thus, to address the problem of travel impacts of highway reconstruction, the analyst must currently choose from a set of analysis and simulation tools to address a specific problem. Since experience is limited on this task, rigid guidelines are not available. Available analysis tools with potential application to the travel impact evaluation process for highway reconstruction projects have been grouped into five categories (Krammes *et al.* 1989):

(a) network-based highway and transit planning models,

(b) quick-response estimation techniques,

(c) highway capacity analysis procedures,

(d) traffic simulation models, and

(e) traffic optimization models.

Figure 2 shows a decision tree for selecting the appropriate model for a given situation. This is illustrative of the general type of application that this article addresses—one where alternative models are available and a particular analysis strategy must be developed. Expert systems are currently being investigated to establish their ability to design modelling strategies.

3.3 Advanced Communications Technology

This section describes a methodology that was recommended to evaluate the proposed Autoguide route guidance system to be tested in London in 1990. A comparative assessment of the London system and that of the technically similar LISB system in Berlin is a further objective (see *Route Guidance, Individual*).

The systems have been described by May and Bonsall (1989) as follows.

> "The LISB and Autoguide systems are controlled by a central computer that controls real time network information using travel time data transmitted from equipped vehicles. The driver requests a destination via the keypad and the in-vehicle computer then selects an optimal route to the destination on the basis of data from the central computer. The driver is then directed along a recommended route via instructions on the visual display or audio device. While the vehicle proceeds along links in the network, the microcomputer stores information on travel conditions and these are transmitted to the central computer in order to update its knowledge of current network conditions."

Data from these demonstrations are being taken to show the traditional, ergonomic and operational performance of different levels of route guidance to enable decisions to be taken to expand these pilots into full-scale systems and to develop systems at other locations.

The simulation model would be required to produce a number of indicators. First, it would need to estimate the distribution of travel time and costs of four user groups (May and Bonsall 1989):

(a) those driving equipped vehicles and following advice carefully,
(b) those driving equipped vehicles but following advice only in part,
(c) those driving equipped vehicles but not following advice at all, and
(d) those driving unequipped vehicles.

Second, it would need to indicate the distribution, in time and space, of traffic flow and congestion on the network, indicating the different impacts on different types of link.

The simulation model should be able to represent the performance of strategies having different specifications (e.g., different beacon densities, different guidance network densities and different travel time estimation procedures). The full range of impact parameters would include the street network, OD matrix, market penetration, guidance system parameters, the route choice behavior of nonuser groups and the behavioral response of equipped drivers.

Initial thought on the model indicates that it must be built for this specific problem but, at the same time, draw from available tools where appropriate. The output should contain a representation of delays at junctions and during special turning movements, rather than simply via link-based speed–flow relationships. The simulation model ought to represent variability and uncertainty in journey times since this will influence the perceived quality of advice from the guidance system. Since the route guidance system is itself dynamic, it will be necessary to build this feature into the simulation model. This implies that drivers will not be assumed to complete the journey along the route to which they have been initially assigned, but that the model will be based on time slices at the end of which the current position of each driver (or group of drivers) will be assigned and recorded prior to a new time slice, during which they will continue their journey in the light of the latest information and changed circumstances. This dynamic model structure should be capable of representing feedback effects whereby the advice affects the road conditions which in turn affect the advice (May and Bonsall 1989).

The simulation model will further need to represent the effect of the type of advice given or the ability or desire of the driver to follow it. It might simply suggest that a given proportion of drivers will fail to follow advice at junctions of a given type. These proportions need to be established with data from demonstration projects.

The real-time routing model is much more complex than current analysis and traffic models, but it can evolve from the current framework.

4. Applications of Expert Systems

4.1 Use of Models

The models used to address the aforementioned problems of analysis and control do not, in many cases,

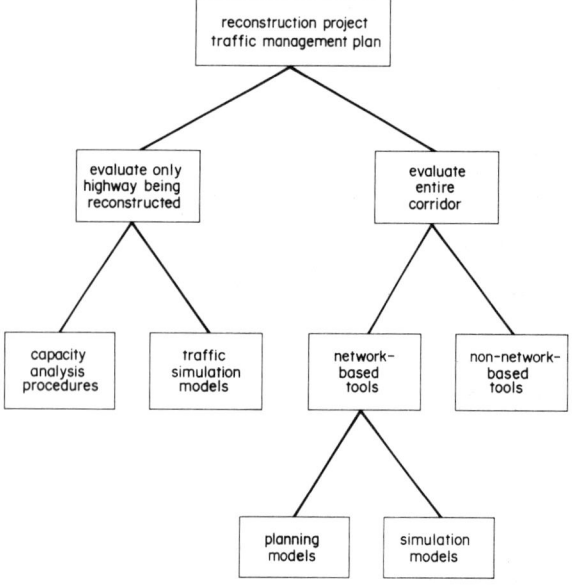

Figure 2
Decision tree for selecting analysis tools for the travel impact evaluation process (after Krammes *et al.* 1989)

provide the decision maker with the direct answer to a problem. This task requires experienced engineers to interpret the results and translate them into design and policy decisions.

Advances in artificial intelligence (AI) offer a potential for solving or lessening these shortcomings and for producing a new generation of traffic and management tools (Faghri 1990). Expert systems is the branch of AI that manages knowledge bases. The knowledge base maintains facts and rules that use the facts for decision making. An intelligent interface is a decision-making program that deals with data preparation, input–output, model interfacing and model selection. For example, for a particular analysis/control problem, an intelligent interface might suggest analytical models for the problem, check the adequacy of the available data and guide data preparation.

The lack of closed-form solutions in most analysis/control problems makes the use of present algorithmic traffic models an art. Knowledge-based expert systems offer a means of analyzing and solving open-ended problems for transportation engineers.

With a knowledge-based expert system, an inexperienced traffic engineer could use sophisticated analytical/control tools. For example, according to Faghri (1990)

> "the engineer will enter some facts about a highway corridor such as peak hour traffic volumes, highway configurations and other relevant information. Then the expert system would recommend feasible transportation management strategies and the necessary procedures and additional resources required to implement the suggested management schemes. The solutions suggested by the traffic management decision system are deduced from knowledge obtained from a variety of experts and based on experiences with similar problems."

4.2 Real-Time Traffic Control

Expert systems are potentially useful in real-time traffic control. Current computerized urban signal control systems are based on minimizing vehicle delays or queues at signalized intersections. These algorithms work well for an undersaturated network but not for congested networks. The expert system permits rules to be added to the algorithm to express different control strategies for congested networks based on principles such as keep intersections clear of spillback vehicles and given the right of way to the direction in which traffic can move. If traffic diversion and simulation capabilities are added, a real-time traffic management decision tool that will suggest traffic congestion or incident diversion schemes will be obtained (see *Expert Systems Approach to Road Traffic Control*).

4.3 Intersection Signal Control

The FHWA is in the process of awarding a contract to develop an expert system for signal design at isolated intersections. This is one of the first nationally (USA) sponsored research projects into expert systems applications in transportation systems analysis/control. If successful, this project should establish a new perspective on state-of-the-art modelling practice. This project was motivated by the fact that there currently exists no truly comprehensive tool for isolated intersection signal design. Capacity analysis procedures do not provide a method for timing signals, while optimization models do not provide the output that the capacity procedures do. In addition, there is no guidance in such methods for considering actuated controllers in the analysis. Accordingly, the solution requires supplementary study to consider such issues as phasing patterns, detector placement and cost–benefit analysis. Thus, an expert system is presumed to be able to enhance isolated intersection signal design by selecting the most appropriate type of controller, estimating optimal signal control parameters, identifying detector configurations and allowing ease of use.

5. Future Developments

The major thrust for transportation systems analysis/control in the coming years will be the combination of current tools into packages that comprehensively address complex problems. The methods will reflect real-time data and generate strategic plans. Sensitivity to advanced guidance systems and other new technologies will be realized. The user will interface with the modelling systems through expert systems which will make highly complex models useful even to novice users.

See also: DRIVE; Traffic Control Systems: Trends

Bibliography

Faghri A 1990 Structuring expert systems and computerized models in transportation. Transportation Research Board Report. TRB, Washington, DC (in press)

Huber M J 1982 Traffic flow theory. *Transportation and Traffic Engineering Handbook,* 2nd edn. Prentice-Hall, Englewood Cliffs, NJ, Chap. 15

Hutchinson B G 1974 *Principles of Urban Transport Systems Planning.* McGraw-Hill, New York

Krammes R A, Ullman G L, Dresser G B, Davis N R 1989 *Application of Analysis Tools to Evaluate the Travel Impacts of Highway Reconstruction with Emphasis on Microcomputer Applications.* Texas Transportation Institute, TX

Manheim M L 1979 *Fundamentals of Transportation Systems Analysis,* Vol. 1. MIT Press, Cambridge, MA

May A D, Bonsall P W 1989 Recommended methodology for evaluation of route guidance in London and the potential for comparability with Berlin. Transport and Road Research Laboratory Contractor Report No. 130. Institute of Transport Studies, Crowthorne, UK

US Department of Transportation 1989 *Discussion Paper on Intelligent Vehicle–Highway Systems.* US Department of Transportation, Washington, DC

Van Aerde M, Voss J, Blum Y 1989a Modelling the Burlington Skyway FTMS during recurring and non-recurring traffic congestion. *Traffic Eng. Control* **30**, 228–40

Van Aerde M, Voss J, Ugge A, Case E R 1989b Managing traffic congestion in combined freeway and traffic signal networks. *ITE J.* **59**(2), 36–42

Yedlin M, Lieberman E B 1988 TRAF—A traffic simulation software system for traffic engineers and planners. Federal Highway Administration Report No. DTFH61-85-C-00005. FHWA, Washington, DC

M. J. Demetsky
[University of Virginia, Charlottesville, Virginia, USA]

Underground Railroad Modelling and Control

This article is devoted to traffic analysis and control design for high-frequency metro lines. Such lines present natural instability; that is, any deviation with respect to the nominal setpoint, affecting a given train, increases with time and disturbs the operation of the other trains. This phenomenon is explained as follows. The number of passengers waiting at a given platform in order to board a train increases with the time elapsed since the departure of the preceding train. If a train is delayed, this time interval, and therefore the number of passengers and the corresponding train staying time, become greater than nominally expected. Hence, the delay of the train increases from one platform to the following one. Conversely, for the next train, the time interval is shorter than expected, the number of passengers and the train staying time are less than their nominal values and the train is in advance. The same argument shows that the next train is delayed, and so on.

Minimal traffic control is implemented, by use of the traffic lights, in order to ensure, according to the security rules, a minimal distance between successive trains and to avoid collisions (see *Automatic Train Control: Protection Principles*). The resulting traffic conditions are, however, not acceptable: the time deviations with respect to the nominal time schedule are important, the passengers repartition between the trains is highly nonuniform and the commercial speed is reduced. A more efficient traffic control strategy is therefore necessary, from the point of view of the passengers as well as of the company.

The purpose of this article is to discuss traffic modelling and regulation design. More precisely,

(a) a linear mathematical model for the metro traffic is proposed, pointing out the natural instability;

(b) using this model, automatic traffic control strategies are designed (either centralized or decentralized) which ensure the stability of the operation around the nominal setpoint; and

(c) the results of simulations under realistic operating conditions show that these control strategies improve the behavior of the system subject to physical security requirements.

1. Traffic Description

Each metro network can be simply seen by the users as a set of platforms where the trains have to stop in order to allow the passengers to get on or to get off. This set of platforms is generally divided into sequenced subsets which constitute metro lines. Such lines are connected together through connection stations which allow the passengers to access the others lines.

Consider an ordered set of trains running on a metro line constituted by a sequence of platforms. Each train and each platform are characterized, respectively, by a number referred to as the train index and the platform index. In this article, only sequential line structures and sequential operation conditions will be considered. This means that:

(a) the sequence of platforms crossed by a given train is ordered and is, therefore, the same for all trains on the line; and

(b) at each platform, the sequence of trains is ordered and is, therefore, the same for all platforms.

2. Traffic Modelling

The sequential operation of the line, as described in Sect. 1, allows a simple control-oriented mathematical model of the traffic to be built, relating the departure instants of the different trains from the different platforms. Another possible approach has been proposed by Levine and Athans (1966) and is based on a continuous description of the vehicle motions. The motion of each train is described in this case by two differential equations relating the position, the velocity, the mass and the applied forces on this train. However, this kind of model is valid only for the description of the motion of the trains between successive platforms and does not take explicitly into account the staying time in station which is, as will be seen, critical for the traffic stability of metro lines.

2.1 Basic Transfer Equation

Consider a section of a metro line with N sequenced platforms and where M trains are operated (see Fig. 1). Throughout this article, a two-indices notation will be used to identify descriptive variables: the upper index refers to the train number and the lower index to the platform number. According to this convention, denote t_k^i as the departure instant of train i from platform k.

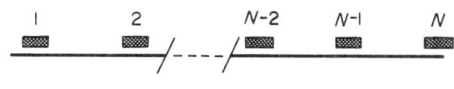

Figure 1
Schematic representation of a metro line

Obviously, the departure instants of train i from two successive stations k and $k+1$ are related by

$$t^i_{k+1} = t^i_k + r^i_k + s^i_{k+1} \quad (1)$$

where r^i_k is the running time of train i from k to $k+1$ and s^i_k is the staying time of train i at platform k. In order to model further the running time r^i_k and the staying time s^i_k, four basic modelling assumptions are now introduced.

(A1) The operation conditions (number of trains, desired interval between successive trains, number of passengers arriving at a given platform per second, etc.) are constant.

(A2) The running time r^i_k of a train between two successive platforms does not depend on the number of passengers on the train.

(A3) The staying time of a train at a platform depends linearly on the number of passengers getting on the train.

(A4) The number of passengers to board train i at platform k is proportional to the interval between the departure instants of the successive trains $i-1$ and i from platform k.

Assumption A1 is introduced only in order to simplify the analysis and can easily be relaxed. The other three assumptions A2, A3 and A4 can be replaced by more sophisticated modelling assumptions. It is possible, for instance, to take into account the load (i.e., the numbers of passengers) of the trains and to relate the staying time not only to the number of embarking passengers (as in A3) but also to the number of passengers getting off the train (e.g., Campion et al. 1985). Simulation results obtained with such more sophisticated models are, however, not significantly different, as far as traffic analysis is concerned.

From A1 and A2, the running time can be expressed as follows:

$$r^i_k = R_k + u^i_k + w1^i_k \quad (2)$$

where R_k is the nominal running time from k to $k+1$, u^i_k is the control action applied to train i between k and $k+1$ in order to increase ($u^i_k \geq 0$) or to decrease ($u^i_k < 0$) the running time and $w1^i_k$ is a disturbance term. From A1, A3 and A4, the staying time s^i_{k+1} can be modelled as

$$s^i_{k+1} = S + c_{k+1}(t^i_{k+1} - t^{i-1}_{k+1}) + w2^i_{k+1} \quad (3)$$

where S is the minimal staying time at a platform, when no passenger gets on the train and the doors are closed as soon as possible, c_{k+1} is the delay rate representing the effect of the time interval between the departure instants of two successive trains (A4) and $w2^i_{k+1}$ is a disturbance term.

In Eqns. (2) and (3), S, R_k and c_k are parameters to be estimated. Parameters S and R_k can be evaluated from the operating conditions while the c_k have to be estimated by linear regression on a large number of observations (t^i_k, s^i_k) at each platform, according to Eqn. (3). Usual values of c_k are in the range 0.01–0.05 (see Campion et al. (1985) for examples relative to Brussels metro lines).

Using Eqns. (2) and (3), Eqn. (1) can be rewritten as

$$(1 - c_{k+1})t^i_{k+1} = t^i_k - c_{k+1}t^{i-1}_{k+1} + S + R_k + u^i_k + w^i_k \quad (4)$$

where $w^i_k = w1^i_k + w2^i_k$. This relation between the departure instants of the trains will be used throughout the subsequent developments. It must be pointed out that the admissible control actions and disturbances are bounded in order to satisfy the security requirements which must prevent any collisions between trains. Equation (4) gives a local description of the traffic behavior relating two successive trains and two successive platforms. The complete set of the departure times t^i_k ($i = 1, \ldots, N$; $k = 1, \ldots, M$) corresponds to a global description of the traffic.

2.2 Time Deviations

It is assumed that ideal traffic planning has been established for the line under consideration. It takes the form of a nominal time schedule which is defined as the set of the nominal departure instants T^i_k of each train at each platform of the line. This nominal time schedule is characterized by a constant time interval H between successive trains (i.e. $H = T^{i+1}_k - T^i_k$). It must be coherent with the natural dynamics of the line; that is, it must satisfy the basic relation given by Eqn. (4) in the absence of control and disturbances ($u^i_k = w^i_k = 0$).

$$T^i_{k+1} = T^i_k + c_{k+1}H + R_k + S \quad (5)$$

Define x^i_k as the time deviation of the actual departure instant, t^i_k, from its nominal value, T^i_k (i.e., $x^i_k \triangleq t^i_k - T^i_k$). Then, the basic dynamical Eqn. (4) is rewritten as follows:

$$(1 - c_{k+1})x^i_{k+1} + c_{k+1}x^{i-1}_{k+1} = x^i_k + u^i_k + w^i_k \quad (6)$$

For given sequences of control actions and disturbances (u^i_k and w^i_k), Eqn. (6) defines completely the set of time deviations $\{x^i_k\}$ which is used in order to evaluate and to improve the quality of the actual operation.

3. Intrinsic Instability of the Traffic Behavior

Define Γ_k as the mean square value of the time deviations of I successive trains at a given platform k:

$$\Gamma_k \triangleq \frac{1}{I} \sum_{i=1}^{I} (x^i_k)^2 \quad (7)$$

This index reflects the quality of the line operation: if all trains are running under nominal operation mode, this index is equal to zero. However, if for some reasons, Γ_k is greater than zero at a platform, it can be proved (Van Breusegem et al. 1991) that Γ_k increases from platform to platform. This means that a metro line is intrinsically an unstable system, whatever the traffic dynamics (i.e., whatever the positive values of the c_ks): any initial time deviation at the first platform is propagated and exponentially amplified along the line.

4. Traffic Regulation

Traffic regulation consists of the development of traffic control strategies ensuring acceptable traffic performance and global traffic stability which means that the operation quality index Γ_k decreases from platform to platform. A method generating a family of such stabilizing traffic control laws, which are designed according to the same principle is presented here: the control action to be applied to a train is computed in order to minimize some adequate performance indexes, minimizing the time deviations with respect to the nominal operation conditions. Depending on the choice of the performance index, the proposed control laws differ in two ways.

(a) They differ by their complexity. The control applied to a train at a given platform can be elaborated by processing only local information (the situation at this platform and possibly at the next one) or the information relative to the complete line. These two strategies are referred to, respectively, as "decentralized" and "centralized" traffic control.

(b) They differ by their performance. A more sophisticated control law (centralized) will ensure better performance of the system, namely as concerns the smoothness of the transient behavior in the presence of disturbances.

4.1 Decentralized Traffic Control

Assume that train i has reached platform k. The control u_k^i is chosen in order to minimize the following quadratic performance index, subject to the linear constraint given by Eqn. (6):

$$J_1 = p(x_{k+1}^i)^2 + q(x_{k+1}^i - x_{k+1}^{i-1})^2 + (u_k^i)^2 \quad (8)$$

The first term penalizes the time deviation at the next platform $k+1$. The second term penalizes the deviation of the time interval at platform $k+1$ between trains $i-1$ and i, with respect to the nominal value H. Indeed, $x_{k+1}^i - x_{k+1}^{i-1} = t_{k+1}^i - t_{k+1}^{i-1} - H$. The third term penalizes the control action and prevents u_k^i becoming too large. The nonnegative weighting coefficients p and q are design parameters at the user's choice and are selected according to the control purpose:

(a) $p=0$ for interval control (regularity);
(b) $q=0$ for time schedule control (punctuality);
(c) $p \neq 0$ and $q \neq 0$ for any trade-off between these two extreme objectives.

The optimal control minimizing the quadratic criterion J_1, under the linear constraint of Eqn. (6), is linear in the time deviations (x) variables:

$$u_k^i = g_k x_k^i + f_k x_{k+1}^{i-1} \quad (9)$$

where $f_k = (q + pc_k)/[(1-c_k)^2 + p + q]$ and $g_k = -(p+q)/[(1-c_k)^2 + p + q]$. This law has a form allowing decentralized traffic control implementation: the control to be applied to train i, at platform k is the linear combination of two time deviations: the deviation of train i at platform k and the deviation of the preceding train $i-1$ at the next station $k+1$.

4.2 Centralized Traffic Control

Consider a generalized criterion J_2 with optimization horizon L:

$$J_2 = \sum_{j=1}^{L} [p_j(x_{k+j}^i)^2 + q_j(x_{k+j}^i - x_{k+j}^{i-1})^2 + r_j(u_{k+j}^i)^2] \quad (10)$$

where L is the horizon and the nonnegative p_j and q_j and positive r_j weighting coefficients are design parameters with the same interpretation as for J_1. The control purpose is to minimize the weighted sum of time deviations and control actions for the L next platforms. Obviously, J_1 is a particular case of J_2, with the choice $L=1$, $p_j=p$, $q_j=q$ and $r_j=1$.

The theoretical optimal control sequence u_j^i of the control actions applied to train i at platform $j(j=k,\ldots,k+L-1)$ is linear in the time deviation variables:

$$u_j^i = \sum_{a=1}^{M} f_{aj} x_{i+j-a}^a \quad (11)$$

The gains f_{aj} are obtained by solving a Riccati equation (e.g., Bryson and Ho (1969) for the complete discussion of this linear quadratic problem). Practically, at each platform k, only the first control of the sequence computed in the window $(j=k,\ldots,k+L-1)$ is applied:

$$u_k^i = \sum_{a=1}^{M} f_{a1} x_{i+k-a}^a \quad (12)$$

The control applied to train i at platform k is therefore the linear combination of the M time deviations characterized by the value $(i+k)$ of the sum of their two indices. It involves information relative to the M trains on the line. This control law therefore requires a centralized implementation.

The advantage of this centralized control over the decentralized one presented before is that the effect of a disturbance affecting a train is immediately reflected on the control actions applied to all the trains on the line. The transient behavior of the disturbances resorption will therefore be smoother with values of the control actions which are smaller, but more uniformly distributed on all the trains, than for a decentralized control strategy. Other performance indexes are discussed by Van Breusegem (1987).

4.3 Practical Implementation

The implementation of the decentralized control law given by Eqn. (9) has been investigated for two lines (A and B) of the Brussels railroad underground transportation system (see Fig. 2). These lines have a long common section where trains of both lines are operated alternately. The standard running times R_k between successive platforms are about 60–90 s, the minimal staying time S is 7 s and the delay rate parameters c_k have been estimated from statistical data. As the passenger flow is not high compared with the maximum embarkment rate, even during the rush hours, the values of the c_k are small (about 0.03). A nominal time schedule has been generated on basis of these values of the parameters.

A program has been implemented for the simulation

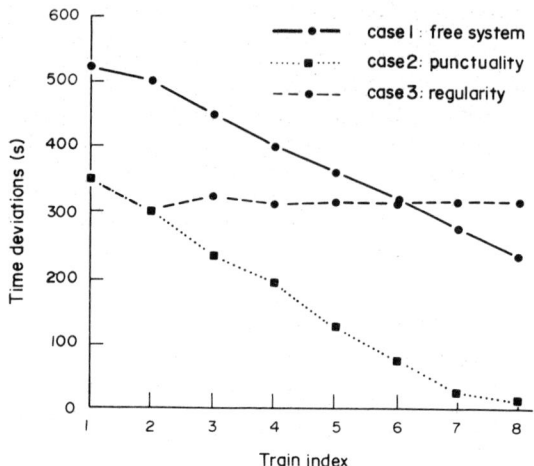

Figure 3
Time deviations at Etangs-Noirs platform (Brussels metro)

of the complete system, taking into account traffic security requirements, minimal distance between trains, and so on. This program generates the absolute departure times t_k^i, in connection with the nominal time schedule. It allows the introduction of control actions u_k^i as well as the disturbance terms for any train at any platform. Equation (9) has been tested on the basis of the simulation program and especially under the following operation conditions.

(a) The constraints on the control are the following: in addition to the respect of the security requirements, the running time modifications may not exceed one-tenth of the standard running time.

(b) For all simulations, a delay of 240 s is imposed on the first train at the first platform of the common section (Mérode Station)

The main results are summarized in Fig. 3, giving the time deviations of the trains at the last platform of the common section (Etangs-Noirs Station). Odd train numbers refer to trains of line A, even numbers to trains of line B. Three different control policies are implemented.

(case 1) Free system—no control is applied. It can be seen that the delays at the last platform become very important (500 s for the first train and 230 s for the eighth train). The delays are propagated because a minimum distance between successive trains has to be ensured.

(case 2) Punctuality—Eqn. (9) with constraints, with $p = 5$ and $q = 0$. It can be seen in Fig. 3 that the delays are smaller than for the free system.

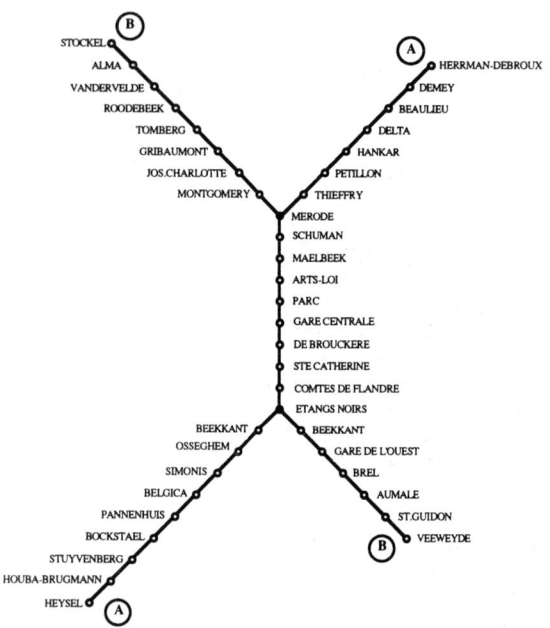

Figure 2
Configuration of lines A and B of the Brussels underground transportation system

(case 3) Regularity—Eqn. (9) with constraints, with $p=0$ and $q=5$. The delays are larger than in case 2, but the interval regularity is nearly perfect.

Clearly, the regulation significantly improves the behavior of the system. A better result can of course be obtained if the severe constraints on the control variables are somewhat relaxed.

5. Conclusions

The generation of control strategies by the minimization of an adequate performance index based on a discrete-time modelling constitutes a very efficient and flexible approach of the traffic control problem on metro lines:

(a) the proposed control laws stabilize the traffic;

(b) by the choice of the criterion a trade-off can be achieved between the complexity of the control (either centralized or decentralized) and the traffic performance; and

(c) the choice of the weighting coefficients involved in the criterion reflects the main control objective: it is possible to favor either the time schedule control (punctuality), or the time interval control (regularity).

For a detailed theoretical study of all these aspects, see Van Breusegem *et al.* (1991).

See also: Railroad Systems: Line Supervision and Control; Underground Railroad Modelling and Control: Discrete-Event Approach

Bibliography

Araya S, Sone S 1984 Traffic dynamics of automated transit systems with pre-established schedules. *IEEE Trans. Syst. Man Cybern.* **14**, 677–87
Bryson A E, Ho Y C 1969 *Applied Optimal Control.* Blaisdell, Waltham, UK
Campion G, Van Breusegem V, Pinson P, Bastin G 1985 Traffic regulation of an underground railway transportation system by state feedback. *Optimal Control Appl. Methods* **6**, 385–402
Levine W S, Athans M 1966 On the optimal error regulation of a string of moving vehicles. *IEEE Trans. Autom. Control* **11**, 355–61
Van Breusegem V 1987 Simulation et régulation de lignes de métro automatisées Ph.D. thesis report. Catholic University of Louvain
Van Breusegem V, Campion G, Bastin G 1991 Traffic modelling and state feedback control for metro lines. *IEEE Trans. Autom. Control* (in press)

V. Van Breusegem, G. Campion and G. Bastin
[Université Catholique de Louvain,
Louvain-la-Neuve, Belgium]

Underground Railroad Modelling and Control: Discrete-Event Approach

In many cases of modelling and designing physical plants, simulation turns out to be a valuable tool. This holds in particular for underground railroad systems, where the intrinsic features of the network, as far as it concerns the transportation model, prevent the use of analytic techniques to study the system behavior. A method of analysis that suits this case well is the discrete-event simulation (Fishman 1978, Banks and Carson 1984). In this simulation environment, the most important entities are the "events" which, on their occurrences, cause the overall state of the system to be updated, some activities to start and new events to be scheduled.

The purpose of this article is just to show how discrete-event modelling and simulation can be used as an effective tool in order to analyze the performances of the considered system and of the applied control strategies. As for the existing literature on models for rail transportation and simulation, see Assad (1980), in particular for optimization, queuing and simulation models. The works by Eichler and Turnheim (1978) and Di Febbraro and Ferrara (1989) are important in showing the benefits that could be achieved whenever discrete-event and continuous simulation are combined while, in Levene *et al.* (1974), an interesting simulation study about the effectiveness of on-line computer control of underground railroads is presented.

1. The Model

For the sake of simplicity, in this article the system to be considered is an underground system that is modelled and simulated neglecting energy consumption and electrical issues, but taking into account the dynamical aspects of the train motions. It is also necessary to introduce some stochastic variables to represent the randomness of real-world traffic situations. Nevertheless, the proposed approach can also be used for modelling more complex underground networks.

The first step in planning a simulation experiment is the model definition, in order to represent the overall system under concern. Such a model is intended to describe the history of the system state variables during the simulative run, by means of the sequence of its state transitions. In the case considered, the following assumptions are made:

(a) the network is modelled as a double-way path;

(b) there are m lines and m stations, a line being the portion of rail track included between two stations;

(c) there are n trains of equal length;

(d) there is a different number of track circuits in each line, depending on the topological aspects of the railroad network;

(e) the normal operating condition of the underground system is established by a timetable and every train is due from the end station according to a previously stated train starting rate; and

(f) stations are modelled by means of a single track circuit under the assumption that a train length is always shorter than that of the track circuit.

Note that the train starting rate is likely to be perturbed during the simulation because of significant deviations (of a single train or more) from the prescribed timetable.

On the basis of real data, the stopping times of trains at the stations are computed taking into account a lower limit based on the prefixed timetable, an additive stochastic term that models the getting in or out of the passengers and, finally, a further time due to the current state of the downstream track circuits. Also, the run times between any two stations are perturbed during the train motion, because of several random effects. However, a lower limit of the run times is fixed for each line, according to safety and comfort requirements.

When modelling an underground railroad system, the necessity of representing in detail the operating functions of the signalling system is apparent. In the considered case of a completely automated network, it works as follows. On the basis of the values of the variables $s_h^{(i)}$ and $s_t^{(i)}$, indicating the position of the head and of the tail of the ith train, respectively, it is possible to determine uniquely the set of track circuits that are presently occupied by the ith train. Then, a set of codes is assigned to a certain set of track circuits that are upstream with respect to the last track circuit occupied by the train. Namely, each of these track circuits is given an ordinal index, to represent its ordinal position with respect to the train tail position. Obviously, there is a precise correspondence between codes and order indexes. Actually, each code represents a maximum speed allowed in a track circuit for a train subsequent to the train that has determined the code imposition. The values of codes depend on the presence of trains in the track circuits ahead of the one considered, and their computation has the function of avoiding collisions among the trains. The signalling system manages the whole line.

During simulation, in absence of correction actions on the train motions due to the application of control strategies, the target speed of each train is computed, when it enters a new track circuit, as the minimum among a reference speed, given *a priori* in accordance with the timetable, a limit speed (i.e., the maximum allowed speed in a track circuit because of its physical aspects) and a code speed, imposed by the signalling system. Besides, a value of acceleration (or deceleration) is imposed, in order to reach the target speed from the actual speed of the train at the entrance of the track circuit. This value is fixed *a priori* purely on the basis of comfort and safety specifications, and is equal for all track circuits. Naturally, depending on the length of the track circuit, the target speed may or may not be reached within the track circuit itself. In any case, the determination of the target speed imposes a nominal speed profile, which allows the nominal run time of that train on that track circuit to be computed easily. Note that the interval between the entrance and the exit of the train head with respect to that track circuit is what is intended by nominal run time. This run time is to be additionally perturbed, in order to take into account such things as mechanical, electrical and random effects not already modelled.

2. Variables and Events

The variables of the model are classified as exogenous (input), state and endogenous (output) variables.

2.1 Exogenous Variables

The exogenous variables are the inputs of the system and are generated according to stochastic distributions deduced from the analysis of the external real environment. To produce the desired distribution, the inverse transform technique or other techniques (Banks and Carson 1984) can be used, all of them based on uniform distribution sampling. Here, the random variables that have to be generated and fed into the system are:

(a) the additive stochastic variables referring to the train stopping times at the stations (these variables are related to the number of passengers waiting and, thus, to the time of the day; moreover, they are also dependent, on the basis of prefixed statistics, on the time that has elapsed since the last passing of a train through that station);

(b) the stochastic speed perturbations relevant to the train motion within the track circuits; and

(c) the inversion times at the end stations.

2.2 State Variables

The state variables, describing at any time the state of the system during the intervals between two consecutive events, take values depending on both the values of the exogenous variables and the one previously taken by the state variables themselves. Of course, state variable transitions are determined by event occurrences. In the model presented, the trains can be in different positions along the lines, move according to fixed speeds and stop. The state variables that are suitable to describe in a right way the system activities, as far as concerns the state transitions, can be divided into train state variables and track circuit state variables.

The train state variables consist of the indication of the state of motion and the position of each train; that is, the number of the track circuit occupied by it, as well as the actual speed of each train when it will enter the next track circuit.

The track circuit state variables consist of a variable indicating if a train is running along the track circuit and a second variable that can take five different values according to the five different codes which can be found in a railroad track. Each code corresponds to a different code speed value, which is communicated to a train by the signalling system.

2.3 Endogenous Variables

The endogenous variables are the result of the evolution of the system when the considered input signals act. Typical variables of interest are:

(a) average or maximum delay of each train; and

(b) global average or maximum delay.

2.4 Description of Events and State Transitions

Events cause state transitions, modifying the overall state of the system. In the model, independent events (i.e., not necessarily coinciding with other events) can be classified as:

(a) a train leaving from a certain station;

(b) a train arrival at a certain station;

(c) the head of a train entering a certain track circuit;

(d) the tail of a train going out of a certain track circuit;

(e) the beginning of the train inversion of the end station.

When an event of type (a) occurs, whether or not the train can really leave is tested. If so, a "train entering a certain track circuit" event is scheduled assuming that such an event is to occur after a certain period, whose length is computed as indicated in Sect. 1. Otherwise, if it is impossible to perform the actual train leaving from the station, the train state is set to a "ready-to-start" state. Then, a next event, which will be able to provoke the train leaving, is awaited.

If an event of type (b) occurs, it has to be determined whether or not the station is an end one. If so, the event that is scheduled is a "beginning of the train inversion at the end station" event. Otherwise, the train state becomes "stopped" and the "train leaving from a certain station" event is scheduled, fixing its occurring time on the basis of a realization of the additive stochastic variable affecting the stopping time.

The occurrence of an event of type (c) causes the target speed to be computed, as described in Sect. 1. Then, the nominal speed profile is computed and further perturbed. On this basis, by means of easy kinematical considerations, it is possible to determine when s_h will reach the end of the present track circuit and when s_t is expected to reach the next track circuit border, if this will happen within the occurrence of the event in point i. Then, a new "train entering a certain track circuit" event and, possibly, a new "train going out of a certain track circuit" event is scheduled.

Whenever an event of type (d) occurs, the codes backward imposed by the signalling system are modified. This can cause an updating of the occurrence times of the previously scheduled events. Besides, a test of possible ready-to-start trains which may not be enabled to start has to be performed, to schedule a "train entering a certain track circuit" even in the affirmative case.

The "beginning of the train inversion at the end station" event (type (e)) causes a "train leaving from a certain station" event to be scheduled. The occurrence time takes into account the time actually needed for inversion, the nominal time necessary for passengers going in and a random delay.

3. Control

An important part of the design of an underground railroad network is the development and testing of some control strategies able to solve problems in train motions. Consequently, it is necessary for the system model to take into account all those real aspects that can cause complex delay situations to occur. For the sake of simplicity, in the application of discrete-event simulation presently considered, it is assumed that perturbations in traffic flow can arise because of significant deviations of one or more trains from the prescribed timetable. A distinction is also made between possible traffic perturbations. Two types are considered:

(a) local perturbations, which involve a single train; and

(b) generalized perturbations, which involve more than a train.

These perurbations can usually be due to two different facts: abnormal passenger crowding at some neighboring stations; and abnormal operating conditions of lines, trains or power supply plants.

Nevertheless, in order to design and evaluate the performances of control strategies, there are issues to be fulfilled. By far the most important requirement is, of course, the passengers' safety, which is assured by the signalling system, but there are also other main requirements, such as adhering to the timetable and guaranteeing the train rate regularity. The normal operating conditions for an underground railroad network are established by a timetable. The purpose of the analyzed control strategies is to take decisions whenever significant perturbations in the system behavior occur. Naturally, the overall system is expected to possess some degrees of freedom, in the sense that some parameters, characteristic of the model, can possibly be manipulated so as to reestablish normal working conditions of the network. For instance, in the present case the following parameters can be changed:

(a) train speed values along the different track circuits;
(b) train leaving times from any station; and
(c) times for train inversion at the end stations.

Generally speaking, the control strategies can be divided as follows:

(a) strategies that aim at restoring the timetable, which are further divided into
   (i) policies involving a single train (timetable regulation), and
   (ii) policies involving more than a train (time interval regulation); and
(b) strategies that lead to the defining of a new timetable, because of the impossibility of bringing the system back to the regular timetable (timetable shift regulation).

In the remainder of this article, some possible regulation strategies belonging to different classes will briefly be mentioned. See Casalino *et al.* (1989) for a more detailed presentation and Van Breusegem and Campion (1989) for the description of an interesting scheme for traffic control design.

3.1 A Timetable Regulation Strategy

A timetable regulation strategy consists of making the best use of the difference existing between the reference speed and the limit speed along the lines ahead of the late train. So, until the delay is eliminated, the train speed is set to the maximum possible value (obviously, not greater than the code speed), while the train stopping times at the stations are as short as possible.

3.2 A Time Interval Regulation Strategy

The purpose of a time interval regulation strategy is also to avoid deviations from the prescribed timetable but it indirectly manages to restore the desired train rate as well. Consider a strategy operating as follows. When a train is late, the train that is immediately downstream is stopped in a certain station for a time longer than the normal one in order to gather as many passengers as possible. As a result, the delay of the late train will not increase because of abnormal crowding of passengers when it stops at the mentioned station.

3.3 Timetable Shift Regulation Strategies

Timetable shift regulation strategies, when implemented, aim at making the trains travel at regular time intervals. From a practical point of view, they can be developed in order to produce an instantaneous shift of the timetable from the moment a perturbation occurs to the end of the simulation experiment.

See also: Railroad Systems: Train Driving Control; Underground Railroad Modelling and Control

Bibliography

Assad A A 1980 Models for rail transportation. *Transp. Res. A* **14**, 205–20
Banks J, Carson J S 1984 *Discrete-Event System Simulation*. Prentice-Hall, Englewood Cliffs, NJ
Casalino G, Di Febbraro A, Ferrara A, Minciardi R, Nicoletti D 1989 Discrete-event simulation and control strategies for underground railways. *Proc. CCCT'89.* AFCET, Paris, pp. 207–11
Di Febbraro A, Ferrara A 1989 Performance analysis of underground railway networks through continuous time and discrete event simulation. *Proc. European Simulation Congress.* SCS, Ghent, Belgium, pp. 653–8
Eichler J, Turnheim A 1978 A combined simulation of a rapid transit system. *Simulation* 155–67
Fishman G S 1978 *Principles of Discrete Event Simulation.* Wiley, New York
Levene S M, Wision J G, Williamson D 1974 An assessment of the application of on-line computers to control an underground railway—A simulation study. *Transp. Res.* **8**, 123–35
Van Breusegem V, Campion G 1989 Traffic modelling and control for circle lines *Proc. CCCT'89.* AFCET, Paris, pp. 173–9

G. Casalino, A. Di Febbraro, A. Ferrara and R. Minciardi
[Università Genova, Genoa, Italy]

D. Nicoletti
[Ansaldo Transporti S.p.A., Genoa, Italy]

Underground Railroad: Organization of Operations

Each metro attempts to meet its specific operating problems in the best way possible, within the constraints of its particular technical, economic and human parameters. However, noticeable differences may be observed between automatic train control (ATC) systems implemented in a new system, where a wide choice of options permits a highly homogeneous and simple design, and ATC systems gradually introduced into an old system, where options are rendered more complex owing to the necessity for compatibility with an entire cultural and technical environment. This requirement acts both as a brake on innovation and as a protection against oversimplified and unrealistic concepts.

However, certain similarities exist between systems and, thus, a simplified description of three such systems, concentrating in each case on an aspect that demonstrates their originality and educational value, provides almost comprehensive knowledge of these systems as a whole.

1. Simplified Description of the Atlanta Metro (USA) ATC

1.1 Automatic Train Protection

Train detection is accomplished by electronically identifying the presence or absence of a train within a certain block of the track system. This information is sent to the station control room and relayed to the central control room.

Table 1
Algorithm for ATC speed regulation

Performance level (PL)	Acceleration rate ($m\ s^{-2}$)	Regulated speed for ATP received speed limits ($km\ h^{-1}$)						
		112	96	80	60	40	24	0
1	1.34	112	96	80	60	40	24	0
2	1.34	101	87	72	60	40	24	0
3	1.34	90	77	64	60	40	24	0
4	0.67	90	77	64	60	40	24	0
5	1.16	101	87	72	60	40	24	0

On each track circuit, the on-board automatic train protection (ATP) equipment (see *Automatic Train Control: Protection Principles*) receives speed-limit information from the fixed installations; if this speed is overstepped, the ATP triggers the train emergency stop, whatever the cause.

Trackside signalling is limited to shunting signals which, on the line, may show red (stop), steady green (line free, diverted track) or flashing green (line free, main line). Other possibilities (yellow and luminous white) are only used at the terminus. Track circuits (welded track, except in the vicinity of points) are at variable frequencies corresponding to different speed levels ($0\ km\ h^{-1}$, $24\ km\ h^{-1}$, $40\ km\ h^{-1}$, $60\ km\ h^{-1}$, $80\ km\ h^{-1}$, $96\ km\ h^{-1}$ and $112\ km\ h^{-1}$ for performance level 1). Acceleration rate and regulated speed in automatic train operation (ATO) can be modified by action on a special controller, giving different levels of performance as indicated in Table 1.

The minimum theoretical interval is 90 s (with a maximum dwell time of 24 s).

1.2 ATO

The ATO (see *Railroad Systems: Train Driving Control*) receives speed-limit information from the ATP and the target performance level from the automatic line supervision (ALS). In line with these two sets of data, the ATO determines the reference speed to which it limits itself in accordance with Table 1.

A system of marker coils installed along the track controls stops in stations. The control of doors, however, remains the responsibility of the driver, who is always present on the train.

In the event of an ATO failure, or for driver training, a manual driving mode is provided with on-board signalling (in the driver's cab) indicating the restricted speed supplied by the ATP system via the running rails, which can thus activate the emergency stop if the driver exceeds the authorized speed.

In the event of an on-board signalling failure, the driver switches to manual driving and respects trackside shunting signals. The only automatic controls are a $40\ km\ h^{-1}$ speed restriction and automatic stopping if the train runs through a stop signal, effected by an automatic train stop system. This system provides a means of preventing a train from entering or leaving the main line without authorization from a transfer or yard area. The ground train stop trip arm is in the upright position and makes contact with the trip cock mounted on the train. This action causes emergency braking of the train by releasing the brake pipe pressure.

Finally, two other manual driving modes are employed, one on the yard and the other at the washing installation, where speeds are limited to $16\ km\ h^{-1}$ and $5\ km\ h^{-1}$, respectively.

1.3 ALS

Between the local ATP–ATO devices and the ALS central control room, an intermediate station control room exists in each station, playing an interface role (see Fig. 1) (see *Railroad Systems: Line Supervision and Control*).

This control room has a small visual control panel making it possible for a technician to be sent to the scene, in the event of a total central control room failure, to control routing in the zone concerned.

The main types of data received by the ALS include the geographical position of the trains, which is given

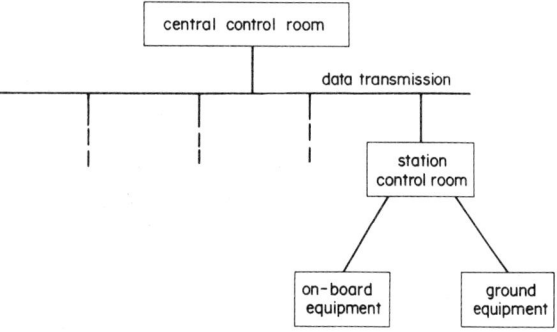

Figure 1
Schematic structure of the Atlanta ATC

by the ATP (occupancy of track circuits). Further information, in the form of train numbers, is initiated in the terminus by the drivers and then transmitted from the train to the station control room via a one-way communication system coupled to the ground equipment at passenger stations and terminal and transfer zones. A receiver picks up the following data transmitted through the running rail: identification number, destination, length and stopped train information. This information is used by the station control room to select an available route through an interlocking, and is relayed to the central control room.

The central control room comprises four workstations: train movement and routing control, power regulation and alarm control, information, and central control room management.

Each of these workstations has its own equipment and the central control room also has two visual control panels visible to all the operators, namely a traffic panel and a power and alarms panel.

The traffic panel, on which both the east–west and north–south lines are schematically represented, carries the following information:

(a) occupancy status of track circuit blocks,

(b) status of signals,

(c) switch points position monitoring, and

(d) monitoring of opening (or closing) of train doors when in station.

The power and alarms panel, on which the two lines are also schematically represented, carries the following information:

(a) monitoring of power supply to elementary track sections, and

(b) alarms corresponding to the opening of a breaker, the reduction in power available at the exit of a rectifier station, and the detection of a fire and an intrusion (i.e., someone entering a technical room without authorization).

(a) *Train movement and routing control.* There are two such workstations for the whole of the two lines. Their function is to check train identification given by drivers, to supplement automatic routing control where necessary and to observe train departures from the terminus and the regularity of traffic. Equipment at these workstations comprises:

(i) a console for line display, either by calling up one zone or by automatic tracking of a train on the line;

(ii) routing control;

(iii) radiotelephone links with drivers;

(iv) eventually, the automatic or manual control of train speeds (variation of acceleration or maximum speed); and

(v) passenger announcements inside the train upon authorization from the driver.

(b) *Power regulation: alarm control workstation.* This is ensured by one person only for the whole of the two lines. Its functions are regulation of traction power, ventilation control and alarm control, including taking a faulty installation out of service (if this is not done automatically) while trying to maintain operation in good safety conditions, alerting the appropriate maintenance service (or the fire service), bringing the repaired installation back into service and providing a maintenance–operation interface (as in the case of various works during operation).

The equipment in such a workstation comprises:

(i) control of power supply via console (able to display three adjoining sectors),

(ii) control of ventilation fans via console,

(iii) radiotelephone links with maintenance staff (special channel), and

(iv) telephone system (in particular an alarm system linked to telephones for passenger use).

(c) *Information workstation.* One person is responsible for information for the whole of the two lines. The functions of the information workstation are to inform passengers (on the platform) of disruptions to service, to ensure coordination with zone centers and, consequently, to operate the maintenance personnel switchboard. Equipment consists of a loudspeaker system on all platforms and the telephone system.

(d) *Central control room manager.* The basic function of the central control room manager is one of coordinating and supervising the central control room sets; for this purpose, the manager has at his disposal the majority of equipment described above.

2. Simplified Description of the Hamburg Metro ATC

2.1 General Information on the Continuous Bilateral Transmission of Telegrams Type A46

Following recommendations from the A46 committee of the International Union of Railways (UIC) Office for Research and Experiments (ORE), a carefully studied track–train transmission principle was defined, to serve as a support for all railway ATC. This principle was applied, along virtually identical lines, by German Railroads and the Munich and Hamburg metros. The same type of transmission has also been used for the intermediate capacity transit system (ICTS) developed in Canada (fully automated transportation systems of Vancouver, Toronto and Detroit).

An ATC system based on this principle can be described as follows. The system incorporates all ATC functions under all traffic conditions. It is characterized by a continuous exchange of information between the track and the train via a cable laid in the middle of the track; binary information is transmitted through 36 kHz and 56 kHz carrier frequencies. All trains running in its control zone are managed at a control center and train equipment ensures the processing of information required to run the train.

The control center receives information:

(a) track occupancy, routing formation, and signals status (from the signalling system and the centers either side), and

(b) weight–braking ratio, maximum speed, length and type of braking, and speed and position of the train (from the train by means of telegrams).

The control center is also aware of the network description (track profile, speed limits). From this information, it sends to each train in turn a telegram containing the following:

(a) the address of the train (its position number);

(b) elements required for calculating the speed limit, carried out on board, which the vehicle must respect;

(c) braking information (braking curve number, distance to cover before braking); and

(d) indicative information (speed and distance to target).

The train equipment picks up and processes the telegram sent to it, gives the driver speed-limit information (with the possibility of automatic implementation via automatic driving), monitors the kinematic behavior of the train, composes and sends a telegram containing information with regard to leaving the position relative to the address received, the exact position within a 100 m section and the actual speed.

2.2 Simplified Description of ATC Architecture

(a) *Technical structure*. The trains are safeguarded and controlled at the lower level by the ATC line section and train equipment. The functions of the middle level are carried out by the line section process control computer.

The ATC line section equipment and line section process control computer are located in section control centers. This is where traffic controllers, train dispatchers and station supervisors work. The traffic controller's job is to monitor train movements and routes, the train dispatcher observes the movement of passengers and gives orders to close the doors and to depart and the station supervisor maintains contact with the passengers and monitors the stations and their technical equipment. The length of line controlled by a section control center is usually 6 km.

The supervisory functions are carried out in the main control center which is equipped with a process control computer operating in a supervisory capacity, directly linked to all the section control centers.

(b) *Main control center*. The functions of the main control center are to monitor and direct the operation of the underground system and to gather and process information on a centralized basis (see Fig. 2). For this purpose, it is equipped with the main process control computer and various communication devices (display units, input and output units and printers). To increase availability, the computer and a few other parts of the system are duplicated.

There are direct data links to the section control centers and to other important facilities (e.g., the control room of the traction power system) to allow the state of operation of the system to be stored and displayed.

The chief supervisor uses the display units to obtain information on the actual state of the system (or, if desired, on a previous state of operation) and to mark deviations from the timetable. This enables appropriate measures to be taken to:

(i) promptly eliminate irregularities and disruptions;

(ii) cope with an exceptionally large volume of traffic;

(iii) provide the necessary assistance in the event of accidents and damage;

(iv) put on special trains and alter train timings in special circumstances; and

(v) coordinate bus connections.

Provision has been made to enter abbreviated inputs into the main control computer to facilitate frequency occurring operations (special trains, altered train timings, information required for the traction power supply). The computer then routes the necessary instructions to the appropriate centers. It also produces the various routine reports, documents and statistical analyses when they are due and passes on indications concerning specific processes to other centers.

The complete timetable is stored in the main computer. To minimize the work required to produce the timetable, the necessary data records are generated off line in an electronic data processing system. The fixed data of a line (mainly travelling and stopping times) and the variable train data (station of origin, destination, length and type of train, time of departure at the station of origin, turnaround information and details about the route) are provided by the timetable office.

The section control centers receive timetable extracts for their own areas directly from the main computer, via the data links.

The software of the line computer (see Fig. 3) consists of the program blocks for processing the timetable and for communication to the staff in the section control center the operating system with dynamic program control for internal computer organization and the program blocks for controlling the operation and supplying information.

This organization has been on successful trial on a short metro line of Hamburg but the Transit Authority

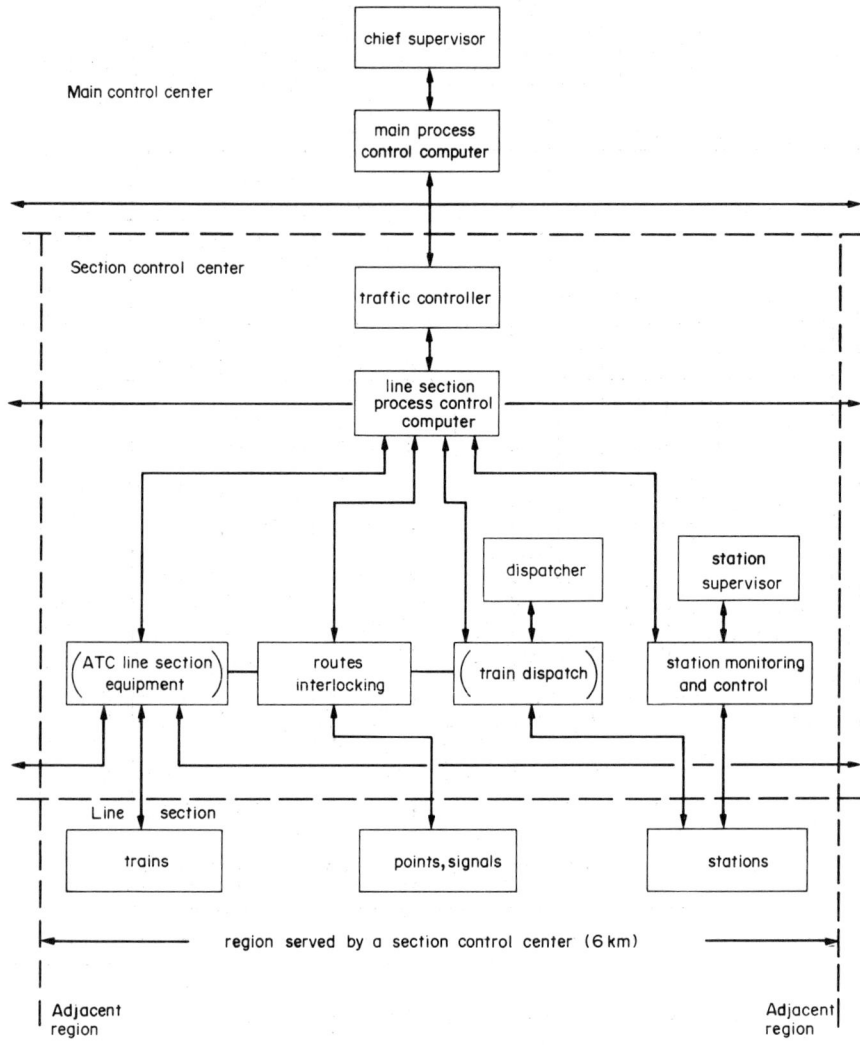

Figure 2
Schematic architecture of the Hamburg ATC

decided (in 1987) that the time has not come to generalize fully automated operations; therefore, the ATO has been put out of service. The U2 line is to be equipped with such a simplified system without the functions in brackets on Fig. 2, in which the driver retains a fundamental task.

3. Simplified Description of the Paris Metro ATC

3.1 ATP and ATO Structure

The ATP consists of track circuits detecting the presence of trains and associated trackside signalling. A signal indicates "track free" (green light) if no spacing restriction (no train down line) and no shunting restriction (appropriate route available for train) are imposed on the train. There is also a permanent speed-limit indication relevant to the track section. If a train is present down line or an appropriate route is not available, a restrictive signal is given, usually the stop signal (red light).

It is on this conventional signalling system that an automatic system is juxtaposed, replacing the driver in a driving capacity as regards train command (ATO) and respect of speed limits and signals (part of ATP) (see Fig. 4).

In this organization, there is a mixing between ATP and ATO functions in the track-to-train transmission and the on-board part.

The transmission of running orders to the train is via a program "written" into the track in the form of segments of varying length. The on-board device

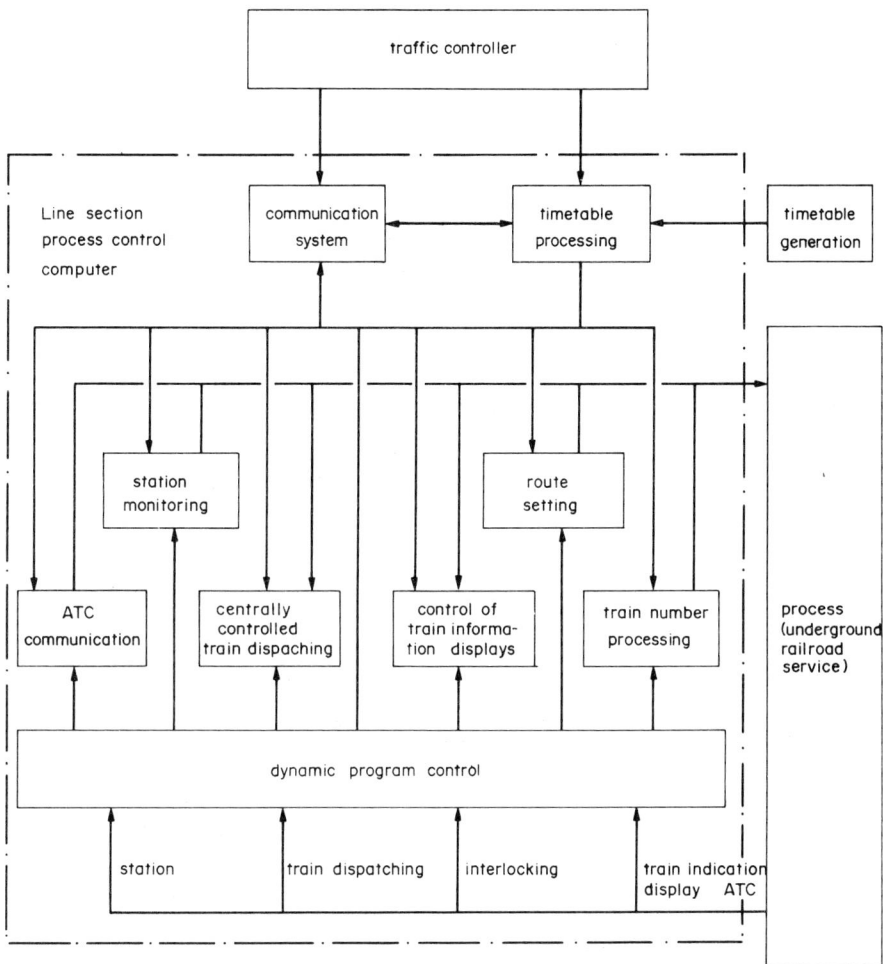

Figure 3
Structure of the line section process control computer of the Hamburg ATC

ensures control of train equipment and of safety systems.

The automatic driving principle is particularly simple:

(a) segment travel time T approaches base time T_0 (300 ms) when the train is travelling at the correct speed,

(b) if segment travel time T deviates from 300 ms, the control system orders an increase in traction ($T > 300$ ms) or braking ($T < 300$ ms); the greater the deviation between T and T_0, the greater the increase.

Therefore, in zones of constant speed, the segments are of constant length; in zones of deceleration or stopping, segment length decreases in an arithmetic progression, giving a parabolic decreasing speed control vs space.

Safety is ensured by continuous monitoring of train speed, emergency braking being triggered on detection of excessive speed compared with that displayed for the point considered by the track program.

3.2 Construction of ATO

A line consisting of two parallel wires is laid along the track. The wires are crossed in places to form segments of varying length (from approximately 0.3 m to 6 m). The segments supply a 135 kHz current modulated with frequencies around 1 kHz.

In front of each signal are installed two separately powered superposed lines running along the track. One is for use when the signal is green (and usually consists of constant length segments), the other when the signal is red (consisting of segments of decreasing length).

On board the train, the vehicle-mounted ATO device both analyzes the characteristics of the current

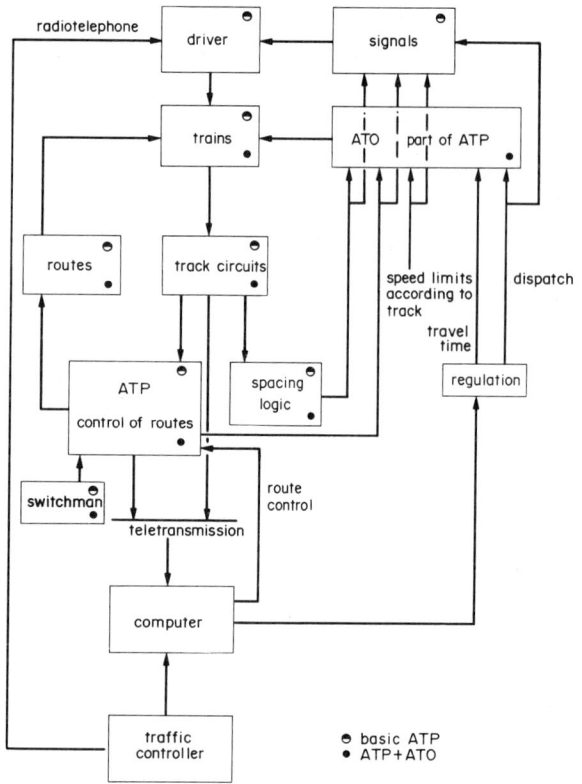

Figure 4
General structure of the Paris Metro ATC

supplying the line running along the tracks and detects where the wires cross, bounding the segments.

An internal clock and a digital counting unit make it possible to measure the travelling time over successive segments. Comparing the travel time and the base time, ATO controls the train power equipment (traction or braking) by means of an internal algorithm.

This reading is obtained with a high degree of accuracy using digital technology. It is not sufficient, however, to ensure safety. Indeed, in the event of failure, it could, for example, wrongly give an acceleration order in a braking area; such failures may occur at track–train level, digital counting and order output level or even at train power equipment level. In order to be prepared for the consequences of such failures, a safety device (part of ATP) monitors both track–train link continuity and train speed. If monitoring results are not satisfactory (loss of track–train link or continued excessive speed), automatic braking is imposed on the train. This monitoring device does not require particularly accurate processing, although it must be reliable. The safety principle adopted is quite specific to railroads and is that of a fail-safe design.

Parallel to the ATO–ATP, the track–train link makes it possible to transmit other information, such as the control of the door-opening side. The high-frequency current carried by the two-wire line is modulated by a dozen or so low frequencies. Each of these frequencies corresponds to a specific order given to the train by the fixed installations.

3.3 ALS

The ALS intervenes on the ATO, for regulation purposes, to modulate the moment of departure from a station by means of a special low frequency of the signal transmitted by the program written into the track. It also modulates the train travel time by means of a time lag in the carrier frequency of the signal transmitted by the program.

The driver is responsible for opening the doors on entering the station and closing them when the exchange of passengers between train and platform has been completed. This takes place either solely by the driver or, generally, at peak hours, upon an order coming from the ALS computer. In addition, the driver fulfills a general surveillance role and has a radiotelephone link with the central control room staff, either to signal a failure or to receive information messages to be passed on to passengers aboard the train.

Furthermore, the driver operates manual drive, either under training, or to cover ATO unavailability. He is then monitored by an automatic system that stops the train if a red signal is run through. Generally, in the Paris metro signalling layout, where a "buffer section" exists between the first red signal encountered and the track circuit containing the preceding train, a train braked in such a way at its maximum speed is able to stop before entering the occupied block.

For each Paris metro line, the ALS has a central control. Owing to the large number of lines in operation, it has been possible to combine all these controls in two large rooms (Fig. 5) with superposition of the visual control panels for two lines. By day, each line is controlled by an operator, but by night, the two lines are monitored by one operator. In addition two work stations, located in the center of each room, are equipped for passenger information.

Technically, two lines are monitored by two computers (one on line, one on standby) whose main job is to follow the progress of the trains and to compare their schedule with the theoretical schedule. Therefore, regulation orders can be sent to the stations (indicators and/or buzzer for the drivers) and to the ATO (maximum speed).

The synoptic layout of each line is carried in the upper part of the visual control panel. This includes two-figure displays for train identification, track circuit monitoring indicator lights, the position of points, shunting signals and emergency rerouting controls, in the event of a section of line being blocked. In the lower part of the panel is a synoptic traction representation of the power supply section and subsection, voltage presence indicator lights, power supply apparatus control buttons (remote-controlled isolating switches and sectioning switches) (see Fig. 6).

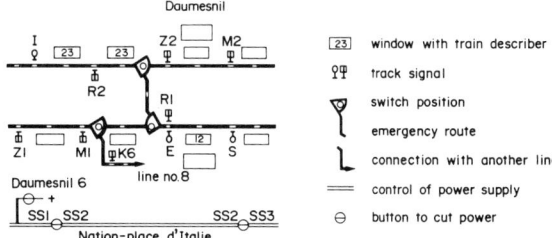

Figure 6
Detail of the visual control panel of the Paris Metro ATC

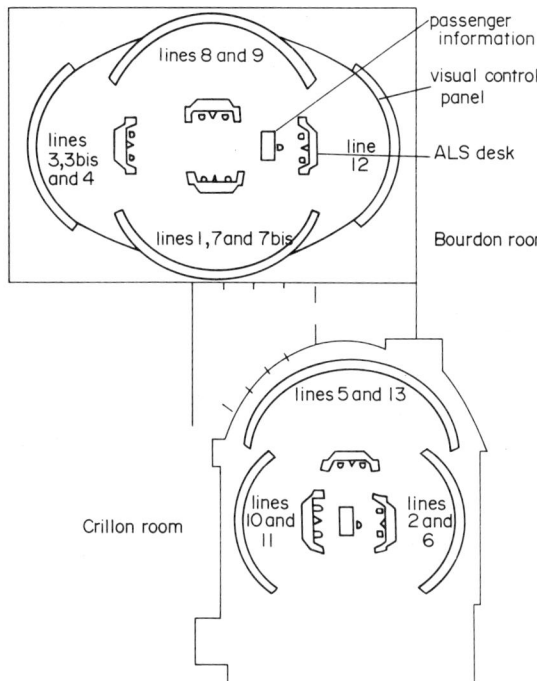

Figure 5
ALS central control rooms of the Paris Metro ATC

It can be noticed that power control at the higher level (from the national company (EDF) sources to the rectifiers or to different low-voltage users) is not managed in the same center. Another room, close to those just described, contains a special power control center (PCE) for the whole RATP network (metro and rapid transit). This is one of the features which differentiates some forms of ALS from others.

On the operator's console are various means of telecommunications (telephone links with stations and fixed installations, radiotelephone links with trains) and some controls for regulation. However, the systematic control of routes at the terminus is supervised by a different operator, with his workstation at the terminus. This local function is necessary owing to the extreme complexity of older terminus designs which make regulation operations (compensations at terminus) complicated. This operator is helped by specific automatism for traffic regulation purposes.

See also: Railroad Systems: Line Supervision and Control; Railroad Systems: Train Driving Control

Bibliography

Frank K H 1980 Automatisierung der Hamburger U Bahn. *Nahverkehrs Praxis* **11**, 12–16
Hooper A 1986 Centralized control of the Jubilee and Metropolitan line. *Prep. IEE February Colloquium.* Institution of Electrical Engineers, London
Perrin J-P (ed.) 1990 *Proc. IFAC Int. Symp. Control, Computers, Communications in Transportation.* Pergamon, Oxford
Perrin J-P, Beuchard P 1984 Automation in rail based urban transit system. *Fr. Railw. Rev.* **2**(2), 101–11
US Congress 1976 *Automatic Train Control in Rail Rapid Transit.* US Government Printing Office, Washington, DC
Yonezawa K, Takemura S 1976 Saporo subway and its computer total system. *Jpn. Railw. Eng.* **16**, 55–126

J.-P. Perrin
[RATP, Paris, France]

Vehicle Guidance by Computer Vision

When vehicles first came into use millennia ago, the task of power generation for locomotion was soon transferred to animals or natural resources (winds for shipping). Higher animals have senses for perception similar to a human, so they are able to learn behavioral patterns when towing carts. Thus, the operator of a tracked vehicle can pay attention to other items without endangering the mission.

Around the eighteenth and nineteenth centuries, the development of technology allowed muscle power to be substituted by engines; with the introduction of this more endurable and more powerful source of propulsion, the capability of partial autonomous navigation was traded against speed and range achievable. Since then, it has been the human operator who has had to provide, in general, both the sensing and the control-actuation capabilities, including the detection of the curvature and surface state of the road, as well as of possible obstacles hindering free passage.

With the advent of powerful, inexpensive digital processing capabilities through microminiaturization of electronic circuitry in the 1970s and 1980s, a new era of transportation may emerge. It will become possible to provide vehicles with sensory systems for light and sound, probably having comparable performance capabilities to biological systems. There is still a long way to go until the introduction of corresponding systems onto the market, but research results being accumulated are very promising.

1. Problem Statement

Observing living beings capable of active locomotion, it is readily noticed that vision is the most important sense for collecting information on the environment and its actual state. It allows remote sensing, in general, even in a purely passive manner, due to the sun providing sufficient light energy on Earth. Visual ranges from centimeters to many kilometers are easily achievable.

However, light rays propagating in a linear fashion map the three-dimensional world onto a two-dimensional visual sensor array. So, the task to be solved is to reconstruct an internal three-dimensional representation of the spatial environment from the two-dimensional image. Fish, land-based animals including humans and birds are all able to learn to orient themselves in a three-dimensional environment early during their lifetime, most of them without any range sensors. Apparently, learning during self-propelled motion is an important step for developing a mature sense of spatial vision.

Therefore, it seems that knowledge about the environment and about the dynamics of motion constitute important components of the capability of real-time vision. During locomotion, the space free of obstacles has to be recognized as well as the relative motion with respect to other objects. Motion prediction, therefore, is an essential part of vision, especially since data processing introduces a time lag between sensing and acting. This knowledge, of course, is specific to the field of application.

2. General Concept for Dynamic Machine Vision

Many efforts have gone into the trial to invert the perspective projection for a sequence of images and, thereby, reconstruct the spatial arrangement and motion of objects. For a survey see Nagel (1983).

The most promising method for real-time machine vision presently known takes a different approach. It tries to solve the analysis task by carrying out an iterative synthesis with prediction error feedback using spatiotemporal world models (see Fig. 1).

In a long-term memory, generic models of objects from the real world are stored as three-dimensional structures carrying very visible features at different spatial positions relative to the center of gravity (cg). Each real object has to occur in space at a certain position and angular orientation relative to the imaging camera, the position and orientation of which relative to the body is assumed to be known. From this knowledge, applying the laws of forward perspective projection (which is done much more easily than the inverse), the position and orientation of visual features in the image can readily be computed.

This world model, however, not only allows the computation of these feature positions but also the computation of how much these positions in the image will change when just one component of the relative state vector is changed. These partial derivatives are computed systematically for each spatial position component and are collected in the Jacobian matrix as detailed information for interpreting the measured image.

In order to introduce continuity conditions in the temporal dimension, knowledge about the behavior of mechanical systems is introduced via dynamic models. Physicists have found differential equations the best-suited means for capturing the motion behavior of mechanical systems. It is assumed here that the objects of interest are rigid. This allows the decomposition of the general motion of points on the object into a translation of the cg and a rotation around the cg, both described by ordinary differential equations containing both spatial position (orientation) and velocity components. These equations comprise the effects of both the state and the control variables.

Knowing these variables at one point in time allows a prediction for the next point to be made when new measurements are going to be taken. Combining the cg state with the shape information, it is possible to determine which measurable features should be visible where and how they should occur in the image. This is a very effective means for attention focusing and for allocating limited computing resources.

If the cycle time of the measurement and control process is small and the state of the object has been well known, the discrepancy between prediction and measurement should be small. Therefore, a linear approximation to the nonlinear equations of the model should be sufficient for capturing the essential interrelationships of the process.

For linear models, however, well-proven methods for recursive state estimation are available. Utilizing the Jacobian matrix mentioned previously, the spatial state estimation through vision can now be transformed to a conventional recursive least-squares estimation process (Kalman or square-root filtering, Maybeck 1979). This provides a link to modern control theory. However, contrary to conventional measurements by dedicated hardware, there is no direct encoding indicating which object the measured data belong to. The correspondence between measured feature data and the object in three-dimensional space also has to be inferred from image analysis. This problem has to be solved through recognition of shape or motion behavior.

Applying this scheme to each object in the environment in parallel, an internal representation of the actual environment can be servomaintained in the interpretation process in the computer by prediction error feedback. Other subprocesses have to monitor the validity of the models used and whether objects leave or enter the scene. In the former case, the spatial movement of this object can still be predicted using reasonable assumptions for its control actuation; in the latter case, a new instance of an object has to be created and installed in short-term memory, including its spatial and temporal properties.

This very general method for dynamic machine vision has been tested in different applications. Dickmanns and Graefe (1988a, b) give more detail and the following applications are discussed:

(a) the docking of vehicles moving in a plane in three independent degrees of freedom (satellite model),

(b) road vehicle guidance,

(c) the autonomous landing approach of an aircraft without any ground support (hardware in the loop simulation).

Only the second application area will be discussed here. Objects of interest in this problem area are roads, the vehicle carrying the camera, and other vehicles or obstacles including humans or animals on the road.

3. Objects

3.1 Roads

The generic model for (high-speed) roads consists of skeletal lines pieced together from clothoids (i.e., arcs with constant curvature change over run length). For each element, the curvature C (the inverse of the radius of curvature R; $C = 1/R$) is linked to the run length l along the skeletal line by

$$C = C_0 + C_l l \qquad (1)$$

where C_0 and C_l are constants. $C_l = 0$ corresponds to circular arcs; $C_l = C_0 = 0$ corresponds to a straight line.

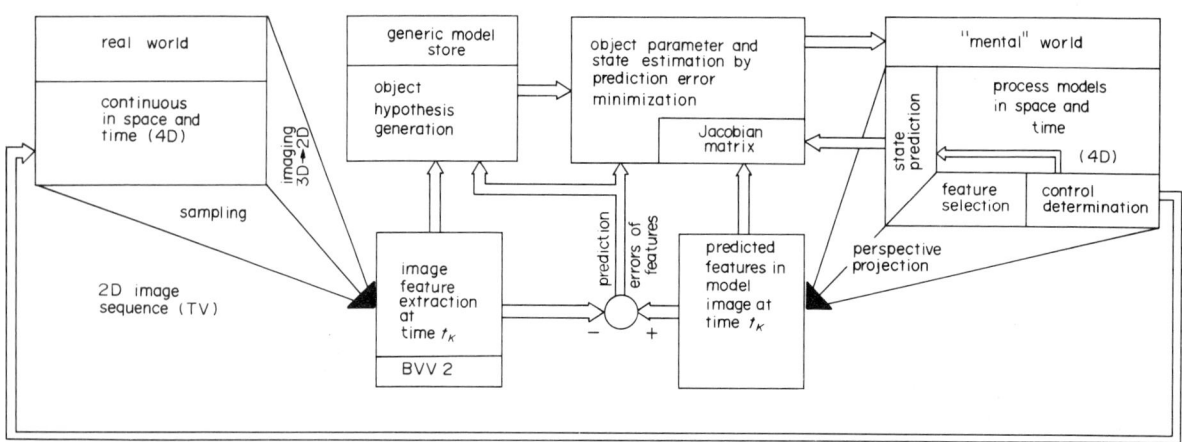

Figure 1
Basic scheme for four-dimensional image sequence understanding by prediction error minimization

Heading angle χ is the integral of curvature

$$\chi = \chi_0 + \int C(l)\,dl$$
$$= \chi_0 + C_0 l + \frac{C_l l^2}{2} \quad (2)$$

Since the radius of curvature on high-speed roads is larger than several hundred meters, for look-ahead distances below 100 m, the heading change of the road remains in a range where the sine can be well approximated by its argument and the cosine by 1. This yields from a second integration of Eqn. (1) for the lateral offset y_c at a distance L relative to a straight line tangential to the road at $l=0$ (see Fig. 2):

$$y_C = \frac{C_0 L^2}{2} + \frac{C_l L^3}{6} \quad (3)$$

and $x = L$ in the direction of the road.

Driving over the road at speed $V = dl/dt$ yields a curvature change with time corresponding to

$$dC/dt = \frac{dC}{dl}\frac{dl}{dt} = C_l V \quad (4)$$

$$\frac{dC_l}{dt} = 0 \text{ or a Dirac impulse at a transition point} \quad (5)$$

The right-hand side of Eqn. (5) is approximated as a noise term $n(t)$ for road parameter estimation taking Eqns. (4) and (5) as the underlying dynamic system.

3.2 Vehicle Carrying the Camera

The position of the camera in the car is assumed to be known: lateral position, longitudinal distance e to the cg and to the axles as well as elevation h above the ground. The viewing direction relative to the body is measured by an angle encoder.

A simplified dynamic model for the lateral motion of the vehicle links the rate of change of the state variables of the vehicle to the magnitude of the state variables and to the control actuations. In this system of differential equations, the knowledge about the dynamics of the system is encoded (Dickmanns and Graefe 1988a, b).

From Fig. 2, the relations between the vehicle position y_v, the viewing direction of the camera ψ_{KV} and the lateral shift of the road boundary due to curvature y_c can be seen to yield for the right road boundary point P at the look-ahead range L

$$y_{gR} \simeq y_v + (L+e)\psi_{VR} + L\psi_{KV} + y_R \simeq b/2 + y_C \quad (6)$$

where y_R is the item measured indirectly through the projected distance y_{BR} in the image plane B. With focal length f, there follows from perspective projection

$$\frac{y_{BR}}{f} = \frac{y_R}{L}, \quad \frac{z_{BR}}{f} = \frac{h}{L} \quad (7)$$

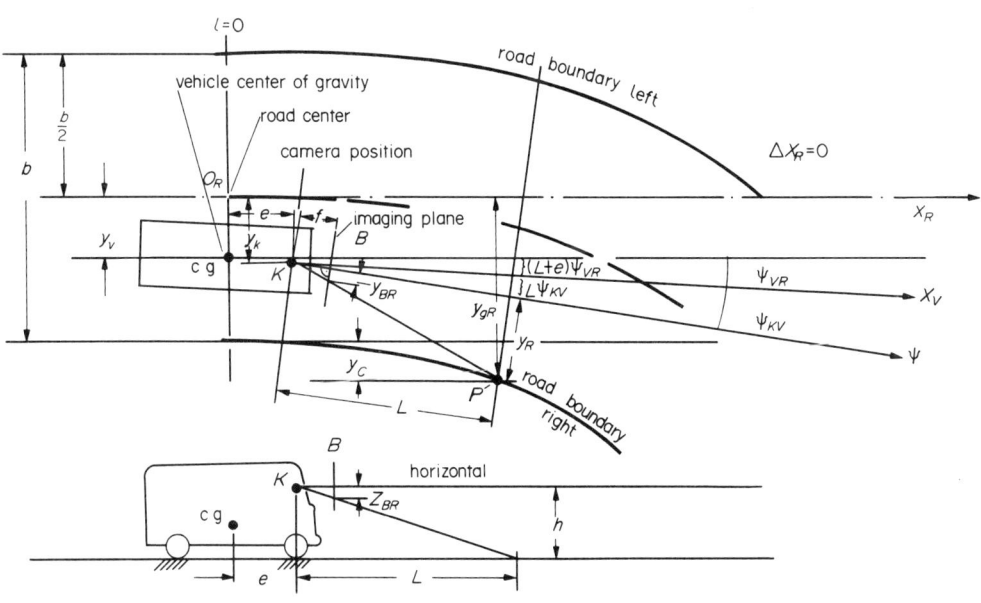

Figure 2
Vehicle/camera/road geometry

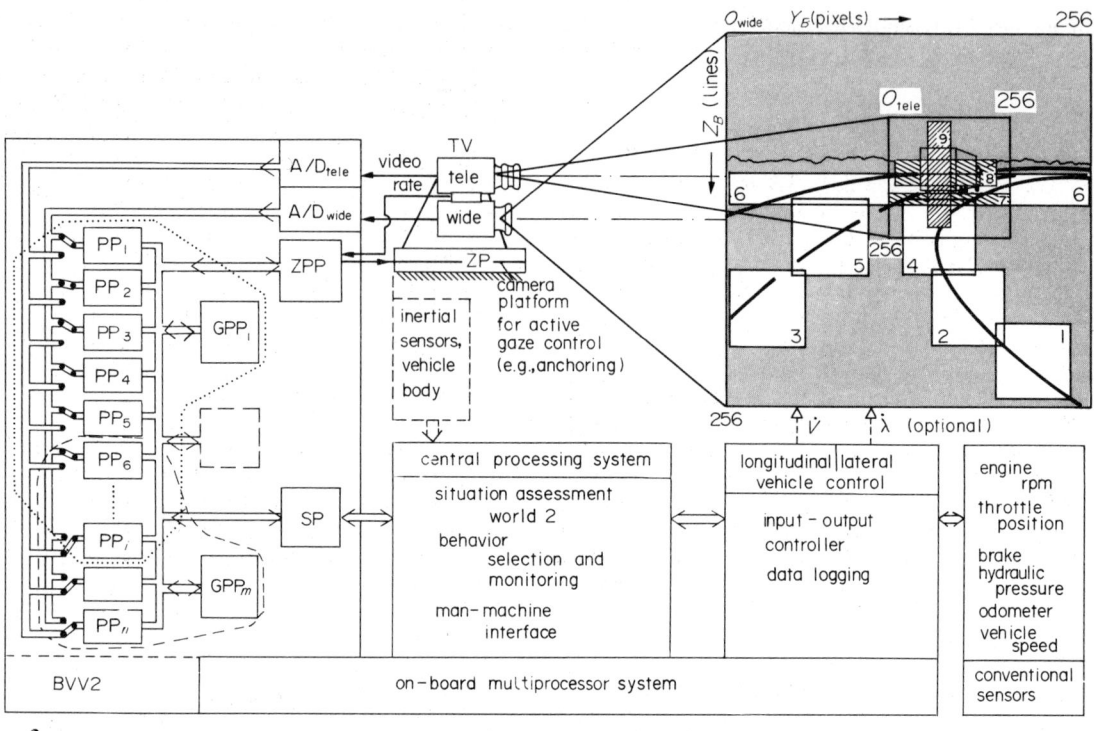

Figure 3
System architecture for active machine vision in road vehicle guidance

Combining with Eqns. (3) and (6) gives

$$y_{BR} = f\left[\frac{b}{2L} - \frac{y_v}{L} - \left(1 + \frac{e}{L}\right)\psi_{VR} - \psi_{KV} + C_0 a_0 + C_l a_l\right] \quad (8)$$

where

$$\left.\begin{array}{l} a_0 = \dfrac{e^2}{2L} + e + \dfrac{L}{2} \\[6pt] a_l = \left[\dfrac{e^3}{L} + 3e(e+L) + L^2\right]\bigg/6 \\[6pt] z_{BR} = \dfrac{fh}{L} \end{array}\right\} \quad (9)$$

which are the perspective mapping equations governing the vision process.

Lateral control is achieved by state vector feedback and a feedforward component proportional to road curvature. Longitudinal control also depends directly on curvature. Since the lateral acceleration a_y is proportional to CV^2, a preset limit value $a_{y,\max}$ determines the speed V_c allowed in the curve of radius $R = 1/C$. If V is larger than V_c, the throttle setting is taken back and the brakes may be applied; if V is smaller than V_c, the throttle setting is increased. The action is chosen proportional to $(V_c - V)$, the speed difference. On straight road segments V_c is set according to other considerations. The style of driving can be influenced by choosing $a_{y,\max}$ appropriately.

3.3 Obstacles

Other objects are considered obstacles only when they are on the road. The reaction to obstacles depends on where they are located relative to the vehicle and to the road. Therefore, both the spatial distance of the object to the vehicle and the lateral position relative to the lane the vehicle is driving in have to be determined. This also makes it clear that good range information alone is not sufficient, since it is the situation that determines the appropriate reaction and not one or two measurement values. Depending on where the obstacle is on the road, the autonomous vehicle might bypass it (left or right) or stop in front of it.

As a very crude first step, the shape of each object may be approximated by its circumscribing rectangular box. This yields grouping of vertical and horizontal contour elements as features indicative of an object (see Fig. 3, windows 8 and 9 forming a cross in the teleimage). Since vertical contour elements close to the ground are invariant with respect to relative aspect angle, they constitute well-suited elements for object detection.

4. Object Recognition by Feature Aggregation

The objects discussed in Sect. 3.3 and their states have to be determined from image sequence evaluation. Figure 3 shows a block diagram of the vision system architecture developed for this purpose. In the upper right of Fig. 3, a typical highway scene is depicted. The image areas of utmost importance to the task of safe and autonomous driving are covered by rectangular windows marked 1 to 9. The scene is observed by two almost coaxial cameras, one of which has a wide-angle lens for acquiring a wide field of view in the near range, the other a variolens in the telerange for good resolution in the far range. The cameras are mounted together on a pan and tilt platform (ZP) with high dynamics for achieving good tracking and stabilization performance. The digital controller for this platform (ZPP) is integrated into the image sequence processing system BVV 2 and has fast access to both inertial and processed image data.

The video signals are digitized into 256×240 pixels of one byte and put onto video busses (upper left in Fig. 3). A number of microprocessors (PP_i) have access to the video busses through a special interface allowing them to fetch up to 4K pixels of data belonging to a rectangular window. These windows may have any size, shape, position and sparseness; parameters controlling these properties can be adjusted by software during run time.

Cooperating groups of processors may be formed by message passing for solving special subtasks in scene recognition; for example, PPs 1 to 6, each individually programmable, are ordered to acquire image data in windows 1 to 6. Each PP searches for linear contour elements under a preferred slope in its window. The slope is prescribed by the higher levels of the system depending on the interpretation context. The positions of candidates for the contour elements are passed to the general-purpose processor (GPP) or the central processing system, which will evaluate them in the light of their specific four-dimensional world model in order to recognize objects or the motion state of the vehicle.

A second group of processors may be assigned the task of tracking another object of relevance for performing the mission (e.g., in Fig. 3, upper right, windows 7 to 9 may have the task of tracking the object a human would readily term a truck). After object detection, for example, by some specialized algorithm in a PP behind window 6, the central processing system (lower center in Fig. 3) may order the processor group marked by the dashed line in the lower left part of Fig. 3 (GPP_m) to acquire and track a combination of features moving in conjunction, which can be interpreted as the image of some moving three-dimensional object.

Each object is represented internally in the computer by shape parameters and a dynamic model with a motion state. This is done individually for each object in a processor group consisting of several PPs and one GPP.

The estimated states of all of these models are gathered in the central processing system which has to produce a situation assessment.

Once the situation is assessed, a decision can be taken whether to continue with the behavioral mode running or whether a mode change—probably with provisions for smooth transition—should be prepared. Behavior is achieved in this four-dimensional approach by state variable feedback to the controls.

Note that image interpretation is not intended to be carried out in image coordinates, but directly in three-dimensional space and time. This immediate transition from image features to spatiotemporal representations of relative positions and velocities of objects is essential for efficient real-time operation. With the moderately powerful multimicroprocessor system described here, image analysis and control cycle times of 0.1 s have been achieved for vehicle guidance using this method.

5. Closed-Loop Performance

The method and hardware described in Sects. 2–4 evolved during many years of experimenting in simulation with real-image sequence processing hardware in the loop. Since 1986 the test vehicle for autonomous mobility and computer vision (Rechnersehen), VaMoRs (see Fig. 4), has been available for real driving experiments. The following results have been demonstrated:

(a) lane following with speed automatically adjusted for a preset maximum acceleration (e.g., 0.1g or 0.2g), maximum speed (limited only by engine

Figure 4
VaMoRs test vehicle of UniBwM

performance) of 96 km/h^{-1}, various weather conditions including sunshine with hard shadows and light rain with wipers operating;

(b) lane changes to the left and to the right;

(c) freeway entry from ending acceleration lane;

(d) driving on unmarked country roads with shadows from trees, speeds up to 60 km h^{-1};

(e) obstacle detection and stopping at a preset distance in front of object from speeds of up to 40 km h^{-1}; and

(f) driving on unsealed dirt roads, speeds up to 20 km h^{-1}.

Image processing hardware under development and further progress with software will steadily push the limits quoted towards higher values in the future.

6. Future Developments

The results achieved show that road vehicle guidance by machine vision is maturing to a state where practical applications in simple environments (e.g., freeways) may come into being, at least for larger-scale tests in order to investigate its potential.

The European car manufacturing companies in conjunction with their governments have, to this end, launched a research program in the framework of the Eureka project called Prometheus. Several vision-guided cars will be developed in the years to come in cooperation between industry and universities. In Japan, similar work has been started. In the USA, autonomous vehicle guidance has been investigated in the framework of the DARPA program on strategic computing since 1982 (autonomous land vehicle (ALV) and Navlab from Carnegie-Mellon University, Pittsburgh).

The integrated spatiotemporal approach in connection with recursive least-squares filtering, on which the outstanding performance of VaMoRs is based, seems to provide a breakthrough in real-time machine vision. The feature-based, servomaintained internal four-dimensional world model is numerically very efficient for fast image sequence processing in a well-structured environment. It allows the bypassing of the nonunique inversion of the perspective projection, exploiting continuity conditions in space and time simultaneously.

Based on broad experience in three application areas, it is safe to state that machine vision will play an important role in the further technical development of transportation systems of any kind. The imaging sensors, of course, are not limited to television but may also be infrared or microwave devices allowing different application regimes. Hybrid systems with complementary properties may be of special interest in some demanding applications. Spacecraft docking and aircraft autonomous navigation including landing without ground support may become possible.

See also: Airborne Collision Avoidance; Video Sensors

Bibliography

Dickmanns E D, Graefe V 1988a Dynamic monocular machine vision. *Int. J. Mach. Vision Appl.* **1**, 223–40

Dickmanns E D, Graefe V 1988b Applications of dynamic monocular machine vision. *Int. J. Mach. Vision Appl.* **1**, 241–61

Hertzberger L O (ed.) 1986 *Proc. Int. Conf. Intelligent Autonomous Systems*. Elsevier, Amsterdam

Maybeck 1979 *Stochastic Models, Estimation and Control*, Vol. 1. Academic Press, New York

Nagel 1983 Overview on image sequence analysis. In: Huang T S (ed.) *Image Sequence Processing and Dynamic Scene Analysis*. Springer, pp. 2–39

E. D. Dickmanns
[Universität der Bundeswehr München, Neubiberg, Germany]

Vehicle Monitoring and Control for Buses, Trolleys and Streetcars

Urban areas require a collective public transport system providing users with a true alternative means of transport that preserves their quality of living. It is essential to improve the quality of the services provided in order to increase network patronage and, eventually attenuate, the increase in operating costs when the offer is increased, by improving resource utilization. Urban public transport networks must, therefore, be able to be equipped with the appropriate technical means. Automatic vehicle monitoring (AVM) systems provide these means. The basic function around which all such systems are designed is vehicle location and communication of this location. Three techniques of vehicle location can be mentioned:

(a) triangulation techniques (Loran C),

(b) satellite systems (Geostar, Locstar), and

(c) vehicle movement integration techniques.

These latest techniques are the most appropriate for vehicles that follow a predetermined itinerary: measurement of distance travelled.

There are two types of families:

(a) continuous location or discrete location, and

(b) real-time link with or without a control center.

This article deals with the most sophisticated systems; that is, continuous location with real-time link with a control center (see Table 1). In France, these systems are called système interactif centralisé à localisation instantanée cyclique (SICLIC), which can be translated as centralized interactive system with cyclic instantaneous location.

1. Objectives

1.1 Regularity and Punctuality

Buses, trolleys and streetcars are subject to random variations in traffic and human behavior. They tend to

Table 1
Worldwide AVM systems with cyclic instantaneous location and real-time link with a central station

Country	Towns
Belgium	Brussels
Canada	Toronto
France	Aix-en-Provence, Angers, Angouleme, Annecy, Belfort, Besancon, Bordeaux, Brest, Caen, Clermont-Ferrand, Evreux, Grenoble, Le Mans, Marseille, Montbelliard, Montpellier, Mulhouse, Nancy, Nantes, Orleans, Paris, Pau, Reims, Saint-Brieux, Saint-Denis-Briétigny, Strasbourg, Toulon, Toulouse (Zelt), Tours, Valenciennes
Germany	Aaken, Amburg, Berlin, Darmstadt, Duisburg, Friedrichshafen, Hannover, Norimberga, Wiesbaden
Ireland	Dublin
Italy	Brescia, Cagliari, Florence, Turin
Japan	Tokyo
Spain	Barcelona, Bilbao, Palma (Majorca)
Sweden	Stockholm
Switzerland	Geneva, Lausanne, Zurich
USA	Baltimore, Chicago, San Francisco
USSR	Moscow

get ahead of schedule when traffic is running well and fall behind schedule during peak hours when roads are congested. It is common to see bunches of vehicles, where the first vehicle is full and the following vehicles are practically empty. Times of departure from the terminus are then not always respected.

The real-time continuous vehicle location AVM system helps the drivers and network supervisors to obtain solutions to these problems. It may

(a) alert the driver when the time of departure is approaching and inform when it is time to depart;

(b) modify departure times, either automatically or on instruction from the controller, to take into account any problems on lines;

(c) supervise vehicle progression, detect problems in real time, inform the network operations supervisor and the vehicle driver by displaying earliness or lateness, thereby relieving the driver of the task of checking adherence to schedule;

(d) control the intervals between vehicles and provide the driver with the information needed to correct speed in order to prevent the formation of bunches of vehicles; and

(e) manage connections between lines by delaying departure of certain vehicles from the connection points if the connecting vehicles are late.

1.2 Giving Priority at Traffic Lights in Order to Increase Commercial Speed

Traffic lights cause significant time losses for public transport vehicles when they are obliged to stop because the lights are red (see *Road Network Control*).

The real-time continuous vehicle location AVM system allows public transport vehicles to obtain priority at traffic lights by extending the green phase or switching a red light to green as the vehicle approaches. This is achieved by direct communication with the central traffic control computer or via the junction traffic light controllers.

1.3 Passenger and Personnel Safety

If a serious incident arises, such as an accident or perhaps even physical assault of the driver or passengers, the driver has to cope with the problem alone, usually by leaving the vehicle in order to signal the problem by telephone.

The real-time continuous vehicle location AVM system makes it possible to have immediate contact with the operations manager at the control center over the voice communication channel. If physical assault is involved, a silent alarm can immediately be transmitted to the control center. The information on vehicle location provided by the system allows help to be sent to the site straight away. The control center can also listen in to what is going on in the vehicle or attempt to dissuade the aggressor by announcing that help is on its way.

1.4 Adaptation to Demand and to External Constraints

The usual methods of finding out about passenger demand (enquiries) and constraints arising from movements of other vehicles (timing running times) are very costly. Consequently, the frequency of vehicles and, above all, the theoretical running times are not always appropriate for each period of the day, for each type of day or for each time of year.

The real-time continuous vehicle location AVM system records actual running times and stopping times at bus stops, counts passengers and may provide statistical analysis in order to optimize theoretical schedules and provide users with the best possible service at the lowest possible cost.

1.5 User Information

Public transport systems are often found to be difficult to use by people who are not familiar with the network (e.g., new users or occasional users).

The real-time continuous vehicle location AVM system may provide users with information to help them:

(a) inside vehicles: location of vehicle on line, announcement of next stop or terminus (audio and/or visual), time of day, real-time display of information messages sent by the controller, general information and advertisements;

(b) on the exterior of vehicles: line number, destination, service number, any special information;

(c) at stops: waiting times till arrival of next buses, per line and per destination, with indications of disruptions (e.g., incidents, temporary diversions), local information, advertisements; and

(d) at home and in public places: interactive consultation of the same information as shown at stops.

1.6 Making Human and Material Resources Pay

On-site supervision tasks take up a large amount of managers' time, thus preventing them from concentrating on improving network operations. The fact that it is difficult to obtain fully exhaustive statistics on network operations makes it difficult to improve performance.

The real-time continuous vehicle location AVM system records the network operating statistics (e.g., schedule monitoring, running times, on-line and dead running kilometers travelled, passenger counts, ticket cancellations, technical alarms, driver's working times).

Statistical analyses can be printed out on request in the appropriate form for utilization (data tables with values, means, standard deviations, graphs).

Improved network service regularity and centralized real-time knowledge of operating problems permit better organization of the working schedules of each driver.

Finally, vehicle management tasks (e.g., recording vehicle running kilometers, fuel consumption, technical incidents) are complicated and can be inaccurate due to lack of information. Automatic monitoring of each vehicle makes it possible to establish precise maintenance planning charts, reduce the number of unjustified interventions, carry out maintenance work "before it is too late" and, thereby, optimize vehicle utilization.

1.7 Improvements of Working Conditions

By having higher running speeds, fewer stops and achieving punctuality, drivers can

(a) start and finish work on time,

(b) benefit from the scheduled rest periods, and

(c) be surer of respecting layover times at terminuses.

By recording vehicle progress on lines on a section-by-section basis, an accurate and incontestable picture is obtained of the points where the vehicle is slowed down and time is lost. Records of actual running times and layover times can provide a basis for objective negotiations between the personnel and management.

Finally, the automation of functions by the system reduces personnel work load while, at the same time, enhancing their functions. This is particularly true for the controllers, who can monitor the position of the vehicles on each line at any moment in time and implement any control actions that might be required to solve daily operating problems.

1.8 Improvement of the Image of the Transport Network

Adhering to schedules, increasing running speed, achieving better vehicle load (passenger) distribution and, above all, providing information to users in vehicles and at home are all factors that give the transport company and the public authorities involved an excellent image.

2. Equipment and Equipment Operation

AVM systems comprise on-board equipment connected by radio to a control station, computer and radio systems at the control station, and ground installations which vary according to the selected options (traffic lights priority, information at stops, etc.).

2.1 On-Board Equipment

A series of modules (see Fig. 1) are usually installed in each vehicle to

(a) accurately determine (to within 20 m) the distance travelled by counting the number of wheel revolutions and recognition of stops;

(b) store the data received from the various sensors (kilometers, number of passengers, ticket cancellations, fuel consumption, alarms, etc.);

(c) transmit information to the driver in clear form (wait/depart, earliness/lateness, wait for connecting vehicle or any other message sent by the controller); and

(d) manage transmission procedures and normal and emergency voice communications.

An overview of these modules is given in Fig. 2 and they often consist of

(a) a wheel revolution counter (odometer);

(b) a door opening detector;

Figure 1
Driver console (courtesy of Compagnie Générale d'Automatisme)

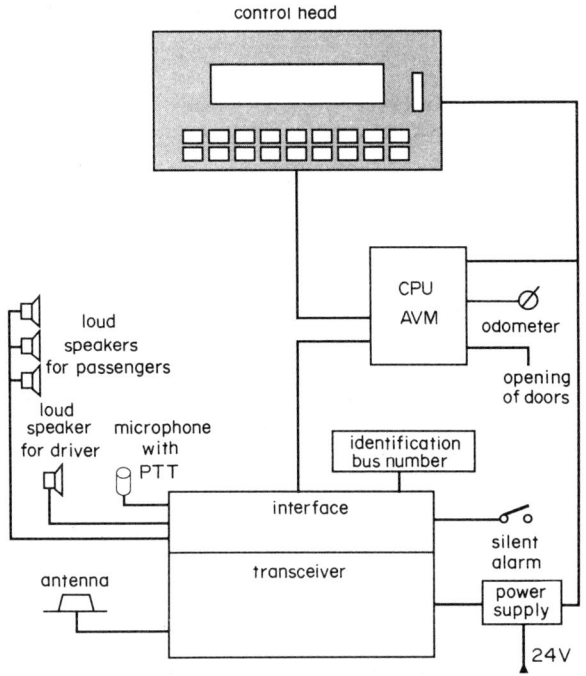

(c) sensors for specific applications (passenger counting, fuel consumption, etc.) with a link to other system sensors (e.g., ticket cancellers);

(d) a microcomputer that manages all on-board functions;

(e) a control console for dialog between the driver and the microcomputer, and an alphanumeric display;

(f) a badge reader for driver identification;

(g) user information equipment which uses visual (screen or alphanumeric display strip) and audio (voice synthesis) techniques;

(h) possibilities for remotely controlling additional items of equipment (ticket cancellers, destination signs, etc.); and

(i) a modem and a two-channel radiotelephone (voice communication and data channels).

2.2 Control Station

The control station (see Fig. 3) usually fulfills the functions of

(a) polling each vehicle every 20 s in order to obtain information concerning vehicle position, counts, incidents and so on, and authorizing immediate transmission of urgent information (e.g., voice calls);

Figure 2
On-board overview

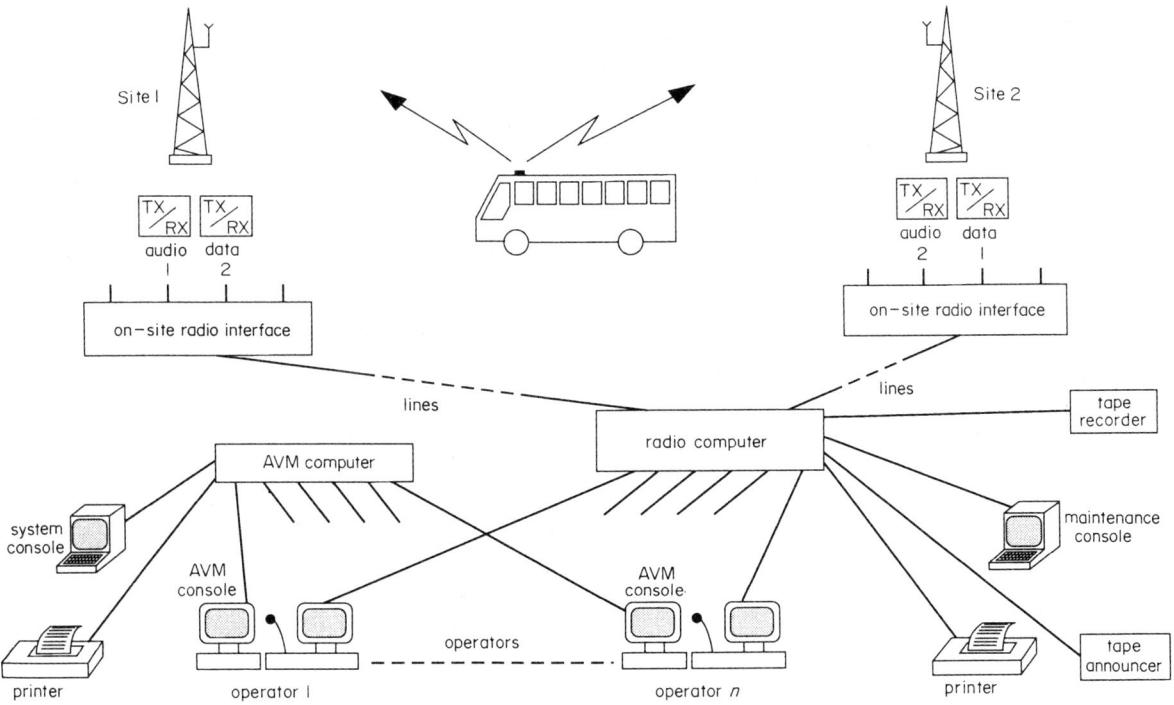

Figure 3
Overview of AVM system

605

(b) applying various algorithms to determine actions to be taken (schedule control, frequency control, interchange management, priority at traffic lights, etc.);

(c) displaying vehicle position and data to the controller;

(d) printing out the log of alarms and main actions requested by the controller;

(e) storing and processing all data acquired in order to print out the statistics requested by the controller;

(f) registering drivers at the start of duty;

(g) managing refuelling operations by recording the vehicle and pump identification; and

(h) exchanging information with other computers, such as
  (i) traffic control computer,
  (ii) management computer,
  (iii) computer for assisting with schedule construction, and
  (iv) Videotex server computer.

The main features of the control station are

(a) computers with fixed disk drives;

(b) mass storage memory (tape unit or disk) for recording statistics;

(c) one or more color monitors for displaying information relating to vehicles and managing operator dialogs and controller commands (switch over to voice communication, image loading, etc.);

(d) monochrome monitors used for creating, consulting and modifying files;

(e) a printer for outputting the daily record, final results and statistics;

(f) an x–y plotter, if necessary; and

(g) equipment for connection to peripheral stations such as fixed radio points, depots and other computers.

2.3 On-Site Equipment

On-site equipment comprises

(a) one or more fixed radio transmission–reception points, essentially comprising two radio transceivers, one for data and one for voice communication;

(b) priority control electrical cabinets at junctions which receive traffic light priority requests;

(c) information stands at stops which are connected to the central system; and

(d) beacons (if needed) to improve the accuracy of vehicle position identification on common route sections and at terminuses.

2.4 Evolution of AVM Systems

In the near future, AVM systems are going to acquire complementary functions such as depot management, which will improve the organization of network operations, and will be interfaced with other decision-aid tools such as expert systems.

3. Results

Overall terminus-to-terminus running times have been reduced on the Strasbourg network as a whole and, in particular, on line No. 4, where the saving in time exceeds 5%.

The French Ministry of Transport carried out a study of effectiveness on the public transport network of the city of Nancy (France), taking measurements both with and without an AVM system. An increase in commercial speed of almost 5.8% was measured in 1985. Using this information, the Nancy district estimated that a complete system would be amortized in 3.5 years, without taking any grants into account. The profitability of this type of system has, therefore, been proved.

Similarly, a reduction of approximately 20% in initial layover times was noted after implementing the system in Strasbourg. This resulted in a 5% saving in work time and, therefore, produced another type of gain.

One of the ratios used to determine productivity is the number of kilometers travelled per year divided by the number of driver duty hours per year. At Nancy, this ratio increased from 12.09 in 1980 to 13.84 in 1985.

An overall gain in excess of 14% was obtained by implementing these measures (system schedule control actions, new running schedules, measures concerning bus lanes).

Energy studies carried out in March 1982 by the French Ministry of Transport highlighted the fuel consumption costs of a stationary vehicle: 1–$5 \, l \, h^{-1}$ for a bus that is stopped at a road junction. The reduction in vehicle stoppages at junctions thanks to traffic light priority control has a direct impact on energy consumption in addition to the gains in productivity.

Improved vehicle and labor utilization is one of the most important aspects of the Toronto AVM system which provides a better service to patrons with fewer vehicles. Comparing AVM routes with non-AVM routes over time before and after AVM installation indicates that savings of 2–3% are obtained at peak periods. Toronto also identified a possible reduction of vehicles of 2–4% during off-peak periods. The introduction of AVM on Wilson routes resulted in a 5% increase in passengers. In spite of a 30% increase in traffic volume and a 13–21% increase in passenger loads per bus, service regularity actually improved.

The Dublin Bus AVM system has resulted in service improvement:

(a) service gaps of more than 15 min reduced by 60%,

(b) service bunches of less than 1 min reduced by 64%, and

(c) lost travelling distance due to congestion down by 30%.

The travel time on the first line of the Grenoble streetcars, which is a distance of 10 km terminal to terminal, is 35 min without traffic priority and 29 min with traffic priority controlled by the AVM system installed.

In addition to the direct economic gains and improved working conditions resulting from the implementation of AVM systems, the beneficial effects of having a more attractive system and a better network image cannot be forgotten.

4. Benefits and Future Developments

The AVM system with cyclic instantaneous location and real-time link with a control center is definitely profitable. Certain benefits have a financial impact: improved vehicle and labor utilization, increased number of passengers and so on. Certain benefits are not easily quantifiable: schedule adherence, service regularity, reductions in complaints, enhancement of inspectors' and operators' working environment, service improvements resulting from better management reports, improved handling of fare disputes, improvement of the image of the transport network and so on.

AVM systems also become the crossroads of all urban transport applications: radio control, ticketing, user information, maintenance aids and so on.

Future research should explore the possible application of artificial intelligence to enhance the service control function.

See also: Underground Railroad: Organization of Operations

Bibliography

Canadian Urban Transit Association 1988 *Proc. Int. Conf. Automatic Vehicle Location in Urban Transit Systems.* CUTA, Ottawa, Canada

Centre d'Etudes des Transports Urbains 1988 *Système d'Aide à l'Exploitation—Les Réalisations Françaises.* CETUR, Bagneux, France

de Kerdaniel P L 1981 Régulation centalisée par ordinateur du traffic automobile et des autobus de Strasbourg. *J. Ecole des Mines de Saint-Etienne* November, 15–18

de Kerdaniel P L 1988 Système d'aide à l'exploitation. *AIMF Int. Conf. Urban Public Transport.* Association Internationale des Mavies des grandes Villes de France, Paris

de Kerdaniel P L 1988 Système d'aide à l'exploitation. *UNIDEC Conf. Traffic and Urban Public Transport.* Union des Maires pour le Développement Economique des Communes, Paris

de Kerdaniel P L 1989 Système d'aide à l'exploitation et a la communication. *Colloque de l'Union des Transports Publics de Suisse.* Union des Transports Publics de Suisse, Berne

Lesne J 1989 Eléments pour une évaluation des effets des systèmes d'aide à l'exploitation. *Note d'Information de la DTT* **109**

Siclop G D 1988 *Recommendations to Automatic Vehicle Location Systems for Small Urban Transit Networks.* Centre d'Etudes des Transports Urbains, Bagneux, France

Texier P Y 1983 *Système d'Aide à l'Exploitation des Transports Publics Urbains—Rapport de Synthèse.* Centre d'Etudes des Transports Urbains, Bagneux, France

P. Y. Texier
[INRETS, Arcueil, France]

P. L. de Kerdaniel
[CGA, Briétigny, France]

Vessel Traffic Services: Management from Shore

Two major issues faced by European coastal states are safety at sea and protection of the maritime environment. In effect, these have become one of the *leitmotivs* (or underlying themes) of their maritime policies.

Such policies must strive for the right balance between the management of sea resources, safety of navigation, protection of the environment and the maintenance of an efficient maritime transport system.

Much effort has been, and continues to be, expended in these areas. However, the many tragic casualties that continue to occur are sad proof that more needs to be done. Important progress can be seen in the development, during recent years, of common initiatives of European scope, aimed at implementing joint approaches to these problems.

The International Maritime Organization (IMO) guidelines for vessel traffic services (VTS) define VTS as follows.

"A VTS is any service implemented by a competent authority, designed to improve safety and efficiency of traffic and the protection of the environment. It may range from the provision of simple information messages to extensive management of traffic within a port or waterway."

The decision in December 1982 of the Council of the European Communities to adopt a concerted action project (known as COST 301) in the field of shore-based aids to marine navigation and, especially, in VTS, was both the natural result of such a context and a catalyst in its further development. This was reinforced in April 1983 by the Community-COST concertation agreement.

1. COST 301 Studies on VTS

The primary objectives of the COST 301 project, as agreed in October 1984, were:

(a) to assess any potential benefits that VTS could bring to the safety and efficiency of traffic and the reduction of pollution risk in European waters;

(b) to make any recommendations on the coordinated European approach to VTS on the basis of the results obtained in (a)

(c) to foster a spirit of European cooperation in the field of research in maritime safety and operational efficiency.

The essential features of a VTS can be described in operational or physical terms.

1.1 Operational Features

A VTS may interact with traffic by implementing one or more of a number of external functions, such as:

(a) general information service,

(b) navigational assistance,

(c) traffic organization,

(d) pollution monitoring, and

(e) search and rescue (SAR).

These and other functions, which show large variations throughout the various VTS centers in Europe, have been more systematically defined within COST 301 as:

(a) primary,

(b) enforcement, and

(c) remedial.

1.2 Physical Features

The physical functions are implemented by a combination of personnel, hardware, software and procedure, under the direction of human VTS operators. Such operators use various technical options to:

(a) acquire information on the traffic;

(b) process this information; and

(c) distribute it in the form of services.

The options for acquisition may or may not require cooperation on the part of the ship; for example, the use of very high frequency (VHF) voice communications, VHF direction finding (DF) and transponders all require some form of cooperation while radar, visual and optoelectronics are essentially noncooperative.

COST 301 has investigated all these aspects as a research project and, as such, it has aimed at producing information useful for decision making.

The main results obtained by the COST 301 project are given by Cutland et al. (1987) and Salvarini et al. (1987). Basic studies were first made on existing VTS. They have provided an improved understanding of how VTS work in managing traffic. These studies are described by Degré et al. (1987) and are summarized in Sect. 2.

2. Existing VTS: Inventory, Functions and Effectiveness

Two basic sets of studies into existing VTS have been carried out. The first produced a comprehensive list of all VTS in European countries, together with their outline features.

In the second set of studies, a limited number of VTS were selected and studied in detail to identify what services they provide, what information they process and how this processing is carried out. The conclusions that can be drawn from these studies are given in Sect. 4.

2.1 Inventory of VTS

As a preliminary activity, a list was produced of all VTS centers in European countries. Questionnaires were produced and sent to centers on the list. The replies to these questionnaires were coded, question by question, for each VTS. Results were presented on a statistical basis.

The preliminary list contained 152 centers which were considered to be VTS, according to the IMO definition. This figure must be treated with caution, in view of the wide interpretation under the IMO definition as to what constitutes a VTS. Replies to the questionnaire were received from 85 VTS. All the results from the questionnaires are expressed as percentages of this figure of 85 VTS.

The following is a selection of these results.

(a) *Type of VTS*. 67% are harbor, 18% river/fairway and 8% coastal.

(b) *Authorities (funding)*. 36% are government funded, 35% are funded by a port authority and 12% are funded by a combination of the two.

(c) *Authorities (operation)*. For 68% of VTS, the authority responsible for operation is a port.

(d) *Pilotage*. Some form of pilotage is compulsory for the areas covered by 82% of VTS.

(e) *Ship reports*. Some form of ship reporting is applicable to 83% of all VTS, and is mandatory for some categories of ships for 53% of all VTS.

(f) *Aims of VTS*. The stated aims were:
  (i) safety of traffic and environment (82%),
  (ii) efficient flow of traffic (78%), and
  (iii) aid to navigation (68%).

(g) *Reasons for establishing a VTS*. The reasons given included:
  (i) efficiency of traffic flow (65%),
  (ii) traffic density (59%),
  (iii) pollution risks (58%),
  (iv) narrow waters (55%), and
  (v) coordination of services (50%).

(h) *VTS functions*. The functions provided were given as:
  (i) information (79%),
  (ii) surveillance (67%),
  (iii) regulation (67%),
  (iv) assistance to navigation (62%),
  (v) SAR (63%),
  (vi) antipollution (71%), and
  (vii) monitoring of aids to navigation (60%).

(i) *Communications.* This is used for VTS–ship communication by 81% of VTS and visual signalling by 33%. Regarding language, English is used by 45% and the national language by 61%. The major problems include interference from other states, congestion of communication traffic and language difficulties.

(j) *Information broadcast.* This is in operation for approximately 50% of VTS and includes information on meteorological and hydrographical aspects, the traffic situation, aids to navigation and unusual events.

(k) *The surveillance function.* Some form of management or regulation of traffic occurs in 55% of VTS. This often takes the form of a time to arrive at a given point.

(l) *The assistance function.* Navigational assistance for ships is provided by 62% of VTS.

(m) *VTS operations.* 81% of VTS provide a 24 h service, normally with the same level of capability throughout the day. Pilots were present in 35% of VTS centers. The duration of a watch in a VTS center varies from 4 h to 24 h.

2.2 Investigations of VTS

In order to investigate how a VTS could improve traffic safety and efficiency, and contribute to the reduction of pollution, a clear understanding is required of the functions of existing VTS, their operation and their effectiveness.

The inventory described above did not give the depth of understanding needed, and it identified some areas where clarification was required—for example, the characterization of VTS and the services supplied. Therefore, a limited number of VTS were selected for detailed study, which was to include descriptions of the services provided, and the associated information content, flow and processing.

It was originally intended that these studies should include the effectiveness of VTS. However, it did not prove feasible to carry out this part during the investigation period and studies were limited to a questionnaire, with the investigations in three parts.

(a) *Preparation of guidelines for investigations.* A guide to the investigations was prepared. This specified the data to be collected, the form of presentation of the results and the methods to be employed. This guide was tested during a visit by a COST 301 team to a Canadian coast guard VTS and some modifications were made.

(b) *Investigations of VTS.* These were carried out by national teams in accordance with the guidelines as follows:

(i) Germany, at the district centres of Cuxhaven and Wilhelmshaven;

(ii) France, at Cross Corsen/Ushant and Gris-Nez;

(iii) UK, at the channel navigation information service (CNIS) and the Harwich harbor operation center;

(iv) Sweden, at the Gothenburg traffic control center; and

(v) The Netherlands, at the Hook of Holland traffic center.

The guidelines specified the collection of data on the following topics:

(i) context of the VTS (the physical area, traffic, casualties and organizational infrastructure); and

(ii) VTS external functions (i.e., services provided, in terms of IMO and COST 301 classifications).

Since a VTS can be considered as an information processing system, the external functions represent data output from a VTS. The investigations were concerned with the three stages of data flow:

(i) data at input (i.e., the content and sources of the data required),

(ii) data at output (i.e., the content of each of the services provided), and

(iii) internal data flow (i.e., how the acquired data are processed).

In addition, the results of a study of operator qualifications and skills in existing VTS are included.

(c) *Analysis and presentation of results.* The results from the various survey were collated and combined, and were presented mostly in tabular form. For details, see Degré *et al.* (1987).

3. Results

3.1 Services Provided by VTS

The detailed investigations on VTS showed large variations in the services provided by different centers, and many of these variations cannot be explained by the context within which they operate. Such variations include different criteria for distributing context-type information (e.g., visibility, ice and sea state) and whether information is provided on rogue vessels, ships at anchor, special operations and major events.

These variations apply to general information broadcasts. Since ships in transit could receive a number of these broadcasts from different authorities, there is a need for harmonizing both the content and the format of such broadcasts. This should be done within a regional or international forum. Similar considerations apply to services provided to individual ships, to pollution surveillance and to SAR activities.

3.2 Internal Operation of VTS

The identification in this study of the internal functions of a VTS could be of considerable value in further studies of VTS. Internal VTS functions are carried out mainly by human operators. However, there is scope for automating many of these, particularly when they involve complex processing or when they are of a routine nature. This could enable VTS operators to concentrate more on higher levels of decision making and on effective communications.

3.3 Levels of Qualification and Training

Since there are large variations in functions and levels of activity from one VTS to another, there will be variations in the levels of skill and expertise required of the VTS operators. However, for the medium-to-large VTS considered, appropriate requirements for VTS operators can be identified.

In Europe, there is a tendency to recruit at a master mariner level, although such qualifications do not always reflect operator tasks.

With some exceptions, there are no formal training programs for VTS operators and, generally, there is no systematic approach to VTS staff and career development. This situation is unlikely to change unless positive moves are made to harmonize VTS external procedures (the services they provide), followed by internal procedures and the organization of human resources to support these procedures.

4. Conclusions

COST 301 studies on existing VTS have been summarized. An inventory of existing VTS in Europe gave an overall view of their outline features.

This study also identified a difficulty which has been present throughout COST 301—namely, what is understood by a VTS. According to the IMO guidelines for VTS, every service implemented by a competent authority, and providing information to maritime traffic, can be considered as a VTS. This could be understood to include, for example, national broadcasting authorities, even though they would not be considered as a VTS by most people.

Using the COST 301 restriction on this definition—that a VTS interacts directly with, and in response to, the state of maritime traffic—national broadcasting authorities would be excluded. However, a small harbor for pleasure craft would be incorporated within this definition if it included the means to observe and communicate with traffic. There would be some reluctance to consider this as a VTS.

Again, within the IMO guidelines for VTS, and the more restricted COST 301 definitions, information systems such as emergency coordination centers for SAR are VTS, although they are not recognized as such by some authorities. COST 301 has provided the framework within which essential characteristics distinguishing a VTS from other information systems could be examined. These characteristics could include the nature of services provided and the intended users of the services.

See also: Marine Fleet Planning and Scheduling

Bibliography

Cutland M J, Degré T, Deutsch C, Glansdorp C C, McAlister K, Salvarini R, Willemse M C 1987 COST 301 Shore-based marine navigation aid systems. Commission of the European Communities. Main Report, EUR 11 304 EN

Degré T, Gerhardt D, Hamer K H, Lyon P R 1987 Existing VTS: Inventory, functions and effectiveness, COST 301 Final Report Annex to Main Report, Vol. 4. Commission of the European Communities. COST 301/FR 3.04/AN 0106

Salvarini R, Cutland M J, Deutsch C 1987 COST 301 Shore-based marine navigation aid systems. Commission of the European Communities. Executive Report, EUR 11250 EN

T. Degré
[INRETS, Arcueil, France]

Video Sensors

The image of a road traffic scene forms information that can be used both for surveillance (more and more cameras are installed to this effect) and for research on traffic theory. As early as 1952, Wardrop clearly stated laws of driver behavior on the basis of car traffic films. Thus, the idea of designing devices capable of automatically processing traffic images for measuring and analyzing road traffic appeared very natural, as soon as data processing techniques proved to be sufficiently efficient.

The basic principle of these devices is to use video cameras placed on road infrastructure in order to record car traffic. The images are numbered and then processed automatically by means of an image processing procedure.

Research in this field started almost simultaneously in Japan and the USA in the 1970s. In Japan, after having been developed at the University of Tokyo, it has now been taken over by private companies. In the USA, a first programme (WADS) was undertaken in the 1970s with public funds by the Jet Propulsion Laboratory, and stated the difficulties of the subject. After a less ambitious intermediate research project (SCAN), the subject is now being dealt with by developing sensor (VIDS) at the University of Minneapolis in Minnesota.

Taking advantage of the important development of methods and materials for image analysis, research and development is also going on in Europe. A system has already been put onto the market by a Belgian society. Other devices, either for different conditions of implementation or for supplementary performances, are being studied in several European institutes (UK, Italy, Germany, Sweden, Spain, France).

1. Possible Usage

The first advantages of this technique appear immediately:

(a) possibilities of fixed or mobile installation, or even *a posteriori* processing (video tape recorders survey scanning), and

(b) access to new traffic measurements (concentration, directional movements and queue lengths).

More specifically, the advantages of this technique compared with traditional techniques concern three aspects of traffic management with which all the people in charge of traffic infrastructure are confronted: measurement, surveillance and control.

1.1 Measurement

Traffic image processing allows the measurement of most parameters accessible by traditional means: traffic volumes, speeds and time occupancy rates. It also allows the access to traffic variables whose measurement is at present impossible except by observation and manual scanning: concentration, directional movements, vehicle store and exit times at junctions, study of conflict points and dangerous zones.

1.2 Surveillance

Systems of remote control generally contain TV cameras for surveillance that it would be possible to use more effectively without doubling the installation by means of conventional detection stations. Automatic processing would allow the detection of incidents, disturbing events, and the selection of the most interesting images to visualize on control screens.

1.3 Control

The main problem that the present urban traffic control systems are incapable of taking into account in a satisfactory way is the blocking of junctions due, in particular, to turning vehicles. From this point of view, the use of video sensors indicating permanently the blocking of junctions could be a considerable step forwards in the management of urban traffic networks, allowing for real-time control to avoid blocking.

1.4 Financial Aspects

The installation of a junction equipped with traffic signals (cabinet and loops) represents a cost of which 60% goes to the civil engineering works necessary for the return of loop information to the cabinet at the junction. A camera covering a much broader field could be installed closer to the junction cabinet, thus allowing savings of civil engineering costs. In a similar way, a camera equipped with an analysis system, installed on a rural freeway, transmitting measurements and compressed images, forms a good measuring means in competition with the traditional measurement station.

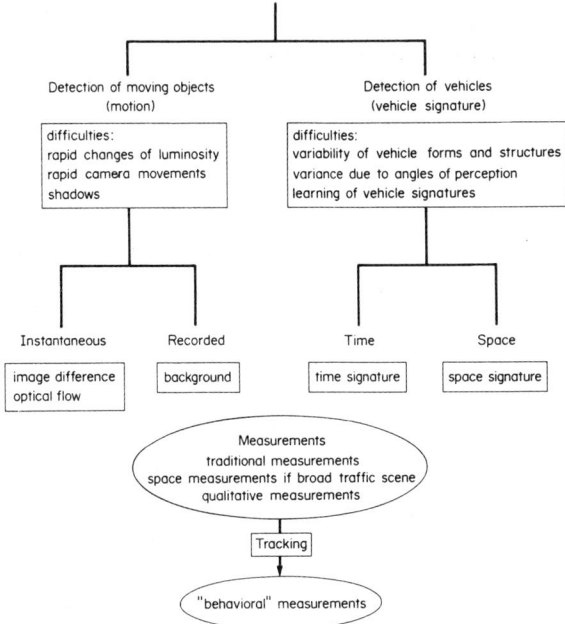

Figure 1
Classification of vehicle detection methods

2. Detection Techniques

Being based on general algorithms for identifying objects in motion or on searching for signatures specific to the vehicles, the detection methods can be classified into two overall categories (see Fig. 1).

2.1 General Methods for Detecting Moving Objects

Using several images, the detection principle is based essentially on the fact that the objects to be searched for are in motion. These methods prioritize the time aspect compared with the space aspect: the detection deals mainly with the analysis of variations in time of one and the same pixel rather than with the information given by the environment of a pixel in one image.

These methods are rather general and suppose here that the only objects in motion are vehicles. The identification process is very rudimentary. In practice, the fact that the objects to be detected have a minimum size is used. Some of these methods can lead to a segregation of the vehicles if the vehicle groups do not have rigid body movements.

In this class, three types of methods, each having a different perception of movement, will be considered.

(a) *Difference between two successive images*. This is the most direct method for making the immobile objects disappear, thus keeping as a consequence the traces of objects in motion between two successive moments. The immediate consequence is that stationary or very slow-moving objects are not detected. In general, this method is not used such as it is. In TITAN

(Blosseville *et al.* 1989) combined with filtering methods, it is used in order to give the images of traffic lanes. In the CRESTA system (Postaire *et al.* 1986), on the basis of three successives images I_{n-1}, I_n and I_{n+1} and the differences $I_{n-1} - I_n$ and $I_n - I_{n+1}$, a gradient algorithm leads to the outlines of the mobile objects in the image I_n being obtained. This system is intended to detect obstacles on the nonprotected track of a tram.

(b) *Optical flow.* Here, use is usually made of the fact that the appearance of a rigid object changes little during motion, in order to proceed to an estimation of the optical flow. In other words, in each point the speed vector $u(x, t)$ is quantified; u is obtained by identifying at the time $t - dt$ the image elements $g(x - u\, dt, t - dt)$ presenting a high resemblance to the elements $g(x, t)$ taken at the time t. Here too, the immobile objects are not detected.

Different attempts to apply this technique to the surveillance of road traffic, obstacle detection, and driving of an autonomous vehicle have been discussed by Nagel (1984), and Enkelman (1989).

(c) *Use of a background image.* In the two preceding methods, the image of motionless objects (background image) is not memorized. On the contrary, this method is based on forming the most precise possible background image. The background image is obtained either manually, in the most primitive applications, taking an image without vehicles, or automatically by making a mathematical or exponential average of successive images as in Fig. 2, an example of an urban image processed by the TITAN system prototype. In Fig. 2, due to the great complexity of the usual urban scene, the camera is almost vertically oriented and only presence or absence measurements can be obtained. Other more elaborate techniques allow more precise control of these background images. The detection is then achieved by means of subtracting the reference image from the current image. Then a thresholding is performed in order to obtain presence/absence information of an object in motion. This is by far the most frequently used technique, and according to the requirements made on calculation time, different approaches can be distinguished dealing with the whole image or part of it.

Thus, an early Japanese system (Onoe *et al.* 1973) and two British ones (Hoose 1989, Houghton *et al.* 1989) process the whole image. Other devices, including Japanese (Onoe and Ohba 1976), Swedish (Abramczuk 1984) and German (Houkes, 1980) ones, process perspective lines along the traffic lanes of freeway images. In other works, there is a preference for processing the images of some lines perpendicular to traffic lanes, including British (Dickinson *et al.* 1985, Dickinson 1986), US (Schlutsmeyer 1982) and Japanese (Takaba and Ooyama 1984) systems.

Reducing the quantity of information processed even more, some researchers (Dods 1984) (Australia) use only a few points of the images.

Most of the devices using one of these three methods based on the *a priori* use of movement limit themselves, in general, to this information in terms of

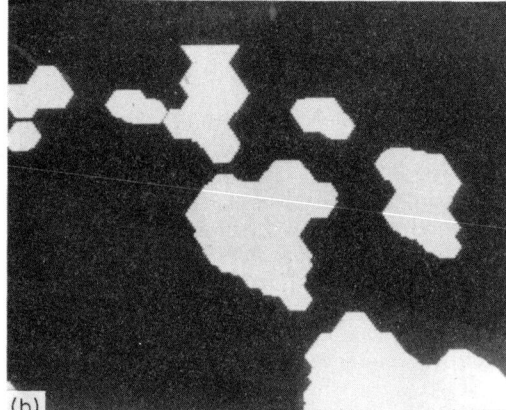

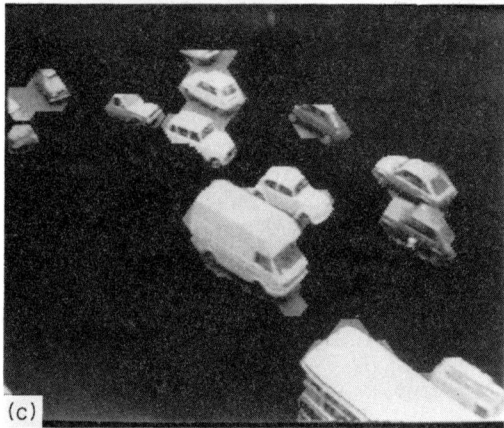

Figure 2
Use of the background image: (a) original image; (b) detection image (the white part of the image corresponds to the presence of moving objects); and (c) detection image superimposed on the original image (only the image of moving vehicles appears)

presence/absence. A second step used by some devices (Dickinson 1986) allows the elimination of artefacts (shadows, false detections) as a refinement of vehicle detection (filling in gaps, etc.).

2.2 Specific Method for Vehicle Detection (Direct)

In the direct detection method, the objects to be identified (vehicles) are described by their characteristics (forms, dimensions, luminosity) allowing their identification in their environment. Time or space signatures are used.

(*a*) *Time signature*. The process developed by Devlonics in 1989 for the CCATS system is capable of measuring traffic on three freeway lanes. Each lane is analyzed by means of a 30-point matrix representing a rectangle on the ground of 5 m × 2 m. The analysis of the time signature recorded on these points is used to derive the presence or absence of vehicles. The time signal of luminosity of each point is analyzed by means of a model with prerecorded and periodically updated characteristics. Space filtering of these binary matrices allows the reinforcement of detection.

(*b*) *Space signature*. This method was used by (Hogg *et al.* 1984) for locating London double-deckers in an image of urban environment. The signatures are then the outlines of the buses reconstituting a certain number of positions in the image thanks to a geometrical model. These *a priori* defined outlines are made up of the sides and hips (gradients) of the images of these vehicles. Then the searching is done by evaluating for each of these preestablished positions in each current image the presence of attributes of an outline.

The method for searching for signatures is also used in TITAN (Blosseville *et al.* 1989) using simpler outlines. The process applies to a front or rear view of a medium–long freeway section (100–300 m). In these conditions, the vehicles appear as a succession of horizontal zones of a certain width, lighter or darker than the carriageway. Thus, the characteristics used are minima and maxima of sufficiently broad images. Then the detected zones are gathered together in order to reconstruct the vehicles. After that, the calculation of trajectories allows some false detections to be removed and traffic measurements to be made, as in Fig. 3 which shows examples of results obtained with the TITAN system prototype (Blosseville *et al.* 1989). Views of 100–300 m long freeway stretches can be monitored (depending on the camera height); the camera can be placed upstream or downstream. Each vehicle is found and tracked while it crosses the scene. Microscopic (vehicle by vehicle) as well as macroscopic measurements (regarding groups of vehicles) can be performed.

Some teams combine the methods using a method based on movement for detection of mobiles, and then a direct method for making a follow-up of the vehicles.

Figure 3
Direct detection method: (a) example of processed image; (b) several markers are used to find the vehicles, white markers indicate the front of vehicles and a vertical view of the freeway is shown on the right; and (c) trajectories of the localized vehicles regarding two axes, the space axis (vertical) and the time axis (horizontal)

Thus, Houghton et al. (1989) use background techniques in order to locate moving objects in a junction and afterwards identify different categories of vehicles based on prerecorded signatures.

3. Difficulties Encountered

Automatic analysis of a road traffic image comes up against a certain number of difficulties due mainly to the great variability of conditions of application. The sources of difficulties can be grouped into three overall themes.

First, there are the factors affecting the quality of the information to be processed:

(a) rapid change of luminosity (cloudy weather),

(b) change in the intensity and direction of light (day/night, shadows etc.),

(c) visibility (fog, snow, rain, etc.), and

(d) rapid camera movements (strong wind).

Second, there are the factors affecting vehicle look:

(a) variability of vehicle forms and structures,

(b) variance in the appearance of vehicles due to different angles of perception (perspective, camera position),

(c) variability of traffic conditions (blocking during peak periods), and

(d) adjustable camera.

Third, there are the factors affecting the complexity level:

(a) dimensions of the field to be processed, and

(b) infrastructure type (freeway/urban).

How can such a level of difficulties be dealt with using the actual computing capacity compatible with the requirements of the sensor market? Each research team tries to answer this question in its own way, for none of the methods known allows all these problems to be solved. An important part of the answer lies in the choice of conditions of application.

Thus, the performance of methods for detection of mobiles will be affected by all the outside conditions that deteriorate the quality and stabilty of the traffic signal. All the parasite variations that cannot be distinguished from the normal variations of the image will produce erroneous detections. The answer given to this type of error consists, in general, of applying the process to images in which the vehicle size is very big compared with the possible errors. Moreover, use of the system is limited to the places of the image in which the vehicles are clearly separated. Thus, a system in which the camera is placed in a very vertical position gives very punctual measurements, like the loops.

The methods using vehicle signatures can be less affected by parasite variations of the traffic signal or the very camera, because the images are processed one after the other. However, they have other drawbacks; they require a definition of the characteristics of each type of scene so that the problem of learning is a crucial one. This may be solved by limiting the system to standard scenes (front or rear scene for TITAN) or by developing techniques for automatic learning such as neuromimetic networks (Dickinson and Wan 1989).

4. Measurements

The choice of a detection method (see Fig. 1) has no direct importance for the nature of the traffic measurements to be produced, but rather for their precision and operating limits. On the contrary, the determining factors are:

(a) the processing frequency,

(b) the extent of the space to be processed, and

(c) the possibilities of follow-up of the vehicles or vehicle groups.

4.1 Traditional

Most of the present systems can reproduce the operating mode of magnetic loops by detecting the presence of a vehicle on a reduced section of the carriageway; the corresponding measurements concern above all:

(a) traffic volumes,

(b) time occupancy rates, and

(c) speeds (corresponding to two successive loops).

Most devices being developed aim at this type of measurements. The closer to the vertical position the camera is and the higher the processing frequency is, the better these measurements are. The results obtained and published are in general good. Large-scale experiments made in Japan by four companies using several methods (Shimizu 1989) were accurate to above 90% for traffic volumes and average speeds.

4.2 Space

As soon as the part of the image processed corresponds to a broad traffic scene (at least 10 m), it becomes possible to have access to lengths and space means of traffic measurements:

(a) vehicle spacing,

(b) queue lengths,

(c) concentration (density), and

(d) space occupancy rates.

Only the devices developed in Sweden (Abramczuk 1984) and France (Blosseville et al. 1989) can really give space measurements on distances of about 100 m. There are no statistics available on performances. The space occupancy rate is, however, often produced on smaller distances: accuracy of above 80% was obtained in the experiment made in Japan.

4.3 Behavioral

Video sensors going as far as to the follow-up of vehicles or vehicle groups, allow the measuring of phenomena relative to behavior:

(a) locating and quantifying lane changes, accelerations and decelerations on freeways; and

(b) analysis of junction crossing by means of measurements such as origin–destination matrices and journey times—a special mention concerns systems for automatic reading of number plates, whose objective is not so much the collection of traffic measurements as automatic toll paying or police surveillance.

4.4 Qualitative

A human operator observing a traffic scene does not need precise measurements of traffic volumes or speeds for evaluating the situation; this statement has given rise to a new approach that consists of extracting relatively qualitative measurements from the image in terms of zones in motion and occupied zones (Hoose 1989). Such a description could prove to be well adapted to the solution of some problems of traffic control and automatic incident detection.

5. Future Developments

Taking into account the relatively old technology used by magnetic loops, there is a real need for a new traffic sensor. Some products based on image processing and capable of making measurements very close to those given by "classical" sensors are available.

However, thanks to very active research in this field and the dynamics of the image processing market, more elaborate sensors capable of describing important traffic phenomena that have been impossible to measure (supervision of large freeway zones, junction centers, queues, etc.) should be available in the short or medium term. The analysis of complex scenes and problems posed by partly hidden vehicles are still, to a great extent, at the research stage. Nevertheless, without waiting for the complete theoretical solution to these problems, application conditions could just be restricted in a way that limits the difficulty. Then the feasibility of a new video sensor dealing with rather complex scenes could be envisaged in the early 1990s.

See also: Detectors for Road Traffic; Vehicle Guidance by Computer Vision

Bibliography

Abramczuk T 1984 A microcomputer based TV-detector for road traffic. *Road Transport Research Program of OECD, Seminar on Micro-Electronics for Road and Traffic Management.* Traffic Bureau, National Police Agency, Tokyo, pp. 87–96

Blosseville J M *et al.* 1989 TITAN: A traffic measurement system using image processing techniques. *IEE 2nd Int. Conf. Road Traffic Monitoring.* Institution of Electrical Engineers, London, pp. 84–8

Dickinson K W 1986 Traffic data capture and analysis using video image processing. Ph.D. thesis, Sheffield University

Dickinson K W *et al.* 1985 Image processing systems for monitoring road traffic. *CEMT/ECMT Int. Seminar Electronics and Traffic on Major Roads.* Organisation for Economic Co-operation and Development, Paris

Dickinson K W, Wan C L 1989 Road traffic monitoring using the TRIP II system. *IEE 2nd Int. Conf. Road Traffic Monitoring.* Institution of Electrical Engineers, London, pp. 56–60

Dods J S 1984 the Australian Road Research Board video based vehicle presence detector. IEE Conference Publication No. 242, Institution of Electrical Engineers, London

Enkelman W 1989 Interpretation of traffic scenes by evaluation of optical flow fields from image sequences. *Proc. IFAC Int. Symp. Control, Computers, Communication on Transportation.* Pergamon, Oxford

Hogg D C *et al.* 1984 Recognition of vehicles in traffic scenes using geometric models. *Proc. Int. Conf. Road Traffic Data Collection.* Institution of Electrical Engineers, London, pp. 115–19

Hoose N 1989 Queue detection using computer image processing. *IEE 2nd Int. Conf. Road Traffic Monitoring.* Institution of Electrical Engineers, London, pp. 94–8

Houghton A D *et al.* 1989 Automatic vehicle recognition. *IEE 2nd Int. Conf. Road Traffic Monitoring*, Institution of Electrical Engineers, London, pp. 71–8

Houkes A 1980 Measurements of speed and time headway of motor vehicles with video camera and computer. *Proc. Int. Conf. Digital Computer Application to Process Control*, Dusseldorf

Hoummady B 1988 Système d'analyse de mouvement dans les images routières. Ph.D. thesis, Université de Saint-Etienne

Nagel H H 1984 New likelihood test methods for change detection in image sequences. *Comput. Vision, Graph. Image Proc.* **26**(1), 72–106

Onoe M, Nobuo M, Ohba K 1973 Computer analysis of traffic flow observed by substractive television. *Comput. Graph. Image Proc.* 377–92

Onoe M, Ohba K 1976 Digital image analysis of traffic flow. *Proc. Int. Joint Conf. Pattern Recognition.* Coronada

Postaire J G, Stelmaszyk P, Bonnet P 1986 A visual surveillance system for traffic collision avoidance control. *IFAC Transportation Symp.* International Federation of Automatic Control, Laxenburg, Austria

Schultsmeyer A P 1982 Wide area detection system (WADS). Report No. FHWA/RD 82/114. Federal Highway Administration, Washington, DC

Shimizu K 1989 Image processing system used cameras for vehicle surveillance. *IEE 2nd Int. Conf. Road Traffic Monitoring.* London, pp. 61–5

Takaba S, Ooyama N 1984 Traffic flow measuring system with image sensors. *Road Transport Research Program of OECD, Seminar on Micro-Electronics for Road and Traffic Management*. Traffic Bureau, National Police Agency, Tokyo, pp. 12–20

Versavel *et al*. Cameras and computer aided traffic sensor. *IEE 2nd Int. Conf. Road Traffic Monitoring*. Institution of Electrical Engineers, London, pp. 94–8

J. M. Blosseville, V. Motyka and S. Espie
[INRETS, Arcueil, France]

Visual and Instrument Flying Rules

It is a considerable problem to manage air traffic for all kinds of air vehicles with a high degree of safety and without too many constraints as to the knowledge to be acquired by pilots.

Before World War I, there were no flying rules at all. When commercial traffic was established on regular schedules, flights in bad visibility or during nighttime became mandatory and rules appeared in the 1940s. In 1945, an international organization, the International Civil Aviation Organisation (ICAO), was created. It currently publishes standard rules and recommendations. Most of these are accepted by countries without any modifications. Supplementary rules are often added to these recommendations. However, the space above a country still remains under the control of the national authority for airborne vehicles (except for satellites). In Europe, Eurocontrol tries to coordinate national activities and is in charge of air traffic control (ATC) above flight level (FL) 195 (FL 195 means an altitude of 19 500 ft (5850 m) read on an altimeter set up at 101.35×10^3 Pa, irrespective of the acutal reference barometric pressure) for a limited but multinational space.

1. Definitions

There are two types of flying rules: visual flying rules (VFR) and instrument flying rules (IFR).

This implies a distinction concerning the type of meteorological conditions. These are similarly divided into two types: visual meteorological conditions (VMC) and instrument meteorological conditions (IMC).

Independently of weather conditions, the airspace is divided into controlled space and uncontrolled space. Figures 1 and 2 give examples of controlled space (Fig. 1 for airways; Fig. 2 for space around an airport, here Toulouse, France). In France, the VMC are defined relative to a height "surface" S (see Fig. 3) for uncontrolled airspace. French VMC are given in Table 1.

Landings are classified as precision approaches if an accurate guidance is provided by instrument landing system (ILS), microwave landing system (ML) or precision approach radar (PAR, a variety of ground control approach (GCA)) equipment and as nonprecision approaches for all the other types of approach. (Nonprecision approach does not mean that the accuracy of the landing (touchdown point on the runway) is degraded or could be degraded; very often nonprecision approaches may require high accuracy in the touchdown point and in alignment because, if the runway is not equipped with radio aids, there is a high probability that it is also a shorter and narrower runway.)

Prior to departure, a flight plan (FPL) which describes the profile of the trajectory (plus additional information concerning the plane) is sent to the ATC for approval. FLPs are not mandatory for VFR flights, but they are recommended.

2. Basic Rules

A VFR flight can be performed only in VMC even in controlled space with, sometimes, restrictions in terminal areas (i.e., around an airport). In any case, it must stay below FL 195. No minimum equipment, such as artificial horizon or radio, is requested. The pilot completely controls safety in respect of collision with other air vehicles or ground obstacles.

An IFR flight can be performed in VMC or IMC. A flight plan is mandatory. Minimum equipment is requested (e.g., radio, navigation aids, receivers, artificial horizons, deicing devices). Prior to takeoff, a first clearance is radioed to the crew: it gives the first legs of the trajectory and the altitude or FL to be achieved over the beacons or the way points (reporting points). It also means that there will be no IFR air vehicles on these legs at the same level or altitude, within the safety distance required for separation. However, it does not mean that some VFR air vehicles will not be present along this trajectory. Consequently, and surprisingly, an IFR flight is safer in IMC (bad weather conditions) than in VMC. Further clearances are regularly given to the crew as the flight progresses. However, IFR flights may occur in noncontrolled space; that is, to join a low traffic airport from an airway and vice versa.

3. Flights in Controlled Airspace

As previously stated, VFR and IFR air vehicles can be present in a controlled airspace. For IFR, authorizations to go on the flight are given by the ATC as the planes evolve on the airways; additional information such as local meteorological conditions or terminal airfield situations are provided both for IFR and VFR. Normally, planes are located by the ATC by reports radioed by the crew when passing over beacons or way points and, in dense areas, by use of transponders (mandatory for IFR flights). (A transponder is an onboard transmitter which is interrogated by secondary surveillance radar (SSR); the reply message includes the aircraft identification and, according to the type of

Visual and Instrument Flying Rules

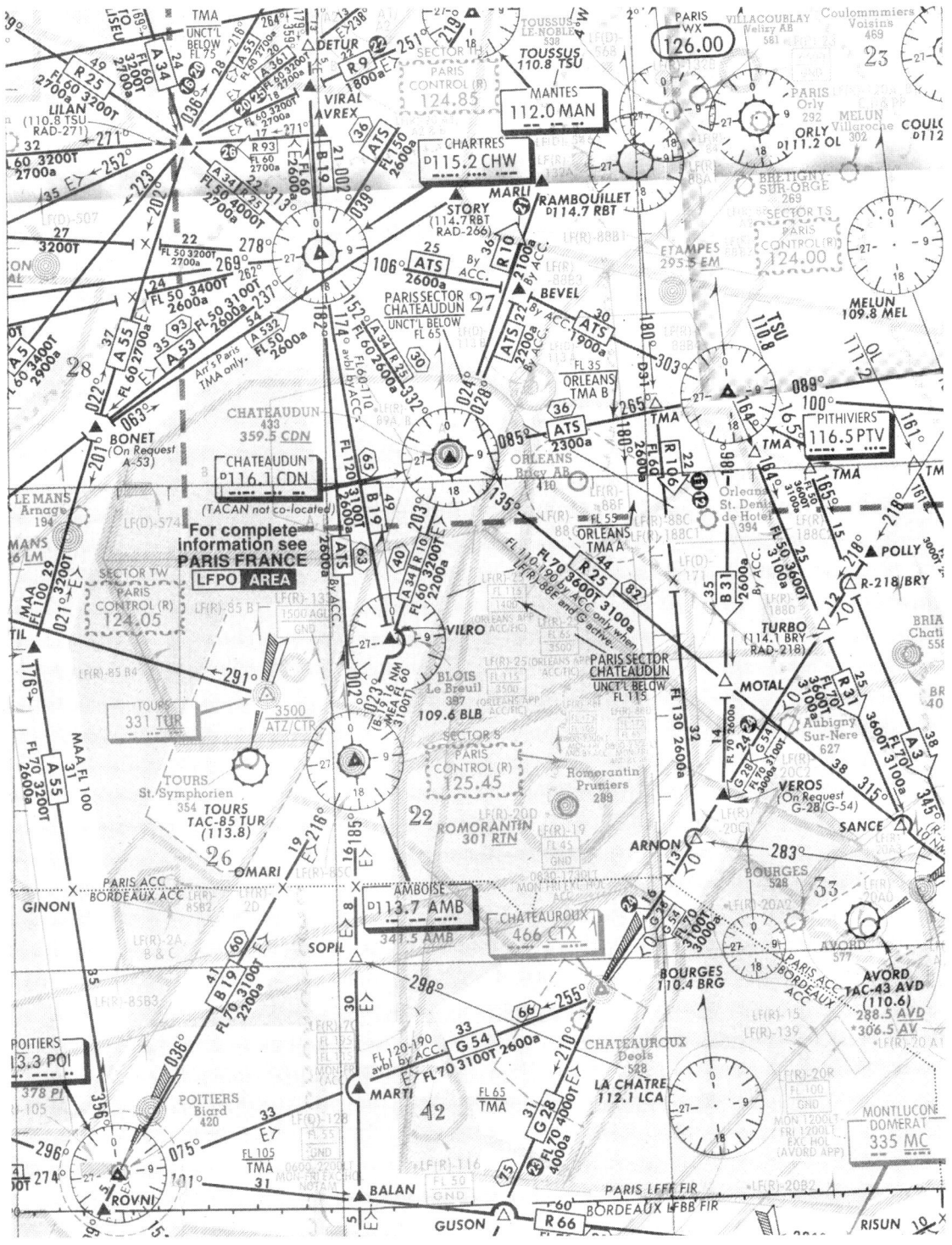

Figure 1
Airways in the Paris area (courtesy Jeppesen Sanderson Inc.)

Visual and Instrument Flying Rules

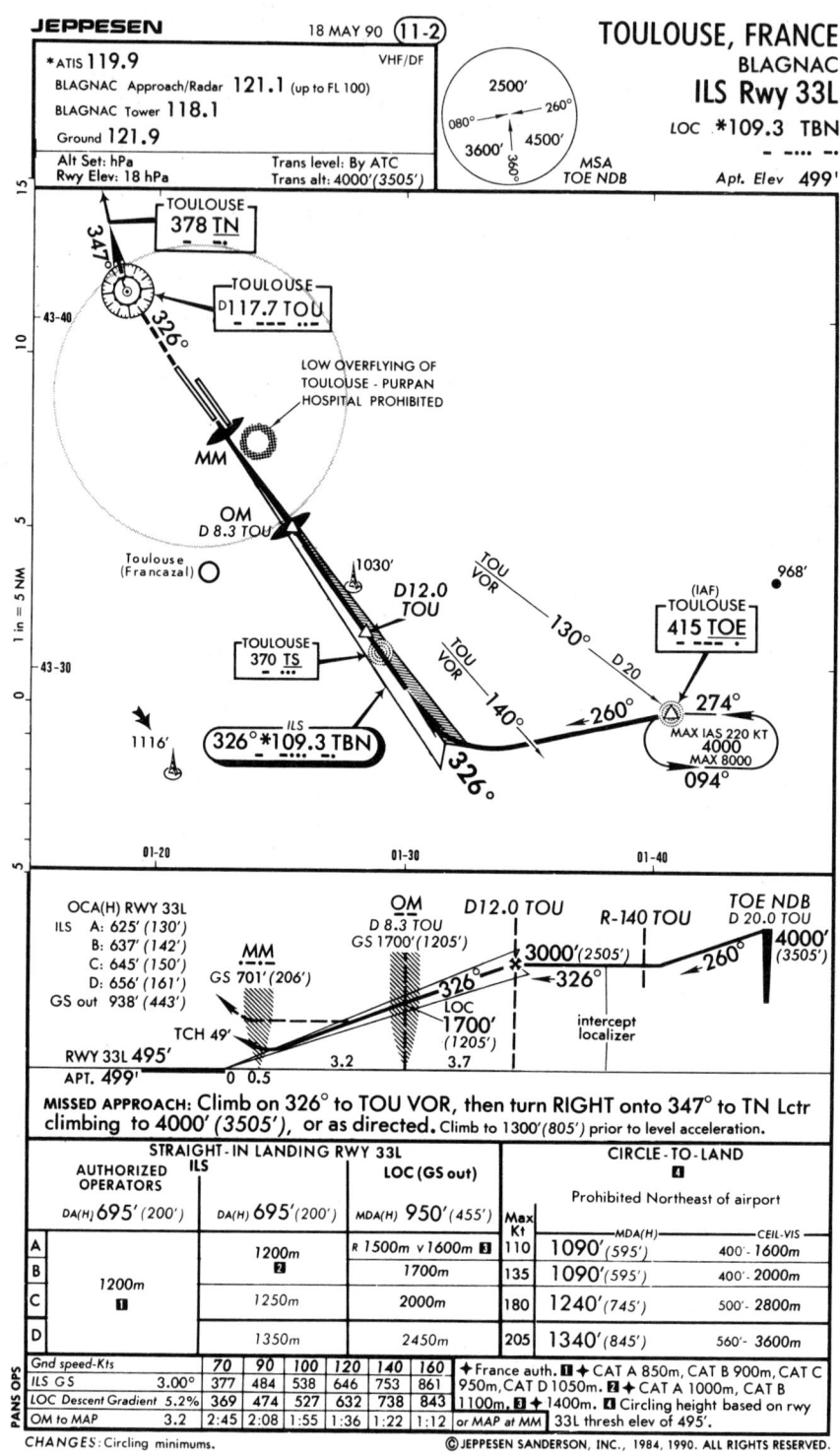

Figure 2
A typical approach pattern for runway 33L with instrument landing system (ILS) at Toulouse, France (courtesy Jeppesen Sanderson Inc.)

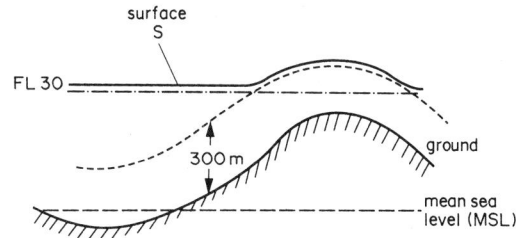

Figure 3
Surface S, defined as the maximum of 300 m above ground level and FL 30.

transponder, it may also indicate the FL of the plane and the airspeeed.)

All that has been stated previously is valid worldwide. This is not true for what follows, but is accepted in many countries.

VRF air vehicles may penetrate into airways (crossing them or using them) if they are flying VMC. If using them, they must fly a prescribed FL, normally "odd plus five levels," when flying routes with bearing comprised between 0° and 179° and "even plus five levels," when flying routes between 180° and 359°. This means that they can use levels such as FL 55, 75, 95, ..., 135, ..., 195 for routes between 0° and 179° and 45, 65, ..., 145, ..., 185 for routes between 180° and 359°.

Note that IFR traffic as well as VFR traffic may change their flight levels. IFR air vehicles will do so after authorization by ATC, but ATC may ignore the presence of VFR planes in the airways or, at least, not attach too much confidence to the data concerning their position or change of FL. In fact, it is highly recommended that VFR planes entering an airway are equipped with radio and inform the ATC about their estimated positon and desired FL; in France, there is no requirement for them to be equipped with a transponder up to FL 120 (it is mandatory above).

In contradiction with what is generally thought, when flying in controlled airspace in VMC (good visibility), collision avoidance is the responsibility of the pilot, whether using VFR or IFR.

Table 1
French VMC in 1981 (km)

	Horizontal visibility	Distance to clouds		Ceiling
		horizontal	vertical	
In uncontrolled airspace				
above S	8	1.5	0.3	
below S	1.5	outside clouds	outside clouds	
In controlled airspace	8	1.5	0.3	
Outside airport terminal areas	8			0.45

4. Classes of Airspace

The controlled space as it currently appears (i.e., with a qualification attached to the geographic situation: airways, terminal areas with a subset of spaces such as TMA-1, TMA-2, CRT, ATZ) will be divided into seven space classes with specific rules in each. Each geographical space will be qualified by one letter A to G. The classes are defined as follows.

(i) *Class A controlled airspace*. This is airspace in which IFR but not VFR flights are permitted. ATC units are responsible for IFR flight spacing.

(ii) *Class B controlled airspace*. Both IFR and VFR flights are permitted. ATC units are responsible for IFR, IFR/VFR and VFR flight spacing.

(iii) *Class C controlled airspace*. Both IFR and VFR flights are permitted. ATC units are responsible for IFR and IFR/VFR flight spacing, as well as for the provision to VFR flights of traffic data on other VFR flights.

(iv) *Class D controlled airspace*. Both IFR and VFR flights are permitted. ATC units are responsible for IFR flights spacing and for the provision to IFR flights of VFR traffic data, as well as the provision to VFR flights of IFR and other VFR traffic data.

(v) *Class E controlled airspace*. Both IFR and VFR flights are permitted. ATC units are responsible for IFR flight spacing.

(vi) *Class F uncontrolled airspace (advisory airspace)*. Both IFR and VFR flights are permitted. ACT units provide advisory services.

(vii) *Class G uncontrolled airspace*. Both IFR and VFR flights are permitted. ATC units provide flight information and alerting services only.

(viii) *Controlled military airspace*. This is airspace of defined dimensions intended for the performance of specific defense activities and the control of military flights requiring special operational and technical (OAT) procedures. The general air traffic services provided to OAT flights are identical to those provided in classes A, B, C, D and E. Clearances are issued by the military ATC units in the light of specific constraints connected with defense activities. (Note that except where otherwise stated, the term controlled airspace also embraces military controlled airspace.)

5. Future Developments

The rapid growth of air traffic needs a revision of the flight rules that are used worldwide. The ICAO recommendations leading to seven classes of airspace with precise regulations for each will maintain the safety level in spite of the traffic growth and it will allow a mixing of both kinds of traffic (VFR and IFR).

However, any real improvement will not emerge until an automatic data link is provided between ground control centers and planes (at least IFR planes). This link has been expected since the early 1970s. Radar mode S (see *Air Traffic Control: An Overview*) and satellites have the capability of transmitting and receiving messages. Compatibility and message content have to be defined in detail. The problem is to reach international agreements on this quickly.

See also: Automatic Landing Systems

M. J. Pélegrin
[ONERA, Toulouse, France]

LIST OF CONTRIBUTORS

Contributors are listed in alphabetical order together with their addresses. Titles of articles that they have authored follow in alphabetical order. Where articles are coauthored, this has been indicated by an asterisk preceding the title.

Akagi, S.
Department of Mechanical Engineering
Faculty of Engineering
Osaka University
2-1 Yamadaoka Suita
Osaka 565
Japan
Fuel Conservation: Sea Transport

Akita, K.
Safety and Telecommunications Laboratory
Railway Technical Research Institute
2-8-38 Hikaricho
Kokubunji-City
Tokyo
Japan
*Railroad Electronic Signalling

Allsop, R. E.
Transport Studies Group
University College London
Gower Street
London WC1E 6BT
UK
Road Traffic: An Introduction
Signal Control at Individual Junctions: Stage-Based Approach

Ambrosino, G.
Automa
Sistemi di Automazione Industriale
Via Al Molo Vecchio
Calata Gadda
I-16126 Genova
Italy
*Container Terminal Management
*Expert Systems Approach to Road Traffic Control

Artynov, A. P.
Far East Department
Institute of Automation and Control Processes
USSR Academy of Sciences
5 Radio Strasse
SU-690032 Vladivostok
USSR
*Marine Fleet Planning and Scheduling

Axhausen, K. W.
Transport Studies Unit
Oxford University
11 Bevington Road
Oxford OX2 6NB
UK
*Transportation Planning: Activity-Based Approach

Balchen, J. G.
Universitetet I Trondheim
Norges Tekniske Høgskole
Institut for Teknisk Kybernetikk
N-7034 Trondheim
Norway
Ship Positioning: Adaptive Control

Balz, W.
Steierwald Schönharting und Partner GmbH
Alexanderstrasse 105
D-7000 Stuttgart 1
Germany
*Freeway Network Control

Barceló, J.
Departament d'Estadistica i Investigació Operativa
Universitat Politecnica de Catalunya
Pau Gargallo 5
E-08028 Barcelona
Spain
Simulation of Urban Traffic: Software Environments
Traffic Management Systems

Bastin, G.
Laboratoire d'Automatique, de Dynamique et d'Analyse des Systèmes
Université Catholique de Louvain
Bâtiment Maxwell
Place du Levant 3
B-1348 Louvain-la-Neuve
Belgium
*Underground Railroad Modelling and Control

Becker, U. J.
Institut für Verkehrswesen
Universität (TH) Karlsruhe
Postfach 6980
Kaiserstrasse 12
D-7500 Karlsruhe
Germany
*Transportation Planning: Microscopic Approach

Belardinelli, F.
Ferrovie dello Stato
Alta Velocità
Via Lamaro 13
I-00173 Roma
Italy
High-Speed Railroad: Systems Approach

Bell, M. C.
Lecturer and Director of Transport Research
Department of Civil Engineering
University of Nottingham
University Park
Nottingham NG7 2RD
UK
Traffic Control Systems: Architecture

Bennett, S.
Department of Control Engineering
University of Sheffield
Mappin Street
Sheffield S1 3JD
UK
Ship Stabilization: History

Benoît, A.
European Organisation for the Safety of Air Navigation
72 Rue de la Loi
B-1040 Bruxelles
Belgium
Fuel Conservation: Air Transport

Betrò, B.
CNR-IAMI
Via Ampère 56
I-20131 Milano
Italy
Discrete-Time Point Processes: Applications to Road Traffic

Bielefeldt, C.
The MVA Consultancy
Victoria Way
Woking
Surrey GU21 1DD
UK
Broadcasting Communication Systems

Bielli, M.
IASI-CNR
Viale Manzoni 30
I-00185 Roma
Italy
Expert Systems Approach to Road Traffic Control

Blosseville, J. M.
INRETS
2 Avenue du Général Malleret
Joinville

BP 34
F-94114 Arcueil Cédex
France
On-Ramp Control, Local
Video Sensors

Boccassi, G.
Automa
Voa Caffaro 8-12
I-16124 Genova
Italy
High-Speed Railroad: Systems Approach

Boero, M.
Automa
Sistemi di Automazione Industriale
Via Al Molo Vecchio
Calata Gadda
I-16126 Genova
Italy
Expert Systems Approach to Road Traffic Control

Burkard, R. E.
Mathematisches Institut
Technische Universität Graz
Kopernikusgasse 24
A-8010 Graz
Austria
Shortest-Path Algorithms

Busch, F.
Department of Traffic Engineering
Siemens AG
Postfach 70 00 70
Hofmannstrasse 51
D-8000 München 70
Germany
Incident Detection

Campion, G.
Laboratoire d'Automatique, de Dynamique et d'Analyse des Systèmes
Université Catholique de Louvain
Bâtiment Maxwell
Place du Levant 3
B-1348 Louvain-la-Neuve
Belgium
Underground Railroad Modelling and Control

Cantarella, G. E.
Istituto di Ingegneria Civile ed Energetica
Facolta di Ingegneria
Università di Reggio Calabria
Via Veneto 69
I-89123 Reggio Calabria
Italy
Road Network Signal Setting: Equilibrium Conditions
Traffic Assignment

Casalino, G.
Department of Communications, Computers and System Sciences
Università Genova
Via Opera Pia 11A
I-16145 Genova
Italy
Underground Railroad Modelling and Control: Discrete-Event Approach

Cascetta, E.
Dipartimento di Ingegneria dei Trasporti
Via Claudio 21
I-80127 Napoli
Italy
Traffic Assignment, Stochastic

Charbonnier, C.
ONERA-CERT/DERA
2 Avenue Edouard Belin
BP 4025
F-31055 Toulouse Cédex
France
Route Guidance, Individual

Chin, S.-M.
Oak Ridge National Laboratory
Oak Ridge, TN 37831
USA
Simulation Programs, Microscopic

Chrisoulakis, J.
146 B, Emm. Benaki Str
GR-11473 Athens
Greece
Simulation Programs, Macroscopic

Cohen, S.
INRETS
2 Avenue du Général Malleret
Joinville
BP 34
F-94114 Arcueil Cédex
France
Flow Variables
Kinematic Wave Theory

Courcoux, L.
Inspection Générale Enseignement Maritime
38 Rue Liancourt
F-75014 Paris
France
Social Issues of Ship Automation

Crämer, W.
Fachhochschule Rosenheim
Lehrgebiet: Mess- und Regelungstechnik
Marienberger Strasse 26
D-8200 Rosenheim
Germany
Magnetic Suspension Railroad Systems

Cremer, M.
TU Hamburg-Harburg
AB Automatisierungstechnik
Lohbrügger Kirchstrasse 65
D-2050 Hamburg 80
Germany
Flow Variables: Estimation
Origin–Destination Matrix: Dynamic Estimation

Davies, P.
Castle Rock Consultants
Heathcoat Building
Highfields Science Park
Nottingham MG7 2QJ
UK
Road Traffic Monitoring Equipment

de Kerdaniel, P.-L.
Compagnie Générale d'Automatisme CGA-HBS
Briétigny
France
Vehicle Monitoring and Control for Buses, Trolleys and Streetcars

de Palma, A.
Department of Civil Engineering
Northwestern University
The Technological Institute
2145 Sheridon Road
Evanston, IL 60208
USA
Road Networks: Dynamic Equilibrium Models

Degre, T.
INRETS
2 Avenue du Général Malleret
Joinville
BP 34
F-94114 Arcueil Cédex
France
Vessel Traffic Services: Management from Shore

Demetsky, M. J.
Department of Civil Engineering
University of Virginia
Thornton Hall
Charlottesville, VA 22903
USA
Transportation Modelling
Transportation Systems: Trends

Desel, U.
Germanweg 23
D-6272 Niedernhausen

Germany
Flight Activity in General Aviation

Di Febbraro, A.
Department of Communications, Computers and System Sciences
Università Genova
Via Opera Pia 11A
I-16145 Genova
Italy
Underground Railroad Modelling and Control: Discrete-Event Approach

Dickmanns, E. D.
Universität der Bundeswehr München
Fakultät für Luft- und Raumfahrttechnik
Institut für Systemdynamik und Flugmechanik
Werner-Heisenberg-Weg 39
D-8014 Neubiberg
Germany
Vehicle Guidance by Computer Vision

Dobias, G.
INRETS
Le Directeur Général
2 Avenue du Général Malleret
Joinville
BP 34
F-94114 Arcueil Cédex
France
Social Issues of Transportation

Donati, F.
Politecnico di Torino
Dipartimento di Automatica e Informatica
Corso Duca degli Abruzzi 24
I-10129 Torino
Italy
Maintenance and Reliability of Traffic Control Systems

Dujardin, C.
Ingénieur Principal des Etudes et de l'Exploitation de la Navigation Aérienne
Service Technique de la Navigation Aérienne
246 Rue Lecourbe
F-75732 Paris
France
Air Traffic Simulators

Eberlein, D.
Deutsche Forschungsanstalt für Luft- und Raumfahrt
Hauptabteilung Verkehrsforschung
Postfach 90 60 58
Linder Hohe
D-5000 Köln 90
Germany
High-Speed Railroad Networks in Europe

Espie, S.
INRETS
2 Avenue du Général Malleret
Joinville
BP 34
F-94114 Arcueil Cédex
France
Video Sensors

Farges, J. L.
ONERA-CERT/DERA
2 Avenue Edouard Belin
BP 4025
F-31055 Toulouse Cédex
France
Route Guidance, Individual

Ferrara, A.
Department of Communications, Computers and System Sciences
Università Genova
Via Opera Pia 11A
I-16145 Genova
Italy
Underground Railroad Modelling and Control: Discrete-Event Approach

Ferrari, P.
Istituto di Strade e Transporti
Facoltà di Ingegneria
Università di Pisa
Via Diotisalvi 2
I-056100 Pisa
Italy
Freeway Capacity: Reliability and Control

Fondacci, R.
Ingénieur Principal des Etudes et de L'Exploitation de la Navigation Aérienne
Service Technique de la Navigation Aérienne
246 Rue Lecourbe
F-75732 Paris Cédex
France
Air Traffic Simulators

Form, P.
Technische Universität Braunschweig
Postfach 33 29
Rebenring 18
D-3300 Braunschweig
Germany
Airborne Collision Avoidance

Frantzeskakis, J. M.
16 Kifissias Ave
GR-15125 Amaroussio

Athens
Greece
Safety of Road Traffic

French, R. L.
R L French & Associates
Suite 201
3815 Lisbon Street
Fort Worth, TX 76107
USA
In-Vehicle Equipment for Future Traffic Control Systems
Mobile Communication

Fukuda, T.
Industrial Electronics and Systems Development
 Laboratory
Mitsubishi Electric Corporation
8-1-1 Tsukaguchi Honmachi
Amagasaki Hyogo 661
Japan
*Expert Systems Approach to Rail Traffic Control

Gabard, J. F.
ONERA-CERT/DERA
2 Avenue Edouard Belin
BP 4025
F-31055 Toulouse Cédex
France
Car-Following Models

Gartner, N. H.
Department of Civil Engineering
University of Lowell
Lowell, MA 01854
USA
Road Traffic Control, Demand-Responsive
Road Traffic Control: Progression Methods

Gazis, D. C.
IBM T.J. Watson Research Center
PO Box 218
Yorktown Heights, NY 10598
USA
*Network Modelling and Control: Store-and-Forward
 Approach*

Genser, R.
Austrian Federal Railways (ÖBB)
GS/Technology Observer
Elisabethstrasse 9
A-1010 Wien
Austria
IFAC Working Group on Transportation Systems
*Transportation Management: Systems Engineering
 Approach*

Goodall, R. M.
Department of Electronic and Electrical Engineering
Loughborough University of Technology
Loughborough
Leicestershire LE11 3TF
UK
*Railroad Systems: Active Control

Gottzein, E.
MBB GmbH
Postfach 80 11 69
D-8000 München 80
Germany
*Magnetic Suspension Railroad Systems

Guiducci, G. P.
Automa
Sistemi di Automazione Industriale
Via Al Molo Verchio
Calata Gadda
I-16124 Genova
Italy
*Container Terminal Management
*Ship Rudder Roll Stabilization

Hadj Salem, H.
INRETS
2 Avenue du Général Malleret
Joinville
BP 34
F-94114 Arcueil Cédex
France
*On-Ramp Control, Local

Haight, F. A.
Institute of Transportation Studies
University of California, Irvine
Irvine, CA 92717
USA
Safety of Road Traffic: Intervention and Evaluation

Haiman, G.
Université de Lille 1
UFR de Matématiques
F-59655 Villeneuve d'Ascq
France
Ship Dynamics: Modelling

Hasegawa, T.
Division of Applied System Engineering
Graduate School of Engineering
Kyoto University
Kyoto
Japan
*On-Ramp Control of Freeway Networks

Hennebert, C.
RATP
127 Avenue Ledru-Rollin

F-75011 Paris
France
Automatic Train Control: Safety and Reliability

Henry, J. J.
ONERA-CERT/DERA
2 Avenue Edouard Belin
BP 4025
F-31055 Toulouse Cédex
France
*Route Guidance, Individual

Hollister, W. M.
Department of Aeronautics and Astronautics
Room 33-117
Massachusetts Institute of Technology
Cambridge, MA 02139
USA
Air Traffic Control: An Overview

Hounsell, N. B.
Department of Civil Engineering
Transportation Research Group
University of Southampton
Southampton S09 5NH
UK
*Road Traffic Control: TRANSYT and SCOOT

Huddart, K. W.
Timbercroft
11 St Leonard's Road
Claygate
Esher
Surrey KT10 0EL
UK
Traffic Office Management

Hurdle, V. F.
Department of Civil Engineering
University of Toronto
Toronto
Canada M5S 1A4
Queuing Theory Applications

Hurrass, K.
DLR
Institut für Flugführung
Postfach 3267
D-3300 Braunschweig
Germany
Navigation Systems, Integrated

Iida, Y.
Department of Transportation Engineering
Faculty of Engineering
Kyoto University

Kyoto
Japan
*On-Ramp Control of Freeway Networks

Imbert, N.
ONERA-CERT/DERA
2 Avenue Edouard Belin
BP 4025
F-31055 Toulouse Cédex
France
Air Traffic Control Near Airports

Improta, G.
Dipartimento di Informatica e Sistemistica
Università di Napoli
Via Claudio 21
I-80125 Napoli
Italy
*Road Network Signal Setting: Equilibrium Conditions
*Signal Control at Individual Junctions: Phase-Based
    Approach*

Inoue, N.
Department of Civil Engineering
Faculty of Engineering
Fukuyama University
Fukuyama 720
Japan
*On-Ramp Control of Freeway Networks

Kamata, J.
Engineering Department of Information
Systems Division
Matsushita Com. Ind. Co. Ltd
4-3-1 Tunashimia-higashi
Kohoku-ku
Yokohama
Japan
*Detectors for Road Traffic

Karamitsos, F.
Commission of the European Communities
Drive Central Office
Rue de la Loi 200
B-1150 Bruxelles
Belgium
DRIVE

Keller, H.
Fachgebiet Verkehrsplanung und Verkehrswesen
Technsche Universität München
Arcisstrasse 21
D-8000 München 2
Germany
Evaluation of Traffic Control Systems

Kidd, P. T.
Research Section
Cheshire Henbury Research & Consultancy
Tamworth House
PO Box 103
Macclesfield SK11 8UW
UK
Marine Propulsion Plants: Control

King, J.
Department of Maritime Studies
University of Wales College of Cardiff
PO Box 907
Cardiff CF1 3YP
UK
Ship Automation and Control

Kitamura, T.
Sumitomo Electric Industries Ltd
3-12 Motoakasa 1-chome
Minato-ku
Tokyo 107
Japan
Road Traffic Control Systems in Japan

Klijnhout, J. J.
Rijkswaterstaat
Dienst Verkeerskunde
PO Box 1031
NL-3000 BA Rotterdam
The Netherlands
Freeway Control: An Overview

Komaya, K.
Industrial Electronics and Systems Development Laboratory
Mitsubishi Electric Corporation
8-1-1 Tsukaguchi Honmachi
Amagasaki Hyogo 661
Japan
*Expert Systems Approach to Rail Traffic Control

Kopitz, D.
European Broadcasting Union
1 Rue de Varembé
1211 Genève 20
Switzerland
*Broadcasting Communication Systems

Kortüm, W.
Deutsche Forschungsanstalt für Luft- und Raumfahrt
Institut für Dynamik der Flugsysteme
Oberpfaffenhofen
Münchner Strasse 20
D-8031 Oberpfaffenhofen
Germany
High-Speed Railroad: Modelling and Simulation
*Railroad Systems: Active Control

Lesort, J. B.
INRETS
2 Avenue du Général Malleret
Joinville
BP 34
F-94114 Arcueil Cédex
France
Prediction of Traffic Flow

Louah, G.
Centre d'Etudes Techniques de l'Equipement de l'Ouest
Rue René Viviani
F-44062 Nantes Cédex
France
Priority Intersection: Modelling

McDonald, M.
Department of Civil Engineering
Transportation Research Group
University of Southampton
Southampton S09 5NH
UK
*Road Traffic Control: TRANSYT and SCOOT

Makigami, Y.
Department of Civil Engineering
Faculty of Science and Engineering
Ritsumeikan University
Kitamachi 56-1
Tojün
Kita-ku
Kyoto 603
Japan
On-Ramp Control: Coordinated Time-of-Day Strategies

Mastretta, M.
Automa
Sistemi di Automazione Industriale
Via Al Molo Vecchio
Calata Gadda
I-16126 Genova
Italy
*Expert Systems Approach to Road Traffic Control

Mauro, V.
MIZAR Automazione S.p.A
Via Vincenzo Monti 48
I-10126 Torino
Italy
Road Network Control

Mensen, H.
Institut fur Luft- und Raumfahrt
Technische Universität Berlin
Marchstrasse 14/F3

D-1000 Berlin 10
Germany
Data Processing in Air Traffic Control

Minciardi, R.
Department of Communications, Computers and System Sciences
Università Genova
Via Opera Pia 11A
I-16145 Genova
Italy
*Underground Railroad Modelling and Control: Discrete-Event Approach

Mochizuki, A.
Head of Speedup Laboratory
Railway Technical Research Institute
Japan Railway Group
2-8-38 Hikari-cho
Kikubunji-shi
Tokyo
Japan
High-Speed Railroad Networks in Japan

Motyka, V.
INRETS
2 Avenue du Général Malleret
Joinville
BP 34
F-94114 Arcueil Cédex
France
*Video Sensors

Nakamura, H.
Safety and Telecommunications Laboratory
Railway Technical Research Institute
2-8-38 Hikaricho
Kokubunji-City
Tokyo
Japan
*Railroad Electronic Signalling

Negre, Y.
Aérospatiale
Toulouse
France
Automatic Landing Systems

Nicoletti, D.
Ansaldo Trasporti S.p.A.
Corso Perrone 25
I-16161 Genova
Italy
*Underground Railroad Modelling and Control: Discrete-Event Approach

Oda, T.
Engineering Department of Information Systems Division
Matsushita Com. Ind. Co. Ltd
4-3-1 Tunashimia-higashi
Kohoku-ku
Yokohama
Japan
*Detectors for Road Traffic

Ozgüner, U.
Department of Electrical Engineering
Ohio State University
2015 Neil Avenue
Columbus, OH 43210
USA
Optimal Routing Applications

Papageorgiou, M.
Lehrstuhl und Laboratorium für Steuerungs- und Regelungstechnik
Technische Universität München
Postfach 20 24 20
D-8000 München 2
Germany
Freeway
*Freeway Traffic Modelling
On-Ramp Control: Coordinated Traffic-Responsive Strategies
*On-Ramp Control, Local
Traffic Control Modes

Pélegrin, M. J.
ONERA-CERT/DERA
2 Avenue Edouard Belin
BP 4025
F-31055 Toulouse Cédex
France
Air Traffic Control: Trends
Data Links in Aeronautics
Visual and Instrument Flying Rules

Perrin, J.-P.
RATP
Département du Développement
7 Square Félix Nadar
F-94684 Vincennes Cédex
France
Automatic Train Control: Protection Principles
Railroad Systems: Line Supervision and Control
Railroad Systems: Train Driving Control
Underground Railroad: Organization of Operations

Polak, J. W.
Transport Studies Unit
Oxford University

11 Bevington Road
Oxford OX2 6NB
UK
Transportation Planning: Activity-Based Approach

Posch, B.
Franziskanerstrasse 26
D-8000 München 80
Germany
Merging Control

Redon, M. S.
Ingénieur Principal des Etudes et de l'Exploitation de la
   Navigation Aérienne
Ecole Nationale de l'Aviation Civile
7 Rue Edouard Belin
F-31055 Toulouse Cédex
France
Air Traffic Simulators

Ross, P.
Traffic Systems Division, HSR-10
Federal Highway Administration
6300 Georgetown Pike
McLean, VA 22101
USA
On-Ramp Control: Realization Principles

Roumégoux, J. P.
INRETS
109 Avenue Salvador-Allende
Case 24
F-69675
Bron Cédex
France
Fuel Conservation: Road Transport

Saridis, G. N.
School of Electrical Computer and Systems Engineering
Rensselaer Polytechnic Institute
Troy, NY 12182
USA
Intelligent Traffic Control Systems

Schmidt, G.
Lehrstuhl und Laboratorium für Steuerungs- und
   Refelungstechnik
Technische Universität München
Postfach 20 24 20
D-8000 München 2
Germany
Freeway Traffic Modelling

Schneider, H.-W.
Heusch/Boesefeldt GmbH

Liebigstrasse 20
D-5100 Aachen
Germany
Parking Control Systems

Schneider, R.
Institut für Verkehrswesen
Universität (TH) Karlsruhe
Postfach 6980
Kaiserstrasse 12
D-7500 Karlsruhe
Germany
Transportation Planning: Microscopic Approach

Schoen, F.
Dipartimento di Scienze dell'Informazione
Via Moretto da Brescia 9
I-20131 Milano
Italy
*Discrete-Time Point Processes: Applications to Road
   Traffic*

Schwarzmann, R.
Institut für Verkehrswesen
Universität (TH) Karlsruhe
Postfach 6980
Kaiserstrasse 12
D-7500 Karlsruhe
Germany
Transportation Planning: Microscopic Approach

Sforza, A.
Instituto di Fisica, Matematica e Informatica
Facolta di Ingegneria
Università di Salerno
F-84084 Fisciano (SA)
Italy
Road Network Signal Setting: Equilibrium Conditions
Traffic Assignment

Soulas, C.
INRETS
2 Avenue du Général Malleret
Joinville
BP 34
F-94114 Arcueil Cédex
France
*Automated Guideway Transit Systems and Personal Rapid
   Transit Systems*
Prediction of Traffic Flow

Speranza, M. G.
Dipartimento di Matematica e Informatica
Via Zanon 6
I-33100 Udine
Italy
*Discrete-Time Point Processes: Applications to Road
   Traffic*

Stathopoulos, A.
Laboratory of Railways and Transport
Department of Transport Planning and Engineering
National Technical University of Athens
5 Iroon Polytecniou Str Zografou
GR-15773 Athens
Greece
Route Guidance, Collective

Swierstra, S.
European Organisation for the Safety of Air Navigation
72 Rue de la Loi
B-1040 Bruxelles
Belgium
Fuel Conservation: Air Transport

Texier, P.-Y.
INRETS
2 Avenue du Général Malleret Joinville
BP 34
F-94114 Arcueil Cédex
France
Vehicle Monitoring and Control for Buses, Trolleys and Streetcars

Vallauri, M.
Politecnico di Torino
Dipartimento di Automatica e Informatica
Corso Duca degli Abruzzi 24
I-10129 Torino
Italy
Maintenance and Reliability of Traffic Control Systems

van Amerongen, J.
Department of Electrical Engineering
University of Twente
PO Box 217
NL-7500 AE Enschede
The Netherlands
Ship Rudder Roll Stabilization
Ship Steering: Model-Reference Adaptive Control

Van Breusegem, V.
Laboratoire d'Automatique, de Dynamique et d'Analyse des Systèmes
Université Catholique de Louvain
Bâtiment Maxwell
Place du Levant 3
B-1348 Louvain-la-Neuve
Belgium
Underground Railroad Modelling and Control

van Vuren, T.
Frank Graham Consulting Engineers
Elgar House
Shrub Hill
Worcester WR4 9EN
UK
Signal Control and Traffic Assignment

Vasilchenko, A. I.
Far East Department
Institute of Automation and Control Processes
USSR Academy of Sciences
5 Radio Strasse
SU-690032 Vladivostok
USSR
Marine Fleet Planning and Scheduling

Wesselink, A. F.
LIPS BV
PO Box 6
NL-5150 BB Drunen
The Netherlands
Navigation Control of Ships

White, J. P. Jr
Department of Systems Engineering
Thornton Hall
University of Virginia
Charlottesville, VA 22901
USA
Transportation Modelling

Wie, B.-W.
School of Travel Industry Management
University of Hawaii
2560 Campus Road
Honolulu, HI 96822
USA
Traffic Assignment, Dynamic

Wilken, D.
DLR
HA Verkehrsforschung
Linder Hohe
D-5000 Köln 90
Germany
Flight Activity: Prevision

Willumsen, L. G.
Steer Davies Gleave
19-21 Conway Street
London W1P 5HL
UK
Origin–Destination Matrix: Static Estimation

Wojnar, A. H.
[deceased, late of Warsaw, Poland]
Cellular Communications Systems

Yagar, S.
Department of Civil Engineering
University of Waterloo
Waterloo
Ontario N2L 3G1
Canada
Corridor Control Systems
Traffic Control Systems: Trends

Zackor, H.
Steierwald Schönharting und Partner GmbH
Alexanderstrasse 105
D-7000 Stuttgart 1
Germany
*Freeway Network Control
Speed Limitation on Freeways: Traffic-Responsive Strategies

SUBJECT INDEX

The Subject Index has been compiled to assist the reader in locating all references to a particular topic in the Encyclopedia. Entries may have up to three levels of heading. Where there is a substantive discussion of the topic, the page numbers appear in ***bold italic*** type. As a further aid to the reader, cross-references have also been given to terms of related interest. These can be found at the bottom of the entry for the first-level term to which they apply. Every effort has been made to make the index as comprehensive as possible and to standardize the terms used.

Accident analysis *425*
  see also Road traffic accidents
Active suspension technology uses *347*
Activity analysis *564*
Activity behavior–choice interaction framework *564*
Activity chain *570*
  simulation models *567*
    German *571*
Activity choice *564*
Activity patterns *565*
Actuators
  active suspension systems *349*
    ride improvement *349*
Adaptive control
  model-based *445*, ***459***
  multivariable gain scheduled algorithms *256*
Adaptive suspensions *347*
Adaptive traffic control systems *545*
  speed limits *508*
Adaptivity
  ship positioning dynamics *447*
Adjustable passive suspensions *348*
Aero engine modification
  marine propulsion systems *250*
Aerodynamics
  high-speed railroad
    infrastructure considerations *210*
    noise *206*
Aeronautic meteorological offices *96*
Aeronautical Fixed Telecommunication Network *90*
Aeronautics
  data links *85*
Aggregate variable models
  freeway traffic simulation *496*
Air charter services *134*
Air passengers *134*
Air pollution monitoring *409*
Air traffic
  controlled *13*, *616*
  density *10*
  objectives *6*, *22*
  safety *6*, *7*, *14*, ***29***
  stacking *21*, *278*
  uncontrolled *13*

Air traffic control (ATC) *1*, ***22***, ***86***
  central databank *87*
  control centers *14*, *22*
    boundaries *2*, *3*
    computerized *94*
    early *1*
    space control units *23*
  control systems *1*
    computer-based *94*
    flight progress strip print system ***90***
    navigation systems development ***16***
    objectives *1*, *6*
    secondary surveillance radar (SSR) *3*, *16*, *85*, *92*
  cost-efficiency *7*, *168*
  current trends *13*
  data processing *23*, ***86***
    meteorological *94*
    radar ***91***
    radiotelephony *96*
  Glonass system *19*
  instrument flying rules ***13***, *616*
  markets *135*
    developmental trends *138*
    German *137*
  mode S radar *4*, *16*, ***85***
  NAVSTAR system *19*
  radar requirements *91*
  routes *13*, ***14***
    Paris area *617*
  simulators ***22***
  training ***22***
  visual flying rules ***13***, *616*
  see also Airports, Automatic landing systems
Air traffic controllers *1*, *2*, *22*
  data handling *90*, *96*
  training simulators *23*
Air traffic simulators ***22***
Air transportation systems *134*
  domestic
    American *135*
    German *137*
  economy strategies ***167***
    zones of convergence concept *173*
  forecasting methodology ***136***

planning 135
Airborne collision warning and avoidance systems 30
   conflict resolution advisories 34
   early developments 31
   mode S related *33*
Aircraft
   collision 29
   collision avoidance 5, 14, *29*
   flight control processes 2, 23, 90
   flight costs minimization 6
      cruise–descent speed combinations 167, 169
      zones of convergence concept 170
   flight levels 14, 619
   flight plans 23, *88*, 616
   flying rules 14, 85, 88, *616*
   fuel conservation *167*
   heights 50, 93
   on-board flight management system 21, 22
   pilot-selected profile transit procedure 172
   safety *6*
   scheduling *6*
   stacking 21, 278
   tracking 3, 10
Aircraft landing 616
   cruise–descent speed profiles 168
      for minimum cost operation 170, 172
   delaying procedures 169
   design heights 50
   navigation 16, 21
   sequencing 7, 11
   see also Automatic landing systems
Aircraft manufacturers
   all-weather automatic landing systems *50*
   American 131
      production rates 132
Aircraft modelling techniques 10, 22, 28
Aircraft navigation
   integrated system accuracy formulations 276
Aircraft transit costs *167*
   pilot-selected profile 172
   transit time control techniques 169
      fuel savings 170
Airline fleet compositions 131, 135
Airports
   air traffic control *6*, 278
      dynamic scheduling of aircraft 7, 13
      inbound flights control 168
      instrument landing system (ILS) 7, *16*
      maximum landing rate development 21
   automated transit links 45
   control simulators 23
   controlled space
      Toulouse, France 618
   descent procedures 7, 21
   infrastructure development
      enviromental issues 138

zones of convergence 6, 7, 11, 169
   aircraft control costs minimization 170
   fuel burn in extended terminal areas 171
Airspace 2, 3, 14, 616
   classes 619
   congestion 6, 20, 138, 278
   control sectors 23
   zones of convergence concept 170, 173
Airways *see* Air traffic
Algebraic volume–density relationship equation 163
Algorithms
   automatic gain controller 451
   automatic incident detection 222
   decision 535
   diagnostic 535
   multivariable gain scheduled adaptive control system 256
   route guidance communication systems 418
   shortest-path 418, *461*
   train dwell time regulation 355
   transportation networks 282
ALINEA strategy
   local traffic ramp metering 296, 298
Alternative technology
   automated guideway transit systems 47
American Mobile Satellite Consortium 267
AMTICS
   Japanese radio data communications system 418
Analog control systems
   limitations 256
Analog propulsion control system 253
Antirolling tanks 456
Arizona Transportation and Traffic Institute traffic simulation model 493
Arrival and departure curves 338
Arterial progression schemes *391*
Artificial intelligence techniques
   air traffic simulation systems 29
   map matching 216
   qualitative modelling 129
   real-world modelling 127
   traffic control research projects 110
   traffic simulation systems 484
   transportation systems analysis 579
Assessment criteria
   automatic incident detection methods 221
Assignment *see* Queuing models
Autarkic route guidance systems 544
Autobahnen *see* Freeways
Autoguide route guidance system 577
Automated container terminal systems *77*
Automated guideway transit systems *36*, 40
   American 37, 41
   Brazilian 38, 47
   Canadian 38, 44
   cable-hauled 39, 45

ropeways 47
command and control systems 40, 44
French 37, 38, *42*
German 38, 45, 46, 48
Japanese 41
linear motor systems 40, 44
personal rapid transit systems 48, *47*
safety 37, 40, 46, 47
specifications 38
Automatic control theory
car-following models *65*, 258
computer simulation evaluation 261
cell-following methods 258
traffic flow metering 297
Automatic gain controller
ship steering machines 451
Automatic landing systems *50*, *54*
demonstration of airworthiness 52
design concepts 51
Automatic line supervision and control system *353*
human involvement 357
underground railroad *589*, *594*
offshore vessels *445*
Automatic road information system 264
Automatic train control *55*, 343
advanced 345
fleet centralized management control 358
French rapid transit lines 57
full 359
Japan Shinkansen railroad 202
safety *58*, 344
fail-safe principles 59, 344
transponders 345
underground railroads *588*
Automatic train operation systems 343, *358*
underground 589, 592, 593
world locations 361
Automatic train protection 343, 588, 592
Autonomous on-board navigation systems 274
Autonomous vehicle guidance 601, 612
Autopilot monitoring 52
Autopilot navigation 271
marine 436
rudder roll stabilization 451
Autoroutes *see* Freeways
Autostrade *see* Freeways
Aviation *see* Civil aviation, General aviation

Beacon-based navigation systems 418, 419
Beaconing
freeway traffic flow 297
Bellman–Moore algorithm 465
Bicycle *see* Pedal cycle strategies
Blackspots
road accidents 426
Bottleneck effects

road traffic flows 233
Bounded rationality concept
driver travel decision-making 377
Branch-and-bound algorithm applications 9, 121, 170, 364
British Rail
advanced passenger train tilting system 351
secondary suspension actuators 349
Broadcasting communication systems *61*, 111, 418
broadcast subcarrier authorization 264
Bus priority systems
freeway access metering 305

CAAD *see* Computer-aided analysis and design
Car-following models *65*, *258*
automatic car following 68
computer simulation evaluation 261
computer simulation modification 489
linear 65
nonlinear 65
Car navigation aids 62
in-vehicle 215, 544
Car parks
guidance system *324*
variable message signs *326*
Car seat belts
road safety 432
enforcement 433
Car sharing scheme model 571
Car transport *505*
Cargo planning
computerized 438
Carminat project
radio data system 418
CASPLAS 244, *247*
CD-ROM
car navigation 62
Cellular communication systems *69*
European 72, 216
radiotelephones 266
Centralization
systems organization 558
Centralized interactive system with cyclic instantaneous location
French continuous location system 602
Change
systems organization
control regulators 559
Channel Tunnel 196
Charter services
aircraft 134
Choice
voluntary route choice schemes 415
Choice models
traffic assignment problems 517, 526, 528
travel behavior analysis 565

Civil aviation 131
Classical feedback techniques
  control design methods 348
Closed-circuit TV system *99*
Closed-loop control
  road intersection control 362
Closed-loop performance
  vehicle guidance by computer vision 601
Cluster analysis
  traffic conditions 223
Collision avoidance
  aircraft 2, 5, 9, 14, *29*
  automated guideway transit systems 40, 42
  marine 439, 441
  trains *55*
  see also Airborne collision warning and avoidance systems
Collision diagram 428
Commercial air transportation system
  global demand 134
    structural characteristics 135
Commercial vehicles
  cellular radio networks 266
  meteor-scatter radio service 267
Communication navigation surveillance *20*, 86
Communication systems
  aeronautical fixed service 90
  computerized 1
  EC research projects 111
  meteor burst 267
  remote 1
  satellite 20
  technologies 111
  see also Broadcasting communication systems, Cellular communication systems, Mobile communication, Radar, Radio communication
Communications architecture
  automatic vehicle identification 412
  EC research project 111
  traffic system 216
    computerized 228, *531*
Commuting problem decisions 372, 375
Compatibility graph and cliques
  road junction 474
Competition
  cooperative 558
Computer-aided analysis and design
  high-speed railroad *187*
  ship propulsion systems 251
Computer-aided planning and scheduling
  marine transportation systems *244*, 438
  see also CASPLAS
Computer-aided traffic control systems 202, 228, *400*
  architecture *531*
  optimal routing applications 306
  signal control methods 537

Computer programming
  data-driven 126
Computer simulation
  road traffic *483*, *546*
    accidents 426
    macroscopic *491*
    microscopic *497*
  ship propulsion systems design 251
  vehicle merging control scheme
    evaluation 261
Computer techniques
  expert systems technology 126
  in simulation systems 27, 192
    graphics displays 488
    hardware architecture 26
    interactive computer graphics interfaces 484
    model building 485, 488
    programming language development 501
    quality requirements 25
    software architecture 26, 193, 485, 489
  learning capabilities 229
Computer vision
  vehicle guidance 597
Computerized data processing 1
  aircraft automatic data link 21, 22
  aircraft tracking 2, 94
  ferry transportation systems 244
  railroad system 203
    electronic data processing (EDP) 203
    fail-safe electronic signalling 343
  ship operational practices 438
  strategic traffic control systems
    processing complex *534*
  train routes 356
  rail traffic 119, 353, 398
  vehicle guidance
    object recognition 601
  see also Automated guideway transit systems
Conflict alert messages 96
Conflict graph and cliques
  road junction 474
Conflict modes
  motion control models 260
Conformance engine
  strategic traffic control system 534
Congestion see Freeway traffic congestion, Road congestion
Connecticut Department of Transportation
  expressway simulation model 494
Consensus formation
  EC DRIVE project 112
Conservation of matter equation 162
Container management optimization 503
Container terminal systems management *75*
  automated *77*
Continuous planning process

marine fleet planning and scheduling 245
CONTRAM traffic assignment model 82, 493, *547*
Control engineering theory 530
    model-based estimation 143, 147
Control law design strategies 348
Control loop strategies
    marine propulsion control system 253
        improvements 256
Control regulators 559
Control theory
    decision making 557, 559
    navigation control process *269*
    network modelling 280
    see also Automatic control theory
Conurbation axes
    Europe 196
    Japan 201
CORQ traffic assignment model 82
Corridors see Transportation corridors
COST 301 project 607
Cost–benefit analysis
    Dutch road signalling policy 156
    public investments *115*
    road safety improvements 429
Cost-effectiveness analysis *116*
Costs
    aircraft global transit costs minimization 6
    air flight economy and transit costs 167
    commuter decisions and personal costs minimization 372, 373
    general aviation flight activity 132
    road traffic accidents 107, *423*, 430
    road traffic delays 217
    shipping industry reduction policies 502
    transportation system 560
    vehicle operating costs 107
Creep force–creepage relations 189, 350
Crews 436, *437*, 502
    reduction policy implications 504

Data collection methods
    travel activity patterns 566
Data-driven programming techniques 126, 129
Data processing systems
    active 78
    flight 23, *87*
    meteorological *94*
    passive 76
    traffic 62, 125, 143, 159, 216, 533
        car parking control 324
        freeway control 510
        road accident 424
        road network control 365
    see also Computerized data processing
Data transmission
    automatic train driving systems 360

automatic vehicle identification systems 411
    road sign systems 161
    train safety systems 345
    vessel traffic services 609
Databases
    CASPLAS 249
    road traffic accident 425
    traffic monitoring 534
        historic data 534
    traffic simulation systems 486
        microscopic traffic simulators 488
Dead-reckoning location 215, 274
Dead-reckoning navigation systems 275, 276, 417
Decentralization
    optimal routing traffic information structures 307
    systems organization 558
Decision analysis *117*
Decision criteria
    driver travel decisions 372, 412
    evaluation methods 117
Decision environment 113
    traffic control systems 118
Decision-making process
    driver travel adjustments 372, 412
    marine fleet planning 245
    multistep process
        transportation management systems 557
    traffic control systems
        stakeholder groups involvement 118
    traffic signal design 127
Delivery tests
    mathematical model validation 442
Demand–capacity strategies 295, 541
Demand-responsive road traffic control *386*, 532
Departure time models
    day-to-day adjustment process 375
        simulation 377
    heterogeneous drivers 374
    homogeneous drivers 373
Design decision process
    expert systems approaches 127
Destination see Origin–destination matrix
Detectors see Road traffic detectors
Deterministic analysis
    road traffic 380
Deterministic equilibrium traffic assignment 515
Diesel engines
    marine use 184, 251, 271
Digital control
    automatic train driving systems 360
    propulsion and engine control systems 255
Digital maps 111, 215, 216
Digraph 462
Dijkstra's algorithm 463
Direct traffic control systems 544, *545*
Direct traffic detection methods 613

Dispatchers *119*
  Japanese railroad system 203, *204*
Distance measuring equipment 274, 276
Doppler effect
  road traffic detectors *96*
Doppler radar
  pulse Doppler processing 93
  train speed detection 345
Dortmuder housing market model 571
DRAFT system
  program generators 486
DRIVE 63, *107*, 218, 224, 429, 490, 506
  consensus formation 112
  private mobile radio 266
  project proposals 109
  workplan *108*
Driver information systems *413*
  in-vehicle 215, 544
Drivers
  behavior
    analysis 615
    speed adjustment 507
    commuters' decisions 372
      day-to-day adjustment process 375
      information factors 373
  voluntary route choice 415
Driving experiments
  test vehicle
    autonomous mobility and computer vision 601
Driving style
  effect on fuel consumption 178
Dynamic equilibrium models 373
  departure time *375*
    simulation 377
Dynamic estimation methods 312
Dynamic machine vision *597*
Dynamic positioning navigational control systems 273, *445*
  simulations 447
Dynamic programming
  road traffic control
    optimization 388
  transportation planning applications 245
Dynamic route guidance 417
  algorithms *420*
Dynamic traffic assignment 521
  mathematical formulations *522*
Dynamic traffic management systems 544
DYNEMO traffic management model 493, 500

Earthquake countermeasures
  railroad systems 205
  warning system 206
Econometric theory
  air transport demand forecasting *135*
Economic factors

  general aviation flight activity 132
  railroad network planning 199
  see also Costs
Effectiveness evaluation 116
  road safety measures 432, 434
  traffic control systems 118
ELECTRA *24*
Electric traction plants
  high-speed railroad system 211
Electricity distribution regulation
  railroad line system 356
Electronic data processing 203
Electronic imaging system
  vehicle identification 410
Electronic position measurement 269
Electronic road pricing systems 545
Electronic roadmap 62, 545
Electronic route guidance systems 417
  signal setting optimization 471
Emergency vehicle priority strategies 385
Energy consumption
  automated guideway transit systems 36
  automated railroad system monitoring 203
  marine power plants 183
Energy-saving measures 179
  marine power plants 184
  waste-heat utilization 185
  wind-power utilization 185
Entropy-maximizing technique
  traffic modelling 317, 319
    limitations 320
Environmental factors
  airport development 138
  automated guideway transit systems 36
  car park locations 323
  high-speed railroad networks 206
  road accidents 424
  ship maneuvering process 270
  vehicle emissions 107, 179
    speed reduction strategies 509
Environmental monitoring systems *409*
Equilibrium network traffic signal setting 368
  global optimization methods 369
  iterative procedures 369
Equilibrium traffic assignment 524
  deterministic 514
  stochastic 518
Error behavior
  navigation systems 275
    dead-reckoning system 276
    INS 277
Estimation techniques
  matrix 316
  model-based 143
    incident detection 147
    traffic flows 145, *312*

travel behavior analysis 565
ESTRAC project *121*
European Broadcasting Union 61, 264
European Community
   air traffic central databank 87
   air traffic levels 20
   aircraft fleets 131
   cellular communication system 72
   common transportation policy 505
   cross-border transport market 195
   dedicated road infrastructure for vehicle safety (DRIVE) *107*, 218, 429, 490, 506
   flying rules 616
   program for a European traffic with highest efficiency and unprecedented safety (Prometheus) 218
European conurbations *197*
   major axes 196
European high-speed rail network scenario *196*
Evaluation
   methodology *114*
   road safety improvements 428, 434
   traffic control systems *113*, 117, 406
   vehicle merger-control computer simulation 260
Expert systems *121*, *124*
   off-line applications 127
   on-line applications 127
      traffic control 128
   rail traffic control *119*
      ESTRAC-III *122*
   road traffic control *125*, 575, 579
      design issues 130
      software technology 130
      system architecture 130
   shipping applications 439
   urban traffic simulation *485*
Expert tuned design 349
Expressways
   Japanese traffic control systems 399

Fail-safe electronic signalling system 343
Feedback control laws
   electronically controlled suspensions design 348
Feedback theory 296
   ALINEA application 298
   road network control application 362
Ferrovie dello Stato high-speed project 208
Ferry transportation
   Russian computer-aided planning 244
   voyage management system 273
Filtering techniques
   mathematical models 146, 223, 275
Fin system
   ship stabilization schemes 457
FLEXSYT traffic simulation model 501
Flight activity

civil aviation *134*
controlled airspace 616
general aviation *131*
   forecasting 133
Flight data links *85*
Flight economy
   transit costs equation 167
   zones of convergence concept 170
Flight levels 14, 619
Flight management systems
   databank *87*
   on-board 21, 22
Flight parameters 167
   costs minimization 168
Flight plans 23, *88*
   data display 91
   data processing *90*
   preflight data 88
      processing 88
Flight progress strip *90*
Flight tests 53
Flight transit costs
   cost–time relationship 168
   minimization 167
Flow equations *see* Traffic flow equations
Floyd's algorithm 466
Flying decline
   recreational 133
Flying rules 14, 85, 88, *616*
Fog accidents 153
Forecasting
   aviation market 133, 135
   ferry transportation external factors 245
   traffic conditions 158
Forecasting methodology
   air transportation demand *136*
      econometric models 137
   travel trip models *561*
      microscopic 570
Frahm's ship antirolling system 456
Free-body diagram 188
Freeway access metering ramps 81, 82, *303*, 537
   French control strategies experiment 297
   merging aids 509
   priority vehicles 305
   sequential ramp closure 302
   signs 304
   see also On-ramp control
Freeway capacity
   control and reliability strategies *149*
   corridor control and traffic demand 289
   demand–capacity strategy 295
   on-ramp control 285, 300
      demand–capacity INRETS 296
   oversaturation problems 154
Freeway control systems *80*, *152*

alternative routes 157
  German system 154
corridor control 289
diversion 299
expert systems techniques 129
hierarchical multilayer control structure *292*
hierarchically intelligent 226
intercity 300
lane reservation 156
objectives 290
priorities 156
queuing theory applications 340
rerouting 154, 157
speed limitation *507*
stabilization techniques 154
system architecture 510
traffic-responsive on-ramp metering strategies *295*
urban 300
see also On-ramp control
Freeway incident detection and control 83, *155*, *219*
  reliability theory 151
Freeway network control 156
  inflow control methods 301
  route guidance model 157
  route guidance system 159
    structure 160
    variable 161
  shortest-path algorithms *461*
  traffic signals 538
Freeway optimization 81, 289
Freeway reliability theory *149*
  traffic control use 150
  traffic instability detection 151
Freeway surveillance systems
  automatic incident detection methods 222
Freeway traffic congestion
  French control strategy experiment 297
  on-ramp control strategies 295
    congestion duration 298
    diversion aspects 299
    downstream control strategy 296
    linear programming 301
    sequential ramp closure 302
  queuing theory applications 340
  speed reduction strategies 509
  traffic density evolution
    simulation 293
Freeway traffic flow process 295
  on-ramp 303
    priority vehicles 305
Freeway traffic modelling *162*
  discrete-time equations 227
  high-order continuous-time equation 227
  linear programming 287, 301
  macroscopic dynamic model 144, *162*
  multilayer control structures 292

objective functions maximization 301
optimization theory *289*
  open-loop optimal control models 291
queuing theory applications 339
simulation models
  macroscopic *491*
  microscopic *497*
total travel time calculations 298
see also Traffic modelling
Freeways *148*
  automatic control and signalling 156
  intercity 300
  lane reservation 156
  ramp-metering installation 305
  urban 300
Freight transport systems
  EC research projects 111
  marine fleets 244
FREQ traffic model *81*
Fuel conservation
  air transit *167*
    zones of convergence concept 170
  maritime scheme 181
  road transportation strategies *173*
  sea transport systems *179*
Fuel consumption
  aircraft assessment techniques 171
  aircraft costs minimization strategies 6, 8, 170
  driving style 178
  gas turbine engines 250
    control 252
    improvements 256
  road bends and slopes equations 176
Function curves
  traffic assignment and signal setting 368
Fundamental diagrams
  traffic problem analysis 142, 158, 231

Gain scheduling
  ship steering control 450
Game theory
  signal control and traffic assignment 468
Gas turbine engines *250*, 271
  control scheme 255
  gas generator control 253
  objectives 252
General aviation
  aircraft fleets 131
    decline 133
    economic development factors 132
    forecasting instruments 133
    market *131*
    size 132
General Aviation Manufacturers Association 131
Georgia freeway traffic model 495
Global air traffic demand 134

Global optimization models
    equilibrium network traffic signal setting problems 369
Global positioning satellite systems 19, 216, 269, 345
Glonass satellite system 19, 20
Goods transportation systems 506
Group rapid transit systems 37
Graphs
    digraphs 462
    undirected 462
Gravity models 316, 562
Gyroscopes *454*

Harbor information systems 76
Hazardous road conditions 153
Headway distributions 139
Heuristic programming 126
Hierarchical structure
    decision-making process 557
    railroad signalling system 343
    SCOOT traffic control system 406
High-frequency oscillation
    vessel motion 445
High-speed railroad systems
    computer-aided analysis and design *187*
    Europe *194*
        economic considerations 199
        estimated demand 198
        network configuration 196
    Japan *201*
        increased speed developments 206
    magnetic suspension *235*
    stability control 350
    systems approach *208*
High-speed trains
    French TGV network 195, 506
    Italian 208
    Japanese 195, *201*
    see also Maglev vehicles
Highways see Freeways
Household interactions model 572
    MASTER simulation model 573
    STARCHILD simulation model *572*
Household scheduling
    travel activity patterns 565
Housing market
    Dortmuder simulation model 571
    dynamic simulation model 572
Human factors
    air traffic control 1, 2, 14, 21
    automatic line supervision 357
    computerized rail traffic control 119, 203, 204
    driving style and fuel consumption 178
    expert systems technique 126
    nonautomatic incident detection 220
    public transport networks optimization 604

road accidents 424
ship crews 436, *437*
traffic signals operation 127
train control 56
transportation planning 569
travel behavior *564*
Hydrodynamic theory *231*
    ship maneuvering process 270
    ship stabilization methods 449, 459
    traffic flow 491
Hydrofoil 183

Ice hazards
    freeway 153
Identification
    system equations 187
Image processing 224, 610
    background image use 612
    problem factors 614
    traffic detection methods 611
Image sensors
    traffic detection *99*, 222, *610*
Image sequence processing system 600, 610
    object recognition 601
In-pavement weight screening systems 410
In-vehicle driver information systems *215*, 398
In-vehicle route guidance 83, 215, 540
Incident detection
    automatic *219*, 539
Inductive loops 265
    traffic flow measurement 222, 363, 404, 411
Inertial navigation systems 274, 277
    error behavior 277
Information networks
    air traffic control systems 2
    computer-aided marine fleet planning 248
    computer-aided railroad systems 202
    container terminal systems *75*
    driver travel decisions 373, 413
    Eurotravel 61
    feedback 558
    freight transportation 244
    hierarchically intelligent traffic control 228
    in-vehicle driver systems *215*
    organization subsystems 558
    road traffic systems 126, 157, 264
        SCOOT 406
        traffic counts 316
    road transportation needs research 111
    route guidance systems 307, 414
    train progress 353
Infrared proximity beacons 265
Infrared systems
    automatic vehicle identification 410
Instrument flying rules *13*, 616
Instrument landing system (ILS) *16*

INTAC traffic simulation model 501
INTEGRATION traffic simulation model 499, *576*
Intelligent plotting aids
    marine collision avoidance 439
Intelligent vehicle–highway systems 278, *283*, 575
Interactive computer graphics interfaces 484
    NETSIM program enhancement 484
International Association for Mathematics and Computers in Simulation 214
International Civil Aviation Organization 1, 20, 33, 616
    airline scheduled services trends 135
International Commission for Air Navigation 1
International Federation for Information Processing 214
International Federation of Automatic Control 213
    Working Group on Transportation Systems *214*
International Federation of Operational Societies 214
International Institute for Applied Systems Analysis 214
International Measurement Confederation 214
International transport links 196
Interrogation modes
    pulse spacing 92
    uses 93
Interrogation signal sequence patterns 3, 92
    discrete 4
Intersections *see* Road junctions
INTRAS traffic simulation model 499
Iterative procedures
    traffic assignment 548
    traffic signal setting 369, 469

Japanese conurbation axes 201
Japanese National Police Agency
    traffic signal control systems 398
Japanese National Railroads (JNR) 201
    secondary suspensions control research 349
    Shinkansen railroad system 202
        computer-aided control system 202
Japanese traffic control systems 399
Journey chain 570

Kalman filter procedures *146*, 223
    dymamic estimation applications 313
    predictive technique applications 330
    time-discrete equations 275
    vessel motion modelling 445
Kármán–Gabrielli diagram *179*
Kinematic wave theory *231*, 291
Knowledge-based simulation systems *486*
Knowledge-based systems *see* Expert systems
Knowledge engineering technology 119, 121

Labor
    robots 440
    shipping crews 436, *437*
Lagrangian variable
    gravity model 318

    optimal solution complexity 309
Land-mobile radio systems 68, 263
    cellular *69*
        European 73
    dedicated 265
    microcellular 73
    specialized 265
Land-use forecasting 560
Landing scheduling of aircraft 7, 10, 16
    costs minimization strategies 170
    for several airports 13
    stacking procedure 21
Law of conservation of mass 231
Leakage coaxial cables 206
Level crossing controller 344
Levitation concepts 235
License plate readers 410
Life-cycle status
    travel behavior 565
Light-emitting-diode-based traffic information boards 398
Linear induction generators 235
Linear induction motors 39, 235
    electrical active guideway systems 45
        French U-shaped motor 45
    passive guideway systems 44
        magnetic levitation systems 44, 236
Linear models
    marine propulsion 252
    marine steering machines 449
    metro traffic 581
    ship dynamics 443
    short-term traffic flow prediction 329
Linear motor car 44
    see also Maglev vehicles
Linear programming
    freeway on-ramp control strategies 287, 301
    oversaturated road intersection 280, 282
    transportation planning problems 245
Linear-quadratic methodology 292
    linear-quadratic Gaussian method 450
        adaptive 451
        optimal 452
Linear state-space techniques
    electronically controlled suspensions design 348
Linear system analysis 192
LIPS stick ship maneuvering 274
Locations
    hazardous location identification 426
    techniques 602
Loop control strategies
    improvements 256
    ship propulsion control 253
Loop detectors *97*, 143, 222, 363, 365, 533
Low-frequency oscillations

vessel motion 445
Lund School of Time–Space Geography 564, 570

Machine vision 597
  active 600
Macroscopic traffic simulation programs *491*
Maglev vehicles 187, 207, *235*
  electrodynamic suspension 235
    control design methods 348
  electromagnetic suspension 235
    vehicles *236*
  German development 238
  high-speed 236
    KOMET 237
  subsystems 237
Magnetic induction speed detectors 98
Magnetic levitation systems 44, 195, *236*
  operational transportation systems 238
Magnetic wheels 237
Man–machine advanced processor
  railroad system data analysis 203
Management information systems
  Shinkansen railroad 204
  ship to shore 504
Map matching *216*
Marine fleets
  planning and scheduling tasks 244
    dynamic programming equations 245
    scheduling process equations 246
  routing 440
Marine personnel
  education and training 504
Marine propulsion machinery 250
Marine propulsion plants 271
  control 253
  fuel consumption improvements 183
  gas turbine powered 250
Marine transportation 502
  computer-aided planning and scheduling 244
    problem formalization 245
  fuel conservation *179*
Marine vessels *see* Ships
Maritime economic crisis 502
Maritime fuel conservation 181
Martingale theory 101
Mass
  law of conservation of mass 231
Mass transit 538
MASTER household processes simulation model 573
Mathematical models
  aircraft scheduling 7, *11*
  car-following theory 65, *258*
  driver travel decisions 373
    dynamic models of departure time 375
  marine transportation planning 245
    fleet scheduling equation 246

time planning equations 246
navigation systems accuracy equations 275
optimization of an objective function 301
road geometric model 599
road traffic flows *101*
  cell-following methods 258
  deterministic models 338, 515
  discrete-time equations 225, 227, 309
  dynamic estimation methods *311*
  filtering techniques 223
  function curves 368
  fundamental diagrams 142
  macroscopic dynamic model 144, 162
  matrix estimation models 316
  optimization theory formulations *289*, 308, 388
  probability density function equations 139
  progression methods *391*
  reliability theory 149
  shortest-path algorithms 418
  store-and-forward network model 307
  validation 165
ship dynamics
  linear models 443, 449, 452
  motion equations 441, 445
ship maneuvering equations *269*
  optimal steering *459*
track–train dynamics 189, 192
traffic assignment problems 515, 519, 522
traffic signal setting *366*
underground railroad
  discrete-time control model *581*
Mathematical programming methodologies
  dynamic system optimal traffic assignment 521
  optimal signal control 475
Matrix algorithm 465
Matrix estimation methods 316
MAXBAND model
  traffic progression maximization 392
Mean down time
  traffic control failures 241, 243
Mean time between failures
  dedicated LSI traffic control system 399
  traffic control system reliability index 241
Mean time to repair
  traffic control system maintainability index 241
MEDYNA interactive multibody software program *192*
META freeway traffic model 492
Meteor burst communications *267*
Meteorological conditions
  all-weather landing categories *51*, 55
  effects
    on air traffic control 1, 10, 21
    on freeway systems *153*, 510
    on fuel consumption 178
    on high-speed railroad networks 205
    on road accidents 424

643

on speed detectors 97
environmental monitoring 409
forecasting 95
instrument (IMC) 14, 616
simulation models 28
visual (VMC) 14, 616
French 619
wind disturbance modelling 449
Meteorological data processing *94*
Metering freeway access 80
corridor control 289
see also On-ramp control
Metro systems *42*, 359
automatic driving process 360
Atlanta 589
Hamburg 592
Paris 592
control center functions 590, 591
human involvement 357
intrinsic instability 582
network sequences 581
safety *55*
simulation
discrete-event model 585
traffic perturbations 587
variables 586
time deviation models 582, 584
traffic modelling *581*
traffic regulation strategies 588
centralized control model 583
decentralized control model 583, 584
evaluation 587
Microcomputer
fail-safe architecture 343
solid-state interlocking structure 344
Microprocessors
automatic driving functions 360
safety functions 41
Microscopic analysis
transportation planning 569
vehicles and pedestrians 380
Microscopic traffic simulation models *497*
car-following models 67
Microscopic transportation simulation models 570
complexities 573
Microwave communications
automatic vehicle identification 411
EC research projects 111
proximity beacon 265
Midwest Research Institute
freeway simulation model 493
mountainous-terrain model 493
Mikhalkin freeway simulation model 495
Military airspace 619
Minorsky's antirolling scheme 457
Minute photoelectron alley system *100*

MISSION traffic simulation model 501
Mobile communication *263*
commercial vehicle applications 267
in-vehicle equipment *215*
land-mobile radio 69, 263, 265
meteor burst 267
microcellular networks 73
satellite 74, 266
see also Broadcasting communication systems, Cellular communication systems, Radio communication
Mobility 2000 218
Mobitex dispatching system 72, 266
Mode-decoupling techniques
electronically controlled suspensions design 348
Mode S selectively addressed SSR 2, 4, 16, *85*
in collision avoidance systems 33
Model-based estimation *143*
Model-reference adaptive control
ship steering optimization *459*
Modelling techniques
aviation 10, 22, *28*
estimation 143
experimental 187
marine 441
microscopic 571
physical 187
vehicle bodies 188
priority intersection 331
qualitative 129
railroad 189
real-world 127
see also Traffic modelling, Transportation modelling
Models see Mathematical models, Queuing models, Simulation models, Traffic assignment
Monitoring
automatic landing systems 52
vehicle *408*
automatic 602
Monopoly 558
Monorails 42, 47
Motion control 258
vehicle merger conflict equations 260
evaluation 261
Motion dynamics 597
Motorways see Freeways
Mountainous-terrain traffic model 493
Moving-cell scheme for traffic junctions 258
Moving-object detection 611
Moving-target indicator 93
Moving-weight scheme
ship stabilization 457
MULTIBAND model
traffic progression maximization 393
Multibody systems 187
MEDYNA software program 192

Multidimensional objective system 116
Multistage stochastic programming 245

Nash noncooperative equilibrium 468
National railroad systems
    France 195
    Japan 201
Naval vessel twin-shaft propulsion 251
Navigation aids and systems
    aviation 13 *16*, 85, 274, 276
        four-dimensional 20, *21*
    integrated *274*
        accuracy formulations 275, 276
        advantages 277
    marine 269, 274, 437, 608
    traffic 61, 398
        on-board 157, 216, 417
    see also Radar, Satellites
Navigation control systems *272*
    maneuvering process *269*
Navigation satellites 14
NAVIGATOR 417
Navstar 216
Negative arc weights 465
Neoclassical control techniques
    control law design 349
NETSIM traffic simulation program 484, *498*
Noise pollution
    high-speed railroad systems 207
Nonlinear control laws
    design techniques 349
Nonlinear dynamics
    marine propulsion 252
    marine steering machine 450
Northwestern University lane-changing simulation model 494

Object recognition by feature aggregation 601
Objectives
    achievement
        organization 558
    automatic train driving systems 358
    automatic vehicle monitoring 602
    identification 117
    multidimensional objective system 116
    social system 555
    traffic control systems 118, 300
        optimization of an objective function 301
    transportation management *556*
Obstacle detector
    train track 345
Offsets
    area traffic control system 398
    traffic signal cycle optimality 472, 532
Offshore oil industry vessels
    control structure 446

dynamic positioning systems *445*
On-line control
    traffic control systems 228
        demand-responsive 387
        dynamic estimation methods 312
On-line learning capability
    hierarchically intelligent traffic control systems 226, 229
On-ramp control *285*, 294, *303*
    beaconing 297
    freeway access 81, 82, 228, 285, 290, *300*
        traffic-responsive 295
    local *294*
    modes 285
        system-mode configuration 286
    rate decrease effect 306
    realization *303*
    strategies 285, 537
        assessment 297
        control strategy parameters 287
        mathematical formulations 287, *290*, *301*
        time-of-day *285*
        total travel time calculations 298
        traffic-responsive *289*, 295
    traffic signals 303, 545
        malfunction 305
        signing 304
    UK motorway 305
    vehicle detectors *304*
One-way communication 412
    algorithms
        shortest-path 418
    systems 418
Open-loop optimal control 291
Optical flow
    traffic surveillance 612
Optical systems
    automatic vehicle identification 411
Optimal control theory
    car-following models 66
    ship steering 459
    traffic networks 279, 282, *291*
Optimality and constraints equations 9
Optimization
    aircraft landing sequencing 9, 170
    complex transportation networks 282
    decision-making process 557
    freight transport plans 244, 248
    layer
        traffic control hierarchy 292
    public transport human resources 604
    road intersection signals 362, 364, *387*, *400*
    road traffic informatics applications 109
    route guidance *306*
    ship steering performance *459*
    traffic corridor 81

traffic flow oversaturation control 280, 301
traffic lights
  fuel economy 175
traffic networks 416
traffic signal sequence setting 468, 481
  global optimization models 369
  mathematical programming methodologies 475
urban traffic control
  SCOOT 404
urban traffic optimization by integrated automation (UTOPIA) 362
Optimization policy for adaptive control flow model 390
Optimization problems
  complex transportation networks algorithm 282
  shortest-path algorithms *461*
  traffic control strategy design *289*
Organization systems 558
Origin–destination matrix
  departure time models 373, 374
  dynamic estimation *310*
  static estimation 310, *315*
Oversaturation of traffic 154
  see also Store-and-forward operation

Pantographs
  servocontrolled 352
Parameter estimation
  origin–destination flows 314
Parking control system *324*
  German 328
    surveys 328
Parking guidance systems *323*
Passenger information
  public transport networks 603
  railroad systems 357
Passenger traffic
  flow variables 337, 603
  Japanese railroad 201
    information dispatchers 204
Passenger transport simulation model 198
Passenger transportation systems 505
Passive suspensions 347
Pathfinder project 218
Pattern recognition
  automatic vehicle identification 410
  traffic flow conditions 223
  see also Image processing
Pedal cycle strategies 385, 406
Pedestrian road crossing delay 384
People movers 37, 41, 48, 49, 237
Percent-occupancy strategy *295*
Personal rapid transit systems 37, *47*
PESASP time–geography model 566, *570*
Piezoelectric sensory monitoring 409
  cable systems 410

Pitch controls
  marine propulsion 253, 256
Planar motion mechanism
  ship model tests 442
Planning
  task of planning and scheduling marine fleet 244
    dynamic programming equations 245
  see also Transportation planning
Platoons 384, 391, 403, 491
Plot extractor data 94
Point processes 101
  discrete-time 101
  filtering for dynamical 102
Poisson test 429
Police
  liaison with traffic management teams 552
  traffic accident reports 425
Pollution monitoring techniques 409
Prediction techniques
  aircraft scheduling 10
  short-term traffic flow *329*
Preview control schemes
  active control suspensions 348
Priority lane model
  macroscopic simulation program 496
Private aviation 131
Probabilistic analysis 59
  priority intersection capacity
    basic gap-acceptance model 331
  probability density function
    traffic flow variables 139, 223
  traffic control systems failures 241
Probabilistic choice models
  stochastic traffic assignment models 518, 526, 528
PROBIT choice model 519
Problem solving
  simulator generators 486
Problem solving equations see Mathematical models
Profit maximization
  marine freight transportation factors 244
Progression methods *391*
Prometheus program 218, 224, 306, 602
Propeller action
  ship maneuvering devices 271
Proportional-plus-integral-plus-derivative (PID) control strategy 448
Proximity beacons 265
Public participation
  DRIVE projects 109
  road safety programs 434
  traffic control systems 118, 554
Public radiotelephony 72
Public transport systems
  automatic vehicle monitoring *602*
    control station functions 605
  EC research projects 111

interaction with traffic light control 243, 603
network operation information 604
park and ride 325
see also Automated guideway transit systems
Public transport vehicles
  maintenance planning 604
  on-board automatic monitoring equipment 604
  priority strategies 362, 366, 385, 603
Pulse Doppler processing 93
Pulse spacing
  radar interrogation signals *92*

Qualitative modelling 129
Queues 337
  arrival and departure curves 338
  traffic junctions *382*
  traffic signals 339
  see also Queuing models, Road congestion, Traffic assignment
Queuing models *82*, 307, 316, 386, 563
  CONTRAM time-dependent method *547*
  deterministic equilibrium models *514*
  dynamic 522
  global criterion 420
  intersection delay 334
  intersection merger strategies 258
  nonproportional 321
  optimal solutions 309, 470
  path choice models 526, 528
  priority intersection capacity formulae 332
  queue-dependent model 420
  SATURN computer simulation 547
  stochastic dynamic models
    day-to-day 525
    within-day 527
  stochastic equilibrium models 518
  traffic signal overflow delay 339
  traffic signal setting links 367, 420, 471
  transportation planning uses 549
  see also Queuing theory
Queuing theory
  traffic modelling applications 337, 373, 374

Radar
  beacon transponders 3, 14, 16, 33, 85
    transmission pulses *92*
  clutter removal 93
  data
    computerized display 94
    conflict warning display 96
  data processing system 23
    air traffic control *91*
    ship maneuvering 274
  mode S 2, 4, 33
    system requirements *85*
  primary (PSR) 1, 3, *91*
    airborne 33
  secondary (SSR) 1, 3, 33, 92, 619
    interrogation modes use 93
Radio communication 1, 13, 34, 61, 85
  automatic vehicle monitoring systems 604
  land-mobile radio 68
    objectives 71
  meteor scatter radio service 267
  roadside traffic information service 398
  traffic radio service 157
Radio data system *61*, 157, 216, 264, 417
  cellular land-mobile radio 69, 266
  traffic message channel *62*
Radio dispatching systems *71*
  French 72
  public 72
  Swedish (Mobitex) 72
Radio-frequency automatic vehicle identification 411
Radio navigation systems 274
Radiotelephony 72, 96, 206
  cellular 266
Rail-guided vehicles
  active control 346
  body tilting systems
    British 351
    Italian 351
    Spanish 351
  infrastructure considerations 210
  modelling strategies 187
  simulation software 192
  wheel–rail interaction 188
    mathematical modelling 190
  see also Railroad vehicles
Rail network configuration 196
Rail operations dispatchers 119, *204*
Rail passenger information monitoring 357
Rail traffic control
  computer-aided *119*, 123, *202*
    human intervention 357
    power regulation 356
    route control 355
    train movement regulation 354
Rail traffic disorders 119, 354
  passenger information control 357
  technical dysfunction management 356
  train schedule adjustment problems *120*
Rail traffic flows
  European networks 199
Rail traffic restoration 119
  expert systems approach 121
Railroad dynamic investigation
  model categories 189
Railroad electronic signalling systems *343*
Railroad–ferry interactive system 244
Railroad services
  international links projects 196

647

Railroad signalling 343
Railroad signalling system
    automatic 343
Railroad suspension systems
    active control concepts 346
        body tilting systems 351
        bogey stability and steering 350
        rail vehicles improvements 349
    magnetic *235*
Railroad systems 194
    automatic driving systems *358*
    automatic line supervision *353*
    see also High-speed railroad systems, Underground
        railroad systems
Railroad track maintenance
    high-speed systems 205, 212
Railroad vehicles
    body tilting systems 351
    bogey steering control 351
    high-speed 210
        bogey stability control 350
        dynamic modelling *187*
    inspection 205
    suspensions 188, 347
    wheel–rail interaction 188
Rain-provoked traffic accidents 154
Ramp metering see On-ramp control
Real-time continuous vehicle location AVM system 602
Real-time information systems 415
Real-time machine vision 597
    test vehicle 601
Real-time traffic control
    expert systems applications 579
Redundancy technique
    automatic landing systems 51
Regression analysis
    aviation forecasting 133
Regulation plants
    high-speed railroad systems 211
Regulators
    systems control 559
Reliability theory
    freeway systems equations 149
        instability detection 151
        traffic control 150
Ringways see Freeways
Risk
    aircraft landing 50
Road automobile communication system
    Japan 265, 418
Road blackspots 426
Road congestion 149, 154, 532, *540*
    control methods 397, 417
    costs 217
    detection 155
    dynamic peak period models 372

expert systems approaches 126, 128
Japanese expressways 399
management 532
    model 576
queue detector 304
queuing in networks 341
research projects 111
speed reduction strategies 509
store-and-forward network 278, 307
traffic assignment models 82, 307, 309
traffic density evolution simulation 293
wave theory approaches 232
    bottlenecks 233
see also Freeway traffic congestion
Road crossings
    pedestrian delay 384
Road databases
    EC research projects 111
    road network database 157
Road geometric model 598
Road infrastructure
    automatic image processing 224
    data integration
        strategic traffic control systems 535
    integrated road transport environment 107
    research projects 110
    origin–destination matrix
        dynamic estimation *310*
    queuing on networks 341
    supported vehicle route guidance 218
    see also DRIVE, Prometheus
Road intersection control *225*, *331*, *361*
    arrival times prediction 362, 363
        variables estimation 363
    optimizing *387*, 401
    oversaturated intersections
        local problem 364
        store-and-forward operation 279
        system of three interacting 281
    priority intersections
        capacity formulae 332
        capacity modelling 331, 402
        delay 334
        impedance effect 333
    queuing theory 337
    signal controls 397, 400
        expert systems application 579
    signal plans 361
        optimal 364, 387
    simulation 489
        TEXAS model 497
    traffic counts 363, 397
    TRANSYT 400, 403
    UTOPIA 364
        Progetto Torino data 365, 366
    vehicle merging *257*

Road junctions *382*, 473
  signal controlled 383
    control variables and parameters 473
    mathematical programming methods 475
      phase-based *476*
      stage-based *478*
  signal cycle 481
  signal timing constraints 481
  signal timing optimization 482
  traffic assignment *468*, *473*
  traffic stream control 479
Road network control
  area level
    controller goals 363, 365
    local observers goals 363, 365
    optimal control problem formulation 365
  communication network 365
  intersection level problems 362
    signal setting *366*
  Japanese *396*
  Progetto Torino UTOPIA strategy 365
  traffic-responsive methods 362
  vehicle operational requirements 385
Road networks 380
  area level model 363
  as series of intersections 362, 402
  capacity *373*
  data input
    traffic control systems 533
  dynamic equilibrium models *372*
  equilibrium conditions signal setting models 367
    global optimization methods 369
    iterative procedures 369, 469
  signal timings and traffic flow 468
    performance monitoring 533
  simulation models 497
  user behavior parameters
    commuting problem strategies 372
    signal setting control strategies 367
  see also Road infrastructure
Road reconstruction projects
  motorists' responses 577
  travel impact evaluation process 577
Road safety 385, *423*, 430
  evaluating improvements 428
  interventions 431
    goals 432
    implementation 433
  speed-spacing variables 140
Road safety programs *430*
Road safety research projects 110
Road system characteristics
  data input
    integrated traffic control systems 533
  effects on fuel consumption 176
  traffic accident factor 424

Road traffic accidents 107
  adverse weather conditions 153, 424
  avoidance 217
  blackspots 426
  causes 424
  control 83, *219*
    estimator models 147
    secondary 155
  countermeasures 431
    evaluation 434
    pilot projects 434
    public support 434
  data collection 424, 433
  detection 83, *219*
    estimator models 147
    secondary 155
  incident definitions 220
  incident detection *219*
  oversaturation 154
  remedial measures determination 428
  secondary 155, 219
  severity index 427
  simulation model 499
  see also Incident detection
Road traffic control
  automatic parking control system *324*
  computer-aided *124*, *531*
  degraded operating conditions 243
  demand-responsive *386*, *464*, 541
  design obstacles 540
  direct 544, *545*
  effectiveness indicators 117, 384
  evaluation methods *113*, 152
    computer simulation *261*, 288
    SCOOT 406
    verification software 534
  expert systems approach *124*, 578
  expressway 399
  failures *239*
    diagnosis 239
  fault control center 239
  features *539*
  freeway controls *80*, *152*, 226
    on-line 228
  fuel conservation effect 175
  hierarchically intelligent *225*
    information flow 227
    system communications network 228
  in-vehicle *215*
  incident detection *219*
  intelligent systems *225*
  international information exchanges 214
  Japan *396*
  junctions 383, 384
  maintenance *239*
    indices 241

mean down time 243
objectives 118, 290, *539*
optimal routing *306*
plans 83, 285, 530
progression methods *391*
real-time database 127
reliability *239*
  automatic backup and recovery 534
  indices 241
rerouting 81, 154
research projects 110
SCOOT 400, *404*
strategic 533
  processing complex 534
  verification 535
theoretical analyses
  extended local traffic-responsive control 288
  extended pretimed traffic-responsive control 287
  optimal control formulations 289, 308
  real-time 576, 579
  roundabout entry and exit 311
traffic-actuated control model *157*
TRANSYT *400*
trends *536*
see also Freeway control systems, Road network control, Urban traffic control, Vehicle merging control
Road traffic detectors 83, *96*, 143, 220, 222, *304*, 397, *408*
  loop detectors 363, 533
  on-line ramp control 228
  remote 533
  siting 406
  video sensors *610*
Road traffic flow variables *139*, 157, 162, *380*, 542
  estimation *143*, 311
Road traffic informatics 107, *143*, 224, 568
  EC research projects *109*, 429
Road traffic modelling *331*, *491*
  see also Traffic modelling
Road traffic monitoring equipment *408*, 533, 611
Road traffic speed–density variation analysis *380*
Road traffic systems 148, *380*
  merging control *257*
    computerized performance assessment 261
  monitoring equipment *408*
  performance measures 384
  roundabout entry and exit flows 311, 383
  simulation of urban traffic *483*
  store-and-forward approach 278
  trip matrices 316
  uncertainties 372
  see also Freeways
Road transportation 107
  arterial road and urban networks control 128
  communication technologies impact 218

corridor control systems 79, 289
  simulation model 495
EC DRIVE initiative *107*
fuel consumption economy strategies *173*
modelling 110
origin–destination matrices *310*, *315*, 373
radio dispatching service 71
safety problems 430
Road users
  road system expenses controls 560
Road vehicles *385*
  fuel conservation *173*
  guidance
    computer vision system 600
  safety campaigns 431
  types 385
Road works 155
  Dutch road maintenance 156
  freeway congestion 340
ROADSIM traffic simulation model 498
Robots
  marine use 440
Rolling-horizon optimization process 388
Route choice
  matrix estimation 316
    route choice proportions 321
  models 378
  prediction
    road network control 363, 365
  voluntary schemes 415
Route diversion strategies *415*
Route guidance *306*, *412*
  controllers 307
  graphical routing plans 326
  information needs *413*
  one-way communication systems 418
  optimality 308
    information structure choice 307
    computational complexity 308
  parking control 323
  store-and-forward model 307
  two-way communication systems 419
Route guidance systems *157*, 218
  automatic 545
  collective *412*
  communications characteristics 417
  dynamic 417
  Japan 218, 306
  in-vehicle 83, 215, 540, 544
    evaluation model 577
  information structures
    decentralization 307
  mathematical models 307
  proximity beacons 265
  routing table 306
Rudder roll stabilization *448*

Safety
  air 1, 6
  automated guideway transit systems 37, 41
  automatic landing 50, 51, 53
  automatic train control 55, *58*, 588
    transponders 345
  campaigns 431
  high-speed railroad networks *205*, 211
  level crossings 344
  Maglev vehicles 237
  public transport vehicle location system 603
  traffic projects 110
  train 344
  see also Road safety
Satellites
  communication 20, 86
  mobile satellite systems 74, 266
    cross-country trucking applications 267
    MSAT North America 266
  navigation 14, *19*, 216
SATURN traffic assignment model *546*
SCATS signal control system 538
Scheduling models
  travel activity patterns 566
SCOOT technique *404*, 538
  evaluation 406
SCOT simulation of corridor traffic model 495
Sea transportation *see* Marine transportation, Ship
Secondary accidents 155
Secondary surveillance radar *see* Radar
Selective addressed SSR *see* Mode S
Sensor technology
  vehicle detection devices *408*
    video sensors *610*
Servohydraulics 349
Shinkansen railroad system *202*
Ship automation *436*, *502*
  maritime training implications 504
Ship control *269*, 436
  computers 438, 504
  expert systems 439
  integrated bridge concept 504
  master's functions 440
  navigation *269*
  robot use 440
  stabilization methods *454*
Ship course control optimization 459
Ship crews 436, 437
  reduction policy implications 437, 503
  training 504
Ship dynamics modelling *441*, *445*, *449*
Ship management 440, 503
  ship-to-shore policy 504
Ship maneuvering process *269*
  automatic steering 436
    model-reference adaptive control *459*

control systems *272*
  dynamic positioning 273
  linear models 443
Ship propulsion systems *250*
  analog propulsion control system 254
  digital control system 255
  efficiency improvement 182
  engine control systems 255, 272
  interconnection of propulsion-system variables 252
  propeller action 270
    rotational speed equation 271
Ship steering
  model-reference adaptive control *459*
Shipping industry changes 502
  Japanese 503
Ships 269
  building 502
  cargo planning 438
  control and separation schemes 441, 448
  cost cutting policy 503
  disturbances 449
  energy consumption characteristics 180, 251
  energy saving 179
    combined diesel gas turbine arrangement 251
    hull form factors 182
  fuel conservation *179*
  fuel demand signal 255
  functional separation 183
  maintenance 438, 440
  performance data 439
  pitch controls 253, 256, 271
  position control
    offshore vessels *445*
  positioning 274
    automatic 445
  rudder 270
  scheduling 246
  spare parts inventory and control 439
  speed control system 254
  stabilization *454*
    rudder roll stabilization *448*
  steering *459*
  steering control loop 444
  transport costs 180
  use of robots 440
  vessel traffic services 607
  voyage management system 273
Shock waves 154, 163, *232*
  traffic flow simulation model 492
Shortest-path problems *462*
Shuttle and loop transit systems 37, 46
Signature searching
  vehicle detection 613
Signs and signals
  energy supply 161
  enhanced freeway 156

road works 155
variable message sign system 160
SIMAUT model
  hydrodynamic theory of traffic flow 491
SIMIR traffic simulation model 501
SIMLA traffic simulation model 501
SIMNET traffic simulation model 500
Simulation environments **483**
Simulation models
  activity chain 567, 571
  commuter day-to-day adjustment patterns 376
  computer-aided design 187
    ship propulsion systems 251
  human travel behavior 569
  passenger transport 198
  railroad system dynamics 192
  road traffic 488
    submodels 489
      TRAF software 575
  road traffic accidents 426
  road traffic flow **491**, **497**, **546**
  route guidance system evaluation 577
  ship dynamics **441**, 447, 449
  traffic density evolution 293
  underground railroad
    discrete-event **585**
  vehicle merging-control scheme 261
    priority intersection 335
Simulation systems
  artificial intelligence approach 485
  functions 24
  knowledge-based 486
  macroscopic **491**
  microscopic **497**
  modelling techniques 28, 489
  technical development 28
  see also Computer simulation, Traffic simulation systems
Simulation techniques 24
  evaluation 261
  partial simulations 122
Simulation tests
  air traffic control 23
  aircraft landing procedures 22, 53
Simulator generators 486
Simulators
  applications 23
  traffic 483
  training
    ELECTRA **24**
Sinha freeway simulation model 494
Sliding window detection technique 94
  operating method 95
Social system objectives 555
Socioeconomic status
  housing market simulation model 571

travel behavior 565
Software architecture
  simulation environments 485
    strategic traffic control **533**, 575
Spatiotemporal world models 597
Spectral analysis
  short-term prediction application 330
Speed
  aircraft descent
    costs 7, 168
  gas turbine engines 253
  high-speed railroad
    environmental considerations 206
    increasing speeds 206
  Maglev transportation systems 238
  ship
    energy consumption ratios 182
  traffic speed–flow relationships 80, 140, 154, 164, **380**
  train
    continuous overspeed protection controls **57**
    high-speed concept 194
  transportation 506
Speed detection
  Doppler radar 345
  image sensors 99
  loop-type 97
  ultrasonic 96
  vehicle profile classication 98
  wheel sensors 345
Speed regulation
  aircraft cruise–descent speed control 170
    fuel consumption savings 172
  automatic train control systems 589
  railroad 202
  ship 254
  traffic 80
    adverse weather conditions 153
    freeway traffic-responsive **507**
    fuel economies 176
  train
    assigned speed system 358
    programmed stop system 359
Sperry's ship stabilizer 456
Spurmann activity chain model 571
Stability
  actively controlled suspension systems 350
  ship
    rudder roll **448**
    stabilization installations 454
Standard traffic code 63
STARCHILD travel activity–household interactions model **572**
State estimators 348
State-feedback laws 348
Stochastic analysis
  road traffic 380

652

Stochastic models
　departure and route choice 379
　dynamic traffic assignment models *524*
　equilibrium approach 518
　process model 526, 529
Stochastic programming
　multistage 245
Stochastic user equilibrium concept 373
Store-and-forward operation 278, 307
　congested system optimal control 279, 308
　　heuristic approach 281
Strain-gauged vehicle screening system 410
Stress
　driver 415
Structural stress
　high-speed railroad systems 211
Supply and demand
　car parking spaces 324
　interactions
　　transport networks 568
　transportation systems *513*
Surface acoustic wave technology 411
Surveillance capability
　air traffic control systems 1, 16, 21, 34, 86
　traffic corridor control systems 83
　traffic intersection signals 128
　video sensors 611
　see also Radar
Suspension systems 188
　actively controlled 346, 352
　　actuators 349
　　design methods 348
　　ride improvements 349
　　schematics 348
　　vehicle considerations 347
System Development Corporation
　diamond interchange model 494
　freeway simulation model 495
Systems analysis
　high-speed railroad *208*
　international liaison 213
　traffic control *113*
Systems dynamics
　traffic flow models 311
Systems engineering
　IFAC Technical Committee on Systems Engineering 213
　transportation systems management *555*
Systems Engineering & Consensus Formation Office(SECFO) 112

Tanner's formula 332, 334
Teleogenic systems 560
Telephone network structure 69
Telepoint system 73

Telesat Mobile Incorporated, Canada 266
Teleterminals 418
Tilting systems
　trains 351
Time
　intersection delay formulae 334
　journey scheduling
　　departure time models 373, 374
　　queuing theory 340
　minimization on freeway system 290
　on-ramp control strategies
　　differences 286
　ratios
　　vehicles parked and in motion 323
　total travel time calculations 298
　travelling 506
Time-dependent queuing theory 383
Time-discrete formulations 275
　metro systems 582
Time-planning equations 246
Time schedule
　objectives realization 558
Time series analyses
　predictive models 329
　　autoregressive integrated moving average formulation 330
Time–space geography 565
　PESASP model 566, *570*
　travel behavior patterns 564
Time variables
　origin–destination estimation procedures 314
Timetable strategies
　metro control strategies 588
　　computerized 592
Toll roads 152
　electronic road pricing systems 545
　Japanese 399
Tourism
　transportation speeds and time 506
Tours
　trip sequence 570
Tracking
　aircraft 3, 10
　containers 75
　radar 94
TRAF simulation software 575
Traffic assignment *513*
　combined signal control–assignment problem 468
　　cycle time and offset changes 472
　　iterative optimization assignment method 470
　deterministic equilibrium
　　graphical representation 516
　　mathematical formulation 515
　　model solution 516
　　Wardrop's principles 514
　dynamic *521*

mathematical models 522
signal control strategies **468**
stochastic dynamic 525
    path choice 526, 528
    stochastic process model 526, 529
stochastic equilibrium
    mathematical formulation 519
    model solution 519
    path choice 518
see also Queuing models, Route choice

Traffic conflict technique 428
Traffic control *see* Road traffic control, Urban traffic control
Traffic control modes **530**
Traffic control systems *see* Road traffic control
Traffic counts 316, 362, 614
    matrix estimation
        nonstructured maximum-entropy model 319
Traffic data centers 216, 533
Traffic engineering knowledge 130, 530
Traffic flow control
    automatic intersection merging control **257**
    expert systems technique **125**
        arterial roads 128
        freeway 129
        intersections control 128
    hydrodynamic theory analogies **231**
    intelligent systems
        decision-making levels 228
        intersections control **225**, 279
        learning capabilities 229
    progression methods 391
        MAXBAND 392
        MULTIBAND 393
    speed limits 507
Traffic flow equations 67, 80, ***103***, ***140***, ***144***
    capacity formulae 332
    density gradient term introduction 163
    discrete-time 225, 227
    equilibrium network traffic signal setting models 366, 469
    linear-quadratic methodology 292
    macroscopic variables 140, ***162***
        changes 221
        validation 166
    maximization models 392
    microscopic variables 139, 381
    motion control 258, 260
        evaluation 261
    on-ramp control strategies **287**
        demand-capacity 295
        optimization theory applications 289, 291
    oversaturated intersections **279**
    saturation 480
    two-dimensional 165
    wave theories 231, 291

Traffic flow estimation equations ***145***, ***311***
    assignment techniques 316
    incident detection capabilities 147
    Kalman filter applications 146, 313
    short-term models 329
    time variables 314
    trip matrices 318
    very-short-term 330
Traffic flow prediction
    short-term 329
    very-short-term 330
Traffic flow variables ***139***, ***162***, 231
    acceleration 140
    concentration 141
    flow 140
    fundamental diagram ***142***, 162, 231, 298
    headway 139
    spacing 140
    speed 140
Traffic fuel consumption variables ***175***
Traffic information services ***61***, 157
    German 64
    Japanese 398, 418
    standard traffic code 63
    see also Traffic signals and signs
Traffic intensity ratio 382
Traffic interactions 542
Traffic jams *see* Freeway traffic congestion, Road congestion
Traffic management ***385***
    bottlenecks 233
    data needs 157
    database 534
    expert systems approaches ***124***
    incident response 83, 126, 220
    intersection 225, 233
    junctions 383
    merging 257
    operations improvements 543
    roundabouts 383
    SCOOT facilities 407
    signal strategies 543
    simulation 500
Traffic management consultants 554
Traffic management systems
    assessment models ***546***
    consultation procedures 554
    dynamic 544
    monitoring 553
    procedural steps 554
    see also Road traffic control, Traffic assignment
Traffic management teams 550
Traffic management techniques 553
Traffic measurement 614
Traffic modelling 65, 81, ***143***, 220, 307, ***316***
    departure time models 373

dynamic approaches 375
    simulated dynamic approaches 377
deterministic models 338
    deterministic queuing model 374
    simulation 339
nonstructured approaches 317
    maximum-entropy matrix estimation 319
predictive *329*
priority intersection *331*
SCOOT technique 405
store-and-forward approach
TRAF simulation software 575
TRANSYT *400*
see also Freeway traffic modelling, Traffic flow
    equations, Traffic simulation systems
Traffic offices 550
    functions 553
    procedural systems 553
Traffic oversaturation 154
    intersection control 279, 397
    merging strategy 263
Traffic radio service 157
Traffic signal control systems 531
    apparent failures 240
        probability analysis 242
    central control 553
    computer-aided *537*
    cycle time 473
        changes 472
        constraints 481
        optimization 482
    dangerous failures 240
        probability analysis 241
    degraded operating conditions 240, 243
    demand-responsive 532
    fault monitoring 533
    fixed-time 295, 479, *532*
    hydrodynamic theory applications 233
    indices of operation 240
        indicative values 241
    integrated 532
    intersection optimization 362
    junctions 397
        phase-based *476*
        stage-based *478*
    mutually consistent 470
    network design problems 469
    offset optimality 472, 532
    optimization for fuel conservation 175
    optimization models *400*, 475
    road infrastructure data determinants 535
    traffic assignment determinants 468, 538
        CONTRAM model 548
    traffic streams 479
        tidal-flow systems 545
    user behavior influences 367

vehicle-actuated 532
Traffic signal design
    enhanced signs 156
    expert systems approaches 127
        intelligent traffic signal control software 128
    freeway ramp metering signing 304
    freeway route guidance display system 160
    queuing theory application 338
    variable guide signs 327
Traffic signals and signs 155
    car parking gudance system 324, *326*
    control methods 397, 543
    cycle 481
    Dutch priority system 156
    energy supply 161
    enhancement 156
    equilibrium network traffic signal setting models *367*
    fault monitoring 533
    light-emitting-diode message boards 398
    queue delays 339
    ramp-metering signs 304
        start up and fail-safe operation 305
    traffic-actuated 339
    traffic assignment *468*
    see also Variable message sign systems
Traffic simulation systems 483
    artificial intelligence approach 485
    knowledge-based 486
    macroscopic models *491*
    microscopic models 66, *497*
        American 497
        British TRAFFICQ 500
        Canadian INTEGRATION 499, 576
        Dutch 501
        French 501
        German 500
        Spanish microscopic simulator project 488
Traffic stabilization techniques *153*
Traffic surveys 552
Traffic weaving phenomena 164
Traffic zones 560
TRAFFICQ simulation model 500
Train diagram 120
Train dispatchers 203, *204*
Train operation restrictions *120*
Train protection *55*
    automatic stopping 57
    solid-state interlocking 344
    see also Automatic train control
Train schedule adjustment *119*
    ESTRAC Project *121*
        ESTRAC-III *122*
    knowledge engineering technology 121
    restrictions *120*
Train supervision 354
Train–track dynamics

655

advanced train control systems 345
automatic driving systems 359
bogey stability 350
dysfunction detection 356
electronic blocking system 344
mathematical model 189
narrow-gauge utiltization 207
power distribution regulation 356
solid-state interlocking structure 344
steering control 350
transmission 360
transponders 345
Train travel
  high-speed concepts 194, 208
Training
  maritime 505
  simulation 23, *24*, 29
  vessel traffic services operators 610
Training flights 131
Trains *see* Rail-guided vehicles
Transit systems 538
Transponders
  air traffic control 616
  train safety systems 345
  see also Radar beacon transponders
Transport markets
  European 195
Transportation
  demand *131*, *134*
  modal choice model 562
  modal split
    France 506
  social aspects *505*
Transportation corridors *79*
  European 196, 198
  freeway
    control strategies 289, 543
    integrated management schemes 540
    reconstruction impact mitigation strategies 576
    road traffic control *79*, *285*, 543
Transportation modelling 110, 196, *560*
  dynamic behavioral models *372*
  expert systems developments 575
  high-speed railroad *187*
  intelligent systems developments 575
  microscopic *570*
  store-and-forward network 278
    optimal operation 280
  see also Origin–destination matrix
Transportation planning 552, 560
  activity-based approach *564*
    limitations 567
  microscopic approach *569*
Transportation supply 513
Transportation systems
  analysis 574

computer-aided planning and control 244
control studies 575
demand management 541
dynamics 555
expenses 560
goods 506
interactions 543
international coordination of automation strategies *213*
planning problems 245
reconstruction impact mitigation strategies 576
  model selection 577
supply and demand relations *513*
trends *574*
Transportation systems management 539
  centralized 558
  decentralized 558
  decision making 557
  information feedback 558
  interactions model 543
  objectives 555
    achievement 558
    realization time schedule 558
  phases 556
Transrapid Versuchanlage Emsland guideway 238
TRANSYT traffic simulation model 339, 370, 391, *400*, 500
Travel behavior analysis *564*
  activity patterns 564
  complexity issues 567
  individual decision-making factors 569
  reconstruction projects impact 576
  traveller groups 565
Travel costs 514, 517
Travel demand 514
  trip-generation model 561
Travel forecasting models *560*
Travel impact evaluation process 576
  analysis tools selection 577
Travel matrices 316
  three-dimensional 322
Travel time calculations 298
  departure time models 373, 374
    day-to-day adjustment process 375
    deterministic queuing model 374
  queuing factors 340
Travel time values 506
Travel trip 570
  assignment model 563
  modes
    modal-choice model 562
  trip-distribution models 561
    gravity model 562
  trip-generation model 561
Traveller groups 567
Travtek project 218
Trilateration techniques 417

Trunking 68
Turbogenerators 185

Ultrasonic Doppler-type speed detector *96*
Underground railroad systems 581
   automatic train control systems *588*
   modelling *581*
      discrete-event simulation *585*
   traffic control strategies models 583
   see also Metro systems
Undirected graphs *462*
United Nations
   UN Educational, Scientific and Cultural Organization 214
   UN Industrial Development Organization 214
Urban magnetic levitation vehicles *see* Maglev vehicles
Urban traffic control *225*, *361*, 386, *400*
   bandwidth maximization problem 392
   demand-responsive *386*
   integrated corridor management 540
   predictive models of traffic flows 329
   traffic-responsive 404
   traffic signals control 545
   urban traffic optimization by integrated automation strategies 362
      intersection control 364
      Progetto Torino 365
   see also Road traffic control, Traffic signal control systems
Urban traffic simulation *483*
   expert systems *485*
Urban transportation
   arterial road and urban networks control
      corridor coordination 289, 540
      expert systems techniques 128
      congestion surveillance 128
         SAGE expert systems technique 129
   supply and demand management 541
   transportation corridors 79
   see also Automated guideway transit systems, Expressways, Freeways
Urban transportation planning modelling system *560*
UTOPIA strategies 362
   intersection control 364
      Progetto Torino data 365, 366

VaMoRs test vehicle
   autonomous mobility and computer vision 601
Variable message sign systems 414, 545
   control model structure 159
   data requirements 157
   display variants 160
   parking control 323
   speed limitations 509
Variable route guidance system *157*
   German 161

Variable speed limits
   freeway traffic
      weather conditions 509
Vehicle acceleration values 140
Vehicle classification 408
Vehicle condition
   road accident factors 424
Vehicle control automatic 541
Vehicle design
   active suspension railroad systems 347
      linearized vehicle model 348
Vehicle detection methods 611
Vehicle engines
   fuel consumption factors 178
Vehicle failure
   rapid transit systems 42
   road accidents 424
Vehicle fuel consumption equations *174*
Vehicle guidance 597
   active machine vision system architecture 600
Vehicle headway 139
Vehicle–highway systems
   American initiatives 218
Vehicle identification methods *410*
   time and space signatures 613
Vehicle identification systems automatic *411*
Vehicle information and communication systems *263*
Vehicle location techniques 602
Vehicle merging control *257*
   car-following methods 258
   cell-following methods 258
   evaluation
      computer simulation 261
   freeway ramp metering 303
      merge detector 304
Vehicle modelling *188*
   linearized 348
Vehicle monitoring *408*, *602*, 614
Vehicle monitoring systems automatic *602*
Vehicle occupancy variables 141
Vehicle priority
   intersection 362, 366
   ramp-metering methods 305
Vehicle profile classification 98
Vehicle profile classifier speed detectors *99*
Vehicle routing 83, 306
   see also Road traffic control, Urban traffic control
Vehicle spacing 614
   concentration equations 141
   traffic flow variable 140
Vehicle speed profiles 140
   mean speed variables 141
Vehicle system dynamics software 191
Vehicle weight screening 410
Velocity
   automatic vehicle merger model 259

high-speed train braking distance 212
Vessel traffic services 607
  inventory 608
  operational features 608
  operators 610
VHF omnidirectional range receivers 274
Video imaging techniques
  traffic control systems 83
  traffic detection 611
Video sensors *610*
Visual flying rules *13*, 616
Visual senses
  vehicle motion 597
  see also Dynamic machine vision
Voice input and synthesis techniques 29
  digital voice transmission 73, 74
  roadside radio systems 398
Volume–density relationship equation 163
Voyage management system *273*

Wardrop's principles 418, 468, 515
Waste-heat utilization
  fuel-saving strategy 185
Water tanks
  ship antirolling systems 456
Watt's antirolling system 456
Wave simulation
  frequency spectrum 449
Wave theory
  traffic flow analogy application 232
Way points 13, *14*, 616
Weather conditions *see* Meteorological conditions
Weighing operations
  heavy vehicle monitoring 410
Wheel–rail interaction
  geometric models 188
  Skytrain 44
Wind-power utilization
  fuel-saving strategy 185

Zone of convergence concept 173
  aircraft control for minimum cost operation 170
  fuel consumption in extended terminal areas 171
Zumkeller activity chain model 571